McGraw-Hill
Dictionary of
ENGINEERING

McGraw-Hill
Dictionary of
ENGINEERING

Sybil P. Parker Editor in Chief

McGraw-Hill Book Company

New York   St. Louis   San Francisco

Auckland   Bogotá   Guatemala   Hamburg
Johannesburg   Lisbon   London   Madrid   Mexico
Montreal   New Delhi   Panama   Paris   San Juan
São Paulo   Singapore   Sydney   Tokyo   Toronto

McGRAW-HILL DICTIONARY OF ENGINEERING
The material in this Dictionary has been published
previously in the McGRAW-HILL DICTIONARY OF
SCIENTIFIC AND TECHNICAL TERMS, Third Edition,
copyright © 1984 by McGraw-Hill, Inc. All rights
reserved. Philippines copyright 1984 by McGraw-Hill,
Inc. Printed in the United States of America. Except as
permitted under the United States Copyright Act of
1976, no part of this publication may be reproduced or
distributed in any form or by any means, or stored in a
data base or a retrieval system, without the prior
written permission of the publisher.

1 2 3 4 5 6 7 8 9 0   DODO   8 9 1 0 9 8 7 6 5 4

ISBN 0-07-045412-4

Library of Congress Cataloging in Publication Data

McGraw-Hill dictionary of engineering.

    "Material in this dictionary has been published
previously in the McGraw-Hill dictionary of scientific
and technical terms, third edition, copyright 1984"—
T.p. verso
    1. Engineering—Dictionaries.   I. Parker, Sybil P.
II. McGraw-Hill Book Company.   III. McGraw-Hill
dictionary of scientific and technical terms.
TA9.M35   1984   620'.003'21   84-12206
ISBN 0-07-045412-4

Preface

In its modern form, engineering is considered an art that involves people, materials, machines, and energy in useful applications of natural phenomena. Over the years, due to the increasing complexity of technology, the field of engineering evolved into various specialized disciplines, which in turn were affected by a number of integrating factors.

The *McGraw-Hill Dictionary of Engineering* is an integrated approach to presenting the specialized vocabularies of the following disciplines: aerospace engineering, civil engineering, design engineering, industrial engineering, materials science, mechanical engineering, metallurgical engineering, mining engineering, petroleum engineering, and systems engineering. For this project, the term "engineering" is used in a restricted sense and excludes the more specialized disciplines, such as chemical engineering, electrical engineering, and food engineering.

Each definition is identified by the field in which it is primarily used; terms may be defined in more than one field. Some definitions, however, are used in several fields, and these are identified by a more general field. For example, a term used in both mechanical and industrial engineering is assigned to engineering.

The 16,000 terms selected for this Dictionary are, in the opinion of the editors, fundamental to understanding the basic principles of engineering. All terms and definitions were drawn from the *McGraw-Hill Dictionary of Scientific and Technical Terms* (3d ed., 1984). Synonyms, acronyms, and abbreviations are given with the definitions and are also listed in the alphabetical sequence as cross-references to the defining terms.

The *McGraw-Hill Dictionary of Engineering* is an important reference tool for professional engineers and engineering students as well as librarians, technicians, and writers who need a fundamental understanding of the language of engineering methods and materials.

Sybil P. Parker
EDITOR IN CHIEF

Editorial Staff

Sybil P. Parker, Editor in Chief
Jonathan Weil, Editor
Betty Richman, Editor
Edward J. Fox, Art director
Ann D. Bonardi, Art production supervisor
Joe Faulk, Editing manager
Ann Jacobs, Editing supervisor
Frank Kotowski, Jr., Editing supervisor
Patricia W. Albers, Senior editing assistant

Consulting and Contributing Editors

from the McGraw-Hill Dictionary of Scientific and Technical Terms

Prof. Eugene A. Avallone—Department of Mechanical Engineering, City University of New York. MECHANICAL ENGINEERING.

Prof. Theodore Baumeister—(Deceased) Consulting Engineer; Stevens Professor of Mechanical Engineering, Emeritus, Columbia University. MECHANICAL ENGINEERING.

Prof. George S. Bonn—Formerly, Graduate School of Library Science, University of Illinois. LIBRARY CONSULTANT.

Waldo G. Bowman—Formerly, Black and Veatch, Consulting Engineers, New York City. CIVIL ENGINEERING.

Dr. John F. Clark—Director, Space Application and Technology, RCA Laboratories, Princeton, New Jersey. SPACE TECHNOLOGY.

Robert L. Davidson—Formerly, Editor in Chief, Business Books and Services, Professional and Reference Division, McGraw-Hill Book Company. PETROLEUM ENGINEERING.

Gerald M. Eisenberg—Senior Engineering Administrator, American Society of Mechanical Engineers. ENGINEERING.

Dr. Gary Judd—Vice Provost, Academic Programs and Budget, and Dean of the Graduate School, Rensselaer Polytechnic Institute. METALLURGICAL ENGINEERING.

Alvin W. Knoerr—Mining Engineer, U.S. Bureau of Mines, Washington, D.C. MINING ENGINEERING.

Dr. Joachim Weindling—Professor of System Engineering and Operations Research, Polytechnic Institute of Brooklyn. INDUSTRIAL AND PRODUCTION ENGINEERING.

How to Use the Dictionary

I. ALPHABETIZATION

The terms in the *McGraw-Hill Dictionary of Engineering* are alphabetized on a letter-by-letter basis; word spacing, hyphen, comma, solidus, and apostrophe in a term are ignored in the sequencing. For example, an ordering of terms would be:

air belt
air-bend die
airblast
air hp
airspeed
AJO breathing apparatus

II. FORMAT

The basic format for a defining entry provides the term in boldface, the field in small capitals, and the single definition in lightface:

term [FIELD] Definition.

A field may be followed by multiple definitions, each introduced by a boldface number:

term [FIELD] **1.** Definition. **2.** Definition. **3.** Definition.

A term may have definitions in two or more fields:

term [MATER] Definition. [MIN ENG] Definition.

A simple cross-reference entry appears as:

term *See* another term.

A cross-reference may also appear in combination with definitions:

term [MATER] Definition. [MIN ENG] *See* another term.

III. CROSS-REFERENCING

A cross-reference entry directs the user to the defining entry. For example, the user looking up "arbor vitae oil" finds:

arbor vitae oil *See* thuja oil.

The user then turns to the "T" terms for the definition.

Cross-references are also made from variant spellings, acronyms, abbreviations, and symbols.

A-5 *See* Vigilante.
aerofoil *See* airfoil.
alt *See* altitude.
ATS *See* applications technology satellite.

IV. ALSO KNOWN AS . . . , etc.

A definition may conclude with a mention of a synonym of the term, a variant spelling, an abbreviation for the term, or other such information, introduced by "Also known as . . . ," "Also spelled . . . ,"

"Abbreviated . . . ," "Symbolized . . . ," "Derived from . . . ." When a term has more than one definition, the positioning of any of these phrases conveys the extent of applicability. For example:

term [MATER] **1.** Definition. Also known as synonym. **2.** Definition. Symbolized T.

In the above arrangement, "Also known as . . ." applies only to the first definition. "Symbolized . . ." applies only to the second definition.

term [MATER] **1.** Definition. **2.** Definition. [MIN ENG] Definition. Also known as synonym.

In the above arrangement "Also known as . . ." applies only to the second field.

term [MATER] Also known as synonym. **1.** Definition. **2.** Definition. [MIN ENG] Definition.

In the above arrangement, "Also known as . . ." applies to both definitions in the first field.

term Also known as synonym. [MATER] **1.** Definition. **2.** Definition. [MIN ENG] Definition.

In the above arrangement, "Also known as . . ." applies to all definitions in both fields.

Field Abbreviations

AERO ENG	aerospace engineering
BUILD	building construction
CIV ENG	civil engineering
DES ENG	design engineering
ENG	engineering
ENG ACOUS	engineering acoustics
IND ENG	industrial engineering
MATER	materials
MECH ENG	mechanical engineering
MET	metallurgy
MIN ENG	mining engineering
PETRO ENG	petroleum engineering
SYS ENG	systems engineering

Scope of Fields

aerospace engineering—Engineering pertaining to the design and construction of aircraft and space vehicles and of power units and dealing with the special problems of flight in both the earth's atmosphere and space, such as in the flight of air vehicles and the launching, guidance, and control of missiles, earth satellites, and space vehicles and probes.

building construction—The art or business of assembling materials into a structure, especially one designated for occupancy.

civil engineering—The planning, design, construction, and maintenance of fixed structures and ground facilities for industry, for transportation, for use and control of water, for occupancy, and for harbor facilities.

design engineering—A branch of engineering concerned with the design of a product or facility according to generally accepted uniform standards and procedures, such as the specification of a linear dimension, or a manufacturing practice, such as the consistent use of a particular size of screw to fasten covers.

engineering—The science by which properties of matter and sources of power in nature are made useful to humans in structures, machines, and products.

engineering acoustics—A field of acoustics that deals with the production, detection, and control of sound by electrical devices, including the study, design, and construction of such things as microphones, loudspeakers, sound recorders and reproducers, and public address systems.

industrial engineering—The application of engineering principles and training and the techniques of scientific management to the maintenance of a high level of productivity at optimum cost in industrial enterprises, as by analytical study, improvement, and installation of methods and systems, operating procedures, quantity and quality measurements and controls, safety measures, and personnel administration.

materials—The study of admixtures of matter or the basic matter from which products are made; includes adhesives, building materials, fuels, paints, leathers, and so on.

mechanical engineering—The branch of engineering concerned with the generation, transmission, and utilization of heat and mechanical power, and with the production and operation of tools, machinery, and their products.

metallurgy—The branch of engineering concerned with the production of metals and alloys, their adaptation to use, and their performance in service; and the study of chemical reactions involved in the processes by which metals are produced, and of the laws governing the physical, chemical, and mechanical behavior of metallic materials.

mining engineering—A branch of engineering concerned with the location and evaluation of coal and mineral deposits, the survey of mining areas, the layout and equipment of mines, the supervision of mining operations, and the cleaning, sizing, and dressing of the product.

petroleum engineering—A branch of engineering concerned with the search for and extraction of oil, gas, and liquefiable hydrocarbons.

systems engineering—The branch of engineering dealing with the design of a complex interconnection of many elements (a system) to maximize an agreed-upon measure of system performance.

McGraw-Hill
Dictionary of
ENGINEERING

A

A-1 *See* Skyraider.

A-3 *See* Skywarrior.

A-5 *See* Vigilante.

abandon [ENG] To stop drilling and remove the drill rig from the site of a borehole before the intended depth or target is reached.

abandoned mine *See* abandoned workings.

abandoned workings [MIN ENG] Deserted excavations, either caved or sealed, in which further mining is not intended, and opening workings which are not ventilated and inspected regularly. Also known as abandoned mine.

abandonment [MIN ENG] Failure to perform work, by conveyance, by absence, and by lapse of time, on a mining claim. [PETRO ENG] *See* abandonment contour.

abandonment contour [PETRO ENG] A graph of actual cumulative yield of an oil well compared with its estimated ultimate yield; useful in determining the most economic time to abandon an oil well. Also known as abandonment.

abate [ENG] **1.** To remove material, for example, in carving stone. **2.** In metalwork, to excise or beat down the surface in order to create a pattern or figure in low relief.

abatement [ENG] **1.** The waste produced in cutting a timber, stone, or metal piece to a desired size and shape. **2.** A decrease in the amount of a substance or other quantity, such as atmospheric pollution.

abat-jour [BUILD] A device that is used to deflect daylight downward as it streams through a window.

abattoir [IND ENG] A building in which cattle or other animals are slaughtered.

abat-vent [BUILD] A series of sloping boards or metal strips, or some similar contrivance, to break the force of wind without being an obstruction to the passage of air or sound, as in a louver or chimney cowl.

abelite [MATER] A substance made of ammonium nitrate and a nitrated aromatic hydrocarbon and used as an explosive.

abherent [MATER] A substance that inhibits a material from adhering to itself or another material.

abietine [MATER] The distillate of the gums of the Jeffrey and digger pines; comprises 96% heptane; used as a cleaning agent, insecticide, and constituent of standard gasolines to measure detonation of engines.

ablation [AERO ENG] The carrying away of heat, generated by aerodynamic heating, from a vital part by arranging for its absorption in a nonvital part, which may melt or vaporize and then pass away, taking the heat with it. Also known as ablative cooling.

ablative agent [MATER] A material from which the surface layer is to be removed, often for the purpose of dissipating extreme heat energy, as in space vehicles reentering the earth's atmosphere. Also known as ablative material.

ablative cooling *See* ablation.

ablative material *See* ablative agent.

ablative shielding [AERO ENG] A covering of material designed to reduce heat transfer to the internal structure through sublimation and loss of mass.

ablatograph [ENG] An instrument that records ablation by measuring the distance a snow or ice surface falls during the observation period.

A block [CIV ENG] A hollow concrete masonry block with one end closed and the other open and with a web between, so that when the block is laid in a wall two cells are produced.

abnormal place [MIN ENG] An area in a coal mine where geological conditions render mining uneconomical.

abort [AERO ENG] **1.** To cut short or break off an action, operation, or procedure with an aircraft, space vehicle, or the like, especially because of equipment failure. **2.** An aircraft, space vehicle, or the like which aborts. **3.** An act or instance of aborting.

abort zone [AERO ENG] The area surrounding the launch within which malperforming missiles will be contained with known and acceptable probability.

aboundikro *See* Sapele mahogany.

about-sledge [MET] A large hammer that is utilized in blacksmithing.

Abrams' law [CIV ENG] In concrete materials, for a mixture of workable consistency the strength of concrete is determined by the ratio of water to cement.

abrasion [ENG] **1.** The removal of surface material from any solid through the frictional action of another solid, a liquid, or a gas or combination thereof. **2.** A surface discontinuity brought about by roughening or scratching.

abrasion resistance [MATER] The ability of a surface to resist wearing due to contact with another surface moving with respect to it.

abrasion resistance index [MATER] In vulcanized material or synthetic rubber compounds, a measure of abrasion resistance relative to a standard rubber compound under defined conditions.

abrasion test [MECH ENG] The measurement of abrasion resistance, usually by the weighing of a material sample before and after subjecting it to a known abrasive stress throughout a known time period, or by reflectance or surface finish comparisons, or by dimensional comparisons.

abrasive [MATER] **1.** A material used, usually as a grit sieved by a specified mesh but also as a solid shape or as a paste or slurry or air suspension, for grinding, honing, lapping, superfinishing, polishing, pressure blasting, or barrel tumbling. **2.** A material sintered or formed into a solid mass such as a hone or a wheel disk, cone, or burr for grinding or polishing other materials. **3.** Having qualities conducive to or derived from abrasion.

abrasive belt [MECH ENG] A cloth, leather, or paper band impregnated with grit and rotated as an endless loop to abrade materials through continuous friction.

abrasive blasting [MECH ENG] The cleaning or finishing of surfaces by the use of an abrasive entrained in a blast of air.

abrasive cloth [MECH ENG] Tough cloth to whose surface an abrasive such as sand or emery has been bonded for use in grinding or polishing.

abrasive cone [MECH ENG] An abrasive sintered or shaped into a solid cone to be rotated by an arbor for abrasive machining.

abrasive disk [MECH ENG] An abrasive sintered or shaped into a disk to be rotated by an arbor for abrasive machining.

abrasive drilling [MIN ENG] A rotary drilling method in which drilling is effected by the abrasive action of the drill steel or drilling medium which rotates while being pressed against the rock.

abrasive jet cleaning [ENG] The removal of dirt from a solid by a gas or liquid jet carrying abrasives to ablate the surface.

abrasive machining [MECH ENG] Grinding, drilling, shaping, or polishing by abrasion.

abrasiveness [MATER] **1.** The property of a material causing wear of a surface by friction. **2.** The quality or characteristic of being able to scratch, abrade, or wear away another material.

abrasive paper [MATER] Tough paper to whose surface an abrasive, such as sand or emery, has been bonded for use in grinding or polishing.

abrasive sand [MATER] Grit used as abrasive, usually graded as to which sieve mesh it will pass through.

abreast milling [MECH ENG] A milling method in which parts are placed in a row parallel to the axis of the cutting tool and are milled simultaneously.

abreuvoir [CIV ENG] A space between stones in masonry to be filled with mortar.

absinthe oil [MATER] A toxic essential oil obtained from the dried leaves of *Artemisia absinthum*; soluble in alcohol; formerly used in medicine. Also known as wormwood oil.

absolute altimeter [ENG] An instrument which employs radio, sonic, or capacitive technology to produce on its indicator the measurement of distance from the aircraft to the terrain below. Also known as terrain clearance indicator.

absolute altitude [ENG] Altitude above the actual surface, either land or water, of a planet or natural satellite.

absolute angle of attack [AERO ENG] The acute angle between the chord of an airfoil at any instant in flight and the chord of that airfoil at zero lift.

absolute blocking [CIV ENG] A control arrangement for rail traffic in which a track is divided into sections or blocks upon which a train may not enter until the preceding train has left.

absolute ceiling [AERO ENG] The greatest altitude at which an aircraft can maintain level flight in a standard atmosphere and under specified conditions.

absolute efficiency [ENG ACOUS] The ratio of the power output of an electroacoustic transducer, under specified conditions, to the power output of an ideal electroacoustic transducer.

absolute instrument [ENG] An instrument which measures a quantity (such as pressure or temperature) in absolute units by means of simple physical measurements on the instrument.

absolute magnetometer [ENG] An instrument used to measure the intensity of a magnetic field without reference to other magnetic instruments.

absolute manometer [ENG] **1.** A gas manometer whose calibration, which is the same for all ideal gases, can be calculated from the measurable physical constants of the instrument. **2.** A manometer that measures absolute pressure.

absolute pressure gage [ENG] A device that measures the pressure exerted by a fluid relative to a perfect vacuum; used to measure pressures very close to a perfect vacuum.

absolute pressure transducer [ENG] A device that responds to absolute pressure as the input and provides a measurable output of a nature different than but proportional to absolute pressure.

absolute roof [MIN ENG] The entire mass of strata overlying a subsurface point of reference.

absolute stop [CIV ENG] A railway signal which indicates that the train must make a full stop and not proceed until there is a change in the signal. Also known as stop and stay.

absolute volume [ENG] The total volume of the particles in a granular material, including both permeable and impermeable voids but excluding spaces between particles.

absorbent [MATER] A material which, in contact with a liquid or gas, extracts one or more substances for which it has an affinity, and is altered physically or chemically during the process.

absorbent cotton [MATER] A cotton fiber that absorbs water because its natural waxes have been removed.

absorbent paper [MATER] Paper capable of absorbing and holding liquids by the capillarity of the pores between or within the closely matted cellulosic fibers.

absorber [MECH ENG] **1.** A device which holds liquid for the absorption of refrigerant vapor or other vapors. **2.** That part of the low-pressure side of an absorption system used for absorbing refrigerant vapor.

absorber oil See absorption oil.

absorbing boom [CIV ENG] A device that floats on the water and is used to stop the spread of an oil spill and aid in its removal.

absorbing well [CIV ENG] A shaft that permits water to drain through an impermeable stratum to a permeable stratum.

absorption bed [CIV ENG] A sizable pit containing coarse aggregate about a distribution pipe system; absorbs the effluent of a septic tank.

absorption cycle [MECH ENG] In refrigeration, the process whereby a circulating refrigerant, for example, ammonia, is evaporated by heat from an aqueous solution at elevated pressure and subsequently reabsorbed at low pressure, displacing the need for a compressor.

absorption dynamometer [ENG] A device for measuring mechanical forces or power in which the mechanical energy input is absorbed by friction or electrical resistance.

absorption-emission pyrometer [ENG] A thermometer for determining gas temperature from measurement of the radiation emitted by a calibrated reference source before and after this radiation has passed through and been partially absorbed by the gas.

absorption field [CIV ENG] Trenches containing coarse aggregate about distribution pipes permitting septic-tank effluent to seep into surrounding soil. Also known as disposal field.

absorption gasoline [MATER] A gasoline obtained by using an oil to absorb the natural or refinery gas containing the gasoline and then distilling it from the oil.

absorption hygrometer [ENG] An instrument with which the water vapor content of the atmosphere is measured by means of the absorption of vapor by a hygroscopic chemical.

absorption loss [CIV ENG] The quantity of water that is lost during the initial filling of a reservoir because of absorption by soil and rocks.

absorption meter [ENG] An instrument designed to measure the amount of light transmitted through a transparent substance, using a photocell or other light detector.

absorption number [ENG] A dimensionless group used in the field of gas absorption in a wetted-wall column; represents the liquid side mass-transfer coefficient.

absorption oil [MATER] A petroleum or coal tar oil that is contacted with a vapor or gas mixture to remove heavy components, as in the recovery of natural gasoline from wet natural gas. Also known as absorber oil; scrubbing oil; wash oil.

absorption refrigeration [MECH ENG] Refrigeration in which cooling is effected by the expansion of liquid ammonia into gas and absorption of the gas by water; the ammonia is reused after the water evaporates.

absorption system [MECH ENG] A refrigeration system in which the refrigerant gas in the evaporator is taken up by an absorber and is then, with the application of heat, released in a generator.

absorption tower [ENG] A vertical tube in which a rising gas is partially absorbed by a liquid in the form of falling droplets. Also known as absorption column.

absorption trench [CIV ENG] A trench containing coarse aggregate about a distribution tile pipe through which septic-tank effluent may move beneath earth.

Abt track [CIV ENG] One of the cogged rails used for railroad tracking in mountains and so arranged that the cogs are not opposite one another on any pair of rails.

abutment [CIV ENG] A surface or mass provided to withstand thrust; for example, end supports of an arch or a bridge.

abutting joint [DES ENG] A joint which connects two pieces of wood in such a way that the direction of the grain in one piece is angled (usually at 90°) with respect to the grain in the other.

abutting tenons [DES ENG] Two tenons inserted into a common mortise from opposite sides so that they contact.

Ac₀ [MET] The temperature at which a magnetic change occurs in cementite; the Curie point of cementite.

Ac₁ [MET] The temperature at which austenite begins to be formed upon heating a steel.

Ac₂ [MET] The Curie point of ferrite.

Ac₃ [MET] The temperature at which the transformation of ferrite to austenite is completed upon heating a steel.

Ac₄ [MET] The temperature at which delta iron is formed from gamma iron upon heating a steel.

Ac_cm [MET] The temperature at which the solution of cementite in austenite is completed upon heating a hypereutectoid steel.

acacia gum *See* gum arabic.

acapau [MATER] A type of ebony obtained from *Voucapapoua americana* in the Amazon valley; used for fine inlay work. Also known as partridge wood.

acaricide [MATER] A pesticide used to destroy mites on domestic animals, crops, and man. Also known as miticide.

accelerated life test [ENG] Operation of a device, circuit, or system above maximum ratings to produce premature failure; used to estimate normal operating life.

accelerated weathering [ENG] A laboratory test used to determine, in a short period of time, the resistance of a paint film or other exposed surface to weathering.

accelerating agent [MATER] **1.** A substance which increases the speed of a chemical reaction. **2.** A compound which hastens and improves the curing of rubber.

accelerating incentive *See* differential piecerate system.

acceleration analysis [MECH ENG] A mathematical technique, often done graphically, by which accelerations of parts of a mechanism are determined.

acceleration feedback [AERO ENG] The use of accelerometers strategically located within the body of a missile so that they sense body accelerations during flight and interact with another device on board the missile or with a control center on the ground or in an airplane to keep the missile's speed within design limits.

acceleration tolerance [ENG] The degree to which personnel or equipment withstands acceleration.

accelerator [MATER] **1.** Any substance added to stucco, plaster, mortar, concrete, cement, and so on to hasten the set. **2.** In the vulcanization process, a substance, added with a curing agent, to speed processing and enhance physical characteristics of a vulcanized material. [MECH ENG] A device for varying the speed of an automotive vehicle by varying the supply of fuel.

accelerator catalyst [MATER] A catalyst that increases the rate of a chemical reaction.

accelerator jet [MECH ENG] The jet through which the fuel is injected into the incoming air in the carburetor of an automotive vehicle with rapid demand for increased power output.

accelerator linkage [MECH ENG] The linkage connecting the accelerator pedal of an automotive vehicle to the carburetor throttle valve or fuel injection control.

accelerator pedal [MECH ENG] A pedal that operates the carburetor throttle valve or fuel injection control of an automotive vehicle.

accelerator pump [MECH ENG] A small cylinder and piston controlled by the throttle of an automotive vehicle so as to provide an enriched air-fuel mixture during acceleration.

accelerogram [ENG] A record made by an accelerograph.

accelerograph [ENG] An accelerometer having provisions for recording the acceleration of a point on the earth during an earthquake or for recording any other type of acceleration.

accelerometer [ENG] An instrument which measures acceleration or gravitational force capable of imparting acceleration.

accent lighting [CIV ENG] Directional lighting which highlights an object or attracts attention to a particular area.

accentuator [MATER] A material that acts to increase the selectivity or intensity of a stain.

acceptability [ENG] State or condition of meeting minimum standards for use, as applied to methods, equipment, or consumable products.

acceptable quality level [IND ENG] The maximum percentage of defects that has been determined tolerable as a process average for a sampling plan during inspection or test of a product with respect to economic and functional requirements of the item. Abbreviated AQL.

acceptable reliability level [IND ENG] The required level of reliability for a part, system, device, and so forth; may be expressed in a variety of terms, for example, number of failures allowable in 1000 hours of operating life. Abbreviated ARL.

acceptance criteria [IND ENG] Standards of judging the acceptability of manufactured items.

acceptance number [IND ENG] The maximum allowable number of defective pieces in a sample of specified size.

acceptance test [IND ENG] A test used to determine conformance of a product to design specifications, as a basis for its acceptance.

access [CIV ENG] Freedom, ability, or the legal right to pass without obstruction from a given point on earth to some other objective, such as the sea or a public highway.

access door [BUILD] A provision for access to concealed plumbing or other equipment without disturbing the wall or fixtures.

accessory [MECH ENG] A part, subassembly, or assembly that contributes to the effectiveness of a piece of equipment without changing its basic function; may be used for testing, adjusting, calibrating, recording, or other purposes.

access road [CIV ENG] A route, usually paved, that enables vehicles to reach a designated facility expeditiously.

access tunnel [CIV ENG] A tunnel provided for an access road.

accident-cause code [IND ENG] Sponsored by the American Standards Association, the code that classifies accidents under eight defective working conditions and nine improper working practices.

accident frequency rate [IND ENG] The number of all disabling injuries per million worker-hours of exposure.

accident severity rate [IND ENG] The number of worker-days lost as a result of disabling injuries per thousand worker-hours of exposure.

accordion door [BUILD] A door that folds and unfolds like an accordion when it is opened and closed.

accordion partition [BUILD] A movable, fabric-faced partition which is fitted into an overhead track and folds like an accordion.

accordion roller conveyor [MECH ENG] A conveyor with a flexible latticed frame which permits variation in length.

accretion [CIV ENG] Artificial buildup of land due to the construction of a groin, breakwater, dam, or beach fill.

accumulated discrepancy [ENG] The sum of the separate discrepancies which occur in the various steps of making a survey.

accumulative timing [IND ENG] A time-study method that allows direct reading of the time for each element of an operation by the use of two stopwatches which operate alternately.

accumulator [AERO ENG] A device sometimes incorporated in the fuel system of a gas-turbine engine to store fuel and release it under pressure as an aid in starting. [MECH ENG] **1.** A device, such as a bag containing pressurized gas, which acts upon hydraulic fluid in a vessel, discharging it rapidly to give high hydraulic power, after which the fluid is returned to the vessel with the use of low hydraulic power. **2.** A device connected to a steam boiler to enable a uniform boiler output to meet an irregular steam demand. **3.** A chamber for storing low-side liquid refrigerant in a refrigeration system. Also known as surge drum; surge header.

accustomization [ENG] The process of learning the techniques of living with a minimum of discomfort in an extreme or new environment.

acetate film [MATER] A cellulose acetate resin sheet that is transparent, airproof, hygienic, and resistant to grease, oil, and dust; used for photographic film, magnetic tapes, and packaging.

acetylated cotton [MATER] Mildewproof cotton made by the chemical conversion of part of the raw cotton fiber to cellulose acetate.

acetylated lanolin [MATER] A clear, nongreasy liquid made by reacting lanolin with polyoxyethylenes; soluble in water, oils, and alcohol; used in cosmetics.

acetylene cutting *See* oxyacetylene cutting.

acetylene generator [ENG] A steel cylinder or tank that provides for controlled mixing of calcium carbide and water to generate acetylene.

acetylene welding *See* oxyacetylene welding.

acicular powder [MET] A metal powder whose grains are needle-shaped.

acid bottom and lining [MET] A melting furnace's inner bottom and lining composed of materials that at operating temperatures of the furnace react with the melt and slag to give an acid reaction; examples of materials are sand, siliceous rock, and silica brick.

acid brittleness [MET] Low ductility of a metal due to its absorption of hydrogen gas, which may occur during an electrolytic process or during cleaning. Also known as hydrogen embrittlement.

acid bronze [MET] A copper-tin alloy containing lead and nickel; used in pumping equipment.

acid dilution [PETRO ENG] Dilution of concentrated hydrochloric acid with water prior to oil-well acidizing.

aciding [ENG] A light etching of a building surface of cast stone.

acidizing [PETRO ENG] Well-stimulation method to increase oil production by injecting hydrochloric acid into the oil-bearing formation; the acid dissolves rock to enlarge the porous passages through which the oil must flow.

acid jetting [PETRO ENG] The jetting, from a device lowered through oil-well tubing, of an acid spray onto bottom-hole rock to clean away mud and scale interfering with oil flow.

acid lead [MET] A 99.9% pure commercial lead made by adding copper to fully refined lead.

acid lining [ENG] In steel production, a silica-brick lining used in furnaces.

acid mine drainage [MIN ENG] Drainage from bituminous coal mines containing a large concentration of acidic sulfates, especially ferrous sulfate.

acid mine water [MIN ENG] Mine water with free sulfuric acid, due to the weathering of iron pyrites.

acid number [ENG] A number derived from a standard test indicating the acid or base composition of lubricating oils; it in no way indicates the corrosive attack of the used oil in service. Also known as corrosion number.

acid open-hearth process [MET] A steelmaking process employing an open-hearth furnace lined with siliceous-type refractories.

acid pickle [MATER] Industrial waste water that is the spent liquor from a chemical process used to clean metal surfaces.

acid polishing [ENG] The use of acids to polish a glass surface.

acid process [MET] A melting process carried out in a furnace lined with acidic materials which combine readily with the oxides in the ore.

acid-proof coating [MATER] Material in liquid form suitable for application, by spraying, to the wall of projectile or bomb cavities to protect the metal from attack by explosives or other shell fillers.

acid refined oils [MATER] A class of linseed oils with the mucilaginous component removed by treatment of raw oil with sulfuric acid.

acid refractory [MATER] A refractory that is composed principally of silica and reacts at high temperatures with bases such as lime, alkalies, and basic oxides.

acid-resistant [MATER] Able to withstand chemical reaction with or degeneration by acids.

acid slag [MET] Furnace slag in which there is more silica and silicates than lime and magnesia.

acid soot [ENG] Carbon particles that have absorbed acid fumes as a by-product of combustion; hydrochloric acid absorbed on carbon particulates is frequently the cause of metal corrosion in incineration.

acid steel [MET] Steel produced in a melting furnace employing siliceous-type refractories.

acid wash [MATER] A solution of phosphoric acid applied to steel parts that removes and neutralizes the alkaline solutions used for grease removal after machining; also leaves a metallic phosphate coating which accepts paint well and, of itself, provides a degree of protection against rust.

acid water pollution [ENG] Industrial waste waters that are acidic; usually appear in effluent from the manufacture of chemicals, batteries, artificial and natural fiber, fermentation processes (beer), and mining.

acieration [MET] Electrolytic coating of a thin metal plate with iron; the iron hardens to steellike strength.

Ackerman linkage See Ackerman steering gear.

Ackerman steering gear [MECH ENG] Differential gear or linkage that turns the two steered road wheels of a self-propelled vehicle so that all wheels roll on circles with a common center. Also known as Ackerman linkage.

acme screw thread [DES ENG] A standard thread having a profile angle of 29° and a flat crest; used on power screws in such devices as automobile jacks, presses, and lead screws on lathes. Also known as acme thread.

acme thread See acme screw thread.

acoubuoy [ENG] An acoustic listening device similar to a sonobuoy, used on land to form an electronic fence that will pick up sounds of enemy movements and transmit them to orbiting aircraft or land stations.

acoustical ceiling [BUILD] A ceiling covered with or built of material with special acoustical properties.

acoustical ceiling system [BUILD] A system for the structural support of an acoustical ceiling; lighting and air diffusers may be included as part of the system.

acoustical door [BUILD] A solid door with gasketing along the top and sides, and usually an automatic door bottom, designed to reduce noise transmission.

acoustical insulation board [MATER] A porous board designed or used for acoustical applications or for sound-insulating construction.

acoustical material [MATER] Any natural or synthetic material that absorbs sound; acoustical tile is an example.

acoustical model [CIV ENG] A model used to investigate certain acoustical properties of an auditorium or room such as sound pressure distribution, sound-ray paths, and focusing effects.

acoustical plaster [MATER] A low-density sound-absorbing plaster applied as a finish coat to provide a uniform finished surface.

acoustical tile [MATER] A sound-absorptive material, usually having unit dimensions of 24 × 24 inches (approximately 61 × 61 centimeters) or less, used to cover an acoustical ceiling.

acoustical treatment [CIV ENG] That part of building planning that is designed to provide a proper acoustical environment; includes the use of acoustical material.

acoustic array [ENG ACOUS] A sound-transmitting or -receiving system whose elements are arranged to give desired directional characteristics.

acoustic center [ENG ACOUS] The center of the spherical sound waves radiating outward from an acoustic transducer.

acoustic clarifier [ENG ACOUS] System of cones loosely attached to the baffle of a loudspeaker and designed to vibrate and absorb energy during sudden loud sounds to suppress these sounds.

acoustic coupler [ENG ACOUS] A device used between the modem of a computer terminal and a standard telephone line to permit transmission of digital data in either direction without making direct connections.

acoustic delay [ENG ACOUS] A delay which is deliberately introduced in sound reproduction by having the sound travel a certain distance along a pipe before conversion into electric signals.

acoustic detection [ENG] Determination of the profile of a geologic formation, an ocean layer, or some object in the ocean by measuring the reflection of sound waves off the object.

acoustic feedback [ENG ACOUS] The reverberation of sound waves from a loudspeaker to a preceding part of an audio system, such as to the microphone, in such a manner as to reinforce, and distort, the original input. Also known as acoustic regeneration.

acoustic filter *See* filter.

acoustic generator [ENG ACOUS] A transducer which converts electrical, mechanical, or other forms of energy into sound.

acoustic hologram [ENG] The phase interference pattern, formed by acoustic beams, that is used in acoustical holography; when light is made to interact with this pattern, it forms an image of an object placed in one of the beams.

acoustic horn *See* horn.

acoustic insulation [MATER] A material used to diminish sound energy that passes through it or strikes its surface.

acoustic jamming [ENG ACOUS] The deliberate radiation or reradiation of mechanical or electroacoustic signals with the objectives of obliterating or obscuring signals which the enemy is attempting to receive and of deterring enemy weapons systems.

acoustic labyrinth [ENG ACOUS] Special baffle arrangement used with a loudspeaker to prevent cavity resonance and to reinforce bass response.

acoustic lens [MATER] Selected materials shaped to refract sound waves in accordance with the principles of geometrical optics, as is done for light. Also known as lens.

acoustic line [ENG ACOUS] The acoustic equivalent of an electrical transmission line, involving baffles, labyrinths, or resonators placed at the rear of a loudspeaker and arranged to help reproduce the very low audio frequencies.

acoustic Mach meter [AERO ENG] A device which registers data on sound propagation for the calculation of Mach number.

acoustic ocean-current meter [ENG] An instrument that measures current flow in rivers and oceans by transmitting acoustic pulses in opposite directions parallel to the flow and measuring the difference in pulse travel times between transmitter-receiver pairs.

acoustic plaster [MATER] Plaster having good acoustic absorbing properties; it contains metal which, upon contact with water, evolves gas to aerate the mass.

acoustic position reference system [ENG] An acoustic system used in offshore oil drilling to provide continuous information on ship position with respect to an ocean-floor acoustic beacon transmitting an ultrasonic signal to three hydrophones on the bottom of the drilling ship.

acoustic radar [ENG] Use of sound waves with radar techniques for remote probing of the lower atmosphere, up to heights of about 1500 meters, for measuring wind speed and direction, humidity, temperature inversions, and turbulence.

acoustic radiator [ENG ACOUS] A vibrating surface that produces sound waves, such as a loudspeaker cone or a headphone diaphragm.

acoustic radiometer [ENG] An instrument for measuring sound intensity by determining the unidirectional steady-state pressure caused by the reflection or absorption of a sound wave at a boundary.

acoustic ratio [ENG ACOUS] The ratio of the intensity of sound radiated directly from a source to the intensity of sound reverberating from the walls of an enclosure, at a given point in the enclosure.

acoustic reflex enclosure [ENG ACOUS] A loudspeaker cabinet designed with a port to allow a low-frequency contribution from the rear of the speaker cone to be radiated forward.

acoustic regeneration *See* acoustic feedback.

acoustic seal [ENG ACOUS] A joint between two parts to provide acoustical coupling with low losses of energy, such as between an earphone and the human ear.

acoustic signature [ENG] In acoustic detection, the profile characteristic of a particular object or class of objects, such as a school of fish or a specific ocean-bottom formation.

acoustic spectrograph [ENG] A spectrograph used with sound waves of various frequencies to study the transmission and reflection properties of ocean thermal layers and marine life.

acoustic spectrometer [ENG ACOUS] An instrument that measures the intensities of the various frequency components of a complex sound wave. Also known as audio spectrometer.

acoustic strain gage [ENG] An instrument used for measuring structural strains; consists of a length of fine wire mounted so its tension varies with strain; the wire is plucked with an electromagnetic device, and the resulting frequency of vibration is measured to determine the amount of strain.

acoustic theodolite [ENG] An instrument that uses sound waves to provide a continuous vertical profile of ocean currents at a specific location.

acoustic theory [AERO ENG] The linearized small-disturbance theory used to predict the approximate airflow past an airfoil when the disturbance velocities caused by the flow are small compared to the flight speed and to the speed of sound.

acoustic tile [MATER] A thin, often decorative tile with sound-absorbing properties, used to cover ceilings and walls.

acoustic transducer [ENG ACOUS] A device that converts acoustic energy to electrical or mechanical energy, such as a microphone or phonograph pickup.

acoustic transformer [ENG ACOUS] A device, such as a horn or megaphone, for increasing the efficiency of sound radiation.

acoustic treatment [BUILD] The use of sound-absorbing materials to give a room a desired degree of freedom from echo and reverberation.

acoustic well logging [ENG] A ground exploration method that uses a high-energy sound source and a receiver, both underground.

acoustoelectronics [ENG ACOUS] The branch of electronics that involves use of acoustic waves at microwave frequencies (above 500 megahertz), traveling on or in piezoelectric or other solid substrates. Also known as pretersonics.

acoustooptical material [MATER] A material in which the refractive index or some other optical property can be changed by an acoustic wave.

acquisition [ENG] The process of pointing an antenna or a telescope so that it is properly oriented to allow gathering of tracking or telemetry data from a satellite or space probe.

acquisition and tracking radar [ENG] A radar set capable of locking onto a received signal and tracking the object emitting the signal; the radar may be airborne or on the ground.

acrometer [ENG] An instrument to measure the density of oils.

acrylic syrup [MATER] Lucite in a liquid form; used as a low-pressure laminating resin; produces stiff, strong, tough laminates that can be adapted to bright or translucent colors.

acrylonitrile-butadiene rubber *See* nitrile rubber.

acrylonitrile rubber *See* nitrile rubber.

actinogram [ENG] The record of heat from a source, such as the sun, as detected by a recording actinometer.

actinograph [ENG] A recording actinometer.

actinometer [ENG] Any instrument used to measure the intensity of radiant energy, particularly that of the sun.

activate [ENG] To set up conditions so that the object will function as designed or required.

activated alumina [MATER] Highly porous, granular aluminum oxide that preferentially absorbs liquids from gases and vapors, and moisture from some liquids; also used as a catalyst or catalyst carrier, as an absorbent to remove fluorides from drinking water, and in chromatography.

activated bauxite *See* filter bauxite.

activated carbon [MATER] A powdered, granular, or pelleted form of amorphous carbon characterized by very large surface area per unit volume because of an enormous number of fine pores. Also known as activated charcoal.

activated charcoal *See* activated carbon.

activated clay [MATER] Bentonite, or other clay, treated with acid to enhance its ability to absorb or bleach.

activated coal plough [MIN ENG] A type of power-operated cutting blade used for coal seams too hard to be sheared by a normal blade.

activated rosin flux [MATER] Soldering flux containing activating agents which promote wetting by the solder. [MET] *See* noncorrosive flux.

activated sintering [MET] Sintering of a metal powder compact in contact with a gaseous atmosphere which reacts with the metal surfaces and enhances the joining of metal particles.

activated sludge [CIV ENG] A semiliquid mass removed from the liquid flow of sewage and subjected to aeration and aerobic microbial action; the end product is dark to golden brown, partially decomposed, granular, and flocculent, and has an earthy odor when fresh.

activated-sludge effluent [CIV ENG] The liquid from the activated-sludge treatment that is further processed by chlorination or by oxidation.

activated-sludge process [CIV ENG] A sewage treatment process in which the sludge in the secondary stage is put into aeration tanks to facilitate aerobic decomposition by microorganisms; the sludge and supernatant liquor are separated in a settling tank; the supernatant liquor or effluent is further treated by chlorination or oxidation.

activating reagent [MATER] Material added to another material or mixture so that a physical or chemical change will take place more rapidly or completely.

activation [MET] **1.** A process of facilitating the separation and collection of ore powders by the use of substances which change the response of the particle surfaces to a flotation fluid. **2.** A process that increases the rate of pressing and heating a metal powder into cohesion.

active communications satellite [AERO ENG] Satellite which receives, regenerates, and retransmits signals between stations.

active controls technology [AERO ENG] The development of special forms of augmentation systems to stabilize airplane configurations and to limit, or tailor, the design loads that the airplane structure must support.

active detection system [ENG] A guidance system which emits energy as a means of detection; for example, sonar and radar.

active earth pressure [CIV ENG] The horizontal pressure that an earth mass exerts on a wall.

active entry [MIN ENG] An entry in which coal is being mined from a portion or from connected sections.

active infrared detection system [ENG] An infrared detection system in which a beam of infrared rays is transmitted toward possible targets, and rays reflected from a target are detected.

active leaf [BUILD] In a door with two leaves, the leaf which carries the latching or locking mechanism. Also known as active door.

active satellite [AERO ENG] A satellite which transmits a signal.

active sludge [CIV ENG] A sludge rich in destructive bacteria used to break down raw sewage.

active sonar [ENG] A system consisting of one or more transducers to send and receive sound, equipment for the generation and detection of the electrical impulses to and from the transducer, and a display or recorder system for the observation of the received signals.

active system [ENG] In radio and radar, a system that requires transmitting equipment, such as a beacon or transponder.

active workings [MIN ENG] All places in a mine that are ventilated and inspected regularly.

activity [SYS ENG] The representation in a PERT or critical-path-method network of a task that takes up both time and resources and whose performance is necessary for the system to move from one event to the next.

activity chart [IND ENG] A tabular presentation of a series of operations of a process plotted against a time scale.

activity duration [SYS ENG] In critical-path-method terminology, the estimated amount of time required to complete an activity.

activity sampling *See* work sampling.

actual cost [IND ENG] Cost determined by an allocation of cost factors recorded during production.

actual exhaust velocity [AERO ENG] **1.** The real velocity of the exhaust gas leaving a nozzle as determined by accurately measuring at a specified point in the nozzle exit plane. **2.** The velocity obtained when the kinetic energy of the gas flow produces actual thrust.

actual horsepower *See* actual power.

actual power [MECH ENG] The power delivered at the output shaft of a source of power. Also known as actual horsepower.

actual time [IND ENG] Time taken by a worker to perform a given task.

actuate [MECH ENG] To put into motion or mechanical action, as by an actuator.

actuated roller switch [MECH ENG] A centrifugal sequence-control switch that is placed in contact with a belt conveyor, immediately preceding the conveyor it controls.

actuator [ENG ACOUS] An auxiliary external electrode used to apply a known electrostatic force to the diaphragm of a microphone for calibration purposes. Also known as electrostatic actuator. [MECH ENG] A device that produces mechanical force by means of pressurized fluid.

adamantine drill [MECH ENG] A core drill with hardened steel shot pellets that revolve under the rim of the rotating tube; employed in rotary drilling in very hard ground.

ada mud [ENG] A conditioning material added to drilling mud to obtain satisfactory cores and samples of formations.

adapter [ENG] A device used to make electrical or mechanical connections between items not originally intended for use together. [MET] A connecting piece, usually made of fireclay, between a horizontal zinc retort and the condenser in which the molten zinc collects.

adapter skirt [AERO ENG] A flange or extension of a space vehicle that provides a ready means for fitting some object to a stage or section.

adaptometer [ENG] An instrument that measures the lowest brightness of an extended area that can barely be detected by the eye.

addendum [DES ENG] The radial distance between two concentric circles on a gear, one being that whose radius extends to the top of a gear tooth (addendum circle) and the other being that which will roll without slipping on a circle on a mating gear (pitch line).

addendum circle [DES ENG] The circle on a gear passing through the tops of the teeth.

adding tape [ENG] A surveyor's tape that is calibrated from 0 to 100 by full feet (or meters) in one direction, and has 1 additional foot (or meter) beyond the zero end which is subdivided in tenths or hundredths.

additive [MATER] **1.** A substance added to another to improve, strengthen, or otherwise alter it; for example, tetraethyllead added to gasoline to prevent engine knock. **2.** See admixture.

additive level [MATER] The total percentage of all the additives in an oil sample.

adherend [MATER] A body attracted to another by means of an adhesive substance.

adhesion [ENG] Intimate sticking together of metal surfaces under compressive stresses by formation of metallic bonds.

adhesive [MATER] A substance used to bond two or more solids so that they act or can be used as a single piece; examples are resins, formaldehydes, glue, paste, cement, putty, and polyvinyl resin emulsions.

adhesive bonding [ENG] The fastening together of two or more solids by the use of glue, cement, or other adhesive.

adhesive strength [ENG] The strength of an adhesive bond, usually measured as a force required to separate two objects of standard bonded area, by either shear or tensile stress.

adhesive tape [MATER] Tape coated with a substance that binds or sticks to a surface.

adiabatic curing [ENG] The curing of concrete or mortar under conditions in which there is no loss or gain of heat.

adiabatic engine [MECH ENG] A heat engine or thermodynamic system in which there is no gain or loss of heat.

adibatic extrusion [ENG] Forming plastic objects by energy produced by driving the plastic mass through an extruder without heat flow.

adit [CIV ENG] An access tunnel used for excavation of the main tunnel. [MIN ENG] A nearly horizontal tunnel for access, drainage, or ventilation of a mine. Also known as side drift.

adjab butter [MATER] A semisolid oil extracted from the nuts of the tree *Bassia toxisperma;* used to make soap. Also known as djave butter.

adjustable base anchor [BUILD] An item which holds a doorframe above a finished floor.

adjustable parallels [ENG] Wedge-shaped iron bars placed with the thin end of one on the thick end of the other, so that the top face of the upper and the bottom face of the lower remain parallel, but the distance between the two faces is adjustable; the bars can be locked in position by a screw to prevent shifting.

adjustable square [ENG] A try square with an arm that is at right angles to the ruler; the position of the arm can be changed to form an L or a T. Also known as double square.

adjustable wrench [ENG] A wrench with one jaw which is fixed and another which is adjustable; the size is adjusted by a knurled screw.

adjutage [ENG] A tube attached to a container of liquid at an orifice to facilitate or regulate outflow.

admiralty brass [MET] An alloy containing 71% copper, 28% zinc, and 0.75–1.0% tin for additional corrosion resistance.

admiralty coal [MIN ENG] A high-quality, smokeless steam coal.

admixture [MATER] A material (other than aggregate, cement, or water) added in small quantities to concrete to produce some desired change in properties. Also known as additive.

adobe brick [MATER] An earth or clay, straw brick of large but varying dimensions, roughly molded and sun-dried.

adobe construction [BUILD] Wall construction with sun-dried blocks of adobe soil.

ADP *See* automatic data processing.

adsorption gasoline [MATER] Gasoline extracted from natural gas or refinery gas.

adsorption system [MECH ENG] A device that dehumidifies air by bringing it into contact with a solid adsorbing substance.

ad valorem tax [PETRO ENG] Property tax for oil-producing properties, assessed at a flat rate for each net barrel of oil produced.

advance [MECH ENG] To effect the earlier occurrence of an event, for example, spark advance or injection advance.

advanced gallery [MIN ENG] A small heading driven in advance of the main tunnel in tunnel excavation.

advance overburden [MIN ENG] Overburden in excess of the average overburden-to-ore ratio that must be removed in open-cut mining.

advance signal [CIV ENG] A signal in a block system up to which a train may proceed within a block that is not completely cleared.

advance slope grouting [ENG] A grouting technique in which the front of the mass of grout is forced to move horizontally through preplaced aggregate.

advance slope method [ENG] A method of concrete placement in which the face of the fresh concrete, which is not vertical, moves forward as the concrete is placed.

advance stripping [MIN ENG] The removal of barren or sub-ore-grade earthy or rock materials to expose the minable grade of ore.

advance wave [MIN ENG] The air pressure wave preceding the flame in a coal dust explosion.

advancing [MIN ENG] Mining outward from the shaft toward the boundary.

advancing longwall [MIN ENG] Mining coal outward from the shaft pillar, with roadways maintained through the worked-out portion of the mine.

adz [DES ENG] A cutting tool with a thin arched blade, sharpened on the concave side, at right angles on the handle; used for rough dressing of timber.

adz block [MECH ENG] The part of a machine for wood planing that carries the cutters.

Ae [MET] The temperature of equilibrium for corresponding phase change Ac; Ae_{cm}, Ae_3, and Ae_4 are similarly related to Ac_{cm}, Ac_3, and Ac_4.

Ae₁ [MET] The temperature, attained without thermal lag, at which cementite-austenite conversion takes place in a hypoeutectoid steel or ferrite-austenite conversion takes place in a hypereutectoid steel.

aerated concrete [MATER] Concrete made by adding substances which will liberate gases by chemical reaction; entrapped gases reduce its density and enhance insulation properties.

aerated flow [ENG] Flowing liquid in which gas is dispersed as fine bubbles throughout the liquid.

aeration [ENG] **1.** Exposing to the action of air. **2.** Causing air to bubble through. **3.** Introducing air into a solution by spraying, stirring, or similar method. **4.** Supplying or infusing with air, as in sand or soil. [MIN ENG] The introduction of air into the pulp in a flotation cell to form air bubbles.

aeration tank [ENG] A fluid-holding tank with provisions to aerate its contents by bubbling air or another gas through the liquid or by spraying the liquid into the air.

aerator [DES ENG] A tool having a roller equipped with hollow fins; used to remove cores of soil from turf. [ENG] **1.** One who aerates. **2.** Equipment used for aeration. **3.** Any device for supplying air or gas under pressure, as for fumigating, welding, or ventilating. [MECH ENG] Equipment used to inject compressed air into sewage in the treatment process. [MET] A device which decreases the density of sand by mixing it with air, thus facilitating the movement of sand particles in packing.

aerial cableway *See* aerial tramway.

aerial photogrammetry [ENG] Use of aerial photographs to make accurate measurements in surveying and mapmaking.

aerial photography [ENG] The making of photographs of the ground surface from an aircraft, spacecraft, or rocket. Also known as aerophotography.

aerial photoreconnaissance [ENG] The obtaining of information by air photography; the three types are strategic, tactical, and survey-cartographic photoreconnaissance. Also known as aerial photographic reconnaissance.

aerial reconnaissance [ENG] The collection of information by visual, electronic, or photographic means while aloft.

aerial ropeway *See* aerial tramway.

aerial sound ranging [AERO ENG] The process of locating an aircraft by means of the sounds it emits.

aerial spud [MECH ENG] A cable for moving and anchoring a dredge.

aerial survey [ENG] A survey utilizing photographic, electronic, or other data obtained from an airborne station. Also known as aerosurvey; air survey.

aerial tramway [MECH ENG] A system for transporting bulk materials that consists of one or more cables supported by steel towers and is capable of carrying a traveling carriage from which loaded buckets can be lowered or raised. Also known as aerial cableway; aerial ropeway.

aerobic-anaerobic interface [CIV ENG] That point in bacterial action in the body of a sewage sludge or compost heap where both aerobic and anaerobic microorganisms participate, and the decomposition of the material goes no further.

aerobic-anaerobic lagoon [CIV ENG] A pond in which the solids from a sewage plant are placed in the lower layer; the solids are partially decomposed by anaerobic bacteria, while air or oxygen is bubbled through the upper layer to create an aerobic condition.

aerobic lagoon [CIV ENG] An aerated pond in which sewage solids are placed, and are decomposed by aerobic bacteria.

aerochlorination [CIV ENG] Treatment of sewage with compressed air and chlorine gas to remove fatty substances.

aerodrome *See* airport.

aeroduct [AERO ENG] A ramjet type of engine designed to scoop up ions and electrons freely available in the outer reaches of the atmosphere or in the atmospheres of other spatial bodies and, by a metachemical process within the engine duct, to expel particles derived from the ions and electrons as a propulsive jetstream.

aerodynamic balance [ENG] A balance used for the measurement of the forces exerted on the surfaces of instruments exposed to flowing air; frequently used in tests made on models in wind tunnels.

aerodynamic center [AERO ENG] A point on a cross section of a wing or rotor blade through which the forces of drag and lift are acting and about which the pitching moment coefficient is practically constant.

aerodynamic characteristics [AERO ENG] The performance of a given airfoil profile as related to lift and drag, to angle of attack, and to velocity, density, viscosity, compressibility, and so on.

aerodynamic chord [AERO ENG] A straight line intersecting or touching an airfoil profile at two points; specifically, that part of such a line between two points of intersection.

aerodynamic configuration [AERO ENG] The form of an aircraft, incorporating desirable aerodynamic qualities.

aerodynamic control [AERO ENG] A control surface whose use causes local aerodynamic forces.

aerodynamic instability [AERO ENG] An unstable state caused by oscillations of a structure that are generated by spontaneous and more or less periodic fluctuations in the flow, particularly in the wake of the structure.

aerodynamic missile [AERO ENG] A missile with surfaces which produce lift during flight.

aerodynamic moment [AERO ENG] The torque about the center of gravity of a missile or projectile moving through the atmosphere, produced by any aerodynamic force which does not act through the center of gravity.

aerodynamic stability [AERO ENG] The property of a body in the air, such as an aircraft or rocket, to maintain its attitude, or to resist displacement, and if displaced, to develop aerodynamic forces and moments tending to restore the original condition.

aerodynamic vehicle [AERO ENG] A device, such as an airplane or glider, capable of flight only within a sensible atmosphere and relying on aerodynamic forces to maintain flight.

aerodyne [AERO ENG] Any heavier-than-air craft that derives its lift in flight chiefly from aerodynamic forces, such as the conventional airplane, glider, or helicopter.

aerofall mill [MECH ENG] A grinding mill of large diameter with either lumps of ore, pebbles, or steel balls as crushing bodies; the dry load is airswept to remove mesh material.

aerofilter [CIV ENG] A filter bed for sewage treatment consisting of coarse material and operated at high speed, often with recirculation.

aerofoil *See* airfoil.

aerograph [ENG] Any self-recording instrument carried aloft by any means to obtain meteorological data.

aerometeorograph [ENG] A self-recording instrument used on aircraft for the simultaneous recording of atmospheric pressure, temperature, and humidity.

aerometer [ENG] An instrument to ascertain the weight or density of air or other gases.

aeromotor [AERO ENG] An engine designed to provide motive power for an aircraft.

aeronaut [AERO ENG] A person who operates or travels in an airship or balloon.

aeronautical engineering [AERO ENG] The branch of engineering concerned primarily with the design and construction of aircraft structures and power units, and with the special problems of flight in the atmosphere.

aerophysics [AERO ENG] The physics dealing with the design, construction, and operation of aerodynamic devices.

aeropulse engine See pulsejet engine.

aeroservoelasticity [AERO ENG] The study of the interaction of automatic flight controls on aircraft and aeroelastic response and stability.

aerosol generator [MECH ENG] A mechanical means of producing a system of dispersed phase and dispersing medium, that is, an aerosol.

aerosol propellant [MATER] Compressed gas or vapor in a container which, upon release of pressure and expansion through a valve, carries another substance from the container; used for cosmetics, household cleaners, and so on; examples are butanes, propane, nitrogen, fluorocarbons, and carbon dioxide.

aerospace engineering [ENG] Engineering pertaining to the design and construction of aircraft and space vehicles and of power units, and to the special problems of flight in both the earth's atmosphere and space, as in the flight of air vehicles and in the launching, guidance, and control of missiles, earth satellites, and space vehicles and probes.

aerospace ground equipment [AERO ENG] Support equipment for air and space vehicles. Abbreviated AGE.

aerospace industry [ENG] Industry concerned with the use of vehicles in both the earth's atmosphere and space.

aerospace vehicle [AERO ENG] A vehicle capable of flight both within and outside the sensible atmosphere.

aerostat [AERO ENG] Any aircraft that derives its buoyancy or lift from a lighter-than-air gas contained within its envelope or one of its compartments; for example, ships and balloons.

aerostatic balance [ENG] An instrument for weighing air.

aerosurvey See aerial survey.

aerozine [MATER] Hydrazine mixed 50:50 with dimethylhydrazine; the most-used bipropellant rocket fuel.

aesthesiometer See esthesiometer.

affreightment [IND ENG] The lease of a vessel for the transportation of goods.

A frame [BUILD] A dwelling whose main frames are in the shape of the letter A. [ENG] Two poles supported in an upright position by braces or guys and used for lifting equipment. Also known as double mast.

African greenheart [MATER] The yellowish-brown wood of *Piptadena africana;* used for shipbuilding and dock timbers. Also known as dahoma.

afterblow [MET] A blow in a Bessemer process, occurring after the flame for the removal of carbon has dropped, to remove phosphorus.

afterbody [AERO ENG] **1.** A companion body that trails a satellite. **2.** A section or piece of a rocket or spacecraft that enters the atmosphere unprotected behind the nose cone or other body that is protected for entry. **3.** The afterpart of a vehicle.

afterbreak [MIN ENG] A phenomenon occurring during mine subsidence; material slides inward after the main break, assumed at right angles to the plane of the seam.

afterburner [AERO ENG] A device for augmenting the thrust of a jet engine by burning additional fuel in the uncombined oxygen in the gases from the turbine.

afterburning [AERO ENG] The function of an afterburner.

afterburnt [AERO ENG] Descriptive of the condition following the complete transformation of the solid propellant to gaseous form.

afterburst [MIN ENG] **1.** A tremor that sometimes follows a rock blast as the ground adjusts to the new stress distribution. **2.** A sudden collapse of rock in an underground mine subsequent to a rock burst.

afterchromed dye [MATER] A dye that is improved in color quality or fastness after the textile is dyed by treatment with sodium dichromate, copper sulfate, or similar materials.

aftercondenser [MECH ENG] A condenser in the second stage of a two-stage ejector; used in steam power plants, refrigeration systems, and air conditioning systems.

aftercooler [MECH ENG] A heat exchanger which cools air that has been compressed; used on turbocharged engines.

aftercooling [MECH ENG] The cooling of a gas after its compression.

afterdamp [MIN ENG] The mixture of gases which remains in a mine after a mine fire or an explosion of firedamp.

afterfilter [MECH ENG] In an air-conditioning system, a high-efficiency filter located near a terminal unit. Also known as final filter.

afterflaming [AERO ENG] With liquid- or solid-propellant rocket thrust chambers, a characteristic low-grade combustion that takes place in the thrust chamber assembly and around its nozzle exit after the main propellant flow has been stopped.

aftergases [MIN ENG] Gases produced by mine explosions or mine fires.

aftertack [MATER] Of a paint film, tackiness or stickiness which remains over a long period of time. Also known as residual tack.

agar [MATER] A gelatinous product extracted from certain red algae and used chiefly as a gelling agent in culture media.

agar attar See oriental linaloe.

agate glass [MATER] Multicolor glass made by blending glasses of two or more colors or by rolling transparent glass into powdered glass of various colors.

AGE See aerospace ground equipment.

age hardening [MET] Increasing the hardness of an alloy by a relatively low-temperature heat treatment that causes precipitation of components or phases of the alloy from the supersaturated solid solution.

agel fiber [MATER] Fiber from the leaves and stem of the gebang palm; used for rope, sailcloth, and fishnets.

Agena rocket [AERO ENG] A liquid-fuel, upper-stage rocket usually used with a first-stage Atlas or Thor booster in certain space satellite projects.

age softening [MET] The loss of strength and hardness which takes place at room temperature in some alloys because of the spontaneous decrease of residual stresses in the strain-hardened structure.

agger [CIV ENG] A material used for road fill over low ground.

agglomeration [MET] Conversion of small pieces of low-grade iron ore into larger lumps by application of heat.

agglomeration test [MIN ENG] A test of a button of coke whose results are used as a measure of the binding qualities of the coal.

aggregate [MATER] The natural sands, gravels, and crushed stone used for mixing with cementing material in making mortars and concretes.

aggregate bin [ENG] A structure designed for storing and dispensing dry granular construction materials such as sand, crushed stone, and gravel; usually has a hopperlike bottom that funnels the material to a gate under the structure.

aggregate interlock [ENG] The projection of aggregate particles or portions thereof from one side of a joint or crack in concrete into recesses in the other side so as to effect load transfer in compression and shear, and to maintain mutual alignment.

aging [ENG] **1.** The changing of the characteristics of a device due to its use. **2.** Operation of a product before shipment to stabilize characteristics or detect early failures. [MATER] **1.** Change in the properties of any substance with time. **2.** Change occurring in powders or slips with the passage of time. **3.** Curing of ceramic materials, such as clays and glazes, by a definite period of time under controlled storage conditions. [MET] **1.** Change in properties of an alloy or metal which generally proceeds slowly at room temperatures and faster at elevated temperatures. **2.** Strain relief, occurring through long storage outdoors under varying temperatures, of iron castings intended for use as toolroom plates or lathe-bed supports. **3.** A second heat treatment of an alloy at a lower temperature, causing precipitation of the unstable phase and increasing hardness, strength, and electrical conductivity.

agitating speed [MECH ENG] The rate of rotation of the drum or blades of a truck mixer or other device used for agitation of mixed concrete.

agitating truck [MECH ENG] A vehicle carrying a drum or agitator body, in which freshly mixed concrete can be conveyed from the point of mixing to that of placing, the drum being rotated continuously to agitate the contents.

agitator [MECH ENG] A device for keeping liquids and solids in liquids in motion by mixing, stirring, or shaking.

agitator body [MECH ENG] A truck-mounted drum for transporting freshly mixed concrete; rotation of internal paddles or of the drum prevents the setting of the mixture prior to delivery.

agricultural chemicals [MATER] Fertilizers, soil conditioners, fungicides, insecticides, weed killers, and other chemicals used to increase farm crop productivity and quality.

agricultural lime [MATER] A hydrated lime which is used to condition soil.

agricultural pipe drain [CIV ENG] A system of porous or perforated pipes laid in a trench filled with gravel or the like; used for draining subsoil.

aided tracking [ENG] A system of radar-tracking a target signal in bearing, elevation, or range, or any combination of these variables, in which the rate of motion of the tracking equipment is machine-controlled in collaboration with an operator so as to minimize tracking error.

aided-tracking mechanism [ENG] A device consisting of a motor and variable-speed drive which provides a means of setting a desired tracking rate into a director or other fire-control instrument, so that the process of tracking is carried out automatically at the set rate until it is changed manually.

aided-tracking ratio [ENG] The ratio between the constant velocity of the aided-tracking mechanism and the velocity of the moving target.

aiguille [ENG] A slender form of drill used for boring or drilling a blasthole in rock.

aileron [AERO ENG] The hinged rear portion of an aircraft wing moved differentially on each side of the aircraft to obtain lateral or roll control moments.

aiming circle [ENG] An instrument for measuring angles in azimuth and elevation in connection with artillery firing and general topographic work; equipped with fine and coarse azimuth micrometers and a magnetic needle.

air-acetylene welding [MET] A gas-welding process in which the heat is obtained from the combustion of acetylene and air.

air adit [MIN ENG] An adit driven for the purpose of ventilating a mine.

air-arc furnace [ENG] An arc furnace designed to power wind tunnels, the air being superheated to 20,000 K and expanded to emerge at supersonic speeds.

air-assist forming [ENG] A plastics thermoforming method in which air pressure is used to partially preform a sheet before it enters the mold.

air atomizing oil burner [ENG] An oil burner in which a stream of fuel oil is broken into very fine droplets through the action of compressed air.

air barrage [MIN ENG] An airtight wall dividing a ventilation gallery in a mine into two parts, so that air is led in through one part and out through the other.

air base [AERO ENG] **1.** In the U.S. Air Force, an establishment, comprising an airfield, its installations, facilities, personnel, and activities, for the flight operation, maintenance, and supply of aircraft and air organizations. **2.** A similar establishment belonging to any other air force. **3.** In a restricted sense, only the physical installation.

air belt [MECH ENG] The chamber which equalizes the pressure that is blasted into the cupola at the tuyeres.

air-bend die [MET] A device for forming metals in which only the two edges of the lower section are in contact with the metal.

air bind [ENG] The presence of air in a conduit or pump which impedes passage of the liquid.

airblast [MIN ENG] A disturbance in underground workings accompanied by a strong rush of air.

airblasting [ENG] A blasting technique in which air at very high pressure is piped to a steel shell in a shot hole and discharged. Also known as air breaking.

air-blown asphalt [MATER] A bituminous product made by reacting the residual oil of petroleum distillation with air at 400–600°F (204–316°C).

airboat *See* seaplane.

airborne [AERO ENG] Of equipment and material, carried or transported by aircraft.

airborne collision warning system [ENG] A system such as a radar set or radio receiver carried by an aircraft to warn of the danger of possible collision.

airborne detector [ENG] A device, transported by an aircraft, whose function is to locate or identify an air or surface object.

airborne electronic survey control [ENG] The airborne portion of very accurate positioning systems used in controlling surveys from aircraft.

airborne intercept radar [ENG] Airborne radar used to track and "lock on" to another aircraft to be intercepted or followed.

airborne magnetometer [ENG] An airborne instrument used to measure the magnetic field of the earth.

airborne profile recorder [ENG] An electronic instrument that emits a pulsed-type radar signal from an aircraft to measure vertical distances between the aircraft and the earth's surface. Abbreviated APR. Also known as terrain profile recorder (TPR).

airborne radar [ENG] Radar equipment carried by aircraft to assist in navigation by pilotage, to determine drift, and to locate weather disturbances; a very important use is locating other aircraft either for avoidance or attack.

airborne waste [ENG] Vapors, gases, or particulates introduced into the atmosphere by evaporation, chemical, or combustion processes; a frequent cause of smog and an irritant to eyes and breathing passages.

air-bound [ENG] Of a pipe or apparatus, containing a pocket of air that prevents or reduces the desired liquid flow.

air brake [MECH ENG] An energy-conversion mechanism activated by air pressure and used to retard, stop, or hold a vehicle or, generally, any moving element.

air breaking *See* airblasting.

air breakup [AERO ENG] The breakup of a test reentry body within the atmosphere.

air-breathing [MECH ENG] Of an engine or aerodynamic vehicle, required to take in air for the purpose of combustion.

air brick [MATER] A brick or brick-sized metal box that is hollow or perforated and used for ventilation.

air cap [MECH ENG] A device used in thermal spraying which directs the air pattern for purposes of atomization.

air carbon arc cutting [MET] A carbon arc cutting process in which the molten metal in the cut is removed by an airblast.

air casing [ENG] A metal casing surrounding a pipe or reservoir and having a space between to prevent heat transmission.

air chamber [MECH ENG] A pressure vessel, partially filled with air, for converting pulsating flow to steady flow of water in a pipeline, as with a reciprocating pump.

air changes [CIV ENG] A measure of the volume of air supplied to or exhausted from a building or room; expressed as the ratio of the volume of air exchanged to the volume of the building or room.

air check [ENG ACOUS] A recording made of a live radio broadcast for filing purposes at the broadcasting facility.

air classifier [MECH ENG] A device to separate particles by size through the action of a stream of air.

air cleaner [ENG] Any of various devices designed to remove particles and aerosols of specific sizes from air; examples are screens, settling chambers, filters, wet collectors, and electrostatic precipitators.

air compressor [MECH ENG] A machine that increases the pressure of air by increasing its density and delivering the fluid against the connected system resistance on the discharge side.

air-compressor unloader [MECH ENG] A device for control of air volume flowing through an air compressor.

air-compressor valve [MECH ENG] A device for controlling the flow into or out of the cylinder of a compressor.

air condenser [MECH ENG] **1.** A steam condenser in which the heat exchange occurs through metal walls separating the steam from cooling air. Also known as air-cooled condenser. **2.** A device that removes vapors, such as of oil or water, from the airstream in a compressed-air line.

air conditioner [MECH ENG] A mechanism primarily for comfort cooling that lowers the temperature and reduces the humidity of air in buildings.

air conditioning [MECH ENG] The maintenance of certain aspects of the environment within a defined space to facilitate the function of that space; aspects controlled include air temperature and motion, radiant heat level, moisture, and concentration of pollutants such as dust, microorganisms, and gases. Also known as climate control.

air content [MATER] The volume of air voids in cement paste, mortar, or concrete, exclusive of pore space in aggregate particles; usually expressed as a percentage of total volume of the mixture.

air conveyor *See* pneumatic conveyor.

air-cooled blast-furnace slag [MET] The material resulting from solidification of molten blast-furnace slag under atmospheric conditions.

air-cooled condenser *See* air condenser.

air-cooled engine [MECH ENG] An engine cooled directly by a stream of air without the interposition of a liquid medium.

air-cooled heat exchanger [MECH ENG] A finned-tube (extended-surface) heat exchanger with hot fluids inside the tubes, and cooling air that is fan-blown (forced draft) or fan-pulled (induced draft) across the tube bank.

air cooling [MECH ENG] Lowering of air temperature for comfort, process control, or food preservation.

air course *See* airway.

aircraft [AERO ENG] Any structure, machine, or contrivance, especially a vehicle, designed to be supported by the air, either by the dynamic action of the air upon the surfaces of the structure or object or by its own buoyancy. Also known as air vehicle.

aircraft axes *See* axes of an aircraft.

aircraft bonding [AERO ENG] Electrically connecting together all of the metal structure of the aircraft, including the engine and metal covering on the wiring.

aircraft detection [ENG] The sensing and discovery of the presence of aircraft; major techniques include radar, acoustical, and optical methods.

aircraft engine [AERO ENG] A component of an aircraft that develops either shaft horsepower or thrust and incorporates design features advantageous for aircraft propulsion.

aircraft fuel [MATER] The material used as a source of energy by an aircraft engine to provide propulsion.

aircraft impactor [ENG] An instrument carried by an aircraft for the purpose of obtaining samples of airborne particles.

aircraft instrumentation [AERO ENG] Electronic, gyroscopic, and other instruments for detecting, measuring, recording, telemetering, processing, or analyzing different values or quantities in the flight of an aircraft.

aircraft instrument panel [AERO ENG] A coordinated instrument display arranged to provide the pilot and flight crew with information about the aircraft's speed, altitude, attitude, heading, and condition; also advises the pilot of the aircraft's response to his control efforts.

aircraft propeller [AERO ENG] A hub-and-multiblade device for transforming the rotational power of an aircraft engine into thrust power for the purpose of moving an aircraft through the air.

aircraft propulsion [AERO ENG] The means, other than gliding, whereby an aircraft moves through the air; effected by the rearward acceleration of matter through the use of a jet engine or by the reactive thrust of air on a propeller.

aircraft pylon [AERO ENG] A suspension device externally installed under the wing or fuselage of an aircraft; it is aerodynamically designed to fit the configuration of specific aircraft so as to create the least amount of drag; it provides a means of attaching fuel tanks, bombs, rockets, torpedoes, rocket motors, or machine-gun pods.

aircraft testing [AERO ENG] The subjecting of an aircraft or its components to simulated or actual flight conditions while measuring and recording pertinent physical phenomena that indicate operating characteristics.

air crossing [MIN ENG] A mine passage in which two airways cross each other.

air current [MIN ENG] The flow of air ventilating the workings of a mine. Also known as airflow.

air curtain [MECH ENG] A stream of high-velocity temperature-controlled air which is directed downward across an opening; it excludes insects, exterior drafts, and so forth, prevents the transfer of heat across it, and permits air-conditioning of a space with an open entrance.

air cushion [MECH ENG] A mechanical device using trapped air to arrest motion without shock.

air-cushion vehicle [MECH ENG] A transportation device supported by low-pressure, low-velocity air capable of traveling equally well over water, ice, marsh, or relatively level land. Also known as ground-effect machine (GEM); hovercraft.

air cycle [MECH ENG] A refrigeration cycle characterized by the working fluid, air, remaining as a gas throughout the cycle rather than being condensed to a liquid; used primarily in airplane air conditioning.

air cylinder [MECH ENG] A cylinder in which air is compressed by a piston, compressed air is stored, or air drives a piston.

air diffuser [BUILD] An air distribution outlet, usually located in the ceiling and consisting of deflecting vanes discharging supply air in various directions and planes, and arranged to promote mixing of the supplied air with the air already in the room.

air-distributing acoustical ceiling [BUILD] A suspended acoustical ceiling in which the board or tile is provided with small, evenly distributed mechanical perforations; designed to provide a desired flow of air from a pressurized plenum above.

air door [MIN ENG] A door placed in a mine roadway to prevent the passage of air.

air drain [CIV ENG] An empty space left around the external foundation wall of a building to prevent the earth from lying against it and causing dampness.

airdraulic [MECH ENG] Combining pneumatic and hydraulic action for operation.

air-dried lumber [MATER] Wood dried by exposure to air under natural conditions; usually has a moisture content not greater than 24%. Also known as air-seasoned lumber; natural-seasoned lumber.

air-dried strength [MET] Tenacity of a sand mixture after a core or mold has been dried in air without application of heat.

air drift [MIN ENG] A roadway, generally inclined, driven in stone for ventilation purposes.

air drill [MECH ENG] A drill powered by compressed air.

air drying [ENG] Removing moisture from a material by exposure to air to the extent that no further moisture is released on contact with air; important in lumber manufacture.

air duct *See* airflow duct.

air ejector [MECH ENG] A device that uses a fluid jet to remove air or other gases, as from a steam condenser.

air eliminator [MECH ENG] In a piping system, a device used to remove air from water, steam, or refrigerant.

air endway [MIN ENG] A narrow roadway driven in a coal seam parallel and close to a winning headway for ventilation.

air engine [MECH ENG] An engine in which compressed air is the actuating fluid.

air-entrained cement [MATER] Cement with improved qualities due to the introduction of air bubbles in its preparation.

air-entraining agent [MATER] An admixture, usually a resin, soap, or grease, for portland cement or concrete to effect air entrainment and, thus, superior properties.

air entrainment [ENG] The inclusion of minute bubbles of air in cement or concrete through the addition of some material during grinding or mixing to reduce the surface tension of the water, giving improved properties for the end product.

air-exhaust ventilator [MECH ENG] Any air-exhaust unit used to carry away dirt particles, odors, or fumes.

air feed [MET] In thermal spraying, transmittal of powdered material by air pressure through the gun into the heat source.

airfield [CIV ENG] The area of an airport for the takeoff and landing of airplanes.

air filter [ENG] A device that reduces the concentration of solid particles in an airstream to a level that can be tolerated in a process or space occupancy; a component of most systems in which air is used for industrial processes, ventilation, or comfort air conditioning.

airfloat clay [MIN ENG] Fine particles of clay obtained by air separation from coarser particles following a grinding operation.

airflow *See* air current.

airflow duct [ENG] A pipe, tube, or channel through which air moves into or out of an enclosed space. Also known as air duct.

airflow orifice [ENG] An opening through which air moves out of an enclosed space.

airflow pipe [ENG] A tube through which air is conveyed from one location to another.

airfoil [AERO ENG] A body of such shape that the force exerted on it by its motion through a fluid has a larger component normal to the direction of motion than along the direction of motion; examples are the wing of an airplane and the blade of a propeller. Also known as aerofoil.

airfoil profile [AERO ENG] The closed curve defining the external shape of the cross section of an airfoil. Also known as airfoil section; airfoil shape; wing section.

airfoil section *See* airfoil profile.

airfoil shape *See* airfoil profile.

airfoil-vane fan [MIN ENG] A centrifugal-type mine fan; the vanes are curved backward from the direction of rotation.

airframe [AERO ENG] The basic assembled structure of any aircraft or rocket vehicle, except lighter-than-air craft, necessary to support aerodynamic forces and inertia loads imposed by the weight of the vehicle and contents.

air furnace [MET] **1.** A furnace using a natural air draft. **2.** A furnace in which the metal is melted by a flame originating from fuel burned at one end, passing over the hearth in the middle, and exiting at the other end.

air gage [ENG] **1.** A device that measures air pressure. **2.** A device that compares the shape of a machined surface to that of a reference surface by measuring the rate of passage of air between the surfaces.

air gap [ENG] **1.** The distance between two components or parts. **2.** In plastic extrusion coating, the distance from the opening of the extrusion die to the nip formed by the pressure and chill rolls. **3.** The unobstructed vertical distance between the lowest opening of a faucet (or the like) which supplies a plumbing fixture (such as a tank or washbowl) and the level at which the fixture will overflow.

air gas [MATER] A gaseous fuel made by blowing air through a coal or coke bed so that CO_2 is reduced to CO.

air grating [BUILD] A fixed metal grille on the exterior of a building through which air is brought into or discharged from the building for purposes of ventilation.

air-handling system [MECH ENG] An air-conditioning system in which an air-handling unit provides part of the treatment of the air.

air-handling unit [MECH ENG] A packaged assembly of air-conditioning components (coils, filters, fan humidifier, and so forth) which provides for the treatment of air before it is distributed.

air-hardening steel [MET] A steel whose content of carbon and other alloying elements is sufficient for the steel to harden fully by cooling in air or any other atmosphere from a temperature above its transformation range. Also known as self-hardening steel.

air heater *See* air preheater.

air-heating system *See* air preheater.

air hoist [MECH ENG] A lifting tackle or tugger constructed with cylinders and pistons for reciprocating motion and air motors for rotary motion, all powered by compressed air. Also known as pneumatic hoist.

airhole [MIN ENG] A small excavation or hole made to improve ventilation by communication with other workings or with the surface.

air horsepower [MECH ENG] The theoretical (minimum) power required to deliver the specified quantity of air under the specified pressure conditions in a fan, blower, compressor, or vacuum pump. Abbreviated air hp.

air hp *See* air horsepower.

air-injection system [MECH ENG] A device that uses compressed air to inject the fuel into the cylinder of an internal combustion engine.

air inlet [MECH ENG] In an air-conditioning system, a device through which air is exhausted from a room or building.

air intake [AERO ENG] An open end of an air duct or similar projecting structure so that the motion of the aircraft is utilized in capturing air to be conducted to an engine or ventilator. [MIN ENG] A device for supplying a compressor with clean air at the lowest possible temperature.

air knife [ENG] A device that uses a thin, flat jet of air to remove the excess coating from freshly coated paper.

air-knife coating [ENG] An even film of coating left on paper after treatment with an air knife.

air-lance [ENG] To direct a pressurized-air stream to remove unwanted accumulations, as in boiler-wall cleaning.

air launch [AERO ENG] Launching from an aircraft in the air.

air leakage [MECH ENG] **1.** In ductwork, air which escapes from a joint, coupling, and such. **2.** The undesired leakage or uncontrolled passage of air from a ventilation system.

airless spraying [ENG] The spraying of paint by means of high fluid pressure and special equipment. Also known as hydraulic spraying.

air lift [MECH ENG] **1.** Equipment for lifting slurry or dry powder through pipes by means of compressed air. **2.** *See* air-lift pump.

airlift [AERO ENG] **1.** To transport passengers and cargo by the use of aircraft. **2.** The total weight of personnel or cargo carried by air.

air-lift hammer [MECH ENG] A gravity drop hammer used in closed die forging in which the ram is raised to its starting point by means of an air cylinder.

air-lift pump [MECH ENG] A device composed of two pipes, one inside the other, used to extract water from a well; the lower end of the pipes is submerged, and air is delivered through the inner pipe to form a mixture of air and water which rises in the outer pipe above the water in the well; also used to move corrosive liquids, mill tailings, and sand. Also known as air lift.

air line [ENG] A fault, in the form of an elongated bubble, in glass tubing. Also known as hairline. [MECH ENG] A duct, hose, or pipe that supplies compressed air to a pneumatic tool or piece of equipment.

air-line lubricator *See* line oiler.

air-line main [MIN ENG] The pipe column supplying air from the compressors to the quarry face.

air lock [ENG] **1.** A chamber capable of being hermetically sealed that provides for passage between two places of different pressure, such as between an altitude chamber and the outside atmosphere, or between the outside atmosphere and the work area in a tunnel or shaft being excavated through soil subjected to water pressure higher than atmospheric. **2.** An air bubble in a pipeline which impedes liquid flow. [MIN ENG] A casing atop an

upcast mine shaft to minimize surface air leakage into the fan.

air-lock strip [BUILD] The weather stripping which is fastened to the edges of each wing of a revolving door.

air log [AERO ENG] A distance-measuring device used especially in certain guided missiles to control range.

airman [MIN ENG] A man who constructs brattices.

air-measuring station [MIN ENG] A place in a mine airway where the volume of air passing is measured periodically.

air meter [ENG] A device that measures the flow of air, or gas, expressed in volumetric or weight units per unit time. Also known as airometer.

air mileage indicator [ENG] An instrument on an airplane which continuously indicates mileage through the air.

air mileage unit [ENG] A device which derives continuously and automatically the air distance flown, and feeds this information into other units, such as an air mileage indicator.

air-mixing plenum [MECH ENG] In an air-conditioning system, a chamber in which the recirculating air is mixed with air from outdoors.

air monitoring [CIV ENG] A practice of continuous air sampling by various levels of government or particular industries.

air motor [MECH ENG] A device in which the pressure of confined air causes the rotation of a rotor or the movement of a piston.

air mover [MIN ENG] A portable compressed-air appliance, used as a blower or exhauster.

airometer [ENG] **1.** An apparatus for both holding air and measuring the quantity of air admitted into it. **2.** *See* air meter.

air outlet [MECH ENG] In an air-conditioning system, a device at the end of a duct through which air is supplied to a space.

air permeability test [ENG] A test for the measurement of the fineness of powdered materials, such as portland cement.

airplane [AERO ENG] A heavier-than-air vehicle designed to use the pressures created by its motion through the air to lift and transport useful loads.

airplane flare [ENG] A flare, often magnesium, that is dropped from an airplane to illuminate a ground area; a small parachute decreases the rate of descent.

air pocket [ENG] An air-filled space that is normally occupied by a liquid. Also known as air trap.

air-pollution control [ENG] A practical means of treating polluting sources to maintain a desired degree of air cleanliness.

airport [CIV ENG] A terminal facility used for aircraft takeoff and landing and including facilities for handling passengers and cargo and for servicing aircraft. Also known as aerodrome.

airport engineering [CIV ENG] The planning, design, construction, and operation and maintenance of facilities providing for the landing and takeoff, loading and unloading, servicing, maintenance, and storage of aircraft.

air preheater [MECH ENG] A device used in steam boilers to transfer heat from the flue gases to the combustion air before the latter enters the furnace. Also known as air heater; air-heating system.

airproof *See* airtight.

air propeller [AERO ENG] A hub-and-multi-blade device for changing rotational power of an aircraft engine into thrust power for the purpose of propelling an aircraft through the air. [MECH ENG] A rotating fan for moving air.

air pump [MECH ENG] A device for removing air from an enclosed space or for adding air to an enclosed space.

air puncher [ENG] A machine consisting essentially of a reciprocating chisel or pick, driven by air.

air purge [MECH ENG] Removal of particulate matter from air within an enclosed vessel by means of air displacement.

air raid shelter [CIV ENG] A chamber, often underground, provided with living facilities and food, for sheltering people against air attacks.

air register [ENG] A device attached to an air-distributing duct for the purpose of controlling the discharge of air into the space to be heated, cooled, or ventilated.

air regulator [MECH ENG] A device for regulating airflow, as in the burner of a furnace. [MIN ENG] An adjustable door installed in permanent air stoppings to control ventilating current.

air reheater [MECH ENG] In a heating system, any device used to add heat to the air circulating in the system.

air release valve [MECH ENG] A valve, usually manually operated, which is used to release air from a water pipe or fitting.

air ring [ENG] In plastics forming, a circular manifold which distributes an even flow of cool air into a hollow tubular form passing through the manifold.

air sampling [ENG] The collection and analysis of samples of air to measure the amounts of various pollutants or other substances in the air, or the air's radioactivity.

air scoop [DES ENG] An air-duct cowl projecting from the outer surface of an aircraft or automobile, which is designed to utilize the dynamic pressure of the airstream to maintain a flow of air.

air screw [MECH ENG] A screw propeller that operates in air.

air-seasoned [ENG] Treated by exposure to air to give a desired quality.

air-seasoned lumber *See* air-dried lumber.

air separator [MECH ENG] A device that uses an air current to separate a material from another of greater density or particles from others of greater size.

air-setting mortar [MATER] Mortar that sets in air at atmospheric pressure and ordinary temperatures.

air shaft [BUILD] An open space surrounded by the walls of a building or buildings to provide ventilation for windows. Also known as air well. [MIN ENG] A usually vertical earth bore or shaft to supply surface air to an underground facility such as a mine.

airship [AERO ENG] A propelled and steered aerial vehicle, dependent on gases for flotation.

air shot [ENG] A shot prepared by loading (charging) so that an air space is left in contact with the explosive for the purpose of lessening its shattering effect.

Airslide conveyor [MECH ENG] An air-activated gravity-type conveyor, of the Fuller Company, using low-pressure air to aerate or fluidize pulverized material to a degree which will permit it to flow on a slight incline by the force of gravity.

air space [ENG] An enclosed space containing air in a wall for thermal insulation.

airspace [AERO ENG] **1.** The space occupied by an aircraft formation or used in a maneuver. **2.** The area around an airplane in flight that is considered an integral part of the plane in order to prevent collision with another plane; the space depends on the speed of the plane.

airspeed [AERO ENG] The speed of an airborne object relative to the atmosphere; in a calm atmosphere, airspeed equals ground speed.

airspeed head [ENG] Any instrument or device, usually a pitot tube, mounted on an aircraft for receiving the static and dynamic pressures of the air used by the airspeed indicator.

airspeed indicator [ENG] A device that computes and displays the speed of an aircraft relative to the air mass in which the aircraft is flying.

air spring [MECH ENG] A spring in which the energy storage element is air confined in a container that includes an elastomeric bellows or diaphragm.

air stack [AERO ENG] A group of planes flying at prescribed heights while waiting to land at an airport. [MIN ENG] A chimney for ventilating a coal mine.

air-standard engine [MECH ENG] A heat engine operated in an air-standard cycle.

airstart [AERO ENG] An act or instance of starting an aircraft's engine while in flight, especially a jet engine after flameout.

air starting valve [MECH ENG] A device that admits compressed air to an air starter.

air strip See landing strip.

air-supply mask See air-tube breathing apparatus.

air surveillance [ENG] Systematic observation of the airspace by visual, electronic, or other means, primarily for identifying all aircraft in that airspace, and determining their movements.

air surveillance radar [ENG] Radar of moderate range providing position of aircraft by azimuth and range data without elevation data; used for air-traffic control.

air survey See aerial survey.

air-suspension system [MECH ENG] Parts of an automotive vehicle that are intermediate between the wheels and the frame, and support the car body and frame by means of a cushion of air to absorb road shock caused by passage of the wheels over irregularities.

air system [MECH ENG] A mechanical refrigeration system in which air serves as the refrigerant in a cycle comprising compressor, heat exchanger, expander, and refrigerating core.

air taxi [AERO ENG] A carrier of passengers and cargo engaged in charter flights, feeder air services to large airline facilities, or contract airmail transportation.

air terminal [CIV ENG] A facility providing a place of assembly and amenities for airline passengers and space for administrative functions.

air thermometer [ENG] A device that measures the temperature of an enclosed space by means of variations in the pressure or volume of air contained in a bulb placed in the space.

airtight [ENG] Not permitting the passage of air. Also known as airproof.

air-to-air resistance [CIV ENG] The resistance provided by the wall of a building to the flow of heat.

air-track drill [MIN ENG] A heavy drilling machine for quarry or opencast blasting, equipped with caterpillar tracks and operated by independent air motors.

air transport [MIN ENG] Movement from one place to another of the filling material in a mine through pneumatic pipelines. Also known as air transportation.

air transportation [AERO ENG] The use of aircraft, predominantly airplanes, to move passengers and cargo. [MIN ENG] See air transport.

air trap [CIV ENG] A U-shaped pipe filled with water that prevents the escape of foul air or gas from such systems as drains and sewers. [ENG] See air pocket.

air-tube breathing apparatus [ENG] A device consisting of a smoke helmet, mask, or mouthpiece supplied with fresh air by means of a flexible tube. Also known as air-supply mask.

air valve [MECH ENG] A valve that automatically lets air out of or into a liquid-carrying pipe when the internal pressure drops below atmospheric.

air vane [AERO ENG] A vane that acts in the air, as contrasted to a jet vane which acts within a jetstream.

air vehicle See aircraft.

air vessel [ENG] **1.** An enclosed volume of air which uses the compressibility of air to minimize water hammer. **2.** An enclosed chamber using the compressibility of air to promote a more uniform flow of water in a piping system.

air void [MATER] A space which is filled with air in cement paste, mortar, or concrete.

air washer [MECH ENG] **1.** A device for cooling and cleaning air in which the entering warm, moist air is cooled below its dew point by refrigerated water so that although the air leaves close to saturation with water, it has less moisture per unit volume than when it entered. **2.** Apparatus to wash particulates and soluble impurities from air by passing the airstream through a liquid bath or spray.

air-water jet [ENG] A jet of mixed air and water which leaves a nozzle at high velocity; used in cleaning the surfaces of concrete or rock.

air-water storage tank [ENG] A water storage tank in which the air above the water is compressed.

airway [BUILD] A passage for ventilation between thermal insulation and roof boards. [MIN ENG] A passage for air in a mine. Also known as air course.

air well See air shaft.

Airy points [ENG] The points at which a horizontal rod is optionally supported to avoid its bending.

aisleway [CIV ENG] A passage or walkway within a factory, storage building, or shop permitting the flow of inside traffic.

Aitken dust counter [ENG] An instrument for determining the dust content of the atmosphere. Also known as Aitken nucleus counter.

Aitken nucleus counter See Aitken dust counter.

Ajax powder [MATER] A high-strength, high-density, gelatinous permitted explosive having good water resistance; contains nitroglycerine, potassium perchlorate, ammonium oxalate, and wood flour.

AJO breathing apparatus [MIN ENG] A breathing device consisting of a Siebe-Gorman mining gas mask with a small oxygen cylinder and a canister which neutralizes mining gases, such as carbon monoxide, sulfureted hydrogen, and nitrous fumes.

ajowan oil [MATER] A yellow essential oil distilled from seeds of the herbaceous plant *Carum copticum* (*Ptychotis ajowan*) and used in pharmaceuticals. Also known as ptychotis oil.

Akins' classifier [MIN ENG] A device for separating fine-size solids from coarser solids in a wet pulp; consists of an interrupted-flight screw conveyor, operating in an inclined trough.

akund [MATER] A silky cotton fiber from the shrub *Calotropis gigantea;* used in combination with kapok fiber for insulation.

alabaster glass [MATER] A glass that contains small inclusions of different diffractive indexes and shows no color reaction to light.

alarm gage [ENG] A device that actuates a signal either when the steam pressure in a boiler is too high or when the water level in a boiler is too low.

alarm system [ENG] A system which operates a warning device after the occurrence of a dangerous or undesirable condition.

alarm valve [ENG] A device that sounds an alarm when water flows in an automatic sprinkler system.

Alaska cedar [MATER] Wood from *Chamaecyparis nootkaensis* or *Cupressus sitkaensis;* has a fine, uniform, straight grain and is light and moderately hard; used for furniture and boat building. Also known as Sitka cypress; yellow cedar; yellow cypress.

albarium [MATER] A white lime used for stucco; made by burning marble.

albedometer [ENG] An instrument used for the measurement of the reflecting power, that is, the albedo, of a surface.

albolite [MATER] A plastic cement composed principally of magnesia and silica.

albumin glue [MATER] A bonding agent composed of soluble dried blood with minor additives and giving strong, durable bonds when coagulated in plywood joints at temperatures of 160–180°F (71–82°C).

alcian blue [MATER] A copper phthalocyanin derivative used as a stain for connective tissue mucins and a number of epithelial mucins and as a gelling agent for lubricating fluids.

alcohol fuel [MATER] A motor fuel of gasoline blended with 5–25% of anhydrous ethyl alcohol; used particularly in Europe.

alcoholimeter *See* alcoholometer.

alcoholmeter *See* alcoholometer.

alcoholometer [ENG] A device, such as a form of hydrometer, that measures the quantity of an alcohol contained in a liquid. Also known as alcoholimeter; alcoholmeter.

alfenol [MET] A permeability alloy that has 16% aluminum and 84% iron; it is brittle and at 572°F (300°C) can be rolled into thin sheets; used for transformer cores and tape recorder heads.

algicide [MATER] A chemical used to kill algae.

alidade [ENG] **1.** An instrument for topographic surveying and mapping by the plane-table method. **2.** Any sighting device employed for angular measurement.

alignment [ENG] Placing of surveying points along a straight line. [MIN ENG] The act of laying out a tunnel or regulating by line; adjusting to a line.

alignment correction [ENG] A correction applied to the measured length of a line to allow for not holding the tape exactly in a vertical plane of the line.

alignment pin [DES ENG] Pin in the center of the base of an octal, loctal, or other tube having a single vertical projecting rib that aids in correctly inserting the tube in its socket.

alignment wire *See* ground wire.

alite [MATER] A constituent of portland cement clinker consisting mostly of calcium silicate.

Aliva concrete sprayer [MIN ENG] A compressed-air machine for spraying concrete on the roof and sides of mine roadways.

alive [MIN ENG] That portion of a lode that is productive.

alizarin yellow [MATER] A dye useful as an acid-base indicator; solutions change color from yellow (acid) to purple (basic) in the pH range 10.1 to 12.0. Also known as *para*-nitrophenylazosalicylate sodium.

alkali cellulose [MATER] Product of wood pulp steeped with sodium hydroxide; first step in manufacture of viscose rayon and other cellulosics.

alkali ion diode [ENG] In testing for leaks, a device which senses the presence of halogen gases by the use of positive ions of alkali metal on the heated diode surfaces.

alkali lead [MET] An alloy of lead hardened with small quantities of alkali metals; used as bearing metals.

alkali lignin [MATER] A type of lignin produced by treating the black liquor from the soda process with acid; used as an extender in the negative plates of storage batteries, in asphalt, and in paperboard products.

alkaline cleaner [MET] An aqueous solution of an alkali used for metal cleaning.

alkali reactivity [MATER] Susceptibility of a concrete aggregate to alkali-aggregate reaction

alkali-resisting paint [MATER] A paint, such as one made with a synthetic resin, that does not undergo saponification when used in such places as bathrooms or on such materials as new concretes.

alkanet [MATER] A chemical indicator made from the root of *Alkanna tinctoria*. Also known as alkanna; anchusa; orcanette.

alkanna *See* alkanet.

alkyd paint [MATER] A paint using an alkyd resin as the vehicle for the pigment.

alkylated gasoline [MATER] A cleaning-burning gasoline with a high-octane rating; prepared by adding neohexane or some other alkylate.

all-burnt time [AERO ENG] The point in time at which a rocket has consumed its propellants.

all-burnt velocity *See* burnout velocity.

allege [BUILD] A part of a wall which is thinner than the rest, especially the spandrel under a window.

Allen red metal [MET] An alloy of copper and lead containing 50% lead and a small quantity of sulfur to hold the lead in solution.

Allen screw [DES ENG] A screw or bolt which has an axial hexagonal socket in its head.

Allen wrench [DES ENG] A wrench made from a straight or bent hexagonal rod, used to turn an Allen screw.

allicin [MATER] An oily liquid extracted from garlic which has a sharp garlic odor; used in medicine as an antibacterial agent.

alligator effect *See* orange peel.

alligatoring [MATER] Cracking of a film of paint or varnish, with broad, deep cracks through one or more coats. Also known as crocodiling. [MET] **1.** A splitting of an end of a rolled steel slab in which the plane of the split is parallel to the rolled surface. Also known as fishmouthing. **2.** The roughening of a sheet-metal surface during forming due to the coarse grain of the metal used.

alligator pear oil *See* avocado oil.

alligator shears *See* lever shears.

alligator squeezer [MET] A tool with a fixed upper jaw and a movable lower jaw used to squeeze a ball of iron produced by the paddling process into a bloom or billet.

alligator wrench [DES ENG] A wrench having fixed jaws forming a V, with teeth on one or both jaws.

allocate [IND ENG] To assign a portion of a resource to an activity.

allowable bearing value [CIV ENG] The maximum permissible pressure on foundation soil that provides adequate safety against rupture of the soil mass or movement of the foundation of such magnitude as to impair the structure imposing the pressure. Also known as allowable soil pressure.

allowable oil [PETRO ENG] The oil an operator is permitted by law to remove from a well in one day.

allowable soil pressure *See* allowable bearing value.

allowance [DES ENG] An intentional difference in sizes of two mating parts, allowing clearance usually for a film of oil, for running or sliding fits.

allowed hour *See* standard hour.

allowed time [IND ENG] Amount of time allowed each employee for personal needs during a work cycle.

alloy [MET] Any of a large number of substances having metallic properties and consisting of two or more elements; with few exceptions, the components are usually metallic elements.

Alloy 750 [MET] A bearing alloy, containing 6.5% tin, 2.5% silicon, 1% copper, 0.5% nickel, and the remainder aluminum; used for automobile engine bearings.

alloying [MET] The addition of a metal or alloy to another metal or alloy.

alloy plating [MET] The codeposition of two or more metals on an electrode by electrolysis.

alloy steel [MET] A steel whose distinctive properties are due to the presence of one or more elements other than carbon.

allspice oil *See* pimenta oil.

alluvial mining [MIN ENG] The exploitation of alluvial deposits by dredging, hydraulicking, or drift mining.

all-weather aircraft [AERO ENG] Aircraft that are designed or equipped to perform by day or night under any weather conditions.

all-weather airport [CIV ENG] An airport with facilities to permit the landing of qualified aircraft and aircrewmen without regard to operational weather limits.

all-weather fighter [AERO ENG] A fighter aircraft equipped with radar and other special devices which enable it to intercept its target in the dark, or in daylight weather conditions that do not permit visual interception; it is usually a multiplace (pilot plus navigator-observer) airplane.

all-weld-metal test specimen [MET] A specimen composed entirely of weld metal used in a weld tension test wherein the axis of the weld from which it is derived is parallel to the axis of the test bar.

almendro [MATER] The wood of tonka bean tree species in Panama *(Coumarouna panamensis)*; resistant to marine borers and used as a substitute for greenheart in marine construction.

almon [MATER] A type of lauan from the tree *Shorea eximia*; wood is fairly strong and hard with a coarse texture.

alnico [MET] One of a series of ferrous alloys containing aluminum, nickel, and cobalt, valued because of their highly retentive magnetic properties; usually designated with a roman-numeral number, such as alnico VII. Also known as aluminum-nickel-cobalt alloy.

aloe wood oil *See* oriental linaloe.

alpha brass [MET] An alloy of copper and zinc containing up to 36% zinc dissolved, rather than chemically combined, with the copper; ductile, easily cold-worked, and corrosion resistant; used for hot-water pipes.

alpha gypsum [MATER] A specially processed gypsum having low consistency and high compressive strength, often exceeding 5000 pounds per square inch (34.5 megapascals).

alpha iron [MET] Iron with a body-centered cubic structure which is stable below 1670°F (910°C).

alpha-ray vacuum gage [ENG] An ionization gage in which the ionization is produced by alpha particles emitted by a radioactive source, instead of by electrons emitted from a hot filament; used chiefly for pressures from 10^{-3} to 10 torrs. Also known as alphatron.

alsifilm [MATER] A bentonite gel in the form of sheets used primarily for electrical insulation, because of its properties of heat and oil resistance.

alt *See* altitude.

altazimuth [ENG] An instrument equipped with both horizontal and vertical graduated circles, for the simultaneous observation of horizontal and vertical directions or angles. Also known as astronomical theodolite; universal instrument.

altazimuth mounting [ENG] A telescope mounting with a vertical axis that determines the azimuth, and a horizontal axis, supported on the vertical axis, that determines the altitude; the telescope must be moved in both altitude and azimuth to follow a celestial object.

alternate immersion test [MET] A corrosion test in which a specimen is repeatedly carried through a cycle of immersion in and removal from a corrosive medium over definite time intervals.

alternate polarity operation [MET] A resistance welding method which utilizes alternating polarity for the progression of weld pulses.

alternating-current welder [ENG] A welding machine utilizing alternating current for welding purposes.

altigraph [ENG] A pressure altimeter that has a recording mechanism to show the changes in altitude.

altimeter [ENG] An instrument which determines the altitude of an object with respect to a fixed level, such as sea level; there are two common types: the aneroid altimeter and the radio altimeter.

altimeter corrections [ENG] Corrections which must be made to the readings of a pressure altimeter to obtain true altitudes; involve horizontal pressure gradient error and air temperature error.

altimeter setting [ENG] The value of atmospheric pressure to which the scale of an aneroid altimeter is set; after United States practice, the pressure that will indicate airport elevation when the altimeter is 10 feet (3 meters) above the runway (approximately cockpit height).

altimeter-setting indicator [ENG] A precision aneroid barometer calibrated to indicate directly the local altimeter setting.

altimetry [ENG] The measurement of heights in the atmosphere (altitude), generally by means of an altimeter.

altitude Abbreviated alt. [ENG] **1.** Height, measured as distance along the extended earth's radius above a given datum, such as average sea level. **2.** Angular displacement above the horizon measured by an altitude curve.

altitude azimuth [ENG] An azimuth determined by solution of the navigational triangle with altitude, declination, and latitude given.

altitude chamber [ENG] A chamber within which the air pressure, temperature, and so on can be adjusted to simulate conditions at different altitudes; used for experimentation and testing.

altitude curve [ENG] The arc of a vertical circle between the horizon and a point on the celestial sphere, measured upward from the horizon.

altitude datum [ENG] The arbitrary level from which heights are reckoned.

altitude difference [ENG] The difference between computed and observed altitudes, or between precomputed and sextant altitudes. Also known as altitude intercept; intercept.

altitude intercept *See* altitude difference.

altitude valve [AERO ENG] A valve that adjusts the composition of the air-fuel mixture admitted into an airplane carburetor as the air density varies with altitude.

altitude wind tunnel [AERO ENG] A wind tunnel in which the air pressure, temperature, and humidity can be varied to simulate conditions at different altitudes.

alum cake [MATER] A material composed of silica and aluminum sulfate produced by the action of sulfuric acid on clay.

alumetized steel *See* aluminized steel.

alumina balls [MATER] Alumina in the form of balls ¼ to ¾ inch (6.4 to 19 mm) in diameter; usually composed of 99% alumina and having high resistance to chemicals and heat; used in reactor and catalytic beds.

alumina brick [MATER] A group of fireclay bricks containing 50, 60, or 70% alumina, used in high temperature applications.

alumina bubble brick [MATER] A lightweight refractory brick used to line kiln walls; manufactured by passing an air jet over molten alumina to produce small hollow bubbles which are pressed into bricks and other shapes.

alumina cement [MATER] A cement made with bauxite and containing a high percentage of aluminate, having the property of setting to high strength in 24 hours.

alumina fibers [MATER] Short, linear crystals of alumina which have a strength of up to 200,000 pounds per square inch (1.38 gigapascals); used in plastics as a filler to improve heat resistance and dielectric properties. Also known as sapphire whiskers.

alumina porcelain [MATER] Porcelain composed principally of alumina; used to make spark plugs.

aluminate cement [MATER] Cement made with bauxite, with high percentage of alumina; sets to a high strength in 24 hours and is used for constructing bank walls and laying roads. Also known as aluminous cement; high-alumina cement; high-speed cement.

aluminize [ENG] To apply a film of aluminum to a material, such as glass. [MET] To form a protective surface alloy on a metal by treatment at elevated temperature with aluminum or an aluminum compound.

aluminized explosive [MATER] An explosive to which aluminum has been added.

aluminized steel [MET] A steel coated with an aluminum-iron alloy coating; prepared by dip-coating and diffusing aluminum into steel at 1600°F (870°C); resists scaling and oxidation up to 1650°F (900°C). Also known as alumetized steel; calorized steel.

aluminothermy [MET] The process of reducing a metallic oxide to the metal and producing great heat by mixing finely divided aluminum with the oxide, which is reduced as the aluminum is oxidized.

aluminous cement *See* aluminate cement.

aluminum alloy [MET] An alloy of aluminum and relatively small amounts of other metals, such as copper, magnesium, or manganese.

aluminum-base grease *See* aluminum-soap grease.

aluminum brass [MET] **1.** A casting brass to which aluminum has been added as a flux to improve the casting qualities and, with the addition of lead, the machining qualities. **2.** A wrought brass to which aluminum has been added to improve the extruding and forging qualities and the oxidation resistance.

aluminum bronze [MET] A copper-aluminum alloy which may also contain iron, manganese, nickel, or zinc.

aluminum coating [MET] A film of aluminum applied to a metallic surface by, for example, spraying, electrolysis, or hot dipping.

aluminum foil [MET] Aluminum in the form of a sheet of thickness not exceeding 0.005 inch (0.127 mm).

aluminum grease *See* aluminum-soap grease.

aluminum-magnesium alloy [MET] An alloy of aluminum, 5–10% magnesium, and sometimes small amounts of other metals, characterized by high resistance to corrosion and high machinability.

aluminum-nickel-cobalt alloy *See* alnico.

aluminum paint [MATER] A mixture of oil varnish and aluminum pigment in the form of thin flakes which overlap in the paint film; reflects the sun's radiation well and retains the heat in hot-air or hot-water pipes or tanks.

aluminum paste [MATER] Aluminum powder finely ground in oil; used in aluminum paints.

aluminum powder [MATER] Small flakes of aluminum metal obtained by stamping or ball-milling foil in the presence of a fatty lubricant, such as stearic acid, which causes the flakes to orient in a pattern to give high brilliance.

aluminum-silicon alloy [MET] An alloy of aluminum, 5–22% silicon, and sometimes small amounts of other metals, characterized by ease of casting and welding, light weight, and high resistance to corrosion.

aluminum-silicon bronze [MET] An alloy consisting chiefly of copper, with aluminum and silicon added to give greater strength and hardness.

aluminum-soap grease [MATER] A lubricating grease consisting of a petroleum oil thickened with aluminum soap. Also known as aluminum-base grease; aluminum grease.

aluminum solder [MET] A solder containing up to 15% aluminum, having a melting point above that of the tin-lead solders, and applied with a brazing torch.

amalgam [MET] An alloy of mercury.

amalgamate [MET] **1.** To unite a metal in an alloy with mercury. **2.** To unite two dissimilar metals. **3.** To cover the zinc elements of a galvanic battery with mercury.

amalgamating table [MET] A sloping wooden table covered with a copper plate on which mercury is spread to amalgamate with precious-metal particles.

amalgamation [MET] Also known as amalgam treatment. **1.** The process of separating metal from ore by alloying the metal with mercury; formerly used for gold and silver recovery, where it has been superseded by the cyanide process. **2.** The formation of an alloy of a metal with mercury.

amalgamation pan [MET] A circular cast-iron pan in which gold or silver ore is ground and the precious metal particles are amalgamated with mercury added to the pan.

amalgamator [MET] A device for bringing pulverized ore into contact with mercury to form an amalgam from which the metal is subsequently recovered.

amalgam barrel [MET] A small batching mill used to grind auriferous concentrates with mercury.

amalgam retort [MET] A retort in which mercury is distilled off from gold, or silver amalgam is obtained in amalgamation.

amalgam treatment See amalgamation.

amatol [MATER] An explosive mixture composed of ammonium nitrate and trinitrotoluene; mixtures with 50% and 80% ammonium nitrate are used for small and large shells, respectively.

amber glass [MATER] A tinted glass made by using different mixtures of sulfur and iron oxide; the color can vary from pale yellow to ruby amber.

amberite [MATER] A smokeless powder composed of guncotton, barium nitrate, and paraffin.

amber oil [MATER] **1.** A yellowish to brown essential oil made by destructive distillation of amber; has an acrid taste. **2.** A light essential oil prepared by destructive distillation of rosin.

ambient [ENG] Surrounding; especially, of or pertaining to the environment about a flying aircraft or other body but undisturbed or unaffected by it, as in ambient air or ambient temperature.

ambrette oil [MATER] A fixative for perfume having a strong musklike odor and distilled from the musk seed of *Hibiscus abelmoschus.*

amebicide [MATER] A chemical used to kill amebas, especially parasitic species.

American basement [BUILD] A basement located above ground level and containing the building's main entrance.

American bond [CIV ENG] A bond in which every fifth, sixth, or seventh course of a wall consists of headers and the other courses consist of stretchers. Also known as common bond; Scotch bond.

American boring system [MIN ENG] A rope system of percussive boring, with a derrick which enables the complete set of boring tools to be raised clear of the hole. Also known as American system.

American caisson See box caisson.

American dillweed oil See dill oil.

American explosive [MATER] One of many explosives that have passed U.S. Bureau of Mines tests and are used under certain conditions.

American filter See disk filter.

American pennyroyal oil See hedeoma oil.

American standard beam [CIV ENG] A type of I beam made of hot-rolled structural steel.

American standard channel [CIV ENG] A C-shaped structural member made of hot-rolled structural steel.

American standard pipe thread [DES ENG] Taper, straight, or dryseal pipe thread whose dimensions conform to those of a particular series of specified sizes established as a standard in the United States. Also known as Briggs pipe thread.

American standard screw thread [DES ENG] Screw thread whose dimensions conform to those of a particular series of specified sizes established as a standard in the United States; used for bolts, nuts, and machine screws.

American system See American boring system.

American system drill See churn drill.

American Table of Distances [ENG] Published data concerning the safe storage of explosives and ammunition.

American wire gage [MET] A particular series of specified diameters and thicknesses established as a standard in the United States and used for nonferrous sheets, rods, and wires. Abbreviated AWG. Also known as Brown and Sharp gage (B and S gage).

American wormseed oil See chenopodium oil.

A-metal [MET] A type of permeability alloy containing 44% nickel and a small amount of copper; used to give nondistortion characteristics upon magnetization in transformers and loudspeakers.

amino plastic [MATER] Any plastic made of compounds derived from ammonia.

ammeter [ENG] An instrument for measuring the magnitude of electric current flow.

ammonal [MATER] A high-explosive mixture, made of ammonium nitrate, trinitrotoluene (TNT), and flaked or powdered aluminum.

ammonia absorption refrigerator [MECH ENG] An absorption-cycle refrigerator which uses ammonia as the circulating refrigerant.

ammoniac [MATER] A gum resin obtained from the stems of the ammoniac plant; used in medicine, perfume, plaster, concrete, and adhesive. Also known as ammoniac gum; ammoniacum; gum ammoniac; Persian ammoniac.

ammoniac gum *See* ammoniac.

ammonia compressor [MECH ENG] A device that decreases the volume of a quantity of gaseous ammonia by the amplification of pressure; used in refrigeration systems.

ammonia condenser [MECH ENG] A device in an ammonia refrigerating system that raises the pressure of the ammonia gas in the evaporating coil, conditions the ammonia, and delivers it to the condensing system.

ammoniacum *See* ammoniac.

ammonia gelatin [MATER] An explosive of the gelatin dynamite class containing ammonium nitrate.

ammonia oil [MATER] Low-pour-test lubricating oil for use in an ammonia compressor in mechanical refrigeration.

ammonia permissible [MATER] A permissible explosive that is an ammonia dynamite.

ammonia valve [ENG] A valve that is resistant to corrosion by ammonia.

ammonite [MATER] An explosive containing 70–95% ammonium nitrate.

ammonium perchlorate explosive [MATER] Any of several compositions consisting of ammonium perchlorate in combination with a high explosive and powdered metal; produces powerful blast.

amorphous film [MATER] A magnetically ordered metallic film that can be deposited on a semiconductor chip or on almost any other material without need for a crystal substrate, for use in magnetic bubble memories.

amortize [IND ENG] To reduce gradually an obligation, such as a mortgage, by periodically paying a part of the principal as well as the interest.

amount limit [IND ENG] In a test for a fixed quantity of work, the time required to complete the work or the total amount of work that can be completed in an unlimited time.

amphibious [MECH ENG] Said of vehicles or equipment designed to be operated or used on either land or water.

amplifier [ENG] A device capable of increasing the magnitude or power level of a physical quantity, such as an electric current or a hydraulic mechanical force, that is varying with time, without distorting the wave shape of the quantity.

amplifier-type meter [ENG] An electric meter whose characteristics have been enhanced by the use of preamplification for the signal input eventually used to actuate the meter.

amplify [ENG ACOUS] To strengthen a signal by increasing its amplitude or by raising its level.

amplitude-frequency response *See* frequency response.

amplitude-modulated indicator [ENG] A general class of radar indicators, in which the sweep of the electron beam is deflected vertically or horizontally from a base line to indicate the existence of an echo from a target. Also known as deflection-modulated indicator; intensity-modulated indicator.

anacardium gum *See* cashew gum.

analemma [CIV ENG] Any raised construction which serves as a support or rest.

analytical aerotriangulation [ENG] Analytical phototriangulation, performed with aerial photographs.

analytical balance [ENG] A balance with a sensitivity of 0.1–0.01 milligram.

analytical centrifugation [ENG] Centrifugation following precipitation to separate solids from solid-liquid suspensions; faster than filtration.

analytical nadir-point triangulation [ENG] Radial triangulation performed by computational routines in which nadir points are utilized as radial centers.

analytical orientation [ENG] The computational steps required to determine tilt, direction of principal line, flight height, angular elements, and linear elements in preparing aerial photographs for rectification.

analytical photogrammetry [ENG] A method of photogrammetry in which solutions are obtained by mathematical methods.

analytical photography [ENG] Photography, either motion picture or still, accomplished to determine (by qualitative, quantitative, or any other means) whether a particular phenomenon does or does not occur.

analytical phototriangulation [ENG] A phototriangulation procedure in which the spatial solution is obtained by computational routines.

analytical radar prediction [ENG] Prediction based on proven formulas, power tables, or graphs; considers surface height, structural and terrain information, and criteria for radar reflectivity together with the aspect angle and range to the target.

analytical radial triangulation [ENG] Radial triangulation performed by computational routines.

analytical three-point resection radial triangulation [ENG] A method of computing the coordinates of the ground principal points of overlapping aerial photographs by resecting on three horizontal control points appearing in the overlap area.

analytical ultracentrifuge [ENG] An ultracentrifuge that uses one of three optical systems (schlieren, Rayleigh, or absorption) for the accurate determination of sedimentation velocity or equilibrium.

analyzer [ENG] A multifunction test meter, measuring volts, ohms, and amperes. Also known as set analyzer. [MECH ENG] The component of an absorption refrigeration system where the mixture of water vapor and ammonia vapor leaving the generator meets the relatively cool solution of ammonia in water entering the generator and loses some of its vapor content.

Anbauhobel [MIN ENG] A rapid plough, traveling at a speed of 75 feet (22.5 meters) per minute, for use on longwall faces.

anchor [CIV ENG] A device connecting a structure to a heavy masonry or concrete object to a metal plate or to the ground to hold the structure in place. [ENG] A device, such as a metal rod, wire, or strap, for fixing one object to another, such as specially formed metal connectors used to fasten together timbers, masonry, or trusses. [MECH ENG] A vehicle used in steam plowing and located on the side of the field opposite that of the engine while maintaining the tension on the endless wire by means of a pulley. [MET] A device that prevents the movement of sand cores in molds.

anchorage [CIV ENG] **1.** An area where a vessel anchors or may anchor because of either suitability or designation. Also known as anchor station. **2.** A device which anchors tendons to the posttensioned concrete member. **3.** In pretensioning, a device used to anchor tendons temporarily during the hardening of the concrete. **4.** *See* deadman.

anchorage deformation [CIV ENG] The shortening of tendons due to their modification or slippage when the prestressing force is transferred to the anchorage device. Also known as anchorage slip.

anchorage slip *See* anchorage deformation.

anchorage zone [CIV ENG] **1.** In posttensioning, the region adjacent to the anchorage for the tendon which is subjected to secondary stresses as a result of the distribution of the prestressing force.

2. In pretensioning, the region in which transfer bond stresses are developed.

anchor and collar [DES ENG] A door or gate hinge whose socket is attached to an anchor embedded in the masonry.

anchor block [BUILD] A block of wood, replacing a brick in a wall to provide a nailing or fastening surface. [CIV ENG] *See* deadman.

anchor bolt [CIV ENG] A bolt used with its head embedded in masonry or concrete and its threaded part protruding to hold a structure or machinery in place. Also known as anchor rod.

anchor buoy [ENG] One of a series of buoys marking the limits of an anchorage.

anchor charge [ENG] A procedure that allows several charges to be preloaded in a seismic shot hole; the bottom charges are fired first, and the upper charges are held down by anchors.

anchored bulkhead [CIV ENG] A bulkhead secured to anchor piles.

anchored-type ceramic veneer [MATER] Ceramic veneer which is attached to a backing by grout and nonferrous metal anchors.

anchor log [CIV ENG] A log, beam, or concrete block buried in the earth and used to hold a guy rope firmly. Also known as deadman.

anchor nut [DES ENG] A nut in the form of a tapped insert forced under steady pressure into a hole in sheet metal.

anchor packer [PETRO ENG] A device used in oil wells to seal the annular space between the tubing and its surrounding casing to help control the oil-producing gas lift.

anchor pile [CIV ENG] A pile that is located on the land side of a bulkhead or pier and anchors it through such devices as rods, cables, and chains.

anchor plate [CIV ENG] A metal or wooden plate fastened to or embedded in a support, such as a floor, and used to hold a supporting cable firmly.

anchor rod *See* anchor bolt.

anchor station *See* anchorage.

anchor tower [CIV ENG] **1.** A tower which is a part of a crane staging or stiffleg derrick and serves as an anchor. **2.** A tower that supports and anchors an overhead transmission line.

anchor wall *See* deadman.

anchusa *See* alkanet.

anechoic chamber [ENG] **1.** A test room in which all surfaces are lined with a sound-absorbing material to reduce reflections of sound to a minimum. Also known as dead room; free-field room. **2.** A room completely lined with a material that absorbs radio waves at a particular frequency or over a range of frequencies;

used principally at microwave frequencies, such as for measuring radar beam cross sections.

anemobiagraph [ENG] A recording pressure-tube anemometer in which the wind scale of the float manometer is linear through the use of springs; an example is the Dines anemometer.

anemoclinometer [ENG] A type of instrument which measures the inclination of the wind to the horizontal plane.

anemogram [ENG] A record made by an anemograph.

anemograph [ENG] **1.** An instrument which records wind velocities. **2.** A recording anemometer.

anemometer [ENG] A device which measures air speed.

anemoscope [ENG] An instrument for indicating the direction of the wind.

anemovane [ENG] A combined contact anemometer and wind vane used in the Canadian Meteorological Service.

aneroid [ENG] **1.** Containing no liquid or using no liquid. **2.** See aneroid barometer.

aneroid altimeter [ENG] An altimeter containing an aneroid barometer that actuates the indicator.

aneroid barograph [ENG] An aneroid barometer arranged so that the deflection of the aneroid capsule actuates a pen which graphs a record on a rotating drum. Also known as aneroidograph; barograph; barometrograph.

aneroid barometer [ENG] A barometer which utilizes an aneroid capsule. Also known as anerroid.

aneroid calorimeter [ENG] A calorimeter that uses a metal of high thermal conductivity as a heat reservoir.

aneroid capsule [ENG] A thin, disk-shaped box or capsule, usually metallic, partially evacuated and sealed, held extended by a spring, which expands and contracts with changes in atmospheric or gas pressure. Also known as bellows.

aneroid diaphragm [ENG] A thin plate, usually metal, covering the end of an aneroid capsule and moving axially as the ambient gas pressure increases or decreases.

aneroid flowmeter [ENG] A mechanism to measure fluid flow rate by pressure of the fluid against a bellows counterbalanced by a calibrated spring.

aneroid liquid-level meter [ENG] A mechanism to measure fluid depth by pressure of the fluid against a bellows which in turn acts on a manometer or signal transmitter.

aneroidograph See aneroid barograph.

aneroid valve [MECH ENG] A valve actuated or controlled by an aneroid capsule.

anerroid See aneroid barometer.

anethum oil See dill oil.

angel echo [ENG] A radar echo from a region where there are no visible targets; may be caused by insects, birds, or refractive index variations in the atmosphere.

angelica oil [MATER] An essential oil with an odor that is strongly aromatic; soluble in alcohol; main ingredients are phellandrene and valeric acid; distilled from the seeds and roots of *Angelica archangelica*; used in medicines, liqueurs, and perfumes.

angle back-pressure valve [MECH ENG] A back-pressure valve with its outlet opening at right angles to its inlet opening.

angle bar [BUILD] An upright bar at the meeting of two faces of a polygonal window, bay window, or bow window.

angle bead [BUILD] A strip, usually of metal or wood, set at the corner of a plaster wall to protect the corner or serve as a guide to float the plaster flush with it.

angle beam [ENG] Ultrasonic waves transmitted for the inspection of a metallic surface at an angle measured from the beam center line to a normal to the test surface.

angle blasting [ENG] Sandblasting, or the like, at an angle of less that 90°.

angle block [ENG] A small block of wood used to fasten adjacent pieces, usually at right angles, or glued into the corner of a wooden frame to stiffen it. Also known as glue block.

angle board [DES ENG] A board whose surface is cut at a desired angle; serves as a guide for cutting or planing other boards at the same angle.

angle bond [CIV ENG] A tie used to bond masonry work at wall corners.

angle brace [ENG] A brace across the interior angle of two members that meet at an angle. Also known as angle tie.

angle brick [ENG] Any brick having an oblique shape to fit an oblique, salient corner.

angle clip [CIV ENG] A short strip of angle iron used to secure structural elements at right angles.

angle closer [ENG] A specially shaped brick used to close the bond at the corner of a wall.

angle collar [DES ENG] A cast-iron pipe fitting which has a socket at each end for joining with the spigot ends of two pipes that are not in alignment.

angle cut [MIN ENG] A drilling pattern in which drill holes converge, so that a core can be blasted out and an open or relieved cavity or free face be left for the following shots, which are timed to ensue with a fractional delay.

angle divider [DES ENG] A square for setting or bisecting angles; one side is an adjustable hinged blade.

Angledozer [MECH ENG] A power-operated machine fitted with a blade, adjustable in height and angle, for pushing, sidecasting, and spreading loose excavated material as for opencast pits, clearing land, or leveling runways. Also known as angling dozer.

angle equation [ENG] A condition equation which expresses the relationship between the sum of the measured angles of a closed figure and the theoretical value of that sum, the unknowns being the corrections to the observed directions or angles, depending on which are used in the adjustment. Also known as triangle equation.

angle fillet [ENG] A wooden strip, triangular in cross section, which is used to cover the internal joint between two surfaces meeting at an angle of less than 180°.

angle fishplates [CIV ENG] Plates which join the rails and prevent the rail joint from sagging where heavy cars and locomotives are used. Also known as angle; angle bar.

angle float [ENG] A trowel having two edge surfaces bent at 90°; used to finish corners in freshly poured concrete and in plastering.

angle gauge [CIV ENG] A template used to set or check angles in building construction.

angle gear See angular gear.

angle iron [CIV ENG] **1.** An L-shaped cleat or brace. **2.** A length of steel having a cross section resembling the letter L.

angle joint [ENG] A joint between two pieces of lumber which results in a change in direction.

angle lacing [CIV ENG] A system of lacing in which angle irons are used in place of bars.

angle of action [MECH ENG] The angle of revolution of either of two wheels in gear during which any particular tooth remains in contact.

angle of advance See angular advance.

angle of approach [CIV ENG] The maximum angle of an incline onto which a vehicle can move from a horizontal plane without interference. [MECH ENG] The angle that is turned through by either of paired wheels in gear from the first contact between a pair of teeth until the pitch points of these teeth fall together.

angle of attack [AERO ENG] The angle between a reference line which is fixed with respect to an airframe (usually the longitudinal axis) and the direction of movement of the body. [MIN ENG] The angle in a mine fan made by the direction of air

approach and the chord of the aerofoil section.

angle of bite See angle of nip.

angle of cant [AERO ENG] In a spin-stabilized rocket, the angle formed by the axis of a venturi tube and the longitudinal axis of the rocket.

angle of climb [AERO ENG] The angle between the flight path of a climbing vehicle and the local horizontal.

angle of departure [AERO ENG] The vertical angle, at the origin, between the line of site and the line of departure. [CIV ENG] The maximum angle of an incline from which a vehicle can move onto a horizontal plane without interference, such as from rear bumpers.

angle of depression [ENG] The angle in a vertical plane between the horizontal and a descending line. Also known as depression angle; descending vertical angle; minus angle.

angle of descent [AERO ENG] The angle between the flight path of a descending vehicle and the local horizontal.

angle of elevation [ENG] The angle in a vertical plane between the local horizontal and an ascending line, as from an observer to an object; used in astronomy, surveying, and so on. Also known as ascending vertical angle; elevation angle.

angle of external friction [ENG] The angle between the abscissa and the tangent of the curve representing the relationship of shearing resistance to normal stress acting between soil and the surface of another material. Also known as angle of wall friction.

angle of glide [AERO ENG] Angle of descent for an airplane or missile in a glide.

angle of nip [MECH ENG] The largest angle that will just grip a lump between the jaws, rolls, or mantle and ring of a crusher. Also known as angle of bite; nip. [MIN ENG] In a rock-crushing machine, the maximum angle subtended by its approaching jaws or roll surfaces at which a specified piece of ore can be gripped.

angle of obliquity See angle of pressure.

angle of pitch [AERO ENG] The angle, as seen from the side, between the longitudinal body axis of an aircraft or similar body and a chosen reference line or plane, usually the horizontal plane.

angle of pressure [DES ENG] The angle between the profile of a gear tooth and a radial line at its pitch point. Also known as angle of obliquity.

angle of recess [MECH ENG] The angle that is turned through by either of two wheels in gear, from the coincidence of the pitch points of a pair of teeth until the last point of contact of the teeth.

angle of repose See angle of rest.

angle of rest [ENG] The maximum slope at which a heap of any loose or fragmented solid material will stand without sliding, or will come to rest when poured or dumped in a pile or on a slope. Also known as angle of repose.

angle of roll [AERO ENG] The angle that the lateral body axis of an aircraft or similar body makes with a chosen reference plane in rolling; usually, the angle between the lateral axis and a horizontal plane.

angle of slide [MIN ENG] The slope, measured in degrees of deviation from the horizontal, on which loose or fragmented solid materials will start to slide.

angle of stall [AERO ENG] The angle of attack at which the flow of air begins to break away from the airfoil, the lift begins to decrease, and the drag begins to increase. Also known as stalling angle.

angle of thread [DES ENG] The angle occurring between the sides of a screw thread, measured in an axial plane.

angle of wall friction See angle of external friction.

angle of yaw [AERO ENG] The angle, as seen from above, between the longitudinal body axis of an aircraft, a rocket, or the like and a chosen reference direction. Also known as yaw angle.

angle paddle [ENG] A hand tool used to finish a plastered surface.

angle plate [DES ENG] An L-shaped plate or a plate having an angular section.

angle post [BUILD] A railing support used at a landing or other break in the stairs.

angle press [MECH ENG] A hydraulic plastics-molding press with both horizontal and vertical rams; used to produce complex moldings with deep undercuts.

angle rafter [BUILD] A rafter, such as a hip rafter, at the angle of the roof.

angle section [CIV ENG] A structural steel member having an L-shaped cross section.

angle set [MIN ENG] **1.** A timber set using an angle brace. **2.** One of a series of sets placed at angles to each other.

angle-stem thermometer [ENG] A device used to measure temperatures in oil-custody tanks; the angle of the calibrated stem may be 90° or greater to the sensitive portion of the thermometer, as needed to fit the tank shell contour.

angle stile [BUILD] A narrow strip of wood used to conceal the joint between a wall and a vertical wood surface which makes an angle with the wall, as at the edge of a corner cabinet.

angle strut [CIV ENG] An angle-shaped structural member which is designed to carry a compression load.

angle tie See angle brace.

angle valve [DES ENG] A manually operated valve with its outlet opening oriented at right angles to its inlet opening; used for regulating the flow of a fluid in a pipe.

angling dozer See Angledozer.

Ångström compensation pyrheliometer [ENG] A pyrheliometer consisting of two identical Manganin strips, one shaded, the other exposed to sunlight; an electrical current is passed through the shaded strip to raise its temperature to that of the exposed strip, and the electric power required to accomplish this is a measure of the solar radiation.

angular accelerometer [ENG] An accelerometer that measures the rate of change of angular velocity between two objects under observation.

angular advance [MECH ENG] The amount by which the angle between the crank of a steam engine and the virtual crank radius of the eccentric exceeds a right angle. Also known as angle of advance; angular lead.

angular aggregate [MATER] An aggregate whose particles possess well-defined edges formed at the intersection of roughly planar faces.

angular bitstalk See angular bitstock.

angular bitstock [MECH ENG] A bitstock whose handles are positioned to permit its use in corners and other cramped areas. Also known as angular bitstalk.

angular clearance [DES ENG] The relieved space located below the straight of a die, to permit passage of blanks or slugs.

angular-contact bearing [MECH ENG] A rolling-contact antifriction bearing designed to carry heavy thrust loads and also radial loads.

angular cutter [MECH ENG] A tool-steel cutter used for finishing surfaces at angles greater or less than 90° with its axis of rotation.

angular error of closure See error of closure.

angular gear [MECH ENG] A gear that transmits motion between two rotating shafts that are not parallel. Also known as angle gear.

angular lead See angular advance.

angular milling [MECH ENG] Milling surfaces that are flat and at an angle to the axis of the spindle of the milling machine.

angular pitch [DES ENG] The angle determined by the length along the pitch circle of a gear between successive teeth.

angular shear [MECH ENG] A shear effected by two cutting edges inclined to each other to reduce the force needed for shearing.

angulator [ENG] An instrument for converting angles measured on an oblique plane to their corresponding projections on a horizontal plane; the rectoblique plotter and the photoangulator are types.

anhydrous calcium sulfate [MATER] Gypsum from which all the water of crystallization has been removed. Also known as dead-burnt gypsum.

anhydrous gypsum plaster [MATER] Plaster which has a greater percentage of the water of crystallization removed than normal gypsum plasters; used as a finish plaster.

aniline ink [MATER] A fast-drying printing ink that is a solution of a coal-tar dye in an organic solvent or a solution of a pigment in an organic solvent or water.

aniline leather [MATER] Leather whose grain pattern has been accentuated by impregnation with aniline dye.

anilol [MATER] An aniline-alcohol mixture used as a blending compound in petroleum products.

animal black See bone black.

animal glue [MATER] A glue made from the bones, hide, horns, and connective tissues of animals.

animal oil See bone oil.

animal power [MECH ENG] The time rate at which muscular work is done by a work animal, such as a horse, bullock, or elephant.

anionic detergent [MATER] A class of detergents having a negatively charged surface-active ion, such as sodium alkylbenzene sulfonate.

anneal [ENG] To treat a metal, alloy, or glass with heat and then cool to remove internal stresses and to make the material less brittle.

annealing furnace [ENG] A furnace for annealing metals or glass. Also known as annealing oven.

annealing oven See annealing furnace.

annual cost comparison [IND ENG] A method of selecting from among several alternative projects or courses of action on the basis of their annual costs, including depreciation.

annual labor See assessment drilling.

annular [MIN ENG] The space between the casing and wall of a hole or between a drill pipe and casing.

annular auger [DES ENG] A ring-shaped boring tool which cuts an annular channel, leaving the core intact.

annular gear [DES ENG] A gear having a cylindrical form.

annular nozzle [DES ENG] A nozzle with a ring-shaped orifice.

annular section [ENG] The open space between two concentric tubes, pipes, or vessels.

annulus [MECH ENG] A plate serving to protect or cover a machine. [MIN ENG] A canvas-covered frame set at such an angle in

the miner's rocker that the gravel and water passing over it are carried to the head of the machine.

annunciator [ENG] A signaling apparatus which operates electromagnetically and serves to indicate visually, or visually and audibly, whether a current is flowing, has flowed, or has changed direction of flow in one or more circuits.

anode copper [MET] Slabs of refined blister copper used as anodes in the electrolytic refining of copper.

anode corrosion [MET] The disintegration of a metal acting as an anode.

anode furnace [MET] A furnace in which blister copper or impure nickel is refined.

anode metal [MET] The metal used as anode in an electroplating process.

anode mud [MET] An insoluble substance or mixture that collects at the anode in an electrolytic refining or plating process. Also known as anode slime.

anode scrap [MET] Portions of anode copper retrieved from electrolytic refining of the metal.

anode slime See anode mud.

anodic cleaning [MET] The removal of a foreign substance from a metallic surface by electrolysis with the metal as the anode. Also known as anodic pickling; reverse-current cleaning.

anodic coating [MET] A film of oxide produced on a metal by electrolysis with the metal as the anode.

anodic pickling See anodic cleaning.

anodic protection [MET] Reduction of the corrosion rate in an anode by polarizing it into a potential region where the dissolution rates low.

anodic reaction [MET] The reaction in the mechanism of electrochemical corrosion in which the metal forming the anode dissolves in the electrolyte in the form of positively charged ions.

anodize [MET] The formation of a decorative or protective passive film on a metal part by making it the anode of a cell and applying electric current.

anodized aluminum [MET] Aluminum coated with a layer of aluminum oxide by an anodic process in a suitable electrolyte such as chromic acid or sulfuric acid solution.

anodized magnesium [MET] An anodic coating on magnesium produced in one of various electrolytes, mainly of fluorides, phosphates, or chromates.

anomaly finder [ENG] A computer-controlled data-plotting system used on ships to measure and record seismic, gravity, magnetic, and other geophysical data and water depth, time, course, and speed.

antemedium [MATER] A substance, such as wax or resin solvent, used in tissue processing prior to infiltration for histological examination.

antenna tilt error [ENG] Angular difference between the tilt angle of a radar antenna shown on a mechanical indicator, and the electrical center of the radar beam.

anthill [MIN ENG] The cuttings around the hole collar in blasthole drilling.

anthracene oil [MATER] A heavy, green oil that is a coal-tar fraction boiling above 270°C; used as a source of anthracene, phenanthrene, and carbazole.

anthracite duff [MATER] In Wales, fine anthracite screenings used in briquets or mixed with bituminous coal to fuel cement kilns on chain grate stokers.

anthracite fines [MIN ENG] Small pieces of material from an anthracite coal preparation plant, usually below 1/8 inch (3 millimeter) diameter.

antiacid additive [MATER] A substance that prevents or retards corrosive acid formation in crankcase oils during use.

antiacid bronze [MET] A high-lead, acid-resistant bronze used for casting chemical machine parts.

anticollision radar [ENG] A radar set designed to give warning of possible collisions during movements of ships or aircraft.

anticorrosive paint [MATER] A paint formulated with a corrosive-resistant pigment (such as lead chromate, zinc chromate, or red lead) and a chemical- and moisture-resistant binder; used to protect iron and steel surfaces.

anticreeper [CIV ENG] A device attached to a railroad rail to prevent it from moving in the direction of its length.

antidesiccant [MATER] Material applied to plants prior to transplanting to reduce the amount of moisture lost by transpiration.

antidetonant *See* antiknock.

antidrag [AERO ENG] **1.** Describing structural members in an aircraft or missile that are designed or built to resist the effects of drag. **2.** Referring to a force acting against the force of drag.

antifogging compound [MATER] A compound of one or more basic chemicals with filler or extenders for preventing condensation of moisture on glass and other transparent material, such as lenses or windshields.

antifouling coating [MATER] A special paint containing copper used on ships' bottoms to prevent marine organisms from attaching themselves.

antifriction [MECH ENG] Employing a rolling contact instead of a sliding contact.

antifriction alloy [MET] An alloy generally having more than 50% tin as a base and cast as facings of machinery bearings,

domestic equipment, and small parts in rolling contact.

antifriction bearing [MECH ENG] Any bearing having the capability of reducing friction effectively.

antifriction material [ENG] A machine element made of Babbitt metal, lignum vitae, rubber, or a combination of a soft, easily deformable metal overlaid on a hard, resistant one.

antifungal agent [MATER] A chemical compound that either destroys or inhibits the growth of fungi.

anti-g suit *See* g suit.

anti-icing [AERO ENG] The prevention of the formation of ice upon any object, especially aircraft, by means of windshield sprayer, addition of antifreeze materials to the carburetor, or heating the wings and tail.

antiknock [MATER] **1.** Resisting detonation or pinging in spark-ignited engines. **2.** A substance, such as tetraethyllead, added to motor and aviation gasolines to increase the resistance of the fuel to knock in spark-ignited engines. Also known as antidetonant.

antiknock blending value [ENG] The numerical improvement by an antiknock additive to gasoline octane, often a greater amount than the additive's own octane value.

antiknock gasoline [MATER] Gasoline containing, for example, tetraethyllead, which in small amounts prevents or lessens knocking in a spark-ignited engine.

antiknock rating [ENG] Measurement of the ability of an automotive gasoline to resist detonation or pinging in spark-ignited engines.

antimagnetic [ENG] Constructed so as to avoid the influence of magnetic fields, usually by the use of nonmagnetic materials and by magnetic shielding.

antimonial lead [MET] A lead alloy containing up to 25% antimony and possessing greater hardness and tensile strength than lead; used for storage-battery plates, pipes, cable coverings, and roofing.

antinoise microphone [ENG ACOUS] Microphone with characteristics which discriminate against acoustic noise.

Antioch process [MET] A method of plaster molding in which a plaster-water mixture is poured over a pattern, after which the mold is steam-treated, allowed to set in air, dried in an oven, and cooled for use in casting certain alloys.

antique finish [MATER] A paper finish, somewhat rougher than eggshell, obtained by operating wet presses and calender stacks at reduced pressures.

antiquing [ENG] **1.** Producing a rich glow on the surface of a leather by applying stain, wax, or oil, allowing it to set, and rubbing or brushing the leather. **2.** A technique of handling wet paint to expose parts of the undercoat, by combing, graining, or marbling. Also known as broken-color work.

antiradar coating [ENG] A surface treatment used to reduce the reflection of electromagnetic waves so as to avoid radar detection.

antireflection coating [ENG] The application of a thin film of dielectric material to a surface to reduce its reflection and to increase its transmission of light or other electromagnetic radiation.

antiresonance [ENG] The condition for which the impedance of a given electric, acoustic, or dynamic system is very high, approaching infinity.

antiskid plate [ENG] A sheet of metal roughed on both sides and placed between piled objects, such as boxes in a freight car, to prevent sliding.

antislip metal [MET] Metal, usually iron, bronze, or aluminum, containing abrasive grains cast with the metal or rolled out to the surface; used for car steps, stair treads, or floor plates.

antislip paint [MATER] A paint with a high coefficient of friction, caused by addition of sand, wood flour, or cork dust; used on steps, porches, and walkways to prevent slipping.

antismudge ring [BUILD] A frame attached around a ceiling-mounted air diffuser, to minimize the formation of rings of dirt on the ceiling.

antistat See antistatic agent.

antistatic agent [MATER] A material used with textiles, plastics, paper products, or wax polishes to reduce static-electrical charges by allowing the charge to leak off. Also known as antistat.

antistripping agent [MATER] An additive used in an asphaltic binder to overcome the affinity of an aggregate for water instead of asphalt; assists the asphalt to adhere to wet surfaces.

AN-TNT slurry [MATER] A mixture of ammonium nitrate and trinitrotoluene; used as an explosive.

anvil [ENG] **1.** The part of a machine that absorbs the energy delivered by a sharp force or blow. **2.** The stationary end of a micrometer caliper. [MET] **1.** A heavy wrought-iron, cast-iron, or steel block upon which metal is hammered in smith forging. **2.** The base of the hammer, holding the die bed and lower die part in drop forging.

anyris oil [MATER] Sandalwood oil from West India sandalwood species.

AOQL See average outgoing quality limit.

apex [ENG] In architecture or construction, the highest point, peak, or tip of any structure.

apitong [MATER] A wood from the Philippine tree *Dipterocarpus grandiflorus*; sold as mahogany although it is not a true mahogany.

Apollo program [AERO ENG] The scientific and technical program of the United States that involved placing men on the moon and returning them safely to earth.

apophorometer [ENG] An apparatus used to identify minerals by sublimation.

apparent density [MET] The weight per unit volume of a metal powder, in contrast to the weight per unit volume of the individual particles.

apparent source See effective center.

appliance [ENG] A piece of equipment that draws electric or other energy and produces a desired work-saving or other result, such as an electric heater, a radio, or an electronic range.

appliance panel [ENG] In electric systems, a metal housing containing two or more devices (such as fuses) for protection against excessive current in circuits which supply portable electric appliances.

applications technology satellite [AERO ENG] Any artificial satellite in the National Aeronautics and Space Administration program for the evaluation of advanced techniques and equipment for communications, meteorological, and navigation satellites. Abbreviated ATS.

applied research [ENG] Research directed toward using knowledge gained by basic research to make things or to create situations that will serve a practical or utilitarian purpose.

applied strategic research [ENG] Research done to provide a basic understanding of a current applied project.

applied trim [BUILD] Supplementary and separate decorative strips of wood or moldings applied to the face or sides of a frame, such as a doorframe.

approach [MECH ENG] The difference between the temperature of the water leaving a cooling tower and the wet-bulb temperature of the surrounding air.

approach signal [CIV ENG] A railway signal warning an engineer of a signal ahead that displays a restrictive indication.

approved flame safety lamp [MIN ENG] A flame safety lamp which has been approved for use in gaseous coal mines.

apricot kernel oil See persic oil.

apron [AERO ENG] A protective device specially designed to cover an area surrounding the fuel inlet on a rocket or spacecraft. [BUILD] **1.** A board on an interior wall beneath a windowsill. **2.** The vertical rear panel of a sink attached to a wall. [CIV ENG] **1.** A covering of a material such as concrete or timber over soil to prevent erosion by flowing water, as at the bottom of a dam. **2.** A concrete or wooden shield that is situated along the bank of a river, along a sea wall, or below a dam.

apron conveyor [MECH ENG] A conveyor used for carrying granular or lumpy material and consisting of two strands of roller chain separated by overlapping plates, forming the carrying surface, with sides 2–6 inches high.

apron feeder [MECH ENG] A limited-length version of apron conveyor used for controlled-rate feeding of pulverized materials to a process or packaging unit. Also known as plate-belt feeder; plate feeder.

apron flashing [BUILD] **1.** The flashing that covers the joint between a vertical surface and a sloping roof, as at the lower edge of a chimney. **2.** The flashing that diverts water from a vertical surface into a gutter.

apron lining [BUILD] The piece of boarding which covers the rough apron piece of a staircase.

apron piece [BUILD] A beam that supports a landing or a series of winders in a staircase.

apron rail [BUILD] A lock rail having a raised ornamental molding.

apron wall [BUILD] In an exterior wall, a panel which extends downward from a windowsill to the top of a window below.

APU *See* auxiliary power unit.

AQL *See* acceptable quality level.

Aquagel [MATER] A hydrous silicate of alumina, of the Silica Products Company, used for waterproofing concrete.

aqualung [ENG] A self-contained underwater breathing apparatus (scuba) of the demand or open-circuit type developed by J.Y. Cousteau.

aqueduct [CIV ENG] An artificial tube or channel for conveying water.

Ar₁ [MET] The temperature at which conversion of austenite to ferrite or to ferrite plus cementite is completed upon cooling a steel.

Ar₃ [MET] The temperature at which austenite begins to convert to ferrite upon cooling a steel.

Ar₄ [MET] The temperature at which delta ferrite is converted to gamma iron (austenite) upon cooling a steel.

Ar_cm [MET] The temperature at which austenite is converted to cementite upon cooling a hypereutectoid steel.

arachis oil *See* peanut oil.

Aramite [MATER] A miticide; used for the control of phytophagous mites.

arbitration [IND ENG] A semijudicial means of settling labor-management disputes in which both sides agree to be bound by the decision of one or more neutral persons selected by some method mutually agreed upon.

arbitration bar [MET] A cast-iron specimen, in the form of a standard-sized bar, to be tested for conformity to specifications of the American Society for Testing and Materials.

arbor [MET] A device which supports sand cores in molds.

arbor collar [ENG] A cylindrical spacer that positions and secures a revolving cutter on an arbor.

arborescent powder *See* dendritic powder.

arbor hole [DES ENG] A hole in a revolving cutter or grinding wheel for mounting it on an arbor.

arbor press [MECH ENG] A machine used for forcing an arbor or a mandrel into drilled or bored parts preparatory to turning or grinding. Also known as mandrel press.

arbor support [ENG] A device to support the outer end or intermediate point of an arbor.

arbor vitae oil *See* thuja oil.

arc [ENG] The graduated scale of an instrument for measuring angles, as a marine sextant; readings obtained on that part of the arc beginning at zero and extending in the direction usually considered positive are popularly said to be on the arc, and those beginning at zero and extending in the opposite direction are said to be off the arc.

arc blow [MET] The shifting of the arc in various directions in electric-arc welding due to the magnetic fields at the arc.

arc brazing [MET] Brazing with the use of an electric arc.

arc cutting [MET] A type of thermal cutting of metal using the temperature generated by an electric arc.

arc furnace [MET] A furnace used to heat materials by the energy from an electric arc. Also known as electric-arc furnace.

arc gouging [MET] The formation of a bevel or groove by an arc cutting process.

arch [CIV ENG] A structure curved and so designed that when it is subjected to vertical loads, its two end supports exert reaction forces with inwardly directed horizontal components; common uses for the arch are as a bridge, support for a roadway or railroad track, or part of a building.

arch band [CIV ENG] Any narrow elongated surface forming part of or connected with an arch.

arch bar [BUILD] **1.** A curved chimney bar. **2.** A curved bar in a window sash.

arch beam [CIV ENG] A curved beam, used in construction, with a longitudinal section bounded by two arcs having different radii and centers of curvature so that the beam cross section is larger at either end than at the center.

arch blocks [MIN ENG] Blocks applied to the wooden voussoirs used in framing a timber support for the roof when driving a tunnel.

arch brace [BUILD] A curved brace, usually used in pairs to support a roof frame and give the effect of an arch.

arch brick [MATER] **1.** A wedge-shaped brick used in arches. **2.** An overburned brick, resulting from contact with the fire in the arch of a kiln.

arch bridge [CIV ENG] A bridge having arches as the main supports.

arch center [CIV ENG] A temporary structure for support of the parts of a masonry or concrete arch during its construction.

arch corner bead [BUILD] A corner bead which is cut on the job; used to form and reinforce the curved portion of arch openings.

arch dam [CIV ENG] A dam having a curved face on the downstream side, the curve being roughly a portion of a cylinder whose axis is vertical.

arc heating [MET] The heating of a material by the heat energy from an electric arc, which has a very high temperature and very high concentration of heat energy. Also known as electric-arc heating.

arched construction [BUILD] A method of construction relying on arches and vaults to support walls and floors.

arch girder [CIV ENG] A normal H-section steel girder bent to a circular shape.

arch-gravity dam [CIV ENG] An arch dam stabilized by gravity due to great mass and breadth of the base.

Archimedes' screw [MECH ENG] A device for raising water by means of a rotating broad-threaded screw or spirally bent tube within an inclined hollow cylinder.

arching [CIV ENG] **1.** The transfer of stress from a yielding part of a soil mass to adjoining less-yielding or restrained parts of the mass. **2.** A system of arches. **3.** The arched part of a structure. [MIN ENG] Curved support for roofs of openings in mines.

architectural acoustics [CIV ENG] The science of planning and building a structure to ensure the most advantageous flow of sound to all listeners.

architectural bronze [MET] An alloy containing 57% copper, 40% zinc, 2.75% lead, 0.25% tin; used for extruded moldings and forgings.

architectural concrete [MATER] Concrete used for ornamentation or finish on exterior or interior surfaces of a building or other structure.

architectural engineering [CIV ENG] The branch of engineering dealing primarily with building materials and components and with the design of structural systems for buildings, in contrast to heavy construction such as bridges.

architectural millwork [CIV ENG] Ready-made millwork especially fabricated to meet the specifications for a particular job, as distinguished from standard or stock items or sizes. Also known as custom millwork.

architectural terra-cotta [MATER] A hard-burnt, glazed or unglazed clay unit used in building construction.

architectural volume [CIV ENG] The cubic content of a building calculated by multiplying the floor area by the height.

architecture [ENG] **1.** The art and science of designing buildings. **2.** The product of this art and science.

arch press [MECH ENG] A punch press having an arch-shaped frame to permit operations on wide work.

arch rib [CIV ENG] One of a set of projecting molded members subdividing the undersurface of an arch.

arch ring [CIV ENG] A curved member that provides the main support of an arched structure.

arch truss [CIV ENG] A truss having the form of an arch or arches.

arc jet engine [AERO ENG] An electromagnetic propulsion engine used to supply motive power for flight; hydrogen and ammonia are used as the propellant, and some plasma is formed as the result of electric-arc heating.

arc melting [MET] Melting and purification of metal in an electric-arc furnace.

arc of action *See* arc of contact.

arc of approach [DES ENG] In toothed gearing, the part of the arc of contact along which the flank of the driving wheel contacts the face of the driven wheel.

arc of contact [MECH ENG] **1.** The angular distance over which a gear tooth travels while it is in contact with its mating tooth. Also known as arc of action. **2.** The angular distance a pulley travels while in contact with a belt or rope.

arc of recess [DES ENG] In toothed gearing, the part of the arc of contact wherein the face of the driving wheel touches the flank of the driven wheel.

arc-seam weld [MET] A linear weld or overlapping spot welds made by an arc welding process.

arc-spot weld [MET] A weld that covers a very small area of the surface in contact and was made by an arc welding process.

arc spraying [MET] Depositing on a surface a metal melted by an electric arc and blown at high speed in an atomized state.

arcticization [ENG] The preparation of equipment for operation in an environment of extremely low temperatures.

arc time [MET] The time interval during which an arc is maintained in the making of an arc weld.

arc triangulation [ENG] A system of triangulation in which an arc of a great circle on the surface of the earth is followed in order to tie in two distant points.

arc voltage [MET] The voltage across a welding arc.

arcwall coal cutter [MIN ENG] A type of electric or compressed-air cutter for under- or overcutting a coal seam in narrow work.

arc welding *See* electric-arc welding.

area coverage [ENG] Complete coverage of an area by aerial photography having parallel overlapping flight lines and stereoscopic overlap between exposures in the line of flight.

area drain [CIV ENG] A receptacle designed to collect surface or rain water from an open area.

area landfill [CIV ENG] A sanitary landfill operation that takes care of the solid waste of more than one municipality in a region.

area light [CIV ENG] **1.** A source of light with significant dimensions in two directions, such as a window or luminous ceiling. **2.** A light used to illuminate large areas.

areal pattern [PETRO ENG] Distribution pattern of oil-production wells and water- or gas-injection wells over a given oil reservoir.

areal sweep efficiency [PETRO ENG] Percentage of the total oil reservoir or pore volume which is within the area being swept of oil by a displacing fluid, as in a natural or artificial gas drive or water injection.

area meter [ENG] A mechanism to measure fluid flow rate through a fixed-area conduit by the movement of a weighted piston or float supported by the flowing fluid; includes rotameters and piston-type meters.

area rule [AERO ENG] A prescribed method of design for obtaining minimum zero-lift drag for a given aerodynamic configuration, such as a wing-body configuration, at a given speed.

area survey [ENG] A survey of areas large enough to require loops of control.

area triangulation [ENG] A system of triangulation designed to progress in every direction from a control point.

area wall [CIV ENG] A retaining wall around an areaway.

areaway [CIV ENG] An open space at subsurface level adjacent to a building, providing access to and utilities for a basement.

Argand lamp [ENG] A gas lamp having a tube-shaped wick, allowing a current of air inside as well as outside the flame.

argentometer [ENG] A hydrometer used to find the amount of silver salt in a solution.

argil *See* potter's clay.

argol [MATER] Any of several manures used as fuel in parts of Asia.

Ariel [AERO ENG] A series of artificial satellites launched for Britain by the United States.

Arkansas stone [ENG] A whetstone made of Arkansas stone, for sharpening edged tools.

ARL *See* acceptable reliability level.

arm [ENG ACOUS] *See* tone arm.

Armco iron [MET] A relatively pure ingot iron which is made by the Armco process in a basic open-hearth furnace in a manner similar to that used for steel.

arm conveyor [MECH ENG] A conveyor in the form of an endless belt or chain to which are attached projecting arms or shelves which carry the materials.

Armco process [MET] The basic open-hearth process used by Armco Steel Corporation to make nearly chemically pure iron for use as a construction material, in electromagnetic cores, and in special steels.

arm elevator [MECH ENG] A chain elevator with protruding arms to cradle fixed-shape objects, such as drums or barrels, as they are moved upward.

armor castings [MET] A type of armor made of high-alloy steel frequently used when complicated shapes are involved.

armored faceplate [DES ENG] A tamper-proof faceplate or lock front, mortised in the edge of a door to cover the lock mechanism.

armored front [DES ENG] A lock front used on mortise locks that consists of two plates, the underplate and the finish plate.

armored wood [MATER] Wood which is faced on one or both sides with metal sheeting.

armor plate [BUILD] A metal plate which protects the lower part of a door from kicks and scratches, covering the door to a height usually 39 inches (1 meter) or more. [MET] Heavy, flat steel, either surface-hardened or hardened throughout, used as a sheathing for warships, tanks, and so forth to resist penetration and deformation from heavy gunfire.

arrastra *See* arrastre.

arrastre [MIN ENG] A mill comprising a circular, rock-lined pit in which broken ore is pulverized by stones, attached to horizontal poles fastened in a central pillar, which stones are dragged around the pit. Also spelled arrastra.

array radar [ENG] A radar incorporating a multiplicity of phased antenna elements.

array sonar [ENG] A sonar system incorporating a phased array of radiating and receiving transducers.

arrester [ENG] A wire screen at the top of an incinerator or chimney which prevents sparks or burning material from leaving the stack.

arrester hook [AERO ENG] A hook in the tail section of an airplane; used to engage the arrester wires on an aircraft carrier's deck.

arrière-voussure [BUILD] **1.** An arch or vault in a thick wall carrying the thickness of the wall, especially one over a door or window frame. **2.** A relieving arch behind the face of a wall.

arris fillet [BUILD] A triangular wooden piece that raises the slates of a roof against a chimney or wall so that rain runs off.

arris gutter [BUILD] A V-shaped wooden gutter fixed to the eaves of a building.

arris hip tile [BUILD] A special roof tile having an L-shaped cross section, made to fit over the hip of a roof. Also known as hip tile.

arris rail [CIV ENG] A rail of triangular section, usually formed by slitting diagonally a strip of square section.

arrissing tool [ENG] A tool similar to a float, but having a form suitable for rounding an edge of freshly placed concrete.

arris tile [BUILD] Any angularly shaped tile.

arrisways [CIV ENG] Diagonally, in respect to the manner of laying tiles, slates, bricks, or timber. Also known as arriswise.

arriswise *See* arrisways.

arrival rate [IND ENG] The mean number of new calling units arriving at a service facility per unit time.

arrow wing [AERO ENG] An aircraft wing of V-shaped planform, either tapering or of constant chord, suggesting a stylized arrowhead.

articulated drop chute [ENG] A drop chute, for a falling stream of concrete, which consists of a vertical succession of tapered metal cylinders, so designed that the lower end of each cylinder fits into the upper end of the one below.

articulated leader [MECH ENG] A wheel-mounted transport unit with a pivotal loading element used in earth moving.

articulated structure [CIV ENG] A structure in which relative motion is allowed to occur between parts, usually by means of a hinged or sliding joint or joints.

articulated train [ENG] A railroad train whose cars are permanently or semipermanently connected.

artificial aging [MET] The heat treatment of an alloy at moderately elevated temperatures to accelerate precipitation of a component from the supersaturated solid solution.

artificial asteroid [AERO ENG] An object made by humans and placed in orbit about the sun.

artificial feel [AERO ENG] A type of force feedback incorporated in the control system of an aircraft or spacecraft whereby a portion of the forces acting on the control surfaces are transmitted to the cockpit controls.

artificial gravity [AERO ENG] A simulated gravity established within a space vehicle by rotation or acceleration.

artificial harbor [CIV ENG] **1.** A harbor protected by breakwaters. **2.** A harbor formed by dredging.

artificial lift [PETRO ENG] Any method of lifting oil out of underground reservoirs, usually by injecting gas into the rock or sand formation to force fluids from wells; an example is a gas lift.

artificial monument [ENG] A relatively permanent object made by humans, such as an abutment or stone marker, used to identify the location of a survey station or corner.

artificial nourishment [CIV ENG] The process of replenishing a beach by artificial means, such as the deposition of dredged material.

artificial recharge [CIV ENG] The recharge of an aquifer depleted by abnormally large withdrawals, by the use of injection wells and other techniques.

artificial satellite [AERO ENG] Any man-made object placed in a near-periodic orbit in which it moves mainly under the gravitational influence of one celestial body, such as the earth, sun, another planet, or a planet's moon.

artificial variable [IND ENG] One type of variable introduced in a linear program model in order to find an initial basic feasible solution; an artificial variable is used for equality constraints and for greater-than or equal inequality constraints.

artificial ventilation [MIN ENG] The inducing of a flow of air through a mine or part of a mine by mechanical or other means.

artificial voice [ENG ACOUS] **1.** Small loudspeaker mounted in a shaped baffle which is proportioned to simulate the acoustical constants of the human head; used for calibrating and testing close-talking microphones. **2.** Synthetic speech produced by a multiple tone generator; used to produce a voice reply in some real-time computer applications.

artificial weathering [ENG] The controlled production of changes in materials under laboratory conditions to simulate actual outdoor exposure.

asbestine [MATER] A material with the properties of asbestos.

asbestos blanket [MATER] Asbestos fibers (alone or in combination with other fibers) stitched, bonded, or woven into flexible blanket form; used for high-temperature insulation or for fire barriers.

asbestos board [MATER] A sheet of fire-resistant material made from asbestos fiber and portland cement.

asbestos cement [MATER] A building material composed of a mixture of asbestos fiber, portland cement, and water made into plain sheets, corrugated sheets, tiles, and piping.

asbestos-cement cladding [BUILD] Asbestos board and component wall systems, directly supported by wall framing, forming a wall or wall facing.

asbestos-cement pipe [MATER] A concrete pipe made of a mixture of portland cement and asbestos fiber and highly resistant to corrosion; used in drainage systems, waterworks systems, and gas lines. Also known by the trade name Transite pipe.

asbestos felt [MATER] A product made by saturating felted asbestos with asphalt or other suitable binder, such as a synthetic elastomer.

asbestos insulation [MATER] A material composed of asbestos fibers bonded with mixtures of clay or sodium silicate; used as thermal insulation for temperatures above 1500°F (816°C).

asbestos joint runner [MATER] An asbestos rope, wrapped around a pipe and then clamped in position; used to hold molten lead which is poured in a caulked joint. Also known as pouring rope.

asbestos plaster [MATER] A fireproof insulating material generally composed of asbestos with bentonite as the binder.

asbestos roofing [MATER] Roofing or wall cladding sheets made of asbestos cement.

asbestos shingle [MATER] A shingle composed of asbestos cement formed under pressure; used on houses for roofing and siding that resist the destructive effects of time, weather, and fire.

as-brazed [MET] A brazement prior to any additional treatment such as thermal, chemical, or mechanical treatments.

ascending vertical angle *See* angle of elevation.

ascent [AERO ENG] Motion of a craft in which the path is inclined upward with respect to the horizontal.

ash [ENG] An undesirable constituent of diesel fuel whose quantitative measure-

ment indicates degree of fuel cleanliness and freedom from abrasive material.

ash collector *See* dust chamber.

ash content [MATER] The mass of incombustible material remaining after burning a given coal sample as a percentage of the orginal mass of the coal.

ash conveyor [MECH ENG] A device that transports refuse from a furnace by fluid or mechanical means.

ash dump [ENG] An opening in the floor of a fireplace or firebox through which ashes are swept to an ash pit below.

ash furnace [ENG] A furnace in which materials are fritted for glassmaking.

ashlar [CIV ENG] Masonry with an exposed side of square or rectangular stones.

ashlar brick [MATER] A brick having rough-hackled faces resembling stone.

ashlar line [BUILD] The outer line of a wall above any projecting base.

ash pan [ENG] A metal receptacle beneath a fireplace or furnace grating for collection and removal of ashes.

ash pit [BUILD] The ash-collecting area beneath a fireplace hearth.

ash pit door [ENG] A cast-iron door providing access to an ash pit for ash removal.

A size [ENG] One of a series of sizes to which trimmed paper and board are manufactured; for size AN, with N equal to any integer from 0 to 10, the length of the longer side is $2^{-(2N-1)/4}$ meters, while the length of the shorter side is $2^{-(2N+1)/4}$ meters, with both lengths rounded off to the nearest millimeter.

aspect [CIV ENG] Of railway signals, what the engineer sees when viewing the blades or lights in their relative positions or colors.

aspect angle [ENG] The angle formed between the longitudinal axis of a projectile in flight and the axis of a radar beam.

aspect ratio [AERO ENG] The ratio of the square of the span of an airfoil to the total airfoil area, or the ratio of its span to its mean chord. [DES ENG] **1.** The ratio of frame width to frame height in television; it is 4:3 in the United States and Britain. **2.** In any rectangular configuration (such as the cross section of a rectangular duct), the ratio of the longer dimension to the shorter.

asphalt [MATER] A brown to black, hard, brittle, or plastic bituminous material composed principally of hydrocarbons; formed in oil-bearing rocks near the Dead Sea, and in Trinidad; prepared by pyrolysis from coal tar, certain petroleums, and lignite tar; melts on heating, insoluble in water but soluble in gasoline; used for paving and roofing and in paints and varnishes.

asphalt base *See* naphthene base.

asphalt base crude [MATER] A crude oil which yields napthenic asphalt residues when processed.

asphalt block [MATER] A paving block composed of a mixture of 88–92% crushed stone with the balance asphalt cement.

asphalt cement [MATER] Fluxed or unfluxed asphalt especially prepared for direct use in making bituminous pavements.

asphalt cutter [MECH ENG] A powered machine having a rotating abrasive blade; used to saw through bituminous surfacing material.

asphalt emulsion [MATER] Asphalt cement in water containing a small amount of emulsifying agent.

asphalt-emulsion slurry seal [MATER] A mixture of slow-setting emulsified asphalt, fine aggregate, and mineral filler, with water added to produce a slurry consistency.

asphaltene [MATER] Any of the dark, solid constituents of crude oils and other bitumens which are soluble in carbon disulfide but insoluble in paraffin naphthas; they hold most of the organic constituents of bitumens.

asphalt flux [MATER] An oil used to reduce the consistency or viscosity of hard asphalt to the point required for use.

asphalt fog seal [MATER] An asphalt surface treatment consisting of a light application of liquid asphalt without a mineral aggregate cover.

asphalt heater [ENG] A piece of equipment for raising the temperature of bitumen used in paving.

asphaltic base oil [MATER] A crude oil that has asphaltum as the predominating solid residual.

asphaltic concrete [MATER] A special concrete consisting of a mixture of graded aggregate and heated asphalt; must be applied and spread while hot.

asphaltic road oil [MATER] A thick, fluid solution of asphalt.

asphalt lamination [MATER] A laminate of sheet material, such as paper or felt, which uses asphalt as the adhesive.

asphalt leveling course [CIV ENG] A layer of an asphalt-aggregate mixture of variable thickness, used to eliminate irregularities in contour of an existing surface, prior to the placement of a superimposed layer.

asphalt macadam [MATER] Pavement made with an asphalt binder rather than tar.

asphalt mastic [MATER] A mixture of asphalt with sand, asbestos, crushed rock, or similar material; used like cement. Also known as mastic asphalt.

asphalt overlay [CIV ENG] One or more layers of asphalt construction on an existing pavement.

asphalt paint [MATER] Asphaltic material dissolved in volatile solvent with or without pigments, drying oils, resins, and so on.

asphalt paper [MATER] A paper that is coated or impregnated with asphalt.

asphalt pavement [CIV ENG] A pavement consisting of a surface layer of mineral aggregate, coated and cemented together with asphalt cement on supporting layers.

asphalt primer [MATER] Low-viscosity, liquid asphaltic material applied to and absorbed by nonbituminous surfaces, as in waterproofing.

asphalt roofing [MATER] A roofing material made by impregnating a dry roofing felt with a hot asphalt saturant, applying asphalt coatings to the weather and reverse sides, and embedding a mineral surfacing in the coating on the weather side.

asphalt shingle [MATER] A roof shingle made of felt impregnated with asphalt and covered with mineral granules.

asphalt soil stabilization [CIV ENG] The treatment of naturally occurring nonplastic or moderately plastic soil with liquid asphalt at normal temperatures to improve the load-bearing qualities of the soil.

asphalt tile [MATER] Floor tile composed of asbestos fibers, mineral coloring pigments, and inert fillers bound together; used on rigid subfloors or hardwood floors.

asphaltum [MATER] Bituminous material in oil of turpentine; used in photomechanical work because of its ability to be rendered insoluble in light.

aspirating *See* dedusting.

aspirating burner [ENG] A burner in which combustion air at high velocity is drawn over an orifice, creating a negative static pressure and thereby sucking fuel into the stream of air; the mixture of air and fuel is conducted into a combustion chamber, where the fuel is burned in suspension.

aspirating screen [MIN ENG] A vibrating screen from which light, liberated particles are removed by suction.

aspiration meteorograph [ENG] An instrument for the continuous recording of two or more meteorological parameters, with the ventilation being provided by a suction fan.

aspiration psychrometer [ENG] A psychrometer in which the ventilation is provided by a suction fan.

aspiration thermograph [ENG] A thermograph in which ventilation is provided by a suction fan.

aspirator [ENG] Any instrument or apparatus that utilizes a vacuum to draw up gases or granular materials. [MIN ENG] A device made of wire gauze, of cloth, or of a fibrous mass held between pieces of meshed material and used to cover the mouth and nose to keep dusts from entering the lungs.

Assam psychrometer [ENG] A psychrometer, shielded from radiation, in which the air is blown over the bulbs of the two thermometers with a small fan.

assault aircraft [AERO ENG] Powered aircraft, including helicopters, which move assault troops and cargo into an objective area and which provide for their resupply.

assay balance [ENG] A sensitive balance used in the assaying of gold, silver, and other precious metals.

assay bar [MET] A bar of pure or nearly pure gold and silver; used by a government as a standard.

assay plan [MIN ENG] A mine map showing the assay, stope, width, and so forth of samples taken from positions marked.

assay pound [MIN ENG] A weight which varies from time to time but is sometimes 0.5 gram, and is used by assayers to proportionately represent a pound.

assay ton [MIN ENG] A unit of weight of ore equal to 29,167 milligrams; the number of milligrams of precious metal in this measure equals the number of troy ounces in a short ton.

assay value [MIN ENG] The amount of gold or silver as shown by assay of any given sample and represented by ounces per ton of ore.

assay walls [MIN ENG] The planes to which an ore body can be profitably mined, the limiting factor being the metal content of the country rock as determined from assays.

assembled stone [MATER] A stone made of two or more gem materials, whether genuine or imitation.

assembling bolt [CIV ENG] A threaded bolt for holding together temporarily the several parts of a structure during riveting.

assembly [MECH ENG] A unit containing the component parts of a mechanism, machine, or similar device.

assembly line [IND ENG] A mass-production arrangement whereby the work in process is progressively transferred from one operation to the next until the product is assembled.

assembly-line balancing [IND ENG] Assigning numbers of operators or machines to each operation of an assembly line so as to meet the required production rate with a minimum of idle time.

assembly method [IND ENG] The technique used to assemble a manufactured product, such as hand assembly, progressive line assembly, and automatic assembly.

assembly time [ENG] **1.** The elapsed time after the application of an adhesive until its strength becomes effective. **2.** The time elapsed in performing an assembly or subassembly operation.

assessment drilling [MIN ENG] Drilling to fulfill the requirement that a prescribed amount of work be done annually on an unpatented mining claim to retain title. Also known as annual labor.

assessment work [MIN ENG] Annual work at an unpatented mining claim in the public domain performed under law to maintain the claim title.

assets [IND ENG] All the resources, rights, and property owned by a person or a company; the book value of these items as shown on the balance sheet.

assignable cause [IND ENG] Any identifiable factor which causes variation in a process outside the predicted limits, thereby altering quality.

assisted takeoff [AERO ENG] A takeoff of an aircraft or a missile by using a supplementary source of power, usually rockets.

assize [CIV ENG] **1.** A cylindrical block of stone forming one unit in a column. **2.** A layer of stonework.

Assmann psychrometer [ENG] A special form of the aspiration psychrometer in which the thermometric elements are well shielded from radiation.

associated gas [PETRO ENG] Gaseous hydrocarbons occurring as a free-gas phase under original oil-reservoir conditions of temperature and pressure.

assumed plane coordinates [ENG] A local plane-coordinate system set up at the convenience of the surveyor.

assured mineral *See* developed reserves.

astatic galvanometer [ENG] A sensitive galvanometer designed to be independent of the earth's magnetic field.

astatic governor *See* isochronous governor.

astatic gravimeter [ENG] A sensitive gravimeter designed to measure small changes in gravity.

astatic magnetometer [ENG] A magnetometer for determining the gradient of a magnetic field by measuring the difference in reading from two magnetometers placed at different positions.

astatic wattmeter [ENG] An electrodynamic wattmeter designed to be insensitive to uniform external magnetic fields.

astatized gravimeter [ENG] A gravimeter, sometimes referred to as unstable, where the force of gravity is maintained in an unstable equilibrium with the restoring force.

astel [MIN ENG] An overhead boarding or arching in a mine gallery.

astern [ENG] To the rear of an aircraft, vehicle, or vessel; behind; from the back.

ASTM-CFR engine [MECH ENG] A special engine developed by the Coordinating Fuel and Equipment Research Committee of the Coordinating Research Council, Inc., to determine the knock tendency of gasolines.

Aston process [MET] A process for making controlled-quality wrought iron synthetically.

astragal [BUILD] **1.** A small convex molding decorated with a string of beads or bead-and-reel shapes. **2.** A plain bead molding. **3.** A member, or combination of members, fixed to one of a pair of doors or casement windows to cover the joint between the meeting stiles and to close the clearance gap.

astragal front [DES ENG] A lock front which is shaped to fit the edge of a door with an astragal molding.

astral dome *See* astrodome.

astral lamp [ENG] An Argand lamp designed so that its light is not prevented from reaching a table beneath it by the flattened annular reservoir holding the oil.

astrodome [AERO ENG] A transparent dome in the fuselage or body of an aircraft or spacecraft intended primarily to permit taking celestial observations in navigating. Also known as astral dome; navigation dome.

astrodynamics [AERO ENG] The practical application of celestial mechanics, astroballistics, propulsion theory, and allied fields to the problem of planning and directing the trajectories of space vehicles.

astrolabe [ENG] An instrument designed to observe the positions and measure the altitudes of celestial bodies.

astronaut [AERO ENG] In United States terminology, a person who rides in a space vehicle.

astronautical engineering [AERO ENG] The engineering aspects of flight in space.

astronautics [AERO ENG] **1.** The art, skill, or activity of operating spacecraft. **2.** The science of space flight.

astronomical instruments [ENG] Specific kinds of telescopes and ancillary equipment used by astronomers to study the positions, motions, and composition of stars and members of the solar system.

astronomical theodolite *See* altazimuth.

astronomical traverse [ENG] A survey traverse in which the geographic positions of the stations are obtained from astronomical observations, and lengths and azimuths of lines are obtained by computation.

as-welded [MET] The condition of a weldment prior to any additional thermal, chemical, or mechanical treatment.

asymmetric rotor [MECH ENG] A rotating element for which the axis (center of rotation) is not centered in the element.

asynchronous timing [IND ENG] A simulation method for queues in which the system model is updated at each arrival or departure, resulting in the master clock being increased by a variable amount.

athermalize [ENG] To make independent of temperature or of thermal effects.

Athey wheel [MECH ENG] A crawler wheel assembly used on tractors for moving over soft terrain.

athodyd [AERO ENG] A type of jet engine, consisting essentially of a duct or tube of varying diameter and open at both ends, which admits air at one end, compresses it by the forward motion of the engine, adds heat to it by the combustion of fuel, and discharges the resulting gases at the other end to produce thrust.

Atlas-Centaur launch vehicle [AERO ENG] A two-stage rocket consisting of an Atlas first stage and a Centaur second stage; used for launching uncrewed spacecraft.

Atlas-Johnson tubing joint [PETRO ENG] A tapered, screw-on joint for connecting lengths of tubing for oil-well casing strings.

atmidometer *See* atmometer.

atmometer [ENG] The general name for an instrument which measures the evaporation rate of water into the atmosphere. Also known as atmidometer; evaporation gage; evaporimeter.

atmospheric braking [AERO ENG] **1.** Slowing down an object entering the atmosphere of the earth or other planet from space by using the drag exerted by air or other gas particles in the atmosphere. **2.** The action of the drag so exerted.

atmospheric cooler [MECH ENG] A fluids cooler that utilizes the cooling effect of ambient air surrounding the hot, fluids-filled tubes.

atmospheric corrosion [MET] The gradual destruction or alteration of a metal or alloy by contact with substances present in the atmosphere, such as oxygen, carbon dioxide, water vapor, and sulfur and chlorine compounds.

atmospheric entry [AERO ENG] The penetration of any planetary atmosphere by any object from outer space; specifically, the penetration of the earth's atmosphere by a crewed or uncrewed capsule or spacecraft.

atmospheric impurity [ENG] An extraneous substance that is mixed as a contaminant with the air of the atmosphere.

atmospheric steam curing [ENG] The steam curing of concrete or cement products at atmospheric pressure, usually at a maximum ambient temperature between 100 and 200°F (40 and 95°C).

atomic hydrogen welding [MET] An arc welding process in which hydrogen gas dissociated by the arc recombines outside the arc to provide intense heat and protection against oxidation for the weld.

atomic moisture meter [ENG] An instrument that measures the moisture content of coal instantaneously and continuously by bombarding it with neutrons and measuring the neutrons which bounce back to a detector tube after striking hydrogen atoms of water.

atomic rocket [AERO ENG] A rocket propelled by an engine in which the energy for the jetstream is to be generated by nuclear fission or fusion. Also known as nuclear rocket.

atomization [MECH ENG] The mechanical subdivision of a bulk liquid or meltable solid, such as certain metals, to produce drops, which vary in diameter depending on the process from under 10 to over 1000 micrometers.

atomizer [MECH ENG] A device that produces a mechanical subdivision of a bulk liquid, as by spraying, sprinkling, misting, or nebulizing.

atomizer burner [MECH ENG] A liquid-fuel burner that atomizes the unignited fuel into a fine spray as it enters the combustion zone.

atomizer mill [MECH ENG] A solids grinder, the product from which is a fine powder.

atomizing humidifier [MECH ENG] A humidifier in which tiny particles of water are introduced into a stream of air.

atom probe [ENG] An instrument for identifying a single atom or molecule on a metal surface; it consists of a field ion microscope with a probe hole in its screen opening into a mass spectrometer; atoms that are removed from the specimen by pulsed field evaporation fly through the probe hole and are detected in the mass spectrometer.

ATS *See* applications technology satellite.

attached thermometer [ENG] A thermometer which is attached to an instrument to determine its operating temperature.

attack plane [AERO ENG] A multiweapon carrier aircraft which can carry bombs, torpedoes, and rockets.

attar of roses *See* rose oil.

attemperation [ENG] The regulation of the temperature of a substance.

attemperation of steam [MECH ENG] The controlled cooling, in a steam boiler, of steam at the superheater outlet or between the primary and secondary stages of the superheater to regulate the final steam temperature.

attenuate [ENG ACOUS] To weaken a signal by reducing its level.

attic [BUILD] The part of a building immediately below the roof and entirely or partly within the roof framing.

attic tank [BUILD] An open tank which is installed above the highest plumbing fixture in a building and which supplies water to the fixtures by gravity.

atticurge [BUILD] Of a doorway, having jambs which are inclined slightly inward, so that the opening is wider at the threshold than at the top.

attic ventilator [BUILD] A mechanical fan located in the attic space of a residence; usually moves large quantities of air at a relatively low velocity.

attitude [AERO ENG] The position or orientation of an aircraft, spacecraft, and so on, either in motion or at rest, as determined by the relationship between its axes and some reference line or plane or some fixed system of reference axes.

attitude control [AERO ENG] **1.** The regulation of the attitude of an aircraft, spacecraft, and so on. **2.** A device or system that automatically regulates and corrects attitude, especially of a pilotless vehicle.

attitude gyro [AERO ENG] Also known as attitude indicator. **1.** A gyro-operated flight instrument that indicates the attitude of an aircraft or spacecraft with respect to a reference coordinate system. **2.** Any gyro-operated instrument that indicates attitude.

attitude indicator *See* attitude gyro.

attitude jet [AERO ENG] **1.** A stream of gas from a jet used to correct or alter the attitude of a flying body either in the atmosphere or in space. **2.** The nozzle that directs this jetstream.

attribute sampling [IND ENG] A quality-control inspection method in which the sampled articles are classified only as defective or nondefective.

attributes testing [ENG] A reliability test procedure in which the items under test are classified according to qualitative characteristics.

attrition [MATER] Wear caused by rubbing or friction; for metal surfaces, also known as scoring; scouring.

attrition mill [MECH ENG] A machine in which materials are pulverized between two toothed metal disks rotating in opposite directions.

Atwood machine [MECH ENG] A device comprising a pulley over which is passed a stretch-free cord with a weight hanging on each end.

audible leak detector [ENG] A device used as an auxiliary to the main leak detector for conversion of the output signal into audible sound.

audio-frequency meter [ENG] One of a number of types of frequency meters usable in the audio range; for example, a resonant-reed frequency meter.

audiometer [ENG] An instrument composed of an oscillator, amplifier, and attenuator and used to measure hearing acuity for pure tones, speech, and bone conduction.

audio-modulated radiosonde [ENG] A radiosonde with a carrier wave modulated by audio-frequency signals whose frequency is controlled by the sensing elements of the instrument.

audio patch bay [ENG ACOUS] Specific patch panels provided to terminate all audio circuits and equipment used in a channel and technical control facility; this equipment can also be found in transmitting and receiving stations.

audio spectrometer *See* acoustic spectrometer.

audio system *See* sound-reproducing system.

audiphone [ENG ACOUS] A device that enables persons with certain types of deafness to hear, consisting of a plate or diaphragm that is placed against the teeth and transmits sound vibrations to the inner ear.

auger [DES ENG] **1.** A wood-boring tool that consists of a shank with spiral channels ending in two spurs, a central tapered feed screw, and a pair of cutting lips. **2.** A large augerlike tool for boring into soil.

auger bit [DES ENG] A bit shaped like an auger but without a handle; used for wood boring and for earth drilling. [MIN ENG] Hard steel or tungsten carbide–tipped cutting teeth used in an auger running on a torque bar or in an auger-drill head running on a continuous-flight auger.

auger boring [ENG] **1.** The hole drilled by the use of auger equipment. **2.** *See* auger drilling.

auger conveyor *See* screw conveyor.

auger drilling [ENG] A method of drilling in which penetration is accomplished by the cutting or gouging action of chisel-type cutting edges forced into the substance by rotation of the auger bit. Also known as auger boring.

auger packer [MECH ENG] A feed mechanism that uses a continuous auger or screw inside a cylindrical sleeve to feed hard-to-flow granulated solids into shipping containers, such as bags or drums.

auget [ENG] A priming tube, used in blasting. Also spelled augette.

augette *See* auget.

augmentation system [AERO ENG] An electronic servomechanism or feedback control system which provides improvements in aircraft performance or pilot handling characteristics over that of the basic unaugmented aircraft.

augmenter tube [AERO ENG] A tube or pipe, usually one of several, through which the exhaust gases from an aircraft reciprocating engine are directed to provide additional thrust.

Augustin process [MET] A silver extraction process by roasting to form a chloride, leaching with a salt solution, and precipitating by metallic copper.

ausforging [MET] The forming of austenitic steel into required shapes by hammering or pressing after cooling.

ausrolling [MET] The working of austenitic steel by passing it, after cooling, between oppositely revolving rollers.

austempering [MET] A process for the heat treatment of austenitic steel.

austenite [MET] Gamma iron with carbon in solution.

austenitic [MET] Composed mainly of austenite.

austenitic cast iron [MET] The product resulting from changing the basic crystalline structure of gray or ductile iron by alloying it with substantial amounts of nickel, manganese, silicon, or other elements.

austenitic manganese steel *See* Hadfield manganese steel.

austenitic stainless steel [MET] Stainless steel composed principally of austenite made stable by alloying with nickel.

austenitic steel [MET] An alloy whose structure is typically that of austenite at room temperature.

austenitize [MET] To heat steel to the temperature range in which the crystalline form of the iron is austenite.

autoclave [ENG] An airtight vessel for heating and sometimes agitating its contents under high steam pressure; used for industrial processing, sterilizing, and cooking with moist or dry heat at high temperatures.

autoclave curing [ENG] Steam curing of concrete products, sand-lime brick, asbestos cement products, hydrous calcium silicate insulation products, or cement in an autoclave at maximum ambient temperatures generally between 340 and 420°F (170 and 215°C).

autoclave molding [ENG] A method of curing reinforced plastics that uses an autoclave with 50–100 psi (345–690 kilopascals) steam pressure to set the resin.

autofrettage [ENG] A process for manufacturing gun barrels; prestressing the metal increases the load at which its permanent deformation occurs.

autogenous grinding [MECH ENG] The secondary grinding of material by tumbling the material in a revolving cylinder, without balls or bars taking part in the operation.

autogenous healing [ENG] A natural process of closing and filling cracks in concrete or mortar while it is kept damp.

autogenous mill *See* autogenous tumbling mill.

autogenous tumbling mill [MECH ENG] A type of ball-mill grinder utilizing as the grinding medium the coarse feed (incoming) material. Also known as autogenous mill.

autogenous volume change [MATER] The change in volume produced by continued hydration of cement, exclusive of effects of external forces or change of water content or temperature.

autogenous welding [MET] A fusion welding process using heat without the addition of filler metal to join two pieces of the same metal.

autogiro [AERO ENG] A type of aircraft which utilizes a rotating wing (rotor) to provide lift and a conventional engine-propeller combination to pull the vehicle through the air.

autoignition [MECH ENG] Spontaneous ignition of some or all of the fuel-air mixture in the combustion chamber of an internal combustion engine.

automanual system [CIV ENG] A railroad signal system in which signals are set manually but are activated automatically to return to the danger position by a passing train.

automata theory [ENG] The theory concerning the operation principles, application, and behavior characteristics of automatic devices.

automatic [ENG] Having a self-acting mechanism that performs a required act at a predetermined time or in response to certain conditions.

automatic batcher [MECH ENG] A batcher for concrete which is actuated by a single starter switch, opens automatically at the start of the weighing operations of each material, and closes automatically when the designated weight of each material has been reached.

automatic brazing [MET] Brazing by the use of either portable or stationary equipment which does not require constant supervision by the operator.

automatic casing hanger [PETRO ENG] Unitized hanger-seal assembly latched at the lower end of an oil-well casing string to support the next smaller string and make a seal between the two strings.

automatic coupling [MECH ENG] A device which couples rail cars when they are bumped together.

automatic dam [MIN ENG] In placer mining, a dam with a gate that automatically discharges the water when it reaches a certain height behind the dam.

automatic data processing [ENG] The machine performance, with little or no human assistance, of any of a variety of tasks involving informational data; examples include automatic and responsive reading, computation, writing, speaking, directing artillery, and the running of an entire factory. Abbreviated ADP.

automatic door bottom [ENG] A movable plunger, in the form of a horizontal bar at the bottom of a door, which drops automatically when the door is closed, sealing the threshold and reducing noise transmission. Also known as automatic threshold closer.

automatic drill [DES ENG] A straight brace for bits whose shank comprises a coarse-pitch screw sliding in a threaded tube with a handle at the end; the device is operated by pushing the handle.

automatic fire pump [MECH ENG] A pump which provides the required water pressure in a fire standpipe or sprinkler system; when the water pressure in the system drops below a preselected value, a sensor causes the pump to start.

automatic flushing system [CIV ENG] A water tank system which functions automatically for the periodic flushing of urinals or other plumbing fixtures, or of pipes having too small a slope to drain effectively.

automatic microfiller [ENG] A device used to place microfilm in jackets at relatively high speeds.

automatic mold [ENG] A mold used in injection or compression molding of plastic objects so that repeated molding cycles are possible, including ejection, without manual assistance.

automatic press [MECH ENG] A press in which mechanical feeding of the work is synchronized with the press action.

automatic record changer [ENG ACOUS] An electric phonograph that automatically plays a number of records one after another.

automatic sampler [MECH ENG] A mechanical device to sample process streams (gas, liquid, or solid) either continuously or at preset time intervals.

automatic screw machine [MECH ENG] A machine designed to automatically produce finished parts from bar stock at high production rates; the term is not an exact, specific machine-tool classification.

automatic shut-off [ENG ACOUS] A switch in some tape recorders which automatically stops the machine when the tape ends or breaks.

automatic spider [MIN ENG] A foot or hydraulically actuated drill-rod clamping device similar to a Wommer safety clamp.

automatic stability [AERO ENG] Stability achieved with the controls operated by automatic devices, as by an automatic pilot.

automatic stabilization equipment [AERO ENG] Apparatus which automatically operates control devices to maintain an aircraft in a stable condition.

automatic stoker [MECH ENG] A device that supplies fuel to a boiler furnace by mechanical means. Also known as mechanical stoker.

automatic tank battery [PETRO ENG] Interconnected system of storage tanks with automatic controls to direct incoming oil to empty tanks in a desired sequence.

automatic test equipment [ENG] Test equipment that makes two or more tests in sequence without manual intervention; it usually stops when the first out-of-tolerance value is detected.

automatic threshold closer *See* automatic door bottom.

automatic time switch [ENG] Combination of a switch with an electric or spring-wound clock, arranged to turn an apparatus on and off at predetermined times.

automatic track shift [ENG ACOUS] A system used with multiple-track magnetic tape recorders to index the tape head, after one track is played, to the correct position for the start of the next track.

automatic-type belt tensioning device [MECH ENG] Any device which maintains a predetermined tension in a conveyor belt.

automatic typewriter [ENG] An electric typewriter that produces a punched paper tape or magnetic recording simultaneously with the conventional typed copy, for subsequent automatic retyping at high speed.

automatic volume compressor *See* volume compressor.

automatic volume expander *See* volume expander.

automatic wagon control [MIN ENG] A mechanism to keep the speed of wagons within certain designed limits; may consist of small hydraulic units fixed at intervals along the inside of the track.

automatic welding [MET] Electric-arc welding with automatic control of the arc movement along the welding line, the electrode feed, and the arc-gap length.

automatic wet-pipe sprinkler system [ENG] A sprinkler system, all of whose parts are filled with water at sufficient pressure to provide an immediate continuous discharge if the system is activated.

automation [ENG] **1.** The replacement of human or animal labor by machines. **2.**

The operation of a machine or device automatically or by remote control.

automobile [MECH ENG] A four-wheeled, trackless, self-propelled vehicle for land transportation of as many as eight people. Also known as car.

automobile chassis [MECH ENG] The automobile frame, together with the wheels, power train, brakes, engine, and steering system.

automotive air conditioning [MECH ENG] A system for maintaining comfort of occupants of automobiles, buses, and trucks, limited to air cooling, air heating, ventilation, and occasionally dehumidification.

automotive body [ENG] An enclosure mounted on and attached to the frame of an automotive vehicle, to contain passengers and luggage, or in the case of commercial vehicles the commodities being carried.

automotive brake [MECH ENG] A friction mechanism that slows or stops the rotation of the wheels of an automotive vehicle, so that tire traction slows or stops the vehicle.

automotive engine [MECH ENG] The fuel-consuming machine that provides the motive power for automobiles, airplanes, tractors, buses, and motorcycles and is carried in the vehicle.

automotive engineering [MECH ENG] The branch of mechanical engineering concerned primarily with the special problems of land transportation by a four-wheeled, trackless, automotive vehicle.

automotive frame [ENG] The basic structure of all automotive vehicles, except tractors, which is supported by the suspension and upon which or attached to which are the power plant, transmission, clutch, and body or seat for the driver.

automotive fuel [MATER] A material, generally a liquid fuel, gasoline, or distillate, whose combustion is used to supply chemical energy to provide the power for an automotive vehicle.

automotive ignition system [MECH ENG] A device in an automotive vehicle which initiates the chemical reaction between fuel and air in the cylinder charge.

automotive steering [MECH ENG] Mechanical means by which a driver controls the course of a moving automobile, bus, truck, or tractor.

automotive suspension [MECH ENG] The springs and related parts intermediate between the wheels and frame of an automotive vehicle that support the frame on the wheels and absorb road shock caused by passage of the wheels over irregularities.

automotive transmission [MECH ENG] A device for providing different gear or drive ratios between the engine and drive wheels of an automotive vehicle, a principal function being to enable the vehicle to accelerate from rest through a wide speed range while the engine operates within its most effective range.

automotive vehicle [MECH ENG] A self-propelled vehicle or machine for land transportation of people or commodities or for moving materials, such as a passenger car, bus, truck, motorcycle, tractor, airplane, motorboat, or earthmover.

autopatrol [MECH ENG] A self-powered blade grader. Also known as motor grader.

autopilot *See* automatic pilot.

autoradiography [ENG] A technique for detecting radioactivity in a specimen by producing an image on a photographic film or plate. Also known as radioautography.

autorail [MECH ENG] A self-propelled vehicle having both flange wheels and pneumatic tires to permit operation on both rails and roadways.

autoranging [ENG] Automatic switching of a multirange meter from its lowest to the next higher range, with the switching process repeated until a range is reached for which the full-scale value is not exceeded. Also known as automatic ranging.

autoreducing tachymeter [ENG] A class of tachymeter by which horizontal and height distances are read simultaneously.

autosled [MECH ENG] A propeller-driven machine equipped with runners and wheels and adaptable to use on snow, ice, or bare roads.

autostoper [MIN ENG] A stoper, or light compressed-air rock drill, mounted on an air-leg support which not only supports the drill but also exerts pressure on the drill bit.

auxanometer [ENG] An instrument used to detect and measure plant growth rate.

auxiliary anode [MET] A supplementary anode that alters the current distribution in electroplating to give a more uniform plating thickness.

auxiliary dead latch [DES ENG] A supplementary latch in a lock which automatically deadlocks the main latch bolt when the door is closed. Also known as auxiliary latch bolt; deadlocking latch bolt; trigger bolt.

auxiliary fan [MIN ENG] A small fan installed underground for ventilating narrow coal drivages or hard headings which are not ventilated by the normal air current.

auxiliary fluid ignition [AERO ENG] A method of ignition of a liquid-propellant rocket engine in which a liquid that is hypergolic with either the fuel or the oxidizer is injected into the combustion chamber to initiate combustion.

auxiliary landing gear [AERO ENG] The part or parts of a landing gear, such as an outboard wheel, which is intended to stabilize the craft on the surface but which bears no significant part of the weight.

auxiliary power plant [MECH ENG] Ancillary equipment, such as pumps, fans, and soot blowers, used with the main boiler, turbine, engine, waterwheel, or generator of a power-generating station.

auxiliary power unit [AERO ENG] A power unit carried on an aircraft or spacecraft which can be used in addition to the main sources of power. Abbreviated APU.

auxiliary rafter [BUILD] A member strengthening the principal rafter in a truss.

auxiliary reinforcement [CIV ENG] In a prestressed structural member, any reinforcement in addition to that whose function is prestressing.

auxiliary rim lock [DES ENG] A secondary or extra lock that is surface-mounted on a door to provide additional security.

auxiliary rope-fastening device [MECH ENG] A device attached to an elevator car, to a counterweight, or to the overhead dead-end rope-hitch support, that automatically supports the car or counterweight in case the fastening for the wire rope (cable) fails.

auxiliary thermometer [ENG] A mercury-in-glass thermometer attached to the stem of a reversing thermometer and read at the same time as the reversing thermometer so that the correction to the reading of the latter, resulting from change in temperature since reversal, can be computed.

auxograph [ENG] An automatic device that records changes in the volume of a body.

auxometer [ENG] An instrument that measures the magnification of a lens system.

available draft [MECH ENG] The usable differential pressure in the combustion air in a furnace, used to sustain combustion of fuel or to transport products of combustion.

available energy [MECH ENG] Energy which can in principle be converted to mechanical work.

available heat [MECH ENG] The heat per unit mass of a working substance that could be transformed into work in an engine under ideal conditions for a given amount of heat per unit mass furnished to the working substance.

aven [MIN ENG] A vertical shaft leading upward from a cave passage, sometimes connecting with passages above.

average acoustic output [ENG ACOUS] Vibratory energy output of a transducer measured by a radiation pressure balance; expressed in terms of watts per unit area of the transducer face.

average assay value [MIN ENG] The weighted result of assays obtained from a number of samples by multiplying the assay value of each sample by the width or thickness of the ore face over which it is taken and dividing the sum of these products by the total width of cross section sampled.

average discount factor *See* discount factor.

average outgoing quality limit [IND ENG] The average quality of all lots that pass quality inspection, expressed in terms of percent defective. Abbreviated AOQL.

averaging device [ENG] A device for obtaining the arithmetic mean of a number of readings, as on a bubble sextant.

averaging pitot tube [ENG] A flowmeter that consists of a rod extending across a pipe with several interconnected upstream holes, which simulate an array of pitot tubes across the pipe, and a downstream hole for the static pressure reference.

avgas *See* aviation gasoline.

aviation [AERO ENG] **1.** The science and technology of flight through the air. **2.** The world of airplane business and its allied industries.

aviation gasoline [MATER] Stable fuel with high volatility and high octane, especially suited for use in aircraft reciprocating engines. Abbreviated avgas.

aviation method [ENG] Determination of knock-limiting power, under lean-mixture conditions, of fuels used in spark-ignition aircraft engines.

aviation mix [MATER] Antiknock fluid containing tetraethyllead, ethylene dibromide, and dye; used in aviation gasoline.

avionics [ENG] The design and production of airborne electrical and electronic devices; term is derived from aviation electronics.

avocado oil [MATER] An oil extracted from ripe fruit of the avocado (*Persea americana*). Also known as alligator pear oil.

avoidable delay [IND ENG] An interruption under the control of the operator during the normal operating time.

AWG *See* American wire gage.

awl [DES ENG] A point tool with a short wooden handle used to mark surfaces and to make small holes, as in leather or wood.

awning window [BUILD] A window consisting of a series of vertically arranged, top-hinged rectangular sections; designed to admit air while excluding rain.

ax [DES ENG] An implement consisting of a heavy metal wedge-shaped head with one or two cutting edges and a relatively long wooden handle; used for chopping wood and felling trees.

axed brick [ENG] A brick, shaped with an ax, that has not been trimmed. Also known as rough-axed brick.

axes of an aircraft [AERO ENG] Three fixed lines of reference, usually centroidal and mutually perpendicular: the longitudinal axis, the normal or yaw axis, and the lateral or pitch axis. Also known as aircraft axes.

axhammer [DES ENG] An ax having one cutting edge and one hammer face.

axial fan [MECH ENG] A fan whose housing confines the gas flow to the direction along the rotating shaft at both the inlet and outlet.

axial-flow compressor [MECH ENG] A fluid compressor that accelerates the fluid in a direction generally parallel to the rotating shaft.

axial-flow jet engine [AERO ENG] **1.** A jet engine in which the general flow of air is along the longitudinal axis of the engine. **2.** A turbojet engine that utilizes an axial-flow compressor and turbine.

axial-flow pump [MECH ENG] A pump having an axial-flow or propeller-type impeller; used when maximum capacity and minimum head are desired. Also known as propeller pump.

axial force diagram [CIV ENG] In statics, a graphical representation of the axial load acting at each section of a structural member, plotted to scale and with proper sign as an ordinate at each point of the member and along a reference line representing the length of the member.

axial hydraulic thrust [MECH ENG] In single-stage and multistage pumps, the summation of unbalanced impeller forces acting in the axial direction.

axial rake [MECH ENG] The angle between the face of a blade of a milling cutter or reamer and a line parallel to its axis of rotation.

axial relief [MECH ENG] The relief behind the end cutting edge of a milling cutter.

axial runout [MECH ENG] The total amount, along the axis of rotation, by which the rotation of a cutting tool deviates from a plane.

axial-type mass flowmeter [ENG] An instrument in which fluid in a pipe is made to rotate at a constant speed by a motor-driven impeller, and the torque required by a second, stationary impeller to straighten the flow again is a direct measurement of mass flow.

axial winding [MATER] A winding used in filament-wound fiberglass-reinforced plastic construction in which the filaments run along the axis at a zero helix angle.

axis of freedom [DES ENG] An axis in a gyro about which a gimbal provides a degree of freedom.

axis of sighting [ENG] A line taken through the sights of a gun, or through the optical center and centers of curvature of lenses in any telescopic instrument.

axis of weld [MET] A line along a weld used to describe the positions of the localized welds.

axle [MECH ENG] A supporting member that carries a wheel and either rotates with the wheel to transmit mechanical power to or from it, or allows the wheel to rotate freely on it.

axle box [ENG] A bushing through which an axle passes in the hub of a wheel.

axle grease [MATER] A lubricating grease containing suspended lime particles and thickened with rosin soap.

axometer [ENG] An instrument that locates the optical axis of a lens, particularly a lens used in eyeglasses.

azimuth-adjustment slide rule [ENG] A circular slide rule by which a known angular correction for fire at one elevation can be changed to the proper correction for any other elevation.

azimuth angle [ENG] An angle in triangulation or in traverse through which the computation of azimuth is carried.

azimuth bar See azimuth instrument.

azimuth circle [DES ENG] A ring calibrated from 0 to 360 degrees over a compass, compass repeater, radar plan position indicator, direction finder, and so on, which provides means for observing compass bearings and azimuths.

azimuth dial [ENG] Any horizontal circle dial that reads azimuth.

azimuth error [ENG] An error in the indicated azimuth of a target detected by radar.

azimuth indicator [ENG] An approach-radar scope which displays azimuth information.

azimuth instrument [ENG] An instrument for measuring azimuths, particularly a device which fits over a central pivot in the glass cover of a magnetic compass. Also known as azimuth bar; bearing bar.

azimuth line [ENG] A radial line from the principal point, isocenter, or nadir point of a photograph, representing the direction to a similar point of an adjacent photograph in the same flight line; used extensively in radial triangulation.

azimuth marker [ENG] **1.** A scale encircling the plan position indicator scope of a radar on which the azimuth of a target from the radar may be measured. **2.** Any of the reference limits inserted electronically at 10 or 15° intervals which extend radially from the relative position of the radar on an off-center plan position indicator scope.

azimuth scale [ENG] A graduated angle-measuring device on instruments, gun carriages, and so forth that indicates azimuth.

azimuth-stabilized plan position indicator [ENG] A north-upward plan position indicator (PPI), a radarscope, which is stabilized by a gyrocompass so that either true or magnetic north is always at the top of the scope regardless of vehicle orientation.

azimuth transfer [ENG] Connecting, with a straight line, the nadir points of two vertical photographs selected from overlapping flights.

azimuth traverse [ENG] A survey traverse in which the direction of the measured course is determined by azimuth and verified by back azimuth.

Azusa [ENG] A continuous-wave, high-accuracy, phase-comparison, single-station tracking system operating at C-band and giving two direction cosines and slant range which can be used to determine space position and velocity of a vehicle (usually a rocket or a missile).

B

B-47 *See* Stratojet.

B-52 *See* Stratofortress.

babassu oil [MATER] A nondrying oil obtained from the kernels of the babassu palm and composed principally of lauric, myristic, and oleic acids.

babbitt metal [MET] Any of the white alloys composed primarily of tin or lead and of lesser amounts of antimony, copper, and perhaps other metals, and used for bearings.

back [MIN ENG] **1.** The upper part of any mining cavity. **2.** A joint, usually a strike joint, perpendicular to the direction of working.

backacter *See* backhoe.

back balance [MIN ENG] **1.** A kind of self-acting incline in a mine. **2.** The means of maintaining tension on a rope transmission or haulage system, consisting of the tension carriage, attached weight, and supporting structure.

backband [BUILD] A piece of millwork used around a rectangular window or door casing as a cover for the gap between the casing and the wall or as a decorative feature. Also known as backbend.

backbend [BUILD] **1.** At the outer edge of a metal door or window frame, the face which returns to the wall surface. **2.** *See* backband.

back boxing *See* back lining.

backbreak *See* overbreak.

backcast stripping [MIN ENG] A stripping method using two draglines; one strips and casts the overburden, and the other recasts a portion of the overburden.

back check [DES ENG] In a hydraulic door closer, a mechanism that slows the speed with which a door may be opened.

backdigger *See* backhoe.

back draft [MET] A reversed taper given to a casting model or pattern to prevent its withdrawal from the mold.

back-draft damper [MECH ENG] A damper with blades actuated by gravity, permitting air to pass through them in one direction only.

back edging [ENG] Cutting through a glazed ceramic pipe by first chipping through the glaze around the outside and then chipping the pipe itself.

back end *See* thrust yoke.

backfill [CIV ENG] Earth refilling a trench or an excavation around a building, bridge abutment, and the like. [MIN ENG] Waste sand or rock used to support the mine roof after removal of ore.

back fillet [BUILD] The return of the margin of a groin, doorjamb, or window jamb when it projects beyond a wall.

backfire [ENG] Momentary backward burning of flame into the tip of a torch. Also known as flashback.

backflap hinge [DES ENG] A hinge having a flat plate or strap which is screwed to the face of a shutter or door. Also known as flap hinge.

backflow [CIV ENG] The flow of water or other liquids, mixtures, or substances into the distributing pipes of a potable supply of water from any other than its intended source.

backflow connection [CIV ENG] Any arrangement of pipes, plumbing fixtures, drains, and so forth, in which backflow can occur.

backflow preventer *See* vacuum breaker.

backflow valve *See* backwater valve.

back gearing [MECH ENG] The technique of using gears on machine tools to obtain an increase in the number of speed changes that can be gotten with cone belt drives.

back gouging [MET] The elimination of excess material from both weld metal and base metal on the opposite side of a partly welded joint; a groove or bevel is formed in order to facilitate complete joint penetration.

background discrimination [ENG] The ability of a measuring instrument, circuit, or other device to distinguish signal from background noise.

background noise [ENG] The undesired signals that are always present in an electronic or other system, independent of whether or not the desired signal is present.

background returns [ENG] Signals on a radar screen from objects which are of no interest.

background signal [ENG] The output of a leak detector caused by residual gas to which the detector element reacts.

back gutter [BUILD] A gutter installed on the uphill side of a chimney on a sloping roof to divert water around the chimney.

backhand welding [MET] A welding technique in which the flame is directed back against the completed weld. Also known as backward welding.

back hearth [BUILD] That part of the hearth (or floor) which is contained within the fireplace itself. Also known as inner hearth.

backhoe [MECH ENG] An excavator fitted with a hinged arm to which is rigidly attached a bucket that is drawn toward the machine in operation. Also known as backacter; backdigger; dragshovel; pullshovel.

back holes [MIN ENG] The holes which are shot last in mine shaft sinking.

backing [MET] *See* backing strip. [MIN ENG] **1.** Timbers across the top of a level, supported in notches cut in the rock. **2.** Rough masonry of a wall faced with finer work. **3.** Earth placed behind a retaining wall.

backing board [BUILD] In a suspended acoustical ceiling, a flat sheet of gypsum board to which acoustical tile is attached by adhesive or mechanical means.

backing brick [CIV ENG] A relatively low-quality brick used behind face brick or other masonry.

backing deals [MIN ENG] Boards, 1–4 inches (2.5–10 centimeters) thick, of sufficient length to bridge the space between timber or steel sets or between rings in skeleton tubbing.

backing off [ENG] Removing excessive body metal from badly worn bits.

backing plate [ENG] A plate used to support the hardware for the cavity used in plastics injection molding.

backing pump [MECH ENG] A vacuum pump, in a vacuum system using two pumps in tandem, which works directly to the atmosphere and reduces the pressure to an intermediate value, usually between 100 and 0.1 pascals. Also known as fore pump.

backing space [ENG] Space between a fore pump and a diffusion pump in a leak-testing system.

backing space technique [ENG] Testing for leaks by connecting a leak detector to the backing space.

backing strip [MET] A piece of metal, asbestos, or other nonflammable material placed behind a joint to facilitate welding. Also known as backing.

backing up [CIV ENG] In masonry, the laying of backing brick.

back jamb *See* back lining.

backjoint [CIV ENG] In masonry, a rabbet such as that made on the inner side of a chimneypiece to receive a slip.

backlash [DES ENG] The amount by which the tooth space of a gear exceeds the tooth thickness of the mating gear along the pitch circles. [ENG] **1.** Relative motion of mechanical parts caused by looseness. **2.** The difference between the actual values of a quantity when a dial controlling this quantity is brought to a given position by a clockwise rotation and when it is brought to the same position by a counterclockwise rotation.

backlining [BUILD] **1.** A thin strip which lines a window casing, next to the wall and opposite the pulley stile, and provides a smooth surface for the working of the weighted sash. Also known as back boxing; back jamb. **2.** That piece of framing forming the back recess for boxing shutters.

back lintel [BUILD] A lintel which supports the backing of a masonry wall, as opposed to the lintel supporting the facing material.

backlog [IND ENG] **1.** An accumulation of orders promising future work and profit. **2.** An accumulation of unprocessed materials or unperformed tasks.

back nailing [BUILD] Nailing the plies of a built-up roof to the substrate to prevent slippage.

back nut [DES ENG] **1.** A threaded nut, one side of which is dished to retain a grommet; used in forming a watertight pipe joint. **2.** A locking nut on the shank of a pipe fitting, tap, or valve.

back off [ENG] **1.** To unscrew or disconnect. **2.** To withdraw the drill bit from a borehole. **3.** To withdraw a cutting tool or grinding wheel from contact with the workpiece.

back order [IND ENG] **1.** An order held for future completion. **2.** A new order placed for previously unavailable materials of an old order.

backout [AERO ENG] An undoing of previous steps during a countdown, usually in reverse order.

backplastering [BUILD] A coat of plaster applied to the back side of lath, opposite the finished surface.

backplate [BUILD] A plate, usually metal or wood, which serves as a backing for a structural member.

backplate lamp holder [DES ENG] A lamp holder, integrally mounted on a plate, which is designed for screwing to a flat surface.

back pressure [MECH ENG] Resistance transferred from rock into the drill stem when the bit is being fed at a faster rate than the bit can cut.

back-pressure curve [PETRO ENG] A graph used to arrive at the capacity of a natural-gas well to deliver gas into a pipeline at a sustained rate; uses data from back-pressure testing.

back-pressure-relief port [ENG] In a plastics extrusion die, an opening for the release of excess material.

back-pressure testing [PETRO ENG] Method of estimating open-flow capacity of natural-gas wells by relating a series of gas-flow rates and their corresponding stabilized pressures at the bottom of the well bore.

back-pressure valve [PETRO ENG] A check valve installed in a natural-gas well bore to shut off gas flow while replacing the blowout preventer (used during drilling) with a christmas tree piping arrangement, which controls gas flow out of the completed well.

back putty [MATER] The bedding of glazing compound which is placed between the face of glass and the frame or sash containing it. Also known as bed glazing.

back rake [DES ENG] An angle on a single-point turning tool measured between the plane of the tool face and the reference plane.

backs [MIN ENG] Ore height available above a given working level.

backsaw [DES ENG] A fine-tooth saw with its upper edge stiffened by a metal rib to ensure straight cuts.

backscatter gage [ENG] A radar instrument used to measure the radiation scattered at 180° to the direction of the incident wave.

backscattering thickness gage [ENG] A device that uses a radioactive source for measuring the thickness of materials, such as coatings, in which the source and the instrument measuring the radiation are mounted on the same side of the material, the backscattered radiation thus being measured.

backset [BUILD] The horizontal distance from the face of a lock or latch to the center of the keyhole, knob, or lock cylinder.

back shot [MIN ENG] A shot used for widening an entry, placed at some distance from the head of an entry.

backsight [ENG] **1.** A sight on a previously established survey point or line. **2.** Reading a leveling rod in its unchanged position after moving the leveling instrument to a different location.

backsight method [ENG] **1.** A plane-table traversing method in which the table orientation produces the alignment of the alidade on an established map line, the table being rotated until the line of sight is coincident with the corresponding ground line. **2.** Sighting two pieces of equipment directly at each other in order to orient and synchronize one with the other in azimuth and elevation.

back siphonage [CIV ENG] The flowing back of used, contaminated, or polluted water from a plumbing fixture or vessel into the pipe which feeds it; caused by reduced pressure in the pipe.

backspace [MECH ENG] To move a typewriter carriage back one space by depressing a back space key.

backstay [ENG] **1.** A supporting cable that prevents a more or less vertical object from falling forward. **2.** A spring used to keep together the cutting edges of purchase shears. **3.** A rod that runs from either end of a carriage's rear axle to the reach. **4.** A leather strip that covers and strengthens a shoe's back seam.

backstep sequence [MET] Sequential deposition of weld beads in the direction opposite to the direction of welding.

back stoping *See* shrinkage stoping.

backup [BUILD] That part of a masonry wall behind the exterior facing. [CIV ENG] Overflow in a drain or piping system, due to stoppage. [ENG] **1.** An item under development intended to perform the same general functions that another item also under development performs. **2.** A compressible material used behind a sealant to reduce its depth and to support the sealant against sag or indentation. [MET] A support used to balance the upsetting force in the workpieces during flash welding. [PETRO ENG] During drilling, the holding of one section of pipe while another is screwed out of it or into it.

backup strip [BUILD] A wood strip which is fixed at the corner of a partition or wall to provide a nailing surface for ends of lath. Also known as lathing board.

backup system [SYS ENG] A system, normally redundant but kept available to replace a system which may fail in operation.

backup tong [ENG] A heavy device used on a drill pipe to loosen the tool joints.

back veneer [MATER] In veneer plywood, the layer of veneer on the side of a plywood sheet which is opposite the face veneer; usually of lower quality.

back vent [CIV ENG] An individual vent for a plumbing fixture located on the downstream (sewer) side of a trap to protect the trap against siphonage.

backward-bladed aerodynamic fan [MECH ENG] A fan that consists of several streamlined blades mounted in a revolving casing.

backward welding *See* backhand welding.

backwater valve [ENG] A type of check valve in a drainage pipe; reversal of flow causes the valve to close, thereby cutting off flow. Also known as backflow valve.

backweld [MET] A weld placed behind a single groove weld.

back work [MIN ENG] Any kind of operation in a mine not immediately concerned with production or transport; literally, work behind the face, such as repairs to roads.

bactericide [MATER] An agent that destroys bacteria.

badger [DES ENG] *See* badger plane. [ENG] A tool used inside a pipe or culvert to remove any excess mortar or deposits.

badger plane [DES ENG] A hand plane whose mouth is cut obliquely from side to side, so that the plane can work close up to a corner. Also known as badger.

bad top [MIN ENG] A weak roof in a coal mine; sometimes develops following a blast.

baffle [ENG] A plate that regulates the flow of a fluid, as in a steam-boiler flue or a gasoline muffler. [ENG ACOUS] A cabinet or partition used with a loudspeaker to reduce interaction between sound waves produced simultaneously by the two surfaces of the diaphragm.

bag [ENG] **1.** A flexible cover used in bag molding. **2.** A container made of paper, plastic, or cloth without rigid walls to transport or store material.

bag filter [ENG] Filtering apparatus with porous cloth or felt bags through which dust-laden gases are sent, leaving the dust on the inner surfaces of the bags.

baghouse [ENG] The large chamber or room for holding bag filters used to filter gas streams from a furnace.

bag molding [ENG] A method of molding plastic or plywood-plastic combinations into curved shapes, in which fluid pressure acting through a flexible cover, or bag, presses the material to be molded against a rigid die.

Bagnold number [ENG] A dimensionless number used in saltation studies.

bag plug [ENG] An inflatable drain stopper, located at the lowest point of a piping system, that acts to seal a pipe when inflated.

bag powder [MATER] An explosive loaded in bags.

bag trap [ENG] An S-shaped trap in which the vertical inlet and outlet pipes are in alignment.

baguette *See* bead molding.

bail [ENG] A loop of heavy wire snap-fitted around two or more parts of a connector or other device to hold the parts together.

bailer [ENG] A long, cylindrical vessel fitted with a bail at the upper end and a flap or tongue valve at the lower extremity; used to remove water, sand, and mud- or cuttings-laden fluids from a borehole. Also known as bailing bucket.

Bailey bridge [CIV ENG] A lattice bridge built of interchangeable panels connected at the corners with steel pins, permitting rapid construction; developed in Britain about 1942 as a military bridge.

Bailey meter [ENG] A flowmeter consisting of a helical quarter-turn vane which operates a counter to record the total weight of granular material flowing through vertical or near-vertical ducts, spouts, or pipes.

bailing [ENG] Removal of the cuttings from a well during cable-tool drilling, or of the liquid from a well, by means of a bailer.

bailing bucket *See* bailer.

bailout [AERO ENG] The exiting from a flying aircraft and descending by parachute in an emergency.

bailout bottle [AERO ENG] A personal supply of oxygen usually contained in a cylinder under pressure and utilized when the individual has left the central oxygen system, as in a parachute jump.

bainite [MET] Steel formed by austempering, having an acicular structure of ferrite and carbides, exhibiting considerable toughness, and combining high strength with high ductility.

baked core [MET] In sand castings, a core which has been heated in core stoves or by dielectric heating to attain uniform physical properties.

baked finish [ENG] A paint or varnish finish obtained by baking, usually at temperatures above 150°F (65°C), thereby developing a tough, durable film.

baked permeability [MET] The property of a molded mass of sand which permits passage of gases through it when molten metal is poured in a mold, baked above 230°F (110°C), and then cooled.

baked strength [MET] The strength of a molded sand mixture when baked above 230°F (110°C) and then cooled to ambient temperature.

bakeout [ENG] The degassing of surfaces of a vacuum system by heating during the pumping process.

baker bell dolphin [CIV ENG] A dolphin consisting of a heavy bell-shaped cap pivoted on a group of piles; a blow from a ship will tilt the bell, thus absorbing energy.

baking [ENG] The use of heat on fresh paint films to speed the evaporation of thinners and to promote the reaction of binder components so as to form a hard polymeric film. Also known as stoving. [MET] Heating metal at low temperatures to remove gases.

baking finish [MATER] Varnish or paint that must be baked at temperatures greater than 150°F (66°C) to develop desired final properties of strength and hardness.

baking varnish [MATER] A chemical-resistant varnish made of synthetic resins that requires baking to be dried.

balance [AERO ENG] **1.** The equilibrium attained by an aircraft, rocket, or the like when forces and moments are acting upon it so as to produce steady flight, especially without rotation about its axes. **2.** The equilibrium about any specified axis that counterbalances something, especially on an aircraft control surface, such as a weight installed forward of the hinge axis to counterbalance the surface aft of the hinge axis. [ENG] An instrument for measuring mass or weight. [MIN ENG] The counterpoise or weight attached by cable to the drum of a winding engine to balance the weight of the cage and hoisting cable and thus assist the engine in lifting the load out of the shaft.

balance arm [BUILD] On a projected window, a side supporting arm which is constructed so that the center of gravity of the sash is not changed appreciably when the window is opened.

balance bar See balance beam

balance beam [CIV ENG] A long beam, attached to a gate (or drawbridge, and such) so as to counterbalance the weight of the gate during opening or closing. Also known as balance bar.

balance car [MIN ENG] In quarrying, a car loaded with iron or stone and connected by means of a steel cable with a channeling machine operating on an inclined track; used to counteract the force of gravity and thus enable the channeling machine to operate with equal ease uphill and downhill.

balanced armature unit [ENG ACOUS] Driving unit used in magnetic loudspeakers, consisting of an iron armature pivoted between the poles of a permanent magnet and surrounded by coils carrying the audio-frequency current; variations in audio-frequency current cause corresponding changes in armature magnetism and corresponding movements of the armature with respect to the poles of the permanent magnet.

balanced construction [BUILD] A plywood or sandwich-panel construction which has an odd number of plies laminated together so that the construction is iden-

tical on both sides of a plane through the center of the panel.

balanced door [BUILD] A door equipped with double-pivoted hardware which is partially counterbalanced to provide easier operation.

balanced draft [ENG] The maintenance of a constant draft in a furnace by monitoring the incoming air and products of combustion.

balanced earthwork [CIV ENG] Cut-and-fill work in which the amount of fill equals the amount of material excavated.

balanced fertilizer [MATER] A material of varying composition added to soil so as to provide essential mineral elements at required levels, improve soil structure, or enhance microbial activity.

balanced gasoline [MATER] An automotive gasoline blended from petroleum hydrocarbons of varying volatilities to provide desired performance for engine starting, warm-up, acceleration, and mileage.

balanced line [IND ENG] A production line for which the time cycles of the operators are made approximately equal so that the work flows at a desired steady rate from one operator to the next.

balanced method [ENG] Method of measurement in which the reading is taken at zero; it may be a visual or audible reading, and in the latter case the null is the no-sound setting.

balanced reinforcement [CIV ENG] An amount and distribution of steel reinforcement in a flexural reinforced concrete member such that the allowable tensile stress in the steel and the allowable compressive stress in the concrete are attained simultaneously.

balanced sash [BUILD] In a double-hung window, a sash which opens by being raised or lowered and which is balanced with counterweights or pretensioned springs so that little force is required to move the sash.

balanced step [BUILD] One of a series of winders arranged so that the width of each winder tread (at the narrow end) is almost equal to the tread width in the straight portion of the adjacent stair flight. Also known as dancing step; dancing winder.

balanced surface [AERO ENG] A control surface that extends on both sides of the axis of the hinge or pivot, or that has auxiliary devices or extensions connected with it, in such a manner as to effect a small or zero resultant moment of the air forces about the hinge axis.

balanced valve [ENG] A valve having equal fluid pressure in both the opening and closing directions.

balance method See null method.

balance shot [MIN ENG] In coal mining, a shot for which the drill hole is parallel to the face of the coal that is to be broken by it.

balance tool [MECH ENG] A tool designed for taking the first cuts when the external surface of a piece in a lathe is being machined; it is supported in the tool holder at an unvarying angle.

balance wheel [MECH ENG] **1.** A wheel which governs or stabilizes the movement of a mechanism. **2.** See flywheel.

balancing a survey [ENG] Distributing corrections through any traverse to eliminate the error of closure and to obtain an adjusted position for each traverse station. Also known as traverse adjustment.

balancing delay [IND ENG] In motion study, idleness of one hand while the other is active to catch up.

balancing plug cock See balancing valve.

balancing valve [ENG] A valve used in a pipe for controlling fluid flow; not usually used to shut off the flow. Also known as balancing plug cock.

balata [MATER] A hard substance, similar to gutta-percha, used mainly in golf balls and belting, which is made by drying the milky juice of the bully tree (*Manilkara bidentata*). Also known as gutta-balata.

balconet [BUILD] A pseudobalcony; a low ornamental railing at a window, projecting only slightly beyond the threshold or sill.

balcony [BUILD] A deck which projects from a building wall above ground level.

balcony outlet [BUILD] In a vertical rainwater pipe that passes through an exterior balcony, a fitting which provides an inlet for the drainage of rainwater from the balcony.

bald [MIN ENG] **1.** Without framing. **2.** A mine timber which has a flat end.

bale [IND ENG] **1.** A large package of material, pressed tightly together, tied with rope, wire, or hoops and usually covered with wrapping. **2.** The amount of material in a bale; sometimes used as a unit of measure, as 500 pounds (227 kilograms) of cotton in the United States.

baler [MECH ENG] A machine which takes large quantities of raw or finished materials and binds them with rope or metal straps or wires into a large package.

baling [CIV ENG] A technique used to convert loose refuse into heavy blocks by compaction; the blocks are then burned and are buried in sanitary landfill.

balk [BUILD] A squared timber used in building construction. [CIV ENG] A low ridge of earth that marks a boundary line. [MIN ENG] A sudden thinning out, for a certain distance, of a bed of coal.

balking [IND ENG] The refusal of a customer to enter a queue for some reason, such as insufficient waiting room.

ball [MECH ENG] In fine grinding, one of the crushing bodies used in a ball mill.

ball-and-race-type pulverizer [MECH ENG] A grinding machine in which balls rotate under an applied force between two races to crush materials, such as coal, to fine consistency. Also known as ball-bearing pulverizer.

ball-and-socket joint [MECH ENG] A joint in which a member ending in a ball is joined to a member ending in a socket so that relative movement is permitted within a certain angle in all planes passing through a line. Also known as ball joint.

ballast [AERO ENG] A relatively dense substance that is placed in the car of a vehicle and can be thrown out to reduce the load or can be shifted to change the center of gravity. [CIV ENG] Crushed stone used in a railroad bed to support the ties, hold the track in line, and help drainage. [MATER] Coarse gravel used as an ingredient in concrete.

ball bearing [MECH ENG] An antifriction bearing permitting free motion between moving and fixed parts by means of balls confined between outer and inner rings.

ball-bearing hinge [MECH ENG] A hinge which is equipped with ball bearings between the hinge knuckles in order to reduce friction.

ball-bearing pulverizer See ball-and-race-type pulverizer.

ball bonding [ENG] The making of electrical connections in which a flame is used to cut a wire, the molten end of which solidifies as a ball, which is pressed against the bonding pad on an integrated circuit.

ball breaker [ENG] **1.** A steel or iron ball that is hoisted by a derrick and allowed to fall on blocks of waste stone to break them or to swing against old buildings to demolish them. Also known as skull cracker; wrecking ball. **2.** A coring and sampling device consisting of a hollow glass ball, 3 to 5 inches (7.5 to 12.5 centimeters) in diameter, held in a frame attached to the trigger line above the triggering weight of the corer; used to indicate contact between corer and bottom.

ball burnishing [MET] A method of giving small stainless steel parts a lustrous finish by rotating them in a wood-lined barrel with water, burnishing soap, and hardened steel balls.

ball bushing [MECH ENG] A type of ball bearing that allows motion of the shaft in its axial direction.

ball catch [DES ENG] A door fastener having a contained metal ball which is under pressure from a spring; the ball engages a striking plate and keeps the door from opening until force is applied.

ball check valve [ENG] A valve having a ball held by a spring against a seat; used to permit flow in one direction only.

ball clay [MATER] A clay used in ceramics that is characterized by strong binding properties, a tendency to ball, and excellent plasticity.

ball float [MECH ENG] A floating device, usually approximately spherical, which is used to operate a ball valve.

ball-float liquid-level meter [ENG] A float which rises and falls with liquid level, actuating a pointer adjacent to a calibrated scale in order to measure the level of a liquid in a tank or other container.

ball grinder *See* ball mill.

ballhead [MECH ENG] That part of the governor which contains flyweights whose force is balanced, at least in part, by the force of compression of a speeder spring.

balling up [MET] Formation of balls of molten brazing filler metal when the base material has not been sufficiently wetted.

ballistic body [ENG] A body free to move, behave, and be modified in appearance, contour, or texture by ambient conditions, substances, or forces, such as by the pressure of gases in a gun, by rifling in a barrel, by gravity, by temperature, or by air particles.

ballistic instrument [ENG] Any instrument, such as a ballistic galvanometer or a ballistic pendulum, that measures an impact or sudden pulse of energy.

ballistic magnetometer [ENG] A magnetometer designed to employ the transient voltage induced in a coil when either the magnetized sample or coil are moved relative to each other.

ballistic pendulum [ENG] A device which uses the deflection of a suspended weight to determine the momentum of a projectile.

ballistic separator [CIV ENG] A device that takes out noncompostable material like stones, glass, metal, and rubber, from solid waste by passing the waste over a rotor that has impellers to fling the material in the air; the lighter organic (compostable) material travels a shorter distance than the heavier (noncompostable) material.

ballistic vehicle [ENG] A nonlifting vehicle; a vehicle that follows a ballistic trajectory.

ballistite [MATER] A smokeless propellant containing nitrocellulose and nitroglycerin; used in some rocket, mortar, and small-arms ammunition.

ball joint *See* ball-and-socket joint.

ball mill [MECH ENG] A pulverizer that consists of a horizontal rotating cylinder, up to three diameters in length, containing a charge of tumbling or cascading steel balls, pebbles, or rods. Also known as ball grinder.

ballonet [AERO ENG] One of the air cells in a blimp, fastened to the bottom or sides of the envelope, which are used to maintain the required pressure in the envelope without adding or valving gas as the ship ascends or descends. Also spelled ballonnet.

ballonnet *See* ballonet.

balloon [AERO ENG] A nonporous, flexible spherical bag, inflated with a gas such as helium that is lighter than air, so that it will rise and float in the atmosphere; a large-capacity balloon can be used to lift a payload suspended from it.

balloon cover [AERO ENG] A cover which fits over a large, inflated balloon to facilitate handling in high or gusty winds. Also known as balloon shroud.

balloon framing [CIV ENG] Framing for a building in which each stud is one piece from roof to foundation.

balloon shroud *See* balloon cover.

balloon-type rocket [AERO ENG] A liquid-fuel rocket, such as the Atlas, that requires the pressure of its propellants (or other gases) within it to give it structural integrity.

ball-peen hammer [ENG] A hammer with a ball at one end of the head; used in riveting and forming metal.

ball pendulum test [ENG] A test for measuring the strength of explosives; consists of measuring the swing of a pendulum produced by the explosion of a weighed charge of material.

ball sealers [PETRO ENG] Balls of rubber, plastic, or metal that are dropped down the well bore to aid the acidizing of impermeable zones of an oil reservoir; they wedge into and plug the bottomhole tubing perforations that are adjacent to the more permeable reservoir zones.

ball sizing [MET] Finishing a hole to a precise diameter and burnishing the surface by forcing a steel ball through it.

ball test [CIV ENG] In a drain, a test for freedom from obstruction and for circularity in which a ball (less than the diameter of the drain by a specified amount) is rolled through the drain.

ballute [AERO ENG] A cross between a balloon and a parachute, used to brake the free fall of sounding rockets.

ball valve [MECH ENG] A valve in which the fluid flow is regulated by a ball moving relative to a spherical socket as a result of fluid pressure and the weight of the ball.

balsam [MATER] An exudate of the balsam tree; a mixture of resins, essential oils, cinnamic acid, and benzoic acid.

balsam of Peru *See* Peru balsam.

Baltimore oil *See* wormseed oil.

baluster [BUILD] A post which supports a handrail and encloses the open sections of a stairway.

balustrade [BUILD] The railing assembly of a stairway consisting of the handrail, balusters, and usually a bottom rail.

Banbury mixer [MECH ENG] Heavy-duty batch mixer, with two counterrotating rotors, for doughy material; used mainly in the plastics and rubber industries.

band [BUILD] Any horizontal flat member or molding or group of moldings projecting slightly from a wall plane and usually marking a division in the wall. Also known as band course; band molding. [DES ENG] A strip or cord crossing the back of a book to which the sections are sewn.

bandage [BUILD] A strap, band, ring, or chain placed around a structure to secure and hold its parts together, as around the springing of a dome.

band brake [MECH ENG] A brake in which the frictional force is applied by increasing the tension in a flexible band to tighten it around the drum.

band chain [ENG] A steel or Invar tape, graduated in feet and at least 100 feet (30.5 meters) long, used for accurate surveying.

band clamp [DES ENG] A two-piece metal clamp, secured by bolts at both ends; used to hold riser pipes.

band clutch [MECH ENG] A friction clutch in which a steel band, lined with fabric, contracts onto the clutch rim.

band course *See* band.

banded structure [MET] The appearance of a metal showing light and dark parallel bands in the direction of rolling or working.

banding [DES ENG] A strip of fabric which is used for bands.

band molding *See* band.

band saw [MECH ENG] A power-operated woodworking saw consisting basically of a flexible band of steel having teeth on one edge, running over two vertical pulleys, and operated under tension.

band wheel [MECH ENG] In a drilling operation, a large wheel that transmits power from the engine to the walking beam.

banister [BUILD] A handrail for a staircase.

bank [AERO ENG] The lateral inward inclination of an airplane when it rounds a curve. [CIV ENG] *See* embankment. [IND ENG] The amount of material allowed to accumulate at a point on a production line where it is not employed or worked upon, to permit reasonable fluctuations in line speed before and after the point. Also known as float. [MIN ENG] **1.** The top of the shaft. **2.** The surface around the mouth of a shaft. **3.** The whole, or sometimes only one side or one end, of a working place underground. **4.** To manipulate materials such as coal, gravel, or sand on a bank. **5.** A terrace-like bench in open-pit mining.

bank and turn indicator [AERO ENG] A device used to advise the pilot that the aircraft is turning at a certain rate, and that the wings are properly banked to preclude slipping or sliding of the aircraft as it continues in flight. Also known as bank indicator.

banker [ENG] The bench or table upon which bricklayers and stonemasons prepare and shape their material.

bank gravel *See* bank-run gravel.

bank height [MIN ENG] The vertical height of a bank as measured between its highest point or crest and its toe at the digging level or bench. Also known as bench height; digging height.

bank indicator *See* bank and turn indicator.

bank material [CIV ENG] Soil or rock in place before excavation or blasting.

bank-run gravel [MATER] Aggregate taken directly from natural deposits; contains both large and small stones. Also known as bank gravel; run-of-bank gravel.

bank slope [MIN ENG] The angle, measured in degrees of deviation from the horizontal, at which the earthy or rock material will stand in an excavated, terracelike cut in an open-pit mine or quarry. Also known as bench slope.

banksman *See* lander.

banks oil *See* cod-liver oil.

bar [MET] An elongated piece of metal of simple uniform cross-section dimensions, usually rectangular, circular, or hexagonal, produced by forging or hot rolling. Also known as barstock. [MIN ENG] *See* bar drill.

Bárány chair [ENG] A chair in which a person is revolved to test his susceptibility to vertigo.

barb bolt [DES ENG] A bolt having jagged edges to prevent its being withdrawn from the object into which it is driven. Also known as rag bolt.

barbed wire [MATER] Two or more wires twisted together with addition of sharp hooks or points (or a single wire furnished with barbs); used for fences. Also known as barbwire.

bar bending [CIV ENG] In reinforced concrete construction, the process of bending reinforcing bars to various shapes.

barberite [MET] A nonferrous alloy with good resistance to sulfuric acid, sea water, and mine waters; 88.5% copper, 5% nickel, 5% tin, 1.5% silicon.

barbwire *See* barbed wire.

bar chair *See* bar support.

bar clamp [DES ENG] A clamping device consisting of a long bar with adjustable clamping jaws; used in carpentry.

bar drawing [MET] An operation in which a metallic bar is pulled through a die so that the cross-sectional area of the bar is reduced.

bar drill [MIN ENG] A small diamond type or other type of rock drill mounted on a bar and used in an underground workplace. Also known as bar.

bareboat charter [IND ENG] An agreement to charter a ship without its crew or stores; the fee for its use for a predetermined period of time is based on the price per ton of cargo handled.

bare electrode [MET] An uncoated electrode used in submerged arc automatic welding with a gas-shielded arc or a granular flux deposited in an elongated mound over a joint.

barefaced tenon [ENG] A tenon having a shoulder cut on one side only.

bar folder [MET] A machine used to bend a metal sheet into a sharp, narrow, and accurate fold, or a rounded fold, along the edge.

barge couple [BUILD] **1.** One of two rafters that support that part of a gable roof which projects beyond the gable wall. **2.** One of the rafters (under the barge course) which serve as grounds for the vergeboards and carry the plastering or boarding of the soffits. Also known as barge rafter.

barge course [BUILD] **1.** The coping of a wall, formed by a course of bricks set on edge. **2.** In a tiled roof, the part of the tiling which projects beyond the principal rafters where there is a gable.

barge rafter *See* barge couple.

barge stone [BUILD] One of the stones, generally projecting, which form the sloping top of a gable built of masonry.

bar hole [ENG] A small-diameter hole made in the ground along the route of a gas pipe in a bar test survey.

bar iron [MET] Wrought iron formed into bars.

barium-base grease [MATER] A lubricating material made from lubricating oil and barium soap.

barium glass [MATER] Glass which differs from ordinary lime-soda glass in that barium oxide replaces part of the calcium oxide.

barium plaster [MATER] A special mill-mixed gypsum plaster containing barium salts; used to plaster walls of x-ray rooms.

barium-sodium niobate [MATER] Synthetic electrooptical crystals used to produce coherent green light in lasers and to manufacture devices such as electrooptical modulators and optical polarimetric oscillators.

bar joist [BUILD] A small steel truss with wire or rod web lacing used for roof and floor supports.

bark [MET] The decarburized layer formed beneath the scale on the surface of steel heated in air.

barker [DES ENG] *See* bark spud. [ENG] A machine, used mainly in pulp mills, which removes the bark from logs.

bark spud [DES ENG] A tool which peels off bark. Also known as barker.

barley coal [MIN ENG] A stream size of anthracite sized on a round punched plate; passes through quarter-inch (6.4 millimeter) holes. Also known as buckwheat no. 3.

bar linkage [MECH ENG] A set of bars joined together at pivots by means of pins or equivalent devices; used to transmit power and information.

bar mining [MIN ENG] The mining of river bars, usually between low and high waters, although the stream is sometimes deflected and the bar worked below water level.

barney [MIN ENG] A small car or truck, attached to a rope or cable, used to push cars up a slope or an inclined plane. Also known as bullfrog; donkey; groundhog; larry; mule; ram; truck.

barogram [ENG] The record of an aneroid barograph.

barograph *See* aneroid barograph.

barometer [ENG] An absolute pressure gage specifically designed to measure atmospheric pressure.

barometric [ENG] Pertaining to a barometer or to the results obtained by using a barometer.

barometric condenser [MECH ENG] A contact condenser that uses a long, vertical pipe into which the condensate and cooling liquid flow to accomplish their removal by the pressure created at the lower end of the pipe.

barometric draft regulator [MECH ENG] A damper usually installed in the breeching between a boiler and chimney; permits air to enter the breeching automatically as required, to maintain a constant overfire draft in the combustion chamber.

barometric elevation [ENG] An elevation above mean sea level estimated from the difference in atmospheric pressure between the point in question and an elevation of known value.

barometric fuel control [AERO ENG] A device that maintains the correct flow of fuel to an engine by adjusting to atmospheric pressure at different altitudes, as well as to impact pressure.

barometric fuse [ENG] A fuse that functions as a result of change in the pressure exerted by the surrounding air.

barometric hypsometry [ENG] The determination of elevations by means of either mercurial or aneroid barometers.

barometric leveling [ENG] The measurement of approximate elevation differences in surveying with the aid of a barometer; used especially for large areas.

barometric switch *See* baroswitch.

barometrograph *See* aneroid barograph.

barometry [ENG] The study of the measurement of atmospheric pressure, with particular reference to ascertaining and correcting the errors of the different types of barometer.

baroscope [ENG] An apparatus which demonstrates the equality of the weight of air displaced by an object and its loss of weight in air.

barostat [ENG] A mechanism which maintains constant pressure inside a chamber.

baroswitch [ENG] **1.** A pressure-operated switching device used in a radiosonde which determines whether temperature, humidity, or reference signals will be transmitted. **2.** Any switch operated by a change in barometric pressure. Also known as barometric switch.

barothermogram [ENG] The record made by a barothermograph.

barothermograph [ENG] An instrument which automatically records pressure and temperature.

barothermohygrogram [ENG] The record made by a barothermohygrograph.

barothermohygrograph [ENG] An instrument that produces graphs of atmospheric pressure, temperature, and humidity on a single sheet of paper.

bar post [CIV ENG] One of the posts driven into the ground to form the sides of a field gate.

barrage [CIV ENG] An artificial dam which increases the depth of water of a river or watercourse, or diverts it into a channel for navigation or irrigation.

barrage-type spillway [CIV ENG] A passage for surplus water with sluice gates across the width of the entrance.

barred-and-braced gate [CIV ENG] A gate with a diagonal brace to reinforce the horizontal timbers.

barred gate [CIV ENG] A gate with one or more horizontal timber rails.

barrel [DES ENG] **1.** A container having a circular lateral cross section that is largest in the middle, and ends that are flat; often made of staves held together by hoops. **2.** A piece of small pipe inserted in the end of a cartridge to carry the squib to the powder. **3.** That portion of a pipe having a constant bore and wall thickness.

barrel bolt [DES ENG] A door bolt which moves in a cylindrical casing; not driven by a key. Also known as tower bolt.

barrel copper [MIN ENG] Copper in lumps small enough to be picked out of the mass of rock and put in the furnace without dressing.

barrel drain [CIV ENG] Any drain which is cylindrical.

barrel fitting [DES ENG] A short length of threaded connecting pipe.

barrelhead [DES ENG] The flat end of a barrel.

barreling *See* tumbling.

barrel plating [MET] An electroplating process by which articles are brought into contact with an electrolyte while rotating in a perforated hardwood barrel.

barrel roof [BUILD] **1.** A roof of semicylindrical section; capable of spanning long distances parallel to the axis of the cylinder. **2.** *See* barrel vault.

barricade [ENG] Structure composed essentially of concrete, earth, metal, or wood, or any combination thereof, and so constructed as to reduce or confine the blast effect and fragmentation of an explosion.

barricade shield [ENG] A type of movable shield made of a material designed to absorb ionizing radiation, for protection from radiation.

barrier curb [CIV ENG] A curb with vertical sides high enough to keep vehicles from crossing it.

barrier material [MATER] Packing material impervious to moisture, vapor, or other liquids and gases.

barrier shield [ENG] A wall or enclosure made of a material designed to absorb ionizing radiation, shielding the operator from an area where radioactive material is being used or processed by remote-control equipment.

barrow *See* handbarrow; wheelbarrow.

barrow run [CIV ENG] A temporary pathway of wood planks or sheets to provide a smooth access for wheeled materials-handling carriers on a building site.

bar sash lift [BUILD] A type of handle, attached to the bottom rail of a sash, for raising or lowering it.

bar screen [MECH ENG] A sieve with parallel steel bars for separating small from large pieces of crushed rock.

bar steel [MET] Steel formed into bars.

barstock *See* bar.

bar strainer [DES ENG] A screening device consisting of a bar or a number or parallel bars; used to prevent objects from entering a drain.

bar support [CIV ENG] A device used to support or hold steel reinforcing bars in proper position before or during the placement of concrete. Also known as bar chair.

bar test survey [ENG] A leakage survey in which bar holes are driven or bored at regular intervals along the way of an underground gas pipe and the atmosphere in the holes is tested with a combustible gas detector or such.

Barth plan [IND ENG] A wage incentive plan intended for a low task and for all efficiency points and defined as: earning = rate per hour × square root of the product (hours standard × hours actual).

bar turret lathe [MECH ENG] A turret lathe in which the bar stock is slid through the headstock and collet on line with the turning axis of the lathe and held firmly by the closed collet.

bar-type grating [CIV ENG] An open grid assembly of metal bars in which the bearing bars (running in one direction) are spaced by rigid attachment to crossbars.

basal tunnel [ENG] A water supply tunnel constructed along the basal water table.

bascule [ENG] A structure that rotates about an axis, as a seesaw, with a counterbalance (for the weight of the structure) at one end.

bascule bridge [CIV ENG] A movable bridge consisting primarily of a cantilever span extending across a channel; it rotates about a horizontal axis parallel with the waterway.

bascule leaf [CIV ENG] The span of a bascule bridge.

base [ENG] Foundation or part upon which an object or instrument rests.

base anchor [BUILD] The metal piece attached to the base of a doorframe for the purpose of securing the frame to the floor.

base apparatus [ENG] Any apparatus designed for use in measuring with accuracy and precision the length of a base line in triangulation, or the length of a line in first- or second-order traverse.

base block [BUILD] **1.** A block of any material, generally with little or no ornament, forming the lowest member of a base, or itself fulfilling the functions of a base, as a member applied to the foot of a door or to window trim. **2.** A rectangular block at the base of a casing or column which the baseboard abuts. **3.** See skirting block.

baseboard [BUILD] A finish board covering the interior wall at the junction of the wall and the floor.

baseboard heater [BUILD] Heating elements installed in panels along the baseboard of a wall.

baseboard radiator [CIV ENG] A heating unit located at the lower portion of a wall, and to which heat is supplied by hot water, warm air, steam, or electricity.

base box [MET] A unit of area used for tin-plated steel sheet; one base box is equivalent to 112 sheets, 14 by 20 inches (35.6 by 50.8 centimeters), or 62,720 square inches of surface, coated on two sides; 1 pound (0.454 kilogram) of tin per base box is equal to a coating of tin 0.000059 inch (0.0014986 millimeter) thick.

base bullion [MET] Crude lead that has enough silver in it to make the extraction of silver worthwhile; gold may be present.

base cap See base molding.

base circle [DES ENG] The circle on a gear such that each tooth-profile curve is an involute of it.

base conditions [PETRO ENG] Standard conditions of 14.65 psia pressure and 60°F (15.6°C) used to calculate the amount of gas contained in oil from a well (the gas-oil ratio).

base correction [ENG] The adjustment made to reduce measurements taken in field exploration to express them with reference to the base station values.

base course [BUILD] The lowest course or first course of a wall. [CIV ENG] The first layer of material laid down in construction of a pavement.

base elbow [DES ENG] A cast-iron pipe elbow having a baseplate or flange which is cast on it and by which it is supported.

base flashing [BUILD] **1.** The flashing provided by upturned edges of a watertight membrane on a roof. **2.** Any metal or composition flashing at the joint between a roofing surface and a vertical surface, such as a wall or parapet.

base fracture [MIN ENG] In quarrying, the broken condition of the base after a blast; it may be a good or bad base fracture.

base line [ENG] **1.** A surveyed line, established with more than usual care, to which surveys are referred for coordination and correlation. **2.** A cardinal line extending east and west along the astronomic parallel passing through the initial point, along which standard township, section, and quarter-section corners are established.

base-line check See ground check.

basement [BUILD] A building story which is wholly or less than half below ground; it is generally used for living space.

basement wall [BUILD] A foundation wall which encloses a usable area under a building.

base metal [MET] **1.** The metal that is to be worked. **2.** The principal metal of an alloy. **3.** Any metal that will oxidize when heated in air. **4.** The metal of parts to be welded. **5.** Metal to which cladding or plating is applied. Also known as basis metal.

base molding [BUILD] Molding used to trim the upper edge of interior baseboard. Also known as base cap.

base net [ENG] A system, in surveying, of quadrilaterals and triangles that include and are quite close to a base line in a triangulation system.

base ore [MIN ENG] Ore in which the gold is associated with sulfides, as contrasted with free-milling ores in which the sulfides have been removed by leaching.

base plate [DES ENG] The part of a theodolite which carries the lower ends of the three foot screws and attaches the theodolite to the tripod for surveying. [ENG] A metal plate that provides support or a foundation.

base screed [ENG] A metal screed with expanded or short perforated flanges that serves as a dividing strip between plaster and cement and acts as a guide to indicate proper thickness of cement or plaster.

base sheet [BUILD] Saturated or coated felt sheeting which is laid as the first ply in a built-up roofing membrane.

base shoe [BUILD] A molding at the base of a baseboard.

base shoe corner [BUILD] A molding piece or block applied in the corner of a room to eliminate the need for mitering the base shoe.

base station [ENG] The point from which a survey begins.

base tee [DES ENG] A pipe tee with a connected baseplate for supporting it.

base tile [BUILD] The lowest course of tiles in a tiled wall.

base time See normal elemental time; normal time.

bashing [MIN ENG] **1.** The building of walls and nonporous stoppings for the complete isolation of a district of a mine in which a fire has occurred. **2.** The complete stowing of old mine workings or roadways after all equipment has been removed.

basic converter See basic-lined converter.

basic dye [MATER] Any of the dyes which are salts of the colored organic bases containing amino and imino groups, combined with a colorless acid, such as hydrochloric or sulfuric.

basic element See elemental motion.

basic feasible solution [IND ENG] A basic solution to a linear program model in which all the variables are nonnegative.

basic-lined [MET] Pertaining to the walls and bottom of a melting furnace made of refractory materials, such as lime or dolomite, that have a basic reaction in the melting process.

basic-lined converter [MET] A converter, such as the Peir-Smith copper converter, which has a basic refractory lining. Also known as basic converter.

basic motion [IND ENG] A single, complete movement of a body member; determined by motion studies.

basic motion-time study [IND ENG] A system of predetermined motion-time standards for basic motions. Abbreviated BMT study.

basic open-hearth process [MET] An open-hearth process for steelmaking under basic slag; used for pig iron and scrap with a phosphorus content too low for the Bessemer process and too high for the acid open-hearth process.

basic refractory [MATER] Any heat-resistant material used for basic linings; examples are dolomite and magnesite.

basic sediment and water [PETRO ENG] Oil, water, and foreign matter that collects in the bottom of petroleum storage tanks. Abbreviated BS&W. Also known as bottoms; bottom settlings; sediment and water.

basic slag [MET] A slag resulting from the steelmaking process; rich in phosphorus, it is ground and used as a nutrient in grasslands.

basic solution [IND ENG] A solution to a linear program model, consisting of m equations in n variables, obtained by solving for m variables in terms of the remaining $(n - m)$ variables and setting the $(n - m)$ variables equal to zero.

basic steel [MET] Steel made by the basic process, in a furnace with a basic lining.

basil [MATER] Sheephide tanned with bark.

basil oil [MATER] Any of the yellow, aromatic essential oils derived from the leaves of sweet basil; used for flavors and perfumes and in medicines. Also known as sweet basil oil.

basin [CIV ENG] **1.** A dock employing floodgates to keep water level constant during tidal variations. **2.** A harbor for small craft. [DES ENG] An open-top vessel with relatively low sloping sides for holding liquids. [MET] The mouth of a sprue in a gating system of castings into which the molten metal is first poured.

basis metal See base metal.

basket [DES ENG] A lightweight container with perforations. [MECH ENG] A type of single-tube core barrel made from thin-wall tubing with the lower end notched into points, which is intended to pick up a sample of granular or plastic rock material by bending in on striking the bottom of the borehole or solid layer; may be used to recover an article dropped into a borehole. Also known as basket barrel; basket tube; sawtooth barrel.

basket barrel *See* basket.

basket tube *See* basket.

basket-weave [BUILD] A checkerboard pattern of bricks, flat or on edge.

bassora gum [MATER] A type of high-colored gum similar to tragacanth gum; includes Indian gum.

bass reflex baffle [ENG ACOUS] A loudspeaker baffle having an opening of such size that bass frequencies from the rear of the loudspeaker emerge to reinforce those radiated directly forward.

bassy [ENG ACOUS] Pertaining to sound reproduction that overemphasizes low-frequency notes.

bastard-cut file [DES ENG] A file that has coarser teeth than a rough-cut file.

bastard pointing *See* bastard tuck pointing.

bastard thread [DES ENG] A screw thread that does not match any standard threads.

bastard tuck pointing [BUILD] An imitation tuck pointing in which the external face is parallel to the wall, but projects slightly and casts a shadow. Also known as bastard pointing.

bat bolt [DES ENG] A bolt whose butt or tang is bashed or jagged.

batch [ENG] **1.** The quantity of material required for or produced by one operation. **2.** An amount of material subjected to some unit chemical process or physical mixing process to make the final product substantially uniform.

batch box [ENG] A container of known volume used to measure and mix the constituents of a batch of concrete, plaster, or mortar, to ensure proper proportions.

batched water [ENG] The mixing water added to a concrete or mortar mixture before or during the initial stages of mixing.

batcher [MECH ENG] A machine in which the ingredients of concrete are measured and combined into batches before being discharged to the concrete mixer.

batching [ENG] Weighing or measuring the volume of the ingredients of a batch of concrete or mortar, and then introducing these ingredients into a mixer.

batch mixer [MECH ENG] A machine which mixes concrete or mortar in batches, as opposed to a continuous mixer.

batch oil [MATER] A pale, lemon-colored, low-viscosity mineral oil used particularly in the manufacture of cordage.

batch plant [ENG] An operating installation of equipment including batchers and mixers as required for batching or for batching and mixing concrete materials.

batch process [ENG] A process that is not in continuous or mass production; operations are carried out with discrete quantities of material or a limited number of items.

batch production *See* series production.

batch-type furnace [MECH ENG] A furnace used for heat treatment of materials, with or without direct firing; loading and unloading operations are carried out through a single door or slot.

batea [MIN ENG] A conical-shaped wood unit (12.3 inches or 31 centimeters in diameter with about 150° apex angle) used to recover valuable metals from river channels and bars.

bathometer [ENG] A mechanism which measures depths in water.

bathtub curve [IND ENG] An equipment failure-rate curve with an initial sharply declining failure rate, followed by a prolonged constant-average failure rate, after which the failure rate again increases sharply.

bathyclinograph [ENG] A mechanism which measures vertical currents in the deep sea.

bathyconductograph [ENG] A device to measure the electrical conductivity of sea water at various depths from a moving ship.

bathygram [ENG] A graph recording the measurements of sonic sounding instruments.

bathymetry [ENG] The science of measuring ocean depths in order to determine the sea floor topography.

bathythermogram [ENG] The record that is made by a bathythermograph.

bathythermograph [ENG] A device for obtaining a record of temperature against depth (actually, pressure) in the ocean from a ship underway. Abbreviated BT. Also known as bathythermosphere.

bathythermosphere *See* bathythermograph.

batt [MATER] A blanket of insulating material usually 16 inches (41 centimeters) wide and 3 to 6 inches (8 to 15 centimeters) thick, used to insulate building walls and roofs. [MIN ENG] A thin layer of coal occurring in the lower part of shale strata that lie above and close to a coal bed.

batted work [ENG] A hand-dressed stone surface scored from top to bottom in narrow parallel strokes (usually 8–10 per inch or 20–25 per centimeter) by use of a batting tool.

batten [AERO ENG] Metal, wood, or plastic panels laced to the envelope of a blimp in the nose cone to add rigidity to the nose and provide a good point of attachment for mooring. [BUILD] **1.** A sawed timber strip of specific dimension—usually 7 inches (18 centimeters) broad, less than 4 inches (10 centimeters) thick, and more than 6 feet (1.8 meters) long—used for outside walls of houses, flooring, and such. **2.** A strip of wood nailed across a door or other structure made of parallel boards to strengthen it and prevent warping.

batten door [BUILD] A wood door without stiles which is constructed of vertical boards held together by horizontal battens on the back side. Also known as ledged door.

battened column [CIV ENG] A column consisting of two longitudinal shafts, rigidly connected to each other by batten plates.

battened wall [BUILD] A wall to which battens have been affixed. Also known as strapped wall.

batten plate [CIV ENG] A rectangular plate used to connect two parallel structural steel members by riveting or welding.

batten roll [BUILD] In metal roofing, a roll joint formed over a triangular-shaped wood piece. Also known as conical roll.

batten seam [BUILD] A seam in metal roofing which is formed around a wood strip.

batter [CIV ENG] A uniformly steep slope in a retaining wall or pier; inclination is expressed as 1 foot horizontally per vertical unit (in feet).

batter board [CIV ENG] Horizontal boards nailed to corner posts located just outside the corners of a proposed building to assist in the accurate layout of foundation and excavation lines.

batter brace [CIV ENG] A diagonal brace which reinforces one end of a truss. Also known as batter post.

batter level [ENG] A device for measuring the inclination of a slope.

batter pile [CIV ENG] A pile driven at an inclination to the vertical to provide resistance to horizontal forces. Also known as brace pile; spur pile.

batter post [CIV ENG] **1.** A post at one side of a gateway or at a corner of a building for protection against vehicles. **2.** See batter brace.

batter stick [CIV ENG] A tapered board which is hung vertically and used to test the batter of a wall surface.

battery amalgamation [MET] Amalgamation by means of mercury placed in the mortar box of a stamp battery.

battery assay [MIN ENG] An assay of samples taken from ore as crushed in a stamp battery.

batting tool [ENG] A mason's chisel usually 3–4½ inches (7.6–11.4 centimeters) wide, used to dress stone to a striated surface.

Bauschinger effect [MET] A phenomenon by which the plastic deformation of a metal increases the tensile yield strength and decreases the compressive yield strength.

bay [AERO ENG] A space formed by structural partitions on an aircraft. [ENG] A housing used for equipment.

bayberry tallow See bayberry wax.

bayberry wax [MATER] Green, bitter-tasting wax derived from boiling of wax myrtle (*Myrica*) berries; used for candles, soaps, medicine. Also known as bayberry tallow; laurel wax; myrtle wax.

Bayer process [MET] A method of producing alumina from bauxite by heating it in a sodium hydroxide solution.

bay oil [MATER] Yellow essential oil with clovelike odor and pungent taste; derived from distillation of West Indian bayberry (*Pimenta acris*) leaves and used in flavors and perfumes. Also known as myrica oil.

bayonet coupling [DES ENG] A coupling in which two or more pins extend out from a plug and engage in grooves in the side of a socket.

bayonet socket [DES ENG] A socket, having J-shaped slots on opposite sides, into which a bayonet base or coupling is inserted against a spring and rotated until its pins are seated firmly in the slots.

bayonet-tube exchanger [MECH ENG] A dual-tube apparatus with heating (or cooling) fluid flowing into the inner tube and out of the annular space between the inner and outer tubes; can be inserted into tanks or other process vessels to heat or cool the liquid contents.

bay rum [MATER] A product originally prepared from the distillation of rum and water mixed with bayberry (*Pimenta acris*) leaves but now made from a mixture of bay oil, orange oil, pimenta oil, alcohol, and water; used in shaving lotions and alcohol rubs.

B bit [MIN ENG] A nonstandard core bit no longer in common use except in drilling deep boreholes to sample gold-bearing deposits in South Africa; the set outside and inside diameters are about $2\frac{1}{16}$ and $1\frac{3}{8}$ inches (52 and 35 millimeters), respectively.

B blasting powder [MATER] A mixture of nitrate of soda, charcoal, and sulfur; used in coal mines. Also known as soda blasting powder.

BDC See bottom dead center.

bd-ft See board-foot.

beach mining [MIN ENG] The mining of the heavy minerals, such as rutile, zircon, monazite, ilmenite, and sometimes gold, which occur in sand dunes, beaches, coastal plains, and deposits located inland from the shoreline.

beacon tracking [ENG] The tracking of a moving object by means of signals emitted from a transmitter or transponder within or attached to the object.

bead [DES ENG] A projecting rim or band. [MET] **1.** The drop of precious metal obtained by cupellation in fire assaying. **2.** *See* weld bead.

bead and butt [BUILD] Framed work in which the panel is flush with the framing and has a bead run on two edges in the direction of the grain; the ends are left plain. Also known as bead butt; bead butt work.

bead-and-flush panel *See* beadflush panel.

bead and quirk *See* quirk bead.

bead and reel [BUILD] A semiround convex molding decorated with a pattern of disks alternating with round or elongated beads. Also known as reel and bead.

bead butt *See* bead and butt.

bead, butt, and square [BUILD] Framed work similar to bead and butt but having the panels flush on the beaded face only, and showing square reveals on the other.

bead butt work *See* bead and butt.

beaded molding [BUILD] A molding or cornice bearing a cast plaster string of beads.

beaded tube end [MECH ENG] The exposed portion of a rolled tube which is rounded back against the sheet in which the tube is rolled.

beadflush panel [BUILD] A panel which is flush with the surrounding framing and finished with a flush bead on all edges of the panel. Also known as bead-and-flush panel.

beading [BUILD] Collectively, the bead moldings used in ornamenting a given surface. [MET] The placing of a bead on a piece of sheet metal for either decorative or strengthening purposes.

beading plane [DES ENG] A plane having a curved cutting edge for shaping beads in wood. Also known as bead plane.

bead-jointed [ENG] Of a carpentry joint, having a bead along the edge of one piece to make the joint less conspicuous.

bead molding [BUILD] A small, convex molding of semicircular or greater profile. Also known as baguette.

beaker sampler [PETRO ENG] A small, cylindrical vessel with a tapered top used to collect crude oil samples; it is made of low-sparking metal or glass, the bottom is weighted, and there is a small stoppered opening at the top.

beaking joint [BUILD] A joint formed by several heading joints occurring in one continuous line; especially used in connection with the laying of floor planks.

beam [CIV ENG] A body, with one dimension large compared with the other dimensions, whose function is to carry lateral loads (perpendicular to the large dimension) and bending movements.

beam-and-girder construction [BUILD] A system of floor construction in which the load is distributed by slabs to spaced beams and girders.

beam-and-slab floor [BUILD] A floor system in which a concrete floor slab is supported by reinforced concrete beams.

Beaman stadia arc [ENG] An attachment to an alidade consisting of a stadia arc on the outer edge of the visual vertical arc; enables the observer to determine the difference in elevation of the instrument and stadia rod without employing vertical angles.

beam-balanced pump [PETRO ENG] An oil well pumping unit having a center-pivoted beam with the sucker rod plunger (pump) at the front end and a counterweight on the rearward extension.

beam bearing plate [CIV ENG] A foundation plate (usually of metal) placed beneath the end of a beam, at its point of support, to distribute the end load at the point.

beam blocking [BUILD] **1.** Boxing-in or covering a joist, beam, or girder to give the appearance of a larger beam. **2.** Strips of wood used to create a false beam.

beam bolster [CIV ENG] A rod which provides support for steel reinforcement in formwork for a reinforced concrete beam.

beam brick [BUILD] A face brick which is used to bond to a poured-in-place concrete lintel.

beam bridge [CIV ENG] A fixed structure consisting of a series of steel or concrete beams placed parallel to traffic and supporting the roadway directly on their top flanges.

beam building [MIN ENG] A process of rock bolting in flat-lying deposits where the bolts are installed in bedded rock to bind the strata together to act as a single beam capable of supporting itself, thus stabilizing the overlying rock.

beam column [CIV ENG] A structural member subjected simultaneously to axial load and bending moments produced by lateral forces or eccentricity of the longitudinal load.

beam-deflection amplifier [MECH ENG] A jet-interaction fluidic device in which the direction of a supply jet is varied by flow from one or more control jets which are oriented at approximately 90° to the supply jet.

beam fill [BUILD] Masonry, brickwork, or cement fill, usually between joists or horizontal beams at their supports; provides increased fire resistance.

beam form [CIV ENG] A form which gives the necessary shape, suppport, and finish to a concrete beam.

beam hanger [PETRO ENG] An attachment at the end of a walking beam above a well casing to lift the pump rods or sacked rods.

beam pattern *See* directivity pattern.

beam pocket [CIV ENG] **1.** In a vertical structural member, an opening to receive a beam. **2.** An opening in the form for a column or girder where the form for an intersecting beam is framed.

beam rider [AERO ENG] A missile for which the guidance system consists of standard reference signals transmitted in a radar beam which enable the missile to sense its location relative to the beam, correct its course, and thereby stay on the beam.

beam riding [AERO ENG] The maneuver of a spacecraft or other vehicle as it follows a beam.

beam splice [CIV ENG] A connection between two lengths of a beam or girder; may be shear or moment connections.

beam spread [ENG] The angle of divergence from the central axis of an electromagnetic or acoustic beam as it travels through a material.

beam test [CIV ENG] A test of the flexural strength (modulus of rupture) of concrete from measurements on a standard reinforced concrete beam.

beam well [PETRO ENG] A well pumped by a walking beam.

bear [MIN ENG] To underhole or undermine; to drive in at the top or side of a working.

bearer [CIV ENG] Any horizontal beam, joist, or member which supports a load.

bearing [CIV ENG] That portion of a beam, truss, or other structural member which rests on the supports. [MECH ENG] A machine part that supports another part which rotates, slides, or oscillates in or on it. [MIN ENG] The direction of a mine drivage, usually given in terms of the horizontal angle turned off a datum direction, such as the true north and south line.

bearing bar [BUILD] A wrought-iron bar placed on masonry to provide a level support for floor joists. [CIV ENG] A load-carrying bar which supports a grating and which extends in the direction of the grating span. [ENG] *See* azimuth instrument.

bearing circle [ENG] A ring designed to fit snugly over a compass or compass repeater, and provided with vanes for observing compass bearings.

bearing cursor [ENG] Of a radar set, the radial line inscribed on a transparent disk which can be rotated manually about an axis coincident with the center of the plan position indicator; used for bearing determination. Also known as mechanical bearing cursor.

bearing distance [CIV ENG] The length of a beam between its bearing supports.

bearing partition [BUILD] A partition which supports a vertical load.

bearing pile [ENG] A vertical post or pile which carries the weight of a foundation, transmitting the load of a structure to the bedrock or subsoil without detrimental settlement.

bearing plate [CIV ENG] A flat steel plate used under the end of a wall-bearing beam to distribute the load over a broader area.

bearing test [ENG] A test of the bearing capacities of pile foundations, such as a field loading test of an individual pile; a laboratory test of soil samples for bearing capacities.

bearing wall [CIV ENG] A wall capable of supporting an imposed load. Also known as structural wall.

bear trap gate [CIV ENG] A type of crest gate with an upstream leaf and a downstream leaf which rest in a horizontal position, one leaf overlapping the other, when the gate is lowered.

beater [ENG] **1.** A tool for packing in material to fill a blasthole containing a charge of powder. **2.** A laborer who shovels or dumps asbestos fibers and sprays them with water in order to prepare them for the beating. [MECH ENG] A machine that cuts or beats paper stock.

beater mill *See* hammer mill.

beating [ENG] A process that reduces asbestos fibers to pulp for making asbestos paper.

beat tone [ENG ACOUS] Musical tone due to beats, produced by the heterodyning of two high-frequency wave trains.

bêche [MECH ENG] A pneumatic forge hammer having an air-operated ram and an air-compressing cylinder integral with the frame.

Beckmann thermometer [ENG] A sensitive thermometer with an adjustable range so that small differences in temperature can be measured.

bed [CIV ENG] **1.** In masonry and bricklaying, the side of a masonry unit on which the unit lies in the course of the wall; the underside when the unit is placed horizontally. **2.** The layer of mortar on which a masonry unit is set. [MECH ENG] The part of a machine having precisely machined ways or bearing surfaces which support or align other machine parts.

Bedaux plan [IND ENG] A wage incentive plan in which work is standardized into man-minute units called bedaux (B); 60 B per hour is 100% productivity, and earnings are based on work units per length of time.

bed charge [MET] The primary charge of fuel in a cupola to initiate the melting process.

bed coke [MET] The primary layer of coke placed in a cupola at a preselected height above the tuyeres for the initial combustion and melting operation.

bed detector [PETRO ENG] Apparatus to detect and measure the extent of underground formations that are potential oil and gas reservoirs; methods include induction logs, gamma-ray logs, and sonic logs.

bedding [CIV ENG] 1. Mortar, putty, or other substance used to secure a firm and even bearing, such as putty laid in the rabbet of a window frame, or mortar used to lay bricks. 2. A base which is prepared in soil or concrete for laying masonry or concrete.

bedding course [CIV ENG] The first layer of mortar at the bottom of masonry.

bedding dot [BUILD] A small spot of plaster built out to the face of a finished wall or ceiling; serves as a screed for leveling and plumbing in the application of plaster.

bede [MIN ENG] A miner's pick.

bed glazing See back putty.

bed joint [CIV ENG] 1. A horizontal layer of mortar on which masonry units are laid. 2. One of the radial joints in an arch.

bed molding [BUILD] 1. The lowest member of a band of moldings. 2. Any molding under a projection, such as between eaves and sidewalls.

beef stearin See oleostearin.

beehive oven [ENG] An arched oven that carbonizes coal into coke by using the heat of combustion of gases that are formed, and of a small part of the coke that is formed, with no recovery of by-products.

beeswax [MATER] Yellow to grayish-brown solid wax obtained from bee honeycombs by boiling and straining; used in floor waxes, waxed paper, and textile finishes and in pharmacy. Also known as yellow wax.

Beethoven exploder [ENG] A machine for the multishot firing of series-connected detonators in tunneling and quarrying.

beetle [MIN ENG] A powerful, cable-hauled propulsion unit, operated under remote control, for moving a train of wagons at the mine surface.

Beilby layer [MET] An amorphous layer formed on a polished metal surface.

Belfast truss [CIV ENG] A bowstring beam for large spans, having the upper member bent and the lower member horizontal; constructed entirely of timber components.

Belgian block [MATER] 1. A stone block used for paving, having the shape of a truncated pyramid, with a depth of 7–8 inches (18–20 centimeters), a base of 5–6 inches (13–15 centimeters) square, and a face opposite the base that is 1 inch (2.5 centimeters) or less smaller than the base. 2. Any stone block used for paving.

Belgian retort process See horizontal retort process.

bell [ENG] 1. A hollow metallic cylinder closed at one end and flared at the other; it is used as a fixed-pitch musical instrument or signaling device and is set vibrating by a clapper or tongue which strikes the lip. 2. See bell tap. [MET] A conical device that seals the top of a blast furnace.

bell-and-spigot joint [ENG] A pipe joint in which a pipe ending in a bell-like shape is joined to a pipe ending in a spigotlike shape.

belled caisson [CIV ENG] A type of drilled caisson with a flared bottom.

bell glass See bell jar.

bell hole [MIN ENG] 1. One of the holes or excavations made at the section joints of a pipeline for the purpose of repairs. 2. A conical cavity in a coal mine roof caused by the falling of a large concretion.

bellite [MATER] An explosive consisting of five parts of ammonium nitrate to one part of metadinitrobenzene with some potassium nitrate.

bell jar [ENG] A bell-shaped vessel, usually made of glass, which is used for enclosing a vacuum, holding gases, or covering objects. Also known as bell glass.

bell jar testing [ENG] A leak testing method in which a vessel is filled with tracer gas and placed in a vacuum chamber; leaks are evidenced by gas drawn into the vacuum chamber.

bell joint clamp [ENG] A clamp applied to a bell-and-spigot joint to prevent leakage.

Bellman's principle of optimality [IND ENG] The principle that an optimal sequence of decisions in a multistage decision process problem has the property that whatever the initial state and decisions are, the remaining decisions must constitute an optimal policy with regard to the state resulting from the first decisions.

bell metal [MET] An alloy of copper and tin, containing 15–25% tin but 20–24% for best tonal quality.

bell mouth [DES ENG] A flared mouth on a pipe opening or other orifice. [ENG] A defect which occurs during metal drilling in which a twist drill produces a hole that is not a perfect circle.

bell nipple [PETRO ENG] A bell-mouthed nipple inserted into the top of oil well casing to allow easy entry of drilling tools and to protect the top of the casing during drilling.

bellows [ENG] **1.** A mechanism that expands and contracts, or has a rising and falling top, to suck in air through a valve and blow it out through a tube. **2.** Any of several types of enclosures which have accordionlike walls, allowing one to vary the volume. **3.** *See* aneroid capsule.

bellows expansion joint [DES ENG] In a run of piping, a joint formed with a flexible metal bellows which compress or stretch to compensate for linear expansion or contraction of the run of piping.

bellows gage [ENG] A device for measuring pressure in which the pressure on a bellows, with the end plate attached to a spring, causes a measurable movement of the plate.

bellows gas meter [ENG] A device for measuring the total volume of a continuous gas flow stream in which the motion of two bellows, alternately filled with and exhausted of the gas, actuates a register.

bellows seal [MECH ENG] A boiler seal in the form of a bellows which prevents leakage of air or gas.

bell screw *See* bell tap.

bell socket *See* bell tap.

bell tap [ENG] A cylindrical fishing tool having an upward-tapered inside surface provided with hardened threads; when slipped over the upper end of lost, cylindrical, downhole drilling equipment and turned, the threaded inside surface of the bell tap cuts into and grips the outside surface of the lost equipment. Also known as bell; bell screw; bell socket; box bill; die; die collar; die nipple.

bell-type manometer [ENG] A differential pressure gage in which one pressure input is fed into an inverted cuplike container floating in liquid, and the other pressure input presses down upon the top of the container so that its level in the liquid is the measure of differential pressure.

belt [CIV ENG] In brickwork, a projecting row (or rows) of bricks, or an inserted row made of a different kind of brick. [MECH ENG] A flexible band used to connect pulleys or to convey materials by transmitting motion and power.

belt conveyor [MECH ENG] A heavy-duty conveyor consisting essentially of a head or drive pulley, a take-up pulley, a level or inclined endless belt made of canvas, rubber, or metal, and carrying and return idlers.

belt course *See* string course.

belt dressing [MATER] A material, usually beeswax, applied to leather belts to increase friction between the belt and pulley surface.

belt drive [MECH ENG] The transmission of power between shafts by means of a belt connecting pulleys on the shafts.

belt feeder [MECH ENG] A short belt conveyor used to transfer granulated or powdered solids from a storage or supply point to an end-use point; for example, from a bin hopper to a chemical reactor.

belt grinding [MET] Grinding with an abrasive-coated continuous belt. Also known as linishing.

belting [MATER] **1.** A sturdy fabric, usually of cotton, used in belts. **2.** A heavy leather, made from hides of cattle, used in power transmission belts. Also known as belting leather.

belting leather *See* belting.

belt sander [MECH ENG] A portable sanding tool having a power-driven abrasive-coated continuous belt.

belt shifter [MECH ENG] A device with fingerlike projections used to shift a belt from one pulley to another or to replace a belt which has slipped off a pulley.

belt slip [MECH ENG] The difference in speed between the driving drum and belt conveyor.

belt tightener [MECH ENG] In a belt drive, a device that takes up the slack in a belt that has become stretched and permanently lengthened.

bench [MIN ENG] **1.** One of two or more divisions of a coal seam, separated by slate and so forth or simply separated by the process of cutting the coal, one bench or layer being cut before the adjacent one. **2.** A long horizontal ledge of ore in an underground working place. **3.** A ledge in an open-pit mine from which excavation takes place at a constant level.

bench assembly [ENG] A technique of fitting and joining parts using a bench as a work surface.

bench blasting [MIN ENG] A mining system used either underground or in surface pits whereby a thick ore or waste zone is removed by blasting a series of successive horizontal layers called benches.

bench check [IND ENG] A workshop or servicing bay check which includes the typical check or actual functional test of an item to ascertain what is to be done to return the item to a serviceable condition or ascertain the item's temporary or permanent disposition.

bench dog [ENG] A wood or metal peg, placed in a slot or hole at the end of a bench; used to keep a workpiece from slipping.

bench height *See* bank height.

bench hook [ENG] Any device used on a carpenter's bench to keep work from moving toward the rear of the bench. Also known as side hook.

benching [CIV ENG] **1.** Concrete laid on the side slopes of drainage channels where the slopes are interrupted by manholes, and so forth. **2.** Concrete laid on sloping sites as a safeguard against sliding. **3.** Concrete laid along the sides of a pipeline to provide additional support. [MIN ENG] A method of working small quarries or opencast pits in steps or benches.

bench lathe [MECH ENG] A small engine or toolroom lathe suitable for attachment to a workbench; bed length usually does not exceed 6 feet (1.8 meters) and workpieces are generally small.

bench mark [ENG] A relatively permanent natural or artificial object bearing a marked point whose elevation above or below an adopted datum—for example, sea level—is known. Abbreviated BM.

bench photometer [ENG] A device which uses an optical bench with the two light sources to be compared mounted one at each end; the comparison is made by a device moved along the bench until matching brightnesses appear.

bench plane [DES ENG] A plane used primarily in benchwork on flat surfaces, such as a block plane or jack plane.

bench sander [MECH ENG] A stationary power sander, usually mounted on a table or stand, which is equipped with a rotating abrasive disk or belt.

bench-scale testing [ENG] Testing of materials, methods, or chemical processes on a small scale, such as on a laboratory worktable.

bench stop [ENG] A bench hook which is used to fasten work in place, often by means of a screw.

bench table [BUILD] A projecting course of masonry at the foot of an interior wall or around a column; generally wide enough to form a seat.

bench vise [ENG] An ordinary vise fixed to a workbench.

benchwork [ENG] Any work performed at a workbench rather than on machines or in the field.

bend [DES ENG] **1.** The characteristic of an object, such as a machine part, that is curved. **2.** A section of pipe that is curved. **3.** A knot formed by a rope fastened to an object or another rope.

bend allowance [DES ENG] Length of the arc of the neutral axis between the tangent points of a bend in any material.

bender See bending machine.

bending [ENG] **1.** The forming of a metal part, by pressure, into a curved or angular shape, or the stretching or flanging of it along a curved path. **2.** The forming of a wooden member to a desired shape by pressure after it has been softened or plasticized by heat and moisture.

bending brake [MECH ENG] A press brake for making sharply angular linear bends in sheet metal.

bending iron [ENG] A tool used to straighten or to expand flexible pipe, especially lead pipe.

bending machine [MECH ENG] A machine for bending a metal or wooden part by pressure. Also known as bender.

bending schedule [CIV ENG] A chart showing the shapes and dimensions of every reinforcing bar and the number of bars required on a particular job for the construction of a reinforced concrete structure.

Bendix-Weiss universal joint [MECH ENG] A universal joint that provides for constant angular velocity of the driven shaft by transmitting the torque through a set of four balls lying in the plane that contains the bisector of, and is perpendicular to, the plane of the angle between the shafts.

bend radius [DES ENG] The radius corresponding to the curvature of a bent specimen or part, as measured at the inside surface of the bend.

bend test [MET] A ductility test in which a specimen is bent through an arc of known radius and angle.

bend wheel [MECH ENG] A wheel used to interrupt and change the normal path of travel of the conveying or driving medium; most generally used to effect a change in direction of conveyor travel from inclined to horizontal or a similar change.

beneficiation [MET] Improving the chemical or physical properties of an ore so that metal can be recovered at a profit. Also known as mineral dressing.

Bengal lights [MATER] Constituent of pyrotechnics made of sulfur, sugar, and potassium nitrate, with barium or strontium salts added for green or red light.

Benioff extensometer [ENG] A linear strainmeter for measuring the change in distance between two reference points separated by 20–30 meters or more; used to observe earth tides.

benjamin gum See benzoin gum.

benne oil See sesame oil.

ben oil [MATER] Oil from seeds of ben (*Moringa*) trees; used in foods, cosmetics, perfumes, and laxatives and as a lubricant.

bent [CIV ENG] A framework support transverse to the length of a structure.

bent bar [CIV ENG] A longitudinal reinforcing bar which is bent to pass from one face of a structural member to the other face.

bentonite slurry [MATER] Slurry composed of water and clay powder consisting chiefly of montmorillonite minerals.

bent-tube boiler [MECH ENG] A water-tube steam boiler in which the tubes terminate in upper and lower steam-and-water drums. Also known as drum-type boiler.

bentwood [ENG] Wood formed to shape by bending, rather than by carving or machining.

benzene-toluene-xylene [MATER] An admixture of three aromatic chemicals formed during the catalytic reforming of petroleum naphthas. Abbreviated BTX.

benzin See petroleum benzin.

benzine See petroleum benzin.

benzoin [MATER] A balsamic resin obtained from trees of the genus *Styrax*; used as an expectorant, as an inhalant in respiratory tract inflammations, and as an antiseptic. Also known as benjamin gum; benzoinam; gum benzoin.

benzoinam See benzoin.

bergamot oil [MATER] A yellow-green essential oil from the rind of bergamot (*Citrus bergamia*) fruit; it is volatile, contains linalyl acetate, limonene, and livalol, and is used in perfumes.

berm [CIV ENG] A horizontal ledge cut between the foot and top of an embankment to stabilize the slope by intercepting sliding earth.

Berthon dynamometer [ENG] An instrument for measuring the diameters of small objects, consisting of two metal straightedges inclined at a small angle and rigidly joined together; a scale on one of the straightedges is used to read the diameters of objects inserted between them.

beryllium [MET] A rare metal, occurring naturally in combinations, with density about one-third of aluminum; used most commonly in the manufacture of beryllium-copper alloys which find numerous industrial and scientific applications.

beryllium alloy [MET] Any dilute alloy of base metals containing a few percent of beryllium in a precipitation-hardening system.

beryllium bronze See beryllium copper.

beryllium copper [MET] **1.** An alloy of copper and beryllium containing not more than about 3% beryllium; used for springs, tools, and plastic molds. Also known as beryllium bronze. **2.** An alloy of copper and beryllium specifically for addition to metals in the foundry.

beryllium detector [ENG] An instrument designed to detect and analyze for beryllium by gamma-ray activation analysis. Also known as berylometer.

beryllium monel [MET] A nickel-copper alloy containing beryllium.

berylometer See beryllium detector.

Bessemer converter [MET] A pear-shaped, basic-lined, cylindrical vessel for producing steel by the Bessemer process.

Bessemer iron [MET] Pig iron with about 1% silicon and a sulfur and a phosphorus content below 0.10%; used to make steel by the Bessemer or the acid open-hearth process. Also known as Bessemer pig iron.

Bessemer matte [MET] Product of the oxidation of furnace matte; contains nickel, copper, cobalt, precious metals, and about 22% sulfur.

Bessemer ore [MET] An iron ore containing very little phosphorus, considered suitable for refining by the Bessemer process.

Bessemer pig iron See Bessemer iron.

Bessemer process [MET] A steelmaking process in which carbon, silicon, phosphorus, and manganese contained in molten pig iron are oxidized by a strong blast of air.

Bessemer steel [MET] Steel manufactured by the Bessemer process.

best commercial practice [ENG] A manufacturing standard for a process vessel which has not been designed according to standard codes, such as the ASME Boiler Code.

beta brass [MET] A type of brass containing nearly equal proportions of copper and zinc.

Bethell process See full-cell process.

Betts' process [MET] A refining process for electrolytically purifying lead.

Betz momentum theory [MECH ENG] A theory of windmill performance that considers the deceleration in the air traversing the windmill disk.

bevel [DES ENG] **1.** The angle between one line or surface and another line or surface, or the horizontal, when this angle is not a right angle. **2.** A sloping surface or line.

bevel gear [MECH ENG] One of a pair of gears used to connect two shafts whose axes intersect.

beveling See chamfering.

Beveloid gear [DES ENG] An involute gear, of the Vinco Corporation, with tapered outside diameter, root diameter, and tooth thickness; used to reduce backlash in the drives of precision instruments.

bezel [DES ENG] **1.** A grooved rim used to hold a transparent glass or plastic window or lens for a meter, tuning dial, or some other indicating device. **2.** A sloping face on a cutting tool.

B-H meter [ENG] A device used to measure the intrinsic hysteresis loop of a sample of magnetic material.

bhp See boiler horsepower; brake horsepower.

bibcock [DES ENG] A faucet or stopcock whose nozzle is bent downward. Also spelled bibb cock.

bicable tramway [MECH ENG] A tramway consisting of two stationary cables on which the wheeled carriages travel, and an endless rope, which propels the carriages.

bid [ENG] An estimate of costs for specified construction, equipment, or services proposed to a customer company by one or more supplier or contractor companies.

bidirectional [ENG] Being directionally responsive to inputs in opposite directions.

bidirectional microphone [ENG ACOUS] A microphone that responds equally well to sounds reaching it from the front and rear, corresponding to sound incidences of 0 and 180°.

Bierbaum scratch hardness test [ENG] A test for the hardness of a solid sample by microscopic measurement of the width of scratch made by a diamond point under preset pressure.

biface tool [DES ENG] A tool, as an ax, made from a coil flattened on both sides to form a V-shaped cutting edge.

bifacial [DES ENG] Of a tool, having both sides alike.

bifilar electrometer [ENG] An electrostatic voltmeter in which two conducting quartz fibers, stretched by a small weight or spring, are separated by their attraction in opposite directions toward two plate electrodes carrying the voltage to be measured.

bifilar micrometer See filar micrometer.

bifilar suspension [ENG] The suspension of a body from two parallel threads, wires, or strips.

big inch pipe [PETRO ENG] A pipeline 24 inches (61 centimeters) in diameter which carries oil or gas, usually for great distances.

bilateral tolerance [DES ENG] The amount that the size of a machine part is allowed to vary above or below a basic dimension; for example, 3.650 ± 0.003 centimeters indicates a tolerance of ± 0.003 centimeter.

bilge block [CIV ENG] A wooden support under the turn of a ship's bilge in dry dock.

bill [DES ENG] One blade of a pair of scissors.

billet [ENG] In a hydraulic extrusion press, a large cylindrical cake of plastic material placed within the pressing chamber. [MET] A semifinished, short, thick bar of iron or steel in the form of a cylinder or rectangular prism produced from an ingot; limited to 1.5 inches (3.8 centimeters) in width and thickness with a cross-

sectional area up to 36 square inches (232 square centimeters).

Billet furnace [MET] A furnace for heating steel in sizes between a bloom and a bar.

billet mill [MET] A rolling mill for making billets from ingots.

bimetal [MATER] A laminate of two dissimilar metals, with different coefficients of thermal expansion, bonded together.

bimetallic corrosion [MET] Corrosion of two different metals exposed to an electrolyte while in electrical contact.

bimetallic strip [ENG] A strip formed of two dissimilar metals welded together; different temperature coefficients of expansion of the metals cause the strip to bend or curl when the temperature changes.

bimetallic thermometer [ENG] A temperature-measuring instrument in which the differential thermal expansion of thin, dissimilar metals, bonded together into a narrow strip and coiled into the shape of a helix or spiral, is used to actuate a pointer. Also known as differential thermometer.

bin [ENG] An enclosed space, box, or frame for the storage of bulk substance.

binary alloy [MET] An alloy composed of two principal metallic components.

binary explosive [MATER] A high explosive composed of a mixture of two high explosives, to secure an explosive which is superior to its components in regard to sensitivity, fragmentation, blast, or loadability.

binary system [ENG] Any system containing two principal components.

binder [MATER] A resin or other cementlike material used to hold particles together and provide mechanical strength or to ensure uniform consistency, solidification, or adhesion to a surface coating; typical binders are resin, glue, gum, and casein.

binder course [CIV ENG] Coarse aggregate with a bituminous binder between the foundation course and the wearing course of a pavement.

binderless briquetting [ENG] The briquetting of coal by the application of pressure without the addition of a binder.

bind-seize See freeze.

biodegradability [MATER] The characteristic of a substance that can be broken down by microorganisms.

bioengineering [ENG] The application of engineering knowledge to the fields of medicine and biology.

bioinstrumentation [ENG] The use of instruments attached to animals and man to record biological parameters such as breathing rate, pulse rate, body temperature, or oxygen in the blood.

biological corrosion [MET] Deterioration of metals as a result of the metabolic activity of microorganisms.

biomedical engineering [ENG] The application of engineering technology to the solution of medical problems; examples are the development of prostheses such as artificial valves for the heart, various types of sensors for the blind, and automated artificial limbs.

bionics [ENG] The study of systems, particularly electronic systems, which function after the manner of living systems.

biopak [ENG] A container for housing a living organism in a habitable environment and for recording biological functions during space flight.

biosatellite [AERO ENG] An artificial satellite designed to contain and support humans, animals, or other living material in a reasonably normal manner for a period of time and to return safely to earth.

biosensor [ENG] A sensor used to provide information about a life process.

biostabilizer [CIV ENG] A component in mechanized composting systems; consists of a drum in which moistened solid waste is comminuted and tumbled for about 5 days until the aeration and biodegradation turns the waste into a fine dark compost.

biotechnology [ENG] The application of engineering and technological principles to the life sciences.

biotelemetry [ENG] The use of telemetry techniques, especially radio waves, to study behavior and physiology of living things.

biotron [ENG] A test chamber used for biological research within which the environmental conditions can be completely controlled, thus allowing observations of the effect of variations in environment on living organisms.

biplane [AERO ENG] An aircraft with two wings fixed at different levels, especially one above and one below the fuselage.

bipolar magnetic driving unit [ENG ACOUS] Headphone or loudspeaker unit having two magnetic poles acting directly on a flexible iron diaphragm.

bipropellant [MATER] A rocket propellant consisting of two unmixed or uncombined chemicals (fuel and oxidizer) fed to the combustion chamber separately.

birch tar oil [MATER] A toxic liquid mixture of guaiacol, phenols, cresol, xylenol, and creosol; derived from distillation of birch tar and by dry distillation of the wood of *Betula alba*; used as a disinfectant, in leather dressing, and in medicine.

Bird centrifuge [MIN ENG] A dewatering machine for fine coal or other fine materials such as potash minerals, clays, and cement rock. Also known as Bird coal filter.

Bird coal filter *See* Bird centrifuge.

Birmingham wire gage [DES ENG] A system of standard sizes of brass wire, telegraph wire, steel tubing, seamless tubing, sheet spring steel, strip steel, and steel plates, bands, and hoops. Abbreviated BWG.

birth-death process [IND ENG] A simple queuing model in which units to be served arrive (birth) and depart (death) in a completely random manner.

biscuit [ENG ACOUS] *See* preform. [MATER] **1.** A clay object that has been fired once prior to glazing. **2.** Pottery that is unglazed in its final form. [MET] An upset blank for drop forging.

biscuit cutter [MIN ENG] A short (6–8 inches or 15–20 centimeters) core barrel that is sharpened at the bottom and forced into the rocks by the jars.

bisilicate [MET] A type of slag whose silicate degree is 2.

bismanol [MET] A magnetic alloy of bismuth and manganese.

bismuth alloy [MET] A group of low-melting alloys (many below 100°C) of bismuth combined with lead, tin, and cadmium; used in automatic sprinklers, special solders, safety plugs in compressed-gas cylinders, automatic shutoffs for water-heating systems, castings, and type metal.

bistable unit [ENG] A physical element that can be made to assume either of two stable states; a binary cell is an example.

bistatic radar [ENG] Radar system in which the receiver is some distance from the transmitter, with separate antennas for each.

bit [DES ENG] **1.** A machine part for drilling or boring. **2.** The cutting plate of a plane. **3.** The blade of a cutting tool such as an ax. **4.** A removable tooth of a saw. **5.** Any cutting device which is attached to or part of a drill rod or drill string to bore or penetrate rocks. [MET] In soldering, the portion of the iron that transfers either heat or solder to the joint involved.

bit blank [DES ENG] A steel bit in which diamonds or other cutting media may be inset by hand peening or attached by a mechanical process such as casting, sintering, or brazing. Also known as bit shank; blank; blank bit; shank.

bit drag [DES ENG] A rotary-drilling bit that has serrated teeth. Also known as drag bit.

bit shank *See* bit blank.

bitter orange oil [MATER] An essential oil from the rind or peel of the orange (*Citrus vulgaris*); insoluble in water; used as a flavoring and in some perfumes.

bitumastic [MATER] A combination of asphalt and filler used mainly to coat metals to protect them against corrosion and weathering.

bitumen [MATER] Naturally occurring or pyrolytically obtained substances of dark to black color consisting almost entirely of carbon and hydrogen, with very little oxygen, nitrogen, or sulfur.

bituminous [MATER] **1.** Containing much organic, or at least carbonaceous, matter, mostly in the form of tarry hydrocarbons which are usually described as bitumen. **2.** Similar to bitumen. **3.** Giving off volatile bituminous substances on heating, as in bituminous coal.

bituminous cement [MATER] A bituminous material suitable for use as a binder, having cementing qualities which are dependent mainly on its bituminous character.

bituminous coating [MATER] A coating made principally of bituminous material and used as a surfacing for roads and as a water-repellent barrier in buildings.

bituminous concrete [MATER] A concrete made with bituminous material as a binder for sand and gravel.

bituminous paint [MATER] Paint with a high proportion of bitumen; usually dark in color.

bivane [ENG] A double-jointed vane which measures vertical as well as horizontal wind direction.

black acids [MATER] Sulfonates in the sludge formed during treatment of petroleum products with sulfuric acid; soluble in water but insoluble in naphtha, benzene, and carbon tetrachloride.

black annealing [MET] A type of box annealing used to impart a black color to the metal surface; first process in tin plating.

black ash [MATER] A carbon product made by furnace heating of black liquor from papermaking processes.

black balsam *See* Peru balsam.

blackboard [MATER] A panel, usually black but sometimes colored, for writing on with chalk. Also known as chalkboard.

black box [ENG] Any component, usually electronic and having known input and output, that can be readily inserted into or removed from a specific place in a larger system without knowledge of the component's detailed internal structure.

black-bulb thermometer [ENG] A thermometer whose sensitive element has been made to approximate a blackbody by covering it with lampblack.

black copper [MET] The more or less impure metallic copper (70–99% copper) produced in blast furnaces when running on oxide ores or roasted sulfide material.

blackdamp [MIN ENG] A nonexplosive mixture of carbon dioxide with other gases, especially with 85–90% nitrogen, which is heavier than air and cannot support flame or life. Also known as chokedamp.

black grease [MATER] Grease which has black coloration due to use of asphalt, either naturally occurring or from residues used in the manufacture of the grease.

blackheart malleable iron *See* whiteheart malleable iron.

blacking [MET] Carbonaceous material, such as powdered graphite, used to coat the inner surfaces of a dry-sand mold to improve the separation and finish of a casting.

black liquor [MATER] **1.** The liquid material remaining from pulpwood cooking in the soda or sulfate papermaking process. **2.** *See* iron acetate liquor.

black mordant *See* iron acetate liquor.

black oil [MATER] Low-grade, black petroleum oil used to lubricate slow-moving or rough-surfaced machinery where high-grade lubricants are impractical or too expensive.

black pepper oil [MATER] An essential oil obtained from the fruit of the black pepper plant; colorless to slightly greenish with a pepper odor and mild taste.

black powder [MATER] A low explosive consisting of an intimate mixture of potassium nitrate or sodium nitrate, charcoal, and sulfur.

black smoke [ENG] A smoke that has many particulates in it from inefficient combustion; comes from burning fossil fuel, either coal or oil.

blacktop [MATER] A black bituminous material that is used to pave roadways; it is spread over a layer of crushed rocks and packed down into a level surface; it may be spread over small areas of roadways in need of repair.

blacktop paver [MECH ENG] A construction vehicle that spreads a specified thickness of bituminous mixture over a prepared surface.

bladder press [MECH ENG] A machine which simultaneously molds and cures (vulcanizes) a pneumatic tire.

blade [ENG] **1.** A broad, flat arm of a fan, turbine, or propeller. **2.** The broad, flat surface of a bulldozer or snowplow by which the material is moved. **3.** The part of a cutting tool, such as a saw, that cuts.

bladed-surface aerator [CIV ENG] A bladed, rotating component of a water treatment plant; used to infuse air into the water.

blade loading [AERO ENG] A rotor's thrust in a rotary-wing aircraft divided by the total area of the rotor blades.

Blake jaw crusher [MECH ENG] A crusher with one fixed jaw plate and one pivoted at the top so as to give the greatest movement on the smallest lump.

blank [DES ENG] *See* bit blank. [ENG] **1.** The result of the final cutting operation on a natural crystal. **2.** *See* blind. [MET] **1.** A semifinished piece of metal to be stamped or forged into a tool or implement. **2.** A semifinished, pressed, compacted mass of powdered metal. **3.** Metal sheet prepared for a forming operation.

blank bit *See* bit blank.

blank carburizing [MET] A simulated carburizing procedure carried out without a carburizing medium. Also known as pseudocarburizing.

blanket [MIN ENG] A textile material used in ore treatment plants for catching coarse free gold and sometimes associated minerals, for example, pyrite.

blanketing [MIN ENG] The material caught on the blankets that are used in concentrating gold-bearing sands or slimes.

blanket insulation [MATER] Insulation in the form of a rolled sheet sometimes having a vapor-barrier treated paper backing.

blank holder [MET] A tool to prevent the edge of a sheet metal blank from wrinkling during deep-drawing operations.

blankholder slide [MECH ENG] The outer slide of a double-action power press; it is usually operated by toggles or cams.

blanking [ENG] **1.** The closing off of flow through a liquid-containing process pipe by the insertion of solid disks at joints or unions; used during maintenance and repair work as a safety precaution. Also known as blinding. **2.** Cutting of plastic or metal sheets into shapes by striking with a punch. Also known as die cutting.

blank nitriding [MET] Simulation of nitriding without the introduction of nitrogen; achieved by use of an inert material or by application of a coating to the piece.

Blasius theorem [AERO ENG] A theorem that provides formulas for finding the force and moment on the airfoil profiler.

blast [ENG] The setting off of a heavy explosive charge.

blast burner [ENG] A burner in which a controlled burst of air or oxygen under pressure is supplied to the illuminating gas used. Also known as blast lamp.

blast chamber [AERO ENG] A combustion chamber, especially in a gas-turbine, jet, or rocket engine.

blast deflector [AERO ENG] A device used to divert the exhaust of a rocket fired from a vertical position.

blast ditching [CIV ENG] The use of explosives to aid in ditch excavation, such as for laying pipelines.

blaster [ENG] A device for detonating an explosive charge; usually consists of a machine by which an operator, by pressing downward or otherwise moving a handle of the device, may generate a powerful transient electric current which is transmitted to an electric blasting cap. Also known as blasting machine.

blast furnace [MET] A tall, cylindrical smelting furnace for reducing iron ore to pig iron; the blast of air blown through solid fuel increases the combustion rate.

blast-furnace coke [MATER] Coke which supplies carbon monoxide to reduce the ore in a blast furnace and supplies heat to melt the iron.

blast-furnace gas [MATER] The gas product from iron ore smelting when hot air passes over coke in blast ovens; contains carbon dioxide, carbon monoxide, hydrogen, and nitrogen and is used as fuel gas.

blast gate [MET] A sliding plate in a cupola blast pipe to channel airflow in the proper proportion.

blasthole [ENG] **1.** A hole that takes a heavy charge of explosive. **2.** The hole through which water enters in the bottom of a pump stock.

blasting [ENG] **1.** Cleaning materials by a blast of air that blows small abrasive particles against the surface. **2.** The act of detonating an explosive.

blasting agent [MATER] A compound or mixture, such as ammonium nitrate or black powder, that detonates as a result of heat or shock; used in mining, blasting, pyrotechnics, and propellants.

blasting barrel [MIN ENG] A piece of iron pipe, usually about one-half inch in diameter, used to provide a smooth passageway through the stemming for the miner's squib; it is recovered after each blast and used until destroyed.

blasting cap [ENG] A copper shell closed at one end and containing a charge of detonating compound, which is ignited by electric current or the spark of a fuse; used for detonating high explosives.

blasting fuse [ENG] A core of gunpowder in the center of jute, yarn, and so on for igniting an explosive charge in a shothole.

blasting gelatin [MATER] A plastic dynamite that contains 5–10% nitrocellulose added to nitroglycerin; used principally in submarine work.

blasting machine *See* blaster.

blasting powder [MATER] A powder containing less nitrate, and in its place more charcoal than black powder; composition is 65–75% sodium or nitrate, potassium nitrate, 10–15% sulfur, and 15–20% charcoal.

blast lamp *See* blast burner; blowtorch.

blast-off [AERO ENG] The takeoff of a rocket or missile.

blast roasting [MIN ENG] The roasting of finely divided ores by means of a blast to maintain internal combustion within the charge. Also known as roast sintering.

blast wall [ENG] A heavy wall used to isolate buildings or areas which contain highly combustible or explosive materials or to protect a building or area from blast damage when exposed to explosions.

bleaching assistant [MATER] A material added to textile bleach baths to cause better penetration by the bleach; includes pine oils, borax, and sulfonated oils.

bleaching clay [MATER] Absorbent clay contacted with petroleum and vegetable and other oils to make decolorized products.

bleaching powder [MATER] A mixture of calcium hydroxide, calcium chloride, and calcium hypochlorite that is used as a bleaching agent. Also known as chloride of lime; chlorinated lime.

bleach liquor [MATER] A water solution of calcium hypochlorite used as a laundry and textile bleach, germicide, and deodorant.

bleed [ENG] To let a fluid, such as air or liquid oxygen, escape under controlled conditions from a pipe, tank, or the like through a valve or outlet.

bleeder [ENG] A connection located at a low place in an air line or a gasoline container so that, by means of a small valve, the condensed water or other liquid can be drained or bled off from the line or container without discharging the air or gas.

bleeder turbine [MECH ENG] A multistage turbine where steam is extracted (bled) at pressures intermediate between throttle and exhaust, for process or feedwater heating purposes.

bleeding [ENG] Natural separation of a liquid from a liquid-solid or semisolid mixture; for example, separation of oil from a stored lubricating grease, or water from freshly poured concrete.

bleeding cycle [MECH ENG] A steam cycle in which steam is drawn from the turbine at one or more stages and used to heat the feedwater. Also known as regenerative cycle.

bleed valve [ENG] A small-flow valve connected to a fluid process vessel or line for the purpose of bleeding off small quantities of contained fluid.

blend [MATER] A mixture so combined as to render the parts indistinguishable from one another.

blended data [ENG] A point that is the combination of scan data and track data to form a vector.

blended fuel oil [MATER] A mixture of petroleum residual and distillate fuel oils.

blending naphtha [MATER] A petroleum distillate used to thin or dilute heavy petroleum stocks (for example, lubricating oil) to simplify handling and processing.

blending problem [IND ENG] A linear programming problem in which it is required to find the least costly mix of ingredients which yields the desired product characteristics.

blending value [ENG] Measure of the ability of an added component (for example, tetraethyllead, isooctane, and aromatics) to affect the octane rating of a base gasoline stock.

blend stop [BUILD] A thin wood strip fastened to the exterior vertical edge of the pulley stile or jamb to hold the sash in position.

blimp [AERO ENG] A name originally applied to nonrigid, pressure-type airships, usually of small size; now applied to airships with volumes of approximately 1,500,000 cubic feet (42,000 cubic meters).

blind [ENG] A solid disk inserted at a pipe joint or union to prevent the flow of fluids through the pipe; used during maintenance and repair work as a safety precaution. Also known as blank.

blind drift [MIN ENG] In a mine, a horizontal passage not yet connected with the other workings.

blind hole [DES ENG] A hole which does not pass completely through a workpiece. [ENG] A type of borehole that does not have the drilling mud or other circulating medium carry the cuttings to the surface.

blinding [ENG] *See* blanking. [MIN ENG] Interference with the functioning of a screen mesh by a matting of fine materials during screening. Also known as blocking; plugging.

blind joint [ENG] A joint which is not visible from any angle.

blind landing [AERO ENG] Landing an aircraft solely by the use of instruments because of poor visibility.

blind nipple [MECH ENG] A short piece of piping or tubing having one end closed off; commonly used in boiler construction.

blind riser [MET] An internal riser that does not extend to the outer surface of a mold.

blind spot [ENG] An area on a filter screen where no filtering occurs. Also known as dead area.

blister [ENG] A raised area on the surface of a metallic or plastic object caused by the pressure of gases developed while the surface was in a partly molten state, or by diffusion of high-pressure gases from an inner surface. [MIN ENG] A protrusion, more or less circular in plan, extending downward into a coal seam.

blister copper [MET] Copper having 96–99% purity and a blistered appearance and formed by forcing air through molten copper matte.

blister furnace [MET] A furnace for smelting ore to produce blister copper.

blister gas [MATER] Any of several war gases, such as lewisite, which cause burning, inflammation, or tissue destruction internally or externally.

blistering [ENG] The appearance of enclosed or broken macroscopic cavities in a body or in a glaze or other coating during firing.

blister steel [MET] Raw steel, made from wrought iron by cementation followed by slow cooling, in which blisters are formed by gas attempting to escape from the metal.

BL method [ENG] Research test for gasoline road octanes; the material is tested at five different speeds in an engine of given compression ratio.

block [DES ENG] **1.** A metal or wood case enclosing one or more pulleys; has a hook with which it can be attached to an object. **2.** *See* cylinder block. [MIN ENG] A division of a mine, usually bounded by workings but sometimes by survey lines or other arbitrary limits. [PETRO ENG] The subdivision of a sea area for the licensing of oil and gas exploration and production rights.

block and fall *See* block and tackle.

block and tackle [MECH ENG] Combination of a rope or other flexible material and independently rotating frictionless pulleys. Also known as block and fall.

block brake [MECH ENG] A brake which consists of a block or shoe of wood bearing upon an iron or steel wheel.

block brazing [MET] The process of joining metals by applying heated blocks to the joint and using a nonferrous filler metal with a melting point above 800°F (427°C).

block caving [MIN ENG] A method of caving where a block, 150–250 feet (46–77 meters) on a side and several hundred feet high, is induced to cave in after it is undercut; the broken ore is drawn off at a bell-shaped draw point.

block coal [MIN ENG] **1.** A bituminous coal that breaks into large lumps or cubical blocks. **2.** Coal that passes over 5-, 6-, and 8-inch (127, 152, and 203 millimeter) block screens; used in smelting iron.

block diagram [ENG] A diagram in which the essential units of any system are drawn in the form of rectangles or blocks and their relation to each other is indicated by appropriate connecting lines.

blocked-out ore *See* developed reserves.

blocked resistance [ENG ACOUS] Resistance of an audio-frequency transducer when its moving elements are blocked so they cannot move; represents the resistance due only to electrical losses.

blocker-type forging [ENG] A type of forging for designs involving the use of large radii and draft angles, smooth contours, and generous allowances.

block grease [MATER] High-melting-point grease that can be handled as a block or stick at normal temperatures; used for journal bearings. Also known as brick grease.

block hole [ENG] A small hole drilled into a rock or boulder into which an anchor bolt or a small charge or explosive may be placed; used in quarries for breaking large blocks of stone or boulders.

blockhouse [ENG] **1.** A reinforced concrete structure, often built underground or half-underground, and sometimes dome-shaped, to provide protection against blast, heat, or explosion during rocket launchings or related activities, and usually housing electronic equipment used in launching the rocket. **2.** The activity that goes on in such a structure.

blocking [ENG] Undesired adhesion between layers of plastic materials in contact during storage or use. [MET] **1.** A preliminary hot-forging operation which imparts an approximate shape to the rough stock. **2.** Reducing the oxygen content of the bath in an open-hearth furnace. [MIN ENG] *See* blinding.

blocking and wedging [MIN ENG] A method of holding mine timber sets in place; blocks of wood are set on the caps directly over the post supports and have a grain of block parallel with the top of the cap; wedges are driven tightly between the blocks and the roof.

block plane [DES ENG] A small type of hand plane, designed for cutting across the grain of the wood and for planing end grains.

block sequence [MET] A procedure by which a continuous multiple-pass joint is completed by alternating the deposition of intermittent cross-sectional buildup and intervening lengths of weld metal.

block system [MIN ENG] A system of pillars in which a series of entries, rooms, and crosscuts are driven to divide the coal into blocks of about equal size which are then extracted on retreat.

blood bank [ENG] A place for storing whole blood or plasma under refrigeration.

bloom [ENG] **1.** Fluorescence in lubricating oils or a cloudy surface on varnished or enameled surfaces. **2.** To apply an antireflection coating to glass. [MET] **1.** A semifinished bar of metal formed from an ingot and having a rectangular cross section exceeding 36 square inches (232 square centimeters). **2.** To hammer or roll metal in order to make its surface bright.

bloomer [MET] A furnace used to shape wrought iron directly from ore.

blooming mill [MET] A rolling mill for making blooms from ingots. Also known as cogging mill.

blotter [ENG] A disk of compressible material used between a grinding wheel and its flanges to avoid concentrated stress.

blotter model [PETRO ENG] An analysis device in which the analogous movement of copper ammonium or zinc ammonium ions in blotting paper or gelatin indicates oil-well injection-fluid movement through porous underground reservoirs.

blotting paper [MATER] An unsized paper used to absorb excess ink from penned letters or signatures; also used for other applications where a soft, spongy paper is required.

blowby [MECH ENG] Leaking of fluid between a cylinder and its piston during operation.

blowdown [MECH ENG] The difference between the pressure at which the safety valve opens and the closing pressure.

blowdown tunnel [AERO ENG] A wind tunnel in which stored compressed gas is allowed to expand through a test section to provide a stream of gas or air for model testing.

blowdown turbine [AERO ENG] A turbine attached to a reciprocating engine which receives exhaust gases separately from each cylinder, utilizing the kinetic energy of the gases.

blower [MECH ENG] A fan which operates where the resistance to gas flow is predominantly downstream of the fan.

blowhole [MET] A pocket of air or gas formed in a metal during solidification.

blow in [MET] To put a blast furnace into operation. [PETRO ENG] Of an oil well, to begin sending forth oil or gas.

blowing See blow molding.

blowing agent [MATER] A chemical added to plastics and rubbers that generates inert gases on heating, causing the resin to assume a cellular structure. Also known as foaming agent.

blowing boundary-layer control [AERO ENG] A technique that is used in addition to purely geometric means to control boundary-layer flow; it consists of re-energizing the retarded flow in the boundary layer by supplying high-velocity flow through slots or jets on the surface of the body.

blowing fan See forcing fan.

blowing pressure [ENG] Pressure of the air or other gases used to inflate the parison in blow molding.

blowing-up furnace [MET] A furnace used for sintering ore and for the volatilization of lead and zinc.

blow molding [ENG] A method of fabricating hollow plastic objects, such as bottles, by forcing a parison into a mold cavity and shaping by internal air pressure. Also known as blowing.

blown asphalt [MATER] Asphalt which is treated or heated with air or steam within a blowing still at relatively high temperature.

blown film [MATER] Film that has been produced by extruding a tube of plastic material which is then expanded in diameter and reduced in thickness by internal air pressure, and then slit along one side.

blown foam [MATER] A cellular plastic consisting of an expanded matrix resembling natural foam.

blown glass [ENG] Glassware formed by blowing air into a ball of liquefied glass until it reaches the desired shape.

blown oil [MATER] A vegetable or animal oil that has been agitated and partially oxidized by a current of warm air or oxygen, including castor, whale, fish, rape, and linseed oils; used in paints, lubricants, and plasticizers.

blown tubing [ENG] A flexible thermoplastic film tube made by applying pressure inside a molten extruded plastic tube to expand it prior to cooling and winding flat onto rolls.

blowoff [AERO ENG] The action of applying an explosive force and separating a package section away from the remaining part of a rocket vehicle or reentry body, usually to retrieve an instrument or to obtain a record made during early flight.

blowoff valves [MECH ENG] Valves in boiler piping which facilitate removal of solid matter present in the boiler water.

blowout [ENG] **1.** The bursting of a container (such as a tube pipe, pneumatic tire, or dam) by the pressure of the contained fluid. **2.** The rupture left by such bursting. **3.** The abrupt escape of air from the working chamber of a pneumatic caisson. [PETRO ENG] A sudden, unplanned escape of oil or gas from a well during drilling.

blowpipe [ENG] **1.** A long, straight tube, used in glass blowing, on which molten glass is gathered and worked. **2.** A small, tapered, and frequently curved tube that leads a jet, usually of air, into a flame to concentrate and direct it; used in flame tests in analytical chemistry and in brazing and soldering of fine work. **3.** See blowtorch.

blow pressure [ENG] Air pressure required for plastics blow molding.

blow rate [ENG] The speed of the cycle at which air or an inert gas is applied intermittently during the forming procedure of blow molding.

blowtorch [ENG] A small, portable blast burner which operates either by having air or oxygen and gaseous fuel delivered through tubes or by having a fuel tank which is pressured by a hand pump. Also known as blast lamp; blowpipe.

blowup ratio [ENG] **1.** In blow molding of plastics, the ratio of the diameter of the mold cavity to the diameter of the parison. **2.** In blown tubing, the ratio of the diameter of the finished product to the diameter of the die.

blubber oil *See* whale oil.

blue annealing [MET] Softening metal sheets by heating in an open furnace and cooling in air; bluish oxide forms on the metal surface.

blue brittleness [MET] Loss of ductility noted for some steels when heated to 400–600°F (204–316°C), the blue heat range.

blue cap [MIN ENG] The characteristic blue halo, or tip, of the flame of a safety lamp when firedamp is present in the air.

blue gas [MATER] A gas consisting chiefly of carbon monoxide and hydrogen, formed by the action of steam upon hot coke; used mainly as a source of hydrogen and in synthesis of other chemical compounds. Also known as blue water gas.

blue glow [MET] Luminescence emitted by certain metallic oxides when heated.

blueing *See* bluing.

blue vitriol [MATER] A hydrous solution of copper sulfate that is applied to the surface of a metal for layout purposes.

blue water gas *See* blue gas.

bluff body [AERO ENG] A body having a broad, flattened front, as in some reentry vehicles.

bluing [MET] Also spelled blueing. **1.** Formation of a bluish oxide film on polished steel; improves appearance and provides some corrosion resistance. **2.** Heating of formed springs to reduce internal stress. **3.** A blue oxide film formed on the polished surface of a metal due to extremely high temperatures.

blunger [ENG] **1.** A large spatula-shaped wooden implement used to mix clay with water. **2.** A vat, containing a rotating shaft with fixed knives, for mixing clay and water into slip.

blunging [ENG] The mixing or suspending of ceramic material in liquid by agitation, to form slip.

blunt file [DES ENG] A file whose edges are parallel.

blunting [DES ENG] Slightly rounding a cutting edge to reduce the probability of edge chipping.

BM *See* bench mark.

BM-AGA coal test [ENG] A laboratory test developed jointly by the United States Bureau of Mines and a committee of the American Gas Association which forecasts the quality of coke producible in commercial practice.

BMT study *See* basic motion-time study.

board drop hammer [MECH ENG] A type of drop hammer in which the ram is attached to wooden boards which slide between two rollers; after the ram falls freely on the forging, it is raised by friction between the rotating rollers. Also known as board hammer.

board-foot [ENG] Unit of volume in measuring lumber; equals 144 cubic inches (2360 cubic centimeters), or the volume of a board 1 foot square and 1 inch thick. Abbreviated bd-ft.

board hammer *See* board drop hammer.

boarding [ENG] **1.** A batch of boards. **2.** Covering with boards.

board measure [ENG] Measurement of lumber in board-feet. Abbreviated bm.

boast [ENG] **1.** To shape stone or curve furniture roughly in preparation for finer work later on. **2.** To finish the face of a building stone by cutting a series of parallel grooves.

boasting chisel [DES ENG] A broad chisel used in boasting stone.

boat spike [DES ENG] A long, square spike used in construction with heavy timbers. Also known as barge spike.

boattail [AERO ENG] Of an elongated body such as a rocket, the rear portion having decreasing cross-sectional area.

bob [MET] A feeding device for providing molten metal to a casting during solidification to prevent shrinkage.

body [AERO ENG] **1.** The main part or main central portion of an airplane, airship, rocket, or the like; a fuselage or hull. **2.** Any fabrication, structure, or other material form, especially one aerodynamically or ballistically designed; for example, an airfoil is a body designed to produce an aerodynamic reaction. [MECH ENG] The part of a drill which runs from the outer corners of the cutting lips to the shank or neck.

body angle [AERO ENG] The angle which the longitudinal axis of the airframe makes with some selected line.

body axis [AERO ENG] Any one of a system of mutually perpendicular reference axes fixed in an aircraft or a similar body and moving with it.

body motion [IND ENG] Motion of parts of a human body requiring a change of posture or weight distribution.

bogey See bogie.

bogie Also spelled bogey; bogy. [AERO ENG] A type of landing-gear unit consisting of two sets of wheels in tandem with a central strut. [ENG] **1.** A supporting and aligning wheel or roller on the inside of an endless track. **2.** A low truck or cart of solid build. **3.** A truck or axle to which wheels are fixed, which supports a railroad car, the leading end of a locomotive, or the end of a vehicle (such as a gun carriage) and which is allowed to swivel under it. **4.** A railroad car or locomotive supported by a bogie. [MECH ENG] The drive-wheel assembly and supporting frame comprising the four rear wheels of a six-wheel truck, mounted so that they can self-adjust to sharp curves and irregularities in the road. [MIN ENG] A small truck or trolley upon which a bucket is carried from the haft to the spoil bank.

bogy See bogie.

Bohemian glass See hard glass.

boiled oil [MATER] A drying oil in which metallic driers are added during the cooking; the reaction releases water, resulting in a boiling of the batch.

boiler [MECH ENG] A water heater for generating steam.

boiler air heater [MECH ENG] A component of a steam-generating unit that transfers heat from the products of combustion after they have passed through the steam-generating and superheating sections to combustion air, which recycles heat to the furnace.

boiler casing [MECH ENG] The gas-tight structure surrounding the component parts of a steam generator.

boiler circulation [MECH ENG] Circulation of water and steam in a boiler, which is required to prevent overheating of the heat-absorbing surfaces; may be provided naturally by gravitational forces, mechanically by pumps, or by a combination of both methods.

boiler cleaning [ENG] A mechanical or chemical process for removal of grease, scale, and other deposits from steam boiler surfaces.

boiler code [MECH ENG] A code, established by professional societies and administrative units, which contains the basic rules for the safe design, construction, and materials for steam-generating units, such as the ASME code.

boiler controls [MECH ENG] Either manual or automatic devices which maintain desired boiler operating conditions with respect to variables such as feedwater flow, firing rate, and steam temperature.

boiler draft [MECH ENG] The difference between atmospheric pressure and some lower pressure existing in the furnace or gas passages of a steam-generating unit.

boiler economizer [MECH ENG] A component of a steam-generating unit that transfers heat from the products of combustion after they have passed through the steam-generating and superheating sections to the feedwater, which it receives from the boiler feed pump and delivers to the steam-generating section of the boiler.

boiler efficiency [MECH ENG] The ratio of heat absorbed in steam to the heat supplied in fuel, usually measured in percent.

boiler feedwater [MECH ENG] Water supplied to a steam-generating unit.

boiler feedwater regulation [MECH ENG] Addition of water to the steam-generating unit at a rate commensurate with the removal of steam from the unit.

boiler fuel [MATER] Natural gas, residual oil, and various solid fuels used to heat a boiler, usually the most economical fuel available.

boiler furnace [MECH ENG] An enclosed space provided for the combustion of fuel to generate steam in a boiler. Also known as steam-generating furnace.

boiler heat balance [MECH ENG] A means of accounting for the thermal energy entering a steam-generating system in terms of its ultimate useful heat absorption or thermal loss.

boiler horsepower [MECH ENG] A measurement of water evaporation rate; 1 boiler horsepower equals the evaporation per hour of 34½ pounds (15.7 kilograms) of water at 212°F (100°C) into steam at 212°F.

boiler hydrostatic test [MECH ENG] A procedure that employs water under pressure, in a new boiler before use or in old equipment after major alterations and repairs, to test the boiler's ability to withstand about 1½ times the design pressure.

boiler layup [MECH ENG] A significant length of time during which a boiler is inoperative in order to allow for repairs or preventive maintenance.

boiler plate [MET] Flat-rolled steel, usually ¼ to ½ inch (6 to 13 millimeters) thick; used mainly for covering ships and making boilers and tanks. Also known as boiler steel.

boiler-plate model [AERO ENG] A metal copy of a flight vehicle, the structure or components of which are heavier than the flight model.

boiler setting [MECH ENG] The supporting steel and gastight enclosure for a steam generator.

boiler steel See boiler plate.

boiler storage [MECH ENG] A steam-generating unit that, when out of service, may be stored wet (filled with water) or dry (filled with protective gas).

boiler superheater [MECH ENG] A boiler component, consisting of tubular elements, in which heat is added to high-pressure steam to increase its temperature and enthalpy.

boiler trim [MECH ENG] Piping or tubing close to or attached to a boiler for connecting controls, gages, or other instrumentation.

boiler tube [MECH ENG] One of the tubes in a boiler that carry water (water-tube boiler) to be heated by the high-temperature gaseous products of combustion or that carry combustion products (fire-tube boiler) to heat the boiler water that surrounds them.

boiler walls [MECH ENG] The refractory walls of the boiler furnace, usually cooled by circulating water and capable of withstanding high temperatures and pressures.

boiler water [MECH ENG] Water in the steam-generating section of a boiler unit.

boil-off assistant [MATER] A material used to facilitate the removal of natural glue from silk during boiling in soap solution (degumming); examples are sulfonated oils, pine oils, and solvents and other wetting agents.

bollard [CIV ENG] A heavy post on a dock or ship used in mooring ships.

boll-weevil hanger [PETRO ENG] A screw-on connector used to connect and seal the lower end of a length of oil-well casing to the next smaller casing string.

bolometer [ENG] An instrument that measures the energy of electromagnetic radiation in certain wavelength regions by utilizing the change in resistance of a thin conductor caused by the heating effect of the radiation. Also known as thermal detector.

bolster [MET] A block of steel to which drop-forging dies are attached.

bolster plate [MECH ENG] A plate fixed on the bed of a power press to locate and support the die assembly.

bolt [DES ENG] A rod, usually of metal, with a square, round, or hexagonal head at one end and a screw thread on the other, used to fasten objects together. [MIN ENG] *See* bolthole.

bolted joint [ENG] The assembly of two or more parts by a threaded bolt and nut or by a screw that passes through one member and threads into another.

bolted rail crossing [CIV ENG] A crossing whose running surfaces are made of rolled rail and whose parts are joined with bolts.

bolthole [MIN ENG] A short, narrow opening made to connect the main working with the airhead or ventilating drift of a coal mine. Also known as bolt.

bolting [ENG] A fastening system using screw-threaded devices such as nuts, bolts, or studs. [MIN ENG] The use of vibrating sieves to separate particles of different sizes.

bolting cloth [MATER] A sieve cloth made of wire, hair, or silk or other thread used to remove lumps from flour or to make screen prints or needlework.

bombardment aircraft *See* bomber.

bomb bay [AERO ENG] The compartment or bay in the fuselage of a bomber where the bombs are carried for release.

bomb calorimeter [ENG] A calorimeter designed with a strong-walled container constructed of a corrosion-resistant alloy, called the bomb, immersed in about 2.5 liters of water in a metal container; the sample, usually an organic compound, is ignited by electricity, and the heat generated is measured.

bomber [AERO ENG] An airplane specifically designed to carry and drop bombs. Also known as bombardment aircraft.

bombproof [ENG] Referring to shelter, building, or other installation resistant or impervious to the effects of bomb explosions.

bomb shelter [CIV ENG] A bomb-proof structure for protection of people.

bomb test [ENG] A leak-testing technique in which the vessel to be tested is immersed in a pressurized fluid which will be driven through any leaks present.

bond [ENG] **1.** A wire rope that fixes loads to a crane hook. **2.** Adhesion between cement or concrete and masonry or reinforcement. [MET] **1.** Material added to molding sand to impart bond strength. **2.** Junction of the base metal and filler metal, or the base metal beads, in a welded joint.

Bond and Wang theory [MECH ENG] A theory of crushing and grinding from which the energy, in horsepower-hours, required to crush a short ton of material is derived.

bond clay [MATER] A type of clay with high plasticity and high dry strength used to bond nonplastic materials; may be refractory.

bonded coating [MATER] A finishing or protecting layer of any compound affixed to a surface.

bonded strain gage [ENG] A strain gage in which the resistance element is a fine wire, usually in zigzag form, embedded in an insulating backing material, such as impregnated paper or plastic, which is cemented to the pressure-sensing element.

bonded transducer [ENG] A transducer which employs a bonded strain gage for sensing pressure.

bonderize [MET] To coat steel with a solution of phosphates for corrosion protection.

bonding [ENG] **1.** The fastening together of two components of a device by means of adhesives, as in anchoring the copper foil of printed wiring to an insulating baseboard. **2.** *See* cladding.

bonding agent [MATER] Any substance that fixes one material to another.

bond paper [MATER] A paper used for writing paper, business forms, and typewriter paper; the less expensive bond papers are made from wood sulfite pulps; rag-content bonds contain 25, 50, 75, or 100% of pulp made from rags, and offer greater permanence and strength.

Bond's law [MECH ENG] A statement that relates the work required for the crushing of solid materials (for example, rocks and ore) to the product size and surface area and the lengths of cracks formed. Also known as Bond's third theory.

Bond's third theory *See* Bond's law.

bondstone [BUILD] A stone joining the coping above a gable to the wall. [CIV ENG] A stone that passes through a masonry wall in order to hold the wall together.

bone black [MATER] A black substance made by carbonizing crushed, defatted bones in closed vessels; used as a paint and varnish pigment, as a decolorizing absorbent in clarifying shellac, in cementation, and in gas masks. Also known as animal black; bone char.

bone char *See* bone black.

bone meal [MATER] A substance made by grinding animal bones; steamed meal, made from pressure-steamed bones, is used as a fertilizer; raw meal is used in animal feed.

bone oil [MATER] Dark brown oil with a disagreeable odor, derived by destructive distillation of bones or other animal substances; used as an alcohol denaturant, an insecticide, or a source of pyrrole and for organic preparations. Also known as animal oil; Dippel's oil; hartshorn oil; Jeppel's oil.

bonus [PETRO ENG] Payment by a lessee of an oil- or gas-production royalty to the landowner at a rate greater than the customary one-eighth of the value of the oil or gas withdrawn.

book mold [MET] A mold made in two halves hinged together like a book.

boom cat [MECH ENG] A tractor supporting a boom and used in laying pipe.

boomer [MIN ENG] In placer mining, an automatic gate in a dam that holds the water until the reservoir is filled, then opens automatically and allows the escape of such a volume that the soil and upper gravel of the placer are washed away.

boomerang sediment corer [ENG] A device, designed for nighttime recovery of a sediment core, which automatically returns to the surface after taking the sample.

boost [AERO ENG] **1.** An auxiliary means of propulsion such as by a booster. **2.** To supercharge. **3.** To launch or push along during a portion of a flight. **4.** *See* boost pressure. [ENG] To bring about a more potent explosion of the main charge of an explosive by using an additional charge to set it off.

booster *See* booster engine; booster rocket; launch vehicle.

booster brake [MECH ENG] An auxiliary air chamber, operated from the intake manifold vacuum, and connected to the regular brake pedal, so that less pedal pressure is required for braking.

booster ejector [MECH ENG] A nozzle-shaped apparatus from which a high-velocity jet of steam is discharged to produce a continuous-flow vacuum for process equipment.

booster engine [AERO ENG] An engine, especially a booster rocket, that adds its thrust to the thrust of the sustainer engine. Also known as booster.

booster fan [MECH ENG] A fan used to increase either the total pressure or the volume of flow.

booster pump [MECH ENG] A machine used to increase pressure in a water or compressed-air pipe.

booster rocket [AERO ENG] Also known as booster. **1.** A rocket motor, either solid- or liquid-fueled, that assists the normal propulsive system or sustainer engine of a rocket or aeronautical vehicle in some phase of its flight. **2.** A rocket used to set a vehicle in motion before another engine takes over.

booster stations [ENG] Booster pumps or compressors located at intervals along a liquid-products or gas pipeline to boost the pressure of the flowing fluid to keep it moving toward its destination.

boost-glide vehicle [AERO ENG] An air vehicle capable of aerodynamic lift which is projected to an extreme altitude by reaction propulsion and then coasts down with little or no propulsion, gliding to increase its range when it reenters the sensible atmosphere.

boost pressure [AERO ENG] Manifold pressure greater than the ambient at atmospheric pressure, obtained by supercharging. Also known as boost.

boot [MIN ENG] **1.** A projecting portion of a reinforced concrete beam acting as a corbel to support the facing material, such as brick or stone. **2.** The lower end of a bucket elevator. [PETRO ENG] *See* surge column.

bootjack [ENG] A fishing tool used in drilling wells.

bootleg [MIN ENG] A hole, shaped somewhat like the leg of a boot, caused by a blast that has failed to shatter the rock properly.

bootstrap [ENG] A technique or device designed to bring itself into a desired state by means of its own action.

bootstrap process [AERO ENG] A self-generating or self-sustaining process; specifically, the operation of liquid-propellant rocket engines in which, during mainstage operation, the gas generator is fed by the main propellants pumped by the turbopump, and the turbopump in turn is driven by hot gases from the gas generator system.

borax glass [MATER] A glassy, transparent solid formed by fusing borax.

Bordeaux mixture [MATER] A fungicide made from a mixture of lime, copper sulfate, and water.

Bordini effect [MET] A phenomenon that takes place when metals having a close-packed crystal structure are subjected to an oscillating stress; there is a peak in the internal friction at a particular frequency of the stress.

bore [DES ENG] Inside diameter of a pipe or tube. [MECH ENG] **1.** The diameter of a piston-cylinder mechanism as found in reciprocating engines, pumps, and compressors. **2.** To penetrate or pierce with a rotary tool. **3.** To machine a workpiece to increase the size of an existing hole in it. [MIN ENG] **1.** A tunnel under construction. **2.** To cut or drill a hole for blasting, water infusion, exploration, or water or firedamp drainage.

borehole [ENG] A hole made by boring into the ground to study stratification, to obtain natural resources, or to release underground pressures.

borehole bit *See* noncoring bit.

borehole mining [PETRO ENG] Extraction of minerals as liquid or gas from the earth's crust by means of boreholes and suction pumps.

borehole survey [ENG] Also known as drillhole survey. **1.** Determining the course of and the target point reached by a borehole, using an azimuth-and-dip recording apparatus small enough to be lowered into a borehole. **2.** The record of the information thereby obtained.

borer [MECH ENG] An apparatus used to bore openings into the earth up to about 8 feet (2.4 meters) in diameter.

borescope [ENG] A straight-tube telescope using a mirror or prism, used to visually inspect a cylindrical cavity, such as the cannon bore of artillery weapons for defects of manufacture and erosion caused by firing.

boresighting [ENG] Initial alignment of a directional microwave or radar antenna system by using an optical procedure or a fixed target at a known location.

boring bar [MECH ENG] A rigid tool holder used to machine internal surfaces.

boring log *See* drill log.

boring machine [MECH ENG] A machine tool designed to machine internal work such as cylinders, holes in castings, and dies; types are horizontal, vertical, jig, and single.

boring mill [MECH ENG] A boring machine tool used particularly for large workpieces; types are horizontal and vertical.

boron alloy [MET] Alloy of boron and iron used to increase the high-temperature strength characteristics of alloy steel.

boron fuel [MATER] The boron compounds alkyl decaborane, diborane, and pentaborane; used for ramjet engines and afterburners.

boron steel [MET] Alloy steel with a small amount (as little as 0.0005%) of boron added to increase hardenability; can be used to replace other alloys in short supply.

borosilicate glass [MATER] A type of glass containing at least 5% boric oxide; used in glassware that resists heat.

borrow [CIV ENG] Earth material such as sand and gravel that is taken from one location to be used as fill at another.

borrow pit [CIV ENG] An excavation dug to provide material (borrow) for fill elsewhere.

Bosch fuel injection pump [MECH ENG] A pump in the fuel injection system of an internal combustion engine, whose pump plunger and barrel are a very close lapped fit to minimize leakage.

Bosch metering system [MECH ENG] A system having a helical groove in the plunger which covers or uncovers openings in the barrel of the pump; most usually applied in diesel engine fuel-injection systems.

bosh [MET] **1.** Tapering lower portion of a blast furnace, from the blast holes of the hearth up to the maximum internal diameter at the bottom of the stack. **2.** Quartz deposited on the furnace lining during the smelting of copper ore.

boss [DES ENG] Protuberance on a cast metal or plastic part to add strength, facilitate assembly, provide for fastenings, or so forth.

Boston ridge [BUILD] A method of applying shingles to the ridge of a house by which the shingles alternate in overlap from one side of the ridge to the other.

bottle [ENG] A container made from pipe or plate with drawn, forged, or spun end closures, and used for storing or transporting gas.

bottle centrifuge [ENG] A centrifuge in which the mixture to be separated is poured into small bottles or test tubes; they are then placed in a rotor assembly which is spun rapidly.

bottled gas [MATER] Butane, propane, or butane-propane mixtures liquefied and bottled under pressure for use as a domestic cooking or heating fuel. Also known as bugas.

bottleneck assignment problem [IND ENG] A linear programming problem in which it is required to assign machines to jobs (or vice versa) so that the efficiency of the least efficient operation is maximized.

bottle thermometer [ENG] A thermoelectric thermometer used for measuring air temperature; the name is derived from the fact that the reference thermocouple is placed in an insulated bottle.

bottom blow [ENG] A type of plastics blow molding machine in which air is injected into the parison from the bottom of the mold.

bottom break [MIN ENG] The break or crack that separates a block of stone from a quarry floor.

bottom chord [CIV ENG] Any of the bottom series of truss members parallel to the roadway of a bridge.

bottom dead center [MECH ENG] The position of the crank of a vertical reciprocating engine, compressor, or pump when the piston is at the end of its downstroke. Abbreviated BDC.

bottom-discharge bit See face-discharge bit.

bottom dump [ENG] A construction wagon with movable gates in the bottom to allow vertical discharge of its contents.

bottomed hole [ENG] A completed borehole, or a borehole in which drilling operations have been discontinued.

bottom flow [ENG] A molding apparatus that forms hollow plastic articles by injecting the blowing air at the bottom of the mold.

bottom-hole cash [PETRO ENG] Cash which is contributed by mineral-rights lessees adjacent to a drilling lessee and which is payable when the well reaches a specified depth, regardless of whether or not the completed well is a producer.

bottom-hole packer [PETRO ENG] An anchored-in-place seal used to provide liquid-proof packing in the annular space between the outside of the oil-producing tubing and the inside of the drill casing.

bottom-hole pressure [PETRO ENG] Gas-drive pressure recorded at the bottom of an oil-well shaft; used to analyze oil-reservoir

performance and evaluate the performance of downhole equipment.

bottom-hole samples [PETRO ENG] Fluid samples from gas-condensate-well reservoirs; used to study the state of the hydrocarbon system under reservoir conditions and to estimate total hydrocarbons in place.

bottoming drill [DES ENG] A flat-ended twist drill designed to convert a cone at the bottom of a drilled hole into a cylinder.

bottom pillar [MIN ENG] A large block of solid coal left unworked around the shaft.

bottoms See basic sediment and water.

bottom sampler [ENG] Any instrument used to obtain a sample from the bottom of a body of water.

bottom sediment [PETRO ENG] A mixture of liquids and solids which form in the bottom of oil storage tanks.

bottom settlings See basic sediment and water.

bottom tap [DES ENG] A tap with a chamfer 1–1½ threads in length. [MET] A hole in the bottom of a furnace for draining out the slag.

Bouin's solution [MATER] A picric acid–acetic acid–formaldehyde fixative and preserving fluid for contractile forms.

boulder buster [ENG] A heavy, pyramidical- or conical-point steel tool which may be attached to the bottom end of a string of drill rods and used to break, by impact, a boulder encountered in a borehole. Also known as boulder cracker.

boulder cracker See boulder buster.

bounce table [MECH ENG] A testing device which subjects devices and components to impacts such as might be encountered in accidental dropping.

bouncing putty [MATER] A silicone polymer in a soft elastic mass; the material's elasticity will increase as the applied force increases.

boundary lubrication [ENG] A lubricating condition that is a combination of solid-to-solid surface contact and liquid-film shear.

boundary monument [ENG] A material object placed on or near a boundary line to preserve and identify the location of the boundary line on the ground.

boundary pillar [MIN ENG] A pillar left in mines between adjoining properties.

boundary survey [ENG] A survey made to establish or to reestablish a boundary line on the ground or to obtain data for constructing a map or plat showing a boundary line.

Bourdon pressure gage [ENG] A mechanical pressure-measuring instrument employing as its sensing element a curved or twisted metal tube, flattened in cross section and closed. Also known as Bourdon tube.

Boussinesq equation [ENG] A relation used to calculate the influence of a concentrated load on the backfill behind a retaining wall.

bow [AERO ENG] The forward section of an aircraft.

Bowden cable [MECH ENG] A wire made of spring steel which is enclosed in a helical casing and used to transmit longitudinal motions over distances, particularly around corners.

Bower-Barff process [MET] A method of coating iron or steel with magnetic oxide, such as Fe_3O_4, in order to minimize atmospheric corrosion.

bowk *See* hoppit.

bowl mill *See* bowl-mill pulverizer.

bowl-mill pulverizer [MECH ENG] A type of pulverizer which directly feeds a coal-fired furnace, in which springs press pivoted stationary rolls against a rotating bowl grinding ring, crushing the coal between them. Also known as a bowl mill.

bowl scraper [MECH ENG] A towed steel bowl hung within a fabricated steel frame, running on four or two wheels; transports soil, in addition to spreading and leveling it.

bowstring beam [CIV ENG] A steel, concrete, or timber beam or girder shaped in the form of a bow and string; the string resists the horizontal forces caused by loads on the arch.

box [DES ENG] *See* boxing. [ENG] A protective covering or housing.

box annealing [MET] Slow heating of metal sheets in a closed metal box to prevent oxidation, followed by cooling; usually limited to iron-base alloys.

box beam *See* box girder.

boxboard [MATER] Paperboard used for making cardboard boxes.

box caisson [CIV ENG] A floating steel or concrete box with an open top which will be filled and sunk at a foundation site in a river or seaway. Also known as American caisson; stranded caisson.

boxcar [ENG] A railroad car with a flat roof and vertical sides, usually with sliding doors, which carries freight that needs to be protected from weather and theft.

box-coking test [ENG] A laboratory test which forecasts the quality of coke producible in commercial practice; uses a specially designed sheet-steel box containing about 60 pounds (27 kilograms) of coal in a commercial coke oven.

box girder [CIV ENG] A hollow girder or beam with a square or rectangular cross section. Also known as box beam.

box-girder bridge [CIV ENG] A fixed bridge consisting of steel girders fabricated by welding four plates into a box section.

box header boiler [MECH ENG] A horizontal boiler with a front header and rear inclined rectangular header connected by tubes.

boxing [DES ENG] The threaded nut for the screw of a mounted auger drill. Also known as box. [ENG] A method of securing shafts solely by slabs and wooden pegs. [MET] Continuing a fillet weld around a corner. Also known as end turning.

boxing shutter [BUILD] A window shutter which can be folded into a boxlike enclosure or recess at the side of the window frame.

box piles [CIV ENG] Pile foundations made by welding together two sections of steel sheet piling or combinations of beams, channels, and plates.

box wrench [ENG] A closed-end wrench designed to fit a variety of sizes and shapes of bolt heads and nuts.

brace [DES ENG] A cranklike device used for turning a bit.

brace [ENG] A diagonally placed structural member that withstands tension and compression, and often stiffens a structure against wind.

brace and bit [DES ENG] A small hand tool to which is attached a metal- or wood-boring bit.

braced framing [CIV ENG] Framing a building with post and braces for stiffness.

braced-rib arch [CIV ENG] A type of steel arch, usually used in bridge construction, which has a system of diagonal bracing.

brace head [ENG] A cross handle attached at the top of a column of drill rods by means of which the rods and attached bit are turned after each drop in chop-and-wash operations while sinking a borehole through overburden. Also known as brace key.

brace key *See* brace head.

brace pile *See* batter pile.

bracing [ENG] The act or process of strengthening or making rigid.

bracket [BUILD] A vertical board to support the tread of a stair. [CIV ENG] A projecting support.

Bradford breaker [MIN ENG] A machine which combines coal crushing and screening.

Bradford preferential separation process [MIN ENG] A flotation process for the treatment of mixed sulfides in which certain mineral salts, such as thiosulfates, are added to the water used in the flotation cells.

Bragg spectrometer [ENG] An instrument for x-ray analysis of crystal structure and measuring wavelengths of x-rays and gamma rays, in which a homogeneous beam of x-rays is directed on the known face of a crystal and the reflected beam is detected in a suitably placed ionization chamber. Also known as crystal spectrometer; crystal diffraction spectrometer; ionization spectrometer.

brainstorming [IND ENG] A procedure used to find a solution for a problem by collecting all the ideas, without regard for feasibility, which occur from a group of people meeting together.

brake [MECH ENG] A machine element for applying friction to a moving surface to slow it (and often, the containing vehicle or device) down or bring it to rest.

brake band [MECH ENG] The contracting element of the band brake.

brake drum [MECH ENG] A rotating cylinder attached to a rotating part of machinery, which the brake band or brake shoe presses against.

brake fluid [MATER] The liquid in the cylinder of an automotive brake which, under the action of a piston, is forced through tubing into cylinders at each car wheel, moving a pair of pistons outward so that the brake shoes are thrust against the revolving brake drums.

brake horsepower [MECH ENG] The power developed by an engine as measured by the force applied to a friction brake or by an absorption dynamometer applied to the shaft or flywheel. Abbreviated bhp.

brake lining [MECH ENG] A covering, riveted or molded to the brake shoe or brake band, which presses against the rotating brake drum; made of either fabric or molded asbestos material.

brake mean-effective pressure [MECH ENG] Applied to reciprocating piston machinery, the average pressure on the piston during the power stroke, derived from the measurement of brake power output.

brake shoe [MECH ENG] The renewable friction element of a shoe brake. Also known as shoe.

brake thermal efficiency [MECH ENG] The ratio of brake power output to power input.

braking ellipses [AERO ENG] A series of ellipses, decreasing in size due to aerodynamic drag, followed by a spacecraft in entering a planetary atmosphere.

braking rocket See retrorocket.

brale [MET] A conical diamond indenter with an angle of 120° used in the Rockwell hardness test.

branch-and-bound technique [IND ENG] A technique in nonlinear programming in which all sets of feasible solutions are divided into subsets, and those having bounds inferior to others are rejected.

branch sewer [CIV ENG] A part of a sewer system that is larger in diameter than the lateral sewer system; receives sewage from both house connections and lateral sewers.

brass [MET] A copper-zinc alloy of varying proportions but typically containing 67% copper and 33% zinc.

brattice [MIN ENG] A temporary board or cloth partition in any mine passage to confine the air and force it into the working places. Also spelled brattish; brettice; brettis.

brattice cloth [MIN ENG] Fire-resistant canvas or duck used to erect a brattice.

brattish See brattice.

braze [MET] To solder metals by melting a nonferrous filler metal, such as brass or brazing alloy (hard solder), with a melting point lower than that of the base metals, at the point of contact. Also known as hard-solder.

brazed joint [MET] The joining of two or more metallic components by brazing or braze welding.

brazed shank tool [MECH ENG] A metal cutting tool made of a material different from the shank to which it is brazed.

braze welding [MET] A method of welding in which coalescence is produced by heating above 800°F (427°C) and by using a nonferrous filler metal having a melting point below that of the base metals; in distinction to brazing, capillary attraction does not distribute the filler metal in the joint.

Brazil wax See carnauba wax.

brazing alloy [MET] **1.** An alloy used as a filler metal for brazing; copper alloys and nickel alloys are used for brazing of steels. **2.** Solder that does not melt below red heat. Formerly known as hard solder.

brazing brass See spelter solder.

brazing metal [MET] A nonferrous metal to be added to a joint in braze welding; can be a pure metal such as copper, zinc, or nickel, or a brazing alloy.

brazing sheet [MET] **1.** Brazing filler metal in sheet form. **2.** Flat-rolled metal clad with brazing filler metal on one or both sides.

breadboard model [ENG] Uncased assembly of an instrument or other piece of equipment, such as a radio set, having its parts laid out on a flat surface and connected together to permit a check or demonstration of its operation.

break [MIN ENG] **1.** A plane of discontinuity in the coal seam such as a slip, fracture, or cleat; the surfaces are in contact or slightly separated. **2.** A fracture or crack in the roof beds as a result of mining operations.

breakaway phenomenon See breakoff phenomenon.

break-bulk cargo [IND ENG] Miscellaneous goods packed in boxes, bales, crates, cases, bags, cartons, barrels, or drums; may also include lumber, motor vehicles, pipe, steel, and machinery.

breakdown [MET] The initial process of rolling and drawing, or a series of such processes, which reduce a casting or extruded shape before its final reduction to desired size.

breaker [MIN ENG] **1.** In anthracite mining, the structure in which the coal is broken, sized, and cleaned for market. Also known as coalbreaker. **2.** One of a row of drill holes above the mining holes in a tunnel face.

breaker cam [MECH ENG] A rotating, engine-driven device in the ignition system of an internal combustion engine which causes the breaker points to open, leading to a rapid fall in the primary current.

breaker plate [ENG] In plastics die forming, a perforated plate at the end of an extruder head; often used to support a screen to keep foreign particles out of the die.

break-even analysis [IND ENG] Determination of the break-even point.

break-even point [IND ENG] The point at which a company neither makes a profit nor suffers a loss from the operations of the business, and at which total costs are equal to total sales volume.

breaking-down rolls [MET] A rolling mill unit used for breakdown operations.

breaking pin device [ENG] A device designed to relieve pressure resulting from inlet static pressure by the fracture of a loaded part of a pin.

breakoff phenomenon [AERO ENG] The feeling which sometimes occurs during high-altitude flight of being totally separated and detached from the earth and human society. Also known as breakaway phenomenon.

breakout [MIN ENG] To pull drill rods or casing from a borehole and unscrew them at points where they are joined by threaded couplings to form lengths that can be stacked in the drill tripod or derrick.

breakout block [PETRO ENG] A heavy plate fitting in a rotary table that holds a drill bit being unscrewed from a drill collar.

breakthrough [MIN ENG] A passage cut through the pillar to allow the ventilating current to pass from one room to another; larger than a doghole. Also known as room crosscut.

breakthrough sweep efficiency [PETRO ENG] The completeness with which an oil-field waterflood sweeps through a reservoir area; related to the critical water saturation (point beyond which oil will not be pushed ahead of the water flow).

breakwater [CIV ENG] A wall built into the sea to protect a shore area, harbor, anchorage, or basin from the action of waves.

breast [MIN ENG] **1.** In coal mines, a chamber driven in the seam from the gangway, for the extraction of coal. **2.** *See* face.

breast boards [CIV ENG] Timber planks used to support the tunnel face when excavation is in loose soil.

breast drill [DES ENG] A small, portable hand drill customarily used by handsetters to drill the holes in bit blanks in which diamonds are to be set; it includes a plate that is pressed against the worker's breast.

breast hole [MET] A hole for raking cinders out of a smelting cupola.

breasting dolphin [CIV ENG] A pile or other structure against which a moored ship rests.

breast wall [CIV ENG] A low wall built to retain the face of a natural bank of earth.

breather pipe [MECH ENG] A pipe that opens into a container for ventilation, as in a crankcase or oil tank.

breath-hold diving [ENG] A form of diving without the use of any artificial breathing mixtures.

breathing [ENG] **1.** Opening and closing of a plastics mold in order to let gases escape during molding. Also known as degassing. **2.** Movement of gas, vapors, or air in and out of a storage-tank vent line as a result of liquid expansions and contractions induced by temperature changes. [MATER] Permeability of plastic sheeting to air, bubbles, voids, or trapped gas globules.

breathing apparatus [ENG] An appliance that enables a person to function in irrespirable or poisonous gases or fluids; contains a supply of oxygen and a regenerator which removes the carbon dioxide exhaled.

breathing line [CIV ENG] A level of 5 feet (1.5 meters) above the floor; suggested temperatures for various occupancies of rooms and other chambers are usually given at this level.

breeching [MECH ENG] A duct through which the products of combustion are transported from the furnace to the stack; usually applied in steam boilers.

Brennan monorail car [MECH ENG] A type of car balanced on a single rail so that when the car starts to tip, a force automatically applied at the axle end is converted gyroscopically into a strong righting moment which forces the car back into a position of lateral equilibrium.

brennschluss [AERO ENG] **1.** The cessation of burning in a rocket, resulting from consumption of the propellants, from deliberate shutoff, or from other cause. **2.** The time at which this cessation occurs.

brettice *See* brattice.

brettis *See* brattice.

brick [MATER] A building material usually made from clay, molded as a rectangular block, and baked or burned in a kiln.

brick grease *See* block grease.

bricking curb [MIN ENG] A curb set in a circular shaft to support the brick walling.

brick walling [MIN ENG] A permanent support for circular shafts by walling or casing.

bridge [CIV ENG] A structure erected to span natural or artificial obstacles, such as rivers, highways, or railroads, and supporting a footpath or roadway for pedestrian, highway, or railroad traffic. [MIN ENG] A piece of timber held above the cap of a set by blocks and used to facilitate the driving of spiling in soft or running ground.

bridge abutment [CIV ENG] The end foundation upon which the bridge superstructure rests.

bridge bearing [CIV ENG] The support at a bridge pier carrying the weight of the bridge; may be fixed or seated on expansion rollers.

bridge cable [CIV ENG] Cable from which a roadway or truss is suspended in a suspension bridge; may be of pencil-thick wires laid parallel or strands of wire wound spirally.

bridge crane [MECH ENG] A hoisting machine in which the hoisting apparatus is carried by a bridgelike structure spanning the area over which the crane operates.

bridge foundation [CIV ENG] The piers and abutments of a bridge, on which the superstructure rests.

bridge pier [CIV ENG] The main support for a bridge, upon which the bridge superstructure rests; constructed of masonry, steel, timber, or concrete founded on firm ground below river mud.

bridge trolley [MECH ENG] Either of the wheeled attachments at the ends of the bridge of an overhead traveling crane, permitting the bridge to move backward and forward on elevated tracks.

bridgewall [MECH ENG] A wall in a furnace over which the products of combustion flow.

bridging [MET] **1.** Formation of arched cavities in a powder compact. **2.** Jamming of the charge in a blast or a cupola furnace due to adherence of fine ore particles to the inner walls. **3.** Formation of solidified metal over the top of the charge in a mold or crucible. [MIN ENG] The obstruction of the receiving opening in a

material-crushing device by two or more pieces wedged together, each of which could easily pass through.

Bridgman sampler [MIN ENG] A mechanical device that automatically selects two samples as the ore passes through.

bridled-cup anemometer [ENG] A combination cup anemometer and pressure-plate anemometer, consisting of an array of cups about a vertical axis of rotation, the free rotation of which is restricted by a spring arrangement; by adjustment of the force constant of the spring, an angular displacement can be obtained which is proportional to wind velocity.

Briggs equalizer [ENG] A breathing device consisting of head harness, mouthpiece, nose clip, corrugated breathing tube, an equalizing device, 120 feet (37 meters) of reinforced air tubes, and a strainer and spike.

Briggs stretcher carriage [MIN ENG] A stretcher used as an ambulance trolley in transporting casualties from underground workings.

bright [MATER] Referring to lubricating oils that are clear, or free from moisture.

bright annealing [MET] Heating and cooling a metal in an inert atmosphere to inhibit oxidation; surface remains relatively bright.

bright dipping [MET] Immersing metal into an acid solution to give a bright, clean surface.

brightener [MET] Any of the agents which are employed in small concentrations in the electrolytic bath for electroplating metal to yield smoother or brighter coatings.

bright plating [MET] An electroplating process resulting in a smooth, lustrous surface without polishing.

bright stock [MATER] High-viscosity refined and dewaxed lubricating oils used in the compounding of motor oils.

Brikollare system [CIV ENG] A method of processing solid waste for composting that is designed to replace windrowing; after metals are separated from solid waste, it is comminuted and sent through a ballistic separator; the organic components are mixed with sewage sludge, compacted into briquets, and stacked; as biodegradation takes place, the briquets are turned; after the process is finished, the bricks are crumbled, bagged, and sold as compost.

brine [MATER] A liquid used in a refrigeration system, usually an aqueous solution of calcium chloride or sodium chloride, which is cooled by contact with the evaporator surface and then goes to the space to be refrigerated.

brine cooler [MECH ENG] The unit for cooling brine in a refrigeration system; the brine usually flows through tubes or pipes surrounded by evaporating refrigerant.

Brinell number [ENG] A hardness rating obtained from the Brinell test; expressed in kilograms per square millimeter.

Brinell test [ENG] A test to determine the hardness of a material, in which a steel ball 1 centimeter in diameter is pressed into the material with a standard force (usually 3000 kilograms); the spherical surface area of indentation is measured and divided into the load; the results are expressed as Brinell number.

briquet [MATER] A block of some compressed substance, such as coal dust, metal powder, or sawdust, used as a fuel. Also spelled briquette.

briquette See briquet.

briquetting [ENG] **1.** The process of binding together pulverized minerals, such as coal dust, into briquets under pressure, often with the aid of a binder, such as asphalt. **2.** A process or method of mounting mineral ore, rock, or metal fragments in an embedding or casting material, such as natural or artificial resins, waxes, metals, or alloys, to facilitate handling during grinding, polishing, and microscopic examination.

brisance index [ENG] The ratio of an explosive's power to shatter a weight of graded sand as compared to the weight of sand shattered by TNT.

Bristol board [MATER] Cardboard with a surface smooth enough for painting or writing, usually at least 0.006 inch (0.15 millimeter) thick.

britannia cell [MIN ENG] In mineral processing, a pneumatic flotation cell 7 to 9 feet (2.1 to 2.7 meters) deep.

britannia metal [MET] A silver-white tin alloy, similar to pewter, containing about 7% antimony, 2% copper, and often some zinc and bismuth; used in domestic utensils.

brittle fracture [MET] A break in a brittle piece of metal which failed because stress exceeded cohesion.

broach [MECH ENG] A multiple-tooth, barlike cutting tool; the teeth are shaped to give a desired surface or contour, and cutting results from each tooth projecting farther than the preceding one.

broaching [ENG] **1.** The restoration of the diameter of a borehole by reaming. **2.** The breaking down of the walls between two contiguous drill holes. [MECH ENG] The machine-shaping of metal or plastic by pushing or pulling a broach across a surface or through an existing hole in a workpiece.

broken ground See loose ground.

broken stone See crushed stone.

bronze [MET] An alloy of copper and tin in varying proportions; other elements such as zinc, nickel, and lead may be added.

bronzing liquid [MATER] A solvent, gloss oil, or varnish containing a bronze powder; used to produce bronze-colored finishes.

Brookhill waffler [MIN ENG] A coal cutter with the ordinary horizontal jib and also a shearing or mushroom jib.

brooming [CIV ENG] A method of finishing uniform concrete surfaces, such as the tops of pavement slabs or floor slabs, by dragging a broom over the surface to produce a grooved texture.

Brown and Sharpe gage See American wire gage.

brown coat [MATER] Mortar with about 1 to 1½ bushels (35 to 53 liters) of hair per 200 pounds (91 kilograms) quicklime; used to make a brown coat of plaster, which is covered with a finish coat, and often covers a scratch coat.

brown petroleum [MATER] A solid or semisolid product formed by air acting on asphalt.

brown smoke [ENG] Smoke with less particulates than black smoke; comes from burning fossil fuel, usually fuel oil.

Brunton See Brunton compass.

Brunton compass [ENG] A compact field compass, with sights and reflector attached, used for geological mapping and surveying. Also known as Brunton; Brunton pocket transit.

Brunton pocket transit See Brunton compass.

brush hopper [IND ENG] A rotating brush that wipes quantities of eyelets, rivets, and other small special parts past shaped openings in a chute.

brushing shot [MIN ENG] A charge fired in the air of a mine to blow out obnoxious gases or to start an air current.

brush plating [MET] Electroplating in which the anode with the solution is in the form of a brush or pad; used for plating equipment too large to be immersed.

brush-shifting motor [ENG] A category of alternating-current motor in which the brush contacts shift to modify operating speed and power factor.

B size [ENG] One of a series of sizes to which trimmed paper and board are manufactured; for size BN, with N equal to any integer from 0 to 10, the length of the shorter side is $2^{-N/2}$ meters, and the length of the longer side is $2^{(1-N)/2}$ meters, with both lengths rounded off to the nearest millimeter.

BS&W See basic sediments and water.

BT See bathythermograph.

BTX See benzene-toluene-xylene.

bubble mold cooling [ENG] In plastics injection molding, cooling by means of a continuous liquid stream flowing into a cavity equipped with an outlet at the end opposite the inlet.

bubble-point reservoir See dissolved-gas-drive reservoir.

bubble test [ENG] Measurement of the largest opening in the mesh of a filter screen; determined by the pressure needed to force air or gas through the screen while it is submerged in a liquid.

bubble tube [ENG] The glass tube in a spirit level containing the liquid and bubble.

bubulum oil See neatsfoot oil.

buck [BUILD] The frame into which the finished door fits. [MIN ENG] **1.** To break up or pulverize ore samples. **2.** A large quartz reef in which there is little or no accessory minerals such as gold. Also known as buck quartz; bull quartz.

bucket [ENG] **1.** A cup on the rim of a Pelton wheel against which water impinges. **2.** A reversed curve at the toe of a spillway to deflect the water horizontally and reduce erosiveness. **3.** A container on a lift pump or chain pump. **4.** A container on some bulk-handling equipment, such as a bucket elevator, bucket dredge, or bucket conveyor. **5.** A water outlet in a turbine. **6.** See calyx.

bucket carrier See bucket conveyor.

bucket conveyor [MECH ENG] A continuous bulk conveyor constructed of a series of buckets attached to one or two strands of chain or in some instances to a belt. Also called bucket carrier.

bucket dredge [MECH ENG] A floating mechanical excavator equipped with a bucket elevator.

bucket drill [MIN ENG] An auger stem drill in which the drill bit is replaced by a bit incorporating a steel cylinder to confine the cutting.

bucket dumper See lander.

bucket elevator [MECH ENG] A bucket conveyor operating on a steep incline or vertical path. Also known as elevating conveyor.

bucket excavator [MECH ENG] An elevating scraper, that is, one that does the work of a conventional scraper but has a bucket elevator mounted in front of the bowl.

bucket ladder See bucket-ladder dredge.

bucket-ladder dredge [MECH ENG] A dredge whose digging mechanism consists of a ladderlike truss on the periphery of which is attached an endless chain riding on sprocket wheels and carrying attached buckets. Also known as bucket ladder; bucket-line dredge; ladder-bucket dredge; ladder dredge.

bucket-ladder excavator See trench excavator.

bucket-line dredge See bucket-ladder dredge.

bucket loader [MECH ENG] A form of portable, self-feeding, inclined bucket elevator for loading bulk materials into cars, trucks, or other conveyors.

bucket temperature [ENG] The surface temperature of ocean water as measured by a bucket thermometer.

bucket thermometer [ENG] A thermometer mounted in a bucket and used to measure the temperature of water drawn into the bucket from the surface of the ocean.

bucket-wheel excavator [MECH ENG] A continuous digging machine used extensively in large-scale stripping and mining. Abbreviated BWE. Also known as rotary excavator.

bucking [MIN ENG] A hand process for crushing ore.

Buckingham's equations [MECH ENG] Equations which give the durability of gears and the dynamic loads to which they are subjected in terms of their dimensions, hardness, surface endurance, and composition.

buckle [MET] An up-and-down wrinkle on the surface of a metal bar or sheet.

buckle plate [CIV ENG] A steel floor plate which is slightly arched to increase rigidity.

Buckley gage [ENG] A device that measures very low gas pressures by sensing the amount of ionization produced in the gas by a predetermined electric current.

buckstay [MECH ENG] A structural support for a furnace wall.

buckwheat no. 3 See barley coal.

buddle [MIN ENG] A device for concentrating ore that uses a circular arrangement from which the finely divided ore is delivered in water from a central point, the heavier particles sinking and the lighter particles overflowing.

Buddy [MIN ENG] A shortwall coal cutter designed for light duty such as stabling on longwall power-loaded faces and for subsidiary developments.

buffeting [AERO ENG] **1.** The beating of an aerodynamic structure or surfaces by unsteady flow, gusts, and so forth. **2.** The irregular shaking or oscillation of a vehicle component owing to turbulent air or separated flow.

buffeting Mach number [AERO ENG] The free-stream Mach number of an aircraft when the local Mach number over the tops of the wings approaches unity.

buffing [ENG] The smoothing and brightening of a surface by an abrasive compound pressed against it by a soft wheel or belt.

buffing wheel [DES ENG] A flexible wheel with a surface of fine abrasive particles for buffing operations.

bug [ENG] **1.** A defect or imperfection present in a piece of equipment. **2.** See bullet.

bugas See bottled gas.

bug dust [MIN ENG] The fine coal or other material resulting from a boring or cutting of a drill, a mining machine, or even a pick.

buggy [ENG] See concrete buggy. [MIN ENG] A four-wheeled steel car used for hauling coal to and from chutes.

buhrstone mill [MECH ENG] A mill for grinding or pulverizing grain in which a flat siliceous rock (buhrstone), generally of cellular quartz, rotates against a stationary stone of the same material.

builder [MATER] A material added to a soap or a synthetic surface-active agent to produce a mixture having enhanced detergency. [MET] A fire-clay brick cull used for bottom construction in kilns or for boxing brick during burning.

building [CIV ENG] A fixed structure for human occupancy and use.

building blocks [MATER] Hollow blocks, either of burned clay or concrete, used in constructing the walls of buildings, and often faced with brick or stone.

building code [CIV ENG] Local building laws to promote safe practices in the design and construction of a building.

building dock [CIV ENG] A type of graving dock or basin, usually built of concrete, in which ships are constructed and then floated out through a caisson gate after flooding the dock.

building line [CIV ENG] A designated line beyond which a building cannot extend.

building paper [MATER] Heavy, waterproof paper used to cover sheathing and subfloors to prevent passage of air and water.

buildup [MET] Deposition of excess metal, either by electroplating or spraying, on worn or undersized machine components to restore required dimensions.

buildup curve [PETRO ENG] Graph of bottom-hole pressure buildup versus shut-in time for a gas or oil well.

buildup pressure [PETRO ENG] The increase in bottom-hole pressure up to an equilibrium value in a shut-in oil or gas well.

built-up beam [ENG] A structural steel member that is fabricated by welding or riveting rather than being rolled.

built-up edge [ENG] Chip material adhering to the tool face adjacent to a cutting edge during cutting.

built-up mica [MATER] Large, laminated plates of mica made by bonding thin splittings of natural mica with shellac, glyptol, or some other suitable adhesive.

built-up roof [BUILD] A roof constructed of several layers of felt and asphalt.

built-up roofing [MATER] A seamless piece of flexible, waterproofed roofing material consisting of plies of felt mopped with asphalt or pitch.

bulb angle [DES ENG] A steel angle iron enlarged to a bulbous thickening at one end.

bulk cargo [IND ENG] Cargo which is loaded into a ship's hold without being boxed, bagged, or hand stowed, or is transported in large tank spaces.

bulk density [ENG] The mass of powdered or granulated solid material per unit of volume.

bulked-down See solid-piled.

bulk factor [ENG] The ratio of the volume of loose powdered or granulated solids to the volume of an equal weight of the material after consolidation into a voidless solid.

bulk flotation [MIN ENG] The rising of a mineralized froth, of more than one mineral, in a single operation.

bulk-handling machine [MECH ENG] Any of a diversified group of materials-handling machines designed for handling unpackaged, divided materials.

bulkhead [AERO ENG] A wall, partition, or such in a rocket, spacecraft, airplane fuselage, or similar structure, at right angles to the longitudinal axis of the structure and serving to strengthen, divide, or help give shape to the structure. [MIN ENG] A tight-seal partition of wood, rock, and mud or concrete in mines that serves to protect against gas, fire, and water.

bulkhead line [CIV ENG] The farthest offshore line to which a structure may be constructed without interfering with navigation.

bulkhead wharf [CIV ENG] A bulkhead that may be used as a wharf by addition of mooring appurtenances, paving, and cargo-handling facilities.

bulking [MATER] The difference in volume of a given mass of sand or other fine material in moist and dry conditions.

bulk insulation [ENG] A type of insulation that retards the flow of heat by the interposition of many air spaces and, in most cases, by opacity to radiant heat.

bulk mining [MIN ENG] Mining in which large quantities of low-grade ore are taken without attempt to segregate the high-grade portions.

bulk molding compound [MATER] A mixture of resin, inert fillers, reinforcements, and other additives which forms a puttylike rope, sheet, or preformed shape; used as a premix in composite manufacture.

bulk plant [PETRO ENG] A wholesale receiving and distributing facility for petroleum products; includes storage tanks, warehouses, railroad sidings, truck loading racks, and related elements. Also known as bulk terminal.

bulk terminal See bulk plant.

bulk transport [MECH ENG] Conveying, hoisting, or elevating systems for movement of solids such as grain, sand, gravel, coal, or wood chips.

bull bit [MIN ENG] A flat drill bit.

bull block [MET] Power-driven machine for drawing wire through a die.

bulldozer [MECH ENG] A wheeled or crawler tractor equipped with a reinforced, curved steel plate mounted in front, perpendicular to the ground, for pushing excavated materials. [MET] A machine for bending, forging, and punching narrow plates and bars, in which a ram is pushed along a horizontal path by a pair of cranks that are linked to two bullwheels with eccentric pins.

bullet [ENG] **1.** A conical-nosed cylindrical weight, attached to a wire rope or line, either notched or seated to engage and attach itself to the upper end of a wire line core barrel or other retrievable or retractable device that has been placed in a borehole. Also known as bug; go-devil; overshot. **2.** A scraper with self-adjusting spring blades, inserted in a pipeline and carried forward by the fluid pressure, clearing away accumulations or debris from the walls of a pipe. Also known as go-devil. **3.** A bullet-shaped weight or small explosive charge dropped to explode a charge of nitroglycerin placed in a borehole. Also known as go-devil. **4.** An electric lamp covered by a conical metal case, usually at the end of a flexible metal shaft. **5.** See torpedo. [MATER] A small, lustrous, nearly spherical industrial diamond.

bullfrog See barney.

bull gear [DES ENG] A bull wheel with gear teeth.

bulling bar [ENG] A bar for ramming clay into cracks containing blasting charges which are about to be exploded.

bullion [MET] Gold or silver in bulk in the shape of bars or ingots. [MIN ENG] A concretion found in some types of coal, composed of carbonate or silica stained by brown humic derivatives; often well-preserved plant structures form the nuclei.

bull ladle [MET] A ladle used in foundry operations for carrying molten metal.

bull nose [BUILD] A rounded external angle, as one used at window returns and doorframes.

bull-nose bit See wedge bit.

bull shaker [MIN ENG] A shaking chute where large coal from the dump is cleaned by hand.

bull wheel [MECH ENG] **1.** The main wheel or gear of a machine, which is usually the largest and strongest. **2.** A cylinder which has a rope wound about it for lifting or hauling. **3.** A wheel attached to the base of a derrick boom which swings the derrick in a vertical plane.

bummer [MIN ENG] The man who runs conveyors in a quarry or mine.

bump-down [PETRO ENG] In an oil well pumping operation, the pump's hitting the bottom in the downstroke because the rod between the pumping jack and the pump seat is too long.

bumper [ENG] **1.** A metal bar attached to one or both ends of a powered transportation vehicle, especially in the front, to prevent damage to the body. **2.** In a drilling operation, the supporting stay between the main foundation sill and the engine block. **3.** In drilling, a fishing tool for loosening jammed cable tools. [MET] A vibrating machine for ramming and consolidating sand into a mold.

bumpiness [AERO ENG] An atmospheric condition causing aircraft to experience sudden jolts.

bumping [AERO ENG] See chugging. [MECH ENG] See chugging. [MET] Forming a dish in metal by many repeated blows.

bumps [MIN ENG] Sudden, violent expulsion of coal from one or more pillars, accompanied by loud reports and earth tremors.

bund [CIV ENG] An embankment or embanked thoroughfare along a body of water; the term is used particularly for such structures in the Far East.

bundling machine [MECH ENG] A device that automatically accumulates cans, cartons, or glass containers for semiautomatic or automatic loading or for shipping cartons by assembling the packages into units of predetermined count and pattern which are then machine-wrapped in paper, film paperboard, or corrugated board.

bunker C fuel oil [MATER] A special grade of bunker fuel oil. Also known as Navy Heavy; No. 6 fuel.

bunker fuel oil [MATER] A heavy residual petroleum oil used as fuel by ships, industry, and large-scale heating and power-production installations.

bunkering [ENG] Storage of solid or liquid fuel in containers from which the fuel can be continuously or intermittently withdrawn to feed a furnace, internal combustion engine, or fuel tank, for example, coal bunkering and fuel-oil bunkering.

Bunsen burner [ENG] A type of gas burner with an adjustable air supply.

Bunsen ice calorimeter [ENG] Apparatus to gage heat released during the melting of a compound by measuring the increase in volume of the surrounding ice-water solution caused by the melting of the ice. Also known as ice calorimeter.

bunton [MIN ENG] A steel or timber element in the lining of a rectangular shaft.

buoy [ENG] An anchored or moored floating object, other than a lightship, intended as an aid to navigation, to attach or suspend measuring instruments, or to mark the position of something beneath the water.

buoyancy pontoons [PETRO ENG] Pontoons that buoy up offshore pipelines during the welding together of sections over bodies of water, after which the pontoons are removed and the pipeline is allowed to sink into position on the bottom.

buoyancy-type density transmitter [ENG] An instrument which records the specific gravity of a flowing stream of a liquid or gas, using the principle of hydrostatic weighing.

buoy sensor [ENG ACOUS] A hydrophone used as a sensor in buoy projects; some hydrophone arrays are designed for telemetering.

burden [MET] **1.** The material which is melted in a direct arc furnace. **2.** In an iron blast furnace, the ratio of iron and flux to coke and other fuels in the charge.

burglar alarm [ENG] An alarm in which interruption of electric current to a relay, caused, for example, by the breaking of a metallic tape placed at an entrance to a building, deenergizes the relay and causes the relay contacts to operate the alarm indicator. Also known as intrusion alarm.

burn [ENG] To consume fuel.

burn cut *See* parallel cut.

burned sand [MET] The dissipated claying portion of a casting sand resulting from the heat of the metal.

burner [ENG] **1.** The part of a fluid-burning device at which the flame is produced. **2.** Any burning device used to soften old paint to aid in its removal. **3.** A worker who operates a kiln which burns brick or tile. **4.** A worker who alters the properties of a mineral substance by burning. **5.** A worker who uses a flame-cutting torch to cut metals. [MECH ENG] A unit of a steam boiler which mixes and directs the flow of fuel and air so as to ensure rapid ignition and complete combustion.

burner fuel oil [MATER] Any of the petroleum distillate and residual oils used to heat homes and buildings.

burner windbox [ENG] A chamber surrounding a burner, under positive air pressure, for proper distribution and discharge of secondary air.

burnettize [ENG] To saturate fabric or wood with a solution of zinc chloride under pressure to keep it from decaying.

burn-in *See* freeze.

burning [ENG] The firing of clay products placed in a kiln. [MET] **1.** Permanent damage to a metal caused by heating beyond temperature limits of the treatment. **2.** Deep pitting of a metal caused by excessive pickling.

burning line [PETRO ENG] A pipeline used to convey refinery fuel gas, as distinguished from gas intended for subsequent processing.

burning oil [MATER] An oil such as kerosine or mineral seal oil suitable for burning.

burning point [ENG] The lowest temperature at which a volatile oil in an open vessel will continue to burn when ignited by a flame held close to its surface; used to test safety of kerosine and other illuminating oils.

burning quality [ENG] Rated performance for a burning oil as determined by specified ASTM tests.

burning-quality index [ENG] Prediction of burning performance of furnace and heater oils; derived from ASTM distillation, API gravity, paraffinicity, and volatility.

burning rate [MATER] The tendency and rate of materials to burn at given temperatures, in contrast to melting or disintegrating.

burning-rate constant [AERO ENG] A constant, related to initial grain temperature, used in calculating the burning rate of a rocket propellant grain.

burnish [ENG] To polish or make shiny. [MET] To develop a smooth, lustrous surface finish by tumbling with steel balls or rubbing with a hard metal pad.

burn-off rate *See* melting rate.

burnout [AERO ENG] **1.** An act or instance of fuel or oxidant depletion or of depletion of both at once. **2.** The time at which this depletion occurs. [ENG] An instance of a device or a part overheating so as to result in destruction or damage.

burnout velocity [AERO ENG] The velocity of a rocket at the time when depletion of the fuel or oxidant occurs. Also known as all-burnt velocity; burnt velocity.

Burnside boring machine [MECH ENG] A machine for boring in all types of ground with the feature of controlling water immediately if it is tapped.

burnt deposit [MET] A dark, powdery deposit obtained by excessive current density in electroplating.

burnt-on sand [MET] A mixture of sand and metal on the surface of a casting due to metal penetration of the sand mold.

burnt plate oil [MATER] A material used to thin etching ink too heavy in body; made by boiling linseed oil and, at a certain temperature, igniting it.

burnt shale [MIN ENG] Carbonaceous shale which has remained for a long period in a colliery tip and undergone spontaneous combustion and converted into a coppery slag material. Also known as oxidized shale.

burnt velocity See burnout velocity.

burr [MET] A thin, ragged fin left on the edge of a piece of metal by a cutting or punching tool.

burrow [MIN ENG] A refuse heap at a coal mine.

burst disk [AERO ENG] A diaphragm designed to burst at a predetermined pressure differential; sometimes used as a valve, for example, in a liquid-propellant line in a rocket. Also known as rupture disk.

burton [MECH ENG] A small hoisting tackle with two blocks, usually a single block and a double block, with a hook block in the running part of the rope.

bus [ENG] A motor vehicle for carrying a large number of passengers.

bushing [MECH ENG] A removable piece of soft metal or graphite-filled sintered metal, usually in the form of a bearing, that lines a support for a shaft.

buster [MET] A pair of dies used in press forging for barreling, for flattening a hot metal billet, or for loosening scale on hot, ferrous forging billets. [MIN ENG] An expanding wedge used to break down coal or rock.

butadiene-styrene rubber [MATER] A synthetic rubber that is formed by copolymerization of butadiene and styrene.

Butler finish See satin finish.

butt [BUILD] The bottom or cover edge of a shingle. [DES ENG] The enlarged and squared-off end of a connecting rod or similar link in a machine. [MIN ENG] Coal exposed at right angles to the face and, in contrast to the face, generally having a rough surface.

butt cable See hand cable.

butter finish [MET] The semilustrous surface produced with a mildly abrasive wheel.

butterfly damper See butterfly valve.

butterfly nut See wing nut.

butterfly valve [ENG] A valve that utilizes a turnable disk element to regulate flow in a pipe or duct system, such as a hydraulic turbine or a ventilating system. Also known as butterfly damper.

buttering [MET] Coating the faces of a weld joint prior to welding to prevent cross contamination of the weld metal and base metal.

Butterworth head [MECH ENG] A mechanical hose head with revolving nozzles; used to wash down shipboard storage tanks.

butt fusion [ENG] The joining of two pieces of plastic or metal pipes or sheets by heating the ends until they are molten and then pressing them together to form a homogeneous bond.

butt gage [ENG] A tool used to mark the outline for the hinges on a door.

butt joint [ENG] A joint in which the parts to be joined are fastened end to end or edge to edge with one or more cover plates (or other strengthening) generally used to accomplish the joining.

buttock [MIN ENG] A corner formed by two coal faces more or less at right angles, such as the end of a working face.

buttock lines [ENG] The lines of intersection of the surface of an aircraft or its float, or of the hull of a ship, with its longitudinal vertical planes. Also known as buttocks.

buttocks See buttock lines.

butt-off [MET] To supplement ramming in the production of castings by either manual or pneumatic jolting.

button [MET] **1.** Mass of metal remaining in a crucible after fusion has been completed. **2.** That part of a weld which tears out in the destructive testing of spot-, seam-, or projection-welded specimens.

button balance [MIN ENG] A small, very delicate balance used for weighing assay buttons.

button die [DES ENG] A mating member, usually replaceable, for a piercing punch. Also known as die bushing.

buttonhead [DES ENG] A screw, bolt, or rivet with a hemispherical head.

buttress [CIV ENG] A pier constructed at right angles to a restraining wall on the side opposite to the restrained material; increases the strength and thrust resistance of the wall.

buttress dam [CIV ENG] A concrete dam constructed as a series of buttresses.

buttress thread [DES ENG] A screw thread whose forward face is perpendicular to the screw axis and whose back face is at an angle to the axis, so that the thread is both efficient in transmitting power and strong.

buttress-thread casing [PETRO ENG] A drill casing in which the ends of the sections are buttressed together and held in place with a short threaded outer sleeve; used where greater than normal clearance, strength, and leak resistance are needed.

butt weld [MET] A weld that joins the ends of two pieces of metal of similar cross section without overlapping.

butyl rubber [MATER] A synthetic rubber made by the polymerization of isoprene and isobutylene.

buy-back crude [PETRO ENG] Oil in which the government of the production territory has a right to a share if it is a stockholder in the oil-producing company. Also known as participation crude.

buzz [AERO ENG] Sustained oscillation of an aerodynamic control surface caused by intermittent flow separation on the surface, or by a motion of shock waves across the surface, or by a combination of flow separation and shock-wave motion on the surface.

BWE *See* bucket-wheel excavator.

BWG *See* Birmingham wire gage.

Byers process [MET] The principal process for manufacturing wrought iron; pig iron is melted in a cupola, desulfurized in a ladle, and refined in a Bessemer converter.

bypass [CIV ENG] A road which carries traffic around a congested district or temporary obstruction. [ENG] An alternating, usually smaller, diversionary flow path in a fluid dynamic system to avoid some device, fixture, or obstruction.

bypass channel [CIV ENG] **1.** A channel built to carry excess water from a stream. Also known as flood relief channel; floodway. **2.** A channel constructed to divert water from a main channel.

bypass valve [ENG] A valve that opens to direct fluid elsewhere when a pressure limit is exceeded.

by-product [ENG] A product from a manufacturing process that is not considered the principal material.

C

cab [ENG] In a locomotive, truck, tractor, or hoisting apparatus, a compartment for the operator.

cabane [AERO ENG] The arrangement of struts used on early types of airplanes to brace the wings.

cabble [MET] To break up into pieces preparatory to the processes of fagoting, fusing, and rolling into bars, as is done with charcoal iron.

cabinet file [DES ENG] A coarse-toothed file with flat and convex faces used for woodworking.

cabinet hardware [DES ENG] Parts for the final trim of a cabinet, such as fastening hinges, drawer pulls, and knobs.

cabinet saw [DES ENG] A short saw, one edge used for ripping, the other for cross-cutting.

cabinet scraper [DES ENG] A steel tool with a contoured edge used to remove irregularities on a wood surface.

cable [DES ENG] A stranded, ropelike assembly of wire or fiber.

cable buoy [ENG] A buoy used to mark one end of a submarine underwater cable during time of installation or repair.

cable conveyor [MECH ENG] A powered conveyor in which a trolley runs on a flexible, torque-transmitting cable that has helical threads.

cable duct [ENG] A pipe, either earthenware or concrete, through which prestressing wires or electric cable are pulled.

cable lacquer [MATER] Black, colored, or clear lacquer made from synthetic resins, having a high dielectric strength and being resistant to oils and heat; used to give a tough, flexible coating.

cable-laid [DES ENG] Consisting of three ropes with a left-hand twist, each rope having three twisted strands.

cableman [ENG] A person who installs, repairs, or otherwise works with cables.

cable paper [MATER] A paper used to insulate electrical cables.

cable railway [MECH ENG] An inclined track on which rail cars trvel, with the cars fixed to an endless steel-wire rope at equal spaces; the rope is driven by a stationary engine.

cable release [ENG] A wire plunger to actuate the shutter of a camera, thus avoiding undesirable camera movement.

cable-stayed bridge [CIV ENG] A modification of the cantilever bridge consisting of girders or trusses cantilevered both ways from a central tower and supported by inclined cables attached to the tower at top or sometimes at several levels.

cable system [MIN ENG] A drilling system involving a heavy string of tools suspended from a flexible cable.

cable-system drill See churn drill.

cable tools [MIN ENG] The bits and other bottom-hole tools and equipment used to drill boreholes by percussive action, using a cable, instead of rods, to connect the drilling bit with the machine on the surface.

cable vault [CIV ENG] A manhole containing electrical cables.

cableway [MECH ENG] A transporting system consisting of a cable extended between two or more points on which cars are propelled to transport bulk materials for construction operations. [MIN ENG] A cable system of material handling in which carriers are supported by a cable and not detached from the operating span.

cableway carriage [MECH ENG] A trolley that runs on main load cables stretched between two or more towers.

caboose [ENG] A car on a freight train, often the last car, usually for use by trainmen.

CABRA numbers [MET] Copper and Brass Research Association number designations for various wrought copper and copper alloy grades.

cabretta [MATER] Tanned sheepskin leather.

cab signal [ENG] A signal in a locomotive that informs the engine operator about conditions affecting train movement.

cacao butter See cocoa butter.

cadastral survey [CIV ENG] A survey made to establish property lines.

cade oil [MATER] A brown, viscous essential oil that has a tar odor and is slightly soluble in water; it is derived by dry distillation of the wood of the European juniper (*Juniperus oxycedrus*); used in antiseptic soaps, perfumes, and pharmaceuticals. Also known as juniper tar oil.

cadmium [MET] A tin-white, malleable, ductile metal capable of high polish; principal use is in the plating of iron and steel, and to a much less extent of copper, brass, and other alloys, to protect them from corrosion and improve solderability and surface conductivity, and as a control absorber and shield in nuclear reactors.

cadmium metallurgy [MET] The extraction of cadmium from zinc ores, or from complex ores as a by-product of zinc, lead, and copper smelting.

cadmium red [MATER] A pigment composed of a mixture of cadmium sulfide, cadmium selenide, and barite; used as a red pigment.

caffetannin [MATER] One of the hydrolyzable tannins obtained from coffee berries and other plant products.

cage [MECH ENG] A frame for maintaining uniform separation between the balls or rollers in a bearing. Also known as separator. [MIN ENG] The car which carries men and materials in a mine hoist.

cage guides [MIN ENG] Directive apparatus used to guide the cages in the mine shaft and to prevent their swinging or colliding with each other.

cage mill [MECH ENG] Pulverizer used to disintegrate clay, press cake, asbestos, packing-house by-products, and various tough, gummy, high-moisture-content or low-melting-point materials.

cage shoes [MIN ENG] Fittings attached on the side of a cage by bolts so that they engage the rigid guides in a shaft.

cairn [ENG] An artificial mound of rocks, stones, or masonry, usually conical or pyramidal, whose purpose is to designate or to aid in identifying a point of surveying or of cadastral importance.

caisson [CIV ENG] 1. A watertight, cylindrical or rectangular chamber used in underwater construction to protect workers from water pressure and soil collapse. 2. A float used to raise a sunken vessel. 3. *See* dry-dock caisson.

caisson foundation [CIV ENG] A shaft of concrete placed under a building column or wall and extending down to hardpan or rock. Also known as pier foundation.

cajeput oil [MATER] An essential oil with a blue-green color and a camphorlike odor; chief constituents are eucalyptol, pinene, and alpha terpineol; distilled from leaves and twigs of species of East Indian trees of the genus *Melaleuca*; used in perfumes and medicines. Also spelled cajuput oil.

cajuput oil *See* cajeput oil.

cake [MIN ENG] 1. Solidified drill sludge. 2. That portion of a drilling mud adhering to the walls of a borehole. 3. To form into a mass, as when ore sinters together in roasting, or coal cakes in coking.

cake of gold [MET] Gold formed into a compact mass (though not melted) by distillation of the mercury from a mercury-gold amalgam. Also known as sponge gold.

caking [ENG] Changing of a powder into a solid mass by heat, pressure, or water.

calamine [MET] An alloy composed of zinc, lead, and tin.

calamus oil [MATER] A yellow essential oil that is slightly soluble in water; composed mainly of eugenol and 2,4,5-trimethoxy-1-propenyl benzene; derived by steam distillation of the roots of calamus (*Acorus calamus*); used in perfumery and medicine.

calcimeter [ENG] An instrument for estimating the amount of lime in soils.

calcimine [MATER] A thin paint, white or colored, made of pigment, glue, and water. Also known as kalsomine.

calcine [ENG] 1. To heat to a high temperature without fusing, as to heat unformed ceramic materials in a kiln, or to heat ores, precipitates, concentrates, or residues so that hydrates, carbonates, or other compounds are decomposed and the volatile material is expelled. 2. To heat under oxidizing conditions. [MATER] Product of calcining or roasting.

calcined clay [MATER] Clay that has been heated to drive out volatile materials; a natural abrasive.

calcined coke [MATER] Coke that has been heated to expel volatile material.

calcined limestone [MATER] Limestone that has been heated in a vertical-shaft kiln to drive off carbon dioxide.

calcinen *See* calcining furnace.

calcining furnace [ENG] A heating device, such as a vertical-shaft kiln, that raises the temperature (but not to the melting point) of a substance such as limestone to make lime. Also known as calciner.

calcium-bomb process [MET] A former process to produce pellets of pure titanium metal by mixing titanium chloride with calcium in a steel bomb and heating.

calefaction [ENG] 1. Warming. 2. The condition of being warmed.

calender [ENG] 1. To pass a material between rollers or plates to thin it into sheets or to make it smooth and glossy. 2. The machine which performs this operation.

calendered paper [MATER] Paper that has passed through the calenders of a paper machine.

calibrated airspeed [AERO ENG] The airspeed as read from a differential-pressure airspeed indicator which has been corrected for instrument and installation errors; equal to true airspeed for standard sea-level conditions.

calibrating tank [ENG] A tank having known capacity used to check the volumetric accuracy of liquid delivery by positive-displacement meters. Also known as meter-proving tank.

calibration curve [ENG] A plot of calibration data, giving the correct value for each indicated reading of a meter or control dial.

calibration markers [ENG] On a radar display, electronically generated marks which provide numerical values for the navigational parameters such as bearing, distance, height, or time.

California sampler [MIN ENG] A drive sampler equipped with a piston that can be retracted mechanically to any desired point within the barrel of the sampler.

California-type dredge [MIN ENG] A single-lift dredge in which closely spaced buckets deliver to a trommel; oversize rocks are piled behind the dredge by a conveyor or stacker; the undersize are washed on gold-saving tables on the deck, and tailings discharge astern through sluices.

caliper [DES ENG] An instrument with two legs or jaws that can be adjusted for measuring linear dimensions, thickness, or diameter.

caliper gage [DES ENG] An instrument, such as a micrometer, of fixed size for calipering.

caliper log [MIN ENG] A graphic record showing the diameter of a drilled hole at each depth; measurements are obtained by drawing a caliper upward through the hole and recording the diameter on quadrile paper.

calite [MET] A heat-resistant alloy of iron, nickel, and aluminum which resists oxidation up to 2200°F (1204°C) and is practically noncorrodible under ordinary conditions of exposure.

Callendar and Barnes' continuous flow calorimeter [ENG] A calorimeter in which the heat to be measured is absorbed by water flowing through a tube at a constant rate, and the quantity of heat is determined by the rate of flow and the temperature difference between water at ends of the tube.

Callendar's thermometer See platinum resistance thermometer.

Callon's rule [MIN ENG] The rule stating that when a pillar is left in an inclined seam for support in a shaft or a structure on the surface, a greater width should be left on the rise side of the shaft or structure than on the dip side.

Callow flotation cell [MIN ENG] Nonmechanical apparatus for separation of floatable solid gangue from pulverized ore, with the mixture suspended in liquid and aerated by air bubbles coming up through a porous medium, so that the lighter gangue floats away from the heavier ore.

Callow process [MIN ENG] A flotation process in which agitation is provided by the forcing of air into the pulp through the canvas-covered bottom of the cell.

Callow screen [MIN ENG] A continuous belt system formed of fine screen wire that is used to separate fine solids from coarse ones.

calorific value [ENG] Quantity of heat liberated on the complete combustion of a unit weight or unit volume of fuel.

calorifier [ENG] A device that heats fluids by circulating them over heating coils.

calorimeter [ENG] An apparatus for measuring heat quantities generated in or emitted by materials in processes such as chemical reactions, changes of state, or formation of solutions.

calorimetric test [ENG] The use of a calorimeter to determine the thermochemical characteristics of propellants and explosives; properties normally determined are heat of combustion, heat of explosion, heat of formation, and heat of reaction.

calorimetry [ENG] The measurement of the quantity of heat involved in various processes, such as chemical reactions, changes of state, and formations of solutions, or in the determination of the heat capacities of substances; fundamental unit of measurement is the joule or the calorie (4.184 joules).

calorize [MET] To treat by a process by which a coating of aluminum and aluminum-iron alloys is produced on iron and steel (and, less commonly, brass, copper, or nickel), which coating protects the metal from burning in temperatures up to 1800°F (982°C).

calorized steel See aluminized steel.

calyx [ENG] A steel tube that is a guide rod and is also used to catch cuttings from a drill rod. Also known as bucket; sludge barrel; sludge bucket.

calyx drill [ENG] A rotary core drill with hardened steel shot for cutting rock. Also known as shot drill.

cam [MECH ENG] A plate or cylinder which communicates motion to a follower by means of its edge or a groove cut in its surface.

cam acceleration [MECH ENG] The acceleration of the cam follower.

camber [AERO ENG] The rise of the curve of an airfoil section, usually expressed as the ratio of the departure of the curve from a straight line joining the extremities of the curve to the length of this straight line. [DES ENG] Deviation from a straight line; the term is applied to a convex, edgewise sweep or curve, or to the increase in diameter at the center of rolled materials.

camber angle [MECH ENG] The inclination from the vertical of the steerable wheels of an automobile.

cam cutter [MECH ENG] A semiautomatic or automatic machine that produces the cam contour by swinging the work as it revolves; uses a master cam in contact with a roller.

cam dwell [DES ENG] That part of a cam surface between the opening and closing acceleration sections.

cam engine [MECH ENG] A piston engine in which a cam-and-roller mechanism seems to convert reciprocating motion into rotary motion.

camera study *See* memomotion study.

cam follower [MECH ENG] The output link of a cam mechanism.

cam mechanism [MECH ENG] A mechanical linkage whose purpose is to produce, by means of a contoured cam surface, a prescribed motion of the output link.

Cammett table [MIN ENG] A side-moving table for concentrating ore.

cam nose [MECH ENG] The high point of a cam, which in a reciprocating engine holds valves open or closed.

cam pawl [MECH ENG] A pawl which prevents a wheel from turning in one direction by a wedging action, while permitting it to rotate in the other direction.

Campbell process [MET] An open-hearth steel manufacturing process in which ore and pig iron are used as raw materials in a tilting furnace.

Campbell-Stokes recorder [ENG] A sunshine recorder in which the time scale is supplied by the motion of the sun and which has a spherical lens that burns an image of the sun upon a specially prepared card.

camp ceiling [BUILD] A ceiling that is flat in the center portion and sloping at the sides.

camphor oil [MATER] An essential oil obtained by steam distillation from the wood of the camphor tree (*Cinnamomum camphora*); used in the manufacture of camphor and safrole.

cam profile [DES ENG] The shape of the contoured cam surface by means of which motion is communicated to the follower. Also known as pitch line.

camshaft [MECH ENG] A rotating shaft to which a cam is attached.

can [DES ENG] A cylindrical metal vessel or container, usually with an open top or a removable cover.

Canada balsam [MATER] A transparent balsam useful for cementing together lenses and other optical elements because its index of refraction is in the same range as that of glass.

canadol [MATER] A light petroleum naphtha fraction having a specific gravity slightly higher than petroleum ether; used as a solvent and an anesthetic.

canal [CIV ENG] An artificial open waterway used for transportation, waterpower, or irrigation. [DES ENG] A groove on the underside of a corona.

canalization [ENG] Any system of distribution canals or conduits for water, gas, electricity, or steam.

cananga oil *See* ilang-ilang oil.

canard [AERO ENG] **1.** An aerodynamic vehicle in which horizontal surfaces used for trim and control are forward of the wing or main lifting surface. **2.** The horizontal trim and control surfaces in such an arrangement.

candelilla wax [MATER] A wax obtained from the wax-coated stems of candelilla shrubs, especially *Euphorbia antisyphilitica*; used for varnishes and furniture and shoe polishes.

candlenut oil *See* lumbang oil.

can hoisting system [MIN ENG] A hoisting method used in shallow lead and zinc mines in which cans are loaded below and hoisted to the surface where they are capsized and the load discharged; the operation is controlled at the top of the shaft.

canned motor [MECH ENG] A motor enclosed within a casing along with the driven element (that is, a pump) so that the motor bearings are lubricated by the same liquid that is being pumped.

canned pump [MECH ENG] A watertight pump that can operate under water.

cannibalize [ENG] To remove parts from one piece of equipment and use them to replace like, defective parts in a similar piece of equipment in order to keep the latter operational.

cannular combustion chambers [AERO ENG] The separate combustion chambers in an aircraft gas turbine. Also known as can-type combustors.

canopy [AERO ENG] **1.** The umbrellalike part of a parachute which acts as its main supporting surface. **2.** The overhead, transparent enclosure of an aircraft cockpit.

cant file [DES ENG] A fine-tapered file with a triangular cross section, used for sharpening saw teeth.

cant hook [DES ENG] A lever with a hooklike attachment at one end, used in lumbering.

cantilever [ENG] A beam or member securely fixed at one end and hanging free at the other end.

cantilever bridge [CIV ENG] A fixed bridge consisting of two spans projecting toward each other and joined at their ends by a suspended simple span.

cantilever footing [CIV ENG] A footing used to carry a load from two columns, with one column and one end of the footing placed against a building line or exterior wall.

cantilever retaining wall [CIV ENG] A type of wall formed of three cantilever beams: stem, toe projection, and heel projection.

cantilever spring [MECH ENG] A flat spring supported at one end and holding a load at or near the other end.

cant strip [BUILD] **1.** A strip placed along the angle between a wall and a roof so that the roofing will not bend sharply. **2.** A strip placed under the edge of the lowest row of tiles on a roof to give them the same slope as the other tiles.

can-type combustors See cannular combustion chambers.

caoutchouc [MATER] Formerly, crude rubber which had been cured over a fire into a solid, dark mass for shipment.

cap [ENG] A detonating or blasting cap.

capacitance altimeter [ENG] An absolute altimeter which determines height of an aircraft aboveground by measuring the variations in capacitance between two conductors on the aircraft when the ground is near enough to act as a third conductor.

capacitance level indicator [ENG] A level indicator in which the material being monitored serves as the dielectric of a capacitor formed by a metal tank and an insulated electrode mounted vertically in the tank.

capacitance meter [ENG] An instrument used to measure capacitance values of capacitors or of circuits containing capacitance.

capacitance-operated intrusion detector [ENG] A boundary alarm system in which the approach of an intruder to an antenna wire encircling the protected area a few feet above ground changes the antenna-ground capacitance and sets off the alarm.

capacitive pressure transducer [ENG] A measurement device in which variations in pressure upon a capacitive element proportionately change the element's capacitive rating and thus the strength of the measured electric signal from the device.

capacitor hydrophone [ENG ACOUS] A capacitor microphone that responds to waterborne sound waves.

capacitor loudspeaker See electrostatic loudspeaker.

capacitor microphone [ENG ACOUS] A microphone consisting essentially of a flexible metal diaphragm and a rigid metal plate that together form a two-plate air capacitor; sound waves set the diaphragm in vibration, producing capacitance variations that are converted into audio-frequency signals by a suitable amplifier circuit. Also known as condenser microphone; electrostatic microphone.

capacitor pickup [ENG ACOUS] A phonograph pickup in which movements of the stylus in a record groove cause variations in the capacitance of the pickup.

capacity correction [ENG] The correction applied to a mercury barometer with a nonadjustable cistern in order to compensate for the change in the level of the cistern as the atmospheric pressure changes.

cap crimper [ENG] A tool resembling a pliers that is used to press the open end of a blasting cap onto the safety fuse before placing the cap in the primer.

cape chisel [DES ENG] A chisel that tapers to a flat, narrow cutting end; used to cut flat grooves.

Capell fan [MIN ENG] A centrifugal type of mine shaft fan.

capillarity correction [ENG] As applied to a mercury barometer, that part of the instrument correction which is required by the shape of the meniscus of the mercury.

capillary collector [ENG] An instrument for collecting liquid water from the atmosphere; the collecting head is fabricated of a porous material having a pore size of the order of 30 micrometers; the pressure difference across the water-air interface prevents air from entering the capillary system while allowing free flow of water.

capillary control [PETRO ENG] Discarded theory of reservoir oil flow to a well hole through capillary pores; it attributed flow resistance to gas bubbles within the capillaries.

capillary drying [ENG] Progressive removal of moisture from a porous solid mass by surface evaporation followed by capillary movement of more moisture to the drying surface from the moist inner region, until the surface and core stabilize at the same moisture concentration.

capillary electrometer [ENG] An electrometer designed to measure a small potential difference between mercury and an electrolytic solution in a capillary tube by measuring the effect of this potential difference on the surface tension between the liquids. Also known as Lippmann electrometer.

capillary equilibrium method [PETRO ENG] Test method to predict oil and gas flow through an oil reservoir core by dethrottling the flow to hold capillary flow in equilibrium between oil and gas within the reservoir.

capillary tube [ENG] A tube sufficiently fine so that capillary attraction of a liquid into the tube is significant.

capillary viscometer [ENG] A long, narrow tube used to measure the laminar flow of fluids.

capital amount factor [IND ENG] Any of 20 common compound interest formulas used to calculate the equivalent uniform annual cost of all cash flows.

capital budgeting [IND ENG] Planning the most effective use of resources to obtain the highest possible level of sustained profits.

capital expenditure [IND ENG] Money spent for long-term additions or improvements and charged to a capital assets account.

cap lamp [MIN ENG] The lamp a miner wears on his safety hat or cap for illumination.

capped fuse [ENG] A length of safety fuse with the cap or detonator crimped on before it is taken to the place of use.

capped steel [MET] Partially deoxidized steel cast in an open-top mold, which is capped to solidify the top metal and enforce internal pressure, resulting in a surface condition similar to that of rimming steel.

cap piece [MIN ENG] A piece of wood fitted over a straight post or timber to provide more bearing surface.

capping [ENG] Preparation of a capped fuse. [MIN ENG] The attachment at the end of a winding rope. [PETRO ENG] **1.** The process of sealing or covering a borehole such as an oil or gas well. **2.** The material or device used to seal or cover a borehole.

cap screw [DES ENG] A screw which passes through a clear hole in the part to be joined, screws into a threaded hole in the other part, and has a head which holds the parts together.

capstan [ENG] A shaft which pulls magnetic tape through a machine at constant speed.

capstan nut [DES ENG] A nut whose edge has several holes, in one of which a bar can be inserted for turning it.

capstan screw [DES ENG] A screw whose head has several radial holes, in one of which a bar can be inserted for turning it.

captive balloon [AERO ENG] A moored balloon, usually by steel cables.

captive fastener [DES ENG] A screw-type fastener that does not drop out after it has been unscrewed.

captive test [ENG] A hold-down test of a propulsion subsystem, rocket engine, or motor.

capture [AERO ENG] The process in which a missile is taken under control by the guidance system.

capture area [ENG ACOUS] The effective area of the receiving surface of a hydrophone, or the available power of the acoustic energy divided by its equivalent plane-wave intensity.

capturing [ENG] The use of a torquer to restrain the spin axis of a gyro to a specified position relative to the spin reference axis.

car *See* automobile.

caramel [MATER] A dark-brown mass formed by heating sugar in the presence of ammonium salts.

caranda wax [MATER] A wax similar to carnauba wax; obtained from the tropical palm caranda (*Copernicia australis*).

caraway oil [MATER] A pale-yellow to colorless liquid that is slightly soluble in water; distilled from the dried fruit of the caraway plant (*Carum carvi*), the main ingredients being carvone and d-limonene; used in flavors, medicines, soaps, and perfumes. Also known as caraway seed oil; carui oil.

caraway seed oil *See* caraway oil.

carbide [MATER] A cemented or compacted mixture of powdered carbides of heavy metals forming a hard material used in metal-cutting tools. Also known as cemented carbide.

carbide lamp [MIN ENG] A lamp that is charged with calcium carbide and water to form acetylene, which it burns.

carbide miner [MIN ENG] An automated coal mining machine; the unit is a continuous miner controlled from outside the coal seam.

carbide tool [DES ENG] A cutting tool made of tungsten, titanium, or tantalum carbides, having high heat and wear resistance.

carbolfuchsin [MATER] A solution of fuchsin, phenol, alcohol, and water; used as a stain in the identification of bacteria.

carbometer [ENG] An instrument for measuring the carbon content of steel by measuring magnetic properties of the steel in a known magnetic field.

carbon arc brazing [MET] Brazing base metals by heating with an electric arc between a carbon electrode and the workpiece.

carbon arc cutting [MET] An arc cutting process which generates heat in order to melt a base metal with a carbon electrode to eventually produce a cut.

carbon arc welding [MET] Welding by maintaining an electric arc between a nonconsumable carbon electrode and the work.

carbon bit [DES ENG] A diamond bit in which the cutting medium is inset carbon.

carbon dioxide indicator [MIN ENG] A detector of carbon dioxide in mines based on the gas's absorption by potassium hydroxide.

carbon dioxide process [MET] A casting process in which the molding material is a mixture of sand and 1.5–6% liquid silicate as a binder, and the mixture is packed around the pattern and hardened by blowing carbon dioxide gas through it.

carbon electrode [MET] A nonfiller-metal electrode consisting of carbon or a graphite rod; sometimes contains copper powder for increased electrical conductivity; used in carbon arc welding.

carbon equivalent [MET] An empirical relationship of the total carbon (TC), silicon (Si) and phosphorus (P) content of gray iron: $CE = \%TC + 0.3(\%Si + \%P)$.

carbon fiber [MATER] Commercial material made by pyrolyzing any spun, felted, or woven raw material to a char at temperatures from 700 to 1800°C.

carbon hydrophone [ENG ACOUS] A carbon microphone that responds to waterborne sound waves.

carbonitrided steel [MET] Steel which is produced by carbonitriding.

carbonitriding [MET] Surface-hardening of low-carbon steel or other solid ferrous alloy by introducing carbon and nitrogen in a gaseous atmosphere containing carbon monoxide or hydrocarbons and ammonia at 800–875°C. Also known as gas cyaniding; nicarbing.

carbon knock [MECH ENG] Premature ignition resulting in knocking or pinging in an internal combustion engine caused when the accumulation of carbon produces overheating in the cylinder.

carbonless copy paper [MATER] A sheet of paper, used to make duplicate copies of written or printed material, whose back is coated with a layer of microcapsules that contain a dye in colorless form in a hydrocarbon solvent; writing or printing pressure breaks the capsules and releases the dye, which reacts with a clay or phenolic resin coating on top of a second paper sheet, located directly below the first, to produce visible color.

carbon microphone [ENG ACOUS] A microphone in which a flexible diaphragm moves in response to sound waves and applies a varying pressure to a container filled with carbon granules, causing the resistance of the microphone to vary correspondingly.

carbon paper [MATER] A paper, coated with dark waxy pigment, used to make duplicate copies while typewriting or handwriting; a sheet of carbon paper is sandwiched between two paper sheets, so that the impression made on the top sheet causes the carbon paper to transfer a pigmented impression onto the bottom sheet.

carbon pile pressure transducer [ENG] A measurement device in which variations in pressure upon a conductive carbon core proportionately change the core's electrical resistance, and thus the strength of the measured electric signal from the device.

carbon potential [MET] Determination of the extent to which an environment containing active carbon can affect the carbon content of a steel.

carbon refractory [MATER] Carbon, generally in the form of graphite, used as a refractory in such equipment as crucibles and stopper nozzles for steel casting.

carbon resistance thermometer [ENG] A highly sensitive resistance thermometer for measuring temperatures in the range 0.05–20 K; capable of measuring temperature changes of the order 10^{-5} degree.

carbon restoration [MET] Carburizing of the surface layer of a material to achieve the original level of carbon content which had been depleted during processing.

carbon steel [MET] Steel containing carbon, to about 2%, as the principal alloying element.

carbon transducer [ENG] A transducer consisting of carbon granules in contact with a fixed electrode and a movable electrode, so that motion of the movable electrode varies the resistance of the granules.

carbonyl process [MET] **1.** A process in powder metallurgy for the production of iron, nickel, and iron-nickel alloy powders for magnetic applications. **2.** A process used in putting a metallic coating on molybdenum tungsten and other metals.

carborundum [MATER] A manufactured crystalline material (silicon carbide), prepared by fusing coke and sand in an electric furnace; used as an abrasive in the grinding of low-tensile-strength materials, and as a semiconductor with a maximum operating temperature of 1300°C, to rectify and detect radio waves.

carbosand [MATER] Sand treated with an organic solution and then roasted; used as a spray to disperse oil slicks.

carbureted water gas [MATER] Water gas that has been enriched by hydrocarbon gases of high fuel value.

carburetion [MECH ENG] The process of mixing fuel with air in a carburetor.

carburetor [MECH ENG] A device that makes and controls the proportions and quantity of fuel-air mixture fed to a spark-ignition internal combustion engine.

carburetor icing [MECH ENG] The formation of ice in an engine carburetor as a consequence of expansive cooling and evaporation of gasoline.

carburize [MET] To surface-harden steel by converting the outer layer of low-carbon steel to high-carbon steel by heating the steel above the transformation range in contact with a carbonaceous material.

cardamon oil [MATER] A pale-yellow, combustible essential oil; insoluble in water, soluble in alcohol and ether; distilled from cardamon seeds, the chief known ingredients being terpinene, borneol, dipentene, limonene, and eucalyptol; used in flavoring and medicine.

Cardan joint See Hooke's joint.

Cardan motion [MECH ENG] The straight-line path followed by a moving centrode in a four-bar centrode linkage.

Cardan shaft [MECH ENG] A shaft with a universal joint at its end to accommodate a varying shaft angle.

Cardan's suspension [DES ENG] An arrangement of rings in which a heavy body is mounted so that the body is fixed at one point; generally used in a gyroscope.

cardboard [MATER] A good quality of chemical pulp or rag pasteboard made by combining two or more webs of paper, either with or without paste, while still wet; used for signs, printed material, and high-quality boxes.

cardioid microphone [ENG ACOUS] A microphone having a heart-shaped, or cardioid, response pattern, so it has nearly uniform response for a range of about 180° in one direction and minimum response in the opposite direction.

cardioid pattern [ENG] Heart-shaped pattern obtained as the response or radiation characteristic of certain directional antennas, or as the response characteristic of certain types of microphones.

car dump [MECH ENG] Any one of several devices for unloading industrial or railroad cars by rotating or tilting the car.

car-following theory [ENG] A mathematical model of the interactions between motor vehicles in terms of relative speed, absolute speed, and separation.

cargo boom [MECH ENG] A long spar extending from the mast of a derrick to support or guide objects lifted or suspended.

cargo mill [IND ENG] A sawmill equipped with docks so the product can be loaded directly onto ships.

cargo winch [MECH ENG] A motor-driven hoisting machine for cargo having a drum around which a chain or rope winds as the load is lifted.

Carinthian furnace [MET] A zinc distillation furnace with small, vertical retorts. [MIN ENG] A small reverberatory furnace with an inclined hearth, in which lead ore is treated by roasting and reaction, wood being the usual fuel.

carnauba wax [MATER] The hardest natural wax, having a melting point of 85°C, exuded from the leaves of the carnauba palm (*Carnauba cerifera*); used for insulating purposes and in making candles, shoe polish, high-luster wax, varnishes, phonograph records, and surface coating of automobiles. Also known as Brazil wax.

Carnot engine [MECH ENG] An ideal, frictionless engine which operates in a Carnot cycle.

Carnoy's solution [MATER] A tissue fixative composed of mercuric salts and acetic acids; used where penetration of hard objects such as seeds is required.

caroba See carob wood.

carob wood [MATER] The wood from a large Brazilian tree, *Jacaranda copaia*. Also known as caroba; jacaranda.

carpenter's level [DES ENG] A bar, usually of aluminum or wood, containing a spirit level.

car pincher [MIN ENG] A worker in a mine who uses a pinch bar to position cars under loading chutes.

carpincho [MATER] Processed capybara skin, noted for its elastic properties.

car retarder [ENG] A device located along the track to reduce or control the velocity of railroad or mine cars.

carriage [ENG] **1.** A device that moves in a predetermined path in a machine and carries some other part, such as a recorder head. **2.** A mechanism designed to hold a paper in the active portion of a printing or typing device, for example, a typewriter carriage.

carriage bolt [DES ENG] A round-head type of bolt with a square neck, used with a nut as a through bolt.

carriage stop [MECH ENG] A device added to the outer way of a lathe bed for accurately spacing grooves, turning multiple diameters and lengths, and cutting off pieces of specified thickness.

Carribel explosive [MATER] A medium-strength explosive used in wet boreholes if immersion does not exceed 2 to 3 hours.

carrier [MECH ENG] Any machine for transporting materials or people. [MIN ENG] **1.**

A piece of timber placed on top of a prop or post in a mine. **2.** The horizontal section of a set of timber that is used as a support in a mine roadway.

carrier gas [MET] The gas used in thermal spraying which transmits powder from the feeder to the spray gun.

carrier pipe [ENG] Pipe used to carry or conduct fluids, as contrasted with an exterior protective or casing pipe.

carrier rocket [AERO ENG] A rocket vehicle used to carry something, as the carrier rocket of the first artificial earth satellite.

carron oil [MATER] A mixture comprising equal volumes of linseed oil and lime water which is used for the relief of burns. Also known as lime liniment.

carrot oil [MATER] A light-yellow oil distilled from the seeds of the carrot (*Daucus carota*), and containing carotene, pinene, palmitic acid, limonene, and butyric or isobutyric acid; used as a flavoring.

car shaker [MECH ENG] A device consisting of a heavy yoke on an open-top car's sides that actively vibrates and rapidly discharges a load, such as coal, gravel, or sand, when an unbalanced pulley attached to the yoke is rotated fast.

car stop [ENG] An appliance used to arrest the movement of a mine or railroad car.

cartesian diver manostat [ENG] Preset, on-off-control manometer arrangement by which a specified low pressure (high vacuum) is maintained via the rise or submergence of a marginally buoyant float within a liquid mercury reservoir.

cartographic satellite [AERO ENG] An applications satellite that is used to prepare maps of the earth's surface and of the culture on it.

cartridge [ENG] A cylindrical, waterproof, paper shell filled with high explosive and closed at both ends; used in blasting. [ENG ACOUS] *See* phonograph pickup; tape cartridge.

cartridge actuated initiator [AERO ENG] An item designed to provide gas pressure for activating various aircraft components such as canopy removers, thrusters, and catapults.

cartridge brass [MET] An alloy containing 70% copper and 30% zinc; uses include cartridge cases, automotive radiator cores and tanks, lighting fixtures, eyelets, rivets, springs, screws, and plumbing products.

cartridge filter [ENG] A filter for the clarification of process liquids containing small amounts of solids; turgid liquid flows between thin metal disks, assembled in a vertical stack, to openings in a central shaft supporting the disks, and solids are trapped between the disks.

cartridge starter [MECH ENG] An explosive device which, when placed in an engine and detonated, moves a piston, thereby starting the engine.

car tunnel kiln [ENG] A long kiln with the fire located near the midpoint; ceramic ware is fired by loading it onto cars which are pushed through the kiln.

carui oil *See* caraway oil.

carved-out payment [PETRO ENG] A proportionate royalty payment based on proceeds from oil or gas production from leased property that has been assigned (carved) out of total payments for the leased property.

caryophyllus oil *See* oil of cloves.

cascade [ENG] An arrangement of separation devices, such as isotope separators, connected in series so that they multiply the effect of each individual device.

cascaded [ENG] Of a series of elements or devices, arranged so that the output of one feeds directly into the input of another, as a series of dynodes or a series of airfoils.

cascade impactor [ENG] A low-speed impaction device for use in sampling both solid and liquid atmospheric suspensoids; consists of four pairs of jets (each of progressively smaller size) and sampling plates working in series and designed so that each plate collects particles of one size range.

cascade pulverizer [MECH ENG] A form of tumbling pulverizer that uses large lumps to do the pulverizing.

cascade sequence [MET] Combined longitudinal and buildup sequence in which weld beads are deposited in overlapping layers.

cascading [MECH ENG] An effect in ball-mill rotating devices when the upper level of crushing bodies breaks clear and falls to the top of the crop load.

cascarilla oil [MATER] An essential oil derived from cascarilla bark (*Croton eluteria*), containing cascarillic acid, $C_{11}H_{20}O_2$; used as a flavoring in foods, medicine, and tobacco.

case. [ENG] An item designed to hold a specific item in a fixed position by virtue of conforming dimensions or attachments; the item which it contains is complete in itself for removal and use outside the container. [MET] Outer layer of a ferrous alloy which has been made harder than the core by case hardening. [MIN ENG] A small fissure admitting water into the mine workings. [PETRO ENG] To line a borehole with steel tubing, such as casing or pipe.

case bay [BUILD] A division of a roof or floor, consisting of two principal rafters and the joists between them.

cased glass [MATER] Glass composed of two or more layers of different colors.

case hardening [MET] Process of carburizing low-carbon steel or other ferrous alloy for making the outer layer (case) harder than the core.

casein glue [MATER] A produce of dried curds of milk, lime, and other chemical ingredients, mixed cold; used for both plywood and assembly work.

casein paint [MATER] A paint with casein substituted for linseed oil.

casein plastic [MATER] A plastic made with casein and used for buttons, beads, knitting needles, and novelties; thin sheets and rods of casein plastic are cured (hardened) in formaldehyde baths.

casement window [BUILD] A window hinged on the side that opens to the outside.

cashew gum [MATER] A gum obtained from the bark of the cashew tree; hard, yellowish-brown substance used for inks, insecticides, pharmaceuticals, varnishes, and bookbinders' gum. Also known as anacardium gum.

cashew nutshell oil [MATER] A toxic, irritating oil obtained from the shell of the cashew nut; contains about 90% anacardic acid; used for varnishes, plasticizers, insecticides, coloring materials, and preservatives.

casing [BUILD] A finishing member around the opening of a door or window. [MECH ENG] A fire-resistant covering used to protect part or all of a steam generating unit. [PETRO ENG] A special steel tubing welded or screwed together and lowered into a borehole to prevent entry of loose rock, gas, or liquid into the borehole or to prevent loss of circulation liquid into porous, cavernous, or crevassed ground.

casing cementing [PETRO ENG] Filling of the area between a casing and the borehole with cement in order to prevent fluid migration between permeable zones and to support the casing.

casing centralizer [PETRO ENG] A device secured around the casing during drilling to hold the pipe in a center position, permitting a uniform cement sheath to be constructed around the pipe.

casing cutter [PETRO ENG] Heavy cylindrical apparatus bearing knives, and run on a string of tubing or drill pipe into a well, to perform cutting on the inner walls of a pipe through rotation, thus freeing a pipe section.

casing hanger See hanger.

casinghead [PETRO ENG] A fitting at the head of an oil or gas well that allows the pumping operation to take place, as well as the separation of oil and gas.

casinghead gas [MATER] The natural gas that is emitted from the mouth or opening of an oil well.

casinghead gasoline [MATER] Liquid hydrocarbon product removed from casinghead gas by absorption, compression, or refrigeration.

casinghead tank [PETRO ENG] Storage tank for natural gasoline or other liquids with vapor pressures between 4 and 40 pounds per square inch gage (28 and 280 kilopascals, gage); intermediate between a general-purpose tank and a compressed-gas tank.

casing joint [PETRO ENG] Joint or union that connects two lengths of pipe used to form an oil-well casing.

casing log [PETRO ENG] Recorded data of a down-hole inspection of an oil or gas well made to determine some characteristic of the formations penetrated by the drill hole; types of logs include resistivity, induction, radioactivity, geologic, temperature, and acoustic.

casing nail [DES ENG] A nail about half a gage thinner than a common wire nail of the same length.

casing pressure [PETRO ENG] Gas pressure which is built up between the casing and tubing in a well.

casing shoe [ENG] A ring with a cutting edge on the bottom of a well casing.

casing spear [PETRO ENG] An instrument used for recovering casing which has accidentally fallen into the well.

casing tester [PETRO ENG] A closely fitting, rubber-flanged bucket or a similar tool let down in a well to determine the location of a leak in the casing.

cassette [ENG] A light-tight container designed to hold photographic film or plates. [ENG ACOUS] A small, compact container that holds a magnetic tape and can be readily inserted into a matching tape recorder for recording or playback; the tape passes from one hub within the container to the other hub.

cassia oil [MATER] An essential oil extracted from the bark of Chinese cinnamon (*Cinnamomum cassia*), and containing cinnamaldehyde. Also known as Chinese oil of cinnamon.

cast [ENG] **1.** To form a liquid or plastic substance into a fixed shape by letting it cool in the mold. **2.** Any object which is formed by placing a castable substance in a mold or form and allowing it to solidify. Also known as casting.

castable [MATER] A refractory aggregate mixed with a bonding agent such as aluminous hydraulic cement which, with addition of water, will develop structural strength and set in a mold.

Castaing-Slodzian mass analyzer *See* direct-imaging mass analyzer.

cast coated paper [MATER] A paper with a high-gloss enamel finish that has been produced by drying coated paper under pressure from a polished cylinder.

castellated bit [DES ENG] **1.** A long-tooth, sawtooth bit. **2.** A diamond-set coring bit with a few large diamonds or hard metal cutting points set in the face of each of several upstanding prongs separated from each other by deep waterways. Also known as padded bit.

castellated nut [DES ENG] A type of hexagonal nut with a cylindrical portion above through which slots are cut so that a cotter pin or safety wire can hold it in place.

caster [ENG] **1.** The inclination of the kingpin or its equivalent in automotive steering, which is positive if the kingpin inclines forward, negative if it inclines backward, and zero if it is vertical as viewed along the axis of the front wheels. **2.** A wheel which is free to swivel about an axis at right angles to the axis of the wheel, used to support trucks, machinery, or furniture.

cast-film extrusion *See* chill-roll extrusion.

Castile soap [MATER] A white, odorless, hard soap made from sodium hydroxide and olive oil.

casting alloy [MET] An alloy which cannot be forged or rolled and can be shaped only as a casting.

casting area [ENG] In plastics injection molding, the moldable area of a thermoplastic material for a given thickness and under given conditions of molding.

casting copper [MET] Copper used for making foundry castings; obtained from copper ores, and inferior to electrolytic copper.

casting ladle [MET] A refractory-lined steel ladle used to transport molten metal from the furnace to a mold.

casting plaster [MATER] A white plaster used for castings and carvings.

castings [ENG] Any term or symbol designating a low-quality drill diamond.

casting shrinkage [MET] **1.** Total reduction in volume of a casting due to partial reductions at each stage of solidification. **2.** Reduction in volume at each stage of solidification of a casting.

casting slip [MATER] A slurry of clay and additives mixed in water with deflocculating agents and used for casting in molds.

casting wheel [MET] A large turntable with molds mounted on the outer edge; used primarily in the base-metal industries for cast ingots, anodes, and so on.

cast iron [MET] Any carbon-iron alloy cast to shape and containing 1.8–4.5% carbon, that is, in excess of the solubility in

austenite at the eutectic temperature. Abbreviated C.I.

castor machine oil [MATER] Petroleum-base lubricating oil thickened with an aluminum-base soap, such as aluminum oleate.

castor oil [MATER] A colorless or greenish nondrying oil extracted from the castor bean; used as a cathartic, in soap, and after processing as a lubricant, and as a leather preservative. Also known as ricinus oil.

cast setting *See* mechanical setting.

cast steel [MET] Steel shaped by casting.

cast stone [MATER] Building stone molded from concrete so that it resembles natural stone.

cast structure [MET] The microstructure of a casting.

cast-weld [MET] Joining parts by pouring molten metal over them in a mold.

Catalan forge [MET] A furnace or forge for making wrought iron from ore, in which the ore is loaded at front and charcoal at rear are covered with fine ore and charcoal dust moistened with water.

catalytically cracked gasoline [MATER] Gasoline produced in a refinery reactor equipped with catalytic cracking equipment.

cat-and-mouse engine [MECH ENG] A type of rotary engine, typified by the Tschudi engine, which is an analog of the reciprocating piston engine, except that the pistons travel in a circular motion. Also known as scissor engine.

catapult [AERO ENG] **1.** A power-actuated machine or device for hurling an object at high speed, for example, a device which launches aircraft from a ship deck. **2.** A device, usually explosive, for ejecting a person from an aircraft.

cataracting [MECH ENG] A motion of the crushed bodies in a ball mill in which some, leaving the top of the crop load, fall with impact to the toe of the load.

catastrophic failure [ENG] **1.** A sudden failure without warning, as opposed to degradation failure. **2.** A failure whose occurrence can prevent the satisfactory performance of an entire assembly or system.

catch basin [CIV ENG] **1.** A basin at the point where a street gutter empties into a sewer, built to catch matter that would not easily pass through the sewer. **2.** A well or reservoir into which surface water may drain off.

catching sample [PETRO ENG] During drilling operations, a sample of a cutting obtained from the drilling fluid as it emerges from the well bore or the bailer of a cable tool.

catch pit [MIN ENG] In mineral processing, a sump in a mill to which the floor slopes gently; spillage gravitates or is hosed to this area.

catchwater [CIV ENG] A ditch for catching water on sloping land.

catenary [MATER] In fiberglass-reinforced plastics, a measure of the sag in an assemblage of a number of strands, which have a minimal amount of twist, at the midpoint of a specified length.

catenary suspension [ENG] Holding a flexible wire or chain aloft by its end points; the wire or chain takes the shape of a catenary.

caterpillar [MECH ENG] A vehicle, such as a tractor or army tank, which runs on two endless belts, one on each side, consisting of flat treads and kept in motion by toothed driving wheels.

caterpillar chain [DES ENG] A short, endless chain on which dogs (grippers) or teeth are arranged to mesh with a conveyor.

caterpillar gate [CIV ENG] A steel gate carried on crawler tracks that is used to control water flow through a spillway.

catgut [MATER] A thin cord made from the submucosa of sheep and other animal intestine; used for sutures and ligatures, for strings of musical instruments, and for tennis racket strings. Also known as gut.

cathead line See catline.

cathedral glass [MATER] Unpolished translucent sheet glass.

cathetometer [ENG] An instrument for measuring small differences in height, for example, between two columns of mercury.

cathode cleaning [MET] Electrolytically removing soil from work connected as the cathode.

cathode copper [MET] Copper deposited at the cathode during electrolytic refining; it is melted and marketed as electrolytic copper.

cathode corrosion [MET] **1.** Corrosion of the cathode of an electrochemical circuit, usually caused by production of alkaline reaction products. **2.** Corrosion of the cathodic member of a galvanic couple.

cathodic coating [MET] Material forming a continuous film on a base metal, deposited by mechanical coating or electroplating.

cathodic protection [MET] Protecting a metal from electrochemical corrosion by using it as the cathode of a cell with a sacrificial anode. Also known as electrolytic protection.

catline [PETRO ENG] In oil well drilling, a heavy line which is used for hoisting. Also known as cathead line.

catwalk [ENG] A narrow, raised platform or pathway used for passage to otherwise inaccessible areas, such as a raised walk-way on a ship permitting fore and aft passage when the main deck is awash, a walkway on the roof of a freight car, or a walkway along a vehicular bridge.

caulking compound [MATER] A heavy paste, such as a synthetic, containing a polysulfide rubber and lead peroxide curing agent, or a natural product such as oakum, used for caulking.

caulking iron [DES ENG] A tool for applying caulking to a seam.

caulk Also spelled calk. [ENG] To make a seam or point airtight, watertight, or steamtight by driving in caulking compound, dry pack, lead wool, or other material. [MATER] Material used to caulk seams.

caustic barley See sabadilla.

caustic cracking See caustic embrittlement.

caustic dip [MET] Immersion of metal into a caustic solution such as sodium hydroxide.

caustic embrittlement [MET] Intercrystalline cracking of steel caused by exposure to caustic solutions above 70°C while under tensile stress; once common in riveted boilers. Also known as caustic cracking.

caustic soda [MATER] Sodium hydroxide that contains 76–78% sodium oxide; the most important of the commercial caustic materials, used in chemical manufacture, petroleum refining, and pulp and paper manufacture.

cave [ENG] A pit or tunnel under a glass furnace for collecting ashes or raking the fire. [MIN ENG] **1.** Fragmented rock materials, derived from the sidewalls of a borehole, that obstruct the hole or hinder drilling progress. Also known as cavings. **2.** The partial or complete failure of borehole sidewalls or mine workings. Also known as cave-in.

cave-in See cave.

Cavendish balance [ENG] An instrument for determining the constant of gravitation, in which one measures the displacement of two small spheres of mass m, which are connected by a light rod suspended in the middle by a thin wire, caused by bringing two large spheres of mass M near them.

caving [MIN ENG] A mining procedure, used when the surface is expendable, in which the ore body is undercut and allowed to fall, breaking into small pieces that are recovered by passages (drifts) driven for that purpose; sublevel caving, block caving, and top slicing are examples.

caving ground [MIN ENG] Rock formation that will not stand in the walls of an underground opening without support such as that offered by cementation, casing, or timber.

cavings See cave.

cavitation [ENG] Pitting of a solid surface such as metal or concrete.

cavitation corrosion See cavitation erosion.

cavitation damage See cavitation erosion.

cavitation erosion [MET] Attack of metal surfaces caused by the collapse of cavitation bubbles on the surface of the liquid and characterized by pitting. Also known as cavitation corrosion; cavitation damage.

cavity resonance [ENG ACOUS] The natural resonant vibration of a loudspeaker baffle; if in the audio range, it is evident as unpleasant emphasis of sounds at that frequency.

cavity wall [BUILD] A wall constructed in two separate thicknesses with an air space between; provides thermal insulation. Also known as hollow wall.

CAVU [AERO ENG] An operational term commonly used in aviation, which designates a condition wherein the ceiling is more than 10,000 feet (3048 meters) and the visibility is more than 10 miles (16 kilometers). Derived from ceiling and visibility unlimited.

C chart [IND ENG] A quality-control chart showing number of defects in subgroups of constant size; gives information concerning quality level, its variability, and evidence of assignable causes of variation.

CD-4 sound See compatible discrete four-channel sound.

cedarwood oil [MATER] An essential alcohol-soluble oil obtained from the heartwood of various cedars, the chief components being cedrol and cedrene.

ceiling [BUILD] The covering made of plaster, boards, or other material that constitutes the overhead surface in a room.

ceiling and visibility unlimited See CAVU.

ceiling balloon [AERO ENG] A small balloon used to determine the height of the cloud base; the height is computed from the ascent velocity of the balloon and the time required for its disappearance into the cloud.

ceiling light [ENG] A type of cloud-height indicator which uses a searchlight to project vertically a narrow beam of light onto a cloud base. Also known as ceiling projector.

ceiling projector See ceiling light.

ceilometer [ENG] An automatic-recording cloud-height indicator.

celery seed oil [MATER] A colorless liquid extracted from celery seeds, containing selinene and apiol.

cell [MIN ENG] A compartment in a flotation machine.

celloidin [MATER] A concentrated solution of pyroxylin used principally in microscopy for embedding specimens or for section-cutting.

cellophane [MATER] A thin, transparent sheeting of regenerated cellulose; it is moisture-proof, and sometimes dyed, and used chiefly as food wrapping or as bags for dialysis.

cellular cofferdam [CIV ENG] A cofferdam consisting of interlocking steel-sheet piling driven as a series of interconnecting cells; cells may be of circular type or of straight-wall diaphragm type; space between lines of pilings is filled with sand.

cellular glass [MATER] Sheets or blocks of thermal insulating material for walls and roofs made from pulverized glass that is heated with a gas-forming chemical at the flow temperature of glass. Also known as cellulated glass; foamed glass.

cellular horn See multicellular horn.

cellular plastic [MATER] A type of plastic with apparent density decreased substantially by the presence of numerous cells disposed throughout its mass.

cellular rubber See rubber sponge.

cellular striation [ENG] Stratum of cells inside a cellular-plastic object that differs noticeably from the cell structure of the remainder of the material.

cellulated glass See cellular glass.

cement [MATER] **1.** A dry powder made from silica, alumina, lime, iron oxide, and magnesia which hardens when mixed with water; used as an ingredient in concrete. **2.** An adhesive for the assembling of surfaces which are not in close contact.

cementation [ENG] **1.** Plugging a cavity or drill hole with cement. Also known as dental work. **2.** Consolidation of loose sediments or sand by injection of a chemical agent or binder. [MET] **1.** High-temperature impregnation of a metal surface with another material. **2.** Conversion of wrought iron into steel by packing layers of bars in charcoal sealed with clay and heating to 1000°C for 7–10 days.

cementation factor [PETRO ENG] Mathematical expression in oil-reservoir analysis for the degree to which precipitated minerals have bound together the grains (for example, of sand).

cementation sinking [MIN ENG] A technique of shaft sinking through strata containing water by injecting liquid cement or chemicals into the ground.

cement brick [MATER] A type of brick made from a mixture of cement and sand, molded under pressure and cured under steam at 200°F (93°C); used as backing brick and where there is no danger of attack from acid or alkaline conditions.

cement copper [MET] A precipitate of copper from copper sulfate solution by the addition of iron.

cemented carbide *See* carbide.

cement gun [MECH ENG] **1.** A machine for mixing, wetting, and applying refractory mortars to hot furnace walls. Also known as cement injector. **2.** A mechanical device for the application of cement or mortar to the walls or roofs of mine openings or building walls.

cement injector *See* cement gun.

cementite [MET] Fe_3C A hard, brittle, crystalline compound occurring as lamellae or plates in steel. Also known as iron carbide.

cementitious material [MATER] Any of various building materials which may be mixed with a liquid, such as water, to form a plastic paste, and to which an aggregate may be added; includes cements, limes, and mortar.

cement kiln [ENG] A kiln used to fire cement to less than complete melting.

cement-lined casing [PETRO ENG] Steel-, oil-, or gas-well casing pipe with internal lining of special cement; used to withstand severe corrosive conditions.

cement log [PETRO ENG] Gamma-ray measurement and logging of the height and condition of cement surrounding down-hole oil-well casing.

cement mill [MECH ENG] A mill for grinding rock to a powder for cement.

cement mortar [MATER] A mixture of approximately four parts of sand to one part of portland cement with a small amount of lime and enough water to make it plastic.

cement paint [MATER] A mixture based on portland cement, with filler, accelerator, and water repellent added, that is combined with water and applied to masonry, concrete, or brickwork; provides a waterproof coating.

cement paste [MATER] A mixture of cement and water, hardened or unhardened.

cement plaster [MATER] A gypsum plaster used in mortar mixtures for plastering interior surfaces.

cement pump [MECH ENG] A piston device used to move concrete through pipes.

cement silo [ENG] A silo used to store dry, bulk cement.

cement valve [MECH ENG] A ball-, flapper-, or clack-type valve placed at the bottom of a string of casing, through which cement is pumped, so that when pumping ceases, the valve closes and prevents return of cement into the casing.

center drill [ENG] A two-fluted tool consisting of a twist drill with a 60° countersink; used to drill countersink center holes in a workpiece to be mounted between centers for turning or grinding.

center gage [DES ENG] A gage used to check angles; for example, the angles of cutting tool points or screw threads, or the angular position of cutting tools.

center-gated mold [ENG] A plastics injection mold with the filling orifice interconnected to the nozzle and the center of the cavity area.

centering [CIV ENG] A curved, temporary support for an arch or dome during a casting or laying operations.

centering machine [MECH ENG] A machine for drilling and countersinking work to be turned on a lathe.

centerless grinder [MECH ENG] A cylindrical metal-grinding machine that carries the work on a support or blade between two abrasive wheels.

center of lift [AERO ENG] The mean of all the centers of pressure on an airfoil.

center of pressure [AERO ENG] The point in the chord of an airfoil section which is at the intersection of the chord (prolonged if necessary) and the line of action of the combined air forces (resultant air force).

center of pressure coefficient [AERO ENG] The ratio of the distance of a center of pressure from the leading edge of an airfoil to its chord length.

center plug [DES ENG] A small diamond-set circular plug, designed to be inserted into the annular opening in a core bit, thus converting it to a noncoring bit.

center punch [DES ENG] A tool similar to a prick punch but having the point ground to an angle of about 90°; used to enlarge prick-punch marks or holes.

center square [DES ENG] A straight edge with a sliding square; used to locate the center of a circle.

central breaker [MIN ENG] A breaker where the coal from several mines in a district is prepared.

central control [AERO ENG] The place, facility, or activity at which the whole action incident to a test launch and flight is coordinated and controlled, from the make-ready at the launch site and on the range, to the end of the rocket flight down-range. [SYS ENG] Control exercised over an extensive and complicated system from a single center.

central gear [MECH ENG] The gear on the central axis of a planetary gear train, about which a pinion rotates. Also known as sun gear.

central heating [CIV ENG] The use of a single steam or hot-water heating plant to serve a group of buildings, facilities, or even a complete community through a system of distribution pipes.

centralized traffic control [CIV ENG] Control of train movements by signal indications given by a train director at a central control point. Abbreviated CTC.

central mix concrete [MATER] A concrete prepared at a concrete mixing plant and transported to the building site.

centrifugal atomizer [MECH ENG] Device that atomizes liquids with a spinning disk; liquid is fed onto the center of the disk, and the whirling motion (3000 to 50,000 revolutions per minute) forces the liquid outward in thin sheets to cause atomization.

centrifugal brake [MECH ENG] A safety device on a hoist drum that applies the brake if the drum speed is greater than a set limit.

centrifugal casting [ENG] A method for casting metals or forming thermoplastic resins in which the molten material solidifies in and conforms to the shape of the inner surface of a heated, rapidly rotating container.

centrifugal clarification [MECH ENG] The removal of solids from a liquid by centrifugal action which decreases the settling time of the particles from hours to minutes.

centrifugal classification [MECH ENG] A type of centrifugal clarification purposely designed to settle out only the large particles (rather than all particles) in a liquid by reducing the centrifuging time.

centrifugal classifier [MECH ENG] A machine that separates particles into size groups by centrifugal force.

centrifugal clutch [MECH ENG] A clutch operated by centrifugal force from the speed of rotation of a shaft, as when heavy expanding friction shoes act on the internal surface of a rim clutch, or a flyball-type mechanism is used to activate clutching surfaces on cones and disks.

centrifugal collector [MECH ENG] Device used to separate particulate matter of 0.1–1000 micrometers from an airstream; some types are simple cyclones, high-efficiency cyclones, and impellers.

centrifugal compressor [MECH ENG] A machine in which a gas or vapor is compressed by radial acceleration in an impeller with a surrounding casing, and can be arranged multistage for high ratios of compression.

centrifugal discharge elevator [MECH ENG] A high-speed bucket elevator from which free-flowing materials are discharged by centrifugal force at the top of the loop.

centrifugal fan [MECH ENG] A machine for moving a gas, such as air, by accelerating it radially outward in an impeller to a surrounding casing, generally of scroll shape.

centrifugal filter [ENG] An adaptation of the centrifugal settler; centrifugal action of a spinning container segregates heavy and light materials but heavy materials escape through nozzles as a thick slurry.

centrifugal filtration [MECH ENG] The removal of a liquid from a slurry by introducing the slurry into a rapidly rotating basket, where the solids are retained on a porous screen and the liquid is forced out of the cake by the centrifugal action.

centrifugal governor [MECH ENG] A governor whose flyweights respond to centrifugal force to sense speed.

centrifugal pump [MECH ENG] A machine for moving a liquid, such as water, by accelerating it radially outward in an impeller to a surrounding volute casing.

centrifugal separation [MECH ENG] The separation of two immiscible liquids in a centrifuge within a much shorter period of time than could be accomplished solely by gravity.

centrifugal switch [MECH ENG] A switch opened or closed by centrifugal force; used on some induction motors to open the starting winding when the motor has almost reached synchronous speed.

centrifugal tachometer [MECH ENG] An instrument which measures the instantaneous angular speed of a shaft by measuring the centrifugal force on a mass rotating with it.

centrifuge [MECH ENG] **1.** A rotating device for separating liquids of different specific gravities or for separating suspended colloidal particles, such as clay particles in an aqueous suspension, according to particle-size fractions by centrifugal force. **2.** A large motor-driven apparatus with a long arm, at the end of which human and animal subjects or equipment can be revolved and rotated at various speeds to simulate the prolonged accelerations encountered in rockets and spacecraft.

cepstrum vocoder [ENG ACOUS] A digital device for reproducing speech in which samples of the cepstrum of speech, together with pitch information, are transmitted to the receiver, and are then converted into an impulse response that is convolved with an impulse train generated from the pitch information.

ceramal *See* cermet.

ceramet *See* cermet.

ceramic [MATER] **1.** A product made by the baking or firing of a nonmetallic mineral, such as tile, cement, plaster refractories, and brick. **2.** Consisting of such a product.

ceramic aggregate [MATER] **1.** Portland cement concrete containing lumps of ceramic material. **2.** Concrete made with porous clay to reduce its weight.

ceramic cartridge [ENG ACOUS] A device containing a piezoelectric ceramic element, used in phonograph pickups and microphones.

ceramic coating [MET] A nonmetallic, inorganic coating made of sprayed aluminum oxide or of zirconium oxide, or a cemented coating of an intermetallic compound, such as aluminum disilicide, of essentially crystalline nature, applied as a protective film on metal to protect against temperatures above 1100°C.

ceramic earphones *See* crystal headphones.

ceramic fiber [MATER] A small-dimension filament or thread composed of a ceramic material, usually alumina and silica, used in lightweight units for electrical, thermal, and sound insulation, filtration at high temperatures, packing, and reinforcing other ceramic materials.

ceramic glaze [ENG] A glossy finish on a clay body obtained by spraying with metallic oxides, chemicals, and clays and firing at high temperature.

ceramic microphone [ENG ACOUS] A microphone using a ceramic cartridge.

ceramic mold casting [MET] A precision casting process using a ceramic body fired to high temperature as the mold, and carbon, low-alloy, or stainless steel as the casting.

ceramic pickup [ENG ACOUS] A phonograph pickup using a ceramic cartridge.

ceramic radiant [ENG] A baked-clay component of a gas heating unit which radiates heat when incandescent from the gas flame.

ceramic rod flame spraying [MET] A method of flame spraying in which the ceramic rod is fed into a gun that utilizes an oxyfuel gas flame to atomize and airblast the rod material to the substrate.

ceramics [ENG] The art and science of making ceramic products.

ceramic tile [MATER] A burned-clay product composed of a clay body with a decorative surface glaze; used principally for decorative and sanitary effects.

ceramic tool [DES ENG] A cutting tool made from metallic oxides.

ceramic transducer *See* electrostriction transducer.

ceramoplastic [MATER] A high-temperature insulating material made by bonding synthetic mica with glass.

cereal binder [MATER] A binding material derived from flour; used for core mixtures in a casting process.

ceresin [MATER] **1.** A hydrocarbon wax refined from veins of wax shale known as ozocerite; used in manufacture of candles, shoe polish, electrical insulation, and floor waxes because of its great compatibility with other substances. Also known as ceresine; ozocerite. **2.** A mixture of paraffin wax and beeswax, or a mixture of ozocerite and paraffin.

ceresine *See* ceresin.

ceresin wax *See* paraffin wax.

Cermak-Spirek furnace [ENG] An automatic reverberatory furnace of rectangular form divided into two sections by a wall; used for roasting zinc and quicksilver ores.

cermet [MATER] Any of a group of composite materials made by mixing, pressing, and sintering metal with ceramic; examples are silicon-silicon carbide and chromium-alumina carbide. Also known as ceramal; ceramet; metal ceramic.

certified color *See* food color.

cesium-ion engine [AERO ENG] An ion engine that uses a stream of cesium ions to produce a thrust for space travel.

cesium magnetometer [ENG] A magnetometer that uses a cesium atomic-beam resonator as a frequency standard in a circuit that detects very small variations in magnetic fields.

cesspit *See* cesspool.

cesspool [CIV ENG] An underground tank for raw sewage collection; used where there is no sewage system. Also known as cesspit.

cevedilla *See* sabadilla.

CFR engine [ENG] Cooperative Fuel Research engine, a standard test engine used to determine the octane number of motor fuels.

chafing corrosion *See* fretting corrosion.

chafing fatigue [MET] Fatigue induced by corrosion damage between metal surfaces in close contact under pressure.

chain [CIV ENG] *See* engineer's chain; Gunter's chain. [DES ENG] **1.** A flexible series of metal links or rings fitted into one another; used for supporting, restraining, dragging, or lifting objects or transmitting power. **2.** A mesh of rods or plates connected together, used to convey objects or transmit power.

chain belt [DES ENG] Belt of flat links to transmit power.

chain block [MECH ENG] A tackle which uses an endless chain rather than a rope, often operated from an overhead track to lift heavy weights especially in workshops. Also known as chain fall; chain hoist.

chain bond [CIV ENG] A masonry bond formed with a chain or bar.

chain breat machine [MIN ENG] A coal-cutting machine, so constructed that a series of cutting points attached to a circulating chain work their way for a certain distance under a seam.

chain coal cutter [MIN ENG] A cutter which makes a groove in the coal by an endless chain moving around a flat plate called a jib.

chain conveyor [MECH ENG] A machine for moving materials that carries the product on one or two endless linked chains with crossbars; allows smaller parts to be added as the work passes.

chain course [CIV ENG] A course of stone held together by iron cramps.

chain drive [MECH ENG] A flexible device for power transmission, hoisting, or conveying, consisting of an endless chain whose links mesh with toothed wheels fastened to the driving and driven shafts.

chain fall *See* chain block.

chain-float liquid-level gage [ENG] Float device to measure the level of liquid in a vessel; the float, suspended from a counterweighted chain draped over a toothed sprocket, rises or falls with the liquid level, and the chain movement turns the sprocket to position a calibrated depth-indicator.

chain gear [MECH ENG] A gear that transmits motion from one wheel to another by means of a chain.

chain grate stoker [MECH ENG] A wide, endless chain used to feed, carry, and burn a noncoking coal in a furnace, control the air for combustion, and discharge the ash.

chain hoist *See* chain block.

chain intermittent fillet welding [MET] **1.** The forming of two lines of intermittent fillet welds in a T joint or lap joint so that the increments of welding in one line are approximately opposite those in the other line. **2.** The forming of two lines of equal-length fillet welds concurrently on opposite sides of the perpendicular member of a T joint at intermittent intervals.

chain radar system [ENG] A number of radar stations located at various sites on a missile range to enable complete radar coverage during a missile flight; the stations are linked by data and communication lines for target acquisition, target positioning, or data-recording purposes.

chain riveting [ENG] Riveting consisting of rivets one behind the other in rows along the seam.

chain saw [MECH ENG] A gasoline-powered saw for felling and bucking timber, operated by one person; has cutting teeth inserted in a sprocket chain that moves rapidly around the edge of an oval-shaped blade.

chain tongs [DES ENG] A tool for turning pipe, using a chain to encircle and grasp the pipe.

chain vise [DES ENG] A vise in which the work is encircled and held tightly by a chain.

chainwall [MIN ENG] A coal mining technique in which the mine roof is supported by coal pillars, between which the coal is mined away.

chairs *See* folding boards.

chalcogenide glass [MATER] A type of glass containing large amounts of one of the chalcogens tellurium, selenium, or sulfur; used in glass switches.

chalk [MATER] Artificially prepared pure calcium carbonate; used as the basis for pastels. Also known as whiting.

chalkboard *See* blackboard.

chalking [MET] Defect of coated metals in which a layer of powder forms between the coating and the base metal.

chamber [CIV ENG] The space in a canal lock between the upper and lower gates. [MIN ENG] **1.** The working place of a miner. **2.** A body of ore with definite boundaries apparently filling a preexisting cavern.

chamber capacity *See* chamber volume.

chambering [MIN ENG] Increasing the size of a drill hole in a quarry by firing a succession of small charges, until the hole can take a proper explosive charge to bring down the face of the quarry.

chamber kiln [ENG] A kiln consisting of a series of adjacent chambers in a ring or oval through which the fire moves, taking several days to make a circuit; waste gas from the fire preheats ware in chambers toward which the fire is moving, while combustion air is preheated by ware in chambers already fired.

chamber pressure [AERO ENG] The pressure of gases within the combustion chamber of a rocket engine.

chamber volume [AERO ENG] The volume of the rocket combustion chamber, including the convergent portion of the nozzle up to the throat. Also known as chamber capacity.

chamfer [ENG] To bevel a sharp edge on a machined part.

chamfer angle [DES ENG] The angle that a beveled surface makes with one of the original surfaces.

chamfering [MECH ENG] Machining operations to produce a beveled edge. Also known as beveling.

chamfer plane [DES ENG] A plane for chamfering edges of woodwork.

chance process [MIN ENG] A method for separating clean coal from slate and other impurities in a mixture of sand and water.

change gear [MECH ENG] A gear used to change the speed of a driven shaft while the speed of the driving remains constant.

changing bag [ENG] An enclosure of light-proof material used for operations such as loading of film holders in daylight.

channel [CIV ENG] A natural or artificial waterway connecting two bodies of water or containing moving water. [ENG] The forming of cavities in a gear lubricant at low temperatures because of congealing. [PETRO ENG] In a drilling operation, a cavity appearing behind the casing because of a defect in the cement.

channeler See channeling machine.

channeling machine [MECH ENG] An electrically powered machine that operates by a chipping action of three to five chisels while traveling back and forth on a track; used for primary separation from the rock ledge in marble, limestone, and soft sandstone quarries. Also known as channeler.

channel iron [DES ENG] A metal strip or beam with a U-shape.

channel sample See groove sample.

channel wing [AERO ENG] A wing that is trough-shaped so as to surround partially a propeller to get increased lift at low speeds from the slipstream.

chaplet [MET] Metal support used to space and hold the core in position within a sand mold.

char See low-temperature coke.

characteristic chamber length [AERO ENG] The length of a straight, cylindrical tube having the same volume as that of the chamber of a rocket engine if the chamber had no converging section.

characteristic exhaust velocity [AERO ENG] Of a rocket engine, a descriptive parameter, related to effective exhaust velocity and thrust coefficient. Also known as characteristic velocity.

characteristic velocity See characteristic exhaust velocity.

charcoal [MATER] Also known as char. **1.** A porous solid product containing 85–98% carbon and produced by heating carbonaceous materials such as cellulose, wood, or peat at 500–600°C in the absence of air. **2.** The residue obtained from the carbonization of a noncoking coal, such as subbituminous coal, lignite, or anthracite. **3.** See low-temperature coke.

charge [ENG] **1.** A unit of an explosive, either by itself or contained in a bomb, projectile, mine, or the like, or used as the propellant for a bullet or projectile. **2.** To load a borehole with an explosive. **3.** The material or part to be heated by induction or dielectric heating. **4.** The measurement or weight of material, either liquid, preformed, or powder, used to load a mold at one time during one cycle in the manufacture of plastics or metal.

[MECH ENG] **1.** In refrigeration, the quantity of refrigerant contained in a system. **2.** To introduce the refrigerant into a refrigeration system. [MET] Material introduced into a furnace for melting.

charging line [PETRO ENG] A pipeline for transporting fresh charging stock of crude oil, gas, oil, and such to a still.

charging stock [PETRO ENG] A product introduced into a still; may be any product recovered through previous distillation such as gas oil or fuel oil, or any product selected for further distillation or refining.

Charlton white See lithopone.

Charpy test [MET] An impact test to determine the ductility of a metal; a freely swinging pendulum is allowed to strike and break a notched specimen laid loosely on a support; the work done by the pendulum is obtained by comparing the position of the pendulum before release with the position to which it swings after breaking the specimen.

charring ablator [MATER] An ablation material characterized by the formation of a carbonaceous layer at the heated surface which impedes heat flow into the material by its insulating and reradiating characteristics.

chart comparison unit [ENG] A device that permits simultaneous viewing of a radar plan position indicator display and a navigation chart so that one appears superimposed on the other. Also known as autoradar plot.

chart datum See datum plane.

chart desk [ENG] A flat surface on which charts are spread out, usually with storage space for charts and other navigating equipment below the plotting surface.

chart recorder [ENG] A recorder in which a dependent variable is plotted against an independent variable by an ink-filled pen moving on plain paper, a heated stylus on heat-sensitive paper, a light beam or electron beam on photosensitive paper, or an electrode on electrosensitive paper. The plot may be linear or curvilinear on a strip chart recorder, or polar on a circular chart recorder.

chart table [ENG] A flat surface on which charts are spread out, particularly one without storage space below the plotting surface, as in aircraft and VPR (virtual PPI reflectoscope) equipment.

chase [BUILD] A vertical passage for ducts, pipes, or wires in a building. [DES ENG] A series of cuts, each having a path that follows the path of the cut before it; an example is a screw thread. [ENG] **1.** The main body of the mold which contains the molding cavity or cavities. **2.** The enclosure used to shrink-fit parts of a mold cavity in place to prevent spreading or distortion, or to enclose an assembly of two or more parts of a split-cavity block. **3.** To straighten and clean threads on screws or pipes.

chase mortise [DES ENG] A mortise with a sloping edge from bottom to surface so that a tenon can be inserted when the outside clearance is small.

chase pilot [AERO ENG] A pilot who flies an escort airplane and advises another pilot who is making a check, training, or research flight in another craft.

chaser [AERO ENG] The vehicle that maneuvers in order to effect a rendezvous with an orbiting object. [ENG] A thread-cutting tool with many teeth.

chase ring [MECH ENG] In hobbing, the ring which restrains the blank from spreading during hob sinking.

chasing tool [DES ENG] A hammer or chisel used to decorate metal surfaces.

chassis [ENG] **1.** A frame on which the body of an automobile or airplane is mounted. **2.** A frame for mounting the working parts of a radio or other electronic device.

chassis lubricant [MATER] A lubricating grease of consistency to be applied with a grease gun through fittings on autos and farm and industrial equipment.

chassis punch [DES ENG] A hand tool used to make round or square holes in sheet metal.

chatter [ENG] An irregular alternating motion of the parts of a relief valve due to the application of pressure where contact is made between the valve disk and the seat. [ENG ACOUS] Vibration of a disk-recorder cutting stylus in a direction other than that in which it is driven.

chaulmoogra oil [MATER] Any of several fixed oils extracted from seeds of trees in the family Flacourtiaceae; widely used at one time to treat leprosy and other diseases.

check [MATER] A lengthwise crack in a board.

check dam [CIV ENG] A low, fixed structure, constructed of timber, loose rock, masonry, or concrete, to control water flow in an erodable channel or irrigation canal.

checkerboard regenerator [ENG] An open-checkerwork arrangement of firebrick in a high-temperature chamber that absorbs heat during a batch processing cycle, then releases it to preheat fresh combustion air during the down cycle; used, for example, in the steel industry with open-hearth and heat-treating furnaces.

checkers [ENG] Open brickwork in a checkerboard regenerator allowing for the passage of hot, spent gases.

check flight [AERO ENG] **1.** A flight made to check or test the performance of an air-

craft, rocket, or spacecraft, or a piece of its equipment, or to obtain measurements or other data on performance. **2.** A familiarization flight in an aircraft, or a flight in which the pilot or the aircrew are tested for proficiency.

checking [MATER] Fine cracks appearing in a ceramic coating. [MET] Temporarily reducing the volume or temperature of the air blast in a blast furnace.

checkout [ENG] A sequence of actions to test or examine a thing as to its readiness for incorporation into a new phase of use or as to the performance of its intended function.

check rail [BUILD] A rail, thicker than the window, that spans the opening between the top and bottom sash; usually beveled and rabbeted. [CIV ENG] See guard rail.

check screen See oversize control screen.

check valve [ENG] A device for automatically limiting flow in a piping system to a single direction.

cheek [MET] Portion of a three-part flask between the cope and the drag.

cheese cement [MATER] A glue made from cheese or milk curd.

cheese head [DES ENG] A raised cylindrical head on a screw or bolt.

Chemag process [MET] A process for oxide-coating iron and steel by electrolytic means, the object to be coated being the anode in an alkaline solution.

chemical agent [MATER] A solid, liquid, or gas employed in three principal categories: war gases, smokes, and incendiaries; through its chemical properties produces lethal, injurious, or irritant effects on humans, makes a screening or colored smoke, or acts as a fire starter.

chemical-cartridge respirator [MIN ENG] An air purification device worn by miners that removes small quantities of toxic gases or vapors from the inspired air; the cartridge contains chemicals which operate by processes of oxidation, absorption, or chemical reaction.

chemical conversion coating [MET] A protective or decorative coating formed on the surface of a metal as the result of chemical reaction of the metal with a selected environment.

chemical engineering [ENG] That branch of engineering serving those industries that chemically convert basic raw materials into a variety of products, and dealing with the design and operation of plants and equipment to perform such work; all products are formed in chemical processes involving chemical reactions carried out under a wide range of conditions and frequently accompanied by changes in physical state or form.

chemical etching [MET] Formation of characteristic surface features when a polished metal surface is etched by suitable reagents.

chemical flux cutting [MET] An oxygen cutting process in which metals are cut by using flux.

chemically foamed plastic [MATER] A foamed plastic having its cellular structure produced by gases generated from chemical reaction of the components.

chemical machining [MET] Making of metal parts to specified dimensions by removing surface metal with chemicals (acids or alkalies). Also known as chemical milling.

chemical metallurgy [MET] The science and technology of extracting metals from ores and refining them. Also known as process metallurgy.

chemical milling *See* chemical machining.

chemical polishing [MET] Smoothing and brightening the surface of a metal by treatment with a chemical agent.

chemical porcelain [MATER] High-purity, nonporous grade of porcelain used to make laboratory analysis utensils, such as crucibles, retorts, and spatulas.

chemical pressurization [AERO ENG] The pressurization of propellant tanks in a rocket by means of high-pressure gases developed by the combustion of a fuel and oxidizer or by the decomposition of a substance.

chemical pulp [MATER] Wood pulp made by separating the fibers of wood chips by the action of alkalies or acids.

chemical pump [PETRO ENG] Skid-mounted pumping unit used to feed chemicals into the power oil (used to operate bottom-hole pumps in oil wells) to reduce corrosion in the system and to assist in water removal when the power oil and well-produced oil reach the ground-level wash tank.

chemical resistance [MATER] Ability of solid materials to resist damage by chemical reactivity or solvent action.

chemical sterilization [ENG] The use of bactericidal chemicals to sterilize solutions, air, or solid surfaces.

chemical stoneware [MATER] Clay pottery material that resists acids and alkalies; used for ball mills, pipes, laboratory sinks and utensils, and so on.

chemical thermometer [ENG] A filled-system temperature-measurement device in which gas or liquid enclosed within the device responds to heat by a volume change (rising or falling of mercury column) or by a pressure change (opening or closing of spiral coil).

chemimechanical pulp [MATER] Plant material treated by the sulfite, soda, or sulfate process for papermaking.

chemonite [MATER] A wood preservative consisting of a water solution of small percentages of copper hydroxide, arsenic trioxide, ammonia, and acetic acid.

chenopodium oil [MATER] An alcohol-soluble, colorless to yellow oil, derived from the herb *Chenopodium ambrosioides*; chief constituents are *para*-cymeme, 1-limonene, and ascaridole; used in medicine. Also known as American wormseed oil; goosefoot oil.

cherry picker [AERO ENG] A crane used to remove the aerospace capsule containing astronauts from the top of the rocket in the event of a malfunction. [MECH ENG] Any of several small traveling cranes, especially one used to hoist a passenger on the end of a boom. [MIN ENG] A small hoist used to facilitate car changing near the loader in a mine tunnel.

Chicago boom [MECH ENG] A hoisting device that is supported on the structure being erected.

Chicago caisson [CIV ENG] A cofferdam about 4 feet (1.2 meters) in diameter lined with planks and sunk in medium-stiff clays to hard ground for pier foundations. Also known as open-well caisson.

chicle [MATER] A gummy exudate obtained from the bark of *Achras zapota*, an evergreen tree belonging to the sapodilla family (Sapotaceae); used as the principal ingredient of chewing gum.

Chile mill [MECH ENG] A crushing mill having vertical rollers running in a circular enclosure with a stone or iron base or die. Also known as edge runner.

chill [MET] **1.** A metal plate inserted in the surface of a sand mold or placed in the mold cavity to rapidly cool and solidify the casting, producing a hard surface. **2.** White or mottled iron occurring on the surface of a rapidly cooled gray iron casting.

chill-block melt spinning [MET] A rapid quenching process in which a jet of molten metal is directed onto a cold moving surface, such as a spinning disk, where the jet is shaped and solidified; quench rates are 1000 to 1,000,000 K/s.

chilled iron [MET] Cast iron made in iron- or steel-faced molds so the surface of the casting cools rapidly, retaining most of the carbon and becoming white and hard.

chilled roll [MET] A roll consisting of an outer hard layer of white (chilled) iron with a middle transitional layer of mottled iron and a core of full gray iron.

chilled shot [MET] Lead shots containing 3–6% antimony.

chilling [MET] Rapidly removing the heat from a casting.

chill roll [ENG] A cored roll used in chill-roll extrusion of plastics.

chill-roll extrusion [ENG] Method of extruding plastic film in which the film is cooled while being drawn around two or more highly polished chill rolls, inside of which there is cooling water. Also known as cast-film extrusion.

chimney [BUILD] A vertical, hollow structure of masonry, steel, or concrete, built to convey gaseous products of combustion from a building.

chimney bar [BUILD] A wrought-iron or steel lintel which is supported by the sidewalls and carries the masonry above the fireplace opening. Also known as turning bar.

chimney core [MECH ENG] The inner section of a double-walled chimney which is separated from the outer section by an air space.

chimney rock [MATER] A porous phosphate rock used principally in chimney construction.

china clay [MATER] A high-grade white kaolin composed principally of the mineral kaolinite, and often occurring as a lenticular-shaped body; used in the manufacture of ceramics, paper, rubber, catalysts, and ink.

China oil *See* Peru balsam.

China wood oil *See* tung oil.

Chinese bean oil *See* soybean oil.

Chinese ink *See* india ink.

Chinese oil *See* Peru balsam.

Chinese oil of cinnamon *See* cassia oil.

Chinese wax [MATER] A white or yellowish crystalline wax formed on certain trees by the secretions of a scale insect, especially *Ceroplastes ceriferus.*

chipboard [MATER] A low-density paper board made from mixed waste paper and used where strength and quality are needed.

chip breaker [DES ENG] An irregularity or channel cut into the face of a lathe tool behind the cutting edge to cause removed stock to break into small chips or curls.

chip cap [DES ENG] A plate or cap on the upper part of the cutting iron of a carpenter's plane designed to give the tool rigidity and also to break up the wood shavings.

chip log [ENG] A line, marked at intervals (commonly 50 feet or 15 meters), that is paid out over the stern of a moving ship and is pulled out by a drag (the chip), to determine the ship's speed.

chipper [ENG] A tool such as a chipping hammer used for chipping.

chipping [MET] Removing seams, surface defects, or other excess fragments from semifinished metal products by using a manual or pneumatic chisel or a continuous machine.

chipping hammer [ENG] A hand or pneumatic hammer with chisel-shaped or pointed faces used to remove rust and scale from metal surfaces.

chip sampling [MIN ENG] Taking small pieces of ore or coal from the width of an ore face exposure; may be done at random or along a line.

chirp radar [ENG] Radar in which a swept-frequency signal is transmitted, received from a target, then compressed in time to give a narrow pulse called the chirp signal.

chisel [DES ENG] A tool for working the surface of various materials, consisting of a metal bar with a sharp edge at one end and often driven by a mallet.

chisel bit *See* chopping bit.

chisel bond [ENG] A thermocompression bond in which a contact wire is attached to a contact pad on a semiconductor chip by applying pressure with a chisel-shaped tool.

chisel-edge angle [DES ENG] The angle included between the chisel edge and the cutting edge, as seen from the end of the drill.

chisel-tooth saw [DES ENG] A circular saw with chisel-shaped cutting edges.

chlorate candle [MATER] A mixture of solid chemical compounds which, when ignited, liberates free oxygen.

chlorate explosive [MATER] A type of explosive with a potassium chlorate base; a substitute for blackpowder in which potassium chlorate is used in place of potassium nitrate. Also known as chlorate powder.

chlorate powder *See* chlorate explosive.

chloride of lime *See* bleaching powder.

chloride paper [MATER] A paper made with an emulsion of silver chloride; usually used in photography as contact paper or very-slow-speed enlarging paper.

chlorinated lime *See* bleaching powder.

chlorinated rubber [MATER] A nonrubbery, incombustible rubber derivative produced by the action of chlorine on rubber in solution; used in corrosion-resistant paints and varnishes, and in inks and adhesives.

chlorine log *See* chlorinolog.

chlorinolog [PETRO ENG] A record of the presence and concentration of chlorine in oil reservoirs, prepared as a method of locating salt-water strata. Also known as chlorine log.

chock [MIN ENG] A square pillar for supporting the roof in a mine, constructed of prop timber laid up in alternate cross layers, in log-cabin style, the center being filled with waste.

choke [MECH ENG] **1.** To increase the fuel feed to an internal combustion engine through the action of a choke valve. **2.** *See* choke valve. [PETRO ENG] A removable nipple inserted in a flow line to control oil or gas flow.

choke crushing [MIN ENG] A recrushing of fine ore.

chokedamp *See* blackdamp

choked neck [DES ENG] Container neck which has a narrowed or constricted opening.

choke valve [MECH ENG] A valve which supplies the higher suction necessary to give the excess fuel feed required for starting a cold internal combustion engine. Also known as choke.

chopper [ENG] Any knife, axe, or mechanical device for chopping or cutting an object into segments.

chopping bit [MECH ENG] A steel bit with a chisel-shaped cutting edge, attached to a string of drill rods to break up, by impact, boulders, hardpan, and a lost core in a drill hole. Also known as chisel bit.

chop-type feeder [MECH ENG] Device for semicontinuous feed of solid materials to a process unit, with intermittent opening and closing of a hopper gate (bottom closure) by a control arm actuated by an eccentric cam.

chord [AERO ENG] **1.** A straight line intersecting or touching an airfoil profile at two points. **2.** Specifically, that part of such a line between two points of intersection. [CIV ENG] The top or bottom, generally horizontal member of a truss.

chordal thickness [DES ENG] The tangential thickness of a tooth on a circular gear, as measured along a chord of the pitch circle.

chord length [AERO ENG] The length of the chord of an airfoil section between the extremities of the section.

Christmas tree [PETRO ENG] An assembly of valves, tees, crosses, and other fittings at the wellhead, used to control oil or gas production and to give access to the well tubing.

chromadizing [MET] Treating the surface of aluminum or aluminum alloys with chromic acid to improve paint adhesion.

chromate treatment [MET] Treatment of metal with a solution of a hexavalent chromium compound to produce a protective coating of metal chromate.

chromating [MET] Performing a chromate treatment.

chrome brick *See* chrome refractory.

chrome leather [MATER] A leather tanned with chromium salts and used in making shoe uppers.

chrome plating [MET] A thin plate of chromium deposited by electrolysis on a corrodible metal, giving a bright, metallic surface which is highly resistant to tarnish; used to coat automobile trimming, bathroom fixtures, and many household and other articles. Also known as chromium coating; chromium plating.

chrome refractory [MATER] A ceramic material made from chrome ore and used to line steel furnaces. Also known as chrome brick.

chrome steel *See* chromium steel.

chrome-vanadium steel *See* chromium-vanadium steel.

chromium [MET] A blue-white, hard, brittle metal used in chrome plating, in chromizing, and in many alloys.

chromium coating *See* chrome plating.

chromium-iron alloy [MET] Any of several acid- and corrosion-resistant alloys containing chromium and iron.

chromium molybdenum steel [MET] Cast steel containing up to 1% carbon, 0.7–1.1% chromium, and 0.2–0.4% molybdenum; characterized by high strength and ductility.

chromium-nickel alloy [MET] Any of several alloys containing chromium and nickel in various proportions together with small amounts of other metals.

chromium plating *See* chrome plating.

chromium steel [MET] Hard, wear-resistant steel containing chromium as the predominating alloying element. Also known as chrome steel.

chromium-vanadium steel [MET] Any of several strong, hard alloy steels containing 0.15–0.25% vanadium, 0.50–1% chromium, and 0.45–0.55% carbon. Also known as chrome-vanadium steel.

chromizing [MET] Surface-alloying of metals in which an alloy is formed by diffusion of chromium into the base metal.

chromoradiometer [ENG] A radiation meter that uses a substance whose color changes with x-ray dosage.

chronocyclegraph [IND ENG] A device used in micromotion studies to record a complete work cycle by taking still pictures with long exposures, the motion paths being traced by small electric lamps fastened to the worker's hands or fingers; time is obtained by interrupting the light circuits with a controlled frequency which produces dots on the film.

chronograph [ENG] An instrument used to register the time of an event or graphically record time intervals such as the duration of an event.

chronometric data [ENG] Data in which the desired quantity is the time of occurrence of an event or the time interval between two or more events.

chronometric radiosonde [ENG] A radiosonde whose carrier wave is switched on and off in such a manner that the interval of time between the transmission of signals is a function of the magnitude of the meteorological elements being measured.

chronometric tachometer [ENG] A tachometer which repeatedly counts the revolutions during a fixed interval of time and presents the average speed during the last timed interval.

chronothermometer [ENG] A thermometer consisting of a clock mechanism whose speed is a function of temperature; automatically calculates the mean temperature.

chuck [DES ENG] A device for holding a component of an instrument rigid, usually by means of adjustable jaws or set screws, such as the workpiece in a metalworking or woodworking machine, or the stylus or needle of a phonograph pickup. [MET] A small bar between flask bars to secure the sand in the upper box (cope) of a flask.

chucking [MECH ENG] The grasping of an outsize workpiece in a chuck or jawed device in a lathe.

chucking lug [MET] A projection forged or cast onto a piece of metal that functions as a location marker when the work is being machined.

chucking machine [MECH ENG] A lathe or grinder in which the outsize workpiece is grasped in a chuck or jawed device.

chuffing *See* chugging.

chugging [AERO ENG] Also known as bumping; chuffing. **1.** A form of combustion instability in a rocket engine, characterized by a pulsing operation at a fairly low frequency, sometimes defined as occurring between particular frequency limits. **2.** The noise that is made in this kind of combustion.

churn drill [MECH ENG] Portable drilling equipment, with drilling performed by a heavy string of tools tipped with a blunt-edge chisel bit suspended from a flexible cable, to which a reciprocating motion is imparted by its suspension from an oscillating beam or sheave, causing the bit to be raised and dropped. Also known as American system drill; cable-system drill.

churn shot drill [MECH ENG] A boring rig with both churn and shot drillings.

chute [ENG] A conduit for conveying free-flowing materials at high velocity to lower levels.

chute conveyor *See* jigging conveyor.

chute spillway [CIV ENG] A spillway in which the water flow passes over a crest into a sloping, lined, open channel; used for earth and rock-fill dams.

chute system [MIN ENG] A method of mining by which ore is broken from the surface downward into chutes and is removed through passageways below. Also known as glory hole system; milling system.

Chworinov rule [MET] The postulation that total freezing time for a casting is a function of the ratio of volume to surface area.

C.I. *See* cast iron.

cinder [MATER] Slag from a metal furnace. [MET] Scale cast off in forging metal.

cinder block [MATER] A hollow block made of cinder concrete. [MET] A block which closes the front of a blast furnace, containing the cinder notch.

cinder concrete [MATER] A concrete containing cinders as the aggregate.

cinder notch [MET] An opening in a blast furnace that allows molten slag to flow out.

cinder pig [MET] Pig iron produced from a mixture of slag in the furnace and crude metal or ore.

cinders [MATER] Incombustible residue from a burning process; in particular, small pieces of clinker from the burning of soft coal.

cinetheodolite [ENG] A surveying theodolite in which 35-millimeter motion picture cameras with lenses of 60- to 240-inch (1.5 to 6.1 meter) focal length are substituted for the surveyor's eye and telescope; used for precise time-correlated observation of distant airplanes, missiles, and artificial satellites.

Cipolletti weir [CIV ENG] Trapezoidal weir in which the sides of the notch slope are one horizontal to four vertical; used to measure water flow in open channels, especially streams and rivers.

circle error probable *See* circle of equal probability.

circle haul [MIN ENG] A haulage system in strip mining; empty units enter the mine over one lateral and leave, loaded, over the lateral nearest the tipple.

circle of equal probability [AERO ENG] A measure of the accuracy with which a rocket or missile can be guided; the radius of the circle at a specific distance in which 50% of the reliable shots land. Also known as circle of probable error; circular error probable.

circle of probable error *See* circle of equal probability.

circle shear [MECH ENG] A shearing machine that cuts circular disks from a metal sheet rolling between the cutting wheels.

circular burner [ENG] A fuel burner having a round opening.

circular channel [ENG] Continuous-length opening with circular cross section through which liquid or gas can be made to flow.

circular-chart recorder [ENG] Graphic pen-and-ink recorder where measured values are drawn onto a rotating circular chart by the backward and forward movement of a pivoted pen actuated by the input signal (such as temperature, pressure, flow, or force) from an instrument transmitter.

circular cutter [MECH ENG] A rotating blade with a square or knife edge used to slit or shear metal.

circular form tool [DES ENG] A round or disk-shaped tool with the cutting edge on the periphery.

circular pitch [DES ENG] The linear measure in inches along the pitch circle of a gear between corresponding points of adjacent teeth.

circular plane [DES ENG] A plane that can be adjusted for convex or concave surfaces.

circular saw [MECH ENG] Any of several power tools for cutting wood or metal, having a thin steel disk with a toothed edge that rotates on a spindle.

circular scanning [ENG] Radar scanning in which the direction of maximum radiation describes a right circular cone.

circular shaft [MIN ENG] A shaft excavated in a round shape.

circulated gas-oil ratio [PETRO ENG] The volume (cubic feet) of gas introduced into a well during gas-lift operations in comparison with the volume (barrel) of oil that is lifted.

circulating fluid [ENG] A fluid pumped into a borehole through the drill stem, the flow of which cools the bit and transports the cuttings out of the borehole.

circulating scrap [MET] At steelworks and founderies, scrap arising during the manufacture of finished iron and steel or of castings.

circulation area [BUILD] The area required for human traffic in a building, including permanent corridors, stairways, elevators, escalators, and lobbies.

circumferentor [ENG] A horizontal compass used in surveying that has arms diametrically placed with vertical slit sights in them.

cistern [CIV ENG] A tank for storing water or other liquid.

cistern barometer [ENG] A pressure-measuring device in which pressure is read by the liquid rise in a vertical, closed-top tube as a result of system pressure on a liquid reservoir (cistern) into which the bottom, open end of the tube is immersed.

citronella oil [MATER] A yellowish oil distilled from the leaves of either of two grasses, *Cymbopogon nardus* or *C. win-terianus*; used as an insect repellent. Also known as Java citronella oil.

civil engineering [ENG] The planning, design, construction, and maintenance of fixed structures and ground facilities for industry, transportation, use and control of water, or occupancy.

cladding [ENG] Process of covering one material with another and bonding them together under high pressure and temperature. Also known as bonding.

clad metal [MET] A metal overlaid on one or both sides with a different metal.

claim *See* mining claim.

clamp [DES ENG] A tool for binding or pressing two or more parts together, by holding them firmly in their relative positions.

clamping coupling [MECH ENG] A coupling with a split cylindrical element which clamps the shaft ends together by direct compression, through bolts or rings, and by the wedge action of conical sections; not considered a permanent part of the shaft.

clamping plate [ENG] A plate on a mold which attaches the mold to a machine.

clamping pressure [ENG] In injection and transfer-molding of plastics, the pressure applied to keep the mold closed in opposition to the fluid pressure of the molding material.

clamp screw [DES ENG] A screw that holds a part by forcing it against another part.

clamp screw sextant [ENG] A marine sextant having a clamp screw for controlling the position of the tangent screw.

clamshell bucket [MECH ENG] A two-sided bucket used in a type of excavator to dig in a vertical direction; the bucket is dropped while its leaves are open and digs as they close. Also known as clamshell grab.

clamshell grab *See* clamshell bucket.

clamshell snapper [MECH ENG] A marine sediment sampler consisting of snapper jaws and a footlike projection which, upon striking the bottom, causes a spring mechanism to close the jaws, thus trapping a sediment sample.

clapboard [MATER] A board, thicker at one edge than the other, used to cover exterior walls.

clapper box [MECH ENG] A hinged device that permits a reciprocating cutting tool (as in a planer or shaper) to clear the work on the return stroke.

clarified oil [MATER] The heavy oil which is taken from the bottom of a fractionator in a catalytic cracking process and from which residual catalyst has been removed.

clarifier [ENG] A device for filtering a liquid.

clarifying centrifuge [MECH ENG] A device that clears liquid of foreign matter by centrifugation.

clarifying filter [ENG] Any filter, such as a sand filter or a cartridge filter, used to purify liquids with a low solid-liquid ratio; in some instances color may be removed as well.

Clarke's soap solution [MATER] Reagent used in standard APHA (American Pharmaceutical Association) method to estimate hardness in water; consists of powdered castile soap in 80% ethyl alcohol solution.

clary sage oil *See* oil of sassafrass.

clasp [DES ENG] A releasable catch which holds two or more objects together.

clasp lock [DES ENG] A spring lock with a self-locking feature.

clasp nut [DES ENG] A split nut that clasps a screw when closed around it.

class A push-pull sound track [ENG ACOUS] Two single photographic sound tracks side by side, the transmission of one being 180° out of phase with the transmission of the other; both positive and negative halves of the sound wave are linearly recorded on each of the two tracks.

class B push-pull sound track [ENG ACOUS] Two photographic sound tracks side by side, one of which carries the positive half of the signal only, and the other the negative half; during the inoperative half-cycle, each track transmits little or no light.

classification [ENG] **1.** Sorting out or categorizing of particles or objects by established criteria, such as size, function, or color. **2.** Stratification of a mixture of various-sized particles (that is, sand and gravel), with the larger particles migrating to the bottom. [IND ENG] *See* grading.

classification track [CIV ENG] A railroad track used to separate cars from a train according to destination.

classification yard [CIV ENG] A railroad yard for separating trains according to car destination.

classifier [MECH ENG] Any apparatus for separating mixtures of materials into their constituents according to size and density.

claw [DES ENG] A fork for removing nails or spikes.

claw clutch [MECH ENG] A clutch consisting of claws that interlock when pushed together.

claw coupling [MECH ENG] A loose coupling having projections or claws cast on each face which engage in corresponding notches in the opposite faces; used in situations in which shafts require instant connection.

claw hammer [DES ENG] A woodworking hammer with a flat working surface and a claw to pull nails.

clay [MATER] A special grade of absorbent clay used as a filtering medium in refineries for removing solids or colorizing matter from lubricating oils.

clay atmometer [ENG] An atmometer consisting of a porous porcelain container connected to a calibrated reservoir filled with distilled water; evaporation is determined by the depletion of water.

clay bit [DES ENG] *See* mud auger. [ENG] A bit designed for use on a clay barrel.

clay brick [MATER] Brick made from diverse types of clays and used for normal constructional purposes.

clay digger [MECH ENG] A power-driven, hand-held spade for digging hard soil or soft rock.

clay press [ENG] A press used to remove excess water from a pottery-clay slurry.

clay slip [MATER] A slurry of clay and water used in glazing pottery.

clay wash [MATER] A light oil such as naphtha or kerosine used to clean fuller's earth after it has been used in a filter.

cleanout auger *See* cleanout jet auger.

cleanout jet auger [ENG] An auger equipped with water-jet orifices designed to clean out collected material inside a driven pipe or casing before taking soil samples from strata below the bottom of the casing. Also known as cleanout auger.

clean room [ENG] A room in which elaborate precautions are employed to reduce dust particles and other contaminants in the air, as required for assembly of delicate equipment.

cleanser [MATER] Any material used to remove dirt, soil, and impurities from surfaces of all kinds.

clean track [ENG ACOUS] A sound track having no leakage from other tracks.

cleanup [AERO ENG] Improving the external shape and smoothness of an aircraft to reduce its drag. [ENG] The time required for a leak-testing system to reduce its signal output to 37% of the signal transmitted at the instant when tracer gases enter the system. [MIN ENG] **1.** The collecting of all the valuable product of a given period of operation in a stamp mill or in a hydraulic or placer mine. **2.** The valuable material resulting from a cleanup.

clearance [ENG] Unobstructed space required for occasional removal of parts of equipment. [MECH ENG] **1.** In a piston-and-cylinder mechanism, the space at the end of the cylinder when the piston is at dead-center position toward the end of the cylinder. **2.** The ratio of the volume of this space to the piston displacement during a stroke. [MIN ENG] The space between the top or side of a car and the roof or wall. [PETRO ENG] The annular space between down-hole drill-string equipment, such as bits, core barrels, and casing, and the walls of the borehole with the down-hole equipment centered in the hole.

clearance angle [MECH ENG] The angle between a plane containing the end surface of a cutting tool and a plane passing through the cutting edge in the direction of cutting motion.

clearance volume [MECH ENG] The volume remaining between piston and cylinder when the piston is at top dead center.

clear gasoline [MATER] Gasoline which is free from antiknock additives.

cleat [CIV ENG] A strip of wood, metal, or other material fastened across something to serve as a batten or to provide strength or support. [DES ENG] A fitting having two horizontally projecting horns around which a rope may be made fast.

cleft weld [MET] A weld in which a V-shaped projection on one piece is joined to a V-shaped groove in the other.

clevis [DES ENG] A U-shaped metal fitting with holes in the open ends to receive a bolt or pin; used for attaching or suspending parts. [MIN ENG] A spring hook or snap hook which, in coal mining, is used to attach the bucket to the hoisting rope. Also known as clivvy.

clevis pin [DES ENG] A fastener with a head at one end, used to join the ends of a clevis.

click [ENG ACOUS] A perforation in a sound track which produces a clicking sound when passed over the projector sound head.

click track [ENG ACOUS] A sound track containing a series of clicks, which may be spaced regularly (uniform click track) or irregularly (variable click track).

climate control See air conditioning.

climb [AERO ENG] The gain in altitude of an aircraft.

climb cutting [MET] A milling technique in which the teeth of a cutting tool advance into the work in the same direction as the feed. Also known as climb milling; down cutting; down milling.

climbing crane [MECH ENG] A crane used on top of a high-rise construction that ascends with the building as work progresses.

climbing irons [DES ENG] Spikes attached to a steel framework worn on shoes to climb wooden utility poles and trees.

climb milling See climb cutting.

clinical thermometer [ENG] A thermometer used to accurately determine the temperature of the human body; the most common type is a mercury-in-glass thermometer, in which the mercury expands from a bulb into a capillary tube past a constriction that prevents the mercury from receding back into the bulb, so that the thermometer registers the maximum temperature attained.

clinker [MATER] An overburned brick.

clinker building [DES ENG] A method of building ships and boilers in which the edge of the wooden planks or steel plates used for the outside covering overlap the edge of the plank or plate next to it; clinched nails fasten the planks together, and rivets fasten the steel plates.

clinograph [ENG] A type of directional surveying instrument that records photographically the direction and magnitude of deviations from the vertical of a borehole, well, or shaft; the information is obtained by the instrument in one trip into and out of the well.

clinometer [ENG] **1.** A hand-held surveying device for measuring vertical angles; consists of a sighting tube surmounted by a graduated vertical arc with an attached level bubble; used in meteorology to measure cloud height at night, in conjunction with a ceiling light, and in ordnance for boresighting. **2.** A device for measuring the amount of roll aboard ship.

clip [DES ENG] A device that fastens by gripping, clasping, or hooking one part to another.

clip and shave [MET] Dual forging operation in which one cutter removes the flash and then another cutter shaves and sizes the piece.

clip bond [CIV ENG] A bond in which the inner edge of face brick is cut off so that bricks laid diagonal to a wall can be joined to those laid parallel to it.

clipping edge [MET] Area of a forging where flash is removed.

clivvy See clevis.

clo [ENG] The amount of insulation which will maintain normal skin temperature of the human body when heat production is 50 kilogram-calories per meter squared per hour, air temperature is 70°F (21°C), and the air is still.

clock drive [ENG] The mechanism that causes an equatorial telescope to revolve about its polar axis so that it keeps the same star in its field of view.

clock valve [ENG] A hinged valve that permits flow in one direction only.

close-control radar [ENG] Ground radar used with radio to position an aircraft over a target that is normally difficult to locate or is invisible to the pilot.

close-coupled pump [MECH ENG] Pump with built-in electric motor (sometimes a steam turbine), with the motor drive and pump impeller on the same shaft.

closed-belt conveyor [MECH ENG] Solids-conveying device with zipperlike teeth that mesh to form a closed tube wrapped snugly around the conveyed material; used with fragile materials.

closed-cell foam [MATER] A cellular plastic in which there is a predominance of non-interconnecting cells.

closed-circuit grinder [MIN ENG] A grinder connected to a size classifier (cyclone or screen) to return oversized particles to the grinding operation in closed-circuit pulverizing.

closed-circuit pulverizing [MIN ENG] A process used in ore dressing in which the material discharged from the pulverizer is passed through an external classifier where the finished product is removed and oversized particles are returned to the pulverizer.

closed-cycle turbine [MECH ENG] A gas turbine in which essentially all the working medium is continuously recycled, and heat is transferred through the walls of a closed heater to the cycle.

closed die [MET] A forming or forging die in which the flow of metal is restricted to the cavity of the die set.

closed ecological system [AERO ENG] A system used in spacecraft that provides for the maintenance of life in an isolated living chamber through complete reutilization of the material available, in particular, by means of a cycle wherein exhaled carbon dioxide, urine, and other waste matter are converted chemically or by photosynthesis into oxygen, water, and food.

closed fireroom system [MECH ENG] A fireroom system in which combustion air is supplied via forced draft resulting from positive air pressure in the fireroom.

closed frame [MIN ENG] A mine support frame that is completely closed; especially in inclined shafts, it is used to protect all sides from rock pressure.

closed-loop telemetry system [ENG] **1.** A telemetry system which is also used as the display portion of a remote-control system. **2.** A system used to check out test vehicle or telemetry performance without radiation of radio-frequency energy.

closed nozzle [MECH ENG] A fuel nozzle having a built-in valve interposed between the fuel supply and combustion chamber.

closed pass [MET] A metal-rolling operation in which the top roll has a collar that fits a groove on the bottom roll, allowing a flash-free shape to be formed in the rolled metal.

closed respiratory gas system [ENG] A self-contained system within a sealed cabin, capsule, or spacecraft that will provide adequate oxygen for breathing, maintain adequate cabin pressure, and absorb the exhaled carbon dioxide and water vapor.

closed rotative gas lift [PETRO ENG] Oil-well control system in which high-pressure compressor gas is injected into a well to force oil fluids from the reservoir, with spent lift gas recompressed for reinjection.

closed shop [IND ENG] An establishment permitting only union members to be employed.

closed steam [ENG] Steam that flows through a heating coil or annulus so that there is no direct contact between the steam and the material being heated.

close-grained [MATER] Consisting of fine, closely spaced particles, crystals, or other elements.

close-talking microphone [ENG ACOUS] A microphone designed for use close to the mouth, so noise from more distant points is suppressed. Also known as noise-canceling microphone.

close-tolerance forging [MET] Forging in which draft angles are on the order of 1–3°, tolerances are less than half of those for commercial designs, and there is little or no allowance for finish.

close work [MIN ENG] Driving a tunnel or drifting between two coal seams.

closing pressure [MECH ENG] The amount of static inlet pressure in a safety relief valve when the valve disk has a zero lift above the seat.

closing rate [AERO ENG] The speed at which two aircraft or missiles come closer together.

cloth wheel [DES ENG] A polishing wheel made of sections of cloth glued or sewn together.

cloudburst hardness test [MET] A procedure in which a shower of steel balls, dropped from a predetermined height, dulls the surface of a hardened part in proportion to its softness and thus reveals defective areas.

cloudburst treatment [MET] Cold-working the surface of a metal by impingement of an avalanche of metal shot; a form of shot-peening.

cloud-detection radar [ENG] A type of weather radar designed specifically for the detection of clouds (rather than precipitation).

cloud-drop sampler [ENG] An instrument for collecting cloud particles, consisting of a sampling plate or cylinder and a shutter, which is so arranged that the sampling surface is exposed to the cloud for a predetermined length of time; the sampling surface is covered with a material which either captures the cloud particles or leaves an impression characteristic of the impinging elements.

cloud height indicator [ENG] General term for an instrument which measures the height of cloud bases.

cloud mirror *See* mirror nephoscope.

clout nail [DES ENG] A nail with a large, thin, flat head used in building.

clove oil *See* oil of cloves.

cloverleaf [CIV ENG] A highway intersection resembling a clover leaf and designed to allow movement and interchange of traffic without direct crossings and left turns.

clusec [MECH ENG] A unit of power used to measure the power of evacuation of a vacuum pump, equal to the power associated with a leak rate of 1 centiliter per second at a pressure of 1 millitorr, or to approximately 1.33322×10^{-6} watt.

cluster [ENG] **1.** A pyrotechnic signal consisting of a group of stars or fireballs. **2.** A grouping of rocket motors fastened together.

cluster mill [MET] A rolling mill in which small-diameter rolls are supported by larger rolls.

clutch [MECH ENG] A machine element for the connection and disconnection of shafts in equipment drives, especially while running.

coach screw [DES ENG] A large, square-headed, wooden screw used to join heavy timbers. Also known as lag bolt; lag screw.

coal auger [MIN ENG] A type of continuous miner which consists of a screw drill of large diameter and which cuts, transports, and loads the coal.

coal bank [MIN ENG] A seam of coal that is exposed.

coal barrier [MIN ENG] A protective pillar composed of coal.

coal blasting [MIN ENG] Breaking coal with explosives.

coalbreaker *See* breaker.

coal chemicals [MATER] Chemicals obtained as by-products in the primary processing of coal to metallurgical coke; the main source of aromatic compounds used as intermediates in the synthesis of dyes, drugs, antiseptics, and solvents.

coal cutter [MIN ENG] A power-operated machine which cuts out a thin strip of coal from the bottom of the seam; it draws itself by rope haulage along the coal face.

coal digger *See* faceman.

coal drill [MIN ENG] Usually, an electric drill of a compact, light design; however, also a light pneumatic drill.

coal dust [MIN ENG] A finely divided coal, sometimes defined as coal that will pass through 100-mesh screens (100 wires to the inch or 40 wires to the centimeter).

coalesced copper [MET] A mass, oxygen-free copper made by compacting and sintering cathode copper at high pressure and temperature.

coalescence [MET] The bonding of welded materials into one body.

coal face [MIN ENG] The mining face from which coal is extracted.

coalfield [MIN ENG] A region containing coal deposits.

coal gas [MATER] **1.** Flammable gas derived from coal either naturally in place, or by induced methods of industrial plants and underground gasification. **2.** Specifically, fuel gas obtained from carbonization of coal.

coal getter *See* faceman.

coal-in-oil suspension [MATER] A fluid mixture of pulverized coal dispersed in either fuel oil or coal tar oil, used as a fuel chiefly in large installations. Also known as colloidal fuel.

coal mining [MIN ENG] The technical and mechanical job of removing coal from the earth and preparing it for market.

coal oil [MATER] **1.** Condensed liquid from coal distillation. **2.** An archaic term for kerosine made from petroleum.

coal planer [MIN ENG] A type of continuous coal mining machine for longwall mining; consists of a heavy steel plow with cutting knives, with power equipment to drag it back and forth across a coal face.

coal plough [MIN ENG] A device with steel blades which shears off coal and pushes it onto the face conveyor.

coal-sensing probe [MIN ENG] A nucleonic instrument to measure the thickness of coal left in the seam floor by means of a gamma-ray backscattering unit.

coal sizes [MIN ENG] The sizes by which anthracite coal is marketed.

coal tar [MATER] A tar obtained from carbonization of coal, usually in coke ovens or retorts, containing several hundred organic chemicals.

coal tar enamel [MATER] Coal tar used as a paintlike coating for petroleum-product pipelines; provides protection from both water and cathodic corrosion.

coal tar light oil *See* light oil.

coal tar pitch [MATER] Dark-brown to black amorphous residue from the redistillation of coal tar; melts at 150°F (66°C); used as a thermoplastic.

coarse *See* low-grade.

coarse aggregate [MATER] Crushed stone or gravel used in concrete; will not, when dry, pass through a sieve with ¼-inch-diameter (6 millimeter) holes.

coarse-grained [MATER] Having a coarse texture.

coarse roll [MIN ENG] A large roll for the preliminary crushing of large pieces of ore, rock, or coal.

coast [ENG] A memory feature on a radar which, when activated, causes the range and angle systems to continue to move in the same direction and at the same speed as that required to track an original target.

coastal engineering [CIV ENG] A branch of civil engineering pertaining to the study of the action of the seas on shorelines and to the design of structures to protect against this action.

coasting flight [AERO ENG] The flight of a rocket between burnout or thrust cutoff of one stage and ignition of another, or between burnout and summit altitude or maximum horizontal range.

coated abrasive [MATER] An abrasive product having the abrasive particles attached to a backing material with glue or a synthetic resin.

coated electrode [MET] A wire covered with metal oxides and silicates and used as a filler-metal electrode in arc welding. Also known as covered electrode.

coated paper [MATER] Paper with a surface coating of clay and other materials to produce a smooth, shiny surface; especially useful for fine, detailed, blur-free reproductions in color or black and white. Also known as enamel paper.

coat hanger die [ENG] A plastics-sheet slot die shaped like a coat hanger on the inside.

coating [MATER] **1.** Any material that will form a continuous film over a surface. **2.** The film formed by the material.

coating density ratio [MET] In thermal spraying, the ratio of actual density to theoretical density of the coating material used.

coaxial [MECH ENG] Mounted on independent concentric shafts.

coaxial speaker [ENG ACOUS] A loudspeaker system comprising two, or less commonly three, speaker units mounted on substantially the same axis in an integrated mechanical assembly, with an acoustic-radiation-controlling structure.

coaxing [MET] Improving fatigue strength of a metal by increasing the stress range, beginning just below the fatigue limit.

cob [MIN ENG] To chip away waste material from an ore, using hand hammers.

cobber [MIN ENG] **1.** A device used to reject waste materials from ore concentrates. **2.** A person who breaks fibers from asbestos rocks or chips low-grade material from ore.

cock [ENG] Any mechanism which starts, stops, or regulates the flow of liquid, such as a valve, faucet, or tap.

cockle finish [MATER] An irregular surface usually produced on rag bond and ledger paper, obtained by coating the paper with sizing and drying it in loop or festoon fashion in heated air.

cockpit [AERO ENG] A space in an aircraft or spacecraft where the pilot sits.

cocoa butter [MATER] A brown fat obtained from cacao seeds; melts at 30–35°C; used in the manufacture of chocolate, cosmetics, and pharmaceuticals. Also known as cacao butter; oleum theobromatis; theobroma oil.

coconut oil [MATER] A nearly colorless or yellow oil from fresh coconut (*Cocos nucifera*) or from copra (dried coconut); used in foods, in making soap, and as a raw material in fatty-acid production.

code-sending radiosonde [ENG] A radiosonde which transmits the indications of the meteorological sensing elements in the form of a code consisting of combinations of dots and dashes. Also known as code-type radiosonde; contracted code sonde.

code-type radiosonde *See* code-sending radiosonde.

cod-liver oil [MATER] A yellow oil, high in vitamin D, extracted from the liver of the Atlantic cod (*Gadus morrhua*); soluble in alcohol. Also known as banks oil; oleum morrhuae.

coelostat [ENG] A device consisting of a clockwork-driven mirror that enables a fixed telescope to continuously keep the same region of the sky in its field of view.

coercimeter [ENG] An instrument that measures the magnetic intensity of a natural magnet or electromagnet.

coextrusion [ENG] Extrusion-forming of plastic or metal products in which two or more compatible feed materials are used in physical admixture through the same extrusion die.

cofferdam [CIV ENG] A temporary damlike structure constructed around an excavation to exclude water.

coffered ceiling [BUILD] An ornamental ceiling constructed of panels that are sunken or recessed.

coffin corner [AERO ENG] The range of Mach numbers between the buffeting Mach number and the stalling Mach number within which an aircraft must be operated.

cog [DES ENG] A tooth on the edge of a wheel.

cog belt [MECH ENG] A flexible device used for timing and for slip-free power transmission.

cogeneration [MECH ENG] The simultaneous on-site generation of electric energy and process steam or heat from the same plant.

cogging mill   See blooming mill.

cog railway   [CIV ENG] A steep railway that employs a cograil that meshes with a cogwheel on the locomotive to ensure traction.

cogwheel   [DES ENG] A wheel with teeth around its edge.

coherent moving-target indicator   [ENG] A radar system in which the Doppler frequency of the target echo is compared to a local reference frequency generated by a coherent oscillator.

coherent noise   [ENG] Noise that affects all tracks across a magnetic tape equally and simultaneously.

coil breaks   [MET] Creases across a metal strip transverse to the direction of coiling and representing areas of reduced thickness.

coil spring   [DES ENG] A helical or spiral spring, such as one of the helical springs used over the front wheels in an automotive suspension.

coil weld   [MET] A butt weld joining the ends of two metal sheets; forms a continuous strip for coiling.

coil winder   [ENG] A manual or motor-driven mechanism for winding coils individually or in groups.

coincidence correction   See dead-time correction.

coin gold   [MET] Gold of the legal fineness for coins.

coining   [MET] 1. A process of forming metals by squeezing between two dies so as to impress well-defined imprints on both surfaces of the work; usually performed cold. 2. Final pressing of a sintered compact in powder metallurgy.

coin silver   [MET] An alloy of 90% silver, 10% copper; has been used for coining American currency.

coir   [MATER] A coarse, brown fiber obtained from the husk of the coconut.

coke   [MATER] A coherent, cellular, solid residue remaining from the dry (destructive) distillation of a coking coal or of pitch, petroleum, petroleum residues, or other carbonaceous materials; contains carbon as its principal constituent, together with mineral matter and volatile matter.

coke breeze   [MECH ENG] Undersized coke screenings passing through a screen opening of approximately ⅝ inch (16 millimeters).

coke knocker   [MECH ENG] A mechanical device used to break loose coke within a drum or tower.

coke-oven gas   [MATER] A gas produced during carbonization of coal to form coke.

cold-air machine   [MECH ENG] A refrigeration system in which air serves as the refrigerant in a cycle of adiabatic compression, cooling to ambient temperature, and adiabatic expansion to refrigeration temperature; the air is customarily reused in a closed superatmospheric pressure system. Also known as dense-air system.

cold bending   [MET] The bending of metal rods, especially concrete-reinforcing rods, without heat.

cold-chamber die casting   [ENG] A die-casting process in which molten metal is ladled either manually or mechanically into a relatively cold cylinder from which it is forced into the die cavity.

cold chisel   [DES ENG] A chisel specifically designed to cut or chip cold metal; made of specially tempered tool steel machined into various cutting edges. Also known as cold cutter.

cold cutter   See cold chisel.

cold differential test pressure   [ENG] The inlet pressure of a pressure-relief valve at which the valve is set to open during testing.

cold drawing   [MET] Drawing a tube or wire through a series of successively smaller dies, without the application of heat, to reduce its diameter.

cold extrusion   [MET] Shaping cold metal by striking a slug in a closed cavity with a punch so that the metal is forced up around the punch. Also known as cold forging; cold pressing; extrusion pressing; impact extrusion.

cold-finished steel   [MET] Steel bars which have been cold-drawn, cold-rolled, centerless-ground, or turned smooth.

cold-flow test   [AERO ENG] A test of a liquid rocket without firing it to check or verify the integrity of a propulsion subsystem, and to provide for the conditioning and flow of propellants (including tank pressurization, propellant loading, and propellant feeding).

cold forging   See cold extrusion.

cold forming   [MET] Any forging operation performed cold, such as cold extrusion, cold drawing, or coining, which enables close dimensional accuracy to be achieved.

cold galvanizing   [MET] Painting iron with a suspension of zinc particles in an organic solvent, so that a zinc coating remains following evaporation of the solvent.

cold heading   [MET] The cold working of metal in order to increase part or all of the cross-sectional area of the stock.

cold inspection   [MET] The inspection of a forging at room temperature by visible or nondestructive means to detect surface conditions or defects at room temperature.

cold joint   [ENG] A soldered connection which was inadequately heated, with the result that the wire is held in place by rosin flux, not solder.

cold molding [ENG] Shaping of an unheated compound in a mold under pressure, followed by heating the article to cure it.

cold nosing *See* wildcat drilling.

cold plate [MECH ENG] An aluminum or other plate containing internal tubing through which a liquid coolant is forced, to absorb heat transferred to the plate by transistors and other components mounted on it. Also known as liquid-cooled dissipator.

cold pressing *See* cold extrusion.

cold rolling [MET] Rolling metal at room temperature to reduce thickness or harden the surface; results in a smooth finish and improved resistance to fatigue.

cold rubber [MATER] Butadiene-styrene type of synthetic rubber produced by polymerization at about 40°F (4°C), instead of the conventional 120°F (49°C); has improved strength and abrasion resistance.

cold saw [MECH ENG] **1.** Any saw for cutting cold metal, as opposed to a hot saw. **2.** A disk made of soft steel or iron which rotates at a speed such that a point on its edge has a tangential velocity of about 15,000 feet per minute (75 meters per second), and which grinds metal by friction.

cold-short [MET] Pertaining to lack of ductility in some metals at temperatures below the recrystallization temperature.

cold shot [MET] Intensely hard, globular portions of the surface of an ingot or casting formed by premature solidification upon first contact with the cold sand during pouring.

cold shut [MET] **1.** A surface defect of a metal casting in the form of a discontinuity where two streams failed to unite. Also known as cold lap. **2.** Freezing of the top surface of an ingot before the mold is full.

cold slug [ENG] The first material to enter an injection mold in plastics manufacturing.

cold slug well [ENG] The area in a plastic injection mold which receives the cold slug from the sprue opening.

cold soldering [MET] Soldering of parts without heat.

cold storage [ENG] The storage of perishables at low temperatures produced by refrigeration, usually above freezing, to increase storage life.

cold-storage locker plant [ENG] A plant with many rental steel lockers, each with a capacity of about 6 cubic feet (0.17 cubic meter) and generally for food storage by an individual family, placed in refrigerated rooms, at about 0°F (−18°C).

cold stretch [ENG] A pulling operation on extruded plastic filaments in which little or no heat is used; improves tensile properties.

cold trap [MECH ENG] A tube whose walls are cooled with liquid nitrogen or some other liquid to condense vapors passing through it; used with diffusion pumps and to keep vapors from entering a McLeod gage.

cold treatment [MET] Subzero cooling, to a temperature of −100°F (−73°C).

cold trimming [MET] The removal of excess metal from a forging at room temperature by means of a trimming press.

cold welding [MET] Welding in which a molecular bond is obtained by a cold flow of metal under extremely high pressures, without heat; widely used for sealing transistors and quartz crystal holders.

cold working [MET] Plastic deformation of a metal below the annealing temperature to cause permanent strain hardening.

coleopter [AERO ENG] An aircraft having an annular (barrel-shaped) wing, the engine and body being mounted within the circle of the wing.

collapse [ENG] Contraction of plastic container walls during cooling; produces permanent indentation.

collar [DES ENG] A ring placed around an object to restrict its motion, hold it in place, or cover an opening. [MIN ENG] The mouth of a mine shaft.

collar beam [BUILD] A tie beam in a roof truss connecting the rafters well above the wall plate.

collar bearing [MECH ENG] A bearing that resists the axial force of a collar on a rotating shaft.

collared hole [ENG] A started hole drilled sufficiently deep to confine the drill bit and prevent slippage of the bit from normal position.

collar locator log [PETRO ENG] Down-hole nuclear-log measurement to locate drill-hole casing collars, usually for precise location of perforating points.

collective bargaining [IND ENG] The negotiation for mutual agreement in the settlement of a labor contract between an employer or his representatives and a labor union or its representatives.

collector [ENG] A class of instruments employed to determine the electric potential at a point in the atmosphere, and ultimately the atmospheric electric field; all collectors consist of some device for rapidly bringing a conductor to the same potential as the air immediately surrounding it, plus some form of electrometer for measuring the difference in potential between the equilibrated collector and the earth itself; collectors differ widely in their speed of response to atmospheric potential changes.

collet [DES ENG] A split, coned sleeve to hold small, circular tools or work in the nose of a lathe or other type of machine. [ENG] **1.** The glass neck remaining on a bottle after it is taken off the glass-blowing iron. **2.** Pieces of glass, ordinarily discarded, that are added to a batch of glass. Also spelled cullet.

colliery [MIN ENG] A whole coal mining plant; generally the term is used in connection with anthracite mining but sometimes to designate the mine, shops, and preparation plant of a bituminous operation.

collimation error [ENG] **1.** Angular error in magnitude and direction between two nominally parallel lines of sight. **2.** Specifically, the angle by which the line of sight of a radar differs from what it should be.

collimation tower [ENG] Tower on which a visual and a radio target are mounted to check the electrical axis of an antenna.

Collins miner [MIN ENG] A type of remote-controlled continuous miner for thin-seam extraction.

collision-avoidance radar [ENG] Radar equipment utilized in a collision-avoidance system.

collision-avoidance system [ENG] Electronic devices and equipment used by a pilot to perform the functions of conflict detection and avoidance.

collision blasting [ENG] The blasting out of different sections of rocks against each other.

collision parameter [AERO ENG] In orbit computation, the distance between a center of attraction of a central force field and the extension of the velocity vector of a moving object at a great distance from the center.

colloidal fuel See coal-in-oil suspension.

colloidal graphite [MATER] Extremely fine flakes of graphite suspended in water, petroleum oil, castor oil, glycerin, or other liquids; used to provide conductive shields on the inside or outside surfaces of electron tubes.

colloidal silver [MATER] Finely divided particles of silver, sometimes used on terminals of electronic components to give a larger surface area for connections.

colloider [CIV ENG] A device that removes colloids from sewage.

colloid mill [MECH ENG] A grinding mill for the making of very fine dispersions of liquids or solids by breaking down particles in an emulsion or paste.

Colmol miner [MIN ENG] A continuous miner, in which the coal is completely augered by two banks of cutting arms fitted with picks; the arms rotate in opposite directions to assist in gathering up the cuttings for the central conveyor.

color-bar code [IND ENG] A code that uses one or more different colors of bars in combination with black bars and white spaces, to increase the density of binary coding of data printed on merchandise tags or directly on products for inventory control and other purposes.

color code [ENG] **1.** Any system of colors used for purposes of identification, such as to identify dangerous areas of a factory. **2.** A system of colors used to identify the type of material carried by a pipe; for example, dangerous materials, protective materials, extra valuable materials.

colored smoke [MATER] Gaseous products of a distinctive color which forms a colored cloud and may be used for signals or target markers.

color lake See lake.

color tempering [MET] Reheating of hardened steel and observing color changes to determine quenching temperature and to obtain the desired hardness.

columbium See niobium.

column [ENG] A vertical shaft designed to bear axial loads in compression.

column crane [MECH ENG] A jib crane whose boom pivots about a post attached to a building column.

column drill [MECH ENG] A tunnel rock drill supported by a vertical steel column.

column formation See trail formation.

column pipe [MIN ENG] The large cast-iron (or wooden) pipe through which the water is conveyed from the mine pumps to the surface.

column splice [CIV ENG] A connection between two lengths of a compression member (column); an erection device rather than a stress-carrying element.

colza oil See rape oil.

combination chuck [DES ENG] A chuck used in a lathe whose jaws either move independently or simultaneously.

combination collar [DES ENG] A collar that has left-hand threads at one end and right-hand threads at the other.

combination cycle See mixed cycle.

combination die [MET] A die having more than one cavity for different castings.

combination-drive reservoir [PETRO ENG] A type of reservoir in which hydrocarbons are swept (displaced) toward the drill hole by injection of water followed by liquefied-petroleum-gas or gas injection.

combination lock [ENG] A lock that can be opened only when its dial has been set to the proper combination of symbols, in the proper sequence.

combination mill [MET] A rolling mill arranged with continuous rolls for roughing and a guide or looping mill for shaping.

combination pliers [DES ENG] Pliers that can be used either for holding objects or for cutting and bending wire.

combination saw [MECH ENG] A saw made in various tooth arrangement combinations suitable for ripping and crosscut mitering.

combination square [DES ENG] A square head and steel rule that when used together have both a 45 and 90° face to allow the testing of the accuracy of two surfaces intended to have these angles.

combination wrench [DES ENG] A wrench that is an open-end wrench at one end and a socket wrench at the other.

combined footing [CIV ENG] A footing, either rectangular or trapezoidal, that supports two columns.

combined moisture [MIN ENG] Moisture in coal that cannot be removed by ordinary drying.

combined sewers [CIV ENG] A drainage system that receives both surface runoff and sewage.

combing [BUILD] In roofing, the topmost row of shingles which project above the ridge line. [ENG] **1.** Using a comb or stiff bristle brush to create a pattern by pulling through freshly applied paint. **2.** Scraping or smoothing a soft stone surface.

comb nephoscope [ENG] A direct-vision nephoscope constructed with a comb (a crosspiece containing equispaced vertical rods) attached to the end of a column 8–10 feet (2.4–3 meters) long and supported on a mounting that is free to rotate about its vertical axis; in use, the comb is turned so that the cloud appears to move parallel to the tips of the vertical rods.

combustible gas [MATER] A gas that burns, including the fuel gases, hydrogen, hydrocarbon, carbon monoxide, or a mixture of these.

combustible loss [ENG] Thermal loss resulting from incomplete combustion of fuel.

combustion chamber [AERO ENG] That part of the rocket engine in which the combustion of propellants takes place at high pressure. Also known as firing chamber. [ENG] Any chamber in which a fuel such as oil, coal, or kerosine is burned to provide heat. [MECH ENG] The space at the head end of an internal combustion engine cylinder where most of the combustion takes place.

combustion chamber volume [MECH ENG] The volume of the combustion chamber when the piston is at top dead center.

combustion deposit [ENG] A layer of ash on the heat-exchange surfaces of a combustion chamber, resulting from the burning of a fuel.

combustion engine [MECH ENG] An engine that operates by the energy of combustion of a fuel.

combustion engineering [MECH ENG] The design of combustion furnaces for a given performance and thermal efficiency, involving study of the heat liberated in the combustion process, the amount of heat absorbed by heat elements, and heat-transfer rates.

combustion furnace [ENG] A furnace whose source of heat is the energy released in the oxidation of fossil fuel.

combustion instability [AERO ENG] Unsteadiness or abnormality in the combustion of fuel, as may occur in a rocket engine.

combustion knock *See* engine knock.

combustion shock [ENG] Shock resulting from abnormal burning of fuel in an internal combustion engine, caused by preignition or fuel-air detonation; or in a diesel engine, the uncontrolled burning of fuel accumulated in the combustion chamber.

combustion turbine *See* gas turbine.

combustor [MECH ENG] The combustion chamber together with burners, igniters, and injection devices in a gas turbine or jet engine.

come-along [DES ENG] **1.** A device for gripping and effectively shortening a length of cable, wire rope, or chain by means of two jaws which close when one pulls on a ring. **2.** *See* puller.

comfort chart [ENG] A diagram showing curves of relative humidity and effective temperature superimposed upon rectangular coordinates of wet-bulb temperature and dry-bulb temperature.

comfort control [ENG] Control of temperature, humidity, flow, and composition of air by using heating and air-conditioning systems, ventilators, or other systems to increase the comfort of people in an enclosure.

comfort curve [ENG] A line drawn on a graph of air temperature versus some function of humidity (usually wet-bulb temperature or relative humidity) to show the varying conditions under which the average sedentary person feels the same degree of comfort; a curve of constant comfort.

comfort zone [ENG] The ranges of indoor temperature, humidity, and air movement, under which most persons enjoy mental and physical well-being. Also known as comfort standard.

command control *See* command guidance.

command guidance [ENG] A type of electronic guidance of guided missiles or other guided aircraft wherein signals or pulses sent out by an operator cause the guided object to fly a directed path. Also known as command control.

command module [AERO ENG] The spacecraft module that carries the crew, the main communication and telemetry equipment, and the reentry capsule during cruising flight.

commercial diesel cycle See mixed cycle.

commercial harbor [CIV ENG] A harbor in which docks are provided with cargo-handling facilities.

commercial lecithin See lecithin.

commercial mine [MIN ENG] A coal mine operated to supply purchasers in general, as contrasted with a captive mine.

commercial ore [MIN ENG] Mineralized material profitable at prevailing metal prices.

comminution [MECH ENG] Breaking up or grinding into small fragments. Also known as pulverization.

comminutor [MECH ENG] A machine that breaks up solids.

commission ore [MIN ENG] Uranium-bearing ore of 0.10% U_3O_8 or higher, for which the U.S. Atomic Energy Commission has an established price.

common bond See American bond.

common brick [MATER] Brick made from natural clay.

common carrier [IND ENG] A company recognized by an appropriate regulatory agency as having a vested interest in furnishing communications services or in transporting commodities or people.

common joist [BUILD] An ordinary floor beam to which floor boards are attached.

common labor [IND ENG] Unskilled workers.

common rafter [BUILD] A rafter which extends from the plate of the roof to the ridge board at right angles to both members, and to which roofing is attached.

common wall [BUILD] A wall that is shared by two dwelling units.

communications [ENG] The science and technology by which information is collected from an originating source, transformed into electric currents or fields, transmitted over electrical networks or space to another point, and reconverted into a form suitable for interpretation by a receiver.

communications satellite [AERO ENG] An orbiting, artificial earth satellite that relays radio, television, and other signals between ground terminal stations thousands of miles apart. Also known as radio relay satellite; relay satellite.

commutator-controlled welding [MET] Spot or projection welding in which several electrodes in contact with the work simultaneously are operated under the control of an electrical commutating device. Also known as ultraspeed welding.

compact [MET] A briquette made by the compression of metal powder, with or without the addition of nonmetallic constituents.

compaction [ENG] Increasing the dry density of a granular material, particularly soil, by means such as impact or by rolling the surface layers.

compactor [MECH ENG] Machine designed to consolidate earth and paving materials by kneading, weight, vibration, or impact, to sustain loads greater than those sustained in an uncompacted state.

companion body [AERO ENG] A nose cone, last-stage rocket, or other body that orbits along with an earth satellite or follows a space probe.

companion flange [DES ENG] A pipe flange that can be bolted to a similar flange on another pipe.

compar [MATER] Generic term for a family of compounded, modified, and plasticized polyvinyl alcohol resins; used for making full- and solvent-resistant tubing and hose, and printing rolls.

comparative rabal [ENG] A rabal observation (that is, a radiosonde balloon tracked by theodolite) taken simultaneously with the usual rawin observation (tracking by radar or radio direction-finder), to provide a rough check on the alignment and operating accuracy of the electronic tracking equipment.

comparator [ENG] A device used to inspect a gaged part for deviation from a specified dimension, by mechanical, electrical, pneumatic, or optical means.

compartment [MIN ENG] A section of a mine shaft separated by framed timbers and planking.

compartment mill [MECH ENG] A multisection pulverizing device divided by perforated partitions, with preliminary grinding at one end in a short ball-mill operation, and finish grinding at the discharge end in a longer tube-mill operation.

compass [ENG] An instrument for indicating a horizontal reference direction relative to the earth.

compass bowl [ENG] That part of a compass in which the compass card is mounted.

compass card [DES ENG] The part of a compass on which the direction graduations are placed; it is usually in the form of a thin disk or annulus graduated in degrees, clockwise from $0°$ at the reference direction to $360°$, and sometimes also in compass points.

compass card axis [DES ENG] The line joining 0° and 180° on a compass card.

compass declinometer [ENG] An instrument used for magnetic distribution surveys; employs a thin compass needle 6 inches (15 centimeters) long, supported on a sapphire bearing and steel pivot of high quality; peep sights serve for aligning the compass box on an azimuth mark.

compass roof [BUILD] A roof in which each truss is in the form of an arch.

compass saw [DES ENG] A handsaw which has a handle with several attachable thin, tapering blades of varying widths, making it suitable for a variety of work, such as cutting circles and curves.

compatibility [SYS ENG] The ability of a new system to serve users of an old system.

compatible discrete four-channel sound [ENG ACOUS] A sound system in which a separate channel is maintained from each of the four sets of microphones at the recording studio or other input location to the four sets of loudspeakers that serve as the output of the system. Abbreviated CD-4 sound.

compensated pendulum [DES ENG] A pendulum made of two materials with different coefficients of expansion so that the distance between the point of suspension and center of oscillation remains nearly constant when the temperature changes.

compensated volume control *See* loudness control.

compensation signals [ENG] In telemetry, signals recorded on a tape, along with the data and in the same track as the data, used during the playback of data to correct electrically the effects of tape-speed errors.

complete fusion [MET] Fusion which has occurred over all surfaces of the base metal exposed for welding.

complete joint penetration [MET] The fusion of weld metal to base metal throughout the entire thickness of the base metal, that is, the deposited weld metal occupies the entire groove.

complete lubrication [ENG] Lubrication taking place when rubbing surfaces are separated by a fluid film, and frictional losses are due solely to the internal fluid friction in the film. Also known as viscous lubrication.

complex frequency [ENG] A complex number used to characterize exponential and damped sinusoidal motion in the same way that an ordinary frequency characterizes simple harmonic motion; designated by the constant s corresponding to a motion whose amplitude is given by Ae^{st}, where A is a constant and t is time.

complex reflector [ENG] A structure or group of structures having many radar-reflecting surfaces facing in different directions.

complex target [ENG] A radar target composed of a number of reflecting surfaces that, in the aggregate, are smaller in all dimensions than the resolution capabilities of the radar.

compo board *See* composition board.

composite [ENG ACOUS] A re-recording consisting of at least two elements. [MATER] A structural material composed of combinations of metal alloys or plastics, usually with the addition of strengthening agents.

composite beam [CIV ENG] Beam action of two materials joined to act as a unit, especially that developed by a concrete slab resting on a steel beam and joined by shear connectors.

composite column [CIV ENG] A concrete column having a structural-steel or cast-iron core with a maximum core area of 20%.

composite compact [MET] A powder compact composed of more than one layer of different components with each layer retaining its identity.

composite electrode [MET] A filler-metal electrode composed of more than one metal.

composite explosive [MATER] A mixture of substances which consume and give off oxygen, together with one or several simple explosives; dynamite is an example.

composite fuel [MATER] A broad class of solid chemical fuels composed of a fuel and oxidizer and used as propellants in rockets; an example of a fuel is phenol formaldehyde, and an oxidizer is ammonium perchlorate. Also known as composite propellant.

composite I-beam bridge [CIV ENG] A beam bridge in which the concrete roadway is mechanically bonded to the I beams by means of shear connectors.

composite joint [MET] A joint connected by welding in conjunction with one or more mechanical means.

composite macromechanics [ENG] The study of composite material behavior wherein the material is presumed homogeneous and the effects of the constituent materials are detected only as averaged apparent properties of the composite.

composite map [MIN ENG] A map in which several levels of a mine are shown on a single sheet.

composite micromechanics [ENG] The study of composite material behavior wherein the constituent materials are studied on a microscopic scale with specific properties being assigned to each constituent; the interaction of the constituent materials is used to determine the properties of the composite.

composite pile [CIV ENG] A pile in which the upper and lower portions consist of different types of piles.

composite plate [MET] A layer of electrodeposited material consisting of at least two different constituents.

composite propellant *See* composite fuel.

composite sampler [ENG] A hydrometer cylinder equipped with sample cocks at regular intervals along its vertical height; used to take representative (vertical composite) samples of oil from storage tanks.

composite steel [MET] Bar steel machined along the entire length which is cast around an insert of tool steel welded to the backing of mild steel; used for shear blades and die parts.

composition board [MATER] A sheet product composed of vegetable fibers mechanically or chemically formed into a pulp which is rolled and pressed. Also known as compo board.

composition metal [MET] A cast copper alloy having a composition of more than 80% copper, with tin, zinc, and lead.

compost [MATER] A mixture of decaying organic matter used to fertilize and condition the soil.

compound angle [ENG] The angle formed by two mitered angles.

compound compact [MET] A powder compact made from a mixture of metals, with each particle retaining its original composition.

compound die [MET] A die designed to perform more than one operation on the work with each stroke of the press.

compound engine [MECH ENG] A multicylinder-type displacement engine, using steam, air, or hot gas, where expansion proceeds successively (sequentially).

compounding [MECH ENG] The series placing of cylinders in an engine (such as steam) for greater ratios of expansion and consequent improved engine economy.

compound lever [MECH ENG] A train of levers in which motion or force is transmitted from the arm of one lever to that of the next.

compound rest [MECH ENG] A principal component of a lathe consisting of a base and an upper part dovetailed together; the base is graduated in degrees and can be swiveled to any angle; the upper part includes the tool post and tool holder.

compound screw [DES ENG] A screw having different or opposite pitches on opposite ends of the shank.

compound shaft [MIN ENG] A shaft in which the upper stage is often a vertical shaft, while the lower stage, or stages, may be inclined and driven into the deposit.

compregnate [ENG] Compression of materials into a dense, hard substance with the aid of heat.

compressed-air blasting [MIN ENG] A method for breaking down coal by compressed-air power.

compressed-air diving [ENG] Any form of diving in which air is supplied under high pressure to prevent lung collapse.

compressed-air loudspeaker [ENG ACOUS] A loudspeaker having an electrically actuated valve that modulates a stream of compressed air.

compressed-air power [MECH ENG] The power delivered by the pressure of compressed air as it expands, utilized in tools such as drills, in hoists, grinders, riveters, diggers, pile drivers, motors, locomotives, and mine ventilating systems.

compression [MECH ENG] *See* compression ratio.

compression coupling [MECH ENG] **1.** A means of connecting two perfectly aligned shafts in which a slotted tapered sleeve is placed over the junction and two flanges are drawn over the sleeve so that they automatically center the shafts and provide sufficient contact pressure to transmit medium loads. **2.** A type of tubing fitting.

compression cup [ENG] A cup from which lubricant is forced to a bearing by compression.

compression failure [ENG] Buckling or collapse caused by compression, as of a steel or concrete column or of wood fibers.

compression ignition [MECH ENG] Ignition produced by compression of the air in a cylinder of an internal combustion engine before fuel is admitted.

compression-ignition engine *See* diesel engine.

compression machine *See* compressor.

compression member [ENG] A beam or other structural member which is subject to compressive stress.

compression mold [ENG] A mold for plastics which is open when the material is introduced and which shapes the material by heat and by the pressure of closing.

compression plant [PETRO ENG] Gas-compression facility used to produce a high-pressure gas stream for injection into reservoir formations to increase oil yield; when the injected gas is that recovered from the well during oil production, the facility is called a gas-cycling plant.

compression pressure [MECH ENG] That pressure developed in a reciprocating piston engine at the end of the compression stroke without combustion of fuel.

compression ratio [MECH ENG] The ratio in internal combustion engines between the volume displaced by the piston plus the clearance space, to the volume of the clearance space. Also known as compression. [MET] Ratio of the volume of loose metal powder to the volume of the compact made from it.

compression refrigeration [MECH ENG] The cooling of a gaseous refrigerant by first compressing it to liquid form (with resultant heat buildup), cooling the liquid by heat exchange, then releasing pressure to allow the liquid to vaporize (with resultant absorption of latent heat of vaporization and a refrigerative effect).

compression release [MECH ENG] Release of compressed gas resulting from incomplete closure of intake or exhaust valves.

compression spring [ENG] A spring, usually a coil spring, which resists a force tending to compress it.

compression stroke [MECH ENG] The phase of a positive displacement engine or compressor in which the motion of the piston compresses the fluid trapped in the cylinder.

compression test [ENG] A test to determine compression strength, usually applied to materials of high compression but low tensile strength, in which the specimen is subjected to increasing compressive forces until failure occurs.

compressor [MECH ENG] A machine used for increasing the pressure of a gas or vapor. Also known as compression machine.

compressor blade [MECH ENG] The vane components of a centrifugal or axial-flow, air or gas compressor.

compressor station [MECH ENG] A permanent facility which increases the pressure on gas to move it in transmission lines or into storage.

compressor valve [MECH ENG] A valve in a compressor, usually automatic, which operates by pressure difference (less than 5 pounds per square inch or 35 kilopascals) on the two sides of a movable, single-loaded member and which has no mechanical linkage with the moving parts of the compressor mechanism.

compromise joint [CIV ENG] 1. A joint bar used for joining rails of different height or section. 2. A rail that has different joint drillings from that of the same section.

compromise rail [CIV ENG] A short rail having different sections at the ends to correspond with the rail ends to be joined, thus providing a transition between rails of different sections.

concave bit [DES ENG] A type of tungsten carbide drill bit having a concave cutting edge; used for percussive boring.

concave fillet weld [MET] A fillet weld having a concave surface.

concentrate [MIN ENG] 1. To separate ore or metal from its containing rock or earth. 2. The clean product recovered in froth flotation or other methods of mineral separation.

concentrating table [MIN ENG] A device consisting of a riffled deck to which a reciprocating motion in a horizontal direction is imparted; the material to be separated is fed in a stream of water, the heavy particles collect between the riffles and are conveyed in the direction of the reciprocating motion, while the lighter particles are borne by the water over the riffles to be discharged laterally from the table.

concentrator [ENG] 1. An apparatus used to concentrate materials. 2. A plant where materials are concentrated.

concentric groove *See* locked groove.

concentric locating [DES ENG] The process of making the axis of a tooling device coincide with the axis of the workpiece.

concentric orifice plate [DES ENG] A fluid-meter orifice plate whose edges have a circular shape and whose center coincides with the center of the pipe.

concession lease [MIN ENG] A lease form that conveys specified national or state permission to a lessee to explore for or produce minerals (such as oil, gas, or uranium) from specified properties.

concrete [MATER] A mixture of aggregate, water, and a binder, usually portland cement; hardens to stonelike condition when dry.

concrete beam [CIV ENG] A structural member of reinforced concrete, placed horizontally over openings to carry loads.

concrete block [MATER] A solid or hollow block of precast concrete.

concrete bridge [CIV ENG] A bridge constructed of prestressed or reinforced concrete.

concrete bucket [ENG] A container with movable gates at the bottom that is attached to power cranes or cables to transport concrete.

concrete buggy [ENG] A cart which carries up to 6 cubic feet (0.17 cubic meter) of concrete from the mixer or hopper to the forms. Also known as buggy; concrete cart.

concrete caisson sinking [CIV ENG] A shaft-sinking method similar to caisson sinking except that reinforced concrete rings are used and an airtight working chamber is not adopted.

concrete cart *See* concrete buggy.

concrete chute [ENG] A long metal trough with rounded bottom and open ends used for conveying concrete to a lower elevation.

concrete column [CIV ENG] A vertical structural member made of reinforced or unreinforced concrete.

concrete dam [CIV ENG] A dam built of concrete.

concrete finish [MATER] The texture or smoothness on the surface of hardened concrete.

concrete form oil [MATER] Nonviscous, neutral mineral oil used on wooden or metal forms to allow easy removal from set concrete.

concrete hardener [MATER] An admixture such as calcium chloride, sodium chloride, or sodium hydroxide that hastens or decreases the hydration rate of cementing material; the concrete takes less time to set and has earlier higher strength.

concrete masonry [MATER] Building units composed of block, brick, or tile laid by masons.

concrete mixer [MECH ENG] A machine with a rotating drum in which the components of concrete are mixed.

concrete pile [CIV ENG] A reinforced pile made of concrete, either precast and driven into the ground, or cast in place in a hole bored into the ground.

concrete pipe [CIV ENG] A porous pipe made of concrete and used principally for subsoil drainage; diameters over 15 inches (38 centimeters) are usually reinforced.

concrete pump [MECH ENG] A device which drives concrete to the placing position through a pipeline of 6-inch (15-centimeter) diameter or more, using a special type of reciprocating pump.

concrete retarder [MATER] A material added to concrete that decreases the hydration rate of cement, thereby increasing the setting time and decreasing the strengthening rate during the early age.

concrete slab [CIV ENG] A flat, reinforced-concrete structural member, relatively sizable in length and width, but shallow in depth; used for floors, roofs, and bridge decks.

concrete steel [MET] Steel used in reinforced concrete, which should comply with standard specifications for prestressed concrete.

concrete vibrator [MECH ENG] Vibrating device used to achieve proper consolidation of concrete; the three types are internal, surface, and form vibrators.

concurrent heating [MET] Application of supplemental heat in metal cutting or welding.

concussion table [MIN ENG] An inclined table, agitated by a series of shocks and operating like a buddle. Also known as percussion table.

condensate [MATER] **1.** The liquid product from a condenser. Also known as condensate liquid. **2.** A light hydrocarbon mixture formed as a liquid product in a gas-recycling plant through expansion and cooling of the gas.

condensate liquid See condensate.

condensate strainer [MECH ENG] A screen used to remove solid particles from the condensate prior to its being pumped back to the boiler.

condensate well [MECH ENG] A chamber into which condensed vapor falls for convenient accumulation prior to removal. [PETRO ENG] Well that produces a natural gas highly saturated with condensable hydrocarbons heavier than methane and ethane.

condenser [MECH ENG] A heat-transfer device that reduces a thermodynamic fluid from its vapor phase to its liquid phase, suchas in a vapor-compression refrigeration plant or in a condensing steam power plant.

condenser-discharge anemometer [ENG] A contact anemometer connected to an electrical circuit which is so arranged that the average wind speed is indicated.

condenser microphone See capacitor microphone.

condenser transducer See electrostatic transducer.

condenser tubes [MECH ENG] Metal tubes used in a heat-transfer device, with condenser vapor as the heat source and flowing liquid such as water as the receiver.

condensing engine [MECH ENG] A steam engine in which the steam exhausts from the cylinder to a vacuum space, where the steam is liquefied.

condensing gas drive [PETRO ENG] Reservoir-oil displacement by gas where hydrocarbon components of the injected gas condense in the oil that it is displacing.

conducting polymer [MATER] A plastic having high conductivity, approaching that of metals.

conduction pump [ENG] A pump in which liquid metal or some other conductive liquid is moved through a pipe by sending a current across the liquid and applying a magnetic field at right angles to current flow.

conductive coating [MATER] A coating used to reduce surface resistance and thus prevent the accumulation of static electric charges.

conductive elastomer [MATER] A rubberlike silicone material in which suspended metal particles conduct electricity.

conductive silver paste [MATER] Silver powder in a suitable vehicle for applying to ceramic or other insulating materials by silk-screening or other methods, then fixing or firing at appropriate temperatures to provide a hard conductive surface or joint.

conductometer [ENG] An instrument designed to measure thermal conductivity; in particular, one that compares the rates at which different rods transmit heat.

conductor pipe [BUILD] A metal pipe through which water is drained from the roof. [PETRO ENG] A short string of large-diameter casing serving primarily to keep the top of a well bore open and to convey upflowing drilling fluid from the well bore to the slush pit.

conduit [ENG] Any channel or pipe for conducting the flow of water or other fluid.

cone [ENG ACOUS] The cone-shaped paper or fiber diaphragm of a loudspeaker. [MET] The part of an oxygen gas flame adjacent to the orifice of the tip.

cone bearing [MECH ENG] A cone-shaped journal bearing running in a correspondingly tapered sleeve.

cone-bottom tank [ENG] Liquids-storage tank with downward-pointing conical bottom to facilitate drainage of bottom, as of water or sludge.

cone brake [MECH ENG] A type of friction brake whose rubbing parts are cone-shaped.

cone classifier [MECH ENG] Inverted-cone device for the separation of heavy particulates (such as sand, ore, or other mineral matter) from a liquid stream; feed enters the top of the cone, heavy particles settle to the bottom where they can be withdrawn, and liquid overflows the top edge, carrying the smaller particles or those of lower gravity over the rim; used in the mining and chemical industries.

cone clutch [MECH ENG] A clutch which uses the wedging action of mating conical surfaces to transmit friction torque.

cone crusher [MECH ENG] A machine that reduces the size of materials such as rock by crushing in the tapered space between a truncated revolving cone and an outer chamber.

conehead rivet [DES ENG] A rivet with a head shaped like a truncated cone.

cone key [DES ENG] A taper saddle key placed on a shaft to adapt it to a pulley with a too-large hole.

cone loudspeaker [ENG ACOUS] A loudspeaker employing a magnetic driving unit that is mechanically coupled to a paper or fiber cone. Also known as cone speaker.

cone mandrel [DES ENG] A mandrel in which the diameter can be changed by moving conical sleeves.

cone nozzle [DES ENG] A cone-shaped nozzle that disperses fluid in an atomized mist.

cone of visibility [AERO ENG] Generally, the right conical space which has its apex at some ground target and within which an aircraft must be located if the pilot is to be able to discern the target while flying at a specified altitude.

cone pulley See step pulley.

cone rock bit [MECH ENG] A rotary drill with two hardened knurled cones which cut the rock as they roll. Also known as roller bit.

cone-roof tank [ENG] Liquids-storage tank with flattened conical roof to allow a vapor reservoir at the top for filling operations.

cone settler [MIN ENG] A conical vessel fed centrally with fine ore pulp, in which the apex discharge carries the larger-sized particles, and the peripheral top overflow carries the finer fraction of the solids.

cone speaker See cone loudspeaker.

cone valve [CIV ENG] A divergent valve whose cone-shaped head in a fixed cylinder spreads water around the wide, downstream end of the cone in spillways of dams or hydroelectric facilities. Also known as Howell-Bunger valve.

confidence level [IND ENG] The probability in acceptance sampling that the quality of accepted lots manufactured will be better than the rejectable quality level (RQL); 90% level indicates that accepted lots will be better than the RQL 90 times in 100.

configuration [AERO ENG] A particular type of specific aircraft, rocket, or such, which differs from others of the same model by the arrangement of its components or by the addition or omission of auxiliary equipment; for example, long-range configuration or cargo configuration. [SYS ENG] A group of machines interconnected and programmed to operate as a system.

confined flow [ENG] The flow of any fluid (liquid or gas) through a continuous container (process vessel) or conduit (piping or tubing).

confinement [ENG] Physical restriction, or degree of such restriction, to passage of detonation wave or reaction zone, for example, that of a resistant container which holds an explosive charge.

congruent transformation [MET] An isothermal or isobaric phase change in an alloy where the integrity of both phases is maintained throughout the process.

conical ball mill [MECH ENG] A cone-shaped tumbling pulverizer in which the steel balls are classified, with the larger balls at the feed end where larger lumps are crushed, and the smaller balls at the discharge end where the material is finer.

conical bearing [MECH ENG] An antifriction bearing employing tapered rollers.

conical refiner [MECH ENG] In paper manufacture, a cone-shaped continuous refiner having two sets of bars mounted on the rotating plug and fixed shell for beating unmodified cellulose fibers.

coniscope *See* koniscope.

connecting rod [MECH ENG] Any straight link that transmits motion or power from one linkage to another within a mechanism, especially linear to rotary motion, as in a reciprocating engine or compressor.

connector [ENG] **1.** A detachable device for connecting electrical conductors. **2.** A metal part for joining timbers.

Conrad machine [MIN ENG] Mechanized pit digger used in checking of alluvial boring; sections of tubing, 5 feet (152 centimeters) long and 2 feet (61 centimeters) in inside diameter, are worked into the ground while spoil is removed by means of a bucket or grab.

conservation gasoline [MATER] Gasoline that is made from charging stocks consisting of vapors collected from distillation and cracking stills, from storage tanks, and from other points where condensation gasoline vapors may be escaping.

consistency [MATER] The degree of solidity or fluidity of a material such as grease, pulp, or slurry.

console [ENG] **1.** A main control desk for electronic equipment, as at a radar station, radio or television station, or airport control tower. Also known as control desk. **2.** A large cabinet for a radio or television receiver, standing on the floor rather than on a table. **3.** A grouping of controls, indicators, and similar items contained in a specially designed model cabinet for floor mounting; constitutes an operator's permanent working position.

consolidation test [MIN ENG] A test in which the specimen is confined laterally in a ring and is compressed between porous plates which are saturated with water.

constant-amplitude recording [ENG ACOUS] A sound-recording method in which all frequencies having the same intensity are recorded at the same amplitude.

constant-distance sphere [ENG ACOUS] The relative response of a sonar projector to variations in acoustic intensity, or intensity per unit band, over the surface of a sphere concentric with its center.

constant-force spring [MECH ENG] A spring which has a constant restoring force, regardless of displacement.

constant-head meter [ENG] A flow meter which maintains a constant pressure differential but varies the orifice area with

flow, such as a rotameter or piston meter.

constant-level balloon [AERO ENG] A balloon designed to float at a constant pressure level. Also known as constant-pressure balloon.

constant-load balance [ENG] An instrument for measuring weight or mass which consists of a single pan (together with a set of weights that can be suspended from a counterpoised beam) that has a constant load (200 grams for the microbalance).

constant-pressure balloon *See* constant-level balloon.

constant-pressure combustion [MECH ENG] Combustion occurring without a pressure change.

constant-pressure gas thermometer [ENG] A thermometer in which the volume occupied by a given mass of gas at a constant pressure is used to determine the temperature.

constant-speed drive [MECH ENG] A mechanism transmitting motion from one shaft to another that does not allow the velocity ratio of the shafts to be varied, or allows it to be varied only in steps.

constant-speed propeller [AERO ENG] A variable-pitch propeller having a governor which automatically changes the pitch to maintain constant engine speed.

constant-velocity recording [ENG ACOUS] A sound-recording method in which, for input signals of a given amplitude, the resulting recorded amplitude is inversely proportional to the frequency; the velocity of the cutting stylus is then constant for all input frequencies having that given amplitude.

constant-velocity universal joint [MECH ENG] A universal joint that transmits constant angular velocity from the driving to the driven shaft, such as the Bendix-Weiss universal joint.

constitution diagram [MET] Graphical representation of phase-stability relationships in an alloy system as a function of temperature. Also known as phase diagram.

constrained mechanism [MECH ENG] A mechanism in which all members move only in prescribed paths.

constraint [ENG] Anything that restricts the transverse contraction which normally occurs in a solid under longitudinal tension.

constricting nozzle [MET] A copper nozzle which has a constricting orifice and envelopes the electrode in plasma arc processes.

constrictor [AERO ENG] The exit portion of the combustion chamber in some designs of ramjets, where there is a narrowing of the tube at the exhaust.

construction [DES ENG] The number of strands in a wire rope and the number of wires in a strand; expressed as two numbers separated by a multiplication sign. [ENG] **1.** Putting parts together to form an integrated object. **2.** The manner in which something is put together.

construction area [BUILD] The area of exterior walls and permanent interior walls and partitions.

construction engineering [CIV ENG] A specialized branch of civil engineering concerned with the planning, execution, and control of construction operations for projects such as highways, dams, utility lines, and buildings.

construction equipment [MECH ENG] Heavy power machines which perform specific construction or demolition functions.

construction joint [CIV ENG] A vertical or horizontal surface in reinforced concrete where concreting was stopped and continued later.

construction paper [MATER] A heavy paper made from mechanical pulp and available in a large range of colors, most of which are not lightfast.

construction survey [CIV ENG] A survey that gives locations for construction work.

construction weight [AERO ENG] The weight of a rocket exclusive of propellant, load, and crew if any. Also known as structural weight.

construction wrench [DES ENG] An open-end wrench with a long handle; the handle is used to align matching rivet or bolt holes.

consumable electrode [MET] A metal electrode that supplies the filler metal for welding.

consumable guide electroslag welding [MET] A modification of the wire process of electroslag welding in which the filler metal is supplied by a welding wire held by a stationary metal tube.

consumable insert [MET] Filler metal which is located at the root of a welded joint and which becomes part of the final weldment.

consumer's risk [IND ENG] The probability that a lot whose quality equals the poorest quality that a consumer is willing to tolerate in an individual lot will be accepted by a sampling plan.

contact [ENG] Initial detection of an aircraft, ship, submarine, or other object on a radarscope or other detecting equipment.

contact aerator [CIV ENG] A tank in which sewage is settled on a bed of stone, cement-asbestos, or other surfaces is treated by aeration with compressed air.

contact anemometer [ENG] An anemometer which actuates an electrical contact at a rate dependent upon the wind speed. Also known as contact-cup anemometer.

contact bed [CIV ENG] A bed of coarse material such as coke, used to purify sewage.

contact ceiling [BUILD] A ceiling in which the lath and construction are in direct contact, without use of furring or runner channels.

contact condenser [MECH ENG] A device in which a vapor, such as steam, is brought into direct contact with a cooling liquid, such as water, and is condensed by giving up its latent heat to the liquid. Also known as direct-contact condenser.

contact corrosion *See* crevice corrosion.

contact-cup anemometer *See* contact anemometer.

contact gear ratio *See* contact ratio.

contact-initiated discharge machining [MECH ENG] An electromachining process in which the discharge is initiated by allowing the tool and workpiece to come into contact, after which the tool is withdrawn and an arc forms.

contact inspection [ENG] A method by which an ultrasonic search unit scans a test piece in direct contact with a thin layer of couplant for transmission between the search unit and entry surface.

contact log [PETRO ENG] Record of electrical-resistivity data pertaining to strata structures along the depth of a drill hole.

contact material [MET] A metal having high electrical and thermal conductivity, low contact resistance, minimum sticking or welding tendencies, and high corrosion resistance.

contact microphone [ENG ACOUS] A microphone designed to pick up mechanical vibrations directly and convert them into corresponding electric currents or voltages.

contact-pressure resin *See* contact resin.

contact ratio [DES ENG] The ratio of the length of the path of contact of two gears to the base pitch, equal to approximately the average number of pairs of teeth in contact. Also known as contact gear ratio.

contact resin [MATER] A liquid resin which thickens or polymerizes on heating and, when used for bonding laminates, requires little or no pressure for adherence. Also known as contact-pressure resin.

contact thermography [ENG] A method of measuring surface temperature in which a thin layer of luminescent material is spread on the surface of an object and is excited by ultraviolet radiation in a darkened room; the brightness of the coating indicates the surface temperature.

contact time [ENG] The length of time a substance is held in direct contact with a treating agent.

contact tube [MET] A device which provides electric current to a continuous electrode in a welding process.

container [IND ENG] A portable compartment of standard, uniform size, used to hold cargo for air, sea, or ground transport.

container car [ENG] A railroad car designed specifically to hold containers.

containerization [IND ENG] The practice of placing cargo in large containers such as truck trailers to facilitate loading on and off ships and railroad flat cars.

contamination [MIN ENG] Separation and accumulation of economic minerals from gangue.

continuity [CIV ENG] Joining of structural members to each other, such as floors to beams, and beams to beams and to columns, so they bend together and strengthen each other when loaded. Also known as fixity.

continuous beam [CIV ENG] **1.** A beam resting upon several supports, which may be in the same horizontal plane. **2.** A beam having several spans in one straight line; generally has at least three supports.

continuous brake [MECH ENG] A train brake that operates on all cars but is controlled from a single point.

continuous bridge [CIV ENG] A fixed bridge supported at three or more points and capable of resisting bending and shearing forces at all sections throughout its length.

continuous bucket elevator [MECH ENG] A bucket elevator on an endless chain or belt.

continuous bucket excavator [MECH ENG] A bucket excavator with a continuous bucket elevator mounted in front of the bowl.

continuous casting [MET] A technique in which an ingot, billet, tube, or other shape is continuously solidified and withdrawn while it is being poured, so that its length is not determined by mold dimensions.

continuous coal cutter [MIN ENG] A coal mining machine that cuts the coal face without being withdrawn from the cut.

continuous dryer [ENG] An apparatus in which drying is accomplished by passing wet material through without interruption.

continuous-flow conveyor [MECH ENG] A totally enclosed, continuous-belt conveyor pulled transversely through a mass of granular, powdered or small-lump material fed from an overhead hopper.

continuous footing [CIV ENG] A footing that supports a wall.

continuous furnace [MET] A type of reheating furnace in which the charge introduced at one end moves continuously through the furnace and is discharged at the other end.

continuous gas lift [PETRO ENG] Oil production in which reservoir gas pressure (natural or injected) is sufficient to provide a continuous upward flow of oil through the well tubing.

continuous industry [IND ENG] An industry in which raw material is subjected to successive operations, turning it into a finished product.

continuous kiln [ENG] **1.** A long kiln through which ware travels on a moving device, such as a conveyor. **2.** A kiln through which the fire travels progressively.

continuous mill [MET] A rolling mill in which metal is successively rolled thinner as it passes through a series of synchronized rolls in tandem.

continuous miner [MIN ENG] Machine designed to remove coal or other soft minerals from the face and to load it into cars or conveyors continuously, without the use of cutting machines, drills, or explosives.

continuous mining [MIN ENG] A type of mining in which the continuous miner cuts or rips coal or other soft minerals from the face and loads it in a continuous operation.

continuous mixer [MECH ENG] A mixer in which materials are introduced, mixed, and discharged in a continuous flow.

continuous operation [ENG] A process that operates on a continuous flow (materials or time) basis, in contrast to batch, intermittent, or sequenced operations.

continuous phase [MET] The matrix or background phase of a multiphasic alloy.

continuous precipitation [MET] Precipitation that is characteristic of certain alloys, from a supersaturated solid solution, involving a gradual change of the lattice parameter of the matrix with aging time.

continuous production [IND ENG] Manufacture of products, such as chemicals or paper, involving a sequence of processes performed by a series of machines receiving the materials through a closed channel of flow.

continuous-rail frog [ENG] A metal fitting that holds continuous welded rail sections to railroad ties.

continuous rating [ENG] The rating of a component or equipment which defines the substantially constant conditions which can be tolerated for an indefinite time without significant reduction of service life.

continuous recorder [ENG] A recorder whose record sheet is a continuous strip or web rather than individual sheets.

continuous sequence [MET] A welding sequence in which each succeeding pass is made longitudinally along the entire length of the joint.

continuous sintering [MET] Sintering process in which materials are moved through the furnace at a fixed rate without interruption.

continuous tube process [ENG] Plastics blow-molding process that uses a continuous extrusion of plastic tubing as feed to a series of blow molds as they clamp in sequence.

continuous-type furnace [MECH ENG] A furnace used for heat treatment of materials, with or without direct firing; pieces are loaded through one door, progress continuously through the furnace, and are discharged from another door.

continuous-wave Doppler radar *See* continuous-wave radar.

continuous-wave radar [ENG] A radar system in which a transmitter sends out a continuous flow of radio energy; the target reradiates a small fraction of this energy to a separate receiving antenna. Also known as continuous-wave Doppler radar.

continuous weld [MET] A weld that is continuous along the entire length of the joint.

contour forming [MET] Shaping a sheet of metal onto a shaped die.

contouring temperature recorder [ENG] A device that records data from temperature sensors towed behind a ship and then plots the vertical distribution of isotherms on a continuous basis.

contour machining [MECH ENG] Machining of an irregular surface.

contour milling [MET] Milling of an irregular surface.

contour plan [MIN ENG] A plan showing surface contours or calculated contours of coal seams to be developed.

contour turning [MECH ENG] Making a three-dimensional reproduction of the shape of a template by controlling the cutting tool with a follower that moves over the surface of a template.

contracted code sonde *See* code-sending radiosonde.

contraction crack [ENG] A crack resulting from restriction of metal in a mold while contracting.

contraction joint [CIV ENG] A break designed in a structure to allow for drying and temperature shrinkage of concrete, brickwork, or masonry, thereby preventing the formation of cracks.

contraction rule [MET] A measuring rule having larger divisions than standard measures to allow for shrinkage of a metal casting. Also known as shrinkage rule; shrink rule.

contraflexure point [CIV ENG] The point in a structure where bending occurs in opposite directions.

contrapropagating ultrasonic flowmeter [ENG] An instrument for determining the velocity of a fluid flow from the difference between the times required for high-frequency sound to travel between two transducers in opposite directions along a path having a component parallel to the flow.

contrarotating propellers [MECH ENG] A pair of propellers on concentric shafts, turning in opposite directions.

contrarotation [ENG] Rotation in the direction opposite to another rotation.

control board [ENG] A panel in which meters and other indicating instruments display the condition of a system, and dials, switches, and other devices are used to modify circuits to control the system. Also known as control panel; panel board.

control chart [IND ENG] A chart in which quantities of data concerning some property of a product or process are plotted and used specifically to determine the variation in the process.

control column [AERO ENG] A cockpit control lever pivoted or sliding in front of the pilot; controls operation of the elevator and aileron.

control diagram *See* flow chart.

control echo [ENG] In an ultrasonic inspection system, consistent reflection from a surface, such as a back reflection, which provides a reference signal.

control feel [AERO ENG] The impression of the stability and control of an aircraft that a pilot receives through the cockpit controls, either from the aerodynamic forces acting on the control surfaces or from forces simulating these aerodynamic forces.

control joint [CIV ENG] An expansion joint in masonry to allow movement due to expansion and contraction.

controllability [AERO ENG] The quality of an aircraft or guided weapon which determines the ease of producing changes in flight direction or in altitude by operation of its controls.

controllable-pitch propeller [MECH ENG] An aircraft or ship propeller in which the pitch of the blades can be changed while the propeller is in motion; five types used for aircraft are two-position, variable-pitch, constant-speed, feathering, and reversible-pitch. Abbreviated CP propeller.

controlled cooling [MET] Process by which an object is cooled from an elevated temperature in a predetermined manner to avoid cracking, internal damage, or hardening, or to produce a desired microstructure.

controlled-leakage system [AERO ENG] A system that provides for the maintenance of life in an aircraft or spacecraft cabin by a controlled escape of carbon dioxide and other waste from the cabin, with replenishment provided by stored oxygen and food.

controlled parameter [ENG] In the formulation of an optimization problem, one of the parameters whose values determine the value of the criterion parameter.

control limits [IND ENG] In statistical quality control, the limits of acceptability placed on control charts; parts outside the limits are defective.

control-moment gyro [AERO ENG] An internal momentum storage device that applies torques to the attitude-control system through large rotating gyros.

control panel *See* control board.

control plane [AERO ENG] An aircraft from which the movements of another craft are controlled remotely.

control rocket [AERO ENG] A vernier engine, retrorocket, or other such rocket used to change the attitude of, guide, or make small changes in the speed of a rocket, spacecraft, or the like.

control room [ENG] A room from which space flights are directed.

control surface [AERO ENG] **1.** Any movable airfoil used to guide or control an aircraft, guided missile, or the like in the air, including the rudder, elevators, ailerons, spoiler flaps, and trim tabs. **2.** In restricted usage, one of the main control surfaces, such as the rudder, an elevator, or an aileron.

control system [ENG] A system in which one or more outputs are forced to change in a desired manner as time progresses.

control track [ENG ACOUS] A supplementary sound track, usually containing tone signals that control the reproduction of the sound track, such as by changing feed levels to loudspeakers in a theater to achieve stereophonic effects.

control valve [ENG] A valve which controls pressure, volume, or flow direction in a fluid transmission system.

control vane [AERO ENG] A movable vane used for control, especially a movable air vane or jet vane on a rocket used to control flight altitude.

convection cooling [ENG] Heat transfer by natural, upward flow of hot air from the device being cooled.

convection section [ENG] That portion of the furnace in which tubes receive heat from the flue gases by convection.

convector [ENG] A heat-emitting unit for the heating of room air; it has a heating element surrounded by a cabinet-type enclosure with openings below and above for entrance and egress of air.

convectron [ENG] An instrument for indicating deviation from the vertical which is based on the principle that the convection from a heated wire depends strongly on its inclination; it consists of a Y-shaped tube, each of whose arms contains a wire forming part of a bridge circuit.

conventional milling [MET] Milling in which the cutter and feed move in opposite directions from the point of contact.

conventional mining [MIN ENG] The cycle which includes cutting the coal, drilling the shot holes, charging and shooting the holes, loading the broken coal, and installing roof support. Also known as cyclic mining.

conventional spinning *See* manual spinning.

convergent die [ENG] A die having internal channels which converge.

convergent-divergent nozzle [DES ENG] A nozzle in which supersonic velocities are attained; has a divergent portion downstream of the contracting section. Also known as supersonic nozzle.

conversion coating [MET] A metal-surface coating consisting of a compound of the base metal.

converter [MET] A type of furnace in which impurities are oxidized out by blowing air through or across a path of molten metal or matte.

convertiplane [AERO ENG] A hybrid form of heavier-than-air craft capable, because of one or more horizontal rotors or units acting as rotors, of taking off, hovering, and landing in a fashion similar to a helicopter; and once aloft and moving forward, capable, by means of a mechanical conversion, of flying purely as a fixed-wing aircraft, especially in higher speed ranges.

conveyor [MECH ENG] Any materials-handling machine designed to move individual articles such as solids or free-flowing bulk materials over a horizontal, inclined, declined, or vertical path of travel with continuous motion.

coolant [MATER] **1.** A cutting fluid for machine operations, which keeps the tool cool to prevent reduction in hardness and resistance to abrasion, and prevents distortion of the work. **2.** A substance, ordinarily fluid, used for cooling any part of a reactor in which heat is generated.

3. In general, any cooling agent, usually a fluid.

cooled-tube pyrometer [ENG] A thermometer for high-temperature flowing gases that uses a liquid-cooled tube inserted in the flowing gas; gas temperature is deduced from the law of convective heat transfer to the outside of the tube and from measurement of the mass flow rate and temperature rise of the cooling liquid.

cooler nail [DES ENG] A thin, cement-coated wire nail.

cooling channel [ENG] A channel in the body of mold through which a cooling liquid is circulated.

cooling coil [MECH ENG] A coiled arrangement of pipe or tubing for the transfer of heat between two fluids.

cooling degree day [MECH ENG] A unit for estimating the energy needed for cooling a building; one unit is given for each degree Fahrenheit that the daily mean temperature exceeds 75°F (24°C).

cooling fin [MECH ENG] The extended element of a heat-transfer device that effectively increases the surface area.

cooling fixture [ENG] A wooden or metal block used to hold the shape or dimensional accuracy of a molding until it cools enough to retain its shape.

cooling load [MECH ENG] The total amount of heat energy that must be removed from a system by a cooling mechanism in a unit time, equal to the rate at which heat is generated by people, machinery, and processes, plus the net flow of heat into the system not associated with the cooling machinery.

cooling power [MECH ENG] A parameter devised to measure the air's cooling effect upon a human body; it is determined by the amount of heat required by a device to maintain the device at a constant temperature (usually 34°C); the entire system should be made to correspond, as closely as possible, to the external heat exchange mechanism of the human body.

cooling-power anemometer [ENG] Any anemometer operating on the principle that the heat transfer to air from an object at an elevated temperature is a function of airspeed.

cooling process [ENG] Physical operation in which heat is removed from process fluids or solids; may be by evaporation of liquids, expansion of gases, radiation or heat exchange to a cooler fluid stream, and so on.

cooling range [MECH ENG] The difference in temperature between the hot water entering and the cold water leaving a cooling tower.

cooling table *See* hotbed.

cooling tower [ENG] A towerlike device in which atmospheric air circulates and cools warm water, generally by direct contact (evaporation).

coolometer [ENG] An instrument which measures the cooling power of the air, consisting of a metal cylinder electrically heated to maintain a constant temperature; the electrical heating power required is taken as a measure of the air's cooling power.

cool time [MET] The period of time between successive heat times in pulsation and seam welding.

cooperative system [ENG] A missile guidance system that requires transmission of information from a remote ground station to a missile in flight, processing of the information by the missile-borne equipment, and retransmission of the processed data to the originating or other remote ground stations, as in azusa and dovap.

coordinating holes [DES ENG] Holes in two parts of an assembly which form a single continuous hole when the parts are joined.

copaiba balsam [MATER] An oleoresin extracted from trees of the genus *Copaifera* of South America; used as a plasticizer and in medicine.

copal [MATER] Hard, resinous substance exuded from certain trees in the East Indies, South America, and Africa and used in varnish and printing ink.

cope [MET] The upper portion of a flask, mold, or pattern.

cope chisel [DES ENG] A chisel used to cut grooves in metal.

coping [BUILD] A covering course on a wall. [MECH ENG] Shaping stone or other nonmetallic substance with a grinding wheel. [MIN ENG] **1.** Process of cutting and trimming the edges of stone slabs. **2.** Process of cutting a stone slab into two pieces.

coping saw [DES ENG] A type of handsaw that has a narrow blade, usually about ⅛ inch (3 millimeters) wide, held taut by a U-shaped frame equipped with a handle; used for shaping and cutout work.

copper [MET] One of the most important nonferrous metals; a ductile and malleable metal found in various ores and used in industry, engineering, and the arts in both pure and alloyed form.

copper alloy [MET] A solid solution of one or more metals in copper.

copper amalgam [MET] An alloy of copper and mercury.

copper brazing [MET] Brazing by using copper as the filler metal.

copper converter [MET] A converter for purifying copper.

copper dirt *See* sour dirt.

copper plating [MET] Coating of a substance with copper by an electrolytic process, to minimize corrosion.

copper powder [MET] A bronzing powder made by saturating nitrous acid with copper and precipitating the latter by the addition of iron.

copper steel [MET] Low-carbon steel containing up to 0.25% copper.

copper strip corrosion [ENG] A qualitative method of determining the corrosivity of a petroleum product by observing its effect on a strip of polished copper suspended or placed in the product.

copperweld [MET] Copper-covered steel, used as a conductor for high-voltage transmission spans where tensile strength is more important than high conductance.

copper wire [MET] Wire commonly made from copper by drawing from a hot-rolled rod without annealing; however, the smaller sizes may involve intermediate anneals.

corbinotron [ENG] The combination of a corbino disk, made of high-mobility semiconductor material, and a coil arranged to produce a magnetic field perpendicular to the disk.

cord [MATER] **1.** A unit of measure for wood stacked for fuel or pulp; equals 4 × 4 × 8, or 128 cubic feet (approximately 3.6246 cubic meters). **2.** A long, flexible, cylindrical construction of natural or synthetic fibers twisted or woven together.

cordage [ENG] Number of cords of lumber per given area. [MATER] Ropes or cords, especially those in the rigging of a ship.

cord foot [ENG] A stack of wood measuring 16 cubic feet (approximately 0.45307 cubic meter).

Cordirie process [MET] The refining of lead by conducting steam through it, while molten, to oxidize certain metallic impurities.

cordite [MATER] A trinitrate cellulose derivative prepared by treating cotton fiber or purified wood pulp with a mixture of nitric and sulfuric acids; an explosive powder.

cord of ore [MIN ENG] A unit of about 7 tons (6.35 metric tons), but measured by wagonloads and not by weight.

cordovan [MATER] Nonporous leather made from split horsehide and tanned with vegetable materials.

cord tire [DES ENG] A pneumatic tire made with cords running parallel to the tread.

cordwood [MATER] Wood stacked and sold in cords.

core [ENG] The inner material of a wall, column, veneered door, or similar structure. [MET] A specially formed part of a mold used to form internal holes in a casting.

core barrel [DES ENG] A hollow cylinder attached to a specially designed bit; used to obtain a continuous section of the rocks penetrated in drilling.

core barrel rod *See* guide rod.

core binder [MATER] A substance for binding core sand together; an example is core oil.

core bit [DES ENG] The hollow, cylindrical cutting part of a core drill.

core blower [MET] A machine using compressed air to blow and pack sand into a core box.

core box [MET] A container for shaping a sand core for a casting.

core catcher *See* split-ring core lifter.

core-catcher case *See* lifter case.

cored ammonium nitrate dynamite [MATER] A class of dynamite that has good water resistance and exhibits velocities higher than straight ammonia dynamite; the gelatin core provides for detonation in the complete explosive column.

cored bar [MET] A bar-shaped powder compact which has been heated by electricity to melt the interior.

cored electrode [MET] An electrode made of metal with a core of flux or other material.

core drier [MET] A light, skeleton cast-iron or aluminum box, whose internal shape conforms closely to the cope portion of a core for molding, used to support, during baking, a core which cannot be placed on a flat plate.

core drill [MECH ENG] A mechanism designed to rotate and to cause an annular-shaped rock-cutting bit to penetrate rock formations, produce cylindrical cores of the formations penetrated, and lift such cores to the surface, where they may be collected and examined.

cored solder [MET] Soldering wire which has a core consisting of flux.

core gripper *See* split-ring core lifter.

core-gripper case *See* lifter case.

core iron [MET] A grade of soft iron suitable for cores of chokes, transformers, and relays.

coreless-type induction heater [ENG] A device in which a charge is heated directly by induction, with no magnetic core material linking the charge. Also known as coreless-type induction furnace.

core lifter *See* split-ring core lifter.

core-lifter case *See* lifter case.

core molding [MET] A molding process which makes use of assembled cores to construct the mold.

core oil [MATER] An oil compound used with sand to make foundry core.

core oven [MET] An oven used for baking cores for molding; the walls are constructed of inner and outer layers of sheet metal separated by rock wool or fiber glass insulation, with interlocked joints.

core print [MET] A projection on a cylindrical casting pattern which supports a core.

corer [ENG] An instrument used to obtain cylindrical samples of geological materials or ocean sediments.

core rod [MET] The part of a die used to make a hole in a compact.

core sand [MATER] Sand used in a core for molding, made from standard molding-sand mixtures or from silica sand, usually with a binder. Also known as foundry core sand.

core-spring case See lifter case.

core wall See cutoff wall.

core wire [MET] Copper wire having a steel core, often used for antennas.

coriander oil [MATER] An essential oil distilled from the fruit of coriander; principal constituents are linalool and pinene; used as a flavoring in gin.

coring [MET] A variable composition of individual crystals across a casting, due to nonequilibrium growth over a range of temperature; the purest material is near the center. [PETRO ENG] The use of a core barrel (hollow length of tubing) to take samples from the underground formation during the drilling operation; used for core analysis.

Coriolis-type mass flowmeter [ENG] An instrument which determines mass flow rate from the torque on a ribbed disk that is rotated at constant speed when fluid is made to enter at the center of the disk and is accelerated radially.

corkboard [MATER] Board made of compressed cork.

cork paint [MATER] A paint containing fine cork particles; used on steel parts on ships to prevent sweating.

cork tile [MATER] Floor tile made of compressed cork bound with phenolic or other resin binders; used on moisture-free rigid subfloors or on plywood or hardboard.

Corliss valve [MECH ENG] An oscillating type of valve gear with a trip mechanism for the admission and exhaust of steam to and from an engine cylinder.

corner bead [BUILD] **1.** Any vertical molding used to protect the external angle of the intersecting surfaces. **2.** A strip of formed galvanized iron, sometimes combined with a strip of metal lath, placed on corners to reinforce them before plastering.

corner chisel [DES ENG] A chisel with two cutting edges at right angles.

corner effect [ENG] In ultrasonic testing, reflection of an ultrasonic beam directed perpendicular to the intersection of two surfaces 90° apart.

corner head [BUILD] A metal molding that is built into plaster in corners to prevent plaster from accidentally breaking off.

cornering tool [DES ENG] A cutting tool with a curved edge, used to round off sharp corners.

cornerite [BUILD] A corner reinforcement for interior plastering.

corner joint [ENG] An L-shaped joint formed by two members positioned perpendicular to each other.

cornerstone [BUILD] An inscribed stone laid at the corner of a building, usually at a ceremony.

cornice brake [MECH ENG] A machine used to bend sheet metal into different forms.

Cornish rolls [MIN ENG] A geared pair of horizontal cylinders, one fixed in a frame and the other held by strong springs; used for grinding.

corn oil [MATER] A semidrying, fatty oil of yellowish color, extracted from germs of corn kernels; used mainly as a salad oil, in soft soaps, and in compounded petroleum lubricants.

coromant cut [MIN ENG] A drill hole pattern; two overlapping holes about 2¼ inches (about 5.5 centimeters) in diameter are drilled in the tunnel center and left uncharged; they form a slot roughly 4 by 2 inches (10 by 5 centimeters) to which the easers can break.

corona [MET] An area sometimes surrounding the nugget at the faying surfaces of a spot weld which provides a degree of bond strength.

coronizing [MET] Process to electroplate zinc on nickel, thermally treated at 375°C; coating is used on ferrous and copper-based alloys to give resistance to sulfur dioxide, SO_2, and sulfur trioxide, SO_3.

correcting wedge [MIN ENG] A deflection wedge used to deflect a crooked borehole back into its intended course.

corrector [ENG] A magnet, piece of soft iron, or device used in the adjustment or compensation of a magnetic compass.

correlated orientation tracking and range See cotar.

correlation detection [ENG] A method of detection of aircraft or space vehicles in which a signal is compared, point to point, with an internally generated reference. Also known as cross-correlation detection.

correlation direction finder [ENG] Satellite station separated from a radar to receive jamming signals; by correlating the signals received from several such stations, range and azimuth of many jammers may be obtained.

correlation tracking and triangulation *See* cotat.

correlation tracking system [ENG] A trajectory-measuring system utilizing correlation techniques where signals derived from the same source are correlated to derive the phase difference between the signals.

correlation ultrasonic flowmeter [ENG] An instrument for determining the velocity of a fluid flow from the time required for discontinuities in the fluid stream to pass between two pairs of transducers that generate and detect high-frequency sound.

correlative rights [PETRO ENG] Legal rights protecting property over a portion of a gas or oil reservoir from excessive or wasteful withdrawal of hydrocarbons by adjoining properties overlying the same reservoir.

corroding lead [MET] Lead that can be corroded to make white lead.

Corrodkote test [MET] An accelerated corrosion test in which the article is coated with a slurry of clay and a salt solution and then exposed to a high humidity for a specified period.

corrosion [MET] Gradual destruction of a metal or alloy due to chemical processes such as oxidation or the action of a chemical agent.

corrosion fatigue [MET] Damage to or failure of a metal due to corrosion combined with fluctuating fatigue stresses.

corrosion fatigue limit [MET] The maximum stress that a corroded material can withstand for a given number of stress reversals.

corrosion potential [MET] The measure of corroding surface potential in an electrolyte in relation to a reference electrode while the circuit is open.

corrosion protection [MET] The minimization of corrosion by coating with a protective metal, with an oxide or phosphide or similar substance, or with a protective paint, or by rendering the metal passive.

corrosion test [MET] Any of various tests to determine the resistance of a metal to chemical attack.

corrosive [MATER] A substance that causes corrosion.

corrosive flux [MET] A soldering flux, usually composed of inorganic salts and acids, which provides oxide removal of the base metal upon application of solder; flux remaining on the base metal is corrosive and should be removed.

corrosiveness [MET] The tendency of a metal to wear away another by chemical attack.

corrugated bar [DES ENG] Steel bar with transverse ridges; used in reinforced concrete.

corrugated fastener [DES ENG] A thin corrugated strip of steel that can be hammered into a wood joint to fasten it.

corrugating [DES ENG] Forming straight, parallel, alternate ridges and grooves in sheet metal, cardboard, or other material.

cosmic-ray telescope [ENG] Any device for detecting and determining the directions of either cosmic-ray primary protons and heavier-element nuclei, or the products produced when these particles interact with the atmosphere.

cosmonaut [AERO ENG] A Soviet astronaut.

Cosmos satellites [AERO ENG] A series of earth satellites launched by the Soviet Union starting in 1962 to conduct geophysical experiments.

Cosslett process [MET] A process in which iron or steel articles immersed for 3 or 4 hours in a boiling solution, made by mixing iron filings with concentrated H_3PO_4 (sufficient to form a paste) and then adding to weak phosphoric acid, become coated with a rust-resisting deposit of basic ferrous phosphate.

cost accounting [IND ENG] The branch of accounting in which one records, analyzes, and summarizes costs of material, labor, and burden, and compares these actual costs with predetermined budgets and standards.

cost analysis [IND ENG] Analysis of the factors contributing to the costs of operating a business and of the costs which will result from alternative procedures, and of their effects on profits.

cost function [SYS ENG] In decision theory, a loss function which does not depend upon the decision rule.

cost-plus contract [ENG] A contract under which a contractor furnishes all material, construction equipment, and labor at actual cost, plus an agreed-upon fee for his services.

cotar [ENG] A passive system used for tracking a vehicle in space by determining the line of direction between a remote ground-based receiving antenna and a telemetering transmitter in the missile, using phase-comparison techniques. Derived from correlated orientation tracking and range.

cotat [ENG] A trajectory-measuring system using several antenna base lines, each separated by large distances, to measure direction cosines to an object; then the object's space position is computed by triangulation. Derived from correlation tracking and triangulation.

cotter [DES ENG] A tapered piece that can be driven in a tapered hole to hold together an assembly of machine or structural parts.

cottered joint [MECH ENG] A joint in which a cotter, usually a flat bar tapered on one side to ensure a tight fit, transmits power by shear on an area at right angles to its length.

cotter pin [DES ENG] A split pin, inserted into a hole, to hold a nut or cotter securely to a bolt or shaft, or to hold a pair of hinge plates together.

cotton oil [MATER] The yellow, viscous fixed oil, containing principally linoleic acid, pressed from the seeds of various *Gossypium* species; the refined oil is colorless and used in foods and some pharmaceutical preparations. Also known as cottonseed oil; oleum gossypii seminis.

cottonseed oil *See* cotton oil.

cotton wax [MATER] The wax composing cottonseed coating.

Cottrell precipitator [ENG] A machine for removing dusts and mists from gases, in which the gas passes through a grounded pipe with a fine axial wire at a high negative voltage, and particles are ionized by the corona discharge of the wire and migrate to the pipe.

Couette viscometer [ENG] A viscometer in which the liquid whose viscosity is to be measured fills the space between two vertical coaxial cylinders, the inner one suspended by a torsion wire; the outer cylinder is rotated at a constant rate, and the resulting torque on the inner cylinder is measured by the twist of the wire. Also known as rotational viscometer.

coulisse [ENG] A piece of wood that has a groove cut in it to enable another piece of wood to slide in it. Also known as cullis.

coulombmeter [ENG] A measuring instrument that measures quantity of electricity in coulombs by integrating a stored charge in a circuit which has very high input impedance.

count [AERO ENG] **1.** To proceed from one point to another in a countdown or plus count, normally by calling a number to signify the point reached. **2.** To proceed in a countdown, for example, T minus 90 and counting.

countdown [AERO ENG] **1.** The process in the engineering definition, used in leading up to the launch of a large or complicated rocket vehicle, or in leading up to a captive test, a readiness firing, a mock firing, or other firing test. **2.** The act of counting inversely during this process. [ENG] A step-by-step process that culminates in a climatic event, each step being performed in accordance with a schedule marked by a count in inverse numerical order.

counter [ENG] A complete instrument for detecting, totalizing, and indicating a sequence of events.

counterbalance *See* counterweight.

counterblow hammer [MECH ENG] A forging hammer in which the ram and anvil are driven toward each other by compressed air or steam.

counterbore [DES ENG] A flat-bottom enlargement of the mouth of a cylindrical bore to enlarge a borehole and give it a flat bottom. [ENG] To enlarge a borehole by means of a counterbore.

countercurrent flow [MECH ENG] A sensible heat-transfer system in which the two fluids flow in opposite directions.

countercurrent spray dryer [ENG] A dryer in which drying gases flow in a direction opposite to that of the spray.

counterflow [ENG] Fluid flow in opposite directions in adjacent parts of an apparatus, as in a heat exchanger.

counterfort [CIV ENG] A strengthening pier perpendicular and bonded to a retaining wall.

counterfort wall [CIV ENG] A type of retaining wall that resembles a cantilever wall but has braces at the back; the toe slab is a cantilever and the main steel is placed horizontally.

counter/frequency meter [ENG] An instrument that contains a frequency standard and can be used to measure the number of events or the number of cycles of a periodic quantity that occurs in a specified time, or the time between two events.

counterlath [BUILD] **1.** A strip placed between two rafters to support crosswise laths. **2.** A lath placed between a timber and a sheet lath. **3.** A lath nailed at a more or less random spacing between two precisely spaced laths. **4.** A lath put on one side of a partition after the other side has been finished.

counterpoise *See* counterweight.

countershaft [MECH ENG] A secondary shaft that is driven by a main shaft and from which power is supplied to a machine part.

countersink [DES ENG] The tapered and relieved cutting portion in a twist drill, situated between the pilot drill and the body.

countersinking [MECH ENG] Drilling operation to form a flaring depression around the rim of a hole.

counterweight [MECH ENG] **1.** A device which counterbalances the original load in elevators and skip and mine hoists, going up when the load goes down, so that the engine must only drive against the unbalanced load and overcome friction. **2.** Any weight placed on a mechanism which is out of balance so as to maintain static equilibrium. Also known as counterbalance; counterpoise.

couplant [ENG] A substance such as water, oil, grease, or paste used to avoid the retarding of sound transmission by air between the transducer and the test piece during ultrasonic examination.

couple [ENG] To connect with a coupling, such as of two belts or two pipes.

coupled engine [MECH ENG] A locomotive engine having the driving wheels connected by a rod.

coupler [ENG] A device that connects two railroad cars.

coupling [ENG] Any device that serves to connect the ends of adjacent parts, as railroad cars. [MECH ENG] The mechanical fastening that connects shafts together for power transmission. Also known as shaft coupling.

course [CIV ENG] A row of stone, block, or brick of uniform height.

coursed rubble [CIV ENG] Masonry in which rough stones are fitted into approximately level courses.

coursing joint [CIV ENG] A mortar joint connecting two courses of brick or pebble.

cover [MIN ENG] The thickness of rock between the mine workings and the surface.

covered electrode *See* coated electrode.

cover half [MET] The stationary portion of a die.

cover hole [MIN ENG] One of a group of boreholes drilled in advance of mine workings to probe for and detect water-bearing fissures or structures.

covering power [ENG] The degree to which a coating obscures the underlying material. [MET] The ability of an electroplating bath to produce a coating at a low current density.

cover plate [ENG] A pane of glass in a welding helmet or goggles which protects the colored lens excluding harmful light rays from damage by weld spatter.

Cowell method [AERO ENG] A method of orbit computation using direct step-by-step integration in rectangular coordinates of the total acceleration of the orbiting body.

cowling [AERO ENG] The streamlined metal cover of an aircraft engine. [ENG] A metal cover that houses an engine.

coyote blasting [MIN ENG] A method of blasting in which large charges are fired in small adits or tunnels driven at the level of the floor, in the face of a quarry or the slope of an open-pit mine. Also known as coyote-hole blasting; gopher-hole blasting; heading blasting.

coyote hole *See* gopher hole.

coyote-hole blasting *See* coyote blasting.

CPM *See* critical path method.

crab locomotive [MIN ENG] A type of trolley locomotive equipped with an electric motor, a drum, and haulage cable mounted on a small truck; used to haul mine cars from workings.

crack [ENG] To open something slightly, for instance, a valve.

cracked [MATER] Applied to those oils produced by the cracking process rather than straight distillation.

cracked gasoline [MATER] Gasoline manufactured by heating crude petroleum distillation fractions or residues under pressure, or by heating with or without pressure in the presence of a catalyst, so that heavier hydrocarbons are broken into others, some of which distill in the gasoline range.

cracking [ENG] Presence of relatively large cracks extending into the interior of a structure, usually produced by overstressing the structural material.

cradle [ENG] A framework or other resting place for supporting or restraining objects.

cradle dump [MIN ENG] A tipple which dumps cars with a rocking motion.

cramp [DES ENG] A metal plate with bent ends used to hold blocks together.

crampon [DES ENG] A device for holding heavy objects such as rock or lumber to be lifted by a crane or hoist; shaped like scissors, with points bent inward for grasping the load. Also spelled crampoon.

crampoon *See* crampon.

crane [MECH ENG] A hoisting machine with a power-operated inclined or horizontal boom and lifting tackle for moving loads vertically and horizontally.

crane hoist [MECH ENG] A mobile construction machine built principally for lifting loads by means of cables and consisting of an undercarriage on which the unit moves, a cab or house which envelops the main frame and contains the power units and controls, and a movable boom over which the cables run.

crane hook [DES ENG] A hoisting fixture designed to engage a ring or link of a lifting chain, or the pin of a shackle or cable socket.

crane truck [MECH ENG] A crane with a jiblike boom mounted on a truck. Also known as yard crane.

crank [MECH ENG] A link in a mechanical linkage or mechanism that can turn about a center of rotation.

crank angle [MECH ENG] **1.** The angle between a crank and some reference direction. **2.** Specifically, the angle between the crank of a slider crank mechanism and a line from crankshaft to the piston.

crank arm [MECH ENG] The arm of a crankshaft attached to a connecting rod and piston.

crank axle [MECH ENG] **1.** An axle containing a crank. **2.** An axle bent at both ends so that it can accommodate a large body with large wheels.

crankcase [MECH ENG] The housing for the crankshaft of an engine, where, in the case of an automobile, oil from hot engine parts is collected and cooled before returning to the engine by a pump.

crankpin [DES ENG] A cylindrical projection on a crank which holds the connecting rod.

crank press [MECH ENG] A punch press that applies power to the slide by means of a crank.

crankshaft [MECH ENG] The shaft about which a crank rotates.

crank throw [MECH ENG] **1.** The web or arm of a crank. **2.** The displacement of a crankpin from the crankshaft.

crank web [MECH ENG] The arm of a crank connecting the crankshaft to crankpin, or connecting two adjacent crankpins.

crater [MECH ENG] A depression in the face of a cutting tool worn down by chip contact. [MET] A depression at the end of the weld head or under the electrode during welding.

crater cuts [MIN ENG] Cuts with one or more fully charged holes in which blasting is conducted toward the face of the tunnel.

crawler [MECH ENG] **1.** One of a pair of an endless chain of plates driven by sprockets and used instead of wheels by certain power shovels, tractors, bulldozers, drilling machines, and such, as a means of propulsion. **2.** Any machine mounted on such tracks.

crawler crane [MECH ENG] A self-propelled crane mounted on two endless tracks that revolve around wheels.

crawler tractor [MECH ENG] A tractor that propels itself on two endless tracks revolving around wheels.

crawler wheel [MECH ENG] A wheel that drives a continuous metal belt, as on a crawler tractor.

crawl space [BUILD] **1.** A shallow space in a building which workers can enter to gain access to pipes, wires, and equipment. **2.** A shallow space located below the ground floor of a house and surrounded by the foundation wall.

crazing [ENG] A network of fine cracks on or under the surface of a material such as enamel, glaze, metal, or plastic. [MET] Development of a network of cracks on a metal surface.

creep [ENG] The tendency of wood to move while it is being cut, particularly when being mitered. [MIN ENG] *See* squeeze.

creeper [ENG] A low platform on small casters that is used for back support and mobility when a person works under a car. [MIN ENG] An endless chain that catches mine car axles on projecting bars.

creep test [ENG] Any one of a number of methods of measuring creep, for example, by subjecting a material to a constant stress or deforming it at a constant rate.

cremone bolt [DES ENG] A fastening for double doors or casement windows; employs vertical rods that move up and down to engage the top and bottom of the frame.

creosote [MATER] A colorless or yellowish oily liquid containing a mixture of phenolic compounds obtained by distillation of tar; commercial creosote is distilled from coal tar, and pharmaceutical creosote is distilled from wood tar.

creosote oil [MATER] A coal tar fraction, boiling between 240 and 270°F (116–132°C); used for producing materials such as creosote and tar acids and used directly as a germicide, insecticide, or pesticide.

crepe paper [MATER] A lightweight, crinkled paper available in many colors and used for displays, floats, and decorations; it has no strength when wet, and the colors run.

crescent beam [ENG] A beam bounded by arcs having different centers of curvature, with the central section the largest.

crest [DES ENG] The top of a screw thread.

crestal injection *See* external gas injection.

crest clearance [DES ENG] The clearance, in a radial direction, between the crest of the thread of a screw and the root of the thread with which the screw mates.

crest gate [CIV ENG] A gate in the spillway of a dam which functions to maintain or change the water level.

cresylic acid [MATER] **1.** A mixture of phenols containing varying amounts of xylenols, cresols, and other high-boiling fractions. **2.** A crude mixture of the three cresol isomers.

crevice corrosion [MET] Corrosive degradation of metal parts at the crevices left at rolled joints or from other forming procedures; common in stainless steel heat exchangers in contact with chloride-containing fluids or other dissolved corrosives. Also known as contact corrosion.

crib [CIV ENG] The space between two successive ties along a railway track. [ENG] **1.** Any structure composed of a layer of timber or steel joists laid on the ground, or two layers across each other, to spread a load. **2.** Any structure composed of frames of timber placed horizontally on top of each other to form a wall.

cricket [BUILD] A device that is used to divert water at the intersections of roofs or at the intersection of a roof and chimney.

crimp [ENG] **1.** To cause something to become wavy, crinkled, or warped, such as lumber. **2.** To pinch or press together, especially a tubular or cylindrical shape, in order to seal or unite.

cripple [BUILD] A structural member, such as a stud above a window, that is cut less than full length.

critical altitude [AERO ENG] The maximum altitude at which a supercharger can maintain a pressure in the intake manifold of an engine equal to that existing during normal operation at rated power and speed at sea level without the supercharger.

critical angle of attack [AERO ENG] The angle of attack of an airfoil at which the flow of air about the airfoil changes abruptly so that lift is sharply reduced and drag is sharply increased. Also known as stalling angle of attack.

critical compression ratio [MECH ENG] The lowest compression ratio which allows compression ignition of a specific fuel.

critical cooling rate [MET] The minimum cooling rate that will suppress undesired transformations in a metal.

critical density [CIV ENG] For a highway, the density of traffic when the volume equals the capacity.

critical flow prover [PETRO ENG] Device used to measure the velocity of gas flow during open-flow testing of gas wells.

critical gas saturation *See* equilibrium gas saturation.

critical humidity [MET] The atmospheric humidity above which the corrosion rate increases rapidly for a particular metal.

critical Mach number [AERO ENG] The free-stream Mach number at which a local Mach number of 1.0 is attained at any point on the body under consideration.

critical path method [SYS ENG] A systematic procedure for detailed project planning and control. Abbreviated CPM.

critical range [MET] The temperature range for the reversible change of austenite to ferrite, pearlite, and cementite.

critical speed [MECH ENG] The angular speed at which a rotating shaft becomes dynamically unstable with large lateral amplitudes, due to resonance with the natural frequencies of lateral vibration of the shaft.

critical strain [MET] The strain at which heating causes rapid growth of large grains in many metals and alloys; phase transformations do not occur.

critical velocity [AERO ENG] In rocketry, the speed of sound at the conditions prevailing at the nozzle throat. Also known as throat velocity.

crochet file [DES ENG] A thin, flat, round-edged file that tapers to a point.

Crockett magnetic separator [MIN ENG] An assembly consisting of a continuous belt submerged in a tank through which ore pulp flows; magnetic solids adhere to the belt, which has a series of flat magnets attached to it, and the solids are dragged clear.

crocodile shears *See* lever shears.

crocodiling *See* alligatoring.

crocus [MATER] Finely powdered oxide of iron, of dark red color, used for buffing and polishing.

croning process [MET] A shell-molding process.

crooked hole [PETRO ENG] A borehole drilled at an angle, often because of steeply dipping formations; not to be confused with holes deliberately deviated from the vertical to avoid obstacles or to tap otherwise unavailable reservoirs.

Crookes glass [MATER] A type of glass that contains cerium and other rare earths and has a high absorption of ultraviolet radiation; used in sunglasses.

crop [MET] Defective end portion of an ingot which is removed for scrap before rolling the ingot.

crop coal [MIN ENG] Coal of inferior quality found near the surface.

cross axle [MECH ENG] **1.** A shaft operated by levers at its ends. **2.** An axle with cranks set at 90°.

crossbar [CIV ENG] In a grating, one of the connecting bars which extend across bearing bars, usually perpendicular to them.

crossbar micrometer [ENG] An instrument consisting of two bars mounted perpendicular to each other in the focal plane of a telescope, and inclined to the east-west path of stars by 45°; used to measure differences in right ascension and declination of celestial objects.

cross-belt drive [DES ENG] A belt drive having parallel shafts rotating in opposite directions.

crossbolt [DES ENG] A lock bolt with two parts which can be moved in opposite directions.

cross bond [CIV ENG] A masonry bond in which a course of alternating lengthwise and endwise bricks (Flemish bond) alternates with a course of bricks laid lengthwise.

cross box [MECH ENG] A boxlike structure for the connection of circulating tubes to the longitudinal drum of a header-type boiler.

cross bracing [BUILD] Boards which are nailed diagonally across studs or other boards so as to impart rigidity to a framework.

cross-correlation detection *See* correlation detection.

cross-country mill [MET] A rolling mill in which the tables of the mill stands are parallel with a crossover table that connects them; used to produce special forms of bar stock.

crosscut [ENG] A cut made through wood across the grain. [MIN ENG] **1.** A small passageway driven at right angles to the main entry of a mine to connect it with a parallel entry of air course. **2.** A passageway in a mine that cuts across the geological structure.

crosscut file [DES ENG] A file with a rounded edge on one side and a thin edge on the other; used to sharpen straight-sided saw teeth with round gullets.

crosscut saw [DES ENG] A type of saw for cutting across the grain of the wood; designed with about eight teeth per inch.

cross drum boiler [MECH ENG] A sectional header or box header type of boiler in which the axis of the horizontal drum is perpendicular to the axis of the main bank of tubes.

crossed belt [MECH ENG] A pulley belt arranged so that the sides cross, thereby making the pulleys rotate in opposite directions.

crossed-field accelerator [AERO ENG] A plasma engine for space travel in which plasma serves as a conductor to carry current across a magnetic field, so that a resultant force is exerted on the plasma.

cross-fade [ENG ACOUS] In dubbing, the overlapping of two sound tracks, wherein the outgoing track fades out while the incoming track fades in.

cross-flow [AERO ENG] A flow going across another flow, as a spanwise flow over a wing.

cross-flow plane [AERO ENG] A plane at right angles to the free-stream velocity.

cross furring ceiling [BUILD] A ceiling in which furring members are attached perpendicular to the main runners or other structural members.

cross gateway *See* cross heading.

cross hair [ENG] An inscribed line or a strand of hair, wire, silk, or the like used in an optical sight, transit, or similar instrument for accurate sighting.

crosshaul [MECH ENG] A device for loading objects onto vehicles, consisting of a chain that is hooked on opposite sides of a vehicle, looped under the object, and connected to a power source and that rolls the object onto the vehicle.

crosshead [MECH ENG] A block sliding between guides and containing a wrist pin for the conversion of reciprocating to rotary motion, as in an engine or compressor. [MET] A device generally employed in wire coating which is attached to the discharge end of the extruder cylinder; designed to facilitate extruding material at an angle. [MIN ENG] A runner or guide positioned just above a sinking bucket to restrict excessive swinging.

cross heading [MIN ENG] Mine passage driven for ventilation from the airway to the gangway, or from one breast through the pillar to the adjoining working. Also known as cross gateway; cross hole; headway.

cross hole *See* cross heading.

crossing plates [CIV ENG] Plates placed between a crossing and the ties to support the crossing and protect the ties.

cross-level [ENG] To level at an angle perpendicular to the principal line of sight.

crossover [CIV ENG] An S-shaped section of railroad track joining two parallel tracks.

crossover flange [ENG] Intermediate pipe flange used to connect flanges of different working pressures.

crossover frequency [ENG ACOUS] **1.** The frequency at which a dividing network delivers equal power to the upper and lower frequency channels when both are terminated in specified loads. **2.** *See* transition frequency.

crossover network [ENG ACOUS] A selective network used to divide the audio-frequency output of an amplifier into two or more bands of frequencies. Also known as dividing network; loudspeaker dividing network.

crossover spiral *See* lead-over groove.

cross-peen hammer [ENG] A hammer with a wedge-shaped surface at one end of the head.

cross-rolling [MET] **1.** Straightening metal sheets by passing them through rolls at right angles to the principal direction of rolling. **2.** Straightening round bars or tubes by passing the work through parallel to the axes of rolls.

cross slide [MECH ENG] A part of a machine tool that allows the tool carriage to move at right angles to the main direction of travel.

crosstalk [ENG ACOUS] *See* magnetic printing.

crosstie [ENG] A timber or metal sill placed transversely under the rails of a railroad, tramway, or mine-car track.

cross turret [MECH ENG] A turret that moves horizontally and at right angles to the lathe guides.

crossvein [MIN ENG] A vein that intersects an older or larger vein.

cross-wire weld [MET] A weld made across wires or bars in order to make wire mesh or other similar products.

croton oil [MATER] A yellow-brown oil obtained from the seeds of the *Croton tiglium;* used as a purgative and as a substitute for castor oil.

crowbar [DES ENG] An iron or steel bar that is usually bent and has a wedge-shaped working end; used as a lever and for prying.

Crowe process [MET] In cyanidation, after extraction of the gold, separation of the solution from the ore tailings by filtration or countercurrent decantation and by passage through a vacuum chamber where deaeration occurs.

crown [CIV ENG] Center of a roadway elevated above the sides. [ENG] **1.** The part of a drill bit inset with diamonds. **2.** The vertex of an arch or arched surface. **3.** The top or dome of a furnace or kiln. [MET] That part of the sheet or roll where the thickness or diameter increases from edge to center. [MIN ENG] A horizontal roof member of a timber up to 16 feet (4.9 meters) long and supported at each end by an upright.

crown block [PETRO ENG] A wooden or steel beam joined to the tops of derrick posts of an oil well to support pulleys.

crown glass [MATER] A soda-lime glass, typically having 72% SiO_2, 13% CaO, and 15% Na_2O, which is hard and will take a simple polish; highly transparent for visible light.

crown saw [DES ENG] A saw consisting of a hollow cylinder with teeth around its edge; used for cutting round holes. Also known as hole saw.

crown sheet [MECH ENG] The structural element which forms the top of a furnace in a fire-tube boiler.

crown wheel [DES ENG] A gear that is light and crown-shaped.

crucible melt extraction [MET] Melt extraction in which the molten metal is contained in a crucible.

crucible steel *See* drill steel.

cruciform wing [AERO ENG] An aircraft wing in the shape of a cross.

crude lecithin *See* lecithin.

crude naphtha [MATER] A light distillate made in the fractionation of crude oil.

crude ore [MIN ENG] The ore as it leaves the mine in an unconcentrated form.

crude scale *See* scale wax.

crude yellow scale [MATER] The trade name for a low-grade paraffin wax.

cruise missile [AERO ENG] A pilotless airplane that can be launched from a submarine, surface ship, ground vehicle, or another airplane; range can be up to 1500 miles (2400 kilometers), flying at a constant altitude that can be as low as 60 meters.

crush [MET] Casting defect caused by damage to the mold before pouring the metal. [MIN ENG] **1.** A general settlement of the strata above a coal mine due to failure of pillars; generally accompanied by numerous local falls of roof in mine workings. **2.** To reduce ore or quartz by stamps, crushers, or rolls.

crushed steel [MATER] An abrasive used in the stone, brick, glass, and metal trades, made by heating high-grade crucible steel to white heat, quenching in a bath of cold water, and crushing the fragments.

crushed stone [MIN ENG] Irregular fragments of rock crushed or ground to smaller sizes after quarrying. Also known as broken stone.

crusher [MECH ENG] A machine for crushing rock and other bulk materials.

crush-forming [ENG] Shaping the face of a grinding wheel by forcing a rotating metal roll into it.

crushing [MIN ENG] The quantity of ore pulverized or crushed at a single operation in processing.

crushing mill *See* stamping mill.

crushing test [ENG] A test of the suitability of stone that might be mined for roads or building use. [MET] A test to determine quality of tubing, especially welded tubing, by applying compression parallel to the axis.

crutter [MIN ENG] **1.** A worker who drills blasting holes and prepares the blasting charge. **2.** A worker who removes blasted rock.

cryogenic engineering [ENG] A branch of engineering specializing in technical operations at very low temperatures (about 200 to 400°R, or −160 to −50°C).

cryogenic gyroscope [ENG] A gyroscope in which a spherical rotor of superconducting niobium spins while in levitation at cryogenic temperatures. Also known as superconducting gyroscope.

cryogenic propellant [MATER] A rocket fuel, oxidizer, or propulsion fluid which is liquid only at very low temperatures.

cryology [MECH ENG] The study of low-temperature (approximately 200°R, or −160°C) refrigeration.

cryometer [ENG] A thermometer for measuring low temperatures.

cryopreservation [ENG] Preservation of food, biologicals, and other materials at extremely low temperatures.

cryoscope [ENG] A device to determine the freezing point of a liquid.

cryosorption pump [MECH ENG] A high-vacuum pump that employs a sorbent such as activated charcoal or synthetic zeolite cooled by nitrogen or some other refrigerant; used to reduce pressure from atmospheric pressure to a few millitorr.

cryostat [ENG] An apparatus used to provide low-temperature environments in which operations may be carried out under controlled conditions.

cryptoclimate [ENG] The climate of a confined space, such as inside a house, barn, or greenhouse, or in an artificial or natural cave; a form of microclimate. Also spelled kryptoclimate.

crystal cartridge [ENG ACOUS] A piezoelectric unit used with a stylus in a phonograph pickup to convert disk recordings into audio-frequency signals, or used with a diaphragm in a crystal microphone to convert sound waves into af signals.

crystal cutter [ENG ACOUS] A cutter in which the mechanical displacements of the recording stylus are derived from the deformations of a crystal having piezoelectric properties.

crystal glass [MATER] A water-clear lead glass which polishes readily and has a high index of refraction.

crystal headphones [ENG ACOUS] Headphones using Rochelle salt or other crystal elements to convert audio-frequency signals into sound waves. Also known as ceramic earphones.

crystal holder [DES ENG] A housing designed to provide proper support, mechanical protection, and connections for a quartz crystal plate.

crystal hydrophone [ENG ACOUS] A crystal microphone that responds to waterborne sound waves.

crystalline alumina [MATER] An abrasive which consists of essentially the same mineral as corundum, but whose physical properties such as crystal structure, size, and shape of grain are so controlled as to produce the most desirable abrasives for specific types of grinding.

crystalline fracture [MET] A break in a polycrystalline metal, with the fractured surface having a grainy appearance.

crystal loudspeaker [ENG ACOUS] A loudspeaker in which movements of the diaphragm are produced by a piezoelectric crystal unit that twists or bends under the influence of the applied audio-frequency signal voltage. Also known as piezoelectric loudspeaker.

crystal microphone [ENG ACOUS] A microphone in which deformation of a piezoelectric bar by the action of sound waves or mechanical vibrations generates the output voltage between the faces of the bar. Also known as piezoelectric microphone.

crystal oven [ENG] A temperature-controlled oven in which a crystal unit is operated to stabilize its temperature and thereby minimize frequency drift.

crystal phase [MET] A crystal structure formed by an alloy over a certain range of values of the relative proportions of its constituents.

crystal pickup [ENG ACOUS] A phonograph pickup in which movements of the needle in the record groove cause deformation of a piezoelectric crystal, thereby generating an audio-frequency output voltage between opposite faces of the crystal. Also known as piezoelectric pickup.

C size [ENG] One of a series of sizes to which trimmed paper and board are manufactured; for size CN, with N equal to any integer, the length of the longer side is $2^{3/8-N/2}$ meters, while the length of the shorter side is $2^{1/8-N/2}$ meters, with both lengths rounded off to the nearest millimeter.

CTC See centralized traffic control.

CTD recorder See salinity-temperature-depth recorder.

C-tube bourdon element [ENG] Hollow tube of flexible (elastic) metal shaped like the arc of a circle; changes in internal gas or liquid pressure flexes the tube to a degree related to the pressure change; used to measure process-stream pressures.

cubeb oil See oil of cubeb.

cubicle [BUILD] Any small, approximately square room or compartment. [ENG] An enclosure for high-voltage equipment.

cul-de-sac [CIV ENG] A dead-end street with a circular area for turning around.

Cullender isochronal method [PETRO ENG] Procedure for back-pressure testing to analyze gas wells that produce from low-permeability reservoirs with a consequent slow approach to stabilized producing conditions.

cullet See collet.

cullis See coulisse.

culm [MIN ENG] Fine, refuse coal, screened and separated from larger pieces.

cultellation [ENG] Transferring a surveyed point from a high level (such as an overhang) to a lower level by dropping a marking pin.

culvert [ENG] A covered channel or a large-diameter pipe that takes a watercourse below ground level.

cumin oil [MATER] A colorless to yellow liquid with a sharp, spicy taste; soluble in alcohol, ether, and chloroform; used in medicine, flavoring, and perfumes.

cumulative compound motor [MECH ENG] A motor with operating characteristics between those of the constant-speed (shunt-wound) and the variable-speed (series-wound) types.

cumulative gas [PETRO ENG] Measurement of total gas produced from a reservoir, usually expressed in graphical relationship to total (cumulative) oil produced from the same reservoir.

cumulative sum chart [IND ENG] A statistical control chart on which the cumulative sum of deviations is plotted over a period of time and which often has a sliding V-shaped mask for comparing the plot with allowable limits. Also known as cusum chart.

cup [DES ENG] A cylindrical part with only one end open. [MET] Sheet metal part formed during the first deep-drawing operation.

cup-and-cone fracture *See* cup fracture.

cup anemometer [ENG] A rotation anemometer, usually consisting of three or four hemispherical or conical cups mounted with their diametral planes vertical and distributed symmetrically about the axis of rotation; the rate of rotation of the cups, which is a measure of the wind speed, is determined by a counter.

cup barometer [ENG] A barometer in which one end of a graduated glass tube is immersed in a cup, both cup and tube containing mercury.

cup-case thermometer [ENG] Total-immersion type of thermometer with a cup container at the bulb end to hold a specified amount and depth of the material whose temperature is to be measured.

cupel [MET] A cup made of bone ash or magnesite, used in assaying precious metals.

cupellation [MET] **1.** Method using a cupel for assaying precious metals. **2.** Process for refining gold and silver by alloying them with lead and then oxidizing the molten lead to separate the base metal from the precious metal.

cup fracture [MET] A break in a ductile material under tensile stress in which the surface of failure on one piece has a central flat area with an exterior extended rim. Also known as cup-and-cone fracture.

cup grease [MATER] A lubricating grease, usually lime base, for many applications.

cupola [MET] A vertical cylindrical furnace for melting gray iron for foundry use; the metal, coke, and flux are put into the top of the furnace onto a bed of coke through which air is blown. Also known as furnace cupola.

cupola drop [MET] The bed and unmelted charges dropped from a cupola at the end of a heat.

cupping [MET] **1.** First operation of a deep-drawing process. **2.** Fracture of a wire or rod in which one fracture surface is conical and the other concave.

cupronickel [MET] A copper-base alloy with 10–30% nickel and small amounts of manganese and iron; used in industrial

and marine installations as condenser and heat-exchanger tubing.

curare [MATER] Poisonous extract from the plant *Strychnos toxifera* containing alkaloids that produce paralysis of the voluntary muscles by acting on synaptic junctions; used as an adjunct to anesthesia in surgery.

curb [CIV ENG] A border of concrete or row of joined stones forming part of a gutter along a street edge. [MIN ENG] A timber frame, circular or square, wedged in a shaft to make a foundation for walling or tubbing, or to support, with or without other timbering, the walls of the shaft.

cure [ENG] A process by which concrete is kept moist for its first week or month to provide enough water for the cement to harden. Also known as mature.

curing time [ENG] Time interval between the stopping of moving parts during thermoplastics molding and the release of mold pressure. Also known as molding time.

curl [MATER] A defect of paper caused by unequal alteration in the dimensions of the top and underside of the sheet.

curling dies [MECH ENG] A set of tools that shape the ends of a piece of work into a form with a circular cross section.

curling machine [MECH ENG] A machine with curling dies; used to curl the ends of cans.

current decay [MET] In certain types of welding operations, controlled reduction of the welding impulse over a predetermined time interval to prevent rapid cooling of the weld nugget.

current drogue [ENG] A current-measuring assembly consisting of a weighted current cross, sail, or parachute, and an attached surface buoy.

current line [ENG] In marine operations, a graduated line attached to a current pole, used to measure the speed of a current; as the pole moves away with the current, the speed of the current is determined by the amount of line paid out in a specified time. Also known as log line.

current meter *See* velocity-type flowmeter.

current pole [ENG] A pole used to determine the direction and speed of a current; the direction is determined by the direction of motion of the pole, and the speed by the amount of an attached current line paid out in a specified time.

current-type flowmeter [ENG] A mechanical device to measure liquid velocity in open and closed channels; similar to the vane anemometer (where moving liquid turns a small windmill-type vane), but more rugged.

cursor [DES ENG] A clear or amber-colored filter that can be placed over a radar screen and rotated until an etched diameter line on the filter passes through a target echo; the bearing from radar to target can then be read accurately on a stationary 360° scale surrounding the filter.

curtain board [BUILD] A fire-retardant partition applied to a ceiling.

curtain wall [CIV ENG] An external wall that is not load-bearing.

Curtis stage [MECH ENG] A velocity-staged impulse turbine using reversing buckets between stages.

Curtis turbine [MECH ENG] A velocity-staged, impulse-type steam engine.

curved beam [ENG] A beam bounded by circular arcs.

cuscus oil *See* vetiver oil.

custodial area [BUILD] Area of a building designated for service and custodial personnel; includes rooms, closets, storage, toilets, and lockers.

custom millwork *See* architectural millwork.

cusum chart *See* cumulative sum chart.

cut [MET] *See* fraction. [MIN ENG] **1.** To intersect a vein or a working. **2.** To excavate coal. **3.** To shear one side of an entry or a crosscut by digging out the coal from floor to roof with a pick.

cut-and-carry method [MET] A die-fabricating method in which the part remains attached to the strip or is forced back into the strip to be fed through the succeeding stations of a progressive die.

cut and fill [CIV ENG] Construction of a road, a railway, or a canal which is partly embanked and partly below ground.

cutback asphalt [MATER] Asphalt which has been softened or liquefied by blending with petroleum distillates.

cutch [MATER] Tannin extracted from mangrove bark.

cut constraint [SYS ENG] A condition sometimes imposed in an integer programming problem which excludes parts of the feasible solution space without excluding any integer solution points.

cut glass [MATER] Flint glass ornamented with patterns cut into its surface.

cut methods [SYS ENG] Methods of solving integer programming problems that employ cut constraints derived from the original problem.

cut nail [DES ENG] A flat, tapered nail sheared from steel plate; it has greater holding power than a wire nail and is generally used for fastening flooring.

cutoff [CIV ENG] **1.** A channel constructed to straighten a stream or to bypass large bends, thereby relieving an area normally subjected to flooding or channel erosion. **2.** An impermeable wall, collar, or other structure placed beneath the base or within the abutments of a dam to prevent or reduce losses by seepage along

otherwise smooth surfaces or through porous strata. [ENG] **1.** A misfire in a round of shots because of severance of fuse owing to rock shear as adjacent charges explode. **2.** The line on a plastic object formed by the meeting of the two halves of a compression mold. Also known as flash groove; pinch-off. [MECH ENG] **1.** The shutting off of the working fluid to an engine cylinder. **2.** The time required for this process. [MIN ENG] **1.** A quarryman's term for the direction along which the granite must be channeled, because it will not split. **2.** The number of feet a bit may be used in a particular type of rock (as specified by the drill foreman). **3.** Minimum percentage of mineral in an ore that can be mined profitably.

cutoff point [MECH ENG] **1.** The point at which there is a transition from spiral flow in the housing of a centrifugal fan to straight-line flow in the connected duct. **2.** The point on the stroke of a steam engine where admission of steam is stopped.

cutoff tool [MECH ENG] A tool used on bar-type lathes to separate the finished piece from the bar stock.

cutoff trench [CIV ENG] A trench which is below the foundation base line of a dam or other structure and is filled with an impervious material, such as clay or concrete, to form a watertight barrier.

cutoff valve [MECH ENG] A valve used to stop the flow of steam to the cylinder of a steam engine.

cutoff wall [CIV ENG] A thin, watertight wall of clay or concrete built up from a cutoff trench to reduce seepage. Also known as core wall.

cutoff wheel [MECH ENG] A thin wheel impregnated with an abrasive used for severing or cutting slots in a material or part.

cut oil [MATER] An oil which has been partially emulsified with water in the presence of air.

cutout *See* horseback.

cutscore [ENG] A knife used in die-cutting processes, designed to cut just partway into the paper or board so that it can be folded.

cut shot [MIN ENG] A shot designed to bring down coal which has been sheared or opened on one side.

cutter [ENG ACOUS] An electromagnetic or piezoelectric device that converts an electric input to a mechanical output, used to drive the stylus that cuts a wavy groove in the highly polished wax surface of a recording disk. Also known as cutting head; head; phonograph cutter; recording head. [MECH ENG] *See* cutting tool. [MIN ENG] **1.** An operator of a coal-cutting or rock-cutting machine, or a worker engaged in underholing by pick or drill. **2.** A joint, usually a dip joint, running in the direction of working; usually in the plural.

cutter bar [MECH ENG] The bar that supports the cutting tool in a lathe or other machine.

cutterhead [MECH ENG] A device on a machine tool for holding a cutting tool.

cutter sweep [MECH ENG] The section that is cut off or eradicated by the milling cutter or grinding wheel in entering or leaving the flute.

cutting angle [MECH ENG] The angle that the cutting face of a tool makes with the work surface back of the tool.

cutting down [MET] Removing surface roughness or irregularities from metal by the use of an abrasive.

cutting drilling [MECH ENG] A rotary drilling method in which cutting occurs through the action of the drill steel rotating while pressed against the rock.

cutting edge [DES ENG] **1.** The point or edge of a diamond or other material set in a drill bit. Also known as cutting point. **2.** The edge of a lathe tool in contact with the work during a machining operation.

cutting fluid [MATER] A fluid flowed over the tool and work in metal cutting to reduce heat generated by friction, lubricate, prevent rust, and flush away chips.

cutting head *See* cutter.

cutting machine [MIN ENG] A power-driven apparatus used to undercut or shear the coal to help in its removal from the face.

cutting-off machine [MECH ENG] A machine for cutting off metal bars and shapes; includes the lathe type using single-point cutoff tools, and several types of saws.

cutting oil [MATER] A type of cutting fluid used in machining metals to lubricate the tool and workpiece, reducing tool wear, increasing cutting speeds, and decreasing power needs; there are two types: active and inactive.

cutting pliers [DES ENG] Pliers with cutting blades on the jaws.

cutting point *See* cutting edge.

cutting process [MET] A process where metal is severed by the application of a gas or an electric arc.

cutting ratio [ENG] As applied to metal cutting, the ratio of depth of cut to chip thickness for a given shear angle.

cutting rule [ENG] A sharp steel rule used in a machine for cutting paper or cardboard.

cuttings [MIN ENG] Rock fragments broken from the penetrated rock during drilling operations.

cutting speed [MECH ENG] The speed of relative motion between the tool and work-piece in the main direction of cutting. Also known as feed rate; peripheral speed.

cutting stylus [ENG ACOUS] A recording stylus with a sharpened tip that removes material to produce a groove in the recording medium.

cutting tip [ENG] The end of the snout of a cutting torch from which gas flows.

cutting tool [MECH ENG] The part of a machine tool which comes into contact with and removes material from the workpiece by the use of a cutting medium. Also known as cutter.

cutting torch [ENG] A torch that preheats metal while the surface is rapidly oxidized by a jet of oxygen issuing through the flame from an additional feed line.

cutwater [CIV ENG] A sharp-edged structure built around a bridge pier to protect it from the flow of water and material carried by the water.

cyanidation *See* cyanide process.

cyanide copper [MET] **1.** An electrolytic solution containing a complex of copper and the cyanide radical. **2.** Copper electrodeposited from the solution.

cyanide process [MET] Process of dissolving powdered gold and silver ores in a weak solution of sodium cyanide or potassium cyanide; the precious metals are precipitated from solution by zinc. Also known as cyanidation.

cyanide pulp [MET] The mixture resulting from grinding of gold and silver ore, then dissolving out the precious-metal content in a solution of sodium cyanide.

cyanide slime [MET] Minute particles of precious metals precipitated from cyanide solutions used in extracting the metals from ore.

cyaniding [MET] Introduction of carbon and nitrogen simultaneously into a ferrous alloy by heating while in contact with molten cyanide; usually followed by quenching to produce a hardened case.

cyanoacrylate adhesive [MATER] An adhesive having a base of an alkyl 2-cyanoacrylate compound and characterized by excellent polymerizing and bonding properties; used for rubber printing plates, tools, and rubber swimming masks.

cybernation [IND ENG] The use of computers in connection with automation.

cycle [ENG] To run a machine through an operating cycle.

cycle annealing [MET] Annealing at a controlled time-temperature cycle to achieve a specific microstructure.

cyclegraph technique [IND ENG] Recording a brief work cycle by attaching small lights to various parts of a worker and then exposing the work motions on a still-film time plate; motion will appear on the plate as superimposed streaks of light constituting a cyclegraph.

cycle skip *See* skip logging.

cycle time [PETRO ENG] In a drilling operation, the time needed for the pump to move the drilling fluid in a bore hole.

cyclic mining *See* conventional mining.

cyclic train [MECH ENG] A set of gears, such as an epicyclic gear system, in which one or more of the gear axes rotates around a fixed axis.

cyclized rubber [MATER] A thermoplastic, nonrubbery, tough or hard rubber derivative formed by the action of certain agents, such as sulfonic acid and chlorostannic acid, on rubber; used in paints and adhesives and for insulation.

cyclograph [ENG] An electronic instrument that produces on a cathode-ray screen a pattern which changes in shape according to core hardness, carbon content, case depth, and other metallurgical properties of a test sample of steel inserted in a sensing coil.

cycloidal gear teeth [DES ENG] Gear teeth whose profile is formed by the trace of a point on a circle rolling without slippage on the outside or inside of the pitch circle of a gear; now used only for clockwork and timer gears.

cyclone [MECH ENG] Any cone-shaped air-cleaning apparatus operated by centrifugal separation that is used in particle collecting and fine grinding operations.

cyclone cellar [CIV ENG] An underground shelter, often built in areas frequented by tornadoes. Also known as storm cellar; tornado cellar.

cyclone classifier *See* cyclone separator.

cyclone furnace [ENG] A water-cooled, horizontal cylinder in which fuel is fired cyclonically and heat is released at extremely high rates.

cyclone separator [MECH ENG] A funnel-shaped device for removing particles from air or other fluids by centrifugal means; used to remove dust from air or other fluids, steam from water, and water from steam, and in certain applications to separate particles into two or more size classes. Also known as cyclone classifier.

cyclopean [MATER] Mass concrete with aggregate larger than 6 inches (15 centimeters); used for thick structures such as dams.

cyclotol [MATER] High explosive composed of RDX (cyclonite) and TNT.

cylinder [CIV ENG] **1.** A steel tube 10–60 inches (25–152 centimeters) in diameter with a wall at least 1/8 inch (3 millimeters) thick that is driven into bedrock, excavated inside, filled with concrete, and used as a pile foundation. **2.** A domed, closed tank for storing hot water to be drawn off at taps. Also known as storage calorifier. [ENG] A container used to hold and transport compressed gas for various pressurized applications. [MECH ENG] *See* engine cylinder.

cylinder actuator [MECH ENG] A device that converts hydraulic power into useful mechanical work by means of a tight-fitting piston moving in a closed cylinder.

cylinder block [DES ENG] The metal casting comprising the piston chambers of a multicylinder internal combustion engine. Also known as block; engine block.

cylinder bore [DES ENG] The internal diameter of the tube in which the piston of an engine or pump moves.

cylinder head [MECH ENG] The cap that serves to close the end of the piston chamber of a reciprocating engine, pump, or compressor.

cylinder liner [MECH ENG] A separate cylindrical sleeve inserted in an engine block which serves as the cylinder.

cylinder machine [ENG] A paper-making machine consisting of one or a series of rotary cylindrical filters on which wet paper sheets are formed.

cylinder oil [MATER] A viscous lubricating oil for the cylinders and valves of steam engines.

cylinder stock [MATER] **1.** Residual material in a still after vaporization of lighter petroleum stock. **2.** Compounded or straight oil used for lubricating steam cylinders.

cylindrical cam [MECH ENG] A cam mechanism in which the cam follower undergoes translational motion parallel to the camshaft as a roller attached to it rolls in a groove in a circular cylinder concentric with the camshaft.

cylindrical cutter [DES ENG] Any cutting tool with a cylindrical shape, such as a milling cutter.

cylindrical grinder [MECH ENG] A machine for doing work on the peripheries or shoulders of workpieces composed of concentric cylindrical or conical shapes, in which a rotating grinding wheel cuts a workpiece rotated from a power headstock and carried past the face of the wheel.

D

dado head [MECH ENG] A machine consisting of two circular saws with one or more chippers in between; used for cutting flat-bottomed grooves in wood.

dado plane [DES ENG] A narrow plane for cutting flat grooves in woodwork.

dahoma *See* African greenheart.

Dall tube [MECH ENG] Fluid-flow measurement device, similar to a venturi tube, inserted as a section of a fluid-carrying pipe; flow rate is measured by pressure drop across a restricted throat.

dalmatian sage oil *See* oil of sage.

Dalmatian wettability [PETRO ENG] Theory of wettability that some in situ reservoir rocks are partly preferentially oil-wet and partly preferentially water-wet.

dam [CIV ENG] **1.** A barrier constructed to obstruct the flow of a watercourse. **2.** A pair of cast-steel plates with interlocking fingers built over an expansion joint in the road surface of a bridge.

dammar [MATER] **1.** A type of hard resin obtained from evergreen trees of the genus *Agathis*. **2.** A type of soft, clear to yellow East Indian resin derived from several trees of the family Dipterocarpaceae. Also known as gum dammar.

dammar varnish [MATER] Varnish made from East Indian dammar.

damp [ENG] To reduce the fire in a boiler or a furnace by putting a layer of damp coals or ashes on the fire bed. [MIN ENG] A poisonous gas in a coal mine.

damp course [CIV ENG] A layer of impervious material placed horizontally in a wall to keep out water.

damp down [MET] To stop the blast in a blast furnace by closing the openings in the furnace.

dampener [ENG] A device for damping spring oscillations after abrupt removal or application of a load.

damper [MECH ENG] A valve or movable plate for regulating the flow of air or the draft in a stove, furnace, or fireplace.

damper loss [ENG] The reduction in rate of flow or of pressure of gas across a damper.

damper pedal [ENG] A pedal that controls the damping of piano strings.

damping [ENG] Reducing or eliminating reverberation in a room by placing sound-absorbing materials on the walls and ceiling. Also known as soundproofing.

damp sheet [MIN ENG] A curtain used in a mine to direct airflow, thus preventing gas accumulation.

dancing step *See* balanced step.

dancing winder *See* balanced step.

dandy roll [MECH ENG] A roll in a Fourdrinier paper-making machine; used to compact the sheet and sometimes to imprint a watermark.

dangler [MET] The flexible electrode used in barrel plating.

Daniell hygrometer [ENG] An instrument for measuring dew point; dew forms on the surface of a bulb containing ether which is cooled by evaporation into another bulb, the second bulb being cooled by the evaporation of ether on its outer surface.

Danjon prismatic astrolabe [ENG] A type of astrolabe in which a Wollaston prism just inside the focus of the telescope converts converging beams of light into parallel beams, permitting a great increase in accuracy.

darby [ENG] A flat-surfaced tool for smoothing plaster.

dark plaster [MATER] A plaster made from calcined, unground gypsum.

dark satellite [AERO ENG] Satellite that gives no information to a friendly ground environment, either because it is controlled or because the radiating equipment is inoperative.

d'Arsonval galvanometer [ENG] A galvanometer in which a light coil of wire, suspended from thin copper or gold ribbons, rotates in the field of a permanent magnet when current is carried to it through the ribbons; the position of the coil is indicated by a mirror carried on it, which reflects a light beam onto a fixed scale. Also known as light-beam galvanometer.

dart configuration [AERO ENG] An aerodynamic configuration in which the control surfaces are at the tail of the vehicle.

dashpot [MECH ENG] A device used to dampen and control a motion, in which an attached piston is loosely fitted to move slowly in a cylinder containing oil.

datum [ENG] **1.** A direction, level, or position from which angles, heights, speeds or distances are conveniently measured. **2.** Any numerical or geometric quantity or value that serves as a base reference for other quantities or values (such as a point, line, or surface in relation to which others are determined).

datum level *See* datum plane.

datum plane [ENG] A permanently established horizontal plane, surface, or level to which soundings, ground elevations, water surface elevations, and tidal data are referred. Also known as chart datum; datum level; reference level; reference plane.

Davis magnetic tester [MIN ENG] An instrument used to determine the magnetic contents of ores.

Davis wing [AERO ENG] A narrow-chord wing that has comparatively low drag and a stable center of pressure and develops lift at relatively small angles of attack.

Davy lamp [MIN ENG] An early safety lamp with a mantle of wire gauze around the flame to dissipate the heat from the flame to below the ignition temperature of methane.

day drift [MIN ENG] A mine passageway that has one end at the surface.

daylight controls [ENG] Special devices which automatically control the electric power to the lamp, causing the light to operate during hours of darkness and to be extinguished during daylight hours.

daylight glass [MATER] A glass that absorbs red light; used in incandescent lamps to remove excess red emission so that the light spectrum resembles natural daylight.

daylighting [CIV ENG] To light an area with daylight.

daylight opening [ENG] The space between two press platens when open.

day wage [IND ENG] A fixed rate of pay per shift or per daily hours of work, irrespective of the amount of work completed.

dc casting *See* direct-chill casting.

deactivation [MET] Chemical removal of the active constituents of a corrosive liquid. [MIN ENG] Treatment of one or more species of mineral particles to reduce floating during froth flotation.

dead [MIN ENG] An area of subsidence that has totally settled and is not likely to move.

dead air [MIN ENG] Air in a mine when it is stagnant or contains carbonic acid.

dead-air space [BUILD] A sealed air space, such as in a hollow wall.

dead axle [MECH ENG] An axle that carries a wheel but does not drive it.

dead band [ENG] The range of values of the measured variable to which an instrument will not effectively respond. Also known as dead zone; neutral zone.

dead block [ENG] A device placed on the ends of railroad passenger cars to absorb the shock of impacts.

dead bolt [DES ENG] A lock bolt that is moved directly by the turning of a knob or key, not by spring action.

dead-bright [MET] Of a metal, polished to remove tool marks.

dead-burn [MIN ENG] A calcination to produce a dense refractory substance; done at a higher temperature and for a longer time than for normal calcination.

dead-burnt gypsum *See* anhydrous calcium sulfate.

dead center [MECH ENG] **1.** A position of a crank in which the turning force applied to it by the connecting rod is zero; occurs when the crank and rod are in a straight line. **2.** A support for the work on a lathe which does not turn with the work.

dead-end tower [CIV ENG] Antenna or transmission line tower designed to withstand unbalanced mechanical pull from all the conductors in one direction together with the wind strain and vertical loads.

deadhead [MET] The portion of a casting that fills up the ingate. [MIN ENG] To begin a new cut without excavating the material from the preceding cut.

deadline [PETRO ENG] The end of the drilling line that is not reeled on the hoisting drum of the rotary rig; usually it is anchored to the derrick substructure and does not move as the traveling block is hoisted.

deadlocking latch bolt *See* auxiliary dead latch.

deadman [CIV ENG] **1.** A buried plate, wall, or block attached at some distance from and forming an anchorage for a retaining wall. Also known as anchorage; anchor block; anchor wall. **2.** *See* anchor log.

deadman's brake [MECH ENG] An emergency device that automatically is activated to stop a vehicle when the driver removes his foot from the pedal.

deadman's handle [MECH ENG] A handle on a machine designed so that the operator must continuously press on it in order to keep the machine running.

dead oil [MATER] An oil, of density greater than water, distilled from tar.

dead-pile *See* solids piled.

dead rail [CIV ENG] One of two rails on a railroad weighing platform that permit an excessive load to leave the platform.

dead-roast [MIN ENG] **1.** A roasting process for driving off sulfur. Also known as sweet roast. **2.** Removing volatiles by roasting within a specified temperature range.

dead room *See* anechoic chamber.

dead soft steel [MET] **1.** Steel very low in carbon. **2.** Steel annealed until it is very soft.

dead stick [AERO ENG] The propeller of an airplane that is not rotating because the engine has stopped.

dead-stroke [MECH ENG] Having a recoilless or nearly recoilless stroke.

dead-stroke hammer [MECH ENG] A power hammer provided with a spring on the hammer head to reduce recoil.

dead time [ENG] The time interval, after a response to one signal or event, during which a system is unable to respond to another. Also known as insensitive time.

dead-time correction [ENG] A correction applied to an observed counting rate to allow for the probability of the occurrence of events within the dead time. Also known as coincidence correction.

dead track [CIV ENG] **1.** Railway track that is no longer used. **2.** A section of railway track that is electrically isolated from the track signal circuits.

deadweight gage [ENG] An instrument used as a standard for calibrating pressure gages in which known hydraulic pressures are generated by means of freely balanced (dead) weights loaded on a calibrated piston.

dead work [MIN ENG] Preparatory work which is for future operations and not directly productive.

dead zone *See* dead band.

deaeration [ENG] Removal of gas or air from a substance, as from feedwater or food.

deaerator [MECH ENG] A device in which oxygen and carbon dioxide are removed from boiler feedwater.

deal [DES ENG] **1.** A face on which numbers are registered by means of a pointer. **2.** A disk usually with a series of markings around its border, which can be turned to regulate the operation of a machine or electrical device.

debriefing [AERO ENG] The relating of factual information by a flight crew at the termination of a flight, consisting of flight weather encountered, the condition of the aircraft, or facilities along the airways or at the airports.

debris dam [CIV ENG] A fixed dam across a stream channel for the retention of sand, gravel, driftwood, or other debris.

debubblizer [ENG] A worker who removes bubbles from plastic rods and tubing.

debug [ENG] To eliminate from a newly designed system the components and circuits that cause early failures.

deburr [MET] To remove burrs, sharp edges, or fins from metal parts by placing them in a revolving barrel containing abrasives suspended in a liquid.

decalescence [MET] Darkening of a metal surface due to isothermal absorption of the latent heat of phase transformation.

decantation [ENG] A method for mechanical dewatering of a wet solid by pouring off the liquid without disturbing underlying sediment or precipitate.

decanter [ENG] Tank or vessel in which solids or immiscible dispersions in a carrier liquid settle or coalesce, with clear upper liquid withdrawn (decanted) as overflow from the top.

decarburize [MET] To remove carbon from the surface of a ferrous alloy, particularly steel, by heating in a medium that reacts with carbon.

decay [MATER] To undergo decomposition.

deceleration parachute *See* drogue.

decelerometer [ENG] An instrument that measures the rate at which the speed of a vehicle decreases.

deceleron [AERO ENG] A lateral control surface of an airplane that is divided so as to combine the functions of an airbrake and an aileron.

decibel meter [ENG] An instrument calibrated in logarithmic steps and labeled with decibel units and used for measuring power levels in communication circuits.

decimal balance [ENG] A balance having one arm 10 times the length of the other, so that heavy objects can be weighed by using light weights.

decision calculus [SYS ENG] A guide to the process of decision-making, often outlined in the following steps: analysis of the decision area to discover applicable elements; location or creation of criteria for evaluation; appraisal of the known information pertinent to the applicable elements and correction for bias; isolation of the unknown factors; weighting of the pertinent elements, known and unknown, as to relative importance; and projection of the relative impacts on the objective, and synthesis into a course of action.

decision rule [SYS ENG] In decision theory, the mathematical representation of a physical system which operates upon the observed data to produce a decision.

decision theory [SYS ENG] A broad spectrum of concepts and techniques which have been developed to both describe and rationalize the process of decision making, that is, making a choice among several possible alternatives.

decision tree [IND ENG] Graphic display of the underlying decision process involved in the introduction of a new product by a manufacturer.

deck [CIV ENG] **1.** A floor, usually of wood, without a roof. **2.** The floor or roadway of a bridge. [ENG] A magnetic-tape transport mechanism.

deck bridge [CIV ENG] A bridge that carries the deck on the very top of the superstructure.

deck charge [MIN ENG] A charge that is separated into several smaller components and placed along a quarry borehole.

decking [CIV ENG] Surface material on a deck. [MIN ENG] Changing tubs on a cage at both ends of a shaft.

deckle [ENG] A detachable wood frame fitted around the edges of a papermaking mold. [MATER] In paper manufacturing, the width of the wet sheet as it comes off the wire of a paper machine.

deckle edge [MATER] The unfinished edge of paper having a characteristic appearance as a result of leakage under the frame (deckle) in which the paper is made; handmade paper has a deckle edge on all four sides, machine-made paper only on two sides.

deckle rod [ENG] A small rod inserted at each end of the extrusion coating die to adjust the die opening length.

deckle strap [ENG] An endless rubber band which runs longitudinally along the wire edges of a paper machine and determines web width.

deck loading [MIN ENG] The method of loading deck charges in a quarry borehole.

deck roof [BUILD] A roof that is nearly flat and without parapet walls.

deck truss [CIV ENG] The frame of a deck.

declination axis [ENG] For an equatorial mounting of a telescope, an axis of rotation that is perpendicular to the polar axis and allows the telescope to be pointed at objects of different declinations.

declination circle [ENG] For a telescope with an equatorial mounting, a setting circle attached to the declination axis that shows the declination to which the telescope is pointing.

declination compass *See* declinometer.

declination variometer [ENG] An instrument that measures changes in the declination of the earth's magnetic field, consisting of a permanent bar magnet, usually about 1 centimeter long, suspended with a plane mirror from a fine quartz fiber 5–15 centimeters in length; a lens focuses to a point a beam of light reflected from the mirror to recording paper mounted on a rotating drum. Also known as D variometer.

declinometer [ENG] A magnetic instrument similar to a surveyor's compass, but arranged so that the line of sight can be rotated to conform with the needle or to any desired setting on the horizontal circle; used in determining magnetic declination. Also known as declination compass.

decompression [ENG] Any procedure for the relief of pressure or compression.

decompression chamber [ENG] **1.** A steel chamber fitted with auxiliary equipment to raise its air pressure to a value two to six times atmospheric pressure; used to relieve a diver who has decompressed too quickly in ascending. **2.** Such a chamber in which conditions of high atmospheric pressure can be simulated for experimental purposes.

deconcentrator [ENG] An apparatus for removing dissolved or suspended material from feedwater.

decontamination [ENG] The removing of chemical, biological, or radiological contamination from, or the neutralizing of it on, a person, object, or area.

decorative stone [MATER] A stone that serves for architectural decoration, as in mantles or store fronts.

decouple [ENG] **1.** To minimize or eliminate airborne shock waves of a nuclear or other explosion by placing the explosives deep under the ground. **2.** To minimize the seismic effect of an underground explosion by setting it off in the center of an underground cavity.

decremeter [ENG] An instrument for measuring the logarithmic decrement (damping) of a train of waves.

dedendum [DES ENG] The difference between the radius of the pitch circle of a gear and the radius of its root circle.

dedendum circle [DES ENG] A circle tangent to the bottom of the spaces between teeth on a gear wheel.

dedusting [MIN ENG] Cleaning ore, using pneumatic means and screening, to remove dust and other fine impurities. Also known as aspirating.

deemphasis [ENG ACOUS] A process for reducing the relative strength of higher audio frequencies before reproduction, to complement and thereby offset the preemphasis that was introduced to help override noise or reduce distortion. Also known as postemphasis; postequalization.

deemphasis network [ENG ACOUS] An *RC* filter inserted in a system to restore preemphasized signals to their original form.

deep-draw [MET] To form shapes with large depth-diameter ratios in sheet or strip metal by considerable plastic distortion in dies.

Deep Space Network [AERO ENG] A spacecraft network operated by NASA which tracks, commands, and receives telemetry for all types of spacecraft sent to explore deep space, the moon, and solar system planets. Abbreviated DSN.

deep-space probe [AERO ENG] A spacecraft designed for exploring space beyond the gravitational and magnetic fields of the earth.

deep underwater muon and neutrino detector [ENG] A proposed device for detecting and determining the direction of extraterrestrial neutrinos passing through a volume of approximately 1 cubic kilometer of ocean water, using an array of several thousand Cerenkov counters suspended in the water to sense the showers of charged particles generated by neutrinos. Abbreviated DUMAND.

deep well [CIV ENG] A well that draws its water from beneath shallow impermeable strata, at depths exceeding 22 feet (6.7 meters).

deep-well pump [MECH ENG] A multistage centrifugal pump for lifting water from deep, small-diameter wells; a surface electric motor operates the shaft. Also known as vertical turbine pump.

defender [IND ENG] A machine or facility which is being considered for replacement.

deferment factor *See* discount factor.

deflashing [ENG] Finishing technique to remove excess material (flash) from a plastic or metal molding.

deflected jet fluidic flowmeter *See* fluidic flow sensor.

deflection [ENG] **1.** Shape change or reduction in diameter of a conduit, produced without fracturing the material. **2.** Elastic movement or sinking of a loaded structural member, particularly of the mid-span of a beam.

deflection bit [DES ENG] A long, cone-shaped, noncoring bit used to drill past a deflection wedge in a borehole.

deflection meter [ENG] A flowmeter that applies the differential pressure generated by a differential-producing primary device across a diaphragm or bellows in such a way as to create a deflection proportional to the differential pressure.

deflection-modulated indicator *See* amplitude-modulated indicator.

deflection ultrasonic flowmeter [ENG] A flowmeter for determining velocity from the deflection of a high-frequency sound beam directed across the flow. Also known as drift ultrasonic flowmeter.

deflection wedge [DES ENG] A wedge-shaped tool inserted into a borehole to direct the drill bit.

deflectometer [ENG] An instrument for measuring minute deformations in a structure under transverse stress.

deflector [ENG] A plate, baffle, or the like that diverts the flow of a forward-moving stream.

defocus [ENG] To make a beam of x-rays, electrons, light, or other radiation deviate from an accurate focus at the intended viewing or working surface.

defoliant [MATER] A chemical sprayed on plants that causes leaves to fall off prematurely.

deformation bands *See* Lüder's lines.

deformation thermometer [ENG] A thermometer with transducing elements which deform with temperature; examples are the bimetallic thermometer and the Bourdon-tube type of thermometer.

deformed bar [CIV ENG] A steel bar with projections or indentations to increase mechanical bonding; used to reinforce concrete.

deformeter [ENG] An instrument used to measure minute deformations in materials in structural models.

defrost [ENG] To keep free of ice or to remove ice.

degas [ENG] To remove gas from a liquid or solid.

degasifier [MET] An alloy added to molten metal to facilitate the removal of dissolved gases.

degradation failure [ENG] Failure of a device because of a shift in a parameter or characteristic which exceeds some previously specified limit.

degras [MATER] **1.** A semioxidized fat obtained from sheep skins by subjecting them to the action of oxidized fish oil and pressing them; used to dress leather. Also known as moellen. **2.** A mixture of this material with other fatty oils or fats or with wool grease. **3.** *See* wool grease.

degrease [MET] To remove grease, oil, or fatty material from a metal surface with fumes from a hot solvent.

degreaser [ENG] A machine designed to clean grease and foreign matter from mechanical parts and like items, usually metallic, by exposing them to vaporized or liquid solvent solutions confined in a tank or vessel. [MATER] A solvent, such as a polyhalogenated hydrocarbon, that removes fat or oil in many industrial processes.

degree-day [MECH ENG] A measure of the departure of the mean daily temperature from a given standard; one degree-day is recorded for each degree of departure above (or below) the standard during a single day; used to estimate energy requirements for building heating and, to a lesser extent, for cooling.

degree of curve [CIV ENG] A measure of the curvature of a railway or highway, equal to the angle subtended by a 100-foot (30-meter) chord (railway) or by a 100-foot arc (highway).

degree of freedom [MECH ENG] Any one of the number of ways in which the space configuration of a mechanical system may change.

dehumidification [MECH ENG] The process of reducing the moisture in the air; serves to increase the cooling power of air.

dehumidifier [MECH ENG] Equipment designed to reduce the amount of water vapor in the ambient atmosphere.

deicer [AERO ENG] Any device to keep the wings and propeller of an airplane free of ice. [MATER] Any substance used to keep a surface free of ice or to rid it of ice; ethylene glycol is used to deice windshields of automobiles and airplanes.

deicing [ENG] The removal of ice deposited on any object, especially as applied to aircraft icing, by heating, chemical treatment, and mechanical rupture of the ice deposit.

delamination [ENG] Separation of a laminate into its constituent layers.

de Laval nozzle [AERO ENG] A converging-diverging nozzle used in certain rockets. Also known as Laval nozzle.

de Lavaud process [MET] A centrifugal casting process employing water-cooled metal molds, used to produce cast iron pipe, gun barrels, and other cylindrical objects.

delay [IND ENG] Interruption of the normal tempo of an operation; may be avoidable or unavoidable.

delay-action detonator *See* delay blasting cap.

delay allowance [IND ENG] A percentage of the normal operating time added to the normal time to allow for delays.

delay blasting cap [ENG] A blasting cap which explodes at a definite time interval after the firing current has been passed by the exploder. Also known as delay-action detonator.

delayed combustion [ENG] Secondary combustion in succeeding gas passes beyond the furnace volume of a boiler.

delayed development well *See* step-out well.

delayed repeater satellite [AERO ENG] Satellite which stores information obtained from a ground terminal at one location, and upon interrogation by a terminal at a different location, transmits the stored message.

delayed yield [MET] Time delay between the sudden application of a yield stress and the appearance of yielding.

delayer [MATER] A substance mixed with solid rocket propellants to decrease the rate of combustion.

delta ferrite *See* delta iron.

delta iron [MET] The nonmagnetic polymorphic form of iron stable between about 1403°C and the melting point, about 1535°C. Also known as delta ferrite.

delta wing [AERO ENG] A triangularly shaped wing of an aircraft.

demagnetization [MIN ENG] Deflocculation in dense-media process using ferrosilicon by passing the fluid through an alternating-current field.

demand meter [ENG] Any of several types of instruments used to determine a customer's maximum demand for electric power over an appreciable time interval; generally used for billing industrial users.

demand system [ENG] A system in an airplane that automatically dispenses oxygen according to the demand of the flyer's body.

demister [MECH ENG] A series of ducts in automobiles arranged so that hot, dry air directed from the heat source is forced against the interior of the windscreen or windshield to prevent condensation.

demolition [CIV ENG] The act or process of tearing down a building or other structure.

dendritic powder [MET] Fine metal particles having a dendritic structure; usually of electrolytic origin. Also known as arborescent powder.

Denison sampler [ENG] A soil sampler consisting of a central nonrotating barrel which is forced into the soil as friction is removed by a rotating external barrel; the bottom can be closed to retain the sample during withdrawal.

dense-air system *See* cold-air machine.

dense-media separator [MIN ENG] A device in which a heavy mineral is dispersed in water, causing heavier ores to sink and lighter ores to float.

densify [ENG] To increase the density of a material such as wood by subjecting it to pressure or impregnating it with another material.

densimeter [ENG] An instrument which measures the density or specific gravity of a liquid, gas, or solid. Also known as densitometer; density gage; density indicator; gravitometer.

densitometer [ENG] 1. An instrument which measures optical density by measuring the intensity of transmitted or reflected light; used to measure photographic density. 2. *See* densimeter.

density [MATER] Closeness of texture or consistency.

density airspeed [AERO ENG] Calibrated airspeed corrected for pressure altitude and true air temperature.

density bottle *See* specific gravity bottle.

density correction [AERO ENG] A correction made necessary because the airspeed indicator is calibrated only for standard air pressure; it is applied to equivalent airspeed to obtain true airspeed, or to calibrated airspeed to obtain density airspeed. [ENG] **1.** The part of the temperature correction of a mercury barometer which is necessitated by the variation of the density of mercury with temperature. **2.** The correction, applied to the indications of a pressure-tube anemometer or pressure-plate anemometer, which is necessitated by the variation of air density with temperature.

density error [AERO ENG] The error in the indications of a differential-pressure-type airspeed indicator due to nonstandard atmospheric density.

density gage *See* densimeter.

density indicator *See* densimeter.

density log [PETRO ENG] Radioactivity logging of reservoir structure densities down an oil-well bore by emission and detection of gamma rays.

density rule [ENG] A grading system for lumber based on the width of annual rings.

density specific impulse [AERO ENG] The product of the specific impulse of a propellant combination and the average specific gravity of the propellants.

density transmitter [ENG] An instrument used to record the density of a flowing stream of liquid by measuring the buoyant force on an air-filled chamber immersed in the stream.

dental coupling [MECH ENG] A type of flexible coupling used to join a steam turbine to a reduction-gear pinion shaft; consists of a short piece of shaft with gear teeth at each end, and mates with internal gears in a flange at the ends of the two shafts to be joined.

dental gold [MET] An alloy composed of 5 to 12% silver, 4 to 10% copper, with the balance gold.

dental work *See* cementation.

Denver cell [MIN ENG] A subaeration type of flotation cell, mechanized and self-aerating.

Denver jig [MIN ENG] A pulsion-suction diaphragm for separating sulfur from coal before flotation; hydraulic water is admitted through a rotary valve.

deodorant [MATER] A substance used to remove, correct, or repress undesirable odors.

deodorized kerosine [MATER] A highly refined petroleum kerosine that has very little odor; used as an illuminant for wick lamps. Also known as refined kerosine.

deorbit [AERO ENG] To recover a spacecraft from earth orbit by providing a new orbit which intersects the earth's atmosphere.

deoxidize [MET] To remove an oxide film from a metal surface.

deoxidized copper [MET] Pure copper deoxidized with phosphorus to reduce cuprous oxide and eliminate porosity.

departure track [CIV ENG] A railroad yard track for combining freight cars into outgoing trains.

dephosphorize [MET] Removal of phosphorus from a molten metal such as steel.

depilation [ENG] Removal of hair from animal skins in processing leather.

depilatory [MATER] A chemical that removes hairs from skin.

depletion drive [PETRO ENG] Displacement mechanism (type of drive) to expel hydrocarbons from porous reservoir formations, that is, to remove more hydrocarbon from the reservoir; types of drives are gas or water (natural or injected) and injected LPG.

depletion-type reservoir [PETRO ENG] Oil reservoir which is initially in (and during depletion remains in) a state of equilibrium between the gas and liquid phases; includes single-phase gas, two-phase bubble-point, and retrograde-gas-condensate (or dew-point) reservoirs.

deposit [MATER] Any material applied to a base by means of vacuum, electrical, chemical, screening, or vapor methods.

deposit attack [MET] Corrosion under or around the edge of a noncontinuous local deposit on a metal surface.

deposited metal [MET] Molten metal used during a welding operation for the fusion of base metals.

deposit gage [ENG] The general name for instruments used in air pollution studies for determining the amount of material deposited on a given area during a given time.

deposition efficiency [MET] The ratio of the weight of deposited metal to the net weight of the consumed electrodes, exclusive of stubs, in welding.

deposition rate [MET] The amount of welding material deposited per unit of time, expressed in pounds per hour.

deposition sequence [MET] The order of deposition of weld-metal increments.

depreciation [IND ENG] Loss of value due to physical deterioration.

depressed center car [ENG] A flat railroad car having a low center section; used to provide adequate tunnel clearance for oversized loads.

depression angle *See* angle of depression.

depropanized material [MATER] Material that has undergone distillation to remove lighter components from butanes and heavier material.

depth finder [ENG] A radar or ultrasonic instrument for measuring the depth of the sea.

depth gage [DES ENG] An instrument or tool for measuring the depth of depression to a thousandth inch.

depth marker [ENG] A thin board or other lightweight substance used as a means of identifying the surface of snow or ice which has been covered by a more recent snowfall.

depth micrometer [DES ENG] A micrometer used to measure the depths of holes, slots, and distances of shoulders and projections.

depth of engagement [DES ENG] The depth of contact, in a radial direction, between mating threads.

depth of fusion [MET] The distance that fusion extends from the original surface into the base metal in a welding operation.

depth of thread [DES ENG] The distance, in a radial direction, from the crest of a screw thread to the base.

depth sounder [ENG] An instrument for mechanically measuring the depth of the sea beneath a ship.

derail [ENG] **1.** To cause a railroad car or engine to run off the rails. **2.** A device to guide railway cars or engines off the tracks to avoid collision or other accident.

derby [MET] A large, usually cylindrical piece of primary metal, whose weight may exceed 100 pounds (45 kilograms), formed by bomb reduction.

derived sound system [ENG ACOUS] A four-channel sound system that is artificially synthesized from conventional two-channel stereo sound by an adapter, to provide feeds to four loudspeakers for approximating quadraphonic sound.

Dermitron [MET] An electrical instrument which applies a high-frequency current and the resulting eddy current series as a measure of thickness of nonmagnetic coatings on nonmagnetic basis metals, provided the conductivity of the coating and that of the basis metal are different.

derrick [MECH ENG] A hoisting machine consisting usually of a vertical mast, a slanted boom, and associated tackle; may be operated mechanically or by hand.

derrick barge [PETRO ENG] A crane barge used in offshore drilling platform construction and suitable for work in rough seas.

derrick crane *See* stiffleg derrick.

descaling [ENG] Removing scale, usually oxides, from the surface of a metal or the inner surface of a pipe, boiler, or other object.

descending node [AERO ENG] That point at which an earth satellite crosses to the south side of the equatorial plane of its primary. Also known as southbound node.

descending vertical angle *See* angle of depression.

descent [AERO ENG] Motion of a craft in which the path is inclined with respect to the horizontal.

deseaming [MET] Removing defects from the surfaces of ingots, blooms, or semifinished products, usually by means of a chipping hammer or an oxy-gas flame.

design engineering [ENG] A branch of engineering concerned with the creation of systems, devices, and processes useful to and sought by society.

design flood [CIV ENG] The flood, either observed or synthetic, which is chosen as the basis for the design of a hydraulic structure.

design gross weight [AERO ENG] The gross weight at takeoff that an aircraft, rocket, or such is expected to have, used in design calculations.

design head [CIV ENG] The planned elevation between the free level of a water supply and the point of free discharge or the level of free discharge surface.

design load [DES ENG] The most stressful combination of weight or other forces a building, structure, or mechanical system or device is designed to sustain.

design pressure [CIV ENG] **1.** The force exerted by a body of still water on a dam. **2.** The pressure which the dam can withstand. [DES ENG] The pressure used in the calculation of minimum thickness or design characteristics of a boiler or pressure vessel in recognized code formulas; static head may be added where appropriate for specific parts of the structure.

design speed [CIV ENG] The highest continuous safe vehicular speed as governed by the design features of a highway.

design standards [DES ENG] Generally accepted uniform procedures, dimensions, materials, or parts that directly affect the design of a product or facility.

design storm [CIV ENG] A storm whose magnitude, rate, and intensity do not exceed the design load for a storm drainage system or flood protection project.

design stress [DES ENG] A permissible maximum stress to which a machine part or structural member may be subjected, which is large enough to prevent failure in case the loads exceed expected values, or other uncertainties turn out unfavorably.

design thickness [DES ENG] The sum of required thickness and corrosion allowance utilized for individual parts of a boiler or pressure vessel.

desilter [MECH ENG] Wet, mechanical solids classifier (separator) in which silt particles settle as the carrier liquid is slowly stirred by horizontally revolving rakes; solids are plowed outward and removed at the periphery of the container bowl.

desilting basin [CIV ENG] A space or structure constructed just below a diversion structure of a canal to remove bed, sand, and silt loads. Also known as desilting works.

desilting works *See* desilting basin.

desilverization [MET] The act or process of removing silver; specifically, the process used to remove silver and gold from ore after softening.

deslimer [MECH ENG] Apparatus, such as a bowl-type centrifuge, used to remove fine, wet particles (slime) from cement rocks and to size pigments and abrasives.

destressing [MIN ENG] Relieving stress on the abutments of an excavation by drilling and blasting to loosen peak stress zones.

destruct [AERO ENG] The deliberate action of destroying a rocket vehicle after it has been launched, but before it has completed its course.

destructive testing [ENG] Intentional operation of equipment until it fails, to reveal design weaknesses.

destruct line [AERO ENG] On a rocket test range, a boundary line on each side of the down-range course beyond which a rocket cannot fly without being destroyed under destruct procedures; or a line beyond which the impact point cannot pass.

detachable bit [ENG] An all-steel drill bit that can be removed from the drill steel, and can be resharpened. Also known as knock-off bit; rip bit.

detailing *See* screening.

det drill *See* fusion-piercing drill.

detector bar [CIV ENG] A device that keeps a railroad switch locked while a train is passing over it.

detector car [ENG] A railroad car used to detect flaws in rails.

detent [MECH ENG] A catch or lever in a mechanism which initiates or locks movement of a part, especially in escapement mechanisms.

detention basin [CIV ENG] A reservoir without control gates for storing water over brief periods of time until the stream has the capacity for ordinary flow plus released water; used for flood regulation.

detergent [MATER] A synthetic cleansing agent resembling soap in the ability to emulsify oil and hold dirt, and containing surfactants which do not precipitate in hard water; may also contain protease enzymes and whitening agents.

detergent additive [MATER] A substance incorporated in lubricating oils which gives them the property of keeping insoluble material in suspension.

detergent oil [MATER] A lubricating oil with special sludge-dispersing properties for use in internal combustion engines.

deterioration [ENG] Decline in the quality of equipment or structures over a period of time due to the chemical or physical action of the environment.

detonating agent [MATER] An explosive, such as PETN, contained in the blasting cap or detonator.

detonating fuse [ENG] A device consisting of a core of high explosive within a waterproof textile covering and set off by an electrical blasting cap fired from a distance by means of a fuse line; used in large, deep boreholes.

detonating primer [MATER] A primer used to fire high explosives that is exploded by a fuse.

detonating relay [ENG] A device used in conjunction with the detonating fuse to avoid short-delay blasting.

detonation [MECH ENG] Spontaneous combustion of the compressed charge after passage of the spark in an internal combustion engine; it is accompanied by knock.

detonation flame spraying [MET] A flame-spraying method in which the combined mixture of fuel gas, oxygen, and powdered coating liquefies and explodes material to the workpiece.

detonation front [ENG] The reaction zone of a detonation.

detonator [ENG] A device, such as a blasting cap, employing a sensitive primary explosive to detonate a high-explosive charge.

detonator safety [ENG] A fuse has detonator safety or is detonator safe when the functioning of the detonator cannot initiate subsequent explosive train components.

detonics [ENG] The study of detonating and explosives performance.

detritus tank [CIV ENG] A tank in which heavy suspended matter is removed in sewage treatment.

Detroit rocking furnace [ENG] An indirect arc type of rocking furnace having graphite electrodes entering horizontally from opposite ends.

DeVecchis process [MIN ENG] A smelting process for pyrites in which the raw material is roasted, concentrated magnetically, and then reduced in a rotary kiln or electric furnace.

developed blank [MET] A blank requiring little or no trimming after being formed.

developed ore *See* developed reserves.

developed planform [AERO ENG] The plan of an airfoil as drawn with the chord lines at each section rotated about the airfoil axis into a plane parallel to the plane of projection and with the airfoil axis rotated or developed and projected into the plane of projection.

developed reserves [MIN ENG] Ore that is exposed on three sides and for which tonnage yield and quality estimates have been made. Also known as assured mineral; blocked-out ore; developed ore; measured ore; ore in sight.

development [MIN ENG] Opening of a coal seam or ore body by sinking shafts or driving levels, as well as installing equipment, for proving ore reserves and exploiting them.

development drift [MIN ENG] A tunnel dug in a mine either from the surface or a point underground to get to coal or ore for exploitation or mining purposes.

development drilling [MIN ENG] Drilling boreholes to locate, identify, and prove an ore body or coal seam.

development rock [MIN ENG] Rock containing both barren and valuable rock, broken during development work.

development well [PETRO ENG] A well drilled to produce oil or gas from a proven productive area.

Devereaux agitator [MIN ENG] An agitator that utilizes an upthrust propeller to stir pulp; used in leach agitation of minerals.

deviation [ENG] The difference between the actual value of a controlled variable and the desired value corresponding to the set point.

deviation hole [PETRO ENG] Drilled hole with deviation from true vertical, usually limited by contract to 3–5°; not to be confused with a crooked hole, resulting from carelessness or a steeply dipping formation.

device [ENG] A mechanism, tool, or other piece of equipment designed for specific uses.

devil's pitchfork [DES ENG] A tool with flexible prongs used in recovery of a bit, underreamer, cutters, or such lost during drilling.

devitrified glass [MATER] A glassy material which has been changed from a vitreous to a brittle crystalline state during manufacture.

dewaterer [MECH ENG] Wet-type mechanical classifier (solids separator) in which solids settle out of the carrier liquid and are concentrated for recovery.

dewatering [ENG] **1.** Removal of water from solid material by wet classification, centrifugation, filtration, or similar solid-liquid separation techniques. **2.** Removing or draining water from an enclosure or a structure, such as a riverbed, caisson, or mine shaft, by pumping or evaporation.

dewaxed oil [MATER] Lubricating oil that has had a portion of the wax removed.

dew cell [ENG] An instrument used to determine the dew point, consisting of a pair of spaced, bare electrical wires wound spirally around an insulator and covered with a wicking wetted with a water solution containing an excess of lithium chloride; an electrical potential applied to the wires causes a flow of current through the lithium chloride solution, which raises the temperature of the solution until its vapor pressure is in equilibrium with that of the ambient air.

dewetting [MET] Flow of solder away from the soldered surface during reheating following initial soldering.

dew-point recorder [ENG] An instrument which gives a continuous recording of the dew point; it alternately cools and heats the target and uses a photocell to observe and record the temperature at which the condensate appears and disappears. Also known as mechanized dew-point meter.

dew-point reservoir [PETRO ENG] A hydrocarbon reservoir in which the temperature lies between the critical temperature and the cricondentherm (maximum temperature and pressure at which two phases can coexist) and in the one-phase region. Also known as retrograde gas-condensate reservoir.

dezincification [MET] Corrosion of brass in which both components of the alloy are dissolved and the copper is redeposited as a porous surface residue.

diagnostics [ENG] Information on what tests a device has failed and how they were failed; used to aid in troubleshooting.

diagonal [CIV ENG] A sloping structural member, under compression or tension or both, of a truss or bracing system.

diagonal bond [CIV ENG] A masonry bond with diagonal headers.

diagonal pitch [ENG] In rows of staggered rivets, the distance between the center of a rivet in one row to the center of the adjacent rivet in the next row.

diagonal pliers [DES ENG] Pliers with cutting jaws at an angle to the handles to permit cutting off wires close to terminals.

diagonal stay [MECH ENG] A diagonal member between the tube sheet and shell in a fire-tube boiler.

diagram factor [MECH ENG] The ratio of the actual mean effective pressure, as determined by an indicator card, to the map of the ideal cycle for a steam engine.

DIAL *See* differential absorption lidar.

dial [DES ENG] A separate scale or other device for indicating the value to which a control is set.

dial cable [DES ENG] Braided cord or flexible wire cable used to make a pointer move over a dial when a separate control knob is rotated, or used to couple two shafts together mechanically.

dial cord [DES ENG] A braided cotton, silk, or glass fiber cord used as a dial cable.

dial feed [MECH ENG] A device that rotates workpieces into position successively so they can be acted on by a machine.

dial indicator [DES ENG] Meter or gage with a calibrated circular face and a pivoted pointer to give readings.

dial press [MECH ENG] A punch press with dial feed.

diameter group [MECH ENG] A dimensionless group, used in the study of flow machines such as turbines and pumps, equal to the fourth root of pressure number 2 divided by the square root of the delivery number.

diameter tape [ENG] A tape for measuring the diameter of trees; when wrapped around the circumference of a tree, it reads the diameter directly.

diametral pitch [DES ENG] A gear tooth design factor expressed as the ratio of the number of teeth to the diameter of the pitch circle measured in inches.

diamond bit [DES ENG] A rotary drilling bit crowned with bort-type diamonds, used for rock boring. Also known as bort bit.

diamond boring [ENG] Boring with a diamond tool.

diamond chisel [DES ENG] A chisel having a V-shaped or diamond-shaped cutting edge.

diamond coring [ENG] Obtaining core samples of rock by using a diamond drill.

diamond count [DES ENG] The number of diamonds set in a diamond crown bit.

diamond crossing [CIV ENG] An oblique railroad crossing that forms a diamond shape between the tracks.

diamond crown [DES ENG] The cutting bit used in diamond drilling; it consists of a steel shell set with black diamonds on the face and cutting edges.

diamond drill [DES ENG] A drilling machine with a hollow, diamond-set bit for boring rock and yielding continuous and columnar rock samples.

Diamond-Hinman radiosonde [ENG] A variable audio-modulated radiosonde used by United States weather services; the carrier signal from the radiosonde is modulated by audio signals determined by the electrical resistance of the humidity- and

temperature-transducing elements and by fixed reference resistors; the modulating signals are transmitted in a fixed sequence at predetermined pressure levels by means of a baroswitch.

diamond indenter [ENG] An instrument that measures hardness by indenting a material with a diamond point.

diamond matrix [DES ENG] The metal or alloy in which diamonds are set in a drill crown.

diamond orientation [DES ENG] The set of a diamond in a cutting tool so that the crystal face will be in contact with the material being cut.

diamond-particle bit [DES ENG] A diamond bit set with small fragments of diamonds.

diamond paste [MATER] An abrasive consisting of diamond dust in a viscous material.

diamond pattern [DES ENG] The arrangement of diamonds set in a diamond crown.

diamond point [DES ENG] A cutting tool with a diamond tip.

diamond-point bit *See* mud auger.

diamond-pyramid hardness number [MET] The quotient of the load applied in the diamond-pyramid hardness test divided by the pyramidal area of the impression.

diamond-pyramid hardness test [MET] An indentation hardness test in which a diamond-pyramid indenter, with a 136° angle between opposite faces, is forced under variable loads into the surface of a test specimen. Also known as Vickers hardness test.

diamond reamer [DES ENG] A diamond-inset pipe behind, and larger than, the drill bit and core barrel that is used for enlarging boreholes.

diamond saw [DES ENG] A circular, band, or frame saw inset with diamonds or diamond dust for cutting sections of rock and other brittle substances.

diamond setter [ENG] A person skilled at setting diamonds by hand in a diamond bit or a bit mold.

diamond size [ENG] In the bit-setting and diamond-drilling industries, the number of equal-size diamonds having a total weight of 1 carat; a 10-diamond size means 10 stones weighing 1 carat.

diamond stylus [ENG ACOUS] A stylus having a ground diamond as its point.

diamond tool [DES ENG] **1.** Any tool using a diamond-set bit to drill a borehole. **2.** A diamond shaped to the contour of a single-pointed cutting tool, used for precision machining.

diamond washer [MIN ENG] An apparatus for shaking and separating rock gravel containing diamonds, utilizing a vertical series of screens with 8-, 4-, 2-, and 1-millimeter mesh.

diamond wheel [DES ENG] A grinding wheel in which synthetic diamond dust is bonded as the abrasive to cut very hard materials such as sintered carbide or quartz.

diaphragm [ENG ACOUS] A thin, flexible sheet that can be moved by sound waves, as in a microphone, or can produce sound waves when moved, as in a loudspeaker.

diaphragm compressor [MECH ENG] Device for compression of small volumes of a gas by means of a reciprocally moving diaphragm, in place of pistons or rotors.

diaphragm gage [ENG] Pressure- or vacuum-sensing instrument in which pressures act against opposite sides of an enclosed diaphragm that consequently moves in relation to the difference between the two pressures, actuating a mechanical indicator or electric-electronic signal.

diaphragm horn [ENG ACOUS] A horn that produces sound by means of a diaphragm vibrated by compressed air, steam, or electricity.

diaphragm jig [MIN ENG] A jig having a flexible diaphragm to pulse water; used in gravity concentration of minerals.

diaphragm meter [ENG] A flow meter which uses the movement of a diaphragm in the measurement of a difference in pressure created by the flow, such as a force-balance-type or a deflection-type meter.

diaphragm pump [MECH ENG] A metering pump which uses a diaphragm to isolate the operating parts from pumped liquid in a mechanically actuated diaphragm pump, or from hydraulic fluid in a hydraulically actuated diaphragm pump.

diaphragm valve [ENG] A fluid valve in which the open-close element is a flexible diaphragm; used for fluids containing suspended solids, but limited to low-pressure systems.

dichromate treatment [MET] Processing technique involving the formation of a corrosion-resistant film on the surface of a magnesium alloy by boiling the alloy in a sodium dichromate solution.

dicing cutter [MECH ENG] A cutting mill for sheet material; sheet is first slit into horizontal strands by blades, then fed against a rotating knife for dicing.

die [DES ENG] A tool or mold used to impart shapes to, or to form impressions on, materials such as metals and ceramics.

die adapter [ENG] That part of an extrusion die which holds the die block.

die blade [ENG] A deformable member attached to a die body which determines the slot opening and is adjusted to pro-

duce uniform thickness across plastic film or sheet.

die block [ENG] **1.** A tool-steel block which is bolted to the bed of a punch press and into which the desired impressions are machined. **2.** The part of an extrusion mold die holding the forming bushing and core.

die body [ENG] The stationary part of an extrusion die, used to separate and form material.

die casting [ENG] A metal casting process in which molten metal is forced under pressure into a permanent mold; the two types are hot-chamber and cold-chamber.

die chaser [ENG] One of the cutting parts of a composite die or a die used to cut threads.

die clearance [ENG] The distance between die members that meet during an operation.

die collar *See* bell tap.

die cushion [ENG] A device located in or under a die block or bolster to provide additional pressure or motion for stamping.

die cutting *See* blanking.

die drawing [MET] Reducing the diameter of wire or tubing by pulling it through a die.

die forging [MET] Shaping metal by plastic deformation in a die.

die forming [MET] Shaping metal by means of a die under pressure.

die gap [ENG] In plastics and metals forming, the distance between the two opposing metal faces forming the opening of a die.

die holder [ENG] A plate or block on which the die block is mounted; it is fastened to the bolster or press bed.

dieing machine [MECH ENG] A vertical press with the slide activated by pull rods attached to the drive mechanism below the bed of the press.

die insert [ENG] A removable part or the liner of a die body or punch.

dielectric [MATER] A material which is an electrical insulator or in which an electric field can be sustained with a minimum dissipation in power.

die lines [ENG] Lines or markings on the surface of a drawn, formed, or extruded product due to imperfections in the surface of the die.

die lubricant [MATER] Any material applied to a die to facilitate movement of the work in the die in certain die-forming operations.

die match [MET] The proper alignment of forging dies in relation to each other in forging equipment.

die opening [MET] The distance between electrodes in flash or upset welding; it is measured with parts in contact but before the beginning or immediately after completion of the weld cycle.

die radius [MET] The radius on the exposed edge of a deep-drawing die.

die scalping [MET] Drawing wire, tubing, bars, or rods through a sharp-edged die to remove surface layers containing defects.

diesel electric locomotive [MECH ENG] A locomotive with a diesel engine driving an electric generator which supplies electric power to traction motors for propelling the vehicle. Also known as diesel locomotive.

diesel electric power generation [MECH ENG] Electric power generation in which the generator is driven by a diesel engine.

diesel engine [MECH ENG] An internal combustion engine operating on a thermodynamic cycle in which the ratio of compression of the air charge is sufficiently high to ignite the fuel subsequently injected into the combustion chamber. Also known as compression-ignition engine.

diesel fuel [MATER] Fuel used for internal combustion in diesel engines; usually that fraction of crude oil that distills after kerosine.

diesel fuel additives [MATER] Compounds added to diesel fuels to improve performance, such as cetane number improvers, metal deactivators, corrosion inhibitors, antioxidants, rust inhibitors, and dispersants.

diesel fuel grades [MATER] Fuels suitable for various classes of engines and service, which must meet specifications for flash point temperature, distillation temperature, cetane number, and concentrations of impurities.

diesel fuel water and sediment [MATER] Undesirable constituents of diesel fuel which should not exceed certain limits to ensure clean fuel.

diesel index [MECH ENG] Diesel fuel rating based on ignition qualities; high-quality fuel has a high index number.

diesel knock [MECH ENG] A combustion knock caused when the delayed period of ignition is long so that a large quantity of atomized fuel accumulates in the combustion chamber; when combustion occurs, the sudden high pressure resulting from the accumulated fuel causes diesel knock.

diesel locomotive See diesel electric locomotive.

diesel oil [MATER] Heavy oil residue used as fuel for certain types of diesel engines.

diesel rig [MECH ENG] Any diesel engine apparatus or machinery.

diesel squeeze [PETRO ENG] A technique of forcing dry cement mixed with diesel oil through casing openings to repair water-bearing areas without affecting the oil-bearing areas.

die set [ENG] A tool or tool holder consisting of a die base for the attachment of a die and a punch plate for the attachment of a punch.

die shoe [MECH ENG] A block placed beneath the lower part of a die upon which the die holder is mounted; spreads the impact over the die bed, thereby reducing wear.

diesinking [ENG] Making a depressed pattern in a die by forming or machining.

die slide [MECH ENG] A device in which the lower die of a power press is mounted; it slides in and out of the press for easy access and safety in feeding the parts.

die steel [MET] Plain carbon steel or alloy steel used in making tools for cutting, machining, shearing, stamping, punching, and chipping.

die swell ratio [ENG] The ratio of the outer parison diameter (or parison thickness) to the outer diameter of the die (or die gap).

Dietert tester [MET] An apparatus for reading Brinell hardness directly from the impression made in the part being tested by means of a depth pin pressed into the depression.

die welding [MET] Forge welding in which the weld is completed under pressure between dies.

difference channel [ENG ACOUS] An audio channel that handles the difference between the signals in the left and right channels of a stereophonic sound system.

differential [MECH ENG] Any arrangement of gears forming an epicyclic train in which the angular speed of one shaft is proportional to the sum or difference of the angular speeds of two other gears which lie on the same axis; allows one shaft to revolve faster than the other, the speed of the main driving member being equal to the algebraic mean of the speeds of the two shafts. Also known as differential gear.

differential absorption lidar [ENG] A technique for the remote sensing of atmospheric gases, in which lasers transmit pulses of radiation into the atmosphere at two wavelengths, one of which is absorbed by the gas to be measured and one is not, and the difference between the return signals from atmospheric backscattering on the absorbed and non-absorbed wavelengths is used as a direct measure of the concentration of the absorbing species. Abbreviated DIAL.

differential air thermometer [ENG] A device for detecting radiant heat, consisting of a U-tube manometer with a closed bulb at each end, one clear and the other blackened.

differential brake [MECH ENG] A brake in which operation depends on a difference between two motions.

differential entrapment [PETRO ENG] Controlling oil and gas migration and accumulation by means of selective trapping or gas flushing in interconnecting reservoirs.

differential flotation [MIN ENG] Separation of a complex ore into two or more mineral components and gangue by flotation.

differential frequency meter [ENG] A circuit that converts the absolute frequency difference between two input signals to a linearly proportional direct-current output voltage that can be used to drive a meter, recorder, oscilloscope, or other device.

differential gear See differential.

differential heating [MET] A thermal gradient caused as heating takes place; can result in a distribution of stress in a material.

differential indexing [MECH ENG] A method of subdividing a circle based on the difference between movements of the index plate and index crank of a dividing engine.

differential instrument [ENG] Galvanometer or other measuring instrument having two circuits or coils, usually identical, through which currents flow in opposite directions; the difference or differential effect of these currents actuates the indicating pointer.

differential leak detector [ENG] A leak detector consisting of two tubes and a trap which directs the tracer gas from the system into the desired tube.

differential leveling [ENG] A surveying process in which a horizontal line of sight of known elevation is intercepted by a graduated standard, or rod, held vertically on the point being checked.

differential manometer [ENG] An instrument in which the difference in pressure between two sources is determined from the vertical distance between the surfaces of a liquid in two legs of an erect or inverted U-shaped tube when each of the legs is connected to one of the sources.

differential microphone See double-button microphone.

differential motion [MECH ENG] A mechanism in which the follower has two driving elements; the net motion of the follower is the difference between the

motions that would result from either driver acting alone.

differential piece-rate system [IND ENG] A wage plan based on a standard task time whereby the worker receives increased or decreased piece rates as his production varies from that expected for the standard time. Also known as accelerating incentive.

differential pressure fuel valve [MECH ENG] A needle or spindle normally closed, with seats at the back side of the valve orifice.

differential pressure gage [ENG] Apparatus to measure pressure differences between two points in a system; it can be a pressured liquid column balanced by a pressured liquid reservoir, a formed metallic pressure element with opposing force, or an electrical-electronic gage (such as strain, thermal-conductivity, or ionization).

differential producing primary device [ENG] An instrument that modifies the flow pattern of a fluid passing through a pipe, duct, or open channel, and thereby produces a difference in pressure between two points, which can then be measured to determine the rate of flow.

differential pulley [MECH ENG] A tackle in which an endless cable passes through a movable lower pulley, which carries the load, and two fixed coaxial upper pulleys having different diameters; yields a high mechanical advantage.

differential scatter [ENG] A technique for the remote sensing of atmospheric particles in which the backscattering from laser beams at a number of infrared wavelengths is measured and correlated with scattering signatures that are uniquely related to particle composition. Abbreviated DISC.

differential screw [MECH ENG] A type of compound screw which produces a motion equal to the difference in motion between the two component screws.

differential steam calorimeter [ENG] An instrument for measuring small specific-heat capacities, such as those of gases, in which the amount of steam condensing on a body containing the substance whose heat capacity is to be measured is compared with the amount condensing on a similar body which is evacuated or contains a substance of known heat capacity.

differential temperature survey [PETRO ENG] Well-temperature logging method that detects very small temperature anomalies; two thermometers, 6 feet (1.8 meters) apart, record the temperature gradient down the well bore, with small difference changes showing up anomalies.

differential thermometer See bimetallic thermometer.

differential windlass [MECH ENG] A windlass in which the barrel has two sections, each having a different diameter; the rope winds around one section, passes through a pulley (which carries the load), then winds around the other section of the barrel.

diffuser [ENG] A duct, chamber, or section in which a high-velocity, low-pressure stream of fluid (usually air) is converted into a low-velocity, high-pressure flow.

diffusion [MECH ENG] The conversion of air velocity into static pressure in the diffuser casing of a centrifugal fan, resulting from increases in the radius of the air spin and in area.

diffusion annealing [MET] Heat treatment of metal to promote homogeneity by diffusion of components.

diffusion bonding [MET] A solid-state process for joining metals by using only heat and pressure to achieve atomic bonding.

diffusion brazing [MET] A process which produces bonding of the faying surfaces by heating them to suitable temperatures; the filler metal is diffused with the base metal and approaches the properties of the base metal.

diffusion coating [MET] An alloy coating produced by allowing the coating material to diffuse into the base at high temperature.

diffusion hygrometer [ENG] A hygrometer based upon the diffusion of water vapor through a porous membrane; essentially, it consists of a closed chamber having porous walls and containing a hygroscopic compound, whose absorption of water vapor causes a pressure drop within the chamber that is measured by a manometer.

diffusion-limited current density [MET] The density corresponding to the maximum transfer rate that a material can sustain due to diffusion limits.

diffusion pump [ENG] A vacuum pump in which a stream of heavy molecules, such as mercury vapor, carries gas molecules out of the volume being evacuated; also used for separating isotopes according to weight, the lighter molecules being pumped preferentially by the vapor stream.

diffusion welding [MET] A welding process which utilizes high temperatures and pressures to coalesce the faying surfaces by solid-state bonding; there is no physical movement, visible deformation of the parts involved, or melting.

digested sludge [CIV ENG] Sludge or thickened mixture of sewage solids with water that has been decomposed by anaerobic bacteria.

digester [CIV ENG] A sludge-digestion tank containing a system of hot water or steam pipes for heating the sludge.

digestion [CIV ENG] The process of sewage treatment by the anaerobic decomposition of organic matter.

digger [ENG] A tool or apparatus for digging in the ground. [MIN ENG] A person who digs in the ground; usually refers to a coal miner.

digging [ENG] A sudden increase in cutting depth of a cutting tool due to an erratic change in load.

digging height See bank height.

digital log [ENG] A well log that has undergone discrete sampling and recording on a magnetic tape preparatory to use in computerized interpretation and plotting.

dihedral [AERO ENG] The upward or downward inclination of an airplane's wing or other supporting surface in respect to the horizontal; in some contexts, the upward inclination only.

dike [CIV ENG] An embankment constructed on dry ground along a riverbank to prevent overflow of lowlands and to retain floodwater.

dilatometer [ENG] An instrument for measuring thermal expansion and dilation of liquids or solids.

dill oil [MATER] A yellowish essential oil, soluble in propylene glycol, slightly soluble in glycerine, obtained by steam distillation of the dill plant *Anethum graveolens;* chief ingredient is carvone. Also known as American dillweed oil; anethum oil.

dilution [MET] The use of a welding filler metal deposit with a base metal or a previously deposited weld material having a lower alloy content.

dimensional stability [MATER] The ability of a material, such as a textile or plastic, to hold its shape over a period of time and under specific conditions.

dimension stone [MATER] Large, sound, relatively flawless blocks of stone used as building stone, monumental stone, paving stone, curbing, and flagging.

dimpling [ENG] Forming a conical depression in a metal surface in order to countersink a rivet head.

Dines anemometer [ENG] A pressure-tube anemometer in which the pressure head on a weather vane is kept facing into the wind, and the suction head, near the bearing which supports the vane, develops a suction independent of wind direction; the pressure difference between the heads is proportional to the square of the wind speed and is measured by a float manometer with a linear wind scale.

dingot [MET] A massive derby, usually a ton or more, produced in a bomb reaction.

Dings magnetic separator [MECH ENG] A device which is suspended above a belt conveyor to pull out and separate magnetic material from burden as thick as 40 inches (1 meter) and at belt speeds up to 750 feet (229 meters) per minute.

dinking [MECH ENG] Using a sharp, hollow punch for cutting light-gage soft metals or nonmetallic materials.

dioctyl phthalate test [ENG] A method used to evaluate air filters to be used in critical air-cleaning applications; a light-scattering technique counts the number of particles of controlled size (0.3 micrometer) entering and emerging from the test filter. Abbreviated DOP test.

dip [ENG] The vertical angle between the sensible horizon and a line to the visible horizon at sea, due to the elevation of the observer and to the convexity of the earth's surface. Also known as dip of horizon.

dip brazing [MET] Soldering by dipping the work into a hot, molten salt or metal bath and by using a nonferrous metal with a melting point above 800°F (427°C).

dip coating [ENG] A coating applied to ceramic ware or metal by immersion into a tank of melted nonmetallic material, such as resin or plastic, then chilling the adhering melt.

diphase cleaning [MET] Removing soilage from metal surfaces in a cleaning tank incorporating a solvent phase and an aqueous phase.

diphead [MIN ENG] A passage that follows the inclination of a coal seam.

dip inductor See earth inductor.

dipmeter [ENG] **1.** An instrument used to measure the direction and angle of dip of geologic formations. **2.** An absorption wavemeter in which bipolar or field-effect transistors replace the electron tubes used in older grid-dip meters.

dip mold [ENG] A one-piece glassmaking mold with an open top; used to mold patterns.

dip needle [ENG] An obsolete type of magnetometer consisting of a magnetized needle that rotates freely in the vertical plane, with an adjustable weight on one side of the pivot.

dip of horizon See dip.

dip oil [MATER] Oil containing about 25% tar acids; used as dip for animals to kill insect parasites.

Dippel's oil See bone oil.

dipper dredge [MECH ENG] A power shovel resembling a grab crane mounted on a flat-bottom boat for dredging under water. Also known as dipper shovel.

dipper shovel See dipper dredge.

dipper stick [MECH ENG] A straight shaft connecting the digging bucket of an excavating machine or power shovel with the boom.

dipper trip [MECH ENG] A device which releases the door of a shovel bucket.

dipping sonar [ENG] A sonar transducer that is lowered into the water from a hovering antisubmarine-warfare helicopter and recovered after the search is complete. Also known as dunking sonar.

dip plating See immersion plating.

dip soldering [MET] A method similar to dip brazing but using a filler metal having a melting point below 800°F (427°C).

dipstick [ENG] A graduated rod which measures depth when dipped in a liquid, used, for example, to measure the oil in an automobile engine crankcase.

direct-acting pump [MECH ENG] A displacement reciprocating pump in which the steam or power piston is connected to the pump piston by means of a rod, without crank motion or flywheel.

direct-acting recorder [ENG] A recorder in which the marking device is mechanically connected to or directly operated by the primary detector.

direct air cycle [AERO ENG] A thermodynamic propulsion cycle involving a nuclear reactor and gas turbine or ramjet engine, in which air is the working fluid. Also known as direct cycle.

direct-arc furnace [ENG] A furnace in which a material in a refractory-lined shell is rapidly heated to pour temperature by an electric arc which goes directly from electrodes to the material.

direct bearing [CIV ENG] A direct vertical support in a structure.

direct-chill casting [MET] A continuous ingot- or billet-casting process in which metal is poured into short molds on a platform and then cooled when the platform is lowered into a water bath. Abbreviated dc casting. Also known as semicontinuous casting.

direct command guidance [ENG] Control of a missile or drone entirely from the launching site by radio or by signals sent over a wire.

direct-connected [MECH ENG] The connection between a driver and a driven part, as a turbine and an electric generator, without intervening speed-changing devices, such as gears.

direct-contact condenser See contact condenser.

direct cost [IND ENG] The cost in goods and labor to produce a product which would not be spent if the product were not made.

direct-coupled [MECH ENG] Joined without intermediate connections.

direct coupling [MECH ENG] The direct connection of the shaft of a prime mover (such as a motor) to the shaft of a rotating mechanism (such as a pump or compressor).

direct-current electrode negative [MET] In direct-current arc welding, the arrangement of leads where the surface to be welded is the positive and the electrode is the negative relative to the welding arc.

direct-current electrode positive [MET] In direct-current arc welding, the arrangement of leads where the surface to be welded is the negative and the electrode is the positive relative to the welding arc.

direct cycle *See* direct air cycle.

direct drive [MECH ENG] A drive in which the driving part is directly connected to the driven part.

direct-drive vibration machine [MECH ENG] A vibration machine in which the vibration table is forced to undergo a displacement by a positive linkage driven by a direct attachment to eccentrics or camshafts.

direct dye [MATER] A group of coal tar dyes that act without mordants, for example, benzidine dyes. Also known as substantive dye.

direct energy conversion [ENG] Conversion of thermal or chemical energy into electric power by means of direct-power generators.

direct-expansion coil [MECH ENG] A finned coil, used in air cooling, inside of which circulates a cold fluid or evaporating refrigerant. Abbreviated DX coil.

direct extrusion [ENG] Extrusion by movement of ram and product in the same direction against a die orifice.

direct-fire [ENG] To fire a furnace without preheating the air or gas.

direct-geared [MECH ENG] Joined by a gear on the shaft of one machine meshing with a gear on the shaft of another machine.

direct-imaging mass analyzer [ENG] A type of secondary ion mass spectrometer in which secondary ions pass through an electrostatic immersion lens which forms an image that bears a point-to-point relation to the ion's place of origin on the sample surface, and then traverse magnetic sectors which effect mass separation. Also known as Castaing-Slodzian mass analyzer.

direction [ENG] The position of one point in space relative to another without reference to the distance between them; may be either three-dimensional or two-dimensional, the horizontal being the usual plane of the latter; usually indicated in terms of its angular distance from a reference direction.

directional control [ENG] Control of motion about the vertical axis; in an aircraft, usually by the rudder.

directional control valve [ENG] A control valve serving primarily to direct hydraulic fluid to the point of application.

directional drilling [ENG] A method of drilling in which the direction of the hole is planned before.

directional gain *See* directivity index.

directional gyro [AERO ENG] A flight instrument incorporating a gyro that holds its position in azimuth and thus can be used as a directional reference. Also known as direction indicator.

directional hydrophone [ENG ACOUS] A hydrophone whose response varies significantly with the direction of sound incidence.

directional log [PETRO ENG] A record of the wellhole drift, from the vertical, and the direction of that drift.

directional microphone [ENG ACOUS] A microphone whose response varies significantly with the direction of sound incidence.

directional property [MET] Any property of a metal whose magnitude varies with the orientation of the test axis to a specific direction within the metal.

directional response pattern *See* directivity pattern.

directional solidification [MET] Controlled solidification of molten metal in a casting so as to provide feed metal to the solidifying front of the casting.

directional stability [AERO ENG] The property of an aircraft, rocket, or such, enabling it to restore itself from a yawing or side-slipping condition. Also known as weathercock stability.

directional well [PETRO ENG] A well drilled at an angle up to 70° from the vertical to avoid obstacles over the reservoir, such as towns, beaches, or bodies of water.

direction cosine [ENG] In tracking, the cosine of the angle between a baseline and the line connecting the center of the baseline with the target.

direction-independent radar [ENG] Doppler radar used in sentry applications.

direction indicator *See* directional gyro.

directivity factor [ENG ACOUS] **1.** The ratio of radiated sound intensity at a remote point on the principal axis of a loudspeaker or other transducer, to the average intensity of the sound transmitted through a sphere passing through the remote point and concentric with the transducer; the frequency must be stated. **2.** The ratio of the square of the voltage produced by sound waves arriving parallel to the principal axis of a microphone or other receiving transducer, to the mean square of the voltage that would be produced if sound waves having the same frequency and mean-square pressure were arriving simultaneously from all directions with random phase; the frequency must be stated.

directivity index [ENG ACOUS] The directivity factor expressed in decibels; it is 10 times the logarithm to the base 10 of the directivity factor. Also known as directional gain.

directivity pattern [ENG ACOUS] A graphical or other description of the response of a transducer used for sound emission or reception as a function of the direction of the transmitted or incident sound waves in a specified plane and at a specified frequency. Also known as beam pattern; directional response pattern.

direct labor [IND ENG] The labor or effort actually producing goods or services.

direct labor standard See standard time.

direct-line drive [PETRO ENG] Waterflood operation involving a network of wells in a direct (straight) line.

direct material [IND ENG] Any raw or semifinished material which will be incorporated into the product.

direct-power generator [ENG] Any device which converts thermal or chemical energy into electric power by methods more direct than the conventional thermal cycle.

direct process [MET] A process in which the metal is produced from the ore in a single step (for example, steel without intermediate pig iron).

direct quenching [MET] Rapid cooling of carburized parts directly from the carburizing process.

direct-radiator speaker [ENG ACOUS] A loudspeaker in which the radiating element acts directly on the air, without a horn.

direct-reading gage [ENG] Gage that records directly (instead of inferentially) measured values, for example, a liquid-level gage pointer actuated by direct linkage with a float.

direct recording [ENG ACOUS] Recording in which a record is produced immediately, without subsequent processing, in response to received signals.

direct-reduction process [MET] Any of several methods for extracting iron ore below the melting point of iron, to produce solid reduced iron that may be converted to steel with little further refining.

direct-writing galvanometer [ENG] A direct-writing recorder in which the stylus or pen is attached to a moving coil positioned in the field of the permanent magnet of a galvanometer.

direct-writing recorder [ENG] A recorder in which the permanent record of varying electrical quantities or signals is made on paper, directly by a pen attached to the moving coil of a galvanometer or indirectly by a pen moved by some form of motor under control of the galvanometer. Also known as mechanical oscillograph.

dirigible [AERO ENG] A lighter-than-air craft equipped with means of propelling and steering for controlled flight.

disappearing filament pyrometer See optical pyrometer.

disappearing stair [BUILD] A stair that is designed to be swung up into a ceiling space.

disassemble [ENG] To take apart into constituent parts.

DISC See differential scatter.

discharge channel [MECH ENG] The passage in a pressure-relief device through which the fluid is released to the outside of the device.

discharge head [MECH ENG] Vertical distance between the intake level of a water pump and the level at which it discharges water freely to the atmosphere.

discharge hydrograph [CIV ENG] A graph showing the discharge or flow of a stream or conduit with respect to time.

discharge tube [MECH ENG] A tube through which steam and water are released into a boiler drum.

discharge-tube leak indicator [ENG] A device which detects the presence of a tracer gas by using a glass tube attached to a high-voltage source; the presence of leaked gas is indicated by the color of the electric discharge.

discharging arch [CIV ENG] A support built over, and not touching, a weak structural member, such as a wooden lintel, to carry the main load. Also known as relieving arch.

disconnect [ENG] To sever a connection.

discontinuity [MET] The place where the structural nature of a weldment is interfered with because of the materials involved or where the mechanical, physical, or metallurgical aspects are not homogeneous.

discontinuous precipitation [MET] Precipitation principally at and away from the grain boundaries in a supersaturated solid solution; diffraction patterns show two lattice parameters, the solute in solution and the precipitate.

discontinuous yielding [MET] The nonuniform plastic deformation of a metal along the length strained in tension.

DISCOS See disturbance compensation system.

discount [IND ENG] A reduction from the gross amount, price, or value.

discount factor [PETRO ENG] The ratio of the present worth of one or a series of future payments to the total undiscounted amount of such future payments. Also known as average discount factor; deferment factor; present-worth factor.

discovery [MIN ENG] Finding of a valuable mineral deposit.

discovery claim [MIN ENG] The first claim for the finding of a mineral deposit.

discovery vein [MIN ENG] The vein on which a mining claim is based.

discovery well [PETRO ENG] A successful exploration well.

discrete sound system [ENG ACOUS] A quadraphonic sound system in which the four input channels are preserved as four discrete channels during recording and playback processes; sometimes referred to as a 4-4-4 system.

disdrometer [ENG] Equipment designed to measure and record the size distribution of raindrops as they occur in the atmosphere.

disengage [ENG] To break the contact between two objects.

dishing [MET] Producing a shallow, concave surface in metal-forming operations.

disinfectant [MATER] A chemical agent that destroys microorganisms but not bacterial spores.

disintegrator [MECH ENG] An apparatus used for pulverizing or grinding substances, consisting of two steel cages which rotate in opposite directions.

disk *See* phonograph record.

disk attrition mill *See* disk mill.

disk brake [MECH ENG] A type of brake in which disks attached to a fixed frame are pressed against disks attached to a rotating axle or against the inner surfaces of a rotating housing.

disk cam [MECH ENG] A disk with a contoured edge which rotates about an axis perpendicular to the disk, communicating motion to the cam follower which remains in contact with the edge of the disk.

disk canvas wheel [DES ENG] A polishing wheel made of disks of canvas sewn together with heavy twine or copper wire, and reinforced by steel side plates and side rings with bolts or screws.

disk centrifuge [MECH ENG] A centrifuge with a large bowl having a set of disks that separate the liquid into thin layers to create shallow settling chambers.

disk clutch [MECH ENG] A clutch in which torque is transmitted by friction between friction disks with specially prepared friction material riveted to both sides and

contact plates keyed to the inner surface of an external hub.

disk coupling [MECH ENG] A flexible coupling in which the connecting member is a flexible disk.

disk engine [MECH ENG] A rotating engine in which the piston is a disk.

disk filter [ENG] A filter in which the substance to be filtered is drawn through membranes stretched on segments of revolving disks by a vacuum inside each disk; the solids left on the membrane are lifted from the tank and discharged. Also known as American filter.

disk grinder [MECH ENG] A grinding machine that employs abrasive disks.

disk grinding [MECH ENG] Grinding with the flat side of a rigid, bonded abrasive disk or segmental wheel.

disk leather wheel [DES ENG] A polishing wheel made of leather disks glued together.

disk loading [AERO ENG] A measure which expresses the design gross weight of a helicopter as a function of the swept areas of the lifting rotor.

disk meter [ENG] A positive displacement meter to measure flow rate of a fluid; consists of a disk that wobbles or nutates within a chamber so that each time the disk nutates a known volume of fluid passes through the meter.

disk mill [MECH ENG] Size-reduction apparatus in which grinding of feed solids takes place between two disks, either or both of which rotate. Also known as disk attrition mill.

disk recording [ENG ACOUS] **1.** The process of inscribing suitably transformed acoustical or electrical signals on a phonograph record. **2.** *See* phonograph record.

disk sander [MECH ENG] A machine that uses a circular disk coated with abrasive to smooth or shape surfaces.

disk signal [CIV ENG] Automatic block signal with colored disks that indicate train movements.

disk spring [MECH ENG] A mechanical spring that consists of a disk or washer supported by one force (distributed by a suitable chuck or holder) at the periphery and by an opposing force on the center or hub of the disk.

disk-wall packer [PETRO ENG] A disklike seal between the outside of the well tubing and the inside of the well casing; used to prevent fluid movement from the pressure differential above and below the sealing point.

disk wheel [DES ENG] A wheel in which a solid metal disk, rather than separate spokes, joins the hub to the rim.

dispenser [ENG] Device that automatically dispenses radar chaff from an aircraft.

dispersal [CIV ENG] The practice of building or establishing industrial plants, government offices, or the like, in separated areas, to reduce vulnerability to enemy attack.

dispersed gas injection [PETRO ENG] Gas-injection pressure maintenance of an oil reservoir in which the injection wells are arranged geometrically to distribute the gas uniformly throughout the oil-productive portions of the reservoir.

disperse dye [MATER] A very slightly water-soluble, colored material for use on cellulose acetate and other synthetic fibers; color is transferred to the fiber as extremely finely divided particles, resulting in a solution of the dye in the solid fiber.

disperser [MATER] Material added to solid-in-liquid or liquid-in-liquid suspensions to separate the individual suspended particles; used in pigment grinding and dye dispersion. Also known as dispersing agent; emulsifier; emulsifying agent.

dispersing agent See disperser.

dispersion [AERO ENG] Deviation from a prescribed flight path; specifically, circular dispersion especially as applied to missiles.

dispersion mill [MECH ENG] Size-reduction apparatus that disrupts clusters or agglomerates of solids, rather than breaking down individual particles; used for paint pigments, food products, and cosmetics.

displacement [ENG] The volume swept out in one stroke by a piston moving in a cylinder as for an engine, pump, or compressor.

displacement compressor [MECH ENG] A type of compressor that depends on displacement of a volume of air by a piston moving in a cylinder.

displacement efficiency [PETRO ENG] In a gas condensate reservoir, the proportion (by volume) of wet hydrocarbons swept out of pores during dry-gas cycling.

displacement engine See piston engine.

displacement gyroscope [ENG] A gyroscope that senses, measures, and transmits angular displacement data.

displacement manometer [ENG] A differential manometer which indicates the pressure difference across a solid or liquid partition which can be displaced against a restoring force.

displacement meter [ENG] A water meter that measures water flow quantitatively by recording the number of times a vessel of known capacity is filled and emptied.

displacement pump [MECH ENG] A pump that develops its action through the alternate filling and emptying of an enclosed volume as in a piston-cylinder construction.

displacer-type meter [ENG] Apparatus to detect liquid level or gas density by measuring the effect of the fluid (gas or liquid) on the buoyancy of a displacer unit immersed within the fluid.

disposable [ENG] Within a manufacturing system, designed to be discarded after use and replaced by an identical item, such as a filter element.

disposal field See absorption field.

disruptive strength [MET] Failure stress caused by hydrostatic tension.

dissipative muffler [ENG] A device which absorbs sound energy as the gas passes through it; a duct lined with sound-absorbing material is the most common type.

dissolved gas See solution gas.

dissolved-gas drive See internal gas drive.

dissolved-gas-drive reservoir [PETRO ENG] Oil reservoir in which the temperature of the liquid phase is below critical, and the liquid is driven from the reservoir by the expansion of dissolved gas. Also known as a bubble-point reservoir.

distance marker [ENG] One of a series of concentric circles, painted or otherwise fixed on the screen of a plan position indicator, from which the distance of a target from the radar antenna can be read directly; used for surveillance and navigation where the relative distances between a number of targets are required simultaneously. Also known as radar range marker; range marker.

distance resolution [ENG] The minimum radial distance by which targets must be separated to be separately distinguishable by a particular radar. Also known as range discrimination; range resolution.

distant signal [CIV ENG] A signal placed at a distance from a block of track to give advance warning when the block is closed.

distillate fuel [MATER] Any one of the wide variety of fuels obtained from fractions boiling above the temperature at which gasoline comes off in the distillation of petroleum.

distillate fuel oil [MATER] A classification for one of the overhead fractions produced from crude oil in conventional distillation operations.

distortion meter [ENG] An instrument that provides a visual indication of the harmonic content of an audio-frequency wave.

distributed-parameter system See distributed system.

distributed system [SYS ENG] A system whose behavior is governed by partial differential equations, and not merely ordinary differential equations. Also known as distributed-parameter system.

distribution amplifier [ENG ACOUS] An audio-frequency power amplifier used to feed a speech or music distribution system and having sufficiently low output impedance so changes in load do not appreciably affect the output voltage.

distribution box [CIV ENG] In sanitary engineering, a box in which the flow of effluent from a septic tank is distributed equally into the lines that lead to the absorption field.

distribution reservoir [CIV ENG] A service reservoir connected with the conduits of a primary water supply; used to supply water to consumers according to fluctuations in demand over short time periods and serves for local storage in case of emergency.

distributor gear [MECH ENG] A gear which meshes with the camshaft gear to rotate the distributor shaft.

district heating [MECH ENG] The supply of heat, either in the form of steam or hot water, from a central source to a group of buildings.

disturbance compensation system [AERO ENG] A system applied to navigational satellites to remove the along-the-track component of drag and radiation forces. Abbreviated DISCOS.

disulfide oil [MATER] One of the oils obtained by oxidizing to disulfides the mercaptans extracted from light petroleum distillates; the disulfides separate from the extract as an oily layer.

ditch [CIV ENG] A small artificial channel cut through earth or rock to carry water for irrigation or drainage.

ditch check [CIV ENG] A small dam positioned at intervals in a road ditch to prevent erosion.

ditcher See trench excavator.

ditching [AERO ENG] A forced landing on water, or the process of making such a landing.

ditching [ENG] The digging of ditches, as around storage tanks or process areas to hold liquids in the event of a spill or along the sides of a roadway for drainage.

dive [AERO ENG] A rapid descent by an aircraft or missile, nose downward, with or without power or thrust. [ENG] To submerge into an underwater environment so that it may be studied or utilized; includes the use of specialized equipment such as scuba, diving helmets, diving suits, diving bells, and underwater research vessels.

dive bomber [AERO ENG] An aircraft designed to release bombs during a steep dive.

dive brake [AERO ENG] An air brake designed for operation in a dive; flaps at the following edge of one wing that can be extended into the airstream to increase drag and hold the aircraft to its "never exceed" dive speed in a vertical dive; used on dive bombers and sailplanes.

divergence speed [AERO ENG] The speed of an aircraft above which no statically stable equilibrium condition exists and the deformation will increase to a point of structural failure.

divergent die [ENG] A die with the internal channels that lead to the orifice diverging, such as the dies used for manufacture of hollow-body plastic items.

divergent nozzle [DES ENG] A nozzle whose cross section becomes larger in the direction of flow.

diverging duct [DES ENG] Fluid-flow conduit whose internal cross-sectional area increases in the direction of flow.

diverging yaw [AERO ENG] In the flight of a projectile, an angle of yaw increasing from the initial yaw, so that the projectile is unstable.

diversion canal [CIV ENG] An artificial channel for diverting water from one place to another.

diversion chamber [ENG] A chamber designed to direct a stream into a channel or channels.

diversion dam [CIV ENG] A fixed dam for diverting stream water away from its course.

diversion gate [CIV ENG] A gate which may be closed to divert water from the main conduit or canal to a lateral or some other channel.

diversion tunnel [CIV ENG] An underground passageway used to divert flowing water around a construction site.

diversity radar [ENG] A radar that uses two or more transmitters and receivers, each pair operating at a slightly different frequency but sharing a common antenna and video display, to obtain greater effective range and reduce susceptibility to jamming.

diverting agent [PETRO ENG] A viscous gel or suspension of graded solids used during acidizing of an oil reservoir to temporarily block off the most permeable sections of the pay zone to force the acid into less permeable sections.

divided lane [CIV ENG] A highway divided into lanes by a median strip.

divided pitch [DES ENG] In a screw with multiple threads, the distance between corresponding points on two adjacent threads measured parallel to the axis.

divider [DES ENG] A tool like a compass, used in metalworking to lay out circles or arcs and to space holes or other dimensions.

dividing network *See* crossover network.

diving bell [ENG] An early diving apparatus constructed in the shape of a box or cylinder without a bottom and connected to a compressed-air hose.

diving suit [ENG] A waterproof outfit designed for diving, especially one with a helmet connected to a compressed-air hose.

divining [MIN ENG] An unscientific method for searching for subsurface water or minerals by means of a divining rod. Also known as dowsing.

divining rod [MIN ENG] An unscientific device in the form of a forked rod or tree branch that is supposed to dip when held over water or minerals, depending on the specialty of the operator, or dowser. Also known as dowsing rod; wiggle stick.

division plate [MECH ENG] A diaphragm which surrounds the piston rod of a crosshead-type engine and separates the crankcase from the lower portion of the cylinder.

division wall [BUILD] A wall used to create major subdivisions in a building.

djave butter *See* adjab butter.

D nickel [MET] A nickel-manganese (4.75%) alloy of medium strength that resists spark erosion and hence is used for sparkplug electrodes and for some electronic applications.

Dobson prop [MIN ENG] A hydraulic supporting post used in mine tunnel construction.

Dobson support system [MIN ENG] A self-advancing unit consisting of three Dobson props used to support longwall faces.

dock [CIV ENG] **1.** The slip or waterway that is between two piers or cut into the land for the berthing of ships. **2.** A basin or enclosure for reception of vessels, provided with means for controlling the water level.

docking [AERO ENG] The mechanical coupling of two or more man-made orbiting objects.

docking block [CIV ENG] A timber used to support a ship in dry dock.

dockyard [CIV ENG] A yard utilized for ship construction and repair.

doctor blade [ENG] A device for regulating the amount of liquid material on the rollers of a spreader. Also known as doctor knife.

doctor knife *See* doctor blade.

dodge chain [DES ENG] A chain with detachable bearing blocks between the links.

Dodge crusher [MIN ENG] A type of jaw crusher with the movable jaw hinged at the bottom, allowing a highly uniform product to be discharged.

Dodge pulverizer [MIN ENG] A hexagonal drum-shaped pulverizer that rotates on a horizontal axis and contains steel balls for reducing rock and ore.

Dodge-Romig tables [IND ENG] Tabular data for acceptance sampling, including lot tolerance and AOQL tables.

dodo [ENG] A rectangular groove cut across the grain of a board.

dog [DES ENG] **1.** Any of various simple devices for holding, gripping, or fastening, such as a hook, rod, or spike with a ring, claw, or lug at the end. **2.** An iron for supporting logs in a fireplace. **3.** A drag for the wheel of a vehicle.

dog clutch [DES ENG] A clutch in which projections on one part fit into recesses on the other part.

dogfish oil *See* shark liver oil.

doghole [MIN ENG] A small opening in a mine.

doghole mine [MIN ENG] A small coal mine employing 15 or less miners.

dog iron [DES ENG] **1.** A short iron bar with ends bent at right angles. **2.** An iron pin that can be inserted in stone or timber in order to lift it.

dogleg [PETRO ENG] Bend or sudden direction change in a wellhole that can cause tubing wear and failure.

dogs *See* folding boards.

dog screw [DES ENG] A screw with an eccentric head; used to mount a watch in its case.

dog's tooth [CIV ENG] A masonry string course in which the brick corner projects.

Dolan equation [PETRO ENG] Empirical equation for reservoir-permeability damage factor by the invasion of drilling mud or other foreign materials.

dolly [ENG] Any of several types of industrial hand trucks consisting of a low platform or specially shaped carrier mounted on rollers or combinations of fixed and swivel casters; used to carry such things as furniture, milk cans, paper rolls, machinery weighing up to 80 tons, and television cameras short distances.

dolphin [CIV ENG] **1.** A group of piles driven close and tied together to provide a fixed mooring in the open sea or a guide for ships coming into a narrow harbor entrance. **2.** A mooring post on a wharf.

dome [ENG ACOUS] An enclosure for a sonar transducer, projector, or hydrophone and associated equipment; designed to have minimum effect on sound waves traveling underwater.

domestic coke [MATER] Coke for residential heating, which must have as low an ash content and as high a softening temperature (preferably above 2300°F, or 1260°C) for the ash as possible.

domestic induction heater [ENG] A cooking utensil heated by current (usually of commercial power line frequency) induced in it by a primary inductor.

domestic refrigerator [MECH ENG] A refrigeration system for household use which typically has a compression machine designed for continuous automatic operation and for conservation of the charges of refrigerant and oil, and is usually motor-driven and air-cooled. Also known as refrigerator.

domestic satellite [AERO ENG] A satellite in stationary orbit 22,300 miles (35,680 kilometers) above the equator for handling up to 12 separate color television programs, up to 14,000 private-line telephone calls, or an equivalent number of channels for other communication services within the United States. Abbreviated DOMSAT.

dome theory [MIN ENG] The theory that the movements of strata resulting from underground excavations are limited by a dome whose base is the area of excavation, and that the movements decrease in intensity as they extend upward from the center of the base.

doming [MIN ENG] Setting up a domelike region above the open space created by stope excavation.

DOMSAT See domestic satellite.

donkey See barney.

donkey engine [MECH ENG] A small auxiliary engine which is usually portable or semiportable and powered by steam, compressed air, or other means, particularly one used to power a windlass to lift cargo on shipboard or to haul logs.

doodlebug [MECH ENG] **1.** A small tractor. **2.** A motor-driven railcar used for maintenance and repair work. [MIN ENG] The treatment plant or washing unit of a dredge which is mounted on a pontoon and can be floated in an excavation dug by a dragline.

door [ENG] A piece of wood, metal, or other firm material pivoted or hinged on one side, sliding along grooves, rolling up and down, revolving, or folding, by means of which an opening into or out of a building, room, or other enclosure is open or closed to passage.

door check See door closer.

door closer [DES ENG] **1.** A device that makes use of a spring for closing, and a compression chamber from which liquid or air escapes slowly, to close a door at a controlled speed. Also known as door check. **2.** In elevators, a device or assembly of devices which closes an open car or hoistway door by the use of gravity or springs.

doorstop [BUILD] A strip positioned on the doorjamb for the door to close against.

dope [MATER] A cellulose ester lacquer used as an adhesive or a coating.

doped solder [MET] Solder having an element added to it to ensure retention of a quality of the base metal on which it is used.

doping [ENG] Coating the mold or mandrel with a substance which will prevent the molded plywood part from sticking to it and will facilitate removal.

Doppler current meter [ENG] An acoustic current meter in which a collimated ultrasonic signal of known frequency is projected into the water and the reverberation frequency is measured; the difference in frequencies (Doppler shift) is proportional to the speed of water traveling past the meter.

Doppler radar [ENG] A radar that makes use of the Doppler shift of an echo due to relative motion of target and radar to differentiate between fixed and moving targets and measure target velocities.

Doppler range See doran.

Doppler sonar [ENG] Sonar based on Doppler shift measurement technique. Abbreviated DS.

Doppler tracking [ENG] Tracking of a target by using Doppler radar.

Doppler ultrasonic flowmeter [ENG] An instrument for determining the velocity of fluid flow from the Doppler shift of high-frequency sound waves reflected from particles or discontinuities in the flowing fluid.

DOP test See dioctyl phthalate test.

doran [ENG] A Doppler ranging system that uses phase comparison of three different modulation frequencies on the carrier wave, such as 0.01, 0.1, and 1 megahertz, to obtain missile range data with high accuracy. Derived from Doppler range.

dore [MET] Gold and silver bullion remaining in a cupeling furnace after removal of the oxidized lead.

dormer window [BUILD] An extension of an attic room through a sloping roof to accommodate a vertical window.

Dorr agitator [MECH ENG] A tank used for batch washing of precipitates which cannot be leached satisfactorily in a tank; equipped with a slowly rotating rake at the bottom, which moves settled solids to the center, and an air lift that lifts slurry to the launders. Also known as Dorr thickener.

Dorr classifier [MECH ENG] A horizontal flow classifier consisting of a rectangular tank with a sloping bottom, a rake mechanism for moving sands uphill along the bottom, an inlet for feed, and outlets for sand and slime.

Dorr thickener See Dorr agitator.

Dosco miner [MIN ENG] A large crawler-tracked cutter-loader that is designed for longwall faces in seams over 4½ feet (1.4 meters) thick and takes a buttock 0.5 foot (0.15 meter) wide; rated at 200 horsepower (1.49 × 10^5 watts); has seven cutter chains mounted on the cutterhead, with a capacity of over 400 tons per machine.

dosing tank [CIV ENG] A holding tank that discharges sewage at a rate required by treatment processes.

double-acting [MECH ENG] Acting in two directions, as with a reciprocating piston in a cylinder with a working chamber at each end.

double-acting compressor [MECH ENG] A reciprocating compressor in which both ends of the piston act in working chambers to compress the fluid.

double-acting hammer [MET] A forging hammer in which the ram is raised and forced down by a charge of air or steam.

double-acting pawl [MECH ENG] A double pawl which can drive in either direction.

double-action die [MET] A die designed to perform more than one operation with each stroke of the press.

double-action forming [MET] Forming in which more than one shape is imparted by each stroke of the press.

double-action mechanical press [MECH ENG] A press having two slides which move one within the other in parallel movements.

double aging [MET] Introduction of a primary (stabilizing) and a secondary aging treatment to control the precipitate formed from a supersaturated alloy and to achieve specific properties in the material.

double arcing [MET] An occurrence in plasma arc welding and cutting where a secondary electric arc displaces the main arc at the outlet of a welding nozzle.

double-base rocket propellant [MATER] A solid rocket propellant using two unstable compounds, such as nitrocellulose and nitroglycerine.

double-bevel groove weld [MET] A type of groove weld in which one member has a joint edge beveled on both sides. Also known as double-V groove weld.

double-block brake [MECH ENG] Two single-block brakes in symmetrical opposition, where the operating force on one lever is the reaction on the other.

double board [PETRO ENG] The platform from which the derrick operator works, located at a height on the derrick equal to two joined pipe lengths.

double-button microphone [ENG ACOUS] A carbon microphone having two carbon-filled buttonlike containers, one on each side of the diaphragm, to give twice the resistance change obtainable with a single button. Also known as differential microphone.

double-cone bit [DES ENG] A type of roller bit having only two cone-shaped cutting members.

double-core barrel drill [DES ENG] A core drill consisting of an inner and an outer tube; the inner member can remain stationary while the outer one revolves.

double-coursed [BUILD] Covered with a material such as shingles in such a way that no area is covered with less than two thicknesses.

double-crank press [MECH ENG] A mechanical press with a single wide slide operated by a crankshaft having two crank pins.

double-cut file [DES ENG] A file covered with two series of parallel ridges crossing at angles to each other.

double-cut planer [MECH ENG] A planer designed to cut in both the forward and reverse strokes of the table.

double-cut saw [DES ENG] A saw with teeth that cut during the forward and return strokes.

double drill column [MIN ENG] Two drill columns connected by a horizontal bar on which a drill machine can be mounted. Also known as double jack.

double-drum hoist [MECH ENG] A hoisting device consisting of two cable drums which rotate in opposite directions and can be operated separately or together.

double-entry method [MIN ENG] A mining arrangement involving twin entries in flat or gently dipping coal, so that rooms can be extended from both entryways.

double floor [BUILD] A floor in which binding joists support the ceiling joists below as well as the floor joists above.

doublehand drilling [ENG] A rock-drilling method performed by two persons, one striking the rock with a long-handled sledge hammer while a second holds the drill and twists it between strokes. Also known as double jacking.

double headings [MIN ENG] A pair of coal headings driven parallel to each other and positioned side by side about 10–20 yards (9–18 meters) apart.

double Hooke's joint [MECH ENG] A universal joint which eliminates the variation in angular displacement and angular velocity between driving and driven shafts, consisting of two Hooke's joints with an intermediate shaft.

double-housing planer [MECH ENG] A planer having two housings to support the cross rail, with two heads on the cross rail and one sidehead on each housing.

double-hung [BUILD] Of a window, having top and bottom sashes which are counterweighted or equipped with a spring on each side for easier raising and lowering.

double jack [DES ENG] A heavy hammer, weighing about 10 pounds (4.5 kilograms), requiring the use of both hands. [MIN ENG] *See* double drill column.

double-J groove weld [MET] A groove weld in which one member has a joint edge in the form of a double J or two half U's, one from each side. Also known as double-U groove weld.

double load [ENG] A charge separated by inert material in a borehole.

double mast *See* A frame.

double-rivet [ENG] To rivet a lap joint with two rows of rivets or a butt joint with four rows.

double-roll crusher [MECH ENG] A machine which crushes materials between teeth on two roll surfaces; used mainly for coal.

double sampling [IND ENG] Inspecting one sample and then deciding whether to accept or reject the lot or to defer action until a second sample is inspected.

double-shot molding [ENG] A means of turning out two-color parts in thermoplastic materials by successive molding operations.

double tempering [MET] A technique for ensuring the stability of the microstructure of a quench-hardened steel by subjecting it to two tempering cycles at approximately the same temperature.

double-theodolite observation [ENG] A technique for making winds-aloft observations in which two theodolites located at either end of a base line follow the ascent of a pilot balloon; synchronous measurements of the elevation and azimuth angles of the balloon, taken at periodic intervals, permit computation of the wind vector as a function of height.

double-track tape recorder [ENG ACOUS] A tape recorder with a recording head that covers half the tape width, so two parallel tracks can be recorded on one tape. Also known as dual-track tape recorder; half-track tape recorder.

double-U groove weld *See* double-J groove weld.

double valves [PETRO ENG] Two valves in series used as subsurface traveling or standing valves in wells, the dual arrangement being more reliable than a single valve.

double-V groove weld *See* double-bevel groove weld.

double-wall cofferdam [CIV ENG] A cofferdam consisting of two lines of steel piles tied to each other, and having the space between filled with sand.

double-wedge cut [MIN ENG] A drill-hole pattern composed of a shallow wedge within a larger, outer wedge.

double-welded joint [MET] Any joint that has been welded from both sides.

doughnut [PETRO ENG] A ring of wedges or a threaded, tapered ring that supports a pipe string.

Douglas fir oil *See* pine needle oil.

douse [MET] To thrust a hot piece of metal into a liquid during the hardening process. [MIN ENG] To locate and delineate subsurface resources such as water, oil, gas, or minerals.

dovetailing [MET] In thermal spraying, roughening the surface by angular cutting prior to the deposit of sprayed material.

dovetail joint [DES ENG] A joint consisting of a flaring tenon in a fitting mortise.

dovetail saw [DES ENG] A short stiff saw with a thin blade and fine teeth; used for accurate woodwork.

dowel [DES ENG] **1.** A headless, cylindrical pin which is sunk into corresponding holes in adjoining parts, to locate the parts relative to each other or to join them together. **2.** A round wooden stick from which dowel pins are cut.

dowel screw [DES ENG] A dowel with threads at both ends.

downcast [MIN ENG] Intake shaft for air in a mine.

down cutting *See* climb cutting.

downdraft carburetor [MECH ENG] A carburetor in which the fuel is fed into a downward current of air.

downhand welding *See* flat-position welding.

downhole drill [MIN ENG] A hammer or percussive drill in which a reciprocating pneumatic piston is located immediately behind the drill bit and can follow and enter the bit down the hole, for minimizing energy losses.

downhole equipment *See* drill fittings.

down lock [AERO ENG] An airplane mechanism that locks the landing gear in a down position after the gear is lowered.

down milling *See* climb cutting.

downrange [AERO ENG] Any area along the flight course of a rocket or missile test range.

downslope time [MET] Time necessary for current decrease when using slope control in resistance welding.

downtime [IND ENG] The lost production time during which a piece of equipment is not operating correctly due to a breakdown, maintenance, necessities, or power failure.

dowsing *See* divining.

dowsing rod *See* divining rod.

Dowty prop [MIN ENG] A self-contained hydraulic supporting post consisting of two telescoping tubes; the upper (inner) tube contains the oil, pump, yield valve, and other accessories.

Draeger breathing apparatus [MIN ENG] A long-service, self-contained oxygen-breathing apparatus with the oxygen feed governed by the lungs; allows the user to do hard work for up to 5 hours and normal work for 7 hours, and can sustain a resting individual for 18 hours.

Draeger escape apparatus [MIN ENG] A portable, self-contained oxygen-breathing apparatus that is carried on the back of the user; protects against poisonous gases or oxygen shortages for 1 hour.

draft Also spelled draught. [CIV ENG] A line of a traverse survey. [ENG] **1.** In molds, the degree of taper on a side wall or the angle of clearance present to facilitate removal of cured or hardened parts from a mold. **2.** The area of a water discharge opening. [MET] **1.** The act or process of drawing, with dies. **2.** The work or quantity of work drawn.

drafter See draftsman.

draft gage [ENG] **1.** A modified U-tube manometer used to measure draft of low gas heads, such as draft pressure in a furnace, or small differential pressures, for example, less than 2 inches (5 centimeters) of water. **2.** A hydrostatic depth indicator, installed in the side of a vessel below the light load line, to indicate amount of submergence.

drafting paper [MATER] A fine white or cream-colored paper that is hard-surfaced and has good erasing characteristics.

draft loss [MECH ENG] A decrease in the static pressure of a gas in a furnace or boiler due to flow resistance.

draftsman [ENG] An individual skilled in drafting, especially of machinery and structures. Also known as drafter.

draft tube [MECH ENG] The piping system for a reaction-type hydraulic turbine that allows the turbine to be set safely above tail water and yet utilize the full head of the site from head race to tail race.

drag [MET] **1.** The bottom part of a flask used in casting. **2.** In thermal cutting, the distance deviating from the theoretical vertical line of cutting measured along the bottom surface of the material. [MIN ENG] Movement of the hanging wall with respect to the foot wall due to the weight of the arch block in an inclined slope.

drag anchor See drogue.

drag-body flowmeter [ENG] Device to meter liquid flow; measures the net force parallel to the direction of flow; the resulting pressure difference is used to solve flow equations.

drag chain [ENG] **1.** A chain dragged along the ground from a motor vehicle chassis

to prevent the accumulation of static electricity. **2.** A chain for coupling rail cars.

drag-chain conveyor [MECH ENG] A conveyor in which the open links of a chain drag material along the bottom of a hard-faced concrete or cast iron trough. Also known as dragline conveyor.

drag chute See drag parachute.

drag classifier [MECH ENG] A continuous belt containing transverse rakes, used to separate coarse sand from fine; the belt moves up through an inclined trough, and fast-settling sands are dragged along by the rakes.

drag-cup generator [ENG] A type of tachometer which uses eddy currents and functions in control systems; it consists of two stationary windings, positioned so as to have zero coupling, and a nonmagnetic metal cup, which is revolved by the source whose speed is to be measured; one of the windings is used for excitation, inducing eddy currents in the rotating cup. Also known as drag-cup tachometer.

drag-cup tachometer See drag-cup generator.

drag cut [ENG] A drill hole pattern for breaking out rock, in which angled holes are drilled along a floor toward a parting, or on a free face and then broken by other holes drilled into them.

drag direction [AERO ENG] In stress analysis of a given airfoil, the direction of the relative wind.

drag-in [MET] Solution carried by the work and handling equipment to another solution.

dragline [MECH ENG] An excavator operated by pulling a bucket on ropes towards the jib from which it is suspended. Also known as dragline excavator.

dragline conveyor See drag-chain conveyor.

dragline excavator See dragline.

dragline scraper [MECH ENG] A machine with a flat, plowlike blade or partially open bucket pulled on rope for withdrawing piled material, such as stone or coal, from a stockyard to the loading platform; the empty bucket is subsequently returned to the pile of material by means of a return rope.

drag link [MECH ENG] A four-bar linkage in which both cranks traverse full circles; the fixed member must be the shortest link.

drag-out [MET] Solution taken from a bath by the work or handling equipment.

drag parachute [AERO ENG] Any of various types of parachutes that can be deployed from the rear of an aircraft, especially during landings, to decrease speed and also, under certain flight conditions, to control and stabilize the aircraft. Also known as drag chute.

drag rope [AERO ENG] A long, heavy rope carried in the basket of a balloon and permitted to hang over the side and drag on the ground in order to lighten the basket.

dragsaw [DES ENG] A saw that cuts on the pulling stroke; used in power saws for cutting felled trees.

dragshovel *See* backhoe.

drag-stone mill [MIN ENG] A mill in which a heavy stone is dragged over ore to grind it.

drag technique [MET] An arc-welding method in which the electrode is in contact with the joint being welded without being in short circuit.

drag truss [AERO ENG] A truss that is positioned horizontally between the wing spars; used to stiffen the wing structure and as a resistance for drag forces acting on the airplane wing.

drag-type tachometer *See* eddy-current tachometer.

drag-weight ratio [AERO ENG] The ratio of the drag of a missile to its total weight.

drag wire [AERO ENG] A part of the truss in an airplane wing and also in the wing support; used to sustain the backward reaction due to the wing's drag.

drain [CIV ENG] 1. A channel which carries off surface water. 2. A pipe which carries off liquid sewage.

drainage [CIV ENG] Removal of groundwater or surface water, or of water from structures, by gravity or pumping.

drainage canal [CIV ENG] An artificial canal built to drain water from an area having no natural outlet for precipitation accumulation.

drainage gallery [CIV ENG] A gallery in a masonry dam parallel to the top of the dam, to intercept seepage from the upstream face and conduct it away from the downstream face.

drainage well [CIV ENG] A vertical shaft in a masonry dam to intercept seepage before it reaches the downstream side.

drain tile [BUILD] A cylindrical tile with holes in the walls used at the base of a building foundation to carry away groundwater.

drape forming [ENG] A method of forming thermoplastic sheet in which the sheet is clamped into a movable frame, heated, and draped over high points of a male mold; vacuum is then applied to complete the forming operation.

draught *See* draft.

draw [ENG] To haul a load. [MET] 1. A fissure or pocket in a casting formed when the supply of molten metal is inadequate during solidification. 2. To remove a

pattern from a foundry flask. [MIN ENG] 1. To remove timber supports, allowing overhanging coal to fall down for collection. 2. To allow ore to run down chutes from stopes, chambers, or ore bins. 3. To collect broken coal in trucks. 4. To hoist coal, rock, ore, or other materials to the surface. 5. The horizontal distance to which creep extends on the surface beyond the stopes.

drawability [MET] The ability of a metal to be deep-drawn.

drawbar [ENG] 1. A bar used to connect a tender to a steam locomotive. 2. A beam across the rear of a tractor for coupling machines or other loads. 3. A clay block submerged in a glassmaking furnace to define the point at which sheet glass is drawn.

drawbar horsepower [MECH ENG] The horsepower available at the drawbar in the rear of a locomotive or tractor to pull the vehicles behind it.

drawbar pull [MECH ENG] The force with which a locomotive or tractor pulls vehicles on a drawbar behind it.

draw bead [MET] A projection on the surface of a metal sheet to control its flow during drawing.

drawbench [MET] A stand on which metal is drawn through dies; used in wire-making, or for drawing of rods and tubing.

drawbridge [CIV ENG] Any bridge that can be raised, lowered, or drawn aside to provide clear passage for ships.

drawdown [PETRO ENG] The difference between the static and the flowing bottom-hole pressure.

drawdown ratio [ENG] The ratio of die opening thickness to product thickness.

drawer [ENG] A box or receptacle that slides or rolls on tracks within a cabinet.

draw-filing [ENG] Filing by pushing and pulling a file sideways across the work.

drawhead [MET] A group of rollers through which strip tubing or solid stock is drawn to form angled sections.

drawhole [MIN ENG] The aperture in a battery through which coal or ore is drawn.

drawing [MET] 1. Pulling a wire or tube through a die to reduce the cross section. 2. Forcing plastic deformation of metal in a die to form recessed parts.

drawing back [MET] Reheating hardened steel to a temperature below the critical temperature in order to change its hardness.

drawing bristol [MATER] A cardboard made of 100% cotton in the higher grades; has good characteristics of permanence, strength, and erasability.

drawing cloth [MATER] A linen cloth that is specially treated to be smooth and translucent so that it may be used for ink tracings.

drawing compound [MET] A material applied to the work during drawing or pressing operations to eliminate draw marks by preventing direct contact between the work and die.

drawing die [MET] A die that forms sheet metal into cuplike, wrinkle-free shapes.

drawing of temper [MET] The process of heating steel to red heat and then letting it cool slowly; opposite of hardening or tempering.

drawing out [MET] Lengthening of a piece of metal through a heating and hammering process, resulting in a proportional reduction in section area.

drawing paper [MATER] One of a wide variety of papers used for pen-and-pencil drawing by artists and architects.

drawing timber [MIN ENG] The act of withdrawing timber and other supports from abandoned or worked-out mines.

drawknife [DES ENG] A woodcutting tool with a long, narrow blade and two handles mounted at right angles to the blade.

draw mark [MET] An impairment of the die or metal surface caused during drawing due to friction or a defect in the die; examples are scoring and die lines.

drawn glass [MATER] Glass made automatically by drawing the molten material through rollers.

draw piece [MET] Any part made by drawing.

drawplate [MET] A circular plate having a central hole through which wire is drawn by a punch. Also known as draw ring.

drawpoint [ENG] A steel point used to scratch lines or to pierce holes.

draw radius [MET] A measure of cutting edge of a die or punch over which the metal is drawn.

draw ring *See* drawplate.

draw works [PETRO ENG] An oil-well drilling mechanism used to supply driving power and to lift heavy objects; consists of a countershaft and drum.

dredge [ENG] A cylindrical or rectangular device for collecting samples of bottom sediment and benthic fauna. [MECH ENG] A floating excavator used for widening or deepening channels, building canals, constructing levees, raising material from stream or harbor bottoms to be used elsewhere as fill, or mining.

dredging [ENG] Removing solid matter from the bottom of a water area.

dress [MECH ENG] **1.** To shape a tool. **2.** To restore a tool to its original shape and sharpness. [MIN ENG] To sort, grind, clean, and concentrate ore.

dresser [ENG] Any tool or apparatus used for dressing something.

Dresser coupling [ENG] A type of coupling for unthreaded pipe.

dressing [ENG] The sharpening, repairing, and replacing of parts, notably drilling bits and tool joints, to ready equipment for reuse.

Dressler kiln [MECH ENG] The first successful muffle-type tunnel kiln.

dribbling [MIN ENG] Fall of debris from the roof of an excavation, usually preceding a heavy fall or cave-in.

drier [ENG] A device to remove water. [MATER] **1.** A substance that absorbs water. **2.** A substance that is used to hasten solidification. **3.** Material, such as salts of lead, manganese, and cobalt, which facilitates the oxidation of oils; used in paints and varnishes to speed drying.

drift [ENG] **1.** A gradual deviation from a set adjustment, such as frequency or balance current, or from a direction. **2.** The deviation, or the angle of deviation, of a borehole from the vertical or from its intended course. [MECH ENG] The water lost in a cooling tower as mist or droplets entrained by the circulating air, not including the evaporative loss. [MIN ENG] A horizontal mine opening which follows a vein or lies within the trend of an ore body.

drift bolt [ENG] **1.** A bolt used to force out other bolts or pins. **2.** A metal rod used to secure timbers.

drifter [MECH ENG] A rock drill, similar to but usually larger than a jack hammer, mounted for drilling holes up to 4½ inches (11.4 centimeters) in diameter. [MIN ENG] **1.** A person who excavates mine drifts. **2.** An air-driven rock drill used for excavating mine drifts and crosscuts.

drift indicator [ENG] Device used to record directional logs; records only the amount of drift (deviation from the vertical), and not the direction.

drifting [MIN ENG] Tunneling along the strike of a lode.

drift mining [MIN ENG] Working of shallow veins or beds through drifts or shafts from the surface.

driftpin [DES ENG] A round, tapered metal rod that is driven into matching rivet holes of two metal parts for stretching the parts and bringing them into alignment.

drift plug [ENG] A plug that can be driven into a pipe to straighten it or to flare its opening.

drill [ENG] A rotating-end cutting tool for creating or enlarging holes in a solid material. Also known as drill bit.

drillability [ENG] Fitness for being drilled, denoting ease of penetration.

drill angle gage *See* drill grinding gage.

drill bit *See* drill.

drill cable [ENG] A cable used to pull up drill rods, casing, and other drilling equipment used in making a borehole.

drill capacity [MECH ENG] The length of drill rod of specified size that the hoist on a diamond or rotary drill can lift or that the brake can hold on a single line.

drill carriage [MECH ENG] A platform or frame on which several rock drills are mounted and which moves along a track, for heavy drilling in large tunnels. Also known as jumbo.

drill chuck [DES ENG] A chuck for holding a drill or other cutting tool on a spindle.

drill collar [DES ENG] A ring which holds a drill bit and gives it radial location with respect to a bearing.

drill column [MIN ENG] A steel pipe that can be wedged across an underground opening in a vertical or horizontal position to serve as a base on which to mount a diamond or rock drill.

drill cuttings [ENG] Cuttings of rock and other subterranean materials brought to the surface during the drilling of wellholes.

drill doctor [MIN ENG] **1.** A person who services drill bits, tools, and steels. **2.** A shop where the mechanic works.

drill drift [ENG] A steel wedge used to remove tapered shank tools from spindles, sockets, and sleeves.

drilled caisson [CIV ENG] A drilled hole filled with concrete and lined with a cylindrical steel casing if needed.

drilled extrusion ingot [MET] A hollow extrusion ingot made from a solid cast extrusion ingot by drilling.

driller [ENG] A person who operates a drilling machine. [MECH ENG] See drilling machine.

drill extractor [ENG] A tool for recovering broken drill pieces or a detached drill from a borehole.

drill feed [MECH ENG] The mechanism by which the drill bit is fed into the borehole during drilling.

drill fittings [ENG] All equipment used in a borehole during drilling. Also known as downhole equipment.

drill floor [ENG] A work area covered with planks around the collar of a borehole at the base of a drill tripod or derrick.

drill footage [ENG] The lineal feet of borehole drilled.

drill gage [DES ENG] A thin, flat steel plate that has accurate holes for many sizes of drills; each hole, identified as to drill size, enables the diameter of a drill to be checked. [ENG] Diameter of a borehole.

drill grinding gage [DES ENG] A tool that checks the angle and length of a twist drill while grinding it. Also known as drill angle gage; drill point gage.

drill hole [ENG] A hole created or enlarged by a drill or auger.

drillhole pattern [ENG] The number, position, angle, and depth of the shot holes forming the round in the face of a tunnel or sinking pit.

drillhole survey See borehole survey.

drill-in [MIN ENG] The act or process of setting casting through overburden by using a drill machine.

drilling [ENG] The creation or enlarging of a hole in a solid material with a drill.

drilling column [ENG] The column of drill rods, with the drill bit attached to the end.

drilling fluid See drilling mud.

drilling machine [MECH ENG] A device, usually motor-driven, fitted with an end cutting tool that is rotated with sufficient power either to create a hole or to enlarge an existing hole in a solid material. Also known as driller.

drilling mud [MATER] A suspension of finely divided heavy material, such as bentonite and barite, pumped through the drill pipe during rotary drilling to seal off porous zones and flush out chippings, and to lubricate and cool the bit. Also known as drilling fluid.

drilling platform [ENG] The structural base upon which the drill rig and associated equipment is mounted during the drilling operation.

drilling rate [MECH ENG] The number of lineal feet drilled per unit of time.

drilling time [ENG] **1.** The time required in rotary drilling for the bit to penetrate a specified thickness (usually 1 foot) of rock. **2.** The actual time the drill is operating.

drilling time log [ENG] Foot-by-foot record of how fast a formation is drilled.

drill jig [MECH ENG] A device fastened to the work in repetition drilling to position and guide the drill.

drill log [ENG] **1.** A record of the events and features of the formations penetrated during boring. Also known as boring log. **2.** A record of all occurrences during drilling that might help in a complete logging of the hole or in determining the cost of the drilling.

drill out [ENG] **1.** To complete one or more boreholes. **2.** To penetrate or remove a borehole obstruction. **3.** To locate and delineate the area of a subsurface ore body or of petroleum by a series of boreholes.

drill-over [ENG] The act or process of drilling around a casing lodged in a borehole.

drill pipe [MIN ENG] A pipe used for driving a revolving drill bit, used especially in drilling wells; consists of a casing within which tubing is run to conduct oil or gas to ground level; drilling mud flows in the annular space between casing and tubing during the drilling operation.

drill point gage *See* drill grinding gage.

drill press [MECH ENG] A drilling machine in which a vertical drill moves into the work, which is stationary.

drill rod [ENG] The long rod that drives the drill bit in drilling boreholes.

drill runner [MIN ENG] A tunnel miner who operates rock drills.

drill sleeve [ENG] A tapered, hollow steel shaft designed to fit the tapered shank of a cutting tool to adapt it to the drill press spindle.

drill socket [ENG] An adapter to fit a tapered shank drill to a taper hole that is larger than that in the drill press spindle.

drill steel [MET] Steel with at least 0.85% carbon content made by the electric furnace process. Formerly known as crucible steel, when made by the crucible process.

drill stem *See* drill string.

drill-stem test [PETRO ENG] Bottom-hole pressure information obtained and used to determine formation productivity.

drill string [MECH ENG] The assemblage of drill rods, core barrel, and bit, or of drill rods, drill collars, and bit in a borehole, which is connected to and rotated by the drill collar of the borehole. Also known as drill stem.

drip [MATER] **1.** Oil which comes through the cloth of a paraffin wax press. **2.** Filter drainings too dark to be included in filter stock. [PETRO ENG] A discharge mechanism installed at a low point in a gas transmission line to collect and remove liquid accumulations.

drip cap [BUILD] A molding on top of the head casing of a window frame to direct water away from the window.

drip edge [BUILD] A metal strip that extends beyond the other parts of the roof and is used to direct rainwater off.

drive [MECH ENG] The means by which a machine is given motion or power (as in steam drive, diesel-electric drive), or by which power is transferred from one part of a machine to another (as in gear drive, belt drive). [MIN ENG] **1.** To excavate in a horizontal or inclined plane. **2.** A horizontal underground tunnel along or parallel to a lode, vein, or ore body.

drive chuck [MECH ENG] A mechanism at the lower end of a diamond-drill drive rod on the swivel head by means of which the motion of the drive rod can be transmitted to the drill string.

drive fit [DES ENG] A fit in which the larger (male) part is pressed into a smaller (female) part; the assembly must be effected through the application of an external force.

drivehead [ENG] A cap fitted over the end of a mechanical part to protect it while it is being driven.

driven caisson [CIV ENG] A caisson formed by driving a cylindrical steel shell into the ground with a pile-driving hammer and then placing concrete inside; the shell may be removed when concrete sets.

driven gear [MECH ENG] The member of a pair of gears to which motion and power are transmitted by the other.

drivepipe [ENG] A thick-walled casing pipe that is driven through overburden or into a deep drill hole to prevent caving.

drive pulley [MECH ENG] The pulley that drives a conveyor belt.

driver [ENG ACOUS] The portion of a horn loudspeaker that converts electrical energy into acoustical energy and feeds the acoustical energy to the small end of the horn.

drive rod [ENG] Hollow shaft in the swivel head of a diamond-drill machine through which energy is transmitted from the drill motor to the drill string. Also known as drive spindle.

drive sampling [ENG] The act or process of driving a tubular device into soft rock material for obtaining dry samples.

drivescrew [DES ENG] A screw that is driven all the way in, or nearly all the way in, with a hammer.

drive shaft [MECH ENG] A shaft which transmits power from a motor or engine to the rest of a machine.

drive shoe [DES ENG] A sharp-edged steel sleeve attached to the bottom of a drivepipe or casing to act as a cutting edge and protector.

driving clock [ENG] A mechanism for driving an instrument at a required rate.

driving pinion [MECH ENG] The input gear in the differential of an automobile.

driving wheel [MECH ENG] A wheel that supplies driving power.

drogue [AERO ENG] **1.** A small parachute attached to a body for stabilization and deceleration. Also known as deceleration parachute. **2.** A funnel-shaped device at the end of the hose of a tanker aircraft in flight, to receive the probe of another aircraft that will take on fuel. [ENG] **1.** A device, such as a sea anchor, usually shaped like a funnel or cone and dragged or towed behind a boat or seaplane for deceleration, stabilization, or speed control. **2.** A current-measuring assembly consisting of a weighted current cross, sail, or parachute and an attached surface buoy. Also known as drag anchor; sea anchor.

drone [AERO ENG] A pilotless aircraft usually subordinated to the controlling influences of a remotely located command station, but occasionally preprogrammed.

drooped ailerons [AERO ENG] Ailerons that are of the hinged trailing-edge type and are so arranged that both the right and left one have a 10 to 15° positive downward deflection with the control column in a neutral position.

droop governor [MECH ENG] A governor whose equilibrium speed decreases as the load on the machinery controlled by the governor increases.

drop [MET] A casting defect due to the falling of a portion of sand from an overhanging section of the mold.

drop ball [ENG] A ball, weighing 3000–4000 pounds (1400–1800 kilograms), dropped from a crane through about 20–33 feet (6–10 meters) onto oversize quarry stones left after blasting; this method is used to avoid secondary blasting.

drop bar [MECH ENG] A bar that guides sheets of paper into a printing or folding machine.

drop black [MATER] Black pigment shaped into droplets.

drop-bottom car [MIN ENG] A mine car designed so that flaps drop open in the bottom to allow the coal to fall out as the car passes over the dump; flaps close as the car leaves.

drop forging [MET] Plastic deformation of hot metal under a falling weight, such as a drop hammer.

drop hammer [MECH ENG] *See* pile hammer. [MET] A hammer used in forging that is raised and then dropped on the metal resting on an anvil or on a die.

drop log [MIN ENG] A timber which can be dropped across a mine track by remote control to derail cars.

dropping test [MET] A chemical method for determining thickness of zinc and cadmium plated coatings on metal in which a reagent is dropped on the surface until the basis metal is exposed.

drop press *See* punch press.

drop siding [BUILD] Building siding with a shiplap joint.

dropsonde [ENG] A radiosonde dropped by parachute from a high-flying aircraft to measure weather conditions and report them back to the aircraft.

dropsonde dispenser [ENG] A chamber from which dropsonde instruments are released from weather reconnaissance aircraft; used only for some models of equipment, ejection chambers being used for others.

drop spillway [CIV ENG] A spillway usually less than 20 feet (6 meters) high having a vertical downstream face, and water drops over the face without touching the face.

drop tank [AERO ENG] A fuel tank on an airplane that may be jettisoned.

drosometer [ENG] An instrument used to measure the amount of dew deposited on a given surface.

dross [MET] An impurity, usually an oxide, formed on the surface of a molten metal.

drossing [MET] A process used in nonferrous pyrometallurgy for removing solid oxide deposits on the surface of a molten metal.

drum [DES ENG] **1.** A hollow, cylindrical container. **2.** A metal cylindrical shipping container for liquids having a capacity of 12–110 gallons (45–416 liters). [MECH ENG] A horizontal cylinder about which rope or wire rope is wound in a hoisting mechanism. Also known as hoisting drum.

drum brake [MECH ENG] A brake in which two curved shoes fitted with heat- and wear-resistant linings are forced against the surface of a rotating drum.

drum cam [MECH ENG] A device consisting of a drum with a contoured surface which communicates motion to a cam follower as the drum rotates around an axis.

drum dryer [MECH ENG] A machine for removing water from substances such as milk, in which a thin film of the product is moved over a turning steam-heated drum and a knife scrapes it from the drum after moisture has been removed.

drum feeder [MECH ENG] A rotating drum with vanes or buckets to lift and carry parts and drop them into various orienting or chute arrangements. Also known as tumbler feeder.

drum filter [MECH ENG] A cylindrical drum that rotates through thickened ore pulp, extracts liquid by a vacuum, and leaves solids, in the form of a cake, on a permeable membrane on the drum end. Also known as rotary filter; rotary vacuum filter.

drum gate [CIV ENG] A movable crest gate in the form of an arc hinged at the apex and operated by reservoir pressure to open and close a spillway.

drum meter *See* liquid-sealed meter.

drummy [MIN ENG] Loose rock or coal, especially in a mine roof, that produces a hollow, weak sound when tapped with a bar.

drum plotter [ENG] A graphics output device that draws lines with a continuously moving pen on a sheet of paper rolled around a rotating drum that moves the paper in a direction perpendicular to the motion of the pen.

drum separator [MIN ENG] A cylindrical vessel which rotates slowly and separates run-of-mine coal into clean coal, middlings, and refuse, can be adjusted for different specific gravities.

drum-type boiler See bent-tube boiler.

dry abrasive cutting [MECH ENG] Frictional cutting using a rotary abrasive wheel without the use of a liquid coolant.

dry assay [MET] Determination of the amount of a desired constituent in ores, metallurgical residues, and alloys by methods other than those involving liquid means of separation.

dry blast cleaning [ENG] Cleaning of metallic surfaces by blasting with abrasive material traveling at a high velocity; abrasive may be accelerated by an air nozzle or a centrifugal wheel.

dry-bulb thermometer [ENG] An ordinary thermometer, especially one with an unmoistened bulb; not dependent upon atmospheric humidity.

dry cargo [IND ENG] Nonliquid cargo, including minerals, grain, boxes, and drums.

dry-cell cap light [MIN ENG] A headlamp with a focusing lens lamp and a dry-cell battery unit clipped to the belt; to prevent explosion in a mine, the bulb is ejected automatically in case of its breakage.

dry cleaning [ENG] To utilize dry-cleaning fluid to remove stains from textile.

dry-cleaning fluid [MATER] An organic solvent such as chlorinated hydrocarbons or petroleum naphtha with narrow, carefully selected boiling points; used in dry cleaning.

dry coloring [ENG] A method to color plastics by tumbleblending colorless plastic particles with dyes and pigments. [MATER] A powdered form of pigment.

dry cooling tower [MECH ENG] A structure in which water is cooled by circulation through finned tubes, transferring heat to air passing over the fins; there is no loss of water by evaporation because the air does not directly contact the water.

dry corrosion [MET] Destruction of a metal or alloy by chemical processes resulting from attack by gases in the atmosphere above the dew point.

dry course [BUILD] An initial roofing course of felt or paper not bedded in tar or asphalt.

dry dock [CIV ENG] A dock providing support for a vessel and a means for removing the water so that the bottom of the vessel can be exposed.

dry-dock caisson [CIV ENG] The floating gate to a dry dock. Also known as caisson.

dry drilling [MIN ENG] Drilling in which chippings and cuttings are lifted out of a borehole by a current of air or gas.

dry gas [MATER] A gas that does not contain fractions which may easily condense under normal atmospheric conditions, for example, natural gas with methane and ethane.

dry grinding [ENG] Reducing particle sizes without a liquid medium.

dry hole [ENG] A hole driven without the use of water. [PETRO ENG] A well in which no oil or gas is found.

drying oil [MATER] Relatively highly unsaturated oil, such as cottonseed, soybean, and linseed oil, that is easily oxidized and polymerized to form a hard, dry film on exposure to air; used in paints and varnish.

drying oven [ENG] A closed chamber for drying an object by heating at relatively low temperatures.

dry kiln [ENG] A heated room or chamber used to dry and season cut lumber.

dry mill [MECH ENG] Grinding device used to powder or pulverize solid materials without an associated liquid.

dry mining [MIN ENG] Mining operation in which there is no moisture in the ventilating air.

dry ore [MIN ENG] An ore of gold or silver which requires added lead and fluxes for treatment.

dry permeability [ENG] A property of dried bonded sand to permit passage of gases while molten material is poured into a mold.

dry pipe [MECH ENG] A perforated metal pipe above the normal water level in the steam space of a boiler which prevents moisture or extraneous matter from entering steam outlet lines.

dry-pipe system [ENG] A sprinkler system that admits water only when the air it normally contains has been vented; used for systems subjected to freezing temperatures.

dry-pit pump [MECH ENG] A pump operated with the liquid conducted to and from the unit by piping.

dry placer [MIN ENG] A gold-bearing alluvial deposit found in arid regions; it cannot be mined due to lack of water.

dry pressing [ENG] Molding clayware by compressing moist clay powder in metal dies.

dry run [ENG] Any practice test or session.

dry sample [MIN ENG] A sample of ore obtained by dry drilling.

dry sand mold [MET] A mold made of greensand and then dried in an oven to increase its strength.

Drysdale ac polar potentiometer [ENG] A potentiometer for measuring alternating-current voltages in which the voltage is applied across a slide-wire supplied with current by a phase-shifting transformer; this current is measured by an ammeter and brought into phase with the unknown voltage by adjustment of the transformer rotor, and the unknown voltage is measured by observation of the slide-wire setting for a null indication of a vibration galvanometer.

dry sieving [ENG] Particle-size distribution analysis of powdered solids; the sample is placed on the top sieve screen of a nest (stack), with mesh openings decreasing in size from the top to the bottom of the nest.

dry sleeve [MECH ENG] A cylinder liner which is not in contact with the coolant.

dry steam drum [MECH ENG] **1.** Pressurized chamber into which steam flows from the steam space of a boiler drum. **2.** That portion of a two-stage furnace that extends forward of the main combustion chamber; fuel is dried and gasified therein, with combustion of gaseous products accomplished in the main chamber; the refractory walls of the Dutch oven are sometimes water-cooled.

dry storage [MECH ENG] Cold storage in which refrigeration is provided by chilled air.

dry strength [ENG] The strength of an adhesive joint determined immediately after drying under specified conditions or after a period of conditioning in the standard laboratory atmosphere.

dry tabling [MIN ENG] A process similar to wet tabling, but without the water; used to separate two or more minerals based on specific gravity differences.

dry test meter [ENG] Gas-flow rate meter with two compartments separated by a movable diaphragm which is connected to a series of gears that actuate a dial; when one chamber is full, a valve switches to the other, empty chamber; used to measure household gas-flow rates and to calibrate flow-measurement instruments.

dry ticket [IND ENG] Tank inspection form signed by shore and ship inspectors before loading and after discharging the ship.

dry wall [BUILD] A wall covered with wallboard, in contrast to plaster. [ENG] A wall constructed of rock without cementing material.

dry washer [MIN ENG] A machine for extracting gold mined from dry placers.

dry well [CIV ENG] **1.** A well that has been completely drained. **2.** An excavated well filled with broken stone and used to receive drainage when the water percolates into the soil. **3.** Compartment of a pumping station in which the pumps are housed.

dry wire drawing [MET] The drawing of dry steel process wire, pretreated by acid cleaning, lime coating, and baking, through a lubricant and wire drawing frame.

Drzwiecki theory [MECH ENG] In theoretical investigations of windmill performance, a theory concerning the air forces produced on an element of the blade.

DS *See* Doppler sonar.

DSN *See* Deep Space Network.

dual-bed dehumidifier [MECH ENG] A sorbent dehumidifier with two beds, one bed dehumidifying while the other bed is reactivating, thus providing a continuous flow of air.

dual-channel amplifier [ENG ACOUS] An audio-frequency amplifier having two separate amplifiers for the two channels of a stereophonic sound system, usually operating from a common power supply mounted on the same chassis.

dual completion well [PETRO ENG] Single well casing containing two production tubing strings, each in a different zone of the reservoir (one higher, one lower) and each separately controlled.

dual-flow oil burner [MECH ENG] An oil burner with two sets of tangential slots in its atomizer for use at different capacity levels.

dual-fuel engine [MECH ENG] Internal combustion engine that can operate on either of two fuels, such as natural gas or gasoline.

dual meter [ENG] Meter constructed so that two aspects of an electric circuit may be read simultaneously.

dual-seal tubing joint [PETRO ENG] Tubing connection joint with two sealing surfaces to assure a leak-free connection between sections.

dual thrust [AERO ENG] A rocket thrust derived from two propellant grains and using the same propulsion section of a missile.

dual-thrust motor [AERO ENG] A solid-propellant rocket engine built to obtain dual thrust.

dual-track tape recorder *See* double-track tape recorder.

dub [ENG ACOUS] **1.** To transfer recorded material from one recording to another, with or without the addition of new sounds, background music, or sound effects. **2.** To combine two or more sources of sound into one record. **3.** To add a new sound track or new sounds to a motion picture film, or to a recorded radio or television production.

duckbill [MECH ENG] A shaking type of combination loader and conveyor whose loading end is generally shaped like a duck's bill.

duct [MECH ENG] A fluid flow passage which may range from a few inches in diameter to many feet in rectangular cross section, usually constructed of galvanized steel, aluminum, or copper, through which air flows in a ventilation system or to a compressor, supercharger, or other equipment at speeds ranging to thousands of feet per minute.

ducted fan [MECH ENG] A propeller or multibladed fan inside a coaxial duct or cowling. Also known as ducted propeller; shrouded propeller.

ducted-fan engine [AERO ENG] An aircraft engine incorporating a fan or propeller enclosed in a duct; especially, a jet engine in which an enclosed fan or propeller is used to ingest ambient air to augment the gases of combustion in the jetstream.

ducted rocket *See* rocket ramjet.

ductile iron *See* nodular cast iron.

ductility [MATER] The ability of a material to be plastically deformed by elongation, without fracture.

duct propulsion [AERO ENG] A means of propelling a vehicle by ducting a surrounding fluid through an engine, adding momentum by mechanical or thermal means, and ejecting the fluid to obtain a reactive force.

Dulong's formula [ENG] A formula giving the gross heating value of coal in terms of the weight fractions of carbon, hydrogen, oxygen, and sulfur from the ultimate analysis.

DUMAND *See* deep underwater muon and neutrino detector.

dumb iron [ENG] **1.** A rod for opening seams prior to caulking. **2.** A rigid connector between the frame of a motor vehicle and the spring shackle.

dumbwaiter [MECH ENG] An industrial elevator which carries small objects but is not permitted to carry people.

Dumet wire [MET] An iron-nickel alloy wire containing 42% nickel, and covered with copper; used to replace platinum as the seal-in wire in incandescent lamps and vacuum tubes; the copper prevents gassing at the seal.

dummy [ENG] Simulating device with no operating features, as a dummy heat coil. [MET] A cathode that undergoes electroplating at low current densities.

dummy block [MET] A thick disk positioned between the ram and billet in extrusion working to prevent the ram from overheating.

dummy joint [ENG] A groove cut into the top half of a concrete slab, sometimes packed with filler, to form a line where the slab can crack with only minimum damage.

dump bailer [ENG] A cylindrical vessel designed to deliver cement or water into a well which otherwise might cave in if fluid was poured from the top.

dump bucket [MECH ENG] A large bucket with movable discharge gates at the bottom; used to move soil or other construction materials by a crane or cable.

dump car [MECH ENG] Any of several types of narrow-gage rail cars with bodies which can easily be tipped to dump material.

dump truck [ENG] A motor or hand-propelled truck for hauling and dumping loose materials, equipped with a body that discharges its contents by gravity.

dump valve [ENG] A large valve located at the bottom of a tank or container used in emergency situations to empty the tank quickly; for example, to jettison fuel from an airplane fuel tank.

dumpy level [ENG] A surveyor's level which has the telescope with its level tube rigidly attached to a vertical spindle and is capable only of horizontal rotary movement.

dunking sonar *See* dipping sonar.

dunnage [IND ENG] **1.** Padding material placed in a container to protect shipped goods from damage. **2.** Loose wood or waste material placed in the ship's hold to protect the cargo from shifting and damage.

Duovac method [MET] Technique for testing for defects in magnetic parts; a moving magnetic field magnetizes the part in many directions, and the part is then sprayed with fluorescent magnetic particles and examined under ultraviolet light so that defects become apparent.

duplex [ENG] Consisting of two parts working together or in a similar fashion.

duplexed system [ENG] A system with two distinct and separate sets of facilities, each of which is capable of assuming the system function while the other assumes a standby status. Also known as redundant system.

duplexing *See* duplex operation; duplex process.

duplex iron [MET] Cast iron heated in an electric furnace after it has been melted in a cupola.

duplex lock [DES ENG] A lock with two independent pin-tumbler cylinders on the same bolt.

duplex operation [ENG] In radar, a condition of operation when two identical and interchangeable equipments are provided, one in an active state and the other immediately available for operation.

duplex practice *See* duplex process.

duplex process [MET] A two-step procedure in which steel is refined by one process (usually the Bessemer process) and finished by another process (usually open-hearth or electric-furnace). Also known as duplexing; duplex practice.

duplex pump [MECH ENG] A reciprocating pump with two parallel pumping cylinders.

duplex tandem compressor [MECH ENG] A compressor having cylinders on two parallel frames connected through a common crankshaft.

duplicate cavity plate [ENG] In plastics molds, the removable plate in which the molding cavities are retained; used in operating where two plates are necessary for insert loading.

Dupont process [MIN ENG] A method for separation of minerals in which organic liquids of high specific gravity and low viscosity are used.

durability [ENG] The quality of equipment, structures, or goods of continuing to be useful after an extended period of time and usage.

durometer [ENG] An instrument consisting of a small drill or blunt indenter point under pressure; used to measure hardness of metals and other materials.

durometer hardness [ENG] The hardness of a material as measured by a durometer.

Durville process [MET] A casting process involving the attachment of an inverted mold to the top of a crucible; the metal is melted in the bottom of the crucible, and then the molten metal is decanted into the mold by inverting the entire apparatus.

dust and fume monitor [MIN ENG] An instrument designed to measure and record concentrations of dust, fume, and gas in mine environments over an extended period of time.

dust chamber [ENG] A chamber through which gases pass to permit deposition of solid particles for collection. Also known as ash collector; dust collector.

dust collector See dust chamber.

dust control system [ENG] System to capture, settle, or inert dusts produced during handling, drying, or other process operations; considered important for safety and health.

dust counter [ENG] A photoelectric apparatus which measures the size and number of dust particles per unit volume of air. Also known as Kern counter.

dust-counting microscope [ENG] A microscope equipped for quantitative dust sample analysis; magnification is usually 100X.

dust explosion [ENG] An explosion following the ignition of flammable dust suspended in the air.

dust filter [ENG] A gas-cleaning device using a dry or viscous-coated fiber or fabric for separation of particulate matter.

dusting [MET] Spontaneous disintegration of a material on cooling due to expansion or inversion.

dusting clay [MATER] Finely pulverized clay used as an extender or carrier in insecticide dust formulations.

dust separator [ENG] Device or system to remove dust from a flowing stream of gas; includes electrostatic precipitators, wet scrubbers, bag filters, screens, and cyclones.

Dutch door [BUILD] A door with upper and lower parts that can be opened and closed independently.

dutchman [ENG] A filler piece for closing a gap between two pipes or between a pipe or fitting and a piece of equipment, if the pipe is too short to achieve closure or if the pipe and equipment are not aligned.

Dutchman's log [ENG] A buoyant object thrown overboard to determine the speed of a vessel; the time required for a known length of the vessel to pass the object is measured, and the speed can then be computed.

Dutch metal [MET] An alloy of 80% copper and 20% zinc that is ductile, is easily drawn, and takes a high polish; used for low-priced jewelry.

duty cycle [ENG] 1. The time intervals devoted to starting, running, stopping, and idling when a device is used for intermittent duty. 2. The ratio of working time to total time for an intermittently operating device, usually expressed as a percent. Also known as duty factor. [MET] The percentage of time that current flows in equipment over a specific period during electric resistance welding.

duty cyclometer [ENG] Test meter which gives direct reading of duty cycle.

duty factor See duty cycle.

D variometer See declination variometer.

dwell [DES ENG] That part of a cam that allows the cam follower to remain at maximum lift for a period of time. [ENG] A pause in the application of pressure to a mold.

Dwight-Lloyd machine [MIN ENG] A continuous sintering machine in which the feed is moved on articulated plates pulled by chains in conveyor-belt fashion.

Dwight-Lloyd process [MIN ENG] Blast roasting, with air currents being drawn downward through the ore.

DX coil See direct-expansion coil.

dyecrete process [ENG] A process of adding permanent color to concrete with organic dyes.

dye penetrant [MET] A dye-containing liquid used for detecting cracks or other surface defects in nonmagnetic materials.

dye-retarding agent [MATER] Materials that decrease the rate of dye absorption, preventing rapid exhaustion of dye baths.

dynamic augment [MECH ENG] Force produced by unbalanced reciprocating parts in a steam locomotive.

dynamic behavior [ENG] A description of how a system or an individual unit functions with respect to time.

dynamic capillary pressure [PETRO ENG] Capillary-pressure saturation curves of a core sample determined by the simultaneous steady-state flow of two fluids through the sample; capillarity pressures are determined by the difference in the pressures of the two fluids.

dynamic check [ENG] Check used to ascertain the correct performance of some or all components of equipment or a system under dynamic or operating conditions.

dynamic compressor [MECH ENG] A compressor which uses rotating vanes or impellers to impart velocity and pressure to the fluid.

dynamic factor [AERO ENG] A ratio formed from the load carried by any airplane part when the airplane is accelerating or subjected to abnormal conditions to the load carried in the conditions of normal flight.

dynamic leak test [ENG] A type of leak test in which the vessel to be tested is evacuated and an external tracer gas is applied; an internal leak detector will respond if gas is drawn through any leaks.

dynamic load [AERO ENG] With respect to aircraft, rockets, or spacecraft, a load due to an acceleration of craft, as imposed by gusts, by maneuvering, by landing, by firing rockets, and so on. [CIV ENG] A force exerted by a moving body on a resisting member, usually in a relatively short time interval. Also known as energy load.

dynamic loudspeaker [ENG ACOUS] A loudspeaker in which the moving diaphragm is attached to a current-carrying voice coil that interacts with a constant magnetic field to give the in-and-out motion required for the production of sound waves. Also known as dynamic speaker; moving-coil loudspeaker.

dynamic microphone [ENG ACOUS] A moving-conductor microphone in which the flexible diaphragm is attached to a coil positioned in the fixed magnetic field of a permanent magnet. Also known as moving-coil microphone.

dynamic model [ENG] A model of an aircraft or other object which has its linear dimensions and its weight and moments of inertia reproduced in scale in proportion to the original.

dynamic noise suppressor [ENG ACOUS] An audio-frequency filter circuit that automatically adjusts its band-pass limits according to signal level, generally by means of reactance tubes; at low signal levels, when noise becomes more noticeable, the circuit reduces the low-frequency response and sometimes also reduces the high-frequency response.

dynamic packing [ENG] Any packing that operates on moving surfaces; in functioning, to retain fluid under pressure, they carry the hydraulic load and therefore operate like bearings.

dynamic sensitivity [ENG] The minimum leak rate which a leak detector is capable of sensing.

dynamic similarity [MECH ENG] A relation between two mechanical systems (often referred to as model and prototype) such that by proportional alterations of the units of length, mass, and time, measured quantities in the one system go identically (or with a constant multiple for each) into those in the other; in particular, this implies constant ratios of forces in the two systems.

dynamic speaker *See* dynamic loudspeaker.

dynamic test [ENG] A test conducted under active or simulated load.

dynamic unbalance [MECH ENG] Failure of the rotation axis of a piece of rotating equipment to coincide with one of the principal axes of inertia due to forces in a single axial plane and on opposite sides of the rotation axis, or in different axial planes.

dynamite [MATER] A generic term covering a class of nitroglycerin-sensitized mixtures of carbonaceous materials (wood, flour, starch) and oxygen-supplying salts, used as explosives for blasting and mining.

dynamometer [ENG] **1.** An instrument in which current, voltage, or power is measured by the force between a fixed coil and a moving coil. **2.** A special type of electric rotating machine used to measure the output torque or driving torque of rotating machinery by the elastic deformation produced.

E

earing [MET] Formation of scallops around the top edge of a deep-drawn product due to differences in the directional properties of the metal sheet.

earliest finish time [IND ENG] The earliest time for completion of an activity of a project; for the entire project, it equals the earliest start time of the final event included in the schedule.

earliest start time [IND ENG] The earliest time at which an activity may begin in the schedule of a project; it equals the earliest time that all predecessor activities can be completed.

earphone [ENG ACOUS] **1.** An electroacoustical transducer, such as a telephone receiver or a headphone, actuated by an electrical system and supplying energy to an acoustical system of the ear, the waveform in the acoustical system being substantially the same as in the electrical system. **2.** A small, lightweight electroacoustic transducer that fits inside the ear, used chiefly with hearing aids.

earplug [ENG] A device made of a pliable substance which fits into the ear opening; used to protect the ear from excessive noise or from water.

ear protector [ENG] A device, such as a plug or ear muff, used to protect the human ear from loud noise that may be injurious to hearing, such as that of jet engines.

earth dam [CIV ENG] A dam having the main section built of earth, sand, or rock, and a core of impervious material such as clay or concrete.

earthenware [ENG] Ceramic products of natural clay, fired at 1742–2129°F (950–1165°C), that is slightly porous, opaque, and usually covered with a nonporous glaze.

earth inductor [ENG] A type of inclinometer that has a coil which rotates in the earth's field and in which a voltage is induced when the rotation axis does not coincide with the field direction; used to measure the dip angle of the earth's magnetic field.

Also known as dip inductor; earth inductor compass; induction inclinometer.

earth inductor compass *See* earth inductor.

earthmover [MECH ENG] A machine used to excavate, transport, or push earth.

earth-nut oil *See* peanut oil.

earth pressure [CIV ENG] The pressure which exists between earth materials (such as soil or sediments) and a structure (such as a wall).

earthquake-resistant [CIV ENG] Of a structure or building, able to withstand lateral seismic stresses at the base.

earth resources technology satellite [AERO ENG] One of a series of satellites designed primarily to measure the natural resources of the earth; functions include mapping, cataloging water resources, surveying crops and forests, tracing sources of water and air pollution, identifying soil and rock formations, and acquiring oceanographic data. Abbreviated ERTS.

earth satellite [AERO ENG] An artificial satellite placed into orbit about the earth.

earth thermometer *See* soil thermometer.

earthwork [CIV ENG] **1.** Any operation involving the excavation or construction of earth embankments. **2.** Any construction made of earth.

easement [CIV ENG] The right held by one person over another person's land for a specific use; rights of tenants are excluded.

easement curve [CIV ENG] A curve, as on a highway, whose degree of curvature is varied to provide a gradual transition between a tangent and a simple curve, or between two simple curves which it connects. Also known as transition curve.

East Indian geranium oil *See* palmarosa oil.

East Indian sandalwood oil *See* sandalwood oil.

eave [BUILD] The border of a roof overhanging a wall.

eaves board [BUILD] A strip nailed along the eaves of a building to raise the end of the bottom course of tile or slate on the roof.

eaves molding [BUILD] A cornicelike molding below the eaves of a building.

Ebert ion counter [ENG] An ion counter of the aspiration condenser type, used for the measurement of the concentration and mobility of small ions in the atmosphere.

EBM *See* electron-beam machining.

ebonite *See* hard rubber.

EC blank fire *See* EC smokeless powder.

EC blank powder *See* EC smokeless powder.

eccentric bit [DES ENG] A modified chisel for drilling purposes having one end of the cutting edge extended further from the center of the bit than the other.

eccentric cam [DES ENG] A cylindrical cam with the shaft displaced from the geometric center.

eccentric gear [DES ENG] A gear whose axis deviates from the geometric center.

eccentric load [ENG] A load imposed on a structural member at some point other than the centroid of the section.

eccentric rotor engine [MECH ENG] A rotary engine, such as the Wankel engine, wherein motion is imparted to a shaft by a rotor eccentric to the shaft.

eccentric signal [ENG] A survey signal whose position is not in a vertical line with the station it is representing.

eccentric station [ENG] A survey point over which an instrument is centered and which is not positioned in a vertical line with the station it is representing.

eccentric valve [ENG] A rubber-lined slurry or fluid valve with an eccentric rotary cutoff body to reduce corrosion and wear on mechanical moving valve parts.

echogram [ENG] The graphic presentation of echo soundings recorded as a continuous profile of the sea bottom.

echograph [ENG] An instrument used to record an echogram.

echo matching [ENG] Rotating an antenna to a position in which the pulse indications of an echo-splitting radar are equal.

echo ranging [ENG] Active sonar, in which underwater sound equipment generates bursts of ultrasonic sound and picks up echoes reflected from submarines, fish, and other objects within range, to determine both direction and distance to each target. Also known as echo location.

echo-ranging sonar [ENG] Active sonar, in which underwater sound equipment generates bursts of ultrasonic sound and picks up echoes reflected from submarines, fish, and other objects within range, to determine both direction and distance to each target.

echo recognition [ENG] Identification of a sonar reflection from a target, as distinct from energy returned by other reflectors.

echo repeater [ENG ACOUS] In sonar calibration and training, an artificial target that returns a synthetic echo by receiving a signal and retransmitting it.

Echo satellite [AERO ENG] An aluminized-surface, Mylar balloon about 100 feet (30 meters) in diameter, placed in orbit as a passive communications satellite for reflecting microwave signals from a transmitter to receivers beyond the horizon.

echosonogram [ENG] A graphic display obtained with ultrasound pulse-reflection techniques; for example, an echocardiogram.

echo sounder *See* sonic depth finder.

echo sounding [ENG] Determination of the depth of water by measuring the time interval between emission of a sonic or ultrasonic signal and the return of its echo from the sea bottom.

echo-splitting radar [ENG] Radar in which the echo is split by special circuits associated with the antenna lobe–switching mechanism, to give two echo indications on the radarscope screen; when the two echo indications are equal in height, the target bearing is read from a calibrated scale.

ECM *See* electrochemical machining.

econometrics [IND ENG] The application of mathematical and statistical techniques to the estimation of mathematical relationships for testing of economic theories and the solution of economic problems.

economic life [IND ENG] The number of years after which a capital good should be replaced in order to minimize the long-run annual cost of operation, repair, depreciation, and capital.

economic lot size [IND ENG] The number of units of a product or item to be manufactured at each setup or purchased on each order so as to minimize the cost of purchasing or setup, and the cost of holding the average inventory over a given period, usually annual.

economic order quantity [IND ENG] The number of orders required to fulfill the economic lot size.

economic purchase quantity [IND ENG] The economic lot size for a purchased quantity.

economics [IND ENG] A social science that deals with production, distribution, and consumption of commodities, or wealth.

economizer [ENG] A reservoir in a continuous-flow oxygen system in which oxygen exhaled by the user is collected for recirculation in the system. [MECH ENG] A forced-flow, once-through, convection-heat-transfer tube bank in which feedwater is raised in temperature on its way to the evaporating section of a steam boiler, thus lowering flue gas temperature, improving boiler efficiency, and saving fuel.

ECR *See* electronic cash register.

EC smokeless powder [MATER] An explosive powder used chiefly in blank cartridges, but also in some .22-caliber and shotgun ammunition, and formerly in fragmentation grenades. Also known as EC blank fire; EC blank powder.

ED$_{50}$ *See* effective dose 50.

eddy-current brake [MECH ENG] A control device or dynamometer for regulating rotational speed, as of flywheels, in which energy is converted by eddy currents into heat.

eddy-current clutch [MECH ENG] A type of electromagnetic clutch in which torque is transmitted by means of eddy currents induced by a magnetic field set up by a coil carrying direct current in one rotating member.

eddy-current damper [AERO ENG] A device used to damp nutation and other unwanted vibration in spacecraft, based on the principle that eddy currents induced in conducting material by motion relative to magnets tend to counteract that motion.

eddy-current heating *See* induction heating.

eddy-current tachometer [ENG] A type of tachometer in which a rotating permanent magnet induces currents in a spring-mounted metal cylinder; the resulting torque rotates the cylinder and moves its attached pointer in proportion to the speed of the rotating shaft. Also known as drag-type tachometer.

edge grain [MATER] The grain pattern produced when soft wood is cut so that the tree's annular rings form an angle of more than 45° with the board's surface.

edge joint [MET] A joint between the edges of welded members which are essentially parallel to each other.

edger [MET] The part of a forging die which portions out the quantity of metal needed for shaping.

edge runner *See* Chile mill.

edging [MET] Controlling the plate width or the edge shape during rolling operations.

EDM *See* electron discharge machining.

eductor [ENG] **1.** An ejectorlike device for mixing two fluids. **2.** *See* ejector.

eductor pump [MIN ENG] A pump which removes slurried material from a hydraulically disseminated subsurface ore matrix.

eff *See* efficiency.

effective angle of attack [AERO ENG] That part of a given angle of attack that lies between the chord of an airfoil and a line

representing the resultant velocity of the disturbed airflow.

effective center [ENG ACOUS] In a sonar projector, the point where lines coincident with the direction of propagation, as observed at different points some distance from the projector, apparently intersect. Also known as apparent source.

effective confusion area [ENG] Amount of chaff whose radar cross-sectional area equals the radar cross-sectional area of the particular aircraft at a particular frequency.

effective decline rate [PETRO ENG] The drop in oil or gas production rate over a period of time; equal to unity (1 month or 1 year) divided by the production rate at the beginning of the period.

effective discharge area [DES ENG] A nominal or calculated area of flow through a pressure relief valve for use in flow formulas to determine valve capacity.

effective exhaust velocity [AERO ENG] A fictitious exhaust velocity that yields the observed value of jet thrust in calculations.

effective molecular weight [PETRO ENG] Empirical relationship of oil graphed against API gravity to give the effective (pseudoaverage) molecular weight of the oil for reservoir calculations.

effective pitch [AERO ENG] The distance traveled by an airplane along its flight path for one complete turn of the propeller.

effective rake [MECH ENG] The angular relationship between the plane of the tooth face of the cutter and the line through the tooth point measured in the direction of chip flow.

effective thrust [AERO ENG] In a rocket motor or engine, the theoretical thrust less the effects of incomplete combustion and friction flow in the nozzle.

efficiency expert [IND ENG] An individual who analyzes procedures, productivity, and jobs in order to recommend methods for achieving maximum utilization of resources and equipment.

efficiency Abbreviated eff. [ENG] **1.** Measure of the degree of heat output per unit of fuel when all available oxidizable materials in the fuel have been burned. **2.** Ratio of useful energy provided by a dynamic system to the energy supplied to it during a specific period of operation.

effluent [CIV ENG] The liquid waste of sewage and industrial processing.

effluent weir [CIV ENG] A dam at the outflow end of a watercourse.

effluvium [IND ENG] By-products of food and chemical processes, in the form of wastes.

effort-controlled cycle [IND ENG] A work cycle which is performed entirely by hand or in which the hand time controls the place. Also known as manually controlled work.

effort rating [IND ENG] Assessing the level of manual effort expended by the operator, based on the observer's concept of normal effort, in order to adjust time-study data. Also known as pace rating; performance rating.

Eggertz's method [MET] A colorimetric estimation of the carbon content of steel by dissolving the metal in nitric acid and comparing the color with that produced by a similar metal of known carbon content.

eicosane [MATER] A mixture of saturated hydrocarbons mostly straight-chained and averaging 20 carbons in the chain; for this reason, the formula $C_{20}H_{42}$ is given to the technical mixture; used in lubricants and plasticizers.

ejection [ENG] The process of removing a molding from a mold impression by mechanical means, by hand, or by compressed air.

ejection capsule [AERO ENG] In an aircraft or spacecraft, a detachable compartment (serving as a cockpit or cabin) or a payload capsule which may be ejected as a unit and parachuted to the ground.

ejection seat [AERO ENG] Emergency device which expels the pilot safely from a high-speed airplane.

ejector [ENG] **1.** Any of various types of jet pumps used to withdraw fluid materials from a space. Also known as eductor. **2.** A device that ejects the finished casting from a mold.

ejector condenser [MECH ENG] A type of direct-contact condenser in which vacuum is maintained by high-velocity injection water; condenses steam and discharges water, condensate, and noncondensables to the atmosphere.

ejector half [MET] The movable half of a casting die.

ejector pin [ENG] A pin driven into the rear of a mold cavity to force the finished piece out. Also known as knockout pin.

ejector plate [ENG] The plate backing up the ejector pins and holding the ejector assembly together.

ejector rod [ENG] A rod that activates the ejector assembly of a mold when it is opened.

Ekman current meter [ENG] A mechanical device for measuring ocean current velocity which incorporates a propeller and a magnetic compass and can be suspended from a moored ship.

Ekman water bottle [ENG] A cylindrical tube fitted with plates at both ends and used for deep-water samplings; when hit by a messenger it turns 180°, closing the plates and capturing the water sample.

elastomer [MATER] A material, such as a synthetic rubber or plastic, which at room temperature can be stretched under low stress to at least twice its original length and, upon immediate release of the stress, will return with force to its approximate original length.

elbow [DES ENG] **1.** A fitting that connects two pipes at an angle, often of 90°. **2.** A sharp corner in a pipe.

elbow meter [ENG] Pipe elbow used as a liquids flowmeter; flow rate is measured by determining the differential pressure developed between the inner and outer radii of the bend by means of two pressure taps located midway on the bend.

electret headphone [ENG ACOUS] A headphone consisting of an electret transducer, usually in the form of a push-pull transducer.

electret microphone [ENG ACOUS] A microphone consisting of an electret transducer in which the foil electret diaphragm is placed next to a perforated, ridged, metal or metal-coated backplate, and output voltage, taken between diaphragm and backplate, is proportional to the displacement of the diaphragm.

electrical blasting cap [ENG] A blasting cap ignited by electric current and not by a spark.

electrical discharge machining *See* electron discharge machining.

electrical disintegration [MET] Removing excess metal by using an electric spark in air.

electrical engineer [ENG] An engineer whose training includes a degree in electrical engineering from an accredited college or university (or who has comparable knowledge and experience), to prepare him for dealing with the generation, transmission, and utilization of electric energy.

electrical engineering [ENG] Engineering that deals with practical applications of electricity; generally restricted to applications involving current flow through conductors, as in motors and generators.

electrical insulating paper *See* insulating paper.

electrical log [ENG] Recorded measurement of the conductivities and resistivities down the length of an uncased borehole; gives a complete record of the formations penetrated.

electrical logging [ENG] The recording in uncased sections of a borehole of the conductivities and resistivities of the penetrated formations; used for geological correlation of the strata and evaluation of possibly productive horizons. Also known as electrical well logging.

electrically suspended gyro [ENG] A gyroscope in which the main rotating element is suspended by an electromagnetic or an electrostatic field.

electrical oil *See* insulating oil.

electrical porcelain *See* insulation porcelain.

electrical pressure transducer *See* pressure transducer.

electrical prospecting [ENG] The use of downhole electrical logs to obtain subsurface information for geological analysis.

electrical resistance meter *See* resistance meter.

electrical-resistance strain gage [ENG] A vibration-measuring device consisting of a grid of fine wire cemented to the vibrating object to measure fluctuating strains.

electrical steel [MET] Low carbon–iron alloy containing 0.5–5% silicon, produced in an electric-arc furnace and used for the cores of transformers, alternators, and other iron-core electric machines.

electrical tape *See* insulating tape.

electrical transcription *See* transcription.

electrical weighing system [ENG] An instrument which weighs an object by measuring the change in resistance caused by the elastic deformation of a mechanical element loaded with the object.

electrical well logging *See* electrical logging.

electric-arc furnace *See* arc furnace.

electric-arc heating *See* arc heating.

electric-arc spraying [MET] A thermal spraying process with an electric arc as a heat source and with compressed gas to propel the material.

electric-arc welding [MET] Welding in which the joint is heated to fusion by an electric arc or by a large electric current. Also known as arc welding.

electric boiler [MECH ENG] A steam generator using electric energy, in immersion, resistor, or electrode elements, as the source of heat.

electric brake [MECH ENG] An actuator in which the actuating force is supplied by current flowing through a solenoid, or through an electromagnet which is thereby attracted to disks on the rotating member, actuating the brake shoes; this force is counteracted by the force of a compression spring. Also known as electromagnetic brake.

electric coupling [MECH ENG] Magnetic-field coupling between the shafts of a driver and a driven machine.

electric detonator [ENG] A detonator ignited by a fuse wire which serves to touch off the primer.

electric drive [MECH ENG] A mechanism which transmits motion from one shaft to another and controls the velocity ratio of the shafts by electrical means.

electric engine [AERO ENG] A rocket engine in which the propellant is accelerated by some electric device. Also known as electric propulsion system; electric rocket.

electric fence [ENG] A fence consisting of one or more lengths of wire energized with high-voltage, low-current pulses, and giving a warning shock when touched.

electric furnace [ENG] A furnace which uses electricity as a source of heat.

electric-furnace steel [MET] Steel produced in an electric furnace.

electric guitar [ENG ACOUS] A guitar in which a contact microphone placed under the strings picks up the acoustic vibrations for amplification and for reproduction by a loudspeaker.

electric hammer [MECH ENG] An electric-powered hammer; often used for riveting or caulking.

electric heating [ENG] Any method of converting electric energy to heat energy by resisting the free flow of electric current.

electric hygrometer [ENG] An instrument for indicating by electrical means the humidity of the ambient atmosphere; usually based on the relation between the electric conductance of a film of hygroscopic material and its moisture content.

electrician [ENG] A skilled worker who installs, repairs, maintains, or operates electric equipment.

electric ignition [MECH ENG] Ignition of a charge of fuel vapor and air in an internal combustion engine by passing a high-voltage electric current between two electrodes in the combustion chamber.

electric instrument [ENG] An electricity-measuring device that indicates, such as an ammeter or voltmeter, in contrast to an electric meter that totalizes or records.

electric locomotive [MECH ENG] A locomotive operated by electric power picked up from a system of continuous overhead wires, or, sometimes, from a third rail mounted alongside the track.

electric meter [ENG] An electricity-measuring device that totalizes with time, such as a watthour meter or ampere-hour meter, in contrast to an electric instrument.

electric power generation [MECH ENG] The large-scale production of electric power for industrial, residential, and rural use, generally in stationary plants designed for that purpose.

electric power meter [ENG] A device that measures electric power consumed, either at an instant, as in a wattmeter, or averaged over a time interval, as in a demand meter. Also known as power meter.

electric power plant [MECH ENG] A power plant that converts a form of raw energy into electricity, for example, a hydro, steam, diesel, or nuclear generating station for stationary or transportation service.

electric power system [MECH ENG] A complex assemblage of equipment and circuits for generating, transmitting, transforming, and distributing electric energy.

electric propulsion [AERO ENG] A general term encompassing all the various types of propulsion in which the propellant consists of electrically charged particles which are accelerated by electric or magnetic fields, or both.

electric propulsion system See electric engine.

electric railroad [MECH ENG] A railroad which has a system of continuous overhead wires or a third rail mounted alongside the track to supply electric power to the locomotive and cars.

electric resistance furnace See resistance furnace.

electric rocket See electric engine.

electric spark machining See electron discharge machining.

electric stacker [MECH ENG] A stacker whose carriage is raised and lowered by a winch powered by electric storage batteries.

electric steel [MET] Steel melted in an electric furnace which permits close control and the addition of alloying elements directly into the furnace.

electric surface-recording thermometer [PETRO ENG] Device to measure temperatures during oil-well temperature surveying; has a thermocouple, resistance wire, or thermistor as the temperature-sensitive element.

electric tachometer [ENG] An instrument for measuring rotational speed by measuring the output voltage of a generator driven by the rotating unit.

electric tank See electrolytic tank.

electric thermometer [ENG] An instrument that utilizes electrical means to measure temperature, such as a thermocouple or resistance thermometer.

electric typewriter [MECH ENG] A typewriter having an electric motor that provides power for all operations initiated by the touching of the keys.

electric vehicle [MECH ENG] Any ground vehicle whose original source of energy is electric power, such as an electric car or electric locomotive.

electroacoustics [ENG ACOUS] The conversion of acoustic energy and waves into electric energy and waves, or vice versa.

electroacoustic transducer [ENG ACOUS] A transducer that receives waves from an electric system and delivers waves to an acoustic system, or vice versa. Also known as sound transducer.

electroceramics [MATER] Ceramic materials having electrical and other properties which make them suitable for use as insulators for power lines and in electrical components.

electrochemical cleaning [MET] Removing soil by the chemical action caused or sustained by a current of electricity in an electrolyte. Also known as electrolytic cleaning.

electrochemical coating [MET] A coating formed by chemical action on the metal surface and effected by a current of electricity through an electrolyte.

electrochemical corrosion [MET] Corrosion of a metal associated with the flow of electric current in an electrolyte. Also known as electrolytic corrosion.

electrochemical machining [MET] Removing excess metal by electrolytic dissolution, effected by the tool acting as the cathode against the workpiece acting as the anode. Abbreviated ECM. Also known as electrolytic machining.

electrochemical power generation [ENG] The direct conversion of chemical energy to electric energy, as in a battery or fuel cell.

electrochemical transducer [ENG] A device which uses a chemical change to measure the input parameter; the output is a varying electrical signal proportional to the measurand.

electrode force [MET] The force that occurs between electrodes during seam, spot, and projection welding. Also known as welding force.

electrodeposition [MET] Electrolytic process in which a metal is deposited at the cathode from a solution of its ions; includes electroplating and electroforming. Also known as electrolytic deposition.

electrode skid [MET] Sliding of an electrode over the work surface in spot, seam, or projection welding.

electrode-type liquid-level meter [ENG] Device that senses liquid level by the effect of the liquid-gas interface on the conductance of an electrode or probe.

electrodrill [MECH ENG] A drilling machine driven by electric power.

electrodynamic ammeter [ENG] Instrument which measures the current passing through a fixed coil and a movable coil connected in series by balancing the torque on the movable coil (resulting from the magnetic field of the fixed coil) against that of a spiral spring.

electrodynamic instrument [ENG] An instrument that depends for its operation on the reaction between the current in one or more movable coils and the current in one or more fixed coils. Also known as electrodynamometer.

electrodynamic loudspeaker [ENG ACOUS] Dynamic loudspeaker in which the magnetic field is produced by an electromagnet, called the field coil, to which a direct current must be furnished.

electrodynamic wattmeter [ENG] An electrodynamic instrument connected as a wattmeter, with the main current flowing through the fixed coil, and a small current proportional to the voltage flowing through the movable coil. Also known as moving-coil wattmeter.

electrodynamometer *See* electrodynamic instrument.

electroerosive machining *See* electron discharge machining.

electroexplosive [ENG] An initiator or a system in which an electric impulse initiates detonation or deflagration of an explosive. [MATER] The explosive substance so detonated or deflagrated.

electroformed mold [MET] A mold made by electroplating metal on the reverse pattern on the cavity; molten steel may be then sprayed on the back of the mold to increase its strength.

electroforming [MET] Shaping components by electrodeposition of the metal on a pattern.

electrogalvanizing [MET] Coating of a metal, especially iron or steel, with zinc by electroplating.

electrogas flux-cored welding [MET] A modification of the flux-cored welding process in which there is an externally supplied source of gas or gas mixture.

electrograph [ENG] Any plot, graph, or tracing produced by the action of an electric current on prepared sensitized paper (or other chart material) or by means of an electrically controlled stylus or pen.

electrokinetograph [ENG] An instrument used to measure ocean current velocities based on their electrical effects in the magnetic field of the earth.

electroless plating [MET] Deposition of a metal coating by immersion of a metal or nonmetal in a suitable bath containing a chemical reducing agent.

electroluminescent phosphor [MATER] Zinc sulfide powder, with small additions of copper or manganese, which emits light when suspended in an insulator in an intense alternating electric field. Also known as electroluminor.

electrolytic brightening *See* electropolishing.

electrolytic cleaning *See* electrochemical cleaning.

electrolytic copper [MET] Metallic copper produced by electrochemical deposition from a copper ion-containing electrolyte.

electrolytic corrosion *See* electrochemical corrosion.

electrolytic deposition *See* electrodeposition.

electrolytic etching [MET] Engraving the surface of a metal by electrolysis.

electrolytic grinding [MECH ENG] A combined grinding and machining operation in which the abrasive, cathodic grinding wheel is in contact with the anodic workpiece beneath the surface of an electrolyte.

electrolytic machining *See* electrochemical machining.

electrolytic model [PETRO ENG] Laboratory simulation of steady-state fluid flow through porous reservoir mediums; depends on the mobility of ions in absorbent mediums (gelatin or blotter), or through a liquid (potentiometric technique). Also known as gelatin model; oil-field model; potentiometric model.

electrolytic pickling [MET] Removal of metal by electrolysis using the metal as an electrode in a suitable electrolyte.

electrolytic polishing *See* electropolishing.

electrolytic powder [MET] Metal powder produced directly or indirectly by electrodeposition.

electrolytic protection *See* cathodic protection.

electrolytic strip *See* humidity strip.

electrolytic tank [ENG] A tank in which voltages are applied to an enlarged scale model of an electron-tube system or a reduced scale model of an aerodynamic system immersed in a poorly conducting liquid, and equipotential lines between electrodes are traced; used as an aid to electron-tube design or in computing ideal fluid flow; the latter application is based on the fact that the velocity potential in ideal flow and the stream function in planar flow satisfy the same equation, Laplace's equation, as an electrostatic potential. Also known as electric tank; potential flow analyzer.

electrolytic tough pitch [MET] Copper which has been refined electrolytically, containing mostly oxygen as an impurity.

electromachining [MECH ENG] The application of electric or ultrasonic energy to a workpiece to effect removal of material.

electromagnetic brake *See* electric brake.

electromagnetic clutch [MECH ENG] A clutch based on magnetic coupling between conductors, such as a magnetic fluid and powder clutch, an eddy-current clutch, or a hysteresis clutch.

electromagnetic crack detector [MET] An instrument that detects cracks in iron or steel objects by applying a strong magnetizing force and measuring the resulting magnetic flux through the object.

electromagnetic flowmeter [ENG] A flowmeter that offers no obstruction to liquid flow; two coils produce an electromagnetic field in the conductive moving fluid; the current induced in the liquid, detected by two electrodes, is directly proportional to the rate of flow. Also known as electromagnetic meter.

electromagnetic log [ENG] A log containing an electromagnetic sensing element extended below the hull of the vessel; this device produces a voltage directly proportional to speed through the water.

electromagnetic logging [ENG] A method of well logging in which a transmitting coil sets up an alternating electromagnetic field, and a receiver coil, placed in the drill hole above the transmitter coil, measures the secondary electromagnetic field induced by the resulting eddy currents within the formation. Also known as electromagnetic well logging.

electromagnetic meter See electromagnetic flowmeter.

electromagnetic mixing [MET] Mixing of molten alloys by exposing the melt to a strong magnetic field while passing direct current between electrodes at opposite ends of the crucible; stirring action results from interaction of the magnetic field of the current-carrying molten alloy with the external transverse magnetic field.

electromagnetic propulsion [AERO ENG] Motive power for flight vehicles produced by electromagnetic acceleration of a plasma fluid.

electromagnetic prospecting See electromagnetic surveying.

electromagnetic surveying [ENG] Underground surveying carried out by generating electromagnetic waves at the surface of the earth; the waves penetrate the earth and induce currents in conducting ore bodies, thereby generating new waves that are detected by instruments at the surface or by a receiving coil lowered into a borehole. Also known as electromagnetic prospecting.

electromagnetic well logging See electromagnetic logging.

electromanometer [ENG] An electronic instrument used for measuring pressure of gases or liquids.

electromechanical [MECH ENG] Pertaining to a mechanical device, system, or process which is electrostatically or electromagnetically actuated or controlled.

electromechanics [MECH ENG] The technology of mechanical devices, systems, or processes which are electrostatically or electromagnetically actuated or controlled.

electrometallurgy [MET] Industrial recovery and processing of metals by electrical and electrolytic procedures.

electrometer [ENG] An instrument for measuring voltage without drawing appreciable current.

electron-beam cutting [MET] A process which uses high-velocity electrons to heat the workpieces to be cut.

electron-beam machining [MET] A machining process in which heat is produced by a focused electron beam at a sufficiently high temperature to volatilize and thereby remove metal in a desired manner; takes place in a vacuum. Abbreviated EBM.

electron-beam magnetometer [ENG] A magnetometer that depends on the change in intensity or direction of an electron beam that passes through the magnetic field to be measured.

electron-beam melting [MET] A melting process in which an electron beam provides the necessary heat.

electron-beam welding [MET] A technique for joining materials in which highly collimated electron beams are used at a pressure below 10^{-3} mmHg (0.1333 pascal) to produce a highly concentrated heat source; used in outer space.

electron compound [MET] Alloy of two metals in which a progressive change in composition is accompanied by a progression of phases, differing in crystal structure. Also known as Hume-Rothery compound; intermetallic compound.

electron discharge machining [MET] A process by which materials that conduct electricity can be removed from a metal by an electric spark; used to form holes of varied shapes in materials of poor machinability. Abbreviated EDM. Also known as electrical discharge machining; electric spark machining; electroerosive machining; electrospark machining.

electronically agile radar [ENG] An airborne radar that uses a phased-array antenna which changes radar beam shapes and beam positions at electronic speeds.

electronic altimeter See radio altimeter.

electronic cash register [ENG] A system for automatically checking out goods from retail food stores, consisting of a device that scans packages and reads symbols imprinted on the label, and a computer that converts the symbol information to tell a cash register the price of the item; the computer can also keep records of sales and inventories. Abbreviated ECR.

electronic dummy [ENG ACOUS] A vocal simulator which is a replica of the head and torso of a person, covered with plastisol flesh that simulates the acoustical and mechanical properties of real flesh, and possessing an artificial voice and two artificial ears. Abbreviated ED.

electronic engineering [ENG] Engineering that deals with practical applications of electronics.

electronic flame safeguard [MECH ENG] An electrode used in a burner system which detects the main burner flame and interrupts fuel flow if the flame is not detected.

electronic fuse [ENG] A fuse, such as the radio proximity fuse, set off by an electronic device incorporated in it.

electronic heating [ENG] Heating by means of radio-frequency current produced by an electron-tube oscillator or an equivalent radio-frequency power source. Also known as high-frequency heating; radio-frequency heating.

electronic humidistat [ENG] A humidistat in which a change in the relative humidity causes a change in the electrical resistance between two sets of alternate metal conductors mounted on a small flat plate with plastic coating, and this change in resistance is measured by a relay amplifier.

electronic logger *See* Geiger-Müller probe.

electronic music [ENG ACOUS] Music consisting of tones originating in electronic sound and noise generators used alone or in conjunction with electroacoustic shaping means and sound-recording equipment.

electronic musical instrument [ENG ACOUS] A musical instrument in which an audio signal is produced by a pickup or audio oscillator and amplified electronically to feed a loudspeaker, as in an electric guitar, electronic carillon, electronic organ, or electronic piano.

electronic packaging [ENG] The technology of packaging electronic equipment; in current usage it refers to inserting discrete components, integrated circuits, and MSI and LSI chips (usually attached to a lead frame by beam leads) into plates through holes on multilayer circuit boards (also called cards), where they are soldered in place.

electronic photometer *See* photoelectric photometer.

electronic speedometer [ENG] A speedometer in which a transducer sends speed and distance pulses over wires to the speed and mileage indicators, eliminating the need for a mechanical link involving a flexible shaft.

electronic thermometer [ENG] A thermometer in which a sensor, usually a thermistor, is placed on or near the object being measured.

electronic voltmeter [ENG] Voltmeter which uses the rectifying and amplifying properties of electron devices and their associated circuits to secure desired characteristics, such as high-input impedance, wide-frequency range, crest indications, and so on.

electron metallography [MET] The study of the microscopic structure of metals employing an electron microscope.

electron stain [MATER] A substance such as phosphotungstic acid or osmic acid which scatters large numbers of electrons and can therefore be used to stain objects to be examined by an electron microscope.

electron vacuum gage [ENG] An instrument used to measure vacuum by the ionization effect that an electron flow (from an incandescent filament to a charged grid) has on gas molecules.

electrooptic radar [ENG] Radar system using electrooptic techniques and equipment instead of microwave to perform the acquisition and tracking operation.

electropainting [ENG] Electrolytic deposition of a thin layer of paint on a metal surface which is made an anode.

electrophoretic coating [MET] A surface coating on a metal deposited by electric discharge of particles from a colloidal solution.

electroplating [MET] Electrodeposition of a metal or alloy from a suitable electrolyte solution; the article to be plated is connected as the cathode in the electrolyte solution; direct current is introduced through the anode which consists of the metal to be deposited.

electropolishing [MET] Smoothing and enhancing the appearance of a metal surface by making it an anode in a suitable electrolyte. Also known as electrolytic brightening; electrolytic polishing.

electropulse engine [AERO ENG] An engine, for propelling a flight vehicle, that is based on the use of spark discharges through which intense electric and magnetic fields are established for periods ranging from microseconds to a few milliseconds; a resulting electromagnetic force drives the plasma along the leads and away from the spark gap.

electrorefining [MET] Purifying metals by electrolysis using an impure metal as anode from which the pure metal is dissolved and subsequently deposited at the cathode. Also known as electrolytic refining.

electroscope [ENG] An instrument for detecting an electric charge by means of the mechanical forces exerted between electrically charged bodies.

electrosensitive paper [MATER] A conductive paper that darkens when electric current is sent through it.

electroslag welding [MET] A welding process in which consumable electrodes are fed into a joint containing flux; the current melts the flux, and the flux in turn melts the faces of the joint and the electrodes, allowing the weld metal to form a continuously cast ingot between the joint faces.

electrospark machining *See* electron discharge machining.

electrostatic actuator *See* actuator.

electrostatic atomization [MECH ENG] Atomization in which a liquid jet or film is exposed to an electric field, and forces leading to atomization arise from either free charges on the surface or liquid polarization.

electrostatic gyroscope [ENG] A gyroscope in which a small beryllium ball is electrostatically suspended within an array of six electrodes in a vacuum inside a ceramic envelope.

electrostatic loudspeaker [ENG ACOUS] A loudspeaker in which the mechanical forces are produced by the action of electrostatic fields; in one type the fields are produced between a thin metal diaphragm and a rigid metal plate. Also known as capacitor loudspeaker.

electrostatic microphone *See* capacitor microphone.

electrostatic painting [ENG] A painting process that uses the particle-attracting property of electrostatic charges; direct current of about 100,000 volts is applied to a grid of wires through which the paint is sprayed to charge each particle; the metal objects to be sprayed are connected to the opposite terminal of the high-voltage circuit, so that they attract the particles of paint.

electrostatic precipitator [ENG] A device which removes dust or other finely divided particles from a gas by charging the particles inductively with an electric field, then attracting them to highly charged collector plates. Also known as precipitator.

electrostatic separation [ENG] Separation of finely pulverized materials by placing them in electrostatic separators. Also known as high-tension separation.

electrostatic separator [ENG] A separator in which a finely pulverized mixture falls through a powerful electric field between two electrodes; materials having different specific inductive capacitances are deflected by varying amounts and fall into different sorting chutes.

electrostatic transducer [ENG ACOUS] A transducer consisting of a fixed electrode and a movable electrode, charged electrostatically in opposite polarity; motion of the movable electrode changes the capacitance between the electrodes and thereby makes the applied voltage change in proportion to the amplitude of the electrode's motion. Also known as condenser transducer.

electrostatic tweeter [ENG ACOUS] A tweeter loudspeaker in which a flat metal diaphragm is driven directly by a varying high voltage applied between the diaphragm and a fixed metal electrode.

electrostatic voltmeter [ENG] A voltmeter in which the voltage to be measured is applied between fixed and movable metal vanes; the resulting electrostatic force deflects the movable vane against the tension of a spring.

electrostatic wattmeter [ENG] An adaptation of a quadrant electrometer for power measurements in which two quadrants are charged by the voltage drop across a noninductive shunt resistance through which the load current passes, and the line voltage is applied between one of the quadrants and a moving vane.

electrostriction transducer [ENG ACOUS] A transducer which depends on the production of an elastic strain in certain symmetric crystals when an electric field is applied, or, conversely, which produces a voltage when the crystal is deformed. Also known as ceramic transducer.

electrothermal ammeter *See* thermoammeter.

electrothermal energy conversion [ENG] The direct conversion of electric energy into heat energy, as in an electric heater.

electrothermal process [ENG] Any process which uses an electric current to generate heat, utilizing resistance, arcs, or induction; used to achieve temperatures higher than can be obtained by combustion methods.

electrothermal propulsion [AERO ENG] Propulsion of spacecraft by using an electric arc or other electric heater to bring hydrogen gas or other propellant to the high temperature required for maximum thrust; an arc-jet engine is an example.

electrothermal voltmeter [ENG] An electrothermal ammeter employing a series resistor as a multiplier, thus measuring voltage instead of current.

electrotinning [MET] Electroplating an object with tin.

electrowinning [MET] Extracting metal from solutions by electrochemical processes.

element [IND ENG] A brief, relatively homogeneous part of a work cycle that can be described and identified.

elemental motion [IND ENG] In time-and-motion study, a fundamental subdivision of the hand movements in manipulating an object. Also known as basic element; fundamental motion; therblig.

element breakdown [IND ENG] Separation of a work cycle into elemental motions.

element time [IND ENG] The time to complete a specific motion element.

elemi [MATER] A soft resin obtained from tropical trees of the family Burseraceae in the Philippines; used as a plasticizer, in cements and printing inks, and for perfumery and waterproofing.

elevate [ENG] To increase the angle of elevation of a gun, launcher, optical instrument, or the like.

elevating conveyor *See* bucket conveyor.

elevating machine *See* elevator.

elevation [ENG] Vertical distance to a point or object from sea level or some other datum.

elevation angle *See* angle of elevation.

elevation meter [ENG] An instrument that measures the change of elevation of a vehicle.

elevation stop [ENG] Structural unit in a gun or other equipment that prevents it from being elevated or depressed beyond certain fixed limits.

elevator [AERO ENG] The hinged rear portion of the longitudinal stabilizing surface or tail plane of an aircraft, used to obtain longitudinal or pitch-control moments. [MECH ENG] Also known as elevating machine. **1.** Vertical, continuous-belt, or chain device with closely spaced buckets, scoops, arms, or trays to lift or elevate powders, granules, or solid objects to a higher level. **2.** Pneumatic device in which air or gas is used to elevate finely powdered materials through a closed conduit. **3.** An enclosed platform or car that moves up and down in a shaft for transporting people or materials. Also known as lift. [PETRO ENG] A clamp gripping a stand or column of casing tubing, drill pipe, or sucker rods so that it can be moved up or down in a borehole being drilled.

elevator angle [AERO ENG] The angular displacement of the elevator from its neutral position; it is positive when the trailing edge of the elevator is below the neutral position, and negative when it is above.

elevator dredge [MECH ENG] A dredge which has a chain of buckets, usually flattened across the front and mounted on a nearly vertical ladder; used principally for excavation of sand and gravel beds under bodies of water.

elevator rod [PETRO ENG] A steel block fitted with an opening and latching device to permit insertion of a sucker rod, and provided with two long links to suspend from the elevator clamp when equipping a well with sucker rods for pumping.

elevon [AERO ENG] The hinged rear portion of an aircraft wing, moved in the same direction on each side of the aircraft to obtain longitudinal control and differentially to obtain lateral control; elevon is a combination of the words elevator and aileron to denote that an elevon combines the functions of aircraft elevators and ailerons.

elinvar [MET] A nickel-chromium steel alloy containing manganese and tungsten in varying amounts and having a low thermal expansion and almost invariable modulus of elasticity; used for chronometer balances and springs for gages and other instruments.

ell [BUILD] A wing built perpendicular to the main section of a building.

elliptical system [ENG] A tracking or navigation system where ellipsoids of position are determined from time or phase summation relative to two or more fixed stations which are the focuses for the ellipsoids.

elliptic gear [MECH ENG] A change gear composed of two elliptically shaped gears, each rotating about one of its focal points.

elliptic spring [DES ENG] A spring made of laminated steel plates, arched to resemble an ellipse.

elutriation [ENG] In a mixture, the separation of finer lighter particles from coarser heavier particles through a slow stream of fluid moving upward so that the lighter particles are carried with it.

elutriator [ENG] An apparatus used to separate suspended solid particles according to size by the process of elutriation.

emanometer [ENG] An instrument for the measurement of the radon content of the atmosphere: radon is removed from a sample of air by condensation or adsorption on a surface, and is then placed in an ionization chamber and its activity determined.

embankment [CIV ENG] **1.** A ridge constructed of earth, stone, or other material to carry a roadway or railroad at a level above that of the surrounding terrain. **2.** A ridge of earth or stone to prevent water from passing beyond desirable limits. Also known as bank.

embossing stylus [ENG ACOUS] A recording stylus with a rounded tip that forms a groove by displacing material in the recording medium.

emergency brake [MECH ENG] A brake that can be set by hand and, once set, continues to hold until released; used as a parking brake in an automobile.

Emerson wage incentive plan [IND ENG] A plan comprising time wages to 66⅔ of standard performance, empiric bonuses from there to standard performance, ending at 120% time wages, and thereafter a straight-line earning which is 20% above and parallel to basic piece rate.

emery [MATER] An abrasive which is composed of pulverized, impure corundum; used in polishing and grinding.

emery cake [MATER] Caked, powdered emery in a binding material.

Emery-Dietz gravity corer [ENG] A tube, with weights attached, which forces sediment samples into its interior as it is dropped on the ocean bottom.

emery paper [MATER] An abrasive paper or cloth with an adherent surface layer of emery powder; used for polishing and cleaning metal.

emery stone [MATER] **1.** A sharpening stone. **2.** A mixture of powdered emery and a binder which can be molded into grinding devices.

emery wheel [DES ENG] A grinding wheel made of or having a surface of emery powder; used for grinding and polishing.

empennage [AERO ENG] The assembly at the rear of an aircraft; it comprises the horizontal and vertical stabilizers. Also known as tail assembly.

empire cloth [MATER] Cotton cloth coated with oxidized oil; used as an electrical insulator.

employment test [IND ENG] Any of a wide variety of tests to measure intelligence, personality traits, skills, interests, aptitudes, or other characteristics; used to supplement interviews, physical examinations, and background investigations before employment.

empty-cell process [ENG] A wood treatment in which the preservative coats the cells without filling them.

emulsified asphalt [MATER] An emulsion of asphalt cement and water with a small quantity of an emulsifying agent.

emulsifier See disperser.

emulsifying agent See disperser.

emulsifying oil See soluble oil.

emulsion paint [MATER] Paint whose vehicle is an emulsion of a binder (oil, resin, latex, and so on) in water.

enamel [MATER] **1.** A finely ground, resin-containing oil paint that dries relatively harder, smoother, and glossier than ordinary paint. **2.** See glaze.

enamel clay [MATER] A ball clay able to float nonplastic enamel slips to make them spray or dip more evenly.

enameled brick [MATER] Brick with a smooth hard surface acquired from the application of a special wash before burning.

enameling [ENG] The application of a vitreous glaze to pottery or metal surfaces, followed by fusing in a kiln or furnace.

enamel kiln [ENG] A kiln in which enamel colors are fired.

enamel oxide [MATER] Any of the mixtures of calcined oxides used to color vitreous enamels used on sheet steel or cast iron.

encastré beam See fixed-end beam.

encrustation [ENG] The buildup of slag or other material inside furnaces and kilns.

end-bearing pile [CIV ENG] A bearing pile that is driven down to hard ground so that it carries the full load at its point. Also known as a point-bearing pile.

end construction [CIV ENG] Structural blocks or tiles laid so that the hollow cells run vertically.

end-construction tile [MATER] A type of structural clay tile designed to receive its principal stress parallel to the axis of the cells.

end cut See heavy fraction.

end-feed centerless grinding [MECH ENG] Centerless grinding in which the piece is fed through grinding and regulating wheels to an end stop.

end item [ENG] A final combination of end products, component parts, or materials which is ready for its intended use; for example, ship, tank, mobile machine shop, or aircraft.

end lap [DES ENG] A joint in which two joining members are made to overlap by removal of half the thickness of each.

end loader [MECH ENG] A platform elevator at the rear of a truck.

end mill [MECH ENG] A machine which has a rotating shank with cutting teeth at the end and spiral blades on the peripheral surface; used for shaping and cutting metal.

end-milled keyway See profiled keyway.

endoradiosonde [ENG] A miniature battery-powered radio transmitter encapsulated like a pill, designed to be swallowed for measuring and transmitting physiological data from the gastrointestinal tract.

end play [MECH ENG] Axial movement in a shaft-and-bearing assembly resulting from clearances between the components.

endurance [ENG] The time that an aircraft, vehicle, or ship can continue operating under given conditions without refueling.

energy beam [ENG] An intense beam of light, electrons, or other nuclear particles; used to cut, drill, form, weld, or otherwise process metals, ceramics, and other materials.

energy conversion efficiency [MECH ENG] The efficiency with which the energy of the working substance is converted into kinetic energy.

energy load See dynamic load.

energy management [AERO ENG] In rocketry, the monitoring of the expenditure of fuel for flight control and navigation.

engaged column [CIV ENG] A column partially built into a wall, and not freestanding.

engine [MECH ENG] A machine in which power is applied to do work by the conversion of various forms of energy into mechanical force and motion.

engine balance [MECH ENG] Arrangement and construction of moving parts in reciprocating or rotating machines to reduce dynamic forces which may result in undesirable vibrations.

engine block See cylinder block.

engine cooling [MECH ENG] Controlling the temperature of internal combustion engine parts to prevent overheating and to maintain all operating dimensions, clearances, and alignment by a circulating coolant, oil, and a fan.

engine cylinder [MECH ENG] A cylindrical chamber in an engine in which the energy of the working fluid, in the form of pressure and heat, is converted to mechanical force by performing work on the piston. Also known as cylinder.

engine displacement [MECH ENG] Volume displaced by each piston moving from bottom dead center to top dead center multiplied by the number of cylinders.

engine distillate [MATER] A heavy naphtha-kerosine distillate fuel of low octane number. Also known as tractor fuel.

engine efficiency [MECH ENG] Ratio between the energy supplied to an engine to the energy output of the engine.

engineer [ENG] An individual who specializes in one of the branches of engineering.

engineering economy [IND ENG] **1.** Application of engineering or mathematical analysis and synthesis to decision making in economics. **2.** The knowledge and techniques concerned evaluating the worth of commodities and services relative to their cost. **3.** Analysis of the economics of engineering alternatives.

engineering geology [CIV ENG] The application of education and experience in geology and other geosciences to solve geological problems posed by civil engineering structures.

engineering plastics [MATER] Plastics that lend themselves to use for engineering design, such as gears and structural members.

engineer's chain [CIV ENG] A surveyor's measuring instrument consisting of 1-foot (30.48-centimeter) steel links joined together by rings, 100 (30.5 meters) or 50 feet long. Also known as chain.

engine fuel [MATER] Any of various substances, usually fluid, which provide heat, chemical, or pressure energy for engine operation.

engine inlet [MECH ENG] A place of entrance for engine fuel.

engine knock [MECH ENG] In spark ignition engines, the sound and other effects associated with ignition and rapid combustion of the last part of the charge to burn, before the flame front reaches it. Also known as combustion knock.

engine lathe [MECH ENG] A manually operated lathe equipped with a headstock of the back-geared, cone-driven type or of the geared-head type.

engine oil [MATER] Oil used for the bearing lubrication of all types of engines, machines, and shafting and for cylinder lubrication in other than steam engines.

engine performance [MECH ENG] Relationship between power output, revolutions per minute, fuel or fluid consumption, and ambient conditions in which an engine operates.

engine sludge [ENG] The insoluble products of degradation of lubricating oils and fuels formed during the operation of an internal combustion engine.

Engler viscometer [ENG] An instrument used in the measurement of the degree Engler, a measure of viscosity; the kinematic viscosity v in stokes for this instrument is obtained from the equation $v = 0.00147t - 3.74/t$, where t is the efflux time in seconds.

English garden-wall bond [CIV ENG] A masonry bond in which there are three courses of stretchers to one of headers.

English red [MATER] Pigment consisting mostly of red iron oxide.

engobe [MATER] A thin layer of fluid clay applied to a piece of earthenware to support a glaze or enamel or to cover blemishes.

enriched gas [MATER] A motor gasoline containing additives to improve combustion characteristics, as by limiting knock.

enrockment [CIV ENG] A grouping of large stones dropped into water to form a base, such as for supporting a pier.

entering angle [MECH ENG] The angle between the side-cutting edge of a tool and the machined surface of the work; angle is 90° for a tool with 0° side-cutting edge angle effective.

entrance [CIV ENG] The seaward end of a channel, harbor, and so on. [ENG] A place of physical entering, such as a door or passage.

entrance angle [ENG] In molding, the maximum angle, measured from the center line of the mandrel, at which molten material enters the land area of a die.

entrance lock [CIV ENG] A lock between the tideway and an enclosed basin made necessary because the levels of the two bodies of water vary; by means of this lock, vessels can pass either way at all states of the tide. Also known as guard lock; tidal lock; tide lock.

entry corridor [AERO ENG] Depth of the region between two trajectories which define the design limits of a vehicle about to enter a planetary atmosphere, or define the desired landing area (footprint).

envelope [ENG] The glass or metal housing of an electron tube or the glass housing of an incandescent lamp. Also known as bulb.

envenomation [MATER] The process by which the surface of a plastic close to or in contact with another surface is deteriorated.

environment [ENG] The aggregate of all natural, operational, or other conditions that affect the operation of equipment or components.

environmental cab [ENG] Operator's compartment in earthmovers equipped with tinted safety glass, soundproofing, air conditioning, and cleaning units.

environmental control [ENG] Modification and control of soil, water, and air environments of man and other living organisms.

environmental control system [ENG] A system used in a closed area, especially a spacecraft or submarine, to permit life to be sustained; the system provides the occupants with a suitably controlled atmosphere to permit them to live and work in the area.

environmental engineering [ENG] The technology concerned with the reduction of pollution, contamination, and deterioration of the surroundings in which humans live.

environmental impact analysis [IND ENG] Predetermination of the extent of pollution or environmental degradation which will be involved in a mining or processing project.

environmental impact statement [ENG] A report of the potential effect of plans for land use in terms of the environmental, engineering, esthetic, and economic aspects of the proposed objective.

environmental protection [ENG] The protection of humans and equipment against stresses of climate and other elements of the environment.

environmental range [ENG] The range of environment throughout which a system or portion thereof is capable of operation at not less than the specified level of reliability.

environmental survey satellite [AERO ENG] One of a series of meteorological satellites which completely photographs the earth each day. Abbreviated ESSA.

environmental test [ENG] A laboratory test conducted to determine the functional performance of a component or system under conditions that simulate the real environment in which the component or system is expected to operate.

environment simulator [ENG] Any machine or artificial device that simulates all or some of the attributes of an environment, such as the solar simulators with artificial suns used in testing spacecraft.

eolian anemometer [ENG] An anemometer which works on the principle that the pitch of the eolian tones made by air moving past an obstacle is a function of the speed of the air.

Eötvös torsion balance [ENG] An instrument which records the change in the acceleration of gravity over the horizontal distance between the ends of a beam; used to measure density variations of subsurface rocks.

EPDM *See* ethylene-propylene terpolymer.

epicyclic gear [MECH ENG] A system of gears in which one or more gears travel around the inside or the outside of another gear whose axis is fixed.

epicyclic train [MECH ENG] A combination of epicyclic gears, usually connected by an arm, in which some or all of the gears have a motion compounded of rotation about an axis and a translation or revolution of that axis.

EP lubricant [MATER] A lubricating oil or grease that contains additives to improve ability to adhere to the surfaces of metals under high bearing pressures. Derived from extreme-pressure lubricant.

epoxy adhesive [MATER] An adhesive material made of epoxy resin.

Eppley pyrheliometer [ENG] A pyrheliometer of the thermoelectric type; radiation is allowed to fall on two concentric silver rings, the outer covered with magnesium oxide and the inner covered with lampblack; a system of thermocouples (thermopile) is used to measure the temperature difference between the rings; attachments are provided so that measurements of direct and diffuse solar radiation may be obtained.

EPT *See* ethylene-propylene terpolymer.

equaling file [DES ENG] A slightly bulging double-cut file used in fine toolmaking.

equalizer [MECH ENG] **1.** A bar to which one attaches a vehicle's whiffletrees to make the pull of draft animals equal. Also known as equalizing bar. **2.** A bar which joins a pair of axle springs on a railway locomotive or car for equalization of weight. Also known as equalizing bar. **3.** A device which distributes braking force among independent brakes of an automotive vehicle. Also known as equalizer brake. **4.** A machine which saws wooden stock to equal lengths.

equalizer brake *See* equalizer.

equalizing bar *See* equalizer.

equalizing reservoir [CIV ENG] A reservoir located between a primary water supply and the consumer for the purpose of maintaining equilibrium between different portions of the distribution system.

equatorial mounting [ENG] The mounting of an equatorial telescope; it has two perpendicular axes, the polar axis (parallel to the earth's axis) that turns on fixed bearings, and the declination axis, supported by the polar axis.

equatorial telescope [ENG] An astronomical telescope that revolves about an axis parallel to the earth's axis and automatically keeps a star on which it has been fixed in its field of view.

equilibristat [ENG] A device for measuring the deviation from equilibrium of a railroad car as it goes around a curve.

equilibrium gas saturation [PETRO ENG] Condition of zero relative permeability of a nonwetting/wetting phase system in a reservoir; relation to the nonwetting phase (for example, oil) to the wetting phase (for example, water) when the nonwetting-phase saturation is so small that relatively few pores contain it. Also known as critical gas saturation.

equilibrium state [IND ENG] A state in which the numbers of customers or items waiting in a queue varies in such a way that the mean and distribution remain constant over a long period.

equipment [ENG] One or more assemblies capable of performing a complete function.

equipment chain [ENG] Group of equipments that are functionally in series; the failure of one or more of the equipments results in loss of the function.

equipment replacement study [IND ENG] A cost analysis based on estimates of operating costs over a stated time for the old facility compared with the new facility.

equipressure contour [PETRO ENG] Within a reservoir, a plot or map of the equal isopressure flow network; used to locate sites for water-injection wells for flood coverage of an areal reservoir pattern.

equity crude [PETRO ENG] Crude produced which belongs to an oil company that owns a concession jointly with a host government.

equivalent airspeed [AERO ENG] The product of the true airspeed and the square root of the density ratio; used in structural design work to designate various design conditions.

equivalent annual rate [IND ENG] A measure used in setting up a monthly rate on a comparable basis for each of the months regardless of their variation in working days, or for making the rate comparable with an annual rate regardless of the variation in working days during each month.

equivalent noise pressure [ENG ACOUS] In an electroacoustic transducer or sound reception system, the root-mean-square sound pressure of a sinusoidal plane progressive wave, which when propagated parallel to the primary axis of the transducer, produces an open-circuit signal voltage equivalent to the root-mean-square of the inherent open-circuit noise voltage of the transducer in a transmission band with a bandwidth of 1 hertz and centered on the frequency of the plane sound wave. Also known as inherent noise pressure.

equivalent orifice [MECH ENG] An expression of fan performance as the theoretical sharp-edge orifice area which would offer the same resistance to flow as the system resistance itself.

equivalent vapor volume [PETRO ENG] The volume occupied by a barrel of oil if all the oil were to become a vapor; expressed as cubic foot per barrel at 60°F (15.6°C).

erection [CIV ENG] Positioning and fixing the frame of a structure.

erection bolt [CIV ENG] A threaded rod with a head at one end, used to temporarily join parts of a structure during construction.

erection tower [CIV ENG] A temporary framework built at a construction site for hoisting equipment.

ergograph [ENG] An instrument with a recording device used to measure work capacity of muscles.

ergometer [ENG] An instrument with a recording device used to measure work performed by muscles under control conditions.

ergonometrics [IND ENG] The application of various procedures for determining the time for an operator to perform a task satisfactorily, using the standard method in the usual environmental conditions.

ergonomics [IND ENG] The study of human capability and psychology in relation to the working environment and the equipment operated by the worker.

Erichsen test [MET] A cupping test to measure the ductility of a piece of sheet metal and to determine its suitability for deep drawing.

Erichsen value [MET] The depth of impression in millimeters required to fracture a cupped sheet metal supported on a ring and deformed at the center by a spherically shaped tool.

erigeron oil [MATER] A volatile oil, whose components are gallic and tannic acid, that is distilled from fleabane and horseweed.

erosion-corrosion [MET] Attack on a metal surface resulting from the combined effects of erosion and corrosion.

error of closure [ENG] Also known as angular error of closure. **1.** The amount by which the measurement of the azimuth of the first line of a traverse, made after completing the circuit, fails to equal the initial measurement. **2.** The amount by which the sum of the angles measured around the horizon differs from 360°.

ERTS *See* earth resources technology satellite.

escalation [IND ENG] Provision in actual or estimated costs for inflational increases in the costs of equipment, materials, labor, and so on, over those specified in an original contract.

escalator [MECH ENG] A continuously moving stairway and handrail.

escape hatch [ENG] A hatch which permits men to escape from a compartment, such as the interior of a submarine or aircraft, when normal means of exiting are blocked.

escapement [MECH ENG] A ratchet device that permits motion in one direction slowly.

escape rocket [AERO ENG] A small rocket engine attached to the leading end of an escape tower, to provide additional thrust to the capsule in an emergency; it helps separate the capsule from the booster vehicle and carries it to an altitude where parachutes can be deployed.

escape tower [AERO ENG] A trestle tower placed on top of a space capsule, connecting the capsule to the escape rocket on top of the tower; used for emergencies.

escape velocity [AERO ENG] In space flight, the speed at which an object is able to overcome the gravitational pull of the earth.

escort fighter [AERO ENG] A fighter designed or equipped for long-range missions, usually to accompany heavy bombers on raids.

escutcheon [DES ENG] An ornamental shield, flange, or border used around a dial, window, control knob, or other panel-mounted part. Also known as escutcheon plate.

esparto wax [MATER] Vegetable wax extracted from esparto grass; it is hard, blends well, and is easy to emulsify; used to give smoothness to polishes.

ESSA *See* environmental survey satellite.

essential oil [MATER] Any of the odoriferous oily products of plant origin which are distillable; the principal constituents are terpenes, but benzenoid and aliphatic compounds may also be present. Also known as ethereal oil.

esthesiometer [ENG] An instrument used to measure tactile sensibility by determining the distance by which two points pressed against the skin must be separated in order that they be felt as separate. Also spelled aesthesiometer.

estimated time [IND ENG] A predicted element or operation time.

estragon oil [MATER] Essential oil, colorless to yellowish green, with aniselike odor and aromatic flavor; distilled from flowering herb tarragon, *Artemisia dracunculus*; used chiefly as flavoring. Also known as tarragon oil.

etch [MET] To corrode the surface of a metal in order to reveal its composition and structure.

etch cleaning [MET] Removing soil by electrolytic or chemical action; removes some of the surface metal along with the dirt.

etch cracks [MET] Shallow cracks in the surface of hardened steel that result from reaction with an acid, causing hydrogen cracking.

etched circuit [ENG] A printed circuit formed by chemical or electrolytic removal of unwanted portions of a layer of conductive material bonded to an insulating base.

etch figures [MET] Minute, faceted pits or surfaces produced on a metal surface by chemical reaction with an etchant.

ethereal oil *See* essential oil.

ethylene-propylene terpolymer [MATER] An elastomer which is based on ethylene and propylene terpolymers with small amounts of a nonconjugated diene and which can be vulcanized; used for automotive parts, cable coating, hose, footwear, and other products. Abbreviated EPDM; EPT.

eucalyptus oil [MATER] Any of various essential oils of various eucalyptus leaves, a yellow liquid miscible with alcohol; varies from peppermint to turpentine in odor; used in perfumery, medicine, and ore flotation.

eucalyptus resin oil *See* eucalyptus tar.

eucalyptus tar [MATER] Residue from caustic treating of oil from distilled eucalyptus leaves; used to perfume soaps and as a disinfectant. Also known as eucalyptus resin oil.

eucaryote *See* eukaryote.

eudiometer [ENG] An instrument for measuring changes in volume during the combustion of gases, consisting of a graduated tube that is closed at one end and has two wires sealed into it, between which a spark may be passed.

European porcelain *See* porcelain.

eutectic [MET] The microstructure that results when a metal of eutectic composition solidifies.

eutectic crystallization [MET] Simultaneous crystallization of the constituents of a eutectic alloy during cooling of the melt.

eutectic melting [MET] Melting of isolated, microscopic areas of an alloy that correspond to the location of the eutectic of the system.

eutectoid [MET] A mixture of phases whose composition is determined by the eutectoid point in the solid region of an equilibrium diagram and whose constituents are formed by the eutectoid reaction.

EV *See* expected value.

evacuate [ENG] To remove something, especially gases and vapors, from an enclosure, such as from the envelope of an electron tube, or from a well. Also known as exhaust.

evaporation gage *See* atmometer.

evaporation pan [ENG] A type of atmometer consisting of a pan, used in the measurement of the evaporation of water into the atmosphere.

evaporation tank [ENG] A tank used to measure the evaporation of water under controlled conditions.

evaporative condenser [MECH ENG] An apparatus in which vapor is condensed within tubes that are cooled by the evaporation of water flowing over the outside of the tubes.

evaporative cooling [ENG] **1.** Lowering the temperature of a large mass of liquid by utilizing the latent heat of vaporization of a portion of the liquid. **2.** Cooling air by evaporating water into it. **3.** *See* vaporization cooling.

evaporative cooling tower *See* wet cooling tower.

evaporator [MECH ENG] Any of many devices in which liquid is changed to the vapor state by the addition of heat, for example, distiller, still, dryer, water purifier, or refrigeration system element where evaporation proceeds at low pressure and consequent low temperature.

evaporimeter *See* atmometer.

evaporite pond [IND ENG] Any containment area for brines or solution-mined effluents constructed to permit solar evaporation and harvesting of dewatered evaporite concentrates.

evapotranspirometer [ENG] An instrument which measures the rate of evapotranspiration; consists of a vegetation soil tank so designed that all water added to the tank and all water left after evapotranspiration can be measured.

Evasé stack [CIV ENG] In tunnel engineering, an exhaust stack for air having a cross section that increases in the direction of airflow at a rate to regain pressure.

even pitch [DES ENG] The pitch of a screw in which the number of threads per inch is a multiple (or submultiple) of the threads per inch of the lead screw of the lathe on which the screw is cut.

event recorder [ENG] A recorder that plots on-off information against time, to indicate when events start, how long they last, and how often they recur.

evolutionary operation [IND ENG] An iterative technique for optimizing a production process by systematically introducing small changes in the process and then observing and evaluating the results.

excavator [MECH ENG] A machine for digging and removing earth.

excelsior [MATER] Fine, curled wood shavings, used as packing material.

excess air [ENG] Amount of air in a combustion process greater than the amount theoretically required for complete oxidation.

excess coefficient [MECH ENG] The ratio $(A-R)/R$, where A is the amount of air admitted in the combustion of fuel and R is the amount required.

exchanger *See* heat exchanger.

exfoliation [MET] Peeling off or separation of metal at its surface in the form of thin, parallel scales or lamellae.

exfoliation corrosion [MET] A type of corrosion that progresses parallel to the metal surface in such a manner that underlying layers are gradually separated.

exhaust [MECH ENG] **1.** The working substance discharged from an engine cylinder or turbine after performing work on the moving parts of the machine. **2.** The phase of the engine cycle concerned with this discharge. **3.** A duct for the escape of gases, fumes, and odors from an enclosure, sometimes equipped with an arrangement of fans.

exhaust deflecting ring [MECH ENG] A type of jetavator consisting of a ring so mounted at the end of a nozzle as to permit it to be rotated into the exhaust stream.

exhaust gas [MECH ENG] Spent gas leaving an internal combustion engine or gas turbine.

exhaust-gas analyzer [ENG] An instrument that analyzes the gaseous products to determine the effectiveness of the combustion process.

exhaust head [ENG] A device placed on the end of an exhaust pipe to remove oil and water and to reduce noise.

exhaust manifold [MECH ENG] A branched system of pipes to carry waste emissions away from the piston chambers of an internal combustion engine.

exhaust nozzle [AERO ENG] The terminal portion of a jet engine tail pipe.

exhaust pipe [MECH ENG] The duct through which engine exhaust is discharged.

exhaust scrubber [ENG] A purifying device on internal combustion engines which removes noxious gases from engine exhaust.

exhaust stream [AERO ENG] The stream of matter or radiation emitted from the nozzle of a rocket or other reaction engine.

exhaust stroke [MECH ENG] The stroke of an engine, pump, or compressor that expels the fluid from the cylinder.

exhaust suction stroke [MECH ENG] A stroke of an engine that simultaneously removes used fuel and introduces fresh fuel to the cylinder.

exhaust valve [MECH ENG] The valve on a cylinder in an internal combustion engine which controls the discharge of spent gas.

exit [ENG] A door, passage, or place of egress.

ex lighterage [IND ENG] Price quoted exclusive of lighterage fees.

exotic fuels [MATER] The hydroborons which have higher calorific values than do the carbon-hydrogen fuels, once proposed as high-energy fuels for aircraft and missiles; include borane (BH_3), borobutane (B_4H_{10}), and borodecane ($B_{10}H_{14}$).

expandable space structure [AERO ENG] A structure which can be packaged in a small volume for launch and then erected to its full size and shape outside the earth's atmosphere.

expanded clay [MATER] A material made from common brick clays by grinding, screening, and then feeding through a gas burner at about 2700°F (1482°C), thus changing the ferric oxide to ferrous oxide and causing the formation of bubbles.

expanded-flow bin [ENG] A bin formed by attaching a mass-flow hopper to the bottom of a funnel-flow bin.

expanded metal [MET] An alloy which has expanded following cooling and solidification.

expanded plastic [MATER] A light, spongy plastic made by introducing pockets of air or gas. Also known as foamed plastic; plastic foam.

expanded slag [MATER] Slag formed by running slag from phosphate rock onto a forehearth at about 2000°F (1093°C) and then treating it with water, high-pressure steam, and air; used to make lightweight concrete blocks.

expanding [MET] A process used to increase the inside diameter of a hollow piece, such as a tube, cup, or shell.

expanding brake [MECH ENG] A brake that operates by moving outward against the inside rim of a drum or wheel.

expansion [MECH ENG] Increase in volume of working material with accompanying drop in pressure of a gaseous or vapor fluid, as in an internal combustion engine or steam engine cylinder.

expansion bolt [DES ENG] A bolt having an end which, when embedded into masonry or concrete, expands under a pull on the bolt, thereby providing anchorage.

expansion chucking reamer [DES ENG] A machine reamer with an expansion screw at the end which increases the diameter.

expansion cooling [MECH ENG] Cooling of a substance by having it undergo adiabatic expansion.

expansion engine [MECH ENG] Piston-cylinder device that cools compressed air via sudden expansion; used in production of pure gaseous oxygen via the Claude cycle.

expansion fit [DES ENG] A condition of optimum clearance between certain mating parts in which the cold inner member is placed inside the warmer outer member and the temperature is allowed to equalize.

expansion joint [MECH ENG] **1.** A joint between parts of a structure or machine to avoid distortion when subjected to temperature change. **2.** A pipe coupling which, under temperature change, allows movement of a piping system without hazard to associated equipment.

expansion opening [ENG] A chamber in line with a pipe or tunnel and of larger diameter than the conduit containing liquid or gas, to allow lowering of pressure within the conduit by expansion of the fluid.

expansion ratio [MECH ENG] In a reciprocating piston engine, the ratio of cylinder volume with piston at bottom dead center to cylinder volume with piston at top dead center.

expansion reamer [ENG] A reamer whose diameter may be adjusted between limits by an expanding screw.

expansion rollers [CIV ENG] Rollers fitted to one support of a bridge or truss to allow for thermal expansion and contraction.

expansion shield [DES ENG] An anchoring device that expands as it is driven into masonry or concrete, pressing against the sides of the hole.

expansion system [PETRO ENG] Gas-liquid recovery system in which the refrigeration effect of rapidly depressurized wellstream effluent through a wellhead choke is used to obtain maximum removal of liquefiable hydrocarbons from the gas stream.

expansion valve [MECH ENG] A valve in which fluid flows under falling pressure and increasing volume.

expansive bit [DES ENG] A bit in which the cutting blade can be set at various sizes.

expansive cement [MATER] A type of hydraulic cement, usually of high sulfate and alumina content, that expands after hardening to compensate for drying shrinkage.

expected utility *See* expected value.

expected value [SYS ENG] In decision theory, a measure of the value or utility expected to result from a given strategy, equal to the sum over states of nature of the product of the probability of the state times the consequence or outcome of the strategy in terms of some value or utility parameter. Abbreviated EV. Also known as expected utility (EU).

expiration date [MATER] The anticipated date when a material may go from usable to unusable.

exploding bridge wire [ENG] An initiator or system in which a very high energy electrical impulse is passed through a bridge wire, literally exploding the bridge wire and releasing thermal and shock energy capable of initiating a relatively insensitive explosive in contact with the bridge wire.

exploitation [MIN ENG] The extraction from the earth and utilization of ore, gas, oil, and minerals found by exploration.

exploration [MIN ENG] The search for economic deposits of minerals, ore, gas, oil, or coal by geological surveys, geophysical prospecting, boreholes and trial pits, or surface or underground headings, drifts, or tunnels.

exploratory well [PETRO ENG] An oil well drilled for purposes of exploration for underlying petroleum.

Explorer program [AERO ENG] A series of earth satellites begun by the U.S. Army and continued by NASA; *Explorer 1*, the first U.S. satellite, went into orbit on January 31, 1958.

explosion door [MECH ENG] A door in a furnace which is designed to open at a predetermined pressure.

explosion rupture disk device [MECH ENG] A protective device used where the pressure rise in the vessel occurs at a rapid rate.

explosion welding [MET] A solid-state process wherein bonding is produced by a controlled detonation, resulting in rapid movement together of the members to be joined.

explosive [MATER] A substance, such as trinitrotoluene, or a mixture, such as gunpowder, that is characterized by chemical stability but may be made to undergo rapid chemical change without an outside source of oxygen, whereupon it produces a large quantity of energy generally accompanied by the evolution of hot gases.

explosive-actuated device [ENG] Any of various devices actuated by means of explosive; includes devices actuated either by high explosives or low explosives, whereas propellant-actuated devices include only the latter.

explosive bolt [AERO ENG] A bolt designed to contain a remote-initiated explosive charge which, upon detonation, will shear the bolt or cause it to fail otherwise; applicable to such uses as stage separation of rockets, jettison of expended fuel tanks, and ejection of parachutes.

explosive cladding [MET] Bonding of a metal coating or metal cladding to a base metal by using the force of an explosive charge.

explosive decompression [AERO ENG] A sudden loss of pressure in a pressurized cabin, cockpit, or the like, so rapid as to be explosive, as when punctured by gunfire.

explosive disintegration [ENG] Explosive shattering when pressure is suddenly released on a pressured, permeable material (wood, mineral, and such) containing gas or liquid; the rupture of wood by this process is used to manufacture Masonite.

explosive echo ranging [ENG] Sonar in which a charge is exploded underwater to produce a shock wave that serves the same purpose as an ultrasonic pulse; the elapsed time for return of the reflected wave gives target range.

explosive filler [MATER] Main explosive charge contained in a projectile, missile, bomb, or the like.

explosive forming [MET] Shaping metal parts in dies by using an explosive charge to generate forming pressure.

explosive fuel [MATER] Any substance which combines with oxygen and other explosive ingredients to produce explosion energy, including aluminum, silicon, carbon, sulfur, glycerol, glucol, paraffin wax, diesel oil, and guar gum.

explosive oxidizer [MATER] Any substance which yields oxygen to combine with fuels or other explosive ingredients to produce explosive energy, such as nitrates, chlorates, and perchlorates.

explosive rivet [ENG] A rivet holding a charge of explosive material; when the charge is set off, the rivet expands to fit tightly in the hole.

exponential horn [ENG ACOUS] A horn whose cross-sectional area increases exponentially with axial distance.

exponential smoothing [IND ENG] A mathematical-statistical method of forecasting used in industrial engineering which assumes that demand for the following period is some weighted average of the demands for the past periods.

export kerosine [MATER] A grade of kerosine once used for export; has the darkest shade of the standard kerosine colors, namely standard white (also called export white). Also known as standard white kerosine.

exposure [BUILD] The distance from the butt of one shingle to the butt of the shingle above it, or the amount of a shingle that is seen.

exposure time [CIV ENG] The time period of interest for seismic hazard calculations such as the design lifetime of a building or the time over which the numbers of casualties should be estimated.

expressway [CIV ENG] A limited-access, high-speed, divided highway having grade separations at points of intersection with other roads. Also known as limited-access highway.

extended area [DES ENG] An engineering surface that has been extended areawise without increasing diameter, as by using pleats (as in filter cartridges) or fins (as in heat exchangers).

extender plasticizer *See* secondary plasticizer.

extension bolt [DES ENG] A vertical bolt that can be slid into place by a long extension rod; used at the top of doors.

extension jamb [BUILD] A jamb that extends past the head of a door or window.

extension ladder [DES ENG] A ladder of two or more nesting sections which can be extended to almost the combined length of the sections.

extension ore *See* possible ore.

extension spring [DES ENG] A tightly coiled spring designed to resist a tensile force.

extensometer [ENG] **1.** A strainometer that measures the change in distance between two reference points separated 20–30 meters or more; used in studies of displacements due to seismic activities. **2.** An instrument designed to measure minute deformations of small objects subjected to stress.

external aileron [AERO ENG] An aileron offset from the wing; that is, not forming a part of the wing.

external brake [MECH ENG] A brake that operates by contacting the outside of a brake drum.

external centerless grinding [MECH ENG] A process by which a metal workpiece is finished on its external surface by supporting the piece on a blade while it is advanced between a regulating wheel and grinding wheel.

external combustion engine [MECH ENG] An engine in which the generation of heat is effected in a furnace or reactor outside the engine cylinder.

external gas injection [PETRO ENG] Pressure-maintenance gas injection with wells located in the structurally higher positions of the reservoir, usually in the primary or secondary gas cap. Also known as crestal injection; gas-cap injection.

external grinding [MECH ENG] Grinding the outer surface of a rotating piece of work.

external header [MECH ENG] Manifold connecting sections of a cast iron boiler.

externally fired boiler [MECH ENG] A boiler that has refractory or cooling tubes surrounding its furnace.

external-mix oil burner [ENG] A burner utilizing a jet stream of air to strike the liquid fuel after it has left the burner orifice.

external shoe brake [MECH ENG] A friction brake operated by the application of externally contracting elements.

external thread [DES ENG] A screw thread cut on an outside surface.

external time [IND ENG] The time used to perform work by the operator outside the machine cycle, resulting in a loss of potential machine operating time.

external upset casing [PETRO ENG] Special oil- or gas-well casing designed for extreme conditions requiring greater than usual strength and leak resistance. Also known as extreme line casing.

extraction parachute [AERO ENG] An auxiliary parachute designed to release and extract cargo from aircraft in flight and to deploy cargo parachutes.

extraction turbine [MECH ENG] A steam turbine equipped with openings through which partly expanded steam is bled at one or more stages.

extractive metallurgy [MET] Extraction of metals from ore by various chemical and mechanical methods.

extract oil [MATER] The less desirable portion of the oil being solvent-refined; it is dissolved in and selectively removed by the solvent.

extractor [ENG] **1.** A machine for extracting a substance by a solvent or by centrifugal force, squeezing, or other action. **2.** An instrument for removing an object.

extravehicular activity [AERO ENG] Activity conducted outside a spacecraft during space flight.

extreme line casing *See* external upset casing.

extreme line tubing [PETRO ENG] Special oil- or gas-well tubing designed for extreme conditions requiring greater than usual strength and leak resistance.

extreme-pressure lubricant *See* EP lubricant.

extrudate [ENG] Ductile metal, plastic, or other semisoft solid material that has been shaped into a continuous form (such as fiber, film, pipe, or wire coating) by forcing the semisolid material through a die opening of appropriate shape.

extruder [ENG] A device that forces ductile or semisoft solids through die openings of appropirate shape to produce a continuous film, strip, or tubing.

extrusion [ENG] A process in which a hot or cold semisoft solid material, such as metal or plastic, is forced through the orifice of a die to produce a continuously formed piece in the shape of the desired product.

extrusion billet [MET] A slug of heated metal that is forced through a die by a hydraulic ram in direct extrusion operations.

extrusion coating [ENG] A process of placing resin on a substrate by extruding a thin film of molten resin and pressing it onto

or into the substrates, or both, without the use of adhesives.

extrusion defect [MET] Impaired flow of an extrusion product due to surface oxidation of the ingot or billet.

extrusion ingot [MET] A cylindrical casting used to form extruded products.

extrusion metal [MET] Any of numerous nonferrous metals, alloys, and other materials used in extrusion operations.

extrusion pressing *See* cold extrusion.

eye assay [MIN ENG] An estimate of the valuable mineral content of a core or ore sample as based on visual inspection. Also known as eyeball assay.

eyeball assay *See* eye assay.

eyebar [DES ENG] A metal bar having a hole or eye through each enlarged end.

eyebolt [DES ENG] A bolt with a loop at one end.

eyelet [DES ENG] A small ring or barrel-shaped piece of metal inserted into a hole for reinforcement.

eyeleting [ENG] Forming a lip around the rim of a hole.

eye screw [DES ENG] A screw with an open loop head.

F

F-6 *See* Skyray.
F-11 *See* Tiger.
F-84 *See* Thunderstreak.
F-86 *See* Sabrejet.
F-89 *See* Scorpion.
F-100 *See* Super Sabre.
F-105 *See* Thunderchief.
Fabian system [MIN ENG] The free-fall drilling system from which all other free-fall systems have originated.
fabrication [ENG] **1.** The manufacture of parts, usually structural or electromechanical parts. **2.** The assembly of parts into a structure.
fabric laminate [MATER] Layers of fabric alternating with plastic, used as insulation in electrical equipment.
fabric-type dust collector [MIN ENG] A collector which removes dust particles from ore by means of a filter made of fabric.
face [MIN ENG] A surface on which mining operations are being performed. Also known as breast.
face area [MIN ENG] The working area toward the interior of the last open crosscut in an entry or room.
face belt conveyor [MIN ENG] A lightweight belt conveyor used at the working face in a mine.
face boss [MIN ENG] A foreman in charge of operations at the working face in a bituminous coal mine.
face brick [MATER] A brick of some esthetic quality to be used on the exposed surface of a building wall or other structure.
face conveyor [MIN ENG] Any type of mine conveyor used at and parallel to a working face.
face-discharge bit [MECH ENG] A liquid-coolant bit designed for drilling in soft formations and for use on a double-tube core barrel, the inner tube of which fits snugly into a recess cut into the inside wall of the bit directly above the inside reaming stones; the coolant flows through the bit and is ejected at the cutting face. Also known as bottom-discharge bit; face-ejection bit.

faced wall [BUILD] A wall whose masonry facing and backing are of different materials.
face-ejection bit *See* face-discharge bit.
face feed [MET] In brazing or soldering, the deposition of filler metal to the joint, usually by hand.
face gear [DES ENG] A gear having teeth cut on the face.
face haulage *See* primary haulage.
face height [MIN ENG] The vertical distance between the top and toe of a quarry or opencast face.
faceman [MIN ENG] A coal miner who performs the duties involved in drilling underground openings into which explosives are charged and set off, to extract coal, slate, and rock. Also known as coal digger; coal getter.
face mechanization [MIN ENG] The use of a cutter-loader on a longwall face.
face milling [MECH ENG] Milling flat surfaces perpendicular to the rotational axis of the cutting tool.
face mold [ENG] A pattern for cutting forms out of sheets of wood, metal, or other material.
face nailing [ENG] Nailing of facing wood to a base, leaving the nailheads exposed.
face of weld [MET] The exposed surface on the welded side of an arc weld or gas weld.
faceplate [ENG] **1.** A disk fixed perpendicularly to the spindle of a lathe and used for attachment of the workpiece. **2.** A protective plate used to cover holes in machines or other devices. **3.** In scuba or skin diving, a glass or plastic window positioned over the face to provide an air space between the diver's eyes and the water.
face sampling [MIN ENG] Taking random samples of ore and rock from exposed faces of ore and waste.
face shield [ENG] A detachable wraparound guard fitted to a worker's helmet to protect the face from flying particles.
face signal [MIN ENG] A wire stretched along the face and connected to a panel near

the main gate to control the running of a face conveyor.

facet [MATER] The plane surface of a crystal, a cut precious stone, or other fractured surface.

face tile [MATER] Tile with one finished surface, intended for use on a face.

face timbering [MIN ENG] Positioning of safety posts at the working portion of a coal face to support the roof of the mine.

face veneer [MATER] Wood veneer selected for its decorative qualities rather than its strength.

facework [CIV ENG] Ornamental or otherwise special material on the front side or outside of a wall.

face worker [MIN ENG] A miner who works regularly at the face.

facing [CIV ENG] A covering or casting of some material applied to the outer face of embankments, buildings, and other structures. [MECH ENG] Machining the end of a flat rotating surface by applying a tool perpendicular to the axis of rotation in a spiral planar path. [MET] A fine molding sand applied to the face of a mold.

facing-point lock [CIV ENG] A lock used on a railroad track, such as a switch track, which contains a plunger that engages a rod on the switch point to lock the device.

facing wall [CIV ENG] Concrete lining against the earth face of an excavation; used instead of timber sheeting.

factor comparison [IND ENG] A quantitative system of job evaluation in which jobs are given relative positions on a rating scale based on a comparison of factors composing the job with certain previously selected key jobs.

factory [IND ENG] A building or group of buildings where goods are manufactured.

factory lumber [MATER] Softwood lumber graded and used in the factory for the manufacture of such items as doors, sashes, moldings, and so on.

Fagergren cell [MIN ENG] A froth-flotation cell in which a squirrel-cage rotor is driven concentrically in a vertical stator, so that air is drawn down the rotor shaft and dispersed into the pulp.

Fagersta cut [MIN ENG] A cut drilled with handheld equipment in two steps, first as a pilot hole and then as an enlargement of this hole.

Fahrenheit's hydrometer [ENG] A type of hydrometer which carries a pan at its upper end on which weights are placed; the relative density of a liquid is measured by determining the weights necessary to sink the instrument to a fixed mark, first in water and then in the liquid being studied.

failed hole [ENG] A drill hole loaded with dynamite which did not explode. Also known as missed hole.

fail-safe system [ENG] A system designed so that failure of power, control circuits, structural members, or other components will not endanger people operating the system or other people in the vicinity.

failure rate [ENG] The probability of failure per unit of time of items in operation; sometimes estimated as a ratio of the number of failures to the accumulated operating time for the items.

faired cable [DES ENG] A trawling cable covered by streamlined surfaces to reduce hydrodynamic drag.

fairing [AERO ENG] A structure or surface on an aircraft or rocket that functions to reduce drag, such as the streamlined nose of a satellite-launching rocket.

fairlead [AERO ENG] A guide through which an airplane antenna or control cable passes. [MECH ENG] A group of pulleys or rollers used in conjunction with a winch or similar apparatus to permit the cable to be reeled from any direction.

fake set See false set.

Falconbridge process [MET] Recovery of nickel from a nickel-copper matte; the matte is first crushed and roasted to remove sulfur, and the copper is acid-leached, filtered off, and electrolyzed; the residual solids are melted, cast as anodes, and refined electrolytically to produce nickel.

Fales-Stuart windmill [MECH ENG] A windmill developed for farm use from the two-blade airfoil propeller. Also known as Stuart windmill.

Falk flexible coupling [MECH ENG] A spring coupling in which a continuous steel spring is threaded back and forth through axial slots in the periphery of two hubs on the shaft ends.

fall [MECH ENG] The rope or chain of a hoisting tackle. [MIN ENG] A mass of rock, coal, or ore which has fallen from the roof or side in any underground working or gallery.

fallaway section [AERO ENG] A section of a rocket vehicle that is cast off from the vehicle during flight, especially such a section that falls back to the earth.

fall block [MECH ENG] A pulley block that rises and falls with the load on a lifting tackle.

faller [MECH ENG] A machine part whose operation depends on a falling action.

falling-ball viscometer See falling-sphere viscometer.

falling-film cooler [ENG] Liquid cooling system in which the cooling liquid flows down vertical tube exterior surfaces in a thin film, and hot process fluid flows upward through the tubes.

falling-film evaporator [ENG] Liquid evaporator system with heated vertical tubes; liquid to be evaporated flows down the inside tube surfaces as a film, evaporating as it flows.

falling-sphere viscometer [ENG] A viscometer which measures the speed of a spherical body falling with constant velocity in the fluid whose viscosity is to be determined. Also known as falling-ball viscometer.

falloff curve [PETRO ENG] Graphical representation of bottom-hole pressure falloff for a shut-in well as the reservoir drainage area expands.

fall of ground [MIN ENG] The fall of rock from the roof into a mine opening.

fallout shelter [CIV ENG] A structure that affords some protection against fallout radiation and other effects of nuclear explosion; maximum protection is in reinforced concrete shelters below the ground. Also known as radiation shelter.

false attic [BUILD] A section under a roof normally occupied by an attic, but which has no windows and does not enclose rooms.

false bottom [CIV ENG] A temporary bottom installed in a caisson to add to its buoyancy. [MET] An insert put in either member of a die set to increase the strength and improve the life of the die. [MIN ENG] A flat, hexagonal or cylindrical iron die upon which ore is crushed in a stamp mill.

false header [CIV ENG] A half brick used to complete a visible bond; it is not a header.

false set [MATER] Rapid hardening of freshly mixed cement paste, mortar, or concrete with minimum evolution of heat; plasticity can be restored by mixing without addition of water. [MIN ENG] A light, temporary lagging set of timber supporting the side and roof lagging until the drive is advanced sufficiently to allow the heavy permanent set to be put, at which time the false set is taken out and used again in advance of the next permanent set. Also known as fake set.

false stull [MIN ENG] A stull so placed as to offer support or reinforcement for a stull, prop, or other timber.

falsework [CIV ENG] A temporary support used until the main structure is strong enough to support itself.

family mold [ENG] A multicavity injection mold where each cavity forms a component part of the finished product.

fan [MECH ENG] **1.** A device, usually consisting of a rotating paddle wheel or an airscrew, with or without a casing, for producing currents in order to circulate, exhaust, or deliver large volumes of air or gas. **2.** A vane to keep the sails of a windmill facing the direction of the wind.

fan brake [MECH ENG] A fan used to provide a load for a driving mechanism.

fan cut [ENG] A cut in which holes of equal or increasing length are drilled in a pattern on a horizontal plane or in a selected stratum to break out a considerable part of the plane or stratum before the rest of the round is fired.

fan drift [MIN ENG] The short tunnel connecting the upcast shaft with the exhaust fan.

fan-drift doors [MIN ENG] Isolation doors for each drift leading to each fan, when there are two fans at a mine.

fan drilling [ENG] **1.** Drilling boreholes in different vertical and horizontal directions from a single-drill setup. **2.** A radial pattern of drill holes from a setup.

fan efficiency [MECH ENG] The ratio obtained by dividing a fan's useful power output by the power input (the power supplied to the fan shaft); it is expressed as a percentage.

fang bolt [DES ENG] A bolt having a triangular nut with sharp projections at its corners; used to attach metal pieces to wood.

fanjet [AERO ENG] A turbojet engine whose performance has been improved by the addition of a fan which operates in an annular duct surrounding the engine.

fan rating [MECH ENG] The head, quantity, power, and efficiency expected from a fan operating at peak efficiency.

fan ring [DES ENG] Circular metallic collar encircling (but spaced away from) the tips of the fan blade in process equipment, such as air-cooled heat exchangers; ring design is critical to the efficiency of fan performance.

fan shaft [DES ENG] The spindle on which a fan impeller is mounted. [MIN ENG] The ventilating shaft to which a mine fan is connected.

fan shooting [ENG] Seismic exploration in which seismometers are placed in a fan-shaped array to detect anomalies in refracted-wave arrival times indicative of circular rock structures such as salt domes.

fan static pressure [MECH ENG] The total pressure rise diminished by the velocity pressure in the fan outlet.

fan test [MECH ENG] Observations of the quantity, total pressure, and power of air circulated by a fan running at a known constant speed.

fan total head [MECH ENG] The sum of the fan static head and the velocity head at the fan discharge corresponding to a given quantity of airflow.

fan total pressure [MECH ENG] The algebraic difference between the mean total pressure at the fan outlet and the mean total pressure at the fan inlet.

fan truss [CIV ENG] A truss with struts arranged as radiating lines.

fan velocity pressure [MECH ENG] The velocity pressure corresponding to the average velocity at the fan outlet.

far-infrared maser [ENG] A gas maser that generates a beam having a wavelength well above 100 micrometers, and ranging up to the present lower wavelength limit of about 500 micrometers for microwave oscillators.

fascia [BUILD] A wide board fixed vertically on edge to the rafter ends or wall which carries the gutter around the eaves of a roof.

fascine [CIV ENG] A cylindrical bundle of brushwood 1–3 feet (30–90 centimeters) in diameter and 10–20 feet (3–6 meters) long, used as a facing for seawalls on riverbanks, as a foundation mat, as a dam in an estuary, or to protect bridge, dike, and pier foundations from erosion.

fast break [MET] Interruption of the current in the magnetizing coil during nondestructive testing of magnetic particles; induces eddy currents and strong magnetization.

fast coupling [MECH ENG] A flexible geared coupling that uses two interior hubs on the shafts with circumferential gear teeth surrounded by a casing having internal gear teeth to mesh and connect the two hubs.

fast-delay detonation [ENG] The firing of blasts by means of a blasting timer or millisecond delay caps.

fastener [DES ENG] **1.** A device for joining two separate parts of an article or structure. **2.** A device for holding closed a door, gate, or similar structure.

fastening [DES ENG] A spike, bolt, nut, or other device to connect rails to ties.

fast-joint [ENG] Pertaining to a joint with a permanently secured pin.

fast line [PETRO ENG] In a drilling operation, the end of the drilling line that is attached to the drum or reel; it travels with greater velocity than any other part of the drilling line.

fast pin [ENG] A pin that fastens immovably, particularly the pin in a fast joint.

fast powder [MATER] Any explosive having a high-speed detonation.

fast-spiral drill See high-helix drill.

fatal accident [MIN ENG] A coal mine accident in which less than five persons are killed and property damage is slight; excludes ignitions and mine fires.

fat dye [MATER] A type of oil-soluble dye used in the coloring of candles and other wax products.

fatigue allowance [IND ENG] An adjustment to normal time to compensate for production time lost due to exhaustion of the worker.

fatigue factor [IND ENG] The element of physical and mental exhaustion in a time-motion study; the multiplier used to add the fatigue allowance to the normal time.

fatigue notch factor [MET] A notch, scratch, or other impairment on the surface of a metal resulting in premature failure of the metal.

fatigue notch sensitivity [MET] A measure of the reduction of fatigue strength of a metal resulting from a notch.

fatigue test [ENG] Test to determine the range of alternating stress which a material can withstand without breaking.

fat liquoring agents [MATER] Oil-in-water emulsions used to replace oils in tanned (deoiled) leather hides.

fat mortar [MATER] Mortar that adheres to the trowel.

fat oil [MATER] Enriched absorber oil that is drawn off from the absorber column after being saturated by hydrocarbon values stripped from a wet natural-gas stream.

fatty-acid pitch See packing house pitch.

faucet [ENG] A fixture through which water is drawn from a pipe or vessel.

Faugeron kiln [ENG] A coal-fired tunnel kiln for firing feldspathic porcelain; the distinctive feature is the separation of the tunnel into a series of chambers by division walls on the cars and drop arches in the roof.

fault finder [ENG] Test set for locating trouble conditions in communications circuits or systems.

fault tolerance [SYS ENG] The capability of a system to perform its functions in accord with design specifications even in the presence of component failures.

Faust jig [MIN ENG] A plunger-type jig, usually built with multiple compartments; distinguished by synchronized plungers on both sides of the screen plate, withdrawal of refuse through kettle valves in each compartment, and discharge of the hutch periodically by means of hand valves.

faying surface [ENG] The surfaces of materials in contact with each other and joined or about to be joined together.

FD&C color See food color.

feasibility study [SYS ENG] **1.** A study of applicability or desirability of any management or procedural system from the standpoint of advantages versus disadvantages in any given case. **2.** A study to determine the time at which it would be practicable or desirable to install such a system when determined to be advantageous. **3.** A study to determine whether a plan is capable of being accomplished successfully.

feasibility test [SYS ENG] A test conducted to obtain data in support of a feasibility study or to demonstrate feasibility.

feasible ground [MIN ENG] Ground that is easy to work and yet will stand without the support of timber or boards.

feather [MECH ENG] To change the pitch on a propeller.

featheredge [CIV ENG] The thin edge of a gravel-surfaced road. [DES ENG] A wood tool with a level edge used to straighten angles in the finish coat of plaster.

feathering [MECH ENG] A pitch position in a controllable-pitch propeller; it is used in the event of engine failure to stop the windmilling action, and occurs when the blade angle is about 90° to the plane of rotation. Also known as full feathering.

feathering propeller [MECH ENG] A variable-pitch marine or airscrew propeller capable of increasing pitch beyond the normal high pitch value to the feathered position.

feather joint [ENG] A joint made by cutting a mating groove in each of the pieces to be joined and inserting a feather in the opening formed when the pieces are butted together. Also known as ploughed-and-tongued joint.

feed [ENG] **1.** Process or act of supplying material to a processing unit for treatment. **2.** The material supplied to a processing unit for treatment. [MECH ENG] Forward motion imparted to the cutters or drills of cutting or drilling machinery.

feed-control valve [MECH ENG] A small valve, usually a needle valve, on the outlet of the hydraulic-feed cylinder on the swivel head of a diamond drill, used to control minutely the speed of the hydraulic piston travel and hence the rate at which the bit is made to penetrate the rock.

feeder [MECH ENG] **1.** A conveyor adapted to control the rate of delivery of bulk materials, packages, or objects, or a control device which separates or assembles objects. **2.** A device for delivering materials to a processing unit. [MET] A runner or riser so placed that it can feed molten metal to the contracting mass of the casting as it cools in its flask, therefore preventing formation of cavities or porous structure.

feeder-breaker [MECH ENG] A unit that breaks and feeds ore or crushed rock to a materials-handling system at a required rate.

feeder canal [CIV ENG] A canal serving to conduct water to a larger canal.

feeder conveyor [MECH ENG] A short auxiliary conveyor designed to transport materials to another conveyor. Also known as stage loader.

feeder road [CIV ENG] A road that feeds traffic to a more important road.

feeder trough [MIN ENG] The trough connected to the conveyor pan line in a duckbill.

feedhead [MET] A reservoir of molten metal that is left above a casting in order to supply additional metal as the casting solidifies and shrinks. Also known as riser; sinkhead.

feeding rod [MET] A rod used by working up and down to keep the passage clear between riser and casting.

feed lines [MET] The pattern produced on the surface of a piece of metal by machine grinding.

feed nut [MECH ENG] The threaded sleeve fitting around the feed screw on a gear-feed drill swivel head, which is rotated by means of paired gears driven from the spindle or feed shaft.

feed off [ENG] To lower the bit continuously or intermittently during a drilling operation by disengaging the drum brake.

feed pipe [MECH ENG] The pipe which conducts water to a boiler drum.

feed pitch [DES ENG] The distance between the centers of adjacent feed holes in punched paper tape.

feed pressure [MECH ENG] Total weight or pressure, expressed in pounds or tons, applied to the drilling stem to make the drill bit cut and penetrate the geologic, rock, or ore formation.

feed pump [MECH ENG] A pump used to supply water to a steam boiler.

feed rate *See* cutting speed.

feed ratio [MECH ENG] The number of revolutions a drill stem and bit must turn to advance the drill bit 1 inch when the stem is attached to and rotated by a screw- or gear-feed type of drill swivel head with a particular pair of the set of gears engaged. Also known as feed speed.

feed reel [ENG] The reel from which paper tape or magnetic tape is being fed.

feed screw [MECH ENG] The externally threaded drill-rod drive rod in a screw- or gear-feed swivel head on a diamond drill; also used on percussion drills, lathes, and other machinery.

feed shaft [MECH ENG] A short shaft or countershaft in a diamond-drill gear-feed swivel head which is rotated by the drill motor through gears or a fractional drive and by means of which the engaged pair of feed gears is driven.

feed speed *See* feed ratio.

feedstock [ENG] The raw material furnished to a machine or process.

feed travel [MECH ENG] The distance a drilling machine moves the steel shank in traveling from top to bottom of its feeding range.

feed trough [MECH ENG] A receptacle into which feedwater overflows from a boiler drum.

feedwater [MECH ENG] The water supplied to a boiler or still.

feedwater heater [MECH ENG] An apparatus that utilizes steam extracted from an engine or turbine to heat boiler feedwater.

feeler pin [MECH ENG] A pin that allows a duplicating machine to operate only when there is a supply of paper.

Feinc filter [MIN ENG] A vacuum-type drum filter in which a system of parallel strings is used to carry the filter cake away from the drum, instead of the usual filter cloth.

Fell system [CIV ENG] A method of traction intended for steep railroad slopes; a central rail is gripped between horizontal wheels on the locomotive.

felt [MATER] A fibrous, watertight heavy paper of organic or asbestos fibers impregnated with asphalt and used as an overlining or an underlining for roofs. Also known as felt paper.

felt side [MATER] The upper side of a sheet of paper which was not in contact with the wire in a papermaking machine.

female fitting [DES ENG] In a paired pipe or an electrical or mechanical connection, the portion (fitting) that receives, contrasted to the male portion (fitting) that inserts.

fence [AERO ENG] A stationary plate or vane projecting from the upper surface of an airfoil, substantially parallel to the airflow, used to prevent spanwise flow. [ENG] **1.** A line of data-acquisition or tracking stations used to monitor orbiting satellites. **2.** A line of radar or radio stations for detection of satellites or other objects in orbit. **3.** A line or network of early-warning radar stations. **4.** A concentric steel fence erected around a ground radar transmitting antenna to serve as an artificial horizon and suppress ground clutter that would otherwise drown out weak signals returning at a low angle from a target. **5.** An adjustable guide on a tool.

fender [CIV ENG] A timber, cluster of piles, or bag of rope placed along dock or bridge pier to prevent damage by docking ships or floating objects. [ENG] A cover over the upper part of a wheel of an automobile or other vehicle. [MIN ENG] A thin pillar of coal adjacent to the gob, left for protection while driving a lift through the mine pillar.

fennel oil [MATER] The essential oil obtained from fennel; a colorless liquid with aromatic scent and bitter taste, insoluble in water and boiling at 160–220°C; used in medicine, perfumes, and liqueurs. Also known as oil of fennel.

fermentation accelerator [MATER] Substance that speeds chemical fermentation (as for wines) without participating in the resulting chemical changes; can be an enzyme or other catalytic agent.

ferment oil [MATER] A volatile oil formed by the fermentation of plant material in which the oil was not present originally.

fernico [MET] An iron-nickel-cobalt alloy used for metal-to-glass seals.

ferrite [MET] Iron that has not combined with carbon in pig iron or steel.

ferrite banding [MET] The formation of faint bands (flow lines) of free ferrite in rolled steel, running in the direction of working. Also known as ferrite ghosts; ferrite streaks; ghost lines; ghost structure.

ferrite ghosts See ferrite banding.

ferrite number [MET] The standard value assigned to austenitic stainless steel to denote a specific ferrite content; used instead of percent ferrite.

ferrite streaks See ferrite banding.

ferritic stainless steel [MET] Any magnetic iron alloy containing more than 12% chromium having a body-centered cubic structure. Also known as stainless iron.

ferroalloy [MET] Any alloy containing iron, usually in major amount. Also known as ferrous alloy.

ferroaluminum [MET] An alloy of iron and aluminum; added to molten steel as a deoxidizer or as an alloying component.

ferroboron [MET] An alloy of iron and boron that is added to steel to form hardened special steels; two grades are used, 10% boron and 17% boron.

ferrocarbon titanium [MET] An alloy of iron with 15–20% titanium and 3–8% carbon; may be added to molten steel as a component of low-alloy steel.

ferrocerium [MET] An alloy of iron with a high percentage of cerium; used to make cigarette lighter flints.

ferrochromium [MET] A crude ferroalloy containing chromium.

ferrocolumbium [MET] An alloy of iron and columbium (niobium); used to add columbium to certain alloy steels.

ferrofluid [MATER] A colloidal suspension of ultramicroscopic magnetic particles in a carrier liquid, used as a lubricant or damping liquid.

ferrograph analyzer [ENG] An instrument used for ferrography; a pump delivers a small sample of the fluid to a microscope slide mounted above a magnet that generates a high-gradient magnetic field, causing particles to be deposited in a gradient of sizes along the slide.

ferrography [ENG] Wear analysis of machine bearing surfaces by collection of ferrous (or nonferrous) wear particles from lubricating oil in a ferrograph analyzer; the method can be applied to human joints by collecting fragments of cartilage, bone, or prosthetic materials from synovial fluid.

ferromanganese [MET] A ferroalloy containing about 80% manganese and used in steelmaking.

ferrometer [ENG] An instrument used to make permeability and hysteresis tests of iron and steel.

ferromolybdenum [MET] A molybdenum-iron alloy produced in the electric furnace or by a thermite process; used to introduce molybdenum into iron or steel alloys, and as a coating material on welding rods.

ferronickel [MET] A crude ferroalloy containing nickel.

ferrophosphorus [MET] A by-product formed in the heating of iron, phosphate rock, silica, and coke; this alloy is used to increase fluidity in steel casting.

ferroprussiate paper [MATER] A paper used in a blueprint process to reproduce plans and drawings.

ferrosilicon [MET] A crude ferroalloy containing 15–95% silicon and used in steelmaking.

ferrosilicon process *See* Pidgeon process.

ferrotitanium [MET] A ferroalloy containing 15–45% titanium and used in steelmaking.

ferrotungsten [MET] A crude ferroalloy containing tungsten and used in steelmaking.

ferrouranium [MET] An alloy of iron and uranium.

ferrous alloy *See* ferroalloy.

ferrovanadium [MET] An iron alloy high in vanadium (35–55%); used to add 0.1–2.5% vanadium during the manufacture of engineering steels and high-strength steels.

ferrule [DES ENG] **1.** A metal ring or cap attached to the end of a tool handle, post, or other device to strengthen and protect it. **2.** A bushing inserted in the end of a boiler flue to spread and tighten it. [ENG] *See* stabilizer.

fertilizer [MATER] Material that is added to the soil to supply chemical elements needed for plant nutrition.

fiber [MET] **1.** The characteristic of wrought metal that indicates directional properties as revealed by etching or by fracture appearance. **2.** The pattern of preferred orientation of metal crystals after a deformation process, usually wiredrawing.

fiberboard [MATER] A hard isotropic board made by compressing wood chips or other vegetable fibers.

fiber grease [MATER] Solid-base lubricating grease; contains soap fibers 1–1000 micrometers long and 0.1–1.0 micrometer wide, which stabilize lubricating action by immobilizing fluid lubricating components.

fiberizer [MIN ENG] A hammer mill which cracks open asbestos-bearing rock to yield a fibrous product.

fiber metal [MET] Any material composed of metal fibers that are pressed or sintered together or infiltrated with resin or other material.

fiber metallurgy [MET] A branch of metallurgy concerned with the study of metal fibers.

fiber plaster [MATER] Gypsum plaster containing hair or wood fiber as a binder.

fibril [MATER] One of the minute threadlike elements of a natural or synthetic fiber.

fibrous composite [MATER] A composite material consisting of fibers embedded in a matrix.

fibrous plaster [MATER] Gypsum plaster reinforced or backed with sisal or canvas.

fibrous structure [MATER] A ropy surface on a fractured material. [MET] **1.** A lamination on an etched section of a forging. **2.** In wrought iron, a structure consisting of slag fibers embedded in ferrite.

field engineer [ENG] **1.** An engineer who is in charge of directing civil, mechanical, and electrical engineering activities in the production and transmission of petroleum and natural gas. **2.** An engineer who operates at a construction site.

field excitation [MECH ENG] Control of the speed of a series motor in an electric or diesel-electric locomotive by changing the relation between the armature current and the field strength, either through a reduction in field current by shunting the field coils with resistance, or through the use of field taps.

field quenching [MET] The quench cooling and tempering of a heated metal object at the site of construction or operation by using portable equipment rather than fixed manufacturing facilities.

field-strength meter [ENG] A calibrated radio receiver used to measure the field strength of radiated electromagnetic energy from a radio transmitter.

field weld [MET] A weld made at the construction site.

FIFO *See* first-in, first-out.

fighter aircraft [AERO ENG] A military aircraft designed primarily to destroy other aircraft in the air; may also be used to bomb military targets; it is maneuverable and has a high rate of climb.

fighter bomber [AERO ENG] A fighter aircraft that is designed to have bombs, or rockets, added to it so that it may be used as a bomber.

fighter interceptor [AERO ENG] A fighter aircraft designed to intercept and shoot down enemy aircraft.

figure [MATER] The natural grain of wood, especially when it is cut as a veneer.

filament [MET] A long, flexible metal wire drawn very fine.

filament drawing [MET] Reducing the cross section of wire by pulling it through a die to form a filament.

filament winding [ENG] A process for fabricating a composite structure in which continuous fiber reinforcement (glass, boron, silicon carbide), either previously impregnated with a matrix material or impregnated during winding, are wound under tension over a rotating core.

filar micrometer [DES ENG] An instrument used to measure small distances in the field of an eyepiece by using two parallel wires, one of which is fixed while the other is moved at right angles to its length by means of an accurately cut screw. Also known as bifilar micrometer.

file [DES ENG] A steel bar or rod with cutting teeth on its surface; used as a smoothing or forming tool.

file hardness [ENG] Hardness of a material as determined by testing with a file of standardized hardness; a material which cannot be cut with the file is considered as hard as or harder than the file.

filiform corrosion [MET] A random threadlike deterioration of a painted or lacquered metal caused by superficial corrosion of the base metal.

fill [CIV ENG] Earth used for embankments or as backfill. [MIN ENG] *See* pack.

filled composite [MATER] Mixture (composite) of thermoplastic or thermosetting resin and granular or short-strand fiber fill.

filled insulation [MATER] A loose insulating material that is poured or blown into walls.

filled stopes [MIN ENG] Stopes filled with barren stone, low-grade ore, sand, or tailings (mill waste) after the ore has been extracted.

filled-system thermometer [ENG] A thermometer which has a bourdon tube connected by a capillary tube to a hollow bulb; the deformation of the bourdon tube depends on the pressure of a gas (usually nitrogen or helium) or on the volume of a liquid filling the system. Also known as filled thermometer.

filled thermometer *See* filled-system thermometer.

filled thermoplastic [MATER] A thermoplastic resin material that has been extended (filled) with an inert filler powder or fibers before curing.

filled thermoset [MATER] Thermosetting resin material that has been extended (filled) with an inert filler powder or fibers before curing.

filler [MATER] **1.** An inert material added to paper, resin, bituminous material, and other substances to modify their properties and improve quality. **2.** A material used to fill holes in wood, plaster, or other surfaces before applying a coating such as paint or varnish. [MET] The rod used to deposit metal in a joint in brazing, soldering, or welding.

filler specks [MATER] In a cast plastic object, visible specks of a filler such as wood flour or asbestos that stand out in color contrast against the surface of the object.

fillet [BUILD] A flat molding that separates rounded or angular moldings. [DES ENG] A concave transition surface between two otherwise intersecting surfaces. [ENG] **1.** Any narrow, flat metal or wood member. **2.** A corner piece at the juncture of perpendicular surfaces to lessen the danger of cracks, as in core boxes for castings.

fillet gage [DES ENG] A gage for measuring convex or concave surfaces.

fillet weld [MET] A weld joining two edges at right angles; cross-sectional configuration is approximately triangular.

fill factor [MECH ENG] The approximate load that the dipper of a shovel is carrying, expressed as a percentage of the rated capacity.

filling [ENG] The loading of trucks with any material. [MIN ENG] Allowing a mine to fill with water.

fill-up work *See* internal work.

film [MATER] A flat section of a thermoplastic resin, a regenerated cellulose derivative, or other material that is extremely thin in comparison to its length and breadth and has a nominal maximum thickness of 0.25 millimeter. [MET] Oxide coating on a metal.

filmogen [MATER] The film-forming material or binder in paint which imparts continuity.

film platen [ENG] A device which holds film in the focal plane during exposure.

film sizing table [MIN ENG] A table used in ore dressing for sorting fine material by means of a film of flowing water.

film strength [MATER] **1.** The measurement of a lubricant's ability to keep an unbroken film over surfaces. **2.** The resistance to disruption by films of all types, such as plastic films or surface-coating films.

film transport [MECH ENG] **1.** The mechanism for moving photographic film through the region where light strikes it in recording film tracks or sound tracks of motion pictures. **2.** The mechanism which moves the film print past the area where light passes through it in reproduction of picture and sound.

film vault [ENG] A place for safekeeping of film.

filter [ENG] A porous article or material for separating suspended particulate matter from liquids by passing the liquid through the pores in the filter and sieving out the solids. [ENG ACOUS] A device employed to reject sound in a particular range of frequencies while passing sound in another range of frequencies. Also known as acoustic filter.

filterability [ENG] The adaptability of a liquid-solid system to filtration; system is not filterable if it is too viscous to be forced through a filter medium, or if the solids are too small to be stopped by the filter medium.

filter aid [MATER] An inert powder or granules such as diatomaceous earth, fly ash, or sand added to a solution that is to be filtered in order to form a porous bed on the filter and increase the rate and improve the quality of filtration.

filter bauxite [MATER] Crushed, screened, and calcined bauxite, in particles that range from 20–60 and from 30–60 mesh grades; used in ore refineries for filtering. Also known as activated bauxite.

filter bed [CIV ENG] A fill of pervious soil that provides a site for a septic field. [ENG] A contact bed used for filtering purposes.

filter cake [MATER] A concentrated solid or semisolid material that is separated from a liquid and remains on the filter after pressure filtration.

filter cloth [MATER] A fabric used as a medium for filtration.

filtered particle testing [ENG] A penetrant method of nondestructive testing by which cracks in porous objects (100 mesh or smaller) are indicated: a fluid containing suspended particles is sprayed on a test object; if a crack exists, particles are filtered out and concentrate at the surface as liquid flows into the crack.

filtered stock [MATER] Lubricating oil which has been filtered or refiltered to improve performance characteristics.

filtering [ENG] The process of interpreting reported information on movements of aircraft, ships, and submarines in order to determine their probable true tracks and, where applicable, heights or depths.

filter medium [MATER] That portion of a filtration system that provides the liquid-solid separation, such as close-woven textiles or metal screens, papers, nonwoven fabrics, granular beds, or porous media.

filter paper [MATER] Porous cellulose paper used for filtering, especially for quantitative purposes.

filter photometer [ENG] A colorimeter in which the length of light is selected by the use of appropriate glass filters.

filter press [ENG] A metal frame on which iron plates are suspended and pressed together by a screw device; liquid to be filtered is pumped into canvas bags between the plates, and the screw is tightened so that pressure is furnished for filtration.

filter pump [MECH ENG] An aspirator or vacuum pump which creates a negative pressure on the filtrate side of the filter to hasten the process of filtering.

filter sand [MATER] Graded sand used for filtering suspended matter from a flowing liquid stream.

filter screen [ENG] A fine-pored medium through which a liquid will pass and on which solids deposit; the medium may be a metal sieve screen or a woven fabric of metal or of natural or synthetic fibers.

filter thickener [ENG] Device that thickens a liquid-solid mixture by removing a portion of the liquid by filtration, rather than by settling.

filter-type respirator [ENG] A protective device which removes dispersoids from the air by physically trapping the particles on the fibrous material of the filter.

fin [AERO ENG] A fixed or adjustable vane or airfoil affixed longitudinally to an aerodynamically or ballistically designed body for stabilizing purposes. [DES ENG] A projecting flat plate or structure, as a cooling fin. [ENG] Material which remains in the holes of a molded part and which must be removed.

final filter *See* afterfilter.

final mass [AERO ENG] The mass of a rocket after its propellants are consumed.

find [IND ENG] The therblig representing the mental reaction which occurs on recognizing an object at the end of the elemental motion search; now seldom used.

fin efficiency [ENG] In extended-surface heat-exchange equations, the ratio of the mean temperature difference from surface-to-fluid divided by the temperature difference from fin-to-fluid at the base or root of the fin.

fine gold [MET] Almost pure gold; the value of bullion gold depends on its percentage of fineness. [MIN ENG] In placer mining, gold in exceedingly small particles.

fine grinding [MECH ENG] Grinding performed in a mill rotating on a horizontal axis in which the material undergoes final size reduction, to -100 mesh.

fineness [MET] Degree of purity of gold or silver in parts per thousand.

fineness modulus [ENG] A number denoting the fineness of a fine aggregate or other fine material such as sand or paint.

fineness ratio [AERO ENG] The ratio of the length of a streamlined body, as that of a fuselage or airship hull, to its maximum diameter.

fines [MATER] **1.** Particles smaller than average in a mixture of particles varying in size. **2.** Fine material which passes through a standard screen on which coarser fragments are retained. [MET] That portion of a metal powder consisting of particles smaller than a specified size.

fine silver [MET] Silver having a minimum fineness of 999; considered to be pure silver.

finger [PETRO ENG] A pair or set of bracketlike projections placed at a strategic point in a drill tripod or derrick to keep a number of lengths of drill rods or casing in place when they are standing in the tripod or derrick.

finger bit [DES ENG] A steel rock-cutting bit having fingerlike, fixed or replaceable steel-cutting points.

finger board [PETRO ENG] A board with projecting dowels or pipe fingers located in the upper part of the drill derrick or tripod to support stands of drill rod, drill pipe, or casing.

finger chute [MIN ENG] Steel rails hinged independently over an ore chute, to control rate of flow of rock.

finger raise [MIN ENG] Steeply sloping openings permitting caved ore to flow down raises through grizzlies to chutes on the haulage level.

finish [MATER] **1.** A chemical or other material applied to surfaces to protect them, to alter their appearances, or to modify their physical properties; finishes can be physically, chemically, or electrolytically applied and have value for fabrics and fibers, metals, paper products, plastics, woods, and so on. **2.** The ultimate quality, condition, or appearance of the surface of a material.

finished goods [IND ENG] Manufactured products in inventory ready for packaging, shipment, or sale.

finished steel [MET] Steel that has undergone final processing and is ready for market.

finisher [CIV ENG] A construction machine used to smooth the freshly placed surface of a roadway, or to prepare the foundation for a pavement.

finish grinding [MECH ENG] The last action of a grinding operation to achieve a good finish and accurate dimensions.

finishing compound [MATER] A substance used to impart surface properties to textiles or leather, such as softness, flexibility, or fire resistance.

finishing hardware [BUILD] Items, such as hinges, door pulls, and strike plates, made in attractive shapes and finishes, and usually visible on the completed structure.

finishing hydrated lime [MATER] Any hydrated lime suitable for use in the finishing coat of plaster; characterized by a high degree of whiteness and plasticity.

finishing mill [MET] A rolling mill in which sheet, plate, and other mill products are subjected to final rolling operations.

finishing nail [DES ENG] A wire nail with a small head that can easily be concealed.

finishing roll [MIN ENG] The last roll, or the one that does the finest crushing in ore dressing.

finishing temperature [MET] The temperature at which hot-working is completed.

finish plate [DES ENG] A plate which covers and protects the cylinder setscrews; it is fastened to the underplate and forms part of the armored front for a mortise lock.

finish turning [MECH ENG] The operation of machining a surface to accurate size and producing a smooth finish.

finite-element method [ENG] An approximation method for studying continuous physical systems, used in structural mechanics, electrical field theory, and fluid mechanics; the system is broken into discrete elements interconnected at discrete node points.

Fink truss [CIV ENG] A symmetrical steel roof truss suitable for spans up to 50 feet (15 meters).

finned surface [MECH ENG] A tubular heat-exchange surface with extended projections on one side.

Fior process [MET] The prereduction of high-grade iron particles or concentrates in a hot gaseous reactor to produce low-oxygen fines for partially metallized briquettes suitable for electric-arc steel-making furnaces.

fire [ENG] To blast with gunpowder or other explosives. [MIN ENG] A warning that a shot is being fired.

fire assay [MET] Analysis of a metal-bearing material, especially gold and silver, by assaying a sample with a suitable flux and measuring the content of resulting metal leads by weighing or atomic absorption techniques.

fire boss [MIN ENG] An individual who examines a mine for gas and other dangers. Also known as mine examiner.

firebox [MECH ENG] The furnace of a locomotive or similar type of fire-tube boiler.

firebreak [MIN ENG] A strip across an area in which either no combustible material is employed or in which, if timber supports are used, sand is filled and packed tightly around them.

firebrick [MATER] A refractory brick, often made of fireclay, that is able to withstand high temperature (up to 1500–1600°C) without fusion; used to line furnaces, fireplaces, and chimneys.

fire bridge [ENG] A low wall separating the hearth and the grate in a reverberatory furnace.

fire crack [ENG] A crack resulting from thermal stress which propagates on the heated side of a shell or header in a boiler or a heat transfer surface.

firecracker [ENG] A cylindrically shaped item containing an explosive and a fuse; used to simulate the noise of an explosive charge.

fire cut [BUILD] An angular cut made at the end of a joist which will rest on a brick wall.

firedamp [MIN ENG] **1.** A gas formed in mines by decomposition of coal or other carbonaceous matter; consists chiefly of methane and is combustible. **2.** An airtight stopping to isolate an underground fire and to prevent the inflow of fresh air and the outflow of foul air. Also known as fire wall.

firedamp alarm [MIN ENG] An instrument which gives a warning signal when the methane content in the mine atmosphere exceeds a known value.

firedamp detector [MIN ENG] A portable device to detect the presence and determine the percentage of firedamp in mine air.

firedamp drainage [MIN ENG] The collection of firedamp from coal strata, generally into pipes, with or without the use of suction. Also known as methane drainage.

firedamp drainage drill [MIN ENG] A heavy, compressed-air-operated, percussive, rotary or rotary-percussive drilling machine for putting up the boreholes in firedamp drainage.

firedamp explosion [MIN ENG] An explosion of a mixture of firedamp and air.

firedamp fringe [MIN ENG] The zone of contact between the coal gases and the ventilation air current at the face of the mine.

firedamp layer [MIN ENG] An accumulation of firedamp under the roof of a mine roadway where the ventilation is insufficient to dilute and remove the gas.

firedamp migration [MIN ENG] The movement of firedamp through the strata or coal of a mine.

firedamp pressure-chamber method [MIN ENG] A method of firedamp drainage in coal mines; pressure chambers built at the intake and the return of a worked-out area are used to trap firedamp, which is drawn off in pipes.

firedamp probe [MIN ENG] A flexible rubber tube connected to a rod, which can be thrust into roof cavities and breaks so that a sample of the air may be transferred to a methanometer and its firedamp content determined.

fire-danger meter [ENG] A graphical aid used in fire-weather forecasting to calculate the degree of forest-fire danger (or burning index); commonly in the form of a circular slide rule, it relates numerical indices of the seasonal stage of foliage, cumulative effect of past precipitation or lack thereof (buildup index), the measured fuel moisture, and the speed of the wind in the woods; the fuel moisture is determined by weighing a special type of wooden stick that has been exposed in the woods, its weight being proportional to its contained water; the calculated burning index falls on a scale of 1 to 100: 1 to 11 is no fire danger; 12 to 35 medium danger; 40 to 100 high danger.

fire detector [ENG] A temperature-sensing device designed to sound an alarm, to turn on a sprinkler system, or to activate some other fire preventive measure at the first signs of fire.

fire door [ENG] **1.** The door or opening through which fuel is supplied to a furnace or stove. **2.** A door that can be closed to prevent the spreading of fire, as through a building or mine.

fired process equipment [ENG] Heaters, furnaces, reactors, incinerators, vaporizers, steam generators, boilers, and other process equipment for which the heat input is derived from fuel combustion (flames); can be direct-fired (flame in contact with the process stream) or indirect-fired (flame separated from the process fluid by a metallic wall).

fire escape [BUILD] An outside stairway usually made of steel and used to escape from a building in case of fire.

fire extinguisher [ENG] Any of various portable devices used to extinguish a fire by the ejection of a fire-inhibiting substance, such as water, carbon dioxide, gas, or chemical foam.

firefinder [ENG] An instrument consisting of a map and a sighting device; used in fire towers to locate forest fires.

fire flooding [PETRO ENG] A method to improve secondary recovery in an oil reservoir; a combustion process is started in the reservoir at an injection well by continued introduction of gas containing oxygen or other material to support combustion, and the combustion wave is driven through the reservoir toward the production well.

fire foam [MATER] A colloidal solution of small gas bubbles produced by chemical reaction or mechanical agitation and used to extinguish hydrocarbon fire.

fire hose [ENG] A collapsible, flameproof hose that can be attached to a hydrant, standpipe, or similar outlet to supply water to extinguish a fire.

fire hydrant [CIV ENG] An outlet from a water main provided inside buildings or outdoors to which fire hoses can be connected. Also known as fire plug; hydrant.

fire load [CIV ENG] The load of combustible material per square foot of floor space.

fire partition [BUILD] A wall inside a building intended to retard fire.

fire plug See fire hydrant.

fireproof [BUILD] Having noncombustible walls, stairways, and stress-bearing members, and having all steel and iron structural members which could be damaged by heat protected by refractory materials. [MATER] The property of being relatively resistant to combustion.

fireproofing compound See fire retardant.

fire protection [CIV ENG] Measures for reducing injury and property loss by fire.

fire pump [MECH ENG] A pump for fire protection purposes usually driven by an independent, reliable prime mover and approved by the National Board of Fire Underwriters.

fire refining [MET] The refining of blister copper by treatment in a furnace under oxidizing conditions to remove the impurities and under reducing conditions to remove the excess oxygen.

fire-resistant [CIV ENG] Of a structural element, able to resist combustion for a specified time under conditions of standard heat intensity without burning or failing structurally.

fire retardant [MATER] A chemical used as a coating for or a component of a combustible material to reduce or eliminate a tendency to burn; used with textiles, plastics, rubbers, paints, and other materials. Also known as fireproofing compound.

fire retardant paint [MATER] A paint applied as a thin coating to reduce the rate of flame spread of a combustible material; based on silicone, casein, polyvinylchloride, or other substance.

fireroom [MECH ENG] That portion of a fossil fuel–burning plant which contains the furnace and associated equipment.

fire scale [MET] Copper oxide remaining below the surface of silver-copper alloys after annealing and pickling.

fire sprinkling system See sprinkler system.

fire standpipe [CIV ENG] A high, vertical pipe or tank that holds water to assure a positive, relatively uniform pressure, particularly to provide fire protection to upper floors of tall buildings.

fire stop [BUILD] An incombustible, horizontal or vertical barrier, as of brick across a hollow wall or across an open room, to stop the spread of fire. Also known as draught stop.

fire tower [BUILD] A fireproof and smokeproof stairway compartment running the height of a building.

fire-tube boiler [MECH ENG] A steam boiler in which hot gaseous products of combustion pass through tubes surrounded by boiler water.

fire wall [CIV ENG] **1.** A fire-resisting wall separating two parts of a building from the lowest floor to several feet above the roof to prevent the spread of fire. **2.** A fire-resisting wall surrounding an oil storage tank to retain oil that may escape and to confine fire. [MIN ENG] See firedamp.

fire welding See forge welding.

firing [ENG] **1.** The act or process of adding fuel and air to a furnace. **2.** Igniting an explosive mixture. **3.** Treating a ceramic product with heat.

firing chamber See combustion chamber.

firing machine [ENG] An electric blasting machine. [MECH ENG] A mechanical stoker used to feed coal to a boiler furnace.

firing mechanism [ENG] A mechanism for firing a primer; the primer may be for initiating the propelling charge, in which case the firing mechanism forms a part of the weapon; if the primer is for the purpose of initiating detonation of the main charge, the firing mechanism is a part of the ammunition item and performs the function of a fuse.

firing pressure [MECH ENG] The highest pressure in an engine cylinder during combustion.

firing rate [MECH ENG] The rate at which fuel feed to a burner occurs, in terms of volume, heat units, or weight per unit time.

firmer chisel [DES ENG] A small hand chisel with a flat blade; used in woodworking.

firm-joint caliper [DES ENG] An outside or inside caliper whose legs are jointed together at the top with a nut and which must be opened and closed by hand pressure.

fir needle oil [MATER] An essential oil distilled from the needles and twigs of some trees of the genus *Abies;* used in perfumery, flavoring, and medicine. Also known as fir oil.

fir oil *See* fir needle oil.

first arrival [ENG] In exploration refraction seismology, the first seismic event recorded on a seismogram; it is noteworthy in that only first arrivals are considered in this usage.

first fire [ENG] The igniter used with pyrotechnic devices, consisting of first fire composition, loaded in direct contact with the main pyrotechnic charge; the ignition of the igniter or first fire is generally accomplished by fuse action.

first fire composition [MATER] A pyrotechnic composition (readily ignitable and easily pressed into a strong, solid mass), compounded to produce a high temperature, preferably with creation of slag to give heat capacity.

first-in, first-out [IND ENG] An inventory cost evaluation method which transfers costs of material to the product in chronological order. Abbreviated FIFO.

first-order leveling [ENG] Spirit leveling of high precision and accuracy in which lines are run first forward to the objective point and then backward to the starting point.

firth *See* estuary.

fished joint [CIV ENG] A structural joint made with fish plates.

fisheye [MATER] A small globular mass which has not blended completely into the surrounding material and is particularly evident in a transparent or translucent material, such as a plastic coating or surface coating. [MET] *See* flake.

fish gelatin *See* isinglass.

fish glue [MATER] An adhesive obtained from the skin of certain fish, principally cod; used in gummed tape, letterpress printing plates, and blueprint paper.

fishing [ENG] In drilling, the operation by which lost or damaged tools are secured and brought to the surface from the bottom of a well or drill hole.

fishing space [CIV ENG] The space between base and head of a rail in which a joint bar is placed.

fishing tool [ENG] A device for retrieving objects from inaccessible locations.

fish ladder [CIV ENG] A type of fishway that carries water around a dam through a series of stepped baffles or boxes and thus facilitates the migration of fish.

fish lead [ENG] A type of sounding lead used without removal from the water between soundings.

fish liver oil [MATER] An oil extracted from certain fish livers and containing vita-min A; high-potency livers are obtained from cod, shark, and halibut; used in medicine and as a dietary supplement.

fishmouthing *See* alligatoring.

fish oil [MATER] Oil obtained from fish such as menhaden, pilchard, herring, and sardine; used as a drying oil in paint and as a raw material for detergents, resins, margarine, and so on.

fish paper [MATER] A type of fiber used in sheet form for insulating purposes where high mechanical strength is required, as in insulating transformer windings from the transformer core.

fish plate [CIV ENG] One of a pair of steel plates bolted to the sides of a rail or beam joint, to secure the joint.

fish screen [CIV ENG] **1.** A screen set across a water intake canal or pipe to prevent fish from entering. **2.** Any similar barrier to prevent fish from entering or leaving a pond.

fishtail [MET] Excess metal trailing on the end of a roll forging.

fishtail bit [DES ENG] A drilling bit shaped like the tail of a fish.

fishtail burner [ENG] A burner in which two jets of gas impinge on each other to form a flame shaped like a fish's tail.

fissure [MET] A small cracklike discontinuity with a slight opening or displacement of the fracture surfaces.

fit [DES ENG] The dimensional relationship between mating parts, such as press, shrink, or sliding fit.

fitter [ENG] One who maintains, repairs, and assembles machines in an engineering shop.

fitting [ENG] A small auxiliary part of standard dimensions used in the assembly of an engine, piping system, machine, or other apparatus.

five-spot well pattern [PETRO ENG] A symmetrical network pattern of five wells (one in center, four equally spaced in a square pattern) as used in water-injection pressure maintenance of reservoirs.

fixative [MATER] **1.** A chemical or a mixture of chemicals used to treat biological specimens before preservation so as to retain a reasonable facsimile of their appearance when alive. **2.** A substance used to increase the durability of another substance; used to fix dye mordants, hold textile dyes and pigments, and slow the rate of perfume evaporation. Also known as fixing agent.

fixed arch [CIV ENG] A stiff arch having rotation prevented at its supports.

fixed bridge [CIV ENG] A bridge having permanent horizontal or vertical alignment.

fixed-charge problem [IND ENG] A linear programming problem in which each variable has a fixed-charge coefficient in addition to the usual cost coefficient; the fixed charge (for example, a setup time charge) is a nonlinear function and is incurred only when the variable appears in the solution with a positive level.

fixed cost [IND ENG] A cost that remains unchanged during short-term changes in production level. Also known as overhead cost.

fixed-electrode method [ENG] A geophysical surveying method used in a self-potential system of prospecting in which one electrode remains stationary while the other is grounded at progressively greater distances from it.

fixed-end beam [CIV ENG] A beam that is supported at both free ends and is restrained against rotation and vertical movement. Also known as encastré beam.

fixed-end column [CIV ENG] A column with the end fixed so that it cannot rotate.

fixed-feed grinding [MECH ENG] Feeding processed material to a grinding wheel, or vice versa, in predetermined increments or at a given rate.

fixed fin [AERO ENG] A nonadjustable vane or airfoil affixed longitudinally to an aerodynamically or ballistically designed body for stabilizing purposes.

fixed mooring berth [CIV ENG] A marine structure consisting of dolphins for securing a ship and a platform to support cargo-handling equipment.

fixed-needle traverse [ENG] In surveying, a traverse with a compass fitted with a sight line which can be moved above a graduated horizontal circle, so that the azimuth angle can be read, as with a theodolite.

fixed oil [MATER] A nonvolatile fatty oil of vegetable origin.

fixed point [ENG] A reproducible value, as for temperature, used to standardize measurements; derived from intrinsic properties of pure substances.

fixed-position welding [MET] A welding operation in which the work is stationary.

fixed rent See minimum rent.

fixed screen [MIN ENG] A stationary panel, commonly of wedge wire, used to remove a large proportion of water and fines from a suspension of coal in water.

fixed sonar [ENG] Sonar in which the receiving transducer is not constantly rotated, in contrast to scanning sonar.

fixing agent See fixative.

fixity See continuity.

fixture [CIV ENG] An object permanently attached to a structure, such as a light or sink. [MECH ENG] A device used to hold and position a piece of work without guiding the cutting tool.

Flade potential [MET] The potential of a passive metal immediately preceding a final steep fall from the passive to the active region.

flag alarm [ENG] A semaphore-type flag in the indicator of an instrument to serve as a signal, usually to warn that the indications are unreliable.

flag float [ENG] A pyrotechnic device that floats and burns upon the water, used for marking or signaling.

flagman [CIV ENG] A range-pole carrier in a surveying party.

flagpole [ENG] A single staff or pole rising from the ground and on which flags or other signals are displayed; on charts the term is used only when the pole is not attached to a building.

flagstaff [ENG] A pole or staff on which flags or other signals are displayed; on charts this term is used only when the pole is attached to a building.

flair [CIV ENG] A gradual widening of the flangeway near the end of a guard line of a track or rail structure.

flake [MATER] 1. Dry, unplasticized, cellulosic plastics base. 2. Plastic chip used as feed in molding operations. [MET] 1. Discontinuous, internal cracks formed in steel during cooling due usually to the release of hydrogen. Also known as fisheye; snowflake; shattercrack. 2. Fishscale, flat particles in powder metallurgy. Also known as flake powder.

flake powder See flake.

flaking [ENG] 1. Reducing or separating into flakes. 2. See frosting. [MIN ENG] Breaking small chips from the face of a refractory, particularly chrome ore containing refractories.

flaking mill [MECH ENG] A machine for converting material to flakes.

flak jacket [ENG] A jacket or vest of heavy fabric containing metal, nylon, or ceramic plates, designed especially for protection against flak; usually covers the chest, abdomen, back, and genitals, leaving the arms and legs free. Also known as flak vest.

flak vest See flak jacket.

flame annealing [MET] The careful heating of a metal part by flames, before or after working.

flame arrester [ENG] An assembly of screens, perforated plates, or metal-gauze packing attached to the breather vent on a flammable-product storage tank.

flame bucket [AERO ENG] A deep, cavelike construction built beneath a rocket launchpad, open at the top to receive the hot gases of the rocket, and open on one or three sides below, with a thick metal fourth side bent toward the open side or sides so as to deflect the exhaust gases.

flame cleaning [MET] Removing scale, rust, and dirt from metal surfaces by using a broad flame.

flame coating *See* flame plating.

flame collector [ENG] A device used in atmospheric electrical measurements for the removal of induction charge on apparatus; based upon the principle that products of combustion are ionized and will consequently conduct electricity from charged bodies.

flame cutting [MET] Use of an oxyacetylene, oxyhydrogen, or oxycoal gas flame to cut thick metal sections.

flame deflector [AERO ENG] **1.** In a vertical launch, any of variously designed obstructions that intercept the hot gases of the rocket engine so as to deflect them away from the ground or from a structure. **2.** In a captive test, an elbow in the exhaust conduit or flame bucket that deflects the flame into the open.

flame detector [MECH ENG] A sensing device which indicates whether or not a fuel is burning, or if ignition has been lost, by transmitting a signal to a control system.

flame gouging [MET] A form of oxygen cutting by means of a cutting torch with a slightly curved tip, enabling the flame to strike the metal surface at a low angle and making shallow cuts possible. Also known as oxygen gouging.

flame hardening [MET] A method for local surface hardening of steel by passing an oxyacetylene or similar flame over the work at a predetermined rate.

flameholder [AERO ENG] A device that sustains combustion in a flowing mixture within the combustion chamber of some types of jet engines.

flameout [AERO ENG] The extinguishing of the flame in a reaction engine, especially in a jet engine.

flame plate [ENG] One of the plates on a boiler firebox which are subjected to the maximum furnace temperature.

flame plating [MET] Coating a thin layer of refractory material on a surface by exploding a mixture of plating powder, oxygen, and acetylene. Also known as flame coating.

flame-retarded resin [MATER] A resin which is compounded with certain chemicals to reduce or eliminate its tendency to burn.

flame spraying [ENG] **1.** A method of applying a plastic coating onto a surface in which finely powdered fragments of the plastic, together with suitable fluxes, are projected through a cone of flame. **2.** Deposition of a conductor on a board in molten form, generally through a metal mask or stencil, by means of a spray gun that feeds wire into a gas flame and drives the molten particles against the work.

flame straightening [MET] Correcting distorted structural metal to a straight form by local application of a gas-flame heat.

flamethrower [ENG] A device used to project ignited fuel from a nozzle so as to cause casualties to personnel or to destroy material such as weeds or insects.

flame trap [ENG] A device that prevents a gas flame from entering the supply pipe.

flame treating [ENG] A method of rendering inert thermoplastic objects receptive to inks, lacquers, paints, or adhesives, in which the object is bathed in an open flame to promote oxidation of the surface.

flammable [MATER] Of a material, capable of supporting combustion.

flammable liquid [MATER] A liquid which gives off combustible vapors.

flanged pipe [DES ENG] A pipe with flanges at the ends; can be bolted end to end to another pipe.

flangeway [CIV ENG] Open way through a rail or track structure that provides a passageway for the flange of a wheel.

flank [CIV ENG] The outer edge of a carriageway. [DES ENG] **1.** The end surface of a cutting tool, adjacent to the cutting edge. **2.** The side of a screw thread.

flank angle [DES ENG] The angle made by the flank of a screw thread with a line perpendicular to the axis of the screw.

flank hole [MIN ENG] **1.** A hole bored in advance of a working place when approaching old workings. **2.** A borehole driven from the side of an underground excavation, not parallel with the center line of the excavation, to detect water, gas, or other danger.

flank wear [ENG] Loss of relief on the flank of a tool behind the cutting edge.

flap [AERO ENG] **1.** Any control surface, such as a speed brake, dive brake, or dive-recovery brake, used primarily to increase the lift or drag on an airplane, or to aid in recovery from a dive. **2.** Any rudder attached to a rocket and acting either in the air or within the jet stream.

flaperon [AERO ENG] A control surface used both as a flap and as an aileron.

flap gate [CIV ENG] A gate that opens or closes by rotation around hinges at the top of the gate. Also known as pivot leaf gate.

flap hinge *See* backflap hinge.

flapping [MET] Striking through the surface of molten copper to hasten oxidation by increasing the exposure to air.

flap valve [MECH ENG] A valve fitted with a hinged flap or disk that swings in one direction only.

flare [AERO ENG] To descend in a smooth curve, making a transition from a relatively steep descent to a direction substantially parallel to the surface, when landing an aircraft. [DES ENG] An expansion at the end of a cylindrical body, as at the base of a rocket. [ENG] A pyrotechnic item designed to produce a single source of intense light for such purposes as target or airfield illumination.

flare chute [ENG] A flare attached to a parachute.

flare factor [ENG ACOUS] Number expressing the degree of outward curvature of the horn of a loudspeaker.

flareout [AERO ENG] That portion of the approach path of an aircraft in which the vertical component is modified to lessen the impact of landing.

flare-type bucket [MIN ENG] A dragline bucket that has flared sides to allow heaped loading.

flare-type burner [ENG] A circular burner which discharges flame in the form of a cone.

flaring [MET] Increasing the diameter at the end of a pipe or tube.

flash [ENG] In plastics or rubber molding or in metal casting, that portion of the charge which overflows from the mold cavity at the joint line. [MET] A fin of excess metal along the mold joint line of a casting, occurring between mating die faces of a forging or expelled from a joint in resistance welding.

flashback *See* backfire.

flashback arrester [ENG] A device which prevents a flashback from passing the point where the arrester is installed in a torch, thereby preventing damage.

flashboard [CIV ENG] A relatively low, temporary barrier constructed of a series of boards along the top of a dam spillway to increase storage capacity.

flash boiler [MECH ENG] A boiler with hot tubes of small capacity; designed to immediately convert small amounts of water to superheated steam.

flash bomb [ENG] A bomb that illuminates the ground for night aerial photography.

flash butt welding [MET] Resistance welding to produce a butt joint by passing an electric current through two pieces of metal in light contact to create an arc which causes flashing and consequent heating; the weld is completed by applying pressure at the joint.

flash coat [MET] A thin coating that is forced from a flash-welded joint after the abutting surfaces are forced together.

flash depressor [MATER] A substance used to reduce the flash from a rocket motor.

flash groove [ENG] **1.** A groove in a casting die so that excess material can escape during casting. **2.** *See* cutoff.

flashing [BUILD] A strip of sheet metal placed at the junction of exterior building sur-

faces to render the joint watertight. [ENG] Burning brick in an intermittent air supply in order to impart irregular color to the bricks. [MET] The violent expulsion of small metal particles due to arcing during flash butt welding.

flashing ring [ENG] A ring around a pipe that holds it in place as it passes through a partition such as a floor or wall.

flash line [ENG] A raised line on the surface of a molding where the mold faces joined.

flash mold [ENG] A mold which permits excess material to escape during closing.

flash plating [MET] Electrodeposition of a thin film of metal.

flash ridge [ENG] The part of a flash mold along which the excess material escapes before the mold is closed.

flash roast [MIN ENG] Rapid removal of sulfur from ore by having finely divided sulfide mineral fall through a heated oxidizing atmosphere. Also known as suspension roast.

flash set [MATER] Rapid hardening of freshly mixed cement paste, mortar, or concrete with considerable evolution of heat; plasticity cannot be restored.

flash smelting [MET] Production of molten metal or matte in a vertical furnace in which concentrates are reacted with hot gases; the molten product is collected in a horizontal refractory-lined accumulator at the base of the furnace.

flash vaporization [ENG] A method used for withdrawing liquefied petroleum gas from storage in which liquid is first flashed into a vapor in an intermediate pressure system, and then a second stage regulator provides the low pressure required to use the gas in appliances.

flash welding [MET] A form of resistance butt welding used to weld wide, thin members or members with irregular faces, and tubing to tubing.

flask [MET] A frame used to hold molding sand in foundry work.

flat [ENG] A nonglossy painted surface.

flat-back stope [MIN ENG] An overhand stoping method in which the ore is broken in slices parallel with the levels.

flatbed plotter [ENG] A graphics output device that draws by moving a pen in both horizontal and vertical directions over a sheet of paper; the overall size of the drawing is limited by the height and width of this bed.

flatbed truck [ENG] A truck whose body is in the form of a platform.

flat belt [DES ENG] A power transmission belt, in the form of leather belting, used where high-speed motion rather than power is the main concern.

flat-belt conveyor [MECH ENG] A conveyor belt in which the carrying run is supported by flat-belt idlers or pulleys.

flat-belt pulley [DES ENG] A smooth, flat-faced pulley made of cast iron, fabricated steel, wood, and paper and used with a flat-belt drive.

flat-blade turbine [MECH ENG] An impeller with flat blades attached to the margin.

flat-bottom crown *See* flat-face bit.

flatcar [ENG] A railroad car without fixed walls or a cover.

flat chisel [DES ENG] A steel chisel used to obtain a flat and finished surface.

flat crank [DES ENG] A crankshaft having one flat bearing journal.

flat-crested weir [CIV ENG] A type of measuring weir whose crest is in the horizontal plane and whose length is great compared with the height of water passing over it.

flat cut [MIN ENG] A manner of placing the boreholes, for the first shot in a tunnel, in which they are started about 2 or 3 feet (60–90 centimeters) above the floor and pointed downward so that the bottom of the hole will be about level with the floor.

flat die forging [MET] Die forging in which the metal is worked between simple contour dies.

flat drill [DES ENG] A type of rotary drill constructed from a flat piece of material.

flat edge trimmer [MECH ENG] A machine designed to trim the notched edges of metal shells.

flat-face bit [DES ENG] A diamond core bit whose face in cross section is square. Also known as flat-bottom crown; flat-nose bit; square-nose bit.

flat fillet weld [MET] A fillet weld having a face that is relatively flat.

flat-flamed burner [ENG] A burner which emits a mixture of fuel and air in a flat stream through a rectangular nozzle.

flat form tool [DES ENG] A tool having a square or rectangular cross section with the form along the end.

flat grain [MATER] The grain pattern formed when soft wood is cut so that the tree's annular rings form an angle of 45° or less with the board's surface.

flathead rivet [DES ENG] A small rivet with a flat manufactured head used for general-purpose riveting.

flat jack [CIV ENG] A hollow steel cushion which is made of two nearly flat disks welded around the edge and which can be inflated with oil or cement under controlled pressure; used at the arch abutments and crowns to relieve the load on the formwork at the moment of striking the formwork.

flat-nose bit *See* flat-face bit.

flat-position welding [MET] Welding above the joint with the face of the weld in the horizontal reference plane. Also known as downhand welding.

flat rope [DES ENG] A steel or fiber rope having a flat cross section and composed of a number of loosely twisted ropes placed side by side, the lay of the adjacent strands being in opposite directions to secure uniformity in wear and to prevent twisting during winding.

flat slab [CIV ENG] A flat plate of reinforced concrete designed to span in two directions.

flat spring *See* leaf spring.

flattening [MET] Straightening of metal sheet by passing it through special rollers which flatten it without changing its thickness. Also known as roll flattening.

flattening test [MET] Quality test performed by flattening metal tubing between parallel plates that are a specified distance from each other.

flatting agent [MATER] Additive substance for paints or varnishes to disperse incident light rays to give the dried surface a nonglossy matte finish.

flat-turret lathe [MECH ENG] A lathe with a low, flat turret on a power-fed cross-sliding headstock.

flat yard [CIV ENG] A switchyard in which railroad cars are moved by locomotives, not by gravity.

flaw [MATER] A discontinuity in a material beyond acceptable established limits.

fleet [MECH ENG] Sidewise movement of a rope or cable when winding on a drum.

fleet angle [MECH ENG] In hoisting gear, the included angle between the rope, in its position of greatest travel across the drum, and a line drawn perpendicular to the drum shaft, passing through the center of the head sheave or lead sheave groove.

Fleming's solution [MATER] A tissue fixative made up of a mixture of osmic, chromic and acetic acids.

Flemish bond [CIV ENG] A masonry bond consisting of alternating stretchers and headers in each course, laid with broken joints.

Flemish garden wall bond [CIV ENG] A masonry bond consisting of headers and stretchers in the ratio of one to three or four in each course, with joints broken to give a variety of patterns.

fleshing machine [ENG] A machine that removes flesh from hides in a tannery.

Fletcher radial burner [ENG] A burner with gas jets arranged radially.

Flettner windmill [MECH ENG] An inefficient windmill with four arms, each consisting of a rotating cylinder actuated by a Savonius rotor.

flexibilizer [MATER] An additive that gives an otherwise rigid plastic flexibility. Also known as plasticizer.

flexible collodion [MATER] A collodion which has two additives (2% camphor and 3% castor oil) to make a pliable film.

flexible coupling [MECH ENG] A coupling used to connect two shafts and to accommodate their misalignment.

flexible glue [MATER] A type of glue used for pliable molds and printers' rollers, for example, a mixture of glue, glycerol, and water.

flexible-joint pipe [ENG] Cast-iron pipe adapted to laying under water and capable of motion through several degrees without leakage.

flexible mold [ENG] A coating mold made of flexible rubber or other elastomeric materials; used mainly for casting plastics.

flexible pavement [CIV ENG] A road or runway made of bituminous material which has little tensile strength and is therefore flexible.

flexible shaft [MECH ENG] **1.** A shaft that transmits rotary motion at any angle up to about 90°. **2.** A shaft made of flexible material or of segments. **3.** A shaft whose bearings are designed to accommodate a small amount of misalignment.

flexible ventilation ducting [MIN ENG] Flexible fabric tubes covered with rubber or polyvinyl chloride, used for auxiliary ventilation.

flexometer [ENG] An instrument for measuring the flexibility of materials.

flicker control [AERO ENG] Control of an aircraft, rocket, or such in which the control surfaces are deflected to their maximum degree with only a slight motion of the controller.

flight [AERO ENG] The movement of an object through the atmosphere or through space, sustained by aerodynamic reaction or other forces. [CIV ENG] A series of stairs between landings or floors.

flight characteristic [AERO ENG] A characteristic exhibited by an aircraft, rocket, or the like in flight, such as a tendency to stall or to yaw, or an ability to remain stable at certain speeds.

flight conveyor [MECH ENG] A conveyor in which paddles, attached to single or double strands of chain, drag or push pulverized or granulated solid materials along a trough. Also known as drag conveyor.

flight deck [AERO ENG] In certain airplanes, an elevated compartment occupied by the crew for operating the airplane in flight.

flight dynamics [AERO ENG] The study of the motion of an aircraft or missile; concerned with transient or short-term effects relating to the stability and control

of the vehicle, rather than to calculating such performance as altitude or velocity.

flight envelope [AERO ENG] The boundary depicting, for a specific aircraft, the limits of speed, altitude, and acceleration which that aircraft cannot safely exceed.

flight feeder [MECH ENG] Short-length flight conveyor used to feed solids materials to a process vessel or other receptacle at a preset rate.

flight instrument [AERO ENG] An aircraft instrument used in the control of the direction of flight, attitude, altitude, or speed of an aircraft, for example, the artificial horizon, airspeed indicator, altimeter, compass, rate-of-climb indicator, accelerometer, turn-and-bank indicator, and so on.

flight level [AERO ENG] A surface of constant atmospheric pressure which is related to the standard pressure datum.

flight path [AERO ENG] The path made or followed in the air or in space by an aircraft, rocket, or such.

flight-path angle [AERO ENG] The angle between the horizontal (or some other reference angle) and a tangent to the flight path at a point. Also known as flight-path slope.

flight-path slope *See* flight-path angle.

flight profile [AERO ENG] A graphic portrayal or plot of the flight path of an aeronautical vehicle in the vertical plane.

flight recorder [ENG] Any instrument or device that records information about the performance of an aircraft in flight or about conditions encountered in flight, for future study and evaluation.

flight science [AERO ENG] The sum total of all knowledge that enables humans to accomplish flight; it is compounded of both science and engineering, and is concerned with airplanes, missiles, and crewed and uncrewed space vehicles.

flight simulator [AERO ENG] A training device or apparatus that simulates certain conditions of actual flight or of flight operations.

flight stability [AERO ENG] The property of an aircraft or missile to maintain its attitude and to resist displacement, and, if displaced, to tend to restore itself to the original attitude.

flight test [AERO ENG] **1.** A test by means of actual or attempted flight to see how an aircraft, spacecraft, space-air vehicle, or missile flies. **2.** A test of a component part of a flying vehicle, or of an object carried in such a vehicle, to determine its suitability or reliability in terms of its intended function by making it endure actual flight.

flinching [IND ENG] In inspection, failure to call a borderline defect a defect.

flint glass [MATER] **1.** Heavy, colorless, brilliant glass that contains lead oxide. **2.** Any high-quality glass.

flint mill [MECH ENG] A mill employing pebbles to pulverize materials (for example, in cement manufacture).

FLIR imager *See* forward-looking infrared imager.

flitch beam [CIV ENG] A built-up beam composed of a metal plate sandwiched between flitches, that is, other strips of metal or other material.

flitch girder [CIV ENG] A girder composed of a metal plate sandwiched between flitches.

flitch plate [CIV ENG] The metal plate in a flitch beam or girder.

float [DES ENG] A file which has a single set of parallel teeth. [ENG] **1.** A flat, rectangular piece of wood with a handle, used to apply and smooth coats of plaster. **2.** A mechanical device to finish the surface of freshly placed concrete paving. **3.** A marble-polishing block. **4.** Any structure that provides positive buoyancy such as a hollow, watertight unit that floats or rests on the surface of a fluid. **5.** *See* plummet. [IND ENG] *See* bank.

floatability [MIN ENG] Response of a specific mineral to the flotation process.

float-and-sink analysis [MIN ENG] Use of a series of heavy liquids diminishing (or increasing) in density by accurately controlled stages in order to divide a sample of crushed coal or other minerals or metals into fractions that are either equal-settling or equal-floating at each stage.

float barograph [ENG] A type of siphon barograph in which the mechanically magnified motion of a float resting on the lower mercury surface is used to record atmospheric pressure on a rotating drum.

float chamber [ENG] A vessel in which a float regulates the level of a liquid.

float control [ENG] Floating device used to transmit a liquid-level reading to a control apparatus, such as an on-off switch controlling liquid flow into and out of a storage tank.

float-cut file [DES ENG] A coarse file used on soft materials.

float finish [CIV ENG] A rough concrete finish, obtained by using a wooden float for finishing.

float gage [ENG] Any one of several types of instruments in which the level of a liquid is determined from the height of a body floating on its surface, by using pullies, levers, or other mechanical devices.

floating action [ENG] Controller action in which there is a predetermined relation between the deviation and the speed of a final control element; a neutral zone, in which no motion of the final control element occurs, is often used.

floating axle [MECH ENG] A live axle used to turn the wheels of an automotive vehicle; the weight of the vehicle is borne by housings at the ends of a fixed axle.

floating block *See* traveling block.

floating chase [ENG] A mold part that can move freely in a vertical plane, which fits over a lower member (such as a cavity or plug) and into which an upper plug can telescope.

floating control [ENG] Control device in which the speed of correction of the control element (such as a piston in a hydraulic relay) is proportional to the error signal. Also known as proportional-speed control.

floating crane [CIV ENG] A crane having a barge or scow for an undercarriage and moved by cables attached to anchors set some distance off the corners of the barge; used for water work and for work on waterfronts.

floating dock [CIV ENG] **1.** A form of dry dock for repairing ships; it can be partly submerged by controlled flooding to receive a vessel, then raised by pumping out the water so that the vessel's bottom can be exposed. Also known as floating dry dock. **2.** A barge or flatboat which is used as a wharf.

floating dry dock *See* floating dock.

floating floor [BUILD] A floor constructed so that the wearing surface is separated from the supporting structure by an insulating layer of mineral wool, resilient quilt, or other material to provide insulation against impact sound.

floating foundation [CIV ENG] **1.** A reinforced concrete slab that distributes the concentrated load from columns; used on soft soil. **2.** A foundation mat several meters below the ground surface when it is combined with external walls.

floating lever [MECH ENG] A horizontal brake lever with a movable fulcrum; used under railroad cars.

floating pan [ENG] An evaporation pan in which the evaporation is measured from water in a pan floating in a larger body of water.

floating platen [ENG] In a multidaylight press, a platen that is between the main head and the press table and can be moved independently of them.

floating plug [MET] A plug or mandrel attached to a rod and used in plug drawing. Also known as a plug die.

floating roof [ENG] A type of tank roof (steel, plastic, sheet, or microballoons) which floats upon the surface of the stored liquid; used to decrease the vapor space and reduce the potential for evaporation.

floating scraper [MECH ENG] A balanced scraper blade that rests lightly on a drum filter; removes solids collected on the rotating drum surface by riding on the drum's surface contour.

floating zone refining [MET] A variation of the zone-refining technique in which the molten zone is held in place by its own surface tension between two collinear rods; since no container is needed, contamination of the pure metal is avoided.

floatless level control [ENG] Any nonfloat device for measurement and control of liquid levels in storage tanks or process vessels; includes use of manometers, capacitances, electroprobes, nuclear radiation, and sonics.

float switch [ENG] A switch actuated by a float at the surface of a liquid.

float-type rain gage [ENG] A class of rain gage in which the level of the collected rainwater is measured by the position of a float resting on the surface of the water; frequently used as a recording rain gage by connecting the float through a linkage to a pen which records on a clock-driven chart.

float valve [ENG] A valve whose on-off action is controlled directly by the fall or rise of a float concurrent with the fall or rise of liquid level in a liquid-containing vessel.

flood [ENG] To cover or fill with fluid. [MECH ENG] To supply an excess of fuel to a carburetor so that the level rises above the nozzle.

flood control [CIV ENG] Use of levees, walls, reservoirs, floodways, and other means to protect land from water overflow.

flood coverage [PETRO ENG] The extent of subterranean coverage within an oil reservoir by the injection of pressure-maintenance (or water-drive) water.

flood dam [CIV ENG] A dam for storing floodwater, or for supplying a flood of water.

flooded system [ENG] A system filled with so much tracer gas that probe testing for leaks suffers from a loss of sensitivity.

floodgate [CIV ENG] **1.** A gate used to restrain a flow or, when opened, to allow a flood flow to pass. **2.** The lower gate of a lock.

flooding [PETRO ENG] Technique of increasing oil (secondary recovery) from a reservoir by injection of water into the formation to drive the oil toward producing wellholes. Also known as waterflooding.

flood-out pattern [PETRO ENG] Pattern of subterranean water penetration and spread in an oil reservoir as a result of water injection.

flood pot test [PETRO ENG] Laboratory simulation of an oil reservoir to appraise the residual reservoir saturation after waterflooding.

flood relief channel See bypass channel.

flood wall [CIV ENG] A levee or similar wall for the purpose of protecting the land from inundation by flood waters.

floodway See bypass channel.

floor [ENG] The bottom, horizontal surface of an enclosed space. [MIN ENG] Boards laid at the heading to receive blasted rocks and to facilitate ore loading.

floor beam [BUILD] A beam used in the framing of floors in buildings. [CIV ENG] A large beam used in a bridge floor at right angles to the direction of the roadway, to transfer loads to bridge supports.

floorboard [MIN ENG] A thick wooden plank constituting part of a drill platform or other work platform.

floor burst [MIN ENG] A type of outburst in longwall faces which is preceded by heavy weighting due to floor lift; gas evolved below the seam collects beneath an impervious layer of rock, and a gas blister forms beneath the face, giving the observed floor lift; later, the floor fractures and the firedamp escapes into the mine atmosphere.

floor cut [MIN ENG] A machine-made cut in the floor dirt just below the coal seam.

floor drain [CIV ENG] A pipe or channel to remove water from under a floor in contact with soil.

floor framing [BUILD] Floor joists together with their strutting and supports.

flooring [MATER] Material suitable for use as a floor.

flooring saw [DES ENG] A pointed saw with teeth on both edges; cuts its own entrance into a material.

floor light [BUILD] A window set in a floor that is adapted for walking on and admitting light to areas below.

floor plate [BUILD] A flat board on a floor used to support wall studs. [ENG] A plate in a floor to which heavy work or machine tools can be bolted.

floor sill [MIN ENG] A large timber laid flat on the ground or in a level, shallow ditch, to which are fastened the drill-platform boards or planking, or which is used as the base for a full timber set.

floor system [CIV ENG] The structural floor assembly between supporting beams or girders in buildings and bridges.

flop gate [MIN ENG] An automatic gate used in placer mining when there is a shortage of water; the gate closes a reservoir until it is filled with water, then automatically opens and allows the water to flow into the sluices; when the reservoir is empty, the gate closes, and the operation is repeated.

Florence oil See olive oil.

flospinning [MET] Power-spinning or flowing metal over a rotating bar for shaping into cylindrical, conical, and curvilinear parts.

floss [MET] Molten or solid slag floating on the surface of a metal melt.

floss hole [MET] A small door or opening of the bottom of a smokestack or flue for removal of ash.

flotation [ENG] A process used to separate particulate solids by causing one group of particles to float; utilizes differences in surface chemical properties of the particles, some of which are entirely wetted by water, others are not; the process is primarily applied to treatment of minerals but can be applied to chemical and biological materials; in mining engineering it is referred to as froth flotation.

flotation cell [MIN ENG] The device in which froth flotation of ores is performed.

flotation collar [ENG] A buoyant bag carried by a spacecraft and designed so that it inflates and surrounds part of the outer surface if the spacecraft lands in the sea.

flotsam [ENG] Floating articles, particularly those that are thrown overboard to lighten a vessel in distress.

flour gold [MET] The finest-size gold dust, much of which will float on water.

flow [ENG] A forward movement in a continuous stream or sequence of fluids or discrete objects or materials, as in a continuous chemical process or solids-conveying or production-line operations.

flow bean [PETRO ENG] A plug containing a small hole placed in the flow line at the well head which serves to maintain oil flow at a proper rate.

flow brazing [MET] A brazing process in which coalescence is produced by the heat of molten filler metal that is poured over a joint.

flow brightening [MET] In a soldering process, the melting of a chemical or mechanical metallic coating on the base metal to be soldered.

flow chart [ENG] A graphical representation of the progress of a system for the definition, analysis, or solution of a data-processing or manufacturing problem in which symbols are used to represent operations, data or material flow, and equipment, and lines and arrows represent interrelationships among the components. Also known as control diagram; flow diagram; flow sheet.

flow-chart symbol [ENG] Any of the existing symbols normally used to represent operations, data or materials flow, or equipment in a data-processing problem or manufacturing-process description.

flow coat [ENG] A coating formed by pouring a liquid material over the object and allowing it to flow over the surface and drain off.

flow coefficient [MECH ENG] A dimensionless number used in studying the power required by fans, equal to the volumetric flow rate through the fan divided by the product of the rate of rotation of the fan and the cube of the impeller diameter.

flow control [ENG] Any system used to control the flow of gases, vapors, liquids, slurries, pastes, or solid particles through or along conduits or channels.

flow control valve [ENG] A valve whose flow opening is controlled by the rate of flow of the fluid through it; usually controlled by differential pressure across an orifice at the valve. Also known as rate-of-flow control valve.

flow diagram *See* flow chart.

flow direction [ENG] The antecedent-to-successor relation, indicated by arrows or other conventions, between operations on a flow chart.

flow graph *See* signal-flow graph.

flowing-film concentration [MIN ENG] A concentration based on the fact that liquid films in laminar flow possess a velocity which is not the same in all depths of the film; by this principle lighter particles of ore may be washed off while the heavier particles accumulate and are intermittently removed.

flowing furnace [MET] A furnace from which molten metal can be tapped or drawn.

flowing pressure [PETRO ENG] Pressure at the bottom of an oil-well bore (bottom-hole pressure) during normal oil production.

flowing-pressure gradient [PETRO ENG] The slope of decreasing pressure plotted against distance measured for upward liquid flow in a continuous-flow gas-lift oil well.

flowing well [PETRO ENG] Oil reservoir in which gas-drive pressure is sufficient to force oil flow up through and out of a wellhole.

flow line [ENG] **1.** The connecting line or arrow between symbols on a flow chart or block diagram. **2.** Mark on a molded plastic or metal article made by the meeting of two input-flow fronts during molding. Also known as weld mark. [PETRO ENG] A pipeline that takes oil from a single well or a series of wells to a gathering center.

flow marks [MATER] Wavy surface marks on a thermoplastic resin molding due to improper flow of material into the mold.

flow measurement [ENG] The determination of the quantity of a fluid, either a liquid, a vapor, or a gas, that passes through a pipe, duct, or open channel.

flowmeter [ENG] An instrument used to measure pressure, flow rate, and discharge rate of a liquid, vapor, or gas flowing in a pipe. Also known as fluid meter.

flow mixer [MECH ENG] Liquid-liquid mixing device in which the mixing action occurs as the liquids pass through it; includes jet nozzles and agitator vanes. Also known as line mixer.

flow nozzle [ENG] A flowmeter in a closed conduit, consisting of a short flared nozzle of reduced diameter inset into the inner diameter of a pipe; used to cause a temporary pressure drop in flowing fluid to determine flow rate via measurement of static pressures before and after the nozzle.

flow process [ENG] System in which fluids or solids are handled in continuous movement during chemical or physical processing or manufacturing.

flow-rating pressure [MECH ENG] The value of inlet static pressure at which the relieving capacity of a pressure-relief device is established.

flow sheet See flow chart.

flow soldering [ENG] Soldering of printed circuit boards by moving them over a flowing wave of molten solder in a solder bath; the process permits precise control of the depth of immersion in the molten solder and minimizes heating of the board. Also known as wave soldering.

flow string [PETRO ENG] Total length of oil- or gas-well tubing made up of a string of interconnected tubing sections.

flow tank [PETRO ENG] A tank which receives oil from the well and where gas and water may be separated before the oil passes into a stock tank.

flow transmitter [ENG] A device used to measure the flow of liquids in pipelines and convert the results into proportional electric signals that can be transmitted to distant receivers or controllers.

flow valve [ENG] A valve that closes itself when the flow of a fluid exceeds a particular value.

flow visualization [ENG] Method of making visible the disturbances that occur in fluid flow, using the fact that light passing through a flow field of varying density exhibits refraction and a relative phase shift among different rays.

flow welding [MET] A welding process in which coalescence is produced by heating with molten filler metal, which is poured over the joint until the welding temperature is attained and the required amount of filler metal is added.

flue [ENG] A channel or passage for conveying combustion products from a furnace, boiler, or fireplace to or through a chimney.

flue dust [MET] Fine particles of metal or alloy emitted with the gases of a smelter or metallurgical furnace.

flue exhauster [ENG] A device installed as part of a vent in order to provide a positive induced draft.

flue gas [ENG] Gaseous combustion products from a furnace.

flue gas analyzer [ENG] A device that monitors the composition of the flue gas of a boiler heating unit to determine if the mixture of air and fuel is at the proper ratio for maximum heat output.

fluid amplifier [ENG] An amplifier in which all amplification is achieved by interaction between jets of fluid, with no electronic circuit and usually no moving parts.

fluid clutch See fluid drive.

fluid coal [MATER] Pulverized coal that, when mixed with air, can be transported through pipes.

fluid coefficient [PETRO ENG] A measure of the flow resistance to the leaking off of reservoir fracturing fluids into the formation during the fracturing operation.

fluid-compressed [MET] Pertaining to steel that has been compressed while still fluid to remove gases and make the material more homogeneous.

fluid-controlled valve [MECH ENG] A valve for which the valve operator is activated by a fluid energy, in contrast to electrical, pneumatic, or manual energy.

fluid coupling [MECH ENG] A device for transmitting rotation between shafts by means of the acceleration and deceleration of a fluid such as oil. Also known as hydraulic coupling.

fluid die [MECH ENG] A die for shaping parts by liquid pressure; a plunger forces the liquid against the part to be shaped, making the part conform to the shape of a die.

fluid distributor [ENG] Device for the controlled distribution of fluid feed to a process unit, such as a liquid-gas or liquid-solids contactor, reactor, mixer, burner, or heat exchanger; can be a simple perforated-pipe sparger, spray head, or such.

fluid drive [MECH ENG] A power coupling operated on a hydraulic turbine principle in which the engine flywheel has a set of turbine blades which are connected directly to it and which are driven in oil, thereby turning another set of blades attached to the transmission gears of the automobile. Also known as fluid clutch; hydraulic clutch.

fluid-energy mill [ENG] A size-reduction unit in which grinding is achieved by collision between the particles being ground and the energy supplied by a compressed fluid entering the grinding chamber at high speed. Also known as jet mill.

fluid-film bearing [MECH ENG] An antifriction bearing in which rubbing surfaces are kept apart by a film of lubricant such as oil.

fluidic device [ENG] A device that operates by the interaction of streams of fluid.

fluidic flow sensor [ENG] A device for measuring the velocity of gas flows in which a jet of air or other selected gas is directed onto two adjacent small openings and is deflected by the flow of gas being measured so that the relative pressure on the two ports is a measure of gas velocity. Also known as deflected jet fluidic flowmeter.

fluidic oscillator meter [ENG] A flowmeter that measures the frequency with which a fluid entering the meter attaches to one of two opposite diverging side walls and then the other, because of the Coanda effect.

fluidics [ENG] A control technology that employs fluid dynamic phenomena to perform sensing, control, information, processing, and actuation functions without the use of moving mechanical parts.

fluid injection [PETRO ENG] The introduction of gases or liquids under pressure into a reservoir to force oil into producing wells.

fluidized bed [ENG] A cushion of air or hot gas blown through the porous bottom slab of a container which can be used to float a powdered material as a means of drying, heating, quenching, or calcining the immersed components.

fluidized-bed coating [ENG] Method for plastic-coating of objects; the heated object is immersed in the fluidized bed of a thermoplastic resin that then fuses into a continuous uniform coating over the immersed object.

fluidized-bed combustion [MECH ENG] A method of burning particulate fuel, such as coal, in which the amount of air required for combustion far exceeds that found in conventional burners; the fuel particles are continually fed into a bed of mineral ash in the proportions of 1 part fuel to 200 parts ash, while a flow of air passes up through the bed, causing it to act like a turbulent fluid.

fluid logic [ENG] The simulation of logical operations by means of devices that employ fluid dynamic phenomena to control the interactions between sets of gases or liquids.

fluid-loss agent [MATER] Material used to thicken or gel crude oil, light oil, and water or acid fracturing fluids to seal off pores and flow channels in the reservoir matrix.

fluid-loss test [PETRO ENG] Measure of fracturing fluid loss versus time (spurt loss) before the fluid-loss agent forms a nonpermeable layer in the reservoir pore matrix.

fluidmeter *See* flow meter.

fluid transmission [MECH ENG] Automotive transmission with fluid drive.

fluid viscosity ratio [PETRO ENG] Ratio of viscosity of a displacing gas to that of oil in a gas-drive reservoir; used in unit displacement efficiency calculations.

flume [ENG] **1.** An open channel constructed of steel, reinforced concrete, or wood and used to convey water to be utilized for power, to transport logs, and so on. **2.** To divert by a flume, as the waters of a stream, in order to lay bare the auriferous sand and gravel forming the bed.

flumed [MIN ENG] In hydraulic mining, pertaining to the transportation of solids by suspension or flotation in flowing water.

fluorescent screen [ENG] A sheet of material coated with a fluorescent substance so as to emit visible light when struck by ionizing radiation such as x-rays or electron beams.

fluoridation [ENG] The addition of the fluorine ion (F^-) to municipal water supplies in a final concentration of 0.8–1.6 ppm (parts per million) to help prevent dental caries in children.

fluorimeter *See* fluorometer.

fluorologging [ENG] A well-logging technique in which well cuttings are examined under ultraviolet light for fluorescence radiation related to trace occurrences of oil.

fluorometer [ENG] An instrument that measures the fluorescent radiation emitted by a sample which is exposed to monochromatic radiation, usually radiation from a mercury-arc lamp or a tungsten or molybdenum x-ray source that has passed through a filter; used in chemical analysis, or to determine the intensity of the radiation producing fluorescence. Also spelled fluorimeter.

fluoroplastics [MATER] A family of plastics based on fluorine replacement of hydrogen atoms in hydrocarbon molecules; includes polytetrafluoroethylene (PTFE), polychlorotrifluoroethylene (PCTFE), polyvinylidene fluoride, and fluorinated ethylene propylene (FEP).

fluoroscope [ENG] A fluorescent screen designed for use with an x-ray tube to permit direct visual observation of x-ray shadow images of objects interposed between the x-ray tube and the screen.

fluoroscopy [ENG] Use of a fluoroscope for x-ray examination.

fluosolids system [MET] In pyrometallurgy, a roasting method for finely divided solids, in which air under pressure is blown through a heated bed of mineral to keep it fluid.

flush [ENG] Pertaining to separate surfaces that are on the same level.

flush coat [CIV ENG] A coating of bituminous material, used to waterproof a surface.

flushed-zone resistivity [PETRO ENG] Electrical resistivity of the reservoir area which surrounds a borehole to a distance of at least 3 inches (7.6 centimeters) and for which the original interstitial fluids have been flushed out by drilling-mud filtrate.

flush gate [CIV ENG] A gate for flushing a channel that lies below the gate of a dam.

flushing [CIV ENG] The removal or reduction to a permissible level of dissolved or suspended contaminants in an estuary or harbor. [ENG] Removing lodged deposits of rock fragments and other debris by water flow at high velocity; used to clean water conduits and drilled boreholes.

flushing oil [MATER] A solvent oil designed to remove used lubricating oil, decomposition products, and accumulated dirt from lubrication passages, crankcase surfaces, and lubricated moving parts of automotive engines.

flush-joint casing [PETRO ENG] Lengths of casing that when connected end to end form a smooth joint flush with the outer diameter of the remainder of the section length.

flushometer [ENG] A valve that discharges a fixed quantity of water when a handle is operated; used to flush toilets and urinals.

flush production [PETRO ENG] First yield from a flowing oil well during its most productive period.

flush tank [CIV ENG] **1.** A tank in which water or sewage is retained for periodic release through a sewer. **2.** A small water-filled tank for flushing a water closet.

flush valve [ENG] A valve used for flushing toilets.

flute [DES ENG] A groove having a curved section, especially when parallel to the main axis, as on columns, drills, and other cylindrical or conical shaped pieces.

fluted chucking reamer [DES ENG] A machine reamer with a straight or tapered shank and with straight or spiral flutes; the ends of the teeth are ground on a slight chamfer for end cutting.

fluted coupling *See* stabilizer.

flute length [DES ENG] On a twist drill, the length measured from the outside corners of the cutting lips to the farthest point at the back end of the flutes.

fluting [MECH ENG] A machining operation whereby flutes are formed parallel to the main axis of cylindrical or conical parts.

flutter [ENG] The irregular alternating motion of the parts of a relief valve due to the application of pressure where no contact is made between the valve disk and the seat.

flutter valve [ENG] A valve that is operated by fluctuations in pressure of the material flowing over it; used in carburetors.

fluvarium [ENG] A large aquarium in which the tanks contain flowing stream water maintained by gravity, not pumps.

flux [MATER] **1.** In soldering, welding, and brazing, a material applied to the pieces to be united to reduce the melting point of solders and filler metals and to prevent the formation of oxides. **2.** A substance used to promote the fusing of minerals or metals. **3.** Additive for plastics composition to improve flow during physical processing. **4.** In enamel work, a substance composed of silicates and other materials that forms a colorless, transparent glass when fired. Also know as fondant.

flux-cored welding [MET] Welding with a metal electrode that has a flux core.

flux factor [MET] A factor for assessing the quality of steelworks-grade silica refractories.

flux gate [ENG] A detector that gives an electric signal whose magnitude and phase are proportional to the magnitude and direction of the external magnetic field acting along its axis; used to indicate the direction of the terrestrial magnetic field.

flux guide [MET] A shaped piece of magnetic material used to guide magnetic flux in induction heating; may be used either to direct the flux to preferred locations or to prevent the flux from spreading beyond definite regions.

fluxing [MET] The development of the liquid phase in a ceramic body under heat treatment by the melting of low-fusion components; used in steel manufacture, metal smelting, and assaying.

fluxing ore [MET] An ore containing usually an appreciable amount of valuable metal, but smelted mainly because it contains fluxing agents which are required in the reduction of other ores.

fluxmeter [ENG] An instrument for measuring magnetic flux.

flux oil [MATER] An oil suitable for blending with bitumen or asphalt to form a product of greater fluidity or softer consistency.

flux oxygen cutting [MET] Oxygen cutting of metal with the aid of a flux to reduce the temperature.

fly [MECH ENG] A fan with two or more blades used in timepieces or light machinery to govern speed by air resistance.

fly ash [ENG] Fine particulate, essentially noncombustible refuse, carried in a gas stream from a furnace.

fly-by-wire system [AERO ENG] A flight control system that uses electric wiring instead of mechanical or hydraulic linkages to control the actuators for the ailerons, flaps, and other control surfaces of an aircraft.

fly cutter [MECH ENG] A cutting tool that revolves with the arbor of a lathe.

fly cutting [MECH ENG] Cutting with a milling cutter provided with only one tooth.

flying angle [AERO ENG] The acute angle between the longitudinal axis of an aircraft and the horizontal axis in normal level flight, or the angle of attack of a wing in normal level flight.

flying boat [AERO ENG] A seaplane with a fuselage that acts as a hull and is the means of the plane's support on water.

flying-crane helicopter [AERO ENG] A heavy-lift helicopter used in rapid loading and unloading of, for example, cargo ships.

flying shear [MET] A machine which cuts lengths of rolled products and allows for continuous production by reciprocating with the product while cutting.

flying switch [ENG] Disconnection of railroad cars from a locomotive while they are moving and switching them to another track under their own momentum.

fly rock [ENG] The fragments of rock thrown and scattered during quarry or tunnel blasting.

flywheel [MECH ENG] A rotating element attached to the shaft of a machine for the maintenance of uniform angular velocity and revolutions per minute. Also known as balance wheel.

foam drilling [MIN ENG] A method of dust suppression in which thick foam is forced through the drill by means of compressed air, and the foam-and-dust mixture emerges from the mouth of the hole in the form of a thick sludge.

foamed glass See cellular glass.

foamed plastic See expanded plastic.

foam glass [MATER] A light, black, opaque, cellular glass made by adding powdered carbon to crushed glass and firing the mixture.

foaming [ENG] Any of various processes by which air or gas is introduced into a liquid or solid to produce a foam material.

foaming agent See blowing agent.

foam-in-place [ENG] The deposition of reactive foam ingredients onto the surface to be covered, allowing the foaming reaction to take place upon that surface, as with polyurethane foam; used in applying thermal insulation for homes and industrial equipment.

foam metal [MET] Cast metal with finely divided gas bubbles evenly distributed throughout the body of the metal; an example is foam aluminum.

foam rubber See rubber sponge.

focal spot [MET] In electron-beam or laser welding, the spot where the beam has the highest concentrated energy level.

focused-current log [ENG] A resistivity log that is obtained by means of a multiple-electrode arrangement.

fogged metal [MET] A metal whose luster has been highly reduced by corrosion products.

fog quenching [MET] Rapid cooling of a metal piece in a fine vapor or mist.

foil [MET] A thin sheet of metal, usually less than 0.006 inch (0.15 millimeter) thick.

foil decorating [ENG] The molding of paper, textile, or plastic foil, printed with compatible inks, into a plastic part so that the foil is visible below the surface of the part as a decoration.

fold See lap.

folded horn [ENG ACOUS] An acoustic horn in which the path from throat to mouth is folded or curled to give the longest possible path in a given volume.

folded-plate roof [BUILD] A roof constructed of flat plates, usually of reinforced concrete, joined at various angles.

folding boards [MIN ENG] A shifting frame on which the cage rests in a mine. Also known as chairs; keeps; keps.

folding door [ENG] A door in sections that can be folded back or can be moved apart by sliding.

folding fin [AERO ENG] A fin hinged at its base to lie flat, especially a fin on a rocket that lies flat until the rocket is in flight.

foliation [MET] Beating metal into thin sheets.

follower [ENG] A drill used for making all but the first part of a hole, the first part being made with a drill of larger gage.

follower rail [MIN ENG] The rail of a mine switch on the other side of the turnout corresponding to the lead rail.

fondant See flux.

food color [MATER] A colorant, either a dye (soluble) or a lake (insoluble), permitted by the Food and Drug Administration for use in foods, drugs, and cosmetics. Also known as certified color; FD&C color.

food engineering [ENG] The technical discipline involved in food manufacturing and processing.

footage [ENG] The extent or length of a material expressed in feet. [MIN ENG] **1.** The number of feet of borehole drilled per unit of time, or that required to complete a specific project or contract. **2.** The payment of miners by the running foot of work.

foot block [ENG] Flat pieces of wood placed under props in tunneling to give a broad base and thus prevent the superincumbent weight from pressing the props down.

foot bridge [CIV ENG] A bridge structure used only for pedestrian traffic.

foot guard [CIV ENG] A filler placed on the space between converging rails to prevent a foot from being wedged between the rails.

foot holes [MIN ENG] Holes cut in the sides of shafts or winzes to enable miners to climb up or down.

footing [CIV ENG] The widened base or substructure forming the foundation for a wall or a column.

foot screw [ENG] **1.** One of the three screws connecting the tribach of a theodolite or other level with the plate screwed to the tripod head. **2.** An adjusting screw that serves also as a foot.

foot section [MECH ENG] In both belt and chain conveyors that portion of the conveyor at the extreme opposite end from the delivery point.

footstock [MECH ENG] A device containing a center which supports the workpiece on a milling machine; usually used in conjunction with a dividing head.

foot valve [MECH ENG] A valve in the bottom of the suction pipe of a pump which prevents backward flow of water.

footwall shaft See underlay shaft.

Foraky boring method [MIN ENG] A percussive boring system; a closed-in derrick contains the crown pulley, over which a steel rope with the boring tools is passed from a drum; the drum moves the tools, which are vibrated by a walking beam.

Foraky freezing process [MIN ENG] A method of shaft sinking through heavily watered sands by freezing the sands.

force-balance meter [ENG] A flowmeter that measures a force, such as that associated with the air pressure in a small bellows, that is required to balance the net force created by the differential pressure, on opposite sides of a diaphragm or diaphragm capsule, generated by a differential-producing primary device.

forced-air heating [MECH ENG] A warm-air heating system in which positive air circulation is provided by means of a fan or a blower.

forced auxiliary ventilation [MIN ENG] A system in which the duct delivers the intake air to the face.

forced-caving system [MIN ENG] A stoping system in which the ore is broken down by large blasts into the stopes that are kept partly full of broken ore.

forced circulation [MECH ENG] The use of a pump or other fluid-movement device in conjunction with liquid-processing equipment to move the liquid through pipes and process vessels; contrasted to gravity or thermal circulation.

forced-circulation boiler [MECH ENG] A once-through steam generator in which water is pumped through successive parts.

forced draft [MECH ENG] Air under positive pressure produced by fans at the point where air or gases enter a unit, such as a combustion furnace.

forced ventilation [MECH ENG] A system of ventilation in which air is forced through ventilation ducts under pressure.

force fit See press fit.

force gage [ENG] An instrument which measures the force exerted on an object.

force main [CIV ENG] The discharge pipeline of a pumping station.

force piece See foreset.

force plate [ENG] A plate that carries the plunger or force plug of a mold and the guide pins on bushings.

force plug [ENG] A mold member that fits into the cavity block, exerting pressure on the molding compound. Also known as piston; plunger.

forceps [DES ENG] A pincerlike instrument for grasping objects.

force pump [MECH ENG] A pump fitted with a solid plunger and a suction valve which draws and forces a liquid to a considerable height above the valve or puts the liquid under a considerable pressure.

forcing fan [MIN ENG] A fan which forces the intake air into mine workings. Also known as blowing fan.

fording depth [ENG] Maximum depth at which a particular vehicle can operate in water.

forebay [CIV ENG] **1.** A small reservoir at the head of the pipeline that carries water to the consumer; it is the last free water surface of a distribution system. **2.** A reservoir feeding the penstocks of a hydro-power plant.

fore drift [MIN ENG] That one of a pair of parallel headings which is kept a short distance in advance of the other.

forehand welding [MET] Welding in which the flame is directed against the base metal ahead of the weld and is moved in the direction of welding. Also known as forward welding.

forehearth [MET] **1.** A bay in front of the hearth of a furnace. **2.** A receptacle in front of a hearth to receive the molten products.

foreign-body locator [ENG] A device for locating foreign metallic bodies in tissue by means of suitable probes that generate a magnetic field; the presence of a magnetic body within this field is indicated by a meter or a sound signal.

foreign element [IND ENG] A work element which is not a part of the normal work cycle, either because it is accidental or because it occurs only occasionally.

forensic engineering [ENG] The application of accepted engineering practices and principles for discussion, debate, argumentative, or legal purposes.

forepoling [MIN ENG] A timbering method for a very weak roof in which a bench of timbers is set and boards or long wedges are placed above the header; as the next bench of timbers is placed at the inbye end of the wedges, other like wedges are driven in under the first wedges and over the second header. Also known as spiling.

fore pump *See* backing pump.

foreset [MIN ENG] **1.** To place a prop under the coal-face end of a bar. **2.** Timber set used for roof support at the working face. Also known as force piece.

foreshaft sinking [MIN ENG] The first 150 feet (46 meters) of shaft sinking from the surface; the plant and services for the main shaft are installed during this step.

foresight [ENG] **1.** A sight or bearing on a new survey point, taken in a forward direction and made in order to determine its elevation. **2.** A sight on a previously established survey point, taken in order to close a circuit. **3.** A reading taken on a level rod to determine the elevation of the point on which the rod rests when read. Also known as minus sight.

forfeiture [MIN ENG] Loss of a mining claim by operation of the law, without regard to the intention of the locator, whenever he fails to preserve his right by complying with the conditions imposed by law.

forge [MET] **1.** To form a metal, usually hot, into desirable shapes by employing compressive forces. **2.** A machine or place in which metal is formed hot, or where iron is produced from its ore.

forgeability [MET] Suitability of a material for forging.

forge delay time [MET] The time between the start of weld time and the time when forging pressure is reached by the electrode force.

forge welding [MET] A group of welding processes in which the parts to be joined, usually iron, are heated to about 1000°C and then hammered or pressed together. Also known as fire welding.

forging [MET] **1.** Using compressive force to shape metal by plastic deformation; dies may be used. **2.** A piece of work made by forging.

forging brass [MET] Brass composed of 60% copper, 38% zinc, and 2% lead, used for hot forgings, hardware, and plumbing supplies; it is extremely plastic when hot, is corrosion-resistant, and has excellent mechanical properties.

forging hammer [MET] A hammer used to pound metal into forgings.

forging plane [MET] The plane of the principal die face when oriented normal to the direction of ram travel.

forging press [MET] A press designed to operate dies in die forging.

forging range [MET] Optimum temperature range in which a metal can be forged.

forging rolls [MET] A machine used in making forgings by rolling the metal.

forging stock [MET] A section or a piece of metal used to make a forging.

forklift [MECH ENG] A machine, usually powered by hydraulic means, consisting of two or more prongs which can be raised and lowered and are inserted under heavy materials or objects for hoisting and moving them.

forklift truck *See* fork truck.

fork truck [MECH ENG] A vehicle equipped with a forklift. Also known as forklift truck.

form [CIV ENG] Temporary boarding, sheeting, or pans of plywood, molded fiber glass, and so forth, used to give desired shape to poured concrete or the like.

formability [MATER] Capability of a material to be shaped by plastic deformation.

formalin [MATER] An aqueous solution of formaldehyde, usually 37% formaldehyde by weight.

formation fracturing [PETRO ENG] Method of applying hydraulic pressure to a reservoir formation to cause the rock to split open, that is to fracture; used to increase oil production.

formation solubility [PETRO ENG] Measure of formation rock solubility in oil-well acidizing solution (hydrochloric acid or hydrochloric-hydrofluoric acids).

formation tester [PETRO ENG] Device for retrieval of samples of fluid from an oil-reservoir formation.

form clamp [CIV ENG] An adjustable metal clamp used to secure planks of wooden forms for concrete columns or beams.

form grinding [MECH ENG] Grinding by use of a wheel whose cutting face is contoured to the reverse shape of the desired form.

forming [ENG] A process for shaping or molding sheets, rods, or other pieces of hot glass, ceramic ware, plastic, or metal by the application of pressure.

forming die [ENG] A die like a drawing die, but without a blank holder.

forming press [MECH ENG] A punch press for forming metal parts.

forming rolls [MECH ENG] Rolls contoured to give a desired shape to parts passing through them.

forming tool [DES ENG] A nonrotating tool that produces its inverse form on the workpiece.

form oil [MATER] An oil utilized on the contact surface of wooden or metal concrete forms to prevent concrete from sticking.

form process chart [IND ENG] A graphic representation of the process flow of paperwork forms. Also known as forms analysis chart; functional forms analysis chart; information process analysis chart.

formwork [CIV ENG] A temporary wooden casing used to contain concrete during its placing and hardening. Also known as shuttering.

Forrester machine [MIN ENG] A pneumatic flotation cell in which pulp is aerated by low-pressure air, delivering a mineralized froth along the overflow, and tailings to the end weir.

Fortin barometer [ENG] A type of cistern barometer; provision is made to increase or decrease the volume of the cistern so that when a pressure change occurs, the level of the cistern can be maintained at the zero of the barometer scale (the ivory point).

forward extrusion [MET] A cold extrusion process in which a formed blank is placed in a die cavity and struck by a punch; the metal is extruded through an annular space between the die and the end of the punch, moving in the same direction as the punch.

forward-looking infrared imager [ENG] An infrared imaging device which employs an optomechanical system to make a two-dimensional scan, and produces a visible image corresponding to the spatial distribution of infrared radiation. Abbreviated FLIR imager. Also known as framing imager.

forward welding See forehand welding.

Foster's formula [MIN ENG] The empirical formula $R = 3\sqrt{DT}$ for determining the radius R of a shaft pillar, where D = depth in feet and T = thickness of lode in feet.

foul bottom [CIV ENG] A hard, uneven, rocky or obstructed bottom having poor holding qualities for anchors, or one having rocks or wreckage that would endanger an anchored vessel.

fouling plates [ENG] Metal plates submerged in water to allow attachment of fouling organisms, which are then analyzed to determine species, growth rate, and growth pattern, as influenced by environmental conditions and time.

fouling point [CIV ENG] 1. The point at a switch or turnout beyond which railroad cars must be placed so as not to interfere with cars on the main track. 2. The location of insulated joints in a turnout on signaled tracks.

foundation [CIV ENG] 1. The ground that supports a building or other structure. 2. The portion of a structure which transmits the building load to the ground.

foundation mat See raft foundation.

foundry [ENG] A building where metal or glass castings are produced.

foundry alloy See master alloy.

foundry core sand See core sand.

foundry engineering [ENG] The science and practice of melting and casting glass or metal.

foundry facing [MET] A material applied to a sand mold to improve the surface quality of a casting.

foundry sand [MET] Sand used in foundries to make molds for the casting of metal shapes.

fourable [PETRO ENG] A section of drill pipe casing or tubing comprising four joints that are screwed together.

fourable board [PETRO ENG] A platform installed in an oil derrick at an elevation of 80–120 feet (24–37 meters) above the derrick floor to support the derrick operator while pipe is being raised or lowered.

four-ball tester [ENG] A machine designed to measure the efficiency of lubricants by driving one ball against three stationary balls clamped together in a cup filled with the lubricant; performance is evaluated by measuring wear-scar diameters on the stationary balls.

four-bar linkage [MECH ENG] A plane linkage consisting of four links pinned tail to head in a closed loop with lower, or closed, joints.

Fourcault process [ENG] A process for forming sheet glass in which the molten glass is drawn vertically upward.

four-channel sound system See quadraphonic sound system.

Fourdrinier machine [MECH ENG] A papermaking machine; a paper web is formed on an endless wire screen; the screen passes through presses and over dryers to the calenders and reels.

Fourier analyzer [ENG] A digital spectrum analyzer that provides push-button or other switch selection of averaging, coherence function, correlation, power spectrum, and other mathematical operations involved in calculating Fourier transforms of time-varying signal voltages for such applications as identification of underwater sounds, vibration analysis, oil prospecting, and brain-wave analysis.

four-piece set [MIN ENG] Squared timber frame used in underground driving to give all-around support to weak ground.

four-stroke cycle [MECH ENG] An internal combustion engine cycle completed in four piston strokes; includes a suction stroke, compression stroke, expansion stroke, and exhaust stroke.

four-track tape [ENG ACOUS] Magnetic tape on which two tracks are recorded for each direction of travel, to provide stereo sound reproduction or to double the amount of source material that can be recorded on a given length of 1/4-inch (0.635 centimeter) tape.

four-way reinforcing [CIV ENG] A system of reinforcing rods in concrete slab construction in which the rods are placed parallel to two adjacent edges and to both diagonals of a rectangular slab.

four-way valve [MECH ENG] A valve at the junction of four waterways which allows passage between any two adjacent waterways by means of a movable element operated by a quarter turn.

four-wheel drive [MECH ENG] An arrangement in which the drive shaft acts on all four wheels of the automobile.

fox lathe [MECH ENG] A lathe with chasing bar and leaders for cutting threads; used for turning brass.

fraction [MET] In powder metallurgy, that portion of sample that lies between two stated particle sizes. Also known as cut.

fractional gas-flow curve [PETRO ENG] Graph of the fraction of free injected gas flowing through a reservoir formation versus the liquid saturation of the gas for various parameter values of oil viscosity; used to calculate displacement efficiency during gas injection.

fractional sampling [MIN ENG] Mechanical selection of samples of uniformly graded material without segregation.

fraction defective [IND ENG] The number of units per 100 pieces which are defective in a lot; expressed as a decimal.

fractography [MET] The microscopic examination of fractured metal surfaces.

fractured formation [PETRO ENG] Reservoir formation in which rock has been split by hydraulic pressure produced by injected fluids.

fracture dome [MIN ENG] The zone of loose or semiloose rock which exists in the immediate hanging or footwall of a stope.

fracture test [ENG] **1.** Macro- or microscopic examination of a fractured surface to determine characteristics such as grain pattern, composition, or the presence of defects. **2.** A test designed to evaluate fracture stress.

fragmentation [MIN ENG] The blasting of coal, ore, or rock into pieces small enough to load, handle, and transport without the need for hand-breaking or secondary blasting.

Frahm frequency meter *See* vibrating-reed frequency meter.

frame [BUILD] The skeleton structure of a building. Also known as framing.

frame set [MIN ENG] The arrangement of the legs and cap or crossbar so as to provide support for the roof of an underground passage. Also known as framing; set.

framework [ENG] The load-carrying frame of a structure; may be of timber, steel, or concrete.

framing [BUILD] *See* frame. [MIN ENG] *See* frame set.

framing imager *See* forward-looking infrared imager.

framing square [DES ENG] A graduated carpenter's square used for cutting off and making notches.

framing table [MIN ENG] An inclined table on which ore or slimes are separated by running water.

Francis turbine [MECH ENG] A reaction hydraulic turbine of relatively medium speed with radial flow of water in the runner.

frankincense *See* olibanum.

frankincense oil *See* olibanum oil.

Franklin equation [ENG ACOUS] An equation for intensity of sound in a room as a function of time after shutting off the source, involving the volume and exposed surface area of the room, the speed of sound, and the mean sound-absorption coefficient.

Frank-Read source [MET] **1.** The creation of dislocations by application of shear stress to an edge dislocation anchored terminally, causing formation of an unstable loop form followed by formation of a closed dislocation line and the establishment of the original condition. **2.** One of the sources of dislocations in a plastically deforming metal.

Frary metal [MET] Metal containing 97–98% lead alloyed with 1–2% barium and calcium; used for bearings.

Frasch process [MIN ENG] A process to remove sulfur from sulfur beds; superheated water is forced under pressure into the sulfur bed, and the molten sulfur is thus forced to the surface.

Fraser's air-sand process [MIN ENG] A process in which dry, specific-gravity separation of coal from refuse is achieved by utilizing a flowing dense medium intermediate in density between coal and refuse.

free ascent [ENG] Emergency ascent by a diver by floating to the surface through natural buoyancy or through assisted buoyancy with a life jacket.

free balloon [AERO ENG] A balloon that ascends without a tether, propulsion or guidance; it is made to descend by the release of gas.

freeboard [CIV ENG] The height between normal water level and the crest of a dam or the top of a flume.

free carbon [MET] Elemental carbon present in a metal in an uncombined state.

free crushing [MIN ENG] Crushing under conditions of speed and feed so that there is ample room for the fine ore to fall away from the coarser material and thereby escape further crushing.

free-cutting steel [MET] Steel that contains a higher percentage of sulfur than carbon steel, making it very easy to machine.

free diving [ENG] Diving with the use of scuba equipment to allow freedom and maneuverability.

freedom to mine [MIN ENG] The law by which anybody has the right to mine certain minerals when he has prospected for them and has filed a proper application for the right to mine them.

free-drop [ENG] To air-drop supplies or equipment without parachute.

free end *See* free face.

free face [MIN ENG] The exposed surface of a mass of rock or of coal. Also known as free end.

free fall [PETRO ENG] In deep drilling, an arrangement by which the bit is permitted to fall freely to the bottom at each drop or down stroke.

free falling [MECH ENG] In ball milling, the peripheral speed at which part of the crop load breaks clear on the ascending side and falls clear to the toe of the charge.

free-fed [MIN ENG] In comminution, pertaining to rolls fed only enough ore to maintain a ribbon of material between them.

free ferrite [MET] Relatively pure metallic iron phase present in steel or cast iron.

free-field room *See* anechoic chamber.

free fit [DES ENG] A fit between mating pieces where accuracy is not essential or where large variations in temperature may occur.

free-flight melt spinning [MET] Rapid-quenching process in which the molten metal is forced through an orifice under pressure and the jet is solidified while in free flight; quench rates reach 1000 K per second.

free gas [PETRO ENG] A hydrocarbon that exists in the gaseous phase at reservoir pressure and temperature and remains a gas when produced under normal conditions.

free-gas saturation [PETRO ENG] Proportion of oil-reservoir pore structure saturated by free (undissolved) gas.

free gold [MET] Gold that is in the free state, that is, not combined with other substances.

free gyroscope [ENG] A gyroscope that uses the property of gyroscopic rigidity to sense changes in altitude of a machine, such as an airplane; the spinning wheel or rotor is isolated from the airplane by gimbals; when the plane changes from level flight, the gyro remains vertical and gives the pilot an artificial horizon reference.

free instruments [ENG] Instruments designed to initially sink to the ocean bottom, release their ballast, and then rise to the surface where they are retrieved with their acquired payload.

free-machining steel [MET] Steel to which impurities have been added to improve machinability.

Freeman-Nichols roaster [MIN ENG] A unit in which pyrite flotation concentrates are flash-roasted.

free milling [MIN ENG] A process applied to ores which contain free gold or silver and can be reduced by crushing and amalgamation (by gravity or on blankets), without roasting or other chemical treatment.

free-milling gold [MET] Gold that has a clean surface so that it readily amalgamates with mercury (by gravity or on blankets) after liberation by comminution.

free-milling ore [MIN ENG] Ore containing gold which can be caught with mercury by a variety of gravity processes or on blankets.

free-piston engine [MECH ENG] A prime mover utilizing free-piston motion controlled by gas pressure in the cylinders.

free port [CIV ENG] An isolated, enclosed, and policed port in or adjacent to a port of entry, without a resident population.

free settling [MIN ENG] In classification, the free fall of particles through fluid media.

free-stream Mach number [AERO ENG] The Mach number of the total airframe (entire aircraft) as contrasted with the local Mach number of a section of the airframe.

free-swelling index [ENG] A test for measuring the free-swelling properties of coal; consists of heating 1 gram of pulverized coal in a silica crucible over a gas flame under prescribed conditions to form a coke button, the size and shape of which are then compared with a series of standard profiles numbered 1 to 9 in increasing order of swelling.

free turbine [MECH ENG] In a turbine engine, a turbine wheel that drives the output shaft and is not connected to the shaft driving the compressor.

free wall [MIN ENG] The wall of an ore vein filling which scales off cleanly from the gouge.

freeze [ENG] **1.** To permit drilling tools, casing, drivepipe, or drill rods to become lodged in a borehole by reason of caving walls or impaction of sand, mud, or drill cuttings, to the extent that they cannot be pulled out. Also known as bind-seize. **2.** To burn in a bit. Also known as burn-in. **3.** The premature setting of cement, especially when cement slurry hardens before it can be ejected fully from pumps or drill rods during a borehole cementation operation. **4.** The act or process of drilling a borehole by utilizing a drill fluid chilled to minus 30–40°F (minus 34–40°C) as a means of consolidating, by freezing, the borehole wall materials or core as the drill penetrates a water-saturated formation, such as sand or gravel.

freeze drying [ENG] A method of drying materials, such as certain foods, that would be destroyed by the loss of volatile ingredients or by drying temperatures above the freezing point; the material is frozen under high vacuum so that ice or other frozen solvent will quickly sublime and a porous solid remain.

freezer [MECH ENG] An insulated unit, compartment, or room in which perishable foods are quick-frozen and stored.

freeze sinking [MIN ENG] A method of shaft sinking in waterlogged strata by the use of cold brine circulating through a system of pipes until an ice wall is formed. Also known as freezing method.

freeze-up [MECH ENG] Abnormal operation of a refrigerating unit because ice has formed at the expansion device.

freezing method *See* freeze sinking.

freezing microtome [ENG] A microtome used to cut frozen tissue.

freight car [ENG] A railroad car in or on which freight is transported.

freighter [ENG] A ship or aircraft used mainly for carrying freight.

freight ton *See* ton.

French chalk [MATER] Finely ground talc.

french coupling [DES ENG] A coupling having both right- and left-handed threads.

French drain [CIV ENG] An underground passage for water, consisting of loose stones covered with earth.

French polish [MATER] Shellac dissolved in methylated spirits.

frequency characteristic *See* frequency-response curve.

frequency meter [ENG] **1.** An instrument for measuring the frequency of an alternating current; the scale is usually graduated in hertz, kilohertz, and megahertz.

2. A device calibrated to indicate frequency of a radio wave.

frequency-modulated radar [ENG] Form of radar in which the radiated wave is frequency modulated, and the returning echo beats with the wave being radiated, thus enabling range to be measured.

frequency-modulation Doppler [ENG] Type of radar involving frequency modulation of both carrier and modulation on radial sweep.

frequency response [ENG] A measure of the effectiveness with which a circuit, device, or system transmits the different frequencies applied to it; it is a phasor whose magnitude is the ratio of the magnitude of the output signal to that of a sine-wave input, and whose phase is that of the output with respect to the input. Also known as amplitude-frequency response; sine-wave response.

frequency-response curve [ENG] A graph showing the magnitude or the phase of the freqency response of a device or system as a function of frequency. Also known as frequency characteristic.

frequency spectrum [SYS ENG] In the analysis of a random function of time, such as the amplitude of noise in a system, the limit as T approaches infinity of $1/2\pi T$ times the ensemble average of the squared magnitude of the amplitude of the Fourier transform of the function from $-T$ to T. Also known as power-density spectrum; power spectrum; spectral density.

frequency study *See* work sampling.

fresh-core technique [PETRO ENG] A method in which a core sample, fresh from the field, is subjected to waterflooding in the laboratory, and the resulting residual oil is determined; used to calculate total waterflood recovery of oil from a reservoir formation.

fretting corrosion [MET] Surface damage usually in an air environment between two surfaces, one or both of which are metals, in close contact under pressure and subject to a slight relative motion. Also known as chafing corrosion.

friability [MATER] The ease with which a material is crumbled, pulverized, or reduced to powder.

friable [MATER] Referring to the property of a substance capable of being easily rubbed, crumbled, or pulverized into powder.

friction bearing [MECH ENG] A solid bearing that directly contacts and supports an axle end.

friction blocks [PETRO ENG] Thin blocks with cylindrical surfaces that drag on the inside of the well casing to prevent rotation of the packer (seal between the outside of tubing and inside of casing).

friction bonding [ENG] Soldering of a semi-conductor chip to a substrate by vibrating the chip back and forth under pressure to create friction that breaks up oxide layers and helps alloy the mating terminals.

friction brake [MECH ENG] A brake in which the resistance is provided by friction.

friction calendering [ENG] Process wherein an elastomeric compound is forced into the interstices of woven or cord fabrics while passing between calender rolls.

friction clutch [MECH ENG] A clutch in which torque is transmitted by pressure of the clutch faces on each other.

friction drive [MECH ENG] A drive that operates by the friction forces set up when one rotating wheel is pressed against a second wheel.

friction feed [MIN ENG] Longitudinal movement or advance of a drill stem and bit accomplished by friction devices in a diamond-drill swivel head, as opposed to a system consisting entirely of meshing gears.

friction gear [MECH ENG] Gearing in which motion is transmitted through friction between two surfaces in rolling contact.

friction horsepower [MECH ENG] Power dissipated in a machine through friction.

friction pile [CIV ENG] A bearing pile surrounded by earth and supported entirely by friction; carries no load at its end.

friction saw [MECH ENG] A toothless circular saw used to cut materials by fusion due to frictional heat.

friction sawing [MECH ENG] A burning process to cut stock to length by using a blade saw operating at high speed; used especially for the structural parts of mild steel and stainless steel.

friction tape [MATER] Cotton tape impregnated with a sticky moisture-repelling compound; used chiefly to hold rubber-tape insulation in position over a joint or splice.

friction-tube viscometer [ENG] Device to determine liquid viscosity by measurement of pressure drop through a friction tube with the liquid in viscous flow; gives direct solution to Poiseuille's equation.

friction welding [ENG] A welding process for metals and thermoplastic materials in which two members are joined by rubbing the mating faces together under high pressure.

friction yielding prop *See* mechanical yielding prop.

frigorimeter [ENG] A thermometer which measures low temperatures.

fringe howl [ENG ACOUS] Squeal or howl heard when some circuit in a receiver is on the verge of oscillation.

Frise aileron [AERO ENG] A type of aileron having its leading edge projecting well ahead of the hinge axis.

frit [MATER] Fusible ceramic mixture used to make glazes and enamels for dinnerware and metallic surfaces, as on stoves and metal-base basins and tubs.

frit seal [ENG] A seal made by fusing together metallic powders with a glass binder, for such applications as hermetically sealing ceramic packages for integrated circuits.

fritting [ENG] Fusing materials for glass by application of heat. [MET] The pasty condition, usually occurring a little below the melting point, of the powdered ore, flux, and other reagents in fire assaying.

frog [DES ENG] A hollow on one or both of the larger faces of a brick or block; reduces weight of the brick or block; may be filled with mortar. [ENG] A device which permits the train or tram wheels on one rail of a track to cross the rail of an intersecting track.

from-to tester [ENG] Test equipment which checks continuity or impedance between points.

frontal-advance performance [PETRO ENG] The theory that, during the waterflood of a formation reservoir, displacement of oil causes a desaturation of the displaced fluids in accordance with relative-permeability relationships, and the displacement is linear.

frontal drive [PETRO ENG] In an oil reservoir with constant pressure (by gas injection or from a large gas-to-oil ratio), the driving of oil fluids into the wellbore by the free gas.

frontal passage [AERO ENG] The transit of an aircraft through a frontal zone.

front-end loader [MECH ENG] An excavator consisting of an articulated bucket mounted on a series of movable arms at the front of a crawler or rubber-tired tractor.

front slagging [ENG] Skimming slag from the mixture of slag and molten metal as it flows through a taphole.

frosted glass [MATER] Glass that has been etched with sand, or appears to have been so treated.

frosting [ENG] Decorating a scraped metal surface with a handscraper. Also known as flaking.

frost line [MATER] In polyethylene film extrusion, a ring-shaped area with a frosty appearance at the point where the film reaches its final diameter.

frost-point hygrometer [ENG] An instrument for measuring the frost point of the atmosphere; air under test is passed continuously across a polished surface whose temperature is adjusted so that a thin deposit of frost is formed which is in equilibrium with the air.

froth flotation [ENG] A process for recovery of particles of ore or other material, in which the particles adhere to bubbles and can be removed as part of the froth.

frothing [ENG] The producing of relatively stable bubbles at an air-liquid interface as the result of agitation, aeration, ebulliation, or chemical reaction; it can be an undesired side effect, but in minerals beneficiation it is the basis of froth flotation.

frothing collector [MIN ENG] An ore collector which in addition produces a stable foam.

frozen pipe [PETRO ENG] A pipe that is immobilized in a borehole because caving has settled around the outside of the pipe.

Frue vanner [MIN ENG] A side-shake type of ore-dressing apparatus consisting of an inclined rubber belt on which the material is washed by a constant flow of water.

fuel [MATER] A material that is burnt to release heat energy, for example, coal, oil, or uranium.

fuel bed [MECH ENG] A layer of burning fuel, as on a furnace grate or a cupola.

fuel filter [ENG] A device, as in an internal combustion engine, that removes particles from the fuel.

fuel gas [MATER] A gaseous fuel used to provide heat energy when burned with oxygen.

fuel injection [MECH ENG] The delivery of fuel to an internal combustion engine cylinder by pressure from a mechanical pump.

fuel injector [MECH ENG] A pump mechanism that sprays fuel into the cylinder of an internal combustion engine at the appropriate part of the cycle.

fuel oil [MATER] A liquid product burned to generate heat, exclusive of oils with a flash point below 100°F (38°C); includes heating oils, stove oils, furnace oils, bunker fuel oils.

fuel pump [MECH ENG] A pump for drawing fuel from a storage tank and delivering it to an engine or furnace.

fuel shutoff [AERO ENG] **1.** The action of shutting off the flow of liquid fuel into a combustion chamber or of stopping the combustion of a solid fuel. **2.** The event or time marking this action.

fuel structure ratio See fuel-weight ratio.

fuel system [MECH ENG] A system which stores fuel for present use and delivers it as needed.

fuel tank [MECH ENG] The operating, fuel-storage component of a fuel system.

fuel-weight ratio [AERO ENG] The ratio of the weight of a rocket's fuel to the weight of the unfueled rocket. Also known as fuel structure ratio.

fugitive air [MIN ENG] Air which moves through the ventilation fan but never reaches the mine workings.

fulchronograph [ENG] An instrument for recording lightning strokes, consisting of a rotating aluminum disk with several hundred steel fins on its rim; the fins are magnetized if they pass between two coils when these are carrying the surge current of a lightning stroke.

fulgurator [ENG] An atomizer used to spray salt solutions into a flame for analysis.

full annealing [MET] Heating steel to a high temperature and then cooling to ambient or near-ambient temperatures.

full automatic plating [MET] Electroplating a piece of work that is carried through the full cycle automatically.

full-cell process [ENG] A process of preservative treatment of wood that uses a pressure vessel and first draws a vacuum on the charge of wood and then introduces the preservative without breaking the vacuum. Also known as Bethell process.

fuller [MET] A die or portion of a die used in preliminary forging operations to reduce the cross section somewhere between the ends of a piece of stock.

full-face firing [MIN ENG] Drilling of small-diameter holes from top to bottom of the face.

full-face tunneling [CIV ENG] A system of tunneling in which the tunnel opening is enlarged to desired diameter before extension of the tunnel face.

full feathering See feathering.

full-gear [MECH ENG] The condition of a steam engine when the valve is operated to the maximum extent by the link motion.

full-mill [BUILD] A type of construction in which all vertical apertures open onto shafts of brick or other fireproof material; used for fire retardance.

full-seam mining [MIN ENG] A mining system in which the entire section is dislodged together and the coal is separated from the rock outside the mine by the cleaning plant.

full subsidence [MIN ENG] The greatest amount of subsidence occurring as a result of mine workings.

full-track vehicle [MECH ENG] A vehicle entirely supported, driven, and steered by an endless belt, or track, on each side; for example, a tank.

full wires [MATER] In wire-mesh cloth, wires running the short way of the cloth as woven. Also known as shute wires.

fully developed mine [MIN ENG] In coal mining, a mine where all development work has reached the boundaries and further extraction will be done on the retreat.

fumble [IND ENG] An unintentional sensory-motor error that may be unavoidable.

fumigating [ENG] The use of a chemical compound in a gaseous state to kill insects, nematodes, arachnids, rodents, weeds, and fungi in confined or inaccessible locations; also used to control weeds, nematodes, and insects in the field.

fundamental motion *See* elemental motion.

fungicide [MATER] An agent that kills or destroys fungi.

fungi-proofing [ENG] Application of a protective chemical coating that inhibits growth of fungi.

fungistat [MATER] A compound that inhibits or prevents growth of fungi.

funicular [ENG] Cable railway up a mountain, with cars simultaneously ascending and descending to counterbalance.

funicular railroad [ENG] A railroad system used primarily to ascend and descend mountains; the weight of the descending train helps to move the ascending train up the mountain.

funnel [DES ENG] A tube with one conical end that sometimes holds a filter; the function is to direct flow of a liquid or, if a filter is present, to direct a flow that was filtered.

funnel-flow bin [ENG] A bin in which solid flows toward the outlet in a channel that forms within stagnant material.

fur [MATER] The dressed pelt of a mammal.

furnace [ENG] An apparatus in which heat is liberated and transferred directly or indirectly to a solid or fluid mass for the purpose of effecting a physical or chemical change.

furnace brazing [MET] Joining two metals by mechanical union of the filler metal and joint, then heating the composite in a furnace.

furnace cupola *See* cupola.

furnace lining [ENG] The interior part of a furnace in contact with a molten charge and hot gases; constructed of heat-resistant material.

furnace oil [MATER] Distillate fuel oil intended primarily for domestic central heating systems; usually No. 1 fuel oil.

furnace refining [MET] Purification of molten metal by treatment in a reverberatory furnace.

furnace soldering [MET] Soldering by heating clamped members to the appropriate temperature in a furnace.

furred ceiling [BUILD] A ceiling in which the furring units are attached directly to the structural units of the building.

furring [BUILD] Thin strips of wood or metal applied to the joists, studs, or wall of a building to level the surface, create an air space, or add thickness.

furring brick [MATER] Hollow brick grooved for plastering.

furring tile [MATER] Non-load-bearing clay tile used for lining interior walls.

furrow [ENG] A trench plowed in the ground.

fuse [ENG] Also spelled fuze. **1.** A device with explosive components designed to initiate a train of fire or detonation in an item of ammunition by an action such as hydrostatic pressure, electrical energy, chemical energy, impact, or a combination of these. **2.** A nonexplosive device designed to initiate an explosion in an item of ammunition by an action such as continuous or pulsating electromagnetic waves or acceleration.

fuse blasting cap [ENG] A small copper cylinder closed at one end and charged with a fulminate.

fuse body [ENG] The part of a fuse contributing the major portion of the total weight, and which houses the majority of the functioning parts, and to which smaller parts are attached.

fused quartz [MATER] A glasslike insulating material made by melting crushed crystals of natural quartz or a certain type of quartz sand.

fused silica *See* silica glass.

fused spray deposit [MET] In thermal spraying, deposit which is sprayed on a preheated substrate and has the capability to coalesce within itself as well as to the substrate.

fuse gage [ENG] An instrument for slicing time fuses to length.

fusehead [ENG] That part of an electric detonator consisting of twin metal conductors, bridged by fine resistance wire, and surrounded by a bead of igniting compound which burns when the firing current is passed through the bridge wire.

fuselage [AERO ENG] In an airplane, the central structure to which wings and tail are attached; it accommodates flight crew, passengers, and cargo.

fuse lighter [ENG] A device for facilitating the ignition of the powder core of a fuse.

fusel oil [MATER] A volatile, poisonous mixture of isoamyl, butyl, propyl, and heptyl alcohols produced as by-products in alcoholic fermentation of starches, grains, or fruits to produce ethyl alcohol.

fusible alloy [MET] A low melting alloy, usually of bismuth, tin, cadmium, and lead, which melts at temperatures as low as 70°C (160°F).

fusible plug *See* safety plug.

fusing disk [MECH ENG] A rapidly spinning disk that cuts metal by melting it.

fusion piercing [ENG] A method of producing vertical blastholes by virtually burning holes in rock. Also known as piercing.

fusion-piercing drill [ENG] A machine designed to use the fusion-piercing mode of producing holes in rock. Also known as det drill; jet-piercing drill; Linde drill.

fusion welding [MET] Any welding oper-

ation involving melting of the base or parent metal.

fusion zone [MET] The volume of base or parent metal melted during a welding operation.

fuze *See* fuse.

fuzz [MATER] Fibers which protrude from the surface of a sheet of paper.

G

gabion [ENG] A bottomless basket of wickerwork or strap iron filled with earth or stones; used in building fieldworks or as revetments in mining. Also known as pannier.

gad [MIN ENG] **1.** A heavy steel wedge, 6 or 8 inches (15 or 20 centimeters) long, with a narrow chisel point used in mining to cut samples, break out pieces of loose rock, and so on. **2.** A small iron punch with a wooden handle used to break up ore.

gadder [MIN ENG] A small car or platform with a drilling machine attached, to make a straight line of holes along its course in getting out dimension stone. Also known as gadding car; gadding machine.

gadding car *See* gadder.

gadding machine *See* gadder.

gage Also spelled gauge. [CIV ENG] The distance between the inner faces of the rails of railway track; standard gage in the United States is 4 feet 8 ½ inches (1.44 meters). [DES ENG] **1.** A device for determining the relative shape or size of an object. **2.** The thickness of a metal sheet, a rod, or a wire. [ENG] The minimum sieve size through which most (95% or more) of an aggregate will pass.

gage block [DES ENG] A chrome steel block having two flat, parallel surfaces with the parallel distance between them being the size marked on the block to a guaranteed accuracy of a few millionths of an inch; used as the standard of precise lineal measurement for most manufacturing processes. Also known as precision block; size block.

gage cock [ENG] A valve located on a water column of a boiler drum.

gaged brick [MATER] Brick which has been ground or otherwise produced to accurate dimensions.

gage glass [ENG] A glass, plastic, or metal tube, usually equipped with shutoff valves, that is connected by a suitable fitting to a tank or vessel, for the measurement of liquid level.

gage length [ENG] Original length of the portion of a specimen measured for strain, length changes, and other characteristics.

gage loss [MIN ENG] The diametrical reduction in the size of a bit or reaming shell caused by wear through use.

gage plate [CIV ENG] A plate inserted between the parallel rails of a railroad track to maintain the gage.

gage point [DES ENG] A point used to position a part in a jig, fixture, or qualifying gage.

gage pressure [MECH ENG] The amount by which the total absolute pressure exceeds the ambient atmospheric pressure.

gager [PETRO ENG] An oil-field worker who gathers oil samples, tests them to determine their gravity and freedom from water, and measures the quantity of oil that is run from the producer's tank to the pipeline.

gagger [MET] An irregular-shaped piece of metal used in a sand mold to reinforce and support a metal casting.

gain [ENG] A cavity in a piece of wood prepared by notching or mortising so that a hinge or other hardware or another piece of wood can be placed on the cavity.

gain sensitivity control *See* differential gain control.

galbanum [MATER] A yellowish to brownish gum resin derived from *Ferula galbaniflua*, a perennial herb of western Asia; used in medicine.

gallery [MIN ENG] A level or drift.

gallery testing [MIN ENG] A method of testing explosives; a test condition is achieved by firing light charges without any stemming, and heavier charges with only 1 inch (2.5 centimeters) of stemming.

galley [ENG] The kitchen of a ship, airplane, or trailer.

galling [MET] Surface damage on mating, moving metal parts due to friction caused by local welding of high spots.

gallium [MET] A silvery-white metal, melting at 29.7°C, boiling at 1983°C.

Galloway sinking and walling stage   *See* sinking and walling scaffold.

Galloway stage [MIN ENG] A platform of several decks suspended near the shaft during the sinking operation.

gallows *See* headframe.

gallows frame *See* headframe.

galvanic corrosion [MET] Electrochemical corrosion associated with the current in a galvanic cell, caused by dissimilar metals in an electrolyte because of the difference in potential (emf) of the two metals.

galvanize [MET] To deposit zinc on the surface of metal by the processes of hot dipping, sherardizing, or sometimes electroplating.

130**galvanometer** [ENG] An instrument for indicating or measuring a small electric current by means of a mechanical motion derived from electromagnetic or electrodynamic forces produced by the current.

galvanometer recorder [ENG ACOUS] A sound recorder in which the audio signal voltage is applied to a coil suspended in a magnetic field; the resulting movements of the coil cause a tiny attached mirror to move a reflected light beam back and forth across a slit in front of a moving photographic film.

gambir [MATER] The yellowish extract from the twigs and leaves of the Malayan wood vine *Uncaria gambir* (Rubiaceae); used for tanning and dyeing and as an astringent. Also known as pale catechu; terra japonica; white cutch.

gambrel roof [BUILD] A roof with two sloping sides stepped at different angles on each side of the center ridge; the lower slope is steeper than the upper slope.

gamene *See* madder.

gamma camera [ENG] An instrument consisting of a large, thin scintillation crystal or array of photomultiplier tubes, a multichannel collimator, and circuitry to analyze the pulses produced by the photomultipliers; used to visualize the distribution of radioactive compounds in the human body.

gamma counter [ENG] A device for detecting gamma radiation, primarily through the detection of fast electrons produced by the gamma rays; it either yields information about integrated intensity within a time interval or detects each photon separately.

gamma iron [MET] Iron having a face-centered cubic lattice structure, stable between 910 and 1400°C.

gamma logging [ENG] Obtaining, by means of a gamma-ray probe, a record of the intensities of gamma rays emitted by the rock strata penetrated by a borehole.

gamma-ray altimeter [ENG] An altimeter, used at altitudes under several hundred feet, that measures the photon backscatter from the earth resulting from the transmission of photons to earth from a cobalt-60 gamma source in the plane.

gamma-ray detector [ENG] An instrument used on ships to identify and measure abnormal concentrations of gamma rays in the oceans.

gamma-ray level indicator [ENG] A level indicator in which the rising level of the liquid or other material reduces the amount of radiation passing from a gamma-ray source through the container to a Geiger counter or other radiation detector.

gamma-ray probe [ENG] A gamma-ray counter built into a watertight case small enough to be lowered into a borehole.

gamma-ray tracking [ENG] Use of three tracking stations, located at the three corners of a triangle centered on a missile about to be launched, to obtain accurate azimuthal tracking of a cobalt-60 gamma source in the tail.

gamma-ray well logging [ENG] Measurement of gamma-ray intensity versus depth down the wellbore; used to identify rock strata, their position, and their thicknesses.

gang chart [IND ENG] A multiple-activity process chart used for groups of men on materials-handling operations.

gang drill [MECH ENG] A set of drills operated together in the same machine; used in rock drilling.

gang milling [ENG] Rolling of material by means of a composite machine with numerous cutting blades.

gang saw [MECH ENG] A steel frame in which thin, parallel saws are arranged to operate simultaneously in cutting logs.

gangway [MIN ENG] **1.** A principal underground haulage road. **2.** A passageway into or out of an underground mine.

ganister [MATER] A fine mixture of quartz and fireclay which is used to line certain furnaces for metallurgical processes.

gantlet [CIV ENG] A stretch of overlapping railroad track, with one rail of one track being between the two rails of another track; used over narrow bridges and passes.

gantry [ENG] A frame erected on side supports so as to span an area and support and hoist machinery and heavy materials.

gantry crane [MECH ENG] A bridgelike hoisting machine having fixed supports or arranged for running along tracks on ground level.

Gantt chart [IND ENG] In production planning and control, a type of bar chart depicting the work planned and done in relation to time; each division of space represents both a time interval and the amount of work to be done during that interval.

Gantt task and bonus plan [IND ENG] A wage incentive plan in which high task efficiency is maintained by providing a percentage bonus as a reward for production in excess of standard.

gap [MET] An opening at the point of closest approach between faces of members in a weld joint.

gap-filler radar [ENG] Radar used to fill gaps in radar coverage of other radar.

gap-framepress [MECH ENG] A punch press whose frame is open at bed level so that wide work or strip work can be inserted.

gap-graded aggregate [MATER] Aggregate in which certain size particles are entirely or substantially absent.

gap lathe [MECH ENG] An engine lathe with a sliding bed providing enough space for turning large-diameter work.

gap scanning [ENG] In ultrasonic testing, a coupling technique in which a sound beam is projected through a short fluid column that flows through a nozzle on an ultrasonic search unit.

garbage pitch [MATER] Dark-brown to black pitch material obtained as a by-product residue from the burning of garbage; properties are analogous to complex hydrocarbons; used to make paints, varnishes, tarred paper, and waterproofing compound.

Garbutt rod [PETRO ENG] A device used to pull the standing valve out of a tubing-type oil-well sucker-rod pump.

Gardner crusher [MIN ENG] A swing-and-hammer crusher; the U-shaped hammers are thrown by a revolving shaft against the feed and a heavy anvil inside the housing.

garland [MIN ENG] A channel fixed around a shaft in order to catch the water draining down the walls and conduct it to a lower level. Also known as water curb; water garland; water ring.

garlic oil [MATER] An essential oil obtained from steam distillation of garlic; contains chiefly a mixture of terpenes, with organic sulfides also present.

garnet hinge [DES ENG] A hinge with a vertical bar and horizontal strap.

garnet paper [MATER] Paper with a layer of crushed garnet on one side; used as an abrasive or polisher.

garret [BUILD] The part of a house just under the roof.

garter spring [DES ENG] A closed ring formed of helically wound wire.

gas *See* gasoline.

gas alarm [MIN ENG] A signal system which warns mine workers of dangerous concentration of firedamp.

gas anchor [PETRO ENG] A downhole gas separator used to reduce gas-in-oil froth before the pump to increase pump efficiency.

gas and mist sampler [MIN ENG] An instrument for automatic collection of one sample per hour of airborne contaminants such as sulfur dioxide or ammonia.

gas bearing [MECH ENG] A journal or thrust bearing lubricated with gas. Also known as gas-lubricated bearing.

gas brazing *See* gas-flame brazing.

gas burner [ENG] A hole or a group of holes through which a combustible gas or gas-air mixture flows and burns.

gas cap [PETRO ENG] Gas occurring above liquid hydrocarbons in a reservoir under such trap conditions as the presence of water which prevents downward migration or the abutment of an impermeable formation against the reservoir.

gas-cap drive [PETRO ENG] Driving liquid hydrocarbons through a porous reservoir and toward well holes by utilizing the pressure of gas overlying the liquid pool.

gas-cap expansion [PETRO ENG] Process of reservoir-liquids displacement by the natural expansion of the reservoir gas cap to fill the voids vacated by recovered liquids.

gas-cap injection *See* external gas injection.

gas-cap reservoir [PETRO ENG] Two-phase reservoir in which a free area of gas (a gas cap) is underlain by an oil or liquid phase.

gas carburizing [MET] Surface hardening by heating a metal in gas of high carbon content in order to introduce carbon into the surface layers.

gas cleaning [ENG] Removing ingredients, pollutants, or contaminants from domestic and industrial gases.

gas-compression cycle [MECH ENG] A refrigeration cycle in which hot, compressed gas is cooled in a heat exchanger, then passes into a gas expander which provides an exhaust stream of cold gas to another heat exchanger that handles the sensible-heat refrigeration effect and exhausts the gas to the compressor.

gas compressor [MECH ENG] A machine that increases the pressure of a gas or vapor by increasing the gas density and delivering the fluid against the connected system resistance.

gas-condensate well [PETRO ENG] A well producing hydrocarbons from a gas-condensate reservoir.

gas coning [PETRO ENG] The tendency of gas in a gas-drive reservoir to push oil downward in an inverse cone contour toward the casing perforations; at the extreme of coning, gas, not oil, will be produced from the well.

gas cut [PETRO ENG] A foamy mixture of gas and drilling mud recovered in testing.

gas cutting [MET] Cutting metal with the heat of an oxyacetylene flame.

gas cyaniding See carbonitriding.

gas cycling [PETRO ENG] A petroleum enhanced-recovery process which injects the gas produced with oil back into the oil sand to help produce more oil.

gas cylinder [MECH ENG] The chamber in which a piston moves in a positive displacement engine or compressor.

gas depletion drive See internal gas drive.

gas detector [MIN ENG] A device which indicates the existence of firedamp or other combustible or noxious gas in a mine.

gas-drive reservoir [PETRO ENG] An oil reservoir in which gas (either natural or reinjected) provides the driving force to sweep liquids through the formation and into the wellbore.

gas emission [MIN ENG] The release of gas from the strata into the mine workings.

gas engine [MECH ENG] An internal combustion engine that uses gaseous fuel.

gaseous conduction analyzer [ENG] A device to detect organic vapors in air by measuring the change in current that flows between a heated platinum anode and a concentric platinum cathode.

gaseous fuel [MATER] A combustible gas that can be burned in a furnace or an engine.

gas etching [ENG] The removal of material from a semiconductor circuit by reaction with a gas that forms a volatile compound.

gas explosion [MIN ENG] An explosion of firedamp in a coal mine; coal dust apparently does not play a significant part.

gas field [PETRO ENG] An area underlain with little or no interruption by one or more reservoirs of commercially valuable gas.

gas-filled thermometer [ENG] A thermometer which uses a gas (usually nitrogen or hydrogen), that approximately follows the ideal gas law.

gas-flame brazing [MET] A brazing process for which the heat is supplied by a gas flame. Also known as gas brazing.

gas furnace [ENG] An enclosure in which a gaseous fuel is burned.

gas generator [MECH ENG] An apparatus that supplies a high-pressure gas flow to drive compressors, airscrews, and other machines.

gas heater [MECH ENG] A unit heater designed to supply heat by forced convection, using gas as a heat source.

gas holder [ENG] Gas storage container with vertically free top section that moves up or down to adjust to the volume of gas held.

gas hole [ENG] A cavity formed in a casting as a result of cavitation.

gasification [MIN ENG] A method for exploiting poor-quality coal and thin coal seams by burning the coal in place to produce combustible gas which can be burned to generate power or processed into chemicals and fuels. Also known as underground gasification.

gas ignition [MIN ENG] The setting on fire of an accumulation of firedamp in a coal mine.

gas injection [MECH ENG] Injection of gaseous fuel into the cylinder of an internal combustion engine at the appropriate part of the cycle. [PETRO ENG] The injection of gas into a reservoir to maintain formation pressure and to drive liquid hydrocarbons toward the wellbores.

gas-injection well [PETRO ENG] A well hole in a reservoir into which pressurized gas is injected to maintain formation pressure or to drive liquid hydrocarbons toward other well holes.

gasket [ENG] A packing made of deformable material, usually in the form of a sheet or ring, used to make a pressure-tight joint between stationary parts. Also known as static seal.

gas lift [PETRO ENG] The injection of gas near the bottom of an oil well to aerate and lighten the column of oil to increase oil production from the well.

gas logging [PETRO ENG] Hot-wire-detector or gas-chromatographic analysis and record of gas contained in the mud stream and cuttings for a well being drilled; a common way of detecting subsurface oil and gas shows.

gas-lubricated bearing See gas bearing.

gasman [MIN ENG] An underground official who examines the mine for firedamp and has charge of its removal.

gas manometer [ENG] A gage for determining the difference in pressure of two gases, usually by measuring the difference in height of liquid columns in the two sides of a U-tube.

gas mask [ENG] A device to protect the eyes and respiratory tract from noxious gases, vapors, and aerosols, by removing contamination with a filter and a bed of adsorbent material.

gas meter [ENG] An instrument for measuring and recording the amount of gas flow through a pipe.

gasogene [MATER] A fuel gas formed by incomplete combustion of charcoal; a European development as a substitute for gasoline. Also spelled gazogene.

gas oil [MATER] A petroleum distillate boiling within the general range 450–800°F (232–426°C); usually includes kerosine, diesel fuel, heating oils, and light fuel oils.

gas-oil ratio [PETRO ENG] Approximation of oil-reservoir composition, expressed in cubic feet of gas per barrel of liquid at 14.7 psia and 60°F (15.6°C).

gas-oil separator [PETRO ENG] An oil-field stock tank or series of tanks in which wellhead pressure is reduced so that the dissolved gas associated with reservoir oil is flashed off or separated as a separate phase. Also known as gas separator; oil-field separator; oil-gas separator; oil separator; separator.

gasoline [MATER] A fuel for internal combustion engines consisting essentially of volatile flammable liquid hydrocarbons; derived from crude petroleum by processes such as distillation reforming, polymerization, catalytic cracking, and alkylation; the common name is gas. Also known as petrol.

gasoline alkylate [MATER] A gasoline component usually formed by union of an olefin with isobutane by a refinery process known as alkylation; the product has high octane value and is blended with motor and aviation gasoline to improve antiknock value.

gasoline engine [MECH ENG] An internal combustion engine that uses a mixture of air and gasoline vapor as a fuel.

gasoline gel *See* gelatinized gasoline.

gasoline locomotive [MIN ENG] A mine locomotive which uses a gaseous fuel, is comparable to the steam locomotive in radius of travel, and has a speed of 3–12 miles per hour (5–19 kilometers per hour).

gasoline pool [MATER] A concept that considers gasolines of different qualities as a single group for the purpose of blending to meet final product specifications.

gasoline pump [MECH ENG] A device that pumps and measures the gasoline supplied to a motor vehicle, as at a filling station.

gasometer [ENG] A piece of equipment that holds and measures gas; may be used in analytical chemistry to measure the quantity of gas evolved in a reaction.

gas packing [IND ENG] Packing a material such as food in an atmosphere consisting of an oxygen-free gas.

gaspar [MATER] A mixture of finely ground glass and quartz; a feldspar substitute in some applications and a hard-rubber filler.

gas pliers [DES ENG] Pliers for gripping round objects such as pipes, tubes, and circular rods.

gas pocket [MET] A cavity in a metal which contains trapped gases.

gas-pressure maintenance [PETRO ENG] The maintenance of oil-reservoir gas pressure, usually by gas injection, to increase hydrocarbon recovery and to improve reservoir production characteristics.

gas rig [MIN ENG] A borehole drill, either rotary or churn type, driven by a combustion-type engine energized by a combustible liquid, such as gasoline, or a combustible gas, such as bottle gas.

gas seal [ENG] A seal which prevents gas from leaking to or from a machine along a shaft.

gas sendout [PETRO ENG] The total gas that is produced, purchased, or withdrawn from underground storage in a certain interval of time.

gas separator *See* gas-oil separator.

gas-shielded arc welding [MET] Use of a gas atmosphere to shield the molten metal from air in arc welding.

gassing [ENG] **1.** Absorption of gas by a material. **2.** Formation of gas pockets in a material. **3.** Evolution of gas from a material during a process or procedure.

gas-solubility factor [PETRO ENG] The number of standard cubic feet of gas liberated under specified gas-oil separator conditions that are in solution in one barrel of stock-tank oil at reservoir temperature and pressure.

gas stimulation [PETRO ENG] The detonation of a nuclear explosive in the strata of a natural-gas field to make the gas flow more freely.

gassy [MIN ENG] A coal mine rating by the U.S. Bureau of Mines, applicable when an ignition occurs or if a methane content exceeding 0.25% can be detected; work must be halted if the methane exceeds 1.5% in a return airway.

gas tank [ENG] A tank for storing gas or gasoline.

gas thermometer [ENG] A device to measure temperature by measuring the pressure exerted by a definite amount of gas enclosed in a constant volume; the gas (preferably hydrogen or helium) is enclosed in a glass or fused-quartz bulb connected to a mercury manometer. Also known as constant-volume gas thermometer.

gas thermometry [ENG] Measurement of temperatures with a gas thermometer; used with helium down to about 1 K.

gas tracer [MIN ENG] Dust clouds, chemical smoke, or gaseous or radioactive tracers used to detect slowly moving air currents in a mine.

gas trap [CIV ENG] A bend or chamber in a drain or sewer pipe that prevents sewer gas from escaping.

gas-tungsten arc welding [MET] Arc welding by means of an electric arc between a single tungsten electrode and the work, with a shield of inert gas, such as argon or helium, around the electrode.

gas turbine [MECH ENG] A heat engine that converts energy of fuel into work by using compressed, hot gas as the working medium and that usually delivers its mechanical output through a rotating shaft. Also known as combustion turbine.

gas-turbine nozzle [MECH ENG] The component of a gas turbine in which the hot, high-pressure gas expands and accelerates to high velocity.

gas valve [ENG] An exhaust valve, held shut by rubber springs, used to discharge gas from the extreme top of a balloon.

gas vent [ENG] A pipe or hole that allows gas to pass off.

gas welding [MET] A welding process in which metals are joined by the heat of an oxyacetylene flame.

gas well [PETRO ENG] A well drilled for extraction of natural gas from a gas reservoir.

gate [CIV ENG] A movable barrier across an opening in a large barrier, a fence, or a wall. [ENG] **1.** A device, such as a valve or door, for controlling the passage of materials through a pipe, channel, or other passageway. **2.** A device for positioning the film in a camera, printer, or projector. [MET] The opening in a casting mold through which molten metal enters the cavity. Also known as in-gate.

gate conveyor [MIN ENG] A conveyor that carries coal from one source or face only, that is, from a single-unit or double-unit face.

gate interlock [MIN ENG] A system that prevents movement of shaft conveyances or transmission of action signals until all shaft gates are closed.

gate road bunker [MIN ENG] An appliance for coal storage from the face conveyors during peaks of production or during a stoppage of the outby transport.

Gates crusher [MECH ENG] A gyratory crusher which has a cone or mantle that is moved eccentrically by the lower bearing sleeve.

gate valve [DES ENG] A valve with a disk-shaped closing element that fits tightly over an opening through which water passes.

gather [MIN ENG] **1.** To assemble loaded cars from several production points and deliver them to main haulage for transport to the surface or pit bottom. **2.** To drive a heading through disturbed or faulty ground so as to meet the seam of coal at a convenient level or point on the opposite side.

gathering area [PETRO ENG] The area, usually down the regional dip from a hydro-

carbon trap, from which the oil or gas may have migrated updip into the trap.

gathering arm loader [MIN ENG] A machine for loading loose rock or coal; has a tractor-mounted chassis and carries a chain conveyor whose front end is built into a wedge-shaped blade.

gathering conveyor [MIN ENG] Any conveyor which gathers coal from other conveyors and delivers it either into mine cars or onto another conveyor.

gathering iron [ENG] A rod used to collect molten glass for glassblowing.

gathering motor [MIN ENG] A lightweight type of electric locomotive used to haul loaded cars from the working places to the main haulage road and to replace them with empties. Also known as electric gathering locomotive; gathering locomotive; gathering mine locomotive.

gathering motorman [MIN ENG] In bituminous coal mining, one who operates a mine locomotive to haul loaded mine cars from working places to sidings.

gathering mule [MIN ENG] The mule used to collect the loaded cars from the separate working places and to return empties.

gathering pump [MIN ENG] A portable or semiportable pump that is required for removing water encountered while opening a new mine, for extending headings or entries in an operating mine, for pump rooms or rib sections lying in the dip, for collecting water from local pools, or for sinking a shaft.

gathering ring [ENG] A clay ring placed on molten glass to collect impurities and thus permit high-quality glass to be taken from the center.

gathering system [PETRO ENG] A pipeline system used to gather gas or oil production from a number of separate wells, bringing the combined production to a central storage, pipelining, or processing terminal.

gating [ENG] A network of connecting channels, including sprues, runners, gates, and cavities, which conduct molten metal to the mold.

gauge See gage.

gaussmeter [ENG] A magnetometer whose scale is graduated in gauss or kilogauss, and usually measures only the intensity, and not the direction, of the magnetic field.

gauze [MATER] **1.** A sheer, loosely woven textile fabric similar to cheesecloth; used for surgical dressings. **2.** A plastic or wire mesh.

Gay-Lussac acid [MATER] Product of a Gay-Lussac tower in the chamber process for sulfuric acid manufacture; a mixture of sulfuric acid and nitrogen oxides.

gazogene See gasogene.

GCA radar *See* ground-controlled approach radar.

GCI radar *See* ground-controlled intercept radar.

gear [DES ENG] A toothed machine element used to transmit motion between rotating shafts when the center distance of the shafts is not too large. [MECH ENG] **1.** A mechanism performing a specific function in a machine. **2.** An adjustment device of the transmission in a motor vehicle which determines mechanical advantage, relative speed, and direction of travel.

gearbox *See* transmission.

gear case [MECH ENG] An enclosure, usually filled with lubricating fluid, in which gears operate.

gear cutter [MECH ENG] A machine or tool for cutting teeth in a gear.

gear cutting [MECH ENG] The cutting or forming of a uniform series of toothlike projections on the surface of a workpiece.

gear down [MECH ENG] To arrange gears so the driven part rotates at a slower speed than the driving part.

gear drive [MECH ENG] Transmission of motion or torque from one shaft to another by means of direct contact between toothed wheels.

geared turbine [MECH ENG] A turbine connected to a set of reduction gears.

gear forming [MECH ENG] A method of gear cutting in which the desired tooth shape is produced by a tool whose cutting profile matches the tooth form.

gear generating [MECH ENG] A method of gear cutting in which the tooth is produced by the conjugate or total cutting action of the tool plus the rotation of the workpiece.

gear grinding [MECH ENG] A gear-cutting method in which gears are shaped by formed grinding wheels and by generation; primarily a finishing operation.

gear hobber [MECH ENG] A machine that mills gear teeth; the rotational speed of the hob has a precise relationship to that of the work.

gearing [MECH ENG] A set of gear wheels.

gearing chain [MECH ENG] A continuous chain used to transmit motion from one toothed wheel, or sprocket, to another.

gearless traction [MECH ENG] Direct drive, without reduction gears.

gear level [MECH ENG] To arrange gears so that the driven part and driving part turn at the same speed.

gear loading [MECH ENG] The power transmitted or the contact force per unit length of a gear.

gear meter [ENG] A type of positive-displacement fluid quantity meter in which

the rotating elements are two meshing gear wheels.

gearmotor [MECH ENG] A motor combined with a set of speed-reducing gears.

gear oil [MATER] A lubricating oil for use in transmissions, most types of differential gears, and gears in gear boxes.

gear pump [MECH ENG] A rotary pump in which two meshing gear wheels contrarotate so that the fluid is entrained on one side and discharged on the other.

gear ratio [MECH ENG] The ratio of the angular speed of the driving member of a gear train or similar mechanism to that of the driven member; specifically, the number of revolutions made by the engine per revolution of the rear wheels of an automobile.

gear shaper [MECH ENG] A machine that makes gear teeth by means of a reciprocating cutter that rotates slowly with the work.

gear-shaving machine [MECH ENG] A finishing machine that removes excess metal from machined gears by the axial sliding motion of a straight-rack cutter or a circular gear cutter.

gearshift [MECH ENG] A device for engaging and disengaging gears.

gear teeth [DES ENG] Projections on the circumference or face of a wheel which engage with complementary projections on another wheel to transmit force and motion.

gear train [MECH ENG] A combination of two or more gears used to transmit motion between two rotating shafts or between a shaft and a slide.

gear up [MECH ENG] To arrange gears so that the driven part rotates faster than the driving part.

gear wheel [MECH ENG] A wheel that meshes gear teeth with another part.

Geco sampler [MIN ENG] Straight-line cutter designed to traverse a falling stream of ore or pulp at regular intervals, so as to divert a representative sample to a holding vessel.

Geiger-Müller probe [ENG] A Geiger-Müller counter in a watertight container, lowered into a borehole to log the intensity of the gamma rays emitted by radioactive substances in traversed rock. Also known as electronic logger; Geiger probe.

Geiger probe *See* Geiger-Müller probe.

Geissler pump [ENG] A type of air pump that uses the principle of the Torricellian vacuum, and in which the vacuum is produced by the flow of mercury back and forth between a vertically adjustable and a fixed reservoir.

gelatin *See* gelatin dynamite.

gelatin dynamite [MATER] A high explosive consisting mainly of a jellylike mass of nitroglycerin, with sodium nitrate, meal, collodion cotton, and sodium carbonate; commonly used by drillers to shatter boulders encountered in driving pipe through overburden, especially in water-filled or saturated ground. Also known as gelatin; gelignite; nitrogelatin.

gelatin extra [MATER] An explosive in which some of the nitroglycerin is replaced with ammonium nitrate.

gelatinized gasoline [MATER] Gasoline treated with a thickening agent; used in napalm bombs and flamethrowers. Also known as gasoline gel; jellied gasoline; thickened fuel.

gelatinizing agent [MATER] In manufacture of propellants, a material which softens the nitrocellulose, permitting the mixture to be processed and formed. Also known as gelling agent.

gelatinobromide [MATER] A preparation of gelatin silver bromide that is light-sensitive and used in photography.

gelation model [PETRO ENG] Electrolytic analog of a reservoir; used to investigate the areal sweep movement of water from multiple injection wells; operates by movement of copper ammonium and zinc ammonium ions through the gelatin in a flat tray that simulates the reservoir.

gel cement [MATER] Cement containing a small percentage of bentonite, which makes the mixture more homogeneous, increases the water-cement ratio, and reduces loss of water to the formation.

gel coat [MATER] A resin applied to the surface of a mold and gelled prior to placing plastic material in the mold in position for production; the gel coat becomes an integral part of the finished laminate and improves surface appearance.

gelignite See gelatin dynamite.

gelling agent See gelatinizing agent.

gel paint [MATER] Paint formulation made thixotropic by the reaction of a small amount of polyamide resin with an alkyd resin vehicle.

general manager [IND ENG] The person of general authority who performs all reasonable tasks in conducting the usual and customary business of the principal head or owner.

generating magnetometer [ENG] A magnetometer in which a coil is rotated in the magnetic field to be measured with the resulting generated voltage being proportional to the strength of the magnetic field.

generating plant See generating station.

generating station [MECH ENG] A stationary plant containing apparatus for large-scale conversion of some form of energy (such as hydraulic, steam, chemical, or nuclear energy) into electrical energy. Also known as generating plant; power station.

generator set [ENG] The aggregate of one or more generators together with the equipment and plant for producing the energy that drives them.

geochemical prospecting [ENG] The use of geochemical and biogeochemical principles and data in the search for economic deposits of minerals, petroleum, and natural gases.

geochemical well logging [ENG] Well logging dependent on geochemical analysis of the data.

geodetic satellite [AERO ENG] An artificial earth satellite used to obtain data for geodetic triangulation calculations.

geodetic survey [ENG] A survey in which the figure and size of the earth are considered; it is applicable for large areas and long lines and is used for the precise location of basic points suitable for controlling other surveys.

geolograph [ENG] A device that records the penetration rate of a bit during the drilling of a well.

geomagnetic electrokinetograph [ENG] An instrument that can be suspended from the side of a ship to measure the direction and speed of ocean currents while the ship is under way by measuring the voltage induced in the moving conductive seawater by the magnetic field of the earth.

geometrical pitch [AERO ENG] The distance a component of an airplane propeller would move forward in one complete turn of the propeller if the path it was moving along was a helix that had an angle equal to an angle between a plane perpendicular to the axis of the propeller and the chord of the component.

geometric construction [ENG] Construction that employs only straightedge and compasses or is carried out by drawing only straight lines and circles.

geometric programming [SYS ENG] A nonlinear programming technique in which the relative contribution of each of the component costs is first determined; only then are the variables in the component costs determined.

geophysical engineering [ENG] A branch of engineering that applies scientific methods for locating mineral deposits.

geophysical prospecting [ENG] Application of quantitative concepts and principles of physics and mathematics in geologic explorations to discover the character of and mineral resources in underground rocks in the upper portions of the earth's crust.

geostationary satellite [AERO ENG] A satellite that orbits the earth from west to east at such a speed as to remain fixed over a given place on the earth's equator at approximately 35,900 kilometers altitude; makes one revolution in 24 hours, synchronous with the earth's rotation. Also known as synchronous satellite.

geotechnics [CIV ENG] The application of scientific methods and engineering principles to civil engineering problems through acquiring, interpreting, and using knowledge of materials of the crust of the earth.

geotechnology [ENG] Application of the methods of engineering and science to exploitation of natural resources.

geotextile [MATER] A woven or nonwoven fabric manufactured from synthetic fibers or yarns that is designed to serve as a continuous membrane between soil and aggregate in a variety of earth structures.

geothermal prospecting [ENG] Exploration for sources of geothermal energy.

geothermal well logging [ENG] Measurement of the change in temperature of the earth by means of well logging.

geothermometer [ENG] A thermometer constructed to measure temperatures in boreholes or deep-sea deposits.

geranium oil [MATER] A pale-yellow or green liquid distilled from the herb of several Pelargonium species; chief known constituents are citronellol and geraniol, both used in perfumery and as flavoring agents.

gerber beam [CIV ENG] A long, straight beam that functions essentially as a cantilevered beam by the insertion of two hinges in alternate spans.

German cupellation [MET] A refining method using a large reverberatory furnace with a fixed bed and a movable roof; the bullion to be cupelled is all charged at once, and the silver is not refined in the same furnace where the cupellation is carried on.

germanium [MET] A rare metal used in semiconductors, alloys, and glass.

German silver See nickel silver.

German tubbing [MIN ENG] A form of tubbing, with internal flanges and bolts, for lining circular shafts sunk through heavily watered strata.

germicide [MATER] An agent that destroys germs.

gesso [MATER] A material made from chalk and gelatin or casein glue; painted on panels to furnish a surface for tempera work or for polymer-based paints.

get [IND ENG] A combination of two or more of the elemental motions of search, select, grasp, transport empty, and transport loaded; applied to time-motion studies.

getter-ion pump [ENG] A high-vacuum pump that employs chemically active metal layers which are continuously or intermittently deposited on the wall of the pump, and which chemisorb active gases while inert gases are "cleaned up" by ionizing them in an electric discharge and drawing the positive ions to the wall, where the neutralized ions are buried by fresh deposits of metal. Also known as sputter-ion pump.

gewel hinge [DES ENG] A hinge consisting of a hook inserted in a loop.

ghatti gum [MATER] A water-soluble gum from the tree Anogeissus latifolia, forming a viscous glue in water; used as an emulsifier.

ghedda wax [MATER] A beeswax that is obtained from African and Indian bees.

ghost lines See ferrite banding.

ghost structure See ferrite banding.

giant See hydraulic monitor.

giant powder [MATER] A blasting powder made of nitroglycerin, sodium nitrate, sulfur, and rosin, sometimes with kieselguhr.

gib [ENG] A removable plate designed to hold other parts in place or act as a bearing or wear surface. [MIN ENG] 1. A temporary support at the face to prevent coal from falling before the cut is complete, either by hand or by machine. 2. A prop put in the holing of a seam while being undercut. 3. A piece of metal often used in the same hole with a wedge-shaped key for holding pieces together.

Gibbs apparatus [ENG] A compressed-oxygen breathing apparatus used by miners in the United States.

Giesler coal test [ENG] A plastometric method for estimating the coking properties of coals.

Gilbreth's micromotion study [IND ENG] A time and motion study based on the concept that all work is performed by using a relatively few basic operations in varying combinations and sequence; basic elements (therbligs) include grasp, search, move, reach, and hold.

gilding metal [MET] A copper alloy (about 90% copper, 10% zinc) used to jacket small-arms bullets, to form detonator or primer cups, and to form rotating bands for artillery projectiles; it can be readily engraved by the lands as the projectile moves down the bore.

gill net [ENG] A net that entangles the gill covers of fish.

Gilmour heat-exchange method [ENG] Thermal design method for heat exchangers by solution of five unique equations containing a minimum number of variables and involving tube-side, shell-side, tube-wall, and dirt resistance.

gimbal [ENG] **1.** A device with two mutually perpendicular and intersecting axes of rotation, thus giving free angular movement in two directions, on which an engine or other object may be mounted. **2.** In a gyro, a support which provides the spin axis with a degree of freedom. **3.** To move a reaction engine about on a gimbal so as to obtain pitching and yawing correction moments. **4.** To mount something on a gimbal.

gimbaled nozzle [MECH ENG] A nozzle supported on a gimbal.

gimbal freedom [ENG] Of a gyro, the maximum angular displacement about the output axis of a gimbal.

gimbal lock [ENG] A condition of a two-degree-of-freedom gyro wherein the alignment of the spin axis with an axis of freedom deprives the gyro of a degree-of-freedom and therefore its useful properties.

gimlet [DES ENG] A small tool consisting of a threaded tip, grooved shank, and a cross handle; used for boring holes in wood.

gimlet bit [DES ENG] A bit with a threaded point and spiral flute; used for drilling small holes in wood.

gin [MECH ENG] A hoisting machine in the form of a tripod with a windlass, pulleys, and ropes.

gingelly oil *See* sesame oil.

ginger-grass oil [MATER] A type of citronella oil derived from sofia grass that contains about 50% geraniol; used in perfumery.

ginger oil [MATER] A thick, yellowish essential oil, soluble in most organic solvents and insoluble in water; distilled from dried ginger; main components are citral, borneol, and phellandrene; used as flavoring in liqueurs and soft drinks.

ginging [MIN ENG] **1.** Lining a shaft with masonry or brick. **2.** The brick or masonry of a shaft lining.

gin pit [MIN ENG] A shallow mine, the hoisting from which is done by a gin.

gin pole [MECH ENG] A hand-operated derrick which has a nearly vertical pole supported by guy ropes; the load is raised on a rope that passes through a pulley at the top and over a winch at the foot. Also known as guyed-mast derrick; pole derrick; standing derrick.

gin tackle [MECH ENG] A tackle made for use with a gin.

gipsy winch [MIN ENG] A small winch that may be attached to a post and operated by a rotary motion or the reciprocating action of a handle having a pair of pawls and a ratchet.

girder [CIV ENG] A large beam made of metal or concrete, and sometimes of wood.

girt [CIV ENG] **1.** A timber in the second-floor corner posts of a house to serve as a footing for roof rafters. **2.** A horizontal member to stiffen the framework of a building frame or trestle. [ENG] A brace member running horizontally between the legs of a drill tripod or derrick. [MIN ENG] In square-set timbering, a horizontal brace running parallel to the drift.

glair [MATER] A sizing liquid made of egg white beaten with vinegar; used to prepare a surface of a book binding for gilding.

gland [ENG] **1.** A device for preventing leakage at a machine joint, as where a shaft emerges from a vessel containing a pressurized fluid. **2.** A movable part used in a stuffing box to compress the packing.

glass [MATER] A hard, amorphous, inorganic, usually transparent, brittle substance made by fusing silicates, sometimes borates and phosphates, with certain basic oxides and then rapidly cooling to prevent crystallization.

glassblowing [ENG] Shaping a mass of viscid glass by inflating it with air introduced through a tube.

glass-bonded mica [MATER] An insulating material made by compressing a mixture of powdered glass and powdered natural or synthetic mica at high temperatures.

glass brick [MATER] A hollow block of translucent glass with patterns molded on the faces; used in partitions.

glass cutter [ENG] A tool equipped with a steel wheel or a diamond point used to cut glass.

glass fiber [MATER] A glass thread less than a thousandth of an inch (25 micrometers) thick, used loosely or in woven form as an acoustic, electrical, or thermal insulating material and as a reinforcing material in laminated plastics.

glass furnace [ENG] A large, covered furnace or tank for melting large batches of glass, in which heat is supplied by a flame playing over the glass surface, and regenerative heating of combustion air and gas is usually employed. Also known as glass tank.

glass heat exchanger [ENG] Any heat exchanger in which glass replaces metal, such as shell-and-tube, cascade, double-pipe, bayonet, and coil exchangers.

glassine [MATER] A thin, dense, transparent, supercalendered paper from highly refined sulfite pulp, used for envelope windows, for sanitary wrapping, and as an insulating paper between layers of iron-core transformer windings.

glass insulator [MATER] An insulator for a power transmission line made of annealed or toughened (tempered) glass.

glassmakers' soap [MATER] A substance such as manganese dioxide added to glass to remove the green color created by iron salts.

glass paper [MATER] **1.** Paper with a layer of pulverized glass; used as an abrasive. **2.** Paper made of glass fibers.

glass pot [ENG] A crucible used for making small amounts of glass.

glass sand [MATER] High-quartz sand used in glassmaking; contains small amounts of aluminum oxide, iron oxide, calcium oxide, and magnesium oxide.

glass seal [ENG] An airtight seal made by molten glass.

glass tank *See* glass furnace.

glass-tube manometer [ENG] A manometer for simple indication of difference of pressure, in contrast to the metallic-housed mercury manometer, used to record or control difference of pressure or fluid flow.

glassware [MATER] Articles, especially tableware, made of glass.

glass wool [MATER] A mass of glass fibers resembling wool and used as insulation, packing, and air filters.

glassy alloy [MET] An alloy having an amorphous or glassy structure. Also known as metallic glass.

glaze [ENG] A glossy coating. Also known as enamel.

glazier's point [ENG] A small piece of sheet metal, usually shaped like a triangle, used to hold a pane of glass in place.

glazing [ENG] **1.** Cutting and fitting panes of glass into frames. **2.** Smoothing the lead of a wiped pipe joint by passing a hot iron over it.

glazing compound [MATER] A caulking compound used to seal and support a glass pane in place.

Gleason bevel gear system [DES ENG] The standard for bevel gear designs in the United States; employs a basic pressure angle of 20° with long and short addenda for ratios other than 1:1 to avoid undercut pinions and to increase strength.

glide [AERO ENG] Descent of an aircraft at a normal angle of attack, with little or no thrust.

glide angle *See* gliding angle.

glide path [AERO ENG] **1.** The flight path of an aeronautical vehicle in a glide, seen from the side. Also known as glide trajectory. **2.** The path used by an aircraft or spacecraft in a landing approach procedure.

glider [AERO ENG] A fixed-wing aircraft designed to glide, and sometimes to soar; usually does not have a power plant.

glide slope *See* gliding angle.

glide trajectory *See* glide path.

gliding angle [AERO ENG] The angle between the horizontal and the glide path of an

aircraft. Also known as glide angle; glide slope.

glitter [MATER] A group of decorative materials consisting of flakes large enough so that each flake produces a plainly visible sparkle or reflection; incorporated into plastic during compounding.

globe valve [ENG] A device for regulating flow in a pipeline, consisting of a movable disk-type element and a stationary ring seat in a generally spherical body.

globular transfer [MET] In electric arc welding, the transfer of weld metal across the arc in large drops.

glory hole [CIV ENG] A funnel-shaped, fixed-crest spillway. [ENG] A furnace for resoftening or fire polishing glass during working, or an entrance in such a furnace. [MIN ENG] An opening formed by the removal of soft or broken ore through an underground passage.

glory hole system *See* chute system.

glossimeter [ENG] An instrument, often photoelectric, for measuring the ratio of the light reflected from a surface in a definite direction to the total light reflected in all directions. Also known as glossmeter.

glossmeter *See* glossimeter.

gloss oil [MATER] Low-grade varnish composed of rosin dissolved in solvent naphtha.

glove box [ENG] A sealed box with gloves attached and passing through openings into the box, so that workers can handle materials in the box; used to handle certain radioactive and biologically dangerous materials and to prevent contamination of materials and objects such as germfree rats or lunar rocks.

glow-discharge microphone [ENG ACOUS] Microphone in which the action of sound waves on the current forming a glow discharge between two electrodes causes corresponding variations in the current.

glow plug [MECH ENG] A small electric heater, located inside a cylinder of a diesel engine, that preheats the air and aids the engine in starting.

glue [MATER] A crude, impure, amber-colored form of commercial gelatin of unknown detailed composition produced by the hydrolysis of animal collagen; gelatinizes in aqueous solutions and dries to form a strong, adhesive layer.

glued-laminated wood [MATER] A wooden member formed by assembling a set of boards or planks with glue so that the grain of all laminations is essentially parallel to the length of the member. Also known as glulam.

glue-joint ripsaw [MECH ENG] A heavy-gage ripsaw used on straight-line or self-feed rip machines; the cut is smooth enough to permit gluing of joints from the saw.

glue-line heating [ENG] Dielectric heating in which the electrodes are designed to give preferential heating to a thin film of glue or other relatively high-loss material located between layers of relatively low-loss material such as wood.

glulam *See* glued-laminated wood.

gnomon [ENG] On a sundial, the inclined plate or pin that casts a shadow. Also known as style.

goaf [MIN ENG] **1.** That part of a mine from which the coal has been worked away and the space more or less filled up. **2.** The refuse or waste left in the mine. Also known as gob.

gob *See* goaf.

gobo [ENG] A panel used to shield a television camera lens from direct light. [ENG ACOUS] A sound-absorbing shield used with a microphone to block unwanted sounds.

gob stink [MIN ENG] **1.** The odor from the burning coal given off by an underground fire. **2.** The odor given off by the spontaneous heating of coal, not necessarily in the gob. Also known as stink.

go-devil [ENG] **1.** A device inserted in a pipe or hole for purposes such as cleaning or for detonating an explosive. **2.** A sled for moving logs or cultivating. **3.** A large rake for gathering hay. **4.** A small railroad car used for transporting workers and materials.

go gage [DES ENG] A test device that just fits a part if it has the proper dimensions (often used in pairs with a "no go" gage to establish maximum and minimum dimensions).

goggles [ENG] Spectaclelike eye protectors having shields at the sides and short, projecting eye tubes.

going [CIV ENG] On a staircase, the distance between the faces of two successive risers.

Golay cell [ENG] A radiometer in which radiation absorbed in a gas chamber heats the gas, causing it to expand and deflect a diaphragm in accordance with the amount of radiation.

gold [MET] The native metallic element; a deep-yellow, very dense, soft, isometric metal, usually found alloyed with silver or copper; used in jewelry, dentistry, gilding, anodes, alloys, and solders.

gold alloy [MET] Any alloy containing gold.

goldbeater's skin [MATER] The treated outside membrane of the large intestine of cattle; used between leaves of metal in goldbeating, and sometimes in hygrometers.

goldbeater's-skin hygrometer [ENG] A hygrometer using goldbeater's skin as the sen-

sitive element; variations in the physical dimensions of the skin caused by its hygroscopic character indicate relative atmospheric humidity.

goldbeating [MET] The process of producing gold leaf.

gold bronze [MET] A powdered alloy of copper used to simulate gold, or an alloy of copper containing 3–5% aluminum.

gold-filled [MET] Covered by a layer of gold alloy.

gold foil [MET] A thin sheet of gold, thicker than gold leaf, formed by rolling or hammering.

gold leaf [MET] Gold beaten or rolled into extremely thin sheets or leaves (10^{-6} inch or 25 nanometers thick); leaves are stored in books (a book consists of 25 leaves), the paper of which is rubbed with chalk to keep the leaves from sticking.

gold metallurgy [MET] The science and technology of gold recovery and refining.

gold plate [MET] Gold which has been electroplated on a material in a thin layer of controlled thickness; used on electric contacts for corrosion resistance and solderability and on jewelry and oraments.

Goldschmidt process [MET] **1.** The thermite process of welding. **2.** A process by which dry chlorine is employed to remove tin from scrap tinplate.

gold size [MATER] An adhesive used to fix gold leaf to a surface.

Gold slide [ENG] A slide rule used on British ships to compute barometric corrections and reduction of pressure to sea level; it includes the effects of temperature, latitude, index correction, and barometric height above sea level.

gold solder [MET] Solder composed of 60% gold, 20% silver, and 20% copper.

golf ball [ENG] A printing element used on some typewriters and serial printers, consisting of a rotating, spherically shape, removable typehead that skims across the printed line while the typewriter or printer carriage does not move.

gondola car [ENG] A flat-bottomed railroad car which has no top, fixed sides, and often removable ends, in which steel, rock, or heavy bulk commodities are transported.

goniometer [ENG] **1.** An instrument used to measure the angles between crystal faces. **2.** An instrument which uses x-ray diffraction to measure the angular positions of the axes of a crystal. **3.** Any instrument for measuring angles.

go/no-go detector [ENG] An instrument having only two operating states, such as a common fuse which is either intact or melted.

go/no-go test [ENG] A test based on the measurement of one or more parameters but which can have only one of two possible results, to pass or reject the device under test.

Goodman duckbill loader [MIN ENG] A gathering and loading assembly for coal, consisting of a shovel trough with a shovel head fitting inside the feeder trough, an operating carrier which controls the connection between these troughs, a sliding shoe which moves to and fro on the floor of the seam, a swivel trough, and a pendulum jack; it loads the coal into the shaker conveyor pan column.

Goodman loader [MIN ENG] **1.** An electrohydraulic power shovel designed for loading coal where the seams are 6 feet (1.8 meters) or more in thickness. **2.** A loader designed for loading coal from thin seams; has a telescoping fan-shaped apron that extends from the entry of the room to the working face.

goop [MATER] A compound in paste form containing finely divided magnesium, used as a constituent of certain incendiary bomb fillings.

goosefoot oil *See* chenopodium oil.

gooseneck [DES ENG] **1.** A pipe, bar, or other device having a curved or bent shape resembling that of the neck of a goose. **2.** *See* water swivel.

gopher hole [ENG] Horizontal T-shaped opening made in rock in preparation for blasting. Also known as coyote hole. [MIN ENG] An irregular pitting hole made during prospecting.

gopher-hole blasting *See* coyote blasting.

gophering [MIN ENG] A method of breaking up a sandy medium-hard overburden where usual blastholes tend to cave in, by firing an explosive charge in each of a series of shallow holes; debris is cleared, and holes are made deeper for further charges, until they are deep enough to take enough explosives to break up the deposit.

Gordon's formula [CIV ENG] An empirical formula which gives the collapsing load of a column in terms of its cross-sectional area, length, and least diameter.

gore [CIV ENG] A small triangular parcel of land.

gouge [DES ENG] A curved chisel for wood, bone, stone, and so on. [MIN ENG] A layer of soft material along the wall of a vein which favors miners by enabling them, after gouging it out with a pick, to attack the solid vein from the side.

gouging [ENG] The removal of material by electrical, mechanical, or manual means for the formation of a groove.

governor [MECH ENG] A device, especially one actuated by the centrifugal force of whirling weights opposed by gravity or by

springs, used to provide automatic control of speed or power of a prime mover.

grab [ENG] An instrument for extricating broken boring tools from a borehole.

grabbing crane [MECH ENG] An excavator made up of a crane carrying a large grab or bucket in the form of a pair of half scoops, hinged to dig into the earth as they are lifted.

grab bucket [MECH ENG] A bucket with hinged jaws or teeth that is hung from cables on a crane or excavator and is used to dig and pick up materials.

grab dredger [MECH ENG] Dredging equipment comprising a grab or grab bucket that is suspended from the jib head of a crane. Also known as grapple dredger.

grabhook [DES ENG] A hook used for grabbing, as in lifting blocks of stone, in which case the hooks are used in pairs connected with a chain, and are so constructed that the tension of the chain causes them to adhere firmly to the rock.

grab sample [MIN ENG] A random mode of sampling; the samples may be taken from the pile broken in the process of mining, or from a truck or car of ore or coal.

Gradall [MECH ENG] A hydraulic backhoe equipped with an extensible boom that excavates, backfills, and grades.

grade [CIV ENG] **1.** To prepare a roadway or other land surface of uniform slope. **2.** A surface prepared for the support of rails, a road, or a conduit. [ENG] The degree of strength of a high explosive. [MIN ENG] **1.** A classification of ore according to recoverable amount of a valuable metal. **2.** To sort and classify diamonds.

gradeability [MECH ENG] The performance of earthmovers on various inclines, measured in percent grade.

grade beam [CIV ENG] A reinforced concrete beam placed directly on the ground to provide the foundation for the superstructure.

grade crossing [CIV ENG] The intersection of roadways, railways, pedestrian walks, or combinations of these at grade.

grade line [CIV ENG] A line or slope used as a longitudinal reference for a railroad or highway.

grade of coal [MIN ENG] A classification of coal based on the amount and nature of the ash and sulfur content.

grader [MECH ENG] A high-bodied, wheeled vehicle with a leveling blade mounted between the front and rear wheels; used for fine-grading relatively loose and level earth.

grade separation [CIV ENG] A grade crossing employing an underpass and overpass.

grade slab [CIV ENG] A reinforced concrete slab placed directly on the ground to provide the foundation for the superstructure.

grade stake [CIV ENG] A stake used as an elevation reference.

gradienter [ENG] An attachment placed on a surveyor's transit to measure angle of inclination in terms of the tangent of the angle.

gradient microphone [ENG ACOUS] A microphone whose electrical response corresponds to some function of the difference in pressure between two points in space.

gradient of equal traction [MIN ENG] The gradient at which the tractive force necessary to pull an empty tram inby (slightly uphill) is equal to that force required to pull a loaded tram outby.

grading [IND ENG] Segregating a product into a number of adjoining categories which often form a spectrum of quality. Also known as classification.

graduated coating [MET] In thermal spraying, a deposit consisting of a number of layers which vary in material composition.

grain [MATER] **1.** The appearance and texture of wood due to the arrangement of constituent fibers. **2.** The woodlike appearance or texture of a rock, metal, or other material. **3.** The direction in which most fibers lie in a sample of paper, which corresponds with the way the paper was made on the manufacturing machine.

grain boundary [MET] Surface between individual grains in a metal.

grainer salt [MATER] Sodium chloride produced by the grainer process.

grain fineness number [MATER] Average grain size of a granular material.

grain flow [MET] The fiber patterns appearing on the surface of a forging and resulting from the alignment of the crystalline structure of the base metal in the direction of working.

grain growth [MET] Enlargement of grains in a metal, usually through heat treatment.

graining [ENG] Simulating a grain such as wood or marble on a painted surface by applying a translucent stain, then working it into suitable patterns with tools such as special combs, brushes, and rags.

grain leather [MATER] Leather made from the grain side of a skin.

grain size [MET] Average size of grains in a metal expressed as average diameter, or grains per unit area or volume.

grain spacing [DES ENG] Relative location of abrasive grains on the surface of a grinding wheel.

Granby car [MIN ENG] An automatically dumped car for hand loading or power-shovel loading; a wheel attached to its side engages an inclined track at the dumping point.

granular-bed separator [ENG] Vessel or chamber in which a bed of granular material is used to remove dust from a dust-laden gas as it passes through the bed.

granular fracture [MET] Grain-like or crystalline surface appearance of a broken metal.

granular powder [MET] Equidimensional metal particles that are not spherical.

granular structure [MATER] Nonuniform appearance of molded or compressed material due to presence of particles of composition, either within the material or on the surface.

granulated metal [MET] Small pellets produced by pouring molten metal through a screen or similar device and chilling the droppings in water.

grapefruit oil [MATER] Pale-yellow, volatile liquid with grapefruit aroma; soluble in oils, insoluble in glycerin; derived from fresh rind of *Citrus paradisi*; used in flavors and toiletries. Also known as oil of grapefruit; oil of shaddock.

grapefruit seed oil [MATER] Reddish-brown oil expressed from grapefruit seeds; has nutlike aroma and bitter taste; solidifies at $-10°C$; used as lubricant for textile fibers and leather.

graphic recording instrument [ENG] An instrument that makes a graphic record of one or more quantities as a function of another variable, usually time.

graphite flake [MET] A curved graphite particle in gray cast iron.

graphite grease [MATER] A lubricating grease that contains 2–10% amorphous graphite; used for bearings, especially in damp places.

graphite oil [MATER] A deflocculated suspension of graphite in oil; used as a lubricant.

graphite rosette [MET] Graphite flakes which extend radially outward from the centers of crystallization.

graphitic carbon [MET] Carbon in iron or steel present in the form of graphite.

graphitic corrosion [MET] Corrosion of gray cast iron in which the metallic iron constituent is converted into corrosion products which cement together the residual graphite.

graphitic steel [MET] Alloy steel containing graphitic carbon.

graphitizing [MET] Annealing cast iron to convert all or some of the combined carbon to graphitic carbon.

grapnel [DES ENG] An implement with claws used to recover a lost core, drill fittings, and junk from a borehole or for other grappling operations. Also known as grapple.

grappier cement [MATER] A cement composed of finely ground lumps of leftover underburned and overburned slaked lime.

grapple dredger See grab dredger.

grapple hook [DES ENG] An iron hook used on the end of a rope to snag lines, to hold one ship alongside another, or as a fishing tool. Also known as grappling iron.

grappling iron See grapple hook.

grasp [IND ENG] A basic element (therblig) in time-motion study; a useful element that accomplishes work.

grasshopper linkage [MECH ENG] A straight-line mechanism used in some early steam engines.

Grassot fluxmeter [ENG] A type of fluxmeter in which a light coil of wire is suspended in a magnetic field in such a way that it can rotate; the ends of the suspended coil are connected to a search coil of known area penetrated by the magnetic flux to be measured; the flux is determined from the rotation of the suspended coil when the search coil is moved.

grass-roots deposit [MIN ENG] A deposit that is discovered in surface croppings, is easily exploited, and can pay for its own development while in progress.

grass-roots mining [MIN ENG] Also known as mining on a shoestring. **1.** Inadequately financed mining operation, with catch-as-catch-can practices. **2.** Mining from surface down to bedrock.

grate [ENG] A support for burning solid fuels; usually made of closely spaced bars to hold the burning fuel, while allowing combustion air to rise up to the fuel from beneath, and ashes to fall away from the burning fuel.

gravel mine [MIN ENG] A mine extracting gold from sand or gravel. Also known as placer mine.

gravel pump [MECH ENG] A centrifugal pump with renewable impellers and lining, used to pump a mixture of gravel and water.

gravel stop [BUILD] Metal flashing placed at the edge of a roof to prevent gravel from falling off.

graveyard shift [IND ENG] The shift of workers that begins at or around midnight; the last shift of the day.

gravimeter [ENG] A highly sensitive weighing device used for relative measurement of the force of gravity by detecting small weight differences of a constant mass at different points on the earth. Also known as gravity meter.

gravimetry [ENG] Measurement of gravitational force.

graving dock [CIV ENG] A form of dry dock consisting of an artificial basin fitted with a gate or caisson, into which a vessel can be floated and the water pumped out to expose the vessel's bottom.

gravitometer See densimeter.

gravity bar [MIN ENG] A 5-foot (1.5-meter) length of heavy half-round rod forming the link between the wedge-oriented coupling and the drill-rod swivel coupling on an assembled Thompson retrievable borehole-deflecting wedge.

gravity bed [ENG] A moving body of solids in which particles (granules, pellets, beads, or briquets) flow downward by gravity through a vessel, while process fluid flows upward; the moving-bed technique is used in blast and shaft furnaces, petroleum catalytic cracking, pellet dryers, and coolers.

gravity chute [ENG] A gravity conveyor in the form of an inclined plane, trough, or framework that depends on sliding friction to control the rate of descent.

gravity classification [MIN ENG] The grading of ores, or the separation of waste from coal, by the differences in specific gravities of the substances.

gravity concentration [ENG] **1.** Any of various methods for separating a mixture of particles, such as minerals, based on the differences in density of the various species and on the resistance to relative motion exerted upon the particles by the fluid or semifluid medium in which separation takes place. **2.** The separation of liquid-liquid dispersions based on settling out of the dense phase by gravity.

gravity conveyor [ENG] Any unpowered conveyor such as a gravity chute or a roller conveyor, which uses the force of gravity to move materials over a downward path.

gravity corer [ENG] Any type of corer that achieves bottom penetration solely as a result of gravitational force acting upon its mass.

gravity dam [CIV ENG] A dam which depends on its weight for stability.

gravity feed [ENG] Movement of materials from one location to another using the force of gravity.

gravity-flow gathering system [PETRO ENG] The use of gravity (downhill flow) through pipelines to transport and collect liquid at a central location; used for gathering of waste water from waterflooding operations for treatment prior to reuse or disposal.

gravity-gradient attitude control [AERO ENG] A device that regulates automatically attitude or orientation of an aircraft or spacecraft by responding to changes in gravity acting on the craft.

gravity haulage [MIN ENG] A type of haulage system in which the set of full cars is lowered at the end of a rope, and gravity force pulls up the empty cars, the rope being passed around a sheave at the top of the incline. Also known as self-acting incline.

gravity incline [MIN ENG] An opening made in the direction of, and along the same gradient as, the dip of the deposit.

gravity meter [ENG] **1.** U-tube-manometer type of device for direct reading of solution specific gravities in semimicro quantities. **2.** An electrical device for measuring variations in gravitation through different geologic formations; used in mineral exploration. **3.** *See* gravimeter.

gravity prospecting [ENG] Identifying and mapping the distribution of rock masses of different specific gravity by means of a gravity meter.

gravity railroad [ENG] A cable railroad in which cars descend a slope by gravity and are hauled back up the slope by a stationary engine, or there may be two tracks with cars so connected that cars going down may help to raise the cars going up and thus conserve energy.

gravity segregation [ENG] Tendency of immiscible liquids or multicomponent granular mixtures to separate into distinct layers in accordance with their respective densities.

gravity separation [ENG] Separation of immiscible phases (gas-solid, liquid-solid, liquid-liquid, solid-solid) by allowing the denser phase to settle out under the influence of gravity; used in ore dressing and various industrial chemical processes.

gravity settling chamber [ENG] Chamber or vessel in which the velocity of heavy particles (solids or liquids) in a fluid stream is reduced to allow them to settle downward by gravity, as in the case of a dust-laden gas stream.

gravity stamp [MIN ENG] Unit in a stamp battery which directs a heavy falling weight onto a die on which rock is crushed.

gravity station [ENG] The site of installation of gravimeters.

gravity survey [ENG] The measurement of the differences in gravity force at two or more points.

gravity wall [CIV ENG] A retaining wall which is kept upright by the force of its own weight.

gravity wheel conveyor [MECH ENG] A downward-sloping conveyor trough with closely spaced axle-mounted wheel units on which flat-bottomed containers or objects are conveyed from point to point by gravity pull.

gravity yard *See* hump yard.

gray casting [MET] A casting of gray iron.

gray iron [MET] Pig or cast iron in which the carbon not contained in pearlite is present in the form of graphitic carbon.

Grayloc tubing joint [PETRO ENG] Special wellbore-tubing joint that has greater leak resistance and strength than standard API tubing joints.

grease [MATER] **1.** Rendered, inedible animal fat that is soft at room temperature and is obtained from lard, tallow, bone, raw animal fat, and other waste products. **2.** A lubricant in the form of a solid to semisolid dispersion of a thickening agent in a fluid lubricant, such as petroleum oil thickened with metallic soap.

grease cup [ENG] A receptacle used to apply a solid or semifluid lubricant to a bearing; the receptacle is packed with grease and the cap forces the grease to the bearing.

greased-deck concentration [MIN ENG] A separation process based on selective adhesion of certain grains (diamonds) to quasi-solid grease.

grease gun [ENG] A small hand-operated device that pumps grease under pressure into bearings.

greasepaint [MATER] Makeup made of melted tallow or grease used by theatrical performers.

grease seal [ENG] **1.** Type of seal used on floating pistons of some hydropneumatic recoil systems to prevent leakage past the piston of gas or oil; also used in cylinders of some hydropneumatic equilibrators. **2.** Seal used to retain grease in a case or housing, as on an axle shaft.

grease table [MIN ENG] An apparatus for concentrating minerals, such as diamonds, which adhere to grease; usually a shaking table coated with grease or wax over which an aqueous pulp is flowed.

grease trap [CIV ENG] A trap in a drain or waste pipe to stop grease from entering a sewer system.

green [MET] Pertaining to an unsintered powder.

Greenburg-Smith impinger [MIN ENG] A dust-sampling apparatus based on the principle of impingement of the dust-carrying air at high velocity upon a wetted glass surface; also involves bubbling the air through a liquid medium.

green glass [MATER] Glass given a blue-green hue by substituting cupric oxide for the chromium compound used in ordinary glass.

green gold [MET] A greenish alloy of gold obtained by using silver, silver and cadmium, or silver and copper as the alloying metal.

greenheart [MATER] Wood from *Octolea rodioei*; resistant to fungi and termites and used for shipbuilding, docks, and marine planking.

green lumber [MATER] Freshly sawed lumber, before drying.

green oil [MATER] In the Scottish shale-oil industry, once-run crude shale oil.

green roof [MIN ENG] A mine roof which has not broken down or which shows no sign of taking weight.

green strength [MET] The mechanical strength which a compacted powder must have in order to withstand mechanical operations to which it is subjected after pressing and before sintering, without damaging its fine details and sharp edges.

greenware [MATER] Ceramic ware that has not yet been fired.

Greenwell formula [MIN ENG] A formula used for calculating the thickness of tubing and involving the required thickness of tubing in feet, the vertical depth in feet, the diameter of the shaft in feet, and an allowance for possible flaws or corrosion.

greyhound [PETRO ENG] A stand of drill pipe comprising one or more lengths with a tool joint at each end; used to fashion lengths of less than one regular stand during drilling.

grid [DES ENG] A network of equally spaced lines forming squares, used for determining permissible locations of holes on a printed circuit board or a chassis. [MIN ENG] Imaginary line used to divide the surface of an area when following a checkerboard placement of boreholes.

grid metal [MET] An alloy of lead with 5–12% antimony and sometimes with a smaller amount of tin; used for grids in storage batteries.

grid nephoscope [ENG] A nephoscope constructed of a grid work of bars mounted horizontally on the end of a vertical column and rotating freely about the vertical axis; the observer rotates the grid and adjusts the position until some feature of the cloud appears to move along the major axis of the grid; the azimuth angle at which the grid is set is taken as the direction of the cloud motion.

grid-rectification meter [ENG] A type of vacuum-tube voltmeter in which the grid and cathode of a tube act as a diode rectifier, and the rectified grid voltage, amplified by the tube, operates a meter in the plate circuit.

Griffith crack [MET] Any small flaw in a metal theorized as creating a low order of fracture strength.

Griffith's white *See* lithopone.

grillage [CIV ENG] A footing that consists of two or more tiers of closely spaced structural steel beams resting on a concrete block, each tier being at right angles to the one below.

grille [ENG ACOUS] An arrangement of wood, metal, or plastic bars placed across the front of a loudspeaker in a cabinet for decorative and protective purposes.

grille cloth [ENG ACOUS] A loosely woven cloth stretched across the front of a loudspeaker to keep out dust and provide protection without appreciably impeding sound waves.

grindability [MATER] Relative ease with which a material can be ground.

grindability index [MATER] A numerical indication of the capacity of a material to be ground.

grinder [MECH ENG] Any device or machine that grinds, such as a pulverizer or a grinding wheel.

grinding [MECH ENG] 1. Reducing a material to relatively small particles. 2. Removing material from a workpiece with a grinding wheel. [MIN ENG] The act or process of continuing to drill after the bit or core barrel is blocked, thereby crushing and destroying any core that might have been produced.

grinding aid [ENG] An additive to the charge in a ball mill or rod mill to accelerate the grinding process.

grinding burn [MECH ENG] Overheating a localized area of the work in grinding operations.

grinding cracks [MATER] Cracks in a workpiece resulting from grinding.

grinding fluid [MATER] Cutting fluid used in metal-grinding operations.

grinding medium [ENG] Any material including balls and rods, used in a grinding mill.

grinding mill [MECH ENG] A machine consisting of a rotating cylindrical drum, that reduces the size of particles of ore or other materials fed into it; three main types are ball, rod, and tube mills.

grinding pebbles [ENG] Pebbles, of chert or quartz, used for grinding in mills, where contamination with iron has to be avoided.

grinding ratio [MECH ENG] Ratio of the volume of ground material removed from the workpiece to the volume removed from the grinding wheel.

grinding relief [MET] A groove at the edge of a metal surface which permits overhang of the corner of the grinding wheel.

grinding sensitivity [MATER] Susceptibility of a material to the formation of grinding cracks.

grinding wheel [DES ENG] A wheel or disk having an abrasive material such as alumina or silicon carbide bonded to the surface.

grindstone [ENG] A stone disk on a revolving axle, used for grinding, smoothing, and shaping.

grit [MATER] An abrasive material composed of angular grains.

grit blasting [ENG] Surface treatment in which steel grit, sand, or other abrasive material is blown against an object to produce a roughened surface or to remove dirt, rust, and scale. Also known as sandblasting.

grit chamber [CIV ENG] A chamber designed to remove sand, gravel, or other heavy solids that have subsiding velocities or specific gravities substantially greater than those of the organic solids in waste water.

grit size [DES ENG] Size of the abrasive particles on a grinding wheel.

grizzly [ENG] **1.** A coarse screen used for rough sizing and separation of ore, gravel, or soil. **2.** A grating to protect chutes, manways, and winzes, in mines, or to prevent debris from entering a water inlet.

grizzly chute [MIN ENG] A chute equipped with grizzlies which separate fine from coarse material as it passes through the chute.

grizzly crusher [MECH ENG] A machine with a series of parallel rods or bars for crushing rock and sorting particles by size.

grizzly worker [MIN ENG] In metal mining, a laborer who works underground at a grizzly.

grog [MATER] Fired refractory material that is used in the manufacture of products which must withstand extreme heat.

groin [CIV ENG] A barrier built out from a seashore or riverbank to protect the land from erosion and sand movements, among other functions. Also known as groyne; jetty; spur dike; wing dam.

grommet [ENG] **1.** A metal washer or eyelet. **2.** A piece of fiber soaked in a packing material and used under bolt and nut heads to preserve tightness.

grommet nut [DES ENG] A blind nut with a round head; used with a screw to attach a hinge to a door.

groove [DES ENG] A long, narrow channel in a surface.

grooved drum [DES ENG] Drum with a grooved surface to support and guide a rope.

groove face [MIN ENG] The portion of a surface of a member that is included in a groove.

groove sample [MIN ENG] A sample of coal or ore obtained by cutting appropriate grooves along or across the road exposures. Also known as channel sample.

groove weld [MET] A weld in the groove between a pair of members.

grooving saw [MECH ENG] A circular saw for cutting grooves.

gross area [BUILD] Sum of the areas of all stories included within the outside face of the exterior walls of a building.

gross porosity [MET] Gas pockets or pores of undesirable size and quantity in a casting or a weld metal.

gross recoverable value [MIN ENG] The part of the total metal recovered from an ore multiplied by the price.

gross unit value [MIN ENG] The weight of metal per long or short ton as determined by assay or analysis, multiplied by the market price of the metal.

gross weight [IND ENG] The weight of a vehicle or container when it is loaded with goods. Abbreviated gr wt.

ground [AERO ENG] To forbid (an aircraft or individual) to fly, usually for a relatively short time.

ground area [BUILD] The area of a building at ground level.

ground check [ENG] **1.** A procedure followed prior to the release of a radiosonde in order to obtain the temperature and humidity corrections for the radiosonde system. **2.** Any instrumental check prior to the ground launch of an airborne experiment. Also known as base-line check.

ground-check chamber [ENG] A chamber that is used to check the sensing elements of radiosonde equipment and that houses sources of heat and water vapor plus instruments for measuring temperature, humidity, and pressure, and in which air circulation is maintained by a motor-driven fan.

ground coal [MIN ENG] The bottom of a coal seam.

ground control [CIV ENG] Supervision or direction of all airport surface traffic, except an aircraft landing or taking off. [ENG] The marking of survey, triangulation, or other key points or system of points on the earth's surface so that they may be recognized in aerial photographs.

ground-controlled approach radar [ENG] A ground radar system providing information by which aircraft approaches may be directed by radio communications. Abbreviated GCA radar.

ground-controlled intercept radar [ENG] A radar system by means of which a controller may direct an aircraft to make an interception of another aircraft. Abbreviated GCI radar.

ground controller [ENG] Aircraft controller stationed on the ground; a generic term, applied to the controller in ground-controlled approach, ground-controlled interception, and so on.

ground data equipment [ENG] Any device located on the ground that aids in obtaining space-position or tracking data (including computation function); reads out data telemetry, video, and so on, from payload instrumentation, or is capable of transmitting command and control signals to a satellite or space vehicle.

ground effect [AERO ENG] Increase in the lift of an aircraft operating close to the ground caused by reaction between high-velocity downwash from its wing or rotor and the ground.

ground-effect machine *See* air-cushion vehicle.

ground environment [ENG] **1.** Environment that surrounds and affects a system or piece of equipment that operates on the ground. **2.** System or part of a system, as of a guidance system, that functions on the ground; the aggregate of equipment, conditions, facilities, and personnel that go to make up a system, or part of a system, functioning on the ground.

ground handling equipment *See* ground support equipment.

groundhog *See* barney.

ground instrumentation *See* spacecraft ground instrumentation.

ground lead *See* work lead.

ground magnetic survey [ENG] A determination of the magnetic field at the surface of the earth by means of ground-based instruments.

groundman [ENG] A person employed in digging or excavating. [MIN ENG] *See* mucker.

ground noise [ENG ACOUS] The residual system noise in the absence of the signal in recording and reproducing; usually caused by inhomogeneity in the recording and reproducing media, but may also include tube noise and noise generated in resistive elements in the amplifier system.

ground sluice [MIN ENG] A channel through which gold-bearing earth is passed in placer mining.

ground support equipment [AERO ENG] That equipment on the ground, including all implements, tools, and devices (mobile or fixed), required to inspect, test, adjust, calibrate, appraise, gage, measure, repair, overhaul, assemble, disassemble, transport, safeguard, record, store, or otherwise function in support of a rocket, space vehicle, or the like, either in the research and development phase or in an operational phase, or in support of the guidance system used with the missile, vehicle, or the like. Abbreviated GSE. Also known as ground handling equipment.

ground surveillance radar [ENG] A surveillance radar operated at a fixed point on the earth's surface for observation and control of the position of aircraft or other vehicles in the vicinity.

ground trace [ENG] The theoretical mark traced upon the surface of the earth by a flying object, missile, or satellite as it passes over the surface, the mark being made vertically from the object making the trace.

ground ways [CIV ENG] Supports, usually made of heavy timbers, which are placed on the ground on either side of the keel of a ship under construction, providing a track for launching, and supporting the sliding ways. Also known as standing ways.

ground wire [CIV ENG] A small-gage, high-strength steel wire used to establish line and grade for air-blown mortar or concrete. Also known as alignment wire; screed wire.

groundwood pulp *See* mechanical pulp.

group incentive [IND ENG] Any wage incentive applied to more than one employee who is engaged in group work characterized by interdependent relationship between operations with consequent physical proximity and unification of interest.

grouser [ENG] A temporary pile or a heavy, iron-shod pole driven into the bottom of a stream to hold a drilling or dredging boat or other floating object in position. Also known as spud.

grout [MATER] **1.** A fluid mixture of cement and water, or a mixture of cement, sand, and water. **2.** Waste material of all sizes obtained in quarrying stone.

grout curtain [ENG] A row of vertically drilled holes filled with grout under pressure to form the cutoff wall under a dam, or to form a barrier around an excavation through which water cannot seep or flow.

grout hole [ENG] **1.** One of the holes in a grout curtain. **2.** Any hole into which grout is forced under pressure to consolidate the surrounding earth or rock.

grouting [ENG] The act or process of applying grout or of injecting grout into grout holes or crevices of a rock.

grout injector [ENG] A machine that mixes the dry ingredients for a grout with water and injects it, under pressure, into a grout hole.

grout pipe [ENG] A pipe that transports grout under pressure for injection into a grout hole or a rock formation.

growth factor [AERO ENG] The additional weight of fuel and structural material required by the addition of 1 pound of payload to the original payload.

groyne *See* groin.

grubbing [CIV ENG] Clearing stumps and roots.

grub screw [DES ENG] A headless screw with a slot at one end to receive a screwdriver.

grubstake [MIN ENG] In the United States, supplies or money furnished to a mining prospector for a share in his discoveries.

grubstake contract [MIN ENG] An agreement between two or more persons to locate mines upon the public domain by their joint aid, effort, labor, or expense, with each to acquire by virtue of the act of location such an interest in the mine as agreed upon in the contract.

GSE *See* ground support equipment.

g suit [ENG] A suit that exerts pressure on the abdomen and lower parts of the body to prevent or retard the collection of blood below the chest under positive acceleration. Also known as anti-g suit.

guaiac [MATER] A resin obtained from the trees *Guaiacum santum* and *G. officinale;* soluble in alcohol, ether, and chloroform; used in medicine and varnish.

guano [MATER] Phosphate- and nitrogen-rich, partially decomposed excrement of seabirds; used as a fertilizer.

guard [ENG] A shield or other fixture designed to protect against injury. [MIN ENG] A support in front of a roll train to guide the bar into the groove.

guard circle [DES ENG] The closed loop at the end of a grooved record.

guard-electrode system [PETRO ENG] System of extra electrodes used during electrical logging of reservoir formations to confine the surveying current from the measuring electrode to a generally horizontal path.

guard lock [CIV ENG] *See* entrance lock. [ENG] An auxiliary lock that must be opened before the key can be turned in a main lock.

guard magnet [MIN ENG] A magnet employed in a crushing system to remove or arrest tramp iron ahead of the machinery.

guardrail [CIV ENG] **1.** A handrail. **2.** A rail made of posts and a metal strip used on a road as a divider between lines of traffic in opposite directions or used as a safety barrier on curves. **3.** A rail fixed close to the outside of the inner rail on railway curves to hold the inner wheels of a railway car on the rail. Also known as check rail; safety rail; slide rail.

guard screen *See* oversize control screen.

guar gum [MATER] A mucilage formed from seeds of the guar plant; light-gray powder dispersible in water; used as a thickening agent in paper, foods, pharmaceuticals, and cosmetics.

gudgeon [ENG] **1.** A pivot. **2.** A pin for fastening stone blocks.

Guggenheim process [CIV ENG] A method of chemical precipitation which employs ferric chloride and aeration to prepare sludge for filtration.

guidance site [ENG] Specific location of high-order geodetic accuracy containing equipment and structures necessary to provide guidance services or a given launch rate; it may be an integrated part of a launch site, or it may be a remote facility.

guidance station equipment [ENG] The ground-based portion of the missile guidance system necessary to provide guidance during missile flight; it specifically includes the tracking radar, the rate measuring equipment, the data link equipment, and the computer, test, and maintenance equipment integral to these items.

guidance system [AERO ENG] The control devices used in guidance of an aircraft or spacecraft.

guide bearing [MECH ENG] A plain bearing used to guide a machine element in its lengthwise motion, usually without rotation of the element.

guide bracket [MIN ENG] A steel bracket fixed to a bunton to secure rigid guides in a shaft.

guide coupling [MIN ENG] A short coupling with a projecting reamer guide or pup to which is attached a reaming bit, which it couples to a reaming barrel.

guided bend test [MET] A bend test in which the specimen is bent to a predetermined shape.

guide frame [MIN ENG] A frame held rigidly in place by roof jacks or timbers, with provisions for attaching a shaker conveyor pan line to the movable portion of the frame; prevents jumping or side movement of the pan line.

guide idler [MECH ENG] An idler roll with its supporting structure mounted on a conveyor frame to guide the belt in a defined horizontal path, usually by contact with the edge of the belt.

guide mill [MET] A hand rolling mill with a series of stands and guides at the entrance to the rolls.

guide pin [ENG] A pin used to line up a tool or die with the work.

guide post [CIV ENG] A post along a road that bears direction signs or guide boards.

guide rail [CIV ENG] A track or rail that serves to guide movement, as of a sliding door, window, or similar element.

guide ring [MIN ENG] A longitudinally grooved, annular ring made almost full borehole size, which is fitted to an extension coupling between the core barrel and the first drill rod.

guide rod [MIN ENG] A heavy drill rod coupled to and having the same diameter as a core barrel on which it is used; gives additional rigidity to the core barrel and helps to prevent deflection of the borehole. Also known as core barrel rod; oversize rod.

guides [MECH ENG] **1.** Pulleys to lead a driving belt or rope in a new direction or to keep it from leaving its desired direction. **2.** Tracks that support and determine the path of a skip bucket and skip bucket bail. **3.** Tracks guiding the chain or buckets of a bucket elevator. **4.** The runway paralleling the path of the conveyor which limits the conveyor or parts of a conveyor to movement in a defined path. [MIN ENG] **1.** Steel, wood, or steel-wire rope conductors in a mine shaft to guide the movement of the cages. **2.** Timber, rope, or metal tracks in a hoisting shaft, which are engaged by shoes on the cage or skip so as to steady it in transit. **3.** The holes in a crossbeam through which the stems of the stamps in a stamp mill rise and fall.

Guinier-Preston zones [MET] The initially formed zones of a precipitate as it comes out of solid solution.

gulch claim [MIN ENG] A claim laid upon and along the bed of an unnavigable stream winding through a canyon, with precipitous, nonmineral, and uncultivable bands wherein have accumulated placer deposits.

gum [MATER] A hydrophilic plant polysaccharide or derivative that swells to produce a viscous dispersion or solution when added to water.

gum ammoniac See ammoniac.

gum arabic [MATER] A water-soluble gum obtained from acacia trees in Africa and Australia; produced commercially as a white powder; used in the manufacture of inks and adhesives, in textile finishing, and as the principal binder in watercolor and gouache. Also known as acacia gum; gum Kordofan; gum Senegal.

gum benzoin See benzoin gum.

gum dammar See dammar.

gum Kordofan See gum arabic.

gummed paper [MATER] **1.** A variety of colored, patterned papers with adhesive on one side for easy application. **2.** Coated or uncoated book paper, gummed on one side, used for stickers, labels, stamps, seals, and tapes.

gum resin [MATER] A group of oleoresinous substances obtained from plants; mixtures of true gums and resins less soluble in alcohol than natural resins; examples are rubber, gutta-percha, gamboge, myrrh, and olibanum; used to make certain pharmaceuticals.

gum Senegal See gum arabic.

gum thus See olibanum.

gun burner [ENG] A burner which sprays liquid fuel into a furnace for combustion.

guncotton [MATER] Any of various nitrocellulose explosives of high nitration (13.35–13.4% nitrogen) made by treating cotton with nitric and sulfuric acids; used principally in the manufacture of single-base and double-base propellants.

gunk See rod dope.

gun-laying radar [ENG] Radar equipment specifically designed to determine range, azimuth, and elevation of a target and sometimes also to automatically aim and fire antiaircraft artillery or other guns.

gunmetal [MET] **1.** Bronze composed of copper and tin in proportions of 9:1, formerly used to make cannons. **2.** Any metal or alloy from which guns are made. **3.** Any metal or alloy treated to give the appearance of black, tarnished copper-alloy gunmetal.

gunner's quadrant [ENG] Mechanical device having scales graduated in mils, with fine micrometer adjustments and leveling or cross-leveling vials; it is a separate, unattached instrument for hand placement on a reference surface.

gun pendulum [ENG] A device used to determine the initial velocity of a projectile fired from a gun in which the gun is mounted as a pendulum and its excursion upon firing is measured.

gunpowder [MATER] A black or brown explosive mixture of potassium nitrate, charcoal, and sulfur; originally, it was made in powder form, now generally in grains of various sizes.

gunpowder paper [MATER] Paper with an explosive on it that is rolled up for use in loading.

Gunter's chain [ENG] A chain 66 feet (20.1168 meters) long, consisting of 100 steel links, each 7.92 inches (20.1168 centimeters) long, joined by rings, which is used as the unit of length for surveying public lands in the United States. Also known as chain.

gun-type burner [ENG] An oil burner that uses a nozzle to atomize the fuel.

gusset [CIV ENG] A plate that is used to strengthen truss joints. [MIN ENG] A V-shaped cut in the face of a heading.

gusset plate [CIV ENG] A rectangular or triangular steel plate that connects members of a truss.

gust-gradient distance [AERO ENG] The horizontal distance along an aircraft flight path from the "edge" of a gust to the point at which the gust reaches its maximum speed.

gustsonde [ENG] An instrument dropped from high altitude by a stable parachute, to measure the vertical component of turbulence aloft; consists of an accelerometer and radio telemetering equipment.

gust tunnel [AERO ENG] A type of wind tunnel that has an enclosed space and is used to test the effect of gusts on an airplane model in free flight to determine how atmospheric gusts affect the flight of an airplane.

gut See catgut.

gutta-balata See balata.

gutta-percha [MATER] A leathery, thermoplastic substance consisting of gutta hydrocarbon with some resin obtained from the latex of certain Malaysian sapotaceous trees; used as insulation for submarine cables, and in golf balls and other products.

gutter [BUILD] A trough along the edge of the eaves of a building to carry off rainwater. [CIV ENG] A shallow trench provided beside a canal, bordering a highway, or elsewhere, for surface drainage. [MET] A groove along the periphery of a die impression to allow for excess flash during forging. [MIN ENG] A drainage trench cut along the side of a mine shaft to conduct the water back into a lodge or sump.

guttering [ENG] A process of quarrying stone in which channels, several inches wide, are cut by hand tools, and the stone block is detached from the bed by pinch bars. [MIN ENG] The process of cutting gutters in a mine shaft.

Gutzkow's process [MET] A modification of the sulfuric acid parting process for bullion containing large amounts of copper; a large excess of acid is used, and the silver sulfate is then reduced with charcoal, or, in the original process, ferrous sulfate.

guy [ENG] A rope or wire securing a pole, derrick, or similar temporary structure in a vertical position.

guy derrick [MECH ENG] A derrick having a vertical pole supported by guy ropes to which a boom is attached by rope or cable suspension at the top and by a pivot at the foot.

guyed-mast derrick See gin pole.

gypsum board [MATER] A plaster board covered with paper.

gypsum cement See gypsum plaster.

gypsum lath [MATER] Lath consisting of a core of set gypsum surfaced with paper that is treated to receive plaster.

gypsum plank [MATER] A structural precast unit consisting of gypsum core rein-

forced with welded galvanized steel mesh and bounded on all four edges with a tongue-and-groove steel form; used as the roof deck of steel-frame buildings, and sometimes for the floor system.

gypsum plaster [MATER] Plaster made principally from gypsum. Also known as gypsum cement.

gypsum wallboard [MATER] Wallboard consisting of a core of set gypsum surfaced with paper or other fibrous material suitable to receive paint or paper.

gyratory breaker See gyratory crusher.

gyratory crusher [MECH ENG] A primary breaking machine in the form of two cones, an outer fixed cone and a solid inner erect cone mounted on an eccentric bearing. Also known as gyratory breaker.

gyratory screen [MECH ENG] Boxlike machine with a series of horizontal screens nested in a vertical stack with downward-decreasing mesh-opening sizes; near-circular motion causes undersized material to sift down through each screen in succession.

gyropendulum [MECH ENG] A gravity pendulum attached to a rapidly spinning gyro wheel.

gyroplane [AERO ENG] A rotorcraft whose rotors are not power-driven.

gyrorepeater [ENG] That part of a remote indicating gyro compass system which repeats at a distance the indications of the master gyro compass system.

gyroscope [ENG] An instrument that maintains an angular reference direction by virtue of a rapidly spinning, heavy mass; all applications of the gyroscope depend on a special form of Newton's second law, which states that a massive, rapidly spinning body rigidly resists being disturbed and tends to react to a disturbing torque by precessing (rotating slowly) in a direction at right angles to the direction of torque. Also known as gyro.

gyroscopic-clinograph method [ENG] A method used in borehole surveying which measures time, temperature, and temperature on 16-millimeter film while a gyroscope maintains the casing on a fixed bearing.

gyroscopic/Coriolis-type mass flowmeter [ENG] An instrument consisting of a C-shape pipe and a T-shaped leaf-spring tuning fork which is excited by an electromagnetic forcer, resulting in an angular deflection of the pipe which is directly proportional to the mass-flow rate within the pipe.

gyroscopic couple [MECH ENG] The turning moment which opposes any change of the inclination of the axis of rotation of a gyroscope.

gyroscopic mass flowmeter [ENG] An instrument in which the torque on a rotating pipe of suitable shape, through which a fluid is made to flow, is measured to determine the mass flow through the pipe.

gyrostabilizer [ENG] A gyroscope used to stabilize ships and airplanes.

gyro wheel [MECH ENG] The rapidly spinning wheel in a gyroscope, which resists being disturbed.

H

Haase system [MIN ENG] Shaft sinking in loose ground or quicksand by piles in the form of iron tubes connected by webs; downward movement is facilitated by water forced down the tubes to wash away loose material beneath their points.

hacksaw [ENG] A hand or power tool consisting of a fine-toothed blade held in tension in a bow-shaped frame; used for cutting metal, wood, and other hard materials.

Hadfield manganese steel [MET] Austenitic steel (face-centered cubic structure) containing 11–14% manganese; resistant to shock and wear. Also known as austenitic manganese steel; manganese steel.

Hadsel mill [MIN ENG] An early autogenous grinding mill in which comminution was caused by the fall of ore or that was rotating in a large-diameter horizontal cylinder.

HAF black *See* high-abrasion furnace black.

hair cracks [MATER] Fine, random cracks in the surface of the top coat of paint or other coating material.

hair felt [MATER] Felt made of cattle hair; used as insulation in buildings.

hair hygrometer [ENG] A hygrometer in which the sensing element is a bundle of human hair, which is held under slight tension by a spring and which expands and contracts with changes in the moisture of the surrounding air or gas.

hairline *See* air line.

hairpin tube [DES ENG] A boiler tube bent into a hairpin, or U, shape.

half-and-half solder [MET] Solder composed of tin and lead in equal parts.

half bat [MATER] One half of a brick, cut across the length.

half-course [MIN ENG] The drift or opening driven at an angle of about 45° to the strike and in the plane of the seam.

half cycle [ENG] The time interval corresponding to half a cycle, or 180°, at the operating frequency of a circuit or device.

half-dog setscrew [DES ENG] A setscrew with a short, blunt point.

half-hard [MET] A rolled-metal product of intermediate hardness or temper.

half-header [MIN ENG] A large cap piece; used by sawing a header in two and placing, generally, two timbers under the half header on one side of the haulage, with the end extending over the haulage.

half nut [DES ENG] A nut split lengthwise so that it can be clamped around a screw.

half-round file [DES ENG] A file that is flat on one side and convex on the other.

half set [MIN ENG] In mine timbering, one leg piece and a collar.

half space [BUILD] A broad step between two half flights of a stair.

half-through arch [CIV ENG] A bridge arch having the roadway running through it at an elevation midway between the base and the crown.

half-tide basin [CIV ENG] A lock of very large size and usually of irregular shape, the gates of which are kept open for several hours after high tide so that vessels may enter as long as there is sufficient depth over the sill; vessels remain in the half-tide basin until the ensuing flood tide, when they may pass through the gate to the inner harbor; if entry to the inner harbor is required before this time, water must be admitted to the half-tide basin from some external source.

half-timbered [BUILD] Pertaining to a timber frame building with brickwork, plaster, or wattle and daub filling the spaces between the timbers.

half-track [MECH ENG] **1.** A chain-track drive system for a vehicle; consists of an endless metal belt on each side of the vehicle driven by one of two inside sprockets and running on bogie wheels; the revolving belt lays down on the ground a flexible track of cleated steel or hard-rubber plates; the front end of the vehicle is supported by a pair of wheels. **2.** A motor vehicle equipped with half-tracks.

half-track tape recorder *See* double-track tape recorder.

halibut liver oil [MATER] Fishy-tasting oil, pale yellow to dark red, with slightly fishy smell; soluble in alcohol, ether, chloroform, and carbon disulfide, insoluble in water; derived from halibut livers; used as medicine, as vitamin A and D source, and to dress leather.

Halimond tube [MIN ENG] A miniature pneumatic flotation cell used for examination of small ore samples under closely controllable conditions.

Hall-effect gaussmeter [ENG] A gaussmeter that consists of a thin piece of silicon or other semiconductor material which is inserted between the poles of a magnet to measure the magnetic field strength by means of the Hall effect.

Hallett table [MIN ENG] A Wilfley-type concentrating table having the tops of the riffles in the same plane as the cleaning planes, and riffles inclined toward the waste water side.

Hallinger shield [MIN ENG] A tunneling shield valuable for working in very soft ground; incorporates a mechanical excavator and does not entail the use of timbering to protect the miners.

Hall process [MET] An electrolytic recovery process for aluminum employing a fused-bauxite (aluminum oxide), cryolite electrolyte.

halo effect [IND ENG] A tendency when rating a person in regard to a specific trait to be influenced by a general impression or by another trait of the person.

halophone [ENG] A device that records patterns in time in a manner analogous to the way that optical holograms record space.

Halsey premium plan [IND ENG] A wage-incentive plan which sets a guaranteed daily rate to an employee and provides for predetermined compensation for superior performance.

hammer [DES ENG] **1.** A hand tool used for pounding and consisting of a solid metal head set crosswise on the end of a handle. **2.** An arm with a striking head for sounding a bell or gong. [MECH ENG] A power tool with a metal block or a drill for the head.

hammer drill [MECH ENG] Any of three types of fast-cutting, compressed-air rock drills (drifter, sinker, and stoper) in which a hammer strikes rapid blows on a loosely held piston, and the bit remains against the rock in the bottom of the hole, rebounding slightly at each blow, but does not reciprocate.

hammer forging [MET] Forging by means of repeated blows of a hammer.

hammerhead [DES ENG] The striking part of a hammer.

hammerhead crane [MECH ENG] A crane with a horizontal jib that is counterbalanced.

hammer mill [MECH ENG] **1.** A type of impact mill or crusher in which materials are reduced in size by hammers revolving rapidly in a vertical plane within a steel casing. Also known as beater mill. **2.** A grinding machine which pulverizes feed and other products by several rows of thin hammers revolving at high speed.

hammer milling [MECH ENG] Crushing or fracturing materials in a hammer mill.

hammer pick [MIN ENG] A pneumatic hand-held machine used to break up the harder rocks in a mine; consists of a pick which is driven by a hammer set in a cylinder which receives compressed air.

hammer test [MET] An impact test conducted by dropping weights from increasing heights until a specified deflection of the weight is produced.

hammer welding [MET] Forge welding by means of a hammer.

Hancock jig [MIN ENG] A moving-screen jig used to treat lead-zinc ores; the material is jigged in a tank of water, and the heavy layer settles through slots.

hand auger [DES ENG] A hand tool resembling a large carpenters' bit or comprising a short cylindrical container with cutting lips attached to a rod; used to bore shallow holes in the soil to obtain samples of it and other relatively unconsolidated near-surface materials.

handbarrow [ENG] A flat, rectangular frame with handles at both ends, carried by two persons to transport objects. Also known as barrow.

hand brake [MECH ENG] A manually operated brake.

hand cable [MIN ENG] A flexible cable for electrical connection between a mining machine and a truck carrying a reel of portable cable. Also known as butt cable; head cable.

handcar [MECH ENG] A small, four-wheeled, hand-pumped car used on railroad tracks to transport men and equipment for construction or repair work; other cars for the same purpose are motor-operated.

hand drill [DES ENG] A small, portable drilling machine which is operated by hand.

hand electric lamp [MIN ENG] In mining, a portable, battery operated hand lamp with a tungsten-filament light source that forms a self-contained unit.

hand feed [ENG] A drill machine in which the rate at which the bit is made to penetrate the rock is controlled by a hand-operated ratchet and lever or a hand-turned wheel meshing with a screw mechanism.

hand forging [MET] Plastic deformation of a metal by manual force.

hand hammer drill [ENG] A hand-held rock drill.

handhole [ENG] A shallow access hole large enough for a hand to be inserted for maintenance and repair of machinery or equipment.

hand jig [MIN ENG] A moving-screen jig operated by hand and used to treat small batches of ore; the jig box is attached to a rocking beam and moved in a tank of water.

hand lance [ENG] A hand-held pipe with a nozzle through which steam or air is discharged; used to remove soot deposits from the external surfaces of boiler tubes.

handle [MECH ENG] The arm connecting the bucket with the boom in a dipper shovel or hoe.

hand lead [ENG] A light sounding lead (7–14 pounds or 3–6 kilograms) usually having a line not more than 25 fathoms (46 meters) in length.

hand level [ENG] A hand-held surveyor's level, basically a telescope with a bubble tube attached so that the position of the bubble can be seen when looking through the telescope.

handling time [IND ENG] The time needed to transport parts or materials to or from a work area.

hand loader [MIN ENG] A miner who uses a shovel, rather than a machine to load coal.

handpicking [MIN ENG] Manual removal of a selected fraction of coarse run-of-mine ore, after washing and screening away waste.

hand punch [DES ENG] A hand-held device for punching holes in paper or cards.

handrail [ENG] A narrow rail to be grasped by a person for support.

hand sampling [MIN ENG] Using manual methods to detach and reduce in size representative samples of ore; one of the major methods in sampling small batches of ore, others being grab sampling, trench or channel sampling, fractional selection, coning and quartering, and pipe sampling.

handsaw [DES ENG] A saw operated by hand, with a backward and forward arm movement.

handset [DES ENG] A combination of a telephone-type receiver and transmitter, designed for holding in one hand.

handset bit [DES ENG] A bit in which the diamonds are manually set into holes that are drilled into a malleable-steel bit blank and shaped to fit the diamonds.

hand time [IND ENG] The time necessary to complete a manual element. Also known as manual time.

hand tool [ENG] Any implement used by hand.

hand tramming [MIN ENG] Pushing of mine cars by workers; limited to small mines with low output.

hand truck [ENG] **1.** A manually operated, two-wheeled truck consisting of a rectangular frame with handles at the top and a plate at the bottom to slide under the load. **2.** Any of various small, manually operated, multiwheeled platform trucks for transporting materials.

hand winch [MECH ENG] A winch that is operated by hand.

hangar [CIV ENG] A building at an airport specially designed in height and width to enable aircraft to be stored or maintained in it.

hanger [CIV ENG] An iron strap which lends support to a joist beam or pipe. [PETRO ENG] **1.** A device to seat in the bowl of a lowermost casing head to suspend the next-smaller casing string and form a seal between the two. Also known as casing hanger. **2.** A device to provide a seal between the tubing and the tubing head. Also known as tubing hanger.

hanger bolt [DES ENG] A bolt with a machine-screw thread on one end and a lag-screw thread on the other.

hangfire [ENG] Delay in the explosion of a charge.

hanging [MET] Sticking or wedging of part of the charge in a blast furnace.

hanging bolt [MIN ENG] A bolt used to suspend wall plates in shaft construction.

hanging coal [MIN ENG] A portion of the coal seam which, by undercutting, has had its natural support removed.

hanging-drop atomizer [MECH ENG] An atomizing device used in gravitational atomization; functions by quasi-static emission of a drop from a wetted surface. Also known as pendant atomizer.

hanging load [MECH ENG] **1.** The weight that can be suspended on a hoist line or hook device in a drill tripod or derrick without causing the members of the derrick or tripod to buckle. **2.** The weight suspended or supported by a bearing.

hanging scaffold [CIV ENG] A movable platform suspended by ropes and pulleys; used by workers for above-ground building construction and maintenance.

hanging sets [MIN ENG] Timbers from which cribs are suspended in working through soft strata.

hanging sheave [MIN ENG] The grooved wheel or pulley which is suspended from the drill tripod clevis or from the roof or side of a haulage road, and over which the hoist line runs to minimize friction.

hanging-wall drift [MIN ENG] A horizontal gallery that is driven in the hanging wall of a vein.

hang-up [ENG] A virtual leak resulting from the release of entrapped tracer gas from a leak detector vacuum system. [MIN ENG] Blockage of the movement of ore by rock in an underground chute.

harbor engineering [CIV ENG] Planning and design of facilities for ships to discharge or receive cargo and passengers.

harbor line [CIV ENG] The line beyond which wharves and other structures cannot be extended.

hard [MATER] Quality of a material that is compact, solid, and difficult to deform.

hard beach [CIV ENG] A portion of a beach especially prepared with a hard surface extending into the water, employed for the purpose of loading or unloading directly into or from landing ships or landing craft.

hardboard [MATER] A fiberboard formed to a density of 30–50 pounds per cubic foot (480–800 kilograms per cubic meter) and having one textured and one smooth face.

hard bottom [MIN ENG] A condition encountered in some opencut mines wherein the rock occasionally does not break down to grade because of an extra-hard streak of ground or because insufficient powder is used.

hard bronze [MET] A high-tensile-strength alloy containing 88% copper, 7% tin, 3% zinc, and 2% lead.

hard-burned brick [MATER] A brick that has been fired and sintered at high temperature.

hardcap [MIN ENG] In bauxite mining the uppermost foot or two (0.3-0.6 meter) deposit of bauxite; usually serves as a roof during mining.

hard chromium [MET] A thick coating of electrodeposited chromium on a base metal; increases wear resistance.

hard-coal plough [MIN ENG] A plough-type cutter-loader consisting of rigid or swiveling kerfing bits which precut the coal in hard-coal seams.

hard-drawn wire [MET] Cold-drawn metal wire, usually of high tensile strength.

hardenability [MET] In a ferrous alloy, the property that determines the depth and distribution of hardness induced by quenching from elevated temperatures.

hardened steel [MET] Steel hardened by quenching from high temperatures.

hardener [MET] A master alloy added to a melt to control hardness.

hardening [MET] 1. Imparting hardness to carbon steel by abrupt cooling (quenching) through a critical temperature range. 2. Heat-treating an age-hardening or precipitation-hardening alloy at intermediate temperatures.

hard-face [MET] To apply a layer of hard, abrasion-resistant metal to a less resistant metal part by plating, welding, spraying, or other techniques. Also known as hard-surface.

hard-fiber [MATER] Indicating vulcanization with zinc chloride; used of paper or boards.

hard glass [MATER] A potash-lime glass with a high silica content, used for making brilliant glassware. Also known as Bohemian glass.

hard grease [MATER] A lubricating grease that flows at a temperature of about 90°C.

hard ground [MIN ENG] Ground that is difficult to work.

Hardgrove grindability index [ENG] The relative grindability of ores and minerals in comparison with standard coal, chosen as 100 grindability, as determined by a miniature ball-ring pulverizer. Also known as Hardgrove number.

Hardgrove number *See* Hardgrove grindability index.

hard hat [ENG] A safety hat usually having a metal crown; used by construction workers and miners.

hardhead [MET] A hard white deposit formed during tin refining by liquation.

Hardinge feeder-weigher [MECH ENG] A pivoted, short belt conveyor which controls the rate of material flow from a hopper by weight per cubic foot.

Hardinge mill [MECH ENG] A tricone type of ball mill; the cones become steeper from the feed end toward the discharge end.

Hardinge thickener [ENG] A machine for removing the maximum amount of liquid from a mixture of liquid and finally divided solids by allowing the solids to settle out on the bottom as sludge while the liquid overflows at the top.

hard iron [MET] Iron or steel which is not readily magnetized by induction, but which retains a high percentage of the magnetism acquired.

hard lac [MATER] Solvent-extracted shellac. Also known as hard-lac resin.

hard-lac resin *See* hard lac.

hard-laid [DES ENG] Pertaining to rope with strands twisted at a 45° angle.

hard landing [AERO ENG] A landing made without deceleration, as by impact on the moon.

hard lead [MET] Lead alloy with reduced malleability due to the presence of impurities, usually antimony.

hard magnetic material [MET] A metal having a high coercive force which gives a large magnetic hysteresis.

hardness [ENG] Property of an installation, facility, transmission link, or equipment that will prevent an unacceptable level of damage. [MATER] Resistance of a metal or other material to indentation, scratching, abrasion, or cutting.

hardness number [ENG] A number representing the relative hardness of a mineral, metal, or other material as determined by any of more than 30 different hardness tests.

hardness test [ENG] A test to determine the relative hardness of a metal, mineral, or other material according to one of several scales, such as Brinell, Mohs, or Shore.

hard paste porcelain See porcelain.

hard porcelain [MATER] A ceramic material having good resistance to thermal shock.

hard-rock driller [MIN ENG] A worker who operates a drill in a mine where the rocks are generally igneous or metamorphosed and considered hard, such as rocks in which coal and salt are generally found.

hard-rock mine [MIN ENG] A mine located in hard rock, especially a mine difficult to drill, blast, and square up.

hard-rock miner [MIN ENG] A worker competent to mine in hard rock, usually an expert miner.

hard-rock tunnel boring [MIN ENG] A tunneling method utilizing a mole to cut out 7-foot-diameter (2.1-meter-diameter) drifts in hard rock at an average rate of 5 feet (1.5 meters) per hour.

hard rubber [MATER] Rubber that has been vulcanized at high temperatures and pressures to give hardness; used as an electrical insulating material and in tool handles. Also known as ebonite.

hard solder See brazing alloy.

hard-solder See braze.

hardstand [CIV ENG] 1. A paved or stabilized area where vehicles or aircraft are parked. 2. Open ground area having a prepared surface and used for storage of material.

hard-surface [CIV ENG] To treat a ground surface in order to prevent muddiness. [MET] See hard-face.

hardware [ENG] Items made of metal, such as tools, fittings, fasteners, and appliances.

Hardwick conveyor loader head [MIN ENG] A dust collector for belt conveyors used at the loading station; a scraper chain runs at the bottom of a coal hopper and collects underbelt fines.

hardwood [MATER] Dense, close-grained wood of an angiospermous tree, such as oak, walnut, cherry, and maple.

hardwood bearing [MECH ENG] A fluid-film bearing made of lignum vitae which has a natural gum, or of hard maple which is impregnated with oil, grease, or wax.

Hardy plankton indicator [ENG] Metal-shrouded net sampler designed to collect specimens of plankton during normal passage of a ship.

Hare's hygrometer [ENG] A type of hydrometer in which the ratio of the densities of two liquids is determined by measuring the heights to which they rise in two vertical glass tubes, connected at their upper ends, when suction is applied.

harmless-depth theory [MIN ENG] Formerly, hypothesis that there was a certain depth below which mining could be carried on without risk of damage to the surface.

harmonic decline [PETRO ENG] One of three types of decline in oil or gas production rate (the others are constant-percentage and hyperbolic), in which the nominal decline in production rate per unit of time expressed as a fraction of the production rate is proportional to the production rate itself.

harmonic speed changer [MECH ENG] A mechanical-drive system used to transmit rotary, linear, or angular motion at high ratios and with positive motion.

harness [AERO ENG] 1. Straps arranged to hold an occupant of a spacecraft or aircraft in the seat. 2. Straps worn by a parachutist or used to suspend a load from a parachute.

harpoon [DES ENG] A barbed spear used to catch whales.

harpoon log [ENG] A log which consists essentially of a rotator and distance registering device combined in a single unit, and towed through the water; it has been largely replaced by the taffrail log; the two types of logs are similar except that the registering device of the taffrail log is located at the taffrail and only the rotator is in the water.

Harris process [MET] A method for refining lead in which the liquid bullion is sprayed through molten caustic soda and molten sodium nitrate; arsenic, antimony, and tin are oxidized, converted into sodium salts, and skimmed from the bath.

Hartford loop [MECH ENG] A condensate return arrangement for low-pressure, steam-heating systems featuring a steady water line in the boiler.

Hartmann lines See Lüders lines.

hartshorn oil See bone oil.

Harz jig [MIN ENG] A device used to separate coal and foreign matter which gives pulsion intermittently with suction.

Hasenclever turntable [MIN ENG] A turntable that is made to rotate by the friction between the positively driven pulley, the car, and the table; used as an alternative to the shunt-back or the traverser for changing the direction of mine cars or tubs, either on the surface or underground.

hatch [ENG] A door or opening, especially on an airplane, spacecraft, or ship.

hatch beam [ENG] A heavy, portable beam which supports a hatch cover.

hatch cover [ENG] A steel or wooden cover for a hatch.

hatchet [DES ENG] A small ax with a short handle and a hammerhead in addition to the cutting edge.

haul [ENG] A single tow of a net or dredge.

haulage [MIN ENG] The movement, in cars or otherwise, of men, supplies, ore, and waste, underground and on the surface.

haulage conveyor [MIN ENG] A conveyor used to transport material between the gathering conveyor and the outside.

haulage curve [MIN ENG] A bend in a haulage road in any direction.

haulage drum [MIN ENG] A cylinder on which steel haulage rope is coiled.

haulage level [MIN ENG] An underground level, either along and inside the ore body or closely parallel to it, and usually in the footwall, in which mineral is loaded into trams and moved out to the hoisting shaft.

haulage stage [MIN ENG] A mine roadway along which a load is moved by one form of haulage without coupling or uncoupling of cars.

haulageway [MIN ENG] The gangway, entry, or tunnel through which loaded or empty mine cars are hauled by animal or mechanical power.

haul-cycle time [MIN ENG] The time required for the scraper to haul a load to the dumping area and to return to its position in the loading area.

haul road [MIN ENG] A road built to carry heavily loaded trucks at a good speed; the grade is limited and usually kept to less than 17% of climb.

Hauzeur furnace [MET] A double furnace for the distillation of zinc wherein waste heat from one set of retorts is utilized for heating the second set.

hawk [ENG] A board with a handle underneath used by a workman to hold mortar.

Hayden process [MET] A method of electrolytic copper refining; anodes of unrefined copper are suspended in an acid electrolyte, and one side of each then acts as an anode and the other as a cathode.

haydite [MATER] Expanded shale, slate, or clay characterized by low unit weight and satisfactory structural properties; used as an aggregate to produce lightweight structural concrete.

Hayward grab bucket [MECH ENG] A clamshell type of grab bucket used for handling coal, sand, gravel, and other flowable materials.

Hayward orange peel [MECH ENG] A grab bucket that operates like the clamshell type but has four blades pivoted to close.

hazard [IND ENG] Any risk to which a worker is subject as a direct result (in whole or in part) of his being employed.

hazardous material [MATER] A poison, corrosive agent, flammable substance, explosive, radioactive chemical, or any other material which can endanger human health or well-being if handled improperly.

haze level See haze line.

H beam [CIV ENG] A beam similar to the I beam but with longer flanges.

H bit [DES ENG] A core bit manufactured and used in Canada having inside and outside diameters of 2.875 and 3.875 inches (73.025 and 98.425 millimeters), respectively; the matching reaming shell has an outside diameter of 3.906 inches (99.2124 millimeters).

head [BUILD] The upper part of the frame on a door or window. [ENG] The end section of a plastics blow-molding machine in which a hollow parison is formed from the melt. [ENG ACOUS] See cutter.

headblock [MIN ENG] **1.** A stop at the head of a slope or shaft to keep cars from going down the shaft or slope. **2.** A cap piece.

headboard [MIN ENG] **1.** A wooden wedge placed against the hanging wall, and against which one end of the stull is jammed. **2.** A board in the roof of a heading, contacting the earth above and supported by a headtree on each side.

head cable See hand cable.

header [BUILD] A framing beam positioned between trimmers and supported at each end by a tail beam. [CIV ENG] Brick or stone laid in a wall with its narrow end facing the wall. [ENG] A pipe, conduit, or chamber which distributes fluid from a series of smaller pipes or conduits; an example is a manifold. [MECH ENG] A machine used for gathering or upsetting materials; used for screw, rivet, and bolt heads. [MIN ENG] **1.** An entry-boring machine that bores the entire section of the entry in one operation. **2.** A rock that heads off or delays progress. **3.** A blasthole at or above the head.

header bond [CIV ENG] A masonry bond consisting of header courses exclusively.

header course [CIV ENG] A masonry course of bricks laid as headers.

headframe [MIN ENG] **1.** The frame at the top of a shaft, on which is mounted the hoisting pulley. Also known as gallows; gallows frame; headstock; hoist frame. **2.** The shaft frame, sheaves, hoisting arrangements, dumping gear, and connected works at the top of a shaft. Also known as headgear.

head gate [CIV ENG] **1.** A gate on the upstream side of a lock or conduit. **2.** A gate at the starting point of an irrigation ditch.

headgear See headframe.

headhouse [MIN ENG] **1.** A timber framing located at the top of a shaft and receiving the shaft guides that carry the cage or elevator. **2.** A structure that houses the headframe.

heading-and-bench mining [MIN ENG] A stoping method used in thicker ore where it is customary first to take out a slice or heading 7 or 8 feet (2.1 or 2.4 meters) high directly under the top of the ore, and then to bench or stope down the ore between the bottom of the heading and the bottom of the ore or floor of the level.

heading blasting *See* coyote blasting.

heading joint [BUILD] **1.** A joint between two pieces of timber which are joined in a straight line, end to end. **2.** A masonry joint formed between two stones in the same course.

heading–overhand bench method [MIN ENG] A tunneling method in which the heading is the lower part of the section and is driven at least a round or two in advance of the upper part (bench), which is taken out by overhand excavating. Also known as inverted heading and bench method.

headline [MIN ENG] In dredging, the line which is anchored ahead of the dredge pond and holds the dredge up to its digging front.

headman [MIN ENG] **1.** A person who brings coal to the tramway from the workings. **2.** One who engages or disengages grips on mine cars at the top of a haulage slope.

head mast [MIN ENG] The tower carrying the working lines of a cable excavator.

head meter [ENG] A flowmeter that is dependent upon change of pressure head to operate.

head motion [MECH ENG] The vibrator on a reciprocating table concentrator which imparts motion to the deck.

headphone [ENG ACOUS] An electroacoustic transducer designed to be held against an ear by a clamp passing over the head, for private listening to the audio output of a communications, radio, or television receiver or other source of audio-frequency signals. Also known as phone.

head pulley [MECH ENG] The pulley at the discharge end of a conveyor belt; may be either an idler or a drive pulley.

head-pulley-drive conveyor [MECH ENG] A conveyor having the belt driven by the head pulley without a snub pulley.

headroom [MIN ENG] **1.** Distance between the drill platform and the bottom of the sheave wheel. **2.** Height between the floor and the roof in a mine opening.

headrope [MIN ENG] In rope haulage, that rope used to pull the loaded transportation device toward the discharge point.

heads [MIN ENG] **1.** Material removed from the ore in the treatment plant and containing the valuable metallic constituents. **2.** The feed to a concentrating system in ore dressing.

head section [ENG] That part of belt conveyor which consists of a drive pulley, a head pulley which may or may not be a drive pulley, belt idlers if included, and the necessary framing.

headset [ENG ACOUS] A single headphone or a pair of headphones, with a clamping strap or wires holding them in position.

head shaft [MECH ENG] The shaft driven by a chain and mounted at the delivery end of a chain conveyor; it serves as the mount for a sprocket which drives the drag chain.

head sheave [MIN ENG] Pulley in the headgear of a winding shaft over which hoisting rope runs.

headsill [BUILD] A horizontal beam at the top of the frame of a door or window.

headstock [MECH ENG] **1.** The device on a lathe for carrying the revolving spindle. **2.** The movable head of certain measuring machines. **3.** The device on a cylindrical grinding machine for rotating the work. [MIN ENG] *See* headframe.

headtree [MIN ENG] The horizontal timber placed at each side of a rectangular heading to support the headboard.

head up [ENG] To tighten bolts on a hatch cover or access hole plate to prevent leakage from or into an operating vessel.

head value [MIN ENG] Assay value of the feed to a concentrating system.

headwall [CIV ENG] A retaining wall at the outlet of a drain or culvert.

headway [MIN ENG] *See* cross heading.

headworks [CIV ENG] Any device or structure at the head or diversion point of a waterway.

heaped capacity [MIN ENG] In scraper loading, the volume of heaped material that a scraper will hold.

heap leaching [MET] A process used for the recovery of copper from weathered ore and material from mine dumps; material is laid to a thickness of 20 feet in alternately fine and coarse beds and treated with water at intervals during which oxidation occurs; liquor that runs off is treated with scrap iron to precipitate copper.

heap roasting [MIN ENG] A process in which ore with a high sulfur content is roasted by the combustion of the sulfur.

heap sampling [MIN ENG] Method of ore sampling in which the material is shoveled into a conical heap which is then flattened with a spade and shoveled into four equal heaps, two of which are retained, crushed, mixed, and formed into another, smaller cone; the process is repeated until the required small sample is produced.

hearing aid [ENG ACOUS] A miniature, portable sound amplifier for persons with impaired hearing, consisting of a microphone, audio amplifier, earphone, and battery.

heart bond [CIV ENG] A masonry bond in which two header stones meet in the middle of the wall, their joint being covered by another stone; no headers stretch across the wall.

hearth [BUILD] 1. The floor of a fireplace or brick oven. 2. The projection in front of a fireplace, made of brick, stone, or cement. [MET] The floor of a reverberatory, open-hearth, cupola, or blast furnace; it is made of refractory material able to support the charge and to collect the molten products.

hearth furnace [MET] A furnace designed to heat the charge, resting on the hearth, by passing hot gases over it.

hearth roasting [MIN ENG] A process in which ore or concentrate enters at the top of a multiple hearth roaster and drops from hearth until it is discharged at the bottom.

heat-affected zone [MET] The zone within a base metal that undergoes structural changes but does not melt during welding, cutting, or brazing.

heat balance [MET] The calculation used in fluidization roasting so that the addition or removal of heat can be controlled to maintain the optimum temperature in the reacting vessel.

heat barrier *See* thermal barrier.

heat check [MET] Parallel surface cracks forming a pattern on the surface of a metal as a result of thermal fatigue.

heat distortion point [ENG] The temperature at which a standard test bar (ASTM test) deflects 0.010 inch (0.254 millimeter) under a load of either 66 or 264 psi (4.55×10^5 or 18.20×10^5 newtons per square meter), as specified.

heat dump *See* heat sink.

heat engine [MECH ENG] A machine that converts heat into work (mechanical energy).

heater [ENG] A contrivance designed to give off heat.

heater oil *See* heating oil.

heat exchanger [ENG] Any device, such as an automobile radiator, that transfers heat from one fluid to another or to the environment. Also known as exchanger.

heating chamber [ENG] The part of an injection mold in which cold plastic feed is changed into a hot melt.

heating fuel *See* heating oil.

heating load [CIV ENG] The quantity of heat per unit time that must be provided to

maintain the temperature in a building at a given level.

heating oil [MATER] No. 2 fuel oil; used in domestic heating units. Also known as heater oil; heating fuel.

heating plant [CIV ENG] The whole system for heating an enclosed space. Also known as heating system.

heating surface [ENG] The surface for the absorption and transfer of heat from one medium to another.

heating system *See* heating plant.

heat insulator [MATER] A substance having relatively low heat conductivity.

heat-loss flowmeter [ENG] Any of various instruments that determine gas velocities or mass flows from the cooling effect of the flow on an electrical sensor such as a thermistor or resistor; a second sensor is used to compensate for the temperature of the fluid. Also known as thermal-loss meter.

heat pipe [ENG] A heat-transfer device consisting of a sealed metal tube with an inner lining of wicklike capillary material and a small amount of fluid in a partial vacuum; heat is absorbed at one end by vaporization of the fluid and is released at the other end by condensation of the vapor.

heat pump [MECH ENG] A device which transfers heat from a cooler reservoir to a hotter one, expending mechanical energy in the process, especially when the main purpose is to heat the hot reservoir rather than refrigerate the cold one.

heat rate [MECH ENG] An expression of the conversion efficiency of a thermal power plant or engine, as heat input per unit of work output; for example, Btu/kWh.

heat-resistant alloy [MET] An oxidation-resistant alloy.

heat-resistant glass [MATER] Glass, such as borosilicate glass, that is heat-treated or leached to remove alkali so that it withstands high heat and sudden cooling without shattering.

heat seal [ENG] A union between two thermoplastic surfaces by application of heat and pressure to the joint.

heat shield [MATER] Any protective layer that gives protection from heat; used on the front of a reentry capsule.

heat-shrinkable tubing [MATER] A type of plastic tubing that can be heated and shrink-fitted over terminals and other objects of varying sizes and shapes, for insulating and other purposes.

heat shunt [MET] A heatsink placed in contact with the lead of a delicate component to prevent overheating during soldering.

heatsink [AERO ENG] 1. A type of protective device capable of absorbing heat and used as a heat shield. **2.** In nuclear propulsion, any thermodynamic device, such as a radiator or condenser, that is designed to absorb the excess heat energy of the working fluid. Also known as heat dump.

heatsink cooling [ENG] Cooling a body or system by allowing heat to be absorbed from it by another body.

heat sterilization [ENG] An act of destroying all forms of life on and in bacteriological media, foods, hospital supplies, and other materials by means of moist or dry heat.

heat time [MET] Duration of a single current impulse in pulsation welding.

heat tinting [MET] Oxidation of a polished metal surface by heating to reveal the microstructure.

heat-transfer oil [MATER] An oil used to transport heat or cold between two areas of process-equipment surface, and especially compounded to avoid heat degradation in the temperature range of application.

heat-treatable alloy [MET] An alloy that can be hardened by thermal treatment.

heat-treating film [MET] An oxide coating formed on a metal surface by heat treating.

heat treatment [MET] Heating and cooling a metal or alloy to obtain desired properties or conditions.

heave [MIN ENG] **1.** A rising of the floor of a mine caused by its being too soft to resist the weight on the pillars. **2.** A predominantly upward movement of the surface of the soil due to expansion or displacement.

heavier-than-air craft [AERO ENG] Any aircraft weighing more than the air it displaces.

heaving [PETRO ENG] Partial or total collapse of drill hole walls resulting from internal pressures mainly due to swelling from hydration or formation gas pressures.

heaving plug [PETRO ENG] A plug at the bottom of an oil well which stops unconsolidated sand from mixing with the oil.

heavy alloy [MET] A tungsten-nickel alloy produced by pressing and sintering the metallic powders; used for screens for x-ray tubes and radioactivity units and for contact surfaces of circuit breakers.

heavy bomber [AERO ENG] Any large bomber considered to be relatively heavy, such as a bomber having a gross weight, including bomb load, of 250,000 pounds (113,000 kilograms) or more, as the B-36 and the B-52.

heavy concrete [MATER] Concrete in which some or all rock aggregate is replaced by metal aggregate.

heavy crude [PETRO ENG] Crude oil having a high proportion of viscous, high-molecular-weight hydrocarbons, and often having a high sulfur content.

heavy-duty [ENG] Designed to withstand excessive strain.

heavy-duty car [MECH ENG] A railway motorcar weighing more than 1400 pounds (635 kilograms), propelled by an engine of 12–30 horsepower (8900–22,400 watts), and designed for hauling heavy equipment and for hump-yard service.

heavy-duty oil [MATER] Lubricating oil with good oxidation stability and corrosion-preventive and detergent-dispersant characterisitics; used in high-speed diesel and gasoline engines under heavy-duty service conditions.

heavy-duty tool block *See* open-side tool block.

heavy ends [MATER] The highest boiling portion of a petroleum fraction.

heavy force fit [DES ENG] A fit for heavy steel parts or shrink fits in medium sections.

heavy fraction [PETRO ENG] The final products retrieved from crude oil during the process of distillation. Also known as end cut.

heavy ground [MIN ENG] Dangerous hanging wall requiring vigilance against possible rock fall.

heavy liquid [MATER] Any of a group of heavy organic liquids, inorganic solutions, and fused salts used for determination of specific gravity of mineral particles or for separation of minerals having lower and higher specific gravities than the liquids; examples are methylene iodide and bromoform.

heavy-liquid separation [MIN ENG] A laboratory technique for separating ore particles by allowing them to settle through, or float above, a fluid of intermediate density.

heavy-media separation [MIN ENG] A series of processes for the concentration of ore developed at one time, but now used in coal cleaning; uses suspensions of magnetic materials such as magnetite.

heavy metal [MET] A metal whose specific gravity is approximately 5.0 or higher.

heavy-mineral prospecting [MIN ENG] Locating the source of an economic mineral by determining the relative amounts of the mineral in stream sediments and tracing the drainage upstream.

heavy naphtha [MATER] A dark amber to red liquid that is a mixture of xylene and higher homologs; it is flammable; used as a solvent for asphalt and in production of coumarone resins. Also known as high-flash naphtha.

heavy oil [MATER] The high-boiling, relatively viscous fractions of petroleum or coal tar oils.

heavy resin oil [MATER] Reddish-brown, high-boiling, heavy coal tar oils.

heavy section car [MECH ENG] A railway motorcar weighing 1200–1400 pounds (544–635 kilograms) and propelled by an 8–12 horsepower (6000–8900 watts) engine.

hedeoma oil [MATER] A yellowish essential oil distilled from the leaves of *Hedeoma pulegioides;* soluble in two or more parts of 70% alcohol, ether, and chloroform; used in medicine and perfumery. Also known as American pennyroyal oil; pulegium oil.

heel *See* heel block.

heel block [MECH ENG] A block or plate that is usually fixed on the die shoe to minimize deflection of a punch or cam. Also known as heel.

heeling adjuster [ENG] A dip needle with a sliding weight that can be moved along one of its arms to balance the magnetic force; used to determine the correct position of a heeling magnet. Also known as heeling error instrument; vertical force instrument.

heeling error instrument *See* heeling adjuster.

heeling magnet [ENG] A permanent magnet placed vertically in a tube under the center of a marine magnetic compass, to correct for heeling error.

heel of a shot [ENG] **1.** In blasting, the front or face of a shot farthest from the charge. **2.** The distance between the mouth of the drill hole and the corner of the nearest free face. **3.** That portion of a drill hole which is filled with the tamping. [MIN ENG] That portion of the coal to be fractured which is outside the powder.

heel plate [CIV ENG] A plate at the end of a truss.

heel post [CIV ENG] A post to which are secured the hinges of a gate or door.

Hegeler furnace [MET] A muffle furnace having seven tiers of hearths; lower hearths are heated by gas burned in flues beneath them.

height finder [ENG] A radar equipment, used to determine height of aerial targets.

height finding [ENG] Determination of the height of an airborne object.

height-finding radar [ENG] A radar set that measures and determines the height of an airborne object.

height gage [ENG] A gage used to measure heights by either a micrometer or a vernier scale.

height of instrument [ENG] **1.** In survey leveling, the vertical height of the line of collimation of the instrument over the station above which it is centered, or above a specified datum level. **2.** In spirit leveling, the vertical distance from datum to line of sight of the instrument. **3.** In stadia leveling the height of center of transit above the station stake. **4.** In differential leveling, the elevation of the line of sight of the telescope when the instrument is leveled.

held in common [MIN ENG] Pertaining to a claim whereof there is more than one owner.

Heliarc welding *See* inert gas-shielded arc welding.

helical conveyor [MECH ENG] A conveyor for the transport of bulk materials which consists of a horizontal shaft with helical paddles or ribbons rotating inside a stationary tube.

helical-flow turbine [MECH ENG] A steam turbine in which the steam is directed tangentially and radially inward by nozzles against buckets milled in the wheel rim; the steam flows in a helical path, reentering the buckets one or more times. Also known as tangential helical-flow turbine.

helical gear [MECH ENG] Gear wheels running on parallel axes, with teeth twisted oblique to the gear axis.

helical milling [MECH ENG] Milling in which the work is simultaneously rotated and translated.

helical rake angle [DES ENG] The angle between the axis of a reamer and a plane tangent to its helical cutting edge; also applied to milling cutters.

helical scanning [ENG] A method of radar scanning in which the antenna beam rotates continuously about the vertical axis while the elevation angle changes slowly from horizontal to vertical, so that a point on the radar beam describes a distorted helix.

helical spline broach [MECH ENG] A broach used to produce internal helical splines having a straight-sided or involute form.

helical spring [DES ENG] A bar or wire of uniform cross section wound into a helix.

helical steel support [MIN ENG] A continuous, screw-shaped steel joist lining used for staple shafts.

helicopter [AERO ENG] An aircraft fitted to sustain itself by motor-driven horizontal rotating blades (rotors) that accelerate the air downward, providing a reactive lift force, or accelerate the air at an angle to the vertical, providing lift and thrust.

heliograph [ENG] An instrument that records the duration of sunshine and gives a qualitative measure of its amount by action of sun's rays on blueprint paper.

heliostat [ENG] A clock-driven instrument mounting which automatically and continuously points in the direction of the sun; it is used with a pyrheliometer when continuous direct solar radiation measurements are required.

heliotrope [ENG] An instrument that reflects the sun's rays over long distances; used in geodetic surveys.

heliox [MATER] A mixture of helium and a few percent of oxygen used for breathing during deep dives.

helipad [CIV ENG] The launch and landing area of a heliport.

heliport [CIV ENG] A place built for helicopter takeoffs and landings.

helium-oxygen diving [ENG] Diving operations employing a breathing mixture of helium and oxygen.

helium refrigerator [MECH ENG] A refrigerator which uses liquid helium to cool substances to temperatures of 4 K or less.

helix angle [DES ENG] That angle formed by the helix of the thread at the pitch-diameter line and a line at right angles to the axis.

helmet [ENG] A globe-shaped head covering made of copper and supplied with air pumped through a hose; attached to the breastplate of a diving suit for deep-sea diving.

Helmholtz resonator [ENG ACOUS] An enclosure having a small opening consisting of a straight tube of such dimensions that the enclosure resonates at a single frequency determined by the geometry of the resonator.

helper grade [MIN ENG] A grade on which helper engines are required to assist road locomotives. Also known as pusher grade.

helper set [MIN ENG] A set of timbers to reinforce the normal set of timbers in a mine.

help-yourself system [IND ENG] A tool-crib system for temporary issue of tools employed in small shops; employees have access to tools in the crib and help themselves.

helve hammer [MET] A belt-driven trip hammer with the hammer face or swage carried on the end of a beam; used for welding, forging, plating, drawing, and other metal-working operations.

hemispherical pyrheliometer [ENG] An instrument for measuring the total solar energy from the sun and sky striking a horizontal surface, in which a thermopile measures the temperature difference between white and black portions of a thermally insulated horizontal target within a partially evacuated transparent sphere or hemisphere.

hemlock oil *See* spruce oil.

hemming [MECH ENG] Forming of an edge by bending the metal back on itself.

hemp-core cable *See* standard wire rope.

hempseed oil [MATER] A fatty oil, light green or brownish yellow when dry, obtained from hempseed; used in soft soap, paints, and varnishes, and in Asia in foods.

Henderson process [MET] The treatment of copper sulfide ores by roasting with salt to form chlorides, which are then leached out and precipitated.

henequen [MATER] A hard plant fiber, obtained from the leaves of the American agave (*Agave fourcroydes*) and other agave species; used to make rope, twine, and cord.

Hepplewhite-Gray lamp [MIN ENG] A lamp which drew its air from the top down through four tubular pillars into the base, where the air fed the flame through a gauze ring; the outlet was through a metal chimney closed by a gauze disk.

Herbert cloudburst test [MET] A hardness test in which a shower of steel balls, dropped from a predetermined height, dulls the surface of a hardened part in proportion to its softness and thus reveals defective areas.

Herbert pendulum method [MET] Hardness testing in which a 1-millimeter steel or jewel ball resting on the surface to be tested acts as the fulcrum for a 4-kilogram compound pendulum of 10-second period; the swinging of the pendulum causes a rolling indentation in the material, and several hardness factors, such as work hardenability, are determined.

herbicide [MATER] A chemical agent that destroys or inhibits plant growth.

herculac lining [MIN ENG] A German method of lining mine roadways subjected to heavy pressures; a closed circular arch of wedge-shaped precast concrete blocks made in two sizes are erected so that alternate blocks offer their wedge action in opposite direction, the larger blocks toward the center of the roadway and the smaller outward.

hermaphrodite caliper [DES ENG] A layout tool having one leg pointed and the other like that of an inside caliper; used to locate the center of irregularly shaped stock or to lay out a line parallel to an edge.

hermetic seal [ENG] An airtight seal.

Herreshoff furnace [MET] **1.** A rectangular-shaft blast furnace for smelting copper ore. **2.** A mechanical, cylindrical, multiple-deck muffle furnace of the McDougall type.

herringbone gear [MECH ENG] The equivalent of two helical gears of opposite hand placed side by side.

herringbone stoping [MIN ENG] Method used in flattish Rand stope panels 500–1000 feet (150–300 meters) long for breaking and moving the ore; the stope is divided into 20-foot (6-meter) panels, and a different gang works each panel; a tramming system delivers the cut rock to a central scraper system.

herringbone timbering [MIN ENG] A method of timber support in a roadway with a weak roof and strong sides, using neither arms nor side uprights; the crossbar is notched into the sides and supported at its center by a bar under it and parallel with the roadway; the bar is supported by struts notched into the sides at about half height.

herring oil [MATER] Pale-yellow to dark-red liquid obtained from herring; soluble in carbon disulfide, chloroform, and ether, insoluble in water; used in soaps, in leather dressing, and as machinery lubricant.

Herschel-type venturi tube [ENG] A type of venturi tube in which the converging and diverging sections are cones, the throat section is relatively short, the diverging cone is long, and the pressures preceding the inlet cone and in the throat are transferred through multiple openings into annular openings, called piezometer rings.

heterodyne analyzer [ENG ACOUS] A type of constant-bandwidth analyzer in which the electric signal from a microphone beats with the signal from an oscillator, and one of the side bands produced by this modulation is then passed through a fixed filter and detected.

Heusler alloy [MET] Any of a group of ferromagnetic nonferrous alloys typically composed of 18–25% manganese, 10–25% aluminum, and the balance copper.

hevea rubber [MATER] Rubber made from latex obtained from the rubber tree (*Hevea brasiliensis*); used for electrical insulation.

hexagonal-head bolt [DES ENG] A standard wrench head bolt with a hexagonal head.

hexagonal nut [DES ENG] A plain nut in hexagon form. Also known as hex nut.

hex nut *See* hexagonal nut.

Heyn's reagent [MET] Double chloride of copper and ammonia; used in microanalysis of carbon steels.

hide [MATER] A raw or dressed pelt, especially from a large, adult animal such as a cow.

hide glue [MATER] A strong, light-brown glue made from animal hides.

hi-fi *See* high fidelity.

high-abrasion furnace black [MATER] A variety of carbon black made by burning oil in a deficiency of air and then quenching to stop the reaction short of equilibrium; particles are 26–30 nanometers in diameter; used in tires and mechanical rubber goods. Abbreviated HAF black.

high-alloy steel [MET] Steel containing large percentages of elements other than carbon.

high-alumina brick [MATER] Refractory brick made from raw materials rich in alumina, such as diaspore and bauxite; when well fired, they contain a large amount of mullite; used for unusually severe temperature or load conditions.

high-alumina cement *See* aluminate cement.

high boiler [MATER] A solvent added to lacquer thinner to slow the rate of evaporation; boiling point 150–200°C.

high brass [MET] The most common commercial wrought brass containing about 35% zinc and 65% copper.

high-carbon chromium [MET] Chromium containing at least 86% chromium, 8–11% carbon, and a maximum of 0.5% each of iron and silicon.

high-carbon steel [MET] A cast or forged steel containing more than 0.5% carbon.

high-energy fuel [MATER] Fuel that upon combustion provides greater energy than that from conventional carbonaceous fuels; specifically, a hydroboron.

high-energy-rate forging [MET] The production of a forging by the use of a machine which utilizes a sudden surge of kinetic energy from compressed gas against a piston, causing a high ram velocity against the work.

higher pair [MECH ENG] A link in a mechanism in which the mating parts have surface (instead of line or point) contact.

high-expansion alloy [MET] An alloy possessing a high coefficient of expansion.

high-expansion foam [MATER] Noncombustible foam made from ammonium lauryl sulfate; used in underground mine fire fighting.

high explosive [MATER] An explosive with a nitroglycerin base requiring a detonator; the explosion is violent and practically instantaneous.

high-explosive plastic [MATER] High-explosive substance or mixture which, within normal ranges of atmospheric temperature, is capable of being molded into desired shapes. Also known as plastic explosive.

high fidelity [ENG ACOUS] Audio reproduction that closely approximates the sound of the original performance. Also known as hi-fi.

high-flash naphtha *See* heavy naphtha.

high-frequency furnace [ENG] An induction furnace in which the heat is generated within the charge, within the walls of the containing crucible, or within both, by currents induced by high-frequency magnetic flux produced by a surrounding coil. Also known as coreless-type induction furnace; high-frequency heater.

high-frequency heater *See* high-frequency furnace.

high-frequency heating *See* electronic heating.

high-frequency welding [MET] Resistance welding in which the heat is produced by the current flow induced by a high-frequency electromagnetic field. Also known as radio-frequency welding.

high-front shovel [MECH ENG] A power shovel with a dipper stick mounted high on the boom for stripping and overburden removal.

high-grade [MIN ENG] To steal or pilfer ore or gold from a mine or miner.

high-grade dynamite [MATER] Dynamite of 40% strength or over.

high-gradient magnetic separation [ENG] A magnetic separation technique applicable to weakly paramagnetic compounds and to particle sizes down to the colloidal domain.

high hat [ENG] A very low tripod head resembling a formal top hat in shape.

high-heat cement [MATER] A type of cement which releases a large amount of heat during curing.

high-helix drill [DES ENG] A two-flute twist drill with a helix angle of 35–40°; used for drilling deep holes in metals, such as aluminum, copper, hard brass, and soft steel. Also known as fast-spiral drill.

high-intensity atomizer [MECH ENG] A type of atomizer used in electrostatic atomization, based on stress sufficient to overcome tensile strength of the liquid.

high-lead bronze [MET] Bronze containing high percentages of lead to give a soft matrix alloy.

high-lift truck [MECH ENG] A forklift truck with a fixed or telescoping mast to permit high elevation of a load.

high-modulus furnace black [MATER] A variety of carbon black made by burning oil in a deficiency of air; particle diameter is 49–60 nanometers; production is now negligible, but it was used in tire carcasses. Abbreviated HMF black.

high-octane [MATER] Having an octane number in the middle or high 90s and therefore having good antiknock properties.

high-pressure gage glass [ENG] A gage glass consisting of a metal tube with thick glass windows.

high-pressure gas injection [PETRO ENG] Oil reservoir pressure maintenance by injection of gas at pressures higher than those used in conventional equilibrium gas drives.

high-pressure laminate [MATER] A plastic-substrate laminate molded and cured at pressures of, commonly, 1200–2000 psi (8–14×10^6 newtons per square meter).

high-pressure separator [PETRO ENG] A horizontal vessel through which a low-temperature, high-pressure gas stream is fed, and in which free liquids separate out from the gas.

high-pressure torch [ENG] A type of torch in which both acetylene and oxygen are delivered to the mixing chamber under pressure.

high-pressure well [PETRO ENG] A well with a shut-in wellhead pressure of more than 2000 psia (1.4×10^7 newtons per square meter, absolute).

high-residual-phosphorus copper [MET] Deoxidized copper having reduced conductivity due to the presence of residual phosphorus, usually less than 0.1%.

high-resolution radar [ENG] A radar system which can discriminate between two close targets.

high-solvency naphtha [MATER] Any of the petroleum-based solvents that boil in the naphtha range (95–650°F; 35–343°C) and have a high aromatic-chemical content to give high-solvency power for nitrocelluloses, dry paints, and certain resins.

high-speed machine [MECH ENG] A diamond drill capable of rotating a drill string at a minimum of 2500 revolutions per minute, as contrasted with the normal maximum speed of 1600–1800 revolutions per minute attained by the average diamond drill.

high-speed steel [MET] An alloy steel that remains hard and tough at red heat.

high-strength alloy *See* high-tensile alloy.

high-strength low-alloy steel [MET] Steel containing small amounts of niobium or vanadium and having higher strength, better low-temperature impact toughness, and, in some grades, better atmospheric corrosion resistance than carbon steel. Abbreviated HSLA steel.

high-strength steel *See* high-tensile steel.

high-temperature alloy [MET] An alloy suitable for use at temperatures of 500°C and above.

high-temperature cement [MATER] A cement that resists fusing, softening, or spalling at elevated temperatures; used to bond refractory materials.

high-temperature coke [MATER] Coke produced at temperatures of 900–1150°C; used mainly for metallurgical purposes.

high-temperature material [MATER] A material with high-temperature capability, including the superalloys, refractory alloys, and ceramics; used in structures such as spacecraft subjected to extreme thermal environments.

high-temperature water boiler [MECH ENG] A boiler which provides hot water, under pressure, for space heating of large areas.

high-tensile alloy [MET] An alloy having a high tensile strength. Also known as high-strength alloy.

high-tensile bolt [ENG] A bolt that is adjusted to a carefully controlled tension by means of a calibrated torsion wrench.

high-tensile steel [MET] Low-alloy steel having a yield strength range of 50,000–100,000 pounds per square inch (3.4 × 10⁸ to 6.9 × 10⁸ newtons per square meter). Also known as high-strength steel.

high-tension detonator [ENG] A detonator requiring an electric potential of about 50 volts for firing.

high-tension separation *See* electrostatic separation.

high-test chain [ENG] Chain made from heat-treatable plain-carbon steel, usually with a carbon content of 0.15–0.20%; used for load binding, tie-downs, and other applications where failure would be costly.

highwall [MIN ENG] The unexcavated face of exposed overburden and coal or ore in an opencast mine or the face or bank of the uphill side of a contour strip-mine excavation.

highway [CIV ENG] A public road where traffic has the right to pass and to which owners of adjacent property have access.

highway engineering [CIV ENG] A branch of civil engineering dealing with highway planning, location, design, and maintenance.

hill-and-dale recording *See* vertical recording.

hill-climbing [MECH ENG] Adjustment, either continuous or periodic, of a self-regulating system to achieve optimum performance.

hillside quarry [MIN ENG] A quarry cut along a hillside.

hindered contraction [MET] Thermal contraction of a casting that is hindered locally due to the particular geometry.

hindered settling [MIN ENG] Settling of particles in a thick suspension in water through which their fall is hindered by rising water.

hindered-settling ratio [MIN ENG] The ratio of the specific gravity of a mineral to that of the suspension of ore raised to a power between one-half and unity.

Hindley screw [DES ENG] An endless screw or worm of hourglass shape that fits a part of the circumference of a worm wheel so as to increase the bearing area and thus diminish wear. Also known as hourglass screw; hourglass worm.

hinge [DES ENG] A pair of metal leaves forming a jointed device on which a swinging part turns.

hinged arch [CIV ENG] A structure that can rotate at its supports or in the center or at both places.

hinged bar [MIN ENG] A steel extension bar placed in contact with the mine roof perpendicular to the longwall face and supported by yielding steel props. Also known as link bar.

hinge moment [AERO ENG] The tendency of an aerodynamic force to produce motion about the hinge line of a control surface.

hip joint [CIV ENG] The junction of an inclined head post and the top chord of a truss.

hip rafter [BUILD] A diagonal rafter extending from the plate to the ridge of a roof.

Hirschback method [MIN ENG] A method for draining firedamp from coal seams by means of superjacent entries located 80–138 feet (24–42 meters) above the seams and supplemented by boreholes drilled at right angles to the entry walls. Also known as superjacent roadway system.

hitch [MIN ENG] **1.** A step cut in the rock face to hold timber support in an underground working. **2.** A hole cut in side rock solid enough to hold the cap of a set of timbers, permitting the leg to be dispensed with.

hitcher [MIN ENG] **1.** The worker who runs trams into or out of the cages, gives the signals, and attends at the shaft when workers are riding in the cage. **2.** A worker at the bottom of a haulage slope or plane who engages the clips or grips by means of which mine cars are attached to a hoisting cable or chain. Also known as hitcher-on.

hitcher-on *See* hitcher.

hitch timbering [MIN ENG] Installing timbers in hitches either cut or drilled in the rock.

HMF black *See* high-modulus furnace black.

H Monel [MET] A Monel containing 3.2% silicon available in cast form; it is harder and stronger than Monel.

hob [DES ENG] A master model made from hardened steel which is used to press the shape of a plastics mold into a block of soft steel. [MECH ENG] A rotary cutting tool with its teeth arranged along a helical thread; used for generating gear teeth.

hobber *See* hobbing machine.

hobbing [DES ENG] In plastics manufacturing, the act of creating multiple mold cavities by pressing a hob into soft metal cavity blanks. [MECH ENG] Cutting evenly spaced forms, such as gear teeth, on the periphery of cylindrical workpieces.

hobbing machine [MECH ENG] A machine for cutting gear teeth in gear blanks or for cutting worm, spur, or helical gears. Also known as hobber.

hobbing steel [MET] A high-speed steel used to make gear teeth cutters.

hobnail [DES ENG] A short, large-headed, sharp-pointed nail; used to attach soles to heavy shoes.

hobo connection [ENG] A parallel electrical connection used in blasting.

hod [CIV ENG] A tray fitted with a handle by which it can be carried on the shoulder for transporting bricks or mortar.

hoe [DES ENG] An implement consisting of a long handle with a thin, flat, straight-edged blade attached transversely to the end; used for cultivating and weeding.

Hoepfner process *See* Hopfner process.

hoe scraper [MIN ENG] A scraper-loader consisting of a box-sided hoe pulled by cables and used in mining to gather and transport severed rock.

hoe shovel [MECH ENG] A revolving shovel with a pull-type bucket rigidly attached to a stick hinged on the end of a live boom.

hogged fuel [MATER] Sawmill refuse that has been fed through a disintegrator, or hog, by which the various sizes and forms are reduced to a practically uniform size of chips or shreds.

hogging [ENG] Mechanical chipping of wood waste for fuel.

hog gum [MATER] A sticky gum, an exudate from various species of *Sterculia* trees, whose chief constituent is galactan; the gum is dried and marketed either as a powder or as white flakes; used in the food, cosmetic, and textile industry.

Hohmann orbit [AERO ENG] A minimum-energy-transfer orbit.

Hohmann trajectory [AERO ENG] The minimum-energy trajectory between two planetary orbits, utilizing only two propulsive impulses.

hoist [MECH ENG] **1.** To move or lift something by a rope-and-pulley device. **2.** A power unit for a hoisting machine, designed to lift from a position directly above the load and therefore mounted to facilitate mobile service.

hoist back-out switch [MECH ENG] A protective switch that permits hoist operation only in the reverse direction in case of overwind.

hoist cable [MECH ENG] A fiber rope, wire rope, or chain by means of which force is exerted on the sheaves and pulleys of a hoisting machine.

hoist frame *See* headframe.

hoist hook [DES ENG] A swivel hook attached to the end of a hoist cable for securing a load.

hoisting [MECH ENG] **1.** Raising a load, especially by means of tackle. **2.** Either of two power-shovel operations: the raising or lowering of the boom, or the lifting or dropping of the dipper stick in relation to the boom.

hoisting compartment [MIN ENG] The section of a shaft used for hoisting the mined mineral to the surface.

hoisting cycle [MIN ENG] The periods of acceleration, uniform speed, retardation, and rest; the deeper the shaft, the greater is the ratio of the time of full-speed hoisting to the whole cycle.

hoisting drum *See* drum.

hoisting machine [MECH ENG] A mechanism for raising and lowering material with intermittent motion while holding the material freely suspended.

hoisting power [MECH ENG] The capacity of the hoisting mechanism on a hoisting machine.

hoistman [ENG] One who operates steam or electric hoisting machinery to lower and raise cages, skips, or instruments into a mine or an oil or gas well. Also known as hoist operator; winch operator.

hoist overspeed device [MECH ENG] A device used to prevent a hoist from operating at speeds greater than predetermined values by activating an emergency brake when the predetermined speed is exceeded.

hoist overwind device [MECH ENG] A device which can activate an emergency brake when a hoisted load travels beyond a predetermined point into a danger zone.

hoist slack-brake switch [MECH ENG] A device that automatically cuts off power to the hoist motor and sets the brake if the links in the brake rigging require tightening or if the brakes require relining.

hoist tower [CIV ENG] A temporary shaft of scaffolding used to hoist materials for building construction.

hold [ENG] The interior of a ship or plane, especially the cargo compartment. [IND ENG] A therblig, or basic operation, in time-and-motion study in which the hand or other body member maintains an object in a fixed position and location.

holdback [MECH ENG] A brake on an inclined-belt conveyor system which is automatically activated in the event of power failure, thus preventing the loaded belt from running downward.

holddown [MET] A device that holds the outer portion of a metal sheet in place during deep-drawing operations in order to keep it from becoming wrinkled. [PETRO ENG] A device to anchor an oil well rod pump in its position.

holddown groove [ENG] A groove in the side wall of the molding surface which assists in holding the molded plastic article in place when the mold opens.

holding furnace [MET] A heated reservoir to hold molten metal preparatory to casting.

hold time [MET] In resistance welding, the time that pressure is applied to the electrodes after the welding current is cut off.

hole deviation [ENG] The change in the course or direction that a borehole follows.

hole director [MIN ENG] A steel framework used in underground tunneling to set the angle at which holes for a blasting round are to be drilled.

hole layout [MIN ENG] In quarrying, an arrangement of vertical and horizontal holes.

hole saw *See* crown saw.

hole-through [MIN ENG] The meeting of two approaching tunnel heads.

holiday [ENG] An undesirable discontinuity or break in the anticorrosion protection on pipe or tubing.

holing [MIN ENG] **1.** The working of a lower part of a bed of coal to bring down the upper mass. **2.** The final act of connecting two workings underground. **3.** The meeting of two mine roadways driven to intersect. Also known as thirling.

hollander [MECH ENG] An elongate tube with a central midfeather and a cylindrical beater roll; formerly used for stock preparation in paper manufacture.

Holland formula [ENG] A formula used to calculate the height of a plume formed by pollutants emitted from a stack in terms of the diameter of the stack exit, the exit velocity and heat emission rate of the stack, and the mean wind speed.

hollow block *See* hollow tile.

hollow drill [DES ENG] A drill rod or stem having an axial hole for the passage of water or compressed air to remove cuttings from a drill hole. Also known as hollow rod; hollow stem.

hollow gravity dam [CIV ENG] A fixed gravity dam, usually of reinforced concrete, constructed of inclined slabs or arched sections supported by transverse buttresses.

hollow mill [MECH ENG] A milling cutter with three or more cutting edges that revolve around the cylindrical workpiece.

hollow-plunger pump [MIN ENG] A pump for mining and quarrying in gritty and muddy water.

hollow reamer [ENG] A tool or bit used to correct the curvature in a crooked borehole.

hollow rod *See* hollow drill.

hollow-rod churn drill [MECH ENG] A churn drill with hollow rods instead of steel wire rope.

hollow-rod drilling [ENG] A modification of wash boring in which a check valve is introduced at the bit so that the churning action may be also used to pump the cuttings up the drill rods.

hollow shafting [MECH ENG] Shafting made from hollowed-out rods or hollow tubing to minimize weight, allow internal support, or permit other shafting to operate through the interior.

hollow stem *See* hollow drill.

hollow tile [MATER] A hollow building block of concrete or burnt clay used for making partitions, exterior walls, or suspended floors or roofs. Also known as hollow block.

hollow wall [BUILD] A masonry wall provided with an air space between the inner and outer wythes.

Holman Airleg [MIN ENG] A drill support consisting of a cylinder in which the piston is actuated by compressed air controlled by a twist-grip valve.

Holman counterbalanced drill rig [MIN ENG] A drill rig consisting of a rail-track carriage with a counterbalanced boom 10 feet (3 meters) long.

Holman dust extractor [MIN ENG] A dust-trapping system in which the dust and chippings from percussive drilling operations are drawn back through the hollow drill rod and along a hose to a metal container with filter elements.

Holman stamp [MIN ENG] A crushing stamp raised by a crank and accelerated in its fall by compressed air.

Holme mud sampler [ENG] A scooplike device which can be lowered by cable to the ocean floor to collect sediment samples.

holystone [MATER] A soft sandstone that is used in scrubbing a ship's deck.

home signal [CIV ENG] A signal at the beginning of a block of railroad track that indicates whether the block is clear.

homing device [ENG] A device incorporated in a guided missile or the like to home in on a target.

homing guidance [ENG] A guidance system in which a missile directs itself to a target by means of a self-contained mechanism that reacts to a particular characteristic of the target.

homogenize [MET] To hold metal at a high temperature long enough to eliminate by diffusion any chemical segregation of the components.

homogenizer [MECH ENG] A machine that blends or emulsifies a substance by forcing it through fine openings against a hard surface.

hone [MATER] A fine-grit stone that is used for sharpening a cutting tool. [MECH ENG] A machine for honing that consists of a holding device containing several oblong stones arranged in a circular pattern.

honeycombing [MATER] **1.** Internal fiber separation in drying timber. **2.** Local roughness and weakening on the face of a concrete wall due to segregation of the concrete, with the result that there is little sand to fill in between the stone aggregate.

honeycomb radiator [MECH ENG] A heat-exchange device utilizing many small cells, shaped like a bees' comb, for cooling circulating water in an automobile.

Honigmann process [MIN ENG] A method of shaft sinking through water-bearing sand; the shaft is bored in stages, increasing in size from the pilot hole, about 4 feet in diameter, to the final size.

honing [MECH ENG] The process of removing a relatively small amount of material from a cylindrical surface by means of abrasive stones to obtain a desired finish or extremely close dimensional tolerance.

hood [DES ENG] An opaque shield placed above or around the screen of a cathode-ray tube to eliminate extraneous light. [ENG] **1.** Close-fitting, rubber head covering that leaves the face exposed; used in scuba diving. **2.** A protective covering, usually providing special ventilation to carry away objectionable fumes, dusts, and gases, in which dangerous chemical, biological, or radioactive materials can be safely handled.

hood test [ENG] A leak detection method in which the vessel under test is enclosed by a metallic casing so that a dynamic leak test may be carried out on a large portion of the external surface.

hoof oil *See* neat's-foot oil.

hook [DES ENG] A piece of hard material, especially metal, formed into a curve for catching, holding, or pulling something.

hookah [ENG] An air supply device used in free diving, comprising a demand regulator worn by the diver and a hose extending to a compressed air supply at the surface.

hook-and-eye hinge [DES ENG] A hinge consisting of a hook (usually attached to a gate post) over which an eye (usually attached to the gate) is placed.

hook bolt [DES ENG] A bolt with a hook or L band at one end and threads at the other to fit a nut.

hooker [MIN ENG] A worker who detaches empty downcoming buckets and hook-loads buckets or cans onto the hoisting rope.

Hooker process [MET] A forming process in which pierced slugs or cups are punched through a die to produce tubing and other shapes.

Hooke's joint [MECH ENG] A simple universal joint; consists of two yokes attached to their respective shafts and connected by means of a spider. Also known as Cardan joint.

hook gage [ENG] An instrument used to measure changes in the level of the water in an evaporation pan; it consists of a pointed metal hook, mounted in the vertical, whose position with respect to its supporting member may be adjusted by means of a micrometer arrangement; the gage is placed on the still well, and a measurement is taken when the point of the hook just breaks above the surface of the water.

hook tender [MIN ENG] In bituminous coal mining, a worker who attaches the hook of a hoisting cable to the link of a trip of cars to be hauled up or lowered down an incline. Also known as rope cutter.

hook-wall packer [PETRO ENG] Fluid-proof seal between the outside of oil well tubing and the inside of the casing; hooks hold it in place.

hook wrench [DES ENG] A wrench with a hook for turning a nut or bolt.

hoop [CIV ENG] A ring-shaped binder placed around the main reinforcement in a reinforced concrete column.

hooped column [CIV ENG] A column of reinforced concrete with hoops around the main reinforcements.

Hooper jig [MIN ENG] A pneumatic jig, used when water is scarce or the ore must be kept dry, to concentrate values from sands.

Hopfner process [MET] A process for the recovery of copper from its sulfide ores by leaching with a solution of cuprous chloride in sodium or calcium chloride and electrolyzing the resulting solution in tanks that are protected by diaphragms. Also spelled Hoepfner process.

hopper [ENG] A funnel-shaped receptacle with an opening at the top for loading and a discharge opening at the bottom for bulk-delivering material such as grain or coal.

hopper car [ENG] A freight car with a permanent roof and a hinged floor sloping to one or more hoppers for discharging contents by gravity.

hopper dryer [ENG] In extrusion and injection molding of plastics, a combined feeding and drying device in which hot air flows through the hopper.

hoppit [MIN ENG] A large bucket, usually up to about 80 cubic feet (2.3 cubic meters), used in shaft sinking for hoisting men, rock, materials, and tools. Also known as bowk; kibble; sinking bucket.

hops oil [MATER] Greenish-yellow essential oil with strong aroma; soluble in alcohol, ether, and chloroform, insoluble in water; distilled from strobiles of the hop (*Humulus lupulus*); main components are humulene, geraniol, and terpenes; used to aromatize beer and tobacco.

horadiam drilling [MIN ENG] The drilling of a number of horizontal boreholes radiating outward from a common center. Also known as horizontal-ring drilling.

horizon mining [MIN ENG] A system of mining suitable for inclined, and perhaps faulted, coal seams; main stone headings are driven, at predetermined levels, from the winding shaft to intersect the seams to be developed.

horizon sensor [ENG] A passive infrared device that detects the thermal discontinuity between the earth and space; used in establishing a stable vertical reference for control of the attitude or orientation of a missile or satellite in space.

horizontal auger [MECH ENG] A rotary drill, usually powered by a gasoline engine, for making horizontal blasting holes in quarries and opencast pits.

horizontal boiler [MECH ENG] A water-tube boiler having a main bank of straight tubes inclined toward the rear at an angle of 5 to 15° from the horizontal.

horizontal borer [MIN ENG] A machine that makes holes 2–6 inches (5–15 centimeters) in diameter, used for drilling at opencut coal mines.

horizontal boring machine [MECH ENG] A boring machine adapted for work not conveniently revolved, for milling, slotting, drilling, tapping, boring, and reaming long holes and for making interchangeable parts that must be produced without jigs and fixtures.

horizontal broaching machine [MECH ENG] A pull-type broaching machine having the broach mounted on the horizontal plane.

horizontal circle [ENG] A graduated disk affixed to the base of a transit or theodolite which is used to measure horizontal angles.

horizontal crusher [MECH ENG] Rotary size reducer in which the crushing cone is supported on a horizontal shaft; needs less headroom than vertical models.

horizontal drilling machine [MECH ENG] A drilling machine in which the drill bits extend in a horizontal direction.

horizontal drive [MIN ENG] An opening with a small inclination directed toward the shaft to drain the water and facilitate hauling of full cars to the shaft. Also known as horizontal crosscut.

horizontal engine [MECH ENG] An engine with horizontal stroke.

horizontal field balance [ENG] An instrument that measures the horizontal component of the magnetic field by means of the torque that the field component exerts on a vertical permanent magnet.

horizontal firing [MECH ENG] The firing of fuel in a boiler furnace in which the burners discharge fuel and air into the furnace horizontally.

horizontal force instrument [ENG] An instrument used to make a comparison between the intensity of the horizontal component of the earth's magnetic field and the magnetic field at the compass location on board a craft; basically, it consists of a magnetized needle pivoted in a horizontal plane, as a dry-card compass; it settles in some position which indicates the direction of the resultant magnetic field; if the needle is started swinging, it damps down with a certain period of oscillation dependent upon the strength of the magnetic field. Also known as horizontal vibrating needle.

horizontal intensity variometer [ENG] Essentially a declination variometer with a larger, stiffer fiber than in the standard model; there is enough torsion in the fiber to cause the magnet to turn 90° out of the magnetic meridian; the magnet is aligned with the magnetic prime vertical to within 0.5° so it does not respond appreciably to changes in declination. Also known as H variometer.

horizontal lathe [MECH ENG] A horizontally mounted lathe with which longitudinal and radial movements are applied to a workpiece that rotates.

horizontal magnetometer [ENG] A measuring instrument for ascertaining changes in the horizontal component of the magnetic field intensity.

horizontal milling machine [MECH ENG] A knee-type milling machine with a horizontal spindle and a swiveling table for cutting helices.

horizontal-position welding [MET] **1.** Making a fillet weld on the upper side of the intersection of a vertical surface and a horizontal surface. **2.** Making a horizontal groove weld on a vertical surface.

horizontal retort [MET] An intermittent unit made from a siliceous fireclay and formerly used for zinc smelting.

horizontal retort process [MET] A zinc-smelting process that employs vast, honeycomblike batteries of fireclay or silicon-carbide retorts set horizontally in a gas- or coal-fired furnace. Also known as Belgian retort process.

horizontal return tubular boiler [MECH ENG] A fire-tube boiler having tubes within a cylindrical shell that are attached to the end closures; products of combustion are transported under the lower half of the shell and back through the tubes.

horizontal-ring drilling *See* horadiam drilling.

horizontal-rolled-position welding [MET] Topside welding of a butt joint connecting two horizontal pieces of rotating pipe.

horizontal scanning [ENG] In radar scanning, rotating the antenna in azimuth around the horizon or in a sector. Also known as searching lighting.

horizontal screen [MECH ENG] Shaking screen with horizontal plates.

horizontal separator [PETRO ENG] Horizontal tank used to separate free oil well gas from liquid hydrocarbons.

horizontal-tube evaporator [MECH ENG] A horizontally mounted tube-and-shell type of liquid evaporator, used most often for preparation of boiler feedwater.

horizontal vibrating needle *See* horizontal force instrument.

horn [ENG ACOUS] A tube whose cross-sectional area increases from one end to the other, used to radiate or receive sound waves and to intensify and direct them. Also known as acoustic horn. [MET] Holding arm for the electrode of a resistance spot-welding machine.

horn loudspeaker [ENG ACOUS] A loudspeaker in which the radiating element is coupled to the air or another medium by means of a horn.

horn socket [DES ENG] A cone-shaped fishing tool especially designed to recover lost collared drill rods, drill pipe, or tools in bored wells.

horn spacing [MET] Unobstructed work space in a resistance-welding machine between horns at right angles to the throat depth.

horn spoon [MIN ENG] A troughlike section cut from a cow horn and scraped thin; used for washing auriferous gravel and pulp when exacting tests are to be performed.

horse *See* horseback.

horseback [MIN ENG] **1.** Shale or sandstone occurring in a channel that was cut by flowing water in a coal seam. Also known as cutout; horse; roll; swell; symon fault; washout. **2.** To move or raise a heavy piece of machinery or timber by using a pinch bar as a lever. Also known as pinch.

horsehead [MIN ENG] Timbers or steel joists used to support planks in tunneling through loose ground.

hose [DES ENG] Flexible tube used for conveying fluids.

hose clamp [DES ENG] Band or brace to attach the raw end of a hose to a water outlet.

hose coupling [DES ENG] Device to interconnect two or more pieces of hose.

hose fitting [DES ENG] Any attachment or accessory item for a hose.

Hoskold formula [MIN ENG] A two-rate valuation formula formerly used to determine present value of mining properties or shares, with redemption of invested capital.

hot acid [PETRO ENG] The use of hot hydrochloric acid (200–300°F; 93–149°C) for oil well acidizing where wellbore scale is slow-dissolving and hard to remove.

hot-air engine [MECH ENG] A heat engine in which air or other gases, such as hydrogen, helium, or nitrogen, are used as the working fluid, operating on cycles such as the Stirling or Ericsson.

hot-air furnace [MECH ENG] An encased heating unit providing warm air to ducts for circulation by gravity convection or by fans.

hot-air soldering [MET] Soldering with a narrow blast of air whose temperature is closely controlled at the value required for soldering individual joints on printed circuit boards.

hot-air sterilization [ENG] A method of sterilization using dry heat for glassware and other heat-resistant materials which need to be dry after treatment; temperatures of 160–165°C are generated for at least 2 hours.

hotbed [MET] An area where hot-rolled metal is placed to cool. Also known as cooling table.

hot-blast stove [MET] A retracting device for preheating the incoming air in an iron blast furnace by using heat from the burning gases of the furnace.

hot-bulb [MECH ENG] Pertaining to an ignition method used in semidiesel engines in which the fuel mixture is ignited in a separate chamber kept above the ignition temperature by the heat of compression.

hot chamber die casting [ENG] A die-casting process in which a piston is driven through a reservoir of molten metal and thereby delivers a quantity of molten metal to the die cavity.

Hotchkiss drive [MECH ENG] An automobile rear suspension designed to take torque reactions through longitudinal leaf springs.

Hotchkiss superdip [ENG] A sensitive dip needle consisting of a freely rotating magnetic needle about a horizontal axis and a nonmagnetic bar with a counterweight at the end which is attached to the pivot point of the needle.

hot-die steel [MET] A high-temperature, shock-resistant alloy steel used in forging machines.

hot dipping [MET] Coating metal components by immersion in a molten metal bath, such as tin or zinc.

hot-draw [ENG] To draw a material while it is hot.

hot extrusion [MET] The process of extruding metal at very high temperatures.

hot forming [MET] Shaping operations performed at temperatures above the recrystallization temperature of the metal.

hot-gas welding [ENG] Joining of thermoplastic materials by softening first with a jet of hot air, then joining at the softened points.

hothouse [ENG] A greenhouse heated to grow plants out of season.

hot patching [ENG] Repair of a hot refractory lining in a furnace, usually by spraying with a refractory slurry.

hot plate [MET] A heated surface on which joints are brought to soldering temperature.

hot press forge [MET] A press in which metal parts are formed by forcing hot metal into dies under high pressure.

hot pressing [ENG] Forming a metal-powder compact or a ceramic shape by applying pressure and heat simultaneously at temperatures high enough for sintering to occur.

hot-pressure welding [MET] A pressure welding process in which macrodeformation of the base material to produce coalescence results from the application of heat and pressure.

hot-quenching [MET] Quenching from high temperatures into a medium of lower but still high temperature.

hot rolling [MET] Rolling of metal bars or sheets when hot.

hot-runner mold [ENG] A plastics mold in which the runners are kept hot by insulation from the chilled cavities.

hot saw [MECH ENG] A power saw used to cut hot metal.

hot shortness [MET] Brittleness, usually of steel or wrought iron, when the metal is hot, due to a high sulfur content.

hotshot wind tunnel [AERO ENG] A wind tunnel in which electrical energy is discharged into a pressurized arc chamber, increasing the temperature and pressure in the arc chamber so that a diaphragm separating the arc chamber from an evacuated chamber is ruptured, and the heated gas from the arc chamber is then accelerated in a conical nozzle to provide flows with mach numbers of 10 to 27 for durations of 10 to 100 milliseconds.

hot-solder coating [ENG] The application of a protective finish to a printed circuit board by dip soldering in a solder bath.

hot stamp [ENG] An impression on a forging made in a heated condition.

hot-swage [MET] To reduce the cross section of a hot metal tube or rod.

hot tear [MET] A separation either internally or externally in a casting due to loadings or internal stresses or both; it results from improper solidification, and shrinkage near the temperature at which the casting is completely solid.

hot trim [MET] Removal of flash in a heated forging.

hot-water heating [MECH ENG] A heating system for a building in which the heat-conveying medium is hot water and the heat-emitting means are radiators, convectors, or panel coils. Also known as hydronic heating.

hot well [MECH ENG] A chamber for collecting condensate, as in a steam condenser serving an engine or turbine.

hot-wire ammeter [ENG] An ammeter which measures alternating or direct current by sending it through a fine wire, causing the wire to heat and to expand or sag, deflecting a pointer. Also known as thermal ammeter.

hot-wire anemometer [ENG] An anemometer used in research on air turbulence and boundary layers; the resistance of an electrically heated fine wire placed in a gas stream is altered by cooling by an amount which depends on the fluid velocity.

hot-wire instrument [ENG] An instrument that depends for its operation on the expansion by heat of a wire carrying a current.

hot-wire microphone [ENG ACOUS] A velocity microphone that depends for its operation on the change in resistance of a hot wire as the wire is cooled by varying particle velocities in a sound wave.

hot working [MET] Plastic deformation of a metal at a rate and temperature such that strain hardening cannot occur.

hourglass screw See Hindley screw.

hourglass worm See Hindley screw.

hour-out line See time front.

house drain [CIV ENG] Horizontal drain in a basement receiving waste from stacks.

house sewer [CIV ENG] Connection between house drain and public sewer.

housing [ENG] A case or enclosure to cover and protect a structure or a mechanical device.

Houskeeper seal [ENG] A vacuum-tight seal made between copper and glass by bringing the copper to a flexible feather edge before fusing it to the glass; the copper then flexes as the glass shrinks during cooling.

hovercraft See air-cushion vehicle.

Howell-Bunger valve See cone valve.

Howe truss [CIV ENG] A truss for spans up to 80 feet (24 meters) having both vertical and diagonal members; made of steel or timber or both.

howl [ENG ACOUS] Undesirable prolonged sound produced by a radio receiver or audio-frequency amplifier system because of either electric or acoustic feedback.

Hoyer method of prestressing *See* pretensioning.

H pile [CIV ENG] A steel pile that is H-shaped in section.

H rod [DES ENG] A drill rod having an outside diameter of 3½ inches (8.89 centimeters).

HSLA steel *See* high-strength low-alloy steel.

hub [DES ENG] **1.** The cylindrical central part of a wheel, propeller, or fan. **2.** A piece in a lock that is turned by the knob spindle, causing the bolt to move. **3.** A short coupling that joins plumbing pipes. [MET] A steel punch used in making a working die for a coin or medal.

hubbing [MET] Forcing a male die into a blank to form a female die.

hubcap [DES ENG] A metal cap fastened or clamped to the end of an axle, as on motor vehicles.

Huggenberger tensometer [ENG] A type of extensometer having a short gage length (10 to 20 millimeters) and employing a compound lever system that gives a magnification of about 1200.

human-factors engineering [ENG] The area of knowledge dealing with the capabilities and limitations of human performance in relation to design of machines, jobs, and other modifications of the human's physical environment.

Humble gage [PETRO ENG] Device to measure oil well bottom-hole pressure; a piston acts through a stuffing box against a helical spring in tension.

Humble relation [PETRO ENG] Equation used by oil companies to estimate porosity of the oil-bearing formation from measurements made with a contact resistivity device, such as a microlog.

Humboldt jig [MIN ENG] An ore jig with a movable screen.

hum-bucking coil [ENG ACOUS] A coil wound on the field coil of an excited-field loudspeaker and connected in series opposition with the voice coil, so that hum voltage induced in the voice coil is canceled by that induced in the hum-bucking coil.

Hume-Rothery compound *See* electron compound.

humidification [ENG] The process of increasing the water vapor content of a gas.

humidifier [MECH ENG] An apparatus for supplying moisture to the air and for maintaining desired humidity conditions.

humidistat [ENG] An instrument that measures and controls relative humidity. Also known as hydrostat.

humidity element [ENG] The transducer of any hygrometer, that is, that part of a hygrometer that quantitatively senses atmospheric water vapor.

humidity strip [ENG] The humidity transducing element in a Diamond-Hinman radiosonde; it consists of a flat plastic strip bounded by electrodes on two sides and coated with a hygroscopic chemical compound such as lithium chloride; the electrical resistance of this coating is a function of the amount of moisture absorbed from the atmosphere and the temperature of the strip. Also known as electrolytic strip.

humidity test [MET] Corrosion test in which a specimen is exposed to an environment of controlled humidity and temperature.

hummer screen [MIN ENG] An electrically vibrated ore screen for sizing moderately small material.

Humpage gears [MECH ENG] A train of bevel gears used for speed reduction.

Humphrey gas pump [MECH ENG] A combined internal combustion engine and pump in which the metal piston has been replaced by a column of water.

Humphrey's spiral [MIN ENG] An ore concentrator consisting of a stationary spiral trough through which ore pulp gravitates; heavy particles stay on the inside and lighter ones climb to the outside.

hump yard [CIV ENG] A switch yard in a railway system that has a hump or steep incline down which freight cars can coast to prescheduled locations. Also known as gravity yard.

hung shot [ENG] A shot whose explosion is delayed after detonation or ignition.

hunt [AERO ENG] **1.** Of an aircraft or rocket, to weave about its flight path, as if seeking a new direction or another angle of attack; specifically, to yaw back and forth. **2.** Of a control surface, to rotate up and down or back and forth without being detected by the pilot.

Hunt and Douglas process [MET] Smelting process involving the roasting of matte carrying copper, lead, gold, and silver to form copper sulfate and oxide (but not silver sulfate); this product is leached with sulfuric acid for copper; the resulting solution is treated with calcium chloride by passing sulfur dioxide through it; the cuprous chloride is then reduced to cuprous oxide by milk of lime, (regenerating calcium chloride), and the cuprous oxide is smelted.

Hunt continuous filter [MIN ENG] A continuous-vacuum filter consisting of a horizontally revolving, annular filter bed on which pulp is washed and then vacuum-dried.

hunting [MECH ENG] Irregular engine speed resulting from instability of the governing device.

Huntington-Heberlein process [MIN ENG] A sink-float process employing a galena medium, which is recovered by froth flotation.

hunting tooth [DES ENG] An extra tooth on the larger of two gear wheels so that the total number of teeth will not be an integral multiple of the number on the smaller wheel.

hurdle sheet [MIN ENG] A brattice-cloth screen across a roadway below a roof cavity or at the ripping lip to divert air current upward, thus diluting and removing firedamp.

hurricane air stemmer [MIN ENG] A mechanical device for rapidly tamping shotholes; consists of a funnel connected to a T piece to the charge tube, with a connection to a compressed-air column; sand is put in the funnel and injected into the shothole as the charge tube is withdrawn.

hurricane beacon [ENG] An air-launched balloon designed to be released in the eye of a tropical cyclone, to float within the eye at predetermined levels, and to transmit radio signals.

hurricane lamp [ENG] An oil lamp with a glass chimney and perforated lid to protect the flame, or a candle with a glass chimney.

hurricane tracking [ENG] Recording of the movement of individual hurricanes by means of airplane sightings and satellite photography.

Hurst formula [PETRO ENG] Relationship used in reservoir material-balance analysis; interrelates field pressure and production data at a number of different times.

Hurst method [PETRO ENG] A calculation method for the bottom-hole static pressure of a well; uses graphical extrapolation of pressure buildup over a short period of time.

hutch [MIN ENG] The bottom compartment of a jig used in ore dressing.

hutch product [MIN ENG] Fine, heavy materials that pass through the screen of a jig and collect in the hutch.

Huwood loader [MIN ENG] A machine consisting of a number of horizontal rotating flight bars working near the floor of the seam which push prepared coal up a ramp onto a low, bottom-loaded coveyor belt.

H variometer See horizontal intensity variometer.

Hybinette process [MET] A process used for refining of crude nickel anodes; anodes are placed in asphalt-lined, reinforced concrete tanks and dissolved electrochemically so that impurities such as copper and iron pass into solution while pure nickel electrolyte is continuously added.

hybrid composite [MATER] A composite material in which two or more high-performance reinforcements are combined.

hybrid inlet noise reduction [ENG ACOUS] A method of reducing the noise from the inlet of a jet engine, which involves the use of both high-Mach-number flows to retard or block the passage of sound waves and acoustic treatment of the walls of the inlet.

hybrid propellant [MATER] A propellant using a combination of liquid and solid materials to provide propulsion energy and working fluid.

hybrid propulsion [AERO ENG] Propulsion utilizing energy released by a liquid propellant with a solid propellant in the same rocket engine.

hybrid rocket [AERO ENG] A rocket with an engine utilizing a liquid propellant with a solid propellant in the same rocket engine.

hydrant See fire hydrant.

hydrated cellulose See hydrocellulose.

hydrated grease [MATER] Grease made with a soap containing a hydrated alkali.

hydraucone [DES ENG] A conical, spreading type of draft tube used on hydraulic turbine installations.

hydraulic [ENG] Operated or effected by the action of water or other fluid of low viscosity.

hydraulic accumulator [MECH ENG] A hydraulic flywheel that stores potential energy by accumulating a quantity of pressurized hydraulic fluid in a suitable enclosed vessel.

hydraulic actuator [MECH ENG] A cylinder or fluid motor that converts hydraulic power into useful mechanical work; mechanical motion produced may be linear, rotary, or oscillatory.

hydraulic air compressor [MECH ENG] A device in which water falling down a pipe entrains air which is released at the bottom under compression to do useful work.

hydraulic backhoe [MECH ENG] A backhoe operated by a hydraulic mechanism.

hydraulic blasting [MIN ENG] Fracturing coal by means of a hydraulic cartridge.

hydraulic bottom-hole pump [PETRO ENG] Liquid power–operated oil production pump; the liquid is oil, piped under pressure to the bottom of the well to operate the engine that drives the pump.

hydraulic brake [MECH ENG] A brake in which the retarding force is applied through the action of a hydraulic press.

hydraulic cartridge [MIN ENG] A device used in mining to split coal or rock and having 8–12 small hydraulic rams in the sides of a steel cylinder.

hydraulic cement [MATER] Cement that hardens underwater.

hydraulic chock [MIN ENG] A steel face-support structure consisting of one to four hydraulic legs mounted in a steel frame with a large head and base plate.

hydraulic circuit [MECH ENG] A circuit whose operation is analogous to that of an electric circuit except that electric currents are replaced by currents of water or other fluids, as in a hydraulic control.

hydraulic circulating system [MIN ENG] A method used to drill a borehole wherein water or a mud-laden liquid is circulated through the drill string.

hydraulic classification [ENG] Classification of particles in a tank by specific gravity, utilizing the action of rising water currents.

hydraulic classifier [MECH ENG] A classifier in which particles are sorted by specific gravity in a stream of hydraulic water that rises at a controlled rate; heavier particles gravitate down and are discharged at the bottom, while lighter ones are carried up and out. Also known as hydrosizer.

hydraulic clutch See fluid drive.

hydraulic conveyor [MECH ENG] A system for handling material, such as ash from a coal-fired furnace; refuse is flushed from a hopper or slag tank to a grinder which discharges to a pump for conveying to a disposal area or a dewatering bin.

hydraulic coupling See fluid coupling.

hydraulic cylinder [MECH ENG] The cylindrical chamber of a positive displacement pump.

hydraulic dredge [MECH ENG] A dredge consisting of a large suction pipe which is mounted on a hull and supported and moved about by a boom, a mechanical agitator or cutter head which churns up earth in front of the pipe, and centrifugal pumps mounted on a dredge which suck up water and loose solids.

hydraulic drill [MECH ENG] A rotary drill powered by hydrodynamic means and used to make shot-firing holes in coal or rock, or to make a well hole.

hydraulic drive [MECH ENG] A mechanism transmitting motion from one shaft to another, the velocity ratio of the shafts being controlled by hydrostatic or hydrodynamic means.

hydraulic ejector [ENG] A pipe for removing excavated material from a pneumatic caisson.

hydraulic elevator [MECH ENG] An elevator operated by water pressure. Also known as hydraulic lift.

hydraulic engineering [CIV ENG] A branch of civil engineering concerned with the design, erection, and construction of sewage disposal plants, waterworks, dams, water-operated power plants, and such.

hydraulic excavation See hydraulicking.

hydraulic excavator digger [MECH ENG] An excavation machine which employs hydraulic pistons to actuate mechanical digging elements.

hydraulic extraction See hydraulicking.

hydraulic filling [MIN ENG] The use of water to wash waste material into stopes in order to prevent failure of rock walls and subsidence.

hydraulic fluid [MATER] A low-viscosity fluid used in operating a hydraulic mechanism.

hydraulic flume transport [MIN ENG] The transport of coal, pulp, or minerals in water flowing in semicircular or rectangular channels.

hydraulic fracturing [PETRO ENG] A method in which sand-water mixtures are forced into underground wells under pressure; the pressure splits the petroleum-bearing sandstone, thereby allowing the oil to move toward the wells more freely.

hydraulic giant See hydraulic monitor.

hydraulic jack [MECH ENG] A jack in which force is applied through the mechanism of a hydraulic press.

hydraulic jetting [ENG] Use of high-pressure water forced through nozzles to clean tube interiors and exteriors in heat exchangers and boilers.

hydraulicking [MIN ENG] Excavating alluvial or other mineral deposits by means of high-pressure water jets. Also known as hydraulic excavation; hydraulic extraction; hydroextraction.

hydraulic lift See hydraulic elevator.

hydraulic lime [MATER] A hydraulic cementitious product that is produced by burning of hydraulic limestone.

hydraulic limestone [MATER] Limestone, containing silica and alumina, which produces lime that hardens in water.

hydraulic loading [MIN ENG] The flushing of coal or other material broken down by jets of water along the mine floor and into flumes.

hydraulic locomotive [MIN ENG] A diesel locomotive in which traction wheels are driven by hydraulic motors powered by a hydraulic system on the unit; used in mine haulage.

hydraulic machine [MECH ENG] A machine powered by a motor activated by the confined flow of a stream of liquid, such as oil or water under pressure.

hydraulic mine [MIN ENG] A placer mine worked by means of a water stream directed against a bank.

hydraulic monitor [MIN ENG] A device for directing a high-pressure jet of water in hydraulicking; essentially, a swivel-mounted, counterweighted nozzle attached to a tripod or other type of stand and so designed that one worker can easily control and direct the vertical and lateral movements of the nozzle. Also known as giant; hydraulic giant; monitor.

hydraulic motor [MECH ENG] A motor activated by water or other liquid under pressure.

hydraulic nozzle [MECH ENG] An atomizing device in which fluid pressure is converted into fluid velocity.

hydraulic packer holddown [PETRO ENG] A pressure-actuated anchor located below a production packer (the seal between tubing and casing) to prevent well pressure from forcing the packer upward in the casing.

hydraulic packing [ENG] Packing material that resists the effects of water even under high pressure.

hydraulic power oil [MATER] Well production oil from which corrosive and abrasive impurities have been removed; used to power downhole hydraulic pumps.

hydraulic power system [MECH ENG] A power transmission system comprising machinery and auxiliary components which function to generate, transmit, control, and utilize hydraulic energy.

hydraulic press [MECH ENG] A combination of a large and a small cylinder connected by a pipe and filled with a fluid so that the fluid pressure created by a small force acting on the small-cylinder piston will result in a large force on the large piston. Also known as hydrostatic press.

hydraulic prop [MIN ENG] A supporting device consisting of two telescoping steel cylinders extended by hydraulic pressure provided by a built-in hand pump.

hydraulic pump See hydraulic ram.

hydraulic ram [MECH ENG] A device for forcing running water to a higher level by using the kinetic energy of flow; the flow of water in the supply pipeline is periodically stopped so that a small portion of water is lifted by the velocity head of a larger portion. Also known as hydraulic pump.

hydraulic rope-geared elevator [MECH ENG] An elevator hoisted by a system of ropes and sheaves attached to a piston in a hydraulic cylinder.

hydraulic scale [MECH ENG] An industrial scale in which the load applied to the load-cell piston is converted to hydraulic pressure.

hydraulic separation [MECH ENG] Mechanical classification using a hydraulic classifier.

hydraulic shovel [MECH ENG] A revolving shovel in which hydraulic rams or motors are substituted for drums and cables.

hydraulic sprayer [MECH ENG] A machine that sprays large quantities of insecticide or fungicide on crops.

hydraulic spraying See airless spraying.

hydraulic stacker [MECH ENG] A tiering machine whose carriage is raised or lowered by a hydraulic cylinder.

hydraulic swivel head [MECH ENG] In a drill machine, a swivel head equipped with hydraulically actuated cylinders and pistons to exert pressure on and move the drill rod string longitudinally.

hydraulic transport [ENG] Movement of material by water.

hydraulic turbine [MECH ENG] A machine which converts the energy of an elevated water supply into mechanical energy of a rotating shaft.

hydride descaling [MET] Removing surface deposits of oxides from a metal by immersion in molten alkali that contains hydrides.

hydrocarbon blending value [ENG] Octane number rating for a 20% blend of a hydrocarbon with a 60:40 mixture of isooctane:n-heptane, which has been recalculated for a hypothetical 100% concentration of the tested hydrocarbon.

hydrocarbon-mud log [PETRO ENG] Record of oil, gas, or cuttings released into mud during rock drilling; used to detect the presence of hydrocarbon-bearing strata.

hydrocarbon pore volume [PETRO ENG] The pore volume in a reservoir formation available to hydrocarbon intrusion.

hydrocarbon stabilization [PETRO ENG] The stepwise pressure reduction of a well stream to allow the release of dissolved gases until the liquid is stable at storage-tank conditions.

hydrocellulose [MATER] A gelatinous mass formed from the reaction of cellulose with water either by grinding cellulose and mixing with water, or by using strong salt solutions, acids, or alkalies; used in the manufacture of artifical fibers such as rayon, mercerized cotton, paper, and vulcanized fiber. Also known as hydrated cellulose.

hydrocyclone [MECH ENG] A cyclone separator in which granular solids are removed from a stream of water and classified by centrifugal force.

hydrodynamic oscillator [ENG ACOUS] A transducer for generating sound waves in fluids, in which a continuous flow through an orifice is modulated by a reciprocating valve system controlled by acoustic feedback.

hydroelectric generator [MECH ENG] An electric rotating machine that transforms mechanical power from a hydraulic turbine or water wheel into electric power.

hydroelectric plant [MECH ENG] A facility at which electric energy is produced by hydroelectric generators. Also known as hydroelectric power station.

hydroelectric power station *See* hydroelectric plant.

hydroextraction *See* hydraulicking.

hydrogen blistering [MET] Cracks or blisters caused when atomic hydrogen penetrates steel via submicroscopic discontinuities or voids and becomes molecular hydrogen and develops internal pressures.

hydrogen brazing [MET] Brazing in an atmosphere rich in hydrogen.

hydrogen damage [MET] Corrosion, common in boilers, caused by diffusion of hydrogen through steel reacting with carbon to form methane, which builds up local stresses at the interfaces between grains, forming voids that ultimately produce failure.

hydrogen embrittlement *See* acid brittleness.

hydrogen loss [MET] Loss of weight by a compact or a metal powder when heated in a hydrogen atmosphere; used as a measure of oxygen content of the sample.

hydrogen overvoltage [MET] An overvoltage occurring at an electrode as a result of the liberation of hydrogen gas.

hydrogen-reduced powder [MET] Metal powder produced by hydrogen-reduction of a metal, metallic compound, or surface-contaminated metal particles.

hydrographic sextant [ENG] A surveying sextant similar to those used for celestial navigation but smaller and lighter, constructed so that the maximum angle that can be read is slightly greater than that on the navigating sextant; usually the angles can be read only to the nearest minute by means of a vernier; it is fitted with a telescope with a large object glass and field of view. Also known as sounding sextant; surveying sextant.

hydrographic sonar [ENG] An echo sounder used in mapping ocean bottoms.

hydrometallurgy [MET] Treatment of metals and metal-containing materials by wet processes.

hydrometer [ENG] A direct-reading instrument for indicating the density, specific gravity, or some similar characteristic of liquids.

hydronic heating *See* hot-water heating.

hydrophone [ENG ACOUS] A device which receives underwater sound waves and converts them to electric waves.

hydropneumatic [ENG] Operated by both water and air power.

hydropneumatic recoil system [MECH ENG] A recoil mechanism that absorbs the energy of recoil by the forcing of oil through orifices and returns the gun to battery by compressed gas.

hydroseparator [MECH ENG] A separator in which solids in suspension are agitated by hydraulic pressure or stirring devices.

hydrosizer *See* hydraulic classifier.

hydrostat *See* humidistat.

hydrostatic bearing [MECH ENG] A sleeve bearing in which high-pressure oil is pumped into the area between the shaft and the bearing so that the shaft is raised and supported by an oil film.

hydrostatic forging [MET] Forging a metal part by using pressure supplied by a liquid.

hydrostatic press *See* hydraulic press.

hydrostatic pressing [ENG] Compacting ceramic or metal powders by packing them in a rubber bag which is subjected to hydrostatic press in a cylinder.

hydrostatic roller conveyor [MECH ENG] A portion of a roller conveyor that has rolls weighted with liquid to control the speed of the moving objects.

hydrostatic test [ENG] Test of strength and leak-resistance of a vessel, pipe, or other hollow equipment by internal pressurization with a test liquid.

hydrowire [ENG] A wire to which equipment is clamped so that it can be lowered over the side of the ship into the water.

hydroxyethylcellulose [MATER] A white powder made from cellulose, used for textile finishes and as a thickener for water-base paints.

hygrodeik [ENG] A form of psychrometer with wet-bulb and dry-bulb thermometers mounted on opposite edges of a specially designed graph of the psychrometric tables, arranged so that the intersections of two curves determined by the wet-bulb and dry-bulb readings yield the relative humidity, dew-point, and absolute humidity.

hygrogram [ENG] The record made by a hygrograph.

hygrograph [ENG] A recording hygrometer.

hygrometer [ENG] An instrument for giving a direct indication of the amount of moisture in the air or other gas, the indication usually being in terms of relative humidity as a percentage which the moisture present bears to the maximum amount of moisture that could be present at the location temperature without condensation taking place.

hygrometry [ENG] The study which treats of the measurement of the humidity of the atmosphere and other gases.

hygrothermograph [ENG] An instrument for recording temperature and humidity on a single chart.

hyperbaric chamber [ENG] A specially equipped pressure vessel used in medicine and physiological research to administer oxygen at elevated pressures.

hyperbolic decline [PETRO ENG] One of three types of decline in oil or gas production rate (the others are constant-percentage and harmonic decline).

hyperbolic flareout [AERO ENG] A flareout obtained by changing the glide slope from a straight line to a hyperbolic curve at an appropriate distance from touchdown at an airport.

hyperbolic horn [ENG] Horn whose equivalent cross-sectional radius increases according to a hyperbolic law.

hyperbolic trajectory [AERO ENG] A trajectory entered by a spacecraft when its velocity exceeds the escape velocity of a planet, satellite, or star.

hypereutectic alloy [MET] Any binary alloy whose composition lies to the right of the eutectic on an equilibrium diagram and which contains some eutectic structure.

hypereutectoid steel [MET] Steel containing more than 0.8% carbon.

hypergolic fuel [MATER] A combination of fuel and oxidizer that ignite spontaneously on contact, such as methanol and hydrogen peroxide; used as rocket propellant.

hyperoid axle [MECH ENG] A type of rear-axle drive gear set which generally carries the pinion 1.5–2 inches (38–51 millimeters) or more below the centerline of the gear.

hypersonic flight [AERO ENG] Flight at speeds well above the local velocity of sound; by convention, hypersonic regime starts at about five times the speed of sound and extends upward indefinitely.

hypersonic glider [AERO ENG] An unpowered vehicle, specifically a reentry vehicle, designed to fly at hypersonic speeds.

hypersonic wind tunnel [ENG] A wind tunnel in which air flows at speeds roughly in the range from 5 to 15 times the speed of sound.

hypervelocity wind tunnel [ENG] A wind tunnel in which higher airspeeds and temperatures can be attained than in a hypersonic wind tunnel.

hypoeutectic alloy [MET] Any binary alloy whose composition lies to the left of the eutectic.

hypoeutectoid steel [MET] Steel containing less than 0.8% carbon.

hypoid gear [MECH ENG] Gear wheels connecting nonparallel, nonintersecting shafts, usually at right angles.

hypoid generator [MECH ENG] A gear-cutting machine for making hypoid gears.

hypsometer [ENG] **1.** An instrument for measuring atmospheric pressure to ascertain elevations by determining the boiling point of liquids. **2.** Any of several instruments for determining tree heights by triangulation.

hypsometric [ENG] Pertaining to hypsometry.

hypsometry [ENG] The measuring of elevation with reference to sea level.

hysteresimeter [ENG] A device for measuring hysteresis.

hysteresis clutch [MECH ENG] A clutch in which torque is produced by attraction between induced poles in a magnetized iron ring and the control field.

I

IAS *See* indicated airspeed.

I beam [CIV ENG] A rolled iron or steel joist having an I section, with short flanges.

ice-accretion indicator [ENG] An instrument used to detect the occurrence of freezing precipitation, usually consisting of a strip of sheet aluminum about 1½ inches (4 centimeters) wide, and is exposed horizontally, face up, in the free air a few meters above the ground.

ice apron [CIV ENG] A wedge-shaped structure which protects a bridge pier from floating ice.

ice buoy [ENG] A sturdy buoy, usually a metal spar, used to replace a more easily damaged buoy during a period when heavy ice is anticipated.

ice calorimeter *See* Bunsen ice calorimeter.

ice load [ENG] The weight of glaze deposited on an overhead wire in a power supply system; standard safety codes require allowance for ½-inch (12.7 millimeters) radial thickness in heavy loading districts and ¼-inch (6.35 millimeters) in medium.

ice pick [DES ENG] A hand tool for chipping ice.

ice tongs [DES ENG] Tongs for handling cubes or blocks of ice.

ichthyocolla *See* isinglass.

icing-rate meter [ENG] An instrument for the measurement of the rate of ice accretion on an unheated body.

ideal productivity index [PETRO ENG] Theoretical straight-line relationship between oil production from a reservoir and the resultant pressure drop within that reservoir.

ideal rocket [AERO ENG] A rocket motor or rocket engine that would have a velocity equal to the velocity of its jet gases.

identification, friend or foe [ENG] A system using pulsed radio transmissions to which equipment carried by friendly forces automatically responds, by emitting a pulse code, thereby identifying themselves from enemy forces; a method of determining the friendly or unfriendly character of aircraft, ships, and army units by other aircraft, ships, or ground force units. Abbreviated IFF.

ID grinding [MET] The grinding of the inner diameter of a piece of tubing or piping.

idle [MECH ENG] To run without a load.

idler gear [MECH ENG] A gear situated between a driving gear and a driven gear to transfer motion, without any change of direction or of gear ratio.

idler pulley [MECH ENG] A pulley used to guide and tighten the belt or chain of a conveyor system.

idler wheel [MECH ENG] **1.** A wheel used to transmit motion or to guide and support something. **2.** A roller with a rubber surface used to transfer power by frictional means in a sound-recording or sound-reproducing system.

idle time [IND ENG] A period of time during a regular work cycle when a worker is not active because of waiting for materials or instruction. Also known as waiting time.

idling jet [MECH ENG] A carburetor part that introduces gasoline during minimum load or speed of the engine.

idling system [MECH ENG] A system to obtain adequate metering forces at low airspeeds and small throttle openings in an automobile carburetor in the idling position.

IFF *See* identification, friend or foe.

Igewsky's solution [MET] Etchant used to prepare carbon steels for microscopic analysis; consists of 5% picric acid in absolute alcohol.

igniter [ENG] **1.** A device for igniting a fuel mixture. **2.** A charge, as of black powder, to facilitate ignition of a propelling or bursting charge.

igniter cord [ENG] A cord which passes an intense flame along its length at a uniform rate to light safety fuses in succession.

ignition charge [MATER] A small quantity of explosive, usually composed of black powder, that facilitates the firing of the main charge.

ignition delay *See* ignition lag.

ignition lag [MECH ENG] In the internal combustion engine, the time interval between the passage of the spark and the inflammation of the air-fuel mixture. Also known as ignition delay.

ignition system [MECH ENG] The system in an internal combustion engine that initiates the chemical reaction between fuel and air in the cylinder charge by producing a spark.

I-head cylinder [MECH ENG] The internal combustion engine construction having both inlet and exhaust valves located in the cylinder head.

ihp *See* indicated horsepower.

ilang-ilang oil [MATER] An oil made synthetically or derived from the flowers of the tree *Canangium odorata*, containing pinene, geraniol, and linalol. Also known as cananga oil; ylang-ylang oil.

illepé fat *See* mowrah fat.

illuminating gas [MATER] Flammable mixture of gases suitable for illuminating purposes; contains hydrogen, methane, ethane, carbon monoxide, and some nitrogen and oxygen.

illuminating oil [MATER] An oil such as kerosine suitable for burning to provide illumination.

illumination design [ENG] Design of sources of lighting and of systems which distribute light in order to effect a comfortable and satisfactory environment for seeing.

imaging radar [ENG] Radar carried on aircraft which forms images of the terrain.

Imhoff cone [CIV ENG] A graduated glass vessel for measuring settled solids in testing the composition of sewage.

Imhoff tank [CIV ENG] A sewage treatment tank in which digestion and settlement take place in separate compartments, one below the other.

immersion cleaning [MET] Removing surface dirt from metal by dipping into a cleaning liquid.

immersion coating [ENG] Applying material to the surface of a metal or ceramic by dipping into a liquid.

immersion plating [MET] Applying an adherent layer of more-noble metal to the surface of a metal object by dipping in a solution of more-noble metal ions; a replacement reaction. Also known as dip plating; metal replacement.

immersion scanning [ENG] Ultrasonic scanning in which the ultrasonic transducer and the object being scanned are both immersed in water or some other liquid that provides good coupling while the transducer is being moved around the object.

immunity [MET] The ability of metal to resist corrosion as a result of thermodynamic stability.

impact bar [ENG] Specimen used to test the relative susceptibility of a plastic material to fracture by shock.

impact breaker [MECH ENG] A device that utilizes the energy from falling stones in addition to power from massive impellers for complete breaking up of stone. Also known as double impeller breaker.

impact crusher [MECH ENG] A machine for crushing large chunks of solid materials by sharp blows imposed by rotating hammers, or steel plates or bars; some crushers accept lumps as large as 28 inches (about 70 centimeters) in diameter, reducing them to ¼ inch (6 millimeters) and smaller.

impact extrusion [MET] A cold extrusion process for producing tubular components by striking a slug of the metal, which has been placed in the cavity of the die, with a punch moving at high velocity.

impact forging [MET] Plastic deformation of a metal using an impactive force.

impact grinding [MECH ENG] A technique used to break up particles by direct fall of crushing bodies on them.

impact load [ENG] A force delivered by a blow, as opposed to a force applied gradually and maintained over a long period.

impact microphone [ENG ACOUS] An instrument that picks up the vibration of an object impinging upon another, used especially on space probes to record the impact of small meteoroids.

impact mill [MECH ENG] A unit that reduces the size of rocks and minerals by the action of rotating blades projecting the material against steel plates.

impact-noise analyzer [ENG] An analyzer used with a sound-level meter to evaluate the characteristics of impact-type sounds and electric noise impulses that cannot be measured accurately with a noise meter alone.

impactometer *See* impactor.

impactor [ENG] A general term for instruments which sample atmospheric suspensoids by impaction; such instruments consist of a housing which constrains the air flow past a sensitized sampling plate. Also known as impactometer. [MECH ENG] A machine or part whose operating principle is striking blows. [MIN ENG] A rotary hammermill which crushes ore by impacting it against crushing plates or elements.

impact predictor [AERO ENG] A device which takes information from a trajectory measuring system and continuously computes the point (in real time) at which the rocket will strike the earth.

impact roll [MECH ENG] An idler roll protected by a covering of a resilient material from the shock of the loading of material onto a conveyor belt, so as to reduce the damage to the belt.

impact screen [MECH ENG] A screen designed to swing or rock forward when loaded and to stop abruptly by coming in contact with a stop.

impact test [ENG] Determination of the degree of resistance of a material to breaking by impact, under bending, tension, and torsion loads; the energy absorbed is measured in breaking the material by a single blow.

impact tube *See* pitot tube.

impact wrench [MECH ENG] A compressed-air or electrically operated wrench that gives a rapid succession of sudden torques.

impeller [MECH ENG] The rotating member of a turbine, blower, fan, axial or centrifugal pump, or mixing apparatus. Also known as rotor.

impeller pump [MECH ENG] Any pump using a mechanical agency to provide continuous power to move liquids.

imperial smelting process [MET] A pyrometallurgical process which treats a complex concentrated feed to a single furnace to recover zinc, copper, lead, cadmium, silver, gold, and other metals in one pass. Abbreviated ISP.

impersonal micrometer [ENG] An instrument consisting of a vertical wire that is mounted in the focal plane of a transit circle and can be moved across the field of view to follow a star, and instrumentation to record the position of the wire as a function of time; used to reduce systematic observational errors.

impervious carbon [MATER] Carbon compressed with bituminous binder, then carbonized by sintering to produce a dense, impervious material; used as brick to line chemical process and storage vessels.

impingement [ENG] Removal of liquid droplets from a flowing gas or vapor stream by causing it to collide with a baffle plate at high velocity, so that the droplets fall away from the stream. Also known as liquid knockout.

impingement attack [MET] Accelerated corrosive attack on a metal by moving liquids, resulting usually from erosion of a protective surface layer.

impinger [ENG] A device used to sample dust in the air that draws in a measured volume of dusty air and directs it through a jet to impact on a wetted glass plate;

the dust particles adhering to the plate are counted.

imposed load [CIV ENG] Any load which a structure must sustain, other than the weight of the structure itself.

impound [CIV ENG] To collect water for irrigation, flood control, or similar purpose.

impounding reservoir [CIV ENG] A reservoir with outlets controlled by gates that release stored surface water as needed in a dry season; may also store water for domestic or industrial use or for flood control. Also known as storage reservoir.

impregnate [ENG] To force a liquid substance into the spaces of a porous solid in order to change its properties, as the impregnation of turquoise gems with plastic to improve color and durability, the impregnation of porous tungsten with a molten barium compound to manufacture a dispenser cathode, or the impregnation of wood with creosote to preserve its integrity against water damage.

impregnated bit [DES ENG] A sintered, powder-metal matrix bit with fragmented bort or whole diamonds of selected screen sizes uniformly distributed throughout the entire crown section.

impregnated timber [MATER] Timber which has been made flame-resistant, fungi-resistant, or insect-proof by forcing into it under vacuum or pressure a flame retardant or a fungal or insect poison.

impression [MET] A machined cavity in a forging die for production of a specific geometric shape in the workpiece.

impression block [PETRO ENG] A block with wax or lead on the bottom run into a well and allowed to rest on a lost tool or other object so that an examination of the resultant impression is revelatory concerning the size, shape, or position of the object.

impulse [MET] A single pulse or several pulses in welding current used in resistance welding.

impulse sealing [ENG] Heat-sealing of plastic materials by applying a pulse of intense thermal energy to the sealing area for a very short time, followed immediately by cooling.

impulse tachometer [ENG] A tachometer in which each rotation of a shaft generates an electric pulse and the time rate of pulses is then measured; classified as capacitory-current, inductory, or interrupted direct-current tachometer.

impulse turbine [MECH ENG] A prime mover in which fluid under pressure enters a stationary nozzle where its pressure (potential) energy is converted to velocity (kinetic) energy and absorbed by the rotor.

in-and-out bond [CIV ENG] Masonry bond composed of vertically alternating stretchers and headers.

inboard [ENG] Toward or close to the longitudinal axis of a ship or aircraft.

inbond [CIV ENG] Pertaining to bricks or stones laid as headers across a wall.

inby [MIN ENG] Away from the shaft or mine entrance and therefore toward the working face.

incentive wage system *See* wage incentive plan.

incinerator [ENG] A furnace or other container in which materials are burned.

inclined cableway [MECH ENG] A monocable arrangement in which the track cable has a slope sufficiently steep to allow the carrier to run down under its own weight.

inclined drilling [ENG] The drilling of blastholes at an angle with the vertical.

inclined orbit [AERO ENG] A satellite orbit which is inclined with respect to the earth's equator.

inclined skip hoist [MIN ENG] A skip hoist that operates on steeply inclined rails placed on a mine pit slope or wall.

inclined-tube manometer [ENG] A glass-tube manometer with the leg inclined from the vertical to extend the scale for more minute readings.

incline shaft [MIN ENG] A shaft which has been dug at an angle to the vertical to follow the depth of the lode.

inclinometer [ENG] **1.** An instrument that measures the attitude of an aircraft with respect to the horizontal. **2.** An instrument for measuring the angle between the earth's magnetic field vector and the horizontal plane. **3.** An apparatus used to ascertain the direction of the magnetic field of the earth with reference to the plane of the horizon.

inclusion [MET] An impure particle, such as sand, trapped in molten metal during solidification.

incompetent rock [ENG] Soft or fragmented rock in which an opening, such as a borehole or an underground working place, cannot be maintained unless artificially supported by casing, cementing, or timbering.

incomplete lubrication [MECH ENG] Lubrication that takes place when the load on the rubbing surfaces is carried partly by a fluid viscous film and partly by areas of boundary lubrication; friction is intermediate between that of fluid and boundary lubrication.

increaser [ENG] An adapter for connecting a small-diameter pipe to a larger-diameter pipe.

incubator oil [MATER] Special grade of long-burning petroleum heating oil used to heat farm incubators.

indelible ink [MATER] An ink that cannot be removed, for example, India ink.

indentation hardness [MET] The resistance of a metal surface to indention when subjected to pressure by a hard pointed or rounded tool. Also known as penetration hardness.

indented bolt [DES ENG] A type of anchor bolt that has indentations to hold better in cemented grout.

independent chuck [DES ENG] A chuck for holding work by means of four jaws, each of which is moved independently of the others.

independent contractor [ENG] One who exercises independent control over the mode and method of operations to produce the results demanded by the contract.

independent footing [CIV ENG] A footing that supports a concentrated load, such as a single column.

independent machine time *See* machine controlled time.

independent suspension [MECH ENG] In automobiles, a system of springs and guide links by which wheels are mounted independently on the chassis.

independent wire-rope core [DES ENG] A core of steel in a wire rope made in accordance with the best practice and design, either bright (uncoated) galvanized or drawn galvanized wire.

indeterminate truss [CIV ENG] A truss having redundant bars.

index center [MECH ENG] One of two machine-tool centers used to hold work and to rotate it by a fixed amount.

index chart [MECH ENG] **1.** A chart used in conjunction with an indexing or dividing head, which correlates the index plate, hole circle, and index crank motion with the desired angular subdivisions. **2.** A chart indicating the arrangement of levers in a machine to obtain desired output speed or fuel rate.

index counter [ENG] A counter indicating revolutions of the tape supply reel, making it possible to index selections within a reel of tape.

index crank [MECH ENG] The crank handle of an index head used to turn the spindle.

index error [ENG] An error caused by the misalignment of the vernier and the graduated circle (arc) of an instrument.

index head [MECH ENG] A headstock that can be affixed to the table of a milling machine, planer, or shaper; work may be mounted on it by a chuck or centers, for indexing.

indexing [MECH ENG] The process of providing discrete spaces, parts, or angles in a workpiece by using an index head.

indexing fixture [MECH ENG] A fixture that changes position with regular steplike movements.

index plate [DES ENG] A plate with circular graduations or holes arranged in circles, each circle with different spacing; used for indexing on machines.

index thermometer [ENG] A thermometer in which steel index particles are carried by mercury in the capillary and adhere to the capillary wall in the high and low positions, thus indicating minimum and maximum inertial scales.

India ink [MATER] A permanent black ink made of lampblack and blue binder; some varieties are waterproof. Also known as Chinese ink; sumi ink.

Indiana limestone *See* spergenite.

Indian balsam *See* Peru balsam.

Indian grass oil *See* palmarosa oil.

Indian gum [MATER] Any of the gums, such as ghatti gum and sterculia gum, with mucilage consistency from trees in the forests in India and Ceylon.

Indian red [MATER] Iron-oxide-base, maroon pigment; used to polish gold, silver, and other metals. Also known as iron saffron.

Indian tragacanth *See* karaya gum.

Indian yellow [MATER] A yellow pigment which may be aureolin, made of cobalt and potassium nitrates; or puree, the impure basic magnesium salt of euxanthic acid; or the synthetic dye primuline.

indicated airspeed [AERO ENG] The airspeed as shown by a differential-pressure airspeed indicator, uncorrected for instrument and installation errors; a simple computation for altitude and temperature converts indicated airspeed to true airspeed. Abbreviated IAS.

indicated altitude [AERO ENG] The uncorrected reading of a barometric altimeter.

indicated horsepower [MECH ENG] The horsepower delivered by an engine as calculated from the average pressure of the working fluid in the cylinders and the displacement. Abbreviated ihp.

indicated ore [MIN ENG] A known mineral deposit for which quantitative estimates are made partly from inference and partly from specific sampling. Also known as probable ore.

indicating gage [ENG] A gage consisting essentially of a case and mounting, a spindle carrying the contact point, an amplifying mechanism, a pointer, and a graduated dial; used to amplify and measure the displacement of a movable contact point.

indication [ENG] In ultrasonic testing, determination of the presence of a flaw by detection of a reflected ultrasonic beam.

indicator [ENG] An instrument for obtaining a diagram of the pressure-volume changes in a running positive-displacement engine, compressor, or pump cylinder during the working cycle.

indicator card [ENG] A chart on which an indicator diagram is produced by an instrument called an engine indicator which traces the real-performance cycle diagram as the machine is running.

indicator diagram [ENG] A pressure-volume diagram representing and measuring the work done by or on a fluid while performing the work cycle in a reciprocating engine, pump, or compressor cylinder.

indicator unit [ENG] An instrument which detects the presence of an electrical quantity without necessarily measuring it.

indirect-arc furnace [ENG] A refractory-lined furnace in which the burden is heated indirectly by the radiant heat from an electric arc.

indirect cost [IND ENG] A cost that is not readily indentifiable with or chargeable to a specific product or service.

indirect extrusion [MET] An extrusion process in which the billet remains stationary while a hollow die stand forces the die back into the cylinder.

indirect labor [IND ENG] Labor not directly engaged in the actual production of the product or performance of a service.

indirect lighting [ENG] A system of lighting in which more than 90% of the light from luminaires is distributed upward toward the ceiling, from which it is diffusely reflected.

indirect material [IND ENG] Any material used in the manufacture of a product which does not itself become a part of the product and whose cost is indirect.

indium [MET] A ductile, silver-white, shiny metal that resists tarnishing and is used in precious-metal alloys for jewelry and dentistry, in glass-sealing alloys, lubricants, and bearing metals, and as an atomic-pile neutron indicator.

induced angle of attack [AERO ENG] The downward vertical angle between the horizontal and the velocity (relative to the wing) of an aircraft) of the airstream passing over the wing.

induced draft [MECH ENG] A mechanical draft produced by suction stream jets or fans at the point where air or gases leave a unit.

induced-draft cooling tower [MECH ENG] A structure for cooling water by circulating air where the load is on the suction side of the fan.

induction brazing [MET] A brazing process in which coalescence is produced by heat generated within the work by an induced electric current.

induction burner [ENG] Fuel-air burner into which the fuel is fed under pressure to entrain needed air into the combustion nozzle area.

induction-electrical survey [ENG] Study of subterranean formations by combined induction and electrical logging.

induction flowmeter [ENG] An instrument for measuring the flow of a conducting liquid passing through a tube, in which the tube is placed in a transverse magnetic field and the induced electromotive force between electrodes at opposite ends of a diameter of the tube perpendicular to the field is measured.

induction furnace [ENG] An electric furnace in which heat is produced in a metal charge by electromagnetic induction.

induction hardening [MET] A quench-hardening technique in which the required elevated temperature is obtained by electromagnetic induction.

induction heating [ENG] Increasing the temperature in a material by induced electric current. Also known as eddy-current heating.

induction inclinometer *See* earth inductor.

induction instrument [ENG] Meter that depends for its operation on the reaction between magnetic flux set up by current in fixed windings, and other currents set up by electromagnetic induction in conducting parts of the moving system.

induction log [ENG] An electric log of the conductivity of rock with depth obtained by lowering into an uncased borehole a generating coil that induces eddy currents on the rocks and these are detected by a receiver coil.

induction loudspeaker [ENG ACOUS] Loudspeaker in which the current which reacts with the steady magnetic field is induced in the moving member.

induction melting [MET] Converting a solid metal to the molten state in an induction furnace.

induction pump [MECH ENG] Any pump operated by electromagnetic induction.

induction salinometer [ENG] A device for measuring salinity by taking voltage readings of the current in seawater.

induction silencer [ENG] A device for reducing engine induction noise, which consists essentially of a low-pass acoustic filter with the inertance of the air-entrance tube and the acoustic compliance of the annular and central volumes providing acoustic filtering elements.

induction soldering [MET] A soldering process in which the metals are heated by an induced electric current.

induction valve *See* inlet valve.

induction welding [MET] A process of welding by means of heat generated within the work by induced electric currents.

inductor microphone [ENG ACOUS] Moving-conductor microphone in which the moving element is in the form of a straight-line conductor.

inductor tachometer [ENG] A type of impulse tachometer in which the rotating member, consisting of a magnetic material, causes the magnetic flux threading a circuit containing a magnet and a pickup coil to rise and fall, producing pulses in the circuit which are rectified for a permanent-magnet, movable-coil instrument.

industrial anthropometry [IND ENG] Application of the knowledge of physical anthropology to the design and construction of equipment for human use, such as automobiles.

industrial car [IND ENG] Any of various narrow-gage railcars used for indoor or outdoor handling of bulk and package materials.

industrial cost control [IND ENG] A specific system or procedure used to keep manufacturing costs in line. Also known as cost control.

industrial engineering [ENG] A branch of engineering concerned with the design, improvement, and installation of integrated systems of people, materials, and equipment. Also known as management engineering.

industrial glass [MATER] Any glass molded into shapes for product parts, for example, lime glass and lead glass.

industrial mobilization [IND ENG] Transformation of industry and other productive facilities and contributory services from their peacetime activities to the fulfillment of the munitions program necessary to support a military effort.

industrial railway [IND ENG] **1.** A usually short feeder line that is either owned or controlled and wholly operated by an industrial firm. **2.** Narrow-gage rail lines used on construction jobs or around industrial plants.

industrial revolution [IND ENG] A widespread change in industrial or production methods, toward production by machine and away from manual labor.

industrial security [IND ENG] The portion of internal security which refers to the protection of industrial installations, resources, utilities, materials, and classified information essential to protection from loss or damage.

industrial waste [ENG] Worthless materials remaining from industrial operations.

inert-gas cutting [MET] Cutting of metal while inert gas flows around the cutting area to prevent oxidation.

inert gas–shielded arc welding [MET] An arc-welding process in which the weld area is shielded by an inert gas to prevent oxidation. Also known as Heliarc welding.

inertia governor [MECH ENG] A speed-control device utilizing suspended masses that respond to speed changes by reason of their inertia.

inertia starter [MECH ENG] A device utilizing inertial principles to start the rotator of an internal combustion engine.

inertia welding [MET] A form of friction welding which utilizes kinetic energy stored in a flywheel system to supply the power required for all of the heating and much of the forging.

inert primer [ENG] A cylinder which enshrouds a detonator but does not interfere with the detonation of the explosive charge.

in-feed centerless grinding [MECH ENG] A metal-cutting process by which a cylindrical workpiece is ground to a prescribed surface smoothness and diameter by the insertion of the workpiece between a grinding wheel and a canted regulating wheel; the rotation of the regulating wheel controls the rotation and feed rate of the workpiece.

inferential flowmeter [ENG] A flowmeter in which the flow is determined by measurement of a phenomenon associated with the flow, such as a drop in static pressure at a restriction in a pipe, or the rotation of an impeller or rotor, rather than measurement of the actual mass flow.

inferential liquid-level meter [ENG] A liquid-level meter in which the level of a liquid is determined by measurement of some phenomenon associated with this level, such as the buoyancy of a solid partly immersed in the liquid, the pressure at a certain level, the conductance of the liquid, or its absorption of gamma radiation, rather than by direct measurement.

inferred ore [MIN ENG] An ore whose estimate of tonnage and grade is based largely on knowledge of the deposit's geological character and to a lesser degree on samples and other data.

infiltration [MET] **1.** Filling the pores of a metal powder compact with metal having a lower melting point. **2.** Movement of molten metal into the pores of a fiber or foam metal.

infiltration gallery [CIV ENG] A large, horizontal underground conduit of porous material or with openings on the sides for collecting percolating water by infiltration.

infinite baffle [ENG ACOUS] A loudspeaker baffle which prevents interaction between the front and back radiation of the loudspeaker.

infinite reservoir [PETRO ENG] In reservoir unsteady-state liquid-diffusion calculations, a reservoir in which the outer boundary is considered to be effectively at an infinite distance from the inner boundary at the well of an aquifer.

inflatable packer [PETRO ENG] A packer (downhole pressure seal between tubing and casing) set and held in place by an element that is inflated with hydraulic pressure.

inflated [ENG] Filled or distended with air or gas.

inflected arch See inverted arch.

influence function [PETRO ENG] Mathematical statement of the influence on pressure and production of each oil reservoir pool in a multipool aquifer.

infrared brazing [MET] A brazing process in which coalescence is produced by heat generated by infrared radiation.

infrared heating [ENG] Heating by means of infrared radiation.

infrared homing [ENG] Homing in which the target is tracked by means of its emitted infrared radiation.

infrared imaging device [ENG] Any device which converts an invisible infrared image into a visible image.

infrared soldering [MET] Soldering in which infrared radiation furnishes the required heat.

infrared-transparent material [MATER] An optical material that transmits infrared radiation; examples include sodium chloride (0.25 to 16 micrometers), cesium iodide (1 to 50 micrometers), and high-density polyethylene (16 to 300 micrometers).

in-gate See gate.

Ingersoll-Rand jumbo columns [MIN ENG] Columns held in place by the pressure of air-operated pistons against the roof; drills are attached to movable arms mounted on the columns.

ingot [MET] **1.** A solid metal casting suitable for remelting or working. **2.** A bar of gold or silver.

ingot iron [MET] Relatively pure iron produced in an open-hearth furnace.

inhabited building distance [ENG] The minimum distance permitted between an ammunition or explosive location and any building used for habitation or where people are accustomed to assemble, except operating buildings or magazines.

inhaler [MIN ENG] An apparatus, of different forms, for permitting the supply of fresh air to a miner.

inhaul cable [MECH ENG] In a cable excavator, the line that pulls the bucket to dig and bring in soil. Also known as digging line.

inherent bursts [MIN ENG] Rock bursts that occur in development.

inherent damping [MECH ENG] A method of vibration damping which makes use of the mechanical hysteresis of such materials as rubber, felt, and cork.

inherent noise pressure *See* equivalent noise pressure.

in-house [IND ENG] Pertaining to an operation produced or carried on within a plant or organization, rather than done elsewhere under contract.

initializing explosive *See* primary explosive.

initial mass [AERO ENG] The mass of a rocket missile at the beginning of its flight.

initial saturation [PETRO ENG] A reservoir's initial relative content (saturation) of water, oil, and gas.

initial set [MATER] The onset of hardening after water has been added to concrete, cement, or plaster.

initiating agent [MATER] An explosive material which has the necessary sensitivity to heat, friction, or percussion to make it suitable for use as the initial element in an explosive train.

injected gas [PETRO ENG] Gas that has been pumped into an oil-producing reservoir to provide a gas-drive for increased oil production.

injected hole [MIN ENG] A borehole into which a cement slurry or grout has been forced by high-pressure pumps and allowed to harden.

injection [AERO ENG] The process of placing a spacecraft into a specific trajectory, such as an earth orbit or an encounter trajectory to Mars. [MECH ENG] The introduction of fuel, fuel and air, fuel and oxidizer, water, or other substance into an engine induction system or combustion chamber. [MIN ENG] The introduction under pressure of a liquid or plastic material into cracks, cavities, or pores in a rock formation.

injection blow molding [ENG] Plastics molding process in which a hollow-plastic tube is formed by injection molding.

injection carburetor [MECH ENG] A carburetor in which fuel is delivered under pressure into a heated part of the engine intake system. Also known as pressure carburetor.

injection fluid [PETRO ENG] Gas or water, depending on the nature of the reservoir and its fluid content, for injection into the formation to increase hydrocarbon production.

injection-fluid front [PETRO ENG] The moving interfacial contact between an injected fluid (gas or water) and the natural fluid content of the reservoir formation.

injection gas-fluid ratio [PETRO ENG] The ratio of gas injected into a reservoir formation to the fluid hydrocarbons produced by the resultant gas lift.

injection mold [ENG] A plastics mold into which the material to be formed is introduced from an exterior heating cylinder.

injection molding [ENG] Molding metal, plastic, or nonplastic ceramic shapes by injecting a measured quantity of the molten material into dies.

injection pressure [PETRO ENG] Pressure of fluid injected into oil formations for waterflood (water) or pressure maintenance (gas).

injection pump [MECH ENG] A pump that forces a measured amount of fuel through a fuel line and atomizing nozzle in the combustion chamber of an internal combustion engine.

injection ram [ENG] In injection molding, the ram that applies pressure to the feed plunger in the process of either injection or transfer molding.

injection signal [ENG ACOUS] The sawtooth frequency-modulated signal which is added to the first detector circuit for mixing with the incoming target signal.

injection well [PETRO ENG] In secondary recovery of petroleum, a well in which a fluid such as gas or water is injected to provide supplemental energy to drive the oil remaining in the reservoir to the vicinity of production wells.

injection-well plugging [PETRO ENG] Plugging of the sand face of an injection well because of lubricant or corrosion-product carryover from surface lines or well equipment.

injectivity index [PETRO ENG] The number of barrels per day of gross liquid pumped into an injection well per psi (pound per square inch) pressure differential between the mean injection pressure and the mean formation pressure.

injectivity test [PETRO ENG] A test series of reservoir water injection rates at different pressures to predict the performance of an injection well.

injector [MECH ENG] **1.** An apparatus containing a nozzle in an actuating fluid which is accelerated and thus entrains a second fluid, so delivering the mixture against a pressure in excess of the actuating fluid. **2.** A plug with a valved nozzle through which fuel is metered to the combustion chambers in diesel- or full-injection engines. **3.** A jet through which feedwater is injected into a boiler, or fuel is injected into a combustion chamber.

injector torch *See* low-pressure torch.

ink [MATER] A dispersion of a pigment or a solution of a dye in a carrier vehicle, yielding a fluid, paste, or powder to be applied to and dried on a substrate; writing, marking, drawing, and printing inks are applied by several methods to paper, metal, plastic, wood, glass, fabric, or other substrate.

inkometer [ENG] An instrument for measuring adhesion of liquids by rotating drums in contact with the liquid.

inlay cladding [MET] A mechanical process in which a groove, 7/100–1/8 inch (1.778–3.175 millimeters) wide, is cut into a base metal and filled with cladding metal; mechanical bonding of the metals is accomplished by passing them through the pressure rolls of a bonding mill.

inlet [ENG] An entrance or orifice for the admission of fluid.

inlet box [MECH ENG] A closure at the fan inlet or inlets in a boiler for attachment of the fan to the duct system.

inlet valve [MECH ENG] The valve through which a fluid is drawn into the cylinder of a positive-displacement engine, pump, or compressor. Also known as induction valve.

in line [ENG] **1.** Over the center of a borehole and parallel with its long axis. **2.** Of a drill motor, mounted so that its drive shaft and the drive rod in the drill swivel head are parallel, or mounted so that the shaft driving the drill-swivel-head bevel gear and the drill-motor drive shaft are centered in a direct line and parallel with each other. **3.** Having similar units mounted together in a line.

in-line assembly machine [IND ENG] An assembly machine that inserts components into a wiring board one at a time as the board is moved from station to station by a conveyor or other transport mechanism.

in-line engine [MECH ENG] A multiple-cylinder engine with cylinders aligned in a row.

in-line equipment [ENG] **1.** A sequence of equipment or processing items mounted along the same vertical or horizontal plane. **2.** Equipment mounted within a process line, such as an in-line pump, pressure-drop flowmeter, or nozzle mixer.

in-line linkage [MECH ENG] A power-steering linkage which has the control valve and actuator combined in a single assembly.

innage [ENG] The volume or the measured height of liquid introduced into a tank or container.

inner tube [MIN ENG] The inside tube which acts as the core container of a double-tube core barrel; used to obtain core

samples for analysis of an ore formation. Also known as inner barrel.

inner tube [ENG] A rubber tube used inside a pneumatic tire casing to hold air under pressure.

inner-tube extension *See* lifter case.

inoculant [MET] A substance which augments a melt, usually in the latter part of the melting operation, thus altering the solidification structure of the cast metal, as in grain refinement of aluminum alloys.

inoculation [MET] Treating a molten material with another material before casting in order to nucleate crystals.

input/output relation [SYS ENG] The relation between two vectors whose components are the inputs (excitations, stimuli) of a system and the outputs (responses) respectively.

inquartation [MET] A step in bullion assay that uses nitric acid to dissolve silver from associated gold. Also known as quartation.

insect attractant [MATER] A chemical agent, usually associated with an insect's sexual drive, which may be used to attract pests to poisoned bait or for insect surveys.

insecticide [MATER] A chemical agent that destroys insects.

insensitive time *See* dead time.

insert [MET] **1.** The part of a die or mold that can be removed. **2.** A part, usually metal, which is placed in a mold and appears as an integral part of the final casting.

insert bit [DES ENG] A bit into which inset cutting points of various preshaped pieces of hard metal (usually a sintered tungsten carbide–cobalt powder alloy) are brazed or hand-peened into slots or holes cut or drilled into a blank bit. Also known as slug bit.

inserted-tooth cutter [DES ENG] A milling cutter in which the teeth can be replaced.

insertion meter [ENG] A type of flowmeter which measures the rotation rate of a small propeller or turbine rotor mounted at right angles to the end of a support rod and inserted into the flowing stream or closed pipe.

insert pump *See* rod pump.

inside caliper [DES ENG] A caliper that has two legs with feet that turn outward; used to measure inside dimensions, as the diameter of a hole.

inside diameter [DES ENG] The length of a line which passes through the center of a hollow cylindrical or spherical object, and whose end points lie on the inner surface of the object. Abbreviated ID.

inside face [DES ENG] That part of the bit crown nearest to or parallel with the inside wall of an annular or coring bit.

inside gage [DES ENG] The inside diameter of a bit as measured between the cutting points, such as between inset diamonds on the inside-wall surface of a core bit.

inside micrometer [DES ENG] A micrometer caliper with the points turned outward for measuring the internal dimensions of an object.

inside work *See* internal work.

in situ combustion [PETRO ENG] A method of driving high-viscosity, low-gravity ore otherwise unrecoverable from a formation by setting fire to the oil sand and thereby heating the oil in the horizon to increase its mobility by reducing its viscosity.

in situ foaming [ENG] Depositing of the ingredients of a foamable plastic onto the location where foaming is to take place; for example, in situ foam insulation on equipment or walls.

inspect [IND ENG] To examine an object to determine whether it conforms to standards; may employ sight, hearing, touch, odor, or taste.

inspection [IND ENG] The critical examination of a product to determine its conformance to applicable quality standards or specifications.

inspector [MIN ENG] One employed to make examinations of and to report upon mines and surface plants relative to compliance with mining laws, rules and regulations, and safety methods.

installation [ENG] Procedures for setting up equipment for use or service.

instantaneous cut [ENG] A cut that is set off by instantaneous detonators to be certain that all charges in the cut go off at the same time; the drilling and ignition are carried out so that all the holes break smaller top angles.

instantaneous detonator [ENG] A type of detonator that does not have a delay period between the passage of the electric current through the detonator and its explosion.

instantaneous fuse [ENG] A fuse with an ignition rate of several thousand feet per minute; an example is PETN.

instantaneous recording [ENG ACOUS] A recording intended for direct reproduction without further processing.

instruction card [IND ENG] A written description of the standard method used by a worker, to guide his activities.

instrument [ENG] A device for measuring and sometimes also recording and controlling the value of a quantity under observation.

instrumental analysis [ENG] The use of an instrument to measure a component, to detect the completion of a quantitative reaction, or to detect a change in the properties of a system.

instrumentation [ENG] Designing, manufacturing, and utilizing physical instruments or instrument systems for detection, observation, measurement, automatic control, automatic computation, communication, or data processing.

instrument correction [ENG] A correction of measurements made on a unit under test for either inaccuracy of the instrument or eroding effect of the instrument.

instrument housing [ENG] A case or enclosure to cover and protect an instrument.

instrument oil [MATER] Special grade of lubricating oil that has been refined to have oxidation resistance and gum resistance, that has compatibility with electrical insulation, and that prevents tarnish or oxidation of contacted metal surfaces; used to lubricate instruments and other intricate mechanisms.

instrument panel [ENG] A panel or board containing indicating meters.

instrument reading time [ENG] The time, after a change in a measured quantity, which it takes for the indication of an instrument to come and remain within a specified percentage of its final value.

instrument shelter [ENG] A boxlike structure designed to protect certain meteorological instruments from exposure to direct sunshine, precipitation, and condensation, while providing adequate ventilation. Also known as thermometer screen; thermometer shelter; thermoscreen.

instrument system [ENG] A system which integrates one or more instruments with auxiliary or associated devices for detection, observation, measurement, automatic control, automatic computation, communication, or data processing.

insulating board [MATER] Any board used in a wall or ceiling to provide insulation.

insulating compound [MATER] A liquid, at low temperatures, which is poured into joint boxes and allowed to solidify; as a poor conductor of heat and electricity, it provides good insulation.

insulating concrete [MATER] Concrete with insulating properties, often made with asbestos fibers and in the form of blocks, corrugated slabs, or sheathing.

insulating oil [MATER] A chlorinated hydrocarbon, such as trichlorobenzene, mixed with fluorinated hydrocarbons, whose high dielectric strength and high flash point allow it to be used in switches, circuit breakers, and transformers as an insulator and cooling medium. Also known as electrical oil.

insulating paper [MATER] A standard material for insulating electrical equipment, usually consisting of bond or kraft paper coated with black or yellow insulating varnish on both sides. Also known as electrical insulating paper; varnish paper.

insulating tape [MATER] Tape impregnated with insulating material, and usually adhesive; used to cover joints in insulated wires or cables. Also known as electrical tape.

insulation [BUILD] Material used in walls, ceilings, and floors to retard the passage of heat and sound.

insulation porcelain [MATER] Any of the various insulating materials consisting of molded silica, molded steatite, or specially compounded ceramics, often containing zirconia or beryllia. Also known as electrical porcelain.

insulation sampler [ENG] A device for collecting deep water which prevents any significant conduction of heat from the water sample so that it maintains its original temperature as it is hauled to the surface.

insulation testing set [ENG] An instrument for measuring insulation resistance, consisting of a high-range ohmmeter having a hand-driven direct-current generator as its voltage source.

insulator [MATER] A material that is a poor conductor of heat, sound, or electricity.

intake [ENG] **1.** An entrance for air, water, fuel, or other fluid, or the amount of such fluid taken in. **2.** A main passage for air in a mine.

intake chamber [CIV ENG] A large chamber that gradually narrows to an intake tunnel; designed to avoid undesirable water currents.

intake gate [CIV ENG] A movable partition for opening or closing a water intake opening.

intake manifold [MECH ENG] A system of pipes which feeds fuel to the various cylinders of a multicylinder internal combustion engine.

intake stroke [MECH ENG] The fluid admission phase or travel of a reciprocating piston and cylinder mechanism as, for example, in an engine, pump, or compressor.

intake valve [MECH ENG] The valve which opens to allow air or an air-fuel mixture to enter an engine cylinder.

integer programming [SYS ENG] A series of procedures used in operations research to find maxima or minima of a function subject to one or more constraints, including one which requires that the values of some or all of the variables be whole numbers.

integral-furnace boiler [MECH ENG] A type of steam boiler which incorporates furnace water-cooling in the circulatory system.

integral-joint casing [PETRO ENG] Oil well casing lengths on whose ends the connection joints are integrally formed.

integral-type flange [DES ENG] A flange which is forged or cast with, or butt-welded to, a nozzle neck, pressure vessel, or piping wall.

integral waterproofing [ENG] Waterproofing concrete by adding the waterproofing material to the cement or to the mixing water.

integraph [ENG] A device used for completing a mathematical integration by graphical methods.

integrated train [MIN ENG] A long string of cars, permanently coupled together, that shuttles endlessly between one mine and one generating plant, not even stopping to load and unload, since rotary couplers permit each car to be flipped over and dumped as the train moves slowly across a trestle.

integrating accelerometer [ENG] A device whose output signals are proportional to the velocity of the vehicle or to the distance traveled (depending on the number of integrations) instead of acceleration.

integrating frequency meter [ENG] An instrument that measures the total number of cycles through which the alternating voltage of an electric power system has passed in a given period of time, enabling this total to be compared with the number of cycles that would have elapsed if the prescribed frequency had been maintained. Also known as master frequency meter.

integrating galvanometer [ENG] A modification of the d'Arsonval galvanometer which measures the integral of current over time; it is designed to be able to measure changes of flux in an exploring coil which last over periods of several minutes.

integrating gyroscope [ENG] A gyroscope that senses the rate of angular displacement and measures and transmits the time integral of this rate.

integrating meter [ENG] An instrument that totalizes electric energy or some other quantity consumed over a period of time.

integrating water sampler [ENG] A water sampling device comprising a cylinder with a free piston whose movement is regulated by the evacuation of a charge of fresh water.

intensifier [PETRO ENG] Hydrofluoric acid added to hydrochloric acid for oil well acidizing; the fluoride destroys silica films that are insoluble by hydrochloric acid.

intensity-modulated indictor *See* ampli-tude-modulated indicator.

intercepting sewer [CIV ENG] A sewer that receives flow from transverse sewers and conducts the water to a treatment plant or disposal point.

intercept method [MET] A method for determining grain size or the quantity of a phase in a microstructure by measuring the number of grains or phase particles per unit length intersected by straight lines.

interceptometer [ENG] A rain gage which is placed under trees or in foliage to determine the rainfall in that location; by comparing this catch with that from a rain gage set in the open, the amount of rainfall which has been intercepted by foliage is found.

interceptor [AERO ENG] A crewed aircraft utilized for the identification or engagement of airborne objects.

interchange [CIV ENG] A junction of two or more highways at a number of separate levels so that traffic can pass from one highway to another without the crossing at grade of traffic streams.

interchangeability [ENG] The ability to re-place the components, parts, or equipment of one manufacturer with those of another, without losing function or suitability.

intercommunicating porosity [MET] The type of porosity in a sintered metal powder compact that allows fluid to pass from pore to pore.

intercommunication [PETRO ENG] Flow inter-connection between the reservoir areas being drained by adjacent wells.

intercondenser [MECH ENG] A condenser be-tween stages of a multistage steam jet pump.

intercooler [MECH ENG] A heat exchanger for cooling fluid between stages of a multistage compressor with consequent saving in power.

intercrystalline *See* transcrystalline.

intercrystalline corrosion [MET] Localized attack occurring along the crystal boundaries of a metal or alloy. Also known as intergranular corrosion.

interdendritic attack *See* interdendritic corrosion.

interdendritic corrosion [MET] Preferential corrosion of the metal immediately sur-rounding dendrites in unworked or slightly worked alloys caused by compo-sition gradients. Also known as inter-dendritic attack.

interference fit [DES ENG] A fit wherein one of the mating parts of an assembly is forced into a space provided by the other part in such a way that the condition of maximum metal overlap is achieved.

interference test [PETRO ENG] Test of pres-sure interrelationships (interference) be-tween wells serving the same formation.

interference time [IND ENG] Idle machine time occurring when a machine opera-tor, assigned to two or more semiauto-matic machines, is unable to service a machine requiring attention.

intergranular corrosion *See* intercrystalline corrosion.

intergranular fracture [MET] Propagation of a crack along the grain boundaries of a metal or alloy.

interlock [ENG] A switch or other device that prevents activation of a piece of equip-ment when a protective door is open or some other hazard exists.

interlocking cutter [DES ENG] A milling cut-ter assembly consisting of two mating sections with uniform or alternate over-lapping teeth.

intermediate annealing [MET] Softening of a metal by heat treatment at one or more stages during cold working and before fi-nal treatment.

intermediate flux [MET] A flux consisting of organic halide compounds whose resi-dues are decomposed by the heat of sol-dering; fluxing action approaches that of corrosive flux.

intermediate gear [MECH ENG] An idler gear interposed between a driver and driven gear.

intermediate haulage *See* relay haulage.

intermediate haulage conveyor [MIN ENG] A type of conveyor, usually 500 to 3000 feet (150 to 900 meters) in length, that trans-ports material between the gathering conveyor and the main haulage con-veyor.

intermediate phase [MET] In an alloy sys-tem, a distinct phase whose composition ranges do not extend to any of the pure constituents of the system.

intermetallic compound *See* electron com-pound.

intermittent defect [ENG] A defect that is not continuously present.

intermittent-duty rating [ENG] An output rating based on operation of a device for specified intervals of time rather than continuous duty. Also known as inter-mittent rating.

intermittent firing [MECH ENG] Cyclic firing whereby fuel and air are burned in a fur-nace for frequent short time periods.

intermittent gas lift [PETRO ENG] A gas-drive oil reservoir that is valved and timed for intermittent activity.

intermittent operation [ENG] Condition in which a device operates normally for a time, then becomes defective for a time, with the process repeating itself at reg-ular or irregular intervals.

intermittent rating *See* intermittent-duty rating.

intermittent weld [MET] A weld in which the continuity is broken by recurring unwelded spaces.

internal brake [MECH ENG] A friction brake in which an internal shoe follows the inner surface of the rotating brake drum, wedging itself between the drum and the point at which it is anchored; used in motor vehicles.

internal broaching [MECH ENG] The removal of material on internal surfaces, by means of a tool with teeth of progressively increasing size moving in a straight line or other prescribed path over the surface, other than for the origination of a hole.

internal combustion engine [MECH ENG] A prime mover in which the fuel is burned within the engine and the products of combustion serve as the thermodynamic fluid, as with gasoline and diesel engines.

internal floating-head exchanger [MECH ENG] Tube-and-shell heat exchanger in which the tube sheet (support for tubes) at one end of the tube bundle is free to move.

internal furnace [MECH ENG] A boiler furnace having a firebox within a water-cooled heating surface.

internal gas drive [PETRO ENG] A primary oil recovery process in which oil is displaced from the reservoir by the expansion of the gas originally dissolved in the liquid. Also known as dissolved-gas drive; gas depletion drive; solution gas drive.

internal gear [DES ENG] An annular gear having teeth on the inner surface of its rim.

internal grinder [MECH ENG] A machine designed for grinding the surfaces of holes.

internally fired boiler [MECH ENG] A fire-tube boiler containing an internal furnace which is water-cooled.

internal mix atomizer [MECH ENG] A type of pneumatic atomizer in which gas and liquid are mixed prior to the gas expansion through the nozzle.

internal oxidation [MET] The subsurface oxidation of components of an alloy due to oxygen diffusion into the metal.

internal spring safety relief valve [ENG] A spring-loaded valve with a portion of the operating mechanism located inside the pressure vessel.

internal thread [DES ENG] A screw thread cut on the inner surface of a hollow cylinder.

internal vibrator [MECH ENG] A vibrating device which is drawn vertically through placed concrete to achieve proper consolidation.

internal work [IND ENG] Manual work done by a machine operator while the machine is automatically operating. Also known as fill-up work; inside work.

international standard annealed copper [MET] An annealed pure copper having a resistivity of 1.7241 microhm-centimeter at 20°C, which is taken as 100% conductivity.

international thread [DES ENG] A standardized metric system in which the pitch and diameter of the thread are related, with the thread having a rounded root and flat crest.

interpass temperature [MET] In a multiple-pass weld, the lowest temperature of the deposited weld metal before the next run is started.

interplanetary flight [AERO ENG] Flight through the region of space between the planets, under the primary gravitational influence of the sun.

interplanetary probe [AERO ENG] An instrumented spacecraft that flies through the region of space between the planets.

interplanetary spacecraft [AERO ENG] A spacecraft designed for interplanetary flight.

interplanetary transfer orbit [AERO ENG] An elliptical trajectory tangent to the orbits of both the departure planet and the target planet.

interpulse time [MET] In resistance welding, the time between successive pulses of an impulse.

interrupted aging [MET] A technique for aging material in several steps; the material is brought to room temperature after each step.

interrupted dc tachometer [ENG] A type of impulse tachometer in which the frequency of pulses generated by the interrupted direct current of an ignition-circuit primary of an internal combustion engine is used to measure the speed of the engine.

interrupted quenching [MET] Quenching in which a material is intermittently removed from the quenching medium while it is still at a higher temperature than the medium.

interrupted screw [DES ENG] A screw with longitudinal grooves cut into the thread, and which locks quickly when inserted into a similar mating part.

intersect [ENG] To find a position by the triangulation method.

intersection [CIV ENG] **1.** A point of junction or crossing of two or more roadways. **2.** A surveying method in which a plane table is used alternately at each end of a measured baseline.

intersection angle [CIV ENG] The angle of deflection at the intersection point between the straights of a railway or highway curve.

intersection point [CIV ENG] That point where two straights or tangents to a railway or road curve would meet if extended.

interspace [BUILD] An air space.

interstellar probe [AERO ENG] An instrumentated spacecraft propelled beyond the solar system to obtain specific information about interstellar environment.

interstellar travel [AERO ENG] Space flight between stars.

interterminal switching [CIV ENG] Moving railroad cars from one line to another within a switching area.

intertube burner [MECH ENG] A burner which utilizes a nozzle that discharges between adjacent tubes.

interval timer [ENG] A device which operates a set of contacts during a preset time interval and, at the end of the interval, returns the contacts to their normal positions. Also known as timer.

in-the-seam mining [MIN ENG] The usual method of mining characterized by the driving of development shafts into the coal seam.

into the solid [MIN ENG] Of a shot, going into the coal beyond the point to which the coal can be broken by the blast. Also known as on the solid.

intraline distance [ENG] The minimum distance permitted between any two buildings within an explosives operating line; to protect buildings from propagation of explosions due to blast effect.

introductory column [MIN ENG] The highest and first column that is inserted in casing a borehole.

intrusion grouting [ENG] A method of placing concrete by intruding the mortar component in position and then converting it into concrete as it is introduced into voids.

intumescence [MATER] The property of a material to swell when heated; intumescent materials in bulk and sheet form are used as fireproofing agents.

invaded zone [PETRO ENG] Transitional downhole area between the area invaded completely by drilling mud and uncontaminated bulk of the reservoir.

invasion efficiency [PETRO ENG] Completeness of invasion of a reservoir formation by a fluid.

inventory [ENG] The amount of plastic in the heating cylinder or barrel in injection molding or extrusion.

inventory control [IND ENG] Systematic management of the balance on hand of inventory items, involving the supply, storage, distribution, and recording of items.

inverse cam [MECH ENG] A cam that acts as a follower instead of a driver.

inverse segregation [MET] Segregation in a cast metal in which an excess of lower-melting metal occurs in the earlier-freezing portions because liquid metal enters cavities developed in the earlier-solidified metal.

inversion [MECH ENG] The conversion of basic four-bar linkages to special motion linkage, slider-crank mechanism, and slow-motion mechanism by successively holding fast, as ground link, members of a specific linkage (as linkages, such as parallelogram drag link).

inversion temperature [ENG] The temperature to which one junction of a thermocouple must be raised in order to make the thermoelectric electromotive force in the circuit equal to zero, when the other junction of the thermocouple is held at a constant low temperature.

invert [CIV ENG] The floor or bottom of a conduit.

inverted arch [CIV ENG] An arch with the crown downward, below the line of the springings; commonly used in tunnels and foundations. Also known as inflected arch.

inverted engine [MECH ENG] An engine in which the cylinders are below the crankshaft.

inverted heading and bench method See heading–overhand bench method.

inverted siphon [CIV ENG] A pressure pipeline crossing a depression or passing under a highway; sometimes called a sag line from its U-shape.

invert level [ENG] The level of the lowest portion at any given section of a liquid-carrying conduit, such as a drain or a sewer, and which determines the hydraulic gradient available for moving the contained liquid.

investment casting [MET] A casting method designed to achieve high dimensional accuracy for small castings by making a mold of refractory slurry, which sets at room temperature, surrounding a wax pattern which is then melted out to leave a mold without joints.

investment compound [MATER] A mixture containing a refractory filler, a binder, and a liquid vehicle which is used to make molds for investment casting.

involute gear tooth [DES ENG] A gear tooth whose profile is established by an involute curve outward from the base circle.

involute spline [DES ENG] A spline having the same general form as involute gear teeth, except that the teeth are one-half the depth and the pressure angle is 30°.

involute spline broach [MECH ENG] A broach that cuts multiple keys in the form of internal or external involute gear teeth.

iodide process [MET] A refining process in which a metal, such as titanium or zirconium, is combined with iodine vapor and then the iodide volatilized and decomposed at high temperatures to yield a pure solid metal.

iodized oil [MATER] A thick, viscous, oily liquid with a garliclike odor that is an iodine addition product of vegetable oil, containing about 40% organically combined iodine; used in medicine as a radiopaque medium for radiography.

ion engine [AERO ENG] An engine which provides thrust by expelling accelerated or high velocity ions; ion engines using energy provided by nuclear reactors are proposed for space vehicles.

ion-exchange resin [MATER] A synthetic resin that can combine or exchange ions with a solution; such a resin produces the exchange of sodium for calcium ions in the softening of hard water.

ion implantation [ENG] A process of introducing impurities into the near-surface regions of solids by directing a beam of ions at the solid.

ion machining [ENG] Use of a high-velocity ion beam to remove material from a surface.

ion microprobe *See* secondary ion mass spectrometer.

ion microprobe mass spectrometer [ENG] A type of secondary ion mass spectrometer in which primary ions are focused on a spot 1–2 micrometers in diameter, mass-charge separation of secondary ions is carried out by a double focusing mass spectrometer or spectrograph, and a magnified image of elemental or isotopic distributions on the sample surface is produced using synchronous scanning of the primary ion beam and an oscilloscope.

ionogram [ENG] A record produced by an ionosonde, that is, a graph of the virtual height of the ionosphere plotted against frequency.

ionophone [ENG ACOUS] A high-frequency loudspeaker in which the audio-frequency signal modulates the radio-frequency supply to an arc maintained in a quartz tube, and the resulting modulated wave acts directly on ionized air to create sound waves.

ionosonde [ENG] A radar system for determining the vertical height at which the ionosphere reflects signals back to earth at various frequencies; a pulsed vertical beam is swept periodically through a fre-quency range from 0.5 to 20 megahertz, and the variation of echo return time with frequency is photographically recorded.

ion-permeable membrane [MATER] A film or sheet of a substance which is preferentially permeable to some species or types of ions.

ion probe *See* secondary ion mass spectrometer.

ion propulsion [AERO ENG] Vehicular motion caused by reaction from the high-speed discharge of a beam of electrically equally charged minute particles ejected behind the vehicle.

iridium [MET] A silver-white, brittle, hard metal used in jewelry, electric contacts, electrodes, resistance wires, and pen tips.

iridosmine [MET] A natural iridium-osmium alloy composed of 10–77% iridium, 17–80% osmium, 0–10% platinum, 0–17% rhodium, 0–9% ruthenium, 0–2% iron, and 0–1% copper; used for surgical needles and compass bearings and for hardening platinum.

iron [MET] A heavy, magnetic, malleable and ductile metal occurring in meteorites and combined in a wide range of ores and most igneous rocks; it is one of the most widely used metals, and plays a role in biological processes.

iron acetate liquor [MATER] Black liquor containing 5–5.5% iron and sometimes copperas or tannin; results from pyroligneous acid attack on iron filings; used for dyeing and printing with logwood, and as a mordant for alizarine and nitroso dyes. Also known as black liquor; black mordant; iron liquor.

iron alloy [MET] An alloy having iron as the principal component.

iron and steel sheet piling [MIN ENG] A technique that uses iron and steel piling instead of wood to drive a shaft through loose, wet ground near the surface.

Ironarc process [MET] An ultra-high-temperature smelting process employing plasma chemistry to process refractory materials, such as zirconia.

iron carbide *See* cementite.

iron castings [MET] Shapes cast in molds from iron.

iron cement [MATER] A mixture of small iron pieces with ammonium chloride, used to join iron or steel surfaces.

iron-Constantan [MET] Dual-metal combination for thermocouple junctions, used for temperature measurement in oxidizing or reducing atmospheres.

iron foundry [MET] A building in which iron castings are made.

ironing [MET] Reducing the wall thickness of a deep-drawn object by reducing the clearance between punch and die.

iron liquor *See* iron acetate liquor.

iron-nickel alloy [MET] An iron alloy containing 20–80% nickel; it has high permeability and low hysteresis losses at low flux densities and is more readily rolled into thin laminations than silicon steels.

iron-ore cement [MATER] A cement in which iron ore is used instead of clay or shale.

iron red [MATER] Any of the pigments made from red varieties of ferric oxide.

iron-retention agent [MATER] Complexing agent that ties dissolved iron into complex ions to prevent reprecipitation and wellbore plugging after well acidizing.

iron saffron See Indian red.

iron scurf [MATER] Glazing material used to color blue bricks; a mixture of stone and iron particles from grinding of gun barrels.

iron soldering [MET] Soldering in which a soldering iron provides the required heat.

iron whiskers [MET] Single-crystal pure iron filaments or fibers.

irradiation [ENG] The exposure of a material, object, or patient to x-rays, gamma rays, ultraviolet rays, or other ionizing radiation.

irreducible saturation [PETRO ENG] In a permeable reservoir, that condition in which the nonwetting phase saturation is so large that the wetting phase can be reduced no more.

irregular element [IND ENG] An element whose frequency of occurrence is irregular but predictable. Also known as incidental element.

irrespirable atmosphere [MIN ENG] Atmosphere in a coal mine requiring workers to wear breathing apparatus because of poisonous gas or insufficient oxygen as a result of an explosion from firedamp or coal dust, or mine fires.

irrigation [CIV ENG] Artificial application of water to arable land for agricultural use.

irrigation canal [CIV ENG] An artificial open channel for transporting water for crop irrigation.

irrigation pipe [CIV ENG] A conduit of connected pipes for transporting water for crop irrigation.

irritant gas [MATER] A nonlethal gas, causing irritation of the skin and flow of tears; any one of the family of tear gases used for training and riot control.

isano oil [MATER] A pale-yellow, viscous drying oil obtained from the nut of an African tree; used as a varnish oil.

isinglass [MATER] A gelatin made from the dried swim bladders of sturgeon and other fishes; used in glues, cements, and printing inks. Also known as fish gelatin; ichthyocolla.

isochronal test [PETRO ENG] Short-time back-pressure test for low-permeability reservoirs that otherwise require excessively long times for pressure stabilization when wells are shut in.

isochronous governor [MECH ENG] A governor that keeps the speed of a prime mover constant at all loads. Also known as astatic governor.

isocrackate [MATER] The liquid products of the process of isocracking.

isogor [PETRO ENG] Constant gas-oil ratio.

isogor map [PETRO ENG] Oil reservoir contour-line map that shows constant gas-oil ratios.

isokinetic sampling [ENG] Any technique for collecting airborne particulate matter in which the collector is so designed that the airstream entering it has a velocity equal to that of the air passing around and outside the collector.

isolated footing [CIV ENG] A concrete slab or block under an individual load or column.

isolation test [ENG] A leak detection method which isolates the evacuated system from the pump, followed by observation of the rate of pressure rise.

isolator [ENG] Any device that absorbs vibration or noise, or prevents its transmission.

isoperm [PETRO ENG] One of the lines of equal (constant) permeability plotted on a reservoir map.

isopotential map [PETRO ENG] A contour-line map to show the initial or calculated daily rate of oil well production in a multiwell field.

isosulfur map [PETRO ENG] A contour-line map to show the percentage of sulfur in underground crude oil.

isothermal annealing [MET] Transformation of an austenitic steel to ferrite and pearlite at constant temperature. Also known as isothermal transformation.

isothermal transformation [MET] See isothermal annealing.

isothermal treatment [MET] Heat treatment of metals at constant temperature.

ISP See imperial smelting process.

ivory [MATER] The ivory-white material composing the tusks and teeth of the elephant; specific gravity is 1.87; takes a high polish and is used for ornamental parts and art objects, and formerly for piano keys.

ivory black [MATER] Animal black made by calcining ivory; used as a pigment.

ivory board [MATER] A highly finished cardboard that is clay-coated on both sides; used for art printing and menu cards.

ivory point [ENG] A small pointer extending downward from the top of the cistern of a Fortin barometer; the level of the mercury in the cistern is adjusted so that it just comes in contact with the end of the pointer, thus setting the zero of the barometer scale.

Izod test [MET] An impact test in which a falling pendulum strikes a fixed, usually notched specimen with 120 foot-pounds (163 joules) of energy at a velocity of 11.5 feet (3.5 meters) per second; the height of the pendulum swing after striking is a measure of the energy absorbed and thus indicates impact strength.

J

jacaranda *See* carob wood.

jack [MECH ENG] A portable device for lifting heavy loads through a short distance, operated by a lever, a screw, or a hydraulic press.

jackbit [DES ENG] A drilling bit used to provide the cutting end in rock drilling; the bit is detachable and either screws on or is taper-fitted to a length of drill steel. Also known as ripbit.

jack chain [DES ENG] **1.** A chain made of light wire, with links arranged in figure-eights with loops at right angles. **2.** A toothed endless chain for moving logs.

jacket [MECH ENG] The space around an engine cylinder through which a cooling liquid circulates. [PETRO ENG] The support structure of a steel offshore production platform; it is fixed to the seabed by piling, and the superstructure is mounted on it.

jackhammer [MECH ENG] A hand-held rock drill operated by compressed air.

jack ladder [ENG] A V-shaped trough holding a toothed endless chain, and used to move logs from pond to sawmill.

jackleg [ENG] A supporting bar used with a jackhammer.

jack line [PETRO ENG] A steel cable or rod that connects the arms of the central pumping engine to the two or more wells that are being pumped.

jack plane [DES ENG] A general-purpose bench plane measuring over 1 foot (30 centimeters) in length.

jack post [MIN ENG] Timber used where a coal seam is separated by a rock band and one bench is loaded out before the other.

jack rafter [BUILD] A short, secondary, or simulated rafter.

jackscrew [MECH ENG] **1.** A jack operated by a screw mechanism. Also known as screw jack. **2.** The screw of such a jack.

jackshaft [MECH ENG] A countershaft, especially when used as an auxiliary shaft between two other shafts.

jack timber [MIN ENG] A timber such as a rafter that is shorter than others with which it is used.

jack truss [BUILD] A minor truss in a hip roof where the roof has a reduced section.

Jacobs taper [DES ENG] A machine tool used for mounting drill chucks in drilling machines.

jag bolt [DES ENG] An anchor bolt with barbs on a flaring shank.

jalap [MATER] An orange or reddish solid or a yellowish to brown powder with acrid taste and slight odor; the dried tuberous root of a Mexican plant (*Exogonium purga*), or the drug prepared from it; used as a cathartic in medicine.

jalousie [BUILD] A window that consists of a number of long, narrow panels, each hinged at the top.

jamb [BUILD] The vertical member on the side of an opening, as a door or window.

jamb brick [MATER] A brick with one rounded corner; used to provide a rounded edge on wall openings.

jamb liner [BUILD] A small strip of wood applied to the edge of a window jamb to increase its width for use in thicker walls.

James concentrator [MIN ENG] A concentration table whose deck is divided into two sections; one section contains riffles for the coarse material, and the other section is smooth to allow settling of the fine particles which will not settle on a riffled surface.

jam nut *See* locknut.

japan [MATER] A glossy, black baking paint or varnish that consists primarily of a hard asphalt base.

Japanese peppermint oil [MATER] An oil distilled from *Mentha arvensis*, grown in Japan, Brazil, and the United States; the oil has less odor than peppermint oil; used for the production of menthol.

japanning [MET] The finishing of metal objects with japan.

Japan paper [MATER] A special paper made with an irregular mottled effect on the surface; used for greeting cards and other decorative applications.

Japan tallow See Japan wax.

Japan wax [MATER] A pale-yellow wax with rancid aroma obtained from the berries of sumac; soluble in benzene and naphtha, insoluble in water; melts at 53°C; used in wax products, polishes, and soaps, and as a beeswax substitute. Also known as Japan tallow; sumac wax.

jar coupling See jars.

jardiniere glaze [MATER] A type of unfritted soft and hard lead glaze; contains oxides of lead, zinc, potassium, calcium, aluminum, and silicon.

jarosite process [MET] A zinc electrometallurgical process in which ferric ions are precipitated from feed solutions in the form of jarosite, a hydrous sulfate of iron and potassium or sodium.

jars [PETRO ENG] A series of links in the drill string to connect drill cables to the drill bit; sets up the uneven motion on the upstroke that helps free the string of tools. Also known as jar coupling.

jasmine oil [MATER] A colorless fragrant essential oil from flowers of a jasmine, as *Jasminum officinale* or *J. grandiflorum*; the oil is extracted from the flowers by enfleurage and is used in perfumery.

JATO engine [AERO ENG] Derived from jet-assisted-takeoff engine. **1.** An auxiliary jet-producing unit or units, usually rockets, for additional thrust. **2.** A JATO bottle or unit; the complete auxiliary power system used for assisted takeoff.

Java citronella oil See citronella oil.

jaw [ENG] A notched part that permits a railroad-car axle box to move vertically.

jawbreaker See jaw crusher.

jaw clutch [MECH ENG] A clutch that provides positive connection of one shaft with another by means of interlocking faces; may be square or spiral; the most common type of positive clutch.

jaw crusher [MECH ENG] A machine for breaking rock between two steel jaws, one fixed and the other swinging. Also known as jawbreaker.

J bolt [DES ENG] A J-shaped bolt, threaded on the long leg of the J.

J box See junction box.

jeep [MECH ENG] A one-quarter-ton, four-wheel-drive utility vehicle in wide use in all United States military services.

Jeffrey crusher [MIN ENG] A crusher to break soft minerals, such as limestone. Also known as whizzer mill.

Jeffrey diaphragm jig [MIN ENG] A plunger-type jig with the plunger beneath the screen.

Jeffrey molveyor [MIN ENG] A string of short conveyors on driven wheels connected together to run alongside a heading or room conveyor; used to keep a continuous miner in operation at all times.

Jeffrey single-roll crusher [MIN ENG] A simple type of crusher for coal, with a drum to which are bolted toothed segments designed to grip the coal, thus forcing it down into the crushing opening.

Jeffrey swing-hammer crusher [MIN ENG] A crusher with swing arms on a revolving shaft for crushing coal, ore, or other material against the iron casing of the crusher; a screen at the bottom allows sufficiently fine pieces to pass through.

Jeffrey-Traylor vibrating feeder [MIN ENG] A feed chute vibrated electromagnetically in a direction oblique to its surface; rate of movement of rock depends on amplitude and frequency of vibration.

Jeffrey-Traylor vibrating screen [MIN ENG] A vibrating screen whose action results from an oscillating armature and a stationary coil.

jellied gasoline See gelatinized gasoline.

jelutong See pontianak gum.

Jenner's stain See May-Grünwald stain.

Jeppel's oil See bone oil.

jerker line [PETRO ENG] A line that radiates from a common point of power to the jack of several wells, permitting the wells to be pumped by a single power unit.

jerk pump [MECH ENG] A pump that supplies a precise amount of fuel to the fuel injection valve of an internal combustion engine at the time the valve opens; used for fuel injection.

jet aircraft [AERO ENG] An aircraft with a jet engine or engines.

jet bit [DES ENG] A modification of a drag bit or a roller bit that utilizes the hydraulic jet principle to increase drilling rate.

jet compressor [MECH ENG] A device, utilizing an actuating nozzle and a combining tube, for the pumping of a compressible fluid.

jet condenser [MECH ENG] A direct-contact steam condenser utilizing the aspirating effect of a jet for the removal of noncondensables.

jet drilling [MECH ENG] A drilling method that utilizes a chopping bit, with a water jet run on a string of hollow drill rods, to chop through soils and wash the cuttings to the surface. Also known as wash boring.

jet engine [AERO ENG] An aircraft engine that derives all or most of its thrust by reaction to its ejection of combustion products (or heated air) in a jet and that obtains oxygen from the atmosphere for the combustion of its fuel. [MECH ENG] Any engine that ejects a jet or stream of gas or fluid, obtaining all or most of its thrust by reaction to the ejection.

jet-flame drill [MIN ENG] A mining drill that utilizes a high-velocity flame to spall out a hole.

jet flap [AERO ENG] A sheet of fluid discharged at high speed close to the trailing edge of a wing so as to induce lift over the whole wing.

jet fuel [MATER] Special grade of kerosine with a flash point of 125°F (52°C), used for jet aircraft; may have methane or naphthene added to produce a 110°F (43°C) flash point, for military aircraft.

jet hole [ENG] A borehole drilled by use of a directed, forceful stream of fluid or air.

jet mill *See* fluid-energy mill.

jet mixer [MECH ENG] A type of flow mixer or line mixer, depending on impingement of one liquid on the other to produce mixing.

jet molding [ENG] Molding method in which most of the heat is applied to the material to be molded as it passes through a nozzle or jet, rather than in a conventional heating cylinder.

jet nozzle [DES ENG] A nozzle, usually specially shaped, for producing a jet, such as the exhaust nozzle on a jet or rocket engine.

jet-piercing drill *See* fusion-piercing drill.

jet propulsion [AERO ENG] The propulsion of a rocket or other craft by means of a jet engine. [ENG] Propulsion by means of a jet of fluid.

jet pump [MECH ENG] A pump in which an accelerating jet entrains a second fluid to deliver it at elevated pressure.

jetsam [ENG] Articles that sink when thrown overboard, particularly those jettisoned for the purpose of lightening a vessel in distress.

jet spinning [ENG] Production of plastic fibers in which a directed blast or jet of hot gas pulls the molten polymer from a die lip; similar to melt spinning.

jet stream [AERO ENG] The stream of gas or fluid expelled by any reaction device, in particular the stream of combustion products expelled from a jet engine, rocket engine, or rocket motor.

jetting [CIV ENG] A method of driving piles or well points into sand by using a jet of water to break the soil. [ENG] During molding of plastics, the turbulent flow of molten resin from an undersized gate or thin section into a thicker mold section, as opposed to laminar, progressive flow.

jetting tool [PETRO ENG] Downhole device that jets a high-pressure, sand-laden fluid stream to clean out wellbore holes, to disintegrate perforating pipe, and to perform other operations.

jettison [ENG] The throwing overboard of objects, especially to lighten a craft in distress.

jetty *See* groin.

jewel [ENG] **1.** A bearing usually made of synthetic corundum and used in precision timekeeping devices, gyros, and other instruments. **2.** A bearing lining of soft metal, used in railroad cars, for example.

jewelry alloy [MET] Any ductile, malleable alloy, usually bronze, of good corrosion resistance, used as a base metal in jewelry.

jib boom [MECH ENG] An extension that is hinged to the upper end of a crane boom.

jib crane [MECH ENG] Any of various cranes having a projecting arm (jib).

jib end [MIN ENG] The delivery end in conveyor systems in which a jib is fitted to deliver the load in advance of and remote from the drive.

jig [ENG] A machine for dyeing piece goods by moving the cloth at full width (open width) through the dye liquor on rollers. [MECH ENG] A device used to position and hold parts for machining operations and to guide the cutting tool. [MIN ENG] A vibrating device in which coal is cleaned and ore is concentrated in water.

jig back [MECH ENG] An aerial ropeway with a pair of containers that move in opposite directions and are loaded or stopped alternately at opposite stations but do not pass around the terminals. Also known as reversible tramway; to-and-fro ropeway.

jig borer [MECH ENG] A machine tool resembling a vertical milling machine designed for locating and drilling holes in jigs.

jigger *See* jigging conveyor.

jigger boss [MIN ENG] A first-line supervisor in some western United States mines.

jiggering [ENG] A mechanization of the ceramic-forming operation consisting of molding the outside of a piece by throwing plastic clay on a plaster of paris mold, placing the mold and clay on a rotating head, and forming the inner surface by forcing a template or jigger knife against the clay; method used in mass-producing dinnerware.

jigging [MIN ENG] A gravity method which separates mineral from gangue particles by utilizing an effective difference in settling rate through a periodically dilated bed.

jigging conveyor [MIN ENG] A series of steel troughs suspended from the roof of the stope or laid on rollers on its floor, and given reciprocating motion mechanically, to move mineral. Also known as chute conveyor; jigger; pan conveyor.

jig grinder [MECH ENG] A precision grinding machine used to locate and grind holes to size, especially in hardened steels and carbides.

jigsaw [MECH ENG] A tool with a narrow blade suitable for cutting intricate curves and lines.

jig washer [MIN ENG] A coal or mineral washer for relatively coarse material; the broken ore is placed on a screen and pulsed vertically with water; the heavy portion passes through the screen and the light portion goes over the sides.

jim crow [DES ENG] A device with a heavy buttress screw thread used for bending rails by hand.

job [IND ENG] **1.** The combination of duties, skills, knowledge, and responsibilities assigned to an individual employee. **2.** A work order.

jobber's reamer [DES ENG] A machine reamer that is solid with straight or helical flutes and taper shanks.

job breakdown [IND ENG] Separation of an operation into elements. Also known as operation breakdown.

job characteristic *See* job factor.

job class [IND ENG] A group of jobs involving a similar type of work, difficulty of performance, or range of pay. Also known as job family; job grade; labor grade.

job classification [IND ENG] Designating job classes on the basis of job factors or level of pay, or on the basis of job evaluation.

job description [IND ENG] A detailed description of the essential activities required to perform a task.

job evaluation [IND ENG] Orderly qualitative appraisal of each job or position in an establishment either by a point system for the specific job characteristics or by comparison of job factors; used for establishing a job hierarchy and wage plans.

job factor [IND ENG] An essential job element which provides a basis for selecting and training employees and establishing the wage plan for the job. Also known as job characteristic.

job family *See* job class.

job grade *See* job class.

job plan [IND ENG] The organized approach to production management involving formal, step-by-step procedures.

joggle [DES ENG] **1.** A flangelike offset on a flat piece of metal. **2.** A projection or notch on a sheet of building material to prevent protrusion. **3.** A dowel for joining blocks of masonry. [MIN ENG] Trusses or sets of timbers joined for taking pressure at right angles.

Johansson block [DES ENG] A type of gage block ground to an accuracy of at least 1/100,000 inch (0.25 micrometer). Also known as Jo block.

Johnson concentrator [MIN ENG] A device used to separate heavy particles such as metallic gold from auriferous pulp; composed of a shell in the shape of a cylinder

that is lined with rubber grooves parallel to the inclined axis.

joint [ENG] The surface at which two or more mechanical or structural components are united.

joint bar [CIV ENG] A rigid steel member used in pairs to join, hold, and align rail ends.

joint buildup sequence [MET] The sequence in which weld beads are deposited relative to the cross section of a multiple-pass joint.

joint clearance [ENG] The distance between mating surfaces of a joint.

joint compound [MATER] A material used primarily to lubricate the threads of pipe joints and secondarily to prevent joint leakage.

joint efficiency [MET] A numerical value expressed as the ratio of the strength of a riveted, welded, or brazed joint to the strength of the parent metal.

jointer [ENG] **1.** Any tool used to prepare, make, or simulate joints, such as a plane for smoothing wood surfaces prior to joining them, or a hand tool for inscribing grooves in fresh cement. **2.** A file for making sawteeth the same height. **3.** An attachment to a plow that covers discarded material. **4.** A worker who makes joints, particularly a construction worker who cuts stone to proper fit.

jointer gage [DES ENG] An attachment to a bench vise that holds a board at any angle desired for planing.

jointing [CIV ENG] Caulking of masonry joints.

joint penetration [MET] The distance extended into a weld joint by the weld metal or fusion zone.

joist [CIV ENG] A steel or wood beam providing direct support for a floor.

Jolly balance [ENG] A spring balance used to measure specific gravity of mineral specimens by weighing a specimen when in the air and when immersed in a liquid of known density.

jolt molding [ENG] A process for shaping refractory blocks in which a mold containing prepared batch is jolted mechanically to consolidate the material.

Jominy end quench test *See* Jominy test.

Jominy test [MET] A hardenability test in which a steel bar is heated to the desired austenitizing temperature and quench-hardened at one end and then measured for hardness along its length, beginning at the quenched end. Also known as Jominy end quench test.

Jones riffle [MIN ENG] An apparatus used to reduce the size of a sample to a desired weight; consists of a hopper which passes samples to a series of open-bottom pockets, each of which divides the sample into two equal parts, and the next pass of each

used

part gives a quarter of the original sample, and so on until the desired sample is obtained.

Jones splitter [MIN ENG] A device used to reduce the volume of a sample, consisting of a belled, rectangular container, the bottom of which is fitted with a series of narrow slots or alternating chutes designed to cast material in equal quantities to opposite sides of the device.

Jones sucker rod [PETRO ENG] Connecting rod between the subsurface pump and the lifting or pumping device on the surface; serves to lift oil out of the cased hole.

jonquil oil [MATER] A colorless oil obtained from flowers of jonquil (*Narcissus jonquilla*); used in perfumes. Also known as narcissus oil.

jordan [MECH ENG] A machine or engine used to refine paper pulp, consisting of a rotating cone, with cutters, that fits inside another cone, also with cutters.

Jordan sunshine recorder [ENG] A sunshine recorder in which the time scale is supplied by the motion of the sun; it consists of two opaque metal semicylinders mounted with their curved surfaces facing each other; each of the semicylinders has a short narrow slit in its flat side; sunlight entering one of the slits falls on light-sensitive paper (blueprint paper) which lines the curved side of the semicylinder.

journal [MECH ENG] That part of a shaft or crank which is supported by and turns in a bearing.

journal bearing [MECH ENG] A cylindrical bearing which supports a rotating cylindrical shaft.

journal box [ENG] A metal housing for a journal bearing.

journal friction [MECH ENG] Friction of the axle in a journal bearing arising mainly from viscous sliding friction between journal and lubricant.

journal oil [MATER] A special grade of lubricating oil for use on journal bearings.

Joy double-ended miner [MIN ENG] A cutter-loader for continuous mining on a longwall face.

Joy extensible conveyor [MIN ENG] A type of belt conveyor consisting of a head section and a tail section, each mounted on crawler tracks and independently driven; used between a loader or continuous miner and the main transport.

Joy extensible steel band [MIN ENG] A hydraulically driven system linking a continuous miner to the main transport; the steel band is coiled on the drivehead.

Joy loader [MIN ENG] A loading machine which uses mechanical arms to collect coal or ore onto an apron that is pushed onto the broken material.

Joy longwall loading machine [MIN ENG] A modified Joy loader comprising a hydraulically elevated loading head fitted with mechanical gathering arms.

Joy microdyne [MIN ENG] A dust collector that wets and traps dust pulled into it and releases the dust as a slurry to be removed by pumps; used at the return end of tunnels or hard headings.

Joy miner [MIN ENG] A continuous miner weighing about 15 tons (13,600 kilograms) and made up of a turntable, a ripper bar, and a discharge boom conveyor; used mainly in coal headings and in extraction of coal pillars.

joystick [AERO ENG] A lever used to control the motion of an aircraft; fore-and-aft motion operates the elevators while lateral motion operates the ailerons. [ENG] A two-axis displacement control operated by a lever or ball, for XY positioning of a device or an electron beam.

Joy-Sullivan hydrodrill rig [MIN ENG] A drill rig set on a jib or boom which can be moved to and locked in any position by hydraulic power controlled from the drill carriage.

Joy transloader [MIN ENG] A rubber-tired, self-propelled machine for loading, transporting, and dumping.

Joy walking miner [MIN ENG] A continuous miner designed to make thin seams; a walking mechanism is used instead of caterpillar tracks for moving the miner.

JP-4 [MATER] Jet engine test fuel made up of 35% light petroleum distillates and 65% gasoline distillates.

JP-5 [MATER] Military jet engine fuel made of specially refined kerosine with specified flash and freezing points.

Judson powder [MATER] An explosive containing sodium nitrate, coal, sulfur, and some nitroglycerin.

jumbo *See* drill carriage.

jumper tube [MECH ENG] A short tube used to bypass the flow of fluid in a boiler or tubular heater.

jumping a claim [MIN ENG] **1.** Taking possession of a mining claim which has been abandoned. **2.** Taking possession of a mining claim that is liable to forfeiture because the requirements of the law are unfulfilled. **3.** Taking possession of a mining claim by stealth, fraud, or force.

junction [CIV ENG] A point of intersection of roads or highways, especially where one terminates.

junction box [ENG] A protective enclosure into which wires or cables are led and connected to form joints. Also known as J box.

juniper berry oil [MATER] Colorless oil that darkens and thickens in air; bitter taste, turpentinelike aroma; derived from the dried ripe fruit of the common juniper (*Juniperus communis*), used in medicine, gins, and liqueurs, and in veterinary practice. Also known as juniper oil.

juniper oil *See* juniper berry oil.

juniper tar oil *See* cade oil.

juniper wood oil [MATER] Oil made by diluting juniper berry oil with turpentine oil or by distilling turpentine oil over juniper wood; used as an external medicine and in veterinary practice.

Junkers engine [MECH ENG] A double-opposed-piston, two-cycle internal combustion engine with intake and exhaust ports at opposite ends of the cylinder.

jury rig [MIN ENG] Any makeshift or temporary device, rig, or piece of equipment.

jute board [MATER] A fiberboard made of jute fiber.

jute paper [MATER] A strong paper composed principally of jute fiber.

K

kadaya gum *See* karaya gum.

kalsomine *See* calcimine.

Kaplan turbine [MECH ENG] A propeller-type hydraulic turbine in which the positions of the runner blades and the wicket gates are adjustable for load change with sustained efficiency.

kapoc oil *See* kapok oil.

kapok oil [MATER] Yellow-green oil with pleasant aroma and taste; soluble in alcohol, ether, and chloroform; derived by pressing seeds of the kapok tree; used in edible oils and soap stock. Also spelled kapoc oil.

karat [MET] A unit for designating the fineness of gold in an alloy; represents a twenty-fourth part; thus, 18-karat gold is 18/24 or 75% pure.

karaya gum [MATER] The exudation from the Indian tree (*Sterculia urens*); white to dark-colored tears; used in pharmaceuticals, textiles, and foods, and as a tragacanth gum substitute. Also known as Indian tragacanth; kadaya gum.

Kata thermometer [ENG] An alcohol thermometer used to measure low velocities in air circulation, by heating the large bulb of the thermometer above 100°F (38°C) and noting the time it takes to cool from 100 to 95°F (38 to 35°C) or some other interval above ambient temperature, the time interval being a measure of the air current at that location.

katchung oil *See* peanut oil.

katharometer [ENG] An instrument for detecting the presence of small quantities of gases in air by measuring the resulting change in thermal conductivity of the air. Also known as thermal conductivity cell.

Kauertz engine [MECH ENG] A type of cat-and-mouse rotary engine in which the pistons are vanes which are sections of a right circular cylinder; two pistons are attached to one rotor so that they rotate with constant angular velocity, while the other two pistons are controlled by a gear-and-crank mechanism, so that angular velocity varies.

kauri gum [MATER] Hard copal resins from kauri pine (*Agathis australis*); used in lacquers and varnishes.

KC-97 *See* Stratofreighter.

KC-135 *See* Stratotanker.

keel block [CIV ENG] docking block used to support a ship's keel. [MET] A simple shape from which a test casting is made in the form of a large head, which is removed and discarded, with a keel on the bottom.

Keene's cement [MATER] An anhydrous calcined gypsum mixed with an accelerator; used as a hard-finish plaster.

keeps *See* folding boards.

kellering [MECH ENG] Three-dimensional machining of a contoured surface by tracer-milling the die block or punch; the cutter path is controlled by a tracer that follows the contours on a die model.

Kelly [PETRO ENG] A pipe attached to the top of a drill string and turned during drilling; transmits twisting torque from the rotary machinery to the drill string and ultimately to the bit.

Kelly ball test [ENG] A test for the consistency of concrete using the penetration of a half sphere; a 1-inch (2.5 centimeters) penetration by the Kelly ball corresponds to about 2 inches (5 centimeters) of slump.

kelly bushing [PETRO ENG] A device added to the rotary table through which the kelly passes, so that the torque of the table is transmitted to the kelly and the drill.

Kennedy key [DES ENG] A square taper key fitted into a keyway of square section and driven from opposite ends of the hub.

kep interlock [MIN ENG] A system designed to prevent the lowering of a shaft conveyance before all keps are fully withdrawn, and to indicate the position of the keps.

keps *See* folding boards.

kerf [ENG] A cut made in wood, metal, or other material by a saw or cutting torch. [MIN ENG] A narrow, deep cut made in the face of coal to facilitate mining.

Kern counter *See* dust counter.

kerosine [MATER] A refined petroleum fraction used as a fuel for heating and cooking, jet engines, lamps, and weed burning and as a base for insecticides; specific gravity is about 0.8; components are mostly paraffinic and naphthenic hydrocarbons in the C_{10} to C_{14} range. Also known as lamp oil.

kerosine distillate [MATER] The distilled cut in the 150–300°C range from petroleum or shale oil; used as lamp, stove, or illuminating oil, as a solvent, and as a component of jet aircraft fuels. Also known as burning oil.

kerosine propellant [MATER] A propellant consisting of highly refined, low-aromatic liquid propellant distillate with a gravity range not exceeding three degrees American Petroleum Institute gravity at 60°F (15.6°C); may contain additives.

Kew barometer [ENG] A type of cistern barometer; no adjustment is made for the variation of the level of mercury in the cistern as pressure changes occur; rather, a uniformly contracting scale is used to determine the effective height of the mercury column.

key [BUILD] **1.** Plastering that is forced between laths to secure the rest of the plaster in place. **2.** The roughening on a surface to be glued or plastered to increase adhesiveness. [CIV ENG] A projecting portion that serves to prevent movement of parts at a construction joint. [DES ENG] **1.** An instrument that is inserted into a lock to operate the bolt. **2.** A device used to move in some manner in order to secure or tighten. **3.** One of the levers of a keyboard. **4.** See machine key. [ENG] The pieces of core causing a block in a core barrel, the removal of which allows the rest of the core in the barrel to slide out. [PETRO ENG] A hooklike wrench fitted to the square of a sucker rod to pull and run each sucker rod of a pumping oil well.

keyboard [ENG] A set of keys or control levers having a systematic arrangement and used to operate a machine or other piece of equipment such as a typewriter, typesetter, processing unit of a computer, or piano.

keyboard perforator [ENG] A typewriterlike device that prepares punched paper tape for communications or computing equipment.

key cut [MIN ENG] In strip mine operations, the section excavated adjacent to the new highwall.

keyhole [DES ENG] A hole or a slot for receiving a key. [MET] A welding method wherein the heat source, because of its concentration, causes a hole through the surface immediately ahead of the molten weld metal in the direction of welding; the hole is filled as the welding progresses, ensuring complete joint penetration.

keyhole saw [DES ENG] A fine compass saw with a blade 11–16 inches (28–41 centimeters) long.

keyhole specimen [MET] A metal specimen containing a keyhole shaped notch and used in certain impact tests.

keying [CIV ENG] Establishing a mechanical bond in a construction joint.

key job [IND ENG] A job that has been evaluated and is considered representative of similar jobs in the same labor market and is used as a benchmark to evaluate the similar jobs and to establish non-key-job wages.

key joint [CIV ENG] A mortar joint with a concave pointing.

key seat See keyway.

keyseater [MECH ENG] A machine for milling beds or grooves in mechanical parts which receive keys.

keystone [MATER] Small crushed stone used as filler for the large aggregate in bituminous bound roads.

keyway [DES ENG] **1.** An opening in a lock for passage of a flat metal key. **2.** The pocket in the driven element to provide a driving surface for the key. **3.** A groove or channel for a key in any mechanical part. Also known as key seat. [ENG] An interlocking channel or groove in a cement or wood joint to provide reinforcement.

kibble See hoppit.

kickback [MECH ENG] A backward thrust, such as the backward starting of an internal combustion engine as it is cranked, or the reverse push of a piece of work as it is fed to a rotary saw.

kickdown [MECH ENG] **1.** Shifting to lower gear in an automotive vehicle. **2.** The device for shifting.

kick over [MECH ENG] To start firing; applied to internal combustion engines.

kickpipe [BUILD] A short pipe protecting an electrical cable at the point where it emerges from a floor.

kickplate [BUILD] A plate used on the bottom of doors and cabinets or on the risers of steps to protect them from shoe marks.

Kick's law [ENG] The law that the energy needed to crush a solid material to a specified fraction of its original size is the same, regardless of the original size of the feed material.

kick wheel [ENG] A potter's wheel worked by a foot pedal.

kill [MATER] To treat in such a way as to destroy undesirable properties; for example, neutralization of an acid by the addition of an alkali. [MET] To add a strong deoxidizer, such as silicon or aluminum, to molten steel in order to stop the reaction between carbon and oxygen forming gaseous carbon monoxide and dioxide during solidification. [PETRO ENG] **1.** In drilling, to prevent well blowout by appropriate measures. **2.** In oil production, to halt well production so that reconditioning of the well may proceed.

killed steel [MET] Thoroughly deoxidized steel, for example, by additions of aluminum or silicon, in which the reaction between carbon and oxygen during solidification is suppressed.

kiln [ENG] A heated enclosure used for drying, burning, or firing materials such as ore or ceramics.

Kind-Chaudron process [MIN ENG] A technique used to sink a large-diameter deep shaft; a pilot shaft of smaller diameter is first dug, then enlarged until the full diameter is reached; a lining with a moss box at the bottom is forced into place when water is found.

kingpin [MECH ENG] The pin for articulation between an automobile stub axle and an axle-beam or steering head. Also known as swivel pin.

king post truss [BUILD] A wooden roof truss having two principal rafters held by a horizontal tie beam, a king post upright between tie beam and ridge, and usually two struts to the rafters from a thickening at the king post foot.

kink [ENG] A tightened loop in a wire rope resulting in permanent deformation and damage to the wire.

Kirkendall effect [MET] The phenomenon whereby a marker placed at the interface between an alloy and a metal moves toward the alloy region when the temperature of the system is raised to the point where diffusion can occur.

Kiruna method [MIN ENG] A borehole-inclination survey method whereby the electrolytic deposition of copper from a solution is used to make a mark on the inside of a metal container.

kish [MET] Free graphite that floats to the surface of molten hypereutectic cast iron as it cools.

kiss-roll coating [ENG] Procedure for coating a substrate web in which the coating roll carries a metered film of coating material; part of the film transfers to the web, part remains on the roll.

klaxon [ENG ACOUS] A diaphragm horn sometimes operated by hand.

Klinkenberg correction [PETRO ENG] Mathematical conversion of laboratory air-permeability measurements (made on formation material) into equivalent liquid-permeability values.

klydonograph [ENG] A device attached to electric power lines for estimating certain electrical characteristics of lightning by means of the figures produced on photographic film by the lightning-produced surge carried over the lines; the size of the figure is a function of the potential and polarity of the lightning discharge.

K Monel [MET] A nonmagnetic, age-hardenable alloy of nickel (28–34%) and copper and 2.75% aluminum that can be heat-treated after finishing to produce a material that is both corrosion-resistant and extra strong.

kneaded eraser [MATER] An eraser made of unvulcanized rubber whose shape can be readily changed by the user for removing pencil marks from paper.

knee [MECH ENG] In a knee-and-column type of milling machine, the part which supports the saddle and table and which can move vertically on the column. [MET] The lower supporting structure for an arm in a resistance welding machine.

knee brace [BUILD] A stiffener between a column and a supported truss or beam to provide greater rigidity in a building frame under transverse loads.

knee pad [ENG] A protective cushion, usually made of sponge rubber, that can be strapped to a worker's knee.

knee rafter [BUILD] A brace placed diagonally between a principal rafter and a tie beam.

knee tool [MECH ENG] A tool holder with a shape resembling a knee, such as the holder for simultaneous cutting and interval operations on a screw machine or turret lathe.

knee wall [BUILD] A partition that forms a side wall or supports roof rafters under a pitched roof.

knife [DES ENG] A sharp-edged blade for cutting.

knife coating [ENG] Procedure for coating a continuous-web substrate in which coating thickness is controlled by the distance between the substrate and a movable knife or bar.

knife-edge [DES ENG] A sharp narrow edge resembling that of a knife, such as the fulcrum for a lever arm in a measuring instrument.

knife-edge bearing [MECH ENG] A balance beam or lever arm fulcrum in the form of a hardened steel wedge; used to minimize friction.

knife-edge cam follower [DES ENG] A cam follower having a sharp narrow edge or point like that of a knife; useful in developing cam profile relationships.

knife file [DES ENG] A tapered file with a thin triangular cross section resembling that of a knife.

knife-line attach [MET] Intergranular corrosion of an alloy adjacent to a weldment after heating the joint above the sensitization temperature.

knob [DES ENG] A component that is placed on a control shaft to facilitate manual rotation of the shaft; sometimes has a pointer or markings to indicate shaft position.

knocker *See* shell knocker.

knock intensity [ENG] The intensity of knock (detonation) recorded when testing a motor gasoline for octane or knock rating.

knockmeter [ENG] A fuels-testing device used to measure the output of the detonation meter used in ASTM knock-test ratings of motor fuels.

knock-off [MECH ENG] **1.** The automatic stopping of a machine when it is operating improperly. **2.** The device that causes automatic stopping.

knockout [ENG] A partially cutout piece in metal or plastic that can be forced out when a hole is needed.

knockout pin *See* ejector pin.

knock rating [ENG] Rating of gasolines according to knocking tendency.

knock suppressor [MATER] A material added to motor fuel to retard or prevent detonation and resultant knock in reciprocating internal combustion engines; an example is tetraethyllead.

Knoop hardness [MET] The relative microhardness of a material, such as metal, determined by the Knoop indentation test.

Knoop indentation test [MET] A diamond pyramid hardness test employing the Knoop indenter; hardness is determined by the depth to which the Knoop indenter penetrates.

Knoop indenter [MET] A diamond indenter which has a rhombic base with diagonals in a 1:7 ratio and included apical angles of $130°$ and $172°30'$; used in the Knoop indentation test.

knot [MATER] A scar on lumber marking a place where a branch grew out of the tree truck.

Knox and Oxborne furnace [MIN ENG] A continuously working shaft furnace for roasting quicksilver ores, having the fireplace built in the masonry at one side; the fuel is wood.

knuckle [MIN ENG] The place on an incline where there is a sudden change in grade.

knuckle joint [DES ENG] A hinge joint between two rods in which an eye on one piece fits between two flat projections with eyes on the other piece and is retained by a round pin.

knuckle joint press [MECH ENG] A short-stroke press in which the slide is actuated by a crank attached to a knuckle joint hinge.

knuckle man [MIN ENG] A worker who connects mine cars to and disconnects them from cables and also couples cars into trains.

knuckle pin [DES ENG] The pin of a knuckle joint.

knuckle post [MECH ENG] A post which acts as the pivot for the steering knuckle in an automobile.

Knudsen gage [ENG] An instrument for measuring very low pressures, which measures the force of a gas on a cold plate beside which there is an electrically heated plate.

Knudsen reversing water bottle [ENG] A type of frameless reversing bottle for collecting water samples; carries reversing thermometers.

Knudsen vacuum gage [ENG] Device to measure negative gas pressures; a rotatable vane is moved by the pressure of heated molecules, proportionately to the concentration of molecules in the system.

knurl [ENG] To provide a surface, usually a metal, with small ridges or knobs to ensure a firm grip or as a decorative feature.

Koch freezing process [MIN ENG] A process used to sink a shaft through a formation such as clay that will not sustain a shaft; magnesium chloride cooled to about $-30°C$ is circulated through pipes sunk in the ground until the ground is frozen.

Koehler lamp [MIN ENG] A naphtha-burning flame safety lamp for use in gaseous mines.

Koepe hoist *See* Koepe winder.

Koepe shear [MIN ENG] A wheel used in place of a winding drum in the Koepe winder; made up of a cast steel hub with steel arms and a welded rim.

Koepe winder brake [MIN ENG] A device that works directly on the Koepe shear to slow or stop the hoist; can be applied by the engineman's brake lever or by safety devices.

Koepe winder [MIN ENG] A hoisting system in which the winding drum is replaced by large wheels or sheaves over which passes an endless rope. Also known as Koepe hoist.

Kollsman window [AERO ENG] A small window on the dial face of an aircraft pressure altimeter in which the altimeter setting in inches of mercury is indicated.

Kondo alloy [MET] A dilute alloy of a magnetic material in a nonmagnetic host which exhibits the Kondo effect.

Kondo effect [MET] The large anomalous increase in the resistance of certain dilute alloys of magnetic materials in nonmagnetic hosts as the temperature is lowered.

Kondo temperature [MET] The temperature below which the Kondo effect predominates for a specified magnetic impurity and host material.

konimeter [ENG] An air-sampling device used to measure dust as in a cement mill or a mine; a measured volume of air drawn through a jet impacts on a glycerin-jelly-coated glass surface; the particles are counted with a microscope.

koniscope [ENG] An instrument which indicates the presence of dust particles in the atmosphere. Also spelled coniscope.

Korfmann arch saver [MIN ENG] A machine that uses a controlled hydraulic system to withdraw steel arches.

Korfmann power loader [MIN ENG] A cutter-loader that is able to cart and load in both directions; its components are four drilling heads and one cutter chain surrounding them.

Kourbatoff's reagents [MET] Etching agents used for microanalysis of carbon steels; there are four different formulations, three with nitric acid, one with hydrochloric acid.

Kozeny's equation [PETRO ENG] Mathematical relationship of flow network permeability to capillary pore dimensions; used for reservoir calculations.

kraft paper [MATER] A strong paper or cardboard made from sulfate-process wood pulp; unbleached varieties are used for wrapping paper and shipping cartons.

K ratio [AERO ENG] The ratio of propellant surface to nozzle throat area.

Krause rolling mill [MET] A type of rolling mill in which the rolls translate as well as rotate, accomplishing a high reduction of thickness for the single passage of a metal sheet.

Kroll process [MET] A reduction process for the production principally of titanium metal sponge from titanium tetrachloride.

Krupp ball mill [MIN ENG] An ore pulverizer in which the grinding is done by chilled iron or steel balls of various sizes moving against each other and the die ring, composed of five perforated spiral plates, each of which overlaps the next; material is discharged through a cylindrical screen.

kryptoclimate *See* cryptoclimate.

K truss [BUILD] A building truss in the form of a K due to the orientation of the vertical member and two oblique members in each panel.

Kullenberg piston corer [MECH ENG] A piston-operated coring device used to obtain 2-inch-diameter (5-centimeter-diameter) core samples.

kytoon [AERO ENG] A captive balloon used to maintain meteorological equipment aloft at approximately a constant height; it is streamlined, and combines the aerodynamic properties of a balloon and a kite.

L

Labarraque's solution [MATER] Aqueous solution of 4–6% sodium hypochlorite and 4–6% sodium chloride with sodium hydroxide or sodium carbonate stabilizer; used as disinfectant.

labdanum oil [MATER] An essential oil, golden yellow with ambergris aroma; soluble in alcohol, chloroform, and ether; derived from gum resin of various rockroses, such as *Cistus ladaniferus;* used in perfumes. Also known as ladanum oil.

labeled cargo [IND ENG] Cargo of a dangerous nature, such as explosives and flammable or corrosive liquids, which is designated by different-colored labels to indicate the requirements for special handling and storage.

labor cost [IND ENG] That part of the cost of goods and services attributable to wages, especially for direct labor.

labor grade *See* job class.

labor relations [IND ENG] The management function that deals with a company's work force; usually the term is restricted to relations with organized labor.

La Bour centrifugal pump [MIN ENG] A self-priming centrifugal pump with a trap which ensures sufficent water for the pump to function, and a separator to remove the entrained air in the water.

labyrinth [ENG ACOUS] A loudspeaker enclosure having air chambers at the rear that absorb rearward-radiated acoustic energy, to prevent it from interfering with the desired forward-radiated energy.

labyrinth seal [ENG] A minimum-leakage seal that offers resistance to fluid flow while providing radial or axial clearance; a labyrinth of circumferential knives or touch points provides for successive expansion of the fluid being piped; used for gas pipes, steam engines, and turbines.

lac [MATER] A resinous material secreted by some insects that live on the sap of certain trees, principally in India; used in the manufacture of shellac.

lacing [CIV ENG] **1.** A lightweight metallic piece that is fixed diagonally to two channels or four angle sections, forming a composite strut. **2.** A course of brick, stone, or tiles in a wall of rubble to give strength. **3.** A course of upright bricks forming a bond between two or more arch rings. **4.** Distribution steel in a slab of reinforced concrete. **5.** A light timber fastened to pairs of struts or walings in the timbering of excavations (including mines).

lacmus *See* litmus.

lacquer [MATER] A material which contains a substantial quantity of a cellulose derivative, most commonly nitrocellulose but sometimes a cellulose ester, such as cellulose acetate or cellulose butyrate, or a cellulose ether such as ethyl cellulose; used to give a glossy finish, especially on brass and other bright metals.

lacquer diluent [MATER] An organic liquid with no solvent power added to lacquer formulations to reduce viscosity and to adjust flow or other properties.

lacrimator *See* tear gas.

LACT *See* lease automatic custody transfer.

lactometer [ENG] A hydrometer used to measure the specific gravity of milk.

lactoprene [MATER] Any of several synthetic rubbers that have good resistance to hydrocarbon oil, ozone, oxygen, and other weather elements excepting cold, and that are polymers or copolymers of an acrylic acid ester.

ladanum oil *See* labdanum oil.

ladder [ENG] A structure, often portable, for climbing up and down; consists of two parallel sides joined by a series of crosspieces that serve as footrests.

ladder-bucket dredge *See* bucket-ladder dredge.

ladder dredge *See* bucket-ladder dredge.

ladder drilling [MECH ENG] An arrangement of retractable drills with pneumatic powered legs mounted on banks of steel ladders connected to a holding frame; used in large-scale rock tunneling, with the advantage that many drills can be worked at the same time by a small labor force.

ladder jack [ENG] A scaffold support which hooks onto a ladder.

ladder track [CIV ENG] A main track that joins successive body tracks in a railroad yard.

ladder trencher [MECH ENG] A machine that digs trenches by means of a bucket-ladder excavator. Also known as ladder ditcher.

ladderway [MIN ENG] Also known as ladder road; manway. **1.** Mine shaft between two main levels, equipped with ladders. **2.** The particular shaft, or compartment of a shaft, containing ladders.

ladle [DES ENG] A deep-bowled spoon with a long handle for dipping up, transporting, and pouring liquids. [MET] A receptacle for transporting and pouring molten metal.

Lafarge cement [MATER] A cement made of plaster of paris, lime, and marble powder; used in mortar for marble and limestone pieces because it is nonstaining.

lagan [ENG] A heavy object thrown overboard and buoyed to mark its location for future recovery.

lag bolt *See* coach screw.

lagging [CIV ENG] **1.** Horizontal wooden strips fastened across an arch under construction to transfer weight to the centering form. **2.** Wooden members positioned vertically to prevent cave-ins in earthworking. [MATER] Asbestos and magnesia plaster that is used as a thermal insulation on process equipment and piping.

lag screw *See* coach screw.

laid paper [MATER] A paper with a pattern of parallel lines spaced so as to give a ribbed effect.

laitance [MATER] Weak material, consisting principally of lime, that is formed on the surface of concrete, especially when excess water is mixed with the cement.

lake [MATER] Any of a large group of dyes that have been combined with or adsorbed by salts of calcium, barium, chromium, aluminum, phosphotungstic acid, or phosphomolybdic acid; used for textile dyeing. Also known as color lake.

lake copper [MET] A pure type of copper produced from ores taken from the Lake Superior region; has high conductivity.

lally column [CIV ENG] A hollow and nearly circular steel column that supports girders or beams.

lambda dispatch [IND ENG] The solution of the problem of finding the most economical use of generators to supply a given quantity of electric power, using the method of Lagrange multipliers, which are symbolized λ.

lamella arch [CIV ENG] An arch consisting basically of a series of intersecting skewed arches made up of relatively short straight members; two members are bolted, riveted, or welded to a third piece at its center.

lamella roof [BUILD] A large span vault built of members connected in a diamond pattern.

lamina [MATER] A flat or curved arrangement of unidirectional or woven fibers in a matrix.

laminar flow control [AERO ENG] The removal of a small amount of boundary-layer air from the surface of an aircraft wing with the result that the airflow is laminar rather than turbulent; frictional drag is greatly reduced.

laminar wing [AERO ENG] A low-drag wing in which the distribution of thickness along the chord is so selected as to maintain laminar flow over as much of the wing surface as possible.

laminate [MATER] A sheet of material made of several different bonded layers.

laminated composite [MATER] A composite material consisting of layers of various materials.

laminated glass *See* nonshattering glass.

laminated metal [MET] A sheet or bar of composite metal composed of two or more bonded layers.

laminated plastic [MATER] A thin sheet made of superposed layers of plastic bonded or impregnated with resin or compressed under heat.

laminated spring [DES ENG] A flat or curved spring made of thin superimposed plates and forming a cantilever or beam of uniform strength.

laminated wood [MATER] Board or timber composed of layers of wood glued together with the grains parallel.

lamination [MATER] One of the thin punchings of iron or steel used in building up a laminated core for a magnetic circuit.

lamp [ENG] A device that produces light, such as an electric lamp.

lampblack [MATER] A grayish-black amorphous, practically pure form of carbon made by burning oil, coal tar, resin, or other carbonaceous substance in an insufficient supply of air; used in making paints, lead pencils, metal polishes, electric brush carbons, crayons, and carbon papers.

lamp-charging rack [MIN ENG] Mine-lamp-charging racks which allow miners to store lamp units for recharging after daily use.

lamphouse [ENG] **1.** The light housing in a motion picture projector, located behind the projector head ordinarily consisting of a carbon arc lamp operating on direct current at about 60 volts, a concave reflector behind the arc which collects the light and concentrates it on the film, and cooling devices. **2.** A box with a small hole containing an electric lamp and a concave mirror behind it, used as a concentrated source of light in a microscope, photographic enlarger, or other instrument.

lamping [MIN ENG] Use of a portable ultraviolet lamp to reveal fluorescent minerals in prospecting.

lampman [MIN ENG] A person responsible for maintaining and servicing miners' lamps.

lamp oil *See* kerosine.

lamp room [MIN ENG] A room or building at the surface of a mine for charging, servicing, and issuing all cap, hand, and flame safety lamps. Also known as lamp cabin; lamp station.

lampshade paper [MATER] Paper that is translucent and either flame-resistant or flame-retardant; often made of wood pulp, vegetable parchment, or laminated glassine.

lamp station *See* lamp room.

Lancashire boiler [MECH ENG] A cylindrical steam boiler consisting of two longitudinal furnace tubes which have internal grates at the front.

lance [MET] To cut into but not through the piece of work.

lance door [MECH ENG] The door to a boiler furnace through which a hand lance is inserted.

Lanchester balancer [MECH ENG] A device for balancing four-cylinder engines; consists of two meshed gears with eccentric masses, driven by the crankshaft.

land [AERO ENG] Of an aircraft, to alight on land or a ship deck. [DES ENG] The top surface of the tooth of a cutting tool, behind the cutting edge. [ENG] **1.** In plastics molding equipment, the horizontal bearing surface of a semipositive or flash mold to allow excess material to escape; or the bearing surface along the top of the screw flight in a screw extruder; or the surface of an extrusion die that is parallel to the direction of melt flow. **2.** The surface between successive grooves of a diffraction grating or phonograph record.

land accretion [CIV ENG] Gaining land in a wet area, such as a marsh or by the sea, by planting maritime plants to encourage silt deposition or by dumping dredged materials in the area. Also known as land reclamation.

land drainage [CIV ENG] The removal of water from land to improve the soil as a medium for plant growth and a surface for land management operations.

lander [MIN ENG] **1.** A worker at one of the levels of a mine shaft to unload rock and to load drilling and blasting supplies to be lowered. **2.** In the quarry industry, one who guides, steadies, and loads trucks or railroad cars with the blocks of stone hoisted from the quarry floor. Also known as top hooker. **3.** In metal mining, one who cleans skips by directing a blast of compressed air into them through a hose, records number of loaded skips hoisted to surface, and loads railroad cars with ore from bins. **4.** In coal mining, one who works with shaft sinking crew at the top of the shaft or at a level immediately above shaft bottom, dumping rock into mine cars from a bucket in which it is raised. Also known as bucket dumper; landing tender; top lander. **5.** The worker who receives the loaded bucket or tub at the mouth of the shaft. Also known as banksman.

landfill [CIV ENG] Disposal of solid waste by burying in layers of earth in low ground.

landing [CIV ENG] A place where boats receive or discharge passengers, freight, and so on. [MIN ENG] **1.** Level stage in a shaft at which cages are loaded and discharged. **2.** The top or bottom of a slope, shaft, or inclined plane.

landing area [AERO ENG] An area intended primarily for landing and takeoff of aircraft.

landing circle [AERO ENG] The approximately circular path flown by an airplane to get into the landing pattern; used particularly with naval aircraft landing on an aircraft carrier.

landing flap [AERO ENG] A movable airfoil-shaped structure located aft of the rear beam or spar of the wing; extends about two-thirds of the span of the wing and functions to substantially increase the lift, permitting lower takeoff and landing speeds.

landing gear [AERO ENG] Those components of an aircraft or spacecraft that support and provide mobility for the craft on land, water, or other surface.

landing light [AERO ENG] One of the floodlights mounted on the leading edge of the wing and below the nose of the fuselage to enable an airplane to land at night.

landing load [AERO ENG] The load on an aircraft's wings produced during landing; depends on descent velocity and landing attitude.

landing stage [CIV ENG] A platform, usually floating and attached to the shore, for the discharge and embarkation of passengers, freight, and so on.

landing strip [AERO ENG] A portion of the landing area prepared for the landing and takeoff of aircraft in a particular direction; it may include one or more runways. Also known as air strip.

landing tee *See* wind tee.

landing tender *See* lander.

landmark [ENG] Any fixed natural or artificial monument or object used to designate a land boundary.

land plaster [MATER] Finely ground gypsum, used as a fertilizer and as a corrective for soil with excess sodium and potassium carbonates.

land reclamation *See* land accretion.

landscape architecture [CIV ENG] The art of arranging and fitting land for human use and enjoyment.

landscape engineer [CIV ENG] A person who applies engineering principles and methods to planning, design, and construction of natural scenery arrangements on a tract of land.

land surveyor [CIV ENG] A specialist who measures land and its natural features and any man-made features such as buildings or roads for drawing to scale as plans or maps.

land tie [CIV ENG] A rod or chain connecting an outside structure such as a retaining wall to a buried anchor plate.

land-use classes [CIV ENG] Categories into which land areas can be grouped according to present or potential economic use.

lane [CIV ENG] An established route, as an air lane, shipping lane, or highway traffic lane.

lang lay [DES ENG] A wire rope lay in which the wires of each strand are twisted in the same direction as the strands.

lanolin [MATER] The hydrous sheep's-wool wax (primarily cholesterol esters of higher fatty acids) derived as a by-product from the preparation of raw wool for the spinner; used as a base for emollients in cosmetics and shampoos.

lantern [ENG] A portable lamp.

lantern pinion [DES ENG] A pinion with bars (between parallel disks) instead of teeth.

lantern ring [DES ENG] A ring or sleeve around a rotating shaft; an opening in the ring provides for forced feeding of oil or grease to bearing surfaces; particularly effective for pumps handling liquids.

lanthanum [MET] A white, soft, malleable metal; tarnishes in moist air; a major component of misch metal.

lanthanum-doped lead zirconate–lead titanate *See* lead lanthanum zirconate titanate.

lap [CIV ENG] The length by which a reinforcing bar must overlap the bar it will replace. [MATER] An abrasive material used for lapping. [MET] A defect caused by folding and then rolling or forging a hot metal fin or corner onto a surface without welding. Also known as fold.

lapel microphone [ENG ACOUS] A small microphone that can be attached to a lapel or pocket on the clothing of the user, to permit free movement while speaking.

lap joint [ENG] A simple joint between two members made by overlapping the ends and fastening them together with bolts, rivets, or welding.

lapping [MET] Polishing with a material such as cloth, lead, plastic, wood, iron, or copper having fine abrasive particles incorporated in or rubbed into the surface.

lap-rivet [MET] To rivet a lap joint.

lap siding [BUILD] Beveled boards used for siding that are similar to clapboards but longer and wider. [CIV ENG] Two railroad sidings, the turnout of one overlapping that of the other.

lap weld [MET] A welded lap joint.

Laray viscometer [ENG] An instrument designed to measure viscosity and other properties of ink.

lard oil [MATER] Yellowish to colorless oil with characteristic aroma and bland taste; melts at $-2°C$; soluble in carbon disulfide, ether, benzene, and chloroform; main components are olein and glycerides of solid fatty acids; used as a lubricant, wool oil, and illuminant, and in soap manufacture.

larry [MIN ENG] **1.** A car with a hopper bottom and adjustable chutes for feeding coke ovens. Also known as lorry. **2.** *See* barney.

Larsen's pile [MIN ENG] A collection of hollow cylinders that increases resistance against bending and crumpling; useful for sinking a shaft in sand or gravel.

Larsen's spiles [MIN ENG] Steel sheet made in various forms to resist bending; used in place of wooden spiles in timbering a weak roof.

larvicide [MATER] A pesticide used to kill larvae.

laryngophone [ENG ACOUS] A microphone designed to be placed against the throat of a speaker, to pick up voice vibrations directly without responding to background noise.

Lasater's bubble-point pressure correction [PETRO ENG] Relation of the gas-oil ratio in a high-pressure oil reservoir to the bubble-point-pressure factor.

laser anemometer [ENG] An anemometer in which the wind being measured passes through two perpendicular laser beams, and the resulting change in velocity of one or both beams is measured.

laser ceilometer [ENG] A ceilometer in which the time taken by a light pulse from a ground laser to travel straight up to a cloud ceiling and be reflected to a receiving photomultiplier is measured and converted into a cathode-ray display that indicates cloud-base height.

laser cutting [MET] A process by which a laser beam impinges on the workpiece in order to heat and sever the piece.

laser earthquake alarm [ENG] An early-warning system proposed for earthquakes, involving the use of two lasers with beams at right angles, positioned across a known geologic fault for continuous monitoring of distance across the fault.

laser glazing [MET] A surface alloying process in which a continuous high-energy carbon dioxide laser traverses the surface of a metal part, creating a thin layer of melt.

laser gyro [ENG] A gyro in which two laser beams travel in opposite directions over a ring-shaped path formed by three or more mirrors; rotation is thus measured without the use of a spinning mass. Also known as ring laser.

laser intrusion-detector [ENG] A photoelectric intrusion detector in which a laser is a light source that produces an extremely narrow and essentially invisible beam around the perimeter of the area being guarded.

laserscope [ENG] A pulsed high-power laser used with appropriate scanning and imaging devices to sense objects over the sea at night or in fog and provide three-dimensional images on a viewing screen.

laser scriber [ENG] A laser-cutting setup used in place of a diamond scriber for dicing thin slabs of silicon, gallium arsenide, and other semiconductor materials used in the production of semiconductor diodes, transistors, and integrated circuits; also used for scribing sapphire and ceramic substrates.

laser seismometer [ENG] A laser interferometer system that detects seismic strains in the earth by measuring changes in distance between two granite piers located at opposite ends of an evacuated pipe through which a helium-neon or other laser beam makes a round trip; movements as small as 80 nanometers (one-eighth the wavelength of the 632.8-nanometer helium-neon laser radiation) can be detected.

laser tracking [ENG] Determination of the range and direction of a target by echoed coherent light.

laser transit [ENG] A transit in which a laser is mounted over the sighting telescope to project a clearly visible narrow beam onto a small target at the survey site.

laser welding [MET] Micro-spot welding with a laser beam.

lashing [ENG] A rope, chain, or wire used for binding, fastening, or wrapping. [MIN ENG] Planks nailed inside of frames or sets in a shaft to keep them in place. Also known as listing.

lashing chain [MIN ENG] A short chain to attach tubs to an overrope in endless rope haulage by wrapping the chain around the rope.

lash-up [ENG] A model or test sample of equipment required in the testing of a new concept or idea which is in the embryo stage.

last in, first out [IND ENG] A method of determining the inventory costs by transferring the costs of material to the product in reverse chronological order. Abbreviated LIFO.

latch [ENG] Any of various closing devices on a door that fit into a hook, notch, or cavity in the frame. [MIN ENG] To make an underground survey with a dial and chain, or to mark out upon the surface, with the same instruments, the position of the workings underneath.

latch bolt [DES ENG] A self-acting spring bolt with a beveled head.

latent defect [IND ENG] A flaw or other imperfection in any article which is discovered after delivery; usually, latent defects are inherent weaknesses which normally are not detected by examination or routine tests, but which are present at time of manufacture and are aggravated by use.

latent load [MECH ENG] Cooling required to remove unwanted moisture from an air-conditioned space.

lateral [ENG] In a gas distribution or transmission system, a pipe branching away from the central, primary part of the system. [MIN ENG] **1.** In horizon mining, a hard heading branching off a horizon along the strike of the seams. **2.** A horizontal mine working.

lateral acceleration [AERO ENG] The component of the linear acceleration of an aircraft or missile along its lateral, or Y, axis.

lateral compliance [ENG ACOUS] That characteristic of a stylus based on the force required to move it from side to side as it follows the grooves of a phonograph record.

lateral controller [AERO ENG] A primary flight control mechanism, generally a part of the longitudinal controller, which controls the ailerons; often resembles an automobile steering wheel but may be a control column.

lateral extensometer [ENG] An instrument used in photoelastic studies of the stresses on a plate; it measures the change in the thickness of the plate resulting from the stress at various points.

lateral flow spillway *See* side-channel spillway.

lateral recording [ENG ACOUS] A type of disk recording in which the groove modulation is parallel to the surface of the recording medium so that the cutting stylus moves from side to side during recording.

lateral search *See* profiling.

lateral sewer [CIV ENG] A sewer discharging into a branch or other sewer and having no tributary sewer.

lateral support [CIV ENG] Horizontal propping applied to a column, wall, or pier across its smallest dimension.

laterlog [ENG] A downhole resistivity measurement method wherein electric current is forced to flow radially through the formation in a sheet of predetermined thickness; used to measure the resistivity in hard-rock reservoirs as a method of determining subterranean structural features.

latex [MATER] **1.** Milky colloid in which natural or synthetic rubber or plastic is suspended in water. **2.** An elastomer product made from latex.

latex cement [MATER] A highly adhesive solvent solution of latex.

latex paint [MATER] A paint consisting of a water suspension or emulsion of latex combined with pigments and additives such as binders and suspending agents. Also known as latex water paint.

latex water paint *See* latex paint.

lath [CIV ENG] **1.** A narrow strip of wood used in making a level base, as for plaster or tiles, or in constructing a light framework, as a trellis. **2.** A sheet of material used as a base for plaster.

lath brick [MATER] A long, narrow brick.

lath crib *See* lath frame.

lath door-set *See* lath frame.

lathe [MECH ENG] A machine for shaping a workpiece by gripping it in a holding device and rotating it under power against a suitable cutting tool for turning, boring, facing, or threading.

lath frame [MIN ENG] A weak construction of laths, surrounding a main crib, the space between being for the insertion of piles. Also known as lath crib; lath door-set.

lathing board *See* backup strip.

latrine [ENG] A toilet facility, either fixed or of a portable nature, such as is maintained underground for use by miners.

latrine cleaner [MIN ENG] A laborer who brings toilet cars in a mine to the surface on a cage and flushes the contents into a sewer. Also called sanitary nipper.

latten [MET] A thin metal sheet, particularly of brass or similar alloy, hot-rolled steel, or tin-covered iron, used for ornamental purposes.

lattice [CIV ENG] A network of crisscrossed strips of metal or wood.

lattice girder [CIV ENG] An open girder, beam, or column built from members joined and braced by intersecting diag-

onal bars. Also known as open-web girder.

lattice truss [CIV ENG] A truss that resembles latticework because of diagonal placement of members connecting the upper and lower chords.

lauan [MATER] Wood from any of several genera of trees of the Philippines, Malaya, and Sarawak; resembles mahogany but shrinks and swells more with changes in moisture.

launch [AERO ENG] **1.** To send off a rocket vehicle under its own rocket power, as in the case of guided aircraft rockets, artillery rockets, and space vehicles. **2.** To send off a missile or aircraft by means of a catapult or by means of inertial force, as in the release of a bomb from a flying aircraft. **3.** To give a space probe an added boost for flight into space just before separation from its launch vehicle.

launch complex [AERO ENG] The composite of facilities and support equipment needed to assemble, check out, and launch a rocket vehicle.

launching [CIV ENG] The act or process of floating a ship after only hull construction is completed; in some cases ships are not launched until after all construction is completed.

launching angle [AERO ENG] The angle between the horizontal plane and the longitudinal axis of a rocket or missile at the time of launching.

launching cradle [CIV ENG] A framework made of wood to support a vessel during launching from sliding ways.

launching ramp [AERO ENG] A ramp used for launching an aircraft or missile into the air.

launching site [AERO ENG] **1.** A site from which launching is done. **2.** The platform, ramp, rack, or other installation at such a site.

launching ways [CIV ENG] Two (or more) sets of long, heavy timbers arranged longitudinally under the bottom of a ship during building and launching, with one set on each side, and sloping toward the water; the lower set, or ground ways, remain stationary and support the upper set, or sliding ways, which carry the weight of the ship after the shores and keel blocks are removed.

launch pad [AERO ENG] The load-bearing base or platform from which a rocket vehicle is launched. Also known as pad.

launch vehicle [AERO ENG] A rocket or other vehicle used to launch a probe, satellite, or the like. Also known as booster.

launch window [AERO ENG] The time period during which a spacecraft or missile must be launched in order to achieve a desired encounter, rendezvous, or impact.

launder [ENG] An inclined channel or trough for the conveyance of a liquid, such as for water in mining and construction engineering or for molten metal.

launder screen [MIN ENG] A screen used in a launder for the sizing and dewatering of small sizes of anthracite.

launder separation process [MIN ENG] A hydraulic process for separating heavy gravity product from the lighter product that flows above it; a stream of fluid carries material down a channel that has draws to separate the heavy material from the light material.

launder washer [MIN ENG] A type of coal washer in which the coal is separated from the refuse by stratification due to hindered settling while being carried in aqueous suspension through the launder.

laundry blue [MATER] Dye-containing solution used to give a blue tint to laundry-yellowed white cottons and linens; usually contains Prussian blue.

laurel wax *See* bayberry wax.

Lauson engine [ENG] Single-cylinder engine used in screening tests prior to the L-series lube oil tests (such as L-1 or L-2 tests).

lautal [MET] A hard aluminum alloy with small percentages of copper and silicon and traces of iron, manganese, or magnesium.

Laval nozzle *See* de Laval nozzle.

lavender oil [MATER] Colorless to yellow or green-yellow essential oil with sweet aroma and bitter taste; distilled from fresh flowers of several species of lavender (*Lavandula*); main components are linalool, linalyl acetate, geraniol, cumarin, furfurol, and borneol; used in perfumery, and in medicine as a stimulant.

lavender spike oil *See* spike oil.

Laves phases [MET] Alloy phases which have the general formula AB_2, and the crystal structures of either $MgCu_2$ (cubic) or $MgZn_2$ or $MgNi_2$ (both hexagonal).

lawnmower [ENG] A helix-type recorder mechanism. [MECH ENG] A machine for cutting grass on lawns.

lay [DES ENG] The direction, length, or angle of twist of the strands in a rope or cable. [MET] The direction of the prevailing surface pattern on a piece of metal after grinding, cutting, lapping, or other processing. [MIN ENG] A share of profit.

lay-by [MIN ENG] Siding in single-track underground tramming road.

layer [MET] The stratum of weld metal consisting of one or more passes and lying parallel to the welding surface.

layering of firedamp [MIN ENG] The formation of a layer of firedamp at the roof of a mine working and above the ventilating air current.

layer loading [MIN ENG] A procedure whereby the coal is placed in the railroad cars in horizontal layers.

lay off [ENG] The process of fairing a ship's lines or an airplane's in a mold loft in order to make molds and templates for structural units.

lay-up [ENG] Production of reinforced plastics by positioning the reinforcing material (such as glass fabric) in the mold prior to impregnation with resin.

lazy jack [ENG] A device that accommodates changes in length of a pipeline or similar structure through the motion of two linked bell cranks.

LCL *See* less-than-carload.

L/D ratio [ENG] Length to diameter ratio, a frequently used engineering relationship.

leaching [MIN ENG] Dissolving soluble minerals or metals out of the ore, as by the use of percolating solutions such as cyanide or chlorine solutions, acids, or water. Also known as lixiviation.

leach ion-exchange flotation process [MET] A method of extraction developed for treatment of copper ores not amenable to direct flotation; the metal is dissolved by leaching, for example, with sulfuric acid, in the presence of an ion-exchange resin; the resin recaptures the dissolved metal and is then recovered in a mineralized froth by the flotation process.

leach material [MIN ENG] Material sufficiently mineralized to be economically recoverable by leaching.

leach pile [MIN ENG] Mineralized materials stacked so as to permit wanted minerals to be effectively and selectively dissolved by leaching.

lead [DES ENG] The distance that a screw will advance or move into a nut in one complete turn. [ENG] A mass of lead attached to a line, as used for sounding at sea. [MET] A soft, heavy metal with a silvery-bluish color; when freshly cut it is malleable and ductile; occurs naturally, mostly in combination; used principally in alloys in pipes, cable sheaths, type metal, and shields against radioactivity.

lead angle [DES ENG] The angle that the tangent to a helix makes with the plane normal to the axis of the helix. [MET] The angle at the point of welding between an electrode and a line perpendicular to the weld axis.

lead-base babbitt [MET] Alloy of 10–15% antimony, 2–10% tin, up to 0.2% copper, sometimes with arsenic, the remainder being lead; used as a bearing metal; a variation used in diesel engine and railway bearings contains alkaline-earth metals. Also known as white-metal bearing alloy.

lead-base grease [MATER] A mixture of soap and mineral oil, usually prepared by the reaction of lead oxide and fatty acid; holds up to extreme pressure and is useful as a gear lubricant.

lead bronze [MET] An alloy of 60–70% copper, up to 2% nickel, and up to 15% tin with the balance lead; used as a bearing metal.

lead curve [CIV ENG] The curve in a railroad turnout between the switch and the frog.

leaded alloy [MET] An alloy, especially of brass, bronze, or steel, to which lead is added to improve machinability and mechanical properties.

leaded gasoline [MATER] Motor gasoline into which a small amount of TEL (tetraethyllead) has been added to increase octane number or rating.

leaded zinc oxide [MATER] A mixture of zinc oxide and basic lead sulfate; used in paints.

lead foil [MET] A foil of lead or of lead alloy containing, for example, 10–12% tin and 1% copper.

lead glass [MATER] Glass into which lead oxide is incorporated to give high refractive index, optical dispersion, and surface brilliance; used in optical glass.

lead-in-air indicator [MIN ENG] An instrument that utilizes reagents to measure the concentration of lead in the air.

leading edge [AERO ENG] The front edge of an airfoil or wing. [DES ENG] The surfaces or inset cutting points on a bit that face in the same direction as the rotation of the bit.

leading edge slat [AERO ENG] A small airfoil attached to the leading edge of a wing of an aircraft that automatically improves airflow at large angles of attack.

leading heading [MIN ENG] **1.** The heading of a pair of parallel headings that is a short distance in front of the other; used to drain a mining area. **2.** A heading dug into the solid coal before the advance of the general face.

lead-in groove [DES ENG] A blank spiral groove at the outside edge of a disk recording, generally of a pitch much greater than that of the recorded grooves, provided to bring the pickup stylus quickly to the first recorded groove. Also known as lead-in spiral.

leading truck [MECH ENG] A swiveling frame with wheels under the front end of a locomotive.

lead-in spiral *See* lead-in groove.

lead joint [ENG] A pipe joint made by caulking with lead wool or molten lead.

lead lanthanum zirconate titanate [MATER] A ferroelectric, ceramic, electrooptical material whose optical properties can be changed by an electric field or by being placed in tension or compression; used in optoelectronic storage and display devices. Abbreviated PLZT. Also known as lanthanum-doped lead zirconate–lead titanate.

lead line *See* sounding line.

lead lining [ENG] Lead sheeting used to line the inside surfaces of liquid-storage vessels and process equipment to prevent corrosion.

lead metallurgy [MET] The science and technology of lead.

lead naphthenate [MATER] Soft, combustible, alcohol-soluble, transparent, resinous material, melting about 100°C; made from addition of lead salt to solution of sodium naphthenate; used as a paint and varnish drier, wood preservative, catalyst, insecticide, and lubricating oil additive.

lead-out groove [DES ENG] A blank spiral groove at the end of a disk recording, generally of a pitch much greater than that of the recorded grooves, connected to either the locked or eccentric groove. Also known as throw-out spiral.

lead-over groove [DES ENG] A groove cut between separate selections or sections on a disk recording to transfer the pickup stylus from one cut to the next. Also known as cross-over spiral.

lead rail [CIV ENG] In an ordinary rail switch, the turnout rail lying between the rails of the main track.

lead screw [MECH ENG] A threaded shaft used to convert rotation to longitudinal motion; in a lathe it moves the tool carriage when cutting threads; in a disk recorder it guides the cutter at a desired rate across the surface of an ungrooved disk.

lead-silver babbitt [MET] Alloy of lead with 10–15% antimony, 2.5–5.1% silicon, up to 5% tin, and up to 0.2% copper; used as a bearing metal.

lead-soap lubricant [MATER] Hard, high-melting-point, extreme-pressure lubricant made of lead salts saponified with fats.

lead solder [MET] Solder composed of a lead alloy.

lead time [IND ENG] The time allowed or required to initiate and develop a piece of equipment that must be ready for use at a given time.

lead track [CIV ENG] A distance measured along a straight railroad track from a switch to a frog.

lead welding [MET] Welding of lead by fusion. Incorrectly known as lead burning.

lead wire [ENG] One of the heavy wires connecting a firing switch with the cap wires.

lead wool [MATER] A coarse lead fiber used to caulk pipe joints.

lead zirconate titanate [MATER] A ferroelectric, ceramic, electrooptic material that has lower optical transparency than lead lanthanum zirconate titanate but similar other properties. Abbreviated PZT.

leaf [BUILD] **1.** A separately movable division of a folding or sliding door. **2.** One of a pair of doors or windows. **3.** One of the two halves of a cavity wall.

leaf spring [DES ENG] A beam of cantilever design, firmly anchored at one end and with a large deflection under a load. Also known as flat spring.

leakage [ENG] Undesired and gradual escape or entry of a quantity, such as loss of neutrons by diffusion from the core of a nuclear reactor, escape of electromagnetic radiation through joints in shielding, flow of electricity over or through an insulating material, and flow of magnetic lines of force beyond the working region. [MIN ENG] An unintentional diversion of ventilation air from its designed path.

leakage intake system [MIN ENG] A ventilation circuit with two adjacent intake roadways leading to the coal face.

leakage rate [ENG] Flow rate of all leaks from an evacuated vessel.

leak detector [ENG] An instrument used for finding small holes or cracks in the walls of a vessel; the helium mass spectrometer is an example.

leak test pressure [MECH ENG] The inlet pressure used for a standard quantitative seat leakage test.

lean [MATER] **1.** Of concrete or mortar, containing little or insufficient cement. **2.** Of clay, deficient in plasticity. **3.** Of coal, having little or no volatile matter. **4.** Of lime, containing impurities. **5.** Of fuel mixture, expecially for internal combustion engines, being low in combustible component. **6.** Of ore, being low-grade.

lean fuel mixture *See* lean mixture.

lean gas [MATER] **1.** In natural-gas absorption processes (such as natural-gasoline recovery from natural gas), gas from which desired liquid components have been removed. **2.** Natural gas poor in butane-and-heavier liquids.

leaning wheel grader [CIV ENG] A grader with skewed wheels to help cut or spread the soil.

lean mixture [MECH ENG] A fuel-air mixture containing a low percentage of fuel and a high percentage of air, as compared with a normal or rich mixture. Also known as lean fuel mixture.

lean oil [MATER] Absorbent oil from which absorbed gas has been stripped; an example is absorber oil in a natural-gaso-

line plant from which absorbed liquids (ethane, propane, butane) have been removed.

lean-to [BUILD] A single-pitched roof whose summit is supported by the wall of a higher structure.

leapfrog system [MIN ENG] A system used in mining coal on a longwall face; self-advancing supports are used, with alternate supports advancing as each web of coal is removed.

lease [IND ENG] **1.** Contract between landowner and another granting the latter the right to use the land, usually upon payment of an agreed rental, bonus, or royalty. **2.** A piece of land that is leased.

lease automatic custody transfer [PETRO ENG] Automatic unattended system to receive and record oil produced from a drilling lease, then to transfer the contents to a pipeline. Abbreviated LACT.

lease-distribution system [PETRO ENG] Any of the electrical distribution systems serving oil-field pump motors and other electrical equipment.

lease tank [PETRO ENG] An oil-field storage tank that stores oil flowing from designated wells.

leather [MATER] Dressed hide or skin of an animal.

lecithin [MATER] **1.** A mixture of phosphatides and oil obtained by drying the separate gums from the degumming of soybean oil; consists of the phosphatides (lecithin), cephalin, other fatlike phosphorus-containing compounds, and 30–35% entrained soybean oil; may be treated to produce more refined grades; used in foods, cosmetics, and paints. Also known as commercial lecithin; crude lecithin; soybean lecithin; soy lecithin. **2.** A waxy mixture of phosphatides obtained by refining commercial lecithin to remove the soybean oil and other materials; used in pharmaceuticals. Also known as refined lecithin.

ledeburite [MET] The eutectic of the iron-carbon system, the constituents being cementite and austenite at high temperatures; cooling decomposes the austenite to ferrite and cementite.

ledge [BUILD] A horizontal timber on the back of a batten door or on a framed and braced door. [ENG] **1.** A raised edge or molding. **2.** A narrow shelf projecting from the side of a vertical structure. **3.** A horizontal timber that supports the putlogs of scaffolding. [MET] Ingate for a foundry mold. [MIN ENG] **1.** A mining quarry exposure. **2.** An outcrop associated with a mineral deposit.

ledged door *See* batten door.

ledger [CIV ENG] A main horizontal member of formwork, supported on uprights and supporting the soffit of the formwork. [ENG] The horizontal support for a scaffold platform.

Ledoux bell meter [ENG] A type of manometer used to measure the difference in pressure between two points generated by any one of several types of flow measurement devices such as a pitot tube; it is equipped with a shaped plug which makes the reading of the meter directly proportional to the flow rate.

Lee-Norse miner [MIN ENG] A continuous miner for driving headings in medium or thick coal seams.

left-hand [DES ENG] Of drilling and cutting tools, screw threads, and other threaded devices, designed to rotate clockwise or cut to the left.

left-handed *See* left-laid.

left-hand screw [DES ENG] A screw that advances when turned counterclockwise.

left-laid [DES ENG] The lay of a wire or fiber rope or cable in which the individual wires or fibers in the strands are twisted to the right and the strands to the left. Also known as left-handed; regular-lay left twist.

leg [ENG] **1.** Anything that functionally or structurally resembles an animal leg. **2.** One of the branches of a forked or jointed object. **3.** One of the main upright members of a drill derrick or tripod. [MECH ENG] The case that encloses the vertical part of the belt carrying the buckets within a grain elevator. [MET] In a fillet weld, the distance between the root and the toe. [MIN ENG] **1.** In mine timbering, a prop or upright member of a set or frame. **2.** A stone that has to be wedged out from beneath a larger one.

leg wire [ENG] One of the two wires forming a part of an electric blasting cap or squib.

Lehigh jig [MIN ENG] A plunger-type jig in which check valves open on the upstroke of the plunger, makeup water is introduced with the feed, the screen plate is at two levels, and the bottom of the discharge end is hinged; used to wash anthracite.

Lehmann process [MIN ENG] A process for treating coal by disintegration and separation of the petrographic constituents.

LEM *See* lunar excursion module.

lemongrass oil [MATER] An essential oil with the odor of lemon, distilled from either of two lemongrasses (*Cymbopogon citratus* or *C. flexuosus*) in the East Indies; contains citral, citronellol, and geraniol; used as a perfume and as a flavoring.

lemon oil [MATER] A yellow essential oil squeezed from lemon rind; it is high in citral, limonene, terpinol, and citronellol; used in soap, perfumes, and flavors.

lengthening joint [ENG] A joint between two members running in the same direction.

length of lay [DES ENG] The distance measured along a line parallel to the axis of the rope in which the strand makes one complete turn about the axis of the rope, or the wires make a complete turn about the axis of the strand.

length of shot [ENG] The depth of the shothole, in which powder is placed, or the size of the block of coal or rock to be loosened by a single blast, measured parallel with the hole. [MIN ENG] In open-pit mining, the distance from the first drill hole to the last drill hole along the bank.

lens *See* acoustic lens.

lens tissue [MATER] Specially prepared paperlike material for cleaning lenses.

Leon firedamp tester [MIN ENG] A device, based on a form of Wheatstone bridge, that is used to detect firedamp.

LES *See* Lincoln experimental satellite.

less-than-carload [IND ENG] Too light to fill a freight car and therefore not eligible for carload rate. Abbreviated LCL.

letdown [AERO ENG] Gradual and orderly reduction in altitude, particularly in preparation for landing.

letters patent *See* patent.

levee [CIV ENG] **1.** A dike for confining a stream. **2.** A pier along a river.

level [CIV ENG] **1.** A surveying instrument with a telescope and bubble tube used to take level sights over various distances, commonly 100 feet (30 meters). **2.** To make the earth surface horizontal. [DES ENG] A device consisting of a bubble tube that is used to find a horizontal line or plane. Also known as spirit level. [MIN ENG] **1.** Mine workings that are at the same elevation. **2.** A gutter for the water to run in.

level crosscut [MIN ENG] A horizontal crosscut.

level drive [MIN ENG] A horizontal shaft which allows access to the length of a deposit and forms the basis for the splitting of the deposit into levels.

leveled elemental time *See* normal elemental time.

leveled time *See* normal time.

leveler [ENG] A back scraper, drag, or other form of device for smoothing land.

level indicator [ENG] An instrument that indicates liquid level. [ENG ACOUS] An indicator that shows the audio voltage level at which a recording is being made; may be a volume-unit meter, neon lamp, or cathode-ray tuning indicator.

leveling [ENG] Adjusting any device, such as a launcher, gun mount, or sighting equipment, so that all horizontal or vertical angles will be measured in the true horizontal and vertical planes. [IND ENG]

A method of performance rating which seeks to rate the principal factors that cause the speed of motions rather than speed itself; it considers that the level at which the operator works is influenced by effort and skill. [MET] Flattening rolled sheet by evening out irregularities, using a roller or tensile straining. [MIN ENG] Measurement of rises and falls, heights, and contour lines.

leveling action [MET] The property exhibited by a plating solution in making the coating smoother than the base metal.

leveling instrument [ENG] An instrument for establishing a horizontal line of sight, usually by means of a spirit level or a pendulum device.

leveling screw [ENG] An adjusting screw used to bring an instrument into level.

level off [AERO ENG] To bring an aircraft to level flight after an ascent or descent.

level-off position [AERO ENG] That position over which a craft ends an ascent or descent and begins relatively horizontal motion.

level rod [ENG] A straight rod or bar, with a flat face graduated in plainly visible linear units with zero at the bottom, used in measuring the vertical distance between a point of the earth's surface and the line of sight of a leveling instrument that has been adjusted to the horizontal position.

level surface [ENG] A surface which is perpendicular to the plumb line at every point.

level valve [MECH ENG] A valve operated by a lever which travels through a maximum arc of 180°.

lever [ENG] A rigid bar, pivoted about a fixed point (fulcrum), used to multiply force or motion; used for raising, prying, or dislodging an object.

lever shears [DES ENG] A shears in which the input force at the handles is related to the output force at the cutting edges by the principle of the lever. Also known as alligator shears; crocodile shears.

levigated abrasive [MATER] A fine abrasive powder for final burnishing of metals or for metallographic polishing; the powder is usually processed to make it chemically neutral.

levisticum oil *See* lovage oil.

levitated vehicle [MECH ENG] A train or other vehicle which travels at high speed at some distance above an electrically conducting track by means of levitation.

levitation [MIN ENG] In froth flotation, raising of particles in a froth to the surface of the pulp, to facilitate separation of selected minerals in the froth.

levitation heating [MET] Providing heat through high-frequency magnetic fields; employed in levitation melting.

levitation melting [MET] Melting metal out of contact with a supporting material by using the induced current provided by a high-frequency surrounding magnetic field to suspend the melt.

lewis [DES ENG] A device for hoisting heavy stones; employs a dovetailed tenon that fits into a mortise in the stone.

lewis bolt [DES ENG] A bolt with an enlarged, tapered head that is inserted into masonry or stone and fixed with lead; used as a foundation bolt.

lewis hole [MIN ENG] A series of two or more holes drilled as closely together as possible, but then connected by knocking out the thin partition between them, thus forming one wide hole having its greatest diameter in a plane with the desired rift.

lewis pin [MIN ENG] A pin used for attachment to a key block; it is placed in a shallow drill hole with a wedge at either side.

L-head engine [MECH ENG] A type of four-stroke cycle internal combustion engine having both inlet and exhaust valves on one side of the engine block which are operated by pushrods actuated by a single camshaft.

licanic oil [MATER] A fatty acid found in drying oils that are in paint formulations.

lichen blue *See* litmus.

lie detector [ENG] An instrument that indicates or records one or more functional variables of a person's body while the person undergoes the emotional stress associated with a lie. Also known as polygraph; psychintegroammeter.

life expectancy [ENG] The predicted useful service life of an item of equipment.

life of mine [MIN ENG] The time in which, through the employment of the available capital, the ore reserves—or such reasonable extension of the ore reserves as conservative geological analysis may justify—will be extracted.

life preserver [ENG] A buoyant device that is used to prevent drowning by supporting a person in the water.

life support system [ENG] A system providing atmospheric control and monitoring, such as a breathing mixture supply system, air purification and filtering system, or carbon dioxide removal system; used in oceanographic submersibles and spacecraft.

life test [ENG] A test in which a device is operated under conditions that simulate a normal lifetime of use, to obtain an estimate of service life.

LIFO *See* last in, first out.

lift [MECH ENG] *See* elevator. [MIN ENG] **1.** The vertical height traveled by a cage in a shaft. **2.** The distance between the first level and the surface or between any two levels. **3.** Any of the various gangways from which coal is raised at a slope colliery.

lift bridge [CIV ENG] A drawbridge whose movable spans are raised vertically.

lift coefficient [AERO ENG] The quantity $C_L = 2L/\rho V^2 S$, where L is the lift of a whole airplane wing, ρ is the mass density of the air, V is the free-stream velocity, and S is the wing area; this is also applicable to other airfoils.

lift-drag ratio [AERO ENG] The lift of an aerodynamic form, such as an airplane wing, divided by the drag.

lifter [MIN ENG] A shothole drilled near the floor when tunneling, and fired subsequent to the cut and relief holes.

lifter case [MIN ENG] The sleeve or tubular part attached to the lower end of the inner tube of M-design core barrels and some other types of core barrels, in which is fitted a core lifter. Also known as core-catcher case; core-gripper case; core-lifter case; core-spring case; inner-tube extension; ring-lifter case; spring-lifter case.

lifter flight [DES ENG] Spaced plates or projections on the inside surfaces of cylindrical rotating equipment (such as rotary dryers) to lift and shower the solid particles through the gas-drying stream during their passage through the dryer cylinder.

lifter roof [ENG] Gas storage tank in which the roof is raised by the incoming gas as the tank fills.

lift fan [AERO ENG] A special turbofan engine used primarily for lift in VTOL/STOL aircraft and often mounted in a wing with vertical thrust axis.

lifting block [MECH ENG] A combination of pulleys and ropes which allows heavy weights to be lifted with least effort.

lifting device [ENG] A device to manually open a pressure relief valve by decreasing the spring loading in order to determine if the valve is in working order.

lifting dog [ENG] **1.** A component part of the overshot assembly that grasps and lifts the inner tube or a wire-line core barrel. **2.** A clawlike hook for grasping cylindrical objects, such as drill rods or casing, while raising and lowering them.

lifting guard [MIN ENG] Fencing placed around the mouth of a shaft and lifted out of the way by the ascending cage.

lifting magnet [ENG] A large circular, rectangular, or specially shaped magnet used for handling pig iron, scrap iron, castings, billets, rails, and other magnetic materials.

lifting reentry [AERO ENG] A reentry into the atmosphere by a space vehicle where aerodynamic lift is used, allowing a more gradual descent, greater accuracy in landing at a predetermined spot; it can accommodate greater errors in the guidance system and greater temperature control.

lifting reentry vehicle [AERO ENG] A space vehicle designed to utilize aerodynamic lift upon entering the atmosphere.

lift-off [AERO ENG] The action of a rocket vehicle as it leaves its launch pad in a vertical ascent.

lift pump [MECH ENG] A pump for lifting fluid to the pump's own level.

lift-slab construction [CIV ENG] Pouring reinforced concrete roof and floor slabs at ground level, then lifting them into position after hardening.

lift truck [MECH ENG] A small hand- or power-operated dolly equipped with a platform or forklift.

lift valve [MECH ENG] A valve that moves perpendicularly to the plane of the valve seat.

ligament [ENG] The section of solid material in a tube sheet or shell between adjacent holes.

ligate tar [MATER] A soft tar produced by destructive distillation of lignite; can be used without further refining as diesel fuel or can be redistilled to yield a substance resembling coal tar distillate.

light-beam galvanometer *See* d'Arsonval galvanometer.

light-beam pickup [ENG ACOUS] A phonograph pickup in which a beam of light is a coupling element of the transducer.

light blasting [ENG] Loosening of shallow or small outcrops of rock and breaking boulders by explosives.

light bomber [AERO ENG] Any bomber with a gross weight of less than 100,000 pounds (45,000 kilograms), including bombs; for example, the A-20 and A-26 bombers in World War II.

light crude [PETRO ENG] Crude oil having a high proportion of low-viscosity, low-molecular-weight hydrocarbons.

light-drawn [MET] Cold-worked very slightly; for copper or copper alloy tubing, drawing entails between 10 and 25% reduction in area.

light ends [MATER] The lower-boiling components of a mixture of hydrocarbons, such as those evaporated or distilled off easily in comparison to the bulk of the mixture; for hydrocarbon mixtures, usually considered to be butane and lighter.

lighterage [IND ENG] **1.** Loading or unloading ships by means of a lighter. **2.** The fee charged for this operation.

lighter-than-air craft [AERO ENG] An aircraft, such as a dirigible, that weighs less than the air it displaces.

lighting-off torch [ENG] A torch used to ignite a fuel oil burner; it consists of asbestos cloth wrapped around an iron rod and soaked with oil.

light-inspection car [MECH ENG] A railway motorcar weighing 400–600 pounds (180–270 kilograms) and having a capacity of 650–800 pounds (295–360 kilograms).

light metal [MET] A metal or alloy of low density, especially aluminum and magnesium alloys.

light meter [ENG] A small, portable device for measuring illumination; an exposure meter is a specific application, being calibrated to give photographic exposures.

light oil [MATER] **1.** A coal tar fraction obtained by distillation; boiling range, 110–210°C; used as a source of benzene, toluene, phenol, and cresols. Also known as coal tar light oil. **2.** Any oil having a boiling range of 110–210°C.

light section car [MECH ENG] A railway motorcar weighing 750–900 pounds (340–408 kilograms) and propelled by 4–6-horsepower (3000–4500-watt) engines.

light water [MATER] A water solution of perfluorocarbon compounds mixed with a polyoxyethylene thickener; used as a fire-fighting agent.

lightweight aggregate [MATER] A lightweight inert material, such as foamed slag, vermiculite, clinker, and perlite, used in unreinforced concrete for making structures of low weight and high insulation.

lightweight concrete [MATER] A type of concrete made with lightweight aggregate.

lignaloe oil *See* linaloe wood oil.

lignite wax *See* montan wax.

ligroin *See* petroleum ether.

Lilly controller [MECH ENG] A device on steam and electric winding engines that protects against overspeed, overwind, and other incidents injurious to workers and the engine.

limb [DES ENG] **1.** The graduated margin of an arc or circle in an instrument for measuring angles, as that part of a marine sextant carrying the altitude scale. **2.** The graduated staff of a leveling rod.

lime-and-cement mortar [MATER] Mortar made of mortar cement, lime putty or hydrated lime, and sand in proportions, by volume, normally of one part cement, one or two lime, and five or six sand; suited for all kinds of masonry.

lime glass [MATER] A type of glass containing a high proportion of lime; used in many commercial glass products, such as bottles.

lime grease [MATER] A type of grease that emulsifies less readily than one made with a soda base and therefore is used in an environment where water may occur.

limelight [ENG] A light source once used in spotlights; it consisted of a block of lime heated to incandescence by means of an oxyhydrogen flame torch.

lime liniment *See* carron oil.

lime mortar [MATER] A mixture of hydrated lime, sand, and water having a compressive strength up to 400 pounds per square inch (2.8×10^6 newtons per square meter); used for interior non-load-bearing walls in buildings.

lime oil [MATER] **1.** An edible essential oil squeezed from the rind of lime and other citrus fruit, whose components are limonene and citral; used in flavorings and perfumes. **2.** The distilled essential oil from citron.

lime putty [MATER] A puttylike cement made from lime slaked in water.

limestone log [ENG] A log that employs an electrical resistivity element in the form of four symmetrically arranged current electrodes to give accurate readings in borehole surveying of hard formations.

limit control [MECH ENG] **1.** In boiler operation, usually a device, electrically controlled, that shuts down a burner at a prescribed operating point. **2.** In machine-tool operation, a sensing device which terminates motion of the workpiece or tool at prescribed points.

limit dimensioning method [DES ENG] Method of dimensioning and tolerancing wherein the maximum and minimum permissible values for a dimension are stated specifically to indicate the size or location of the element in question.

limited-access highway *See* expressway.

limited-pressure cycle *See* mixed cycle.

limited-rotation hydraulic actuator [MECH ENG] A type of hydraulic actuator that produces limited reciprocating rotary force and motion; used for lifting, lowering, opening, closing, indexing, and transferring movements; examples are the piston-rack actuator, single-vane actuator, and double-vane actuator.

limit governor [MECH ENG] A mechanical governor that takes over control from the main governor to shut the machine down when speed reaches a predetermined excess above the allowable rated. Also known as topping governor.

limit lines [IND ENG] Lines on a chart designating specification limits.

limit-load design *See* ultimate-load design.

limits [DES ENG] In dimensioning, the maximum and minimum values prescribed for a specific dimension; the limits may be of size if the dimension concerned is a size dimension, or they may be of location if the dimension concerned is a location dimension.

limnimeter [ENG] A type of tide gage for measuring lake level variations.

limnograph [ENG] A recording made on a limnimeter.

linaloe wood oil [MATER] An essential oil derived from fruit and wood of *Bursera* species; a colorless to yellow liquid, soluble in fixed oils and alcohol; used in perfumery and for flavoring. Also known as lignaloe oil; Mexican linaloe oil.

Lincoln experimental satellite [AERO ENG] A series of military communication satellites initiated in 1965; carried out successful X band experiments. Abbreviated LES.

Linde drill *See* fusion-piercing drill.

Lindemann glass [MATER] A lithium borate–beryllium oxide glass having no element higher in atomic number than oxygen; used as window material for low-voltage x-ray tubes because it will pass x-rays of extremely long wavelength, such as Grenz rays.

line-and-staff organization [IND ENG] A form of organization structure which combines functional subunits with staff officers in line functions.

linear actuator [MECH ENG] A device that converts some kind of power, such as hydraulic or electric power, into linear motion.

linear meter [ENG] A meter in which the deflection of the pointer is proportional to the quantity measured.

linear scanning [ENG] Radar beam which moves with constant angular velocity through the scanning sector, which may be a complete 360°.

line brattice [MIN ENG] A partition in an opening to divide it into intake and return airways.

line clinometer [ENG] A clinometer designed to be inserted between rods at any point in a string of drill rods.

line drilling [MIN ENG] The combined methods of drilling and broaching for the primary cut in quarrying; deep, closely spaced holes are drilled in a straight line by means of a reciprocating drill, and webs between holes are removed by a drill or a flat broaching tool.

line functions [IND ENG] Organizational functions having direct authority and responsibility.

line hydrophone [ENG ACOUS] A directional hydrophone consisting of one straight-line element, an array of suitably phased elements mounted in line, or the acoustic equivalent of such an array.

line loss [ENG] The quantity of gas that is lost in a distribution system or pipeline.

line lubricator *See* line oiler.

line microphone [ENG ACOUS] A highly directional microphone consisting of a single straight-line element or an array of small parallel tubes of different lengths, with one end of each abutting a microphone element. Also known as machine-gun microphone.

line of action [MECH ENG] The locus of contact points as gear teeth profiles go through mesh.

line of balance [IND ENG] A production planning system that schedules key events leading to completion of an assembly on the basis of the delivery date for the completed system. Abbreviated LOB.

line of tunnel [ENG] The width marked by the exterior lines or sides of a tunnel.

line oiler [MECH ENG] An apparatus inserted in a line conducting air or steam to an air- or steam-activated machine that feeds small controllable amounts of lubricating oil into the air or steam. Also known as air-line lubricator; line lubricator.

line pack [ENG] The actual amount of gas in a pipeline or distribution system.

liner [DES ENG] A replaceable tubular sleeve inside a hydraulic or pump-pressure cylinder in which the piston travels. [ENG] A string of casing in a borehole. [MET] 1. The cylindrical chamber that holds the billet for extrusion. 2. The slab of coating metal that is placed on the core alloy and is subsequently rolled down to form a clad composite. [MIN ENG] 1. A foot piece for uprights in timber sets. 2. Timber supports erected to reinforce existing sets which are beginning to collapse due to heavy strata pressure. 3. A bar put up between two other bars to assist in carrying the roof. 4. Replaceable facings inside a grinding mill.

liner bushing [DES ENG] A bushing, provided with or without a head, that is permanently installed in a jig to receive the renewable wearing bushings. Also known as master bushing.

line rod *See* range rod.

liner plate cofferdam [CIV ENG] A cofferdam made from steel plates about 16 inches (41 centimeters) high and 3 feet (91 centimeters) long, and corrugated for added stiffness.

line scanner [ENG] An infrared imaging device which utilizes the motion of a moving platform, such as an aircraft or satellite, to scan infrared radiation from the terrain. Also known as thermal mapper.

line shafting [MECH ENG] One or more pieces of assembled shafting to transmit power from a central source to individual machines.

linesman [ENG] **1.** A worker who sets up and repairs communication and power lines. **2.** An assistant to a surveyor.

line space lever [MECH ENG] A lever on a typewriter used to move the carriage to a new line.

line timbers [MIN ENG] Timbers placed along the sides of the track of a working place in rows according to a predetermined plan.

line up [MIN ENG] **1.** A command signifying that the drill runner wants the hoisting cable attached to the drill stem, threaded through the sheave wheel, or wound on the hoist drum. **2.** To reposition a drill so that the drill stem is centered over the parallel to a newly collared drill hole.

lining [MATER] A material used to protect inner surfaces, as of tunnels, pipes, or process equipment.

lining bar [DES ENG] A crowbar with a pinch, wedge, or diamond point at its working end.

lining pole *See* range rod.

lining sight [MIN ENG] An instrument consisting of a plate with a slot in the middle, and the means of suspending it; used with a plumbline for directing the courses of underground drifts or headings.

linishing *See* belt grinding.

link [CIV ENG] A standardized part of a surveyor's chain, which is 7.92 inches (20.1168 centimeters) in the Gunter's chain and 1 foot (30.48 centimeters) in the engineer's chain. [DES ENG] **1.** One of the rings of a chain. **2.** A connecting piece in the moving parts of a machine.

linkage [MECH ENG] A mechanism that transfers motion in a desired manner by using some combination of bar links, slides, pivots, and rotating members.

link bar *See* hinged bar.

link V belt [DES ENG] A V belt composed of a large number of rubberized-fabric links joined by metal fasteners.

linoleum [MATER] A floor covering made by applying a mixture of gelled linseed oil, pigments, fillers, and other materials to a burlap backing, and curing to produce a hard, resilient sheet.

linseed cake [MATER] The residue formed during pressing of commercial linseed oil; used for cattle feed and fertilizer.

linseed oil [MATER] A product made from the seeds of the flax plant by crushing and pressing either with or without heat; formulated in various grades and with various drying agents and used as a vehicle in oil paints and as a component of oil varnishes.

lint [MATER] During the first stage of processing cotton, the fiber that is separated from the seeds in a cotton gin.

lintel [BUILD] A horizontal member over an opening, such as a door or window, usually carrying the wall load.

linter [MECH ENG] A machine for removing fuzz linters from ginned cottonseed.

linters [MATER] Short residual fibers that adhere to ginned cottonseed; used for making fabrics that do not require long fibers, as plastic fillers, and in the manufacture of cellulosic plastics.

lip [CIV ENG] A parapet placed on the downstream margin of a millrace or apron in order to minimize scouring of the river bottom. [DES ENG] Cutting edge of a fluted drill formed by the intersection of the flute and the lip clearance angle, and extending from the chisel edge at the web to the circumference.

Lippmann electrometer *See* capillary electrometer.

liquation [MET] **1.** Separation of fusible metals from less fusible ones by applying heat. **2.** The partial melting of an alloy.

liquefied gas [MATER] A gaseous compound or mixture converted to the liquid phase by cooling or compression; examples are liquefied petroleum gas (LPG), liquefied natural gas (LNG), liquid oxygen, and liquid ammonia.

liquefied natural gas [MATER] A product of natural gas which consists primarily of methanes; its critical temperature is about $-100°F$ ($-73°C$), and thus it must be liquefied by cooling to cryogenic temperatures and must be well insulated to be held in the liquid state; used as a domestic fuel. Abbreviated LNG.

liquefied petroleum gas [MATER] A product of petroleum gases; principally propane and butane, it must be stored under pressure to keep it in a liquid state; it is often stored in metal cylinders (bottled gas) and used as fuel for tractors, trucks, and buses, and as a domestic cooking or heating fuel in rural areas. Abbreviated LPG.

liquefier [ENG] Equipment or system used to liquefy gases; usually employs a combination of compression, heat exchange, and expansion operations.

liquid asphalt *See* residual oil.

liquid blast cleaning [MET] Cleaning metal surfaces with a suspension of abrasive in water accelerated to high velocities by compressed air, or by a centrifugal wheel.

liquid blocking [PETRO ENG] The blocking or plugging of the sand around an injection-well borehole, usually caused by lubricant carryover from compressors.

liquid bright gold [MATER] Any of several gold compounds applied to ceramics in the form of varnish which is dried and heated to redness, decomposing the compound and leaving a thin film of gold firmly attached to the underlying ceramic; used in decorating china and for the production of printed electrical circuits on ceramics.

liquid carburizing [MET] Surface hardening of steel by immersion into a molten bath consisting of cyanides and other salts, for example, at 1600–1750°F (850–950°C).

liquid-column gage *See* U-tube manometer.

liquid compass [ENG] A compass in a bowl filled with liquid.

liquid-cooled dissipator *See* cold plate.

liquid-cooled engine [MECH ENG] An internal combustion engine with a jacket cooling system in which liquid, usually water, is circulated to maintain acceptable operating temperatures of machine parts.

liquid cooling [ENG] Use of circulating liquid to cool process equipment and hermetically sealed components such as transistors.

liquid fuel [MATER] A rocket fuel which is liquid under the conditions in which it is utilized in the rocket. Also known as liquid propellant.

liquid grease [MATER] Lubricating oil of light or medium grade that is thickened with calcium soap.

liquid honing *See* vapor blasting.

liquid-in-glass thermometer [ENG] A thermometer in which the thermally sensitive element is a liquid contained in a graduated glass envelope; the indication of such a thermometer depends upon the difference between the coefficients of thermal expansion of the liquid and the glass; mercury and alcohol are liquids commonly used in meteorological thermometers.

liquid-in-metal thermometer [ENG] A thermometer in which the thermally sensitive element is a liquid contained in a metal envelope, frequently in the form of a Bourdon tube.

liquid insulator [MATER] A liquid with a resistivity greater than about 10^{14} ohmcentimeters, such as a petroleum oil, silicone oil, or halogenated aromatic hydrocarbon.

liquid knockout *See* impingement.

liquid level control [ENG] Regulation of the linear vertical distance between the surface of a liquid and some reference point.

liquid-metal embrittlement [MET] The rapid loss of mechanical properties of a metal or an alloy due to contact with certain liquid metals.

liquid-oxygen explosive [MATER] Sawdust or other carbonaceous material formed into a cartridge and dipped into liquid oxygen, to use in blasting. Abbreviated LOX.

liquid penetrant test [ENG] A penetrant method of nondestructive testing used to locate defects open to the surface of nonporous materials; penetrating liquid is applied to the surface, and after 1–30 minutes excess liquid is removed, and a developer is applied to draw the penetrant out of defects, thus showing their location, shape, and size.

liquid petrolatum *See* white mineral oil.

liquid piston rotary compressor [MECH ENG] A rotary compressor in which a multiblade rotor revolves in a casing partly filled with liquid, for example, water.

liquid rosin *See* tall oil.

liquid-sealed meter [ENG] A type of positive-displacement meter for gas flows consisting of a cylindrical chamber that is more than half filled with water and divided into four rotating compartments formed by trailing vanes; gas entering through the center shaft into one compartment after another forces rotation that allows the gas then to exhaust out the top as it is displaced by the water. Also known as drum meter.

liquid-sorbent dehumidifier [MECH ENG] A sorbent type of dehumidifier consisting of a main circulating fan, sorbent-air contactor, sorbent pump, and reactivator; dehumidification and reactivation are continuous operations, with a small part of the sorbent constantly bled off from the main circulating system and reactivated to the concentration required for the desired effluent dew point.

liquor finish [MET] A bright, smooth finish on wet-drawn wire achieved by using fermented-grain mash liquor as a lubricant.

list [ENG] To lean to one side, or deviate from the vertical.

listening station [ENG] A radio or radar receiving station that is continuously manned for various purposes, such as for radio direction finding or for gaining information about enemy electronic devices.

listing *See* lashing.

litharge-glycerin cement [MATER] Mixture of glycerin, water, and litharge (lead monoxide) to give, when cured, an acid-resistant cement.

lithium grease [MATER] Heat-stable, water-resistant lubricating grease with lithium salts of higher fatty acids (or lithium soaps of fatty glycerides) as a base; used for low-temperature service in aircraft.

lithol red [MATER] Any of various pigments derived from combination of β-naphthol and Tobias acid; available as sodium, barium, and calcium toners and lakes; used in outside, drum, and toy enamels.

lithophone *See* lithopone.

lithopone [MATER] A white pigment produced as a filtered, heated, quenched precipitate from reaction of barium sulfide and zinc sulfide; used as a pigment for paint, ink, filled leather, paper, linoleum, oilcloth, and cosmetics. Also known as Charlton white; Griffith's white; lithophone; Orr's white; zinc baryta white; zinc sulfide white.

litmus [MATER] Blue, water-soluble powder from various lichens, especially *Variolaria lecanora* and *V. rocella*; turns red in solutions at pH 4.5, and blue at pH 8.3; used as an acid-base indicator. Also known as lacmus; lichen blue.

litmus paper [MATER] White, unsized paper saturated by litmus in water; used as a pH indicator.

little giant [MIN ENG] A jointed iron nozzle used in hydraulic mining.

live axle [MECH ENG] An axle to which wheels are rigidly fixed.

live center [MECH ENG] A lathe center that fits into the headstock spindle.

live oil [MATER] An oil containing dissolved gas.

liver [MATER] Intermediate layer of dark-colored, oily material formed by hydrolyzation of acid sludge from sulfuric acid treatment of petroleum oil; insoluble in weak acid and oil.

live-roller conveyor [MECH ENG] Conveying machine which moves objects over a series of rollers by the application of power to all or some of the rollers.

Livingstone sphere [ENG] A clay atmometer in the form of a sphere; evaporation indicated by this instrument is supposed to be somewhat representative of that from plant growth.

Ljungström heater [MECH ENG] Continuous, regenerative, heat-transfer air heater (recuperator) made of slow-moving rotors packed with closely spaced metal plates or wires with a housing to confine the hot and cold gases to opposite sides.

Ljungström steam turbine [MECH ENG] A radial outward-flow turbine having two opposed rotation rotors.

LNG *See* liquefied natural gas.

load [ENG] **1.** To place ammunition in a gun, bombs on an airplane, explosives in a missile or borehole, fuel in a fuel tank, cargo or passengers into a vehicle, and the like. **2.** The quantity of gas delivered or required at any particular point on a gas supply system; develops primarily at gas-consuming equipment. [MIN ENG] Unit of weight of ore used in the South African diamond mines; equal to 1600 pounds (725 kilograms); the equivalent of about 16 cubic feet (0.453 cubic meter) of broken ore.

load-and-carry equipment [MECH ENG] Earth-moving equipment designed to load and transport material.

load-bearing tile [MATER] A tile with the capacity to support superimposed loads.

load chart [IND ENG] A graph showing the amount of work still to be performed by a factory producing unit such as a machine or assembly group.

load controller [MIN ENG] A device to control the load and prevent spillage on a gathering conveyor receiving coal or mineral from several loading points or subsidiary conveyors; it is a simplified weightometer.

load diagram [CIV ENG] A diagram showing the distribution and intensity of loads on a structure.

loaded concrete [MATER] Concrete to which elements of high atomic number or capture cross section have been added to increase its effectiveness as a radiation shield in nuclear reactors.

loaded wheel [ENG] A grinding wheel that is dull as a result of becoming filled with particles from the material being ground.

loader [MECH ENG] A machine such as a mechanical shovel used for loading bulk materials.

loading [ENG] **1.** Buildup on a cutting tool of the material removed in cutting. **2.** Filling the pores of a grinding wheel with material removed in the grinding process. [ENG ACOUS] Placing material at the front or rear of a loudspeaker to change its acoustic impedance and thereby alter its radiation. [MET] Filling of a die cavity with powdered metal.

loading density [ENG] The number of pounds of explosive per foot length of drill hole.

loading head [MECH ENG] The part of a loader which gathers the bulk materials.

loading pan [MIN ENG] A box or scoop into which broken rock is shoveled in a sinking shaft while the hoppit is traveling in the shaft.

loading rack [ENG] The shelter and associated equipment for the withdrawal of liquid petroleum or a chemical product from a storage tank and loading it into a railroad tank car or tank truck.

loading space [ENG] Space in a compression mold for holding the plastic molding material before it is compressed.

loading station [MECH ENG] A device which receives material and puts it on a conveyor; may be one or more plates or a hopper.

loading tray [ENG] A tray with a sliding bottom used to simultaneously load the plastic charge into the cavities of a multicavity mold.

loading weight [ENG] Weight of a powder put into a container.

loam [MET] Molding material consisting of sand, silt, and clay used over backup material for producing massive castings, usually of iron or steel.

LOB *See* line of balance.

lobe [DES ENG] A projection on a cam wheel or a noncircular gear wheel. [ENG ACOUS] A portion of the directivity pattern of a transducer representing an area of increased emission or response.

lobed impeller meter [ENG] A type of positive displacement meter in which a fluid stream is separated into discrete quantities by rotating, meshing impellers driven by interlocking gears.

local action [MET] Electrochemical corrosion resulting from the action of local cells.

local Mach number [AERO ENG] The Mach number of an isolated section of an airplane or its airframe.

local preheating [MET] The heating of a specific portion of a material or structure prior to the performance of a joining or fabrication process.

locate [MIN ENG] To mark out the boundaries of a mining claim and establish the right of possession.

locating [MECH ENG] A function of tooling operations accomplished by designing and constructing the tooling device so as to bring together the proper contact points or surfaces between the workpiece and the tooling.

locating hole [MECH ENG] A hole used to position the part in relation to a cutting tool or to other parts and gage points.

location damages [MIN ENG] Compensation by an operator to the surface owner for injury to the surface or to growing crops, resulting from the drilling of a well.

location dimension [DES ENG] A dimension which specifies the position or distance relationship of one feature of an object with respect to another.

location fit [DES ENG] The characteristic wherein mechanical sizes of mating parts are such that, when assembled, the parts are accurately positioned in relation to each other.

location notice [MIN ENG] A written sign placed prominently on a claim, showing the locator's name and describing the claim's extent and boundaries.

location plan [MIN ENG] A scale map of the projected mine development indicating, among other things, proposed shafts and works in relation to existing surface features.

location work [MIN ENG] Labor required by law to be done on mining claims within 60 days of location, in order to establish ownership.

locator [ENG] A radar or other device designed to detect and locate airborne aircraft.

lock [CIV ENG] A chamber with gates on both ends connecting two sections of a canal or other waterway, to raise or lower the water level in each section. [DES ENG] A fastening device in which a releasable bolt is secured. [MET] A condition in forging in which the flash line is in more than one plane.

lockalloy [MET] A beryllium-base alloy composed of 62% beryllium and 38% aluminum; used as a structural aerospace alloy because of low density and high (47,000 pounds per square inch or 3.2×10^8 newtons per square meter) yield strength.

lock bolt [ENG] **1.** The bolt of a lock. **2.** A bolt equipped with a locking collar instead of a nut. **3.** A bolt for adjusting and securing parts of a machine.

lock chamber [CIV ENG] A compartment between lock gates in a canal.

locked-coil rope [DES ENG] A completely smooth wire rope that resists wear, made of specially formed wires arranged in concentric layers about a central wire core. Also known as locked-wire rope.

locked groove [DES ENG] A blank and continuous groove placed at the end of the modulated grooves on a disk recording to prevent further travel of the pickup. Also known as concentric groove.

locked-wire rope *See* locked-coil rope.

lock front [DES ENG] On a door lock or latch, the plate through which the latching or locking bolt (or bolts) projects.

lock gate [CIV ENG] A movable barrier separating the water in an upper or lower section of waterway from that in the lock chamber.

locking [ENG] Automatic following of a target by a radar antenna.

locking fastener [DES ENG] A fastening used to prevent loosening of a threaded fastener in service, for example, a seating lock, spring stop nut, interference wedge, blind, or quick release.

lock joint [DES ENG] A joint made by interlocking the joined elements, with or without other fastening.

locknut [DES ENG] **1.** A nut screwed down firmly against another or against a washer to prevent loosening. Also known as jam nut. **2.** A nut that is self-locking when tightened. **3.** A nut fitted to the end of a pipe to secure it and prevent leakage.

lock rail [BUILD] An intermediate horizontal structural member of a door, between the vertical stiles, at the height of the lock.

lockset [ENG] **1.** A complete lock including the lock mechanism, keys, plates, and other parts. **2.** A jig or template for making cuts in a door for holding a lock.

lock washer [DES ENG] A solid or split washer placed underneath a nut or screw to prevent loosening by exerting pressure.

locomotive [MECH ENG] A self-propelling machine with flanged wheels, for moving loads on railroad tracks; utilizes fuel (for steam or internal combustion engines), compressed air, or electric energy.

locomotive boiler [MECH ENG] An internally fixed horizontal fire-tube boiler with integral furnace; the doubled furnace walls contain water which mixes with water in the boiler shell.

locomotive crane [MECH ENG] A crane mounted on a railroad flatcar or a special chassis with flanged wheels. Also known as rail crane.

locomotive gradient [MIN ENG] The gradient set by law for a locomotive haulage; maximum is 1 in 15, but the limit for practical purposes is 1 in 25.

locomotive haulage [MIN ENG] The use of locomotive-hauled mine cars to carry coal ore, workers, and materials in a mine.

lode claim [MIN ENG] That portion of a vein or lode, and of the adjoining surface, which has been acquired by a compliance with the law, both Federal and state.

Lofco car feeder [MIN ENG] A carrying chain running between the rails that controls mine cars at loading points, for marshaling trains and for loading cars into cages and tipplers.

loft [BUILD] **1.** An upper part of a building. **2.** A work area in a factory or warehouse.

loft building [BUILD] A building with a large open floor area.

log [ENG] The record of, or the act or process of recording, events or the type and characteristics of the rock penetrated in drilling a borehole as evidenced by the cuttings, core recovered, or information obtained from electronic devices. [MATER] Unshaped timber either rough or squared.

Logan slabbing machine [MIN ENG] A machine that has three cutting chains; two are horizontal—one at the base of the coal seam, the other at a distance from the floor; the third is mounted vertically and shears off the coal at the back of the cut; a short conveyor transfers the coal to the face conveyor.

logger [ENG] A recorder that automatically scans measured quantities at specified times and records, or logs, their values on a chart.

logging [ENG] Continuous recording versus depth of some characteristic datum of the formations penetrated by a drill hole; for example, resistivity, spontaneous potential, conductivity, fluid content, radioactivity, or density.

log line *See* current line.

long clay [MATER] A clay used in ceramics that has a high degree of plasticity.

long column [CIV ENG] A column so slender that bending is the primary deformation, generally having a slenderness ratio greater than 120–150.

longeron [AERO ENG] A principal longitudinal member of the structural framework of a fuselage, nacelle, or empennage boom.

long-flame burner [ENG] A burner in which fuel and air do not readily mix, resulting in a comparatively long flame.

long hole [MIN ENG] An underground borehole and blasthole exceeding 10 feet (3 meters) in depth or requiring the use of two or more lengths of drill steel or rods coupled together to attain the desired depth.

long-hole drill [MIN ENG] A rotary- or a percussive-type drill used to drill long holes.

long-hole jetting [MIN ENG] A hydraulic mining system consisting essentially of drilling a hole down the pitch of the vein, replacing the drilling head with a jet cutting head, and then retracting the drill column with the jets in operation to remove the coal.

Longhurst-Hardy plankton sampler [ENG] A nonquantitative metal-shrouded net for trapping plankton.

longitudinal controller [AERO ENG] A primary flight control mechanism which controls pitch attitude; located in the cockpit, this may be a control column or a side stick.

longitudinal direction [MET] The principal direction of flow in a plastically deformed metal.

longitudinal drum boiler [MECH ENG] A boiler in which the axis of the horizontal drum is parallel to the tubes, both lying in the same plane.

longitudinal resistance seam welding [MET] The performance of resistance seam welding parallel to the throat depth of the welding machine.

longitudinal sequence [MET] The sequence in which successive welds are deposited along the length of a continuous weld.

longitudinal stability [ENG] The ability of a ship or aircraft to recover a horizontal position after a vertical motion of its ends about a horizontal axis perpendicular to the centerline.

long-nose pliers [DES ENG] Small pincer with long, tapered jaws.

long oil [MATER] Varnish containing a large percentage of oil.

long-pillar work [MIN ENG] A coal-winning technique used in underground mining in which large pillars of coal are left as the face is advanced; finally all the pillars are removed together.

long-playing record [ENG ACOUS] A 10- or 12-inch (25.4- or 30.48-centimeter) phonograph record that operates at a speed of 33⅓ rpm (revolutions per minute) and has closely spaced grooves, to give playing times up to about 30 minutes for one 12-inch side. Also known as LP record; microgroove record.

long residuum [MATER] Residue from crude-oil distillation in a petroleum refinery when a relatively small proportion of the feed is distilled overhead.

long span [ENG] Span of open wire exceeding 250 feet (76 meters) in length.

long-span steel framing [BUILD] Framing system used when there is a greater clear distance between supports than can be spanned with rolled beams; girders, simple trusses, arches, rigid frames, and cantilever suspension spans are used in this system.

long string [PETRO ENG] The last string of casing that is set in a well, through the producing zone. Also known as oil string.

long-time burning oil [MATER] Carefully refined kerosine used in railway semaphore signal lamps; heavy ends are removed so that the oil will burn without charring the wick.

long tom [MIN ENG] A trough, longer than a rocker, for washing gold-bearing earth.

longwall coal cutter [MIN ENG] Compact machine driven by electricity or compressed air which cuts into the coal on relatively long faces, with its jib at right angles to its body.

longwall peak stoping [MIN ENG] An underland stoping method in which the rapid advance of the face goes on, and by working the faces at a 60° angle to the strike, the peak travels down the dip at twice the face advance rate.

longwall pillar working [MIN ENG] A technique used to extract coal pillars left behind in long-pillar working.

longwall retreating [MIN ENG] A longwall working system in which all the roadways are in the solid coal seam and the waste areas are left behind; developing headings are driven close to the boundary or limit, and the coal is taken out by the longwall retreating toward the shaft.

longwall system [MIN ENG] A method of mining in which the faces are advanced from the shaft toward the boundary, and the roof is allowed to cave in behind the miners as work progresses.

look angle [AERO ENG] The elevation and azimuth at which a particular satellite is predicted to be found at a specified time. [ENG] The solid angle in which an instrument operates effectively, generally used to describe radars, optical instruments, and space radiation detectors.

lookout [BUILD] A horizontal wood framing member that extends out from the studs to the end of rafters and overhangs a part of a roof, such as a gable.

lookout station [ENG] A structure or place on shore at which personnel keep watch of events at sea or along the shore.

lookout tower [ENG] In marine operations, any tower surmounted by a small house in which a watch is habitually kept, as distinguished from an observation tower in which no watch is kept.

loop [AERO ENG] A flight maneuver in which an airplane flies a circular path in an approximately vertical plane, with the lateral axis of the airplane remaining horizontal, that is, an inside loop. [ENG] **1.** A reel of motion picture film or magnetic tape whose ends are spliced together, so that it may be played repeatedly without interruption. **2.** A closed circuit of pipe in which materials and components may be placed to test them under different conditions of temperature, irradiation, and so forth.

looped pipeline [PETRO ENG] A pipeline that is paralleled (looped) by a second pipeline, both of which serve the same liquid or gas source and destination.

looping [ENG] Laying a parallel pipeline along another, or along just a section of it, to increase capacity.

looping mill [MET] An arrangement of mills such that a hot bar discharged from one mill is fed into a second mill in the opposite direction.

loop tunnel [ENG] A tunnel which is looped or folded back on itself to gain grade in a tunnel location.

loop-type pit bottom [MIN ENG] An arrangement at the pit bottom in which loaded cars are fed to the cage on one side only, and the empties are returned by a loop roadway to the same place.

loopway [MIN ENG] A double-track loop in a main single-track haulage plane at which mine cars may pass.

loose-detail mold [ENG] A plastics mold with parts that come out with the molded piece.

loose fit [DES ENG] A fit with enough clearance to allow free play of the joined members.

loose ground [MIN ENG] Broken, fragmented, or loosely cemented bedrock material that tends to slough from sidewalls into a borehole and must be supported, as with timber sets. Also known as broken ground.

loose-joint butt [DES ENG] A knuckle hinge in which the pin on one half slides easily into a slot on the other half.

loose pulley [MECH ENG] In belt-driven machinery, a pulley which turns freely on a shaft so that the belt can be shifted from the driving pulley to the loose pulley, thereby causing the machine to stop.

lopping shears [DES ENG] Long-handled shears used for pruning branches.

lorry *See* larry.

lose returns *See* lost circulation.

lose water *See* lost circulation.

loss [ENG] Power that is dissipated in a device or system without doing useful work. Also known as internal loss.

loss-in-weight feeder [MECH ENG] A device to apportion the output of granulated or powdered solids at a constant rate from a feed hopper; weight-measured decrease in hopper content actuates further opening of the discharge chute to compensate for flow loss as the hopper overburden decreases; used in the chemical, fertilizer, and plastics industries.

loss-of-head gage [ENG] A gage on a rapid sand filter, which indicates loss of head for a filtering operation.

lost circulation [PETRO ENG] A condition that occurs when the drilling fluid escapes into crevices or porous sidewalls of a borehole and does not return to the collar of the drill hole. Also called lose returns; lose water; lost returns; lost water.

lost motion [MECH ENG] The delay between the movement of a driver and the movement of a follower.

lost returns *See* lost circulation.

lost time [ENG ACOUS] The period in a frequency-modulation sonar, just after flyback, during which the sound field must be reestablished; its duration equals travel time of the signal to and from the target.

lost water *See* lost circulation.

lost-wax process [MET] A method used in investment casting in which a wax pattern between a two-layered mold is removed by melting and replaced with molten metal; used for casting bronze statues and in jewelry casting.

lot [CIV ENG] A piece of land with fixed boundaries. [IND ENG] A quantity of material, such as propellant, the units of which were manufactured under identical conditions.

lot line [CIV ENG] The legal boundary line of a piece of property.

lot number [IND ENG] Identification number assigned to a particular quantity or lot of material from a single manufacturer.

lot plot method [IND ENG] A variables acceptance sampling plan based on the frequency plot of a random sample of 50 items taken from a lot.

lot tolerance percent defective [IND ENG] The percent of defectives in a lot which is considered bad and should be rejected for some specified fraction, usually 90%, of the time.

loudness control [ENG ACOUS] A combination volume and tone control that boosts bass frequencies when the control is set for low volume, to compensate automatically for the reduced response of the ear to low frequencies at low volume levels. Also known as compensated volume control.

loudspeaker [ENG ACOUS] A device that converts electrical signal energy into acoustical energy, which it radiates into a bounded space, such as a room, or into outdoor space. Also known as speaker.

loudspeaker dividing network *See* crossover network.

loudspeaker voice coil *See* voice coil.

louver [BUILD] An opening in a wall or ceiling with slanted or sloping slats to allow sun and ventilation and exclude rain; may be fixed or adjustable, and may be at the opening of a ventilating duct. [ENG] Any arrangement of fixed or adjustable slatlike openings to provide ventilation. [ENG ACOUS] An arrangement of concentric or parallel slats or equivalent grille members used to conceal and protect a loudspeaker while allowing sound waves to pass.

lovage oil [MATER] A yellow-brown essential oil, soluble in alcohol and fixed oils, obtained from the root and fruit of *Levisticum officinale*; used for flavors and perfumes. Also known as levisticum oil.

low-alloy steel [MET] A hardenable carbon steel generally containing not more than about 1% carbon and one or more of the following alloyed components: < (less than) 2% manganese, < 4% nickel, < 2% chromium, < 0.6% molybdenum, and < 0.2% vanadium.

low boiler [MATER] A fast-evaporating solvent used in lacquer thinner to give a rapid initial set; boiling point is 70–100°C.

low brass [MET] Brass containing 20% zinc, 80% copper.

low-carbon steel [MET] Steel containing 0.15% or less of carbon.

low-density dynamite [MATER] Any dynamite containing up to 80% ammonium nitrate as the principal explosive ingredient.

Lowenhertz thread [DES ENG] A screw thread that differs from U.S. Standard form in that the angle between the flanks measured on an axial plane is 53°8'; height equals 0.75 times the pitch, and width of flats at top and bottom equals 0.125 times the pitch.

lower chord [CIV ENG] The bottom member of a truss.

lower pair [MECH ENG] A link in a mechanism in which the mating parts have surface (instead of line or point) contact.

lower punch [MET] In powder metallurgy, the portion of the die forming the bottom of the die cavity.

lower yield point [MET] In annealed carbon steels, the lowest value of stress after the initial dropoff and before the load begins to rise continuously.

lowest safe waterline [MECH ENG] The lowest water level in a boiler drum at which the burner may safely operate.

low-expansion alloy [MET] An alloy whose dimensions do not vary appreciably with temperature.

low explosive [MATER] An explosive which when used in its normal manner deflagrates or burns rather than detonates; used when shattering must be prevented.

low-freezing dynamite [MATER] A dynamite designed to work under freezing conditions; part of the nitroglycerin of the straight dynamite is replaced by nitrated sugar, nitrotoluene, nitrated polymerized glycerin, or ethylene glycol dinitrate.

low-frequency cycle [MET] In resistance welding, one positive- and one negative-current pulse within a heat time at a lower frequency than the electrical power source.

low-frequency induction furnace [ENG] An induction furnace in which current flow at the commercial power-line frequency is induced in the charge to be heated.

low-grade [MATER] An arbitrary designation of dynamites of less strength than 40. [MIN ENG] Pertaining to ores that have a relatively low content of minerals. Also known as coarse; lean.

low-heat cement [MATER] A chemically altered portland cement with a low initial heat liberation.

low-helix drill [DES ENG] A two-flute twist drill with a lower helix angle than a conventional drill. Also known as slow-spiral drill.

low-hydrogen electrode [MET] A covered electrode used in arc welding that provides an atmosphere low in hydrogen.

low-intensity atomizer [MECH ENG] A type of electrostatic atomizer operating on the principle that atomization is the result of Rayleigh instability, in which the presence of charge in the surface counteracts surface tension.

low-level condenser [MECH ENG] A direct-contact water-cooled steam condenser that uses a pump to remove liquid from a vacuum space.

low-lift truck [MECH ENG] A hand or powered lift truck that raises the load sufficiently to make it mobile.

low-melting glass [MATER] Glass to which selenium, thallium, arsenic, or sulfur is added to give melting points of 260–660°F (127–349°C).

low population zone [ENG] An area of low population density sometimes required around a nuclear installation; the number and density of residents is of concern in providing, with reasonable probability, that effective protection measures can be taken if a serious accident should occur.

low-pressure area [MECH ENG] The point in a bearing where the pressure is the least and the area or space for a lubricant is the greatest.

low-pressure laminate [MATER] A plastic laminate molded and cured at pressures in general of 400 psi (approximately 27 atmospheres or 2.8×10^6 newtons per square meter).

low-pressure torch [ENG] A type of torch in which acetylene enters a mixing chamber, where it meets a jet of high-pressure oxygen; the amount of acetylene drawn into the flame is controlled by the velocity of this oxygen jet. Also known as injector torch.

low-pressure well [PETRO ENG] An oil or gas well with a shut-in wellhead pressure of less than 2000 psia (1.38×10^7 newtons per square meter).

low-residual-phosphorus copper [MET] Deoxidized copper with a 0.004–0.012% residual phosphorus content.

Lowry process [ENG] A system for wood preservation which uses atmospheric pressure at the start and then introduces preservative into the wood in a vacuum.

low-shaft furnace [MET] A blast furnace having a short shaft; used to produce pig iron, ferroalloys, alumina, and other products from low-grade ores using low-grade fuel.

low-speed wind tunnel [ENG] A wind tunnel that has a speed up to 300 miles (480 kilometers) per hour and the essential features of most wind tunnels.

low-temperature coke [MATER] Coke produced at temperatures of 500–750°C, used chiefly for house heating, particularly in England. Also known as char.

low-temperature hygrometry [ENG] The study that deals with the measurement of water vapor at low temperatures; the techniques used differ from those of conventional hygrometry because of the extremely small amounts of moisture present at low temperatures and the difficulties imposed by the increase of the time constants of the standard instruments when operated at these temperatures.

low-water fuel cutoff [MECH ENG] A float device which shuts off fuel supply and burner when boiler water level drops below the lowest safe waterline.

LOX *See* liquid-oxygen explosive.

lozenge file [DES ENG] A small file with four sides and a lozenge-shaped cross section; used in forming dies.

L pad [ENG ACOUS] A volume control having essentially the same impedance at all settings.

LPF process [MIN ENG] Recovery of metals from tailings by a sequence of leaching, precipitation, and flotation.

LPG *See* liquefied petroleum gas.

LP record *See* long-playing record.

L-1 test [ENG] A 480-hour engine test in a single-cylinder Caterpillar diesel engine to determine the detergency of heavy-duty lubricating oils.

L-2 test [ENG] An engine test made in a single-cylinder Caterpillar diesel engine to determine the oiliness of an engine oil. Also known as scoring test.

L-3 test [ENG] An engine test in a four-cylinder Caterpillar engine to determine stability of crankcase oil at high temperatures and under severe operating conditions.

L-4 test [ENG] An engine test in a six-cylinder spark-ignition Chevrolet engine to evaluate crankcase oil oxidation stability, bearing corrosion, and engine deposits.

L-5 test [ENG] An engine test in a General Motors diesel engine to determine detergency, corrosiveness, ring sticking, and oxidation stability properties of lubricating oils.

lube cut [MATER] The distilled fraction of crude oil with suitable boiling range and viscosity to yield a lubricating oil when it is completely refined. Also known as lube-oil distillate; lube stock.

lube oil *See* lubricating oil.

lube-oil distillate *See* lube cut.

lube stock *See* lube cut.

lubricant [MATER] A substance used to reduce friction between parts or objects in relative motion.

lubricant additive [MATER] Any material added to lubricants (greases or oils) to give the product special properties, such as resistance to extremes of pressure, cold, or heat, improved viscosity, and detergency.

lubricated gasoline [MATER] A motor gasoline into which a lubricant has been added.

lubricating film [MATER] A thin layer of oil or grease applied between rubbing surfaces.

lubricating grease [MATER] A solid or semisolid lubricant consisting of a thickening agent (soap or other additives) in a fluid lubricant (usually petroleum lubricating oil).

lubricating oil [MATER] Selected fractions of refined petroleum or other oils (with or without additives) used to lessen friction between moving surfaces. Also known as lube oil.

lubrication action [MATER] The ability of the lubricant to maintain a fluid film between solid surfaces and to prevent their physical contact.

lubricator [ENG] A device for applying a lubricant.

lubricity [MATER] The ability of a material to lubricate.

lucca oil *See* olive oil.

Luckiesh-Moss visibility meter [ENG] A type of photometer that consists of two variable-density filters (one for each eye) that are adjusted so that an object seen through them is just barely discernible; the reduction in visibility produced by the filters is read on a scale of relative visibility related to a standard task.

Lüders' bands *See* Lüders' lines.

Lüders' lines [MET] Surface markings on a metal caused by flow of the material strained beyond its elastic limit. Also known as deformation bands; Hartmann lines; Lüders' bands; Piobert lines; stretcher strains.

lug [DES ENG] A projection or head on a metal part to serve as a cap, handle, support, or fitting connection.

lug bolt [DES ENG] **1.** A bolt with a flat extension or hook instead of a head. **2.** A bolt designed for securing a lug.

lug brick [MATER] A brick with lugs for spacing adjacent bricks.

lumbang oil [MATER] Colorless or yellow liquid with pleasant aroma and bland taste; soluble in alcohol, ether, chloroform, and carbon disulfide; expressed from candlenut; used as an illuminant and wood preservative, and in paints, calking, and soap manufacture. Also known as candlenut oil.

lumber [MATER] Logs that have been sawed and prepared for market.

luminescent dye [MATER] A dye that is made luminous by excitation with an outside energy source; used in luminous paint.

luminous paint [MATER] A type of paint in which luminous pigments are used.

luminous pigment [MATER] A pigment that absorbs light energy and radiates visible light when exposed to ultraviolet light;

made of phosphors such as strontium, zinc, and cadmium sulfides.

lump coal [MIN ENG] Bituminous coal that passes through a 6-inch (15-centimeter) round mesh in initial screening.

Luna program [AERO ENG] A series of Soviet space probes launched for flight missions to the moon.

lunar excursion module [AERO ENG] A manned spacecraft designed to be carried on top of the Apollo service module and having its own power plant for making a manned landing on the moon and a return from the moon to the orbiting Apollo spacecraft. Abbreviated LEM. Also known as lunar module (LM).

lunar flight [AERO ENG] Flight by a spacecraft to the moon.

lunar module *See* lunar excursion module.

lunar orbit [AERO ENG] Orbit of a spacecraft around the moon.

lunar probe [AERO ENG] Any space probe launched for flight missions to the moon.

lunar satellite [AERO ENG] A satellite making one or more revolutions about the moon.

lunar spacecraft [AERO ENG] A spacecraft designed for flight to the moon.

lung-governed breathing apparatus [ENG] A breathing apparatus in which the oxygen that is supplied to the wearer is governed by the wearer's demand.

lusterless paint [MATER] Paint which absorbs light rays so that no shine or polish appears on its surface; used extensively on U.S. Army vehicles.

lute [MATER] A substance, such as cement or clay, for packing a joint or coating a porous surface to produce imperviousness to gas or liquid.

lycopodium [MATER] A yellow powder prepared from the spores of *Lycopodium clavatum*; used as a desiccant and absorbent.

lyddite [MATER] An explosive composed chiefly of picric acid.

lysimeter [ENG] An instrument for measuring the water percolating through soils and determining the materials dissolved by the water.

M

M_f [MET] In a carbon steel, the temperature at which martensite formation finishes during cooling of austenite.

M_s [MET] In a carbon steel, the temperature at which martensite formation begins during cooling of austenite.

macadam [CIV ENG] Uniformly graded stones consolidated by rolling to form a road surface; may be bound with water or cement, or coated with tar or bitumen.

macanilla oil [MATER] An oil obtained from the nuts of the palm *Guilielma garipaes* of Venezuela and Central America.

Macarthur and Forest cyanide process [MET] A process for recovering gold by leaching the pulped gold ore with a solution of 0.2–0.8% potassium cyanide and next with water; the gold is then obtained by precipitation on zinc or aluminum or by electrolysis.

Macassar oil [MATER] **1.** A fat obtained from seeds of kusam, used in hair dressing, cooking, and illumination. **2.** Any of several fatty oils or oily preparations that have similar properties and are used in hair preparations.

mace oil [MATER] An essential oil obtained by distillation from mace and containing pinene and dipentene; used in flavoring.

machete [DES ENG] A knife with a broad blade 2 to 3 feet (60 to 90 centimeters) long.

machinability [MET] **1.** The ability of a metal to be machined. **2.** The difficulty or ease with which a metal can be machined.

machinability index [MET] A numerical value that designates the degree of difficulty or ease with which a particular material can be machined; originally based on turning B1112 steel at 180 feet per minute (0.9144 meter per second) with a high-speed tool for an index of 100; with replacement of high-speed steels with carbides in turning operations, it has been found that the machinability index of a given material changes with the type of operation and the tool material.

machinable carbide [MET] Titanium carbide in a matrix of Ferro-Tic C tool steel.

Mach indicator *See* Machmeter.

machine [MECH ENG] A combination of rigid or resistant bodies having definite motions and capable of performing useful work.

machine attention time [IND ENG] Time during which a machine operator must observe the machine's functioning and be available for immediate servicing, while not actually operating or servicing the machine.

machine bolt [DES ENG] A heavy-weight bolt with a square, hexagonal, or flat head used in the automotive, aircraft, and machinery fields.

machine controlled time [IND ENG] The time necessary for a machine to complete the automatic portion of a work cycle. Also known as independent machine time; machine element; machine time.

machine cut [MIN ENG] A groove or slot made horizontally or vertically in a coal seam by a coal cutter, as a step to shot firing.

machine design [DES ENG] Application of science and invention to the development, specification, and construction of machines.

machine drill [MECH ENG] Any mechanically driven diamond, rotary, or percussive drill.

machine element [DES ENG] Any of the elementary mechanical parts, such as gears, bearings, fasteners, screws, pipes, springs, and bolts used as essentially standardized components for most devices, apparatus, and machinery. [IND ENG] *See* machine controlled time.

machine file [DES ENG] A file that can be clamped in the chuck of a power-driven machine.

machine forging [MET] Forging operations performed in and by certain machines.

machine-gun microphone *See* line microphone.

machine-hour [IND ENG] A unit representing the operation of one machine for 1 hour; used in the determination of costs and economics.

machine idle time [IND ENG] Time during a work cycle when a machine is idle, awaiting completion of manual work.

machine key [DES ENG] A piece inserted between a shaft and a hub to prevent relative rotation. Also known as key.

machine loading [IND ENG] **1.** Feeding work into a machine. **2.** Planning the amount of use of a unit of equipment during a given time period.

machine mining [MIN ENG] The use of power machines and equipment in the excavation and extraction of ore or coal.

machine oil [MATER] Medium-density lubricating oil used for machine parts.

machine-paced operation [IND ENG] The proportion of an operation cycle during which the machine controls the speed of work progress.

machinery [MECH ENG] A group of parts or machines arranged to perform a useful function.

machine screw [DES ENG] A blunt-ended screw with a standardized thread and a head that may be flat, round, fillister, or oval, and may be slotted, or constructed for wrenching; used to fasten machine parts together.

machine setting See mechanical setting.

machine shop [ENG] A workshop in which work, metal or other material, is machined to specified size and assembled.

machine shot capacity [ENG] In injection molding, the maximum weight of a given thermoplastic resin which can be displaced by a single stroke of the injection ram.

machine steel [MET] Plain carbon steel with a 0.2–0.3% carbon content.

machine taper [MECH ENG] A taper that provides a connection between a tool, arbor, or center and its mating part to ensure and maintain accurate alignment between the parts; permits easy separation of parts.

machine time See machine controlled time.

machine tool [MECH ENG] A stationary power-driven machine for the shaping, cutting, turning, boring, drilling, grinding, or polishing of solid parts, especially metals.

machine welding [MET] Welding with a machine under the control and observation of an operator; may be loaded and unloaded either manually or mechanically.

machining [MET] Performing various cutting or grinding operations on a piece of work.

machining stress [MET] Residual stress in the work caused by machining.

machinist's file [DES ENG] A type of double-cut file that removes metal fast and is used for rough metal filing.

Machmeter [ENG] An instrument that measures and indicates speed relative to the speed of sound, that is, indicates the Mach number. Also known as Mach indicator.

Mack's cement [MATER] Cement made of dehydrated gypsum with a small amount of calcined sodium sulfate and potassium sulfate.

Mac-Lane system [MIN ENG] A means of conveying dirt to the top of a heap; an inclined rail track goes from the loading station at the bottom of the heap to an extending frame at the top; the rope that hauls the tubs with dirt up the rail track passes around a return sheave in the extending frame; a tub is tipped over at the top by a gear and the dirt discharged.

macroanalytical balance [ENG] A relatively large type of analytical balance that can weigh loads of up to 200 grams to the nearest 0.1 milligram.

macroscopic anisotropy [ENG] Phenomenon in electrical downhole logging wherein electric current flows more easily along sedimentary strata beds than perpendicular to them.

macroscopic stress [MET] Residual stress in a metal in a distance comparable to the gage length of strain measurement specimens and therefore detectable by x-ray or dissection techniques. Also known as macrostress.

macrostructure [MET] Structure of an etched metal visible to the naked eye or at magnifications up to 10 diameters.

macrotome [ENG] A device for making large anatomical sections.

madder [MATER] The root of the madder plant (*Rubia tinctorium*), pulverized and used as source of glucosides to produce alizarin by fermentation. Also known as gamene.

madder lake [MATER] Bluish-red, transparent pigment produced from alizarin red; used to make stains and inks, and as a component of artists' oil colors.

madia oil [MATER] An oil that is made by crushing the seeds of the melosa plant; used as a substitute for olive oil.

Madsen impedance meter [ENG] An instrument for measuring the acoustic impedance of normal and deaf ears, based on the principle of the Wheatstone bridge.

mafura tallow [MATER] A bitter-tasting, heavy vegetable fat obtained from the nuts of the mafura tree (*Trichilia emetica*); used for soaps, candles, and ointments.

magazine [ENG] **1.** A storage area for explosives. **2.** A building, compartment, or structure constructed and located for the storage of explosives or ammunition.

magnesia brick [MATER] A type of refractory brick composed of magnesium oxide with about 15% of other oxides. Also known as magnesite brick.

magnesia cement *See* magnesium oxychloride cement.

magnesia refractory [MATER] Heat- and corrosion-resistant material made of magnesium oxide; used in cement or brick form to line high-temperature process vessels or furnaces.

Magnesil [MET] A magnetic alloy used in making cores for magnetic amplifiers, similar to Hypersod and Selectron.

magnesite brick *See* magnesia brick.

magnesite wheel [ENG] A grinding wheel made with magnesium oxychloride as the bonding agent.

magnesium [MET] A silvery-white, lightweight, malleable, ductile metal, used in metallurgical and chemical processes, photography, pyrotechny, and light alloys.

magnesium dust [MET] Magnesium metal powder; flammable; used in photographic flash lights and pyrotechnics.

magnesium lime [MATER] Lime containing more than 20% magnesium oxide; slakes more slowly, evolves less heat, expands less, sets more rapidly, and produces higher-strength mortars than does high-calcium quicklime.

magnesium oxychloride cement [MATER] Cement made by adding a magnesium chloride solution to magnesia; used for interior flooring. Also known as magnesia cement.

magnet alloy [MET] An alloy such as Alnico or Alcomax having strong magnetic properties; used in making permanent magnets.

magnet grate [MIN ENG] A series of magnetized bars used to trap and remove tramp iron from a flow of pulverized or granulated dry solids passing over the grate; used to protect crushing or grinding equipment.

magnetically hard alloy [MET] A ferromagnetic alloy that can be permanently magnetized after the removal of an externally applied magnetic field.

magnetically soft alloy [MET] A ferromagnetic alloy which is capable of being magnetized upon application of an external magnetic field, but which returns to a nonmagnetic condition when the field is removed.

magnetic analysis inspection [MET] A nondestructive inspection method to determine the presence of variations in magnetic flux in ferromagnetic materials of constant cross section caused by defects, variations in hardness, discontinuities, or other irregularities.

magnetic annealing [MET] Annealing and cooling in a strong magnetic field.

magnetic brake [MECH ENG] A friction brake under the control of an electromagnet.

magnetic cartridge *See* variable-reluctance pickup.

magnetic chuck [MECH ENG] A chuck in which the workpiece is held by magnetic force.

magnetic clutch *See* magnetic fluid clutch; magnetic friction clutch.

magnetic crack detection *See* magnetic-particle test.

magnetic cutter [ENG ACOUS] A cutter in which the mechanical displacements of the recording stylus are produced by the action of magnetic fields.

magnetic earphone [ENG ACOUS] An earphone in which variations in electric current produce variations in a magnetic field, causing motion of a diaphragm.

magnetic element [ENG] That part of an instrument producing or influenced by magnetism.

magnetic fluid [MATER] A mixture of iron particles in oil or other liquid; viscosity increases sharply in a strong magnetic field.

magnetic fluid clutch [MECH ENG] A friction clutch that is engaged by magnetizing a liquid suspension of powdered iron located between pole pieces mounted on the input and output shafts. Also known as magnetic clutch.

magnetic-force welding [MET] A welding process in which the mechanical force is exerted by a magnetic field.

magnetic forming [MET] The forming of metal into desired shapes by using strong magnetic fields, produced by charging a large capacitor bank and then discharging it into an induction coil in less than 10^{-6} second, to push the metal against a forming die.

magnetic friction clutch [MECH ENG] A friction clutch in which the pressure between the friction surfaces is produced by magnetic attraction. Also known as magnetic clutch.

magnetic hardness comparator [ENG] A device for checking the hardness of steel parts by placing a unit of known proper hardness within an induction coil; the unit to be tested is then placed within a similar induction coil, and the behavior of the induction coils compared; if the standard and test units have the same magnetic properties, the hardness of the two units is considered to be the same.

magnetic induction gyroscope [ENG] A gyroscope without moving parts, in which alternating- and direct-current magnetic fields act on water doped with salts which exhibit nuclear paramagnetism.

magnetic ink [MATER] Ink containing magnetic particles to permit reading of printed characters by a magnetic character reader as well as by humans.

magnetic inspection oil [MET] A light petroleum oil, such as kerosine or naphtha, to which has been added fine ferromagnetic particles (usually colored black or red for contrast) to form an inspection penetrant; when the penetrant is applied to a metal surface being inspected, the ferrous particles accumulate in any surface cracks by magnetic attraction, thereby permitting the cracks to be discernible.

magnetic inspection paste [MET] A paste containing ferromagnetic particles designed to be added to a light distilled petroleum oil, such as kerosine or naphtha, to form an inspection penetrant; when the inspection penetrant is applied to a metal, the ferrous particles accumulate in any surface cracks by magnetic attraction, thereby permitting the cracks to be discerned.

magnetic inspection powder [MET] A dry powder containing ferromagnetic particles colored gray, black, or red for contrast, designed to be dusted on metal parts being inspected by a magnetic inspection machine; the ferrous powder accumulates in any surface cracks (flaws) by magnetic attraction, thereby permitting the cracks to be readily discerned; if the ferrous particles are fluorescent, surface cracks will be brilliantly illuminated under black light.

magnetic loudspeaker [ENG ACOUS] Loudspeaker in which acoustic waves are produced by mechanical forces resulting from magnetic reactions. Also known as magnetic speaker.

magnetic microphone [ENG ACOUS] A microphone consisting of a diaphragm acted upon by sound waves and connected to an armature which varies the reluctance in a magnetic field surrounded by a coil. Also known as reluctance microphone; variable-reluctance microphone.

magnetic-particle test [MET] A nondestructive test to determine the existence and extent of macrodefects such as cracks in ferromagnetic materials; discontinuities in the material create variations of magnetic field which are outlined by fine magnetic particles. Also known as magnetic crack detection.

magnetic pickup See variable-reluctance pickup.

magnetic potentiometer [ENG] Instrument that measures magnetic potential differences.

magnetic pressure transducer [ENG] A type of pressure transducer in which a change of pressure is converted into a change of magnetic reluctance or inductance when one part of a magnetic circuit is moved by a pressure-sensitive element, such as a bourdon tube, bellows, or diaphragm.

magnetic prospecting [ENG] Carrying out airborne or ground surveys of variations in the earth's magnetic field, using a magnetometer or other equipment, to locate magnetic deposits of iron, nickel, or titanium, or nonmagnetic deposits which either contain magnetic gangue minerals or are associated with magnetic structures.

magnetic pulley [ENG] Magnetized pulley device for a conveyor belt; removes tramp iron from dry products being moved by the belt.

magnetic recording paper [MATER] A particle-oriented paper in which both machine-readable and visible traces can be produced by a magnetic recording head; reusable because the trace can be erased by a combination of alternating-current and direct-current magnetic fields.

magnetic rubber [MATER] Synthetic rubber to which magnetic metal powder is added; produced in sheets or strips.

magnetic separator [ENG] A machine for separating magnetic from less magnetic or nonmagnetic materials by using strong magnetic fields; used for example, in tramp iron removal, or concentration and purification.

magnetic sound track [ENG ACOUS] A magnetic tape, attached to a motion picture film, on which a sound recording is made.

magnetic speaker See magnetic loudspeaker.

magnetic wire [MATER] A wire made from magnetic material suitable for magnetic recording.

magneto anemometer [ENG] A cup anemometer with its shaft mechanically coupled to a magnet; both the frequency and amplitude of the voltage generated are proportional to the wind speed, and may be indicated or recorded by suitable electrical instruments.

magnetohydrodynamic arcjet [AERO ENG] An electromagnetic propulsion system utilizing a plasma that is heated in an electric arc and then adiabatically expanded through a nozzle and further accelerated by a crossed electric and magnetic field.

magnetometer [ENG] An instrument for measuring the magnitude and sometimes also the direction of a magnetic field, such as the earth's magnetic field.

magnetostrictive loudspeaker [ENG ACOUS] Loudspeaker in which the mechanical forces result from the deformation of a material having magnetostrictive properties.

magnetostrictive microphone [ENG ACOUS] Microphone which depends for its operation on the generation of an electromotive force by the deformation of a material having magnetostrictive properties.

magno [MET] An alloy of 95.5% nickel and 4.5% manganese, used in the manufacture of incandescent lamps and radio tubes.

maguey [MATER] A fiber obtained from the agave (*Agave cantala*); maguey fibers are white, stiff, brilliant, and light in weight, and are used chiefly for binder twine.

mahogany [MATER] The hard wood of these trees, especially the red or yellow-brown wood of the West Indies mahogany tree (*Swietenia mahagoni*).

mahogany acid [MATER] A dark-colored mixture of sulfonic acid derived from petroleum; the salts are used as emulsifying agents and in lubricants.

mahogany soap [MATER] The sodium salt of crude or refined petroleum sulfonic acids, used as flotation agents and to increase the oil absorption of mineral pigments in paint.

mahubarana fat [MATER] A pale-yellow solid oil, melting point 40–44°C, obtained from seeds of trees of the genus *Boldoa;* used for soaps and candles.

main [CIV ENG] A duct or pipe that supplies or drains ancillary branches.

main-and-tail haulage [MIN ENG] A single-track haulage system that is operated by a haulage engine with two drums, each with a separate rope.

main crosscut [MIN ENG] A crosscut that traverses the entire mining field and penetrates all deposits.

main entry [MIN ENG] The principal entry or set of entries driven through the coalbed from which cross entries, room entries, or rooms are turned.

main fans [MIN ENG] Fans that produce the general ventilating current of the mine, being of large capacity and permanently installed.

main firing [ENG] The firing of a round of shots by means of current supplied by a transformer fed from a main power supply.

main haulage [MIN ENG] The section of the haulage system which moves the coal from the secondary or intermediate haulage system to the shaft or mine opening.

main haulage conveyor [MIN ENG] A conveyor used to transport material in the main haulage section of the mine, between the intermediate haulage conveyor and a car-loading point or the outside.

main-line locomotive [MIN ENG] A large, high-powered locomotive which hauls trains of cars over the main haulage system.

main return [MIN ENG] The main return airway of a mine.

main rope *See* pull rope.

main-rope haulage system [MIN ENG] A system of haulage for hauling loaded trains of tubs or cars up, or lowering them down, a comparatively steep gradient which is not steep enough, in the latter case, for a self-acting incline.

main shaft [MECH ENG] The line of shafting receiving its power from the engine or motor and transmitting power to other parts.

maintainability [ENG] **1.** The ability of equipment to meet operational objectives with a minimum expenditure of maintenance effort under operational environmental conditions in which scheduled and unscheduled maintenance is performed. **2.** Quantitatively, the probability that an item will be restored to specified conditions within a given period of time when maintenance action is performed in accordance with prescribed procedures and resources.

maintenance [IND ENG] The upkeep of industrial facilities and equipment.

maintenance engineering [IND ENG] The function of providing policy guidance for maintenance activities, and of exercising technical and management review of maintenance programs.

maintenance kit [ENG] A collection of items not all having the same basic name, which are of a supplementary nature to a major component or equipment; the items within the collection may provide replacement parts and facilitate such functions as inspection, test repair, or preventive types of maintenance, for the specific purpose of restoring and improving the operational status of a component or equipment comparable to its original capacity and efficiency.

maintenance vehicle [ENG] Vehicle used for carrying parts, equipment, and personnel for maintenance or evacuation of vehicles.

Majac mill [MIN ENG] A mill for dry-grinding mica by means of fluid energy; consists of a chamber which contains two horizontal, directly opposing jets and into which mica is fed continuously from a screw conveyor.

major assembly [ENG] A self-contained unit of individual identity; a completed assembly of component parts ready for operation, but utilized as a portion of, and intended for further installation in, an end item or major item.

major defect [IND ENG] Defect which causes serious malfunctioning of a product.

major diameter [DES ENG] The largest diameter of a screw thread, measured at the crest for an external (male) thread and at the root for an internal (female) thread.

major repair [ENG] Repair work on items of material or equipment that need complete overhaul or substantial replacement of parts, or that require special tools.

making hole [PETRO ENG] During a drilling operation, deepening the well bore, with reference to progress at a given time.

malleable [MET] Capable of undergoing plastic deformation without rupture; a property characteristic of metals.

malleable iron [MET] White cast iron which has been rendered malleable by heat treatment.

malleableize [MET] To render a material malleable, such as by heat-treating white cast iron.

malleable pig iron [MET] A grade of pig iron suitable for production of white cast iron from which malleable iron is made.

mallet [DES ENG] An implement with a barrel-shaped head made of wood, rubber, or other soft material; used for driving another tool, such as a chisel, or for striking a surface without causing damage.

malmstone [MATER] A name applied to chert when it is used in building and paving.

malodorant See odorant.

Malotte's metal [MET] A fusible alloy composed of 46% bismuth, 20% lead, and 34% tin; melts at 96–123°C.

mamarron oil [MATER] A cream-colored fat high in lauric acid and similar to coconut oil in characteristics and odor; obtained from a species of *Attalea* palm.

management game [IND ENG] A training exercise in which prospective decision makers act out managerial decision-making roles in a simulated environment. Also known as business game; operational game.

man cage [MIN ENG] A special cage for raising and lowering workers in a mine shaft.

man car [MIN ENG] A kind of car for transporting miners up and down the steeply inclined shafts of some mines.

mandarin oil [MATER] Golden-yellow or olive-green essential oil with refreshing aroma, obtained from the peel of mandarin oranges; chief constituents are limonene and esters; used in medicine and as flavoring. Also known as tangerine oil.

mandrel [MECH ENG] A shaft inserted through a hole in a component to support the work during machining. [MET] A metal bar serving as a core around which other metals are cast, forged, or extruded, forming a true central hole.

mandrel hanger [PETRO ENG] A device used to provide a liquid- or gas-tight seal (blowout preventer) between oil-well tubing and the tubing head.

mandrel press [MECH ENG] A press for driving mandrels into holes.

manganese [MET] A hard, brittle, grayish-white metal used chiefly in making steel.

manganese-aluminum [MET] A hardener alloy employed for making additions of manganese to aluminum alloys such as Duralumin; typical composition is 25% manganese, 75% aluminum.

manganese-boron [MET] Manganese alloyed with boron; used as an ingredient for hardening and deoxidizing bronze.

manganese brass [MET] A brass containing about 70% copper, 29% zinc, and 1.3% manganese.

manganese bronze [MET] A type of brass or bronze containing about 59% copper, 39% zinc, 1.5% iron, 1% tin, and 0.1% manganese; another composition by the same name contains about 66% copper, 23% zinc, 3% iron, 4.5% aluminum, and 3.7% manganese.

manganese-silicon [MET] An alloy that contains 73–78% silicon, 20–25% manganese, a maximum of 1.5% iron, and a maximum of 0.25% carbon; used for adding manganese and silicon to metals.

manganese steel See Hadfield manganese steel.

manganese-titanium [MET] An alloy usually composed of 38% manganese, 29% titanium, 8% aluminum, 3% silicon, 22% iron, and no carbon; used as a deoxidizer for high-grade steels and for nonferrous alloys.

mangle gearing [MECH ENG] Gearing for producing reciprocating motion; a pinion rotating in a single direction drives a rack with teeth at the ends and on both sides.

mangrove [MATER] Liquid derived from the mangrove tree *Rhizophora mucronata* and used in the leather tanning industry.

Manhattan Project [ENG] A United States project lasting from August 1942 to August 1946, which developed the atomic energy program, with special reference to the atomic bomb.

manhead See manhole.

manhole [ENG] An access hole to a tank or boiler, to underground passages, or in a deck or bulkhead of a ship; usually covered with a cast iron or steel plate. Also known as manhead.

man-hour [IND ENG] A unit of measure representing one person working for one hour.

manifold [ENG] The branch pipe arrangement which connects the valve parts of a multicylinder engine to a single carburetor or to a muffler.

manifolding [ENG] The gathering of multiple-line fluid inputs into a single intake chamber (intake manifold), or the division of a single fluid supply into several outlet streams (distribution manifold).

manifold paper [MATER] An extremely thin paper used for making duplicate copies, such as onionskin paper.

manifold pressure [MECH ENG] The pressure in the intake manifold of an internal combustion engine.

manikin [ENG] A correctly proportioned doll-like figure that is jointed and will assume any human position and hold it; useful in art to draw a human figure in action, or in medicine to show the relations of organs by means of movable parts.

Manila copal *See* Manila resin.

Manila paper [MATER] Yellowish paper or Bristol board; the term originally referred to paper manufactured from Manila hemp.

Manila resin [MATER] A type of resin extracted from trees of the genus *Agathis* in the Philippines that is soluble in ethyl and methyl alcohol, insoluble in water; used in printing ink, varnishes, paints, and linoleum. Also known as Manila copal.

manipulators [ENG] Mechanical devices used for safe handling of dangerous materials of any kind, especially radioactive materials; frequently, they are remotely operated from behind a protective shield.

manjak [MATER] A variety of grahamite, manjak is the blackest of the asphalts; used for insulation and varnishes.

manketti oil [MATER] A light-yellow, viscous varnish oil, obtained from the nuts of the African tree *Ricinodendron rautonemii*.

man-machine chart [IND ENG] A two-column, multiple-activity process chart listing the steps performed by an operator and the operations performed by a machine and showing the corresponding idle times for each.

man-machine system [ENG] A system in which the functions of the worker and the machine are interrelated and necessary for the operation of the system.

manna [MATER] The concrete, yellowish, saccharine exudation of the flowering ash (*Fraxinus ornus*); contains mannitol, sugar, mucilage, and resin and has been used as a mild laxative.

manned orbiting laboratory [AERO ENG] An earth-orbiting satellite containing instrumentation and personnel for continuous measurement and surveillance of the earth, its atmosphere, and space. Abbreviated MOL.

manned spacecraft [AERO ENG] A vehicle capable of sustaining a person above the terrestrial atmosphere.

Mannesmann mill [MET] A mill consisting of two rolls mounted with their axes slightly inclined.

Mannesmann process [MET] A process for making seamless tubing by forcing a billet between the rolls of a Mannesmann mill so as to pierce the center, and then forcing the metal over a mandrel to form the central bore.

manometer [ENG] A double-leg liquid-column gage used to measure the difference between two fluid pressures.

manometry [ENG] The use of manometers to measure gas and vapor pressures.

manostat [ENG] Fluid-filled, upside-down manometer-type device used to control pressures within an enclosure, as for laboratory analytical distillation systems.

man process chart [IND ENG] A graphic representation of the activities performed or to be performed by a worker. Abbreviated MPC.

M-A-N scavenging system [MECH ENG] A system for removing used oil and waste gases from a cylinder of an internal combustion engine in which the exhaust ports are located above the intake ports on the same side of the cylinder, so that gases circulate in a loop, leaving a dead spot in the center of the loop.

mantle [ENG] A lacelike hood or envelope (sack) of refractory material which, when positioned over a flame and heated to incandescence, gives light. [MET] That part of the outer wall and casing of a blast furnace located above the hearth.

man trip [MIN ENG] A trip made by mine cars and locomotives to take workers, rather than coal, to and from the working places.

manual casing hanger [PETRO ENG] A device in the bowl of the lowermost (or intermediate) casing head to suspend the next smaller casing string and to provide a seal between the suspended casing and the casing-head bowl.

manual element [IND ENG] An element completed by hand.

manually controlled work *See* effort-controlled cycle.

manual spinning [MET] A sheet-metal forming process that forms the material over a rotating mandrel with little or no change in the thickness of the original blank. Also known as conventional spinning.

manual tracking [ENG] System of tracking a target in which all the power required is supplied manually through the tracking handwheels.

manual welding [MET] A welding method in which the operator manually guides an electrode, clamped in a hand-held electrode holder.

manufactured gas [MATER] A gaseous fuel that is manufactured from soft coal or from various petroleum products; the gas mixture is composed of producer gas or carbureted water gas.

manufacturer's part number [IND ENG] Identification number of symbol assigned by the manufacturer to a part, subassembly, or assembly.

manure [MATER] Animal excreta collected from stables and barnyards with or without litter; used to enrich the soil.

maple [MATER] The hard, light-colored, close-grained wood, especially from sugar maple (*Acer saccharum*).

maraging steel [MET] High-strength, low-carbon iron-nickel alloy in which a martensitic structure is formed on cooling; contains 7–6% nickel, 0–11% cobalt, 0–5% molybdenum, and small percentages of titanium, aluminum, and columbium; hardening is accomplished by heating the quenched alloy at 400–500°C.

marble dust *See* marble flour.

marble flour [MATER] Finely divided marble chips; used as a filler or abrasive in hand soaps and for casting. Also known as marble dust.

marble shot [PETRO ENG] An explosive shot in open-hole well completions in which glass marbles are packed around the explosive in the wellbore; the marbles become projectiles that help break up the formation.

marbling [ENG] The use of antiquing techniques to achieve the appearance of marble in a paint film.

Marconaflo slurry transport [MIN ENG] A system which recovers solids from ore or tailings piles and mixes them with water to produce a transportable slurry in pipelines.

Marcy mill [MIN ENG] A ball mill with a vertical grate diaphragm placed near the discharge end; screens for sizing the material are located between the diaphragm and the end of the tube.

marginal cost [IND ENG] The extra cost incurred for an extra unit of output.

marginal product [IND ENG] The extra unit of output obtained by one extra unit of some factor, all other factors being held constant.

marginal revenue [IND ENG] The extra revenue achieved by selling an extra unit of output.

margin of safety [DES ENG] A design criterion, usually the ratio between the load that would cause failure of a member or structure and the load that is imposed upon it in service.

margosa oil *See* neem oil.

Marietta miner [MIN ENG] Trade name for a continuous miner mounted on caterpillar treads; has two cutter arms and two cutter chains at the working face, and a conveyor system to carry the broken coal to be loaded onto cars.

marigraph [ENG] A self-registering gage that records the heights of the tides.

marina [CIV ENG] A harbor facility for small boats, yachts, and so on, where supplies, repairs, and various services are available.

marine engineering [ENG] The design, construction, installation, operation, and maintenance of main power plants, as well as the associated auxiliary machinery and equipment, for the propulsion of ships.

marine glue [MATER] An adhesive that is insoluble in water; usually made of rubber and shellac, sometimes with resins.

marine railway [CIV ENG] A type of dry dock consisting of a cradle of wood or steel with rollers on which the ship may be hauled out of the water along a fixed inclined track leading up the bank of a waterway.

Mariner program [AERO ENG] A United States program, begun in 1962, to send a series of uncrewed, solar-powered spacecraft to the vicinity of Venus and Mars, to carry out observations of cosmic rays and solar wind in interplanetary space, to investigate interactions of the solar wind with planets, and to make atmospheric and surface measurements of planets, including photographic scanning of the surface of Mars.

marine terminal [CIV ENG] That part of a port or harbor with facilities for docking, cargo-handling, and storage.

MARISAT [AERO ENG] A geostationary communication satellite equipped with a repeater operating at microwave frequencies, for ship-to-shore communication by satellite. Derived from maritime satellite.

maritime satellite *See* MARISAT.

marjoram oil [MATER] A colorless essential liquid whose chief components are terpenes, obtained from marjoram plants of the genus *Origanum*; used as a perfume in soaps, and in flavorings.

market analysis [IND ENG] The collection and evaluation of data concerned with the past, present, or future attributes of potential consumers for a product or service.

Marseilles soap [MATER] A castile soap to which is added olive oil and soda.

Mars probe [AERO ENG] A United States unmanned spacecraft intended to be sent to the vicinity of the planet Mars, such as in the Mariner or Viking programs.

martempering [MET] Quenching austenitized steel to a temperature just above, or in the upper part of, the martensite range, holding it at this point until the temperature is equalized throughout, and then cooling in air to room temperature.

martensite [MET] A metastable transitional structure formed by a shear process during a phase transformation, characterized by an acicular or needlelike pattern; in carbon steel it is a hard, supersaturated solid solution of carbon in a body-centered tetragonal lattice of iron.

martensite range [MET] The temperature interval between the temperature (M_s) at which formation of martensite initiates and the temperature (M_f) at which the formation is complete.

martensitic stainless steel [MET] A hard, quenched magnetic martensitic steel containing principally 11–18% chromium and 0.1–1.2% carbon.

martensitic steel [MET] Quenched carbon steel composed chiefly of martensite.

martensitic structure [MET] Of, pertaining to, or having the structure of martensite, that is, an interstitial, supersaturated solid solution of carbon in iron having a body-centered tetragonal lattice; the microstructure is characterized by an acicular or needlelike pattern.

martensitic transformation [MET] A phase transformation which occurs in some metals, resulting in the formation of martensite. Also known as shear transformation.

Martin's cement [MATER] A gypsum cement made with potassium carbonate instead of alum.

martonite [MATER] A poison gas composed of 20% chloroacetone and 80% bromoacetone; acts as a powerful lacrimator.

Marvin sunshine recorder [ENG] A sunshine recorder in which the time scale is supplied by a chronograph, and consisting of two bulbs (one of which is blackened) that communicate through a glass tube of small diameter, which is partially filled with mercury and contains two electrical contacts; when the instrument is exposed to sunshine, the air in the blackened bulb is warmed more than that in the clear bulb; the warmed air expands and forces the mercury through the connecting tube to a point where the electrical contacts are shorted by the mercury; this completes the electrical circuit to the pen on the chronograph.

mash seam weld [MET] A seam weld at a lap joint in which the overall lap thickness is reduced plastically to the approximate thickness of one of the lapped parts.

mask [DES ENG] A frame used in front of a television picture tube to conceal the rounded edges of the screen. [MET] A protective device in thermal spraying against blasting or coating effects which are reflected from the substrate surface.

masking [ENG] Preventing entrance of a tracer gas into a vessel by covering the leaks.

masonry [CIV ENG] A construction of stone or similar materials such as concrete or brick.

masonry cement [MATER] A blended cement, made by combining either natural or portland cements with fattening materials such as hydrated lime and, sometimes, with air-entraining mixtures; used in the mortar of brick and block masonry.

masonry dam [CIV ENG] A dam constructed of stone or concrete blocks set in mortar.

masonry drill [DES ENG] A drill tipped with cemented carbide for drilling in concrete or masonry.

masonry nail [DES ENG] Spiral-fluted nail designed to be driven into mortar joints in masonry.

mason's hydrated lime [MATER] Any hydrated lime suitable for use in mortars, base-coat plasters, and concrete.

mass concrete [CIV ENG] Concrete set without structural reinforcement.

mass-distance [ENG] The mass carried by a vehicle multiplied by the distance it travels.

mass-flow bin [ENG] A bin whose hopper walls are sufficiently steep and smooth to cause flow of all the solid, without stagnant regions, whenever any solid is withdrawn.

mass flowmeter [ENG] An instrument that measures the mass of fluid that flows through a pipe, duct, or open channel in a unit time.

mass-haul curve [CIV ENG] A curve showing the quantity of excavation in a cutting which is available for fill.

mass law of sound insulation [CIV ENG] The rule stating that sound insulation for a single wall is determined almost wholly by its weight per unit area; doubling the weight of the partition increases the insulation by 5 cibels.

mass ratio [AERO ENG] The ratio of the mass of the propellant charge of the rocket to the total mass of the rocket when charged with the propellant.

mass spectrograph [ENG] A mass spectroscope in which the ions fall on a photographic plate which after development shows the distribution of particle masses.

mass spectrometer [ENG] A mass spectroscope in which a slit moves across the paths of particles with various masses, and an electrical detector behind it records the intensity distribution of masses.

mass spectroscope [ENG] An instrument used for determining the masses of atoms or molecules, in which a beam of ions is sent through a combination of electric and magnetic fields so arranged that the ions are deflected according to their masses.

mast [ENG] **1.** A vertical metal pole serving as an antenna or antenna support. **2.** A slender vertical pole which must be held in position by guy lines. **3.** A drill, derrick, or tripod mounted on a drill unit, which can be raised to operating position by mechanical means. **4.** A single pole, used as a drill derrick, supported in its upright or operating position by guys.

master [ENG] **1.** A device which controls subsidiary devices. **2.** A precise workpiece through which duplicates are made. [ENG ACOUS] *See* master phonograph record.

master alloy [MET] An alloy of selected elements that can be added to a charge of molten metal to provide a desired composition or texture or to deoxidize the material. Also known as foundry alloy.

masterbatch [MATER] A plastic, rubber, or elastomer mixture in which there is a high additives concentration, such as rubber with carbon black, or plastic with color pigment; used to proportion additives accurately into large bulks of plastic, rubber, or elastomer.

master bushing *See* liner bushing.

master cylinder [MECH ENG] The container for the fluid and the piston, forming part of a device such as a hydraulic brake or clutch.

master frequency meter *See* integrating frequency meter.

master gage [DES ENG] A locating device with fixed hole locations or part positions; locates in three dimensions and generally occupies the same space as the part it represents.

master layout [DES ENG] A permanent template record laid out in reference planes and used as a standard of reference in the development and coordination of other templates.

master mechanic [ENG] The supervisor, as at the mine, in charge of the maintenance and installation of equipment.

master phonograph record [ENG ACOUS] The negative metal counterpart of a disk recording, produced by electroforming as one step in the production of phonograph records. Also known as master.

master/slave manipulator [ENG] A mechanical, electromechanical, or hydromechanical device which reproduces the hand or arm motions of an operator, enabling the operator to perform manual motions while separated from the site of the work.

mastic [MATER] **1.** A glasslike, brittle, yellow to greenish yellow resinous exudation of the mastic tree (*Pistacia lentiscus*); used in medicine, condiments, adhesive, incense, and lacquer. Also

known as mastiche; mastix; pistachia galls. **2.** Mixture of finely powdered rock and asphaltic material used for highway construction.

mastic asphalt *See* asphalt mastic.

mastiche *See* mastic.

mastic oil [MATER] Colorless essential oil with balsamic odor; the chief constituents are pinenes; used in medicine.

mastix *See* mastic.

mat [CIV ENG] **1.** A steel or concrete footing under a post. **2.** Mesh reinforcement in a concrete slab. **3.** A heavy steel-mesh blanket used to suppress rock fragments during blasting. [MATER] Randomly distributed felt or glass fibers used in reinforced-plastics lay-up molding. [MIN ENG] An accumulation of broken mine timbers, rock, earth, and other debris coincident with the caving system of mining.

match [ENG] **1.** A charge of gunpowder put in a paper several inches long and used for igniting explosives. **2.** A short flammable piece of wood, paper, or other material tipped with a combustible mixture that bursts into flame through friction.

matched edges [ENG] Die face edges machined at right angles to each other to provide for alignment of the dies in machining equipment.

matched-metal molding [ENG] Forming of reinforced-plastic articles between two close-fitting metal molds mounted in a hydraulic press.

match plate [MET] A plate on which metal-casting patterns are mounted or formed as an integral part, to facilitate the molding operation.

materials control [IND ENG] Inventory control of materials involved in manufacturing or assembly.

materials handling [ENG] The loading, moving, and unloading of materials.

materials science [ENG] The study of the nature, behavior, and use of materials applied to science and technology.

mat foundation [CIV ENG] A large, thick, usually reinforced concrete mat which transfers loads from a number of columns, or columns and walls, to the underlying rock or soil. Also known as raft foundation.

Matheson joint [DES ENG] A wrought-pipe joint made by enlarging the end of one pipe length to receive the male end of the next length.

matico oil [MATER] An alcohol-soluble, yellowish-brown volatile oil distilled from the leaves or flowers of matico; chief constituents are asarone and methyl eugenol.

mat packs [MIN ENG] Timbers laid side by side into a solid mass 2 by 2½ feet (61 by 76 centimeters) square and 4–6 inches (10–15 centimeters) thick, kept together by wires through holes drilled through the edges of the mass, carried into a mine and built up to make effective roof supports.

matrix [ENG] A recessed mold in which something is formed or cast. [MATER] A binding agent used to make an agglomerate mass. [MET] **1.** The principal component of an alloy. **2.** The precisely shaped form used as the cathode in electroforming.

matrix sound system [ENG ACOUS] A quadraphonic sound system in which the four input channels are combined into two channels by a coding process for recording or for stereo frequency-modulation broadcasting and decoded back into four channels for playback of recordings or for quadraphonic stereo reception.

matte [MATER] Dull, as applied to appearance of a surface. [MET] An impure metallic sulfide mixture produced by smelting the sulfide ores of such metals as copper, lead, or nickel.

matte dip [MET] An etching solution which reacts with the surface of a metal, giving a dull finish.

matte smelting [MET] Smelting of copper-bearing materials in a reverberatory furnace.

mattock [DES ENG] A tool with the combined features of an adz, an ax, and a pick.

max-flow min-cut theorem [IND ENG] In the analysis of networks, the concept that for any network with a single source and sink, the maximum feasible flow from source to sink is equal to the minimum cut value for any of the cuts of the network.

maximal flow [IND ENG] Maximum total flow from the source to the sink in a connected network.

maximum allowable working pressure [MECH ENG] The maximum gage pressure in a pressure vessel at a designated temperature, used for the determination of the set pressure for relief valves.

maximum angle of inclination [MECH ENG] The maximum angle at which a conveyor may be inclined and still deliver an amount of bulk material within a given time.

maximum belt slope [MECH ENG] A slope beyond which the material on the belt of a conveyor tends to roll downhill.

maximum belt tension [MECH ENG] The total of the starting and operating tensions in a conveyor.

maximum continuous load [MECH ENG] The maximum load that a boiler can maintain for a designated length of time.

maximum gradability [MECH ENG] Steepest slope a vehicle can negotiate in low gear;

usually expressed in precentage of slope, namely, the ratio between the vertical rise and the horizontal distance traveled; sometimes expressed by the angle between the slope and the horizontal.

maximum producible oil index [PETRO ENG] An approximation of the maximum amount of oil per bulk formation volume that is producible with water drive.

maximum production life [MECH ENG] The length of time that a cutting tool performs at cutting conditions of maximum tool efficiency.

maximum thermometer [ENG] A thermometer that registers the maximum temperature attained during an interval of time.

maximum working area [IND ENG] That portion of the working area that is readily accessible to the hands of a worker when in his normal operating position.

Maxwellian distribution See Maxwell distribution.

May-Grünwald stain [MATER] A saturated solution of methylene blue eosinate in methyl alcohol; used to stain blood. Also known as Jenner's stain.

McCaa breathing device [MIN ENG] A self-contained, compressed-oxygen breathing apparatus designed by the U.S. Bureau of Mines for mine rescue work; it is carried on the back in a protective aluminum cover.

McGinty [MIN ENG] Three sheaves over which a rope is passed so as to take a course somewhat like that of the letter M.

McLuckie gas detector [MIN ENG] A portable nonautomatic means of analyzing air that can be used underground or aboveground if the air sample is brought out of the mine in small rubber bladders.

McNally-Carpenter centrifuge [MIN ENG] A machine that removes water from fine coal; it has a cone-shaped rotating vertical element, into the top of which the wet feed is put by gravity; a distributing disk forces the material onto the screen or basket lining the cone; as the material approaches the cone bottom, the drying action increases until the dry coal is discharged from the bottom.

McNally-Norton jig [MIN ENG] A device used to clean raw coal by carrying it to a wash box and using water whose level rises and falls as a result of air pulsations; the incoming coal is suspended by the pulsating water while the heavier refuse sinks to the bottom; the coal then spills over into the next wash box where the process is repeated, and the clean coal is discharged by the dewatering screens.

McNally-Vissac dryer [MIN ENG] A coal dryer that operates on the convection from the heavy forced draft from a coal-fired furnace; it consists of an inclined reciprocating screen over which coal moves; moisture is removed by the hot air from the furnace passing down through the bed of coal.

McQuaid-Ehn test [MET] A test for determining the grain-size characteristics of a steel in which a sample is carburized for 8 hours at 1700°F (927°C) and cooled slowly; the high-carbon case on slow cooling will reject cementite at the austenite grain boundaries and, by polishing and etching, the grains will clearly be seen under a microscope.

M-design bit [DES ENG] A long-shank, box-threaded core bit made to fit M-design core barrels.

M-design core barrel [DES ENG] A double-tube core barrel in which a 2½°-taper core lifter is carried inside a short tubular sleeve coupled to the bottom end of the inner tube, and the sleeve extends downward inside the bit shank to within a very short distance behind the face of the core bit.

meadow [ENG] Range of air-fuel ratio within which smooth combustion may be had.

mean camber line [AERO ENG] A line on a cross section of a wing of an aircraft which is equidistant from the upper and lower surfaces of the wing.

mean chord [AERO ENG] That chord of an airfoil that is equal to the sum of all the airfoil's chord lengths divided by the number of chord lengths; equivalently, that chord whose length is equal to the area of the airfoil section divided by the span.

mean effective pressure [MECH ENG] A term commonly used in the evaluation for positive displacement machinery performance which expresses the average net pressure difference in pounds per square inch on the two sides of the piston in engines, pumps, and compressors. Abbreviated mep; mp. Also known as mean pressure.

mean pressure *See* mean effective pressure.

mean time to failure [ENG] A measure of reliability of a piece of equipment, giving the average time before the first failure.

mean time to repair [ENG] A measure of reliability of a piece of repairable equipment, giving the average time between repairs.

measured daywork [IND ENG] Work done for an hourly wage on which specific productivity levels have been determined but which provides no incentive pay.

measured drilling depth [ENG] The apparent depth of a borehole as measured along its longitudinal axis.

measured mile [CIV ENG] The distance of 1 mile (1609.344 meters), the units of which have been accurately measured and marked.

measured ore *See* developed reserves.

measured relieving capacity [DES ENG] The measured amounts of fluid which can be exhausted through a relief device at its rated operating pressure.

measurement ton *See* ton.

measuring chute [MIN ENG] An ore bin or coal bin installed adjacent to the shaft bottom in skip winding and having a capacity equal to that of the skip used; ensures rapid, correct loading of skips without spillage. Also known as measuring pocket.

measuring day [MIN ENG] The day when work is measured and recorded for assessing the wages.

measuring machine [ENG] A device in which an astronomical photographic plate is viewed through a fixed low-power microscope with cross-hairs and which is mounted on a carriage that is moved by micrometer screws equipped with scales, in order to measure the relative positions of images on the plate.

measuring pocket *See* measuring chute.

mechanical [ENG] Of, pertaining to, or concerned with machinery or tools.

mechanical advantage [MECH ENG] The ratio of the force produced by a machine such as a lever or pulley to the force applied to it.

mechanical analysis [MECH ENG] Mechanical separation of soil, sediment, or rock by sieving, screening, or other means to determine particle-size distribution.

mechanical area [BUILD] The areas in a building that include equipment rooms, shafts, stacks, tunnels, and closets used for heating, ventilating, air conditioning, piping, communication, hoisting, conveying, and electrical services.

mechanical bearing cursor *See* bearing cursor.

mechanical classification [MECH ENG] A sorting operation in which mixtures of particles of mixed sizes, and often of different specific gravities, are separated into fractions by the action of a stream of fluid, usually water.

mechanical classifier [MECH ENG] Any of various machines that are commonly used to classify mixtures of particles of different sizes, and sometimes of different specific gravities; the Dorr classifier is an example.

mechanical comparator [ENG] A contact comparator in which movement is amplified usually by a rack, pinion, and pointer or by a parallelogram arrangement.

mechanical damping [ENG ACOUS] Mechanical resistance which is generally associated with the moving parts of an electromechanically transducer such as a cutter or a reproducer.

mechanical dewaxing [PETRO ENG] A dewaxing procedure in which cooled oil is forced through wax presses, separating the solid wax particles from the oil.

mechanical draft [MECH ENG] A draft that depends upon the use of fans or other mechanical devices; may be induced or forced.

mechanical-draft cooling tower [MECH ENG] Cooling tower that depends upon fans for introduction and circulation of its air supply.

mechanical efficiency [MECH ENG] In an engine, the ratio of brake horsepower to indicated horsepower.

mechanical engineering [ENG] The branch of engineering that deals with the generation, transmission, and utilization of heat and mechanical power and with the production of tools, machines, and their products.

mechanical equation of state [MET] An equation that expresses the relation of stress, strain, strain rate, and temperature for a metal.

mechanical filter [PETRO ENG] Granule-packed steel shell used to filter suspended floc or undissolved solids out of treated waterflood water; granules can be graded sand and gravel, anthracite coal, graphitic ore, or aluminum-oxide plates with granular filter medium.

mechanical flotation cell [MIN ENG] A device that separates minerals from ore water pulp; it consists of a cell in which the pulp is kept mixed and moving by an impeller at the bottom of the cell; the impeller pulls air down the standpipe and disperses it as bubbles through the pulp; the floatable minerals concentrate in the froth above, and the pulp is removed by a scraper.

mechanical hygrometer [ENG] A hygrometer in which an organic material, most commonly a bundle of human hair, which expands and contracts with changes in the moisture in the surrounding air or gas is held under slight tension by a spring, and a mechanical linkage actuates a pointer.

mechanical lift dock [CIV ENG] A type of dry dock or marine elevator in which a vessel, after being placed on the keel and bilge blocks in the dock, is bodily lifted clear of the water so that work may be performed on the underwater body.

mechanical linkage [MECH ENG] A set of rigid bodies, called links, joined together at pivots by means of pins or equivalent devices.

mechanical loader [MECH ENG] A power machine for loading mineral, coal, or dirt.

mechanically foamed plastic [MATER] A foamed plastic having its cellular structure produced by gases that are physically incorporated.

mechanical metallurgy [MET] The science and technology of the behavior of metals relating to mechanical forces imposed on them; includes rolling, extruding, deep drawing, bending, and other processes.

mechanical mucking [ENG] Loading of dirt or stone in tunnels or mines by machines.

mechanical oil valve [PETRO ENG] A float-operated liquid-level control valve used to control liquid flow out of oil-lease gas-oil separator tank systems.

mechanical oscillograph *See* direct-writing recorder.

mechanical patent [ENG] A patent granted for an inventive improvement in a process, manufacture, or machine.

mechanical plating [MET] Deposition of one metal on another by a cold-peening process, such as tumbling.

mechanical press [MECH ENG] A press whose slide is operated by mechanical means.

mechanical puddling *See* vibration puddling.

mechanical pulp [MATER] Wood pulp produced by grinding and soaking the wood fibers. Also known as groundwood pulp.

mechanical pulping [MECH ENG] Mechanical, rather than chemical, recovery of cellulose fibers from wood; unpurified, finely ground wood is made into newsprint, cheap Manila papers, and tissues.

mechanical pump [MECH ENG] A pump through which fluid is conveyed by direct contact with a moving part of the pumping machinery.

mechanical refrigeration [MECH ENG] The removal of heat by utilizing a refrigerant subjected to cycles of refrigerating thermodynamics and employing a mechanical compressor.

mechanical scale [ENG] A weighing device that incorporates a number of levers with precisely located fulcrums to permit heavy objects to be balanced with counterweights or counterpoises.

mechanical seal [MECH ENG] Mechanical assembly that forms a leakproof seal between flat, rotating surfaces to prevent high-pressure leakage.

mechanical separation [MECH ENG] A group of industrial operations by means of which particles of solid or drops of liquid are removed from a gas or liquid, or are separated into individual fractions, or both, by gravity separation (settling), centrifugal action, and filtration.

mechanical setting [MECH ENG] Producing bits by setting diamonds in a bit mold into which a cast or powder metal is placed, thus embedding the diamonds and forming the bit crown; opposed to hand setting. Also known as cast setting; machine setting; sinter setting.

mechanical shovel [MECH ENG] A loader limited to level or slightly graded drivages; when full, the shovel is swung over the machine, and the load is discharged into containers or vehicles behind.

mechanical splice [ENG] A splice made to terminate wire rope by pressing one or more metal sleeves over the rope junction.

mechanical spring *See* spring.

mechanical stage [ENG] A stage on a microscope provided with a mechanical device for positioning or changing the position of a slide.

mechanical stoker *See* automatic stoker.

mechanical torque converter [MECH ENG] A torque converter, such as a pair of gears, that transmits power with only incidental losses.

mechanical twin [MET] A twin formed in a metal crystal by plastic deformation, involving shear of the lattice.

mechanical working [MET] Formation of a desired shape and physical properties of a metal by subjecting it to pressure by rolls, presses, or hammers.

mechanical yielding prop [MIN ENG] A steel prop in which yield is controlled by friction between two sliding surfaces or telescopic tubes. Also known as friction yielding prop.

mechanism [MECH ENG] That part of a machine which contains two or more pieces so arranged that the motion of one compels the motion of the others.

mechanize [MECH ENG] **1.** To substitute machinery for human or animal labor. **2.** To produce or reproduce by machine.

mechanized dew-point meter *See* dew-point recorder.

mechanooptical vibrometer [ENG] A vibrometer in which the motion given to a probe by a surface whose vibration amplitude is to be measured is used to rock a mirror; a light beam reflected from the mirror and focused onto a scale provides an indication of the vibration amplitude.

median strip [CIV ENG] A paved or planted section dividing a highway into lanes according to direction of travel.

medicinal oil [MATER] A highly refined, colorless, tasteless and odorless petroleum oil used medicinally as an internal lubricant and for the manufacture of salves

and ointments. Also known as mineral oil.

medium boiler [MATER] A solvent, intermediate in volatility between high and low boilers, which is added to lacquer thinner and influences flow and freedom from an orange peel condition during drying; boiling point is 115–145°C.

medium carbon steel [MET] Steel containing 0.15–0.30% carbon.

medium-curing asphalt [MATER] Liquid product composed of asphalt cement and a kerosine-type diluent. Also known as M-C asphalt.

melamine resin [MATER] An amino resin made from formaldehyde and melamine; it is used as a molding compound with fillers added to it; it may also be used for laminating.

melt [MET] A charge of molten metal.

melter [ENG] A chamber used for melting.

melt extraction [MET] A rapid quenching process in which the molten metal is brought into contact with the periphery of a rotating heat-extracting disk; quench rates exceed 1,000,000 K per second.

melt extractor [ENG] A device used to feed an injection mold, separating molten feed material from partially molten pellets.

melt-fabricable [MATER] Referring to a plastic material that can be shaped as a melt without decomposing, and is capable of being extruded.

melt index [ENG] Number of grams of thermoplastic resin at 190°C that can be forced through a 0.0825-inch (2.0955-millimeter) orifice in 10 minutes by a 2160-gram force.

melting furnace [ENG] A furnace in which the frit for glass is melted.

melting loss [MET] Weight loss due to volatilization or oxidation during metal melting in a foundry.

melting rate [MET] In electric arc welding, the weight or length of electrode melted in a specified unit of time. Also known as burn-off rate; melt-off rate.

melting ratio [MET] The ratio of metal weight to fuel weight in a melting process.

melt-off rate *See* melting rate.

member [CIV ENG] A structural unit such as a wall, column, beam, or tie, or a combination of any of these.

membrane [BUILD] In built-up roofing, a weather-resistant (flexible or semiflexible) covering consisting of alternate layers of felt and bitumen, fabricated in a continuous covering and surfaced with aggregate or asphaltic material.

membrane waterproofing [CIV ENG] Curing concrete, especially in pavements, by spraying a liquid material over the surface to form a solid, impervious layer which holds the mixing water in the concrete. Also known as membrane curing.

memomotion study [IND ENG] A technique of work measurement and methods analysis using a motion picture camera operated at less than normal camera speed. Also known as camera study.

menhaden oil [MATER] A combustible drying oil that is soluble in benzene or ether, and is derived by cooking or pressing the menhaden fish.

meniscus [MET] In reference to a solder joint, the minimum angle at which the solder tapers from the joint to the flat area.

menstruum [MATER] A solvent, commonly one that extracts certain principles from entire plant or animal tissues.

mep *See* mean effective pressure.

Mercer engine [MECH ENG] A revolving-block engine in which two opposing pistons operate in a single cylinder with two rollers attached to each piston; intake ports are uncovered when the pistons are closest together, and exhaust ports are uncovered when they are farthest apart.

mercerizing assistant [MATER] A wetting agent, such as cresylic acid and derivatives or oils, used to increase the penetration of textile mercerization baths.

merchant mill [MET] A mill, consisting of a group of stands of three rolls each, used to roll rounds, flats, or squares of smaller dimensions than could be rolled on a bar mill.

mercury barometer [ENG] An instrument which determines atmospheric pressure by measuring the height of a column of mercury which the atmosphere will support; the mercury is in a glass tube closed at one end and placed, open end down, in a well of mercury. Also known as Torricellian barometer.

mercury manometer [ENG] A manometer in which the instrument fluid is mercury; used to record or control difference of pressure or fluid flow.

Mercury program [AERO ENG] First United States program to use manned spacecraft, carried out in 1961–1963; the craft carried one man.

mercury thermometer [ENG] A liquid-in-glass thermometer or a liquid-in-metal thermometer using mercury as the liquid.

meridian circle *See* transit circle.

meridian transit *See* transit; transit circle.

merit pay plan [IND ENG] Work performed for a set hourly wage that varies from one pay period to another as a function of the worker's productivity, but never declines below a guaranteed minimum wage.

Merrick Weightometer [MECH ENG] An instrument which measures the weight of material being carried by a belt conveyor, in which the weight of the load on a portion of the conveyor is balanced, using a system of levers, by the buoyancy of a cylindrical steel float partly immersed in a bath of mercury.

Merrill-Crowe process [MET] Removal of gold from cyanide solution by deoxygenation followed by precipitation on zinc dust, the work being completed by filtration to give the resultant auriferous gold slimes.

Merton nut [DES ENG] A nut whose threads are made of an elastic material such as cork, and are formed by compressing the material into a screw.

mesh [DES ENG] A size of screen or of particles passed by it in terms of the number of openings occurring per linear inch. Also known as mesh size. [MECH ENG] Engagement or working contact of teeth of gears or of a gear and a rack. [MIN ENG] **1.** A closed path traversed through the network in ventilation surveys. **2.** The size of diamonds as determined by sieves.

mesh weld [MET] A seam weld in which the finished weld is only slightly thicker than the sheets, and the lap disappears.

messenger [ENG] A small, cylindrical metal weight that is attached around an oceanographic wire and sent down to activate the tripping mechanism on various oceanographic devices.

metachromatic stain [MATER] A stain which changes apparent color when absorbed by certain cell constituents.

metal [MATER] An opaque crystalline material usually of high strength with good electrical and thermal conductivities, ductility, and reflectivity; properties are related to the structure, in which the positively charged ions are bonded through a field of free electrons which surrounds them forming a close-packed structure.

metal arc cutting [MET] Cutting metal with the heat of an arc between a metal electrode and the base metal.

metal arc welding [MET] Arc welding using covered metal electrodes.

metal ceramic *See* cermet.

metal coating [MET] A thin film of metal bonded to a base material in order to add specific surface properties, such as corrosion or oxidation resistance, color, wear resistance, or optical characteristics.

metal distribution ratio [MET] In plating operations, the ratio of the thickness of metal deposited on a near portion of a cathode to that deposited on a far portion of the cathode.

metal dye [MATER] Any of the special dyes, such as alizarin cyanin RR or alizarin green S, used to color oxided surfaces of aluminum or steel.

metal-foil paper [MATER] Paper backed with metal foils, manufactured in a number of vivid colors.

metal forming [MET] Any manufacturing process by which parts or components are fabricated by shaping or molding a piece of metal stock.

metaliding [MET] A process of depositing a metal as an alloy on a substrate from a fused complex metal salt.

metal lath [ENG] A mesh of metal used to provide a base for plaster.

metal leaf [MET] Metal sheet, thinner than foil, formed by beating.

metallic corrosion [MET] Destruction of a metal by dissolution, oxidation, or other chemical reaction of the metal with its environment.

metallic glass See glassy alloy.

metallic mortar [MATER] Mortar made with ceramic oxide binders and containing a high percentage of lead powder; mixed with water to form plasters or for casting sections and blocks; used for x-ray and nuclear installation shielding.

metallic paint [MATER] **1.** Paint used for covering metal surfaces; the pigment is commonly iron oxide. **2.** Paint with a metal pigment.

metallic paper [MATER] **1.** A paper coated with zinc white, or with clay and other materials; the surface can be marked on with metal points (of silver, aluminum, or gold, for example), but it cannot be erased. **2.** Paper coated with finely flaked metal.

metallic pigment [MATER] Thin, opaque aluminum or copper alloy flakes that are incorporated into plastic masses to produce metallike effects.

metallize [ENG] To coat or impregnate a metal or nonmetal surface with a metal, as by metal spraying or by vacuum evaporation.

metallized slurry blasting [ENG] The breaking of rocks by using slurried explosive medium containing a powdered metal, such as powdered aluminum.

metallized wood [MATER] Wood impregnated with molten metal, filling the cells in the wood to increase hardness, compressive strength, and flexural strength; the wood becomes an electrical conductor lengthwise of the grain.

metallographic tests [MET] Tests to determine the structural composition of a metal as shown at low and high magnification and by x-ray diffraction methods; tests include macroexamination, microexamination, and x-ray diffraction studies.

metallography [MET] The study of the structure of metals and alloys by various methods, especially by the optical and the electron microscope, and by x-ray diffraction.

metallurgical balance sheet [MET] Material balance of a metallurgical process.

metallurgical coke [MATER] Coke resulting from high-temperature retorting of suitable coal; a dense, crush-resistant fuel for use in shaft furnaces.

metallurgical dust [MET] A mixture of particles of elements and nonmetallic and metallic compounds.

metallurgical engineer [ENG] A person who specializes in metallurgical engineering.

metallurgical engineering [ENG] Application of the principles of metallurgy to the engineering sciences.

metallurgical fume [MET] A mixture of fine particles of elements and metallic and nonmetallic compounds either sublimed or condensed from the vapor state.

metallurgical microscope [ENG] A microscope used in the study of metals, usually optical.

metal mining [MIN ENG] The industry that supplies the various metals and associated products.

metal pointing See pointing.

metal positive See mother.

metal powder [MET] A finely divided metal or alloy.

metal replacement See immersion plating.

metal rolling See rolling.

metal-slitting saw [MECH ENG] A milling cutter similar to a circular saw blade but sometimes with side teeth as well as teeth around the circumference; used for deep slotting and sinking in cuts.

metal spinning See spinning.

metal spraying [ENG] Coating a surface with droplets of molten metal or alloy by using a compressed gas stream.

meteorogram [ENG] A record obtained from a meteorograph.

meteorograph [ENG] An instrument that measures and records meteorological data such as air pressure, temperature, and humidity.

meteorological balloon [ENG] A balloon, usually of high-quality neoprene, polyethylene, or Mylar, used to lift radiosondes to high altitudes.

meteorological instrumentation [ENG] Apparatus and equipment used to obtain quantitative information about the weather.

meteorological radar [ENG] Radar which is used to study the scattering of radar waves by various types of atmospheric phenomena, for making weather observations and forecasts.

meteorological rocket [ENG] Small rocket system used to extend observation of atmospheric character above feasible limits for balloon-borne observing and telemetering instruments. Also known as rocketsonde.

meteorological satellite [AERO ENG] Earth-orbiting spacecraft carrying a variety of instruments for measuring visible and invisible radiations from the earth and its atmosphere.

meter [ENG] A device for measuring the value of a quantity under observation; the term is usually applied to an indicating instrument alone.

meter bar [ENG] A metal bar for mounting a gas meter, having fittings at the ends for the inlet and outlet connections of the meter.

meter factor [ENG] A factor used with a meter to correct for ambient conditions, for example, the factor for a fluid-flow meter to compensate for such conditions as liquid temperature change and pressure shrinkage.

metering installation [PETRO ENG] Oil-production receiving system that includes with the tank battery a metering separator, metering treater, or other type of meter used in conjunction with test separators or emulsion treaters.

metering pin *See* metering rod.

metering rod [ENG] A device consisting of a long metallic pin of graduated diameters fitted to the main nozzle of a carburetor (on an internal combustion engine) or passage leading thereto in such a way that it measures or meters the amount of gasoline permitted to flow by it at various speeds. Also known as metering pin.

metering screw [MECH ENG] An extrusion-type screw feeder or conveyor section used to feed pulverized or doughy material at a constant rate.

metering separator [PETRO ENG] Oil-field process vessel that performs the dual functions of gas-oil separation and liquids metering.

meter oil [MATER] High-purity grade of oil used to lubricate the moving elements of meters.

meter prover [ENG] A device that determines the accuracy of a gas meter; a quantity of air is collected over water or oil in a calibrated cylindrical bell, and then the bell is allowed to sink into the liquid, forcing the air through the meter; the calibrated measurement is then compared with the reading on the meter dial.

meter-proving tank *See* calibrating tank.

meter run [ENG] The length of straight, unobstructed fluid-flow conduit preceding an orifice or venturi meter.

meter sensitivity [ENG] The accuracy with which a meter can measure a voltage, current, resistance, or other quantity.

meter wheel [ENG] A special block used to support the oceanographic wire paid out over the side of a ship; attached directly or connected by means of a speedometer cable to a gearbox which measures the length of wire.

methane drainage *See* firedamp drainage.

methane indicator [MIN ENG] A portable analytical instrument that can determine the methane content in the mine air at the place where the sample is taken; air is brought into the instrument through an aspirator bulb and passed through a cartridge filter to remove moisture.

methane monitoring system [MIN ENG] A system that samples methane content in mine air continuously and feeds this information into an electrical device that cuts off power in each mining machine when the methane content rises above a predetermined level.

method of images [PETRO ENG] Method of calculating the interference between reservoirs by assuming a mirror image of one reservoir on the far side of a geologic fault.

method of joints [ENG] Determination of stresses for joints at which there are not more than two unknown forces by the methods of the stress polygon, resolution, or moments.

methods design [IND ENG] Design for a new, more efficient method of job performance.

methods engineering [IND ENG] A technique used by management to improve working methods and reduce labor costs in all areas where human effort is required.

methods study [IND ENG] An analysis of the methods in use, of the means and potentials for their improvement, and of reducing costs.

methyl acetone [MATER] Flammable, water-white liquid, a mixture of acetone, methanol, and methyl acetate in various proportions; miscible with water, oils, and hydrocarbons; used as a solvent.

metric thread gearing [DES ENG] Gears that may be interchanged in change-gear systems to provide feeds suitable for cutting metric and module threads.

Mexican linaloe oil *See* linaloe wood oil.

micellar flooding [PETRO ENG] A two-step enhanced oil recovery process in which a surfactant slug is injected into the well followed by a larger slug of water containing a high-molecular-weight polymer which pushes the chemicals through the field and improves mobility and sweep efficiency. Also known as microemulsion flooding; surfactant flooding.

Michaelson actinograph [ENG] A pyrheliometer of the bimetallic type used to measure the intensity of direct solar radiation; the radiation is measured in terms of the angular deflection of a blackened bimetallic strip which is exposed to the direct solar beams.

Michigan cut [MIN ENG] A technique used to break off ore at a heading; a large hole or series of small holes at the center of the heading are drilled parallel to the tunnel direction but not charged with explosive; other holes are drilled in the heading and charged so that upon detonation they break out toward the uncharged holes.

Michigan tripod [MIN ENG] A support for a drilling outfit; consists of three debarked pine or fir timber poles about 25 feet (7.6 meters) long whose butt ends are about 12 inches (30 centimeters) in diameter; a sheave suspended from a clevis at the top of the tripod is aligned over the hoisting drum and the borehole; there is a minimum of 22 feet (6.7 meters) of headroom above the drill floor.

microbalance [ENG] A small, light type of analytical balance that can weigh loads of up to 0.1 gram to the nearest microgram.

microbarogram [ENG] The record or trace made by a microbarograph.

microbicide [MATER] An agent that kills microbes.

microcaliper log [PETRO ENG] A detailed and accurate record of drill-hole diameter; used to detect caved sections and to verify the presence of mud cake.

microcrystalline wax [MATER] A petroleum wax containing small, indistinct crystals, and having a higher molecular weight, melting point, and viscosity than paraffin wax; used in laminated paper and electrical coil coating.

microemulsion flooding See micellar flooding.

microfissure [MET] A crack of microscopic dimensions. Also known as microcrack.

microfluoroscope [ENG] A fluoroscope in which a very fine-grained fluorescent screen is optically enlarged.

microforge [ENG] In micromanipulation techniques, an optical-mechanical device for controlling the position of needles or pipets in the field of a low-power microscope by a simple micromanipulator.

micrograph [ENG] An instrument for making very tiny writing or engraving.

microgroove record See long-playing record.

microhardness [MET] Hardness of microscopic areas of a metal or alloy.

microlaterolog [PETRO ENG] Modification of the downhole microlog in which extra electrodes focus electric current into a trumpet-shaped area; gives greater resistivity-measurement resolution than does the microlog.

microlog [PETRO ENG] A drill-hole resistivity log recorded with electrodes mounted at short distances from each other in the face of a rubber-padded microresistivity sonde.

micromanipulator [ENG] A device for holding and moving fine instruments for the manipulation of microscopic specimens under a microscope.

micromanometer [ENG] Any manometer that is designed to measure very small pressure differences.

micrometeorite penetration [AERO ENG] Penetration of the thin outer shell (skin) of space vehicles by small particles traveling in space at high velocities.

micrometer [ENG] **1.** An instrument attached to a telescope or microscope for measuring small distances or angles. **2.** A caliper for making precise measurements; a spindle is moved by a screw thread so that it touches the object to be measured; the dimension can then be read on a scale. Also known as micrometer caliper.

micrometer caliper See micrometer.

micromotion study [IND ENG] Study of the fundamental elements of an operation by means of a motion picture camera operated at normal or faster than normal speed, and a timing device to measure element motion time.

micronized clay [MATER] A pure kaolin pulverized to a fineness of 400 to 800 mesh; used as a filler material in rubber.

micronized mica [MATER] Powdered mica of a fineness of 400 to 1000 mesh; used as a filler.

microphone [ENG ACOUS] An electroacoustic device containing a transducer which is actuated by sound waves and delivers essentially equivalent electric waves.

microphone transducer [ENG ACOUS] A device which converts variation in the position or velocity of some body into corresponding variations of some electrical quantity, in a microphone.

microphotometer [ENG] A photometer that provides highly accurate illumination measurements; in one form, the changes in illumination are picked up by a phototube and converted into current variations that are amplified by vacuum tubes.

micropipet [ENG] **1.** A pipet with capacity of 0.5 milliliter or less, to measure small volumes of liquids with a high degree of accuracy; types include lambda, straight-bore, and Lang-Levy. **2.** A fine-pointed pipette used for microinjection.

microporosity [MET] Extremely fine porosity, visible only with the aid of a microscope.

micropycnometer [ENG] A small-volume pycnometer with a capacity from 0.25 to 1.6 milliliters; weighing precision is 1 part in 10,000, or better.

microresistivity survey [PETRO ENG] General term for downhole resistivity surveys of oil-bearing formations; includes microlog and microlaterolog surveys.

microscopic anisotropy [PETRO ENG] Phenomenon in electrical downhole logging wherein electric current flows most easily along the water-filled interstices, usually parallel to sedimentary bed strata.

microscopic stress [MET] Residual stress ranging from compression to tension in a metal within a distance often comparable to the grain size. Also known as microstress.

microsegregation [MET] Segregation within a grain, crystal, or particle of microscopic size.

microseismic instrument [MIN ENG] An instrument for the study of roof strata and supports; it is inserted in holes, drilled at selected points, for listening to subaudible vibrations that precede rock failure.

microshrinkage [MET] A casting defect consisting of interdendritic voids, visible only at magnifications over 10 diameters.

microstress *See* microscopic stress.

microtome [ENG] An instrument for cutting thin sections of tissues or other materials for microscopical examination.

microwave early warning [ENG] High-power, long-range radar with a number of indicators, giving high resolution, and with a large traffic-handling capacity; used for early warning of missiles.

microwave oven [ENG] An oven that uses microwave heating for fast cooking of meat and other foods.

Midas [AERO ENG] A two-object trajectory-measuring system whereby two complete cotar antenna systems and two sets of receivers at each station, with the multiplexing done after phase comparison, are utilized in tracking more than one object at a time.

midcourse correction [AERO ENG] A change in the course of a spacecraft some time between the end of the launching phase and some arbitrary point when terminal guidance begins.

middle-third rule [CIV ENG] The rule that no tension is developed in a wall or foundation if the resultant force lies within the middle third of the structure.

middling [MIN ENG] An ore product intermediate in mineral content between a concentrate and a tailing.

midget impinger [MIN ENG] A dust-sampling impinger requiring only a 12-inch (30-centimeter) head of water for its operation.

migration [MET] The uncontrolled movement of certain metals, particularly silver, from one location to another, usually with associated undesirable effects such as oxidation or corrosion.

mild abrasive [MATER] An abrasive material, such as chalk or talc, having a hardness of 1–2 on Mohs scale; used in silver polishes and window cleaners.

mild steel [MET] Carbon steel containing 0.05–0.25% carbon.

milepost [CIV ENG] **1.** A post placed a mile away from a similar post. **2.** A post indicating mileage from a given point.

military aircraft [AERO ENG] Aircraft that are designed or modified for highly specialized use by the armed services of a nation.

military engineering [ENG] Science, art, and practice involved in design and construction of defensive and offensive military works as well as construction and maintenance of transportation systems.

military geology [ENG] The application of the earth sciences to such military concerns as terrain analysis, water supply, foundations, and construction of roads and airfields.

military satellite [AERO ENG] An artificial earth satellite used for military purposes; the six mission categories are communication, navigation, geodesy, nuclear test detection, surveillance, and research and technology.

military technology [ENG] The technology needed to develop and support the armament used by the military.

milk glass [MATER] A white, and sometimes colored, opaque glass made by adding calcium fluoride and alumina to soda-lime glass.

mill [IND ENG] **1.** A machine that manufactures paper, textiles, or other products by the continuous repetition of some simple process or action. **2.** A building that houses machinery for manufacturing processes. [MIN ENG] **1.** An excavation made in the country rock, by a crosscut from the workings on a vein, to obtain waste for filling; it is left without timber so that the roof can fall in and furnish the required rock. **2.** A passage connecting a stope or upper level with a lower level intended to be filled with broken ore that can then be drawn out at the bottom as desired for further transportation.

millboard [MATER] Hard, strong paperboard; used for furniture panels.

mill building [CIV ENG] A steel-frame building in which roof trusses span columns in the outside wall; originally, this type of building housed milling machinery, as for wood or metal, hence the name.

milled soap [MATER] Soap, such as toilet soap, made by adding color and perfume to soap chips and then passing the mixture through milling rollers and pressing in molds.

miller *See* milling machine.

mill finish [MET] The characteristic surface finish on a rolled metal product.

mill-head ore *See* run-of-mill.

milling [MECH ENG] Mechanical treatment of materials to produce a powder, to change the size or shape of metal powder particles, or to coat one powder mixture with another. [MIN ENG] A combination of open-cut and underground mining, wherein the ore is mined in open cut and handled underground.

milling cutter [DES ENG] A rotary tool-steel cutting tool with peripheral teeth, used in a milling machine to remove material from the workpiece through the relative motion of workpiece and cutter.

milling machine [MECH ENG] A machine for the removal of metal by feeding a workpiece through the periphery of a rotating circular cutter. Also known as miller.

milling ore *See* second-class ore.

milling planer [MECH ENG] A planer that uses a rotary cutter rather than single-point tools.

milling system *See* chute system.

milling width [MIN ENG] Width of lode designated for treatment in the mill, as calculated with regard to daily tonnage.

millisecond delay cap [ENG] A delay cap with an extremely short (20–500 thousandths of a second) interval between passing of current and explosion. Also known as short-delay detonator.

mill ore [MIN ENG] An ore that must be given some preliminary treatment before a marketable grade or a grade suitable for further treatment can be obtained.

millrace [CIV ENG] A canal filled with water that flows to and from a waterwheel acting as the power supply for a mill.

mill run [MIN ENG] **1.** A given quantity of ore tested for its quality by actual milling. **2.** The yield of such a test.

mill scale [MET] A surface layer of ferric oxide (Fe_3O_4) that forms on steel or iron during hot rolling.

Mills-Crowe process [MIN ENG] Method of regeneration of cyanide liquor from the gold leaching process; the barren solution is acidified, liberating hydrocyanic acid which is separated and reabsorbed in alkaline solution.

millsite [MIN ENG] A plot of ground suitable for the erection of a mill, or reduction works, to be used in connection with mining operations.

millwork [MATER] Ready-made products which are manufactured at a wood-planing mill or woodworking plant, such as moldings, doors, doorframes, window sashes, stairwork, and cabinets.

millwright [ENG] **1.** A person who plans, builds, or sets up the machinery for a mill. **2.** A person who repairs milling machines.

Milorganite [MATER] Granular form of sewage-plant activated sludge; used as fertilizer.

minable [MIN ENG] Material that can be mined under present-day mining technology and economics.

mine [MIN ENG] An opening or excavation in the earth for extracting minerals.

mine captain [MIN ENG] **1.** The director of work in a mine, with or without superior officials or subordinates. **2.** *See* mine superintendent.

mine car [MECH ENG] An industrial car, usually of the four-wheel type, with a low body; the door is at one end, pivoted at the top with a latch at the bottom used for hauling bulk materials.

mine characteristic [MIN ENG] The relation between pressure p and volume Q in the ventilation of a mine of resistance R, expressed as $p = RQ^2$.

mine characteristic curve [MIN ENG] A graph derived by plotting the static or total mine head, or both, against the quantity; used to solve problems in mine ventilation.

mine development [MIN ENG] The operations involved in preparing a mine for ore extraction, including tunneling, sinking, crosscutting, drifting, and raising.

mine dust [MIN ENG] Dust from drilling, blasting, or handling rock.

mined volume [MIN ENG] A statistic used in mine subsidence, computed by multiplying the mined area by the mean thickness of the bed or of that part of the bed which has been dug out.

mine fan signal system [MIN ENG] A system which indicates by electric light or electric audible signal, or both, the slowing down or stopping of a mine ventilating fan.

mine fire truck [MIN ENG] A low-slung railcar designed to fight fires in mines; has a water supply and a pump to supply high pressure to the fire hoses.

mine foreman [MIN ENG] The worker charged with the general supervision of the underground workings of a mine and the persons employed therein.

Mine Gel [MATER] Brand name for a series of semigelatin dynamites.

mine hoist [MIN ENG] A device for raising and lowering ore, rock, or coal in a mine, and men and supplies.

mine inspector [MIN ENG] Generally, the state mine inspector, as contrasted to the Federal mine inspector; inspects mines to find fire and dust hazards and inspects the safety of working areas, electric circuits, and mine equipment.

mine jeep [MIN ENG] An electrically driven car for underground transportation of officials, inspectors, rescue workers, and repair, maintenance, and surveying crews.

minelite [MATER] An explosive made by mixing a chlorate compound with paraffin wax.

mine locomotive [MIN ENG] A low, heavy haulage engine designed for underground operation; usually propelled by electricity, gasoline, or compressed air.

mine props [MIN ENG] Sections of wood used for holding up pieces of rock in the roof of mines.

mine radio telephone system [MIN ENG] A communication system between the dispatcher and locomotive operators; radio impulses travel along the trolley wire and down the trolley pole to the radio telephone.

mineral additive [MATER] A mineral-derived substance added during grease manufacture, particularly to heavy-duty grease.

mineral black [MATER] Black pigment made from ground slate, shale, coke, coal, or slaty coal; used in inks, plastics, and coatings.

mineral cotton *See* mineral wool.

mineral dressing *See* beneficiation.

mineral dye [MATER] A natural dyestuff made from minerals; examples are ochre, chrome yellow, and Prussian blue.

mineral economics [MIN ENG] Study and application of the processes used in management and finance connected with the discovery, exploitation, and marketing of minerals.

mineral engineering *See* mining engineering.

mineral fuel [MATER] A carbonaceous fuel mined or stripped from the earth, such as petroleum, coal, peat, shale oil, or tar sands.

mineral jelly [MATER] A viscous, thick material derived from petroleum; used as a stabilizer in explosives.

mineral land [MIN ENG] Land which is worth more for mining than for agriculture or other use.

mineral lard oil [MATER] A mixture of refined mineral oil with lard oil, having a fatty content of 25–30, and a flash point about 300°F (149°C).

mineral lease *See* mining lease.

mineral monument [MIN ENG] A permanent monument established in a mining district to provide for an accurate description of mining claims and their location.

mineral oil *See* medicinal oil.

mineral processing [MIN ENG] Procedures, such as dry and wet crushing and grinding of ore or other products containing minerals, to raise the concentration of the substance being mined.

mineral right *See* mining right.

mineral rubber [MATER] Asphaltine minerals (such as gilsonite and grahamite) and blown asphalts; used in compounding of rubber, coatings, and paints.

mineral seal oil [MATER] A petroleum distillate with a boiling point higher than that of kerosine; used as a solvent oil and illuminant.

mineral wool [MATER] A fibrous substance, technically a glass, made from molten slag, rock, glass, or a selected combination of these ingredients; produced by blowing, drawing, or other means of fabricating into fine fibers; used for insulation, fireproofing, and as a filter medium. Also known as mineral cotton; rock wool; silicate cotton; slag wool.

mine rescue apparatus [MIN ENG] Certain types of apparatus worn by workers and permitting them to perform in noxious or irrespirable atmospheres, such as during mine fires or following mine explosions.

mine rescue crew [MIN ENG] A crew consisting usually of five to eight persons who are thoroughly trained in the use of mine rescue apparatus, which they wear in rescue or recovery work in a mine following an explosion or during a fire.

mine rescue lamp [MIN ENG] A particular type of electric safety hand lamp used in rescue operations; equipped with a lens for concentrating or diffusing the light.

mine resistance [MIN ENG] Resistance by a mine to the passage of an air current.

miner's friend [MIN ENG] **1.** The Davy safety lamp. **2.** A steam engine once used to pump water from underground.

miner's hammer [MIN ENG] A hammer used to break ore.

miner's hand lamp [MIN ENG] A self-contained mine lamp with handle for carrying.

miner's helmet [MIN ENG] A hat designed for miners to provide head protection and to hold the cap lamp.

miner's horn [MIN ENG] A metal spoon or horn to collect ore particles in gold washing.

miner's inch [MIN ENG] The quantity of water that will escape from an aperture 1-inch (2.54-centimeter) square through a 2-inch-thick (5.08-centimeter-thick) plank, with a steady flow of water standing 6 inches (15.24 centimeters) above the top of the escape aperture, the quantity so discharged amounting to 2274 cubic feet (64.39 cubic meters) in 24 hours.

miner's lamp [MIN ENG] Any one of a variety of lamps used by a miner to furnish light, such as oil lamps, carbide lamps, flame safety lamps, and cap lamps.

miner's right [MIN ENG] An annual permit from the government to occupy and work mineral land.

miners' rules [MIN ENG] Rules and regulations proclaimed by the miners of any district, relating to the location of, recording of, and the work necessary to hold possession of a mining claim.

miner's self-rescuer [MIN ENG] A pocket gas mask effective against carbon monoxide; air passes through a cannister containing fused calcium chloride before entering the mouth.

mine run [MIN ENG] The unscreened output of a mine. Also known as run-of-mine.

mine sample [MIN ENG] Coal or mineral extracted at underground exposures for analysis.

mine signal system [MIN ENG] Signal lights installed at individual track switches immediately indicating to the motorman whether or not he can safely proceed.

mine skips [MIN ENG] Skips used to bring mined ore to the surface of a shaft.

mine superintendent [MIN ENG] A mine manager or group manager. Also known as mine captain.

mine surveyor [MIN ENG] The official who periodically surveys the mine workings and prepares plans for the manager.

mine track devices [MIN ENG] Track devices to provide maximum safety for haulage trains in mines.

mine tractor [MIN ENG] A trackless, self-propelled vehicle used to transport equipment and supplies.

mine valuation [MIN ENG] Properly weighing the financial considerations to place a present value on mineral reserves.

mine ventilating fan [MIN ENG] A motor-driven disk, propeller, or wheel for blowing or exhausting air to provide ventilation of a mine; large units are used for stationary systems, while small portable types provide fresh air in inaccessible locations, such as dead ends.

mine ventilation system [MIN ENG] A combination of connecting airways with air pressure sources and governing devices that are instrumental in making and controlling airflow.

mine water [MIN ENG] Water pumped from mines.

minimum bend radius [MET] The minimum radius through which a piece of metal can be bent to form a given angle without fracturing.

minimum flight altitude [AERO ENG] The lowest altitude at which aircraft may safely operate.

minimum metal condition [DES ENG] The condition corresponding to the removal of the greatest amount of material permissible in a machined part.

minimum rent [MIN ENG] The right to work coal acquired by a mine owner by the payment of an annual rent and a royalty to the landowner (the coal or mineral owner). Also known as fixed rent.

minimum thermometer [ENG] A thermometer that automatically registers the lowest temperature attained during an interval of time.

minimum turning circle [ENG] The diameter of the circle described by the outermost projection of a vehicle when the vehicle is making its shortest possible turn.

mining [MIN ENG] The technique and business of mineral discovery and exploitation.

mining camp [MIN ENG] A term loosely applied to any mining town.

mining claim [MIN ENG] That portion of the public mineral lands which a miner, for mining purposes, takes and holds in accordance with mining laws. Also known as claim.

mining engineer [MIN ENG] One qualified by education, training, and experience in mining engineering.

mining engineering [ENG] Engineering concerned with the discovery, development, and exploitation of coal, ores, and minerals, as well as the cleaning, sizing, and dressing of the product. Also known as mineral engineering.

mining geology [MIN ENG] The study of the structure and occurrence of mineral deposits and the geologic aspects of mine planning.

mining ground [MIN ENG] Land from which a mineral substance is extracted by the process of mining.

mining hazard [MIN ENG] Any of the dangers unique to the winning and working of minerals and coal.

mining lease [MIN ENG] A contract to work a mine and extract mineral or other deposits from it under specified conditions. Also known as mineral lease.

mining machine truck [MIN ENG] A truck used to transport shortwall mining machines.

mining on a shoestring See grass-roots mining.

mining partnership [MIN ENG] The arrangement whereby two or more persons acquire a mining claim and actually engage in working it.

mining property [MIN ENG] Property valued for its mining possibilities.

mining retreating [MIN ENG] A process of mining by which the ore or coal is untouched until after all the gangways and such are driven, at which time the work of extraction begins at the boundary and progresses toward the shaft.

mining right [MIN ENG] A right to enter upon and occupy a specific piece of ground for the purpose of working it, either by underground excavations or open workings, to obtain the mineral ores which may be deposited therein. Also known as mineral right.

mining shield [MIN ENG] A canopy or cover for the protection of workers and machines at the face of a mechanized coal heading.

mining title [MIN ENG] A claim, exclusive prospecting license, right, concession, or lease.

mining town [MIN ENG] A town that has arisen next to a mine or mines.

mining width [MIN ENG] The minimum width needed to extract ore regardless of the actual width of ore-bearing rock.

minitrack [AERO ENG] A satellite tracking system consisting of a field of separate antennas and associated receiving equipment interconnected so as to form interferometers which track a transmitting beacon in the payload itself.

Minofor Alloy [MET] A white alloy containing 68% tin, 18% antimony, 3% copper, and 10% zinc; properties are similar to those of britannia metal.

minor defect [IND ENG] A defect which reduces the effectiveness of the product, without causing serious malfunctioning.

minor diameter [DES ENG] The diameter of a cylinder bounding the root of an external thread or the crest of an internal thread.

minus angle See angle of depression.

minus sieve [MET] In powder metallurgy, the portion of a powder sample that passes through a specified standard sieve.

minus sight See foresight.

minute pressure [PETRO ENG] A measurement of gas well capacity, determined by shutting the gate valve and recording the gage pressure each minute, with the pressure at the end of the first minute providing an estimate of the gas volume.

mirror interferometer [ENG] An interferometer used in radio astronomy, in which the sea surface acts as a mirror to reflect radio waves up to a single antenna, where the reflected waves interfere with the waves arriving directly from the source.

mirror nephoscope [ENG] A nephoscope in which the motion of a cloud is observed by its reflection in a mirror. Also known as cloud mirror; reflecting nephoscope.

mirror scale [ENG] A scale with a mirror used to align the eye perpendicular to the scale and pointer when taking a reading; improves accuracy by eliminating parallax.

mirror transit circle [ENG] A development of the conventional transit circle in which light from a star is reflected into fixed horizontal telescopes pointing due north and south by a plane mirror that is mounted on a horizontal east-west axis and attached to a large circle with accurately calibrated markings to determine the mirror's position.

misch metal [MET] An alloy consisting of a crude mixture of cerium, lanthanum, and other rare-earth metals obtained by electrolysis of the mixed chlorides of the metals dissolved in fused sodium chloride; used in making aluminum alloys, in some steels and irons, and in coating the cathodes of glow-type voltage regulator tubes.

miscible-phase displacement [PETRO ENG] Method of increasing reservoir oil recovery by displacement with an oil-miscible driving fluid, such as gas or liquefied petroleum gas.

miscible-slug process [PETRO ENG] A miscible-phase displacement in which reservoir oil recovery is increased by displacement with liquefied petroleum gas as the driving fluid.

miser [PETRO ENG] A well-boring bit that is tubular with a valve at the bottom, and has a screw for forcing the earth upward. Also spelled mizer.

mismatch [MET] Failure to match forged surfaces formed in opposite dies.

misrun [MET] Incompletely formed casting due to premature solidification of metal before the mold is filled.

missed hole See failed hole.

missed round [ENG] A round in which all or part of the explosive has failed to detonate.

missile site radar [ENG] Phased array radar located at a missile launch area to provide a guidance link with interceptor missiles enroute to their targets.

mist extractor [ENG] A device that removes liquid mist or droplets from a gas stream via impingement, flow-direction change, velocity change, centrifugal force, filters, or coalescing packs.

mist projector [MIN ENG] An appliance to form a mist spray to allay dust and fume during blasting operations in a tunnel.

miter bend [DES ENG] A pipe bend made by mitering (angle cutting) and joining pipe ends.

miter box [ENG] A troughlike device of metal or wood with vertical slots set at various angles in the upright sides, for guiding a handsaw in making a miter joint.

miter gate [CIV ENG] Either of a pair of canal lock gates that swing out from the side walls and meet at an angle pointing toward the upper level.

miter gear [DES ENG] A bevel gear whose bevels are in 1:1 ratio.

miter joint [DES ENG] A joint, usually perpendicular, in which the mating ends are beveled.

miter saw [DES ENG] A hollow-ground saw in diameters from 6 to 16 inches (15.24 to 40.64 centimeters), used for cutting off and mitering on light stock such as moldings and cabinet work.

miter valve [DES ENG] A valve in which a disk fits in a seat making a 45° angle with the axis of the valve.

miticide [MATER] An agent that kills mites. Also known as acaricide.

mixed-base oil [MATER] A crude oil that has both paraffin and asphaltum as predominating solid residuals.

mixed cycle [MECH ENG] An internal combustion engine cycle which combines the Otto cycle constant-volume combustion and the Diesel cycle constant-pressure combustion in high-speed compression-ignition engines. Also known as combination cycle; commercial Diesel cycle; limited-pressure cycle.

mixed-flow fan [MIN ENG] A mine fan in which the flow is both radial and axial.

mixed-flow impeller [MECH ENG] An impeller for a pump or compressor which combines radial- and axial-flow principles.

mixing chamber [ENG] The space in a welding torch in which the gases are mixed.

mixing valve [ENG] Multi-inlet valve used to mix two or more fluid intakes to give a mixed product of desired composition.

mixture ratio [AERO ENG] The ratio of the weight of oxidizer used per unit of time to the weight of fuel used per unit of time.

M meter [ENG] A class of instruments which measure the liquid water content of the atmosphere.

mobile crane [MECH ENG] **1.** A cable-controlled crane mounted on crawlers or rubber-tired carriers. **2.** A hydraulic-powered crane with a telescoping boom mounted on truck-type carriers or as self-propelled models.

mobile drill [MIN ENG] A drill unit mounted on wheels or crawl-type tracks to facilitate moving.

mobile filling [MIN ENG] Filling which is supplemented only from above, and which sinks, filling the mined-out rooms.

mobile hoist [MECH ENG] A platform hoist mounted on a pair of pneumatic-tired road wheels, so it can be towed from one site to another.

mobile loader [MECH ENG] A self-propelling power machine for loading coal, mineral, or dirt.

mobility ratio [PETRO ENG] The ratio of the mobility of the driving fluid (such as water) to that of the driven fluid (such as gas) in a petroleum reservoir.

mocaya oil [MATER] Oil from kernels of the Paraguayan palm (*Acrocomia sclerocarpa*), found in South America and the West Indies; used in manufacture of tinplate, soaps, candles, and margarine.

mock silver [MET] **1.** An aluminum alloy containing 5% copper and 10% tin, or 5% copper and 5% silver. **2.** A white brass containing 55% zinc and 45% copper.

mockup [ENG] A model, often full-sized, of a piece of equipment, or installation, so devised as to expose its parts for study, training, or testing.

model basin [ENG] A large basin or tank of water where scale models of ships can be tested. Also known as model tank; towing tank.

model tank *See* model basin.

modification [ENG] A major or minor change in the design of an item, effected in order to correct a deficiency, to facilitate production, or to improve operational effectiveness. [MET] Treatment of molten aluminum alloys containing 8–13% silicon with small amounts of a sodium fluoride or sodium chloride mixture; improves mechanical properties.

modification kit [ENG] A collection of items not all having the same basic name which are employed individually or conjunctively to alter the design of a component or equipment.

modified asphalt [MATER] Asphalt modified by addition of a rosin ester or synthetic resin.

modified gunmetal [MET] Gunmetal containing about 2.5% lead; used for gears and bearings.

modifier [MATER] In flotation, any of the chemicals which increase the specific attraction between collector agents and particle surfaces or which increase the wettability of those surfaces.

modular structure [BUILD] A building that is constructed of preassembled or presized units of standard sizes; uses a 4-inch (10.16-centimeter) cubical module as a reference.

modulation [MECH ENG] Regulation of the fuel-air mixture to a burner in response to fluctuations of load on a boiler.

modulation meter [ENG] Instrument for measuring the degree of modulation (modulation factor) of a modulated wave train, usually expressed in percent.

modulation transformer [ENG ACOUS] An audio-frequency transformer which matches impedances and transmits audio frequencies between one or more plates of an audio output stage and the grid or plate of a modulated amplifier.

module [AERO ENG] A self-contained unit which serves as a building block for the overall structure in space technology; usually designated by its primary function, such as command module or lunar landing module. [ENG] A unit of size used as a basic component for standardizing the design and construction of buildings, building parts, and furniture.

modulus of strain hardening See rate of strain hardening.

moellen See degras.

moil [MIN ENG] A long steel wedge with a rounded point used for breaking up rocks in a mine.

moist-heat sterilization [ENG] Sterilization with steam under pressure, as in an autoclave, pressure cooker, or retort; most bacteriological media are sterilized by autoclaving at 121°C, with 15 pounds (103 kilopascals) of pressure, for 20 minutes or more.

moisture barrier [MATER] A material that retards the passage of moisture into walls.

moisture loss [MECH ENG] The difference in heat content between the moisture in the boiler exit gases and that of moisture at ambient air temperature.

MOL See manned orbiting laboratory.

molasses/A.N. explosive [MATER] An explosive mixture consisting of ammonium nitrate mixed with molasses and water.

mold [ENG] **1.** A pattern or template used as a guide in construction. **2.** A cavity which imparts its form to a fluid or malleable substance. [ENG ACOUS] The metal part derived from the master by electroforming in reproducing disk recordings; has grooves similar to those of the recording.

moldability [MATER] The capability of being molded.

mold base [ENG] The assembly of all parts of an injection mold except the cavity, cores, and pins.

molded-fabric bearing [DES ENG] A bearing composed of laminations of cotton or other fabric impregnated with a phenolic resin and molded under heat and pressure.

molded lines [ENG] Full-size lines of a ship or airplane which are laid out in a mold loft.

mold efficiency [ENG] In a multimold blow-molding system, the percentage of the total turn-around time actually required for the forming, cooling, and ejection of the formed objects.

molding cycle [ENG] **1.** The time required for a complete sequence of molding oper-

ations. **2.** The combined operations required to produce a set of moldings.

molding machine [MET] A machine that compacts sand around a pattern to form a mold.

molding powder [MATER] Powdered plastic-material ingredients (such as resin, filler, pigments, and plasticizers) ready for compression in molding.

molding press [MET] A press used to form compacts in powder metallurgy.

molding pressure [ENG] Pressure needed to force softened plastic to fill a mold cavity.

molding shrinkage [ENG] Difference in dimensions between the molding and the mold cavity, measured at normal room temperature.

mold loft [ENG] A large building with a smooth wooden floor where full-size lines of a ship or airplane are laid down and templates are constructed from them to lay off the steel for cutting.

mold release See release agent.

mold shift [MET] The mismatch of mold halves at the parting line, resulting in a casting defect.

mold steel [MET] Steel of tool-steel quality used to make molds for plastics; properties include uniform texture, good machinability with die-sinking tools, and lack of microscopic porosity.

mold wash [MET] An aqueous or alcoholic suspension or emulsion used to coat the surfaces of a mold cavity.

mole [CIV ENG] A breakwater or berthing facility, extending from shore to deep water, with a core of stone or earth. [MECH ENG] A mechanical tunnel excavator.

molecular drag pump [ENG] A vacuum pump in which pumping is accomplished by imparting a high momentum to the gas molecules by impingement of a body rotating at very high speeds, as much as 16,000 revolutions per minute; such pumps achieve a vacuum as high as 10^{-6} torr.

molecular pump [MECH ENG] A vacuum pump in which the molecules of the gas to be exhausted are carried away by the friction between them and a rapidly revolving disk or drum.

mole drain [CIV ENG] A subsurface channel for water drainage; formed by pulling a solid object, usually a solid cylinder having a wedge-shaped point at one end, through the soil at the proper slope and depth.

mole mining [MIN ENG] A method of working coal seams about 30 inches (75 centimeters) thick; a small continuous-miner type of machine is used, which is remote-controlled from the roadway, without associated supports.

Moll thermopile [ENG] A thermopile used in some types of radiation instruments; alternate junctions of series-connected manganan-constantan thermocouples are embedded in a shielded nonconducting plate having a large heat capacity; the remaining junctions, which are blackened, are exposed directly to the radiation; the voltage developed by the thermocouple is proportional to the intensity of radiation.

molluscicide [MATER] An agent that kills mollusks.

molybdate orange [MATER] Pigment that is a solid solution of lead chromate, molybdate, and sulfate; used in paints, inks, and plastics.

molybdenum [MET] A silvery-gray metal used in iron-based alloys.

molybdenum cast iron [MET] Cast iron containing small amounts of molybdenum, added as ferromolybdenum or calcium molybdenum; increases strength, toughness, and wear resistance.

molybdenum silicide [MET] A mixture of molybdenum, silicon, and iron in the proportion 60:30:10; used to introduce molybdenum into steel melts.

molybdenum steel [MET] **1.** A carbon steel containing usually less than 0.5% molybdenum to aid hardenability. **2.** A tool steel containing up to 10% molybdenum, up to 1.5% carbon, and varying amounts of chromium, vanadium tungsten, and sometimes cobalt.

moly-blacks [MATER] Lustrous, black, molybdenum-containing decorative coatings used to blacken zinc or zinc-base alloys.

moment coefficient [AERO ENG] The coefficients used for moment are similar to coefficients of lift, drag, and thrust, and are likewise dimensionless; however, these must include a characteristic length, in addition to the area; the span is used for rolling or yawing moment, and the chord is used for pitching moment.

Momertz-Lentz system [MIN ENG] Placement of two winding engines alongside the top of the mine shaft using the shaft collar as a common foundation; results in practically vertical ropes and less rope oscillation.

monaural sound [ENG ACOUS] Sound produced by a system in which one or more microphones are connected to a single transducing channel which is coupled to one or two earphones worn by the listener.

Mond process [MET] A process for extracting and purifying nickel whereby nickel carbonyl is first formed by reaction of the reduced metal with carbon monoxide, and then the nickel carbonyl is decomposed thermally, resulting in deposition of nickel.

monitor [ENG] **1.** An instrument used to measure continuously or at intervals a condition that must be kept within prescribed limits, such as radioactivity at some point in a nuclear reactor, a variable quantity in an automatic process control system, the transmissions in a communication channel or bank, or the position of an aircraft in flight. **2.** To use meters or special techniques to measure such a condition. **3.** A person who watches a monitor. [MIN ENG] See hydraulic monitor.

monkey [MIN ENG] **1.** An appliance for mechanically gripping or releasing the rope in rope haulage. **2.** An airway in an anthracite mine.

monkey drift [MIN ENG] A small drift driven in for prospecting purposes, or a crosscut driven to an airway above the gangway.

monkey heading [MIN ENG] A narrow, low passage in the coal, providing refuge for miners while coal is blasted.

monkey ladder [MIN ENG] A ladder of saplings; the widely separated steps rest in the coal.

monkey winch [MIN ENG] A device for exerting a strong pull; consists of a framework containing a hand-operated drum, around which a steel rope 50 feet (15 meters) long is wound.

monkey wrench [DES ENG] A wrench having one jaw fixed and the other adjustable, both of which are perpendicular to a straight handle.

monocable [MECH ENG] An aerial ropeway that uses one rope to both support and haul a load.

monocoque [AERO ENG] A type of construction, as of a rocket body, in which all or most of the stresses are carried by the skin.

monofuel propulsion [AERO ENG] Propulsion system which obtains its power from a single fuel; in rocket units, the fuel furnishes both oxygen supply and the hydrocarbon for combustion.

monolith [MATER] A large concrete block.

monolithic [CIV ENG] Pertaining to concrete construction which is cast in one jointless piece.

monophonic sound [ENG ACOUS] Sound produced by a system in which one or more microphones feed a single transducing channel which is coupled to one or more loudspeakers.

monopropellant [MATER] A rocket propellant consisting of a single substance, especially a liquid, capable of creating rocket thrust without the addition of a second substance.

monopulse radar [ENG] Radar in which directional information is obtained with high precision by using a receiving antenna system having two or more partially overlapping lobes in the radiation patterns.

Mono pump [MIN ENG] A pump designed and manufactured of special materials to meet mining conditions; consists of a rubber stator shaped like a double internal helix and a single helical rotor which travels in the stator with a slightly eccentric motion.

monorail [CIV ENG] A single rail used as a track; usually elevated, with cars straddling or hanging from it.

monostat [ENG] Fluid-filled, upside-down manometer-type device used to control pressures within an enclosure, as for laboratory analytical distillation systems.

monostatic radar [ENG] Conventional radar, in which the transmitter and receiver are at the same location and share the same antenna; in contrast to bistatic radar.

monotron [MET] A machine for determining indentation hardness by measuring the load required to dent a specimen to a constant depth with a diamond 5/8 millimeter in diameter.

monotype metal [MET] A type metal typically composed of 76% lead, 16% antimony, and 8% tin, with good wear resistance and compressive strength.

montan wax [MATER] A hard mineral wax with a melting point of 80–90°C; white after purification and brown in the crude form; soluble in carbon tetrachloride, benzene, and chloroform; used for shoe and furniture polishes, adhesive pastes, roofing paints, and phonograph records. Also known as lignite wax.

monument [ENG] A natural or artificial (but permanent) structure that marks the location on the ground of a corner or other survey point.

Moody formula [MECH ENG] A formula giving the efficiency e' of a field turbine, whose runner has diameter D', in terms of the efficiency e of a model turbine, whose runner has diameter D; $e' = 1 - (1 - e)(D/D')^{1/5}$.

moon shot [AERO ENG] The launching of a rocket intended to travel to the vicinity of the moon.

moor [ENG] Securing a ship or aircraft by attaching it to a fixed object or a mooring buoy with chains or lines, or with anchors or other devices.

mooring buoy [ENG] A buoy secured to the bottom by permanent moorings and provided with means for mooring a vessel by use of its anchor chain or mooring lines;

in its usual form a mooring buoy is equipped with a ring.

mooring mast [AERO ENG] A mast or pole with fittings at the top to secure any lighter-than-air craft, such as a dirigible or blimp.

Mo-Permalloy [MET] A magnetic alloy having very high permeability and low saturation, consisting of 79% nickel, 17% iron, and 4% molybdenum.

mora fat *See* mowrah fat.

mordant dye [MATER] Textile dye that requires a mordant (third substance) to bind the dye onto the fiber.

mordanting assistant [MATER] A chemical used with textile dye mordants to cause decomposition of the mordant and uniform deposition on the fibers; examples are sulfuric, oxalic, and lactic acids.

mordant rouge [MATER] Aluminum acetate–acetic acid solution used in dyeing and calico printing. Also known as red acetate; red liquor.

Morisette expansion reamer [MIN ENG] A reaming device with three tapered lugs or cutters designed so that drilling pressure necessary to penetrate rock with a noncoring pilot bit forces the diamond-faced cutters of the reamer to expand outward, thereby enlarging the pilot hole sufficiently to allow the casing to follow the reamer as drilling progresses.

morning glory spillway *See* shaft spillway.

Morse taper reamer [DES ENG] A machine reamer with a taper shank.

mortar [MATER] A mixture of cement, lime, and sand used for laying bricks or masonry.

mortise [ENG] A groove or slot in a timber for holding a tenon.

mortise and tenon [DES ENG] A type of joint, principally used for wood, in which a hole, slot, or groove (mortise) in one member is fitted with a projection (tenon) from the second member.

mortise lock [DES ENG] A lock designed to be installed in a mortise rather than applied to a door's surface.

mortising machine [MECH ENG] A machine employing an auger and a chisel to produce a square or rectangular mortise in wood.

mossy zinc [MET] Zinc granules made by pouring molten zinc into water.

mother [ENG ACOUS] A mold derived by electroforming from a master; used to produce the stampers from which disk records are molded in large quantities. Also known as metal positive.

motion analysis [IND ENG] Detailed study of the motions used in a work task or at a given work area.

motion cycle [IND ENG] The complete sequence of motions and activities required to complete one work cycle.

motion economy [IND ENG] Simplification and reduction of body motions to simplify and reduce work content.

motion picture projector [ENG] An optical and mechanical device capable of flashing pictures taken by a motion picture camera on a viewing screen at the same frequency the action was photographed, thus producing an image that appears to move.

motive column [MIN ENG] The ventilating pressure in a mine in units of feet of air column; the height of a column of air whose density is the same as the air in the downcast shaft, which exerts a pressure equal to the ventilating pressure.

motor benzol [MATER] Grade of benzene used in fuels for internal combustion engines.

motorcycle [MECH ENG] An automotive vehicle, essentially a motorized bicycle, with two tandem and sometimes three rubber wheels.

motor element [ENG ACOUS] That portion of an electroacoustic receiver which receives energy from the electric system and converts it into mechanical energy.

motor meter [ENG] An integrating meter which has a rotor, one or more stators, a retarding element which makes the speed of the rotor proportional to the quantity (such as power or current) whose integral over time is being measured, and a register which counts the total number of revolutions of the rotor.

motor reducer [MECH ENG] Speed-reduction power transmission equipment in which the reducing gears are integral with drive motors.

motortruck [MECH ENG] An automotive vehicle which is used to transport freight.

motor vehicle [MECH ENG] Any automotive vehicle that does not run on rails, and generally having rubber tires.

mottled iron [MET] A cast iron showing gray areas that contain graphite, perlite, and sometimes ferrite, and white areas containing primarily cementite.

mount [ENG] **1.** Structure supporting any apparatus, as a gun, searchlight, telescope, or surveying instrument. **2.** To fasten an apparatus in position, such as a gun on its support.

Mount Rose snow sampler [ENG] A particular pattern of snow sampler having an internal diameter of 1.485 inches (3.7727 centimeters), so that each inch of water in the sample weighs 1 ounce (28.349 grams).

mouse hole [PETRO ENG] A drill hole under the derrick floor cased with a suspended drill pipe, to be connected later on the drill string.

mouse trap [ENG] A cylindrical fishing tool having the open bottom end fitted with an inward opening valve.

mousse de chêne *See* oakmoss resin.

mouth [ENG ACOUS] The end of a horn that has the larger cross-sectional area. [MIN ENG] **1.** The end of a shaft, adit, drift, entry, or tunnel emerging at the surface. **2.** The collar of a borehole.

movable bridge [CIV ENG] A bridge in which either the horizontal or vertical alignment can be readily changed to permit the passage of traffic beneath it. Often called drawbridge (an anachronism).

movable platen [ENG] The large platen at the back of an injection-molding machine to which the back half of the mold is fastened.

movable-point crossing [CIV ENG] A small-angle rail crossing with two center frogs, each of which consists essentially of a knuckle rail and two opposed movable center points.

moving-coil galvanometer [ENG] Any galvanometer, such as the d'Arsonval galvanometer, in which the current to be measured is sent through a coil suspended or pivoted in a fixed magnetic field, and the current is determined by measuring the resulting motion of the coil.

moving-coil loudspeaker *See* dynamic loudspeaker.

moving-coil microphone *See* dynamic microphone.

moving-coil voltmeter [ENG] A voltmeter in which the current, produced when the voltage to be measured is applied across a known resistance, is sent through coils pivoted in the magnetic field of permanent magnets, and the resulting torque on the coils is balanced by control springs so that the deflection of a pointer attached to the coils is proportional to the current.

moving-coil wattmeter *See* electrodynamic wattmeter.

moving-conductor loudspeaker [ENG ACOUS] A loudspeaker in which the mechanical forces result from reactions between a steady magnetic field and the magnetic field produced by current flow through a moving conductor.

moving-iron meter [ENG] A meter that depends on current in one or more fixed coils acting on one or more pieces of soft iron, at least one of which is movable.

moving-iron voltmeter [ENG] A voltmeter in which a field coil is connected to the voltage to be measured through a series resistor; current in the coil causes two vanes, one fixed and one attached to the shaft carrying the pointer, to be similarly magnetized; the resulting torque on the shaft is balanced by control springs.

moving-magnet voltmeter [ENG] A voltmeter in which a permanent magnet aligns itself with the resultant magnetic field produced by the current in a field coil and another permanent control magnet.

moving sidewalk [CIV ENG] A sidewalk constructed on the principle of an endless belt, on which pedestrians are moved.

mowa fat *See* mowrah fat.

mowrah fat [MATER] A vegetable fat extracted from the seeds of *Bassia longifolia;* used to make soap. Also known as illepé fat; mora fat; mowa fat.

mp *See* mean effective pressure; melting point.

MPC *See* man process chart.

M synchronization [ENG] A linking arrangement between a camera lens and the flashbulb unit to allow a 15-millisecond delay of the shutter so that the bulb burns to its brightest point before the shutter opens.

mucilage [MATER] **1.** A sticky material employed as an adhesive. **2.** A gummy material derived from plants.

muck [CIV ENG] Rock or earth removed during excavation. [MIN ENG] *See* waste rock.

mucker [MIN ENG] A worker who loads broken mineral into trams or pushes them from stope chute to shaft. Also known as groundman.

mucking [ENG] Clearing and loading broken rock and other excavated materials, as in tunnels or mines.

mud *See* slime.

mud auger [DES ENG] A diamond-point bit with the wings of the point twisted in a shallow augerlike spiral. Also known as clay bit; diamond-point bit; mud bit.

mud berth [CIV ENG] A berth where a vessel rests on the bottom at low water.

mud bit *See* mud auger.

mud blasting [ENG] The detonation of sticks of explosive stuck on the side of a boulder with a mud covering, so that little of the explosive energy is used in breaking the boulder.

mud cake [ENG] A caked layer of clay adhering to the walls of a well or borehole, formed where the water in the drilling mud filtered into a porous formation during rotary drilling.

mud-cake resistivity [PETRO ENG] Resistivity of drilling mud cake pressed from a sample of the mud; important in mud log interpretation.

mudcap [ENG] A quantity of wet mud, wet earth, or sand used to cover a charge of dynamite or other high explosive fired in contact with the surface of a rock in mud blasting.

mud conditioning [PETRO ENG] In a well drilling operation, the treatment and control of drilling mud to ensure proper gel strength, viscosity, density and so on.

mud drilling [MIN ENG] Drilling operations in which a mud-laden circulation fluid is used. Also known as mud flush drilling.

mud flush test [MIN ENG] A test carried out at the boring site to determine whether the mud solution is of the correct viscosity and density.

mud log [PETRO ENG] A continuous record of changes in oil or gas contents of circulating drilling mud while drilling a well.

mud pit *See* slushpit.

mud-removal acid [PETRO ENG] Mixture of hydrochloric and hydrofluoric acids with inhibitors, surfactants, and demulsifiers; used to dissolve drilling-mud clays away from the drillhole face.

mudsill [CIV ENG] The lowest sill of a structure, usually embedded in the earth.

mud still [ENG] An instrument used to separate oil, water, and other volatile materials in a mud sample by distillation, permitting determination of the quantities of oil, water, and total solid contents in the original sample.

muffle furnace [ENG] A furnace with an externally heated chamber; the walls of which radiantly heat the contents of the chamber.

muffler [ENG] A device to deaden the noise produced by escaping gases or vapors.

mulch [MATER] A mixture of organic material, such as straw, peat moss, or leaves, that is spread over soil to prevent evaporation, maintain an even soil temperature, prevent erosion, control weeds, and enrich soil.

mule *See* barney.

mule skinner [MIN ENG] A mule driver.

mull [ENG] To mix thoroughly or grind.

muller [ENG] A foundry sand-mixing machine.

mulling [ENG] The combining of clay, water, and sand, prior to molding, by compressing with a roller to ensure development of optimum sand properties by the adequate distribution of ingredients.

mullion [BUILD] A vertical bar separating two windows in a multiple window.

multicellular horn [ENG ACOUS] A combination of individual horn loudspeakers having individual driver units or joined in groups to a common driver unit. Also known as cellular horn.

multideck cage [MIN ENG] A cage with two or more compartments or platforms to hold the mine cars.

multideck clarifiers [ENG] Extraction units which remove pollutants from recycled plant waste water.

multideck screen [MIN ENG] A screen with two or more superimposed screening surfaces mounted within a common frame.

multideck sinking platform [MIN ENG] A sinking platform of several decks so that various shaft-sinking operations may be performed simultaneously.

multideck table [MIN ENG] A type of shaking table that is double-decked; while each deck is fed and discharged independently, one mechanism vibrates both.

multifuel burner [ENG] A burner which utilizes more than one fuel simultaneously for combustion.

multifunction array radar [ENG] Electronic scanning radar which will perform target detection and identification, tracking, discrimination, and some interceptor missile tracking on a large number of targets simultaneously and as a single unit.

multifuse igniter [ENG] A black powder cartridge that allows several fuses to be fired at the same time by lighting a single fuse.

multilayer bit [DES ENG] A bit set with diamonds arranged in successive layers beneath the surface of the crown.

multimeter See volt-ohm-milliammeter.

multiple [MET] A piece of stock cut from bar for use in a forging which provides the exact length needed for a single workpiece.

multiple-activity process chart [IND ENG] A chart showing the coordinated synchronous or simultaneous activities of a work system comprising one or more machines or men; each machine or man is shown in a separate, parallel column indicating its or his activities as related to the other parts of the work system.

multiple-arch dam [CIV ENG] A dam composed of a series of arches inclined at about 45° and carried on parallel buttresses or piers.

multiple-completion packer [PETRO ENG] A device used to provide a seal between the outside surfaces of the two or more parallel tubing strings in a multiple-completion well, and the inside surface of the common wellbore casing.

multiple-completion well [PETRO ENG] Oil well in which there is production from more than one oil-bearing zone (different depths) with parallel tubing strings within a single wellbore casing string. Also known as multicompletion well.

multiple connector [ENG] A flow chart symbol that indicates the merging of several flow lines into one line or the dispersal of a flow line into several lines.

multiple-entry system [MIN ENG] A system of access or development openings generally in bituminous coal mines involving more than one pair of parallel entries, one for haulage and fresh-air intake and the other for return air.

multiple-factor incentive plan [IND ENG] A wage incentive plan based on productivity and other factors such as yield, material usage, and reduction of scrap.

multiple firing [ENG] Electrically firing with delay blasting caps in a number of holes at one time.

multiple hydraulic pump [PETRO ENG] Oil-well pump arrangement by which a single pump can be used in alternating operation to lift oil from two producing zones within a single multiple completion well.

multiple-impulse welding [MET] Spot, upset, or projection welding in which more than one current impulse is generated during a single machine cycle. Also known as pulsation welding.

multiple openings [MIN ENG] Any series of underground openings separated by rib pillars or connected at frequent intervals to form a system of rooms and pillars.

multiple-parallel-tubing string [PETRO ENG] Two or more parallel and closely packed oil-well tubing strings used in multiple-completion wells. Also known as multiple-tubing string.

multiple-pass weld [MET] A weld made by depositing filler metal with two or more passes in succession.

multiple piece rate plan [IND ENG] A wage incentive plan wherein increasingly higher unit pay rates are given to the worker as his productivity increases.

multiple-purpose tester See volt-ohm-milliammeter.

multiple-row blasting [ENG] The drilling, charging, and firing of rows of vertical boreholes.

multiple sampling [IND ENG] A plan for quality control in which a given number of samples from a group are inspected, and the group is either accepted, resampled, or rejected, depending on the number of failures found in the samples.

multiple series [ENG] A method of wiring a large group of blasting charges by connecting small groups in series and connecting these series in parallel. Also known as parallel series.

multiple shooting [ENG] The firing of an entire face at one time by means of connecting shot holes in a single series and shooting all holes at the same instant.

multiple-shot survey [PETRO ENG] The determining and recording of drill-hole direction.

multiple-slide press [MECH ENG] A press with individual adjustable slides built into the main slide or connected independently to the main shaft.

multiple spot welding [MET] Spot welding in which several spots are welded during a single machine cycle.

multiple-stage rocket *See* multistage rocket.

multiple-stage separator [PETRO ENG] Oil-well gas-oil separator in which wellhead pressure is reduced in several stages, with the flashing off of gas at each pressure reduction.

multiple-strand conveyor [MECH ENG] A conveyor with two or more spaced strands of chain, belts, or cords as the supporting or propelling medium.

multiple-tubing string *See* multiple-parallel-tubing string.

multiplex [ENG] Stereoscopic device to project aerial photographs onto surfaces so that the images may be viewed in three dimensions by using anaglyphic spectacles; used to prepare topographic maps.

multiple x-y recorder [ENG] Recorder that plots a number of independent charts simultaneously, each showing the relation of two variables, neither of which is time.

multiple zone [PETRO ENG] Two or more discrete oil or gas reservoirs in the same geographic area, but at different depths.

multiport burner [ENG] A burner having several nozzles which discharge fuel and air.

multirope friction winder [MECH ENG] A winding system in which the drive to the winding ropes is the frictional resistance between the ropes and the driving sheaves.

multistage [ENG] Functioning or occurring in separate steps.

multistage compressor [MECH ENG] A machine for compressing a gaseous fluid in a sequence of stages, with or without intercooling between stages.

multistage pump [MECH ENG] A pump in which the head is developed by multiple impellers operating in series.

multistage queuing [IND ENG] A situation involving two or more sequential stages in a process, each of which involves waiting in line.

multistage rocket [AERO ENG] A vehicle having two or more rocket units, each unit firing after the one in back of it has exhausted its propellant; normally, each unit, or stage, is jettisoned after completing its firing. Also known as multiple-stage rocket; step rocket.

multistatic radar [ENG] Radar in which successive antenna lobes are sequentially engaged to provide a tracking capability

without physical movement of the antenna.

multitrack recording system [ENG] Recording system which provides two or more recording paths on a medium, which may carry either related or unrelated recordings in common time relationship.

multiwell gas-lift system [PETRO ENG] An installation to allow gas-lift production from the various tubing strings involved in a multiple completion well.

Mumetal [MET] An alloy of high magnetic permeability, containing 14% iron, 5% copper, 1.5% chromium, and the balance nickel.

municipal engineering [CIV ENG] Branch of engineering dealing with the form and functions of urban areas.

Muntz metal [MET] A 60/40 type of brass composed of 58–61% copper, up to 1% lead, and remainder zinc. Also known as malleable brass.

murumuru oil [MATER] An oil obtained from nuts of the palm *Attalea orbignya*, containing 40% lauric acid, 35% myristic acid, and some palmitic, stearic, linoleic, and oleic acids.

music wire [MET] High-quality, high-carbon steel wire used for making mechanical springs.

Muskat equation [PETRO ENG] Equation used to calculate oil reservoir permeability from the pressure buildup curve recorded when a producing well is shut in after a flow test.

Myklestad method [AERO ENG] A method of determining the mode shapes and frequencies of the lateral bending modes of space vehicles, taking into account secondary effects of shear and rotary inertia, in which one imagines masses to be concentrated at a finite number of points along the beam, with elastic properties remaining constant between consecutive mass points.

myotome [ENG] An instrument used to divide a muscle, particularly through its belly.

myrica oil *See* bay oil.

myristica oil *See* nutmeg oil.

myrrh [MATER] A gum resin of species of myrrh (*Commiphora*); partially soluble in water, alcohol, and ether; used in dentifrices, perfumery, and pharmaceuticals.

myrtle oil [MATER] Light-yellow liquid distilled from the flowers and leaves of the European myrtle (*Myrtus communis*); aromatic aroma; formerly used in medicine; now used for flavors and as perfume fixative.

myrtle wax *See* bayberry wax.

not terminate at the origin of the first leg.

open-web girder See lattice girder.

open well [CIV ENG] **1.** A well whose diameter is great enough (1 meter or more) for a person to descend to the water level. **2.** An artificial pond filling a large excavation in the zone of saturation up to the water table.

open-well caisson See Chicago caisson.

open workings [MIN ENG] Surface workings, for example, a quarry or open-cast mine.

operating pressure [ENG] The system pressure at which a process is operating.

operating water level [MECH ENG] The water level in a boiler drum which is normally maintained above the lowest safe level.

operation [IND ENG] A job, usually performed in one location, and consisting of one or more work elements.

operational [ENG] Of equipment such as aircraft or vehicles, being in such a state of repair as to be immediately usable.

operational game See management game.

operational maintenance [ENG] The cleaning, servicing, preservation, lubrication, inspection, and adjustment of equipment; it includes that minor replacement of parts not requiring high technical skill, internal alignment, or special locative training.

operation analysis [IND ENG] An analysis of all procedures concerned with the design or improvement of production, the purpose of the operation, inspection standards, materials used and the manner of handling them, the setup, tool equipment, and working conditions and methods.

operation analysis chart [IND ENG] A form that lists all the essential factors influencing the effectiveness of an operation.

operation breakdown See job breakdown.

operation process chart [IND ENG] A graphic representation that gives an overall view of an entire process, including the points at which materials are introduced, the sequence of inspections, and all operations not involved in material handling.

operator [ENG] A person whose duties include the operation, adjustment, and maintenance of a piece of equipment.

operator process chart [IND ENG] A chart of the time relationship of the movements made by the body members of a workman performing an operation.

operator productivity [IND ENG] The ratio of standard hours to actual hours for a given task.

operator training [IND ENG] The process used to prepare the employee to make his expected contribution to his employer, usually involving the teaching of specialized skills.

operator utilization [IND ENG] The ratio of working time to total clock time; a ratio of 1.00 (or 100) indicates full utilization of the operator's work time.

opisometer [ENG] An instrument for measuring the length of curved lines, such as those on a map; a wheel on the instrument is traced over the line.

opposed engine [MECH ENG] A reciprocating engine having the pistons on opposite sides of the crankshaft, with the piston strokes on each side working in a direction opposite to the direction of the strokes on the other side.

optical amplifier [ENG] An optoelectronic amplifier in which the electric input signal is converted to light, amplified as light, then converted back to an electric signal for the output.

optical bench [ENG] A rigid horizontal bar or track for holding optical devices in experiments; it allows device positions to be changed and adjusted easily.

optical comparator [ENG] Any comparator in which movement of a measuring plunger tilts a small mirror which reflects light in an optical system. Also known as visual comparator.

optical fluid-flow measurement [ENG] Any method of measuring the varying densities of a fluid in motion, such as schlieren, interferometer, or shadowgraph, which depends on the fact that light passing through a flow field of varying density is retarded differently through the field, resulting in refraction of the rays, and in a relative phase shift among different rays.

optical gage [ENG] A gage that measures an image of an object, and does not touch the object itself.

optical glass [MATER] A type of glass which is free from imperfections, such as unmelted particles, bubbles, and chemical inhomogeneities, which would affect its transmission of light.

optical indicator [ENG] An instrument which makes a plot of pressure in the cylinder of an engine as a function of piston (or volume) displacement, making use of magnification by optical systems and photographic recording; for example, the small motion of a pressure diaphragm may be transmitted to a mirror to deflect a beam of light.

optical lantern [ENG] A device for projecting positive transparent pictures from glass or film onto a reflecting screen; it consists of a concentrated source of light, a condenser system, a holder (or changer) for the slide, a projection lens, and (usually) a blower for cooling the slide. Also known as slide projector.

N

naftalan *See* naphthalan.

nail [DES ENG] A slender, usually pointed fastener with a head, designed for insertion by impact. [ENG] To drive nails in a manner that will position and hold two or more members, usually of wood, in a desired relationship.

nailer [ENG] A wood strip or block which serves as a backing into which nails can be driven.

nailhead [DES ENG] Flat protuberance at the end of a nail opposite the point.

naked light [MIN ENG] Open flame, such as a match or a burning cigarette, that is a fire risk in mines.

naked-light mine [MIN ENG] A coal mine that is nongassy, where naked lights can be used by miners.

Nansen bottle [ENG] A bottlelike water-sampling device with valves at both ends that is lowered into the water by wire; at the desired depth it is activated by a messenger which strikes the reversing mechanism and inverts the bottle, closing the valves and trapping the water sample inside. Also known as Petterson-Nansen water bottle; reversing water bottle.

napalm [MATER] **1.** Aluminum soap in powder form, used to gelatinize oil or gasoline for use in napalm bombs or flame throwers. **2.** The resultant gelatinized substance.

naphtha [MATER] **1.** Petroleum fraction with volatility between gasoline and kerosine; used as a gasoline ingredient, solvent for paints and rubber, and cleaning solvent. **2.** Aromatic solvent from coal tar, either solvent naphtha or heavy naphtha.

naphtha gas [MATER] Illuminating gas charged with a low-boiling-point fraction of distilled naphtha.

naphthalan [MATER] Soft, greenish-black mass distilled from Armenian naphtha; soluble in ether and hydrocarbons, insoluble in water; melts at 70°C; used in medicine. Also known as naftalan.

naphthene base [MATER] Crude oil with a high carbon and low oxygen content that leaves an asphaltic residue after refining. Also known as asphalt base.

naphthenic crude [MATER] Crude petroleum containing a significant proportion of naphthenic compounds.

napped leather *See* suede.

narcissus oil *See* jonquil oil.

narrow-band pyrometer [ENG] A pyrometer in which light from a source passes through a color filter, which passes only a limited band of wavelengths, before falling on a photoelectric detector. Also known as spectral pyrometer.

narrow gage [CIV ENG] A railway gage narrower than the standard gage of 4 feet 8½ inches (143.51 centimeters).

natural aging [MET] Spontaneous aging at room temperature of a supersaturated metallic solid solution.

natural cement [MATER] Hydraulic cement made from pulverized and heated limestone containing clay, magnesia, and iron.

natural-draft cooling tower [MECH ENG] A cooling tower that depends upon natural convection of air flowing upward and in contact with the water to be cooled.

natural gas [MATER] A combustible, gaseous mixture of low-molecular-weight paraffin hydrocarbons, generated below the surface of the earth; contains mostly methane and ethane with small amounts of propane, butane, and higher hydrocarbons, and sometimes nitrogen, carbon dioxide, hydrogen sulfide, and helium.

natural gasoline [MATER] The liquid paraffin hydrocarbon contained in natural gas and recovered by compression, distillation, and absorption.

natural laminar flow [AERO ENG] Airflow over a portion of the wing such that local pressure decreases in the direction of flow and flow in the boundary layer is laminar rather than turbulent; frictional drag on the aircraft is greatly reduced.

natural pressure cycle [MET] A cycle in which pressure buildup conforms proportionately to the buildup of stresses due to forming.

natural resource [MATER] A deposit of minerals, water, or other materials furnished by nature.

natural-seasoned lumber *See* air-dried lumber.

natural splitting [MIN ENG] In mine ventilation, a flow of air dividing among the branches, of its own accord and without regulation, in inverse relation to the resistance of each airway.

natural steel [MET] **1.** Steel made directly from cast iron. **2.** Steel, such as wootz, made directly from the ore.

natural ventilation [MIN ENG] The weak and varying ventilation in a mine caused by the difference in air density between shafts.

natural ventilation pressure [MIN ENG] A pressure difference across the shaft-bottom doors caused by a lack of balance in the two vertical air columns.

naval architecture [ENG] The study of the physical characteristics and the design and construction of buoyant structures, such as ships, boats, barges, submarines, and floats, which operate in water; includes the construction and operation of the power plant and other mechanical equipment of these structures.

naval brass [MET] Brass composed of 60–62% copper, 37–39% zinc, and 0.75–1% tin; relatively resistant to corrosion by seawater. Also known as naval bronze.

naval bronze *See* naval brass.

naval stores [MATER] **1.** Pitch and rosin formerly used in the construction of wooden ships. **2.** All pine wood products, including rosin, turpentine, and pine oils.

navigation [ENG] The process of directing the movement of a craft so that it will reach its intended destination; subprocesses are position fixing, dead reckoning, pilotage, and homing.

navigational satellite [AERO ENG] An artificial earth-orbiting satellite designed for use in at least four widely different navigational systems.

navigation dam [CIV ENG] A structure designed to raise the level of a stream to increase the depth for navigation purposes.

navigation dome *See* astrodome.

Navy Heavy *See* bunker C fuel oil.

NBR *See* nitrile rubber.

neat cement grout [MATER] Grout made from a mixture of cement and water.

neat line [CIV ENG] The line to which a masonry wall should generally conform.

neat plaster [MATER] A base-coat plaster, having sand added at the job location.

neatsfoot oil [MATER] Pale-yellow oil with unusual odor; soluble in organic solvents and kerosine; obtained by boiling shinbones and hoofless feet of cattle; used to treat leather, as a lubricant, and to oil wool. Also known as bubulum oil; hoof oil.

neat soap [MATER] Soap in the molten state formed during manufacture, especially after fitting and settling out of nigre and lye.

neck [MET] In a tensile test, that portion of the metal at which fracture is imminent during the later stages of plastic deformation in a tensile test.

neck-down [MET] **1.** A thin core used for restricting the riser neck; facilitates cutting off the riser from the casting. **2.** Localized area reduction of a test piece during plastic deformation.

neck-in [ENG] When coating by extrusion, the width difference between the extruded web leaving the die and that of the coating on the surface.

necking [MET] Reducing the diameter or cross-sectional area of a tube or other piece of metal by stretching.

necking down [MET] Localized reduction in cross-sectional area of a specimen during tensile deformation.

needle [DES ENG] **1.** A device made of steel pointed at one end with a hole at the other; used for sewing. **2.** A device made of steel with a hook at one end; used for knitting. [ENG] **1.** A piece of copper or brass about ½ inch (13 millimeters) in diameter and 3 or 4 feet (90 or 120 centimeters) long, pointed at one end, thrust into a charge of blasting powder in a borehole and then withdrawn, leaving a hole for the priming, fuse, or squib. Also known as pricker. **2.** A thin pointed indicator on an instrument dial. [ENG ACOUS] *See* stylus.

needle beam [CIV ENG] A temporary member thrust under a building or a foundation for use in underpinning.

needle bearing [DES ENG] A roller-type bearing with long rollers of small diameter; the rollers are retained in a flanged cup, have no retainer, and bear directly on the shaft.

needle blow [ENG] A blow-molding technique in which air is injected into the plastic article through a hollow needle inserted in the parison.

needle dam [CIV ENG] A barrier made of horizontal bars across a pass through a dam or of planks that can be removed in case of flooding.

needle file [DES ENG] A small file with an extended tang that serves as a needle.

needle nozzle [MECH ENG] A streamlined hydraulic turbine nozzle with a movable element for converting the pressure and kinetic energy in the pipe leading from the reservoir to the turbine into a smooth jet of variable diameter and discharge but practically constant velocity.

needle tubing [ENG] Stainless steel tubing with outside diameters from 0.014 to 0.203 inch (0.36 to 5.16 millimeters); used for surgical instruments and radon implanters.

needle valve [DES ENG] A slender, pointed rod fitting in a hole or circular or conoidal seat; used in hydraulic turbines and hydroelectric systems.

needle weir [CIV ENG] A type of frame weir in which the wooden barrier is constructed of vertical square-section timbers placed side by side against the iron frames.

needling [CIV ENG] Underpinning the upper part of a building with horizontally placed timber or steel beams.

neem oil [MATER] An aromatic oil from the seeds and fruit of the neem tree (*Melia azadirachta*); contains sulfur compounds; used as an anthelmintic and as an alcohol denaturant. Also known as margosa oil; nim oil.

negative easement [CIV ENG] An easement that can be exercised to prevent the owner of a piece of land from using his property in certain ways that he would otherwise be entitled to.

negative rake [MECH ENG] The orientation of a cutting tool whose cutting edge lags the surface of the tooth face.

negotiated contract [IND ENG] A purchase or sales agreement made by a United States government agency without normally employing techniques required by formal advertising.

nematicide [MATER] A chemical used to kill plant-parasitic nematodes. Also spelled nematocide.

nematocide *See* nematicide.

neodymium glass [MATER] A glass containing small amounts of neodymium oxide; used for color television filter plates since it transmits 90% of the blue, green, and red light rays and no more than 10% of the yellow.

neophane glass [MATER] A glass containing neodymium oxide to reduce glare; used for yellow sunglasses or for windshield glass.

neoprene [MATER] A synthetic rubber with outstanding resistance to ozone, weathering, various chemicals, oil, and flame, made by polymerization of chloroprene (2-chlorobutadiene-1,3); varies from amber to silver to cream in color; used in paints, putties, adhesives, shoe soles, tank linings, and rubber products.

nepheloscope [ENG] An instrument for the production of clouds in the laboratory by condensation or expansion of moist air.

nephometer [ENG] A general term for instruments designed to measure the amount of cloudiness; an early type consists of a convex hemispherical mirror mapped into six parts; the amount of cloud coverage on the mirror is noted by the observer.

nephoscope [ENG] An instrument for determining the direction of cloud motion.

neptune powder [MATER] An explosive consisting of nitroglycerin with a more or less explosive dope.

Nernst-Lindemann calorimeter [ENG] A calorimeter for measuring specific heats at low temperatures, in which the heat reservoir consists of a metal of high thermal conductivity such as copper, to promote rapid temperature equalization; none of the material under study is more than a few millimeters from a metal surface, and the whole apparatus is placed in an evacuated vessel and heated by current through a platinum heating coil.

neroli oil *See* oil of orange blossoms.

nerve gas [MATER] Chemical agent which is absorbed into the body by breathing, by ingestion, or through the skin, and affects the nervous and respiratory systems and various body functions; an example is isopropylphosphonofluoridate.

net [ENG] **1.** Threads or cords tied together at regular intervals to form a mesh. **2.** A series of surveying or leveling stations that have been interconnected in such a manner that closed loops or circuits have been formed, or that are arranged so as to provide a check on the consistency of the measured values. Also known as network.

net floor area [BUILD] Gross floor area of a building, excluding the area occupied by walls and partitions, the circulation area (where people walk), and the mechanical area (where there is mechanical equipment).

net flow area [DES ENG] The calculated net area which determines the flow after the complete bursting of a rupture disk.

net positive suction head [MECH ENG] The minimum suction head required for a pump to operate; depends on liquid characteristics, total liquid head, pump speed and capacity, and impeller design. Abbreviated NPSH.

net thrust [AERO ENG] The gross thrust of a jet engine minus the drag due to the momentum of the incoming air.

network *See* net.

network structure [MET] A crystal structure in a metal in which one constituent occurs primarily at grain boundaries enveloping the grains made up of other constituents.

Neuberg blue [MATER] Pigment made up of a mixture of copper blue and iron blue.

Neumann bands *See* Neumann lines.

Neumann lines [MET] Mechanical deformation twins seen as straight, serrated narrow bands parallel to preferred planes in the crystals of an etched metal which has been strained, usually by sudden impact; most often observed along the 112 planes of body-centered-cubic ferrite. Also known as Neumann bands.

neutral atmosphere [ENG] An atmosphere which neither oxidizes nor reduces immersed materials.

neutrally buoyant float *See* swallow float.

neutral oil [MATER] Light oil from dewaxed petroleum; medium or low viscosity, flash point 143–160°C.

neutral point [MET] In rolling mills, the point at which the speed of the work is equal to the peripheral speed of the rolls.

neutral zone *See* dead band.

neutron-gamma well logging [ENG] Neutron well logging in which the varying intensity of gamma rays produced artificially by neutron bombardment is recorded.

neutron logging *See* neutron well logging.

neutron shield [ENG] A shield that protects personnel from neutron irradiation.

neutron soil-moisture meter [ENG] An instrument for measuring the water content of soil and rocks as indicated by the scattering and absorption of neutrons emitted from a source, and resulting gamma radiation received by a detector, in a probe lowered into an access hole.

neutron well logging [ENG] Study of formation fluid-content properties down a wellhole by neutron bombardment and detection of resultant radiation (neutrons or gamma rays). Also known as neutron logging.

nevyanskite [MET] A tin-white variety of iridosmine containing 35–50% osmium or more than 50% iridium; occurs in flat scales.

new blue [MATER] Any of several iron blue types of pigments; varieties are called mendola blue and prussian blue.

newel post [CIV ENG] **1.** A pillar at the end of an oblique retaining wall of a bridge. **2.** The post about which a circular staircase winds. **3.** A large post at the foot of a straight stairway or on a landing.

New Jersey retort process *See* vertical retort process.

Newson's boring method [MIN ENG] A method of boring small shafts, up to 5½ feet (1.7 meters) in diameter using the principle of chilled-shot drilling on a large scale.

newsprint [MATER] The paper used in the publication of newspapers; an impermanent material made from mechanical wood pulp, with some chemical wood pulp.

Newton's alloy [MET] A fusible alloy made up of 50% bismuth, 31% lead, and 19% tin; melts at 95°C; used in applications where it is required to fall away at predetermined temperatures, as for automatic sprinkler links. Also known as Newton's metal.

Newton's metal *See* Newton's alloy.

NH propellant [MATER] A propellant which, by reason of its formulation or method of manufacture, does not absorb moisture from the air.

nib [ENG] A small projecting point.

nibbling [MECH ENG] Contour cutting of material by the action of a reciprocating punch that takes repeated small bites as the work is passed beneath it.

nicarbing *See* carbonitriding.

Nicholson's hydrometer [ENG] A modification of Fahrenheit's hydrometer in which the lower end of the instrument carries a scale pan to permit the determination of the relative density of a solid.

Nichols radiometer [ENG] An instrument, used to measure the pressure exerted by a beam of light, in which there are two small, silvered glass mirrors at the ends of a light rod that is suspended at the center from a fine quartz fiber within an evacuated enclosure.

nickel [MET] A silver-gray, ductile, malleable, tough metal; used in alloys, plating, coins (to replace silver), ceramics, and electronic circuits.

nickel-aluminum bronze [MET] An alloy composed of an 8–10% aluminum bronze with nickel added to increase strength, corrosion resistance, and heat resistance; used for dies, molds, cast propellers, and valve seats.

nickel brass *See* nickel silver.

nickel bronze [MET] Bronze containing nickel; a common type contains 88% copper, 5% tin, 5% nickel, and 2% zinc.

nickel cast iron [MET] An improved-strength alloy cast iron containing a small percentage of nickel (2–5%); in larger amounts (15–36%) nickel primarily imparts corrosion resistance.

nickel-chromium steel [MET] Steel containing nickel (0.2–3.75%) and chromium (0.3–1.5%) as alloying elements.

nickel-molybdenum iron [MET] An alloy containing 20–40% molybdenum and up to 60% nickel with some carbon added; has high acid resistance.

nickel-molybdenum steel [MET] Steel containing 0.2–0.3% molybdenum and 1.65–3.75% nickel as alloying elements.

nickel plating [MET] Electrolytic deposition of a metallic nickel coating.

nickel rhodium [MET] Nickel alloy with 25–80% rhodium; can also contain other metals, such as platinum or molybdenum; used for pen points, reflectors, electrodes, and chemical equipment.

nickel silver [MET] A silver-white alloy composed of 52–80% copper, 10–35% zinc, and 5–35% nickel; sometimes also contains a few percent of lead and tin. Also known as German silver; nickel brass.

nickel steel [MET] Carbon steel containing up to 9% nickel as a major alloying element.

nickel-vanadium steel [MET] A nickel steel containing about 1.5% nickel, 1% manganese, 0.28% carbon, and 0.10% vanadium; used for high-strength cast parts.

niello [MATER] Mixture of sulfides of copper, silver, and lead, with black metallic appearance; used in ornamental inlays engraved on metals such as silver.

nigrite [MATER] A mixture of rubber with ozocerite distillation residue; used as a substitute for gutta-percha.

nigrosine [MATER] Any of a group of blue or black azine dyes used for coloring inks, shoe polish, leather, and wood; can be water-, alcohol-, or oil-soluble.

nikkel oil [MATER] A bright-yellow liquid with a lemon and cinnamon odor obtained from the leaves and twigs of a laurel tree, *Cinnamomum zeylanicum*; contains citral and cineol; used in perfumery.

nim oil *See* neem oil.

nine-light indicator [ENG] A remote indicator for wind speed and direction used in conjunction with a contact anemometer and a wind vane; the indicator consists of a center light, connected to the contact anemometer, surrounded by eight equally spaced lights which are individually connected to a set of similarly spaced electrical contacts on the wind vane; wind speed is determined by counting the number of flashes of the center light during an interval of time; direction, indicated by the position of illuminated outer bulbs, is given to points of the compass.

niobium [MET] A platinum-gray, ductile metal with brilliant luster; used in alloys, especially stainless steels. Also known as columbium.

nip *See* angle of nip; squeeze.

Nipher shield [ENG] A conically shaped, copper, rain-gage shield; used to prevent the formation of vertical wind eddies in the vicinity of the mouth of the gage, thereby making the rainfall catch a representative one.

nippers [DES ENG] Small pincers or pliers for cutting or gripping.

nipple [DES ENG] A short piece of tubing, usually with an internal or external thread at each end, used to couple pipes.

nipple chaser [ENG] A member of a drilling crew who procures and delivers the tools and equipment necessary for an operation.

Nissen stamp [MIN ENG] **1.** Machine used in crushing rock to sand sizes. **2.** An individual stamp worked in its own circular mortar box.

niter balls [MATER] A pellet form of potassium nitrate, used as a fertilizer. Also known as sal prunella; throat balls.

nitriding [MET] Surface hardening of steel by formation of nitrides; nitrogen is introduced into the steel usually by heating in gaseous ammonia.

nitrile-butadiene rubber *See* nitrile rubber.

nitrile rubber [MATER] A synthetic rubber formed by polymerization of acrylonitrile with butadiene; the structure of the polymer is —$CH_2CH=CHCH_2CH_2CH(CN)$—. Also known as acrylonitrile-butadiene rubber; acrylonitrile rubber; NBR; nitrile-butadiene rubber; NR.

nitrocellulose propellant [MATER] A single-base propellant whose main constituent is nitrocellulose, with only minor percentages of additives for stabilizing and other purposes.

nitrogelatin *See* gelatin dynamite.

nitrogenated oil [MATER] A class of essential oils containing carbon, hydrogen, oxygen, and nitrogen; an example is oil of bitter almonds.

nitrogenous fertilizer [MATER] Fertilizer materials, natural or synthesized, containing nitrogen available for fixation by vegetation, such as potassium nitrate, KNO_3, or ammonium nitrate, NH_4NO_3.

nitroglycerin powder [MATER] Any explosive characterized by a low nitroglycerin content, up to 10, and a high ammonium nitrate content of 80–85, with carbonaceous material forming the remainder.

***para*-nitrophenylazosalicylate sodium** *See* alizarin yellow.

nitrophosphate [MATER] A nitrogen-phosphorus fertilizer made by reacting nitric acid with phosphate rock.

NLGI number [ENG] One of a series of numbers developed by the National Lubricating Grease Institute and used to classify the consistency range of lubricating greases; NLGI numbers are based on the ASTM cone penetration number.

no-atmospheric control [AERO ENG] Any device or system designed or set up to control a guided rocket missile, rocket craft, or the like outside the atmosphere or in regions where the atmosphere is of such tenuity that it will not affect aerodynamic controls.

noble metal [MET] A metal, or alloy, such as gold, silver, or platinum having high resistance to corrosion and oxidation; used in the construction of thin-film circuits, metal-film resistors, and other metal-film devices.

no-bottom sounding [ENG] A sounding in the ocean in which the bottom is not reached.

no-cut rounds [MIN ENG] Set of holes drilled straight into the face for blasting underground.

no-draft forging [MET] A forging designed with little or no taper for removal from dies, and with extremely fine tolerances for closer control of grain flow during production of the final part.

nodular cast iron [MET] Cast iron treated in the molten state with a master alloy containing an element such as magnesium which favors formation of spheroidal graphite. Also known as ductile iron; spheroidal graphite cast iron.

nodular powder [MET] Irregularly shaped metal powder particles.

nodulizing [ENG] Creation of spherical lumps from powders by working them together, coalescing them with binders, drying fluid-solid mixtures, heating, or chemical reaction.

no-fines concrete [MATER] Concrete made without sand and therefore containing a high proportion of communicating pores which provide thermal insulation and drainage.

nog [MIN ENG] **1.** Roof support for stopes, formed of rectangular piles of logs squared at the ends and filled with waste rock. **2.** A wood block wedged tightly into the cut in a coal seam after the coal cutter has passed; it forms a temporary support.

no-go gage [ENG] A limit gage designed not to fit a part being tested; usually employed with a go gage to set the acceptable maximum and minimum dimension limits of the part.

noise-canceling microphone *See* close-talking microphone.

noise radial [ENG] The brightening of all range points on a particular plan position indicator bearing on a radar screen caused by noise reception from the indicated direction.

noise reduction [ENG ACOUS] A process whereby the average transmission of the sound track of a motion picture print, averaged across the track, is decreased for signals of low level; since background noise introduced by the sound track is less at low transmission, this process reduces noise during soft passages.

noise-type flowmeter [ENG] A flowmeter that measures the noise generated in a selected frequency band.

nominal bandwidth [ENG] The difference between the nominal upper and lower cutoff frequencies of an acoustic or electric filter.

nominal decline rate [PETRO ENG] The negative slope of the curve representing the hydrocarbon production rate versus time for an oil gas reservoir.

nominal pass-band center frequency [ENG] The geometric mean of the nominal upper and lower cutoff frequencies of an acoustic or electric filter.

nominal size [DES ENG] Size used for purposes of general identification; the actual size of a part will be approximately the same as the nominal size but need not be exactly the same; for example, a rod may be referred to as 1/4 inch, although the actual dimension on the drawing is 0.2495 inch, and in this case 1/4 inch is the nominal size.

nominal stress [MET] The stress calculated by simple elasticity theory, ignoring stress raisers and plastic flow; in tensile testing of a notched specimen, the load applied at the notch divided by the initial cross-sectional area at the notch.

nonasphaltic road oil [MATER] Any of the nonhardening petroleum distillates or residual oils used to lay road dust; they have viscosities low enough to be applied without prior heating.

nonassociated-gas reservoir [PETRO ENG] Formation in which gaseous hydrocarbons exist as a free phase in a reservoir that is not commercially productive of crude oil.

nonbearing wall [CIV ENG] A wall that bears no vertical weight other than its own.

noncarbon oil [MATER] **1.** Oil in which little or no free carbon is suspended. **2.** Oil which, upon decomposition, contributes little so-called carbon deposits.

noncondensable gas [MATER] A gas from chemical or petroleum processing units (such as distillation columns or steam ejectors) that is not easily condensed by cooling; consists mostly of nitrogen, light hydrocarbons, carbon dioxide, or other gaseous materials.

nonconsumable electrode [MET] An electrode, such as of carbon or tungsten, that is not consumed during a welding or melting operation.

noncoring bit [ENG] A general type of bit made in many shapes which does not produce a core and with which all the rock cut in a borehole is ejected as sludge; used mostly for blasthole drilling and in the unmineralized zones in a borehole where a core sample is not wanted. Also known as borehole bit; plug bit.

noncorrosive flux [MET] A soldering flux composed of rosin or of rosin in a volatile solvent; the residue is nonhygroscopic, noncorrosive, and nonconducting; suitable for soldering electronic components. Also known as activated rosin flux.

nondeforming steel [MET] A group of alloy steels which do not easily deform when heat-treated. Also known as nonshrinking steel.

nondestructive testing [ENG] Any testing method which does not involve damaging or destroying the test sample; includes use of x-rays, ultrasonics, radiography, magnetic flux, and so on.

nondissipative muffler See reactive muffler.

nonexpendable [ENG] Pertaining to a supply item or piece of equipment that is not consumed, and does not lose its identity, in use, as a weapon, vehicle, machine, tool, piece of furniture, or instrument.

nonferrous metal [MET] Any metal other than iron and its alloys.

nonferrous metallurgy [MET] A branch of metallurgy that deals with metals other than iron and iron-base alloys.

nonflowing well [ENG] A well that yields water at the land surface only by means of a pump or other lifting device.

nonfreezing explosive [MATER] An explosive to which 15–20% of nitroethylene glycol has been added.

nonimpinging injector [AERO ENG] An injector used in rocket engines which employs parallel streams of propellant usually emerging normal to the face of the injector.

nonionic detergent [MATER] A detergent with molecules that do not ionize in aqueous solution, for example, detergents derived from condensation products of long-chain glycols and octyl or nonyl phenols.

nonlinear distortion [ENG ACOUS] The ratio of the total root-mean-square (rms) harmonic distortion output of a microphone to the rms value of the fundamental component of the output.

non-load-bearing tile [MATER] Tile unable to carry superimposed loads.

nonmagnetic steel [MET] A steel alloy containing about 12% manganese and sometimes a small quantity of nickel; it is practically nonmagnetic at ordinary temperatures.

nonnitroglycerin explosive [MATER] An explosive which contains TNT instead of nitroglycerin to sensitize ammonium nitrate, and a little aluminum powder may also be added to increase the power and sensitivity.

nonreclosing pressure relief device [MECH ENG] A device which remains open after relieving pressure and must be reset before it can operate again.

nonrecording rain gage [ENG] A rain gage which indicates but does not record the amount of precipitation.

nonrigid plastic [MATER] A plastic with modulus of elasticity not greater than 50,000 psi (3.45×10^8 newtons per square meter) at 25°C, according to standard ASTM test procedures.

nonsegregated reservoir [PETRO ENG] Solution-gas-drive oil reservoir in which the gas does not separate from the oil as a function of height or upward movement.

nonselective mining [MIN ENG] Mining methods permitting low cost, generally by using a cheap stoping method combined with large-scale operations; can be used in deposits where the individual stringers, bands, or lenses of high-grade ore are numerous and so irregular in occurrence and separated by such thin lenses of waste that a selective method cannot be employed.

nonshattering glass [MATER] Two sheets of plate glass with a sheet of transparent resinoid between, the whole molded together under heat and pressure; it will crack without shattering. Also known as laminated glass; shatterproof glass.

nonshrinking steel See nondeforming steel.

nonskid [CIV ENG] Pertaining to a surface that is roughened to reduce slipping, as a concrete floor treated with iron filings or carborundum powder, or indented while wet.

nonslip concrete [MATER] Rough-surface concrete made by applying oxide grains to the mixture before it hardens; used for steps.

nonsoap grease [MATER] Mineral oil thickened with solid lubricants such as graphite, mica, talc, molybdenum sulfide, asbestos fiber, uncombined fats, or rosin oils; used as a lubricant.

nonspinning rope See nonstranded rope.

nonstranded rope [DES ENG] A wire rope with the wires in concentric sheaths instead of in strands, and in opposite directions in the different sheaths, giving the rope nonspinning properties. Also known as nonspinning rope.

nonsymmetrical aquifer [PETRO ENG] In an oil reservoir formation, a water-containing part that is irregular, being neither radial nor linear.

nonsynchronous initiation [MET] In resistance welding, random starting and stopping of transformer primary current relative to the voltage wave.

nontransferred arc [MET] In arc welding and cutting, an arc made between the electrode and constricting nozzle, excluding the workpiece from the circuit.

nonviscous neutral [MATER] A neutral petroleum-derived oil with viscosity less than

135 SSU (Saybolt Seconds Universal) at 100°F (38°C).

nonwetting [MET] Of a metal or alloy, when it is molten, not adhering to or wetting the surface of a diamond which is to be set.

nonwetting phase [PETRO ENG] The oil phase contained in a reservoir pore structure when the reservoir fluid is two-phase (oil and water) or three-phase (oil, water, and gas).

normal barometer [ENG] A barometer of such accuracy that it can be used for the determination of pressure standards; an instrument such as a large-bore mercury barometer is usually used.

normal benzine [MATER] A mixture of hydrocarbons; clear, colorless, water-insoluble liquid distilled from petroleum; boils at 65–95°C; density 0.695–0.705. Also known as benzoline.

normal effort [IND ENG] The effort expended by the average operator in performing manual work with average skill and application.

normal elemental time [IND ENG] The selected or average elemental time adjusted to obtain the elemental time used by an average qualified operator. Also known as base time; leveled elemental time.

normal-incidence pyrheliometer [ENG] An instrument that measures the energy in the solar beam; it usually measures the radiation that strikes a target at the end of a tube equipped with a shutter and baffles to collimate the beam.

normalize [MET] To heat a ferrous alloy to some temperature above the transformation range, followed by air cooling.

normal operation [MECH ENG] The operation of a boiler or pressure vessel at or below the conditions of coincident pressure and temperature for which the vessel has been designed.

normal pace [IND ENG] The manual pace achieved by normal effort.

normal pitch [MECH ENG] The distance between working faces of two adjacent gear teeth, measured between the intersections of the line of action with the faces.

normal-plate anemometer [ENG] A type of pressure-plate anemometer in which the plate, restrained by a stiff spring, is held perpendicular to the wind; the wind-activated motion of the plate is measured electrically; the natural frequency of this system can be made high enough so that resonance magnification does not occur.

normal segregation [MET] Segregation of the lower-melting-point constituents of an alloy primarily near the center of a casting (last portion to solidify).

normal time [IND ENG] **1.** The time required by a trained worker to perform a task at a normal pace. **2.** The total of all the normal elemental times constituting a cycle or operation. Also known as base time; leveled time.

northerly turning error [AERO ENG] An acceleration error in the magnetic compass of an aircraft in a banked attitude during a turn, so called because it was first noted and is most pronounced during turns made from initial north-south courses; during a turn the magnetic needle is tilted from the horizontal, due to acceleration and the banking of the aircraft; in this position the compass needle will be acted upon by the vertical as well as the horizontal component of the earth's magnetic field; in addition, the compass needle is mechanically restricted in movement, due to tilt. Also known as turning error.

north-stabilized plan-position indicator [ENG] A heading-upward plan-position indicator; this term is deprecated because it may be confused with azimuth-stabilized plan-position indicator, a north-upward plan-position indicator.

north-upward plan position indicator [ENG] A plan position indicator on which north is maintained at the top of the indicator, regardless of the heading of the craft.

nose [ENG] The foremost point or section of a bomb, missile, or something similar.

nose cone [AERO ENG] A protective cone-shaped case for the nose section of a missile or rocket; may include the warhead, fusing system, stabilization system, heat shield, and supporting structure and equipment.

nose-heavy [AERO ENG] Pertaining to an airframe in which the nose tends to sink when the longitudinal control is released in any attitude of normal flight.

nose radius [MECH ENG] The radius measured in the back rake or top rake plane of a cutting tool.

nose sill [ENG] A short timber located under the end of the main sill of a standard rig front of a well.

nosing [BUILD] Projection of a tread of a stair beyond the riser below it.

notch [ENG] A V-shaped indentation or cut in a surface or edge.

notch acuity [MET] The severity of the stress concentration produced by a given notch in a structure; it is expressed as the ratio of the notch depth to the notch radius (depth is small compared to width, or diameter, of the narrowest cross section).

notch brittleness [MET] Susceptibility of a material to brittle fracture at areas of stress concentration; in notch tensile testing, a material has notch brittleness if the notch strength lies below the tensile strength.

notch depth [MET] The distance from the surface of a metal specimen to the bottom of the notch.

notch ductility [MET] Percentage reduction in area of the specimen after failure in a notched tensile test.

notched-bar test [MET] Test in which a notched metal specimen is bent with the notch in tension.

notching [MECH ENG] Cutting out various shapes from the ends or edges of a workpiece.

notching press [MECH ENG] A mechanical press for notching straight or rounded edges.

notch sensitivity [MET] A measure of the reduction in strength of a metal caused by the presence of a notch.

notch strength [MET] The ratio of maximum tensional load required to fracture a notched specimen to the original minimum cross-sectional area.

notch test [MET] A tensile or creep test of a metal to determine the effect of a surface notch.

novolac resin [MATER] Any of the thermoplastic phenol-formaldehyde resins made with an excess of phenol in the reaction; used in varnishes.

nozzle [DES ENG] A tubelike device, usually streamlined, for accelerating and directing a fluid, whose pressure decreases as it leaves the nozzle.

nozzle blade [AERO ENG] Any one of the blades or vanes in a nozzle diaphragm. Also known as nozzle vane.

nozzle-contraction-area ratio [DES ENG] Ratio of the cross-sectional area for gas flow at the nozzle inlet to that at the throat.

nozzle efficiency [MECH ENG] The efficiency with which a nozzle converts potential energy into kinetic energy, commonly expressed as the ratio of the actual change in kinetic energy to the ideal change at the given pressure ratio.

nozzle exit area [DES ENG] The cross-sectional area of a nozzle available for gas flow measured at the nozzle exit.

nozzle-expansion ratio [DES ENG] Ratio of the cross-sectional area for gas flow at the exit of a nozzle to the cross-sectional area available for gas flow at the throat.

nozzle-mix gas burner [ENG] A burner in which injection nozzles mix air and fuel gas at the burner tile.

nozzle throat [DES ENG] The portion of a nozzle with the smallest cross section.

nozzle throat area [DES ENG] The area of the minimum cross section of a nozzle.

nozzle thrust coefficient [AERO ENG] A measure of the amplification of thrust due to gas expansion in a particular nozzle as compared with the thrust that would be exerted if the chamber pressure acted only over the throat area. Also known as thrust coefficient.

nozzle vane See nozzle blade.

NPSH See net positive suction head.

NR See nitrile rubber.

N rod bit [DES ENG] A Canadian standard noncoring bit having a set diameter of 2.940 inches (74.676 millimeters).

nuclear-electric propulsion [AERO ENG] A system of propulsion utilizing a nuclear reactor to generate electricity which is then used in an electric propulsion system or as a heat source for the working fluid.

nuclear-electric rocket engine [AERO ENG] A rocket engine in which a nuclear reactor is used to generate electricity that is used in an electric propulsion system or as a heat source for the working fluid.

nuclear excavation [ENG] The use of nuclear explosions to remove earth for constructing harbors, canals, and other facilities.

nuclear gyroscope [ENG] A gyroscope in which the conventional spinning mass is replaced by the spin of atomic nuclei and electrons; one version uses optically pumped mercury isotopes, and another uses nuclear magnetic resonance techniques.

nuclear magnetic resonance flowmeter [ENG] A flowmeter in which nuclei of the flowing fluid are resonated by a radio-frequency field superimposed on an intense permanent magnetic field, and a detector downstream measures the amount of decay of the resonance, thereby sensing fluid velocity.

nuclear magnetometer [ENG] Any magnetometer which is based on the interaction of a magnetic field with nuclear magnetic moments, such as the proton magnetometer. Also known as nuclear resonance magnetometer.

nuclear power plant [MECH ENG] A power plant in which nuclear energy is converted into heat for use in producing steam for turbines, which in turn drive generators that produce electric power.

nuclear resonance magnetometer See nuclear magnetometer.

nuclear rocket See atomic rocket.

nuclear snow gage [ENG] Any type of gage using a radioactive source and a detector to measure, by the absorption of radiation, the water-equivalent mass of a snowpack.

nucleated glass [MATER] Glass treated with a nucleating agent to transform it into a crystalline material.

nucleonics [ENG] The technology based on phenomena of the atomic nucleus such as radioactivity, fission, and fusion; includes nuclear reactors, various applications of radioisotopes and radiation, particle accelerators, and radiation-detection devices.

nucleus counter [ENG] An instrument which measures the number of condensation nuclei or ice nuclei per sample volume of air.

nugget [MET] A weld bead.

null-balance recorder [ENG] An instrument in which a motor-driven slide wire in a measuring circuit is continuously adjusted so that the voltage or current to be measured will be balanced against the voltage or current from this circuit; a pen linked to the slide wire makes a graphical record of its position as a function of time.

null detector *See* null indicator.

null indicator [ENG] A galvanometer or other device that indicates when voltage or current is zero; used chiefly to determine when a bridge circuit is in balance. Also known as null detector.

null method [ENG] A method of measurement in which the measuring circuit is balanced to bring the pointer of the indicating instrument to zero, as in a Wheatstone bridge, and the settings of the balancing controls are then read. Also known as balance method; zero method.

Number six fuel *See* bunker C fuel oil.

nut [DES ENG] An internally threaded fastener for bolts and screws.

nutating antenna [ENG] An antenna system used in conical scan radar, in which a dipole or feed horn moves in a small circular orbit about the axis of a paraboloi-dal reflector without changing its polarization.

nutating-disk meter [ENG] An instrument for measuring flow of a liquid in which liquid passing through a chamber causes a disk to nutate, or roll back and forth, and the total number of rolls is mechanically counted.

nutator [ENG] A mechanical or electrical device used to move a radar beam in a circular, conical, spiral, or other manner periodically to obtain greater air surveillance than could be obtained with a stationary beam.

nutmeg oil [MATER] A pale-yellow or colorless essential oil with spicy taste and nutmeg aroma; obtained from nutmegs; soluble in alcohol, carbon bisulfide, and glacial acetic acid; chief constituents are myristin, pinene, and dipentene; used in flavors, perfumes, and medicines. Also known as myristica oil.

nylon [MATER] Generic name for long-chain polymeric amide molecules in which recurring amide groups are part of the main polymer chain; used to make fibers, fabrics, sheeting, and extruded forms.

nylon 6 [MATER] Nylon made by polycondensation of caprolactam.

nylon 6,6 [MATER] Nylon made by condensation of hexamethylene diamine with adipic acid.

nylon 6,10 [MATER] Nylon made by the condensation of hexamethylene diamine with sebacic acid.

O

oakmoss resin [MATER] Concrete oleoresin from the oakmoss lichens *Evernia prunastri* and *E. furfuracea*; used as a perfume fixative. Also known as mousse de chêne.

oakum [MATER] Old hemp or jute fiber, loosely twisted and impregnated with tar or a tar derivative, used to caulk sides and decks of ships and to pack joints of pipes and caissons.

oatmeal paper [MATER] A paper in which fine sawdust is added to produce a sheet with a coarse texture; it can be used as an inexpensive sketching paper for work in pastels and charcoal, or as wallpaper.

OBA *See* octave-band analyzer.

obliterated corner [CIV ENG] In surveying, a corner for which visible evidence of the previous surveyor's work has disappeared, but whose original position can be established from other physical evidence and testimony.

obscure glass [MATER] Translucent glass.

observation well [PETRO ENG] A special well drilled in a selected location for the purpose of observing parameters such as fluid levels and pressure changes (for example, within an oil reservoir) as production proceeds.

obsolescence [ENG] Decreasing value of functional and physical assets or value of a product or facility from technological changes rather than deterioration.

obsolete [ENG] No longer satisfactory for the purpose for which obtained, due to improvements or revised requirements.

occluded gases [MIN ENG] Gases entering the mine atmosphere from feeders and blowers, and also from blasting operations.

occlusion [ENG] The retention of undissolved gas in a solid during solidification.

occupy [ENG] To set a surveying instrument over a point for the purpose of making observations or measurements.

ocean engineering [ENG] A subfield of engineering involved with the development of new equipment concepts and the methodical improvement of techniques which allow humans to operate successfully beneath the ocean surface in order to develop and utilize marine resources.

oceanographic dredge [ENG] A device used aboard ship to bring up large samples of deposits and sediments from the ocean bottom.

oceanographic platform [ENG] A human-made structure with a flat horizontal surface higher than the water, on which oceanographic equipment is suspended or installed.

ocean thermal energy conversion [MECH ENG] The conversion of energy arising from the temperature difference between warm surface water of oceans and cold deep-ocean current into electrical energy or other useful forms of energy.

ocotea oil [MATER] A volatile essential oil obtained from the wood of a Brazilian tree, *Ocotea cymbarum*; a source of safrole; used to make heliotropin and other technical preparations.

octane number [ENG] A rating that indicates the tendency to knock when a fuel is used in a standard internal combustion engine under standard conditions; *n*-heptane is 0, isooctane is 100; different test methods give research octane, motor octane, and road octane.

octane requirement [MECH ENG] The fuel octane number needed for efficient operation (without knocking or spark retardation) of an internal combustion engine.

octane scale [ENG] Series of arbitrary numbers from 0 to 120.3 used to rate the octane number of a gasoline; *n*-heptane is 0 octane, isooctane is 100, and isooctane + 6 milliliters TEL (tetraethyllead) is 120.3.

octave-band analyzer [ENG ACOUS] A portable sound analyzer which amplifies a microphone signal, feeds it into one of several band-pass filters selected by a switch, and indicates the magnitude of sound in the corresponding frequency band on a logarithmic scale; all the bands except the highest and lowest span an octave in frequency. Abbreviated OBA.

octave-band filter [ENG ACOUS] A band-pass filter in which the upper cutoff frequency is twice the lower cutoff frequency.

octoid [DES ENG] Pertaining to a gear tooth form used to generate the teeth in bevel gears; the octoid form closely resembles the involute form.

ocuba wax [MATER] A waxy fat obtained from the fruit of the myrtle *Myristica ocuba;* melting point, 40°C; used in candles.

odd-leg caliper [DES ENG] A caliper in which the legs bend in the same direction instead of opposite directions.

odograph [ENG] An instrument installed in a vehicle to automatically plot on a map the course and distance traveled by the vehicle.

odometer [ENG] **1.** An instrument for measuring distance traversed, as of a vehicle. **2.** The indicating gage of such an instrument. **3.** A wheel pulled by surveyors to measure distance traveled.

O'Donahue's theory [MIN ENG] A mine subsidence theory, with subsidence regarded as taking place in two stages: first, a breaking of the rocks in which the lines of fracture tend to run at right angles to the stratification; then, an inward movement from the sides, resulting in a pull or draw beyond the edges of the workings.

odorant [MATER] Material added to odorless fuel gases to give them a distinctive odor for safety purposes; usually a sulfur- or mercaptan-containing compound. Also known as malodorant; stench; warning agent.

Oehman's survey instrument [ENG] A drill-hole surveying apparatus that makes a photographic record of the compass and clinometer readings.

Oetling freezing method [MIN ENG] A method of shaft sinking by freezing the wet ground in sections as the sinking proceeds.

off [ENG] Designating the inoperative state of a device, or one of two possible conditions (the other being "on") in a circuit.

off-airways [AERO ENG] Pertaining to any aircraft course or track that does not lie within the bounds of prescribed airways.

off-highway truck [MIN ENG] A truck of such size, weight, or dimensions that is cannot be used on public highways.

officials' inspection lamp [MIN ENG] A portable combined electric lamp and battery, fitted with a reflector to provide directional illumination.

off-line [ENG] **1.** A condition existing when the drive rod of the drill swivel head is not centered and parallel with the borehole being drilled. **2.** A borehole that has

deviated from its intended course. **3.** A condition existing wherein any linear excavation (shaft, drift, borehole) deviates from a previously determined or intended survey line or course. [IND ENG] State in which an equipment or subsystem is in standby, maintenance, or mode of operation other than on-line.

offset [ENG] **1.** A short perpendicular distance measured to a traverse course or a surveyed line or principal line of measurement in order to locate a point with respect to a point on the course or line. **2.** In seismic prospecting, the horizontal distance between a shothole and the line of profile, measured perpendicular to the line. **3.** In seismic refraction prospecting, the horizontal displacement, measured from the detector, of a point for which a calculated depth is relevant. **4.** In seismic reflection prospecting, the correction of a reflecting element from its position on a preliminary working profile to its actual position in space. [MIN ENG] **1.** A short drift or crosscut driven from a main gangway or level. **2.** The horizontal distance between the outcrops of a dislocated bed.

offset cab [ENG] Operator's cab positioned to one side of earthmoving equipment for greater visibility and safety.

offset cylinder [MECH ENG] A reciprocating part in which the crank rotates about a center off the centerline.

offset drilling [PETRO ENG] The drilling of a well on property under which oil is being drained away by a well on adjacent property to make up for the loss of oil from the first property.

offset line [ENG] A secondary line established close to and roughly parallel with the primary survey line to which it is referenced by measured offsets.

offset paper [MATER] Paper with a certain degree of porosity as a result of coating with an alkali-swelling resin; used for offset printing.

offset screwdriver [DES ENG] A screwdriver with the blade set perpendicular to the shank for access to screws in otherwise awkward places.

offset well [PETRO ENG] A well made by offset drilling.

offshore drilling [PETRO ENG] The drilling of oil or gas wells into water-covered locations, usually on submerged continental shelves.

offshore gas [PETRO ENG] Neutral gas produced from reservoirs under the offshore continental shelves.

offshore oil [PETRO ENG] Oil produced from reservoirs under the offshore continental shelves.

offshore survey [PETRO ENG] Seismic geophysical survey procedures conducted over water-covered continental-shelf areas in the search for possible oil reservoirs.

off-the-road equipment [MIN ENG] Tires and earthmoving equipment designed for off-highway duty in surface mines and quarries.

off-the-road hauling [MIN ENG] Hauling off the public highways, and generally on the mining site or excavation site.

off-the-shelf [IND ENG] Available for immediate shipment.

off time [MET] In resistance welding, usually in repetitive cycles, the time that the electrode is not in contact with the work.

Ohio sampler [MIN ENG] A single tube or pipe with a thread on top, and the bottom beveled and hardened for driving into the ground to obtain a soil sample.

ohmmeter [ENG] An instrument for measuring electric resistance; scale may be graduated in ohms or megohms.

ohms per volt [ENG] Sensitivity rating for measuring instruments, obtained by dividing the resistance of the instrument in ohms at a particular range by the full-scale voltage value at that range.

OHV engine *See* overhead-valve engine.

oil [MATER] Any of various viscous, combustible, water-immiscible liquids that are soluble in certain organic solvents, as ether and naphtha; may be of animal, vegetable, mineral, or synthetic origin; examples are fixed oils, volatile or essential oils, and mineral oils.

oil ammoniac [MATER] A yellow liquid distilled from gum ammoniac; boiling point is 275°C; soluble in alcohol and benzene.

oil asphalt [MATER] Water-insoluble, heavy black residue left after removing the tar tailings during the distillation of petroleum; used in roofing, paints, and coatings.

oil-base mud [PETRO ENG] Drilling mud made with oil as the solvent carrier for the solids content.

oil bath [ENG] **1.** Oil, in a container, within which a mechanism works or into which it dips. **2.** Oil in which a piece of apparatus is submerged. **3.** Oil that is poured on a cutting tool. [MET] Oil used in tempering.

oil burner [ENG] Liquid-fuel burner device using a mixture of air and vaporized or atomized oil for combustion.

oil cake [MATER] Solid residue after removal of vegetable oils from oil-bearing seeds (such as soya beans) by expression or solvent extraction; used as fertilizer and animal feed.

oil core [MET] A core in which sand is held together by an oil binder.

oil cup [ENG] A permanently mounted cup used to feed lubricant to a gear, usually with some means of regulating the flow.

oil cut [PETRO ENG] A mixture of oil and drilling mud that is recovered during oil exploration.

oil derrick [PETRO ENG] Tower structure used during oil well drilling to aid in raising and lowering of drill and piping strings.

oil dilution valve [MECH ENG] A valve used to mix gasoline with engine oil to permit easier starting of the gasoline engine in cold weather.

oil emulsion [MATER] Suspension of oil droplets in another liquid in which the oil is insoluble.

oil-extended rubber [MATER] Synthetic rubber into which 25–50% of a petroleum oil emulsion has been incorporated to decrease cost and increase low-temperature flexibility and resilience.

oil field [PETRO ENG] The surface boundaries of an area from which petroleum is obtained; may correspond to an oil pool or may be circumscribed by political or legal limits.

oil field emulsion [PETRO ENG] A crude oil that reaches the surface as an oil-water emulsion.

oil field model [PETRO ENG] Laboratory simulation of steady-state fluid flow through porous reservoir media by electrical (Ohm's-law system), electronic (graphite-impregnated cloth), or electrolytic (gelatin, blotter, potentiometric-liquid) models.

oil field separator *See* gas-oil separator.

oil filter [ENG] Cartridge-type filter used in automotive oil-lubrication systems to remove metal particles and products of heat decomposition from the circulating oil.

oil fogging [ENG] Spraying a fine oil mist into the gas stream of a distribution system to alleviate the drying effects of gas on certain kinds of distribution and utilization equipment.

oil furnace [MECH ENG] A combustion chamber in which oil is the heat-producing fuel.

oil gas [MATER] A heating gas made by interaction of petroleum oil vapors and steam in a process similar to the water-gas reaction.

oil-gas separator *See* gas-oil separator.

oil gas tar [MATER] A type of tar produced during the oil gas process by cracking the oil vapors at high temperatures.

oil groove [DES ENG] One of the grooves in a bearing which distribute and collect lubricating oil.

oil hardening [MET] Quenching of carbon steel in an oil bath; the steel cools slowly and a more uniform and desirable hardness is attained.

oil hole [ENG] A small hole for injecting oil for a bearing.

oil-hole drill [DES ENG] A twist drill containing holes through which oil can be fed to the cutting edges.

oil horizon [PETRO ENG] The upper surface of oil in a well, or the stratum in which the oil surface is located.

oiliness [ENG] The effect of a lubricant to reduce friction between two solid surfaces in contact; the effect is more than can be accounted for by viscosity alone.

oil in place [PETRO ENG] The total volume of oil estimated to be present in an oil reservoir.

oil isoperms [PETRO ENG] Reservoir-map plotting areas of equal oil permeability.

oil length [MATER] The ratio of oil to resin in varnish; expressed as gallons of oil per 100 pounds (45.3 kilograms) of resin.

oilless bearing [MECH ENG] A self-lubricating bearing containing solid or liquid lubricants in its material.

oil lift [MECH ENG] Hydrostatic lubrication of a journal bearing by using oil at high pressure in the area between the bottom of the journal and the bearing itself so that the shaft is raised and supported by an oil film whether it is rotating or not.

oil of amber [MATER] Brown essential oil distilled from amber; miscible with alcohol; has balsamic aroma.

oil of bitter almond [MATER] A fatty, nondrying oil obtained from bitter almond kernels and used in cosmetics and perfumes.

oil of cloves [MATER] A thin, colorless to pale-yellow liquid distilled from cloves; thickens and darkens with time; boils at 250–260°C; has aromatic aroma and pungent taste; soluble in ether and chloroform; chief component is eugenol; used in medicine, perfumery, flavoring, and soaps. Also known as caryophyllus oil; clove oil.

oil of coriander [MATER] A colorless or pale-yellow essential oil extracted from the dried seed of the coriander plant (*Coriandrum sativum*); used in medicines, beverages, and flavoring extracts.

oil of cubeb [MATER] A camphorous tasting, colorless, or pale green or yellow liquid with a peppery odor; distilled from cubebs; soluble in ether, alcohol, and chloroform; boils at 175–180°C; chief components are sesquiterpenes, cadinene, and dipentane; used in medicine. Also known as cubeb oil.

oil of fennel See fennel oil.

oil of grapefruit See grapefruit oil.

oil of orange blossoms [MATER] Bitter-tasting, fluorescent, pale-yellow essential oil with orange aroma; soluble in alcohol; main components are limonene, linalool, and geraniol; used for perfumes and flavors. Also known as neroli oil; orange flower oil.

oil of peppermint [MATER] Colorless or slightly yellow essential oil with minty aroma and taste; soluble in ether, alcohol, and chloroform; derived from leaves and flowering tops of the peppermint plant (*Mentha piperita*); has high menthol content; used in medicines, flavors, perfumes, and liqueurs. Also known as peppermint oil.

oil of sage [MATER] **1.** An alcohol-soluble yellow oil with sage aroma; obtained from leaves of the common sage (*Salvia officinalis*); used chiefly in flavors. Also known as dalmatian sage oil; sage oil; salvia oil. **2.** A pale-yellow oil with ambergris aroma obtained from the flowers of *Salvia sclarea*; used chiefly in perfumes.

oil of sassafras [MATER] A pungent, aromatic, yellowish or reddish-yellow liquid obtained from the bark of American sassafras (*Sassafras albidum*); soluble in organic solvents, glacial acetic acid, and carbon disulfide; used for flavors, perfumes, and medicines. Also known as clary sage oil; sassafras oil.

oil of shaddock See grapefruit oil.

oil of turpentine [MATER] Water-insoluble, colorless, volatile essential oil distilled from turpentine; contains pinene, sylvestrene, and dipentene; used as a carminative, solvent, paint vehicle, and disinfectant.

oil paint [MATER] A paint made with a vegetable oil as the filmogen.

oil pump [MECH ENG] A pump of the gear, vane, or plunger type, usually an integral part of the automotive engine; it lifts oil from the sump to the upper level in the splash and circulating systems, and in forced-feed lubrication it pumps the oil to the tubes leading to the bearings and other parts.

oil reclaiming [ENG] **1.** A process in which oil is passed through a filter as it comes from equipment and then returned for reuse, in the same manner that crank case oil is cleaned by an engine filter. **2.** A method in which solids are removed from oil by treatment in settling tanks.

oil ring [MECH ENG] **1.** A ring located at the lower part of a piston to prevent an excess amount of oil from being drawn up onto the piston during the suction stroke. **2.** A ring on a journal, dipping into an oil bath for lubrication.

oil saturation [PETRO ENG] Measurement of the degree of saturation of reservoir pore structure by reservoir oil.

oil seal [ENG] **1.** A device for preventing the entry or return of oil from a chamber. **2.**

A device using oil as the sealing medium to prevent the passage of fluid from one chamber to another.

oil seed [MATER] Seeds of plants from which oil can be derived by expression or solvent extraction, such as soya beans.

oil separator *See* gas-oil separator.

oil-soluble resin [MATER] A resin that, at moderate temperatures, dissolves in, disperses in, or reacts with drying oils to produce a homogeneous film of modified characteristics.

oil stain [MATER] Thin oil paint with very little pigment, used to stain wood surfaces.

oilstone [MATER] A whetstone used with oil.

oil varnish [MATER] A varnish composed of resins dissolved in oil.

oil well [PETRO ENG] A hole drilled (usually vertically) into an oil reservoir for the purpose of recovering the oil trapped in porous formations.

oil-well cement [MATER] A type of hydraulic cement which has a slow setting rate under the high temperatures obtained in oil wells; uses include support of tubing and bypassing of unwanted zones.

oil-well drive *See* reservoir drive mechanism.

oil-well pump [PETRO ENG] Device for artificial or secondary (non-gas-lift) oil production; about 85% of the production is by ground-level sucker-rod pumps, the remainder by downhole hydraulic lift pumps.

oil white [MATER] Mixture of lithopone and zinc white or white lead; used as a house-paint pigment.

oiticica oil [MATER] A light-yellowish oil obtained from seeds of the Brazilian oiticica tree (*Licania rigida*); raw oil becomes buttery unless heat-treated (semipolymerized); used principally in paint and varnish as a drying oil as a substitute for tung oil or with tung oil.

okonite [MATER] Insulating material made from the vulcanization of ozokerite and resin with rubber and sulfur.

Oldham-Wheat lamp [MIN ENG] A cap lamp designed for full self-service.

old workings [MIN ENG] Mines which have been abandoned, allowed to collapse, and sometimes sealed off.

olemeter [ENG] **1.** A device for measuring specific gravity of oils. **2.** An instrument for determining the proportion of oil in a substance.

oleo oil [MATER] Yellow liquid fat used to make oleomargarine; consists of liquid olein and palmitin from cold-pressed tallow.

oleoresin [MATER] A resin-essential oil mixture with pungent taste; extracted from various plants; used in pharmaceutical

preparations; examples are Peru, tulu, and styrax balsams.

oleoresinous varnish [MATER] A varnish made by compounding the resin with oxidizable oil, such as linseed oil.

oleostearin [MATER] Edible solid fat from tissues of cattle (genus *Bos*); the solid remaining after oleo oil or tallow oil is removed from tallow. Also known as beef stearin.

oleo strut [MECH ENG] A shock absorber consisting of a telescoping cylinder that forces oil into an air chamber, thereby compressing the air; used on aircraft landing gear.

oleum gossypii seminis *See* cotton oil.

oleum morrhuae *See* cod-liver oil.

oleum theobromatis *See* cocoa butter.

olibanum [MATER] A gum resin distilled from the dried exudation of African and Arabian trees of the genus *Boswellia*; used as a perfume, fixative, in incense, in fumigants, and in pharmacy. Also known as frankincense; gum thus.

olibanum oil [MATER] Colorless oil with balsamic aroma; distilled from olibanum; soluble in ether, chloroform, and carbon disulfide; main components are pinene, phellandrene, and dipentene; used in medicine. Also known as frankincense oil.

olive infused oil [MATER] A synthetic olive oil made by infusing corn oil with a paste of finely ground, partly dehydrated ripe olives; contains carotene.

olive oil [MATER] Pale- or greenish-yellow edible oil; main components are olein and palmitin; soluble in ether, chloroform, and carbon disulfide; derived from the pulp of olive tree fruit; used in foods, ointments, linaments, and soaps, as a lubricant, and for tanning. Also known as Florence oil; lucca oil; sweet oil.

Oliver filter [MIN ENG] A continuous-type filter made in the form of a cylindrical drum with filter cloth stretched over the convex surface of the drum.

Olsen ductility test [MET] A cupping test in which a piece of sheet metal is deformed at the center by a steel ball until fracture occurs; ductility is measured by the height of the cup at the time of failure.

ombroscope [ENG] An instrument consisting of a heated, water-sensitive surface which indicates by mechanical or electrical techniques the occurrence of precipitation; the output of the instrument may be arranged to trip an alarm or to record on a time chart.

omnibearing converter [ENG] An electromechanical device which combines an omnirange signal with heading information to furnish electrical signals for the operation of the pointer of a radio magnetic indicator.

omnibearing indicator [ENG] An instrument providing automatic and continuous indication of omnibearing.

omnibearing selector [ENG] A device capable of being set manually to any desired omnibearing, or its reciprocal, to control a course-line deviation indicator. Also known as radial selector.

omnidirectional hydrophone [ENG ACOUS] A hydrophone whose response is fundamentally independent of the incident sound wave's angle of arrival.

omnigraph [ENG] An automatic acetylene cutter controlled by a mechanical pointer that traces a pattern; capable of cutting several duplicates simultaneously.

omnimeter [ENG] A theodolite with a microscope that can be used to observe vertical angular movement of the telescope.

on [ENG] Designating the operating state of a device or one of two possible conditions (the other being "off") in a circuit.

on center [BUILD] The measurement made between the centers of two adjacent members.

once-through boiler [MECH ENG] A boiler in which water flows, without recirculation, sequentially through the economizer, furnace wall, and evaporating and superheating tubes.

one-hundred-percent premium plan [IND ENG] A wage incentive plan wherein each unit produced by an employee in excess of standard is compensated at the same rate paid for each unit of standard production. Also known as straight piecework system; straight proportional system.

one-piece set [MIN ENG] A single stick of timber used as a post, stull, or prop.

one-shot molding [ENG] Production of urethane-plastic foam in which the isocynate, polyol, and catalyst and other additives are mixed directly together and a foam is produced immediately.

one-sided acceptance sampling test [IND ENG] A test against a single specification only, in which permissible values in one direction are not limited.

one-way slab [CIV ENG] A concrete slab in which the reinforcing steel runs perpendicular to the supporting beams, that is, one way.

on grade [CIV ENG] **1.** At ground level. **2.** Supported directly on the ground.

onion oil [MATER] A sharp-odored yellow liquid; derived from the bulb of the onion *Allium cepa*; soluble in ether, chloroform, and carbon disulfude; main component is allyl propyl disulfide; used in flavorings.

onionskin paper [MATER] A lightweight, durable bond paper; usually quite translucent, resembling the dry outer skin of an onion; used for duplicate typewriter copies and in interleaving order books.

on-line analyzer [MIN ENG] An instrument which monitors the content of materials at various stages in flotation or other mineral-processing flow sheets.

onsetter [MIN ENG] The worker in charge of hoisting coal, ore, men, or materials in a mine shaft.

on the run [MIN ENG] Manner of working a seam of coal when there is sufficient inclination to cause the coal, as worked toward the rise, to fall by gravity to the gangways for loading into cars.

on the solid [MIN ENG] **1.** Pertaining to the practice of blasting heavy charges of explosives, in lieu of undercutting or channeling. **2.** *See* into the solid.

on-top flight [AERO ENG] Flight above an overcast.

opacifier [MATER] A substance, used to treat a solid rocket propellant, that absorbs light and heat and protects the propellant from deterioration until ready for use.

opal glass [MATER] Translucent or opaque glass, often milky white, made by adding impurities such as fluorine compounds to the melt; it appears white by reflected light but shows color images through thin sections; used for ornamental glass and as an efficient light diffuser.

OPDAR [ENG] A laser system for measuring elevation angle, azimuth angle, and slant range of a missile during its firing period. Derived from optical direction and ranging. Also known as optical radar.

open-arc furnace [MET] An electrosmelting furnace in which the arc is generated above the level of the furnace feed.

open-belt drive [DES ENG] A belt drive having both shafts parallel and rotating in the same direction.

open berth [CIV ENG] An anchorage berth in an open roadstead.

open caisson [CIV ENG] A caisson in the form of a cylinder or shaft that is open at both ends; it is set in place, pumped dry, and filled with concrete.

open-cell foam [MATER] Foamed material, natural or synthetic, rigid or flexible, organic or metallic, in which there is interconnection between the cells.

open-center plan position indicator [ENG] A plan position indicator on which no signal is displayed within a set distance from the center.

open-circuit grinding [MECH ENG] Grinding system in which material passes through the grinder without classification of product and without recycle of oversize lumps; in contrast to closed-circuit grinding.

open-circuit scuba [ENG] The simplest type of scuba equipment, in which all exhaled gas is discharged directly into the water and the utilization of gas is therefore equal to the mass exhaled.

opencut [CIV ENG] An open trench, such as across a hill. [MIN ENG] **1.** To drive headings out, or to commence working in the coal after sinking the shafts. **2.** To commence longwall working. **3.** To increase the size of a shaft when it intersects a drift.

open-cycle engine [MECH ENG] An engine in which the working fluid is discharged after one pass through boiler and engine.

open-cycle gas turbine [MECH ENG] A gas turbine prime mover in which air is compressed in the compressor element, fuel is injected and burned in the combustor, and the hot products are expanded in the turbine element and exhausted to the atmosphere.

open die [MET] A forming or forging die in which there is little or no restriction to the lateral flow of metal within the die set.

open-die forging [MET] Forging performed with open dies.

open-end method [MIN ENG] A technique of mining pillars in which no stump is left.

open-end wrench [DES ENG] A wrench consisting of fixed jaws at one or both ends of a handle.

open fire [MIN ENG] A fire at a roadway or at the coal face in a mine.

open-flow capacity [PETRO ENG] Fluid flow rate from a gas well flowing open to the atmosphere.

open-flow potential [PETRO ENG] Gas flow rate in thousands of cubic feet of gas per 24 hours that would be produced by a well if the only pressure against the face of the producing formation wellbore were atmospheric pressure.

open-flow test [PETRO ENG] The flowing of wells wide open to the atmosphere with simultaneous measurement of gas-flow rate and pressure drop; used to analyze the potential of a reservoir to deliver gas to the wellbore.

open-flow well [PETRO ENG] Gas well flowing open to the atmosphere.

open-hearth furnace [MET] A reverberatory melting furnace with a shallow hearth and a low roof, in which the charge is heated both by direct flame and by radiation from the roof and walls of the furnace.

oven-hearth process [MET] A steel-making process carried out in an open-hearth furnace in which selected pig iron and malleable scrap iron are melted, with the addition of pure iron ore.

open hole [ENG] **1.** A well or borehole, or a portion thereof, that has not been lined with steel tubing at the depth referred to. **2.** An unobstructed borehole. **3.** A borehole being drilled without cores. [PETRO ENG] In an oil-drilling operation, the unprotected hole which lies below the shoes of the last landed string of casing.

opening [MIN ENG] **1.** A widening of a crevice, in consequence of a softening or decomposition of the adjacent rock, so as to leave a vacant space. **2.** A short heading driven between two or more parallel headings or levels for ventilation. **3.** An area in a coal mine between pillars, or between pillars and ribs.

opening die [MECH ENG] A die head for cutting screws that opens automatically to release the cut thread.

opening material [MATER] A material added to plastic clay in ceramic making to speed drying and reduce shrinking.

opening pressure [MECH ENG] The static inlet pressure at which discharge is initiated.

open-pit mining [MIN ENG] Extracting metal ores and minerals that lie near the surface by removing the overlying material and breaking and loading the ore. Also known as open-cast mining; opencut mining.

open-pit quarry [MIN ENG] A quarry in which the opening is the full size of the excavation.

open plan [BUILD] Arrangement of the interior of a building without distinct barriers such as partitions.

open shop [IND ENG] A shop in which employment is not restricted to members of a labor union.

open-side planer [DES ENG] A planer constructed with one upright or housing to support the crossrail and tools.

open-side tool block [DES ENG] A toolholder on a cutting machine consisting of a T-slot clamp, a C-shaped block, and two or more tool clamping screws. Also known as heavy-duty tool block.

open stope [MIN ENG] Underground working place that is unsupported, or supported by timbers or pillars of rock.

open-stope method [MIN ENG] Stoping in which no regular artificial method of support is employed, although occasional props or cribs may be used to hold local patches of insecure ground.

open-timbered roof [BUILD] A roof in which the supporting timbers are left uncovered, forming part of the ceiling.

open timbering [MIN ENG] A method of supporting the ground in a mine shaft or tunnel; supports are several feet apart, with the ground between them secured by struts.

open traverse [ENG] A surveying traverse in which the last leg, because of error, does

optically pumped magnetometer [ENG] A type of magnetometer that measures total magnetic field intensity by observation of the precession frequency of magnetic atoms, usually gaseous rubidium, cesium, or helium, which are magnetized by irradiation with circularly polarized light of a suitable wavelength.

optical material [MATER] A material which is transparent to light or to infrared, ultraviolet, or x-ray radiation, such as glass and certain single crystals, polycrystalline materials (chiefly for the infrared), and plastics.

optical plastic [MATER] A plastic which is transparent to light, occasionally used in optical systems for reasons of economy, special index-dispersion relation, light weight, and nonbrittleness.

optical pyrometer [ENG] An instrument which determines the temperature of a very hot surface from its incandescent brightness; the image of the surface is focused in the plane of an electrically heated wire, and current through the wire is adjusted until the wire blends into the image of the surface. Also known as disappearing filament pyrometer.

optical radar *See* OPDAR.

optical rangefinder [ENG] An optical instrument for measuring distance, usually from its position to a target point, by measuring the angle between rays of light from the target, which enter the rangefinder through the windows spaced apart, the distance between the windows being termed the baselength of the rangefinder; the two types are coincidence and stereoscopic.

optical recording [ENG] Production of a record by focusing on photographic paper a beam of light whose position on the paper depends on the quantity to be measured, as in a light-beam galvanometer.

optical reflectometer [ENG] An instrument which measures on surfaces the reflectivity of electromagnetic radiation at wavelengths in or near the visible region.

optical square [ENG] A surveyor's hand instrument used for laying of right angles; employs two mirrors at a 45° angle.

optical tracking [ENG] The determination of spatial positions of distant airplanes, missiles, and artificial satellites as a function of time, or the recording of engineering events, by precise time-correlated observations with various types of telescopes or ballistic cameras.

optician [ENG] A maker of optical instruments or lenses.

optimization [SYS ENG] **1.** Broadly, the efforts and processes of making a decision, a design, or a system as perfect, effective, or functional as possible. **2.** Narrowly, the specific methodology, techniques, and procedures used to decide on the one specific solution in a defined set of possible alternatives that will best satisfy a selected criterion. Also known as system optimization.

optimum flight [AERO ENG] An aircraft flight so planned and navigated that it is completed under the optimum conditions of minimum time and minimum exposure to dangerous flying weather.

optimum separation point [PETRO ENG] In extraction of natural gasoline, the pressure and temperature conditions necessary for maximum condensation in the separators under field conditions.

optoelectronic amplifier [ENG] An amplifier in which the input and output signals and the method of amplification may be either electronic or optical.

optoelectronic shutter [ENG] A shutter that uses a Kerr cell to modulate a beam of light.

optophone [ENG ACOUS] A device with a photoelectric cell to convert ordinary printed letters into a series of sounds; used by the blind.

orange flower oil *See* oil of orange blossoms.

orange lake [MATER] Any of various transparent orange pigments from the precipitation of an orange dyestuff on aluminum hydrate or other base; used to produce transparent coatings for metal cans and bottle caps.

orange mineral [MATER] A bright orange-red lead oxide pigment used in printing inks and primers.

orange oil *See* bitter orange oil; sweet orange oil.

orange peel [MATER] A pebbled film surface, resembling an orange skin, on lacquer or enamel as a result of too rapid drying after spraying, or failure to exhibit the desired leveling effects. [MET] A rough, pebble-grained metal surface resulting from either plastic deformation or electropolishing. Also known as alligator effect; pebbling.

orange-peel bucket [DES ENG] A type of grab bucket that is multileaved and generally round in configuration.

orbital curve [AERO ENG] One of the tracks on a primary body's surface traced by a satellite that orbits about it several times in a direction other than normal to the primary body's axis of rotation; each track is displaced in a direction opposite and by an amount equal to the degrees of rotation between each satellite orbit.

orbital decay [AERO ENG] The lessening of the eccentricity of the elliptical orbit of an artificial satellite.

orbital direction [AERO ENG] The direction that the path of an orbiting body takes; in the case of an earth satellite, this path may be defined by the angle of inclination of the path to the equator.

orbital rendezvous [AERO ENG] **1.** The meeting of two or more orbiting objects with zero relative velocity at a preconceived time and place. **2.** The point in space at which such an event occurs.

orbit point [AERO ENG] A geographically defined reference point over land or water, used in stationing airborne aircraft.

orcanette *See* alkanette.

orchil [MATER] Dark-brownish-red coloring matter derived from lichens as paste or aqueous extract; main components are orcin and orcein; used as carpet-yarn dye. Also known as orseille.

order point [IND ENG] The inventory level at which a replenishment order must be placed.

order quantity [IND ENG] The number of pieces ordered to replenish the inventory.

ordinary cut [MIN ENG] The arrangement of drill holes in a mine in which the drill holes are symmetrical with respect to the vertical center-line of the section, extend horizontally, and make a large angle with the working face.

ordinary gear train [MECH ENG] A gear train in which all axes remain stationary relative to the frame.

ordinary sheathed explosive [MATER] A permitted explosive (one passing certain safety tests) whose safety has been further increased by a sheath of sodium bicarbonate; when the explosive is detonated, the sheath forms carbon dioxide, which tends to extinguish the flame around the detonator wave.

ordnance [ENG] Military materiel, such as combat weapons of all kinds, with ammunition and equipment for their use, vehicles, and repair tools and machinery.

ore bin [MIN ENG] A receptacle for ore awaiting treatment or shipment.

ore block [MIN ENG] A vein of ore which is bound above, below, and at one or both ends; it is ready for excavation.

ore blocked out [MIN ENG] Ore exposed on three sides within a reasonable distance of each other.

ore bridge [MIN ENG] A gantry crane used to load and unload stockpiles of ore.

ore car [MIN ENG] A mine car for carrying ore or waste rock.

ore chute [MIN ENG] An inclined passage for the transfer of ore to a lower level.

ore crusher [MIN ENG] A machine for breaking up masses of ore, usually previous to passing through other size-reduction equipment.

ore developed [MIN ENG] Ore exposed on four sides in blocks variously prescribed.

ore dressing [MIN ENG] The cleaning of ore by the removal of certain valueless portions, as by jigging, cobbing, or vanning.

ore expectant [MIN ENG] The whole or any part of the ore below the lowest level or beyond the range of vision.

ore faces [MIN ENG] Those ore bodies that are exposed on one side, or show only one face.

ore flotation promoter [MATER] Material that gives a water-repellent surface to mineral particles so that air bubbles will adhere and cause selective flotation.

ore grader [MIN ENG] In metal mining, a person who directs the storage of iron ores in bins at shipping docks so that the various grades in each bin will contain approximate percentages of iron.

oreide bronze [MET] A series of brass compositions containing 68–87% copper, 10–32% zinc, and sometimes small amounts of tin; used for hardware.

ore in sight [MIN ENG] **1.** Ore exposed on at least three sides within reasonable distance of each other. **2.** Ore which may be reasonably assumed to exist, though not actually blocked out. **3.** *See* developed reserves.

ore intersection [MIN ENG] **1.** The point at which a borehole, crosscut, or other underground opening encounters an ore vein or deposit. **2.** The thickness of the ore-bearing deposit so traversed.

ore pass [MIN ENG] A vertical or inclined passage for the downward transfer of ore.

ore pocket [MIN ENG] **1.** Excavation near the hoisting shaft into which ore from stopes is moved, preliminary to hoisting. **2.** An unusual concentration of ore in the lode.

ore reduction [MIN ENG] The size reduction of solids by crushing and grinding in mineral processing plants.

ore reserve [MIN ENG] The total tonnage and average value of proved ore, plus the total tonnage and value (assumed) of the probable ore.

ore sampling [MIN ENG] The process in which a portion of ore is selected so that its composition will represent the average composition of the entire bulk of ore.

organic bonded wheel [DES ENG] A grinding wheel in which organic bonds are used to hold the abrasive grains.

organic coating [MATER] Material used to protect metal surfaces from chemical or atmospheric attack; includes latex paints, plastics, asphaltic materials, rubbers, and elastomers.

organic glass [MATER] An amorphous, solid, glasslike material made of transparent plastic.

organic semiconductor [MATER] An organic material having unusually high conductivity, often enhanced by the presence of certain gases, and other properties commonly associated with semiconductors; an example is anthracene.

organisol [MATER] A dispersion of very finely divided resin particles that are suspended in an organic-liquid mixture which cannot dissolve the resin at normal temperatures.

organization chart [IND ENG] Graphic representation of the interrelationships within an organization, depicting lines of authority and responsibility and provisions for control.

organosol [MATER] Finely divided or colloidal suspension of insoluble material in a suspending organic liquid; known as plastisol when the solid is a synthetic resin suspended in an organic liquid; used for coatings, moldings, and casting of films.

orient [ENG] **1.** To place or set a map so that the map symbols are parallel with their corresponding ground features. **2.** To turn a transit so that the direction of the 0° line of its horizontal circle is parallel to the direction it had in the preceding or initial setup, or parallel to a standard reference line.

oriental linaloe [MATER] A rosewood oil distilled from highly perfumed parts of *Aquilaria agollocha* trees of Burma, eastern India, and Java. Also known as agar attar; aloe wood oil.

orientation [ENG] Establishment of the correct relationship in direction with reference to the points of the compass.

oriented core [ENG] A core that can be positioned on the surface in the same way that it was arranged in the borehole before extraction.

orifice gas [MET] The torch gas in a plasma arc welding or cutting process which becomes ionized in the arc to form plasma and is ejected from the orifice in a jet stream.

orifice meter [ENG] An instrument that measures fluid flow by recording differential pressure across a restriction placed in the flow stream and the static or actual pressure acting on the system.

orifice mixer [MECH ENG] Arrangement in which two or more liquids are pumped through an orifice constriction to cause turbulence and consequent mixing action.

orifice plate [DES ENG] A disk, with a hole, placed in a pipeline to measure flow.

orifice well tester [PETRO ENG] Velocity-type meter used to measure gas flow quantity from a gas well; static pressure differences before and after a sharp-edged orifice are converted to flow values.

origanum oil [MATER] A light-yellow essential oil obtained from herbs of the genus *Origanum*; contains carvacrol and cymene plus other components; used in flavors and pharmaceuticals.

O ring [DES ENG] A flat ring made from synthetic rubber, used as an airtight seal or a seal against high pressures.

orograph [ENG] A machine that records both distance and elevations as it is pushed across land surfaces; used in making topographic maps.

orometer [ENG] A barometer with a scale that indicates elevation above sea level.

orris [MATER] The fragrant powder from the root of the plants *Iris florentinea, I. germanica,* and *I. pallida*; used in perfume, medicine, and tooth powder. Also known as orrisroot.

orris oil [MATER] A yellow, fatty, semisolid, fragrant essential oil obtained from roots of the Florentine iris; melts at 44–50°C; soluble in ether, chloroform, and alcohol; main components are myristic acid, oleic acid, and irone; used in flavoring and perfumes.

orrisroot *See* orris.

Orr's white *See* lithopone.

orseille *See* orchil.

orthometric correction [ENG] A systematic correction that must be applied to a measured difference in elevation since level surfaces at varying elevations are not absolutely parallel.

orthometric height [ENG] The distance above sea level measured along a plumb line.

Orthonik [MET] A magnetic alloy composed of 45–50% nickel, with the remainder iron, having a grainy structure, high permeability, and a rectangular hysteresis loop; used in magnetic cores.

orthotropic deck [CIV ENG] A bridge deck constructed typically of flat steel plate and longitudinal and transverse ribs; functions in carrying traffic and acting as top flanges of floor beams.

oscillating conveyor [MECH ENG] A conveyor on which pulverized solids are moved by a pan or trough bed attached to a vibrator or oscillating mechanism. Also known as vibrating conveyor.

oscillating granulator [MECH ENG] Solids size-reducer in which particles are broken by a set of oscillating bars arranged in cylindrical form over a screen of suitable mesh.

oscillating screen [MECH ENG] Solids separator in which the sifting screen oscillates at 300 to 400 revolutions per minute in a plane parallel to the screen.

oscillogram [ENG] The permanent record produced by an oscillograph, or a photograph of the trace produced by an oscilloscope.

oscillograph [ENG] A measurement device for determining waveform by recording the instantaneous values of a quantity such as voltage as a function of time.

osmium [MET] A hard white metal of rare natural occurrence.

osseine [MATER] The organic residue formed when bone is dissolved in hydrochloric acid; used to make glue and gelatin.

Ostwald viscometer [ENG] A viscometer in which liquid is drawn into the higher of two glass bulbs joined by a length of capillary tubing, and the time for its meniscus to fall between calibration marks above and below the upper bulb is compared with that for a liquid of known viscosity.

otter *See* paravane.

Otto engine [MECH ENG] An internal combustion engine that operates on the Otto cycle, where the phases of suction, compression, combustion, expansion, and exhaust occur sequentially in a four-stroke-cycle or two-stroke-cycle reciprocating mechanism.

otto of rose oil *See* rose oil.

ounce metal [MET] An alloy composed of 1 ounce each of lead, tin, and zinc to 1 pound of copper. Also known as composition metal.

ouricury wax [MATER] A hard brown wax obtained from leaves of the ouricury palm (*Cocos coronapa*); similar to carnauba wax in use and properties.

outage [PETRO ENG] The difference between the full or rated capacity of a barrel, tank, or tank car as compared to actual content.

outage method [PETRO ENG] Deduction of the liquid content of a tank by measurement of the distance from the top of the tank to the surface of the liquid; in contrast to the innage method.

outburst [MIN ENG] The sudden issue of gases, chiefly methane (sometimes accompanied by coal dust), from the working face of a coal mine.

outby [MIN ENG] Toward the mine entrance or shaft and therefore away from the working face.

outfall [CIV ENG] The point at which a sewer or drainage channel discharges to the sea or to a river.

outgassing [ENG] The release of adsorbed or occluded gases or water vapor, usually by heating, as from a vacuum tube or other vacuum system.

output indicator [ENG] A meter or other device that is connected to a radio receiver to indicate variations in output signal strength for alignment and other pur-

poses, without indicating the exact value of output.

output-limited [ENG] Restricted by the need to await completion of an output operation, as in process control or data processing.

output meter [ENG] An alternating-current voltmeter connected to the output of a receiver or amplifier to measure output signal strength in volume units or decibels.

output-meter adapter [ENG] Device that can be slipped over the plate prong of the output tube of a radio receiver to provide a conventional terminal to which an output meter can be connected during alignment.

output shaft [MECH ENG] The shaft that transfers motion from the prime mover to the driven machines.

output standard *See* standard time.

outrigger [ENG] A steel beam or lattice girder extending from a crane to provide stability by widening the base.

outside caliper [DES ENG] A caliper having two curved legs which point toward each other; used for measuring outside dimensions of a workpiece.

outside diameter [DES ENG] The outer diameter of a pipe, including the wall thickness; usually measured with calipers. Abbreviated OD.

oven [ENG] A heated enclosure for baking, heating, or drying.

overaging [MET] Aging at a higher temperature or for a longer time than is required to produce maximum or optimum properties.

overall drilling time [MIN ENG] The total time for rock drilling including time for setting up, withdrawing, and moving drills, time for mechanical delays, and the time for the actual drilling.

overall efficiency [AERO ENG] The efficiency of a jet engine, rocket engine, or rocket motor in converting the total heat energy of its fuel first into available energy for the engine, then into effective driving energy.

overarching weight [MIN ENG] The pressure of the rocks over the active mine workings.

overarm [MECH ENG] One of the adjustable supports for the end of a milling-cutter arbor farthest from the machine spindle.

overbending [MET] Compensation for springback in metalforming by bending the material through a greater arc than that required for the finished part.

overbreak [CIV ENG] Rock excavated in excess of the neat lines of a tunnel or cutting. Also known as backbreak.

overburden [MIN ENG] To charge in a furnace too much ore and flux in proportion to the amount of fuel.

overcast [MIN ENG] **1.** An enclosed airway to permit one air current to pass over another without interruption. **2.** To move overburden removed from coal mined from surface mines to an area from which the coal has been mined.

overcoating [ENG] Extruding a plastic web beyond the edge of the substrate web in extrusion coating.

overcut [MIN ENG] A machine cut made along the top or near the top of a coal seam; sometimes used in a thick seam or a seam with sticky coal.

overcutting machine [MIN ENG] A coal-cutting machine designed to make the cut at a desired place in the coal seam some distance above the floor.

overdraft [MET] Upward curving of a piece of metal after leaving the rolls during forming, due to higher speed of the lower roll.

overdrilling [ENG] The act or process of drilling a run or length of borehole greater than the core-capacity length of the core barrel, resulting in loss of the core.

overdrive [MECH ENG] An automobile engine device that lowers the gear ratio, thereby reducing fuel consumption.

overfall dam *See* overflow dam.

overfeed [MIN ENG] To attempt to make a diamond- or rock-drill bit penetrate rock at a rate in excess of that at which the optimum economical performance of the bit is attained, needlessly damaging the bit and shortening its life.

overfire draft [MECH ENG] The air pressure in a boiler furnace during occurrence of the main flame.

overflow [CIV ENG] Any device or structure that conducts excess water or sewage from a conduit or container.

overflow capacity [ENG] Capacity of a container measured to its top, or to the point of overflow.

overflow channel [CIV ENG] An artificial waterway for conducting water away from an overflowing structure such as a reservoir or canal.

overflow dam [CIV ENG] A dam built with a crest to allow the overflow of water. Also known as overfall dam; spillway dam.

overflow groove [ENG] Small groove on a plastics mold that allows material to flow freely, to prevent weld lines and low density in the finished product and to dispose of excess material.

overflow pipe [ENG] Open pipe protruding above the surface of a liquid in a container, such as a distillation or absorption column or a toilet tank, to control the height of the liquid; excess liquid enters the pipe's open end and drains away.

overgear [MECH ENG] A gear train in which the angular velocity ratio of the driven shaft to driving shaft is greater than unity, as when the propelling shaft of an automobile revolves faster than the engine shaft.

overglaze color [MATER] Any of the mixtures of ground pigment and low-melting glass melting at 704–816°C; used for decorative designs fired onto china and ceramics.

overgrinding [MIN ENG] Grinding an ore to a smaller particle size than that necessary to free the desired mineral from other materials.

overhand cut and fill [MIN ENG] A method of mining ore in which material removed from the roof of a drive drops through chutes to a lower drive, from which the material is removed.

overhand stope [MIN ENG] A stope in which the ore above the point of entry is attacked, so that severed ore tends to gravitate toward discharge chutes, and the stope is self-draining.

overhand stoping [MIN ENG] A method of mining in which the ore is blasted from a series of ascending stepped benches; both horizontal and vertical holes may be employed.

overhang [BUILD] The distance measured horizontally that a roof projects beyond a wall.

overhaul [ENG] A maintenance procedure for machinery involving disassembly, the inspecting, refinishing, adjusting, and replacing of parts, and reassembly and testing.

overhauling [MET] Removing scale and surface defects from metal castings or slabs by cutting away surface layers.

overhead cableway [MIN ENG] A type of equipment for the removal of soil or rock, consisting of a strong overhead cable which is usually attached to towers at either end, and on which a car or traveler may run back and forth; from this car a pan or bucket may be lowered to the surface, then raised and locked to the car and transported to any position on the cable where it is desired to dump.

overhead camshaft [MECH ENG] A camshaft mounted above the cylinder head.

overhead position [MET] In welding, the position by which the deposit is made from the underside of the joint.

overhead shovel [MECH ENG] A tractor which digs with a shovel at its front end, swings the shovel rearward overhead, and dumps the shovel at its rear end.

overhead-traveling crane [MECH ENG] A hoisting machine with a bridgelike structure moved on wheels along overhead trackage which is usually fixed to the building structure.

overhead-valve engine [MECH ENG] A four-stroke-cycle internal combustion engine having its valves located in the cylinder head, operated by pushrods that actuate rocker arms. Abbreviated OHV engine. Also known as valve-in-head engine.

overheat [MET] To heat a metal or alloy to such high temperatures that its physical properties are impaired.

overlap [MET] **1.** Projection of the weld metal beyond the bond at the toe of the weld. **2.** Extension of one sheet over another in spot, seam, or projection welding.

overlap radar [ENG] Radar located in one sector whose area of useful radar coverage includes a portion of another sector.

overlay [CIV ENG] A repair topping of asphalt or concrete placed on a worn roadway. [ENG] **1.** Nonwoven fibrous mat (glass or other fiber) used as the top layer in a cloth or mat lay-up to give smooth finish to plastic products or to minimize the fibrous pattern on the surface. Also known as surfacing mat. **2.** An ornamental covering, as of wood or metal.

overlay plywood [MATER] Plywood with a resin-treated fiber surface.

overloader [MIN ENG] A loading machine which digs with a bucket, raises the bucket, and swings it in a wide horizontal arc to the dumping point.

overpass [CIV ENG] **1.** A grade separation in which traffic at the higher level is raised, and traffic at the lower level moves at approximately its original level. **2.** The upper part of a grade crossing.

overrun [CIV ENG] A cleared area extending beyond the end of a runway.

overrunning clutch [MECH ENG] A clutch that allows the driven shaft to turn freely only under certain conditions; for example, a clutch in an engine starter that allows the crank to turn freely when the engine attempts to run.

overshoot [ENG] **1.** An initial transient response to a unidirectional change in input which exceeds the steady-state response. **2.** The maximum amount by which this transient response exceeds the steady-state response.

overshot [ENG] **1.** A fishing tool for recovering lost drill pipe or casing. **2.** See bullet.

overshot wheel [MECH ENG] A horizontal-shaft waterwheel with buckets around the circumference; the weight of water pouring into the buckets from the top rotates the wheel.

oversize control screen [MIN ENG] A screen used to prevent the entry into a machine of coarse particles which might interfere with its operation. Also known as check screen; guard screen.

oversize powder [MET] A metal powder having coarser particles than the maximum permitted.

oversize rod See guide rod.

overspeed governor [MECH ENG] A governor that stops the prime mover when speed is excessive.

oversquare engine [MECH ENG] An engine with bore diameter greater than the stroke length.

overstressed area [MIN ENG] In strata control, an area where the force is concentrated on pillars.

overstressing [ENG] Cyclically stressing a material at a level higher than that used at the end of a fatigue test.

overtub system [MIN ENG] An endless-rope system in which the rope runs over the tubs or cars in the center of the rails.

overwind [ENG] To wind a spring, rope, or cable too tightly or too far.

oxidation pond [CIV ENG] A shallow lagoon or basin in which wastewater is purified by sedimentation and aerobic and anaerobic treatment.

oxidized cellulose See oxycellulose.

oxidized microcrystalline wax [MATER] Refined, oxidized wax from bottoms of storage tanks for solvent-extracted petroleum; used in floor polishes.

oxidized shale See burnt shale.

oxyacetylene cutting [ENG] The flame cutting of ferrous metals in which the preheating of the metal is accomplished with a flame produced by an oxyacetylene torch. Also known as acetylene cutting.

oxyacetylene torch [ENG] An acetylene gas-mixing and burning tool that produces a hot flame for the welding or cutting of metal. Also known as acetylene torch.

oxyacetylene welding [MET] A welding process in which the heat is supplied by an oxyacetylene flame. Also known as acetylene welding.

oxycellulose [MATER] Cellulose mixed with reaction products from oxidation of cellulose in the presence of steam or alkalies or by strong sunlight. Also known as oxidized cellulose.

oxychloride cement [MATER] A strong, hard cement composed of magnesium chloride and calcined magnesia; used for floors and stucco. Also known as Sorel cement.

oxygenated oil [MATER] A class of essential oils containing carbon, hydrogen, and oxygen; an example is oil of cassia.

oxygen bomb calorimeter [ENG] Device to measure heat of combustion; the sample is burned with oxygen in a closed vessel, and the temperature rise is noted.

oxygen corrosion [MET] The reaction of oxygen with metallic surfaces to form an oxide of the metal or alloy.

oxygen cutting [ENG] Any of several types of cutting processes in which metal is removed with or without a flux by a chemical reaction of the base metal with oxygen at high temperatures.

oxygen-free copper [MET] Pure copper having a conductivity greater than that of copper containing impurities such as cuprous oxide; used for the construction of high-power electron tubes because it does not release appreciable gas when hot.

oxygen furnace steel [MET] Steel made by a process in which oxygen under pressure is directed onto or into the molten metal.

oxygen gouging *See* flame gouging.

oxygen-kerosine burner [ENG] Liquid-fuel device using a mixture of oxygen and vaporized or atomized kerosine for combustion.

oxygen lance [MET] A pipe used to direct oxygen under pressure into a bath of molten steel.

oxygen mask [ENG] A mask that covers the nose and mouth and is used to administer oxygen.

oxygen propellant [MATER] A propellant having a minimum assay by volume of 99.9% oxygen when gasified.

oxygen steelmaking [MET] The manufacture of steel from molten pig iron and steel scrap by methods which employ pure oxygen gas (99 + %) and suitable fluxes to remove carbon and phosphorus (and in part, sulfur) without introducing nitrogen or hydrogen.

oxyhydrogen welding [MET] Welding with an oxyhydrogen flame.

ozocerite *See* ceresin.

ozone generator [ENG] Apparatus that converts oxygen, O_2, into ozone, O_3, by subjecting the oxygen to an electric-brush discharge. Also known as ozonizer.

ozonizer *See* ozone generator.

P

pace rating *See* effort rating.

pachymeter [ENG] An instrument used to measure the thickness of a material, for example, a sheet of paper.

pack [IND ENG] To provide protection for an article or group of articles against physical damage during shipment; packing is accomplished by placing articles in a shipping container, and blocking, bracing, and cushioning them when necessary, or by strapping the articles or containers on a pallet or skid. [MIN ENG] **1.** A pillar built in the waste area or roadside within a mine to support the mine roof; constructed from loose stones and dirt. **2.** Waste rock or timber used to support the roof or underground workings or used to fill excavations. Also known as fill.

package freight [IND ENG] Freight shipped in lots insufficient to fill a complete car; billed by the unit instead of by the carload.

pack builder [MIN ENG] **1.** One who builds packs or pack walls. **2.** In anthracite and bituminous coal mining, one who fills worked-out rooms, from which coal has been mined, with rock, slate, or other waste to prevent caving of walls and roofs, or who builds rough walls and columns of loose stone, heavy boards, timber, or coal along haulageways and passageways and in rooms where coal is being mined to prevent caving of roof or walls during mining operations. Also known as packer; pillar man; timber packer; waller.

pack carburizing [MET] A method of surface hardening of steel in which parts are packed in a steel box with the carburizing compound and heated to elevated temperatures.

packer *See* pack builder; production packer.

packer fluid [PETRO ENG] Fluid inserted in the annulus between the tubing and casing above a packer in order to reduce pressure differentials between the formation and the inside of the casing and across the packer.

packfong [MET] Chinese name for an alloy of nickel, zinc, and copper, which resembles nickel silver.

pack hardening [MET] A process of heat treating in which the workpiece is packed in a metal box together with carbonaceous material; carbon penetration is proportional to the length of heating; after treatment the workpiece is reheated and quenched.

packing [ENG] *See* stuffing. [ENG ACOUS] Excessive crowding of carbon particles in a carbon microphone, produced by excessive pressure or by fusion particles due to excessive current, and causing lowered resistance and sensitivity. [MET] In powder metallurgy, a material in which compacts are embedded during presintering or sintering operations.

packing house pitch [MATER] Dark-brown to black by-product residue from manufacturing soap and candle stock or from refining vegetable oils, refuse, or wool grease; soluble in naphtha and carbon disulfide; used to make paints, varnishes, and tar paper, and in marine caulking and waterproofing. Also known as fatty-acid pitch.

packing ring *See* piston ring.

pack rolling [MET] Hot rolling of two or more sheets of metal packed together; a thin surface oxide film prevents their welding.

pack wall [MIN ENG] A wall of dry stone built along the side of a roadway, or in the waste area, of a coal or metal mine to help support the roof and to retain the packing material and prevent its spreading into the roadway.

pad [AERO ENG] *See* launch pad. [ENG] **1.** A layer of material used as a cushion or for protection. **2.** A projection of excess metal on a casting forging, or welded part. **3.** A takeoff or landing point for a helicopter or space vehicle. [MET] The brickwork that is beneath the molten iron at the base of a blast furnace.

padded bit *See* castellated bit.

paddle [AERO ENG] A large, flat, paddle-shaped support for solar cells, used on some satellites. [DES ENG] Any of various implements consisting of a shaft with a broad, flat blade or bladelike part at one or both ends.

paddle wheel [MECH ENG] **1.** A device used to propel shallow-draft vessels, consisting of a wheel with paddles or floats on its circumference, the wheel rotating in a plane parallel to the ship's length. **2.** A wheel with paddles used to move leather in a processing vat.

padlock [DES ENG] An unmounted lock with a shackle that can be opened and closed; the shackle is usually passed through an eye, then closed to secure a hasp.

pail [DES ENG] A cylindrical or slightly tapered container.

paint [MATER] A mixture of a pigment and a vehicle, such as oil or water, that together form a liquid or paste that can be applied to a surface to provide an adherent coating that imparts color to and often protects the surface.

paint clay [MATER] A light-yellow to dark-reddish-brown iron- or manganese-bearing clay that mixes easily with linseed oil.

paint remover [MATER] Liquid or paste formulation used to remove dried paint, varnish, enamel, or lacquer; contains solvents such as methanol, ethyl alcohol, acetone, toluene, benzene, and ethyl acetate.

paint vehicle [MATER] The liquid constituent of paint; consists of volatile solvent or thinner and a film-forming component.

pair [MECH ENG] Two parts in a kinematic mechanism that mutually constrain relative motion; for example, a sliding pair composed of a piston and cylinder.

pairing element [MECH ENG] Either of two machine parts connected to permit motion.

palau [MET] A palladium-gold alloy; used as a platinum substitute in analytical chemistry.

palba wax [MATER] A grayish-yellow wax from older green leaves of the palm tree *Copernicia cerifera*.

pale catechu *See* gambir.

pale oil [MATER] A petroleum lubricating or process oil refined until its color (measured by transmitted light) is straw to pale yellow.

palladium [MET] A white, ductile malleable metal that resembles platinum and follows it in abundance and importance of applications; does not tarnish at normal temperatures.

palladium barrier leak detector [ENG] A type of leak detector in which hydrogen is diffused through a barrier of hot palladium into an evacuated vacuum gage.

pallet [BUILD] A flat piece of wood laid in a wall to which woodwork may be securely fastened. [ENG] **1.** A lever that regulates or drives a ratchet wheel. **2.** A hinged valve on a pipe organ. **3.** A tray or platform used in conjunction with a fork lift for lifting and moving materials. [MECH ENG] One of the disks or pistons in a chain pump.

palletize [IND ENG] To package material for convenient handling on a pallet or lift truck.

palmarosa oil [MATER] Colorless to light-yellow, volatile essential oil with roselike aroma obtained from a rosha grass (*Cymbopogon martinii* var. *motia*); soluble in alcohol and mineral oils; main component is geraniol; used in soaps and perfumes. Also known as East Indian geranium oil; Indian grass oil; Rusa oil; Turkish geranium oil.

palm butter [MATER] A reddish-yellow edible fatty oil expressed from putrid or fermented fruit pulp of the African oil palm (*Elaesis quineensis*); soluble in alcohol, ether, carbon disulfide, and chloroform; main components are palmitic acid, stearic acid, and glycerides of palmitic and oleic acids; melts at 27–42°C; used to make soaps and candles, as a lubricant, a color for butter substitutes, and an emollient. Also known as palm grease; palm oil.

palm grease *See* palm butter.

palm kernel oil *See* palm nut oil.

palm nut oil [MATER] Yellowish fatty oil; soluble in alcohol, either, carbon disulfide, and chloroform; main components are triolein and triglycerides of stearic, palmitic, myristic, lauric, and other fatty acids; used to make soap, chocolate products, margarine, cosmetics, and candles, and as an illuminant and a color for butter substitutes. Also known as palm kernel oil.

palm oil *See* palm butter.

palm wax [MATER] A yellow wax from the Ecuadoran palm (*Ceroxylon andicola*); used as a beeswax substitute.

pan [MIN ENG] **1.** A shallow, circular, concave steel or porcelain dish in which drillers or samplers wash the drill sludge to gravity-separate the particles of heavy, dense minerals from the lighter rock powder as a quick visual means of ascertaining if the rocks traversed by the borehole contain minerals of value. **2.** The act or process of performing the above operation.

pan-amalgamation process [MIN ENG] A process for extracting gold or silver from their ores; the ore is crushed and mixed with salt, copper sulfate, and mercury, and the gold or silver amalgamize with the mercury.

Pan-American jig [MIN ENG] Mineral jig developed to treat alluvial sands; the jig cell is pulsated vertically on a flexible diaphragm seated above the stationary hutch.

pan bolt [DES ENG] A bolt with a head resembling an upside-down pan.

pancake [MIN ENG] A concrete disk employed in stope support.

pancake auger [DES ENG] An auger having one spiral web, 12 to 15 inches (30 to 38 centimeters) in diameter, attached to the bottom end of a slender central shaft; used as removable deadman to which a drill rig or guy line is anchored.

pancake engine [MECH ENG] A compact engine with cylinders arranged radially.

pancake forging [MET] A rough, flat, forged shape made quickly with a minimum of tooling.

pancake landing [AERO ENG] Landing of an aircraft at a low forward speed and at a very high rate of descent.

pan conveyor [MECH ENG] A conveyor consisting of a series of pans. [MIN ENG] *See* jigging conveyor.

pan crusher [MECH ENG] Solids-reduction device in which one or more grinding wheels or mullers revolve in a pan containing the material to be pulverized.

pane [BUILD] A sheet of glass in a window or door. [DES ENG] One of the sides on a nut or on the head of a bolt.

panel [CIV ENG] **1.** One of the divisions of a lattice girder. **2.** A sheet of material held in a frame. **3.** A distinct, usually rectangular, raised or sunken part of a construction surface or a material. [ENG] A metallic or nonmetallic sheet on which operating controls and dials of an electronic unit or other equipment are mounted. [MIN ENG] **1.** A system of coal extraction in which the ground is laid off in separate districts or panels, pillars of extra-large size being left between. **2.** A large rectangular block or pillar of coal.

panel board [ENG] A drawing board with an adjustable outer frame that is forced over the drawing paper to hold and strain it. [MATER] A rigid paperboard used for paneling in buildings and automobile bodies.

panel coil *See* plate coil.

panel cooling [CIV ENG] A system in which the heat-absorbing units are in the ceiling, floor, or wall panels of the space which is to be cooled.

panel heating [CIV ENG] A system in which the heat-emitting units are in the ceiling, floor, or wall panels of the space which is to be heated.

panel length [CIV ENG] The distance between adjacent joints on a truss, measured along the upper or lower chord.

panel point [CIV ENG] The point in a framed structure where a vertical or diagonal member and a chord intersect.

panel system [BUILD] A wall composed of factory-assembled units connected to the building frame and to each other by means of anchors.

panel wall [BUILD] A nonbearing partition between columns or piers.

pan head [DES ENG] The head of a screw or rivet in the shape of a truncated cone.

pannier *See* gabion.

panoramic radar [ENG] Nonscanning radar which transmits signals over a wide beam in the direction of interest.

pan out [MIN ENG] To give a result, especially as compared with expectations; for example, in mining, the gravel may be said to pan out.

pantograph [ENG] A device that sits on the top of an electric locomotive or cars in an electric train and picks up electricity from overhead wires to run the train.

pantography [ENG] System for transmitting and automatically recording radar data from an indicator to a remote point.

pantometer [ENG] An instrument that measures all the angles necessary for determining distances and elevations.

pan-type car [MIN ENG] A vehicle for removing material from quarries; it is doorless, is reversible in direction, and can be dumped from either side.

Panzer-Forderer snaking conveyor [MIN ENG] An armored conveyor that is moved forward behind the coal plough by means of a traveling wedge pulled along by the plough or by means of jacks or compressed-air-operated rams attached at intervals to the conveyor structure.

paper [MATER] Felted or matted sheets of cellulose fibers, formed on a fine-wire screen from a dilute water suspension, and bonded together as the water is removed and the sheet is dried.

paperboard [MATER] A composition board available in varying thicknesses and degrees of rigidity.

paper clay [MATER] A special-grade clay that is mixed with paper pulp to add body, weight, and finish to paper products.

paper coating [MATER] Surface coating for paper; made from suspension of clays, starches, casein, rosin, polymers, wax, or various combinations; used to give strength and special surface qualities.

paper cutter [DES ENG] A hand-operated paper cutter and trimmer, consisting of a cutting blade bolted at one end to a ruled board; when the blade is drawn flush with the board, which has a metal strip at the cutting edge, a shearing action takes place which cuts the paper cleanly and evenly.

paper insulation [MATER] Electrical insulation made of paper, chiefly from coniferous woods but also from rags, rope, and other materials, which are chemically treated, beaten into a dispersed pulp, formed into a loose sheet by filtering on a moving wire screen, and compacted into paper by calendering with heated rolls.

paper machine [MECH ENG] A synchronized series of mechanical devices for transforming a dilute suspension of cellulose fibers into a dry sheet of paper.

paper mill [IND ENG] A building or complex of buildings housing paper machines.

paperoid [MATER] A heavy composition board, generally made of rope pulp and having a reddish color; used for large expandible filing envelopes.

papier maché [MATER] A lightweight molding material made from paper pulped with glue and other additives; dries to a hard finish that can be drilled, sanded, or painted.

papite [MATER] A poison gas composed of acrolein with stannic chloride.

papyrus [MATER] A paperlike material made by pressing the pith of the papyrus plant in water.

parabolic flight [AERO ENG] A space flight occurring in a parabolic orbit.

parabolic microphone [ENG ACOUS] A microphone used at the focal point of a parabolic sound reflector to give improved sensitivity and directivity, as required for picking up a band marching down a football field.

paraboloid [ENG] A reflecting surface which is a paraboloid of revolution and is used as a reflector for sound waves and microwave radiation.

parabomb [ENG] An equipment container with a parachute which is capable of opening automatically after a delayed drop.

paracentric [DES ENG] Pertaining to a key and keyway with longitudinal ribs and grooves that project beyond the center, as used in pin-tumbler cylinder locks to deter lockpicking.

parachute [AERO ENG] **1.** A contrivance that opens out somewhat like an umbrella and catches the air so as to retard the movement of a body attached to it. **2.** The canopy of this contrivance. [MIN ENG] A kind of safety catch for mine shaft cages.

parachute flare [ENG] Pyrotechnic device attached to a parachute and designed to provide intense illumination for a short period; it may be discharged from aircraft or from the surface.

parachute-opening shock [AERO ENG] The shock or jolt exerted on a suspended parachute load when the parachute fully catches the air.

parachute weather buoy [ENG] A general-purpose automatic weather station which can be air-dropped; it is 10 feet (3 meters) long and 22 inches (56 centimeters) in diameter, and is designed to operate for 2 months on a 6-hourly schedule, transmitting station identification, wind speed, wind direction, barometric pressure, air temperature, and sea-water temperature.

paracrate [ENG] Rigid equipment container for dropping equipment from an airplane by parachute.

paraffin See paraffin wax.

paraffin-base crude [MATER] Crude petroleum oil that contains predominately paraffin hydrocarbons, as contrasted with asphaltic- or naphthenic-base crudes; used as a source of fuels and high-grade lubricating oils.

paraffin distillate [MATER] In a petroleum refinery, the distillate oils ready for pressing to produce crystalline paraffin wax and paraffin oil.

paraffin jelly [MATER] A light, amber-colored petrolatum; used for medicinal purposes.

paraffin oil [MATER] A viscous, pale to yellow oil made from petroleum; used as a lubricant, medicine, and leather dressing.

paraffin press [ENG] A filter press used during petroleum refining for the separation of paraffin oil and crystallizable paraffin wax from distillates.

paraffin scale [MATER] Unrefined paraffin wax remaining in the chamber after oil has been removed from a mixture of oil and paraffin by sweating.

paraffinum liquidum See white mineral oil.

paraffin wax [MATER] A solid, crystalline hydrocarbon mixture derived from the paraffin distillate portion of crude petroleum; used in paper coating, candles, creams, emollients, and lipsticks. Also known as ceresin wax; paraffin.

parallel baffle muffler [DES ENG] A muffler constructed of a series of ducts placed side by side in which the duct cross section is a narrow but long rectangle.

parallel cut [ENG] A group of parallel holes, not all charged with explosive, to create the initial cavity to which the loaded holes break in blasting a development round. Also known as burn cut.

parallel drum [DES ENG] A cylindrical form of drum on which the haulage or winding rope is coiled.

parallel entry [MIN ENG] An intake airway parallel to the haulageway.

parallel firing [ENG] A method of connecting together a number of detonators which are to be fired electrically in one blast.

parallel laminate [MATER] A laminate in which all the layers of material are set approximately parallel with respect to a particular characteristic, such as the grain or the direction of tension.

parallel linkage [MECH ENG] A linkage system in which reciprocating motion is amplified.

parallel reliability [SYS ENG] Property of a system composed of functionally parallel elements in such a way that if one of the elements fails, the parallel units will continue to carry out the system function.

parallels [ENG] **1.** Spacers located between steam plate and press platen of the mold to prevent bending of the middle section. **2.** Spacers or pressure pods located between steam plates of a mold to regulate height and prevent crushing of mold parts.

parallel series *See* multiple series.

parallel shot [ENG] In seismic prospecting, a test shot which is made with all the amplifiers connected in parallel and activated by a single geophone so that lead, lag, polarity, and phasing in the amplifier-to-oscillograph circuits can be checked.

parallel wire method [MIN ENG] An electrical prospecting method employing equipotential lines or curves in searching for ore bodies.

paramagnetic alloy [MET] An alloy whose permeability is slightly greater than that of vacuum and is independent of the magnetic field strength, such as intermetallic compounds of nickel and titanium.

paramagnetic iron [MET] Iron which has been transformed from a ferromagnetic to a paramagnetic substance by application of a pressure somewhat greater than 10^5 bars (10^{10} newtons per square meter).

parametric excitation [ENG] The method of exciting and maintaining oscillation in either an electrical or mechanical dynamic system, in which excitation results from a periodic variation in an energy storage element in a system such as a capacitor, inductor, or spring constant.

parametrized voice response system [ENG ACOUS] A voice response system which first extracts informative parameters from human speech, such as natural resonant frequencies (formants) of the speaker's vocal tract and the fundamental frequency (pitch) of the voice, and which later reconstructs speech from such stored parameters.

parapack [ENG] A package or bundle with a parachute attached for dropping from an aircraft.

parasheet [AERO ENG] A simple form of parachute in which the canopy is a single piece of material or two or more pieces sewed together; it may have any geometrical form, such as square or hexagonal, and the hem may be gathered to assist in the development of a crown when the parasheet is opened.

parathene [MATER] Any of a group of high-grade hydrocarbons that are extracted from lubricating oil stocks by the solvent process or refining.

para toner [MATER] Water-insoluble red pigment made from β-naphthol and paranitroaniline; used in paints, in printing, and to make para lakes.

paravane [ENG] A torpedo-shaped device with sawlike teeth along its forward end, towed with a wire rope underwater from either side of the bow of a ship to cut the cables of anchored mines. Also known as otter.

parchment [MATER] **1.** The skin of a goat or sheep that has been treated so that it can be used to write upon. **2.** A drawing or written text on this material.

parchment paper [MATER] Paper that has been manufactured so that its appearance resembles parchment.

parging [CIV ENG] A thin coating of mortar or plaster on a brick or stone surface.

Parian cement [MATER] Gypsum plaster containing borax, which dries to a hard finish.

parison [ENG] A hollow plastic tube from which a bottle or other hollow object is blow-molded.

parison swell [ENG] In blow molding, the ratio of the cross-sectional area of the parison to that of the die opening.

parkerizing [MET] Trade name for a process for the production of phosphate coating on steel articles by immersion in an aqueous solution of manganese or zinc acid with phosphate.

Parkes process [MET] A process for recovering precious metals from lead by stirring about 2% zinc into the melt to form zinc compounds with gold and silver which can then be skimmed off the surface.

parking apron [CIV ENG] A hard-surfaced area used for parking aircraft.

parking lot [CIV ENG] An outdoor lot for parking automobiles.

parking orbit [AERO ENG] A temporary earth orbit during which the space vehicle is checked out and its trajectory carefully measured to determine the amount and time of increase in velocity required to send it into a final orbit or into space in the desired direction.

parkway [CIV ENG] A broad landscaped expressway which is not open to commercial vehicles.

parquet flooring [BUILD] Wood flooring made of strips laid in a pattern to form designs.

Parshall flume [ENG] A calibrated device for measuring the flow of liquids in open conduits by measuring the upper and lower beads at a specified distance from an obstructing sill.

parsley oil [MATER] Colorless or pale-greenish-yellow liquid with parsley aroma; soluble in alcohol, ether, and chloroform; distilled from parsley seeds; used in medicine.

Parsons-stage steam turbine [MECH ENG] A reaction-type steam-turbine stage in which the pressure drop occurs partially across the stationary nozzles and partly across the rotating blades.

part [ENG] An element of a subassembly, not normally useful by itself and not amenable to further disassembly for maintenance purposes.

partial pressure maintenance [PETRO ENG] The partial replacement of produced gas in an oil reservoir by gas injection to maintain a portion of the initial reservoir pressure.

participation crude *See* buy-back crude.

particle board [MATER] Construction board made with wood particles impregnated with low-molecular-weight resin and then cured.

particle-oriented paper [MATER] A chart paper that has a magnetic coating which is produced by combining microscopic magnetic flakes with oil to form droplets and then forming these particles into an emulsion that can be applied to the surface of ordinary bond paper or to a clear plastic substrate; the magnetic field of a small-diameter recording head rotates the magnetic flakes so that they absorb or scatter incident light to give a visible dark trace that can also be read magnetically.

particle size [MET] The average and controlling lineal dimension of an individual particle of metal powder as determined by suitable screens or other methods of analysis.

particle-size analysis [ENG] Determination of the proportion of particles of a specified size in a granular or powder sample.

particle-size distribution [ENG] The percentages of each fraction into which a granular or powder sample is classified, with respect to particle size, by number or weight.

particulate composite [MATER] A composite material composed of particles embedded in a matrix.

particulate mass analyzer [ENG] A unit which measures dust concentrations in emissions from furnaces, kilns, cupolas, and scrubbers.

particulates [MATER] Fine solid particles which remain individually dispersed in gases and stack emissions.

parting [MET] **1.** Recovery of gold (or occasionally another metal) from its alloys by a corrosion process. **2.** Zone of separation between cope and drag portions of mold or flask in sand casting. **3.** In sand molding, a composition to facilitate removal of the pattern. **4.** A shearing operation to produce two or more parts from a stamping.

parting agent *See* release agent.

parting compound [MET] A material, such as silica or graphite, used to facilitate the separation of the cope and drag parting surfaces.

parting line [MET] **1.** The line along which a mold is separated. **2.** A line or seam on a casting corresponding to the joint of mold parts.

parting sand [MET] Fine, dry sand applied to the faces of a sand mold to allow disassembly.

parting stop [BUILD] A thin strip of wood that separates the sashes in a double-hung window.

partition [BUILD] An interior wall having a height of one story or less, which divides a structure into sections. [IND ENG] A slotted sheet of paperboard that can be assembled with similar sheets to form cells for holding goods during shipment.

partridgewood *See* acapau.

parts kit [ENG] A group of parts, not all having the same basic name, used for repair or replacement of the worn broken parts of an item; it may include instruction sheets and material, such as sandpaper, tape, cement, and gaskets.

parts list [ENG] One or more printed sheets showing a manufacturer's parts or assemblies of an end item by illustration or a numerical listing of part numbers and names; it does not outline any assembly, maintenance, or operating instructions, and it may or may not have a price list cover sheet.

party wall [BUILD] A wall providing joint service between two buildings.

pass [AERO ENG] **1.** A single circuit of the earth made by a satellite; it starts at the time the satellite crosses the equator from the Southern Hemisphere into the Northern Hemisphere. **2.** The period of time in which a satellite is within telemetry range of a data acquisition station. [MECH ENG] **1.** The number of times that combustion gases are exposed to heat transfer surfaces in boilers (that is, single-pass, double-pass, and so on). **2.** In metal rolling, the passage in one direction of metal deformed between rolls. **3.** In metal cutting, transit of a metal cutting tool past the workpiece with a fixed tool setting. [MET] **1.** Passage of a metal bar between rolls. **2.** Open space between two grooved rolls through which metal is processed. **3.** Weld metal deposited in one trip along the axis of a weld. [MIN ENG] **1.** A mine opening through which coal or ore is delivered from a higher to a lower level. **2.** A passage left in old workings for workers to travel as they move from one level to another. **3.** A treatment of the whole ore sample in a sample divider. **4.** A passage of an excavation or grading machine. **5.** In surface mining, a complete excavator cycle in removing overburden.

pass-by [ENG] The double-track part of any single-track system of rail transport. [MIN ENG] A passage around the working part of a shaft.

passenger car [ENG] **1.** A railroad car in which passengers are carried. **2.** An automobile for carrying as many as nine passengers.

passing point [MIN ENG] The point at which two vehicles, such as coal cars or mine elevators, pass each other while going in opposite directions.

passing track [ENG] A sidetrack with switches at both ends.

passivation [MET] To render passive; to reduce the reactivity of a chemically active metal surface by electrochemical polarization or by immersion in a passivating solution.

passive-active cell [MET] An electrochemical corrosion cell established between passive and active areas on a metal surface.

passive AND gate [ENG] A fluidic device which achieves an output signal, by stream interaction, only when both of two control signals appear simultaneously.

passive communications satellite [AERO ENG] A satellite that reflects communications signals between stations, without providing amplification; an example is the Echo satellite. Also known as passive satellite.

passive earth pressure [CIV ENG] The maximum value of lateral earth pressure exerted by soil on a structure, occurring when the soil is compressed sufficiently to cause its internal shearing resistance along a potential failure surface to be completely mobilized.

passive metal [MET] A metal on which a surface film forms by natural process or by immersion in a passivating solution, making the metal resistant to corrosion.

passive method [CIV ENG] A construction method in permafrost areas in which the frozen ground near the structure is not disturbed or altered, and the foundations are provided with additional insulation to prevent thawing of the underlying ground.

passive radar [ENG] A technique for detecting objects at a distance by picking up the microwave electromagnetic energy that is both radiated and reflected by all bodies.

passive sonar [ENG] Sonar that uses only underwater listening equipment, with no transmission of location-revealing pulses.

passivity [MET] The property of a metal that has been made passive.

paste [MATER] An adhesive mixture with a characteristic plastic consistency, a high order of yield value, and a low bond strength; for example, a paste prepared by heating a starch and water mixture, then cooling the hydrolyzed product. [MET] Finely divided particles of ferromagnetic material in paste form used in the wet method of magnetic particle inspection.

pasteboard [MATER] A type of thin cardboard made from gluing together two or more sheets of paper.

paste ink [MATER] A pastelike mixture of pigment or dye with oil and other additives, such as resins, driers, tackifiers, and adhesives; used in paper and textile printing and ballpoint pens.

pastel fixative [MATER] A thin varnish material for the simple and even application to drawings; only enough fixative should be used to keep the pastel particles from falling off the paper.

paste mixer [ENG] Device for the blending together of solid particles and a liquid, with the final formation of a single paste phase.

paste resin [MATER] Solventless, fluid or semisolid mixture of powdered resin and plasticizer.

paste solder [MET] Finely powdered solder metal combined with a flux.

pasteurizer [ENG] An apparatus used for pasteurization of fluids.

patch bolt [DES ENG] A bolt with a countersunk head having a square knob that twists off when the bolt is screwed in tightly; used to repair boilers and steel ship hulls.

patchouli oil [MATER] A brownish essential oil with strong camphor aroma; soluble in ether, chloroform, alcohol, and oils; derived from dried leaves of the patchouli (*Pogostemon patchouly*); main components are patchouli alcohol, eugenol, and cinnamic aldehyde; used to perfume toiletries.

patent [IND ENG] A certificate of grant by a government of an exclusive right with respect to an invention for a limited period of time. Also known as letters patent.

patented claim [MIN ENG] A mining claim to which a patent has been secured from the government by compliance with the laws relating to such claims.

patenting [MET] A process used in the production of high-strength steel wire containing 0.35–0.85% carbon, in which the wire is heated to above the transformation temperature, then quenched in molten lead or molten salt, or cooled in air.

Patera process [MET] A method used in the 19th century for extracting silver from ore: the ore was roasted with sodium chloride, a solution of sodium thiosulfate was then added to leach out the silver chloride, and sodium sulfide precipitated the silver as silver sulfide.

patina [MET] The greenish product, usually basic copper sulfate, formed on copper and copper-rich alloys as a result of prolonged atmospheric corrosion.

patio process [MET] A crude chemical method of reducing silver from its ores, followed by amalgamation in low heaps with the aid of salt and copper sulfate.

pattern [AERO ENG] The flight path flown by an aircraft, or prescribed to be flown, as in making an approach to a landing. [ENG] A form designed and used as a model for making things.

pattern flood [PETRO ENG] Waterflood of a petroleum reservoir in which there is an areal pattern of injection wells located so as to sweep oil from the area toward the bores of producing wells.

pattern shooting [ENG] In seismic prospecting, firing of explosive charges arranged in geometric pattern.

Pattinson process [MET] A method for separating silver from its alloys rich in lead by slow cooling of the melt so that silver-poor lead crystals separate out and are removed.

paulin [MATER] A fabricated textile item generally used as a weather protection cover for various items or materials during storage or transit.

pavement [BUILD] A hard floor of concrete, brick, tiles, or other material. [CIV ENG] A paved surface.

pavement light [CIV ENG] A window built into the surface of a pavement to admit daylight to a space below ground level.

paver [MECH ENG] Any of several machines which, moving along the road, carry and lay paving material.

paving brick [MATER] A vitrified clay brick used in the construction of pavements.

paving-brick clay [MATER] Impure shale or fire clay with good tensile strength and plasticity; used to make paving bricks.

pawl [MECH ENG] The driving link or holding link of a ratchet mechanism, permits motion in one direction only.

pay dirt [MIN ENG] Profitable mineral-rich earth or ore.

payload [AERO ENG] That which an aircraft, rocket, or the like carries over and above what is necessary for the operation of the vehicle in its flight. [MIN ENG] The weight of coal, ore, or mineral handled, as distinct from dirt, stone, or gangue.

payload–mass ratio [AERO ENG] Of a rocket, the ratio of the effective propellant mass to the initial vehicle mass.

pay ore [MIN ENG] Ore which can be mined, concentrated, or smelted at current cost of exploitation profitably at ruling market value of products.

payout time [IND ENG] A measurement of profitability or liquidity of an investment, being the time required to recover the original investment in depreciable facilities from profit and depreciation; usually, but not always, calculated after income taxes.

pay sand [PETRO ENG] That portion of an oil or gas sand in which the oil or gas is found in commercial quantity.

pay streak [MIN ENG] A layer of oil, ore, or other mineral that can be mined profitably.

pay zone [PETRO ENG] The reservoir rock in which oil and gas are found in exploitable quantities.

p chart [IND ENG] A chart of the fraction defective, either observed in the sample or in some production period.

PDR See precision depth recorder.

peach kernel oil See persic oil.

peakless pumping [MIN ENG] Spreading the pumping load over the entire day in a mine.

peak load [ENG] The maximum quantity of a specified material to be carried by a conveyor per minute in a specified period of time.

peanut oil [MATER] A yellow to greenish-yellow fatty oil obtained from peanuts; soluble in ether, chloroform, benzine, and carbon disulfide; main components are glycerides of oleic and linoleic acids; used as an edible substitute for olive oil, and in soaps and medicine. Also known as arachis oil; earth-nut oil; katchung oil.

pearl [MATER] A dense, more or less round, white or light-colored concretion having various degrees of luster formed within or beneath the mantle of various mollusks by deposition of thin concentric layers of nacre about a foreign particle.

pearl ash [MATER] An impure substance derived from potash following partial purification from wood ash.

pearl essence [MATER] A brilliant, translucent, lustrous material obtained from fish scales; used in pearl lacquers and to make artificial pearls. Also known as pearl white.

pearlite [MET] A lamellar aggregate of ferrite (almost pure iron) and cementite (Fe_3C) often occurring in carbon steels and in cast iron.

pearl white *See* pearl essence.

peat coal [MATER] Artificially carbonized peat that is used as a fuel.

peat tar [MATER] A peat distillate containing 2–6% tar.

peat wax [MATER] A hard, waxy material extracted from peat; it is similar to, and a substitute for, montan wax.

Peaucellier linkage [MECH ENG] A mechanical linkage to convert circular motion exactly into straight-line motion.

pebble mill [MECH ENG] A solids size-reduction device with a cylindrical or conical shell rotating on a horizontal axis, and with a grinding medium such as balls of flint, steel, or porcelain.

pebbles [MATER] Grinding media for pebble mills, usually balls of hard flint or hard burned white porcelain.

pebbling *See* orange peel.

pedal [DES ENG] A lever operated by foot.

pedestal [CIV ENG] **1.** The support for a column. **2.** A metal support carrying one end of a bridge truss or girder and transmitting any load to the top of a pier or abutment. [ENG] A supporting part or the base of an upright structure, such as a radar antenna.

pedestal pile [CIV ENG] A concrete pile with a bulbous enlargement at the bottom.

pedigree mud [MATER] A high-chemical-content drilling mud that includes barium sulfate, caustic soda, soda ash, sodium bicarbonate, and phosphates.

pedometer [ENG] **1.** An instrument for measuring and weighing a newborn child. **2.** An instrument that registers the number of footsteps in walking.

peel-back [ENG] The separation of two bonded materials, one or both of which are flexible, by stripping or pulling the flexible material from the mating surface at a 90 or 180° angle to the plane in which it is adhered.

peeling [MATER] Stripping or detaching a rubber coating from a metal, cloth, or other material.

peel-off time [ENG] In seismic prospecting, the time correction applied to observed data to adjust them to a depressed reference datum.

peel test [ENG] A test to ascertain the adhesive strength of bonded strips of metals by peeling or pulling the metal strips back and recording the adherence values.

peen [DES ENG] The end of a hammer head with a hemispherical, wedge, or other shape; used to bend, indent, or cut.

peening [MET] Surface-hardening a piece of metal by hammering or by bombarding with hard shot.

peepdoor [MECH ENG] A small door in a furnace with a glass opening through which combustion may be observed.

peg [ENG] **1.** A small pointed or tapered piece, often cylindrical, used to pin down or fasten parts. **2.** A projection used to hang or support objects. [MET] *See* plug.

peg count meter [ENG] A meter or register that counts the number of trunks tested, the number of circuits passed busy, the number of test failures, or the number of repeat tests completed.

pegging rammer [MET] A rod with an oblong piece of iron at its end; used to compact sand in a mold.

pelleting [ENG] Method of accelerating solidification of cast explosive charges by blending precast pellets of the explosives into the molten charge.

pelletization [MIN ENG] Forming aggregates of about 1/2 inch (13 millimeter) diameter from finely divided ore or coal.

pellet mill [MECH ENG] Device for injecting particulate, granular or pasty feed into holes of a roller, then compacting the feed into a continuous solid rod to be cut off by a knife at the periphery of the roller.

Pelton turbine *See* Pelton wheel.

Pelton wheel [MECH ENG] An impulse hydraulic turbine in which pressure of the water supply is converted into velocity by a few stationary nozzles, and the water jets then impinge on the buckets mounted on the rim of a wheel; usually limited to high head installations, exceeding 500 feet (150 meters). Also known as Pelton turbine.

pen [ENG] **1.** A small place for confinement, storage, or protection. **2.** A device for writing with ink.

pencil [ENG] An implement for writing or making marks with a solid substance; the three basic kinds are graphite, carbon, and colored.

pencil cave [ENG] A driller's term for hard, closely jointed shale that caves into a well in pencil-shaped fragments.

pendant-drop melt extraction [MET] Melt extraction in which the molten metal is produced by heating the end of a rod above a disk.

pendant post [BUILD] A post on a solid support and set against a wall to support a collar beam or other part of a roof.

pendular ring [PETRO ENG] Distribution of two nonmiscible liquids in a porous system; the pendular ring is a state of reservoir saturation in which the wetting phase is not continuous and the nonwetting phase is in contact with some of the solid surface.

pendulum anemometer [ENG] A pressure-plate anemometer consisting of a plate which is free to swing about a horizontal axis in its own plane above its center of gravity; the angular deflection of the plate is a function of the wind speed; this instrument is not used for station measurements because of the false reading which results when the frequency of the wind gusts and the natural frequency of the swinging plate coincide.

pendulum level [ENG] A leveling instrument in which the line of sight is automatically kept horizontal by a built-in pendulum device (such as a horizontal arm and a plumb line at right angles to the arm).

pendulum press [MECH ENG] A punch press actuated by a swinging treadle operated by the foot.

pendulum saw [MECH ENG] A circular saw that swings in a vertical arc for crosscuts.

pendulum scale [ENG] Weight-measurement device in which the load is balanced by the movement of one or more pendulums from vertical (zero weight) to horizontal (maximum weight).

pendulum seismograph [ENG] A seismograph that measures the relative motion between the ground and a loosely coupled inertial mass; in some instruments, optical magnification is used whereas others exploit electromagnetic transducers, photocells, galvanometers, and electronic amplifiers to achieve higher magnification.

penetrant [MATER] A liquid with low surface tension, usually containing a dye or fluorescent chemical; when flowed over a metal surface, it is used to determine the existence and extent of cracks and other discontinuities.

penetrating oil [MATER] Low-viscosity oil that can penetrate between closely fitted parts, such as the leaves of springs and screw threads; used to loosen rusted parts.

penetration [AERO ENG] That phase of the letdown from high altitude to a specified approach altitude. [MET] **1.** The distance from the original surface of the base metal to that point at which weld fusion ends. **2.** A surface defect on a casting caused by molten metal filling voids in the sand mold.

penetration depth [ENG] The greatest depth in an ultrasonic test piece at which indications can be measured.

penetration hardness *See* indentation hardness.

penetration log [ENG] A record of the speed with which a drill penetrates a bore, including such factors as hole size, bit size, mud pressure, rotation speed, and force on the bit; used to determine the thickness of coal and dirt bands in the bore.

penetration macadam [MATER] A paving material consisting of crushed stone in two sizes bound together by asphalt or tar.

penetration number [ENG] The consistency of greases, waxes, petrolatum, and asphalt or other bituminous materials expressed as the distance that a standard needle penetrates the sample under specified ASTM test conditions.

penetration of fractures [PETRO ENG] The depth to which artificially produced fractures penetrate the fractured reservoir formation.

penetration rate [MECH ENG] The actual rate of penetration of drilling tools.

penetration speed [MECH ENG] The speed at which a drill can cut through rock or other material.

penetration test [ENG] A test to determine the relative values of density of noncohesive sand or silt at the bottom of boreholes.

penetrometer [ENG] **1.** An instrument that measures the penetrating power of a beam of x-rays or other penetrating radiation. **2.** An instrument used to determine the consistency of a material by measurement of the depth to which a standard needle penetrates into it under standard conditions.

Pennsylvania-base crude [MATER] A type of crude petroleum produced in Pennsylvania, New York, West Virginia, and parts of Ohio; contains a high percentage of paraffin-base lube-oil stock.

Pennsylvania truss [CIV ENG] A truss characterized by subdivided panels, curved top chords on through trusses, and curved bottom chords on deck spans; used on long bridge spans.

pennyroyal oil [MATER] An essential oil distilled from the dried leaves and tops of the small pennyroyal plants *Hedeoma pulegioides* or *Mentha pulegium;* used as a counterirritant in liniments, in insect repellents, and for the production of methanol.

pen recorder [ENG] A device in which the varying inputs (electrical, pneumatic, mechanical) are marked by a signal-controlled pen onto a continuous recorder chart (circular or roll chart).

penstock [CIV ENG] A valve or sluice gate for regulating water or sewage flow. [ENG] A closed water conduit controlled by valves and located between the intake and the turbine in a hydroelectric plant.

pentacite [MATER] Alkyd resin in which pentaerythritol is the polyhydric alcohol; used in coatings and printing inks.

pentane lamp [ENG] A pentane-burning lamp formerly used as a standard for photometry.

penthouse [BUILD] **1.** An enclosed space built on a flat roof to cover a stairway, elevator, or other equipment. **2.** A dwelling built on top of the main roof. **3.** A sloping shed or roof attached to a wall or building.

pentice [MIN ENG] **1.** A rock pillar left, or a heavy timber bulkhead placed, in the bottom of a deep shaft of two or more compartments; the shaft is then further sunk through the pentice. **2.** In shaft sinking, a solid rock pillar left in the bottom of the shaft for overhead protection of miners while the shaft is being extended by sinking.

pentolite [MATER] A high explosive composed of pentaerythritol and trinitrotoluene.

peppermint oil See oil of peppermint.

pepper sludge [MATER] Fine particles of sludge produced during the acid treatment of lubricating oils and other petroleum products.

peptized fuel [MATER] Thickened flamethrower fuel to which water or other chemical is added before mixing, to reduce mixing time and to increase storage stability.

percentage depletion [PETRO ENG] Oil- or gas-reservoir depletion allowance calculated on the basis of unit sales and initial depletable leasehold cost.

percentage extraction [MIN ENG] The proportion of a coal seam or other ores which is removed from a mine.

percentage log [ENG] A sample log in which the percentage of each type of rock (except obvious cavings) present in each sample of cuttings is estimated and plotted.

percentage map [PETRO ENG] Contoured map in which the percentage of one component (such as sand) is compared with the total unit (such as sand plus shale).

percentage subsidence [MIN ENG] The measured amount of subsidence expressed as a percentage of the thickness of coal extracted.

percentage support [MIN ENG] The percentage of the total wall area which will actually be covered by supports.

percent compaction [ENG] The ratio, expressed as a percentage, of dry unit weight of a soil to maximum unit weight obtained in a laboratory compaction test.

percent defective [IND ENG] The ratio of defective pieces per lot or sample, expressed as a percentage.

percolation [MIN ENG] Gentle movement of a solvent through an ore bed in order to extract a mineral.

percolation leaching [MIN ENG] The selective removal of a mineral by causing a suitable solvent to seep into and through a mass or pile of material containing the desired soluble mineral.

percolation test [CIV ENG] A test to determine the suitability of a soil for the installation of a domestic sewage-disposal system, in which a hole is dug and filled with water and the rate of water-level decline is measured.

percussion bit [MECH ENG] A rock-drilling tool with chisellike cutting edges, which when driven by impacts against a rock surface drills a hole by a chipping action.

percussion drill [MECH ENG] A drilling machine usually using compressed air to drive a piston that delivers a series of impacts to the shank end of a drill rod or steel and attached bit.

percussion drilling [MECH ENG] A drilling method in which hammer blows are transmitted by the drill rods to the drill bit.

percussion powder [MATER] Powder so composed as to ignite by a slight percussion.

percussion side-wall sampling [PETRO ENG] A method of side-wall core sampling from wellbores in softer reservoir formations.

percussion table See concussion table.

percussion welding [MET] Resistance welding with arc heat and simultaneously applied pressure from a hammerlike blow.

perfect lubrication [ENG] A complete, unbroken film of liquid formed over each of two metal surfaces moving relatively to one another with no contact.

perforated metal [MATER] Sheet metal with round, square, diamond, or rectangular perforations; used for screens and for construction.

perforating [PETRO ENG] Special oil-well downhole procedure to make holes in tubing walls and surrounding cement; used to allow formation oil or gas to enter the wellbore tubing, or to allow water to be forced out into the formation to cause fracturing (hydraulic fracturing).

perforations [PETRO ENG] Downwell holes made in well tubing, usually by shot-and-explosive or shaped-charge techniques; used for oil or gas production from desired horizons, or for injection of acidizing or fracturing fluids into the formation at predetermined depths.

performance bond [ENG] A bond that guarantees performance of a contract.

performance characteristic [ENG] A characteristic of a piece of equipment, determined during its test or during its operation.

performance chart [ENG] A graph used in evaluating the performance of any device, for example, the performance of an electrical or electronic device, such as a graph of anode voltage versus anode current for a magnetron.

performance curves [ENG] Graphical representations showing the abilities of rotating equipment at various operating conditions; for example, the performance curve for a compressor would include rotor speed for various intake and outlet pressures versus gas flow rate adjusted for temperature, density, viscosity, head, and other factors.

performance data [ENG] Data on the manner in which a given substance or piece of equipment performs during actual use.

performance evaluation [IND ENG] The analysis in terms of initial objectives and estimates, and usually made on site, of accomplishments using an automatic data-processing system, to provide information on operating experience and to identify corrective actions required, if any.

performance index [IND ENG] The ratio of standard hours to the hours of work actually used; a ratio exceeding 1.00 (or 100%) indicates standard output is being exceeded.

performance number [ENG] One of a series of numbers (constituting the PN, or performance-number, scale) used to convert fuel antiknock values in terms of a reference fuel into an index which is an indication of relative engine performance; used mostly to rate aviation gasolines with octane values greater than 100.

performance rating *See* effort rating.

performance sampling [IND ENG] A technique in work measurement used to determine the leveling factor to be applied to an operator or a group of operators by short, randomly spaced observations of the performance index.

perfume base [MATER] Any natural or synthetic material used by perfumers as a starting point for perfume manufacture.

perfume oil [MATER] Any volatile oil distilled or extracted from the leaves, flowers, gums, or woods of plant life (but occasionally of animal origin) and used in making perfume; examples are linalyl acetate (from citral), crab apple, wisteria, lavendar, and attar of rose.

peridynamic loudspeaker [ENG ACOUS] Box-type loudspeaker baffle designed to give good bass response by minimizing acoustic standing.

perilla oil [MATER] A light-yellow drying oil derived from seeds of mints of the genus *Perilla*; soluble in alcohol, benzene, carbon disulfide, ether, and chloroform; used as a substitute for linseed oil, as an edible oil, and in manufacture of varnishes and artificial leather.

perimeter blasting [MIN ENG] A method of blasting in tunnels, drifts, and raises, designed to minimize overbreak and leave clean-cut solid walls; the outside holes are loaded with very light continuous explosive charges and fired simultaneously, so that they shear from one hole to the other.

perimeter of airway [MIN ENG] In mine ventilation, the linear distance in feet of the airway perimeter rubbing surface at right angles to the direction of the airstream.

periodic kiln [ENG] A kiln in which the cycle of setting ware in the kiln, heating up, "soaking" or holding at peak temperature for some time, cooling, and removing or "drawing" the ware is repeated for each batch.

peripheral milling [MET] Removing metal from a surface parallel to the axis of a milling cutter.

peripheral speed *See* cutting speed.

perm [PETRO ENG] A unit indicating the degree of permeability of a porous reservoir structure; the unit is expressed as bbl day^{-1} ft^{-2} psi^{-1} ft cp or ft^3 day^{-1} ft^{-2} psi^{-1} ft cp.

permafil [MATER] Polymerizable mixture that cures without any evaporation.

permafrost drilling [ENG] Boreholes drilled in subsoil and rocks in which the contained water is permanently frozen.

permalloy [MET] A trade name for any of several highly magnetically permeable iron-base alloys containing about 45–80% nickel.

permanent anode [MET] A very-corrosion-resistant anode, made of a material such as a carbon, aluminum, or lead alloy, or 14.5% silicon iron; used in cathodic protection against corrosion.

permanent benchmark [ENG] A readily identifiable, relatively permanent, recoverable benchmark that is intended to maintain its elevation without change over a long period of time with reference to an adopted datum, and is located where disturbing influences are believed to be negligible.

permanent-completion packer [PETRO ENG] A packer able to withstand large pressure differentials to allow for its permanent installation in a producing well.

permanent ink [MATER] Ink that contains up to 1% dissolved iron to prevent fading or washing away when dried.

permanent-magnet dynamic loudspeaker *See* permanent-magnet loudspeaker.

permanent-magnet loudspeaker [ENG ACOUS] A moving-conductor loudspeaker in which the steady magnetic field is produced by a permanent magnet. Also known as permanent-magnet dynamic loudspeaker.

permanent-magnet moving-coil instrument [ENG] An ammeter or other electrical instrument in which a small coil of wire, supported on jeweled bearings between the poles of a permanent magnet, rotates when current is carried to it through spiral springs which also exert a restoring torque on the coil; the position of the coil is indicated by an attached pointer.

permanent-magnet moving-iron instrument [ENG] A meter that depends for its operation on a movable iron vane that aligns itself in the resultant magnetic field of a permanent magnet and adjacent current-carrying coil.

permanent mold [MET] A reusable metal mold for the production of many castings of the same kind.

permanent monument [MIN ENG] A monument of a lasting character for marking a mining claim; it may be a mountain, hill, or ridge.

permanent pump [MIN ENG] A pump on which the mine depends for the final disposal of its drainage.

permanent starch [MATER] An emulsion of polyvinyl acetate used for starching clothing and textiles; it is not removed by washing.

permeability alloy [MET] An iron-nickel alloy having greater magnetic susceptibility than iron.

permeability-block method [PETRO ENG] Calculation method for oil recovery from water-drive oil fields in which there are variable-permeability distributions.

permeability number [ENG] A numbered value assigned to molding materials indicating the relative ease of passage of gases through them.

permeability profile [PETRO ENG] A graphical plot of porous reservoir permeability versus distance down the wellbore.

permeameter [ENG] **1.** Device for measurement of the average size or surface area of small particles; consists of a powder bed of known dimension and degree of packing through which the particles are forced; pressure drop and rate of flow are related to particle size, and pressure drop is related to surface area. **2.** A device for measuring the coefficient of permeability by measuring the flow of fluid through a sample across which there is a pressure drop produced by gravity. **3.** An instrument for measuring the magnetic flux or flux density produced in a test specimen of ferromagnetic material by a given magnetic intensity, to permit computation of the magnetic permeability of the material.

Permendur [MET] A magnetic alloy which is composed of equal parts of iron and cobalt and has an extremely high permeability when saturated.

permenorm alloy [MET] An alloy containing 50% nickel and 50% iron; used as magnet core material and in magnetic amplifiers.

permissible [MIN ENG] Said of equipment completely assembled and conforming in every respect with the design formally approved by the U.S. Bureau of Mines for use in gassy and dusty mines.

permissible explosive [MATER] An explosive approved by the U.S. Bureau of Mines as safe for blasting in gassy and dusty mines.

permissible lamp [MIN ENG] A lamp that meets the standards of the U.S. Bureau of Mines.

permissible machine [MIN ENG] A machine, such as a drill, mining machine, loading machine, conveyor, or locomotive, that meets the standards of the U.S. Bureau of Mines.

permissible velocity [CIV ENG] The highest velocity at which water is permitted to pass through a structure or conduit without excessive damage.

permissive stop [CIV ENG] A railway signal indicating the train must stop but can proceed slowly and cautiously after a specified interval, usually 1 minute.

perm-plug method [PETRO ENG] Laboratory method of measuring the permeability of reservoir core samples (or plugs) by the measurement of airflow through the sample at several flow rates.

pernetti [ENG] **1.** Small iron pins or tripods that support ware while it is being fired in a kiln. **2.** The marks left on baked pottery by these supporting feet.

Persian ammoniac *See* ammoniac.

persic oil [MATER] A pale-yellow to red fatty oil, soluble in ether, chloroform, and carbon disulfide; taste and aroma are similar to almond oil; the oil is expressed from blanched seeds of peaches or apricots; used as a flavoring, in medicine, and as a nutrient similar to olive and almond oils. Also known as apricot kernel oil; peach kernel oil.

persistent war gas [MATER] War gas that is normally effective in the open at the point of dispersion for more than 10 minutes.

Persoz's reagent [MATER] Chemical reagent used to detect the presence of silk with wool (only silk dissolves); consists of zinc chloride and zinc oxide in water.

Pers sunshine recorder [ENG] A type of sunshine recorder in which the time scale is supplied by the motion of the sun.

PERT [SYS ENG] A management control tool for defining, integrating, and interrelating what must be done to accomplish a desired objective on time; a computer is used to compare current progress against planned objectives and give management the information needed for planning and decision making. Derived from program evaluation and review technique.

Peru balsam [MATER] A dark, viscous liquid with bitter taste and pleasant aroma; soluble in alcohol, ether, acetone, chloroform, benzene, and glacial acetic acid; derived from the tropical American tree *Myroxylon pereirae;* main components are esters of cinnamic and benzoic acids; used in perfumery, medicine, and chocolate manufacture. Also known as balsam of Peru; black balsam; China oil; Chinese oil; Indian balsam; Peruvian balsam.

Peruvian balsam *See* Peru balsam.

pesticide [MATER] A chemical agent that destroys pests. Also known as biocide.

peter out [ENG] To fail gradually in size, quantity, or quality; for example, a mine may be said to have petered out.

Petersen grab [ENG] A bottom sampler consisting of two hinged semicylindrical buckets held apart by a cocking device which is released when the grab hits the ocean floor.

petitgrain oil [MATER] A yellowish oil extracted from the leaves and twigs of the bitter orange tree; soluble in 70% alcohol, ether, and chloroform; used in perfumes for soaps and skin cream, and for flavoring.

petrol *See* gasoline.

petrolatum [MATER] A smooth, semisolid blend of mineral oil with waxes crystallized from the residual type of petroleum lubricating oil; the wax molecules contain 30–70 carbon atoms and are straight chains with a few branches or naphthene rings; used as a lubricant, as a carrier in polishes and cosmetics, and as a rust preventive.

petroleum asphalt [MATER] Asphalt recovered or made from petroleum.

petroleum benzin [MATER] Colorless, volatile, flammable, toxic petroleum distillate with boiling range of 35–80°C; a special grade is ligroin (petroleum ether; boiling range 20–135°C), used as an extractive solvent, especially for drugs. Also known as benzin; benzine (archaic usage).

petroleum coke [MATER] A carbonaceous solid material made by the destructive heating of high-molecular-weight petroleum-refining residues.

petroleum engineer [PETRO ENG] An engineer whose primary objective is to find and produce oil or gas from petroleum reserves.

petroleum engineering [ENG] The application of almost all types of engineering to the drilling for and production of oil, gas, and liquefiable hydrocarbons.

petroleum ether [MATER] A volatile fraction of petroleum consisting chiefly of pentanes and hexanes. Also known as ligroin.

petroleum products [MATER] Materials derived from petroleum, natural gas, or asphalt deposits; includes gasolines, diesel and heating fuels, liquefied petroleum gases (LPG and bugas), lubricants, waxes, greases, petroleum coke, petrochemicals, and (from sour crudes and natural gases) sulfur.

petroleum secondary engineering [PETRO ENG] The process of removing oil from its native reservoirs by the use of supplemental energies after the natural energies causing oil production have been depleted.

petroleum sulfonates [MATER] Sulfonated petroleum products made as by-products of SO_3 treatment of white oil or lube-stock; used as lube-oil additives, textile-processing emulsifiers, and rust preventives.

petroleum tailings *See* wax tailings.

petroleum tar [MATER] A viscous, black or dark-brown product of petroleum refining; yields substantial quantity of solid residue when partly evaporated or fractionally distilled.

petroleum wax [MATER] A wax occurring naturally in various fractions of crude petroleum; there are two groups: paraffin wax and microcrystalline wax.

petrous [MATER] Referring to a material whose hardness resembles that of stone.

Petterson-Nansen water bottle *See* Nansen bottle.

Pettit truss [CIV ENG] A bridge truss in which the panel is subdivided by a short diagonal and a short vertical member, both intersecting the main diagonal at its midpoint.

pewter [MET] An alloy that typically contained tin as the principal component and some antimony and copper; older produced pewter typically contains lead along with the other components.

phase [MET] A constituent of an alloy that is physically distinct and is homogeneous in chemical composition.

phase-angle meter *See* phase meter.

phase behavior [PETRO ENG] The equilibrium relationships between water, liquid hydrocarbons, and dissolved or free gas, either in reservoirs or as liquids and gases are separated above ground in gas-oil separator systems.

phase diagram *See* constitution diagram.

phase-locked system [ENG] A radar system, having a stable local oscillator, in which information regarding the target is gained by measuring the phase shift of the echo.

phase meter [ENG] An instrument for the measurement of electrical phase angles. Also known as phase-angle meter.

phenolic laminate [MATER] Canvas, linen, kraft paper, glass fiber, or other substrate impregnated with 30% or more of thermosetting phenolic resin and cured; used for structural, mechanical, and electrical purposes.

phenolic plastic [MATER] A thermosetting plastic material available in many combinations of phenol and formaldehyde, often with added fillers to provide a broad range of physical, electrical, chemical, and molding properties.

Philips Hot-air engine [MECH ENG] A compact hot-air engine that is a Philips Research Lab (Holland) design; it uses only one cylinder and piston, and operates at 3000 revolutions per minute, with hot-chamber temperature of 1200°F (650°C), maximum pressure of 50 atmospheres (5.07 megapascals), and mean effective pressure of 14 atmospheres (1.42 megapascals).

Phillips screw [DES ENG] A screw having in its head a recess in the shape of a cross; it is inserted or removed with a Phillips screwdriver that automatically centers itself in the screw.

phleger corer [ENG] A device for obtaining ocean bottom cores up to about 4 feet (1.2 meters) in length; consists of an upper tube, main body weight, and tailfin assembly with a check valve that prevents the flow of water into the upper section and a consequent washing out of the core sample while hoisting the corer.

pH meter [ENG] An electronic voltmeter using a pH-responsive electrode that gives a direct conversion of voltage differences to differences of pH at the temperature of the measurement.

phone *See* headphone.

phonemic synthesizer [ENG ACOUS] A voice response system in which each word is abstractly represented as a sequence of expected vowels and consonants, and speech is composed by juxtaposing the expected phonemic sequence for each word with the sequences for the preceding and following words.

phonograph [ENG ACOUS] An instrument for recording or reproducing acoustical signals, such as voice or music, by transmission of vibrations from or to a stylus that is in contact with a groove in a rotating disk.

phonograph cartridge *See* phonograph pickup.

phonograph cutter *See* cutter.

phonograph needle *See* stylus.

phonograph pickup [ENG ACOUS] A pickup that converts variations in the grooves of a phonograph record into corresponding electric signals. Also known as cartridge; phonograph cartridge.

phonograph record [ENG ACOUS] A shellac-composition or vinyl-plastic disk, usually 7 or 12 inches (18 or 30 centimeters) in diameter, on which sounds have been recorded as modulations in grooves. Also known as disk; disk recording.

phonotelemeter [ENG] A device consisting essentially of a stopwatch, for estimating the distance of guns in action by measuring the interval between the flash and the arrival of the sound waves from the discharge.

phosphate coating [MET] A conversion coating on metal, usually steel, produced by dipping in an aqueous solution of zinc or manganese acid phosphate; used to furnish a black finish to small arms, artillery, or automotive components to provide resistance to corrosion.

phosphate fertilizer [MATER] Fertilizer compound or mixture containing available (soluble) phosphate; examples are phosphate rock (phosphorite), superphosphates or triple superphosphates, nitrophosphate, potassium phosphates, or N-P-K mixtures.

phosphate glass [MATER] Glass in which phosphorus pentoxide is a major component; resistant to hydrofluoric acid.

phosphate recovery process [MIN ENG] A process developed by the U.S. Bureau of Mines for recovering phosphate from low-grade phosphorus-bearing shales.

phosphate rock [MATER] **1.** A rock that is naturally high enough in phosphorus to be used directly in fertilizer manufacturing. **2.** The beneficiated concentrate of a phosphate deposit.

phosphating [MET] Forming a phosphate coating on a metal. Also known as phosphatizing.

phosphor bronze [MET] A hard copper-base alloy containing several percent tin, and sometimes smaller percentages of lead, deoxidized with phosphorus.

phosphorescent paint [MATER] A luminous paint containing phosphors or phosphorogens which requires activation from an outside source of light, depending upon the ability of the chemical to absorb light energy, and to emit it in the form of photons of light.

phosphorized copper [MET] A phosphorus deoxidized copper.

phosphor tin [MET] A master alloy of tin and phosphorus, usually containing up to 5% phosphorus; used to make phosphor bronze.

photoalidade [ENG] A photogrammetric instrument which has a telescopic alidade, a plateholder, and a hinged ruling arm and is mounted on a tripod frame; used for plotting lines of direction and measuring vertical angles to selected features appearing on oblique and terrestrial photographs.

photochromic glass [MATER] A glass that darkens when exposed to light but regains its original transparency a few minutes after light is removed; the rate of clearing increases with temperature.

photoclinometer [ENG] A directional surveying instrument which records photographically the direction and magnitude of well deviations from the vertical.

photodraft [DES ENG] A photographic reproduction of a master layout or design on a specially prepared emulsion-coated piece of sheet metal; used as a master in a tool-construction department.

photoecology [ENG] The application of air photography to ecology, integrated land resource studies, and forestry.

photoelectric colorimeter [ENG] A colorimeter that uses a phototube or photocell, a set of color filters, an amplifier, and an indicating meter for quantitative determination of color.

photoelectric densitometer [ENG] An electronic instrument used to measure the density or opacity of a film or other material; a beam of light is directed through the material, and the amount of light transmitted is measured with a photocell and meter.

photoelectric fluorometer [ENG] Device using a photoelectric cell to measure fluorescence in a chemical sample that has been excited (one or more electrons have been raised to higher energy level) by ultraviolet or visible light; used for analysis of chemical mixtures.

photoelectric liquid-level indicator [ENG] A level indicator in which rising liquid interrupts the light beam of a photoelectric control system; used in a tank or process vessel.

photoelectric photometer [ENG] A photometer that uses a photocell, phototransistor, or phototube to measure the intensity of light. Also known as electronic photometer.

photoelectric pyrometer [ENG] An instrument that measures high temperatures by using a photoelectric arrangement to measure the radiant energy given off by the heated object.

photoelectric reflectometer [ENG] A reflectometer that uses a photocell or phototube to measure the diffuse reflection of surfaces, powders, pastes, and opaque liquids.

photoelectric transmissometer [ENG] A device to measure the runway visibility at an airport by measuring the degree to which a light beam falling on a photocell is obscured by clouds or fog.

photoelectric turbidimeter [ENG] Device for measurement of solution turbidity by use of photocells to detect the loss of intensity of light beamed through the solution.

photoemissive tube photometer [ENG] A photometer which uses a tube made of a photoemissive material; it is highly accurate, but requires electronic amplification, and is used mainly in laboratories.

photoengraving zinc [MET] Pure zinc mixed with a small amount of iron to reduce grain size, and alloyed with a maximum of 0.2% each of cadmium, manganese, and magnesium; used for printing plates.

photoflash bomb [ENG] A missile dropped from aircraft; it contains a photoflash mixture and a means for ignition at a distance above the ground, to produce a brilliant light of short duration for photographic purposes.

photoflash composition [MATER] A pyrotechnic material which, when loaded in a suitable casing and ignited, produces a flash of sufficient intensity and duration for photographic purposes; used as the filler in photoflash bombs and cartridges.

photogoniometer [ENG] A goniometer that uses a phototube or photocell as a sensing device for studying x-ray spectra and x-ray diffraction effects in crystals.

photogrammetry [ENG] **1.** The science of making accurate measurements and maps from aerial photographs. **2.** The practice of obtaining surveys by means of photography.

photographic barograph [ENG] A mercury barometer arranged so that the position of the upper or lower meniscus may be measured photographically.

photographic interpretation See photointerpretation.

photographic surveying [ENG] Photographing of plumb bobs, clinometers, or magnetic needles in borehole surveying to provide an accurate permanent record.

photointerpretation [ENG] The science of identifying and describing objects in a photograph, such as deducing the topographic significance or the geologic structure of landforms on an aerial photograph. Also known as photographic interpretation.

photometer [ENG] An instrument used for making measurements of light or electromagnetic radiation, in the visible range.

photon curve [PETRO ENG] A graphical plot of depth versus gamma radiation (photon) scatter during the radioactive logging of a well bore; used to detect differences in density at various reservoir depths.

photonephelometer [ENG] A nephelometer that uses a photocell or phototube to measure the amount of light transmitted by a suspension of particles.

photon sail *See* solar sail.

photoscanner [ENG] A scanner used to make a film record of gamma rays passing through tissue from an injected radioactive material.

phototheodolite [ENG] A ground-surveying instrument used in terrestrial photogrammetry which combines the functions of a theodolite and a camera mounted on the same tripod.

phototopography [ENG] The science of mapping and surveying in which details are plotted entirely from photographs taken at suitable ground stations.

phototriangulation [ENG] The extension of horizontal or vertical control points, or both, by photogrammetric methods, whereby the measurements of angles and distances on overlapping photographs are related into a spatial solution using the perspective principles of the photographs.

phototube current meter [ENG] A device for measuring the speed of water currents in which a perforated disk, which rotates with the current by means of a propeller, is placed in the path of a beam of light that is then reflected from a mirror onto a phototube.

phugoid [AERO ENG] Pertaining to variations in the longitudinal motion or course of the center of mass of an aircraft.

physical compatibility [ENG] The ability of two or more materials, substances, or chemicals to be used together without ill effect.

physical metallurgy [MET] The branch of metallurgy concerned with physical and mechanical properties of metals as affected by composition, mechanical working, and heat treatment.

physical testing [ENG] Determination of physical properties of materials based on observation and measurement.

phytometer [ENG] A device for measuring transpiration, consisting of a vessel containing soil in which one or more plants are rooted and sealed so that water can escape only by transpiration from the plant.

piano wire [MET] High-tensile-strength, 0.75 to 0.85% carbon steel wire colddrawn to uniform thickness.

Picatinny test [ENG] An impact test used in the United States for evaluating the sensitivity of high explosives; a small sample of the explosive is placed in a depression in a steel die cup and capped by a thin brass cover, a cylindrical steel plug is placed in the center of the cover, and a 2-kilogram weight is dropped from varying heights on the plug; the reported sensitivity figure is the minimum height, in inches, at which at least 1 firing results from 10 trials.

Piche evaporimeter [ENG] A porous-paperwick atmometer.

pick [DES ENG] **1.** The steel cutting points used on a coal-cutter chain. **2.** A miner's steel or iron digging tool with sharp points at each end. [ENG] **1.** To dress the sides of a shaft or other excavation. **2.** To remove shale, dirt, and such from coal.

pick-a-back conveyor [MIN ENG] A short conveyor that advances with a loader or continuous miner at the face of a mine and loads coal on the main haulage system.

Pickard core barrel [MIN ENG] A type of double-tube core barrel; the distinguishing feature of the barrel is that when blocked the inner barrel slides upward into the head, closing the water ports and stopping the flow of the circulating liquid, without irreparably damaging the bit until the barrel is pulled and the blocked inner tube cleared.

pickax [DES ENG] A pointed steel or iron tool mounted on a wooden handle and used for breaking earth and stone.

picker [MIN ENG] **1.** An employee who picks or discards slate and other foreign matter from the coal in an anthracite breaker or at a picking table. **2.** A mechanical arrangement for removing slate from coal.

pick hammer [DES ENG] A hammer with a point at one end of the head and a blunt surface at the other end.

picking [MIN ENG] **1.** Removal of waste material from an ore. **2.** Extraction of the lightest-grade ore from a mine. **3.** Emission of particles from the roof of a mine on the verge of collapse.

picking conveyor [MIN ENG] A continuous belt or apron conveyor used to carry a relatively thin bed of material past pickers who hand-sort or pick the material being conveyed.

picking table [MIN ENG] A flat or slightly inclined platform on which the coal or ore is run to be picked free from slate or gangue.

pick lacing [DES ENG] The pattern to which the picks are set in a cutter chain.

pickle liquor [MET] A spent pickling solution.

pickle patch [MET] A coating of oxide or scale that remains adherent after pickling.

pickle stain [MET] Discoloration of a metal surface due to chemical cleaning but without adequate washing and drying.

pickling [MET] Preferential removal of oxide or mill scale from the surface of a metal by immersion usually in an acidic or alkaline solution.

pick miner [MIN ENG] **1.** In anthracite and bituminous coal mining, one who uses hand tools to extract coal in underground working places. **2.** One who cuts out a channel under the bottom of the working face of coal with a pick.

pickoff [MECH ENG] A mechanical device for automatic removal of the finished part from a press die.

pickup [AERO ENG] A potentiometer used in an automatic pilot to detect the motion of the airplane around the gyro and initiate corrective adjustments. [MET] Transfer of metal from the work to the tool, or from the tool to the work, during a forming operation.

picoammeter [ENG] An ammeter whose scale is calibrated to indicate current values in picoamperes.

picratol [MATER] A binary explosive composed of 52% ammonium picrate and 48% TNT (trinitrotoluene); it can be melt-loaded; less sensitive than TNT, it was developed for use in armor-piercing bombs.

Pictet's liquid [MATER] Liquid mixture of carbon dioxide and sulfur dioxide; used to produce low temperatures.

picture window [BUILD] A large window framing an exterior view.

Pidgeon process [MET] A method for producing magnesium from calcined dolomite by reduction with ferrosilicon. Also known as ferrosilicon process; silicothermic process.

piece mark [ENG] Identification number for an individual part, subassembly, or assembly; shown on the drawing, but not necessarily on the part.

piece rate [IND ENG] Wages paid per unit of production.

piecework [IND ENG] Work paid for in accordance with the amount done rather than the hours taken.

pier [BUILD] A concrete block that supports the floor of a building. [CIV ENG] **1.** A vertical, rectangular or circular support for concentrated loads from an arch or bridge superstructure. **2.** A structure with a platform projecting from the shore into navigable waters for mooring vessels.

piercing See fusion piercing.

pier foundation See caisson foundation.

pierhead line [CIV ENG] The line in navigable waters beyond which construction is prohibited; open-pier construction may extend outward from the bulkhead line to the pierhead line.

piezoelectric ceramic [MATER] A ceramic material that has piezoelectric properties similar to those of some natural crystals.

piezoelectric detector [ENG] A seismic detector constructed from a stack of piezoelectric crystals with an inertial mass mounted on top and intervening metal foil to collect the charges produced on the crystal faces when the crystals are strained.

piezoelectric gage [ENG] A pressure-measuring gage that uses a piezoelectric material to develop a voltage when subjected to pressure; used for measuring blast pressures resulting from explosions and pressures developed in guns.

piezoelectric loudspeaker See crystal loudspeaker.

piezoelectric microphone See crystal microphone.

piezoelectric pickup See crystal pickup.

piezometer [ENG] **1.** An instrument for measuring fluid pressure, such as a gage attached to a pipe containing a gas or liquid. **2.** An instrument for measuring the compressibility of materials, such as a vessel that determines the change in volume of a substance in response to hydrostatic pressure.

piezometer opening See pressure tap.

pig [ENG] In-line scraper (brush, blade cutter, or swab) forced through pipelines by fluid pressure; used to remove scale, sand, water, and other foreign matter from the interior surfaces of the pipe. [MET] A crude metal casting prepared for storage, transportation, or remelting.

pig iron [MET] **1.** Crude, high-carbon iron produced by reduction of iron ore in a blast furnace. **2.** Cast iron in the form of pigs.

pigment [MATER] A solid that reflects light of certain wavelengths while absorbing light of other wavelengths, without producing appreciable luminescence; used to impart color to other materials.

pilaster [CIV ENG] A vertical rectangular architectural member that is structurally a pier and architecturally a column.

pilchard oil [MATER] Pale-yellow oil expressed from pickled pilchards (of the herring family); used to make paints and potash soft soap.

pile [ENG] A long, heavy timber, steel, or reinforced concrete post that has been driven, jacked, jetted, or cast vertically into the ground to support a load.

pile bent [CIV ENG] A row of timber or concrete bearing piles with a pile cap forming that part of a trestle which carries the adjacent ends of timber stringers or concrete slabs.

pile cap [CIV ENG] A mass of reinforced concrete cast around the head of a group of piles to ensure that they act as a unit to support the imposed load.

pile dike [CIV ENG] A dike consisting of a group of piles braced and lashed together along a riverbank.

pile driver [MECH ENG] A hoist and movable steel frame equipped to handle piles and drive them into the ground.

pile extractor [MECH ENG] **1.** A pile hammer which strikes the pile upward so as to loosen its grip and remove it from the ground. **2.** A vibratory hammer which loosens the pile by high-frequency jarring.

pile foundation [CIV ENG] A substructure supported on piles.

pile hammer [MECH ENG] The heavy weight of a pile driver that depends on gravity for its striking power and is used to drive piles into the ground. Also known as drop hammer.

pile shoe [CIV ENG] A cast-iron point on the foot of a timber or concrete driven pile to facilitate penetration of the ground.

Pilger tube-reducing process [MET] A tube-reducing process in which pierced billets are forced over a mandrel between two rolls with inclined axes and given a rotary forging treatment; the tube is advanced during the gap in each revolution.

pillar [CIV ENG] A column for supporting part of a structure. [MIN ENG] An area of coal or ore left to support the overlying strata or hanging wall in a mine.

pillar-and-breast system [MIN ENG] A system of coal mining in which the working places are rectangular rooms usually five or ten times as long as they are broad, opened on the upper side of the gangway.

pillar-and-room system [MIN ENG] A system of mining whereby solid blocks of coal are left on either side of working places to support the roof until first-mining has been completed, when the pillar coal is then recovered.

pillar-and-stall system [MIN ENG] A system of working coal and other minerals where the first stage of excavation is accomplished with the roof sustained by coal or ore.

pillar bolt [DES ENG] A bolt projecting from a part so as to support it.

pillar burst [MIN ENG] A failure of a pillar, by crushing.

pillar crane [MECH ENG] A crane whose mechanism can be rotated about a fixed pillar.

pillar drive [MIN ENG] A wide irregular drift or entry, in firm dry ground, in which the roof is supported by pillars of the natural earth, or by artificial pillars of stone, no timber being used.

pillar extraction [MIN ENG] Removal of the pillars of coal left over from mining by the pillar-and-stall method. Also known as pillar mining.

pillaring [MIN ENG] The process of extracting pillars. Also called pillar robbing; pulling pillars; robbing pillars.

pillar line [MIN ENG] Air currents which have definitely coursed through an inaccessible abandoned panel or area or which have ventilated a pillar line or a pillar area, regardless of the methane content, or absence of methane, in such air.

pillar man See pack builder.

pillar mining See pillar extraction.

pillar press [MECH ENG] A punch press framed by two upright columns; the driving shaft passes through the columns, and the slide operates between them.

pillar robbing See pillaring.

pillar split [MIN ENG] An opening or crosscut driven through a pillar in the course of extraction of ore.

pilot [AERO ENG] **1.** A person who handles the controls of an aircraft or spacecraft from within the craft, and guides or controls the craft in flight. **2.** A mechanical system designed to exercise control functions in an aircraft or spacecraft. [MECH ENG] A cylindrical steel bar extending through, and about 8 inches (20 centimeters) beyond the face of, a reaming bit; it acts as a guide that follows the original unreamed part of the borehole and hence forces the reaming bit to follow, and be concentric with, the smaller-diameter, unreamed portion of the original borehole.

pilot balloon [ENG] A small balloon whose ascent is followed by a theodolite in order to obtain data for the computation of the speed and direction of winds in the upper air.

pilot bit [DES ENG] A noncoring bit with a cylindrical diamond-set plug of somewhat smaller diameter than the bit proper, set in the center and projecting beyond the main face of the bit.

pilot channel [CIV ENG] One of a series of cutoffs for converting a meandering stream into a straight channel of greater slope.

pilot chute [AERO ENG] A small parachute canopy attached to a larger canopy to actuate and accelerate the opening of the load-bearing canopy.

pilot drill [MECH ENG] A small drill to start a hole to ensure that a larger drill will run true to center.

pilot flood [PETRO ENG] A test waterflood operation (water-injection well) designed to evaluate procedures and to give advance information prior to instituting an extensive, multipoint waterflood.

pilot hole [ENG] A small hole drilled ahead of a larger borehole.

pilotless aircraft [AERO ENG] An aircraft adapted to control by or through a preset self-reacting unit or a radio-controlled unit, without the benefit of a human pilot.

pilot light [ENG] A small, constantly burning flame used to ignite a gas burner.

pilot line operation [IND ENG] Minimum production of an item in order to preserve or develop the art of its production.

pilot materials [IND ENG] A minimum quantity of special materials, partially finished components, forgings, and castings, identified with specific production equipment and processes and required for the purpose of proofing, tooling, and testing manufacturing processes to facilitate later reactivation.

pilot model [IND ENG] An early production model of a product used to debug the manufacturing process.

pilot plant [IND ENG] A small version of a planned industrial plant, built to gain experience in operating the final plant.

pilot production [PETRO ENG] Limited or test production of oil or gas from a field to determine reservoir and product characteristics before commencing full-scale recovery operations.

pilot tunnel [ENG] A small tunnel or shaft excavated in advance of the main drivage in mining and tunnel building to gain information about the ground, create a free face, and thus simplify the blasting operations.

pimenta oil [MATER] A yellow to brownish essential oil with spicy aroma and pungent taste; derived from allspice (*Pimenta officinalis*); main components are eugenol, cineol, and phellandrene; used in medicine and flavors. Also known as allspice oil; pimento oil.

pimento oil *See* pimenta oil.

pimple [MATER] A small, conical elevation on the surface of a plastic.

pin [DES ENG] **1.** A cylindrical fastener made of wood, metal, or other material used to join two members or parts with freedom of angular movement at the joint. **2.** A short, pointed wire with a head used for fastening fabrics, paper, or similar materials.

pin bar [MET] A small-diameter, case-hardened steel rod used for making dowel pins.

pinch [ENG] The closing-in of borehole walls before casing is emplaced, resulting from rock failure when drilling in formations having a low compressional strength. [MIN ENG] *See* horseback; squeeze.

pinch bar [DES ENG] A pointed lever, used somewhat like a crowbar, to roll heavy wheels.

pinch-off *See* cutoff.

pinch-off blades [ENG] In blow molding, the part that compresses the parison to seal it prior to blowing, and to allow easy cooling and removal of flash.

pinch pass [MET] A cold rolling of sheet metal to effect a very small reduction in thickness and to produce a piece of accurate dimensions.

pinch trimming [MET] Trimming a tubular or hollow part by pinching the flange or lip over the cutting edge of a punch.

pinch-tube process [ENG] A plastics blow-molding process in which the extruder drops a tube between mold halves, and the tube is pinched off when the mold closes.

pine needle oil [MATER] An essential oil derived from various pines; colorless to yellowish oil with balsamic aroma; soluble in alcohol; used in perfumes and medicines. Also known as Douglas fir oil; fir wood oil.

pine oil [MATER] Any of a group of volatile essential oils with pinaceous aromas distilled from cones, needles, or stumps of various pine or other conifer species; used as solvents, emulsifying agents, wetting agents, deodorants, germicides, and sources of chemicals.

pine oleoresin [MATER] Fused solid blend of turpentine and rosin.

pine tar [MATER] A viscous black mass obtained as a by-product in the distillation of pine wood; used for roofing.

pine tar pitch [MATER] Viscous residue resulting from distillation of volatile oils from pine tar.

pin expansion test [MET] A test for determining tube expandability or for revealing longitudinal weaknesses by forcing a tapered pin into the open end of the tube.

pinger [ENG ACOUS] A battery-powered, low-energy source for an echo sounder.

pinhole [MET] A material fault resulting from small blisters that have burst in a casting or that have formed during electroplating.

pinhole detector [ENG] A photoelectric device that detects extremely small holes and other defects in moving sheets of material.

pinion [MECH ENG] The smaller of a pair of gear wheels or the smallest wheel of a gear train.

pin joint [DES ENG] A joint made with a pin hinge which has a removable pin.

pin metal [MET] Brass with a composition of 63% copper and 37% zinc, used as cold-drawn wire for making ordinary dressmaking pins.

pinpoint gate [ENG] In plastics molding, an orifice of 0.030 inch (0.76 millimeter) or less in diameter through which molten resin enters a mold cavity.

pin rod [DES ENG] A rod designed to connect two parts so they act as one.

pin timbering [MIN ENG] A method of mine roof support in which bolts are driven up into strong material, thus supporting lower weak layers.

pintle [DES ENG] A vertical pivot pin, as on a rudder or a gun carriage.

pintle chain [DES ENG] A chain with links held together by pivot pins; used with sprocket wheels.

pin-type mill [MECH ENG] Solids pulverizer in which protruding pins on high-speed rotating disk provide the breaking energy.

Piobert lines *See* Lüders lines.

pioneer tunnel [MIN ENG] A small tunnel parallel to but ahead of a main tunnel and used to make crosscuts to the path that the main tunnel will follow.

pipe [DES ENG] A tube made of metal, clay, plastic, wood, or concrete and used to conduct a fluid, gas, or finely divided solid. [MET] **1.** The central cavity in an ingot or casting formed by contraction of the metal during solidification. **2.** An extrusion defect caused by the oxidized surface of the billet flowing toward the center of the rod at the back end.

pipe bit [DES ENG] A bit designed for attachment to standard coupled pipe for use in socketing the pipe in bedrock.

pipe clamp [DES ENG] A device similar to a casing clamp, but used on a pipe to grasp it and facilitate hoisting or suspension.

pipe culvert [CIV ENG] A buried pipe for carrying a watercourse below ground level.

pipe cutter [DES ENG] A hand tool consisting of a clamplike device with three cutting wheels which are forced inward by screw pressure to cut into a pipe as the tool is rotated around the pipe circumference.

pipe elbow meter [ENG] A variable-head meter for measuring flow around the bend in a pipe.

pipe fitting [ENG] A piece, such as couplings, unions, nipples, tees, and elbows for connecting lengths of pipes.

pipe flow [ENG] Conveyance of fluids in closed conduits.

pipe laying [ENG] The placing of pipe into position in a trench, as with buried pipelines for oil, water, or chemicals.

pipeline [ENG] A line of pipe connected to valves and other control devices, for conducting fluids, gases, or finely divided solids.

pipe pile [CIV ENG] A steel pipe 6–30 inches (15–76 centimeters) in diameter, usually filled with concrete and used for underpinning.

pipe scale [ENG] Rust and corrosion products adhering to the inner surfaces of pipes; serve to decrease ability to transfer heat and to increase the pressure drop for flowing fluids.

pipe tap [ENG] A small threaded hole or entry made into the wall of a pipe; used for sampling of pipe contents, or connection of control devices or pressure-drop-measurement devices.

pipe tee [DES ENG] A T-shaped pipe fitting with two outlets, one at 90° to the connection to the main line.

pipe thread [DES ENG] Most commonly, a 60° thread used on pipes and tubes, characterized by flat crests and roots and cut with 3/4-inch taper per foot (about 1.9 centimeters per 30 centimeters). Also known as taper pipe thread.

pipe tongs [ENG] Heavy tongs that are hung on a cable and used for screwing pipe and tool joints.

pipe train [ENG] In the extrusion of plastic pipe, the entire equipment assembly used to fabricate the pipe (such as the extruder, die, cooling bath, haul-off, and cutter).

pipe wrench [DES ENG] A tool designed to grip and turn a pipe or rod about its axis in one direction only.

piping [ENG] A system of pipes provided to carry a fluid.

pistachia galls *See* mastic.

piston [ENG] *See* force plug. [MECH ENG] A sliding metal cylinder that reciprocates in a tubular housing, either moving against or moved by fluid pressure.

piston blower [MECH ENG] A piston-operated, positive-displacement air compressor used for stationary, automobile, and marine duty.

piston corer [MECH ENG] A steel tube which is driven into the sediment by a free fall and by lead attached to the upper end, and which is capable of recovering undistorted vertical sections of sediment.

piston displacement [MECH ENG] The volume which a piston in a cylinder displaces in a single stroke, equal to the distance the piston travels times the internal cross section of the cylinder.

piston drill [MECH ENG] A heavy percussion-type rock drill mounted either on a horizontal bar or on a short horizontal arm fastened to a vertical column; drills holes to 6 inches (15 centimeters) in diameter. Also known as reciprocating drill.

piston engine [MECH ENG] A type of engine characterized by reciprocating motion of pistons in a cylinder. Also known as displacement engine; reciprocating engine.

piston head [MECH ENG] That part of a piston above the top ring.

piston meter [ENG] A variable-area, constant-head fluid-flow meter in which the position of the piston, moved by the buoyant force of the liquid, indicates the flow rate. Also known as piston-type area meter.

piston pin [MECH ENG] A cylindrical pin that connects the connecting rod to the piston. Also known as wrist pin.

piston pump [MECH ENG] A pump in which motion and pressure are applied to the fluid by a reciprocating piston in a cylinder. Also known as reciprocating pump.

piston ring [DES ENG] A sealing ring fitted around a piston and extending to the cylinder wall to prevent leakage. Also known as packing ring.

piston rod [MECH ENG] The rod which is connected to the piston, and moves or is moved by the piston.

piston skirt [MECH ENG] That part of a piston below the piston pin bore.

piston speed [MECH ENG] The total distance a piston travels in a given time; usually expressed in feet per minute.

piston-type area meter *See* piston meter.

piston valve [MECH ENG] A cylindrical type of steam engine slide valve for admission and exhaust of steam.

piston viscometer [ENG] A device for the measurement of viscosity by the timed fall of a piston through the liquid being tested.

pit [MET] A small hole in the surface of a metal: usually caused by corrosion or formed during electroplating operations. [MIN ENG] **1.** A coal mine; the term is not commonly used by the coal industry, except in reference to surface mining where the workings may be known as a strip pit. **2.** Any quarry, mine, or excavation area worked by the open-cut method to obtain material of value.

pitch [DES ENG] The distance between similar elements arranged in a pattern or between two points of a mechanical part, as the distance between the peaks of two successive grooves on a disk recording or on a screw. [MATER] A dark heavy liquid or solid substance obtained as a residue after distillation of tar, oil, and such materials; occurs naturally as asphalt.

pitch circle [DES ENG] In toothed gears, an imaginary circle concentric with the gear axis which is defined at the thickest point

on the teeth and along which the tooth pitch is measured.

pitch coke [MATER] Coke made from coal tar pitch, characterized by high carbon and low ash content, and used mainly for production of electrode carbon.

pitch cone [DES ENG] A cone representing the pitch surface of a bevel gear.

pitch cylinder [DES ENG] A cylinder representing the pitch surface of a spur gear.

pitch diameter [DES ENG] The diameter of the pitch circle of a gear.

pitch indicator [AERO ENG] An instrument for indicating the existence and approximate magnitude of the angular velocity about the lateral axis of an airframe.

pitch line *See* cam profile.

pitch mining [MIN ENG] Mining coal beds with steep slopes.

pitchover [AERO ENG] The programmed turn from the vertical that a rocket under power takes as it describes an arc and points in a direction other than vertical.

pit furnace [MET] A low-temperature furnace in which steel is tempered.

pit limits [MIN ENG] The vertical and lateral extent to which the mining of a mineral deposit by open pitting may be carried economically.

pitman [ENG] A worker in or near a pit, as in a quarry, mine, garage, or foundry.

pitometer [ENG] Reversed pitot-tube-type flow-measurement device with one pressure opening facing upstream and the other facing downstream.

pitometer log [ENG] A log consisting essentially of a pitot tube projecting into the water, and suitable registering devices.

pitot tube [ENG] An instrument that measures the stagnation pressure of a flowing fluid, consisting of an open tube pointing into the fluid and connected to a pressure-indicating device. Also known as impact tube.

pitot-tube anemometer [ENG] A pressure-tube anemometer consisting of a pitot tube mounted on the windward end of a wind vane and a suitable manometer to measure the developed pressure, and calibrated in units of wind speed.

pitot-venturi flow element [ENG] Liquid-flow measurement device in which a pair of concentric venturi elements replaces the pitot-tube probe.

pit sampling [MIN ENG] Using small untimbered pits to gain access to shallow alluvial deposits or ore dumps for purpose of testing or valuation.

pit slope [MIN ENG] The angle at which the wall of an open pit or cut stands as measured along an imaginary plane extended along the crests of the berms or from the slope crest to its toe.

pitting [MET] Selective localized formation of rounded cavities in a metal surface due to corrosion or to nonuniform electroplating. [MIN ENG] The act of digging or sinking a pit.

pitting potential [MET] The electrochemical potential in a given environment above which, but not below, a corrosion pit initiates in a metal surface.

pivot bridge [CIV ENG] A bridge in which a span can open by pivoting about a vertical axis.

pivot-bucket conveyor-elevator [MECH ENG] A bucket conveyor having overlapping pivoted buckets on long-pitch roller chains; buckets are always level except when tripped to discharge materials.

pivoted window [BUILD] A window having a section which is pivoted near the center so that the top of the section swings in and the bottom swings out.

placer claim [MIN ENG] A mining claim located upon gravel or ground whose mineral contents are extracted by the use of water, as by sluicing, or hydraulicking.

placer dredge [MIN ENG] A dredge for mining metals from placer deposits; it consists of a chain of closely connected buckets passing over an idler tumbler and an upper or driving tumbler, mounted on a structural-steel ladder which carries a series of rollers.

placer location [MIN ENG] Location of a tract of land for the sake of loose mineral-bearing or other valuable deposits on or near its surface, rather than within lodes or veins in rock in place.

placer mine *See* gravel mine.

placer mining [MIN ENG] **1.** The extraction and concentration of heavy metals from placers. **2.** Mining of gold by washing the sand, gravel, or talus.

plain concrete [CIV ENG] Concrete without reinforcement but often with light steel to reduce shrinkage and temperature cracking.

plain-laid [DES ENG] Pertaining to a rope whose strands are twisted together in a direction opposite to that of the twist in the strands.

plain milling cutter [DES ENG] A cylindrical milling cutter with teeth on the periphery only; used for milling plain or flat surfaces. Also known as slab cutter.

plain turning [MECH ENG] Lathe operations involved when machining a workpiece between centers.

planar process [ENG] A silicon-transistor manufacturing process in which a fractional-micrometer-thick oxide layer is grown on a silicon substrate; a series of etching and diffusion steps is then used to produce the transistor inside the silicon substrate.

planchet [ENG] A small metal container or sample holder; usually used to hold radioactive materials that are being checked for the degree of radioactivity in a proportional counter or scintillation detector. [MATER] A milled metal disk ready for coining.

plane [DES ENG] A tool consisting of a smooth-soled stock from the face of which extends a wide-edged cutting blade for smoothing and shaping wood.

plane correction [ENG] A correction applied to observed surveying data to reduce them to a common reference plane.

plan equation [MECH ENG] The mathematical statement that horsepower = $plan/33,000$, where p = mean effective pressure (pounds per square inch), l = length of piston stroke (feet), a = net area of piston (square inches), and n = number of cycles completed per minute.

planer [MECH ENG] A machine for the shaping of long, flat, or flat contoured surfaces by reciprocating the workpiece under a stationary single-point tool or tools. [MIN ENG] A fixed-blade device for continuous longwall mining of narrow seams of friable coal; the machine is operated along the coal face, planing a narrow cut from the solid coal as it travels.

plane surveying [ENG] Measurement of areas on the assumption that the earth is flat.

plane table [ENG] A surveying instrument consisting of a drawing board mounted on a tripod and fitted with a compass and a straight-edge ruler; used to graphically plot survey lines directly from field observations.

plane-table method [MIN ENG] A method of measuring areas of mine roadways; a drawing board is set up on a tripod in the plane of the mine section to be measured; the distance from a central point on the board to the perimeter of the roadway is measured with a tape along various offsets; the distance measured is scaled on the drawing board along the proper offset line.

planetary gear train [MECH ENG] An assembly of meshed gears consisting of a central gear, a coaxial internal or ring gear, and one or more intermediate pinions supported on a revolving carrier.

planet gear [MECH ENG] A pinion in a planetary gear train.

planform [AERO ENG] The shape or form of an object, such as an airfoil, as seen from above, as in a plan view.

planimeter [ENG] A device used for measuring the area of any plane surface by tracing the boundary of the area.

planimetric method [MET] A method of measuring grain size by counting the number of grains in a given area.

planing [ENG] Smoothing or shaping the surface of wood, metal, or plastic workpieces.

planishing [MECH ENG] Smoothing the surface of a metal by a rapid series of overlapping, light hammerlike blows or by rolling in a planishing mill.

plank [MATER] A heavy board with thickness of 2–4 inches (5–10 centimeters) and a width of at least 8 inches (20 centimeters).

plankton net [ENG] A net for collecting plankton.

plant [IND ENG] The land, buildings, and equipment used in an industry.

plant layout [IND ENG] The location of equipment and facilities in a manufacturing plant.

plant protection [IND ENG] That portion of industrial security which concerns the safeguarding of industrial installations, resources, utilities, and materials by physical measures such as guards, fences, and lighting designation of restricted areas.

plasma-arc cutting [ENG] Metal cutting by melting a localized area with an arc followed by removal of metal by high-velocity, high-temperature ionized gas.

plasma-arc welding [MET] Welding metal in a gas stream heated by a tungsten arc to temperatures approaching 60,000°F (33,315°C).

plasma engine [AERO ENG] An engine for space travel in which neutral plasma is accelerated and directed by external magnetic fields that interact with the magnetic field produced by current flow through the plasma. Also known as plasma jet.

plasma jet See plasma engine.

plasma propulsion [AERO ENG] Propulsion of spacecraft and other vehicles by using electric or magnetic fields to accelerate both positively and negatively charged particles (plasma) to a very high velocity.

plasma rocket [AERO ENG] A rocket that is accelerated by means of a plasma engine.

plasma spraying [MET] In thermal spraying, melting and transference of a metal coating to a workpiece by use of a nontransferred arc.

plasma torch [ENG] A torch in which temperatures as high as 50,000°C are achieved by injecting a plasma gas tangentially into an electric arc formed between electrodes in a chamber; the resulting vortex of hot gases emerges at very high speed through a hole in the negative electrode, to form a jet for welding, spraying of molten metal, and cutting of hard rock or hard metals.

plaster [MATER] A plastic mixture of various materials, such as lime or gypsum, and water which sets to a hard, coherent solid.

plasterboard [MATER] A large, thin sheet of pulpboard, paper, or felt bonded to a hardened gypsum plaster core and used as a wall backing or as a substitute for plaster.

plaster coat [BUILD] A thin layer of plaster lining walls in buildings.

plaster ground [BUILD] A piece of wood used as a gage to control the thickness of a plaster coat placed on a wall; usually put around windows and doors and at the floor.

plaster shooting [ENG] A surface blasting method used when no rock drill is necessary or one is not available; consists of placing a charge of gelignite, primed with safety fuse and detonator, in close contact with the rock or boulder and covering it completely with stiff damp clay.

plastic [MATER] A polymeric material (usually organic) of large molecular weight which can be shaped by flow; usually refers to the final product with fillers, plasticizers, pigments, and stabilizers included (versus the resin, the homogeneous polymeric starting material); examples are polyvinyl chloride, polyethylene, and urea-formaldehyde.

plasticate [ENG] To soften a material by heating or kneading. Also known as plastify.

plastic bonding [ENG] The joining of plastics by heat, solvents, adhesives, pressure, or radio frequency.

plastic bronze [MET] A copper alloy containing lead, usually on the order of 30%, of sufficient plasticity to make a good bearing.

plastic cement [MATER] A plastic material used to seal narrow openings in buildings.

plastic clay [MATER] Fireclay which forms a moldable mass when mixed with water.

plastic design See ultimate-load design.

plastic dielectric [MATER] A plastic used in an application in which its high resistance, dielectric strength, or other electrical properties are important, such as for electrical insulation or in a capacitor.

plastic explosive See high-explosive plastic.

plastic film [MATER] Film with thickness from 0.0015 to 0.006 inch (3.8×10^{-3} to 15×10^{-3} centimeter); made from polyvinyl chloride, polyethylene, polypropylene, polystyrene, Mylar, and other resins; used for wrapping, sealing, garment waterproofing, and coating wood, paper, or fabric.

plastic foam See expanded plastic.

plasticize [ENG] To soften a material to make it plastic or moldable by adding a plasticizer or by using heat.

plasticizer *See* flexibilizer.

plasticizing oil [MATER] Coal tar distillate or solvent naphthas distilling in a wide range above 300°C; used with plastics as a plasticizer.

plasticorder [ENG] Laboratory device used to predict the performance of a plastic material by measurement of temperature, viscosity, and shear-rate relationships. Also known as plastigraph.

plastic paint [MATER] Paint composed of a plastic (such as vinyl or nitrocellulose) in a solvent.

plastic semiconductor [MATER] An organic plastic resin with a conjugated double-bond structure, such as polyacetylene; the material is a semiconductor due to resistance of electrons to transfer from one molecule to another.

plastic wood [MATER] Wood flour or wood cellulose compounded with a synthetic resin of high molecular weight; it is adhesive but does not penetrate wood, and is used to fill cavities or seams in wood products.

plastify *See* plasticate.

plastigel [MATER] A plastisol with gellike flow properties achieved by adding a thixotropic agent (such as bentonite) to the plastisol.

plastigraph *See* plasticorder.

plastisol [MATER] A vinyl resin dissolved in a plasticizer to make a pourable liquid.

plastometer [ENG] Instrument used to determine the flow properties of a thermoplastic resin by forcing molten resin through a specified die opening or orifice at a given pressure and temperature.

plate [BUILD] **1.** A shoe or base member, such as of a partition or other kind of frame. **2.** The top horizontal member of a row of studs used in a frame wall. [DES ENG] A rolled, flat piece of metal of some arbitrary minimum thickness and width depending on the type of metal. [MET] A thick flat particle of metal powder.

plate amalgamation [MET] Use of copper-alloy plates or copper coated with mercury in order to trap gold from crushed ore pulp as it flows over the plates.

plate anemometer *See* pressure-plate anemometer.

plateau level [PETRO ENG] The peak production level reached by an oil field.

plate bearing test [ENG] Former method to estimate the bearing capacity of a soil; a rigid steel plate about 1 foot (30 centimeters) square was placed on the foundation level and then loaded until the foundation failed, as evidenced by rapid sinking of the plate.

plate-belt feeder *See* apron feeder.

plate cam [MECH ENG] A flat, open cam that imparts a sliding motion.

plate coil [MECH ENG] Heat-transfer device made from two metal sheets held together, one or both plates embossed to form passages between them for a heating or cooling medium to flow through. Also known as panel coil.

plate conveyor [MECH ENG] A conveyor with a series of steel plates as the carrying medium; each plate is a short trough, all slightly overlapped to form an articulated band, and attached to one center chain or to two side chains; the chains join rollers running on an angle-iron framework and transmit the drive from the driveheads, installed at intermediate points and sometimes also at the head or tail ends.

plate cut [BUILD] The cut made in a rafter to rest on the plate.

plate feeder *See* apron feeder.

plate-fin exchanger [MECH ENG] Heat-transfer device made up of a stack or layers, with each layer consisting of a corrugated fin between flat metal sheets sealed off on two sides by channels or bars to form passages for the flow of fluids.

plate girder [CIV ENG] A riveted or welded steel girder having a deep vertical web plate with a pair of angles along each edge to act as compression and tension flanges.

plate girder bridge [CIV ENG] A fixed bridge consisting, in its simplest form, of two flange plates welded to a web plate in the overall shape of an I.

plate glass [MATER] Flat, high-quality glass with plane, parallel surfaces.

platen [ENG] **1.** A flat plate against which something rests or is pressed. **2.** The rubber-covered roller of a typewriter against which paper is pressed when struck by the typebars. [MECH ENG] A flat surface for exchanging heat in a boiler or heat exchanger which may have extended heat transfer surfaces.

plate-shear test [ENG] A method used to get true shear data on a honeycomb core by bonding the core between two thick steel plates and subjecting the core to shear by displacing the plates relative to each other by loading in either tension or compression.

plate-type exchanger [MECH ENG] Heat-exchange device similar to a plate-and-frame filter press; fluids flow between the frame-held plates, transferring heat between them.

platform [MIN ENG] A wooden floor on the side of a gangway at the bottom of an inclined seam, to which the coal runs by gravity, and from which it is shoveled into mine cars.

platform balance [ENG] A weighing device with a flat plate mounted above a balanced beam.

platform blowing [ENG] Special technique for blow-molding large parts made of plastic without sagging of the part being formed.

platform conveyor [MECH ENG] A single- or double-strand conveyor with plates of steel or hardwood forming a continuous platform on which the loads are placed.

platform framing [BUILD] A construction method in which each floor is framed independently by nailing the horizontal framing member to the top of the wall studs.

platina [MET] A white brittle brass containing 75% zinc and 25% copper; used for jewelry.

plating [MET] Forming a thin, adherent layer of metal on an object. Also known as metal plating.

plating rack [MET] A fixture that holds, and conducts current to, a piece of work during electrodeposition.

platinoid [MET] **1.** Resembling or related to platinum. **2.** A copper-nickel-zinc alloy used for electrical resistance wire.

platinum [MET] A soft, ductile, malleable, grayish white noble metal with relatively high electric resistance; used in alloys, in electrical and electronic devices, and in jewelry.

platinum black [MET] Black-colored, finely divided metallic platinum; soluble in aqua regia; used as a catalyst, as an absorbent for gases (hydrogen, oxygen), and for gas ignition. Also known as platinum Mohr.

platinum-iridium alloy [MET] An alloy with 1–30% iridium; as concentration of iridium increases, so do hardness, chemical resistance, and melting point; used in jewelry, electrical contacts, and hypodermic needles.

platinum Mohr See platinum black.

platinum resistance thermometer [ENG] The basis of the International Practical Temperature Scale of 1968 from 259.35° to 630.74°C; used in industrial thermometers in the range 0 to 650°C; capable of high accuracy because platinum is noncorrosive, ductile, and nonvolatile, and can be obtained in a very pure state. Also known as Callendar's thermometer.

platinum-rhodium alloy [MET] An alloy with up to 40% rhodium; as concentration of rhodium increases, so do chemical resistance and hardness (although less hard than for platinum-iridium alloys); used as a catalyst to make nitric acid, in thermocouples, and in rayon spinnerets.

platinum sponge [MET] Porous, grayish-black mass of finely divided platinum; soluble in aqua regia; used as a catalyst, and for ignition of combustible gases.

Plattner's process [MET] A process for extracting gold in which a charge of gold-bearing pulp is placed in a revolving iron drum lined with lead, and a stream of chlorine gas is conducted through the pulp, producing chloride of gold, which is soluble in water.

playback [ENG ACOUS] Reproduction of a sound recording.

pleated cartridge [DES ENG] A filter cartridge made into a convoluted form that resembles the folds of an accordion.

plenum [ENG] A condition in which air pressure within an enclosed space is greater than that in the outside atmosphere.

plenum chamber [ENG] An enclosed space in which a plenum condition exists; air is forced into it for slow distribution through ducts.

plenum system [MECH ENG] A heating or air conditioning system in which air is forced through a plenum chamber for distribution to ducts.

pliers [DES ENG] A small instrument with two handles and two grasping jaws, usually long and roughened, working on a pivot; used for holding small objects and cutting, bending, and shaping wire.

plinth block See skirting block.

plot [CIV ENG] A measured piece of land.

plotter [ENG] A visual display or board on which a dependent variable is graphed by an automatically controlled pen or pencil as a function of one or more variables.

plotting board [ENG] The surface portion of a plotter, on which graphs are recorded. Also known as plotting table.

plotting table See plotting board.

plough [ENG] A groove cut lengthwise with the grain in a piece of wood.

plough [MIN ENG] **1.** A continuous mining machine in which cutting blades, moved over the face being worked, bite into the coal as they are pulled along and discharge it on an accompanying conveyor. **2.** A V-shaped scraper that presses against the return belt of a conveyor, removing coal and debris from it.

plough deflector [MIN ENG] A steel plate at the end of a cutter-loader to deflect cut coal onto the face conveyor.

ploughed-and-tongued joint See feather joint.

plowshare [DES ENG] The pointed part of a plow moldboard, which penetrates and cuts the soil first.

plow steel [MET] High-quality, high-strength steel with 0.5 to 0.95% carbon content, used for wire rope.

plug [MET] **1.** A rod or mandrel over which a pierced tube is forced, or that fills a tube as it is drawn through a die. **2.** A punch or mandrel over which a cup is drawn. **3.** A protruding portion of a die impression for forming a corresponding recess in the forging. **4.** A false bottom in a die. Also known as peg. [MIN ENG] A watertight seal in a shaft formed by removing the lining and inserting a concrete dam, or by placing a plug of clay over ordinary debris used to fill the shaft up to the location of the plug.

plug-and-feather hole [ENG] A hole drilled in quarries for the purpose of splitting a block of stone by the plug-and-feather method.

plug-and-feather method [MIN ENG] A method of breaking large quarry stones into smaller blocks; a row of holes is drilled in the stone along a line where the break is desired; a pair of feathers (semicircular cross-section rods) is inserted in each hole; a plug (steel wedge) is inserted between each feather pair; the plugs are hammered in succession until the stone fractures.

plug back [PETRO ENG] To place cement or a mechanical plug in a well bottom for the purpose of excluding bottom water, sidetracking, or producing from a formation already drilled through.

plug bit *See* noncoring bit.

plug cock *See* plug valve.

plug die *See* floating plug.

plug drawing [MET] Drawing tubing over a plug or mandrel and through a die simultaneously to reduce diameter and thickness and to produce a smooth symmetrical bore surface.

plug forming [ENG] Thermoforming process for plastics molding in which a plug or male mold is used to partially preform the part before forming is completed, using vacuum or pressure.

plug gage [DES ENG] A steel gage that is used to test the dimension of a hole; may be straight or tapered, plain or threaded, and of any cross-sectional shape.

plugging [ENG] The formation of a barrier (plug) of solid material in a process flow system, such as a pipe or reactor. [MIN ENG] *See* blinding. [PETRO ENG] The act or process of stopping the flow of water, oil, or gas in strata penetrated by a borehole or well so that fluid from one stratum will not escape into another or to the surface; especially the sealing up of a well that is dry and is to be abandoned.

plugging agent [PETRO ENG] A chemical used to plug or block off selected permeable zones of a reservoir formation; used during formation acidizing to direct the acid to the tighter (less permeable) zones; examples are viscous gels, suspensions of graded solids, and finely ground vegetable material.

plughole [MIN ENG] **1.** A passageway left open while an old portion of a mine is sealed off, to help maintain normal ventilation; it is sealed when the work is finished. **2.** A hole for an explosive charge or for a bolt.

plug meter [ENG] A variable-area flowmeter in which a tapered plug, located in an orifice and raised until the resulting opening is sufficient to handle the fluid flow, is used to measure the flow rate.

plug valve [DES ENG] A valve fitted with a plug that has a hole through which fluid flows and that is rotatable through 90° for operation in the open or closed position. Also known as plug cock.

plug weld [MET] A circular fusion weld made in the hole of a slotted lap or tee joint.

plumb [ENG] Pertaining to an object or structure in true vertical position as determined by a plumb bob.

plumb bob [ENG] A weight suspended on a string to indicate the direction of the vertical.

plumb bond [CIV ENG] A masonry bond in which corresponding joints (for example, on alternate courses) are aligned.

plumbing [CIV ENG] The system of pipes and fixtures concerned with the introduction, distribution, and disposal of water in a building.

plumb line [ENG] The string on which a plumb bob hangs.

plummet [ENG] A loose-fitting metal plug in a tapered rotameter tube which moves upward (or downward) with an increase (or decrease) in fluid flow rate upward through the tube. Also known as float.

plunge [ENG] **1.** To set the horizontal cross hair of a theodolite in the direction of a grade when establishing a grade between two points of known level. **2.** *See* transit.

plunge grinding [MECH ENG] Grinding in which the wheel moves radially toward the work.

plunger [DES ENG] A wooden shaft with a large rubber suction cup at the end, used to clear plumbing traps and waste outlets. [ENG] *See* force plug. [MECH ENG] The long rod or piston of a reciprocating pump.

plunger jig washer [MIN ENG] A machine for washing ore, coal, or stones in which water is forced alternately up or down by a plunger.

plunger lift [PETRO ENG] A method of lifting oil by using compressed gas to drive a free piston from the lower end of the tubing string to the surface.

plunger overtravel [PETRO ENG] Excessive upward or downward movement in a reciprocating-plunger-type sucker-rod oil-well pump.

plunger pump [MECH ENG] A reciprocating pump where the packing is on the stationary casing instead of the moving piston.

plunger-type instrument [ENG] Moving-iron instrument in which the pointer is attached to a long and specially shaped piece of iron that is drawn into or moved out of a coil carrying the current to be measured.

plus sieve [MET] That portion of a powder sample that is retained by a standard sieve of a specified number.

ply [MATER] A thin sheet of wood or other material bonded to one or more additional thin sheets, as in plywood.

plymetal [MATER] **1.** A material consisting of layers of dissimilar metals bonded together. **2.** Plywood faced with aluminum on both sides.

plywood [MATER] A material composed of thin sheets of wood glued together, with the grains of adjacent sheets oriented at right angles to each other.

PLZT *See* lead lanthanum zirconate titanate.

pneumatic [ENG] Pertaining to or operated by air or other gas.

pneumatic atomizer [MECH ENG] An atomizer that uses compressed air to produce drops in the diameter range of 5–100 micrometers.

pneumatic caisson [CIV ENG] A caisson having a chamber filled with compressed air at a pressure equal to the pressure of the water outside.

pneumatic controller [MECH ENG] A device for the mechanical movement of another device (such as a valve stem) whose action is controlled by variations in pneumatic pressure connected to the controller.

pneumatic control valve [MECH ENG] A valve in which the force of compressed air against a diaphragm is opposed by the force of a spring to control the area of the opening for a fluid stream.

pneumatic conveyor [MECH ENG] A conveyor which transports dry, free-flowing, granular material in suspension, or a cylindrical carrier, within a pipe or duct by means of a high-velocity airstream or by pressure of vacuum generated by an air compressor. Also known as air conveyor.

pneumatic drill [MECH ENG] Compressed-air drill worked by reciprocating piston, hammer action, or turbo drive.

pneumatic drilling [MECH ENG] Drilling a hole when using air or gas in lieu of conventional drilling fluid as the circulating medium; an adaptation of rotary drilling.

pneumatic filling [MIN ENG] A filling method using compressed air to blow filling material into the mined-out stope.

pneumatic hammer [MECH ENG] A hammer in which compressed air is utilized for producing the impacting blow.

pneumatic hoist *See* air hoist.

pneumatic injection [MIN ENG] A method for fighting underground coal fires, developed by the U.S. Bureau of Mines; this air-blowing technique involves the injection of incombustible mineral, like rock wool or dry sand, through 6-inch (15 centimeter) boreholes drilled from the surface to intersect underground passageways in the mines.

pneumatic lighting [MIN ENG] Lighting of underground chambers by a compressed-air turbomotor driving a small dynamo.

pneumatic loudspeaker [ENG ACOUS] A loudspeaker in which the acoustic ouput results from controlled variation of an airstream.

pneumatic method [MIN ENG] A method of flotation in which gas is introduced near the bottom of the flotation vessel.

pneumatic riveter [MECH ENG] A riveting machine having a rapidly reciprocating piston driven by compressed air.

pneumatic steelmaking [MET] Any steelmaking process which employs air or oxygen and for which all heat is derived from the initial heat content of the charge materials and from the thermal energy of the refining reactions.

pneumatic stowing [MIN ENG] A method of filling used mine cavities with crushed rock; which is forced by compressed air into the cavity.

pneumatic tank switcher [PETRO ENG] Pneumatic actuated valving system for oil-field tanks to shut off crude-oil flow to a filled tank and then to direct incoming crude flow to the next available empty tank.

pneumatic telemetering [ENG] The transmission of a pressure impulse by means of pneumatic pressure through a length of small-bore tubing; used for remote transmission of signals from primary process-unit sensing elements for pressure, temperature, flow rate, and so on.

pneumatic test [ENG] Pressure testing of a process vessel by the use of air pressure.

pneumatic weighing system [ENG] A system for weight measurement in which the load is detected by a nozzle and balanced by modulating the air pressure in an opposing capsule.

pocket [MIN ENG] A receptacle from which coal, ore, or waste is loaded into wagons or cars.

pod [AERO ENG] An enclosure, housing, or detachable container of some kind on an airplane or space vehicle, as an engine pod. [DES ENG] **1.** The socket for a bit in a brace. **2.** A straight groove in the barrel of a pod auger.

Poetsch process [MIN ENG] Shaft sinking in which brine at subzero temperature is circulated through boreholes to freeze running water through which a shaft or tunnel is to be driven, during development of a waterlogged mine.

Pohlé air lift pump [MECH ENG] A pistonless pump in which compressed air fills the annular space surrounding the uptake pipe and is free to enter the rising column at all points of its periphery.

poidometer [ENG] An automatic weighing device for use on belt conveyors.

point angle [DES ENG] The angle at the point or edge of a cutting tool.

pointer [ENG] The needle-shaped rod that moves over the scale of a meter.

pointing [CIV ENG] **1.** Finishing a mortar joint. **2.** Pressing mortar into a raked joint. [MATER] The material (mortar) used in pointing. [MET] Reducing the diameter and tapering a short length at the end of a rod, wire, or tube. Also known as metal pointing.

pointing trowel [ENG] A tool used to apply pointing to the joints between bricks.

point initiation [ENG] Application of the initial impulse from the detonator to a single point on the main charge surface; for a cylindrical charge this point is usually the center of one face.

point of control [IND ENG] Fraction defective in those lots that have a probability of .50 of acceptance according to a specific sampling acceptance plan.

point of frog [CIV ENG] The place of intersection of the gage lines of the main track and a turnout.

point of intersection [CIV ENG] The point at which two straight sections or tangents to a road curve or rail curve meet when extended.

point of switch [CIV ENG] That place in a track where a car passes from the main track to a turnout.

point of tangency [CIV ENG] The point at which a road curve or railway curve becomes straight or changes its curvature. Also known as tangent point.

point system [IND ENG] **1.** A system of job evaluation wherein job requirements are rated according to a scale of point values. **2.** A wage incentive plan based on points instead of man-minutes.

poison [MATER] A substance that in relatively small doses has an action that either destroys life or impairs seriously the functions of organs or tissues.

poison gas [MATER] A substance employed in chemical warfare to disable enemy troops; may be a gas, or a liquid or solid that gives off a gas. An example is mustard gas.

poke welding *See* push welding.

polarized ceramics [MATER] A substance, such as lead zirconate and barium titanate, having high electromechanical conversion efficiency and used as a transducer element in an ultrasonic system.

polarized meter [ENG] A meter having a zero-center scale, with the direction of deflection of the pointer depending on the polarity of the voltage or the direction of the current being measured.

polarized-vane ammeter [ENG] An ammeter of only moderate accuracy in which the current to be measured passes through a small coil, distorting the field of a circular permanent magnet, and an iron vane aligns itself with the axis of the distorted field, the deflection being roughly proportional to the current.

polar orbit [AERO ENG] A satellite orbit running north and south, so the satellite vehicle orbits over both the North Pole and the South Pole.

polar radiation pattern [ENG ACOUS] Diagram showing the strength of sound waves radiated from a loudspeaker in various directions in a given plane, or a similar response pattern for a microphone.

polar solvent [MATER] A solvent in whose molecules there is either a permanent separation of positive and negative charges, or the centers of positive and negative charges do not coincide; these solvents have high dielectric constants, are chemically active, and form coordinate covalent bonds; examples are alcohols and ketones.

polar timing diagram [MECH ENG] A diagram of the events of an engine cycle relative to crankshaft position.

polder [CIV ENG] Land reclaimed from the sea or other body of water by the construction of an embankment to restrain the water.

pole derrick *See* gin pole.

pole-dipole array [ENG] An electrode array used in a lateral search conducted during a resistivity or induced polarization survey, or in drill hole logging, in which one current electrode is placed at infinity while another current electrode and two potential electrodes in proximity are moved across the structure to be investigated.

pole lathe [MECH ENG] A simple lathe in which the work is rotated by a cord attached to a treadle.

pole-pole array [ENG] An electrode array, used in lateral search or in logging, in which one current electrode and the other potential electrode are kept in proximity and traversed across the structure.

polestar recorder [ENG] An instrument used to determine approximately the amount of cloudiness during the dark hours; consists of a fixed long-focus camera positioned so that Polaris is permanently within its field of view; the apparent motion of the star appears as a circular arc on the photograph and is interrupted as clouds come between the star and the camera.

poling [MET] A technique used in the refining of copper that consists of the thrusting of green-wood poles into the molten metal in order to generate the reducing gases that react with the oxides in the metal. [MIN ENG] The act or process of temporarily protecting the face of a level, drift, or cut by driving poles or planks along the sides of the yet unbroken ground.

poling back [MIN ENG] Carrying out excavation behind timbering already in place.

poling board [CIV ENG] A timber plank driven into soft soil to support the sides of an excavation.

polish [MATER] A powder, liquid, or semiliquid used to give smoothness, surface protection, or decoration to finishes; for example, finely ground red oxide (rouge) is used to polish plate glass, mirror backs, and optical glass; solvent-wax liquids and pastes are used to protect and enhance leather and wood surfaces; nitrocellulose lacquers are used to paint finger- and toenails.

polished-joint hanger [PETRO ENG] A type of tubing hanger that is slipped over or assembled around the top tubing joint in an oil well tubing string.

polishing [MECH ENG] Smoothing and brightening a surface such as a metal or a rock through the use of abrasive materials.

polishing roll [MECH ENG] A roll or series of rolls on a plastics mold; has highly polished chrome-plated surfaces; used to produce a smooth surface on a plastic sheet as it is extruded.

polishing wheel [DES ENG] An abrasive wheel used for polishing.

polyblend [MATER] A mechanical mixture of two or more polymers, such as polystyrene and rubber.

polycrystal [MATER] A polycrystalline solid.

polycrystalline [MATER] **1.** Pertaining to a material composed of aggregates of individual crystals. **2.** Characterized by variously oriented crystals.

polyester film [MATER] Thin film made of polyester resin; used for packaging food and other products.

polyester laminate [MATER] Glass fabric or fiber mat impregnated with a polyester resin slurry, and cured; used to make sheets, bars, and structural shapes.

polyester-reinforced urethane [MATER] A poromeric material which may have a urethane impregnation or a silicone coating; used for shoe uppers and as a substitute for industrial leathers.

polygonal method [MIN ENG] A method of estimating ore reserves in which it is assumed that each drill hole has an area of influence extending halfway to the neighboring drill holes.

polygraph *See* lie detector.

polyliner [ENG] A perforated sleeve with longitudinal ribs that is used inside the cylinder of an injection-molding machine.

polymer gasoline [MATER] A product of polymerization of normally gaseous hydrocarbons to form high-octane liquid hydrocarbons boiling in the gasoline range.

polymer paint [MATER] A paint made of acrylic resin or vinyl resin, or a combination of both resins, in a liquid form with water as the base; it spreads out in a layer, and the water evaporates to leave a continuous, flexible, and waterproof film of plastic.

polymer plastic [MATER] The product of a high polymer with or without additives, such as plasticizers, autooxidants, colorants, or fillers; can be sprayed, shaped, molded, extruded, cast or foamed, depending on whether it is thermoplastic or thermosetting.

polyolefin fiber [MATER] Continuous-strand fiber made from a polyolefin.

polyphase meter [ENG] An instrument which measures some electrical quantity, such as power factor or power, in a polyphase circuit.

polyphase wattmeter [ENG] An instrument that measures electric power in a polyphase circuit.

polysulfone resin [MATER] A thermoplastic polymer containing the sulfone linkage

$(O=S=O)$; has exceptional high temperature, low-creep, and arc-resistance properties, and is self-extinguishing.

polyunsaturated fat [MATER] A fat or oil based on fatty acids such as linoleic or linolenic acids which have two or more double bonds in each molecule; corn oil and safflower oil are examples.

polyurethane foam [MATER] A solid or spongy cellular material produced by the reaction of a polyester (such as glycerin) with a diisocyanate (such as toluene diisocyanate) while carbon dioxide is liberated by the reaction of a carboxyl with the isocyanate; used for thermal insulation, soundproofing, and padding.

PONA analysis [ENG] ASTM analysis of paraffins (P), olefins (O), naphthenes (N) and aromatics (A) in gasolines.

ponding [BUILD] An accumulation of water on a flat roof because of clogged or inadequate drains. [CIV ENG] The impoundment of stream water to form a pond.

pontianak gum [MATER] A grayish-white copal from various jelutongs of the genus *Dyera*, indigenous to Malacca and Borneo; used in rubber manufacture, chewing gum, adhesives, paints, and varnishes. Also known as jelutong.

pontoon [AERO ENG] A float on an airplane.

pontoon bridge [CIV ENG] A fixed floating bridge supported by pontoons.

pontoon-tank roof [ENG] A type of floating tank roof, supported by buoyant floats on the liquid surface of a tank; the roof rises and falls with the liquid level in the tank; used to minimize vapor space above the liquid, thus reducing vapor losses during tank filling and emptying.

pony set [MIN ENG] A small timber set or frame incorporated in the main sets of a haulage level to accommodate an ore chute or other equipment from above or below.

pony truss [CIV ENG] A truss too low to permit overhead braces.

pool [CIV ENG] A body of water contained in a reservoir, by a dam, or by the gates of a lock. [MIN ENG] **1.** To wedge for splitting in quarrying or mining. **2.** To undermine or undercut.

pop [MIN ENG] A drill hole blasted to reduce larger pieces of rock or to trim a working face. Also known as pop hole; pop shot.

pop action [MECH ENG] The action of a safety valve as it opens under steam pressure when the valve disk is lifted off its seat.

pop hole *See* pop.

poppet [CIV ENG] One of the timber and steel structures supporting the fore and aft ends of a ship for launching from sliding ways. [DES ENG] A spring-loaded ball engaging a notch; a ball latch. [MIN ENG] A pulley frame or the headgear over a mine shaft.

poppet valve [MECH ENG] A cam-operated or spring-loaded reciprocating-engine mushroom-type valve used for control of admission and exhaust of working fluid; the direction of movement is at right angles to the plane of its seat.

popping [MIN ENG] Exploding a stick of dynamite on a boulder so as to break it for easy removal from a quarry or opencast mine.

popping pressure [MECH ENG] In compressible fluid service, the inlet pressure at which a safety valve disk opens.

poppy oil [MATER] Golden-yellow drying oil with pleasant taste and aroma; soluble in carbon disulfide, ether, and chloroform; expressed from poppy seeds, especially of the opium poppy; used as a food oil; in artist's colors, soaps, varnishes, and lubricants. Also known as poppy-seed oil.

poppy-seed oil *See* poppy oil.

porcelain [MATER] A high-grade ceramic ware characterized by high strength, a white color, very low absorption, good translucency, and a hard glaze. Also known as European porcelain; hard paste porcelain; true porcelain.

porcelain cement [MATER] A cement for bonding porcelain to porcelain, such as a mixture of gutta-percha and shellac.

porcelain clay [MATER] A clay suitable for use in the manufacture of porcelain, specifically kaolin. Also known as porcelain earth.

porcelain earth *See* porcelain clay.

porcelain enamel *See* vitreous enamel.

porcelain insulator [MATER] An electrical insulator made from porcelain; the porcelain is often made in a one-fire process, the glaze being applied to the green or unfired ware, in contrast to the two-fire process used in making ordinary porcelain.

porcupine boiler [MECH ENG] A boiler having dead end tubes projecting from a vertical shell.

pore [MET] A minute cavity in a powder compact, metal casting, or electroplated coating.

pore diameter [DES ENG] The average or effective diameter of the openings in a membrane, screen, or other porous material.

porosimeter [ENG] Laboratory compressed-gas device used for measurement of the porosity of reservoir rocks.

porosity feet [PETRO ENG] Reservoir porosity fraction multiplied by net pay in feet, where porosity fraction is the portion of the reservoir that is porous, and net pay is the depth and areal extent of the hydrocarbons-containing reservoir.

porous [MATER] **1.** Filled with pores. **2.** Capable of absorbing liquids.

porous bearing [DES ENG] A bearing made from sintered metal powder impregnated with oil by a vacuum treatment.

porous carbon [MATER] Plates, tubes, or disks of uniform carbon particles pressed together without a binder; used for the filtration of corrosive liquids and gases.

porous graphite [MATER] Plates, tubes, or disks of uniform graphite particles pressed together without a binder; more resistant to oxidation but lower in strength than porous carbon.

porous metals [MET] Metals, made by powder metallurgy, having uniformly distributed controlled pore sizes, in the form of sheets, tubes, and shapes; used for filtering liquids and gases at elevated temperatures.

porous mold [ENG] A plastic-forming mold made from bonded or fused aggregates (such as powdered metal or coarse pellets) so that the resulting mass contains numerous open interstices through which air or liquids can pass.

porous reservoir model [PETRO ENG] Scaled laboratory model of porous reservoir used for the study of reservoir areal waterflood efficiencies.

porous wheel [DES ENG] A grinding wheel having a porous structure and a vitrified or resinoid bond.

porpoise oil [MATER] A pale-yellow fatty oil obtained from blubber of the brown porpoise; soluble in ether, benzene, carbon disulfide, and chloroform; used as a lubricant, leather dressing, and illuminating oil, and in soap stock.

port [ENG] The side of a ship or airplane on the left of a person facing forward. [ENG ACOUS] An opening in a bass-reflex enclosure for a loudspeaker, designed and positioned to improve bass response.

portable [ENG] Capable of being easily and conveniently transported.

portal [ENG] A redundant frame consisting of two uprights connected by a third member at the top. [MIN ENG] **1.** An entrance to a mine. **2.** The rock face at which a tunnel is started.

portal crane [MECH ENG] A jib crane carried on a four-legged portal built to run on rails.

Porterfield [ENG] A circular opening in the side of a ship or airplane, usually serving as a window and containing one or more panes of glass.

porthole [DES ENG] The opening or passageway connecting the inside of a bit or core barrel to the outside and through which the circulating medium is discharged.

porthole die [MET] An extrusion die having two or more sections in which metal is extruded separately in each section and welded before leaving the die to form intricate hollow shapes.

portland cement [MATER] A hydraulic cement resembling portland stone when hardened; made of pulverized, calcined argillaceous and calcareous materials; the proper name for ordinary cement.

portland-pozzolana cement [MATER] Portland cement to which pozzolana has been added, in the amount of about 20%, to reduce the liability of leaching.

port of entry [CIV ENG] A location for clearance of foreign goods and citizens through a customhouse.

position blocks [MIN ENG] Unproved mining claims that are in a position to contain a lode if the lode continues in the direction in which it has been proved in other claims.

positioned weld [MET] A weld made in a joint in which members have been positioned to facilitate welding.

position indicator [ENG] An electromechanical dead-reckoning computer, either an air-position indicator or a ground-position indicator.

positioning [MECH ENG] A tooling function concerned with manipulating the workpiece in relationship to the working tools.

position sensor [ENG] A device for measuring a position and converting this measurement into a form convenient for transmission. Also known as position transducer.

position telemetering [ENG] A variation of voltage telemetering in which the system transmits the measurand by positioning a variable resistor or other component in a bridge circuit so as to produce relative magnitudes of electrical quantities or phase relationships.

position transducer See position sensor.

positive clutch [MECH ENG] A clutch designed to transmit torque without slip.

positive derail [MIN ENG] A device installed in or on a mine track to derail runaway cars or trips.

positive-displacement compressor [MECH ENG] A compressor that confines successive volumes of fluid within a closed space in which the pressure of the fluid is increased as the volume of the closed space is decreased.

positive-displacement meter [ENG] A fluid quantity meter that separates and captures definite volumes of the flowing stream one after another and passes them downstream, while counting the number of operations.

positive-displacement pump [MECH ENG] A pump in which a measured quantity of liquid is entrapped in a space, its pressure is raised, and then it is delivered; for example, a reciprocating piston-cylinder or rotary-vane, gear, or lobe mechanism.

positive draft [MECH ENG] Pressure in the furnace or gas passages of a steam-generating unit which is greater than atmospheric pressure.

positive drive belt See timing belt.

positive mold [ENG] A plastics mold designed to trap all of the molding resin when the mold closes.

positive motion [MECH ENG] Motion transferred from one machine part to another without slippage.

positive ore [MIN ENG] Ore exposed on four sides in blocks of a size variously prescribed.

positron camera [ENG] An instrument that uses photomultiplier tubes in combination with scintillation counters to detect oppositely directed gamma-ray pairs resulting from the annihilation with electrons of positrons emitted by short-lived radioisotopes used as tracers in the human body.

possible ore [MIN ENG] A class of ore whose existence is a reasonable possibility, based upon geologic-mineralogic relationships and the extent of ore bodies already developed. Also known as extension ore.

possible reserves [PETRO ENG] Primary petroleum reserves that may exist, but available data do not confirm their presence.

post [CIV ENG] **1.** A vertical support such as a pillar, upright, or fence stake. **2.** A pole used as a boundary marker. [MIN ENG] **1.** A mine timber, or any upright timber, more commonly the uprights which support the roof crosspieces. **2.** The support fastened between the roof and floor of a coal seam, used with certain types of mining machines or augers. **3.** A pillar of coal or ore.

post-and-beam construction [BUILD] A type of wall construction using posts instead of studs.

post brake [MECH ENG] A brake occasionally fitted on a steam winder or haulage, and consisting of two upright posts mounted on either side of the drum that operate on brake paths bolted to the drum cheeks.

postcard paper [MATER] Lightweight Bristol board, made from soda and sulfite pulps, that has a smooth, firm surface for writing with pen or pencil, or for ordinary printing.

postcure bonding [ENG] A method of postcuring at elevated temperatures of parts previously subjected to autoclave or press in order to obtain higher heat-resistant properties of the adhesive bond.

post drill [ENG] An auger or drill supported by a post.

postemphasis See deemphasis.

postequalization See deemphasis.

poster board [MATER] A stiff cardboard used for show cards, posters, display advertising, and signs; it may be white or colored on one side.

poster paint [MATER] A water paint with a gum binder, which is brilliant, opaque, and fast-drying, and usually sold in jars. Also known as show-card color.

poster paper [MATER] A strong, waterproofed paper used for billboard poster work; it is white or colored with nonfading pigments, and does not curl when paste is applied to it.

postforming [ENG] Forming, bonding, or shaping of heated, flexible thermoset laminates before the final thermoset reaction has occurred; upon cooling, the formed shape is held.

postheating [MET] Application of heat after thermal spraying, brazing, or welding to control cooling rate.

posthole [CIV ENG] A hole bored in the ground to hold a fence post.

posttensioning [ENG] Compressing of cast concrete beams or other structural members to impart the characteristics of prestressed concrete.

postweld interval [MET] The elapsed time between the end of a resistance welding current and the start of hold time.

potash regulations [MIN ENG] Rules governing the prospecting and exploitation of land containing potash.

pot clay [MATER] Refractory clay used to make the pots in which glass is produced.

pot die forming [MECH ENG] Forming sheet or plate metal through a hollow die by the application of pressure which causes the workpiece to assume the contour of the die.

pot earth See potter's clay.

potential flow analyzer See electrolytic tank.

potential ore [MIN ENG] **1.** As yet undiscovered mineral deposits. **2.** A known mineral deposit for which recovery is not yet economically feasible.

potentiometer [ENG] A device for the measurement of an electromotive force by comparison with a known potential difference.

potentiometric model study [PETRO ENG] Analogic electrical-resistance model (electrolyte in a contoured container) of an underground reservoir based on Darcy's law that is, the steady-state flow of liquids through porous media is analogous to the flow of current through an electrical conductor; used to predict conditions in gas-condensate oil reservoirs.

potentiostat [ENG] An automatic laboratory instrument that controls the potential of a working electrode to within certain limits during coulometric (electrochemical reaction) titrations.

pot furnace [ENG] **1.** A furnace containing several pots in which glass is melted. **2.** A furnace in which the charge is contained in a pot or crucible.

pothole [CIV ENG] A pot-shaped hole in a pavement surface.

potometer [ENG] A device for measuring transpiration, consisting of a small vessel containing water and sealed so that the only escape of moisture is by transpiration from a leaf, twig, or small plant with its cut end inserted in the water.

potomology [CIV ENG] The systematic study of the factors affecting river channels to provide the basis for predictions of the effects of proposed engineering works on channel characteristics.

pot plunger [ENG] A plunger used to force softened plastic molding material into the closed cavity of a transfer mold.

potter's clay [MATER] A plastic clay, free from iron and devoid of fissility, suitable for modeling or pottery making or adapted for use on a potter's wheel. Also known as argil; pot earth; potter's earth.

potter's earth *See* potter's clay.

potter's wheel [ENG] A revolving horizontal disk that turns when a treadle is operated; used to shape clay by hand.

pottery [MATER] Objects made of clay which may be nonvitreous, porous, opaque, and glazed or unglazed; also included is earthenware such as stoneware.

pounce [MATER] Pumice in the form of a very fine powder, used for preparing parchment and tracing cloth.

pouncing paper [MATER] Paper coated with pumice for polishing felt hats.

pounds per square inch differential [ENG] The difference in pressure between two points in a fluid-flow system, measured in pounds per square inch. Abbreviated psid.

pour depressant *See* pour-point depressant.

pouring [MET] Transferring molten metal to a mold or ladle.

pouring rope *See* asbestos joint runner.

pour point [MET] Temperature at which a molten alloy is cast.

pour-point depressant [MATER] An additive that lowers the pour point of a wax-containing petroleum-base lubricating oil by reducing the tendency of the wax to solidify. Also known as pour depressant; pour-point inhibitor.

pour-point inhibitor *See* pour-point depressant.

pour reversion [MATER] The difference between the original ASTM pour point of a lubricating oil and the relatively high solidification temperature observed in the field.

pour stability [MATER] Ability of a pour-depressant-treated petroleum lubricating oil to maintain its original ASTM pour point at low temperatures approximating winter conditions.

pour test [ENG] The chilling of a liquid under specified test conditions to determine the ASTM pour point.

powder [MATER] **1.** A general term for explosives. **2.** A loose grouping or aggregation of solid particles, usually smaller than 1000 micrometers.

powder box [MIN ENG] A wooden box used by miners to store explosive powder and blasting caps.

powder clutch [MECH ENG] A type of electromagnetic disk clutch in which the space between the clutch members is filled with dry, finely divided magnetic particles; application of a magnetic field coalesces the particles, creating friction forces between clutch members.

powder flowmeter [ENG] A device used to measure the flow rate of a metal powder.

powder house [CIV ENG] A magazine for the temporary storage of explosives.

powder insulation [MATER] Thermal insulation material made up of a finely divided solid held between two surfaces (one hot, one cold); the powder reduces both convection and radiation heat flow between the surfaces.

powder keg [ENG] A small metal keg for black blasting powder.

powder lubricant [MET] An agent mixed with a powder metal to facilitate formation and ejection of a compact.

powderman [MIN ENG] A person in charge of explosives in an operation of any nature requiring their use.

powder metal [MET] Finely divided particles of a metal.

powder metallurgy [MET] The production of massive materials and shaped objects by pressing, binding, and sintering powdered metal.

powder mine [MIN ENG] An excavation filled with powder for the purpose of blasting rocks.

powder-moisture test [ENG] Determination of moisture in a propellant by drying under prescribed conditions; expressed as percentage by weight.

powder molding [ENG] Generic term for plastics-molding techniques to produce objects of varying sizes and shapes by melting polyethylene powder, usually against the heated inside of a mold.

powder train [ENG] **1.** Train, usually of compressed black powder, used to obtain time action in older fuse types. **2.** Train of explosives laid out for destruction by burning.

power-actuated pressure relief valve [MECH ENG] A pressure relief valve connected to and controlled by a device which utilizes a separate energy source.

power brake [MECH ENG] An automotive brake with engine-intake-manifold vacuum used to amplify the atmospheric pressure on a piston operated by movement of the brake pedal.

power car [AERO ENG] A suspended structure on an airship that houses an engine. [MECH ENG] **1.** A railroad car with equipment for furnishing heat and electric power to a train. **2.** A railroad car with controls, which can be operated by itself or as part of a train.

power control valve [MECH ENG] A safety relief device operated by a power-driven mechanism rather than by pressure.

power dam [CIV ENG] A dam designed to raise the level of a stream to create or concentrate hydrostatic head for power purposes.

power-density spectrum See frequency spectrum.

power drill [MECH ENG] A motor-driven drilling machine.

power-driven [MECH ENG] Of a component or piece of equipment, moved, rotated, or operated by electrical or mechanical energy, as in a power-driven fan or power-driven turret.

power equation [MIN ENG] The relationship indicating that the natural ventilating power plus the power required to force air through a mine is equal to the power used in lifting water out of the mine plus the power lost in the kinetic energy of air leaving the mine plus the power converted to heat in overcoming friction.

power-factor meter [ENG] A direct-reading instrument for measuring power factor.

power grizzly [MIN ENG] Power-operated machine for removing dirt and fine particles from ore before it is crushed.

power loader [MIN ENG] Power-operated machine for loading ore, coal, or other material into a car, conveyor, or other collector.

power meter See electric power meter.

power oil [MATER] Fluid used to actuate (power) hydraulic pumps, motors, and other power-type equipment; can be based on petroleum oils or synthetic materials.

power package [MECH ENG] A complete engine and its accessories, designed as a single unit for quick installation or removal.

power plant [MECH ENG] Any unit that converts some form of energy into electrical energy, such as a hydroelectric or steam-generating station, a diesel-electric engine in a locomotive, or a nuclear power plant. Also known as electric power plant.

power saw [MECH ENG] A power-operated woodworking saw, such as a bench or circular saw.

power shovel [MECH ENG] A power-operated shovel that carries a short boom on which rides a movable dipper stick carrying an open-topped bucket; used to excavate and remove debris.

power-shovel mining [MIN ENG] A technique utilizing power shovels to mine ores by mining or stripping and taking away overburden.

power spectrum See frequency spectrum.

power station See generating station.

power steering [MECH ENG] A steering control system for a propelled vehicle in which an auxiliary power source assists the driver by providing the major force required to direct the road wheels.

power stroke [MECH ENG] The stroke in an engine during which pressure is applied to the piston by expanding steam or gases.

power train [MECH ENG] The part of a vehicle connecting the engine to propeller or driven axle; may include drive shaft, clutch, transmission, and differential gear.

pozzolan [MATER] Cement made by mixing and grinding together slaked lime and pozzolan without burning; sometimes used for concrete not exposed to the air.

Pratt truss [CIV ENG] A truss having both vertical and diagonal members between the upper and lower chords, with the diagonals sloped toward the center.

preassembled [ENG] Assembled beforehand.

prebreaker [MECH ENG] Device used to break down large masses of solids prior to feeding them to a crushing or grinding device.

precast concrete [MATER] Concrete components which are cast and partly matured in a factory or on the site before being lifted into their final position on a structure.

precharge [MET] The pressure introduced into the cavity of a mold before forming a part.

prechlorination [CIV ENG] Chlorination of water before filtration.

precious metal [MET] A relatively scarce, valuable metal, such as gold, silver, and members of the platinum group.

precipitation gage [ENG] Any device that measures the amount of precipitation; principally, a rain gage or snow gage.

precipitator See electrostatic precipitator.

precision block See gage block.

precision casting [MET] A metal casting of accurately reproducible dimensions.

precision depth recorder [ENG] A machine that plots sonar depth soundings on electrosensitive paper; can plot variations in depth over a range of 400 fathoms (730 meters) on a paper 18.85 inches (47.9 centimeters) wide. Abbreviated PDR. Also known as precision graphic recorder (PGR).

precision graphic recorder *See* precision depth recorder.

precision grinding [MECH ENG] Machine grinding to specified dimensions and low tolerances.

precoat [MET] In casting, thin coating of refractory slurry applied to an expendable wax or plastic pattern as a base for the application of the main slurry.

precoat filter [ENG] A device designed to filter solid particles from a liquid-solid slurry after a precoat of builtup solid material (filter aid or filtered solid) has been applied to the inner surface of the filter medium.

precoating [ENG] The depositing of an inert material, such as filter aid, onto the filter medium prior to the filtration of suspended solids from a solid-liquid slurry.

precombustion chamber [MECH ENG] A small chamber before the main combustion space of a turbine or reciprocating engine in which combustion is initiated.

precooler [MECH ENG] A device for reducing the temperature of a working fluid before it is used by a machine.

preferential shop [IND ENG] An establishment in which preference is given to union members in hiring, layoffs, and dismissals, with the understanding that nonunion workers may be employed without being required to join the union when the union cannot supply workers.

prefilter [ENG] Filter used to remove gross solid contaminants before the liquid stream enters a separator-filter.

preform [ENG ACOUS] The small slab of record stock material that is loaded into a press to be formed into a disk recording. Also known as biscuit (deprecated usage).

preforming [MET] 1. Initial pressing of a powder metal to form a compact. 2. Preliminary shaping of a refractory metal compact after presintering.

preheat current [MET] Current impulses which occur prior to and apart from the electric current in a resistance welding process.

preheater [MECH ENG] A device for preliminary heating of a material, substance, or fluid that will undergo further use or treatment by heating.

preheat roll [ENG] In plastic-extrusion coating, the heated roll between the pressure roll and the unwind roll; used to heat the substrate before it is coated.

preignition [MECH ENG] Ignition of the charge in the cylinder of an internal combustion engine before ignition by the spark.

preimpregnation [ENG] The mixing of a plastic resin with reinforcing material or substrate before molding takes place.

preloading [ENG] For back-pressure-control gas valves, a weight or spring device to control the gas pressure at which the valve will open or close.

premium motor oil [MATER] A lubricating oil with improved oxidation stability and having corrosion-preventive and detergent properties; used in internal combustion engines operating under severe conditions.

premix [ENG] In plastics molding, materials in which the resin, reinforcement, extenders, fillers, and so on have been premixed before molding.

premix gas burner [ENG] Fuel (gas or oil) burner in which fuel and air are premixed prior to ignition in the combustion chamber.

preparatory work [MIN ENG] Various excavations within a deposit so that actual mining can begin; includes inclines, drives between levels, crosscuts, and chutes.

preplastication [ENG] Premelting of injection-molding powders in a chamber separate from the injection cylinder.

prepolymer molding [ENG] A urethane-foam-producing system in which a portion of the polyol is prereacted with the isocyanate to form a liquid prepolymer with a pumpable viscosity; when combined with a second blend containing more polyol, catalyst, or blowing agent, the two components react and a foamed plastic results.

prepreg [ENG] A reinforced-plastics term for the reinforcing material that contains or is combined with the full complement of resin before the molding operation.

present value [MIN ENG] The sum of money which, if expended on a mine for purchase, development, and equipment, would produce over the life of the mine a return of the original investment plus a commensurate profit.

present-worth factor *See* discount factor.

preservative [MATER] A chemical added to foodstuffs to prevent oxidation, fermentation, or other deterioration, usually by inhibiting the growth of bacteria.

preset guidance [ENG] Guidance in which a predetermined path is set into the guidance mechanism of a craft, drone, or missile and is not altered after launching.

presintering [MET] Heating a compact to a temperature lower than the final sintering temperature to facilitate handling or to remove a binder or lubricant.

press [MECH ENG] Any of various machines by which pressure is applied to a workpiece, by which a material is cut or shaped under pressure, by which a substance is compressed, or by which liquid is expressed.

press bonding   [ENG] A method of bonding structures or materials through the application of pressure by a platen press or other tool.

press drip   [MATER] Oil that drips from the wax press after pressed petroleum distillate has been removed.

pressed brick   [MATER] Brick subjected to pressure before burning to eliminate imperfections of shape and texture.

pressed density   [MET] The density of a metal powder compact before sintering.

pressed distillate   [MATER] Oil recovered when refinery paraffin distillate is pressed to separate the liquid from the solid wax.

pressed glass   [MATER] Glass shaped by being poured into a mold under pressure or pressed into a mold in a plastic state.

pressed loading   [ENG] A loading operation in which bulk material, such as an explosive in granular form, is reduced in volume by the application of pressure.

press fit   [ENG] An interference or force fit assembled through the use of a press. Also known as force fit.

press forging   [MET] Forging hot metal between dies in a press.

pressing   [ENG ACOUS] A phonograph record produced in a record-molding press from a master or stamper.   [MET] **1.** Shallow-drawing metal sheet or plate. **2.** Using compressive force to form a metal powder compact.

press polish   [ENG] High-sheen finish on plastic sheet stock produced by contact with a smooth metal under heat and pressure.

press slide   [MECH ENG] The reciprocating member of a power press on which the punch and upper die are fastened.

pressure altimeter   [ENG] A highly refined aneroid barometer that precisely measures the pressure of the air at the altitude an aircraft is flying, and converts the pressure measurement to an indication of height above sea level according to a standard pressure-altitude relationship. Also known as barometric altimeter.

pressure angle   [MECH ENG] The angle that the line of force makes with a line at right angles to the center line of two gears at the pitch points.

pressure bag   [ENG] A bag made of rubber, plastic, or other impermeable material that provides a flexible barrier between the pressure medium and the part being bonded.

pressure bar   [MECH ENG] A bar that holds the edge of a metal sheet during press operations, such as punching, stamp-

ing, or forming, and prevents the sheet from buckling or becoming crimped.

pressure block   [MIN ENG] The pressure on pillars, walls, and other supports in a mine caused by removal of surrounding formations from masses of rocks or by natural geological formations.

pressure bomb   [PETRO ENG] Pipe-and-valve device used to capture downhole pressurized gas samples from oil wells; used to measure downhole pressure.

pressure bulb   [CIV ENG] The zone in a loaded soil mass bounded by an arbitrarily selected isobar of stress.

pressure bump   [MIN ENG] Sudden failure of a coal pillar overloaded by the weight of the rock above it.

pressure burst   [MIN ENG] A rockburst produced under stresses exceeding the elastic strength of the rock.

pressure carburetor   See injection carburetor.

pressure casting   [MET] Making castings of molten or plastic metal in metal molds under applied pressures.

pressure chamber   [ENG] A chamber in which an artificial environment is established at low or high pressures to test equipment under simulated conditions of operation.   [MIN ENG] An enclosed space that seals off a part of a mine and in which the air pressure can be raised or lowered.

pressure-containing member   [MECH ENG] The part of a pressure-relieving device which is in direct contact with the pressurized medium in the vessel being protected.

pressure control   [ENG] Any device or system able to maintain, raise, or lower the pressure in a vessel or processing system as desired.

pressure cooker   [ENG] An autoclave designed for high-temperature cooking.

pressure decline   [PETRO ENG] The loss or decline in reservoir pressure resulting from pressure drawdown during the production of gas or oil. Also known as pressure depletion.

pressure deflection   [ENG] In a Bourdon or bellows-type pressure gage, the deflection or movement of the primary sensing element when pressured by the fluid being measured.

pressure depletion   See pressure decline.

pressure distillate   [MATER] Light, gasoline-bearing distillate product from petroleum-refinery pressure stills; the product is cracked, as contrasted to virgin or straight-run stock; includes pressure naphtha.

pressure distribution   [PETRO ENG] The relative pressures (pressure gradients) between various portions of a producing reservoir zone; lowest pressures are nearest the producing wellbores.

pressure drawdown [PETRO ENG] The drop in reservoir pressure related to the withdrawal of gas from a producing well; for low-permeability formations, pressures near the wellbores can be much lower than in the main part of the reservoir; leads to pressure decline in the reservoir and ultimate pressure depletion.

pressure-drop manometer [ENG] Manometer device (liquid-filled U tube) open at both ends, each end connected by tubing to a different location in a flow system (such as fluid- or gas-carrying pipe) to measure the drop in system pressure between the two points.

pressure dye test [ENG] A leak detection method in which a pressure vessel is filled with liquid dye and is pressurized under water to make possible leakage paths visible.

pressure elements [ENG] Those portions of a pressure-measurement gage which are moved or temporarily deformed by the gas or liquid of the system to which the gage is connected; the amount of movement or deformation is proportional to the pressure and is indicated by the position of a pointer or movable needle.

pressure fan [MIN ENG] A fan that forces fresh air into a mine as distinguished from one that exhausts air from the mine.

pressure forming [ENG] A plastics thermoforming process using pressure to push the plastic sheet to be formed against the mold surface, as opposed to using vacuum to suck the sheet flat against the mold.

pressure gage [ENG] An instrument having metallic sensing element (as in a Bourdon pressure gage or aneroid barometer) or a piezoelectric crystal (as in a quartz pressure gage) to measure pressure.

pressure hydrophone [ENG ACOUS] A pressure microphone that responds to waterborne sound waves.

pressure interface [PETRO ENG] The interrelation of several individual reservoir pressures whose productions are supported by water influx from a common aquifer; pressure depletion from withdrawal of oil from one reservoir will affect the position of the common aquifer and thus affect the pressures and gas-oil or water-oil contacts in the other reservoirs.

pressure maintenance [PETRO ENG] The maintenance of gas pressure in a reservoir by an active water drive, water injection, gas injection, or a combination of the foregoing.

pressure measurement [ENG] Measurement of the internal forces of a process vessel, tank, or piping caused by pressurized gas or liquid; can be for a static or dynamic pressure, in English or metric units,

either absolute (total) or gage (absolute minus atmospheric) pressure.

pressure microphone [ENG ACOUS] A microphone whose output varies with the instantaneous pressure produced by a sound wave acting on a diaphragm; examples are capacitor, carbon, crystal, and dynamic microphones.

pressure naphtha [MATER] Petroleum naphtha made by cracking, as contrasted to virgin or straight-run naphtha; a special grade of pressure distillate.

pressure pad [ENG] A steel reinforcement in the face of a plastics mold to help the land absorb the closing pressure. [ENG ACOUS] A felt pad mounted on a spring arm, used to hold magnetic tape in close contact with the head on some tape recorders.

pressure pillow [ENG] A mechanical-hydraulic snow gage consisting of a circular rubber or metal pillow filled with a solution of antifreeze and water, and containing either a pressure transducer or a riser pipe to record increase in pressure of the snow.

pressure plate [MECH ENG] The part of an automobile disk clutch that presses against the flywheel.

pressure-plate anemometer [ENG] An anemometer which measures wind speed in terms of the drag which the wind exerts on a solid body; may be classified according to the means by which the wind drag is measured. Also known as plate anemometer.

pressure radius [PETRO ENG] The effective radius of increased reservoir pressure surrounding a water-injection well.

pressure rating [ENG] The operating (allowable) internal pressure of a vessel, tank, or piping used to hold or transport liquids or gases.

pressure-regulating valve [ENG] A valve that releases or holds process-system pressure (that is, opens or closes) either by preset spring tension or by actuation by a valve controller to assume any desired position between full open and full closed.

pressure regulator [ENG] Open-close device used on the vent of a closed, gas-pressured system to maintain the system pressure within a specified range.

pressure relief [ENG] A valve or other mechanical device (such as a rupture disk) that eliminates system overpressure by allowing the controlled or emergency escape of liquid or gas from a pressured system.

pressure relief device [MECH ENG] 1. In pressure vessels, a device designed to open in a controlled manner to prevent the internal pressure of a component or system from increasing beyond a specified value, that is, a safety valve. **2.** A spring-loaded machine part which will yield, or deflect, when a predetermined force is exceeded.

pressure relief valve [MECH ENG] .A valve which relieves pressure beyond a specified limit and recloses upon return to normal operating conditions.

pressure-retaining member [MECH ENG] That part of a pressure-relieving device loaded by the restrained pressurized fluid.

pressure ring [MIN ENG] A ring about a large excavated area, evidenced by distortion of the openings near the main excavation.

pressure roll [ENG] In plastics-extrusion coating, the roll that with the chill roll applies pressure to the substrate and the molten extruded web.

pressure seal [ENG] A seal used to make pressure-proof the interface (contacting surfaces) between two parts that have frequent or continual relative rotational or translational motion.

pressure-sensitive adhesive [MATER] An adhesive that develops maximum bonding power when applied by a light pressure only.

pressure-stabilized [AERO ENG] Referring to membrane-type structures that require internal pressure for maintenance of a stable structure.

pressure storage [ENG] The storage of a volatile liquid or liquefied gas under pressure to prevent evaporation.

pressure survey [MIN ENG] A study to determine the pressure distribution or pressure losses along consecutive lengths or sections of a ventilation circuit. [PETRO ENG] The measurement of static bottom-hole pressures in an oil field with producing wells shut in for a time interval sufficient for reservoir pressure buildup to stabilize.

pressure system [ENG] Any system of pipes, vessels, tanks, reactors, and other equipment, or interconnections thereof, operating with an internal pressure greater than atmospheric.

pressure tank [CIV ENG] An airtight water tank in which air is compressed to exert pressure on the water and which is used in connection with a water distribution system.

pressure tap [ENG] A small perpendicular hole in the wall of a pressurized, fluid-containing pipe or vessel; used for connection of pressure-sensitive elements for the measurement of static pressures. Also known as piezometer opening; static pressure tap.

pressure thrust [AERO ENG] In rocketry, the product of the cross-sectional area of the exhaust jet leaving the nozzle exit and the difference between the exhaust pressure and the ambient pressure.

pressure transducer [ENG] An instrument component that detects a fluid pressure and produces an electrical signal related to the pressure. Also known as electrical pressure transducer.

pressure traverse [PETRO ENG] Measurement of reservoir pressures at progressive depths.

pressure-tube anemometer [ENG] An anemometer which derives wind speed from measurements of the dynamic wind pressures; wind blowing into a tube develops a pressure greater than the static pressure, while wind blowing across a tube develops a pressure less than the static; this pressure difference, which is proportional to the square of the wind speed, is measured by a suitable manometer.

pressure tunnel [CIV ENG] A waterway tunnel under pressure because the hydraulic gradient lies above the tunnel crown.

pressure vessel [ENG] A metal container generally cylindrical or spheroid, capable of withstanding bursting pressures.

pressure welding [MET] Welding of metal surfaces by the application of pressure; examples are percussion welding, resistance welding, seam welding, and spot welding.

pressurization [ENG] **1.** Use of an inert gas or dry air, at several pounds above atmospheric pressure, inside the components of a radar system or in a sealed coaxial line, to prevent corrosion by keeping out moisture, and to minimize high-voltage breakdown at high altitudes. **2.** The act of maintaining normal atmospheric pressure in a chamber subjected to high or low external pressure.

pressurize [ENG] To maintain normal atmospheric pressure in a chamber subjected to high or low external pressures.

pressurized blast furnace [ENG] A blast furnace operated under pressure above the ambient; pressure is obtained by throttling the off-gas line, which permits a greater volume of air to be passed through the furnace at a lower velocity, and results in increase in smelting rate.

pressurized stoppings [MIN ENG] Stoppings which are erected in the intake and return roadways of a district to isolate an open fire or spontaneous heating and in which the pressures on both sides of each stopping are made equal by the use of auxiliary fans.

prestress [ENG] To apply a force to a structure to condition it to withstand its working load more effectively or with less deflection.

prestressed concrete [MATER] Concrete compressed with heavily loaded wires or bars to reduce or eliminate cracking and tensile forces.

pretensioning [ENG] Process of precasting concrete beams with tensioned wires embedded in them. Also known as Hoyer method of prestressing.

pretersonics *See* acoustoelectronics.

preventive maintenance [ENG] A procedure of inspecting, testing, and reconditioning a system at regular intervals according to specific instructions, intended to prevent failures in service or to retard deterioration.

preweld interval [MET] In resistance spot welding, elapsed time between the end of squeeze time and the beginning of welding current.

Price meter [ENG] The ocean current meter in use in the United States: six conical cups, mounted around a vertical axis, rotate and cause a signal in a set of headphones with each rotation; tail vanes and a heavy weight stabilize the instrument.

pricker *See* needle.

prick punch [DES ENG] A tool that has a sharp conical point ground to an angle of 30–60°C; used to make a slight indentation on a workpiece to locate the intersection of centerlines.

prill [MATER] Spherical particles about the size of buckshot. [MIN ENG] **1.** The best ore after cobbing. **2.** A circular particle about the size of buckshot. **3.** Compressed and sized explosives such as ammonium nitrate.

primary [MET] Of a metal, obtained directly from ore.

primary air [MECH ENG] That portion of the combustion air introduced with the fuel in a burner.

primary breaker [MECH ENG] A machine which takes over the work of size reduction from blasting operations, crushing rock to maximum size of about 2-inch (5 centimeter) diameter; may be a gyratory crusher or jaw breaker. Also known as primary crusher.

primary crusher *See* primary breaker.

primary detector *See* sensor.

primary drilling [ENG] The process of drilling holes in a solid rock ledge in preparation for a blast by means of which the rock is thrown down.

primary excavation [ENG] Digging performed in undisturbed soil.

primary explosive [MATER] Explosive or explosive mixture sensitive to shock and friction; used in primers and detonators to initiate explosion. Also known as initiating explosive.

primary haulage [MIN ENG] A short haul in which there is no secondary- or main-line haulage. Also known as face haulage.

primary high explosive [MATER] An explosive which is extremely sensitive to heat and shock and is normally used to initiate a secondary high explosive; examples are mercury fulminate, lead azide, lead styphnate, and tetracene.

primary instrument [ENG] A measuring instrument that can be calibrated without reference to another instrument.

primary lead [MET] Lead recovered from ore, as contrasted with recycled scrap (secondary) lead.

primary measuring element [ENG] The portion of a measuring or sensing device that is in direct contact with the variables being measured (such as temperature, pressure, pH, or velocity).

primary plasticizer [MATER] A plasticizer material for plastics formulations that has sufficient affinity to a polymer or resin so that it is considered compatible and therefore may be used as the sole plasticizer.

primary radar [ENG] Radar in which the incident beam is reflected from the target to form the return signal. Also known as primary surveillance radar (PSR).

primary reference fuel [MATER] **1.** Gasoline; isooctane, *n*-heptane, or mixtures thereof used in the ASTM-CFR gasoline test engine to determine the octane rating of commercial gasoline. **2.** Diesel fuel; cetane, α-methylnaphthalene, or mixtures thereof used in ASTM-CFR diesel test engines to rate the cetane number of commercial diesel fuels.

primary reserve [PETRO ENG] Petroleum reserve recoverable commercially at current prices and costs by conventional methods and equipment as a result of the natural energy inherent in the reservoir.

primary sewage sludge [CIV ENG] A semiliquid waste resulting from sedimentation with no additional treatment.

primary structure [AERO ENG] The main framework, of an aircraft including fittings and attachments; any structural member whose failure would seriously impair the safety of the missile is a part of the primary structure.

primary surveillance radar *See* primary radar.

primary treatment [CIV ENG] Removal of floating solids and suspended solids, both fine and coarse, from raw sewage.

prime [ENG] **1.** Main or primary, as in prime contractor. **2.** In blasting, to place a detonator in a cartridge or charge of explosive. **3.** To treat wood with a primer or penetrant primer.

prime coat *See* primer.

prime contractor [ENG] A contractor having a direct contract for an entire project, who may in turn assign portions of the work to subcontractors.

prime mover [MECH ENG] The component of a power plant that transforms energy from the thermal or the pressure form to the mechanical form.

primer [ENG] In general, a small, sensitive initial explosive train component which on being actuated initiates functioning of the explosive train, and will not reliably initiate high explosive charge; classified according to the method of initiation, for example, percussion primer, electric primer, or friction primer. [MATER] A prefinishing coat applied to a wood surface that is to be painted or otherwise finished. Also known as prime coat.

primer cup [ENG] A small metal cup, into which the primer mixture is loaded.

primer-detonator [ENG] A unit, in a metal housing, in which are assembled a primer, a detonator, and when indicated, an intervening delay charge.

primer leak [ENG] Defect in a cartridge which allows partial escape of the hot propelling gases in a primer, caused by faulty construction or an excessive charge.

primer mixture [MATER] An explosive mixture containing a sensitive explosive and other ingredients, used in a primer.

primes [MET] High-quality metal products, particularly sheet and plate, that are free from visible defects.

prime-white kerosine [MATER] A kerosine with an off-white color, between water-white and standard-white kerosine.

priming [MECH ENG] In a boiler, the excessive carryover of fine water particles along with the steam because of insufficient steam space, faulty boiler design, or faulty operating conditions.

priming composition [MATER] A physical mixture of materials that is very sensitive to impact or percussion and, when so exploded, undergoes very rapid auto-combustion, producing hot gases and incandescent solid particles; priming compositions are used for the ignition of primary high explosives, black powder igniter charges, and propellants in small arms ammunition.

priming pump [MECH ENG] A device on motor vehicles and tanks, providing a means of injecting a spray of fuel into the engine to facilitate starting.

principal axis [ENG ACOUS] A reference direction for angular coordinates used in describing the directional characteristics of a transducer; it is usually an axis of structural symmetry or the direction of maximum response.

principal item [ENG] Item which, because of its major importance, requires detailed analysis and examination of all factors affecting its supply and demand, as well as an unusual degree of supervision; its selection is based upon such criteria as strategic importance, high monetary value, unusual complexity of issue, and procurement difficulties.

principal meridian [CIV ENG] One of the meridians established by the United States government as a reference for subdividing public land.

principle of reciprocity See reciprocity theorem.

printing ink [MATER] Ink generally made from carbon black, lampblack, or other pigment suspended in an oil vehicle, with a resin, solvent, adhesive, and drier; available in many variations.

prior-art search [ENG] **1.** A search for prior art which may possibly anticipate an invention which is being considered for patentability. **2.** A similar search but for the purpose of determining what the status of existing technology is before going ahead with new research; it is done to avoid unwittingly retracing new steps taken by other workers in the field.

prismatic astrolabe [ENG] A surveying instrument that makes use of a pan of mercury forming an artificial horizon, and a prism mounted in front of a horizontal telescope to determine the exact times at which stars reach a fixed altitude, and thereby to establish an astronomical position.

prismatic compass [ENG] A hand compass used by surveyors which is equipped with a prism that allows the compass to be read while the site is being taken.

prismatic plane [MET] Any plane parallel to the principal c axis in noncubic metals.

prism level [ENG] A surveyor's level with prisms that allow the levelman to view the level bubble without moving his eye from the telescope.

probable ore [MIN ENG] **1.** A mineral deposit adjacent to a developed ore but not yet proved by development. **2.** See indicated ore.

probable reserves [PETRO ENG] Primary petroleum reserves based on limited evidence, but not proved by a commercial oil-production rate.

probe [AERO ENG] An instrumented vehicle moving through the upper atmosphere or space or landing upon another celestial body in order to obtain information about the specific environment. [ENG] A small tube containing the sensing element of electronic equipment, which can be lowered into a borehole to obtain measurements and data.

probe gas [ENG] Tracer gas emitted from a small orifice for impingement on a restricted area being tested for leaks.

probe-type liquid-level meter [ENG] Device to sense or measure the level of liquids in storage or process vessels by means of an immersed electrode or probe.

process [ENG] A system or series of continuous or regularly occurring actions taking place in a predetermined or planned manner; for example, as oil refining process or chemicals manufacturing process.

process annealing [MET] Softening a ferrous alloy by heating to a temperature close to but below the lower limit of the transformation range and then cooling.

process chart [IND ENG] A graphic representation of events occurring during a series of actions or operations.

process control [ENG] Manipulation of the conditions of a process to bring about a desired change in the output characteristics of the process.

process control chart [IND ENG] A tabulated graphical arrangement of test results and other pertinent data for each production assembly unit, arranged in chronological sequence for the entire assembly.

process control engineering [ENG] A field of engineering dealing with ways and means by which conditions of continuous processes are automatically kept as close as possible to desired values or within a required range.

process dynamics [ENG] The dynamic response interrelationships between components (units) of a complex system, such as in a chemical process plant.

process engineering [ENG] A service function of production engineering that involves selection of the processes to be used, determination of the sequence of all operations, and requisition of special tools to make a product.

processing [ENG] The act of converting material from one form into another desired form.

process metallurgy [MET] The branch of metallurgy concerned with the extraction of metals from ore, and with the refining of metals; usually synonymous with extractive metallurgy.

process planning [IND ENG] Determining the conditions necessary to convert material from one state to another.

process time [IND ENG] **1.** Time needed for completion of the machine-controlled portion of a work cycle. **2.** Time required for completion of an entire process.

producer gas [MATER] Fuel gas high in carbon monoxide and hydrogen, produced by burning a solid fuel with a deficiency of air or by passing a mixture of air and steam through a bed of incandescent fuel; used as a cheap, low-Btu industrial fuel.

producer's risk [IND ENG] The probability that in an acceptance sampling plan, material of an acceptable quality level will be rejected.

producing gas-oil ratio [PETRO ENG] The ratio of gas to oil (GOR, or gas-oil ratio) from a producing well; an increase in GOR is a danger signal in the efficient control of reservoir performance.

producing horizon [PETRO ENG] A reservoir bed within the stratigraphic series of an oil province from which gas or liquid hydrocarbons can be obtained by drilling a well.

producing reserves [PETRO ENG] Developed (proved) petroleum reserves to be produced by existing wells in that portion of a reservoir subjected to full-scale secondary-recovery operations.

product [IND ENG] **1.** An item or goods made by an industrial firm. **2.** The total of such items or goods.

product design [DES ENG] The determination and specification of the parts of a product and their interrelationship so that they become a unified whole.

production [ENG] Output, such as units made in a factory, oil from a well, or chemicals from a processing plant.

production control [IND ENG] The procedure for planning, routing, scheduling, dispatching, and expediting the flow of materials, parts, subassemblies, and assemblies within a plant, from the raw state to the finished product, in an orderly and efficient manner.

production-decline curve [PETRO ENG] A graphical means to estimate the ultimate recovery (oil or gas) from a reservoir; cumulative production is plotted against time, the curve being extrapolated to an end point (that is, ultimate recovery).

production engineering [IND ENG] The planning and control of the mechanical means of changing the shape, condition, and relationship of materials within industry toward greater effectiveness and value.

production model [IND ENG] A model in its final mechanical and electrical form of final production design made by production tools, jigs, fixtures, and methods.

production packer [PETRO ENG] A downhole tool used to assist in the efficient production of oil and gas from a well having one or more productive horizons; the function is to provide a seal between the outside of the tubing and the inside of the casing to prevent movement of fluids past that point. Also known as packer.

production requirements [IND ENG] The sum of authorized stock levels and pipeline needs less stocks expected to become available, stock on hand, stocks due in, returned stocks, and stocks from salvage, reclamation, rebuild, and other sources.

production standard *See* standard time.

production track [ENG ACOUS] A sound track which is either prerecorded or recorded directly on the set, and which exists in the film at that time when the music breakdown for scoring is about to begin.

production tubing [PETRO ENG] The final string of pipe that is placed in an oil well after the oil has stopped flowing naturally.

productive time [IND ENG] Time during which useful work is performed in an operation or process.

productivity [IND ENG] **1.** The effectiveness with which labor and equipment are utilized. **2.** The production output per unit of effort. [PETRO ENG] Measure of an oil well's ability to produce liquid or gaseous hydrocarbons; categories include relative, specific, ultimate, and fractured-well productivity.

productivity index [PETRO ENG] The number of barrels of oil produced per day per decline in well bottom-hole pressure in pounds per square inch.

productivity ratio [PETRO ENG] **1.** The amount of damage or improvement to reservoir formation permeability adjacent to the borehole (due to invasion or reduction of drilling mud present, drilling-fluid filtrate water, swollen clay particles, or salt or wax deposition). **2.** The ratio of permeability calculated from the productivity index to the permeability calculated from reservoir buildup pressure.

productivity test [PETRO ENG] Graphical relation of bottomhole static pressure (calculated or measured) versus producing pressure for various gas flow rates; used to predict future oil well behavior.

product line [IND ENG] **1.** The range of products offered by a firm. **2.** A group of basically similar products, differentiated only by such characteristics as color, style, or size.

profile die [ENG] A plastics extrusion die used to produce continuous shapes, but not tubes or sheets.

profiled keyway [DES ENG] A keyway for a straight key formed by an end-milling cutter. Also known as end-milled keyway.

profile thickness [AERO ENG] The maximum distance between the upper and lower contours of an airfoil, measured perpendicular to the mean line of the profile.

profiling [ENG] Electrical exploration wherein the transmitter and receiver are moved in unison across a structure to obtain a profile of mutual impedance between transmitter and receiver. Also known as lateral search.

profiling machine [MECH ENG] A machine used for milling irregular profiles; the cutting tool is guided by the contour of a model.

profilograph [ENG] An instrument for measuring and recording roughness of the surface over which it travels.

profilometer [ENG] An instrument for measuring the roughness of a surface by means of a diamond-pointed tracer arm attached to a coil in an electric field; movement of the arm across the surface induces a current proportional to surface roughness.

profit in sight [MIN ENG] Probable gross profit from a mine's ore reserves, as distinct from the ground that is still to be blocked out.

profit sharing [IND ENG] Sharing of company profits with the employees.

program [AERO ENG] In missile guidance, the planned flight path events to be followed by a missile in flight, including all the critical functions, preset in a program device, which control the behavior of the missile.

program level [ENG ACOUS] The level of the program signal in an audio system, expressed in volume units.

programmed turn [AERO ENG] The automatically controlled turn of a ballistic missile into the curved path that will lead to the correct velocity and vector for the final portion of the trajectory.

programming [ENG] In a plastics process, extruding a parison whose thickness differs longitudinally in order to equalize wall thickness of the blown container.

progress chart [IND ENG] A graphical representation of the degree of completion of work in progress.

progressive aging [MET] Aging of metals achieved by increasing the temperature in stages or by continuous elevation of the temperature.

progressive block sequence [MET] A welding sequence in which the joint is completed in sections from one end to the other or from the center alternately to both ends.

progressive bonding [ENG] A method of curing a resin adhesive wherein heat and pressure are applied in successive steps.

progressive die [MET] A die in which two or more operations are performed sequentially at different positions.

progressive forming [MET] Sequential forming at consecutive stations either with a single die or with separate dies.

progressive granulation [MATER] Propellant granulation in which the surface area of a grain increases during burning.

progressive powder [MATER] A slow-burning explosive.

project [ENG] A specifically defined task within a research and development field, which is established to meet a single requirement, either stated or anticipated, for research data, an end item of material, a major component, or a technique.

projected planform [AERO ENG] The contour of the planform as viewed from above.

projected window [BUILD] A window having one or more rotatable sashes which swing either inward or outward.

project engineering [ENG] **1.** The engineering design and supervision (coordination) aspects of building a manufacturing facility. **2.** The engineering aspects of a specific project, such as development of a product or solution to a problem.

projection thermography [ENG] A method of measuring surface temperature in which thermal radiation from a surface is imaged by an optical system on a thin screen of luminescent material, and the pattern formed corresponds to the heat radiation of the surface.

projection welding [MET] Resistance welding in which the welds are localized at projections, intersections, and overlaps on the parts.

projector [ENG ACOUS] **1.** A horn designed to project sound chiefly in one direction from a loudspeaker. **2.** An underwater acoustic transmitter.

prony brake [MECH ENG] An absorption dynamometer that applies a friction load to the output shaft by means of wood blocks, a flexible band, or other friction surface.

proof [ENG] Reproduction of a die impression by means of a cast.

proof load [ENG] A predetermined test load, greater than the service load, to which a specimen is subjected before acceptance for use.

prop [MIN ENG] Underground supporting post set across the lode, seam, bed, or other opening.

propagated blast [ENG] A blast of a number of unprimed charges of explosives plus one hole primed, generally for the purpose of ditching, where each charge is detonated by the explosion of the adjacent one, the shock being transmitted through the wet soil.

prop-crib timbering [MIN ENG] Shaft timbering with cribs kept apart at the proper distance by means of props.

propellant [MATER] A combustible substance that produces heat and supplies ejection particles as in a rocket engine.

propellant-actuated device [ENG] A device that employs the energy supplied by the gases produced by burning propellants to accomplish or initiate a mechanical action other than propelling a projectile.

propellant additive [MATER] Any material added to the basic formulation of a solid propellant to accomplish some special purpose such as to increase or decrease the rate of burning.

propellant binder [MATER] An elastomeric fuel used in a composite propellant so that the propellant may be cast directly into the combustion chamber, where the binder cures to a rubber and the propellant grain is then supported by adhesion to the walls; used in missiles.

propellant grain [MATER] An elongated molding or extrusion, often of intricate shape, of solid propellant for a rocket, regardless of size.

propellant injector [AERO ENG] A device for injecting propellants, which include fuel and oxidizer, into the combustion chamber of a rocket engine.

propellant mass ratio [AERO ENG] Of a rocket, the ratio of the effective propellant mass to the initial vehicle mass. Also known as propellant mass fraction.

propellant powder [MATER] A low explosive of fine granulation which, through burning, produces gases at a controlled rate to provide the energy for propelling a projectile.

propellant weight fraction [AERO ENG] The weight of the solid propellant charge divided by weight of the complete solid propellant propulsion unit.

propeller [MECH ENG] A bladed device that rotates on a shaft to produce a useful thrust in the direction of the shaft axis.

propeller anemometer [ENG] A rotation anemometer which is encased in a strong glass outer shell that protects it against hydrostatic pressure.

propeller blade [DES ENG] One of two or more plates radiating out from the hub of a propeller and normally twisted to form part of a helical surface.

propeller boss [DES ENG] The central portion of the screw propeller which carries the blades, and forms the medium of attachment to the propeller shaft. Also known as propeller hub.

propeller efficiency [MECH ENG] The ratio of the thrust horsepower delivered by the propeller to the shaft horsepower as delivered by the engine to the propeller.

propeller fan [MECH ENG] An axial-flow blower, with or without a casing, using a propeller-type rotor to accelerate the fluid.

propeller hub See propeller boss.

propeller meter [ENG] A quantity meter in which the flowing stream rotates a propellerlike device and revolutions are counted.

propeller pump *See* axial-flow pump.

propeller shaft [MECH ENG] A shaft, carrying a screw propeller at its end, that transmits power from an engine to the propeller.

propeller slip angle [MECH ENG] The angle between the plane of the blade face and its direction of motion.

propeller tip speed [MECH ENG] The speed in feet per minute swept by the propeller tips.

propeller turbine [MECH ENG] A form of reactive-type hydraulic turbine using an axial-flow propeller rotor.

propeller windmill [MECH ENG] A windmill that extracts wind power from horizontal air movements to rotate the blades of a propeller.

prop-free [MIN ENG] A face with no posts between the coal and the conveyor used to remove it in longwall mining of a coal seam.

prop-free front [MIN ENG] In coal mining, longwall working in which support to the roof is given by roof beams cantilevered from behind the working face.

proplatinum [MET] A nickel-silver-bismuth alloy used as a substitute for platinum.

proportional dividers [DES ENG] Dividers with two legs, pointed at both ends, and an adjustable pivot; distances measured by the points at one end can be marked off in proportion by the points at the other end.

proportional-speed control *See* floating control.

proportioning probe [ENG] A leak-testing probe capable of changing the air–tracer gas ratio without changing the amount of flow it transmits to the testing device.

propped cantilever [CIV ENG] A beam having one built-in support and one simple support.

propping agent [PETRO ENG] A granular substance, for example, sand grains or walnut shells, suspended in the drilling fluid during the fracturing portion of the drilling operation to keep the fracture open when the fluid is withdrawn.

propulsion system [MECH ENG] For a vehicle moving in a fluid medium, such as an airplane or ship, a system that produces a required change in momentum in the vehicle by changing the velocity of the air or water passing through the propulsive device or engine; in the case of a rocket-propelled vehicle operating without a fluid medium, the required momentum change is produced by using up some of the propulsive device's own mass, called the propellant.

prospect [MIN ENG] **1.** To search for minerals or oil by looking for surface indications, by drilling boreholes, or both. **2.** A plot of ground believed to be mineral-

ized enough to be of economic importance.

prospecting seismology [PETRO ENG] The application of seismology to the exploration for natural resources, especially gas and oil.

prospector [MIN ENG] A person engaged in exploring for valuable minerals, or in testing supposed discoveries of the same.

Prospector [AERO ENG] A specific unmanned spacecraft designed to make a soft landing on the moon to take measurements, photographs, and soil samples, and then return to earth.

prospect pit [MIN ENG] A pit excavated for the purpose of prospecting mineral-bearing ground.

prospect shaft [MIN ENG] A shaft constructed for the purpose of excavating mineral-bearing ground.

protected thermometer [ENG] A reversing thermometer which is encased in a strong glass outer shell that protects it against hydrostatic pressure.

protective atmosphere [MET] A substance such as inert or combusted fuel gas which surrounds a workpiece to be heat-treated, welded, brazed, or thermally sprayed under controlled conditions.

protective finish [ENG] A coating applied to equipment to protect it from corrosion and wear; many substances, including metals, glass, and ceramics, are used.

proteinaceous [MATER] **1.** Pertaining to any material having a protein base. **2.** Pertaining to adhesive materials having a protein base such as animal glue, casein, and soya.

protobitumen [MATER] Any of the fats, oils, waxes, or resins which are present as unaltered or nearly unaltered plant and animal products from which fossil bitumens are formed.

protore [MIN ENG] **1.** A primary mineral deposit which, through enrichment, can be modified to form an economic ore. **2.** A deposit which could become economically workable if technological change occurred or prices were increased.

prototype [ENG] A model suitable for use in complete evaluation of form, design, and performance.

protractor [ENG] An instrument used to construct and measure angles formed by lines of a plane; the midpoint of the diameter of the semicircle is marked and serves as the vertex of angles contructed or measured.

proved ore [MIN ENG] Ore in which there is practically no risk of failure of continuity.

proved reserves [PETRO ENG] Reserves (primary or secondary) that have been proved by production at commercial flow rates.

proving ring [DES ENG] A ring used for calibrating test machines; the diameter of the ring changes when a force is applied along a diameter.

proximity detector [ENG] A sensing device that produces an electrical signal when approached by an object or when approaching an object.

prudent limit of endurance [AERO ENG] The time during which an aircraft can remain airborne and still retain a given safety margin of fuel.

pseudocarburizing See blank carburizing.

pseudoliquid density [PETRO ENG] For reservoir studies, the calculated value of a pseudodensity for reservoir liquid at atmospheric conditions, followed by application of suitable correction factors to obtain an approximate value for actual liquid density in the reservoir.

pseudostatic SP See pseudostatic spontaneous potential.

pseudostatic spontaneous potential [PETRO ENG] Theoretical maximum spontaneous potential current that can be measured in a downhole, mud-column log in shaly sand. Abbreviated pseudostatic SP; PSP.

pseudo steady-state pressure distribution [PETRO ENG] The condition when the declining pressure distribution within a reservoir system closed at one boundary is declining at a uniform rate everywhere (or, there is constant pressure gradient within the reservoir).

psid See pounds per square inch differential.

psophometer [ENG] An instrument for measuring noise in electric circuits; when connected across a 600-ohm resistance in the circuit under study, the instrument gives a reading that by definition is equal to half of the psophometric electromotive force actually existing in the circuit.

PSP See pseudostatic spontaneous potential.

psychogalvanometer [ENG] An instrument for testing mental reaction by determining how skin resistance changes when a voltage is applied to electrodes in contact with the skin.

psychointegroammeter See lie detector.

psychosomatograph [ENG] An instrument for recording muscular action currents or physical movements during tests of mental–physical coordination.

psychrometer [ENG] A device comprising two thermometers, one a dry bulb, the other a wet or wick-covered bulb, used in determining the moisture content or relative humidity of air or other gases. Also known as wet and dry bulb thermometer.

psychrometric calculator [ENG] A device for quickly computing certain psychrometric data, usually the dew point and the relative humidity, from known values of the dry- and wet-bulb temperatures and the atmospheric pressure.

psychrometry [ENG] The science and techniques associated with measurements of the water vapor content of the air or other gases.

ptychotis oil See ajowan oil.

public address system See sound reinforcement system.

public area [BUILD] The total nonrentable area of a building, such as public conveniences and rest rooms.

public utility [IND ENG] A business organization considered by law to be vested with public interest and subject to public regulation.

public works [IND ENG] Government-owned and financed works and improvements for public enjoyment or use.

puckering [MET] Corrugations in metal parts resulting from pressing or drawing.

puddle [ENG] To apply water in order to settle loose dirt. [MET] A batch of molten iron within the puddling furnace.

puddling [MET] A process for the production of wrought iron by agitation of a bath of molten pig iron with iron oxide in order to reduce the carbon, silicon, phosphorus, and manganese content.

puddling furnace [MET] A coal-fired reverberatory furnace for puddling pig iron.

puff [MECH ENG] A small explosion within a furnace due to combustion conditions.

puffer [MIN ENG] A small stationary engine used in coal mines for hoisting material.

pug mill [MECH ENG] A machine for mixing and tempering a plastic material by the action of blades revolving in a drum or trough.

pulegium oil See hedeoma oil.

pull crack [MET] A crack in a casting caused by contraction strains during cooling and resulting from the shape of the casting.

puller [MECH ENG] A lever-operated chain or wire-rope hoist for lifting or pulling at any angle, which has a reversible ratchet mechanism in the lever permitting short-stroke operation for both tensioning and relaxing, and which holds the loads with a Weston-type friction brake or a releasable ratchet. Also known as come-along.

pulley [DES ENG] A wheel with a flat, round, or grooved rim that rotates on a shaft and carries a flat belt, V-belt, rope, or chain to transmit motion and energy.

pulley lathe [MECH ENG] A lathe for turning pulleys.

pulley stile [BUILD] The upright part of a window frame which holds the pulley and guides the sash.

pulley top [MECH ENG] A top with a long shank used to tap setscrew holes in pulley hubs.

pulling [PETRO ENG] Withdrawing sucker rods and production tubing from a pumping well prior to cleaning out or replacing parts of the pump.

pulling pillars See pillaring.

pull-in torque [MECH ENG] The largest steady torque with which a motor will attain normal speed after accelerating from a standstill.

pull-out torque [MECH ENG] The largest torque under which a motor can operate without sharply losing speed.

pull rope [MIN ENG] **1.** The rope that pulls a journey of loaded cars on a haulage plane. **2.** The rope that pulls the loaded scoop or bucket in a scraper loader layout. Also known as main rope.

pullshovel See backhoe.

pull tube [PETRO ENG] Tube used in rod-type, traveling-barrel oil well pumps to connect the pump plunger with the seating anchor.

pulp [ENG] See slime. [MATER] The cellulosic material produced by reducing wood mechanically or chemically and used in making paper and cellulose products. Also known as wood pulp.

pulpboard [MATER] Chipboard to which is added a percentage of mechanical wood pulp.

pulper [MECH ENG] A machine that converts materials to pulp, for example, one that reduces paper waste to pulp.

pulping [ENG] Reducing wood to pulp.

pulp molding [ENG] A plastics-industry process in which a resin-impregnated pulp material is preformed by application of a vacuum, after which it is oven-cured and molded.

pulpstone [MATER] A block of sandstone cut into wheels for grinding, especially wood pulp in paper manufacture.

pulpwood [MATER] Any wood that can be reduced to pulp.

pulsating flow [ENG] Irregular fluid flow in a piping system often resulting from the pressure variations of reciprocating compressors or pumps within the system.

pulsation dampening [ENG] Device installed in a fluid piping system (gas or liquid) to eliminate or even out the fluid-flow pulsations caused by reciprocating compressors, pumps, and such.

pulsation welding See multiple-impulse welding.

pulse altimeter [ENG] A device which is used to measure the distance of an aircraft above the ground by sending out radar signals in short pulses and measuring the time delay between the leading edge

of the transmitted pulse and that of the pulse returned from the ground.

pulse-amplitude discriminator [ENG] Electronic instrument used to investigate the amplitude distribution of the pulses produced in a nuclear detector.

pulse-compression radar [ENG] A radar system in which the transmitted signal is linearly frequency-modulated or otherwise spread out in time to reduce the peak power that must be handled by the transmitter; signal amplitude is kept constant; the receiver uses a linear filter to compress the signal and thereby reconstitute a short pulse for the radar display.

pulsed-light ceilometer See pulsed-light cloud-height indicator.

pulsed-light cloud-height indicator [ENG] An instrument used for the determination of cloud heights; it operates on the principle of pulse radar, employing visible light rather than radio waves. Also known as pulsed-light ceilometer.

pulse-Doppler radar [ENG] Pulse radar that uses the Doppler effect to obtain information about the velocity of a target.

pulse dot soldering iron [ENG] A soldering iron that provides heat to the tip for a precisely controlled time interval, as required for making a good soldered joint without overheating adjacent parts.

pulse-echo method [MET] A nondestructive test in which pulses of energy are directed into a part, and the time for the echo to return from one or more surfaces is measured.

pulse hardening [MET] A surface-hardening process performed by heating to the required temperature in a span of several milliseconds by using an energy and time-controlled pulse of very high power at a very high frequency, about 27 megahertz.

pulsejet engine [AERO ENG] A type of compressorless jet engine in which combustion occurs intermittently so that the engine is characterized by periodic surges of thrust; the inlet end of the engine is provided with a grid to which are attached flap valves; these can be sucked inward by a negative differential pressure to allow a regulated amount of air to flow inward to mix with the fuel. Also known as aeropulse engine.

pulse-modulated radar [ENG] Form of radar in which the radiation consists of a series of discrete pulses.

pulse radar [ENG] Radar in which the transmitter sends out high-power pulses that are spaced far apart in comparison with the duration of each pulse; the receiver is active for reception of echoes in the interval following each pulse.

pulse-time-modulated radiosonde [ENG] A radiosonde which transmits the indications of the meteorological sensing elements in the form of pulses spaced in time; the meteorological data are evaluated from the intervals between the pulses. Also known as time-interval radiosonde.

pulse tracking system [ENG] Tracking system which uses a high-energy, short-duration pulse radiated toward the target from which the velocity, direction, and range are determined by the characteristics of the reflected pulse.

pulsometer [MECH ENG] A simple, lightweight pump in which steam forces water out of one of two chambers alternately.

pulverizer [MECH ENG] Device for breaking down of solid lumps into a fine material by cleavage along crystal faces.

pump [MECH ENG] A machine that draws a fluid into itself through an entrance port and forces the fluid out through an exhaust port.

pumpability [MATER] **1.** The property of a lubricating grease that causes it to flow under pressure through lines, nozzles, and fittings. **2.** The ability of any liquid, slurry, or suspension to be moved through a flow conduit by pressure from a pump.

pumpability test [ENG] Standard test to ascertain the lowest temperature at which a petroleum fuel oil may be pumped.

pump bob [MECH ENG] A device such as a crank that converts rotary motion into reciprocating motion.

pump-down time [ENG] The length of time required to evacuate a leak-tested vessel.

pumphouse [CIV ENG] A building in which are housed pumps that supply an irrigation system, a power plant, a factory, a reservoir, a farm, a home, and so on.

pumping loss [MECH ENG] Power consumed in purging a cylinder of exhaust gas and sucking in fresh air instead.

pumping pressure [PETRO ENG] Pressure required to inject (pump) water, gas, or acid into a pressurized petroleum reservoir.

pumping station [CIV ENG] A building in which two or more pumps operate to supply fluid flowing at adequate pressure to a distribution system.

pumping well [PETRO ENG] A producing oil well in which liquid products are recovered from the reservoir by means of a pump, rather than by gas lift.

pump off [PETRO ENG] In an oil well, to pump so rapidly that the oil level drops below the pump's standing valve.

punch [MECH ENG] A tool that forces metal into a die for extrusion or similar operations.

punched-plate screen [ENG] Flat, perforated plate with round, square, hexagonal, or elongated openings; used for screening (size classification) of crushed or pulverized solids.

punching [ENG] **1.** A piece removed from a sheet of metal or other material by a punch press. **2.** A method of extrusion, cold heading, hot forging, or stamping in a machine for which mating die sections determine the shape or contour of the work.

punch press [MECH ENG] **1.** A press consisting of a frame in which slides or rams move up and down, of a bed to which the die shoe or bolster plate is attached, and of a source of power to move the slide. Also known as drop press. **2.** Any mechanical press.

punch radius [DES ENG] The radius on the bottom end of the punch over which the metal sheet is bent in drawing.

purge meter interlock [MECH ENG] A meter to maintain airflow through a boiler furnace at a specific level for a definite time interval; ensures that the proper air-fuel ratio is achieved prior to ignition.

purging [ENG] Replacing the atmosphere in a container by an inert substance to prevent formation of explosive mixtures.

purify [ENG] To remove unwanted constituents from a substance.

purlin [BUILD] A horizontal roof beam, perpendicular to the trusses or rafters; supports the roofing material or the common rafters.

purple lakes [MATER] A class of lake (pigment) used in printing inks; derived from a combination of such compounds as β-hydroxynaphthoic acid and 2-diazonaphthalene-1-sulfonic acid.

purse seine [ENG] A net that can be dropped by two boats to encircle a school of fish, then pulled together at the bottom and raised, thereby catching the fish.

push-bar conveyor [MECH ENG] A type of chain conveyor in which two endless chains are cross-connected at intervals by push bars which propel the load along a stationary bed or trough of the conveyor.

push bench [MECH ENG] A machine used for drawing tubes of moderately heavy gage by cupping metal sheet and applying pressure to the inside bottom of the cup to force it through a die.

pusher grade *See* helper grade.

pusher tractor [MIN ENG] A bulldozer exerting pressure on the rear of a scraper-loader while the loader is digging and loading unconsolidated ground during opencast mining.

push fit [DES ENG] A hand-tight sliding fit between a shaft and a hole.

push nipple [MECH ENG] A short length of pipe used to connect sections of cast iron boilers.

push-pull sound track [ENG ACOUS] A sound track having two recordings so arranged that the modulation in one is 180° out of phase with that in the other.

push rod [MECH ENG] A rod, as in an internal combustion engine, which is actuated by the cam to open and close the valves.

push-up [ENG] Concave bottom contour of a plastic container; allows an even bearing surface on the outer edge and prevents the container from rocking.

push welding [MET] Spot or projection welding in which the force is applied manually by one electrode; the work takes the place of the other electrode. Also known as poke welding.

putlog [CIV ENG] A crosspiece in a scaffold or formwork; supports the soffits and is supported by the ledgers.

putty [MATER] A cement of dough consistency made of whiting and boiled linseed oil and used in fastening glass in sashes and sealing crevices in woodwork.

putty knife [DES ENG] A knife with a broad flexible blade, used to apply and smooth putty.

putty oil [MATER] Petroleum oil that is added to putty; serves to lubricate the putty and keep it soft after the linseed oil dries.

pycnometer [ENG] A container whose volume is precisely known, used to determine the density of a liquid by filling the container with the liquid and then weighing it. Also spelled pyknometer.

pyknometer *See* pycnometer.

pylon [AERO ENG] A suspension device externally installed under the wing or fuselage of an aircraft; it is aerodynamically designed to fit the configuration of specific aircraft, thereby creating an insignificant amount of drag; it includes means of attaching to accommodate fuel tanks, bombs, rockets, torpedoes, rocket motors, or the like. [CIV ENG] **1.** A massive structure, such as a truncated pyramid, on either side of an entrance. **2.** A tower supporting a wire over a long span. **3.** A tower or other structure marking a route for an airplane.

pyramidal horn [ENG] Horn whose sides form a pyramid.

pyranometer [ENG] An instrument used to measure the combined intensity of incoming direct solar radiation and diffuse sky radiation; compares heating produced by the radiation on blackened metal strips with that produced by an electric current. Also known as solarimeter.

pyrethrum [MATER] A toxicant obtained in the form of dried powdered flowers of the plant of the same name; mixed with petroleum distillates, it is used as an insecticide.

pyrgeometer [ENG] An instrument for measuring radiation from the surface of the earth into space.

pyrheliometer [ENG] An instrument for measuring the total intensity of direct solar radiation received at the earth.

pyrite roasting [MIN ENG] Thermal processing of iron pyrite (FeS_2, iron disulfide) in the presence of air to produce iron oxide sinter (used in steel mills) and elemental sulfur.

pyroceram [MATER] Hard, strong, opaque-white nucleated glass with nonporous crystalline structure; softens at 2460°F (1350°C); has high flexural strength and shock resistance; used for molded mechanical and electrical parts, heat-exchanger tubes, and coatings.

pyrometallurgy [MET] High-temperature process metallurgy.

pyrometer [ENG] Any of a broad class of temperature-measuring devices; they were originally designed to measure high temperatures, but some are now used in any temperature range; includes radiation pyrometers, thermocouples, resistance pyrometers, and thermistors.

pyrophoric alloy [MET] **1.** An alloy such as ferrocerium that produces a spark when struck with metal (steel) at an angle; used for automatic cigarette lighters. **2.** An alloy in powder form that spontaneously oxidizes in air, reaching high temperatures.

pyrophoric propellant [MATER] A propellant combination of a liquid fuel and a fluid oxidizer (usually air) that will quickly react when brought into intimate contact and achieve ignition temperature.

pyrostat [ENG] **1.** A sensing device that automatically actuates a warning or extinguishing mechanism in case of fire. **2.** A high-temperature thermostat.

pyrotechnic pistol [ENG] A single-shot device designed specifically for projecting pyrotechnic signals.

pyrotechnics [ENG] Art and science of preparing and using fireworks. [MATER] Items which are used for both military and nonmilitary purposes to produce a bright light for illumination, or colored lights or smoke for signaling, and which are consumed in the process.

pyroxylin cement [MATER] A solution of nitrocellulose in a chemical solvent, compounded with a resin, or plasticized with a gum or synthetic; dries by evaporation of the solvent.

PZT *See* lead zirconate titanate.

Q

Q meter [ENG] A direct-reading instrument which measures the Q of an electric circuit at radio frequencies by determining the ratio of inductance to resistance, and which has also been developed to measure many other quantities. Also known as quality-factor meter.

Q point [ENG] Data point describing the position and movement of a radar target, based on two or more radar observations.

quadrangle [CIV ENG] **1.** A four-cornered, four-sided courtyard, usually surrounded by buildings. **2.** The buildings surrounding such a courtyard. **3.** A four-cornered, four-sided building.

quadrant [ENG] **1.** An instrument for measuring altitudes, used, for example, in astronomy, surveying, and gunnery; employs a sight that can be moved through a graduated 90° arc. **2.** A lever that can move through a 90° arc. [MECH ENG] A device for converting horizontal reciprocating motion to vertical reciprocating motion.

quadrant electrometer [ENG] An instrument for measuring electric charge by the movement of a vane suspended on a wire between metal quadrants; the charge is introduced on the vane and quadrants in such a way that there is a proportional twist to the wire.

quadraphonic sound system [ENG ACOUS] A system for reproducing sound by means of four loudspeakers properly situated in the listening room, usually at the four corners of a square, with each loudspeaker being fed its own identifiable segment of the program signal. Also known as four-channel sound system.

quadrille paper [MATER] A good-quality white ledger paper with light-blue lines ruled on it.

quadruple thread [DES ENG] A multiple thread having four separate helices equally spaced around the circumference of the threaded member; the lead is equal to four times the pitch of the thread.

qualification test [ENG] A formally defined series of tests by which the functional, environmental, and reliability performance of a component or system may be evaluated in order to satisfy the engineer, contractor, or owner as to its satisfactory design and construction prior to final approval and acceptance.

quality analysis [IND ENG] Examination of the quality goals of a product or service.

quality assurance [IND ENG] Testing and inspecting all of or a portion of the final product to ensure that the desired quality level of product reaches the consumer.

quality control [IND ENG] Inspection, analysis, and action applied to a portion of the product in a manufacturing operation to estimate overall quality of the product and determine what, if any, changes must be made to achieve or maintain the required level of quality.

quality control chart [IND ENG] A control chart used to indicate and control the quality of a product.

quality-factor meter *See* Q meter.

quantity meter [ENG] A type of fluid meter used to measure volume of flow.

quarantine anchorage [CIV ENG] An area where a vessel anchors when satisfying quarantine regulations.

quarry [ENG] An open or surface working or excavation for the extraction of building stone, ore, coal, gravel, or minerals.

quarry bar [ENG] A horizontal bar with legs at each end, used to carry machine drills.

quarry face [MIN ENG] The freshly split face of ashlar, squared off for the joints only and used for massive work.

quarrying [ENG] The surface exploitation and removal of stone or mineral deposits from the earth's crust.

quarrying machine [MECH ENG] Any machine used to drill holes or cut tunnels in native rock, such as a gang drill or tunneling machine; most commonly, a small locomotive bearing rock-drilling equipment operating on a track.

quarry powder [MATER] Ammonium nitrate dynamites used in quarrying where blasts of several tons of explosives are needed.

quarry sap *See* quarry water.

quarry water [ENG] Subsurface water retained in freshly quarried rock. Also known as quarry sap.

quartation *See* inquartation.

quartering machine [MECH ENG] A machine that bores parallel holes simultaneously in such a way that the center lines of adjacent holes are 90° apart.

quarter-sawed [MATER] The grain pattern that is produced when hardwood is cut so that the annular rings are at an angle of 45° or less with the board's surface.

quarter-turn drive [MECH ENG] A belt drive connecting pulleys whose axes are at right angles.

quartz claim [MIN ENG] In the United States, a mining claim containing ore in veins or lodes, as contrasted with placer claims carrying mineral, usually gold, in alluvium.

quartz horizontal magnetometer [ENG] A type of relative magnetometer used as a geomagnetic field instrument and as an observatory instrument for routine calibration of recording equipment.

quartz mine [MIN ENG] A mine in which a valuable constituent, such as gold, is found in veins rather than in placers; so named because quartz is the chief accessory mineral in such deposits.

quartz pressure gage [ENG] A pressure gage that uses a highly stable quartz crystal resonator whose frequency changes directly with applied pressure.

quartz resonator force transducer [ENG] A type of accelerometer which measures the change in the resonant frequency of a small quartz plate with a longitudinal slot, forming a double-ended tuning fork, when a longitudinal force associated with acceleration is applied to the plate.

quartz thermometer [ENG] A thermometer based on the sensitivity of the resonant frequency of a quartz crystal to changes in temperature.

quaternary alloy [MET] An alloy containing four principal elements apart from accidental impurities.

quay [CIV ENG] A solid embankment or structure parallel to a waterway; used for loading and unloading ships.

quebracho [MATER] A drilling-fluid additive used for thinning or dispersing in order to control viscosity and thixotropy; made from an extract of the quebracho tree and consisting essentially of tannic acid.

queen post [CIV ENG] Either of two vertical members, one on each side of the apex of a triangular truss.

queen's metal [MET] An alloy consisting principally of tin to which antimony, zinc, and lead or copper are added.

quench aging [MET] Aging of metal induced by rapid cooling from solution heat-treatment temperatures.

quench annealing [MET] Annealing an austenitic ferrous alloy by heating followed by quenching from solution temperatures.

quench bath [ENG] A liquid medium, such as oil, fused salt, or water, into which a material is plunged for heat-treatment purposes.

quench hardening [MET] The hardening of a ferrous alloy by quenching from a temperature above the transformation range.

quenching [ENG] Shock cooling by immersing liquid or molten material into a cooling medium (liquid or gas); used in metallurgy, plastics forming, and petroleum refining. [MET] Rapid cooling from solution temperatures.

quenching oil [MET] Animal, vegetable, or mineral oil, such as fish oil, cottonseed oil, or lard, used in quenching baths for carbon and alloy steels; removes heat from the steel more slowly and uniformly than water.

quenching stress [MET] Internal stresses set up in a metal as a result of quenching.

quench-tank extrusion [ENG] Plastic-film or metal extrusion that is cooled in a quenching medium.

quench temperature [ENG] The temperature of the medium used for quenching.

queue *See* waiting line.

queuing [ENG] The movement of discrete units through channels, such as programs or data arriving at a computer, or movement on a highway of heavy traffic.

quick [MIN ENG] Referring to an economically valuable or productive mineral deposit.

quick-change gearbox [MECH ENG] A cluster of gears on a machine tool, the arrangement of which allows for the rapid change of gear ratios.

quick malleable iron [MET] Malleable iron containing 2.2% carbon, 1.5% silicon, 0.30–0.60% manganese, and 0.75–1% copper.

quickmatch [ENG] Fast-burning fuse made from a cord impregnated with black powder.

quick return [MECH ENG] A device used in a reciprocating machine to make the return stroke faster than the power stroke.

quicksand [MATER] A loose sand mixture with a high proportion of water, thus having a low bearing pressure.

quiescent [ENG] State of a body at rest, or inactive, such as an undisturbed liquid in a storage or process vessel.

quill [DES ENG] A hollow shaft into which another shaft is inserted in mechanical devices.

quill drive [MECH ENG] A drive in which the motor is mounted on a nonrotating hollow shaft surrounding the driving-wheel axle; pins on the armature mesh with spokes on the driving wheels, thereby transmitting motion to the wheels; used on electric locomotives.

quill gear [MECH ENG] A gear mounted on a hollow shaft.

quire [MATER] Twenty-five sheets of paper, or one-twentieth of a 500-sheet ream.

quirk [BUILD] **1.** An indentation separating one element from another, as between moldings. **2.** A V groove in the finish-coat plaster where it abuts the return on a door or window.

quirk bead [BUILD] **1.** A bead with a quirk on one side only, as on the edge of a board.

Also known as bead and quirk. **2.** A bead that is flush with the adjoining surface and separated from it by a quirk on each side. Also known a bead and quirk; double-quirked bead; flush bead; recessed bead. **3.** A bead located at a corner with quirks at either side at right angles to each other. Also known as bead and quirk; return bead. **4.** A bead with a quirk on its face. Also known as bead and quirk.

quitclaim [MIN ENG] Legal release of a claim, right, title, or interest by one person or estate to another.

quoin [BUILD] One of the members forming an outside corner or exterior angle of a building, and differentiated from the wall by color, texture, size, or projection.

quoin post [CIV ENG] The vertical member at the jointed end of a gate in a navigation lock.

R

rabbet [ENG] **1.** A groove cut into a part. **2.** A strip applied to a part as, for example, a stop or seal. **3.** A joint formed by fitting one member into a groove, channel, or recess in the face or edge of a second member.

rabbet plane [DES ENG] A plane with the blade extending to the outer edge of one side that is open.

rabbit [PETRO ENG] A small plug driven by pressure through a flow line to clean the line or to check that it is unobstructed.

rabbit ear [MET] A recess in the corner of a die allowing wrinkling or folding of the blank.

rabbling [ENG] Stirring a molten charge, as of metal or ore.

race [DES ENG] Either of the concentric pair of steel rings of a ball bearing or roller bearing. [ENG] A channel transporting water to or away from hydraulic machinery, as in a power house.

rack [AERO ENG] A suspension device permanently fixed to an aircraft; it is designed for attaching, arming, and releasing one or more bombs; it may also be utilized to accommodate other items such as mines, rockets, torpedoes, fuel tanks, rescue equipment, sonobuoys, and flares. [CIV ENG] A fixed screen composed of parallel bars placed in a waterway to catch debris. [DES ENG] See relay rack. [ENG] A frame for holding or displaying articles. [MECH ENG] A bar containing teeth on one face for meshing with a gear. [MIN ENG] An inclined trough or table for washing or separating ore.

racking [CIV ENG] Setting back the end of each course of brick or stone from the end of the preceding course. [MET] Suspending work from a frame that holds and conducts current to one or more electrodes for electroplating and other electrochemical operations. [PETRO ENG] During oil well drilling, placing stands of pipe in an orderly fashion in the derrick while hoisting pipe out of the well bore.

rack railway [CIV ENG] A railway with a rack between the rails which engages a gear on the locomotive; used on steep grades.

radar [ENG] **1.** A system using beamed and reflected radio-frequency energy for detecting and locating objects, measuring distance or altitude, navigating, homing, bombing, and other purposes; in detecting and ranging, the time interval between transmission of the energy and reception of the reflected energy establishes the range of an object in the beam's path. Derived from radio detection and ranging. **2.** See radar set.

radar bombsight [ENG] An airborne radar set used to sight the target, solve the bombing problem, and drop bombs.

radar command guidance [ENG] A missile guidance system in which radar equipment at the launching site determines the positions of both target and missile continuously, computes the missile course corrections required, and transmits these by radio to the missile as commands.

radar contact [ENG] Recognition and identification of an echo on a radar screen; an aircraft is said to be on radar contact when its radar echo can be seen and identified on a PPI (plan-position indicator) display.

radar coverage [ENG] The limits within which objects can be detected by one or more radar stations.

radar coverage indicator [ENG] Device that shows how far a given aircraft should be tracked by a radar station, and also provides a reference (detection) range for quality control; takes into account aircraft size, altitude, screening angle, site elevation, type radar, antenna radiation pattern, and antenna tilt.

radar dome [ENG] Weatherproof cover for a primary radiating element of a radar or radio device which is transparent to radio-frequency energy, and which permits active operation of the radiating element, including mechanical rotation or other movement as applicable.

radar gun-layer [ENG] A radar device which tracks a target and aims a gun or guns automatically.

radar homing [ENG] Homing in which a missile-borne radar locks onto a target and guides the missile to that target.

radar marker [ENG] A fixed facility which continuously emits a radar signal so that a bearing indication appears on a radar display.

radar netting [ENG] The linking of several radars to a single center to provide integrated target information.

radar netting station [ENG] A center which can receive data from radar tracking stations and exchange these data among other radar tracking stations, thus forming a radar netting system.

radar paint [MATER] A coating that absorbs radar waves.

radar picket [ENG] A ship or aircraft equipped with early-warning radar and operating at a distance from the area being protected, to extend the range of radar detection.

radar prediction [ENG] A graphic portrayal of the estimated radar intensity, persistence, and shape of the cultural and natural features of a specific area.

radar range marker See distance marker.

radar relay [ENG] **1.** Equipment for relaying the radar video and appropriate synchronizing signal to a remote location. **2.** Process or system by which radar echoes and synchronization data are transmitted from a search radar installation to a receiver at a remote point.

radar scanning [ENG] The process or action of directing a radar beam through a space search pattern for the purpose of locating a target.

radarscope overlay [ENG] A transparent overlay placed on a radarscope for comparison and identification of radar returns.

radar set [ENG] A complete assembly of radar equipment for detecting and ranging, consisting essentially of a transmitter, antenna, receiver, and indicator. Also known as radar.

radarsonde [ENG] **1.** An electronic system for automatically measuring and transmitting high-altitude meteorological data from a balloon, kite, or rocket by pulse-modulated radio waves when triggered by a radar signal. **2.** A system in which radar techniques are used to determine the range, elevation, and azimuth of a radar target carried aloft by a radiosonde.

radar station [ENG] The place, position, or location from which, or at which, a radar set transmits or receives signals.

radar surveying [ENG] Surveying in which airborne radar is used to measure accurately the distance between two ground radio beacons positioned along a baseline; this eliminates the need for measuring distance along the baseline in inaccessible or extremely rough terrain.

radar telescope [ENG] A large radar antenna and associated equipment used for radar astronomy.

radar theodolite [ENG] A theodolite that uses radar to obtain azimuth, elevation, and slant range to a reflecting target, for surveying or other purposes.

radar threshold limit [ENG] For a given radar and specified target, the point in space relative to the focal point of the antenna at which initial detection criteria can be satisfied.

radar tracking [ENG] Tracking a moving object by means of radar.

radar tracking station [ENG] A radar facility which has the capability of tracking moving targets.

radar triangulation [ENG] A radar system of locating targets, usually aircraft, in which two or more separate radars are employed to measure range only; the target is located by automatic trigonometric solution of the triangle composed of a pair of radars and the target in which all three sides are known.

radar wind system [ENG] Apparatus in which radar techniques are used to determine the range, elevation, and azimuth of a balloon-borne target, and hence to compute upper-air wind data.

radial bearing [MECH ENG] A bearing with rolling contact in which the direction of action of the load transmitted is radial to the axis of the shaft.

radial draw forming [MECH ENG] A metal-forming method in which tangential stretch and radial compression are applied gradually and simultaneously.

radial drill [MECH ENG] A drilling machine in which the drill spindle can be moved along a horizontal arm which itself can be rotated about a vertical pillar.

radial drilling [ENG] The drilling of several holes in one plane, all radiating from a common point.

radial engine [MECH ENG] An engine characterized by radially arranged cylinders at equiangular intervals around the crankshaft.

radial-flow [ENG] Having the fluid working substance flowing along the radii of a rotating tank. [PETRO ENG] Pertaining to spokelike flow of reservoir fluids radially inward toward a wellbore focal area.

radial-flow turbine [MECH ENG] A turbine in which the gases flow primarily in a radial direction.

radial force [MECH ENG] In machining, the force acting on the cutting tool in a direction opposite to depth of cut.

radial gate *See* tainter gate.

radial load [MECH ENG] The load perpendicular to the bearing axis.

radial locating [MECH ENG] One of the three locating problems in tooling to maintain the desired relationship between the workpiece, the cutter, and the body of the machine tool; the other two locating problems are concentric and plane locating.

radial percussive coal cutter [MIN ENG] A heavy coal cutter having a percussive drill, with extension rods; used in headings and rooms in pillar methods of working and for drilling shot-firing holes.

radial-ply [DES ENG] Pertaining to the construction of a tire in which the cords run straight across the tire, and an additional layered belt of fabric is placed around the circumference between the plies and the tread.

radial rake [MECH ENG] The angle between the cutter tooth face and a radial line passing through the cutting edge in a plane perpendicular to the cutter axis.

radial saw [MECH ENG] A power saw that has a circular blade suspended from a transverse head mounted on a rotatable overarm.

radial selector *See* omnibearing selector.

radiant-energy thermometer *See* radiation pyrometer.

radiant heating [ENG] Any system of space heating in which the heat-producing means is a surface that emits heat to the surroundings by radiation rather than by conduction or convection.

radiant superheater [MECH ENG] A superheater designed to transfer heat from the products of combustion to the steam primarily by radiation.

radiant-type boiler [MECH ENG] A water-tube boiler in which boiler tubes form the boundary of the furnace.

radiation [ENG] A method of surveying in which points are located by knowledge of their distances and directions from a central point.

radiation corrosion [MET] Accelerated corrosion of a metal caused by radiation.

radiation hardening [ENG] Improving the ability of a device or piece of equipment to withstand nuclear or other radiation; applies chiefly to dielectric and semiconductor materials.

radiation loss [MECH ENG] Boiler heat loss to the atmosphere by conduction, radiation, and convection.

radiation oven [ENG] Heating chamber relying on tungsten-filament infrared lamps with reflectors to create temperatures up to 600°F (315°C); used to dry sheet and granular material and to bake surface coatings.

radiation pyrometer [ENG] An instrument which measures the temperature of a hot object by focusing the thermal radiation emitted by the object and making some observation on it; examples include the total-radiation, optical, and ratio pyrometers. Also known as radiant-energy thermometer; radiation thermometer.

radiation shield [ENG] A shield or wall of material interposed between a source of radiation and a radiation-sensitive body, such as a person, radiation-detection instrument, or photographic film, to protect the latter.

radiation thermometer *See* radiation pyrometer.

radiation vacuum gage [ENG] Vacuum (reduced-pressure) measurement device in which gas ionization from an α-source of radiation varies measurably with changes in the density (molecular concentration) of the gas being measured.

radiation well logging *See* radioactive well logging.

radiator [ENG] Any of numerous devices, units, or surfaces that emit heat, mainly by radiation, to objects in the space in which they are installed.

radiator temperature drop [MECH ENG] In internal combustion engines, the difference in temperature of the coolant liquid entering and leaving the radiator.

radioacoustic position finding *See* radioacoustic ranging.

radioacoustic ranging [ENG] A method for finding the position of a vessel at sea; a bomb is exploded in the water, and the sound of the explosion transmitted through water is picked up by the vessel and by shore stations, other vessels, or buoys whose positions are known; the received sounds are transmitted instantaneously by radio to the surveying vessel, and the elapsed times are proportional to the distances to the known positions. Abbreviated RAR. Also known as radioacoustic position finding; radioacoustic sound ranging.

radioacoustic sound ranging *See* radioacoustic ranging.

radioactive paint [MATER] A luminous paint that gives off light without being activated.

radioactive snow gage [ENG] A device which automatically and continuously records the water equivalent of snow on a given surface as a function of time; a small sample of a radioactive salt is placed in the ground in a lead-shielded collimator which directs a beam of radioactive particles vertically upward; a Geiger-Müller counting system (located above the snow level) measures the amount of depletion of radiation caused by the presence of the snow.

radioactive well logging [ENG] The recording of the differences in radioactive content (natural or neutron-induced) of the various rock layers found down an oil well borehole; types include γ-ray, neutron, and photon logging. Also known as radiation well logging; radioactivity prospecting.

radioactivity log [ENG] Record of radioactive well logging.

radioactivity prospecting See radioactive well logging.

radio altimeter [ENG] An absolute altimeter that depends on the reflection of radio waves from the earth for the determination of altitude, as in a frequency-modulated radio altimeter and a radar altimeter. Also known as electronic altimeter; reflection altimeter.

radio atmometer [ENG] An instrument designed to measure the effect of sunlight upon evaporation from plant foliage; consists of a porous-clay atmometer whose surface has been blackened so that it absorbs radiant energy.

radioautography See autoradiography.

radio autopilot coupler [ENG] Equipment providing means by which an electrical navigational signal operates an automatic pilot.

radio detection [ENG] The detection of the presence of an object by radiolocation without precise determination of its position.

radio detection and location [ENG] Use of an electronic system to detect, locate, and predict future positions of earth satellites.

radio Doppler [ENG] Direct determination of the radial component of the relative velocity of an object by an observed frequency change due to such velocity.

radio echo observation [ENG] A method of determining the distance of objects in the atmosphere or outer space, in which a radar pulse is directed at the object and the time that elapses from transmission of the pulse to reception of a reflected pulse is measured.

radio engineering [ENG] The field of engineering that deals with the generation, transmission, and reception of radio waves and with the design, manufacture, and testing of associated equipment.

radio-frequency head [ENG] Unit consisting of a radar transmitter and part of a radar receiver, the two contained in a package for ready removal and installation.

radio-frequency heating See electronic heating.

radio-frequency preheating [ENG] Preheating of plastics-molding materials by radio frequencies of 10–100 megahertz per second to facilitate the molding operation or to reduce the molding-cycle time. Abbreviated rf preheating.

radio-frequency welding See high-frequency welding.

radiogoniometry [ENG] Science of locating a radio transmitter by means of taking bearings on the radio waves emitted by such a transmitter.

radio-inertial guidance system [ENG] A command type of missile guidance system consisting essentially of a radar tracking unit; a computer that accepts missile position and velocity information from the tracking system and furnishes to the command link appropriate signals to steer the missile; the command link, which consists of a transmitter on the ground and an antenna and receiver on the missile; and an inertial system for partial guidance in case of radio guidance failure.

radio interferometer [ENG] Radiotelescope or radiometer employing a separated receiving antenna to measure angular distances as small as 1 second of arc; records the result of interference between separate radio waves from celestial radio sources.

radiolocation [ENG] Determination of relative position of an object by means of equipment operating on the principle that propagation of radio waves is at a constant velocity and rectilinear.

radio mast [ENG] A tower, pole, or other structure for elevating an antenna.

radiometer [ENG] An instrument for measuring radiant energy; examples include the bolometer, microradiometer, and thermopile.

radiopasteurization [ENG] Pasteurization by surface treatment with low-energy irradiation.

radio position finding [ENG] Process of locating a radio transmitter by plotting the intersection of its azimuth as determined by two or more radio direction finders.

radio prospecting [ENG] Use of radio and electric equipment to locate mineral or oil deposits.

radio relay satellite See communications satellite.

radiosensitive [MATER] Sensitive to damage by radiant energy.

radiosonde [ENG] A balloon-borne instrument for the simultaneous measurement and transmission of meteorological data; the instrument consists of transducers for the measurement of pressure, temperature, and humidity, a modulator for the conversion of the output of the transducers to a quantity which controls a property of the radio-frequency signal, a selector switch which determines the sequence in which the parameters are to be transmitted, and a transmitter which generates the radio-frequency carrier.

radiosonde balloon [AERO ENG] A balloon used to carry a radiosonde aloft; it is considerably larger than a pilot balloon or a ceiling balloon.

radiosonde-radio-wind system [ENG] An apparatus consisting of a standard radiosonde and radiosonde ground equipment to obtain upper-air data on pressure, temperature, and humidity, and a self-tracking radio direction finder to provide the elevation and azimuth angles of the radiosonde so that the wind vectors may be obtained.

radiosonde set [ENG] A complete set for automatically measuring and transmitting high-altitude meteorological data by radio from such carriers as a balloon or rocket.

radio sonobuoy *See* sonobuoy.

radio telescope [ENG] An astronomical instrument used to measure the amount of radio energy coming from various directions in the sky, consisting of a highly directional antenna and associated electronic equipment.

radio tracking [ENG] The process of keeping a radio or radar beam set on a target and determining the range of the target continuously.

radius cutter [MECH ENG] A formed milling cutter with teeth ground to produce a radius on the workpiece.

radius of action [ENG] The maximum distance a ship, aircraft, or other vehicle can travel away from its base along a given course with normal load and return without refueling, but including the fuel required to perform those maneuvers made necessary by all safety and operating factors.

radius of protection [ENG] The radius of the circle within which a lightning discharge will not strike, due to the presence of an elevated lightning rod at the center.

radius rod [ENG] A rod which restricts movement of a part to a given arc.

raft [ENG] A quantity of timber or lumber secured together by means of ropes, chains, or rods and used for transportation by floating.

rafter [BUILD] A roof-supporting member immediately beneath the roofing material.

rafter dam [CIV ENG] A dam made of horizontal timbers that meet in the center of the stream like rafters in a roof.

raft foundation [CIV ENG] A continuous footing that supports an entire structure, such as a floor. Also known as foundation mat.

rag bolt *See* barb bolt.

rag papers [MATER] The most expensive papers, made wholly or partly from cotton or linen rags; rag content is expressed as 25, 50, 75, or 100%; pure rag papers are the strongest and most resistant to the discoloration and deterioration due to age.

rail [ENG] **1.** A bar extending between posts or other supports as a barrier or guard. **2.** A steel bar resting on the crossties to provide track for railroad cars and other vehicles with flanged wheels. [MECH ENG] A high-pressure manifold in some fuel injection systems.

rail anchor [CIV ENG] A device that prevents tracks from moving longitudinally and maintains the proper gap between sections of rail.

rail bender [ENG] A portable appliance for bending rails for track or for straightening bent or curved rails.

rail capacity [CIV ENG] The maximum number of trains which can be planned to move in both directions over a specified section of track in a 24-hour period.

rail clip [CIV ENG] **1.** A plate that holds a rail at its base. **2.** A device used to fasten a derrick or crane to the rails of a track to prevent tipping. **3.** A support on a track rail, used for holding a detector bar.

rail crane *See* locomotive crane.

railhead [CIV ENG] **1.** The topmost part of a rail, supporting the wheels of railway vehicles. **2.** A point at which railroad traffic originates and terminates. **3.** The temporary ends of a railroad line under construction.

railing [CIV ENG] A barrier consisting of a rail and supports.

rail joint [CIV ENG] A rigid connection of the ends of two sections of railway track.

railroad [CIV ENG] A permanent line of rails forming a route for freight cars and passenger cars drawn by locomotives.

railroad engineering [CIV ENG] That part of transportation engineering involved in the planning, design, development, operation, construction, maintenance, use, or economics of facilities for transportation of goods and people in wheeled units of rolling stock running on, and guided by, rails normally supported on crossties and held to fixed alignment. Also known as railway engineering.

railroad jack [MECH ENG] **1.** A hoist used for lifting locomotives. **2.** A portable jack for lifting heavy objects. **3.** A hydraulic jack, either powered or lever-operated.

railroad thermit [MET] Red thermit to which is added up to 16% nickel, manganese, and steel.

rail steel [MET] Steel used to make rail track.

railway dry dock [CIV ENG] A railway dock consisting of tracks built on an incline on a strong foundation, and extending from a sufficient distance in shore to allow a vessel to be hauled out of the water.

railway end-loading ramp [CIV ENG] A sloping platform situated at the end of a track and rising to the level of the floor of the railcars (wagons).

rainbow [PETRO ENG] Chromatic iridescence observed in drilling fluid that has been circulated in a well, indicating contamination or contact with fresh hydrocarbons.

rain gage [ENG] An instrument designed to collect and measure the amount of rain that has fallen. Also known as ombrometer; pluviometer. Also known as statement; udometer.

rain-gage shield [ENG] A device which surrounds a rain gage and acts to maintain horizontal flow in the vicinity of the funnel so that the catch will not be influenced by eddies generated near the gage. Also known as wind shield.

rain-intensity gage [ENG] An instrument which measures the instantaneous rate at which rain is falling on a given surface. Also known as rate-of-rainfall gage.

raise [MIN ENG] A shaftlike mine opening, driven upward from a level to connect with a level above, or to the surface.

raise boring machine [MIN ENG] A machine that is used to drill pilot holes between levels in a mine, and to ream the pilot hole to the finished dimension of the raise.

raise drill [MIN ENG] A circular raise driving machine which bores a pilot hole and reams it to finished raise diameter.

raise driller [MIN ENG] A person who works in a raise. Also known as raiseman.

Rajakaruna engine [MECH ENG] A rotary engine that uses a combustion chamber whose sides are pin-jointed together at their ends.

rake [BUILD] The exterior finish and trim applied parallel to the sloping end walls of a gabled roof. [DES ENG] A hand tool consisting of a long handle with a row of projecting prongs at one end; for example, the tool used for gathering leaves or grass on the ground. [ENG] The angle between an inclined plane and the vertical. [MECH ENG] The angle between the tooth face or a tangent to the tooth face of a cutting tool at a given point and a reference plane or line.

raked joint [CIV ENG] A mortar, or masonry, joint from which the mortar has been scraped out to about ¾ inch (20 millimeters).

ram [AERO ENG] The forward motion of an air scoop or air inlet through the air. [MECH ENG] A plunger, weight, or other guided structure for exerting pressure or drawing something by impact. [MIN ENG] See barney.

ram effect [MECH ENG] The increased air pressure in a jet engine or in the manifold of a piston engine, due to ram.

ramjet engine [AERO ENG] A type of jet engine with no mechanical compressor, consisting of a specially shaped tube or duct open at both ends, the air necessary for combustion being shoved into the duct and compressed by the forward motion of the engine; the air passes through a diffuser and is mixed with fuel and burned, the exhaust gases issuing in a jet from the rear opening.

ramjet exhaust nozzle [AERO ENG] The discharge nozzle in a ramjet engine; hot gas is ejected rearward through this nozzle.

ramming [ENG] Packing a powder metal or sand into a compact mass.

ramoff [MET] A defect in a casting due to improper ramming of the sand.

ramp [ENG] **1.** A uniformly sloping platform, walkway, or driveway. **2.** A stairway which gives access to the main door of an airplane. [MIN ENG] A slope between levels in open-pit mining.

ram penetrometer See ramsonde.

ramp mining [MIN ENG] The development of moderately inclined accessways from the surface to mining levels for haulage of ore, materials, waste, men, and equipment.

RAMPS See resource allocation in multiproject scheduling.

ramp weight [AERO ENG] The static weight of a mission aircraft determined by adding operating weight, payload, flight plan fuel load, and fuel required for ground turbine power unit, taxi, runup, and takeoff.

ram recovery See recovery.

ram rocket [AERO ENG] **1.** A rocket motor mounted coaxially in the open front end of a ramjet, used to provide thrust at low speeds and to ignite the ramjet fuel. **2.** The entire unit or power plant consisting of the ramjet and such a rocket.

Ramsbottom safety valve [ENG] A steam safety valve with provision for manually checking the pressure adjustment.

ramsonde [ENG] A cone-tipped metal rod or tube that is driven downward into snow to measure its hardness. Also known as ram penetrometer.

ram travel [ENG] In injection or transfer molding, the distance moved by the injection ram when filling the mold.

ram-type turret lathe [MECH ENG] A horizontal turret lathe in which the turret is mounted on a ram or slide which rides on a saddle.

random line [ENG] A trial surveying line that is directed as closely as circumstances permit toward a fixed terminal point that cannot be seen from the initial point. Also known as random traverse.

random-sampling voltmeter [ENG] A sampling voltmeter which takes samples of an input signal at random times instead of at a constant rate; the synchronizing portions of the instrument can then be simplified or eliminated.

random sequence [MET] A longitudinal sequence of weld beads deposited in random increments.

random traverse *See* random line.

Raney nickel [MET] A nickel powder prepared from an alloy of nickel and aluminum in equal parts by preferentially dissolving the aluminum in a warm solution of sodium hydroxide.

range [CIV ENG] Any series of contiguous townships of the U.S. Public Land Survey system. [ENG] **1.** The distance capability of an aircraft, missile, gun, radar, or radio transmitter. **2.** A line defined by two fixed landmarks, used for missile or vehicle testing and other test purposes.

range calibration [ENG] Adjustment of a radar set so that when on target the set will indicate the correct range.

range coding [ENG] Method of coding a radar transponder beacon response so that it appears as a series of illuminated bars on a radarscope; the coding provides identification.

range control [AERO ENG] The operation of an aircraft to obtain the optimum flying time.

range control chart [AERO ENG] A graph kept in flight on which actual fuel consumption is plotted against distance flown for comparison with planned fuel consumption.

range corrector setting [ENG] Degree to which the range scale of a position-finding apparatus must be adjusted before use.

range discrimination *See* distance resolution.

rangefinder *See* optical rangefinder.

range-height indicator [ENG] A scope which simultaneously indicates range and height of a radar target; this presentation is commonly used by height finders.

range marker *See* distance marker.

range oil [MATER] Kerosine similar in properties to a No. 1 distillate heating oil; used for space heating.

range pole *See* range rod.

range recorder [ENG] An item which makes a permanent representation of distance, expressed as range, versus time. [ENG ACOUS] A display used in sonar in which a stylus sweeps across a paper moving at a constant rate and chemically treated so that it is darkened by an electrical signal from the stylus; the stylus starts each sweep as a sound pulse is emitted so that the distance along the trace at which the echo signal appears is a measure of the range to the target.

range resolution *See* distance resolution.

range rod [ENG] A long (6–8 feet or 1.8–2.4 meters) rod fitted with a sharp-pointed metal shoe and usualy painted in 1-foot (30-centimeter) bands of alternate red and white; used for sighting points and lines in surveying or for showing the position of a ground point. Also known as line rod; lining pole; range pole; ranging rod; sight rod.

Ranger program [AERO ENG] A series of nine spacecraft, launched in 1961–1965, designed to transmit photographs back to earth while on a collision course with the moon; the first six Rangers failed, but *Rangers 7, 8,* and *9* successfully transmitted high-resolution television pictures of the lunar surface up to the instant of impact.

range surveillance [ENG] Surveillance of a missile range by means of electronic and other equipment.

ranging rod *See* range rod.

Rankine efficiency [MECH ENG] The efficiency of an ideal engine operating on the Rankine cycle under specified conditions of steam temperature and pressure.

ranking method [IND ENG] A system of job evaluation wherein each job as a whole is given a rank with respect to all the other jobs, and no attempt is made to establish a measure of value.

Ranney oil-mining system [PETRO ENG] A method used to get oil from oil sands that involves driving mine galleries from shafts communicating to the surface in impermeable strata above and below the oil strata; holes drilled at short intervals along the galleries into the oil sands drain the oil or gas through pipes sealed in the drill holes into tanks from which the gas or oil is pumped to the surface.

Ranney well [CIV ENG] A well that has a center caisson with horizontal perforated pipes extending radially into an aquifer; particularly applicable to the development of thin aquifers at shallow depths.

rape oil [MATER] A fatty, nondrying or semidrying, viscous, dark-brown to yellow oil with unpleasant taste and aroma, obtained from the seed of rape and turnip; soluble in ether, carbon disulfide, and chloroform; solidifies at -2 to $-10°C$; used to make lubricants and rubber substitutes, as an illuminant, and in steel heat treatment. Also known as colza oil; rape-seed oil.

rape-seed oil See rape oil.

rapid-curing asphalt [MATER] A liquid asphalt composed of asphalt cement and a gasoline- or naphtha-type diluent. Abbreviated RC asphalt.

rapid quenching [MET] Superfast cooling ($1-5 \times 10^6$ K per second) of a molten metal to produce new and amorphous alloys and new crystalling material with improved properties.

rapid sand filter [CIV ENG] A system for purifying water, which is forced through layers of sand and gravel under pressure.

rapid traverse [MECH ENG] A machine tool mechanism which rapidly repositions the workpiece while no cutting takes place.

RAR See radioacoustic ranging.

rare-earth alloy [MET] An alloy containing rare-earth materials.

rare-earth garnet [MATER] A synthetic garnet having the general structure of grossularite, but with calcium replaced by a rare-earth metal, and aluminum and silicon replaced by iron; used for electronic applications.

rare metal [MET] Any metal that is difficult to extract from ore and is rare and expensive commercially; includes masurium, alabamine, and virginium.

rash [MIN ENG] Very impure coal, so mixed with waste material as to be unsalable.

rashing [MIN ENG] A soft, friable, and flaky or scaly rock (shale or clay) immediately beneath a coal seam, often containing much carbonaceous material and readily mixed with the coal in mining.

RA size [ENG] One of a series of sizes to which untrimmed paper is manufactured; for reels of paper, the standard sizes in millimeters are 430, 610, 860, and 1220; for sheets of paper, the sizes are RA0, 860 × 1220; RA1, 610 × 860; RA2, 430 × 610; RA sizes correspond to A sizes when trimmed.

rasp [DES ENG] A metallic tool with a rough surface of small points used for shaping and finishing metal, plaster, stone, and wood; designed in a number of useful curved shapes.

ratchet [DES ENG] A wheel, usually toothed, operating with a catch or a pawl so as to rotate in only a single direction.

ratchet coupling [MECH ENG] A coupling between two shafts that uses a ratchet to allow the driven shaft to be turned in one direction only, and also to permit the driven shaft to overrun the driving shaft.

ratchet jack [DES ENG] A jack operated by a ratchet mechanism.

ratchet tool [DES ENG] A tool in which torque or force is applied in one direction only by means of a ratchet.

rate climb [AERO ENG] The climb of an aircraft to higher altitudes at a constant rate.

rated capacity [MECH ENG] The maximum capacity for which a boiler is designed, measured in pounds of steam per hour delivered at specified conditions of pressure and temperature.

rated engine speed [MECH ENG] The rotative speed of an engine specified as the allowable maximum for continuous reliable performance.

rate descent [AERO ENG] An aircraft descent from higher altitudes at a constant rate.

rated flow [ENG] **1.** Normal operating flow rate at which a fluid product is passed through a vessel or piping system. **2.** Flow rate for which a vessel or process system is designed.

rated horsepower [MECH ENG] The normal maximum, allowable, continuous power output of an engine, turbine motor, or other prime mover.

rated load [MECH ENG] The maximum load a machine is designed to carry.

rated relieving capacity [DES ENG] The measured relieving capacity for which the pressure relief device is rated in accordance with the applicable code or standard.

rate gyroscope [MECH ENG] A gyroscope that is suspended in just one gimbal whose bearings form its output axis and which is restrained by a spring; rotation of the gyroscope frame about an axis perpendicular to both spin and output axes produces precession of the gimbal within the bearings proportional to the rate of rotation.

rate integrating gyroscope [MECH ENG] A single-degree-of-freedom gyro having primarily viscous restraint of its spin axis about the output axis; an output signal is produced by gimbal angular displacement, relative to the base, which is proportional to the integral of the angular rate of the base about the input axis.

rate of approach [AERO ENG] The relative speed of two aircraft when the distance between them is decreasing.

rate of climb [AERO ENG] Ascent of aircraft per unit time, usually expressed as feet per minute.

rate-of-climb indicator [AERO ENG] A device used to indicate changes in the vertical position of an aircraft by comparing the actual outside air pressure to a reference volume that lags the outside pressure because a calibrated restrictor imposes a lag-time constant to the reference pressure volume. Also known as rate-of-descent indicator; vertical speed indicator.

rate of departure [AERO ENG] The relative speed of two aircraft when the distance between them is increasing.

rate-of-descent indicator See rate-of-climb indicator.

rate-of-flow control valve See flow control valve.

rate-of-rainfall gage See rain-intensity gage.

rate of return [AERO ENG] Aircraft relative to its base, either fixed or moving.

rate of rise [ENG] The time rate of pressure increase during an isolation test for leaks.

rate of strain hardening [MET] Rate of change of true stress with respect to true strain in the plastic range. Also known as modulus of strain hardening.

rate response [ENG] Quantitative expression of the output rate of a control system as a function of its input signal.

rathole [MIN ENG] A shallow, small-diameter, auxiliary hole alongside the main borehole, drilled at an angle to the main hole; after core drilling is completed, the rathole is reamed out and the larger-size hole is advanced, usually by some noncoring method.

rating [ENG] A designation of an operating limit for a machine, apparatus, or device used under specified conditions.

ratio delay study See work sampling.

ratio meter [ENG] A meter that measures the quotient of two electrical quantities; the deflection of the meter pointer is proportional to the ratio of the currents flowing through two coils.

ratio of expansion [MECH ENG] The ratio of the volume of steam in the cylinder of an engine when the piston is at the end of a stroke to that when the piston is in the cutoff position.

ratio of reduction [ENG] The ratio of the maximum size of the stone which will enter a crusher, to the size of its product.

rato [AERO ENG] A rocket system providing additional thrust for takeoff of an aircraft. Derived from rocket-assisted takeoff.

rattail [MET] A small irregular line marking a minor buckle on the surface of a casting.

rattail file [DES ENG] A round tapering file used for smoothing or enlarging holes.

Rauschelback rotor [ENG] A free-turning S-shaped propeller used to measure ocean currents; the number of rotations per unit time is proportional to the flow.

rawhide [MATER] Untanned animal hide that has been dried or treated with a preservative.

raw material [IND ENG] A crude, unprocessed or partially processed material used as feedstock for a processing operation; for example, crude petroleum is the raw material from which naphtha is obtained; naphtha is the raw material from which benzene-toluene-xylene aromatics are obtained.

raw sewage [CIV ENG] Untreated waste materials.

raw sludge [CIV ENG] Sewage sludge preliminary to primary and secondary treatment processes.

raw water [CIV ENG] Water that has not been purified.

Raykin fender [CIV ENG] Sandwich-type fender buffer to protect docks from the impact of mooring ships; made of a connected series of steel plates cemented to layers of rubber.

Raymond concrete pile [CIV ENG] A pile made by driving a thin steel shell into the ground with a tapered mandrel and filling it with concrete.

rayon coning oil [MATER] Lubricant oil used to reduce static in yarns being wound on cones; composed of low-viscosity mineral oils emulsifiable in water.

RC asphalt See rapid-curing asphalt.

reach [CIV ENG] A portion of a waterway between two locks or gages. [ENG] The length of a channel, uniform with respect to discharge, depth, area, and slope.

reach rod [MECH ENG] A rod motion in a link used to transmit motion from the reversing rod to the lifting shaft.

reactant ratio [AERO ENG] The ratio of the weight flow of oxidizer to fuel in a rocket engine.

reaction engine [AERO ENG] An engine that develops thrust by its reaction to a substance ejected from it; specifically, such an engine that ejects a jet or stream of gases created by the addition of energy to the gases in the engine. Also known as reaction motor.

reaction flux [MATER] Soldering flux which reacts chemically with the base metal and has a rapid fluxing action when heated.

reaction motor See reaction engine.

reaction propulsion [AERO ENG] Propulsion by means of reaction to a jet of gas or fluid projected rearward, as by a jet engine, rocket engine, or rocket motor.

reaction turbine [MECH ENG] A power-generation prime mover utilizing the steady-flow principle of fluid acceleration, where nozzles are mounted on the moving element.

reaction wheel [MECH ENG] A device capable of storing angular momentum which may be used in a space ship to provide torque to effect or maintain a given orientation.

reactive dye [MATER] Dye that reacts with the textile fiber to produce both a hydroxyl and an oxygen linkage, the chlorine combining with the hydroxyl to form a strong ether linkage; gives fast, brilliant colors.

reactive fluid [PETRO ENG] Any fluid that alters the internal geometry of a reservoir's porosity; for example, water is a reactive fluid when it causes swelling of clays and consequent changes in porosity.

reactive muffler [ENG] A muffler that attenuates by reflecting sound back to the source. Also known as nondissipative muffler.

readiness time [ENG] The length of time required to obtain a stabilized system ready to perform its intended function (readiness time includes warm-up time); the time is measured from the point when the system is unassembled or uninstalled to such time as it can be expected to perform as accurately as at any later time; maintenance time is excluded from readiness time.

reading [ENG] **1.** The indication shown by an instrument. **2.** Observation of the readings of one or more instruments.

ready-mixed concrete [MATER] Concrete mixed away from the construction site and delivered in readiness for placing.

ream [ENG] To enlarge or clean out a hole. [MATER] **1.** A layer of nonhomogeneous material in flat glass. **2.** Five hundred sheets of paper; a printer's ream consists of 516 sheets.

reamed extrusion ingot [MET] A hollow extrusion ingot whose original inside surface has been removed by reaming.

reamer [DES ENG] A tool used to enlarge, shape, smooth, or otherwise finish a hole.

reaming bit [DES ENG] A bit used to enlarge a borehole. Also known as broaching bit; pilot reaming bit.

rear response [ENG ACOUS] The maximum pressure within 60° of the rear of a transducer in decibels relative to the pressure on the acoustic axis.

rebound clip [DES ENG] A clip surrounding the back and one or two other leaves of a leaf spring, to distribute the load during rebounds.

rebound leaf [DES ENG] In a leaf spring, a leaf placed over the master leaf to limit the rebound and help carry the load imposed by it.

rebreather [ENG] A closed-loop oxygen supply system consisting of gas supply and face mask.

rebuild [ENG] To restore to a condition comparable to new by disassembling the item to determine the condition of each of its component parts, and reassembling it, using serviceable, rebuilt, or new assemblies, subassemblies, and parts.

recalescence [MET] Brightening (reglowing) of iron on cooling through the γ- to α-phase transformation temperature caused by liberation of the latent heat of transformation.

recarburize [MET] **1.** To increase the carbon content of molten steel or cast iron. **2.** To carburize a metal part, making up for any loss of carbon during processing.

receiver [MECH ENG] An apparatus placed near the compressor to equalize the pulsations of the air as it comes from the compressor to cause a more uniform flow of air through the pipeline and to collect moisture and oil carried in the air.

receiving gage [ENG] A fixed gage designed to inspect a number of dimensions and also their reaction to each other.

receiving station [MECH ENG] The location or device on conveyor systems where bulk material is loaded or otherwise received onto the conveyor.

recess [ENG] A surface groove or depression.

recessed tube wall [MECH ENG] A boiler furnace wall which has openings to partially expose waterwall tubes to the radiant combustion gases.

recharge basin [CIV ENG] A basin constructed in sandy material to collect water, as from storm drains, for the purpose of replenishing groundwater supply.

reciprocal leveling [CIV ENG] A variant of straight differential leveling applied to long distances in which levels are taken on two points, and the average of the two elevation differences is the true difference.

reciprocating compressor [MECH ENG] A positive-displacement compressor having one or more cylinders, each fitted with a piston driven by a crankshaft through a connecting rod.

reciprocating drill *See* piston drill.

reciprocating engine *See* piston engine.

reciprocating flight conveyor [MECH ENG] A reciprocating beam or beams with hinged flights that advance materials along a conveyor trough.

reciprocating-plate feeder [MECH ENG] A back-and-forth shaking tray used to feed abrasive materials, such as pulverized coal, into process units.

reciprocating pump *See* piston pump.

reciprocating screen [MECH ENG] Horizontal solids-separation screen (sieve) oscillated back and forth by an eccentric gear; used for solids classification.

reciprocity calibration [ENG ACOUS] A measurement of the projector loss and hydrophone loss of a reversible transducer by means of the reciprocity theorem and comparisons with the known transmission loss of an electric network, without knowing the actual value of either the electric power or the acoustic power.

reciprocity theorem Also known as principle of reciprocity. [ENG ACOUS] The sensitivity of a reversible electroacoustic transducer when used as a microphone divided by the sensitivity when used as a source of sound is independent of the type and construction of the transducer.

recirculator [ENG] A self-contained underwater breathing apparatus that recirculates an oxygen supply (mix-gas or pure) to the diver until the oxygen is depleted.

reclaimed oil [MATER] Used lubricating oil that is collected, reprocessed, and sold for reuse. Also known as recovered oil.

reclaimed rubber [MATER] Scrap rubber (natural or synthetic) prepared for reuse; fragmented scrap is digested in hot caustic solution to which reclaiming agents have been added; reclaimed rubber is used to blend with virgin rubber, or for low-grade rubber products.

reclaim rinse [MET] A nonflowing rinse used to recover dragout in electroplating operations.

reclamation [CIV ENG] **1.** The recovery of land or other natural resource that has been abandoned because of fire, water, or other cause. **2.** Reclaiming dry land by irrigation.

recoil oil [MATER] A neutral, constant-viscosity oil used in hydropneumatic and hydrospring recoil systems.

recompletion [PETRO ENG] Redrilling an oil well to a new producing zone (new depth) when the current zone is depleted.

reconditioning [ENG] Restoration of an object to a good condition.

reconnaissance [ENG] A mission to secure data concerning the meteorological, hydrographic, or geographic characteristics of a particular area.

reconnaissance drone [AERO ENG] An unmanned aircraft guided by remote control, with photographic or electronic equipment for providing information about an enemy or potential enemy.

reconnaissance spacecraft [AERO ENG] A satellite put into orbit about the earth and containing electronic equipment designed to pick up and transmit back to earth information pertaining to activities such as military.

reconnaissance survey [ENG] A preliminary survey, usually executed rapidly and at relatively low cost, prior to mapping in detail and with greater precision.

reconstituted mica [MATER] Mica sheets or shaped objects made by breaking up scrap natural mica, combining with a binder,

and pressing into forms suitable for use as electrical insulating material.

reconstructed coal [MATER] Coal formed from crushed or powdered, briquetted lignite or coal, waterproofed with a coating of pitch.

record changer [ENG ACOUS] A record player that plays a number of records automatically in succession.

recorder *See* recording instrument.

recording head *See* cutter.

recording instrument [ENG] An instrument that makes a graphic or acoustic record of one or more variable quantities. Also known as recorder.

recording optical tracking instrument [ENG] Optical system used for recording data in connection with missile flights.

recording rain gage [ENG] A rain gage which automatically records the amount of precipitation collected, as a function of time. Also known as pluviograph.

recording thermometer *See* thermograph.

record player [ENG ACOUS] A motor-driven turntable used with a phonograph pickup to obtain audio-frequency signals from a phonograph record.

recovered oil *See* reclaimed oil.

recovery [AERO ENG] **1.** The procedure or action that obtains when the whole of a satellite, or a section, instrumentation package, or other part of a rocket vehicle, is retrieved after a launch. **2.** The conversion of kinetic energy to potential energy, such as in the deceleration of air in the duct of a ramjet engine. Also known as ram recovery. **3.** In flying, the action of a lifting vehicle returning to an equilibrium attitude after a nonequilibrium maneuver. [MET] **1.** The percentage of valuable material obtained from a processed ore. **2.** Reduction or elimination of work-hardening effects, usually by heat treatment. [MIN ENG] The proportion or percentage of coal or ore mined from the original seam or deposit. [PETRO ENG] The removal (recovery) of oil or gas from reservoir formations.

recovery area [AERO ENG] An area in which a satellite, satellite package, or spacecraft is recovered after reentry.

recovery capsule [AERO ENG] A space capsule designed to be recovered after reentry.

recovery factor [PETRO ENG] The ratio of recoverable oil reserves to the oil in place in a reservoir.

recovery package [AERO ENG] A package attached to a reentry or other body designed for recovery, containing devices intended to locate the body after impact.

recovery vehicle [MECH ENG] A special-purpose vehicle equipped with winch, hoist, or boom for recovery of vehicles.

recrystallization [MET] A process which takes place in metals and alloys following distortion and fragmentation of constituent crystals by severe mechanical deformation, in which some fragments grow at the expense of others, so that larger, strain-free grains are formed; it progresses slowly at room temperature, but is greatly speeded by annealing.

recrystallization annealing [MET] Producing a new grain structure without phase change by annealing cold-worked metal.

recrystallization temperature [MET] The minimum temperature at which complete recrystallization occurs in an annealed cold-worked metal within a specified time.

rectangular weir [CIV ENG] A weir with a rectangular notch at top for measurement of water flow in open channels; it is simple, easy to make, accurate, and popular.

rectification [CIV ENG] A new alignment to correct a deviation of a stream channel or bank.

rectifier instrument [ENG] Combination of an instrument sensitive to direct current and a rectifying device whereby alternating current (or voltages) may be rectified for measurement.

recuperative air heater [ENG] An air heater in which the heat-transferring metal parts are stationary and form a separating boundary between the heating and cooling fluids.

recuperator [ENG] An apparatus in which heat is conducted from the combustion products to incoming cooler air through a system of thin-walled ducts.

recurring demand [IND ENG] A request made periodically or anticipated to be repetitive by an authorized requisitioner for material for consumption or use, or for stock replenishment.

recycling [ENG] The extraction and recovery of valuable materials from scrap or other discarded materials.

red acetate *See* mordant rouge.

red brass [MET] Brass containing 85% copper, 5% zinc, 5% tin, and 5% lead.

red charcoal [MATER] An impure charcoal made by heating wood to 300°C; much of the oxygen and hydrogen is retained.

red glass [MATER] Soda-zinc glass to which small amounts of cadmium and selenium are added.

red-hardness [MET] In reference to high-speed steel and other cutting tool materials, the property of being hard enough to cut metals even when heated to a dull-red color.

red lake C pigment [MATER] An organic azo pigment; made by coupling the diazonium salt (barium or sodium) of *ortho*-chloro-*meta*-toluidine-*para*-sulfonic acid with β-naphthol; used to color inks, plastics, and rubbers.

Redler conveyor [MECH ENG] A conveyor in which material is dragged through a duct by skeletonized or U-shaped impellers which move the material in which they are submerged because the resistance to slip through the element is greater than the drag against the walls of the duct.

red liquor *See* mordant rouge.

red metal [MET] A copper matte having a copper content of about 48%.

red mud [MET] An iron oxide–rich residue obtained in purifying bauxite in the Bayer process.

red nitric acid [MATER] A type of liquid bipropellant with boiling point 104°F (40°C), freezing point $-80°F$ ($-2°C$), and density 1.58 grams per milliliter; used to supply power for jet propulsion.

red oil [MATER] **1.** Any intermediate-grade petroleum lubricating oil that is red in color by transmitted light; includes so-called red engine oils, bearing oils, and machinery oils. **2.** *See* oleic acid.

red thermit [MET] Thermit made with red iron oxide.

reduced crude [MATER] In petroleum refining, a residual product remaining after removal by distillation or other means of an appreciable quantity of the more volatile components of crude oil.

reduced nickel [MET] Nickel obtained by the precipitation of nickel hydroxide or nickel carbonate onto kieselguhr, then reducing the precipitate by heating it with hydrogen.

reduced oil [MATER] **1.** Oil rerun in a vacuum or steam still from an oil that is already distilled. **2.** An oil made from the residue in the still after another product has been distilled from the crude oil.

reduced viscosity [ENG] In plastics processing, the ratio of the specific viscosity to concentration.

reducing coupling [ENG] A coupling used to connect a smaller pipe to a larger one.

reductant [MET] Coal or other reducing materials introduced in a smelting process to remove oxygen from ores or concentrates.

reduction gear [MECH ENG] A gear train which lowers the output speed.

reduction of area [MET] In tensile testing, the percentage of decrease in cross-sectional area of a specimen at the point of rupture.

reduction ratio [ENG] Ratio of feed size to product size for a mill (crushing or grinding) operation; measured by lump and sieve sizes.

reduction roll [MET] A roller used to reduce the thickness of a piece of metal.

reduction to sea level [ENG] The application of a correction to a measured horizontal length on the earth's surface, at any altitude, to reduce it to its projected or corresponding length at sea level.

redundant system *See* duplexed system.

Redwood viscometer [ENG] A standard British-type viscometer in which the viscosity is determined by the time, in seconds, required for a certain quantity of liquid to pass out through the orifice under given conditions; used for determining viscosities of petroleum oils.

reed [ENG] A thin bar of metal, wood, or cane that is clamped at one end and set into transverse elastic vibration, usually by wind pressure; used to generate sound in musical instruments, and as a frequency standard, as in a vibrating-reed frequency meter.

reed frequency meter *See* vibrating-reed frequency meter.

reed horn [ENG ACOUS] A horn that produces sound by means of a steel reed vibrated by air under pressure.

reeding [ENG] Corrugating or serrating, as in coining or embossing.

reef [MIN ENG] A major ore trend or ore body.

reel [DES ENG] A revolving spool-shaped device used for storage of hose, rope, cable, wire, magnetic tape, and so on.

reel and bead *See* bead and reel.

reel locomotive [MIN ENG] A trolley locomotive with a wire-rope reel for drawing mining cars out of rooms.

reentrant [ENG] Having one or more sections directed inward, as in certain types of cavity resonators.

reentry [AERO ENG] The event when a spacecraft or other object comes back into the sensible atmosphere after being in space.

reentry angle [AERO ENG] That angle of the reentry body trajectory and the sensible atmosphere at which the body reenters the atmosphere.

reentry body [AERO ENG] That part of a space vehicle that reenters the atmosphere after flight above the sensible atmosphere.

reentry nose cone [AERO ENG] A nose cone designed especially for reentry, consists of one or more chambers protected by a heat sink.

reentry trajectory [AERO ENG] That part of a rocket's trajectory that begins at reentry and ends at the target or at the surface.

reentry vehicle [AERO ENG] Any payload-carrying vehicle designed to leave the sensible atmosphere and then return through it to earth.

reentry window [AERO ENG] The area, at the limits of the earth's atmosphere, through which a spacecraft in a given trajectory can pass to accomplish a successful reentry for a landing in a desired region.

reference dimension [DES ENG] In dimensioning, a dimension without tolerance used for informational purposes only, and does not govern machining operations in any way; it is indicated on a drawing by writing the abbreviation REF directly following or under the dimension.

reference fuel [MATER] One of the standardized laboratory engine fuels, blends of which are used to determine the octane numbers of motor gasoline and the cetane numbers of diesel fuels.

reference level [ENG] *See* datum plane. [ENG ACOUS] The level used as a basis of comparison when designating the level of an audio-frequency signal in decibels or volume units. Also known as reference signal level.

reference lot [IND ENG] A lot of select components, used as a standard.

reference plane [ENG] *See* datum plane. [MECH ENG] The plane containing the axis and the cutting point of a cutter.

reference range [ENG] Range obtained from the radar coverage indicator for a given penetrating aircraft.

reference seismometer [ENG] In seismic prospecting, a detector placed to record successive shots under similar conditions, to permit overall time comparisons.

reference signal level *See* reference level.

reference tone [ENG] Stable tone of known frequency continuously recorded on one track of multitrack signal recordings and intermittently recorded on signal track recordings by the collection equipment operators for subsequent use by the data analysts as a frequency reference.

referencing [ENG] The process of measuring the horizontal (or slope) distances and directions from a survey station to nearby landmarks, reference marks, and other permanent objects which can be used in the recovery or relocation of the station.

refine [ENG] To free from impurities, as the separation of petroleum, ores, or chemical mixtures into their component parts.

refined kerosine *See* deodorized kerosine.

refined lecithin *See* lecithin.

refined oil [MATER] A class of petroleum oil used for home lighting and cooking purposes. Also known as burning oil.

refined paraffin wax [MATER] A grade of paraffin wax; a hard, crystalline hydrocarbon wax derived from mixed-base or paraffin-base crude oils.

refined tar [MATER] A tar from which water has been extracted by evaporation or distillation.

refinery [MET] System of process units used to convert nonferrous-metal ores into pure metals, such as copper or zinc.

refinery gas [MATER] Gas produced in petroleum refineries by cracking, reforming, and other processes; principally methane, ethane, ethylene, butanes, and butylenes.

refining temperature [MET] The temperature just above the transformation range employed in the heat treatment of steel in order to refine grain size.

reflected signal indicator [ENG] Pen recorder which presents the radar signals within frequency gates; these recordings enable the operator to determine that an airborne object has penetrated the Doppler link and its direction of penetration.

reflecting nephoscope *See* mirror nephoscope.

reflecting sign [CIV ENG] A road sign painted with reflective paint so as to be easily visible in the light of a headlamp.

reflection altimeter *See* radio altimeter.

reflection goniometer [ENG] A goniometer that measures the angles between crystal faces by reflection of a parallel beam of light from successive crystal faces.

reflection profile [ENG] A seismic profile obtained by designing the spread geometry in such a manner as to enhance reflected energy.

reflection seismology *See* reflection shooting.

reflection shooting [ENG] A procedure in seismic prospecting based on the measurement of the travel times of waves which, originating from an artificially produced disturbance, have been reflected to detectors from subsurface boundaries separating media of different elastic-wave velocities; used primarily for oil and gas exploration. Also known as reflection seismology.

reflection survey [ENG] Study of the presence, depth, and configuration of underground formations; a ground-level explosive charge (shot) generates vibratory energy (seismic rays) that strike formation interfaces and are reflected back to ground-level sensors. Also known as seismic survey.

reflection x-ray microscopy [ENG] A technique for producing enlarged images in which a beam of x-rays is successively reflected at grazing incidence, from two crossed cylindrical surfaces; resolution is about 0.5–1 micrometer.

reflective insulation [MATER] An insulating material used to retard the flow of heat by reflecting heat radiation; usually made of aluminum foil or sheets, although coated steel sheets, aluminized paper, gold and silver surfaces, and refractory metals at higher temperatures are also used.

reflectometer [ENG] A photoelectric instrument for measuring the optical reflectance of a reflecting surface.

reflector microphone [ENG ACOUS] A highly directional microphone which has a surface that reflects the rays of impinging sound from a given direction to a common point at which a microphone is located, and the sound waves in the speech-frequency range are in phase at the microphone.

reflector satellite [AERO ENG] Satellite so designed that radio or other waves bounce off its surface.

reflex baffle [ENG ACOUS] A loudspeaker baffle in which a portion of the radiation from the rear of the diaphragm is propagated forward after controlled shift of phase or other modification, to increase the overall radiation in some portion of the audio-frequency spectrum. Also known as vented baffle.

reflowing [ENG] Melting and resolidifying an electrodeposited or other type coating.

reformate [MATER] Product from a petroleum-refinery reforming process; types are thermal reformate (from thermal reforming), and cat or catalytic reformate (from catalytic reforming).

reformed gas [MATER] A lower-thermal-value fuel gas made by pyrolysis and steam decomposition of high-thermal-value natural and refinery gases.

reformed gasoline [MATER] Gasoline made by a catalytic or thermal reforming process.

refraction process [ENG] Seismic (reflection) survey in which the distance between the explosive shot and the receivers (sensors) is large with respect to the depths to be mapped.

refraction profile [ENG] A seismic profile obtained by designing the spread geometry in such a manner as to enhance refracted energy.

refraction shooting [ENG] A type of seismic shooting based on the measurement of seismic energy as a function of time after the shot and of distance from the shot, by determining the arrival times of seismic waves which have traveled nearly parallel to the bedding in high-velocity layers, in order to map the depth of such layers.

refractometer [ENG] An instrument used to measure the index of refraction of a substance in any one of several ways, such as measurement of the refraction produced by a prism, measurement of the critical angle, observation of an interference pattern produced by passing light through the substance, and measurement of the substance's dielectric constant.

refractory [MATER] **1.** A material of high melting point. **2.** The property of resisting heat.

refractory cement [MATER] Any of a variety of mixtures, such as fireclay-silica-ganister mixture, or fireclay mixed with crushed brick, or fireclay and silica sand, with a refractory range of 2600–2800°F (1412–1523°C); used for furnace and oven linings and for fillers.

refractory clay [MATER] Clay with a melting point above 1600°C; used to make firebrick and linings for furnaces and reactors.

refractory coating [MATER] A coating composed of a refractory material.

refractory concrete [MATER] Heat-resistant concrete made with high-alumina or calcium-aluminate cement and a refractory aggregate.

refractory enamel [MATER] An enamel for coating and protecting metals against attack by hot gases.

refractory-lined firebox boiler [MECH ENG] A horizontal fire-tube boiler with the front portion of the shell located over a refractory furnace; the rear of the shell contains the first-pass tubes, and the second-pass tubes are located in the upper part of the shell.

refractory metal [MET] A metal or alloy that is heat-resistant, having a high melting point.

refractory sand [MATER] Sand used for refractory which is capable of resisting high temperatures.

refrigerant [MATER] A substance that by undergoing a change in phase (liquid to gas, gas to liquid) releases or absorbs a large latent heat in relation to its volume, and thus effects a considerable cooling effect; examples are ammonia, sulfur dioxide, ethyl or methyl chloride (these are no longer widely used), and the fluorocarbons, such as Freon, Ucon, and Genetron.

refrigerated truck [MECH ENG] An insulated truck equipped and used as a refrigerator to transport fresh perishable or frozen products.

refrigeration [MECH ENG] The cooling of a space or substance below the environmental temperature.

refrigeration condenser [MECH ENG] A vapor condenser in a refrigeration system, where the refrigerant is liquefied and discharges its heat to the environment.

refrigeration oil [MATER] A mineral oil with all moisture and wax removed; used for lubricating refrigerating machinery.

refrigeration system [MECH ENG] A closed-flow system in which a refrigerant is compressed, condensed, and expanded to produce cooling at a lower temperature level and rejection of heat at a higher temperature level for the purpose of extracting heat from a controlled space.

refrigerator [MECH ENG] An insulated, cooled compartment.

refrigerator car [MECH ENG] An insulated freight car constructed and used as a refrigerator.

regenerated cellulose [MATER] **1.** Rayon in which the raw cellulose is changed physically but not chemically, such as viscose, cuprammonium, and nitrocellulose rayons. **2.** A transparent cellulose plastic material made by mixing cellulose xanthate with a dilute sodium hydroxide solution to form a viscose.

regenerative air heater [MECH ENG] An air heater in which the heat-transferring members are alternately exposed to heat-surrendering gases and to air.

regenerative cooling [ENG] A method of cooling gases in which compressed gas is cooled by allowing it to expand through a nozzle, and the cooled expanded gas then passes through a heat exchanger where it further cools the incoming compressed gas.

regenerative cycle *See* bleeding cycle.

regenerative engine [AERO ENG] **1.** A jet or rocket engine that utilizes the heat of combustion to preheat air or fuel entering the combustion chamber. **2.** Specifically, to a type of rocket engine in which one of the propellants is used to cool the engine by passing through a jacket prior to combustion.

regenerative pump [MECH ENG] Rotating-vane device that uses a combination of mechanical impulse and centrifugal force to produce high liquid heads at low volumes. Also known as turbine pump.

regenerator [MECH ENG] A device used with hot-air engines and gas-burning furnaces which transfers heat from effluent gases to incoming air or gas.

regional migration [PETRO ENG] Horizontal movement of gas or oil through a reservoir formation as a result of artificial pressure differences created by withdrawal of gas or oil at well sites.

register [ENG] Also known as registration. **1.** The accurate matching or superimposition of two or more images, such as the three color images on the screen of a color television receiver, or the patterns on opposite sides of a printed circuit board, or the colors of a design on a printed sheet. **2.** The alignment of positions relative to a specified reference or coordinate, such as hole alignments in punched cards, or positioning of images in an optical character recognition device. [MECH ENG] The portion of a burner which directs the flow of air used in the combustion process.

register mark [ENG] A mark or line printed or otherwise impressed on a web of material for use as a reference to maintain register.

registration *See* register.

regular element [IND ENG] An element that occurs with a fixed frequency in each work cycle. Also known as repetitive element.

regular lay [DES ENG] The lay of a wire rope in which the wires in the strand are twisted in directions opposite to the direction of the strands.

regular-lay left twist *See* left-laid.

regular motor oil [MATER] A petroleum lubricating oil suitable for use in internal combustion engines under normal operating conditions.

regular sampling [MIN ENG] The continuous or intermittent sampling of the same coal or coke received regularly at a given point.

regular ventilating circuit [MIN ENG] All places in the mine through which there is a positive natural flow of air.

regulated split [MIN ENG] In mine ventilation, a split where it is necessary to control the volumes in certain low-resistance splits to cause air to flow into the splits of high resistance.

regulating reservoir [CIV ENG] A reservoir that regulates the flow in a water-distributing system.

regulator [MIN ENG] An opening in a wall or door in the return airway of a district to increase its resistance and reduce the volume of air flowing.

regulus [MET] Impure metal formed beneath the slag during smelting or reduction of ores.

rehabilitation engineering [ENG] The use of technology to make disabled persons as independent as possible by providing assistive devices to compensate for disability.

Reid equation [PETRO ENG] Relation of gas-well flow rate to pitot-tube readings for various impact pressures.

Reid vapor pressure [ENG] A measure in a test bomb of the vapor pressure in pounds pressure of a sample of gasoline at 100°F (37.8°C).

reinforced beam [CIV ENG] A concrete beam provided with steel bars for longitudinal tension reinforcement and sometimes compression reinforcement and reinforcement against diagonal tension.

reinforced brickwork [CIV ENG] Brickwork strengthened by expanded metal, steel-wire mesh, hoop iron, or thin rods embedded in the bed joints.

reinforced column [CIV ENG] **1.** A long concrete column reinforced with longitudinal bars with ties or circular spirals. **2.** A composite column. **3.** A combination column.

reinforced concrete [CIV ENG] Concrete containing reinforcing steel rods or wire mesh.

reinforced molding compound [MATER] A compound containing polymer or resin and a reinforcing filler, supplied in the form of ready-to-use material as distinguished from premix.

reinforced plastic [MATER] High-strength filled plastic product used for mechanical, construction, and electrical products, automotive components, and ablative coatings; filling can be whiskers of glass, metal, boron, or other materials.

reinforcement [CIV ENG] Strengthening concrete, plaster, or mortar by embedding steel rods or wire mesh in it.

reinforcement of weld [MET] Weld metal that extends beyond the surface or plane of the weld joint.

reinforcing bars [CIV ENG] Steel rods that are embedded in building materials such as concrete for reinforcement.

relative compaction [ENG] The percentage ratio of the field density of soil to the maximum density as determined by standard compaction.

relative-density bottle *See* specific-gravity bottle.

relative force [ENG] Ratio of the force of a test propellant to the force of a standard propellant, measured at the same initial temperature and loading density in the same closed chamber.

relative gravity instrument [ENG] Any device for measuring the differences in the gravity force or acceleration at two or more points.

relative interference effect [ENG ACOUS] Of a single-frequency electric wave in an electroacoustic system, the ratio, usually expressed in decibels, of the amplitude of a wave of specified reference frequency to that of the wave in question when the two waves are equal in interference effects.

relative magnetometer [ENG] Any magnetometer which must be calibrated by measuring the intensity of a field whose strength is accurately determined by other means; opposed to absolute magnetometer.

relative pressure response [ENG ACOUS] The amount, in decibels, by which the acoustic pressure induced by a projector under some specified condition exceeds the pressure induced under a reference condition.

relative transmitting response [ENG ACOUS] In a sonar projector, the ratio of the transmitting response for a given bearing and frequency to the transmitting response for a specified bearing and frequency.

relaxation test [ENG] A creep test in which the decrease of stress with time is measured while the total strain (elastic and plastic) is maintained constant.

relay haulage [MIN ENG] Single-track, high-speed mine haulage from one relay station to another. Also known as intermediate haulage.

relay rack [DES ENG] A standardized steel rack designed to hold 19-inch (48.26-centimeter) panels of various heights, on which are mounted radio receivers, amplifiers, and other units of electronic equipment. Also known as rack.

relay satellite *See* communications satellite.

release [MECH ENG] A mechanical arrangment of parts for holding or freeing a device or mechanism as required.

release agent [MATER] A lubricant, such as wax or silicone oil, used to coat a mold cavity to prevent the molded piece from sticking when removed. Also known as mold release; parting agent.

release altitude [AERO ENG] Altitude of an aircraft above the ground at the time of release of bombs, rockets, missiles, tow targets, and so forth.

reliability [ENG] The probability that a component part, equipment, or system will satisfactorily perform its intended function under given circumstances, such as environmental conditions, limitations as to operating time, and frequency and thoroughness of maintenance for a specified period of time.

relief [MECH ENG] **1.** A passage made by cutting away one side of a tailstock center so that the facing or parting tool may be advanced to or almost to the center of the work. **2.** Clearance provided around the cutting edge by removal of tool material.

relief angle [MECH ENG] The angle between a relieved surface and a tangential plane at a cutting edge.

relief frame [MECH ENG] A frame placed between the slide valve of a steam engine and the steam chest cover; reduces pressure on the valve and thereby reduces friction.

relief hole [ENG] Any of the holes fired after the cut holes and before the lifter holes in breaking ground for tunneling or shaft sinking.

relief well [CIV ENG] A well that drains a pervious stratum, to relieve waterlogging at the surface. [PETRO ENG] A directional well which is drilled to intersect a well that is blowing out, and down which heavy drilling fluid is pumped to kill the blow-out well.

relieving [MECH ENG] Treating an embossed metal surface with an abrasive to reveal the base-metal color on the elevations or highlights of the surface.

relieving arch *See* discharging arch.

relieving platform [CIV ENG] A deck on the land side of a retaining wall to transfer loads vertically down to the wall.

relighter flame safety lamp [MIN ENG] A locked spirit-burning lamp fitted with an internal relighting device.

relish [ENG] The shoulder of a tenon, used in a mortise and tenon system.

reluctance microphone *See* magnetic microphone.

reluctance pickup *See* variable-reluctance pickup.

reluctance pressure transducer [ENG] Pressure-measurement transducer in which pressure changes activate equivalent magnetic-property changes.

remoistening adhesive [MATER] Any adhesive material, such as dextrin, animal glue, or gum arabic, which is reactivated with the application of water upon the adhesive surface.

remotely piloted vehicle [AERO ENG] A robot aircraft, controlled over a two-wave radio link from a ground station or mother aircraft that can be hundreds of miles away; electronic guidance is generally supplemented by remote control television cameras feeding monitor receivers at the control station. Abbreviated RPV.

remote manipulation [ENG] Use of mechanical equipment controlled from a distance to handle materials, such as radioactive materials.

remote manipulator [ENG] A mechanical, electromechanical, or hydromechanical device which enables a person to perform manual operations while separated from the site of the work.

remote metering *See* telemetering.

remote sensing [ENG] The gathering and recording of information without actual contact with the object or area being investigated.

rendezvous [AERO ENG] **1.** The event of two or more objects meeting with zero relative velocity at a preconceived time and place. **2.** The point in space at which such an event takes place, or is to take place.

Renn-Walz process [MET] A method of reclaiming iron and other metals from the waste materials produced in the smelting of zinc and lead ores; this material is brought up to 1000°C in the preheating zone of the kiln by the countercurrent gases, and the oxidized metal vapors are carried off in the flue gases, from which they are subsequently filtered.

reorder cycle [IND ENG] The interval between successive reorder (procurement) actions.

reorder point [IND ENG] An arbitrary level of stock on hand plus stock due in, at or below which routine requisitions for replenishment purposes are submitted in accordance with established requisitioning schedules.

repair [ENG] To restore that which is unserviceable to a serviceable condition by replacement of parts, components, or assemblies.

repair cycle [ENG] The period that elapses from the time the item is removed in a reparable condition to the time it is returned to stock in a serviceable condition.

repair dock [CIV ENG] A graving dock or floating dry dock built primarily for ship repair.

repair forecast [ENG] The quantity of items estimated to be repaired or rebuilt for issue during a stated future period.

repair kit [ENG] A group of parts and tools, not all having the same basic name, used for repair or replacement of the worn or broken parts of an item; it may include instruction sheets and material, such as sandpaper, tape, cement, gaskets, and the like.

repair parts list [ENG] List approved by designated authorities, indicating the total quantities of repair parts, tools, and equipment necessary for the maintenance of a specified number of end items for a definite period of time.

reperforation [PETRO ENG] Creation of new perforations (holes) in oil well tubing opposite to oil-bearing reservoir zones; creates more opportunity for fluid to drain from the formation into the wellbore.

repetitive element See regular element.

repetitive time method [IND ENG] A technique where the stopwatch is read and simultaneously returned to zero at each break point. Also known as snapback method.

replacement bit See reset bit.

replacement demand [ENG] A demand representing replacement of items consumed or worn out.

replacement factor [ENG] The estimated percentage of equipment or repair parts in use that will require replacement during a given period.

replacement study [IND ENG] An economic analysis involving the comparison of an existing facility and a proposed replacement facility.

replica [ENG] A thin plastic or inorganic film which is formed on a surface and then removed from it for study in an electron microscope.

representative sample [MIN ENG] In testing or valuation of a mineral deposit, a sample so large and average in composition as to be considered representative of a specified volume of the surrounding ore body.

repressing [MET] Applying pressure to a pressed and sintered compact to improve some physical property.

repressuring [PETRO ENG] Forcing gas or water under pressure into an oil reservoir with the intention of increasing the recovery of crude oil.

reproducing stylus See stylus.

reproducing system See sound-reproducing system.

required thickness [DES ENG] The thickness calculated by recognized formulas for boiler or pressure vessel construction before corrosion allowance is added.

rerailer [ENG] A small, lightweight Y-shaped device, used to retrack railroad cars and locomotives; as the car is pulled across the device, the derailed wheels are channeled back onto the tracks. Also known as retracker.

resaw [ENG] To cut lumber to boards of final thickness.

resealing pressure [MECH ENG] The inlet pressure at which leakage stops after a pressure relief valve is closed.

research method [ENG] A standard test to determine the research octane number (or rating) of fuels for use in spark-ignition engines.

research octane number [ENG] An expression for the antiknock rating of a motor gasoline as a guide to how vehicles will operate under mild conditions associated with low engine speeds.

research rocket [AERO ENG] A rocket-propelled vehicle used to collect scientific data.

resection [ENG] **1.** A method in surveying by which the horizontal position of an occupied point is determined by drawing lines from the point to two or more points of known position. **2.** A method of determining a plane-table position by orienting along a previously drawn foresight line and drawing one or more rays through the foresight from previously located stations.

reserve aircraft [AERO ENG] Those aircraft which have been accumulated in excess of immediate needs for active aircraft and are retained in the inventory against possible future needs.

reserved minerals [MIN ENG] Economic minerals that belong to the state, which confers the right to prospect for and to mine them on any applicant.

reserves [MIN ENG] The quantity of workable mineral or of gas or oil which is calculated to lie within given boundaries.

reserves-decline relationship [PETRO ENG] Relationship between production-rate decline over a period of time to the total remaining hydrocarbon reserves in a reservoir.

reservoir [CIV ENG] A pond or lake built for storage of water, usually by the construction of a dam across a river.

reservoir cycling [PETRO ENG] Repressuring of an oil reservoir by reinjection of dry gas (gas with liquids stripped out) into the formation.

reservoir drive mechanism [PETRO ENG] The physical action by which hydrocarbons (gas or liquid) are moved through the porous reservoir structure; for example, gas drive or water drive. Also known as oil-well drive.

reservoir dynamics [PETRO ENG] Fluid-flow performance within an oil or gas reservoir.

reset bit [DES ENG] A diamond bit made by reusing diamonds salvaged from a used bit and setting them in the crown attached to a new bit blank. Also known as replacement bit.

reset rate [ENG] The number of times per minute that the effect of the proportional-position action upon the final control element is repeated by the proportional-speed floating action.

resid *See* residual oil.

residual elements [MET] Elements present in small amounts in a metal or alloy, not added intentionally.

residual free gas [PETRO ENG] Free gas-cap gas in equilibrium with residual liquid hydrocarbons in a depleted reservoir, such as a reservoir at the end of its primary or economic producing life.

residual fuel oil [MATER] Topped crude petroleum or viscous residuums from refinery operations; commercial grades of burner oils Nos. 5 and 6 are residual oils, and include the bunker fuels.

residual method [MET] Magnetic particle inspection in which particles are supplied to a specimen after the magnetizing force has been removed.

residual oil [MATER] Petroleum-refinery term for combustible, viscous, or semiliquid bottoms product from crude oil distillation; used in adhesives, roofing compounds, asphalt manufacture, low-grade fuel oils, and sealants. Also known as liquid asphalt; resid; residuum; tailings.

residual tack *See* aftertack.

residuum *See* residual oil.

resin-anchored bolt [ENG] A bolt is anchored in the resin placed at the back of the hole in a glass cartridge, which ruptures when the bolt is inserted.

resin emulsion [MATER] Stable emulsion of a resin in a solvent carrier, such as the latex emulsions used in water-based latex paints.

resinoid wheel [DES ENG] A grinding wheel bonded with a synthetic resin.

resinol [MATER] Heat- and oxidation-sensitive, benzene-soluble coal tar fraction containing phenols; insoluble in light petroleum.

resinous cement [MATER] An acid-proof cement with a base of synthetic resin.

resin roof bolting [MIN ENG] The fixation of metal roof bolts in rock holes with a bonding resin.

resist [MATER] An acid-resistant nonconducting coating used to protect desired portions of a wiring pattern from the action of the etchant during manufacture of printed wiring boards. [MET] An insulating material, for example lacquer, applied to the surface of work to prevent electroplating or electrolytic action at the coated area. Also known as stopoff.

resistance brazing [MET] Brazing employing the heat developed by an electric current, the joint being part of the electric circuit.

resistance furnace [ENG] An electric furnace in which the heat is developed by the passage of current through a suitable internal resistance that may be the charge itself, a resistor embedded in the charge, or a resistor surrounding the charge. Also known as electric resistance furnace.

resistance magnetometer [ENG] A magnetometer that depends for its operation on variations in the electrical resistance of a material immersed in the magnetic field to be measured.

resistance meter [ENG] Any instrument which measures electrical resistance. Also known as electrical resistance meter.

resistance methanometer [ENG] A catalytic methanometer, with platinum used as the filament, which both heats the detecting element and acts as a resistance-type thermometer.

resistance pyrometer *See* resistance thermometer.

resistance-rate flowmeter *See* resistive flowmeter.

resistance seam welding [MET] Resistance welding process which produces a series of individual spot welds, overlapping spot welds, or a continuous nugget weld made by circular or wheel-type electrodes.

resistance spot welding [MET] Resistance welding process in which the parts are lapped and held in place under pressure; the size and shape of the electrodes (usually circular) control the size and shape of the welds.

resistance thermometer [ENG] A thermometer in which the sensing element is a resistor whose resistance is an accurately known function of temperature. Also known as electrical resistance thermometer; resistance pyrometer.

resistance welding [MET] Joining metals together under pressure by making use of heat developed by an electric current, the work being part of the electrical circuit.

resistance wire [MET] Wire made from a metal or alloy having high resistance per unit length, such as Nichrome; used in wire-wound resistors and heating elements.

resistive flowmeter [ENG] Liquid flow-rate measurement device in which flowrates are read electrically as the result of the rise or fall of a conductive differential-pressure manometer fluid in contact with a resistance-rod assembly. Also known as resistance-rate flowmeter.

resistivity index [PETRO ENG] Ratio of the true electrical resistivity of a rock system at a specified water saturation, to the resistivity of the rock itself; used for calculation of electrical well-logging data.

resistivity method [ENG] Any electrical exploration method in which current is introduced in the ground by two contact electrodes and potential differences are measured between two or more other electrodes.

resistivity well logging [PETRO ENG] The measurement of subsurface electrical resistivities (normal and lateral to the borehole) during electrical logging of oil wells.

resistor bulb [ENG] A temperature-measurement device inside of which is a resistance winding; changes in temperature cause corresponding changes in resistance, varying the current in the winding.

resistor furnace [ENG] An electric furnace in which heat is developed by the passage of current through distributed resistors (heating units) mounted apart from the charge.

resistor oven [ENG] Heating chamber relying on an electrical-resistance element to create temperatures of up to 800°F (430°C); used for drying and baking.

resolution in azimuth [ENG] The angle by which two targets must be separated in azimuth in order to be distinguished by a radar set when the targets are at the same range.

resolution in range [ENG] Distance by which two targets must be separated in range in order to be distinguished by a radar set when the targets are on the same azimuth line.

resolving time [ENG] Minimum time interval, between events, that can be detected; resolving time may refer to an electronic circuit, to a mechanical recording device, or to a counter tube.

resonance method [ENG] In ultrasonic testing, a method of measuring the thickness of a metal by varying the frequency of the beam transmitted to excite a maximum amplitude of vibration.

resonant jet [AERO ENG] A pulsejet engine, exhibiting intensification of power under the rhythm of explosions and compression waves within the engine.

resource allocation in multiproject scheduling [IND ENG] A system that employs network analysis as an aid in making the best assignment of resources which must be stretched over a number of projects. Abbreviated RAMPS.

respirator [ENG] A device for maintaining artificial respiration to protect the respiratory tract against irritating and poisonous gases, fumes, smoke, and dusts, with or without equipment supplying oxygen or air; some types have a fitting which covers the nose and mouth.

respirometer [ENG] **1.** An instrument for studying respiration. **2.** A diver's helmet containing a compressed air supply for replenishing oxygen used by the diver.

restart [AERO ENG] The act of firing a stage of a rocket after a previous powered flight.

restraint of loads [ENG] The process of binding, lashing, and wedging items into one unit onto or into its transporter in a manner that will ensure immobility during transit.

restricted adhesive [MATER] An adhesive which for any reason cannot satisfactorily pass its evaluation test; as a result, the maximum time required for curing, that is, its usable life, cannot be assigned; it cannot be used for structural bonding.

restricted air cargo [IND ENG] Cargo which is not highly dangerous under normal conditions, but which possesses certain qualities which require extra precautions in packing and handling.

restricted gate [ENG] Small opening between runner and cavity in an injection or transfer mold which breaks cleanly when the piece is ejected.

restricted work [IND ENG] Manual or machine work where the work pace is only partially under the control of the worker.

resultant rake [MECH ENG] The angle between the face of a cutting tooth and an axial plane through the tooth point measured in a plane at right angles to the cutting edge.

resupply [IND ENG] The act of replenishing stocks in order to maintain required levels of supply.

resuscitator [ENG] A device for supplying oxygen to and inducing breathing in asphyxiation victims.

retainer [ENG] A device that holds a mechanical component in place.

retainer plate [ENG] The plate on which removable mold parts (such as a cavity or ejector pin) are mounted during molding.

retainer wall [ENG] A wall, usually earthen, around a storage tank or an area of storage tanks (tank farm); used to hold (retain) liquid in place if one or more tanks begin to leak.

retaining ring [DES ENG] **1.** A shoulder inside a reaming shell that prevents the core lifter from entering the core barrel. **2.** A steel ring between the races of a ball bearing to maintain the correct distribution of the balls in the races.

retaining wall [CIV ENG] A wall designed to maintain differences in ground elevations by holding back a bank of material.

retard [CIV ENG] A permeable bank-protection structure, situated at and parallel to the toe of a slope and projecting into a stream channel, designed to check stream velocity and induce silting or accretion.

retardant *See* retarder.

retarded acid [PETRO ENG] Oil well acidizing solution whose reactivity is slowed by addition of artificial gums and thickening agents, so that the acid penetrates deeper into the formation before being spent.

retarder [MATER] A material that inhibits the action of another substance, such as flameproofing agents or substances added to cement to retard setting time. Also known as retardant. [MECH ENG] **1.** A braking device used to control the speed of railroad cars moving along the classification tracks in a hump yard. **2.** A strip inserted in a tube of a fire-tube boiler to increase agitation of the hot gases flowing therein.

retarding basin [CIV ENG] A basin designed and operated to provide temporary storage and thus reduce the peak flood flows of a stream.

retarding conveyor [MECH ENG] Any type of conveyor used to restrain the movement of bulk materials, packages, or objects where the incline is such that the conveyed material tends to propel the conveying medium.

reticulated glass [MATER] Ornamental glassware containing interlacing sets of lines.

retracker *See* rerailer.

retreat [MIN ENG] Workings in the opposite direction of advance work which, when completed, will permit the area to be abandoned as finished.

retreater [ENG] A defective maximum thermometer of the liquid-in-glass type in which the mercury flows too freely through the constriction; such a thermometer will indicate a maximum temperature that is too low.

retrievable inner barrel [ENG] The inner barrel assembly of a wire-line core barrel, designed for removing core from a borehole without pulling the rods.

retroactive refit *See* retrofit.

retrofire time [AERO ENG] The computed starting time and duration of firing of retrorockets to decrease the speed of a recovery capsule and make it reenter the earth's atmosphere at the correct point for a planned landing.

retrofit [ENG] A modification of equipment to incorporate changes made in later production of similar equipment; it may be done in the factory or field. Derived from retroactive refit.

retrograde gas-condensate reservoir *See* dew-point reservoir.

retrorocket [AERO ENG] A rocket fitted on or in a spacecraft, satellite, or the like to produce thrust opposed to forward motion. Also known as braking rocket.

return [BUILD] The continuation of a molding, projection, member, cornice, or the like, in a different direction, usually at a right angle.

return bead *See* quirk bead.

return bend [DES ENG] A pipe fitting, equal to two ells, used to connect parallel pipes so that fluid flowing into one will return in the opposite direction through the other.

return connecting rod [MECH ENG] A connecting rod whose crankpin end is located on the same side of the crosshead as the cylinder.

return-flow burner [MECH ENG] A mechanical oil atomizer in a boiler furnace which regulates the amount of oil to be burned by the portion of oil recirculated to the point of storage.

return idler [MECH ENG] The idler or roller beneath the cover plates on which the conveyor belt rides after the load which it was carrying has been dumped.

return wall [BUILD] An interior wall of about the same height as the outside wall of a building; distinct from a partition or a low wall.

return water [PETRO ENG] In a water-injection operation (waterflood) for an oil reservoir, the reinjection of salt water that is produced along with the oil.

reveal [BUILD] **1.** The side of an opening for a door or window, doorway, or the like, between the doorframe or window frame and the outer surface of the wall. **2.** The distance from the face of a door to the face of the frame on the pivot side.

reverberatory furnace [ENG] A furnace in which heat is supplied by burning of fuel in a space between the charge and the low roof.

reversal speed [AERO ENG] The speed of an aircraft above which the aeroelastic loads will exceed the control surface loading of a given flight control system; the resultant load will act in the reverse direction from the control surface loading, causing the control system to act in a direction opposite to that desired.

reverse circulation drilling [MIN ENG] **1.** A variation of the rotary drilling method in which the cuttings are pumped up and out of the drill pipe, an advantage in certain large diameter holes. **2.** Diamond core drilling in which the water is injected through a stuffing box into the annular space around the drill rods and thus forced up special large drill rods.

reverse-current cleaning See anodic cleaning.

reverse drawing [MET] Drawing for a second time, in a direction opposite to the original drawing.

reverse flange [ENG] A flange made by shrinking.

reverse lay [DES ENG] The lay of a wire rope with strands alternating in a right and left lay.

reverse pitch [MECH ENG] A pitch on a propeller blade producing thrust in the direction opposite to the normal one.

reverse polarity [MET] An arc-welding circuit in which the electrode is connected to the positive terminal.

reverse-printout typewriter [ENG] An automatic typewriter that eliminates conventional carriage return by typing one line from left to right and the next line from right to left.

reverse-roll coating [ENG] Substrate coating that is premetered between rolls and then wiped off on the web; amount of coating is controlled by the metering gap and the rotational speed of the roll.

reversible-pitch propeller [MECH ENG] A type of controllable-pitch propeller; of either controllable or constant speed, it has provisions for reducing the pitch to and beyond the zero value, to the negative pitch range.

reversible steering gear [MECH ENG] A steering gear for a vehicle which permits road shock and wheel deflections to come through the system and be felt in the steering control.

reversible tramway See jig back.

reversible transit circle [ENG] A transit circle that can be lifted out of its bearings and

rotated through 180°, enabling systematic errors in both orientations to be determined.

reversing mill [MET] A rolling mill in which the workpiece is passed forward and backward through a given pair of rolls.

reversing thermometer [ENG] A mercury-in-glass thermometer which records temperature upon being inverted and thereafter retains its reading until returned to the first position.

reversing water bottle See Nansen bottle.

revetment [CIV ENG] A facing made on a soil or rock embankment to prevent scour by weather or water.

revolution counter [ENG] An instrument for registering the number of revolutions of a rotating machine. Also known as revolution indicator.

revolution indicator See revolution counter.

revolving-block engine [MECH ENG] Any of various engines which combine reciprocating piston motion with rotational motion of the entire engine block.

revolving door [BUILD] A door consisting of four leaves that revolve together on a central vertical axis within a circular vestibule.

revolving shovel [MECH ENG] A digging machine, mounted on crawlers or on rubber tires, that has the machinery deck and attachment on a vertical pivot so that it can swing freely.

Reynier's isolator [ENG] A mechanical barrier made of steel that surrounds the area in which germ-free vertebrates and accessory equipment are housed; has electricity for light and power, an exit-entry opening with a steam barrier, a means for sterile air exchange, glass viewing port, and neoprene gloves which allow handling of the animals.

rf preheating See radio-frequency preheating.

rheocasting [MET] A process in which a liquid metal is vigorously agitated during initial stages of solidification to produce a globular semisolid structure which remains highly fluid when more than 60% solidification has occurred.

rheostatic braking [ENG] A system of dynamic braking in which direct-current drive motors are used as generators and convert the kinetic energy of the motor rotor and connected load to electrical energy, which in turn is dissipated as heat in a braking rheostat connected to the armature.

rheotaxial growth [ENG] A chemical vapor deposition technique for producing silicon diodes and transistors on a fluid layer having high surface mobility.

rheotropic brittleness [MET] A low-temperature or high-strain-rate brittleness that may be eliminated by prestraining under milder conditions.

rhinestone [MATER] A clear, colorless imitation of diamond, made of glass, paste, or gem quartz, backed with metallic foil.

rhodamine toner [MATER] Rhodamine dye and phosphotungstic or phosphomolybdic acid; red to maroon; used in printing inks.

rhodinol [MATER] Colorless, combustible liquid mixture of terpene alcohols with rose scent; soluble in mineral oil and alcohol; derived from geranium oil; used in perfumes and flavors.

rhodinyl acetate [MATER] Terpene-alcohol-acetates mixture; colorless-to-yellow, combustible liquid with rose scent; soluble in mineral oil, alcohol, and glycerin; used in perfumes and flavors.

rhodium [MET] A silver-white metal in the platinum family; sometimes alloyed with platinum for thermocouples or used as a tarnish-resistant electrode posit.

rhythmic driving [MIN ENG] Driving carried out between two shifts; that is, the drilling, loading, and blasting are carried out in one shift and the mucking and transportation in the following one.

RIAA curve [ENG ACOUS] **1.** Recording Industry Association of America curve representing standard recording characteristics for long-play records. **2.** The corresponding equalization curve for playback of long-play records.

rib [AERO ENG] A transverse structural member that gives cross-sectional shape and strength to a portion of an airfoil. [MIN ENG] **1.** A solid pillar of coal or ore left for support. **2.** A thin stratum in a seam of coal.

rib arch [CIV ENG] An arch consisting of ribs placed side by side and extending from the springings on one end to those on the other end.

ribbed-clamp coupling [DES ENG] A rigid coupling which is split longitudinally and bored to shaft diameter, with a shim separating the two halves.

ribbon [BUILD] A horizontal piece of wood nailed to the face of studs; usually used to support the floor joists.

ribbon conveyor [MECH ENG] A type of screw conveyor which has an open space between the shaft and a ribbon-shaped flight, used for wet or sticky materials which would otherwise build up on the spindle.

ribbon microphone [ENG ACOUS] A microphone whose electric output results from the motion of a thin metal ribbon mounted between the poles of a permanent magnet and driven directly by sound waves; it is velocity-actuated if open to sound waves on both sides, and pressure-actuated if open to sound waves on only one side.

ribbon mixer [MECH ENG] Device for the mixing of particles, slurries, or pastes of solids by the revolution of an elongated helicoid (spiral) ribbon of metal.

ribbon parachute [AERO ENG] A type of parachute having a canopy consisting of an arrangement of closely spaced tapes; this parachute has high porosity with attendant stability and slight opening shock.

rib hole [MIN ENG] One of the final holes fired in blasting ground at the sides of a shaft or tunnel. Also known as trimmer.

rib pillar [MIN ENG] A pillar whose length is large compared with its width.

rice bran oil [MATER] Clear, combustible liquid, derived by solvent-extraction of oil from fresh rice bran; used to make soaps and animal feeds, salad and cooking oils, and hydrogenated shortening.

rice glue [MATER] A paste made from ground rice boiled in soft water; used in molded objects such as statuary.

rice paper [MATER] **1.** A product, not a true paper and not made from rice, but manufactured from the pith of a tree grown in Taiwan; tissue-thin sheets of the pith are peeled away as a cylindrical section of the wood rotates against a knife. **2.** Any of various oriental papers used in block printing.

Richardson automatic scale [ENG] An automatic weighing and recording machine for flowable materials carried on a conveyor; weighs batches from 200 to 1000 pounds (90 to 450 kilograms).

Richard's solder [MET] A yellow brass containing 3% aluminum and 3% phosphor tin.

rich concrete [MATER] Concrete with a high cement content.

rich oil [MATER] Natural-gasoline-plant absorption oil containing dissolved natural-gasoline fractions.

rich ore [MIN ENG] Relatively high grade ore.

ricin [MATER] White, poisonous powder derived from pressed castor oil bran.

ricinus oil *See* castor oil.

riddle [DES ENG] A sieve used for sizing or for removing foreign material from foundry sand or other granular materials.

rider [MIN ENG] A steel or iron crossbeam which slides between the guides in a sinking shaft; it is carried by the hoppit and serves to guide and steady the hoppit during its movement up and down the shaft.

ridge board [BUILD] A horizontal board placed on edge at the apex of the roof.

ridge cap [BUILD] Wood or metal cap which is placed over the angle of the ridge.

ridge pole [BUILD] The horizontal supporting member placed along the ridge of a roof.

riffler [DES ENG] A small, curved rasp or file for filing interior surfaces or enlarging holes.

rifle [DES ENG] A drill core that has spiral grooves on its outside surface. [ENG] A borehole that is following a spiral course.

rifling [MECH ENG] The technique of cutting helical grooves inside a rifle barrel to impart a spinning motion to a projectile around its long axis.

rift saw [DES ENG] **1.** A saw for cutting wood radially from the log. **2.** A circular saw divided into toothed arms for sawing flooring strips from cants.

rig [MECH ENG] A tripod, derrick, or drill machine complete with auxiliary and accessory equipment needed to drill.

rigging [AERO ENG] The shroud lines attached to a parachute.

right-and-left-hand chart [IND ENG] A graphic symbolic representation of the motions made by one hand in relation to those made by the other hand.

right-cut tool [DES ENG] A single-point lathe tool which has the cutting edge on the right side when viewed face up from the point end.

right-hand cutting tool [DES ENG] A cutter whose flutes twist in a clockwise direction.

right-handed [DES ENG] **1.** Pertaining to screw threads that allow coupling only by turning in a clockwise direction. **2.** See right-laid.

right-hand screw [DES ENG] A screw that advances when turned clockwise.

right-laid [DES ENG] Rope or cable construction in which strands are twisted counterclockwise. Also known as right-handed.

right lang lay [DES ENG] Rope or cable in which the individual wires or fibers and the strands are twisted to the right.

right-of-way [CIV ENG] **1.** Areas of land used for a road and along the side of the roadway. **2.** A thoroughfare or path established for public use. **3.** Land occupied and used by a railroad or a public utility.

rigid coupling [MECH ENG] A mechanical fastening of shafts connected with the axes directly in line.

rigid frame [BUILD] A steel skeleton frame in which the end connections of all members are rigid so that the angles they make with each other do not change.

rigidizer [ENG] A supporting structure providing ridigity to an instrument that might otherwise be subject to undesirable vibrations.

rigid pavement [CIV ENG] A thick portland cement pavement on a gravel base and subbase, with steel reinforcement and often with transverse joints.

rigid PVC [MATER] Polyvinyl chloride or a polyvinyl chloride–acetate copolymer with a relatively high hardness; may be formulated with or without a small percentage of plasticizer; a rigid resin.

rigid resin [MATER] A resin with a modulus of 10,000 psi (6.895×10^7 newtons per square meter) or greater.

rim [DES ENG] **1.** The outer part of a wheel, usually connected to the hub by spokes. **2.** An outer edge or border, sometimes raised or projecting.

rim clutch [MECH ENG] A frictional contact clutch having surface elements that apply pressure to the rim either externally or internally.

rim drive [ENG ACOUS] A phonograph or sound recorder drive in which a rubber-covered drive wheel is in contact with the inside of the rim of the turntable.

rimmed steel [MET] Low-carbon steel, partially deoxidized, which on cooling continuously, evolves sufficient carbon monoxide to form a case or rim of metal virtually free of voids.

ring [DES ENG] A tie member or chain link; tension or compression applied through the center of the ring produces bending moment, shear, and normal force on radial sections.

ring and circle shear [DES ENG] A rotary shear designed for cutting circles and rings where the edge of the metal sheet cannot be used as a start.

ringbolt [DES ENG] An eyebolt with a ring passing through the eye.

ring crusher [MECH ENG] Solids-reduction device with a rotor having loose crushing rings held outwardly by centrifugal force, which crush the feed by impact with the surrounding shell.

Ringelmann chart [ENG] A chart used in making subjective estimates of the amount of solid matter emitted by smoke stacks; the observer compares the grayness of the smoke with a series of shade diagrams formed by horizontal and vertical black lines on a white background.

ring gage [DES ENG] A cylindrical ring of steel whose inside diameter is finished to gage tolerance and is used for checking the external diameter of a cylindrical object.

ring gate [CIV ENG] A type of gate used to regulate and control the discharge of a morning-glory spillway; like a drum gate, it offers a minimum of interference to the passage of ice or drift over the gate and requires no external power for operation.

ring gear [MECH ENG] The ring-shaped gear in an automobile differential that is driven by the propeller shaft pinion and transmits power through the differential to the line axle.

ring holes [MIN ENG] The group of bore-holes radially drilled from a common-center setup.

ringing time [ENG] In an ultrasonic testing unit, the length of time that the vibrations in a piezoelectric crystal remain after the generation of ultrasonic waves ceases.

ring jewel [DES ENG] A type of jewel used as a pivot bearing in a time-keeping device, gyro, or instrument.

ring job [MECH ENG] Installation of new piston rings on a piston.

ring laser *See* laser gyro.

ring lifter *See* split-ring core lifter.

ring-lifter case *See* lifter case.

ringlock nail [DES ENG] A nail ringed with grooves to provide greater holding power.

ring-oil [MECH ENG] To oil (a bearing) by conveying the oil to the point to be lubricated by means of a ring, which rests upon and turns with the journal, and dips into a reservoir containing the lubricant.

ring-roller mill [MECH ENG] A grinding mill in which material is fed past spring-loaded rollers that apply force against the sides of a revolving bowl. Also known as roller mill.

ring-rolling [MET] Producing a thin, large-diameter ring from a thicker, smaller-diameter ring by placing the ring between two rotating rolls.

ring stress [MIN ENG] The zone of stress in rock surrounding all development excavations.

riometer [ENG] An instrument that measures changes in ionospheric absorption of electromagnetic waves by determining and recording the level of extraterrestrial cosmic radio noise. Derived from relative ionospheric opacity meter.

riot-control agent [MATER] A chemical that produces temporary irritating or disabling effects when in contact with the eyes or when inhaled.

Rio Tinto process [MIN ENG] Heap leaching of curiferous sulfides that have been oxidized to sulfates by prolonged atmospheric weathering.

rip [ENG] To saw wood with the grain. [MIN ENG] To break down the roof in mine roadways to increase the headroom for haulage, traffic, and ventilation.

rip panel [AERO ENG] A part of a manned free balloon; it is the panel to which the ripcord is attached and extends about ¼ to ⅕ of the circumference of the balloon along one of its meridians; it is torn open when the ripcord is pulled so that all the gas in the balloon escapes.

ripping bar [DES ENG] A steel bar with a chisel at one end and a curved claw for pulling nails at the other.

ripping face support [MIN ENG] A timber or steel support at the ripping lip.

ripping lip [MIN ENG] The end of the enlarged roadway section where work is proceeding.

ripping punch [DES ENG] A tool with a rectangular cutting edge, used in a punch press to crosscut metal plates.

riprap [CIV ENG] A foundation or revetment in water or on soft ground made of irregularly placed stones or pieces of boulders; used chiefly for river and harbor work, for roadway filling, and on embankments.

ripsaw [MECH ENG] A heavy-tooth power saw used for cutting wood with the grain.

riser [CIV ENG] **1.** A board placed vertically beneath the tread of a step in a staircase. **2.** A vertical steam, water, or gas pipe. [MET] *See* feedhead. [PETRO ENG] In an offshore drilling facility, a system of piping extending from the hole and terminating at the rig.

riser plate [CIV ENG] A plate used to support a tapering switch rail above the base of the rail; used with a railroad gage or tie plate to maintain minimum gage.

rising hinge [BUILD] A hinge that raises a door slightly as it is opened.

Rittinger's law [MECH ENG] The law that energy needed to reduce the size of a solid particle is directly proportional to the resultant increase in surface area.

river engineering [CIV ENG] A branch of transportation engineering consisting of the physical measures which are taken to improve a river and its banks.

river gage [ENG] A device for measuring the river stage; types in common use include the staff gage, the water-stage recorder, and wire-weight gage. Also known as stream gage.

river mining [MIN ENG] Mining or excavating beds of existing rivers after deflecting their course, or by dredging without changing the flow of water.

rivet [DES ENG] A short rod with a head formed on one end; it is inserted through aligned holes in parts to be joined, and the protruding end is pressed or hammered to form a second head.

riveting [ENG] The permanent joining of two or more machine parts or structural members, usually plates, by means of rivets.

riveting hammer [MECH ENG] A hammer used for driving rivets.

rivet pitch [ENG] The center-to-center distance of adjacent rivets.

rivet weld [MET] A weld shaped like a countersunk rivet.

road [CIV ENG] An open way for travel and transportation. [MIN ENG] Any mine passage or tunnel.

roadbed [CIV ENG] The earth foundation of a highway or a railroad.

road capacity [CIV ENG] The maximum traffic flow obtainable on a given roadway, using all available lanes, usually expressed in vehicles per hour or vehicles per day.

road grade [CIV ENG] The level and gradient of a road, measured along its center way.

road net [CIV ENG] The system of roads available within a particular locality or area.

road octane [ENG] A numerical value for automotive antiknock properties of a gasoline; determined by operating a car over a stretch of level road or on a chassis dynamometer under conditions simulating those encountered on the highway.

road oil [MATER] A heavy residual petroleum oil, usually one of the slow-curing grades of liquid asphalt.

road test [ENG] A motor-vehicle test conducted on the highway or on a chassis dynamometer to determine the performance of fuels or lubricants or the performance of the vehicle.

roadway [CIV ENG] The portion of the thoroughfare over which vehicular traffic passes.

roast [MET] To heat ore to effect some chemical change that will facilitate smelting.

roaster [ENG] Equipment for the heating of materials, such as in pyrite roasting; a furnace.

roast sintering See blast roasting.

rob [MIN ENG] To take out ore or coal from a mine with a view to immediate product, and not to subsequent working.

robber [MET] An extra cathode that reduces current density at local areas of the work being electroplated for the purpose of producing a more uniform thickness coating.

robbing pillars See pillaring.

Roberts' linkage [MECH ENG] A type of approximate straight-line mechanism which provided, early in the 19th century, a practical means of making straight metal guides for the slides in a metal planner.

Robins-Messiter system [MECH ENG] A stacking conveyor system in which material arrives on a conveyor belt and is fed to one or two wing conveyors.

Robitzsch actinograph [ENG] A pyranometer whose design utilizes three bimetallic strips which are exposed horizontally at the center of a hemispherical glass bowl; the outer strips are white reflectors, and the center strip is a blackened absorber; the bimetals are joined in such a manner that the pen of the instrument deflects in proportion to the difference in temperature between the black and white strips, and is thus proportional to the

intensity of the received radiation; this instrument must be calibrated periodically.

rockair [AERO ENG] A high-altitude sounding system consisting of a small solid-propellant research rocket carried aloft by an aircraft; the rocket is fired while the aircraft is in vertical ascent.

rock-a-well [PETRO ENG] The procedure of bleeding pressure alternately from the casing of a well and from the tubing until the well starts flowing.

rock bit [ENG] Any one of many different types of roller bits used on rotary-type drills for drilling large-size holes in soft to medium-hard rocks.

rockbolt [ENG] A bar, usually constructed of steel, which is inserted into predrilled holes in rock and secured for the purpose of ground control.

rock bolting [ENG] A method of securing or strengthening closely jointed or highly fissured rocks in mine workings, tunnels, or rock abutments by inserting and firmly anchoring rock bolts oriented perpendicular to the rock face or mine opening.

rock bump [MIN ENG] The sudden release of the weight of the rocks over a coal seam or of enormous lateral stresses.

rockburst [MIN ENG] A sudden and violent rock failure around a mining excavation on a sufficiently large scale to be considered a hazard.

rock channeler [MECH ENG] A machine used in quarrying for cutting an artificial seam in a mass of stone.

rock drill [MECH ENG] A machine for boring relatively short holes in rock for blasting purposes; motive power may be compressed air, steam, or electricity.

rock dust distributor See rock duster.

rock duster [MIN ENG] A machine that distributes rock dust over the interior surfaces of a coal mine by means of air to prevent coal dust explosions. Also known as rock dust distributor.

rocker [CIV ENG] A support at the end of a truss or girder which permits rotation and horizontal movement to allow for expansion and contraction. [MIN ENG] A small digging bucket mounted on two rocker arms in which auriferous alluvial sands are agitated by oscillation, in water, to collect gold.

rocker arm [MECH ENG] In an internal combustion engine, a lever that is pivoted near its center and operated by a pushrod at one end to raise and depress the valve stem at the other end.

rocker bearing [CIV ENG] A bridge support that is free to rotate but cannot move horizontally.

rocker bent [CIV ENG] A bent used on a bridge span; hinged at one or both ends to provide for the span's expansion and contraction.

rocker cam [MECH ENG] A cam that moves with a rocking motion.

rocker dump car [MIN ENG] A small-capacity mining car; the most popular and most widely used are the gravity dump types, designed so that the weight of the load tips the body when a locking latch is released by hand.

rocker panel [ENG] The part of the paneling on a passenger vehicle located below the passenger compartment doorsill.

rocker shovel [MIN ENG] A digging and loading machine consisting of a bucket attached to a pair of semicircular runners; lifts and dumps the bucket load into a car or another materials-transport unit behind the machine.

rocket [AERO ENG] **1.** Any kind of jet propulsion capable of operating independently of the atmosphere. **2.** A complete vehicle driven by such a propulsive system.

rocket airplane [AERO ENG] An airplane using a rocket or rockets for its chief or only propulsion.

rocket assist [AERO ENG] An assist in thrust given an airplane or missile by use of a rocket motor or rocket engine during flight or during takeoff.

rocket chamber [AERO ENG] A chamber for the combustion of fuel in a rocket; in particular, that section of the rocket engine in which combustion of propellants takes place.

rocket engine [AERO ENG] A reaction engine that contains within itself, or carries along with itself, all the substances necessary for its operation or for the consumption or combustion of its fuel, not requiring intake of any outside substance and hence capable of operation in outer space. Also known as rocket motor.

rocket fuel [MATER] Any of the substances or mixtures of substances that can burn rapidly with controlled combustion to produce large volumes of gas at high pressures and temperatures; includes monopropellants (hydrogen peroxide and hydrazine), liquid bipropellant fuels (organic fuel and oxidizer), and solid propellants (mixed oxidizer-fuel in a propellant grain).

rocket igniter [AERO ENG] An igniter for the propellant in a rocket.

rocket launcher [AERO ENG] A device for launching a rocket, wheel-mounted, motorized, or fixed for use on the ground; rocket launchers are mounted on aircraft, as under the wings, or are installed below or on the decks of ships.

rocket motor *See* rocket engine.

rocket nose section [AERO ENG] The extreme forward portion of a rocket, designed to contain instrumentation, spotting charges, fusing or arming devices, and the like, but does not contain the payload.

rocket propellant [MATER] **1.** Any agent which is used for consumption or combustion in a rocket, and from which the rocket derives its thrust, such as a fuel, oxidizer, and additive. **2.** The ejected fluid in a nuclear rocket.

rocket propulsion [AERO ENG] Reaction propulsion by a rocket engine.

rocket ramjet [AERO ENG] A ramjet engine having a rocket mounted within the ramjet duct, the rocket being used to bring the ramjet up to the necessary operating speed. Also known as ducted rocket.

rocketry [AERO ENG] **1.** The science or study of rockets, embracing theory, research, development, and experimentation. **2.** The art and science of using rockets, especially rocket ammunition.

rocket sled [AERO ENG] A sled that runs on a rail or rails and is accelerated to high velocities by a rocket engine; the sled is used in determining g tolerances and for developing crash survival techniques.

rocket-sled testing [AERO ENG] A method of subjecting structures and devices to high accelerations or decelerations and aerodynamic flow phenomena under controlled conditions; the test object is mounted on a sled chassis running on precision steel rails and accelerated by rockets or decelerated by water scoops.

rocketsonde *See* meteorological rocket.

rocket staging [AERO ENG] The use of successive rocket sections or stages, each having its own engine or engines; each stage is a complete rocket vehicle in itself.

rocket station [ENG] A life-saving station equipped with line-carrying rocket apparatus.

rocket thrust [AERO ENG] The thrust of a rocket engine.

rocket tube [AERO ENG] **1.** A launching tube for rockets. **2.** A tube or nozzle through which rocket gases are ejected.

rock-fill [CIV ENG] Composed of large, loosely placed rocks.

rock-fill dam [CIV ENG] A dam constructed of loosely placed rock or stone.

rocking furnace [MECH ENG] A horizonal, cylindrical melting furnace that is rolled back and forth on a geared cradle.

rocking pier [CIV ENG] A pier that is hinged to allow for longitudinal expansion or contraction of the bridge.

rocking valve [MECH ENG] An engine valve in which a disk or cylinder turns in its seat to permit fluid flow.

rocklath [MATER] A sheet of gypsum used as a base for plaster.

rock loader [MIN ENG] Any device or machine used for loading slate or rock inside a mine.

rockoon [AERO ENG] A high-altitude sounding system consisting of a small solid-propellant research rocket carried aloft by a large plastic balloon.

rock saw [MIN ENG] A type of mechanical miner that is used to remove large blocks of material; cuts narrow slots or channels by the action of a moving steel band or blade and a slurry of abrasive particles (sometimes diamonds) rather than teeth; small flame jets are also used.

rockshaft [MIN ENG] A shaft through which rock can be brought into a mine for filling, stopes, or other excavations.

Rocksite program [MIN ENG] A U.S. Navy program concerned with undersea mining or consolidated mineral deposits; studied direct sea-floor access at remote sites through shafts drilled in the sea floor.

Rockwell hardness [ENG] A measure of hardness of a material as determined by the Rockwell hardness test.

Rockwell hardness test [ENG] One of the arbitrarily defined measures of resistance of a material to indentation under static or dynamic load; depth of indentation of either a steel ball or a 120° conical diamond with rounded point, $\frac{1}{16}$, $\frac{1}{8}$, $\frac{1}{4}$, or $\frac{1}{2}$ inch (1.5875, 3.175, 6.35, 12.7 millimeters) in diameter, called a brale, under prescribed load is the basis for Rockwell hardness; 60, 100, 150 kilogram load is applied with a special machine, and depth of impression under initial minor load is indicated on a dial whose graduations represent hardness number.

rock wool *See* mineral wool.

rod [DES ENG] **1.** A bar whose end is slotted, tapered, or screwed for the attachment of a drill bit. **2.** A thin, round bar of metal or wood.

rod bit [DES ENG] A bit designed to fit a reaming shell that is threaded to couple directly to a drill rod.

rod coupling [DES ENG] A double-pin-thread coupling used to connect two drill rods together.

rod dope [MATER] Grease or other material used to protect or lubricate drill rods. Also called gunk; rod grease.

rodenticide [MATER] A chemical agent used to kill rodents.

rod grease *See* rod dope.

rod level [ENG] A spirit level attached to a level rod or stadia rod to ensure the vertical position of the rod prior to instrument reading.

rod mill [MECH ENG] A pulverizer operated by the impact of heavy metal rods. [MET] A mill for making metal rods.

rod pump [PETRO ENG] Type of oil well sucker-rod pump that can be inserted into or removed from oil well tubing without moving or disturbing the tubing itself. Also known as insert pump.

rod slide *See* slide.

rod string [MECH ENG] Drill rods coupled to form the connecting link between the core barrel and bit in the borehole and the drill machine at the collar of the borehole.

rod stuffing box [ENG] An annular packing gland fitting between the drill rod and the casing at the borehole collar; allows the rod to rotate freely but prevents the escape of gas or liquid under pressure.

Rogallo wing [AERO ENG] A glider folded inside a spacecraft; to be deployed during the spacecraft's reentry like a parachute, gliding the spacecraft to a landing.

Rohrback solution [MATER] Toxic, clear, yellow liquid used for specific-gravity separation of minerals and microchemical detection of alkaloids.

roily oil [MATER] Crude petroleum oil that is more or less emulsified with water.

rolamite mechanism [MECH ENG] An elemental mechanism consisting of two rollers contained by two parallel planes and bounded by a fixed S-shaped band under tension.

roll [MECH ENG] A cylinder mounted in bearings; used for such functions as shaping, crushing, moving, or printing work passing by it. [MIN ENG] *See* horseback.

roll compacting [MET] Compacting a metal powder by using a rolling mill.

roll control [ENG] The exercise of control over a missile so as to make it roll to a programmed degree, usually just before pitchover.

roll crusher [MECH ENG] A crusher having one or two toothed rollers to reduce the material.

rolled glass [MATER] Thick flat glass made by passing a roller over the molten glass.

rolled gold [MET] Same as gold-filled except that the proportion of gold alloy to total weight of the article may be less than 1/20; fineness of the gold alloy may not be less than 10 karat.

rolled joint [ENG] A joint made by expanding a tube in a tube sheet hole by use of an expander.

roller [DES ENG] A cylindrical device for transmitting motion and force by rotation.

roller analyzer [ENG] Device for quantitative separation of fine particles (down to 5 micrometers) by use of the graduated lift of a variable-rate pneumatic stream.

roller bearing [MECH ENG] A shaft bearing characterized by parallel or tapered steel rollers confined between outer and inner rings.

roller bit *See* cone rock bit.

roller cam follower [MECH ENG] A follower consisting of a rotatable wheel at the end of the shaft.

roller chain [MECH ENG] A chain drive assembled from roller links and pin links.

roller coating [ENG] The application of paints, lacquers, or other coatings onto raised designs or letters by means of a roller.

roller conveyor [MECH ENG] A gravity conveyor having a track of parallel tubular rollers set at a definite grade, usually on antifriction bearings, at fixed locations, over which package goods which are sufficiently rigid to prevent sagging between rollers are moved by gravity or propulsion.

roller gate [CIV ENG] A cylindrical, usually hollow crest gate that is raised and lowered by large toothed wheels running on sloping racks.

roller-hearth kiln [ENG] A type of tunnel kiln through which the ware is conveyed on ceramic rollers.

roller leveling [MECH ENG] Leveling flat stock by passing it through a machine having a series of rolls whose axes are staggered about a mean parallel path by a decreasing amount.

roller mill *See* ring-roller mill.

roller pulverizer [MECH ENG] A pulverizer operated by the crushing action of rotating rollers.

roller stamping die [MECH ENG] An engraved roller used for stamping designs and other markings on sheet metal.

roll flattening *See* flattening.

roll forging [MET] Forging metal by using grooved rotating dies.

roll forming [MET] Metal forming by using contoured rolls.

rolling [MET] Reducing or changing the cross-sectional area of a workpiece by the compressive forces exerted by rotating rolls. Also known as metal rolling.

rolling-contact bearing [MECH ENG] A bearing composed of rolling elements interposed between an outer and inner ring.

rolling door [ENG] A door that moves up and down or from side to side by means of wheels moving along a track.

rolling lift bridge [CIV ENG] A bridge having on the shore end of the lifting portion a segmental bearing that rolls on a flat surface.

roll mill [MECH ENG] A series of rolls operating at different speeds for grinding and crushing.

roll resistance spot welding [MET] Resistance spot welding using rotating circular electrodes.

roll roofing [MATER] Composition sheet roofing supplied in rolls from which it is laid in overlapping strips.

roll straightening [ENG] Unbending of metal stock by passing it through staggered rolls in different planes.

roll threading [MECH ENG] Threading a metal workpiece by rolling it either between grooved circular rolls or between grooved straight lines.

roll welding [MET] Forge welding by heating in a furnace and applying pressure with rolls.

roof [BUILD] The cover of a building or similar structure. [MIN ENG] The rock immediately above a coal seam; corresponds to a hanging wall in metal mining.

roof beam [BUILD] A load-bearing member in the roof structure.

roof bolt [MIN ENG] One of the long steel bolts driven into walls or roofs of underground excavations to strengthen the pinning of rock strata.

roof control [MIN ENG] The study of rock behavior when undermined by mining operations, and the most effective measures to control movements.

roof cut [MIN ENG] A machine cut made with a turret coal cutter in the roof immediately above the seam.

roofing [MATER] Material used in roof construction, such as tar, tar paper, shingles, slate, and tin.

roofing copper [MET] Copper that has been hot-rolled to sheets in 14- to 32-ounce (400- to 900-gram) weights.

roofing felt [MATER] Thick asphalt-impregnated paper used for roofing.

roofing granules [MATER] Graded particles of crushed rock, slate, slag, porcelain, or tile, used as surfacing on asphalt roofing and shingles.

roofing nail [DES ENG] A nail used for attaching paper or shingle to roof boards; usually short with a barbed shank and a large flat head.

roofing putty [MATER] Heavy consistency asphalt solution with asbestos fibers; used for caulking metal roofs.

roofing slate [MATER] Hard varieties of slate varying in size from 12 × 6 inches (30 × 15 centimeters) to 24 × 14 inches (60 × 35 centimeters), and from ⅛ to ¾ inch (3 to 19 millimeters) in thickness.

roof jack [MIN ENG] A screw- or pump-type extension post used as a temporary roof support.

roof stringer [MIN ENG] A lagging bar running parallel with the working place above the header in a weak or scaly top in narrow rooms or entries which have short life.

roof truss [BUILD] A truss used in roof construction; it carries the weight of roof deck and framing and of wind loads on the upper chord; an example is a Fink truss.

room [BUILD] A partitioned-off area inside a building or dwelling. [MIN ENG] **1.** Space driven off an entry in which coal is produced. **2.** Working place in a flat mine.

room-and-pillar [MIN ENG] A system of mining in which the coal or ore is mined in rooms separated by narrow ribs or pillars; pillars are subsequently worked.

room conveyor [MIN ENG] Any conveyor which carries coal from the face of a room toward the mouth.

room crosscut *See* breakthrough.

root [CIV ENG] The portion of a dam which penetrates into the ground where the dam joins the hillside. [DES ENG] The bottom of a screw thread.

root circle [DES ENG] A hypothetical circle defined at the bottom of the tooth spaces of a gear.

root crack [MET] A crack in the weld or in the heat-affected zone at the root of the weld.

rooter [ENG] A heavy plowing device equipped with teeth and used for breaking up the ground surface; a towed scarifier.

root fillet [DES ENG] The rounded corner at the angle of a gear tooth flank and the bottom land.

root of joint [MET] The area of closest proximity between members of a joint to be welded.

root of weld [MET] The points at which the bottom of the weld and the base metal surfaces intersect.

root opening [MET] In welding, the distance between members at the root of the joint.

root pass [MET] The first weld bead deposited in a multiple pass weld. Also known as root sealer bead.

root penetration [MET] The depth of penetration of the weld metal into the root of a joint.

Roots blower [MECH ENG] A compressor in which a pair of hourglass-shaped members rotate within a casing to deliver large volumes of gas at relatively low pressure increments.

rope [MATER] A long, flexible object which consists of many strands of wire, plastic, or vegetable fiber such as manila.

rope-and-button conveyor [MECH ENG] A conveyor consisting of an endless wire rope or cable with disks or buttons attached at intervals.

rope boring [ENG] A method similar to rod drilling except that rigid rods are replaced by a steel rope to which the boring tools are attached and allowed to fall by their own weight.

rope cutter *See* hook tender.

rope drive [MECH ENG] A system of ropes running in grooved pulleys or sheaves to transmit power over distances too great for belt drives.

rope rider *See* trip rider.

rope sheave [DES ENG] A grooved wheel, usually made of cast steel or heat-treated alloy steel, used for rope drives.

rope socket [DES ENG] A drop-forged steel device, with a tapered hole, which can be fastened to the end of a wire cable or rope and to which a load may be attached.

ropewalk [ENG] A long walkway down which a worker carries and lays rope in a manufacturing plant.

ropeway [ENG] One or a pair of steel cables between several supporting towers which serve as tracks for transporting materials in mountainous areas or at sea.

rose absolute [MATER] First filtrate from the cooled alcohol solvent solution of rose concrete (after removal of waxes) during perfume manufacture; pure oil of rose.

rose bit [DES ENG] A hardened steel or alloy noncore bit with a serrated face to cut or mill out bits, casing, or other metal objects lost in the hole.

rose chucking reamer [DES ENG] A machine reamer with a straight or tapered shank and a straight or spiral flute; cutting is done at the ends of the teeth only; produces a rough hole since there are few teeth.

rose concrete [MATER] Semisolid mixture of essential oils and waxes solvent-extracted from rose flower petals, leaves, and bark.

rose flower oil *See* rose oil.

rosemary oil [MATER] Pungent, combustible, colorless to yellow essential oil with camphorlike aroma; soluble in ether, glacial acetic acid, and alcohol; derived from flowers of rosemary (*Rosmarinus officinalis*); used in flavors, perfumes, and medicines.

rose oil [MATER] Transparent, combustible, yellow-to-green or red essential oil with fragrant scent and sweet taste; solidifies at 18–37°C; steam-distilled from rose flowers; used in flavors, perfumes, and medicines. Also known as attar of roses; otto of rose oil; rose flower oil.

rose reamer [DES ENG] A reamer designed to cut on the beveled leading ends of the teeth rather than on the sides.

Rose's metal [MET] An alloy of bismuth tin and lead; melts at 94°C.

rosette [MET] **1.** Rounded constituents in a microstructure arranged in whorls. **2.** Strain gages arranged to indicate at a single position the strains in three different directions.

rose water [MATER] Aqueous solution with rose scent (from steam disillation of fresh flowers of rose plants); used in lotions, flavors, and perfumes.

rosin [MATER] A translucent yellow, umber, or reddish resinous residue from the distillation of crude turpentine from the sap of pine trees (gum rosin) or from an extract of the stumps and other parts of the tree (wood rosin); used in varnishes, lacquers, printing inks, adhesives, and soldering fluxes, in medical ointments, and as a preservative.

rosin-core solder [MATER] Solder made up in tubular or other hollow form, with the inner space filled with noncorrosive rosin flux.

rosin essence [MATER] That part of rosin that can be distilled off at a temperature below 360°C.

rosin-extended rubber [MATER] Cold rubber with up to 50% rosin.

rosin oil [MATER] Viscous, water-insoluble, white-to-brown liquid; soluble in ether, chloroform, carbon disulfide, and fatty oils; distilled from rosin; used as a lubricant and in adhesives, inks, and linoleum.

rosin size [MATER] An alkali-treated rosin used as a dry powder or emulsion to surface-size paper products.

Ross feeder [MECH ENG] A chute for conveying bulk materials by means of a screen of heavy endless chains hung on a sprocket shaft; rotation of the shaft causes materials to slide.

Rossman drive [ENG] A method used to provide speed control of alternating-current motors; an induction motor stator is mounted on trunnion bearings and driven with an auxiliary motor, to provide the desired change in slip between the stator and rotor.

rot See curl.

rotameter [ENG] A variable-area, constant-head, rate-of-flow volume meter in which the fluid flows upward through a tapered tube, lifting a shaped weight to a position where upward fluid force just balances its weight.

rotary [MECH ENG] **1.** A rotary machine, such as a rotary printing press or a rotary well-drilling machine. **2.** The turntable and its supporting and rotating assembly in a well-drilling machine.

rotary abutment meter [ENG] A type of positive displacement meter in which two displacement rotating vanes interleave with cavities on an abutment rotor in such a way that the three elements are geared together.

rotary actuator [MECH ENG] A device that converts electric energy into controlled rotary force; usually consists of an electric motor, gear box, and limit switches.

rotary air heater [MECH ENG] A regenerative air heater in which heat-transferring members are moved alternately through the gas and air streams.

rotary annular extractor [MECH ENG] Vertical, cylindrical shell with an inner, rotating cylinder; liquids to be contacted flow countercurrently through the annular space between the rotor and shell; used for liquid-liquid extraction processes.

rotary atomizer [MECH ENG] A hydraulic atomizer having the pump and nozzle combined.

rotary belt cleaner [MECH ENG] A series of blades symmetrically spaced about the axis of rotation and caused to scrape or beat against the conveyor belt for the purpose of cleaning.

rotary blasthole drilling [MIN ENG] A term applied to two types of drilling: in quarrying and open pit mining it implies rotary drilling with roller-type bits, using compressed air for cuttings removal, either conventional rotary table drive or hydraulic motor to produce rotation, with hydraulic or wire-line mechanisms to add part of the weight of the drill to the weight of the tools to increase bit pressure; and in underground mining and sometimes aboveground, it implies the drilling of small-diameter blastholes with a diamond drill, using either coring or noncoring diamond bits.

rotary blower [MECH ENG] Positive-displacement, rotating-impeller, air-movement device; can be straight-lobe, screw, sliding-vane, or liquid-piston type.

rotary boring [MECH ENG] A system of boring in which rock penetration is achieved by the rotation of the hollow cutting tool.

rotary breaker [MIN ENG] A breaking machine for coal or ore; consists of a trommel screen with a heavy steel shell fitted with lifts which raise and convey the coal and stone forward and break it; as the material is broken, the undersize passes through the apertures.

rotary bucket [MECH ENG] A 12- to 96-inch-diameter (30- to 244-centimeter-diameter) posthole augerlike device, the bottom end of which is equipped with cutting teeth used to rotary-drill large-diameter shallow holes to obtain samples of soil lying above the groundwater level.

rotary compressor [MECH ENG] A positive-displacement machine in which compression of the fluid is effected directly by a rotor and without the usual piston, connecting rod, and crank mechanism of the reciprocating compressor.

rotary crane [MECH ENG] A crane consisting of a boom pivoted to a fixed or movable structure.

rotary crusher [MECH ENG] Solids-reduction device in which a high-speed rotating cone on a vertical shaft forces solids against a surrounding shell.

rotary-cup oil burner [ENG] Oil burner that uses centrifugal force to spray fuel oil from a rotary fuel atomizing cup into the combustion chamber.

rotary cutter [MECH ENG] Device used to cut tough or fibrous materials by the shear action between two sets of blades, one set on a rotating holder, the other stationary on the surrounding casing.

rotary drill [MECH ENG] Any of various drill machines that rotate a rigid, tubular string of rods to which is attached a rock cutting bit, such as an oil well drilling apparatus.

rotary drilling [MECH ENG] The act or process of drilling a borehole by means of a rotary-drill machine, such as in drilling an oil well.

rotary dryer [MECH ENG] A cylindrical furnace slightly inclined to the horizontal and rotated on suitable bearings; moisture is removed by rising hot gases.

rotary dump car [MIN ENG] A small mine car in which the car body is mounted on a rotary dumper.

rotary dumper [MIN ENG] A steel structure on which a mine car revolves and discharges the contents, usually sideways.

rotary engine [MECH ENG] A positive displacement engine (such as a steam or internal combustion type) in which the thermodynamic cycle is carried out in a mechanism that is entirely rotary and without the more customary structural elements of a reciprocating piston, connecting rods, and crankshaft.

rotary excavator See bucket-wheel excavator.

rotary feeder [MECH ENG] Device in which a rotating element or vane discharges powder or granules at a predetermined rate.

rotary filter See drum filter.

rotary furnace [MECH ENG] A heat-treating furnace of circular construction which rotates the workpiece around the axis of the furnace during heat treatment; workpieces are transported through the furnace along a circular path.

rotary kiln [ENG] A long cylindrical kiln lined with refractory, inclined at a slight angle, and rotated at a slow speed.

rotary-percussive drill [MECH ENG] Drilling machine which operates as a rotary machine by the action of repeated blows to the bit.

rotary pump [MECH ENG] A displacement pump that delivers a steady flow by the action of two members in rotational contact.

rotary rig [PETRO ENG] The collective equipment used with a rotary drill; includes prime movers or engines, derrick or mast, hoisting and rotating equipment, drill pipe, drill collars and bit, and the mud system used to circulate drilling fluid.

rotary roughening [MECH ENG] A metal preparation technique in which the workpiece surface is roughened by a cutting tool.

rotary shear [MECH ENG] A sheet-metal cutting machine having two rotary-disk cutters mounted on parallel shafts and driven in unison.

rotary shot drill [MECH ENG] A rotary drill used to drill blastholes.

rotary swager [MECH ENG] A machine for reducing diameter or wall thickness of a bar or tube by delivering hammerlike blows to the surface of the work supported on a mandrel.

rotary table [MECH ENG] A milling machine attachment consisting of a round table with T-shaped slots and rotated by means of a handwheel actuating a worm and worm gear. [PETRO ENG] A circular unit on the floor of a derrick which rotates the drill pipe and bit.

rotary vacuum filter See drum filter.

rotary valve [MECH ENG] A valve for the admission or release of working fluid to or from an engine cylinder where the valve member is a ported piston that turns on its axis.

rotary-vane meter [ENG] A type of positive-displacement rate-of-flow meter having spring-loaded vanes mounted on an eccentric drum in a circular cavity; each time the drum rotates, a fixed volume of fluid passes through the meter.

rotary vibrating tippler [MIN ENG] A tippler designed to overcome the tendency for coal or dirt to stick to the bottom of the tubs so that when the tippler is inverted, the car rests upon a vibrating frame which frees any material tending to stick to the bottom.

rotary voltmeter [ENG] Type of electrostatic voltmeter used for measuring high voltages.

rotating-beam ceilometer [ENG] An electronic, automatic-recording meteorological device which determines cloud height by means of triangulation.

rotating meter See velocity-type flowmeter.

rotating spreader [ENG] Plastics-molding injection device consisting of a finned torpedo that is rotated by a shaft extending through a tubular cross-section injection ram behind it.

rotating viscometer vacuum gage [ENG] Vacuum (reduced-pressure) measurement device in which the torque on a spinning armature is proportional to the viscosity (and the pressure) of the rarefied gas being measured; sensitive for absolute pressures of 1 millimeter of mercury (133.32 newtons per square meter), down to a few tens of micrometers.

rotational casting [ENG] Method to make hollow plastic articles from plastisols and lattices using a hollow mold rotated in one or two planes; the hot mold fuses the plastisol into a gel, which is then chilled and the product stripped out. Also known as rotational molding.

rotational molding *See* rotational casting.

rotational viscometer *See* Couette viscometer.

rotation anemometer [ENG] A type of anemometer in which the rotation of an element serves to measure the wind speed; rotation anemometers are divided into two classes: those in which the axis of rotation is horizontal, as exemplified by the windmill anemometer; and those in which the axis is vertical, such as the cup anemometer.

rotation firing [ENG] Setting off explosions so that each hole throws its burden toward the space made by the preceding explosions.

rotor [AERO ENG] An assembly of blades designed as airfoils that are attached to a helicopter or similar aircraft and rapidly rotated to provide both lift and thrust. [MECH END] *See* impeller.

rotor balancing *See* shaft balancing.

rottenstone [MATER] A soft, decomposed limestone, light gray to olive in color; used in powder form as a polishing material for metal and wood.

rouge [MATER] Finely divided, hydrated iron oxide, used in polishing glass, metal, or gems, and as a pigment.

rough air [AERO ENG] An aviation term for turbulence encountered in flight.

rough-axed brick *See* axed brick.

rough burning [AERO ENG] Pressure fluctuations frequently observed at the onset of burning and at the combustion limits of a ramjet or rocket.

roughcast [CIV ENG] A rough finish on a surface; in particular, a plaster made of lime and shells or pebbles, applied by throwing it against a wall with a trowel.

rough cut [ENG] A heavy cut (or cuts) made before the finish cut, the primary object of which is the rapid removal of material.

rougher cell [MIN ENG] Flotation cells in which the bulk of the gangue is removed from the ore.

rough grinding [MECH ENG] Preliminary grinding without regard to finish.

rough hardware [ENG] Utility items such as nails, sash balances, and studs, without attractive finished appearance.

roughing [ENG] The start of evacuation of a vacuum system under test for leaks.

roughing stand [MET] The first stand of rolls, or the last stand before the finishing rolls, through which a preheated billet is passed.

roughing tool [ENG] A single-point cutting tool having a sharp or small-radius nose, used for deep cuts and rapid material removal from the workpiece.

rough lumber [MATER] Sawed, undressed lumber.

rough machining [MECH ENG] Preliminary machining without regard to finish.

roughness-width cutoff [MECH ENG] The maximum width of surface irregularities included in roughness height measurements.

rough threading [ENG] **1.** Rapid removal of the bulk of the material in a threading operation. **2.** Roughening a surface prior to hot-metal spraying to enhance adhesions.

rough turning [MECH ENG] The removal of excess stock from a workpiece as rapidly and efficiently as possible.

round [ENG] A series of shots fired either simultaneously or with delay periods between them.

round-face bit [DES ENG] Any bit with a rounded cutting face.

round file [DES ENG] A file having a circular cross section.

round-head bolt [DES ENG] A bolt having a rounded head at one end.

round-head buttress dam [CIV ENG] A mass concrete dam built of parallel buttresses thickened at the upstream end until they meet.

roundnose chisel [DES ENG] A chisel having a rounded cutting edge.

roundnose tool [DES ENG] A large-radius-nose cutting tool generally used in finishing operations.

round strand rope [DES ENG] A rope composed generally of six strands twisted together or laid to form the rope around a core of hemp, sisal, or manila, or, in a wire-cored rope, around a central strand composed of individual wires.

round trip [ENG] The combined operations of entering and leaving a hole during drilling operations.

route locking [CIV ENG] Electrically locking in position switches, movable point frogs, or derails on the route of a train, after the train has passed a proceed signal.

router plane [DES ENG] A plane for cutting grooves and smoothing the bottom of grooves.

route survey [CIV ENG] A survey for the design and construction of linear works, such as roads and pipelines.

roving [MATER] Fibrous glass in which spun strands are woven into a tubular rope.

row shooting [MIN ENG] Setting off a row of holes nearest the face first, and then other rows in succession behind it.

royal jelly [MATER] A protein complex high in vitamin B secreted by bees to nourish the larva of the queen bee; used in face creams.

rubber adhesive [MATER] An adhesive made with a rubber base by using natural or synthetic rubber in an evaporative solvent; a tacky mixture of rubber and filler material, as used on pressure-sensitive tapes; or rubber-solvent-catalyst mixtures (usually two-part) that cure in place.

rubber-base paint [MATER] A paint in which chlorinated rubber or synthetic latex is the nonvolatile vehicle.

rubber belt [DES ENG] A conveyor belt that consists essentially of a rubber-covered fabric; fabric is cotton, or nylon or other synthetic fiber, with steel-wire reinforcement.

rubber blanket [ENG] A rubber sheet used as a functional die in rubber forming.

rubber cement [MATER] An adhesive composed of unvulcanized rubber in an organic solvent.

rubber-covered steel conveyor [DES ENG] A steel conveyor band with a cover of rubber bonded to the steel.

rubber fiber [MATER] A fiber composed of natural or synthetic rubber; used to make elastic yarn for clothing.

rubber foam See rubber sponge.

rubber plating [ENG] The laying down of a rubber coating onto metals by electrodeposition or by ionic coagulation.

rubber solvent [MATER] Fast-evaporating petroleum distillate used as a solvent for tackifying rubber during plying (laminating) operations, and in compounding rubber cements.

rubber sponge [MATER] Foamed, flexible rubber; produced by beating air into unvulcanized latex, or by incorporating a gas-producing ingredient (such as sodium bicarbonate) into a strongly masticated rubber stock; used for comfort cushioning, packaging, and shock insulation. Also known as cellular rubber; foam rubber; rubber foam; sponge rubber.

rubber wheel [DES ENG] A grinding wheel made with rubber as the bonding agent.

rubbing oil [MATER] **1.** A low-viscosity petroleum oil used either with or without an abrasive to polish dried surfaces, such as paint. **2.** A nonviscous oil used for polishing wood furniture.

rubble [CIV ENG] **1.** Rough, broken stones and other debris resulting from the deterioration and destruction of a building. **2.** Rough stone or brick used in coarse masonry or to fill the space in a wall between the facing courses.

rubble-mound structure [CIV ENG] A mound of nonselectively formed and placed stones which are protected with a covering layer of selected stones or of specially shaped concrete armored elements.

rubidium magnetometer See rubidium-vapor magnetometer.

rubidium-vapor magnetometer [ENG] A highly sensitive magnetometer in which the spin precession principle is combined with optical pumping and monitoring for detecting and recording variations as small as 0.01 gamma (0.1 microoersted) in the total magnetic field intensity of the earth. Also known as rubidium magnetometer.

ruby glass [MATER] Glass of a rich red color produced by adding selenium or cadmium sulfide, or copper oxide to the glass.

rudder [ENG] **1.** A flat, usually foil-shaped movable control surface attached upright to the stern of a boat, ship, or aircraft, and used to steer the craft. **2.** See rudder angle.

rudder angle [ENG] The acute angle between a ship or plane's rudder and its fore-and-aft line. Also known as rudder.

rue oil [MATER] Yellow, combustible essential oil, soluble in most fixed and mineral oils; derived from blooming plants of genus *Ruta*; used in perfumery and veterinary medicine, and as a chemical intermediate.

rule of approximation [MIN ENG] A rule applicable to placer mining locations and entries upon surveyed lands, to be applied on the basis of 10-acre (40,469-square-meter) legal subdivisions.

ruler [ENG] A graduated strip of wood, metal, or other material, used to measure lines or as a guide in drawing lines.

rumble See turntable rumble.

run [ENG] A portion of pipe or fitting lying in a straight line in the same direction of flow as the pipe to which it is connected. [MIN ENG] See slant.

run a line of soundings [ENG] To obtain a series of soundings along a course line.

runaround [MIN ENG] A bypass driven in the shaft pillar to permit safe passage from one side of the shaft to the other.

runback [ENG] **1.** To retract the drill feed mechanism to its starting position. **2.** To drill slowly downward toward the bottom of the hole when the drill string has been lifted off-bottom for rechucking.

run in [ENG] To lower the assembled drill rods and auxiliary equipment into a borehole.

runner [ENG] In a plastics injection or transfer mold, the channel (usually circular) that connects the sprue with the gate to the mold cavity. [MET] **1.** The part of a casting between itself and the gate assembly of the mold. **2.** A channel through which molten metal flows from one receptacle to another. [MIN ENG] A vertical timber sheet pile used to prevent collapse of an excavation.

runner box [MET] A box that divides the molten metal into several streams before it enters the cavity of the mold.

running block *See* traveling block.

running bond [CIV ENG] A masonry bond involving the placing of each brick as a stretcher and overlapping the bricks in adjoining courses.

running fit [DES ENG] The intentional difference in dimensions of mating mechanical parts that permits them to move relative to each other.

running gate [MET] A gate through which molten metal enters a mold.

running gear [MECH ENG] The means employed to support a truck and its load and to provide rolling-friction contact with the running surface.

running ground [MIN ENG] Insecure, incoherent ground that may be semiplastic or plastic and deforms readily under pressure.

running-in [ENG] The process of operating new or repaired machinery or equipment in order to detect any faults and to ensure smooth, free operation of parts before delivery.

run-of-bank gravel *See* bank-run gravel.

runoff [MIN ENG] Collapse of a coal pillar in a mine.

runoff pit [MIN ENG] Catchment area to which spillage from classifiers, thickeners, and slurry pumps can gravitate if it becomes necessary to dump their contents. Also known as spill pit.

run-of-mill [MIN ENG] Ore accepted for treatment, after waste and dense media rejection. Also known as mill-head ore.

run-of-mine *See* mine run.

runout [MET] **1.** Escape of molten metal from a casting mold, crucible, or furnace. **2.** Defect in a casting caused by escape of metal from a mold.

runout table [MET] A roll table used to receive a rolled or extruded member.

run-out time [IND ENG] Time required by machine tools after cutting time is finished before tool and material are completely free of interference and before the start of the next sequence of operation.

runway [CIV ENG] A straight path, often hard-surfaced, within a landing strip, normally used for landing and takeoff of aircraft.

Rüping process [ENG] A system for preservative treatment of wood by using positive initial pressure, followed by introduction of the preservative and release of air, creating a vacuum.

rupture disk *See* burst disk.

rupture disk device [MECH ENG] A nonreclosing pressure relief device which relieves the inlet static pressure in a system through the bursting of a disk.

Rusa oil *See* palmarosa oil.

Russell flask [PETRO ENG] Device for volumetric determination of the true volume of sand grains within a unit bulk volume of grains plus voids.

rust [MET] The iron oxides formed on corroded ferrous metals and alloys.

rusting [MET] The formation of rust on ferrous metals and alloys.

rust joint [ENG] A joint to which some oxidizing agent is applied either to cure a leak or to withstand high pressure.

rust prevention [ENG] Surface protection of ferrous structures or equipment to prevent formation of iron oxide; can be by coatings, surface treatment, plating, chemicals, cathodic arrangements, or other means.

rust preventive [MATER] One of a group of products, often with petroleum thinners, used to prevent corrosion to metal surfaces.

rusty gold [MET] Native gold that has a thin coat of iron oxide or silica that prevents it from amalgamating readily.

ruthenium [MET] A hard, brittle, grayish-white metal used as a catalyst; workable only at high temperatures.

Rzeppa joint [MECH ENG] A special application of the Bendix-Weiss universal joint in which four large balls are transmitting elements, while a center ball acts as a spacer; it transmits constant angular velocity through a single universal joint.

S

S-2 *See* Tracker.

sabadilla [MATER] Ripe seeds of the sabadilla plant (*Schoenocaulon officinale*) that have been dried; used as an insecticide on cattle. Also known as caustic barley; cevedilla.

Sabathé's cycle [MECH ENG] An internal combustion engine cycle in which part of the combustion is explosive and part at constant pressure.

saber saw [MECH ENG] A portable saw consisting of an electric motor, a straight saw blade with reciprocating mechanism, a handle, baseplate, and other essential parts.

sabrejet [AERO ENG] An airplane widely used by the U.S. Air National Guard and free world countries; it has a range of beyond 1000 miles (1600 kilometers) and a speed of over 650 miles (1050 kilometers) per hour, carries two 1000-pound (450-kilogram) bombs or sixteen 5-inch (12.7-centimeter) rockets, or a combination plus two additional 1000-pound bombs in lieu of fuel tanks; has a crew of one. Designated F-86.

saccharimeter [ENG] An instrument for measuring the amount of sugar in a solution, often by determining the change in polarization produced by the solution.

saccharometer [ENG] An instrument for measuring the amount of sugar in a solution, by determining either the specific gravity or the gases produced by fermentation.

saddle [DES ENG] A support shaped to fit the object being held.

saddle leather [MATER] **1.** Tanned cattlehide used in furnishings for saddle horses. **2.** Leather resembling saddle leather used in handbags and other leather goods.

saddle-type turret lathe [MECH ENG] A turret lathe designed without a ram and with the turret mounted directly on a support (saddle) which slides on the bedways of the lathe.

saddling [MET] Forming a seamless ring by forging a pierced disk over a mandrel (or saddle).

SAE EP lubricant tester [ENG] A machine that tests the extreme pressure (EP) properties of a lubricant under a combined rolling and sliding action, in which the revolving members are two bearing cups that rotate at different speeds.

SAE number [ENG] A classification of motor, transmission, and differential lubricants to indicate viscosities, standardized by the Society of Automotive Engineers; SAE numbers do not connote quality of the lubricant.

safety [ENG] Methods and techniques of avoiding accident or disease.

safety belt [ENG] A strong strap or harness used to fasten a person to an object, such as the seat of an airplane or automobile.

safety board [PETRO ENG] A board placed in a derrick for a man to stand on when handling drill rods at single, double, triple, or quadruple levels; the boards are placed at suitable heights to handle a stand of drill rods for that number of joints.

safety bolt [CIV ENG] A bolt that can be opened from only one side of the door or gate it fastens.

safety cable [MIN ENG] A mining machine cable designed to cut off power when the positive conductor insulation is damaged.

safety cage [MIN ENG] A cage, box, or platform used for lowering and hoisting miners, tools, and equipment into and out of mines.

safety can [ENG] A cylindrical metal container used for temporary storage or handling of flammable liquids, such as gasoline, naphtha, and benzine, in buildings not provided with properly constructed storage rooms; these cans are also used to transport such liquids for filling and supply purposes within local areas.

safety car [MIN ENG] Any mine car or hoisting cage provided with safety stops, catches, or other precautionary devices.

safety chain [MIN ENG] A chain connecting the first and last cars of a trip to prevent separation, if a coupling breaks.

safety chuck [DES ENG] Any drill chuck on which the heads of the set screws do not protrude beyond the outer periphery of the chuck.

safety door [MIN ENG] An extra door ready for use in the event of damage to the existing ventilation door or in any emergency, for example, explosion or fire.

safety engineer [IND ENG] A person who inspects all possible danger spots in a factory, mine, or other industrial building or plant.

safety engineering [IND ENG] The testing and evaluating of equipment and procedures to prevent accidents.

safety explosive [MATER] An explosive which may be handled safely under ordinary conditions; it requires a powerful detonating force.

safety flange [DES ENG] A type of flange with tapered sides designed to keep a wheel intact in the event of accidental breakage.

safety fuse [ENG] A train of black powder which is enclosed in cotton, jute yarn, and waterproofing compounds, and which burns at the rate of 2 feet (60 centimeters) per minute; it is used mainly for small-scale blasting.

safety gate [MIN ENG] An automatically operated gate at the top of a mine shaft or at landings both to guard the entrance and to prevent falling into the shaft.

safety glass [MATER] **1.** A glass that resists shattering (such as a glass containing a net of wire or constructed of sheets separated by plastic film). **2.** A glass that has been tempered so that when it shatters, it breaks up into grains instead of jagged fragments.

safety hoist [MECH ENG] A hoisting gear that does not continue running when tension is released.

safety hook [DES ENG] A hoisting hook with a spring-loaded latch that prevents the load from accidentally slipping off the hook.

safety lamp [MIN ENG] In coal mining, a lamp that is relatively safe to use in atmospheres which may contain flammable gas.

safety level of supply [IND ENG] The quantity of material, in addition to the operating level of supply, required to be on hand to permit continuous operations in the event of minor interruption of normal replenishment or unpredictable fluctuations in demand.

safety match [ENG] A match that can be ignited only when struck against a specially made friction surface.

safety plug [ENG] A protective device used on a heated pressure vessel (for example, a steam boiler), and containing a fusible element that melts at a predetermined safe temperature to prevent the buildup of excessive pressure. Also known as fusible plug.

safety post [MIN ENG] A timber placed near the face of workings to protect the workmen. Also known as safety prop.

safety prop See safety post.

safety rail See guard rail.

safety relief valve See safety valve.

safety shoe [ENG] A special shoe without spark-producing nails or plates, worn by personnel working around explosives.

safety stop [MECH ENG] **1.** On a hoisting apparatus, a device by which the load may be prevented from falling. **2.** An automatic device on a hoisting engine designed to prevent overwinding.

safety valve [ENG] A spring-loaded, pressure-actuated valve that allows steam to escape from a boiler at a pressure slightly above the safe working level of the boiler; fitted by law to all boilers. Also known as safety relief valve.

safe yield [CIV ENG] The maximum dependable draft that can be made continuously upon a source of water supply over a given period of time during which the probable driest period, and therefore period of greatest deficiency in water supply, is likely to occur.

safflower oil [MATER] Nonyellowing oil derived from safflower seed and similar to linseed oil; contains high proportion of linoleic acid; used as a drying oil and in food and medicine.

sag bolt [MIN ENG] A device to measure roof sag; a 12-foot (3.7-meter) unit installed without a bearing plate and securely anchored with the aid of a heavy nut, extending about 2 inches (5 centimeters) from the hole; three ½-inch (1.3 centimeter) strips of colored tape wrapped around the extending section of the bolt, green at the roof line followed by yellow and then red, help detect roof sag at a glance.

sage oil See oil of sage.

saggar clay [MATER] A fire clay of which the case is made that is used for the firing of porcelain and pottery. Also spelled sagger clay.

sagger clay See saggar clay.

sago [MATER] A starch obtained from the trunks of certain tropical palms, such as the sago; used as a thickening agent in food and as textile stiffening.

Saint Joseph retort process [MET] An electrothermic retort process for processing zinc ore and zinc from secondary sources into zinc; heat of reaction between the sintered zinc concentrate and the coke mixture is supplied by passage of heavy electric current through the resistance of the charge.

salable coal [MIN ENG] Total output of a coal mine, less the tonnage rejected or consumed during preparation for market.

salamander stove [ENG] A small portable stove used for temporary or emergency heat; for example, on construction sites or in greenhouses.

salimeter [ENG] A hydrometer graduated to read directly the percentage of salt in a solution such as brine.

salina *See* saltworks.

salinity logging [PETRO ENG] Technique for measurement and recording of salt-water-bearing zones in an oil or gas reservoir; uses a combination of neutron logging with a chlorine curve.

salinity-temperature-depth recorder [ENG] An instrument consisting of sensing elements usually lowered from a stationary ship, and a recorder on board which simultaneously records measurements of temperature, salinity, and depth. Also known as CTD recorder; STD recorder.

salinometer [ENG] An instrument that measures water salinity by means of electrical conductivity or by a hydrometer calibrated to give percentage of salt directly.

salmon oil [MATER] Combustible, pale golden-yellow liquid with sweet taste; soluble in alcohol, ether, chloroform, and carbon disulfide; obtained from the waste in the canning of salmon; used in pet foods, soaps, and leather dressing.

sal prunella *See* niter balls.

salt [ENG] To add an accelerator or retardant to cement. [MIN ENG] **1.** To introduce extra amounts of a valuable or waste mineral into a sample to be assayed. **2.** To artificially enrich, as a mine, usually with fraudulent intent.

salt bath [MET] Molten salts in which steel is heated for hardening and tempering.

saltern *See* salt garden; saltworks.

salt-fog test [MET] An accelerated corrosion test in which a piece of metal is subjected to a fine spray of sodium chloride solution. Also known as salt-spray test.

salt garden [ENG] A large, shallow basin or pond where sea water is evaporated by solar heat. Also known as saltern.

salt glaze [ENG] Glaze formed on the surface of stoneware by putting salt into the kiln during firing.

salt mine [MIN ENG] A mine containing deposits of rock salt.

salt-mud combination log [PETRO ENG] Record of electrical logging of the mud in oil well boreholes in the presence of sodium chloride incursions from adjacent formations.

salt-spray test *See* salt-fog test.

salt velocity meter [ENG] A rate-of-flow volume meter used to find the transit time of passage between two fixed points of a small quantity of salt or radioactive isotope in a flowing stream by measuring electrical conductivity or radiation level at those points.

salt-water well [PETRO ENG] A well from which salt water flows after the petroleum contents are depleted.

salt well [ENG] A bored or driven well from which brine is obtained.

saltworks [ENG] A building or group of buildings where salt is produced commercially, as by extraction from sea water or from the brine of salt springs. Also known as salina; saltern.

salvage procedure [ENG] The recovery, evacuation, and reclamation of damaged, discarded, condemned, or abandoned allied or enemy materiel, ships, craft, and floating equipment for reuse, repair, refabrication, or scrapping.

salvage value [ENG] The net worth of diamonds recovered from a used diamond-inset tool.

salvia [MATER] The dried leaves of the sage, *Salvia officinalis;* contains volatile oil, resin, and tannin; used in food engineering as a flavoring agent and condiment, and in medicine as an antisecretory agent.

salvia oil *See* oil of sage.

SAMOS program [AERO ENG] A series of military reconnaissance satellites carrying cameras and other surveillance equipment; each satellite has a jettisonable camera package that is ejected from orbit and recovered in midair.

sampled grade [MIN ENG] The amount of valuable metal in the ore in place as determined by underground, surface, or drill-hole sampling.

sample log [ENG] Record of core samples or drill cuttings; gives geological, visual, and hydrocarbon-content record versus depth of drilling.

sampler [ENG] A mechanical or other device designed to obtain small samples of materials for analysis; used in biology, chemistry, and geology.

sample splitter [ENG] An instrument, generally constructed of acrylic resin, designed to subdivide a total sample of marine plankton while maintaining a quantitatively correct relationship between the various phyla in the sample.

sampling [ENG] Process of obtaining a sequence of instantaneous values of a wave.

sampling area ratio [MIN ENG] The volume of the soil displaced in proportion to the volume of the sample; a well-designed tool has an area ratio of about 20 percent.

sampling bottle [ENG] A cylindrical container, usually closed at a chosen depth, to trap a water sample and transport it to the surface without introducing contamination.

sampling pipe [MIN ENG] A small pipe built into a seal to take air samples in a sealed area.

sampling plan [IND ENG] A plan stating sample sizes and the criteria for accepting or rejecting items or taking another sample during inspection of a group of items.

sampling probe [ENG] A leak-testing probe which collects tracer gas from the test area of an object under pressure and feeds it to the leak detector at reduced pressure.

sampling process [ENG] The process of obtaining a sequence of instantaneous values of some quantity that varies continuously with time.

sampling rate [ENG] The rate at which measurements of physical quantities are made; for example, if it is desired to calculate the velocity of a missile and its position is measured each millisecond, then the sampling rate is 1000 measurements per second.

sampling risk [IND ENG] In inspection procedure, the probability, under the sampling plan used, that acceptable material will be rejected or that unsatisfactory material will be accepted.

sampling spoon [MIN ENG] A cylinder with a spoonlike cutting edge for taking soil samples.

sampling time [ENG] The time between successive measurements of a physical quantity.

sampling voltmeter [ENG] A special type of voltmeter that detects the instantaneous value of an input signal at prescribed times by means of an electronic switch connecting the signal to a memory capacitor; it is particularly effective in detecting high-frequency signals (up to 12 gigahertz) or signals mixed with noise.

sandal oil *See* sandalwood oil.

sandalwood oil [MATER] Pale-yellow essential oil with harsh taste and faint aromatic scent; soluble in fixed oils; insoluble in glycerin; derived from the sandalwood *Santalum album*; used in medicine, perfumes, and flavors. Also known as East Indian sandalwood oil; sandal oil; santal oil; santalwood oil.

sandarac gum [MATER] Yellow, brittle, water-insoluble, natural resin obtained from the African sandarac tree of Morocco; used in varnishes and lacquers.

sandbag [ENG] A bag filled with sand; used to build temporary protective walls.

sandblasting *See* grit blasting.

sand-cast [MET] Made by pouring molten metal into a mold made of sand.

sand control [MET] A process to regulate the properties of foundry sand to produce defect-free castings.

sand count [PETRO ENG] Determination of the total thickness of an oil or gas reservoir's permeable section (excluding shale streaks and other impermeable zones); can be derived from electrical logs.

sand drain [CIV ENG] A vertical boring through a clay or silty soil filled with sand or gravel to facilitate drainage.

sander [MECH ENG] **1.** An electric machine used to sand the surface of wood, metal, or other material. **2.** A device attached to a locomotive or electric rail car which sands the rails to increase friction on the driving wheels.

sand filter [CIV ENG] A filter consisting of graded layers of sand and aggregate for purifying domestic water.

sand finish [ENG] A smooth finish on a plaster surface made by rubbing the sand or mortar coat.

sand-grain volume [PETRO ENG] In oil reservoir porosity calculations, the actual volume filled by sand grains, without allowance for spaces (voids) between the grains.

sandhog [ENG] A worker in compressed-air environments, as in driving tunnels by means of pneumatic caissons.

sanding [ENG] **1.** Covering or mixing with sand. **2.** Smoothing a surface with sandpaper or other abrasive paper or cloth.

sand-lime brick [MATER] A clay building material made by molding a mixture of sand with 6% hydrated lime and water.

sand line [ENG] A wire line used to raise and lower a bailer or sand pump to remove cuttings from a borehole.

sand mill [MECH ENG] Variation of a ball-type size-reduction mill in which grains of sand serve as grinding balls.

sandpaper [MATER] Paper with abrasive glued to the surface.

sand pile [CIV ENG] A compacted filling of sand in a deep round hole formed by ramming the sand with a pile; used for foundations in soft soil.

sandpit [CIV ENG] An excavation dug in sand, especially as a source of sand for construction materials.

sand pump [MECH ENG] A pump, usually a centrifugal type, capable of handling sand- and gravel-laden liquids without clogging or wearing unduly; used to extract mud and cuttings from a borehole. Also known as sludge pump.

sand reel [MECH ENG] A drum, operated by a band wheel, for raising or lowering the sand pump or bailer during drilling operations.

sand return [PETRO ENG] The return of injected sand to the wellbore following formation fracturing; it constitutes a problem.

sands [MIN ENG] The coarser and heavier particles of crushed ore, of such size that they settle readily in water and may be leached by allowing the solution to percolate.

sand slinger [MECH ENG] A machine which delivers sand to and fills molds at high speed by centrifugal force.

sand trap [ENG] A device in a conduit for trapping sand or soil particles carried by the water.

sand wheel [MECH ENG] A wheel fitted with steel buckets around the circumference for lifting sand or sludge out of a sump to stack it at a higher level.

sandwich braze [MET] A technique by which a shim is placed between materials to be brazed as a transition layer to decrease thermal stress.

sandwich construction [DES ENG] Composite construction of alloys, plastics, wood, or other materials consisting of a foam or honeycomb layer laminated and glued between two hard outer sheets. Also known as sandwich laminate.

sandwich heating [ENG] Method for heating both sides of a thermoplastic sheet simultaneously prior to forming or shaping.

sandwich rolling [MET] Rolling strips of metal together to form a metallurgically bonded composite sheet.

sanitary engineering [CIV ENG] A field of civil engineering concerned with works and projects for the protection and promotion of public health.

sanitary landfill [CIV ENG] The disposal of garbage by spreading it in layers covered with soil or ashes to a depth sufficient to control rats, flies, and odors.

sanitary nipper *See* latrine cleaner.

sanitary sewer [CIV ENG] A sewer which is restricted to carrying sewage and to which storm and surface waters are not admitted.

sanitation [CIV ENG] The act or process of making healthy environmental conditions.

sanitizer [MATER] Disinfectant formulated to clean food-processing equipment and dairy and eating utensils.

santal oil *See* sandalwood oil.

santalwood oil *See* sandalwood oil.

Sapele mahogany [MATER] A figured wood from *Entandrophragma cylindricum*, a big tree growing on the Ivory Coast, Ghana, and Nigeria. Also known as aboundikro; scented mahogany; West African cedar.

sapphire whiskers *See* alumina fibers.

sardine oil [MATER] Combustible, yellow liquid obtained from sardines; soluble in alcohol, ether, and chloroform; solidifies at about 30°C; used in soaps and pet foods, and as a lubricant.

sarking [BUILD] A layer of boards or bituminous felt placed beneath tiles or other roofing to provide thermal insulation or to prevent ingress of water.

SAS *See* stability augmentation system.

sash [BUILD] A frame for window glass.

sash bars [BUILD] Strips of wood which separate the panes in a window composed of several panes. Also known as muntins.

sash cord [BUILD] A cord or chain used to attach a counterweight to the window sash.

satellite *See* artificial satellite.

satellite and missile surveillance [ENG] The systematic observation of aerospace for the purpose of detecting, tracking, and characterizing objects, events, and phenomena associated with satellites and inflight missiles, friendly and enemy.

satellite tracking [AERO ENG] Determination of the positions and velocities of satellites through radio and optical means.

satelloid [AERO ENG] A vehicle that revolves about the earth or other celestial body, but at such altitudes as to require sustaining thrust to balance drag.

satin finish [MET] A finish involving soft scratch-brushing of polished metal surfaces to produce a soft sheen. Also known as Butler finish; scratch-brush finish.

saturable-core magnetometer [ENG] A magnetometer that depends for its operation on the changes in permeability of a ferromagnetic core as a function of the magnetic field to be measured.

saturator [ENG] A device, equipment, or person that saturates one material with another; examples are a tank in which vapors become saturated with ammonia from coal (in carbonization of coal), a humidifier, and the operator of a machine for impregnating roofing felt with asphalt.

Saturn [AERO ENG] One of the very large launch vehicles built primarily for the Apollo program; begun by Army Ordnance but turned over to the National Aeronautics and Space Administration for the manned space flight program to the moon.

Saunders air-lift pump [MECH ENG] A device for raising water from a well by the introduction of compressed air below the water level in the well.

sauterelle [ENG] A device used by masons for tracing and forming angles.

savin oil [MATER] Pale-yellow, volatile oil, soluble in alcohol; derived from the fresh tops of the savin (*Juniperus sabina*); used in medicine.

Savonius rotor [MECH ENG] A rotor composed of two offset semicylindrical elements rotating about a vertical axis.

Savonius windmill [MECH ENG] A windmill composed of two semicylindrical offset cups rotating about a vertical axis.

savory oil [MATER] A yellow to brown essential oil derived from the whole dried plant *Satureia hortensis*; used for flavoring. Also known as summer savory oil.

saw [DES ENG] Any of various tools consisting of a thin, usually steel, blade with continuous cutting teeth on the edge.

saw bit [DES ENG] A bit having a cutting edge formed by teeth shaped like those in a handsaw.

sawdust [MATER] Wood fragments made by a saw in cutting.

sawdust concrete [MATER] Concrete containing sawdust as the principal aggregate.

saw gumming [MECH ENG] Grinding away the punch marks in the spaces between the teeth in saw manufacture.

sawhorse [ENG] A wooden rack used to support wood that is being sawed.

sawing [ENG] Cutting with a saw.

sawmill [IND ENG] A plant that houses sawing machines. [MECH ENG] A machine for cutting logs with a saw or a series of saws.

sawtooth barrel *See* basket.

sawtooth blasting [MIN ENG] The cutting of a series of slabs which, in plan, resemble sawteeth by blasting oblique, horizontal holes along a face.

sawtooth crusher [MECH ENG] Solids crusher in which feed is broken down between two sawtoothed shafts rotating at different speeds.

sawtooth floor channeling [MIN ENG] A method of channeling inclined beds of marble by removing right-angle blocks in succession from the various beds, thus giving the floor a zigzag or sawtooth appearance.

sawtooth stoping [MIN ENG] In the United States, overhand stoping in which the line of advance is up the dip, and benches are advanced in a line parallel with the drift.

sax [DES ENG] A tool for chopping away the edges of roof slates; it has a pick at one end for making nail holes.

Saybolt color [ENG] A color standard for petroleum products determined with a Saybolt chromometer.

Saybolt Furol viscosimeter [ENG] An instrument for measuring viscosity of very thick fluids, for example, heavy oils; similar to the Saybolt Universal viscosimeter, but with a larger-diameter tube so that the efflux time is about one-tenth that of the Universal instrument.

Saybolt Universal viscosimeter [ENG] An instrument for measuring viscosity by the time it takes a fluid to flow through a calibrated tube; used for the lighter petroleum products and lubricating oils.

SBR *See* styrene-butadiene rubber.

scab [MET] A defect consisting of a flat, partially detached piece of metal joined to the surface of a casting or piece of rolled metal.

scaffold [CIV ENG] A temporary or movable platform supported on the ground or suspended; used for working at considerable heights above the ground.

scale [ENG] **1.** A series of markings used for reading the value of a quantity or setting. **2.** To change the magnitude of a variable in a uniform way, as by multiplying or dividing by a constant factor, or the ratio of the real thing's magnitude to the magnitude of the model or analog of the model. **3.** A weighing device. **4.** A ruler or other measuring stick. **5.** An indication of represented to actual distances on a map, chart or drawing. [MET] A thick metallic oxide coating formed usually by heating metals in air.

scaleboard [MATER] Thin sheets of wood used as veneer.

scale effect [AERO ENG] The necessary corrections applied to measurements of a model in a wind tunnel to ascertain corresponding values for a full-sized object.

scale factor [ENG] The factor by which the reading of an instrument or the solution of a problem should be multiplied to give the true final value when a corresponding scale factor is used initially to bring the magnitude within the range of the instrument or computer.

scale-up [DES ENG] Design process in which the data of an experimental-scale operation (model or pilot plant) is used for the design of a large (scaled-up) unit, usually of commercial size.

scale wax [MATER] The paraffin wax derived by sweating the greater part of the oil from slack wax; contains up to 6% oil. Also known as crude scale; paraffin scale.

scaling [ENG] Removing scale (rust or salt) from a metal or other surface. [MET] **1.** Forming of a thick layer of metallic oxide on metals at high temperatures. **2.** Depositing of solid inorganic solutes from water on a metal surface, such as a cooling tube or boiler. [MIN ENG] Removing loose rocks and coal from the roof, walls, or face after blasting.

scaling factor [ENG] Factor used in heat-exchange calculations to allow for the loss in heat conductivity of a material because of the development of surface scale, as inside pipelines and heat-exchanger tubes.

scaling ratio [ENG] The ratio of a certain property of a laboratory model to the same property in the natural prototype.

scalp [MET] To remove surface layers, and thereby defects, from ingots, billets, or slabs by machining. [MIN ENG] To remove undesirable fine material from broken ore, stone, or gravel.

scalped extrusion ingot [MET] A cast, solid, or hollow extrusion ingot which has been machined to remove outside surface layers.

scalpel [DES ENG] A small, straight, very sharp knife (or detachable blade for a knife), used for dissecting.

scan [ENG] **1.** To examine an area, a region in space, or a portion of the radio spectrum point by point in an ordered sequence; for example, conversion of a scene or image to an electric signal or use of radar to monitor an airspace for detection, navigation, or traffic control purposes. **2.** One complete circular, up-and-down, or left-to-right sweep of the radar, light, or other beam or device used in making a scan.

scanner [ENG] **1.** Any device that examines an area or region point by point in a continuous systematic manner, repeatedly sweeping across until the entire area or region is covered; for example, a flying-spot scanner. **2.** A device that automatically samples, measures, or checks a number of quantities or conditions in sequence, as in process control.

scanning proton microprobe [ENG] An instrument used for determining the spatial distribution of trace elements in samples, in which a beam of energetic protons is focused on a narrow spot which is swept over the sample, and the characteristic x-rays emitted from the target are measured.

scanning radiometer [ENG] An image-forming system consisting of a radiometer which, by the use of a plane mirror rotating at 45° to the optical axis, can see a circular path normal to the instrument.

scanning sequence [ENG] The order in which the points in a region are scanned; for example, in television the picture is scanned horizontally from left to right and vertically from top to bottom.

scanning sonar [ENG] Sonar in which all targets of interest are shown simultaneously, as on a radar PPI (plan position indicator) display or sector display; the sound pulse may be transmitted in all directions simultaneously and picked up by a rotating receiving transducer, or transmitted and received in only one direction at a time by a scanning transducer.

scantlings [BUILD] Sections of timber measuring less than 8 inches (20 centimeters) wide and from 2 to 6 inches (5.1 to 15 centimeters) thick; used for studding.

scarfing [MET] **1.** Cutting away of surface defects on metals by use of a gas torch. **2.** A forging process in which the ends of two pieces to be joined are tapered to avoid an enlarged joint.

scarf joint [DES ENG] A joint made by the cutting of overlapping mating parts so that the joint is not enlarged and the patterns are complementary, and securing them by glue, fasteners, welding, or other joining method.

scarifier [ENG] An implement or machine with downward projecting tines for breaking down a road surface 2 feet (60 centimeters) or less.

SC asphalt *See* slow-curing liquid asphaltic material.

scatterometer [ENG] A microwave sensor that is essentially a radar without ranging circuits, used to measure only the reflection or scattering coefficient while scanning the surface of the earth from an aircraft or a satellite.

scavenger [MET] A reactive metal added to a molten metal to combine with and remove dissolved gases.

scavenging [MECH ENG] Removal of spent gases from an internal combustion engine cylinder and replacement by a fresh charge or air. [MET] Removal of dissolved gases from molten metal.

scend [ENG] **1.** The upward motion of the bow and stern of a vessel associated with pitching. **2.** The lifting of the entire vessel by waves or swell. Also known as send.

scene paint [MATER] A paint used in theatrical scene painting; it is a dry pigment mixed with a glue-water mixture called size water.

scented mahogany *See* Sapele mahogany.

Scheffel engine [MECH ENG] A type of multirotor engine that uses nine approximately equal rotors turning in the same clockwise sense.

schlanite [MATER] The soluble resin extracted from anthracoxene by ether.

schlempe *See* vinasse.

Schlumberger dipmeter [ENG] An instrument that measures both the amount and direction of dip by readings taken in the borehole; it consists of a long, cylindrical body with two telescoping parts and three long, springy metal strips, arranged symmetrically round the body, which press outward and make contact with the walls of the hole.

Schlumberger photoclinometer [ENG] An instrument that measures simultaneously the amount and direction of the deviation of a borehole; the sonde, designed to lie exactly parallel to the axis of the borehole, is fitted with a small camera on the axis of a graduated glass bowl, in which a steel ball rolls freely and a compass is mounted in gimbals; the camera is electrically operated from the surface and takes a photograph of the bowl, the steel ball marks the amount of deviation, the position in relation to the image of the compass needle gives the direction of deviation.

Schmidt field balance [ENG] An instrument that operates as both a horizontal and vertical field balance and consists of a permanent magnet pivoted on a knife edge.

Schneider recoil system [MECH ENG] A recoil system for artillery, employing the hydropneumatic principle without a floating piston.

Schoop process [ENG] A process for coating surfaces by spraying with high-velocity molten metal particles.

Schuler tuning [ENG] The designing of gyroscopic devices so that their periods of oscillation will be about 84.4 minutes.

Schultze powder [MATER] A smokeless powder propellant, consisting of wood pellets impregnated with barium nitrate and potassium nitrate.

Schweydar mechanical detector [ENG] A seismic detector that senses and records refracted waves; a lead sphere is suspended by a flat spring, the sphere's motion is magnified by an aluminum cone that moves a bow around a spindle carrying a mirror, and this motion is then photographically recorded.

scissor engine See cat-and-mouse engine.

scissor jack [MECH ENG] A lifting jack driven by a horizontal screw; the linkages of the jack are parallelograms whose horizontal diagonals are lengthened or shortened by the screw.

scissors bridge [CIV ENG] A light metal bridge that can be folded and carried by a military tank.

scissors crossover [CIV ENG] A scissor-shaped junction between two parallel railway tracks. Also called double crossover.

scissors truss [BUILD] A roof truss in which the braces cross like scissors blades.

sclerometer [ENG] An instrument used to determine the hardness of a material by measuring the pressure needed to scratch or indent a surface with a diamond point.

scleroscope [ENG] An instrument used to determine the hardness of a material by measuring the height to which a standard ball rebounds from its surface when dropped from a standard height.

scleroscope hardness test See Shore scleroscope hardness test.

scoop [DES ENG] **1.** Any of various ladle-, shovel-, or bucketlike utensils or containers for moving liquid or loose materials. **2.** A funnel-shaped opening for channeling a fluid into a desired path. [MECH ENG] A large shovel with a scoop-shaped blade.

scoopfish See underway sampler.

scorching [ENG] **1.** Burning an exposed surface so as to change color, texture, or flavor without consuming. **2.** Destroying by fire.

scoria [MATER] Refuse after melting metals or reducing ore.

scorification [MET] Concentration of precious metals, such as gold and silver, in molten lead by oxidation employing appropriate fluxes.

scoring [ENG] Scratching the surface of a material. [MATER] See attrition.

scoring test See L-2 test.

Scorpion [AERO ENG] An all-weather interceptor aircraft with twin turbojet engines; its armament consists of air-to-air rockets with nuclear or nonnuclear warheads. Designated F-89.

scotch [ENG] A wooden stopblock or iron catch placed under a wheel or other curved object to prevent slipping or rolling.

scotch boiler [MECH ENG] A fire-tube boiler with one or more cylindrical internal furnaces enveloped by a boiler shell equipped with five tubes in its upper part; heat is transferred to water partly in the furnace area and partly in passage of hot gases through the tubes. Also known as dry-back boiler; scotch marine boiler (marine usage).

Scotch bond See American bond.

Scotch derrick See stiffleg derrick.

Scotch yoke [MECH ENG] A type of four-bar linkage; it is employed to convert a steady rotation into a simple harmonic motion.

scotophor [MATER] A solid that exhibits reversible darkening and bleaching actions of tenebrescence under suitable irradiation.

Scott-Darey process [CIV ENG] A chemical precipitation method used for fine solids removal in sewage plants; employs ferric chloride solution made by treating scrap iron with chlorine.

scouring [ENG] Physical or chemical attack on process equipment surfaces, as in a furnace or fluid catalytic cracker. [MATER] See attrition. [MECH ENG] Mechanical finishing or cleaning of a hard surface by using an abrasive and low pressure.

scouring basin [CIV ENG] A basin containing impounded water which is released at about low water in order to maintain the desired depth in the entrance channel. Also known as sluicing pond.

scout [ENG] An engineer who makes a preliminary examination of promising oil and mining claims and prospects.

Scout [AERO ENG] A four-stage all-solid-propellant rocket, used as a space probe and orbital test vehicle; first launched July 1, 1960, with a 150-pound (68-kilogram) payload.

scout boring [MIN ENG] A bore made to test a geologic formation being prospected.

scout hole [MIN ENG] **1.** A borehole penetrating only the uppermost part of an ore body in order to delineate the surface configuration. **2.** A shallow borehole used to ascertain the presence of ore or to explore an area in a preliminary manner.

scramble [AERO ENG] To take off as quickly as possible (usually followed by course and altitude instructions).

scram drive [MIN ENG] Underground drive above the tramming level, along which scrapers move ore to a discharge chute.

scramjet [AERO ENG] Essentially a ramjet engine, intended for flight at hypersonic speeds. Derived from supersonic combustion ramjet.

scrap [ENG] Any solid material cutting or reject of a manufacturing operation, which may be suitable for recycling as feedstock to the primary operation; for example, scrap from plastic or glass molding or metalworking.

scraper conveyor [MECH ENG] A type of flight conveyor in which the element (chain and flight) for moving materials rests on a trough.

scraper hoist [MECH ENG] A drum hoist that operates the scraper of a scraper loader.

scraper loader [MECH ENG] A machine used for loading coal or rock by pulling a scoop through the material to an apron or ramp, where the load is discharged onto a car or conveyor.

scraper ring [MECH ENG] A piston ring that scrapes oil from a cylinder wall to prevent it from being burnt.

scraper ripper [MIN ENG] A piece of strip-mine equipment with teeth on the lip to rip or break the coal and with a flight conveyor to remove the broken coal.

scraper trap [ENG] A device for the insertion or recovery of pigs, or scrapers, that are used to clean the inside surfaces of pipelines.

scrap mica [MATER] Mica whose size, color, or quality is below specifications for sheet mica.

scratch-brush finish *See* satin finish.

scratch coat [ENG] The first layer of plaster applied to a surface; the surface is scratched to improve the bond with the next coat.

scratch filter [ENG ACOUS] A low-pass filter circuit inserted in the circuit of a phonograph pickup to suppress higher audio frequencies and thereby minimize needle-scratch noise.

scratch hardness [MATER] A measure of the resistance of minerals or metals to scratching; for minerals it is defined by comparison with 10 selected minerals which are numbered in order of increasing hardness according to the Mohs scale.

scratch hardness test [MET] A hardness test in which a cutting point under given pressure is drawn across the surface of a metal, and the width of the scratch is measured.

screaming [AERO ENG] A form of combustion instability, especially in a liquid-propellant rocket engine, of relatively high frequency, characterized by a high-pitched noise.

screeching [AERO ENG] A form of combustion instability, especially in an afterburner, of relatively high frequency, characterized by a harsh, shrill noise.

screed [CIV ENG] **1.** A straight-edged wood or metal template, fixed temporarily to a surface as a guide when plastering or concreting. **2.** An oscillating metal bar mounted on wheels and spanning a freshly placed road slab, used to strike off and smooth the surface.

screed wire *See* ground wire.

screen analysis [ENG] A method for finding the particle-size distribution of any loose, flowing, conglomerate material by measuring the percentage of particles that pass through a series of standard screens with holes of various sizes.

screen cloth [MATER] A woven material suitable for use in a screen deck.

screen deck [DES ENG] A surface provided with apertures of specified size, used for screening purposes.

screening [ENG] **1.** The separation of a mixture of grains of various sizes into two or more size-range portions by means of a porous or woven-mesh screening media. **2.** The removal of solid particles from a liquid-solid mixture by means of a screen. **3.** The material that has passed through a screen. [IND ENG] The elimination of defective pieces from a lot by inspection for specified defects. Also known as detailing.

screen mesh [ENG] A wire network or cloth mounted in a frame for separating and classifying materials.

screen pipe [ENG] Perforated pipe with a straining device in the form of closely wound wire coils wrapped around it to admit well fluids while excluding sand.

screen size [MIN ENG] A standard for determining the size of diamond particles; the size of the screened particle is determined by the size of the opening through which the diamond particle will not pass.

screw [DES ENG] **1.** A cylindrical body with a helical groove cut into its surface. **2.** A fastener with continuous ribs on a cylindrical or conical shank and a slotted, recessed, flat, or rounded head. Also known as screw fastener.

screw compressor [MECH ENG] A rotary-element gas compressor in which compression is accomplished between two intermeshing, counterrotating screws.

screw conveyor [MECH ENG] A conveyor consisting of a helical screw that rotates upon a single shaft within a stationary trough or casing, and which can move bulk material along a horizontal, inclined, or vertical plane. Also known as auger conveyor; spiral conveyor; worm conveyor.

screwdriver [DES ENG] A tool for turning and driving screws in place; a thin, wedge-shaped or fluted end enters the slot or recess in the head of the screw.

screw elevator [MECH ENG] A type of screw conveyor for vertical delivery of pulverized materials.

screw fastener *See* screw.

screwfeed [MECH ENG] A system or combination of gears, ratchets, and friction devices in the swivel head of a diamond drill, which controls the rate at which a bit penetrates a rock formation.

screw feeder [MECH ENG] A mechanism for handling bulk (pulverized or granulated solids) materials, in which a rotating helicoid screw moves the material forward, toward and into a process unit.

screw machine [MECH ENG] A lathe for making relatively small, turned metal parts in large quantities.

screw pile [CIV ENG] A pile having a wide helical blade at the foot which is twisted into position, for use in soft ground or other location requiring a large supporting surface.

screw plasticating injection molding [ENG] A plastic-molding technique in which plastic is converted from pellets to a viscous (plasticated) melt by an extruder screw that is an integral part of the molding machine.

screw press [MECH ENG] A press having the slide operated by a screw mechanism.

screw propeller [MECH ENG] A marine and airplane propeller consisting of a streamlined hub attached outboard to a rotating engine shaft on which are mounted two to six blades; the blades form helicoidal surfaces in such a way as to advance along the axis about which they revolve.

screw pump [MECH ENG] A pump that raises water by means of helical impellers in the pump casing.

screw rivet [DES ENG] A short rod threaded along the length of the shaft that is set without access to the point.

screw spike [DES ENG] A large nail with a helical thread on the upper portion of the shank; used to fasten railroad rails to the ties.

screwstock [MECH ENG] Free-machining bar, rod, or wire.

screw thread [DES ENG] A helical ridge formed on a cylindrical core, as on fasteners and pipes.

screw-thread gage [DES ENG] Any of several devices for determining the pitch, major, and minor diameters, and the lead, straightness, and thread angles of a screw thread.

screw-thread micrometer [DES ENG] A micrometer used to measure pitch diameter of a screw thread.

scriber [DES ENG] A sharp-pointed tool used for drawing lines on metal workpieces.

scroll gear [DES ENG] A variable gear resembling a scroll with teeth on one face.

scroll saw [ENG] A saw with a narrow blade, used for cutting curves or irregular designs.

scrub [AERO ENG] To cancel a scheduled firing, either before or during countdown.

scrubber [ENG] A device for the removal, or washing out, of entrained liquid droplets or dust, or for the removal of an undesired gas component from process gas streams. Also known as wet collector. [MIN ENG] A device, such as a wash screen, wash trommel, log washer, and hydraulic jet or monitor, in which a coarse and sticky material, for example, ore or clay, is either washed free of adherents or mildly disintegrated.

scrubbing oil *See* absorption oil.

scruff [MET] A mixture of tin oxide and iron-tin alloy formed as dross on a molten tin-coating bath.

scuba diving [ENG] Any of various diving techniques using self-contained underwater breathing apparatus.

scuffing [ENG] The dull mark, sometimes the result of abrasion, on the surface of glazed ceramic or glassware.

scuffle hoe [DES ENG] A hoe having two sharp edges so that it can be pushed and pulled.

scum [MATER] **1.** A film of impurities that rises to or is formed on the surface of a liquid. **2.** A slimy film formed on the surface of a solid object.

scum chamber [CIV ENG] An enclosed compartment in an Imhoff tank, in which gas escapes from the scum which rises to the surface of sludge during sewage digestion.

scutch [ENG] A small, picklike tool which has flat cutting edges for trimming bricks.

scutcheon plate *See* scutcheon.

scuttle [BUILD] An opening in the ceiling to provide access to the attic or roof.

scythe [DES ENG] A tool with a long curved blade attached at a more or less right angle to a long handle with grips for both hands; used for cutting grass as well as grain and other crops.

sea bank *See* seawall.

Sea Bat [AERO ENG] An antisubmarine warfare helicopter equipped with active-passive sonar, acoustic homing torpedoes, and instrument–night flight capability. Designated SH-34G.

seadrome [CIV ENG] **1.** A designated area for landing and takeoff of seaplanes. **2.** A platform at sea for landing and takeoff of land planes.

sea gate [CIV ENG] A gate which serves to protect a harbor or tidal basin from the sea, such as one of a pair of supplementary gates at the entrance to a tidal basin exposed to the sea.

seal [ENG] **1.** Any device or system that creates a nonleaking union between two mechanical or process-system elements; for example, gaskets for pipe connection seals, mechanical seals for rotating members such as pump shafts, and liquid seals to prevent gas entry to or loss from a gas-liquid processing sequence. **2.** A tight, perfect closure or joint.

seal coat [MATER] A layer of bituminous material applied to bituminous macadam or concrete to seal the surface.

sealed cabin [AERO ENG] The occupied space of an aircraft or spacecraft characterized by walls which do not allow gaseous exchange between the inner atmosphere and its surrounding atmosphere, and containing its own mechanisms for maintenance of the inside atmosphere.

sealer [MATER] A preliminary coating applied to seal the pores in a porous, uncoated surface, such as wood.

Seale rope [DES ENG] A wire rope with six or eight strands, each having a large wire core covered by nine small wires, which, in turn, are covered by nine large wires.

sea-level datum [ENG] A determination of mean sea lvel that has been adopted as a standard datum for heights or elevations, based on tidal observations over many years at various tide stations along the coasts.

sealing [MET] **1.** Impregnation of porous castings with resins to overcome porosity. **2.** Reducing porosity of an anodic oxide film on aluminum and aluminum alloys by immersion in boiling water.

sealing tape [MATER] Gummed tape for sealing packages.

sealing wax [MATER] A colored, scented mixture of resins and shellac; used for sealing containers and documents.

seal off [ENG] To close off, as a tube or borehole, by using a cement or other sealant to eliminate ingress or egress. [PETRO ENG] Penetration of a drilling fluid into a formation so that the formation is prevented from producing.

seal oil [MATER] A yellowish, liquid, fatty oil obtained from seal blubber; soluble in ether and chloroform; melts at 22–33°C; used in soapmaking, in dressing animal skins, and as a lubricant.

seal weld [MET] A weld designed primarily for preventing leakage.

seam [ENG] **1.** A mechanical or welded joint. **2.** A mark on ceramic or glassware where matching mold parts join. [MET] An unwelded fold or lap which appears as a crack on the surface of a casting or wrought product.

sea marker [ENG] A patch of color on the ocean surface produced by releasing dye; used, for example, to attract the attention of the crew of a rescue airplane.

seam blast [MIN ENG] A blast made by placing powdered or other explosive along and in a seam or crack between the solid wall and the stone or coal to be removed.

seaming [MET] The joining of the edges of sheet-metal parts by interlocking folds.

seamless ring rolling [MET] The hot-rolling of a circular blank, with a hole in the center, to form a seamless ring.

seamless tubing [MET] A tubing made by extrusion or by piercing and rolling a billet.

seam weld [MET] **1.** A longitudinal weld joining of sheet-metal parts or in making tubing. **2.** Arc or resistance welding in which a series of overlapping spot welds is produced.

seaplane [AERO ENG] An airplane that takes off from and alights on the water; it is supported on the water by pontoons, or floats, or by a hull which is a specially designed fuselage. Also known as airboat.

seaport [CIV ENG] A harbor or town that has facilities for seagoing ships and is active in marine activities.

search [ENG] To explore a region in space with radar.

search and rescue [ENG] The use of aircraft, surface craft, submarines, specialized rescue teams and equipment to search for and rescue personnel in distress on land or at sea.

searching control [ENG] A mechanism that changes the azimuth and elevation settings on a searchlight automatically and constantly, so that its beam is swept back and forth within certain limits.

searching lighting *See* horizontal scanning.

searchlight-control radar [ENG] A ground-based radar used to direct searchlights at aircraft.

searchlight-type sonar [ENG] A sonar system in which both transmission and reception are effected by the same narrow beam pattern.

search radar [ENG] A radar intended primarily to cover a large region of space and to display targets as soon as possible after they enter the region; used for early warning, in connection with ground-controlled approach and interception, and in air-traffic control.

search unit [ENG] The portion of an ultrasonic testing system which incorporates sending and in some cases receiving transducers to scan the workpiece.

season check [MATER] A longitudinal crack in wood, caused by uneven seasoning.

season crack [MET] A stress-corrosion crack produced in a copper-base alloy subject to a residual or applied tensile stress and exposed to a specific environment such as moist air containing traces of ammonia.

seasoned lumber [MATER] Lumber which has been cured by drying to ensure a uniform moisture content.

sea surveillance [ENG] The systematic observation of surface and subsurface sea areas by all available and practicable means primarily for the purpose of locating, identifying, and determining the movements of ships, submarines, and other vehicles, friendly and enemy, proceeding on or under the surface of seas and oceans.

seat [MECH ENG] The fixed, pressure-containing portion of a valve which comes into contact with the moving portions of that valve.

seating-lock locking fastener [DES ENG] A locking fastener that locks only when firmly seated and is therefore free-running on the bolt.

sea van [IND ENG] Commercial or government-owned (or leased) shipping containers which are moved via ocean transportation; since wheels are not attached, they must be lifted on and off the ship.

seawall [CIV ENG] A concrete, stone, or metal wall or embankment constructed along a shore to reduce wave erosion and encroachment by the sea. Also known as sea bank.

seawater thermometer [ENG] A specially designed thermometer to measure the temperature of a sample of seawater; an instrument consisting of a mercury-in-glass thermometer protected by a perforated metal case.

Secchi disk [ENG] An opaque white disk used to measure the transparency or clarity of seawater by lowering the disk into the water horizontally and noting the greatest depth at which it can be visually detected.

secondary air [MECH ENG] Combustion air introduced over the burner flame to enhance completeness of combustion.

secondary circuit [MET] The part of a welding machine which conducts secondary current between transformer and electrodes or between electrodes and workpiece.

secondary crusher [MECH ENG] Any of a group of crushing and pulverizing machines used after the primary treatment to further reduce the particle size of shale or other rock.

secondary drilling [MIN ENG] The process of drilling the so-called popholes for the purpose of breaking the larger masses of rock thrown down by the primary blast.

secondary grinding [MECH ENG] A further grinding of material previously reduced to sand size.

secondary hardening [MET] The hardening of certain alloy steels at moderate temperatures (250–650°C) by the precipitation of carbides; the resultant hardness is greater than that obtained by tempering the steel at some lower temperature for the same time.

secondary haulage [MIN ENG] That portion of the haulage system which collects coal from gathering-haulage delivery points and delivers it to the main portion of the system.

secondary high explosive [MATER] A high explosive which is relatively insensitive to heat and shock and is usually initiated by a primary high explosive; used for boosters and bursting charges.

secondary ion mass analyzer [ENG] A type of secondary ion mass spectrometer that provides general surface analysis and depth-profiling capabilities.

secondary ion mass spectrometer [ENG] An instrument for microscopic chemical analysis, in which a beam of primary ions with an energy in the range 5–20 kiloelectronvolts bombards a small spot on the surface of a sample, and positive and negative secondary ions sputtered from the surface are analyzed in a mass spectrometer. Abbreviated SIMS. Also known as ion microprobe; ion probe.

secondary metal [MET] Metal recovered from scrap by remelting and refining.

secondary oil recovery [PETRO ENG] Procedures used to increase the flow of oil from depleted or nearly depleted wells; includes fracturing, acidizing, waterflood, and gas injection.

secondary plasticizer [MATER] A plastics plasticizer that has insufficient affinity for a resin for it to be the sole plasticizer, and must be blended with a primary plasticizer. Also known as extender plasticizer.

secondary port [CIV ENG] A port with one or more berths, normally at quays, which can accommodate oceangoing ships for discharge.

secondary reference fuel [MATER] A commercially produced internal combustion engine fuel which is acceptable for knock testing or cetane rating, and which has been calibrated against primary reference fuels by engine tests.

secondary rescue facilities [ENG] Local airbase-ready aircraft, crash boats, and other air, surface, subsurface, and ground elements suitable for rescue missions, including government and privately operated units and facilities.

secondary reserves [PETRO ENG] Reserves recoverable commercially at current prices and costs as a result of artificial supplementation of the reserve's natural (gas-drive) energy.

secondary sewage sludge [CIV ENG] Sludge that includes activated sludge, mixed sludge, and chemically precipitated sludge.

secondary shaft [MIN ENG] The shaft which extends a mine downward from the bottom of, but not in line with, the primary shaft.

secondary splits [MIN ENG] Splits formed by separation of the main air splits.

secondary tide station [ENG] A place at which tide observations are made over a short period to obtain data for a specific purpose.

second-class ore [MIN ENG] An ore that needs some preliminary treatment before it is of a sufficiently high grade to be acceptable for market. Also known as milling ore.

second-order leveling [ENG] Spirit leveling that has less stringent requirements than those of first-order leveling, in which lines between benchmarks established by first-order leveling are run in only one direction.

second pilot [AERO ENG] A pilot, not necessarily qualified on type, who is responsible for assisting the first pilot to fly the aircraft and is authorized as second pilot.

section [CIV ENG] A piece of land usually 1 mile square (640 acres or approximately 2.58999 square kilometers) with boundaries conforming to meridians and parallels within established limits; 1 of 36 units of subdivision of a township in the U.S. Public Land survey system.

sectional conveyor [MECH ENG] A belt conveyor that can be lengthened or shortened by the addition or the removal of interchangeable sections.

sectional core barrel [DES ENG] A core barrel whose length can be increased by coupling unit sections together.

sectional header boiler [MECH ENG] A horizontal boiler in which tubes are assembled in sections into front and rear headers; the latter, in turn, are connected to the boiler drum by vertical tubes.

section house [CIV ENG] A building near a railroad section for housing railroad workers, or for storing maintenance equipment for the section.

section line [CIV ENG] A line representing the boundary of a section of land.

sector [CIV ENG] A clearly defined area or airspace designated for a particular purpose.

sector gate [CIV ENG] A horizontal gate with a pie-slice cross section used to regulate the level of water at the crest of a dam; it is raised and lowered by a rack and pinion mechanism.

sector gear [DES ENG] **1.** A toothed device resembling a portion of a gear wheel containing the center bearing and a part of the rim with its teeth. **2.** A gear having such a device as its chief essential feature. [MECH ENG] A gear system employing such a gear as a principal part.

secular [ENG] Of or pertaining to a long indefinite period of time.

securite explosive [MATER] A type of plastic explosive with a balanced oxygen content; it is built up on a nonexplosive, hydrophilic gel and contains oxygen-emitting salts, solid high explosive, and water.

sediment and water *See* basic sediment and water.

sedimentation [MET] Classification of metal powders by the rate of settling in a fluid.

sedimentation tank [ENG] A tank in which suspended matter is removed either by quiescent settlement or by continuous flow at high velocity and extended retention time to allow deposition.

sediment bulb [ENG] A bulb for holding sediment that settles from the liquid in a tank.

sediment corer [ENG] A heavy coring tube which punches out a cylindrical sediment section from the ocean bottom.

sediment trap [ENG] A device for measuring the accumulation rate of sediment on the floor of a body of water.

seep [PETRO ENG] An oil spring whose daily yield ranges from a few drops to several barrels of oil; usually located at low elevations where water has accumulated.

Seger cone [MATER] Any of a series of conical shaped thermometric devices made of materials that deform at specified temperatures; consists of mixtures of clay, salt, and other materials in such proportions that their softening temperatures vary progressively through the series; used to indicate temperatures of furnaces, particularly in ceramic industries. Also known as pyrometric cone.

segmental meter [ENG] A variable head meter whose orifice plate has an opening in the shape of a half circle.

segment die [MET] A die made of parts that can be disassembled to facilitate removal of the workpiece. Also known as split die.

segment saw [MECH ENG] A saw consisting of steel segments attached around the edge of a flange and used for cutting veneer.

segregation [ENG] **1.** The keeping apart of process streams. **2.** In plastics molding, a close succession of parallel, relatively narrow, and sharply defined wavy lines of color on the surface of a plastic that differ in shade from surrounding areas and create the impression that the components have separated. [MET] The nonuniform distribution of alloying elements, impurities, or microphases, resulting in localized concentrations.

seine net [ENG] A net used to catch fish by encirclement, usually by closure of the two ends and the bottom.

seismic constant [CIV ENG] In building codes dealing with earthquake hazards, an arbitrarily set quantity of steady acceleration, in units of acceleration of gravity, that a building must withstand.

seismic detector [ENG] An instrument that receives seismic impulses.

seismic exploration [ENG] The exploration for economic deposits by using seismic techniques, usually involving explosions, to map subsurface structures.

seismic profiler [ENG] A continuous seismic reflection system used to study the structure beneath the sea floor to depths of 10,000 feet (3000 meters) or more, using a rotating drum to record reflections.

seismic shooting [ENG] A method of geophysical prospecting in which elastic waves are produced in the earth by the firing of explosives.

seismic survey *See* reflection survey.

seismochronograph [ENG] A chronograph for determining the time at which an earthquake shock appears.

seismogram [ENG] The record made by a seismograph.

seismograph [ENG] An instrument that records vibrations in the earth, especially earthquakes.

seismometer [ENG] An instrument that detects movements in the earth.

seismoscope [ENG] An instrument for recording only the occurrence or time of occurrence (not the magnitude) of an earthquake.

Sejournet process [MET] During hot extrusion, the lubrication and insulation of a metal billet with molten glass. Also known as Ugine Sejournet process.

selected time [IND ENG] An observed actual time value for an element, measured by time study, which is identified as being the most representative of the situation observed.

selective acidizing [PETRO ENG] Oil-reservoir acid treatment (acidizing) in which the acid is injected into specific reservoir zones; contrasted with uncontrolled acidizing in which the acid solution is simply pumped down the casing and is forced into adjacent rock.

selective filling [MIN ENG] Filling by hand so that stone or dirt is rejected and only clean coal or ore is loaded.

selective flotation [MIN ENG] The surface or froth selecting of the valuable minerals rather than the gangue.

selective fracturing [PETRO ENG] Procedures for obtaining multiple formation fractures in a specific reservoir zone by plugging casing perforations or by isolating (with packers) the desired zone prior to fracturing operations.

selective heating [MET] Heating only certain portions of a workpiece to impart desired properties.

selective mining [MIN ENG] A method of mining whereby ore of unwarranted high value is mined in such manner as to make the low-grade ore left in the mine incapable of future profitable extraction; in other words, the best ore is selected in order to make good mill returns, leaving the low-grade ore in the mine.

selective plating [MET] An electrochemical process in which the base metal is masked, except the area to receive the plate, with a nonconductive material; the masked metal, with an electric current running through it, is then sprayed with a solution of plating metal which adheres only to the unmasked section.

selective quenching [MET] Quenching only certain portions of a piece of metal.

selective transmission [MECH ENG] A gear transmission with a single lever for changing from one gear ratio to another; used in automotive vehicles.

selector [CIV ENG] A device that automatically connects the appropriate railroad signal to control the track selected. [ENG] **1.** A device for selecting objects or materials according to predetermined properties. **2.** A device for starting or stopping at predetermined positions. [MECH

ENG] **1.** The part of the gearshift in an automotive transmission that selects the required gearshift bar. **2.** The lever with which a driver operates an automatic gearshift. [MET] A converter that separates purified copper from residue in a single operation.

selenium stainless steel [MET] Stainless steel to which about 0.1 percent or more selenium is added to improve machinability.

selenotrope [ENG] A device used in geodetic surveying for reflecting the moon's rays to a distant point, to aid in long-distance observations.

self-acting door [MIN ENG] A ventilation door that is constructed of two halves which move on small pulleys and which are forced apart centrally as the trams come in contact with the converging beams that operate the door.

self-acting incline See gravity haulage.

self-adapting system [SYS ENG] A system which has the ability to modify itself in response to changes in its environment.

self-advancing supports [MIN ENG] An assembly of hydraulically operated steel hydraulic supports, on a long-wall face, which are moved forward as a unit. Also known as walking props.

self-centering chuck [MECH ENG] A drill chuck that, when closed, automatically positions the drill rod in the center of the drive rod of a diamond-drill swivel head.

self-cleaning [ENG] Pertaining to any device that is designed to clean itself without disassembly, for example, a filter in which accumulated filter cake or sludge is removed by an internal scraper or by a blowdown or backwash action.

self-contained breathing apparatus [ENG] A portable breathing unit which permits freedom of movement.

self-contained portable electric lamp [MIN ENG] An electric lamp which is operated by an electric battery and is specifically designed to be carried about by its user.

self-contained range finder [ENG] Instrument used for measuring range by direct observation, without using a base line; the two types are the coincidence range finder and the stereoscopic range finder.

self-dumping cage [MIN ENG] A cage in which the deck is pivoted so that as the cage is lifted, toward the end of the lift, the deck tilts and the end door is lifted, discharging the coal.

self-dumping car [MIN ENG] A mine car which can be side-tipped while in motion on the rail track; it is fitted with a spherically contoured wheel which engages a ramp structure and gradually tilts the car.

self-energizing brake [MECH ENG] A brake designed to reinforce the power applied to it, such as a hand brake.

self-extinguishing [MATER] The ability of a material to cease burning once the source of the flame has been removed.

self-fluxing alloy [MET] Any alloy used in thermal spraying which does not require the addition of a flux in order to wet the substrate and coalesce when heated.

self-hardening steel See air-hardening steel.

self-locking nut [DES ENG] A nut having an inherent locking action, so that it cannot readily be loosened by vibration.

self-organizing system [SYS ENG] A system that is able to affect or determine its own internal structure.

self-propelled [MECH ENG] Pertaining to a vehicle given motion by means of a self-contained motor.

self-rescuer [MIN ENG] A small filtering device carried by a miner underground to provide immediate protection against carbon monoxide and smoke in case of a mine fire or explosion; used for escape purposes only, because it does not sustain life in atmospheres containing deficient oxygen.

self-sealing [ENG] A fluid container, such as a fuel tank or a tire, lined with a substance that allows it to close immediately over any small puncture or rupture.

self-starter [MECH ENG] An attachment for automatically starting an internal combustion engine.

self-tapping screw [DES ENG] A screw with a specially hardened thread that makes it possible for the screw to form its own internal thread in sheet metal and soft materials when driven into a hole. Also known as sheet-metal screw; tapping screw.

self-timer [ENG] A device that delays the tripping of a camera shutter so that the photographer can be included in the photograph.

sellers hob [MECH ENG] A hob that turns on the centers of a lathe, the work being fed to it by the lathe carriage.

Selwood engine [MECH ENG] A revolving-block engine in which two curved pistons opposed 180° run in toroidal tracks, forcing the entire engine block to rotate.

semiautomatic transmission [MECH ENG] An automobile transmission that assists the driver to shift from one gear to another.

semiautomatic welding [MET] An arc-welding method in which the electrode, a long length of small-diameter bare wire, usually in coil form, is positioned and advanced by the operator from a hand-held welding gun which feeds the electrode through the nozzle.

semichemical pulp [MATER] Wood which has been pulped by the process of semichemical pulping.

semiclosed-cycle gas turbine [MECH ENG] A heat engine in which a portion of the expanded gas is recirculated.

semicompreg [MATER] Resin-impregnated wood compressed to a density not exceeding 1.25.

semiconductive loading tube [ENG] A loading tube for blasthole explosives which dissipates static electric charges to prevent premature blasts.

semicontinuous casting *See* direct-chill casting.

semidiesel engine [MECH ENG] **1.** An internal combustion engine of a type resembling the diesel engine in using heavy oil as fuel but employing a lower compression pressure and spraying it under pressure, against a hot (uncooled) surface or spot, or igniting it by the precombustion or supercompression of a portion of the charge in a separate member or uncooled portion of the combustion chamber. **2.** A true diesel engine that uses a means other than compressed air for fuel injection.

semifinishing [MET] The preliminary finishing operations.

semifloating axle [MECH ENG] A supporting member in motor vehicles which carries torque and wheel loads at its outer end.

semigel [MATER] A cohesive powder used as an explosive.

semigloss [MATER] Pertaining to a surface finish intermediate between flat and glossy; used especially of paint and varnish.

semikilled steel [MET] Incompletely deoxidized steel containing enough dissolved oxygen to react with the carbon it contains to form carbon monoxide, the latter offsetting solidification shrinkage.

semilive skid [ENG] A platform having two fixed legs at one end and two wheels at the other; used for moving bulk materials.

semimat [MATER] Intermediate between glossy and mat, as photographic paper.

semimember [CIV ENG] A part in a frame or truss that ceases to bear a load when the stress in it starts to reverse.

semimonocoque [AERO ENG] A fuselage structure in which longitudinal members (stringers) as well as rings or frames which run circumferentially around the fuselage reinforce the skin and help carry the stress. Also known as stiffened-shell fuselage.

semipermanent mold [MET] A reusable metal mold with expendable sand cores.

semiplastic explosive [MATER] An explosive in which the quantities of liquid products are insufficient to render the mixture compressible.

semipositive mold [ENG] A plastics mold that allows a small amount of excess material to escape when it is closed.

semirefined wax [MATER] Commercial grades of petroleum wax which are inferior to fully refined grades but which meet specified requirements as to color and oil content.

semirigid plastic [MATER] A plastic that has a stiffness or apparent modulus of elasticity of between 10,000 and 100,000 psi (6.895×10^7 and 6.895×10^8 newtons per square meter) under prescribed test conditions.

semisilica refractory [MATER] A silica refractory made from clay with a high silica (sand) content (over 70% total silica); characterized by dimensional stability when heated or fired.

semisteel [MET] Low-carbon steel made by replacing about one-fourth of the pig iron in the cupola with steel scrap.

semitrailer [ENG] A cargo-carrying piece of equipment that has one or two axles at the rear; the load is carried on these axles and on the fifth wheel of the tractor that supplies motive power to the semitrailer.

semivitreous [MATER] Pertaining to ceramics whose glassy content is not sufficient to reduce porosity below 0.2%.

sems [DES ENG] A preassembled screw and washer combination.

send *See* scend.

Sendust [MET] A magnetic alloy composed of 85% iron, 9.5% silicon, and 5.5% aluminum; used in high-frequency powders.

Sendzimir mill [MET] A mill having small-diameter working rolls, each backed by a pair of supporting rolls, and each pair of these supported by a cluster of three rolls; used for cold-rolling wide sheets of metal to close tolerance.

sensation level *See* level above threshold.

sense [ENG] To determine the arrangement or position of a device or the value of a quantity.

sensing element *See* sensor.

sensitive altimeter [ENG] An aneroid altimeter constructed to respond to pressure changes (altitude changes) with a high degree of sensitivity; it contains two or more pointers to refer to different scales, calibrated in hundreds of feet, thousands of feet, and so on.

sensitivity [ENG] **1.** A measure of the ease with which a substance can be caused to explode. **2.** A measure of the effect of a change in severity of engine-operating conditions on the antiknock performance of a fuel; expressed as the difference between research and motor octane numbers. Also known as spread.

sensitometer [ENG] An instrument for measuring the sensitivity of light-sensitive materials.

sensor [ENG] The generic name for a device that senses either the absolute value or a change in a physical quantity such as temperature, pressure, flow rate, or pH, or the intensity of light, sound, or radio waves and converts that change into a useful input signal for an information-gathering system; a television camera is therefore a sensor, and a transducer is a special type of sensor. Also known as primary detector; sensing element.

separated aggregate [MATER] Aggregate for concrete that has been separated into fine and coarse constituents.

separate sewage system [CIV ENG] A drainage system in which sewage and groundwater are carried in separate sewers.

separation [AERO ENG] The action of a fallaway section or companion body as it casts off from the remaining body of a vehicle, or the action of the remaining body as it leaves a fallaway section behind it. [ENG] **1.** The action segregating phases, such as gas-liquid, gas-solid, liquid-solid. **2.** The segregation of solid particles by size range, as in screening. [ENG ACOUS] The degree, expressed in decibels, to which left and right stereo channels are isolated from each other. [MIN ENG] The removal of gangue from raw ores, as in frothing.

separator [ENG] **1.** A machine for separating materials of different specific gravity by means of water or air. **2.** Any machine for separating materials, as the magnetic separator. [MECH ENG] *See* cage. [PETRO ENG] *See* gas-oil separator.

separator-filter [ENG] A vessel that removes solids and entrained liquid from a liquid or gas stream, using a combination of a baffle or coalescer with a screening (filtering) element.

sepia [MATER] A brown pigment prepared from the dried, inky exudation of a cuttlefish; used as a dye and in watercolors and ink.

septic tank [CIV ENG] A settling tank in which settled sludge is in immediate contact with sewage flowing through the tank while solids are decomposed by anaerobic bacterial action.

sequence [ENG] An orderly progression of items of information or of operations in accordance with some rule.

sequencer [ENG] A mechanical or electronic device that may be set to initiate a series of events and to make the events follow in a given sequence.

sequence timer [MET] A device used in resistance welding to control the sequence and duration of all elements of the weld cycle, except weld time or heat time.

sequence weld timer [MET] A sequence timer which also controls weld time or heat time.

sequencing [IND ENG] Designating the order of performance of tasks to assure optimal utilization of available production facilities.

sequential collation of range [ENG] Spherical, long-baseline, phase-comparison trajectory-measuring system using three or more ground stations, time-sharing a single transponder, to provide nonambiguous range measurements to determine the instantaneous position of a vehicle in flight.

sequential sampling [IND ENG] A sampling plan in which an undetermined number of samples are tested one by one, accumulating the results until a decision can be made.

serial [IND ENG] An element or a group of elements within a series which is given a numerical or alphabetical designation for convenience in planning, scheduling, and control.

series firing [ENG] The firing of detonators in a round of shots by passing the total supply current through each of the detonators.

series-parallel firing [ENG] The firing of detonators in a round of shots by dividing the total supply current into branches, each containing a certain number of detonators wired in series.

series production [IND ENG] The manufacture of a product or service by a group of operations sequenced so that all materials will be routed successively through each production state. Also known as batch production.

series reliability [SYS ENG] Property of a system composed of elements in such a way that failure of any one element causes a failure of the system.

series shots [ENG] The connecting and firing of a number of loaded holes one after the other.

series ventilation [MIN ENG] A system of ventilating a number of faces consecutively by the same air current.

series welding [MET] Making two or more resistance welds simultaneously by using a single welding transformer with three or more electrodes forming a series circuit.

service [ENG] To perform services of maintenance, supply, repair, installation, distribution, and so on, for or upon an instrument, installation, vehicle, or territory.

serviceability [IND ENG] The reliability of equipment according to some objective criterion such as serviceability ratio, utilization ratio, or operating ratio.

serviceability ratio [IND ENG] The ratio of up time to the sum of up time and down time.

service agreement [ENG] A contract which agrees to provide mechanical maintenance of a machine for a fixed period of time at a stated charge.

service brake [MECH ENG] The brake used for ordinary driving in an automotive vehicle; usually foot-operated.

service ceiling [AERO ENG] The height at which, under standard atmospheric conditions, an aircraft is unable to climb faster than a specified rate (100 feet or 30 meters per minute in the United States, Great Britain, and Canada).

service compartment [MIN ENG] The section of a mine shaft that houses water pipes, compressed-air pipeline, cables and telephone wires, and signaling and similar arrangements.

service engineering [ENG] The function of determining the integrity of material and services in order to measure and maintain operational reliability, approve design changes, and assure their conformance with established specifications and standards.

service factor [ENG] For a chemical or a petroleum processing plant or its equipment, the measure of the continuity of an operation, computed by dividing the time on-stream (actual running time) by the total elapsed time.

service life [ENG] The length of time during which a machine, tool, or other apparatus or device can be operated or used economically or before breakdown.

service pipe [CIV ENG] A pipe linking a building to a main pipe.

service road [CIV ENG] A small road parallel to the main road for convenient access to shops and houses.

service shaft [MIN ENG] A shaft used only for hoisting men and materials to and from underground.

servo brake [MECH ENG] **1.** A brake in which the motion of the vehicle is used to increase the pressure on one of the shoes. **2.** A brake in which the force applied by the operator is augmented by a power-driven mechanism.

servonoise [ENG] Hunting action of the tracking servomechanism of a radar, which results from backlash and compliance in the gears, shafts, and structures of the mount.

sesame oil [MATER] A combustible, yellow, optically active, semidrying fatty oil obtained from sesame seeds; soluble in ether, benzene, and carbon disulfide, slightly soluble in alcohol; melts at 20–25°C; used in edible food products, such as shortenings, salad oils, and margarine. Also known as benne oil; gingelly oil; teel oil.

sessile dislocation [MET] A dislocation in a metal lattice that is relatively immobile, offering an obstacle to the movement of other dislocations.

set [ENG] **1.** A combination of units, assemblies, and parts connected or otherwise used together to perform an operational function, such as a radar set. **2.** In plastics processing, the conversion of a liquid resin or adhesive into a solid state by curing or evaporation of solvent or suspending medium, or by gelling. **3.** Saw teeth bent out of the plane of the saw body, resulting in a wide cut in the workpiece. [MATER] **1.** The hardening or firmness displayed by some materials when left undisturbed. **2.** Permanent deformation of a material, such as metal or plastic, when stressed beyond the elastic limit. [MIN ENG] *See* frame set.

set analyzer *See* analyzer.

set bit [DES ENG] A bit insert with diamonds or other cutting media.

set casing [ENG] Introducing cement between the casing and the wall of the hole to seal off intermediate formations and prevent fluids from entering the hole.

set copper [MET] An intermediate copper product obtained at the end of the oxidizing portion of the fire-refining cycle and containing about 3–4% cuprous oxide.

set grease soap [MATER] A soap made at room temperature by saponifying slack lime with resin oil; used to make a cold-set grease such as axle grease.

set hammer [DES ENG] **1.** A hammer used as a shaping tool by blacksmiths. **2.** A hollow-face tool used in setting rivets.

setover [ENG] A device which helps move a lathe tailstock or headstock on its base so that a taper on a turned piece can be obtained.

set pressure [MECH ENG] The inlet pressure at which a relief valve begins to open as required by the code or standard applicable to the pressure vessel to be protected.

set screw [DES ENG] A small headless machine screw, usually having a point at one end and a recessed hexagonal socket or a slot at the other end, used for such purposes as holding a knob or gear on a shaft.

setting angle [MECH ENG] The angle, usually 90°, between the straight portion of the tool shank of the machined portion of the work.

setting circle [ENG] A coordinate scale on an optical pointing instrument, such as a telescope or surveyor's transit.

setting gage [ENG] A standard gage for testing a limit gage or setting an adjustable limit gage.

setting temperature [ENG] The temperature at which a liquid resin or adhesive, or an assembly involving them, will set, that is, harden, gel, or cure.

setting time [ENG] The length of time that a resin or adhesive must be subjected to heat or pressure to cause them to set, that is, harden, gel, or cure.

settleable solids test [CIV ENG] A test used in examination of sewage to help determine the sludge-producing characteristics of sewage; a measurement of the part of the suspended solids heavy enough to settle is made in an Imhoff cone.

settled ground [MIN ENG] Ground which has ceased to subside over the waste area of a mine, having reached a state of full subsidence.

settlement [CIV ENG] The gradual downward movement of an engineering structure, due to compression of the soil below the foundation. [MIN ENG] The gradual lowering of the overlying strata in a mine, due to extraction of the mined material.

settler [ENG] A separator, such as a tub, pan, vat, or tank in which the partial separation of a mixture is made by density difference; used to separate solids from liquid or gas, immiscible liquid from liquid, or liquid from gas.

settling [ENG] The gravity separation of heavy from light materials; for example, the settling out of dense solids or heavy liquid droplets from a liquid carrier, or the settling out of heavy solid grains from a mixture of solid grains of different densities.

settling basin [CIV ENG] An artificial trap designed to collect suspended stream sediment before discharge of the stream into a reservoir. [IND ENG] A sedimentation area designed to remove pollutants from factory effluents.

settling chamber [ENG] A vessel in which solids or heavy liquid droplets settle out of a liquid carrier by gravity during processing or storage.

settling pond [MIN ENG] A natural or artificial pond for recovering the solids from an effluent.

settling reservoir [CIV ENG] A reservoir consisting of a series of basins connected in steps by long weirs; only the clear top layer of each basin is drawn off.

settling tank [ENG] A tank into which a two-phase mixture is fed and the entrained solids settle by gravity during storage.

setup [IND ENG] The preparation of a facility or a machine for a specific work method, activity, or process.

setup time [MATER] The time required for a cement or a gelatin to harden.

sewage [CIV ENG] The fluid discharge from medical, domestic, and industrial sanitary appliances.

sewage disposal plant [CIV ENG] The land, building, and apparatus employed in the treatment of sewage by chemical precipitation or filtration, bacterial action, or some other method.

sewage sludge [CIV ENG] A semiliquid waste with a solid concentration in excess of 2500 parts per million, obtained from the purification of municipal sewage. Also known as sludge.

sewage system [CIV ENG] Any of several drainage systems for carrying surface water and sewage for disposal.

sewage treatment [CIV ENG] A process for the purification of mixtures of human and other domestic wastes; the process can be aerobic or anaerobic.

sewer [CIV ENG] An underground pipe or open channel in a sewage system for carrying water or sewage to a disposal area.

sewerage [CIV ENG] The sewage system in a particular district.

sewer gas [MATER] The gas evolved from the decomposition of municipal sewage; it has a high content of methane and hydrogen sulfide, and can be used as a fuel gas.

sewing machine [MECH ENG] A mechanism that stitches cloth, leather, book pages, or other material by means of a double-pointed or eye-pointed needle.

SFC *See* specific fuel consumption.

shackle [DES ENG] An open or closed link of various shapes with extended legs; each leg has a transverse hole to accommodate a pin, bolt, or the like, which may or may not be furnished.

shackle bolt [DES ENG] A cylindrically shaped metal bar for connecting the ends of a shackle.

shading ring [ENG ACOUS] A heavy copper ring sometimes placed around the central pole of an electrodynamic loudspeaker to serve as a shorted turn that suppresses the hum voltage produced by the field coil.

shaft [MECH ENG] A cylindrical piece of metal used to carry rotating machine parts, such as pulleys and gears, to transmit power or motion. [MIN ENG] An excavation of limited area compared with its depth, made for finding or mining ore or coal, raising water, ore, rock, or coal, hoisting and lowering men and material, or ventilating underground workings; the term is often specifically applied to approximately vertical shafts as distinguished from an incline or an inclined shaft.

shaft allowance [MIN ENG] The extra space between the excavation diameter and the finished diameter to accommodate the permanent shaft lining.

shaft balancing [DES ENG] The process of redistributing the mass attached to a rotating body in order to reduce vibrations arising from centrifugal force. Also known as rotor balancing.

shaft cable [MIN ENG] A specially armored cable of great mechanical strength running down a mine shaft.

shaft capacity [MIN ENG] The output of ore or coal that can be expected to be raised regularly and in normal circumstances.

shaft column [MIN ENG] A length of pipes installed in a mine shaft for pumping, for hydraulic stowing, or for compressed air.

shaft coupling See coupling.

shaft crusher [MIN ENG] A hard-rock crusher in a shaft, set to reduce large lumps of ore to a convenient size for delivery to the skip.

shaft deformation bar [MIN ENG] A length of 1½-inch (3.8-centimeter) pipe fitted at one end with a micrometer and at the other end with a hard-steel cone for measuring the deformation in the cross section of a shaft.

shaft drilling [MIN ENG] The drilling of small shafts up to about 5 feet (1.5 meters) in diameter with the shot drill.

shaft furnace [ENG] A vertical, refractory-lined cylinder in which a fixed bed (or descending column) of solids is maintained, and through which an ascending stream of hot gas is forced; for example, the pig-iron blast furnace and the phosphors-from-phosphate-rock furnace.

shaft hopper [MECH ENG] A hopper that feeds shafts or tubes to grinders, threaders, screw machines, and tube benders.

shaft horsepower [MECH ENG] The output power of an engine, motor, or other prime mover; or the input power to a compressor or pump.

shaft house [MIN ENG] A building at the mouth of a shaft, where ore or rock is received from the mine.

shafting [MECH ENG] The cylindrical machine element used to transmit rotary motion and power from a driver to a driven element; for example, a steam turbine driving a ship's propeller.

shaft kiln [ENG] A kiln in which raw material fed into the top, moves down through hot gases flowing up from burners on either side at the bottom, and emerges as a product from the bottom; used for calcining operations.

shaft lining [MIN ENG] The timber, steel, brick, or concrete structure fixed around a shaft to support the walls.

shaft pillar [MIN ENG] A large area of a coal or ore seam which is left unworked around the shaft bottom to protect the shaft from damage by subsidence.

shaft plumbing [MIN ENG] The operation of orienting two plumb bobs, both at surface and at depth in order to transfer the bearing underground.

shaft pocket [MIN ENG] Ore storage pocket, of one or more compartments, cut into the wall on one or both sides of a vertical shaft or in the hanging wall of an inclined shaft.

shaft siding [MIN ENG] The station or landing place arranged for buckets or tubs at the bottom of the winding shaft.

shaft signaling [MIN ENG] The transmission of visible and audible signals between the onsetter or hitcher at the pit bottom and the banksman or hoistman at the pit top.

shaft sinking [MIN ENG] Excavating a shaft downward, usually from the surface, to the workable coal or ore.

shaft sinking drill [MIN ENG] A large-diameter drill with multiple rotary cones or cutting bits, used for shaft sinking.

shaft spillway [CIV ENG] A vertical shaft which has a funnel-shaped mouth and ends in an outlet tunnel, providing an overflow duct for a reservoir. Also known as morning glory spillway.

shaft station [MIN ENG] An enlargement of a level near a shaft from which ore, coal, or rock may be hoisted and supplies unloaded.

shagreen [MATER] **1.** A leather made by pressing grains into the hide to create indentations. **2.** The skin of certain sharks and rays containing small knobs.

shake [MATER] **1.** Separation between adjoining layers of wood, due to causes other than drying. **2.** A thick hand-cut shingle.

shakedown test [ENG] An equipment test made during the installation work.

shakeout [MET] Removing a casting from a sand mold.

shaker conveyor [MIN ENG] A conveyor consisting of a length of metal troughs, with suitable supports, to which a reciprocating motion is imparted by drives.

shake table See vibration machine.

shake-table test [ENG] A laboratory test for vibration tolerance, in which the device to be tested is placed on a shake table.

shaking screen [MECH ENG] A screen used in separating material into desired sizes; has an eccentric drive or an unbalanced rotating weight to produce shaking.

shaking table See Wilfley table.

shale clay [MATER] A clay made from ground shale.

shale naphtha [MATER] Naphtha derived from shale oil, usually containing 60–70% olefins and other hydrocarbons.

shale oil [MATER] Liquid obtained from the destructive distillation of kerogens in oil shale; further processing is needed to convert shale oil into products similar to petroleum oils.

shale shaker [PETRO ENG] A vibrating screen over which drilling fluid is passed to trap the drill cuttings as the fluid passes through.

shank [DES ENG] **1.** The end of a tool which fits into a drawing holder, as on a drill. **2.** *See* bit blank.

shank-type cutter [DES ENG] A cutter having a shank to fit into the machine tool spindle or adapter.

shape coding [DES ENG] The use of special shapes for control knobs, to permit recognition and sometimes also position monitoring by sense of touch.

shaped-chamber manometer [ENG] A flow measurement device that measures differential pressure with a uniform flow-rate scale with a specially shaped chamber.

shaper [MECH ENG] A machine tool for cutting flat-on-flat, contoured surfaces by reciprocating a single-point tool across the workpiece.

shaping [MECH ENG] A machining process in which a reciprocating single-point tool moves over the work in straight, parallel lines to produce a flat surface.

shaping dies [MECH ENG] A set of dies for bending, pressing, or otherwise shaping a material to a desired form.

shapometer [ENG] A device used to measure the shape of sedimentary particles.

shark liver oil [MATER] A yellow to brown oil with strong aroma, obtained from the livers of various sharks; insoluble in water, soluble in ether, benzene, and carbon disulfide; used as a vitamin A source, in biochemical research. Also known as dogfish oil; shark oil.

shark oil *See* shark liver oil.

sharp-crested weir [CIV ENG] A weir in which the water flows over a thin, sharp edge.

sharpen [ENG] To give a thin keen edge or a sharp acute point to.

sharpening stone [ENG] A device such as a whetstone used for sharpening by hand.

sharp iron [ENG] A tool used to open seams for caulking.

sharp V thread [DES ENG] A screw thread having a sharp crest and root; the included angle is usually 60°.

shattercrack *See* flake.

shatterproof glass *See* nonshattering glass.

shave hook [DES ENG] A plumber's or metalworker's tool composed of a sharp-edged steel plate on a shank; used for scraping metal.

shaving [ENG ACOUS] Removing material from the surface of a disk recording medium to obtain a new recording surface. [MECH ENG] **1.** Cutting off a thin layer from the surface of a workpiece. **2.** Trimming uneven edges from stampings, forgings, and tubing.

Shaw process [MET] A foundry molding process which makes use of wood or metal patterns and a refractory mold bonded with an ethyl silicate base material.

shear [DES ENG] A cutting tool having two opposing blades between which a material is cut. [MIN ENG] To make vertical cuts in a coal seam that has been undercut.

shear angle [MECH ENG] The angle made by the shear plane with the work surface.

shear burst [MIN ENG] The explosive breaking of wall rock in a deep mining field by the occurrence of a single shear crack parallel to the face in one of the walls, causing rock behind the shear plane to expand freely into the stope and then to disrupt and fill the place with debris.

shearing [MECH ENG] Separation of material by the cutting action of shears. [MIN ENG] The vertical side cutting which, together with holing or horizontal undercutting, constitutes the attack upon a face of coal.

shearing die [MECH ENG] A die with a punch for shearing the work from the stock.

shearing machine [MECH ENG] A machine for cutting cloth or bars, sheets, or plates of metal or other material.

shearing punch [MECH ENG] A punch that cuts material by shearing it, with minimal crushing effect.

shearing tool [DES ENG] A cutting tool (for a lathe, for example) with a considerable angle between its face and a line perpendicular to the surface being cut.

shear lip [MET] An area or ridge at the edge of a shear fracture surface.

shear mark [ENG] A crease on a piece of pressed glass; results when the piece is sheared off for pressing.

shear pin [DES ENG] **1.** A pin or wire provided in a fuse design to hold parts in a fixed relationship until forces are exerted on one or more of the parts which cause shearing of the pin or wire; the shearing is usually accomplished by setback or set forward (impact) forces; the shear member may be augmented during transportation by an additional safety device. **2.** In a propellant-actuated device, a locking member which is released by shearing. **3.** In a power train, such as a winch, any pin, as through a gear and shaft, which is designed to fail at a predetermined force in order to protect a mechanism.

shear spinning [MECH ENG] A sheet-metal-forming process which forms parts with rotational symmetry over a mandrel with the use of a tool or roller in which deformation is carried out with a roller in such a manner that the diameter of the original blank does not change but the thickness of the part decreases by an amount dependent on the mandrel angle.

shear steel [MET] A cutlery steel made from short sheared lengths of blister steel; the lengths are heated, joined by rolling or hammering, and finished by hammering.

shear test [ENG] Any of various tests to determine shear strength of soil samples.

shear transformation *See* martensitic transformation.

sheathed explosive [ENG] A permitted explosive enveloped by a sheath containing a noncombustible powder which reduces the temperature of the resultant gases of the explosion and, therefore, reduces the risk of these hot gases causing a firedamp ignition.

sheathing board [MATER] A composition board (for example, of fiber or gypsum cement) used instead of wood sheathing.

sheathing paper [MATER] A paper that is heavier and of better quality than the usual building paper.

sheave [DES ENG] A grooved wheel or pulley.

SHED *See* solar heat exchanger drive.

sheepsfoot roller [DES ENG] A cylindrical steel drum to which knob-headed spikes are fastened; used for compacting earth.

sheepskin wheel [DES ENG] A polishing wheel made of sheepskin disks or wedges either quilted or glued together.

sheet [MATER] A material in a configuration similar to a film except that its thickness is greater than 0.25 millimeter.

sheet asphalt [MATER] Asphalt which provides a smooth surface and is used for continuous road surfacing.

sheet copper [MET] Copper rolled into sheets; for roofing sometimes used as it leaves the rolls, but for other purposes it is commonly employed after it has been cold-rolled to increase hardness and strength.

sheet forming [ENG] The process of producing thin, flat sections of solid materials; for example, sheet metal, sheet plastic, or sheet glass.

sheet glass [MATER] Flat sections of glass made by drawing a continuous thin film of glass from a molten bath, then cooling and cutting it; used for common glazing.

sheeting [MATER] **1.** A continuous film of a material such as plastic. **2.** Steel or wood members used to face the walls of an excavation such as a basement or a trench.

sheeting caps [MIN ENG] A row of caps put on blocks about 14 inches (36 centimeters) high which are placed on top of the drift sets when constructing the permanent floor in the stope.

sheet metal [MET] Thin sections of metal formed by rolling hot metal and usually less than 0.25 inch (6.35 millimeters) thick; when thicker than 0.25 inch, called plate.

sheet-metal gage [MET] A standard for expressing the thickness of metal sheets; some manufacturers, for example, Brown & Sharpe (B&S), Birmingham (BG), and Imperial, use code numbers with actual thickness in inches or millimeters.

sheet mica [MATER] Mica that is relatively flat and sufficiently free from structural defects to enable it to be punched or stamped into specified shapes for use by the electronic and electrical industries.

sheet piling [CIV ENG] Closely spaced piles of wood, steel, or concrete driven vertically into the ground to obstruct lateral movement of earth or water, and often to form an integral part of the permanent structure.

sheet plastic [MATER] Flat sections of extruded, molded, or cast plastic, with a thickness greater than that for film, that is, greater than 0.05 inch (1.3 millimeters).

Sheetrock [MATER] A plasterboard, usually made of two sheets of heavy paper separated by a layer of gypsum.

sheet rubber [MATER] Latex that has been rolled into sheets, either smooth or ribbed.

sheet separation [MET] The gap between faying surfaces surrounding the weld in spot, seam, or projection welding.

sheet steel [MET] Steel rolled in the form of sheet, usually used for deep-drawing applications.

sheet train [ENG] The entire assembly needed to produce plastic sheet; includes the extruder, die, polish rolls, conveyor, draw rolls, cutter, and stacker.

Sheffield plate [MET] A cladding of silver rolled and fused on both sides of a copper sheet.

Shelby tube [ENG] A thin-shelled tube used to take deep-soil samples; the tube is pushed into the undisturbed soil at the bottom of the casting of the borehole driven into the ground.

shelf angle [CIV ENG] A mild steel angle section, riveted or welded to the web of an I beam to support the formwork for hollow tiles or the floor or roof units, or to form a seat for precast concrete.

shelf life [ENG] The time that elapses before stored food, chemicals, batteries, and other materials or devices become inoperative or unusable due to age or deterioration.

shell [BUILD] A building without internal partitions or furnishings. [DES ENG] **1.** The case of a pulley block. **2.** A thin hollow cylinder. **3.** A hollow hemispherical structure. **4.** The outer wall of a vessel or tank. [MET] **1.** The outer wall of a metal mold. **2.** The hard layer of sand and thermosetting plastic formed over a pattern and used as a mold wall in shell molding. **3.** The metal sleeve remaining when a billet is extruded with a dummy block at smaller diameter. **4.** A tubular casting used in preparing seamless drawn tubes. **5.** A pierced forging.

shellac [MATER] A natural, alcohol-soluble, water-insoluble, flammable resin; made from lac resin deposited on tree twigs in India by the lac insect (*Laccifer lecca*) used as an ingredient of wood coatings.

shellac varnish [MATER] A solution of shellac in denatured alcohol; used in wood finishing where a fast-drying, light-colored, hard finish is desired.

shellac wax [MATER] A hard wax with 3% shellac, from which it is extracted, and used in polishes and insulating materials.

shellac wheel [DES ENG] A grinding wheel having the abrasive bonded with shellac.

shell-and-tube exchanger [ENG] A device for the transfer of heat from a hot fluid to a cooler fluid; one fluid passes through a group (bundle) of tubes, the other passes around the tubes, through a surrounding shell.

shell capacity [ENG] The amount of liquid that a tank car or tank truck will hold when the liquid just touches the underside of the top of the tank shell.

shell clearance [DES ENG] The difference between the outside diameter of a bit or core barrel and the outside set or gage diameter of a reaming shell.

shell core [MET] A sand core formed by shell molding.

shell innage [ENG] The depth of a liquid in a tank car or tank truck shell.

shell knocker [ENG] A device to strike the external surface of a horizontally rotating process vessel (for example, a kiln or a dryer) to loosen accumulations of solid materials from the inner walls or flights of the shell. Also known as knocker.

shell molding [MET] Forming a rigid, porous, self-supporting refractory mold by sprinkling molding sand blended with thermosetting plastic or resin over a pre-heated metal pattern and then curing in an oven.

shell outage [ENG] The unfilled portion of a tank car or tank truck shell; the distance from the underside of the top of the shell to the level of the liquid in the shell.

shell pump [MECH ENG] A simple pump for removing wet sand or mud; consists of a hollow cylinder with a ball or clack valve at the bottom.

shell reamer [DES ENG] A machine reamer consisting of two parts, the arbor and the replaceable reamer, with straight or spiral flutes; designed as a sizing or finishing reamer.

shell roof [BUILD] A roof made of a thin, curved, platelike structure, usually of concrete but lumber and steel are also used.

Shelton loader [MIN ENG] A modified coal-cutting machine in which the picks of the cutter chain are replaced by loading flights, which push the prepared coal up a ramp on to the face conveyor.

sherardizing [MET] Coating iron with zinc by tumbling the article in powdered zinc at about 250–375°C.

SH-34G *See* Sea Bat.

shield [ENG] An iron, steel, or wood framework to support the ground ahead of the lining in tunneling and mining.

shielded arc welding [MET] Arc welding in which the electric arc and the weld metal are protected by gas, decomposition products of the electrode covering, or a blanket of fusible flux.

shielded metal-arc welding [MET] Arc welding in which heating with an electric arc between the electrode and the work produces fusion of the electrode covering which shields the work.

shielding [MET] Placing a nonconducting object in an electrolytic bath during plating to alter the current distribution.

shielding gas [MET] Gas, such as nitrogen, oxygen, and carbon dioxide, used in shielded arc welding to protect molten weld from contamination and damage by the atmosphere.

shift [IND ENG] The number of hours or the part of any day worked. Also known as tour. [MECH ENG] To change the ratio of the driving to the driven gears to obtain the desired rotational speed or to avoid overloading and stalling an engine or a motor. [MET] A casting defect caused by malalignment of the mold parts.

shift joint [BUILD] A vertical joint placed on a solid member of the course below.

shift work [IND ENG] Work paid for by day wage.

shim [ENG] **1.** In the manufacture of plywood, a long, narrow patch glued into the panel or cemented into the lumber core itself. **2.** A thin piece of material placed between two surfaces to obtain a proper fit, adjustment, or alignment.

shingle [MATER] A rectangular piece of wood, metal, or other material that is used like a tile and arranged in overlapping rows for covering roofs and walls.

shingle lap [DES ENG] A lap joint of tapered sections, the bottom of each section overlapping the top of the section below it.

shingle nail [DES ENG] A nail about a half to a full gage thicker than a common nail of the same length.

ship auger [DES ENG] An auger consisting of a spiral body having a single cutting edge, with or without a screw; there is no spur at the outer end of the cutting edge.

shipbuilding [CIV ENG] The construction of ships.

shipfitter [CIV ENG] A worker who builds the steel structure of a ship, including laying-off and fabricating the individual members, subassembly, and erection on the shipway.

ship motion [ENG] Translational and rotational motions of a ship in a wave system which cause the center of gravity to deviate from simple straight-line motion; these motions are heave, surge, sway, roll, pitch, and yaw.

shipping and storage container [IND ENG] A reusable noncollapsible container of any configuration designed to provide protection for a specific item against impact, vibration, climatic conditions, and the like, during handling, shipment, and storage.

shipping document [IND ENG] A document listing the items in a shipment, and showing other supply and transportation information that is required by agencies concerned in the movement of material.

shipping time [ENG] The time elapsing between the shipment of material by the supplying activity and receipt of material by the requiring activity.

shipping ton *See* ton.

shipway [CIV ENG] **1.** The ways on which a ship is constructed. **2.** The supports placed underneath a ship in dry dock.

shipwright [CIV ENG] A worker whose responsibility is to ensure that the structure of a ship is straight and true and to the designed dimensions; the work starts with the laying down of the keel blocks and continues throughout the steelwork; applicable also to wood ship builders.

shipyard [CIV ENG] A facility adjacent to deep water where ships are constructed or repaired.

Shirley-Ferranti viscometer [ENG] An instrument used to determine an ink's resistance to flow.

shivering [MATER] Cracks and scales on a pottery glaze caused by unequal contraction during cooling.

shock absorber [MECH ENG] A spring, a dashpot, or a combination of the two, arranged to minimize the acceleration of the mass of a mechanism or portion thereof with respect to its frame or support.

shock bump [MIN ENG] A rock bump resulting from the sudden collapse of a strong deposit.

shock isolation [MECH ENG] The application of isolators to alleviate the effects of shock on a mechanical device or system.

shock mount [MECH ENG] A mount used with sensitive equipment to reduce or prevent transmission of shock motion to the equipment.

shock resistance [ENG] The property which prevents cracking or general rupture when impacted.

shock strut [AERO ENG] The primary working part of any landing gear, which supplies the force as the airplane sinks toward the ground, turning the flight path from one intersecting the ground to one parallel to the ground.

shock test [ENG] The test to determine whether the armor sample will crack or spall under impact by kinetic energy or high-explosive projectiles.

shock tunnel [ENG] A hypervelocity wind tunnel in which a shock wave generated in a shock tube ruptures a second diaphragm in the throat of a nozzle at the end of the tube, and gases emerge from the nozzle into a vacuum tank with Mach numbers of 6 to 25.

shoe [ENG] In glassmaking, an open-ended crucible placed in a furnace for heating the blowing irons. [MECH ENG] **1.** A metal block used as a form or support in various bending operations. **2.** A replaceable piece used to break rock in certain crushing machines. **3.** *See* brake shoe. [MIN ENG] **1.** Pieces of steel fastened to a mine cage and formed to fit over the guides to guide it when it is in motion. **2.** The bottom wedge-shaped piece attached to tubbing when sinking through quicksand. **3.** A trough to convey ore to a crusher. **4.** A coupling of rolled, cast, or forged steel to protect the lower end of the casting or drivepipe in overburden, or the bottom end of a sampler when pressed into a formation being sampled.

shoe brake [MECH ENG] A type of brake in which friction is applied by a long shoe, extending over a large portion of the rotating drum; the shoe may be external or internal to the drum.

shoofly *See* slant.

shoot [ENG] To detonate an explosive, used to break coal loose from a seam or in blasting operation or in a borehole.

shooting board [ENG] **1.** A fixture used as a guide in planing boards; it is more accurate than a miter. **2.** A table and plane used for trimming printing plates.

shop fabrication [ENG] Making parts and materials in the shop rather than at the work site.

shop lumber [MATER] Softwood lumber graded and used in the factory for general cut-up purposes; similar to factory lumber but of a lower grade.

shop supplies [ENG] Expendable items consumed in operation and maintenance (for example, waste, oils, solvents, tape, packing, flux, or welding rod).

shop weld [ENG] A weld made in the workshop prior to delivery to the construction site.

shore [ENG] Timber or other material used as a temporary prop for excavations or buildings; may be sloping, vertical, or horizontal.

Shore hardness [ENG] A method of rating the hardness of a metal or of a plastic or rubber material.

shore protection [CIV ENG] Preventing erosion of the ground bordering a body of water.

Shore scleroscope [ENG] A device used in rebound hardness testing of rubber, metal, and plastic; consists of a small, conical hammer fitted with a diamond point and acting in a glass tube.

Shore scleroscope hardness test [MET] A rebound hardness test in which a metal body is dropped vertically down a glass tube onto the surface of the material being tested; the height of the rebound is a measure of the hardness. Also known as scleroscope hardness test.

shore tank [PETRO ENG] A shoreside storage tank for liquid petroleum products discharged by tankers.

shoring [ENG] Providing temporary support with shores to a building or an excavation.

short [ENG] In plastics injection molding, the failure to fill the mold completely. Also known as short shot.

short-baseline system [AERO ENG] A trajectory measuring system using a baseline the length of which is very small compared with the distance of the object being tracked.

short-circuiting transfer [ENG] Transfer of melted material from a consumable electrode during short circuits.

short column [CIV ENG] A column in which both compression and bending is significant, generally having a slenderness ratio between 30 and 120–150.

shortcoming [DES ENG] An imperfection or malfunction occurring during the life cycle of equipment, which should be reported and which must be corrected to increase efficiency and to render the equipment completely serviceable.

short-delay blasting [ENG] A method of blasting by which explosive charges are detonated in a given sequence with short time intervals.

short-delay detonator *See* millisecond delay cap.

short fuse [ENG] **1.** Any fuse that is cut too short. **2.** The practice of firing a blast, the fuse on the primer of which is not sufficiently long to reach from the top of the charge to the collar of the borehole; the primer, with fuse attached, is dropped into the charge while burning.

short leg [ENG] One of the wires on an electric blasting cap, which has been shortened so that when placed in the borehole, the two splices or connections will not come opposite each other and make a short circuit.

shortness [MET] A form of brittleness in metal, designated as hot, cold, or red to indicate the temperature range in which brittleness occurs.

short oil [MATER] Varnish containing a small percentage of oil.

short-range radar [ENG] Radar whose maximum line-of-sight range, for a reflecting target having 1 square meter of area perpendicular to the beam, is between 50 and 150 miles (80 and 240 kilometers).

short run [MET] Pertaining to a mold or casting filled only partially with molten metal.

shorts [ENG] Oversize particles held on a screen after sieving the fines through the screen.

short shipment [ENG] Freight listed or manifested but not received.

short supply [IND ENG] An item is in short supply when the total of stock on hand and anticipated receipts during a given period is less than the total estimated demand during that period.

short takeoff and landing [AERO ENG] The ability of an aircraft to clear a 50-foot (15-meter) obstacle within 1500 feet (450 meters) of commencing takeoff, or in landing, to stop within 1500 feet after passing over a 50-foot obstacle. Abbreviated STOL.

shortwall [MIN ENG] **1.** A method of mining in which comparatively small areas are worked separately. **2.** A length of coal face between about 5 and 30 yards (4.6 and

27 meters), generally employed in pillar methods of working.

shortwall coal cutter [MIN ENG] A machine for undercutting coal which has a long, rigid chain jib fixed in relation to the main body of the machine and which cuts across a heading from right to left, being drawn across by means of a steel-wire rope.

shot [AERO ENG] An act or instance of firing a rocket, especially from the earth's surface. [ENG] **1.** A charge of some kind of explosive. **2.** Small spherical particles of steel. **3.** Small steel balls used as the cutting agent of a shot drill. **4.** The firing of a blast. **5.** In plastics molding, the yield from one complete molding cycle, including scrap. [MIN ENG] Coal broken by blasting or other methods.

shot bit [DES ENG] A short length of heavy-wall steel tubing with diagonal slots cut in the flat-faced bottom edge.

shot blasting [MET] Cleaning and descaling metal by shot peening or by means of a stream of abrasive powder blown through a nozzle under air pressure in the range 30–150 pounds per square inch (2×10^5 to 1.0×10^6 newtons per square meter).

shot boring [ENG] The act or process of producing a borehole with a shot drill.

shot break [ENG] In seismic prospecting, the electrical impulse which records the instant of explosion.

shot capacity [ENG] The maximum weight of molten resin that an accumulator can push out with one forward stroke of the ram during plastics forming operations.

shotcreting [ENG] A process of conveying mortar or concrete through a hose at high velocity onto a surface; the material bonds tenaciously to a properly prepared concrete surface and to a number of other materials.

shot depth [ENG] The distance from the surface to the charge.

shot elevation [ENG] Elevation of the dynamite charge in the shot hole.

shot feed [MECH ENG] A device to introduce chilled-steel shot, at a uniform rate and in the proper quantities, into the circulating fluid flowing downward through the rods or pipe connected to the core barrel and bit of a shot drill.

shot-firing curtain [MIN ENG] A steel frame with chains about 6 inches (15 centimeters) apart suspended from the roof about 9 to 12 feet (2.7 to 3.7 meters) from the face of an advancing tunnel to intercept flying debris when shot-firing at the face.

shothole [ENG] The borehole in which an explosive is placed for blasting.

shothole casing [ENG] A lightweight pipe, usually about 4 inches (10 centimeters) in diameter and 10 feet (3 meters) long,

with threaded connections on both ends, used to prevent the shothole from caving and bridging.

shothole drill [MECH ENG] A rotary or churn drill for drilling shotholes.

shot peening [MET] Shot blasting with small steel balls driven by a blast of air.

shot point [ENG] The point at which an explosion (such as in seismic prospecting) originates, generating vibrations in the ground.

shot rock [ENG] Blasted rock.

shotting [MET] Making shot by pouring molten metal in finely divided streams; the particles solidify during descent and are cooled in a tank of water.

shoulder [DES ENG] The portion of a shaft, a stepped object, or a flanged object that shows an increase of diameter. [ENG] A projection made on a piece of shaped wood, metal, or stone, where its width or thickness is suddenly changed.

shoulder harness [ENG] A harness in a vehicle that fastens over the shoulders to prevent a person's being thrown forward in the seat.

shoulder screw [DES ENG] A screw with an unthreaded cylindrical section, or shoulder, between threads and screwhead; the shoulder is larger in diameter than the threaded section and provides an axis around which close-fitting moving parts operate.

shovel [DES ENG] A hand tool having a flattened scoop at the end of a long handle for moving soil, aggregate, cement, or other similar material. [MECH ENG] A mechanical excavator.

shovel dozer See tractor loader.

shovel loader [MECH ENG] A loading machine mounted on wheels, with a bucket hinged to the chassis which scoops up loose material, elevates it, and discharges it behind the machine.

show-card color See poster paint.

shrinkage [ENG] **1.** Contraction of a molded material, such as metal or resin, upon cooling. **2.** Contraction of a plastics casting upon polymerizing.

shrinkage cavity [MET] A cavity resulting from shrinkage during casting.

shrinkage crack [MET] An irregular interdendritic crack in a casting caused by unequal contraction or inadequate feeding.

shrinkage rule See contraction rule.

shrinkage stoping [MIN ENG] A modification of overhead stoping, involving the use of a part of the ore for the purpose of support and as a working platform. Also known as back stoping.

shrink fit [DES ENG] A tight interference fit between mating parts made by shrinking-on, that is, by heating the outer member to expand the bore for easy assembly and then cooling so that the outer member contracts.

shrink forming [DES ENG] Forming metal wherein the piece undergoes shrinkage during cooling following the application of heat, cold upset, or pressure.

shrink-mixed concrete [MATER] Concrete that is partially mixed before being put in a truck mixer.

shrink ring [DES ENG] A heated ring placed on an assembly of parts, which on subsequent cooling fixes them in position by contraction.

shrink rule *See* contraction rule.

shrink wrapping [ENG] A technique of packaging with plastics in which the strains in plastics film are released by raising the temperature of the film, causing it to shrink-fit over the object being packaged.

shroud [ENG] A protective covering, usually of metal plate or sheet.

shunt [MIN ENG] To shove or turn off to one side, as a car or train from one track to another.

shunt valve [ENG] A valve that gives a fluid under pressure a more readily available escape route than the normal route.

shut-down circuit [ENG] An electronic, electric, or pneumatic system designed to shut off and close down process systems or equipment; can be used for routine or emergency situations.

shut height [MECH ENG] The distance in a press between the bottom of the slide and the top of the bed, indicating the maximum die height that can be accommodated.

shut-in pressure [PETRO ENG] The equilibrated reservoir pressure measured when all the gas or oil outflow has been shut off.

shut-in well [PETRO ENG] An oil or gas well that is closed off; the well is shut so that it does not produce a fluid product of any kind.

shutoff [AERO ENG] In rocket propulsion, the intentional termination of burning by command from the ground or from a self-contained guidance system.

shutoff head [MECH ENG] The pressure developed in a centrifugal or axial flow pump when there is zero flow through the system.

shutter dam [CIV ENG] A dam consisting of a series of pieces that can be lowered or raised by revolving them about their horizontal axis.

shuttle [MECH ENG] A back-and-forth motion of a machine which continues to face in one direction.

shuttle car [MIN ENG] An electrically propelled vehicle on rubber tires or caterpillar treads used to transfer raw materials, such as coal and ore, from loading machines in trackless areas of a mine to the main transportation system.

shuttle conveyor [MECH ENG] Any conveyor in a self-contained structure movable in a defined path parallel to the flow of the material.

shuttling [ENG] A movement involving two or more trips or partial trips by the same motor vehicles between two points.

Siamese blow [ENG] In the plastics industry, the blow molding of two or more parts of a product in a single blow, then cutting them apart.

sickle [DES ENG] A hand tool consisting of a hooked metal blade with a short handle, used for cutting grain or other agricultural products.

side bar [ENG] A bar on which molding pins are carried; operated from outside the mold.

side-channel spillway [CIV ENG] A dam spillway in which the initial and final flow are approximately perpendicular to each other. Also known as lateral flow spillway.

side-construction tile [MATER] A type of structural clay tile designed to receive its principal stress at right angles to the axis of the cells.

side-cut brick [MATER] Brick cut by taut wire along the long side, as opposed to the edge.

side-discharge shovel [MIN ENG] A shovel loader having a 21-cubic-foot (0.59-cubic-meter) bucket, hinged to the chassis to dig, lift, and discharge the material sideways onto a scraper or a belt conveyor.

side draw pin [ENG] Projection used to core a hole in a molded article at an angle other than the line of mold closing; must be withdrawn before the article is ejected.

side drift *See* adit.

side dumper [MIN ENG] An ore, rock, or coal car that can be tilted sidewise and thus emptied.

side-end lines [MIN ENG] Limits of a mining claim looked on as boundary lines, especially in the case of veins originating within but extending outside the claim.

side-facing tool [ENG] A single-point cutting tool having a nose angle of less than 60° and used for finishing the tailstock end of work being machined between centers or the face of a workpiece mounted in a chuck.

sidehill bit [DES ENG] A drill bit which is set off-center so that it cuts a hole of larger diameter than that of the bit.

side hook *See* bench hook.

side-looking radar [ENG] A high-resolution airborne radar having antennas aimed to the right and left of the flight path; used to provide high-resolution strip maps with photographlike detail, to map unfriendly territory while flying along its perimeter, and to detect submarine snorkels against a background of sea clutter.

side milling [MECH ENG] Milling with a side-milling cutter to machine one vertical surface.

side-milling cutter [DES ENG] A milling cutter with teeth on one or both sides as well as around the periphery.

side plates [MIN ENG] In timbering, where both a cap and a sill are used, and the posts act as spreaders, the cap and the sill are termed side plates.

side rake [MECH ENG] The angle between the tool face and a reference plane for a single-point turning tool.

side relief angle [DES ENG] The angle that the portion of the flanks of a cutting tool below the cutting edge makes with a plane normal to the base.

side rod [MECH ENG] **1.** A rod linking the crankpins of two adjoining driving wheels on the same side of a locomotive; distributes power from the main rod to the driving wheels. **2.** One of the rods linking the piston-rod crossheads and the side levers of a side-lever engine.

siderograph [ENG] An instrument that keeps the time of the Greenwich longitude; consists of a clock and a navigation instrument.

side shot [ENG] A reading or measurement from a survey station to locate a point that is off the traverse or that is not intended to be used as a base for the extension of the survey.

side slope [ENG] A test course used to determine lateral stability of a vehicle as well as steering, carburetion, and other functions.

side-tipping loader [MIN ENG] A front-end loading machine which discharges the bucket load by tipping it sideways.

sidetrack [CIV ENG] **1.** To move railroad cars onto a siding. **2.** *See* siding.

sidetracking [ENG] The deliberate act or process of deflecting and drilling a borehole away from a normal, straight course. [PETRO ENG] Drilling procedure to bypass a broken drill or casing permanently lodged in the hole being drilled for an oil well, usually by using a whipstock.

sidewalk [CIV ENG] **1.** A walkway for pedestrians on the side of a street or road. **2.** A foot pavement.

side-wall sampling [PETRO ENG] A technique for taking rock or sand samples from the sides of oil well boreholes.

siding [CIV ENG] A short railroad track connected to the main track at one or more points and used to move railroad cars in order to free traffic on the main line or for temporary storage of cars. Also known as sidetrack. [MATER] Any wall cladding, except masonry or brick.

sienna [MATER] Any of various yellowish-brown earthy substances consisting of hydrated iron oxide occurring in limonite; becomes orange-brown when burnt and is generally darker and more transparent in oils than is ocher; used as pigment for oil paints and stains.

sieve [ENG] **1.** A meshed or perforated device or sheet through which dry loose material is refined, liquid is strained, and soft solids are comminuted. **2.** A meshed sheet with apertures of uniform size used for sizing granular materials.

sieve analysis [ENG] The size distribution of solid particles on a series of standard sieves of decreasing size, expressed as a weight percent. Also known as sieve classification; sieving.

sieve classification *See* sieve analysis.

sieve diameter [ENG] The size of a sieve opening through which a given particle will just pass.

sieve fraction [ENG] That portion of solid particles which pass through a standard sieve of given number and is retained by a finer sieve of a different number.

sieve mesh [DES ENG] The standard opening in sieve or screen, defined by four boundary wires (warp and woof); the laboratory mesh is square and is defined by the shortest distance between two parallel wires as regards aperture (quoted in micrometers or millimeters), and by the number of parallel wires per linear inch as regards mesh; 60-mesh equals 60 wires per inch.

sieving *See* sieve analysis.

sight-feed [ENG] Pertaining to piping in which the flowing liquid can be observed through a transparent tube or wall.

sight glass [ENG] A glass tube or a glass-faced section on a process line or vessel; used for visual reading of liquid levels or of manometer pressures.

sighting tube [ENG] A tube, usually ceramic, inserted into a hot chamber whose temperature is to be measured; an optical pyrometer is sighted into the tube to observe the interior end of the tube to give a temperature reading.

sight rod *See* range rod.

sigma phase [MET] A brittle, nonmagnetic phase of tetragonal structure occurring in many transition-metal alloys; frequently encountered in high chromium stainless steels.

signal correction [ENG] In seismic analysis, a correction to eliminate the time differences between reflection times, resulting from changes in the outgoing signal from shot to shot.

signal effect [ENG] In seismology, variation in arrival times of reflections recorded with identical filter settings, as a result of changes in the outgoing signal.

signal flare [ENG] A pyrotechnic flare of distinct color and character used as a signal.

signal-flow graph [SYS ENG] An abbreviated block diagram in which small circles, called nodes, represent variables of the system, and the nodes are connected by lines, called branches, which represent one-way signal multipliers; an arrow on the line indicates direction of signal flow, and a letter near the arrow indicates the multiplication factor. Also known as flow graph.

signal generator [ENG] An electronic test instrument that delivers a sinusoidal output at an accurately calibrated frequency that may be anywhere from the audio to the microwave range; the frequency and amplitude are adjustable over a wide range, and the output usually may be amplitude- or frequency-modulated. Also known as test oscillator.

signal light [ENG] A signal, illumination, or any pyrotechnic light used as a sign.

signal tower [CIV ENG] A switch tower from which railroad signals are displayed or controlled.

Silastic [MATER] A trade name for any of several heat-resistant silicone rubbers.

silent speed [ENG] The speed at which silent motion pictures are fed through a projector, equal to 16 frames per second (sound-film speed is 24 frames per second).

silent stock support [MECH ENG] A flexible metal guide tube in which the stock tube of an automatic screw machine rotates; it is covered with a casing which deadens sound and prevents transfer of noise and vibration.

silex [MATER] Heat- and shock-resistant glass containing about 98% quartz.

silica aerogel [MATER] A colloidal silica powder whose grains have small pores; used as a low-temperature insulator.

silica brick [MATER] A type of refractory brick formed of at least 90% silica cemented with, for example, slurried lime; used to line furnace roofs.

silica cement [MATER] A mortar used with silica cement; it is a refractory material.

silica flour [MET] A sand additive for casting produced by pulverizing quartz sand.

silica glass [MATER] A translucent or transparent vitreous material consisting almost entirely of silica. Also known as fused silica; vitreous silica.

silicate cement [MATER] The silicate of soda glue, used as an adhesive in cardboard and plywood boxes.

silicate cotton *See* mineral wool.

silicate grinding wheel [DES ENG] A mild-acting grinding wheel where the abrasive grain is bonded with sodium silicate and fillers.

silicate paint [MATER] A paint in which the vehicle is water-soluble sodium silicate; used for painting mortar.

silicatization [MIN ENG] The sealing off of water by the injection of calcium silicate under pressure; sometimes used to reduce the leakage of water through defective lengths of tubing in a shaft.

siliceous dust [MIN ENG] The dust arising from the dry-working of sand, sandstone, trap, granite, and other igneous rocks; the dust is not soluble in the body fluids, and often results in a form of pneumoconiosis, known as silicosis.

silicomanganese [MET] A crude alloy made up of 65–70% manganese, 16–25% silicon, and 1–2.5% carbon; used in the manufacture of low-carbon steel.

silicon bronze [MET] An alloy of copper with 1–5% silicon; it is corrosion-resistant and has good mechanical properties.

silicon copper [MET] An alloy containing 70–80% copper and 20–30% silicon, used as an addition to molten copper or brass.

silicone [MATER] A fluid, resin, or elastomer; can be a grease, a rubber, or a foamable powder; the group name for heat-stable, water repellent, semiorganic polymers of organic radicals attached to the silicones, for example, dimethyl silicone; used in adhesives, cosmetics, and elastomers.

siliconizing [MET] Diffusing silicon into solid metal at an elevated temperature.

silicon steel [MET] A steel that contains 0.5–4.5% silicon, used in electric transformer coils.

silicospiegel [MET] A spiegeleisen pig iron containing 15–20% manganese and 8–15% silicon and up to 4% carbon with the balance iron; used in making steel.

silicothermic process *See* Pidgeon process.

silk paper [MATER] **1.** A paper containing a small amount of silk fibers which give a mottled appearance. **2.** A safety paper sometimes used for postage and revenue stamps.

silky fracture [MET] A metal fracture in which the broken surface is fine in texture and dull in appearance; characteristic of tough, strong metals.

sill [BUILD] The lowest horizontal member of a framed partition or of a window or door frame. [CIV ENG] **1.** A timber laid across the foot of a trench or a heading under the side truss. **2.** The horizontal overflow line of a dam spillway or other weir structure. **3.** A horizontal member on which a lift gate rests when closed. **4.** A low concrete or masonry dam in a small stream to retard bottom erosion. [MIN ENG] **1.** A piece of wood laid across a drift to constitute a frame to support uprights of timber sets and to carry the track of the tramway. **2.** The floor of a gallery or passage in a mine.

sill anchor [BUILD] A fastener projecting from a foundation wall or foundation slab to secure the sill to the foundation.

silo [AERO ENG] A missile shelter that consists of a hardened vertical hole in the ground with facilities either for lifting the missile to a launch position, or for direct launch from the shelter. [CIV ENG] A large vertical, cylindrical structure, made of reinforced concrete, steel, or timber, and used for storing grain, cement, or other materials.

silting [CIV ENG] The filling up or raising of the bed of a body of water by depositing silt.

silting index [ENG] The measurement of the tendency of a solids- or gel-carrying fluid to cause silting in close-tolerance devices, such as valves or other process-line flow constrictions.

silver [MET] A sonorous, ductile, malleable metal that is capable of a high degree of polish and that has high thermal and electric conductivity.

silver alloy [MET] A metal consisting of silver and one or more additional metallic components.

silver brazing [MET] Brazing in which silver-base alloys are used as the filler metal.

silver-brazing alloy *See* silver solder.

silver coating *See* silver plating.

silver-disk pyrheliometer [ENG] An instrument used for the measurement of direct solar radiation; it consists of a silver disk located at the lower end of a diaphragmed tube which serves as the radiation receiver for a calorimeter; radiation falling on the silver disk is periodically intercepted by means of a shutter located in the tube, causing temperature fluctuations of the calorimeter which are proportional to the intensity of the radiation.

silver foil [MET] Silver or a silver-colored metal in very thin sheets.

silver metallurgy [MET] The art and science of extracting silver metal economically from ores, and the reclamation of silver from industrial processes or from scrap metal.

silver plating [MET] Electrolytically depositing a coating of metallic silver on a base metal. Also known as silver coating.

silver solder [MET] A solder composed of silver, copper, and zinc, having a melting point lower than silver but higher than lead-tin solder. Also known as silver-brazing alloy.

silvery iron [MET] A variety of cast iron with a high silicon content, a light-gray color, and a fine grain.

similitude [ENG] A likeness or resemblance; for example, the scale-up of a chemical process from a laboratory or pilot-plant scale to a commercial scale.

simmer [ENG] The detectable leakage of fluid in a safety valve below the popping pressure.

simo chart [IND ENG] A basic motion-time chart used to show the simultaneous nature of motions; commonly a therblig chart for two-hand work with motion symbols plotted vertically with respect to time, showing the therblig abbreviation and a brief description for each activity, and individual times values and body-member detail. Also known as simultaneous motion-cycle chart.

Simon's theory [ENG] A theory of drilling which includes the effects of drilling by percussion and by vibration with a rotary (oil well) bit, cable tool, and pneumatic hammer; the rate of penetration of a chisel-shaped bit into brittle rock may be defined as follows: $R = NAf_v/\pi D$, where R equals the rate of advance of bit, N equals the number of wings of bit, f_v equals the number of impacts per unit time, D equals the diameter of the bit, and A equals the cross-sectional area of the crater at the periphery of the drill hole.

simple balance [ENG] An instrument for measuring weight in which a beam can rotate about a knife-edge or other point of support, the unknown weight is placed in one of two pans suspended from the ends of the beam and the known weights are placed in the other pan, and a small weight is slid along the beam until the beam is horizontal.

simple engine [MECH ENG] An engine (such as a steam engine) in which expansion occurs in a single phase, after which the working fluid is exhausted.

simple machine [MECH ENG] Any of several elementary machines, one or more being incorporated in every mechanical machine; usually, only the lever, wheel and axle, pulley (or block and tackle), inclined plane, and screw are included, although the gear drive and hydraulic press may also be considered simple machines.

simplex concrete pile [CIV ENG] A molded-in-place pile made by using a hollow cylindrical mandrel which is filled with concrete after having been driven to the desired depth and raised a few feet at a time, the concrete flowing out at the bottom and filling the hole in the earth.

simplex pump [MECH ENG] A pump with only one steam cylinder and one water cylinder.

Simpson's rule [PETRO ENG] A mathematical relationship for calculating the oil- or gas-bearing net-pay volume of a reservoir; uses the contour lines from a subsurface geological map of the reservoir, including gas-oil and gas-water contacts.

SIMS *See* secondary ion mass spectrometer.

simulate [ENG] To mimic some or all of the behavior of one system with a different, dissimilar system, particularly with computers, models, or other equipment.

simulator [ENG] A computer or other piece of equipment that simulates a desired system or condition and shows the effects of various applied changes, such as a flight simulator.

simultaneous motion-cycle chart *See* simo chart.

sine bar [DES ENG] A device consisting of a steel straight edge with two cylinders of equal diameter attached near the ends with their centers equidistant from the straightedge; used to measure angles accurately and to lay out work at a desired angle in relationship to a surface.

sine galvanometer [ENG] A type of magnetometer in which a small magnet is suspended in the center of a pair of Helmholtz coils, and the rest position of the magnet is measured when various known currents are sent through the coils.

sine-wave response *See* frequency response.

single acting [MECH ENG] Acting in one direction only, as a single-acting plunger, or a single-acting engine (admitting the working fluid on one side of the piston only).

single-action press [MECH ENG] A press having a single slide.

single axis gyroscope [ENG] A gyroscope suspended in just one gimbal whose bearings form its output axis; an example is a rate gyroscope.

single-base powder [MATER] An explosive or propellant powder in which nitrocellulose is the only active ingredient.

single-bevel groove weld [MET] A groove weld in which one member has a joint edge beveled from one side.

single-block brake [MECH ENG] A friction brake consisting of a short block fitted to the contour of a wheel or drum and pressed up against the surface by means of a lever on a fulcrum; used on railroad cars.

single-button carbon microphone [ENG ACOUS] Microphone having a carbon-filled buttonlike container on only one side of its flexible diaphragm.

single completion [PETRO ENG] An oil or gas well drilled to produce fluids from a single reservoir level or zone, and using a single tubing string.

single compound explosive [MATER] Explosive composed of a single chemical compound.

single-cut file [DES ENG] A file with one set of parallel teeth, extending diagonally across the face of the file.

single-ended spread [ENG] A spread of geophones in which the shot point is located at one end of the arrangement.

single-hand drilling [ENG] A method of rock drilling in which the drill steel, which is held in the hand, is struck with a 4-pound (1.8-kilogram) hammer, the drill being turned between the blows.

single-impulse welding [MET] Spot, projection, or upset welding by means of a single current impulse.

single-J groove weld [MET] A groove weld in which one member has a joint edge in the form of a J from one side.

single-layer bit *See* surface-set bit.

single packing [MIN ENG] Strip packing on a longwall face in which the widest pack is along the roadside.

single-pass weld [MET] A weld made by depositing the filler metal with a single pass.

single-perforated grain [MATER] A cylindrical propellant grain with a single perforation located in its axis; this type of granulation is used in propelling charges for several calibers of guns, and in rockets.

single-phase meter [ENG] A type of power-factor meter that contains a fixed coil that carries the load current, and crossed coils that are connected to the load voltage; there is no spring to restrain the moving system, which takes a position to indicate the angle between the current and voltage.

single-piece milling [MECH ENG] A milling method whereby one part is held and milled in one machine cycle.

single-point tool [ENG] A cutting tool having one face and one continuous cutting edge.

single sampling [IND ENG] A sampling inspection in which the lot is accepted or rejected on the basis of one sample.

single-shot exploder [ENG] A magneto exploder operated by the twist action given by a half turn of the firing key.

single-shot survey [PETRO ENG] An oil well directional log or record with a single-reading device that is either run down into the drill pipe or positioned in a nonmagnetic drill collar.

single-sized aggregate [MATER] Aggregate in which most of the particles lie between narrow size limits.

single-stage compressor [MECH ENG] A machine that effects overall compression of a gas or vapor from suction to discharge conditions without any sequential multiplicity of elements, such as cylinders or rotors.

single-stage pump [MECH ENG] A pump in which the head is developed by a single impeller.

single-stage rocket [AERO ENG] A rocket or rocket missile to which the total thrust is imparted in a single phase, by either a single or multiple thrust unit.

single-stand mill [MET] A rolling mill in which the product is in contact with only two rolls at a time.

single thread [DES ENG] A screw thread having a single helix in which the lead and pitch are equal.

single-U groove weld [MET] A groove weld in which the joint edge of both members is prepared in the form of a J from one side, giving a final U form to the completed weld.

single-V groove weld [MET] A groove weld in which the joint edge of each member is beveled from the same side.

single welded joint [MET] A joint welded from one side only.

sink [MIN ENG] **1.** To excavate strata downward in a vertical line for the purpose of winning and working minerals. **2.** To drill or put down a shaft or borehole.

sinker [MIN ENG] **1.** A person who sinks mine shafts and puts in framing. **2.** A special movable pump used in shaft sinking. **3.** See sinker drill.

sinker bar [MIN ENG] A short, heavy rod placed above the drill jars to increase the effect of the upward sliding jars in well-drilling with cable tools.

sinker drill [MIN ENG] A jackhammer type of rock drill used in shaft sinkings. Also known as sinker.

sink-float separation process [ENG] A simple gravity process used in ore dressing that separates particles of different sizes or composition on the basis of differences in specific gravity.

sinkhead See feedhead.

sinking and walling scaffold [MIN ENG] A platform designed for use in shaft sinking to enable sinking and walling to be performed simultaneously. Also known as Galloway sinking and walling stage.

sinking bucket See hoppit.

sinking fund [IND ENG] A fund established by periodically depositing funds at compound interest in order to accumulate a given sum at a given future time for some specific purpose.

sinking pump [MIN ENG] A long, narrow, electrically driven centrifugal-type pump designed for keeping a shaft dry during sinking operations.

sinking tubing [MET] Drawing tubing through a die or passing it through rolls without the use of a tool in the bore to control the inside diameter.

sink mark [ENG] A shallow depression or dimple on the surface of an injection-molded plastic part due to collapsing of the surface following local internal shrinkage after the gate seals.

sinter [MET] **1.** The product of a sintering operation. **2.** A shaped body composed of metal powders and produced by sintering with or without previous compacting.

sintered copper [MET] Copper prepared by heating a compressed powder of the metal to form a solid mass.

sintered steel [MET] Steel prepared by heating compressed iron powder and graphite to form a solid.

sintering [MET] Forming a coherent bonded mass by heating metal powders without melting; used mostly in powder metallurgy.

sintering furnace [MET] A furnace in which presintering and sintering operations are carried out.

sinter setting See mechanical setting.

siphon [ENG] A tube, pipe, or hose through which a liquid can be moved from a higher to a lower level by atmospheric pressure forcing it up the shorter leg while the weight of the liquid in the longer leg causes continuous downward flow.

siphon barograph [ENG] A recording siphon barometer.

siphon barometer [ENG] A J-shaped mercury barometer in which the stem of the J is capped and the cusp is open to the atmosphere.

siphon recorder [ENG] A recorder in which a small siphon discharges ink to make the record; used in submarine telegraphy.

siphon spillway [CIV ENG] An enclosed spillway passing over the crest of a dam in which flow is maintained by atmospheric pressure.

siporex [MATER] A building material composed of sand, lime or cement, and aluminum powder which are mixed and cast into molds to be made into roof slabs, door lintels, and wall blocks which give excellent heat and sound insulation.

siren [ENG ACOUS] An apparatus for generating sound by the mechanical interruption of the flow of fluid (usually air) by a perforated disk or cylinder.

sisal-hemp wax [MATER] A hard wax derived from sisal waste; melts at 63°C, decomposes at 95°C. Also known as sisal wax.

sister hook [DES ENG] **1.** Either of a pair of hooks which can be fitted together to form a closed ring. **2.** A pair of such hooks.

site [ENG] Position of anything; for example, the position of a gun emplacement.

Sitka cypress *See* Alaska cedar.

Six's thermometer [ENG] A combination maximum thermometer and minimum thermometer; the tube is shaped in the form of a U with a bulb at either end; one bulb is filled with creosote which expands or contracts with temperature variation, forcing before it a short column of mercury having iron indexes at either end; the indexes remain at the extreme positions reached by the mercury column, thus indicating the maximum and minimum temperatures; the indexes can be reset with the aid of a magnet.

six-tenths factor [IND ENG] An empirical relationship between the cost and the size of a manufacturing facility; as size increases, cost increases by an exponent of six-tenths, that is $\text{cost}_1/\text{cost}_2 = (\text{size}_1/\text{size}_2)^{0.6}$.

size [MATER] Materials used to surface-treat textiles, papers, and leathers; examples are starch, gelatins, casein, water-soluble gums, and waxes.

size block *See* gage block.

size classification *See* sizing.

size dimension [DES ENG] In dimensioning, a specified value of a diameter, width, length, or other geometrical characteristic directly related to the size of an object.

size effect [MET] The effect of the size of a piece of metal on its properties and manufacturing variables; in general, mechanical properties are lower for a larger size.

size of weld [MET] **1.** The joint penetration in a groove weld. **2.** The lengths of the nominal legs of a fillet weld.

size reduction [MECH ENG] The breaking of large pieces of coal, ore, or stone by a primary breaker, or of small pieces by grinding equipment.

sizing [ENG] **1.** Separating an aggregate of mixed particles into groups according to size, using a series of screens. Also known as size classification. **2.** *See* sizing treatment. [MECH ENG] A finishing operation to correct surfaces and shapes to meet specified dimensions and tolerances. [MET] Final pressing of a metal powder compact after sintering.

sizing screen [DES ENG] A mesh sheet with standard-size apertures used to separate granular material into classes according to size; the Tyler standard screen is an example.

sizing treatment [ENG] Also known as sizing; surface sizing. **1.** Application of material to a surface to fill pores and thus reduce the absorption of subsequently applied adhesive or coating; used for textiles, paper, and other porous materials. **2.** Surface-treatment applied to glass fibers used in reinforced plastics.

skeleton framing [BUILD] Framing in which steel framework supports all the gravity loading of the structure; this system is used for skyscrapers.

skelp [MET] A strip or sheet of steel which will be rolled and welded to form a tube.

skew [MECH ENG] Gearing whose shafts are neither interesecting nor parallel.

skew back [CIV ENG] The beveled or inclined support at each end of a segmental arch.

skew bridge [CIV ENG] A bridge which spans a gap obliquely and is therefore longer than the width of the gap.

skewed bridge [CIV ENG] A bridge for which the deck in plan is a parallelogram.

skew level gear [DES ENG] A level gear whose axes are not in the same place.

skid [AERO ENG] The metal bar or runner used as part of the landing gear of helicopters and planes. [ENG] **1.** A device attached to a chain and placed under a wheel to prevent its turning when descending a steep hill. **2.** A timber, bar, rail, or log placed under a heavy object when it is being moved over bare ground. **3.** A wood or metal platform support on wheels, legs, or runners used for handling and moving material. Also known as skid platform. [MECH ENG] A brake for a power machine. [MIN ENG] An arrangement upon which certain coal-cutting machines travel along the working faces.

skid-mounted [ENG] Equipment or processing systems mounted on a portable platform.

skid platform *See* skid.

skim coat [BUILD] A finish coat of plaster composed of lime putty and fine white sand.

skim gate [MET] A gate used to prevent slag and other undesirable materials from passing into the casting.

skin [AERO ENG] The covering of a body, such as the covering of a fuselage, a wing, a hull, or an entire aircraft. [BUILD] The exterior wall of a building. [ENG] In flexible bag molding, a protective covering for the mold; it may consist of a thin piece of plywood or a thin hardwood. [MET] A thin outside layer of metal differing in composition, structure, or other characteristics from the main mass of metal but not formed by bonding or electroplating.

skin diving [ENG] Diving without breathing apparatus, using fins and faceplate only.

skin effect [PETRO ENG] The restriction to fluid flow through a reservoir adjacent to the borehole; calculated as a factor of reservoir pressure, product rate, formation volume and thickness, porosity, and other related parameters.

skin lamination [MET] Surface rupture in flat-rolled metals due to exposure of a subsurface lamination.

skintle [CIV ENG] To set bricks in an irregular fashion so that they are out of alignment with the face by ¼ inch (6 millimeters) or more.

skip *See* skip hoist.

skip distance [ENG] In angle-beam ultrasonic testing, the distance between the point of entry on the workpiece and the point of first reflection.

skip hoist [MECH ENG] A basket, bucket, or open car mounted vertically or on an incline on wheels, rails, or shafts and hoisted by a cable; used to raise materials. Also known as skip.

skip logging [ENG] A phenomenon during acoustical (sonic) logging in which the acoustical energy is attenuated by low-elasticity formations and lacks the energy to trip the second sonic receiver (skips a cycle). Also known as cycle skip.

skip shaft [MIN ENG] A mine shaft prepared for hauling a skip.

skip vehicle [AERO ENG] A reentry body which climbs after striking the sensible atmosphere in order to cool the body and to increase its range.

skirting block [BUILD] Also known as base block; plinth block. **1.** A corner block where a base strip and vertical enframement meet. **2.** A concealed block to which a baseboard is anchored.

skirt roof [BUILD] A false band of roofing projecting from between the stories of a building.

skiver [MATER] Thin, soft leather made from the grain side of a split sheepskin.

skiving [MECH ENG] **1.** Removal of material in thin layers or chips with a high degree of shear or slippage of the cutting tool. **2.** A machining operation in which the cut is made with a form tool with its face at an angle allowing the cutting edge to progress from one end of the work to the other as the tool feeds tangentially past ten rotating workpieces.

skull [MET] A layer of solidified metal or dross left in the pouring vessel after the molten metal has been poured.

skull cracker [ENG] A heavy iron or steel ball that can be swung freely or dropped by a derrick to raze buildings or to compress bulky scrap. Also known as wrecking ball.

skull crucible [MET] A consumable-electrode vacuum arc melting and casting furnace; used in the production of turbine buckets for aircraft jet engines using a nickel-base high-temperature alloy.

Skyhawk [AERO ENG] A United States single-engine, turbojet attack aircraft designed to operate from aircraft carriers, and capable of delivering nuclear or nonnuclear weapons, providing troop support, or conducting reconnaissance missions; it can act as a tanker, and can itself be air refueled; it possesses a limited all-weather attack capability, and can operate from short, unprepared fields.

skyhook [MIN ENG] To drive bolts into the overhead rock of a mine in order to reinforce the ceiling.

skyhook balloon [AERO ENG] A large plastic constant-level balloon for duration flying at very high altitudes, used for determining wind fields and measuring upper-atmospheric parameters.

Skylab [AERO ENG] The first United States space station, attended by three separate flight crews of three astronauts each in 1973 and 1974.

skylight [ENG] An opening in a roof or ship deck that is covered with glass or plastic and designed to admit daylight.

Skyraider [AERO ENG] A United States single reciprocating-engine, general-purpose attack aircraft designed to operate from aircraft carriers; it is capable of relatively long-range, low-level nuclear and nonnuclear weapons delivery, minelaying, reconnaissance, torpedo delivery, and troop support. Designated A-1.

Skyray [AERO ENG] A United States single-engine, single-pilot, supersonic, limited all-weather jet fighter designed for operating from aircraft carriers for interception and destruction of enemy aircraft; armament includes the Sidewinder. Designated F-6.

skyscraper [BUILD] A very tall, multistory building.

Skywarrior [AERO ENG] A United States twin-engine, turbojet, tactical, all-weather attack aircraft designed to operate from aircraft carriers, and capable of delivering nuclear or nonnuclear weapons, conducting reconnaissance, or minelaying missions; its range can be extended by in-flight refueling, and it has a crew of four. Designated A-3.

slab [CIV ENG] That part of a reinforced concrete floor, roof, or platform which spans beams, columns, walls, or piers. [ENG] The outside piece cut from a log when sawing it into boards. [MATER] A thin piece of concrete or stone. [MET] A piece of metal, intermediate between ingot and plate, with the width at least twice the thickness. [MIN ENG] A slice taken off the rib of an entry or room in a mine.

slabbing cut *See* slipping cut.

slabbing cutter [MECH ENG] A face-milling cutter used to make wide, rough cuts.

slabbing machine [MIN ENG] A coal-cutting machine designed to make cuts in the side of a room or entry pillar preparatory to slabbing.

slabbing method [MIN ENG] A method of mining pillars in which successive slabs are cut from one side or rib of the pillar after a room is finished, until as much of the pillar is removed as can safely be recovered.

slabbing mill [MET] A steel rolling mill for making slabs.

slab cutter *See* plain milling cutter.

slab entry [MIN ENG] An entry which is widened or slabbed to provide a working place for a second miner.

slab oil [MATER] White petroleum-based oil used by candymakers and bakers to oil the slab or surface on which the candy or pastry is worked.

slack [ENG] Looseness or play in a mechanism, as the play in the trigger of a small-arms weapon.

slack barrel [PETRO ENG] A petroleum-industry container used for shipment of petroleum paraffin; generally contains 235 to 245 pounds (107 to 111 kilograms) net, and is of lighter construction than the ordinary oil barrel but of the same general shape.

slackline cableway [MECH ENG] A machine, widely used in sand-and-gravel plants, employing an open-ended dragline bucket suspended from a carrier that runs upon a track cable, which can dig, elevate, and convey materials in one continuous operation.

slack quenching [MET] Formation of transformation products other than martensite as a result of quenching at a rate slower than the critical cooling rate.

slack time [ENG] For an activity in a PERT or critical-path-method network, the difference between the latest possible completion time of each activity which will not delay the completion of the overall project, and the earliest possible completion time, based on all predecessor activities.

slack wax [MATER] A soft, oily, crude wax obtained from the pressing of petroleum paraffin distillate or wax distillate.

slag [MET] A nonmetallic product resulting from the interaction of flux and impurities in the smelting and refining of metals.

slag cement [MATER] Cement produced by grinding blast-furnace slag and mixing it with lime, portland cement, or dehydrated gypsum.

slagging [MET] Freeing from or converting into slag.

slag inclusion [MET] Slag entrapped in solidified metal.

slag sand [MATER] Slag that has been finely crushed for use in mortar and concrete.

slag wool *See* mineral wool.

slamming stile [BUILD] The vertical strip that a closed door abuts; it receives the bolt when the lock engages.

slant [MIN ENG] **1.** Any short, inclined crosscut connecting the entry with its air course to facilitate the hauling of coal. Also known as shoofly. **2.** A heading driven diagonally between the dip and the strike of a coal seam. Also known as run.

slant depth [DES ENG] The distance between the crest and root of a screw thread measured along the angle forming the flank of the thread.

slant drilling [ENG] The drilling of a borehole or well at an angle to the vertical.

slat [AERO ENG] A movable auxiliary airfoil running along the leading edge of a wing, remaining against the leading edge in normal flight conditions, but lifting away from the wing to form a slot at certain angles of attack.

slat conveyor [MECH ENG] A conveyor consisting of horizontal slats on an endless chain.

sled [ENG] An item equipped with runners and a suitable body designed to transport loads over ice and snow.

sledgehammer [DES ENG] A large heavy hammer that is usually wielded with two hands; used for driving stakes or breaking stone.

sleeper [CIV ENG] A timber, steel, or precast concrete beam placed under rails to hold them at the correct gage.

sleeve [ENG] A cylindrical part designed to fit over another part.

sleeve bearing [MECH ENG] A machine bearing in which the shaft turns and is lubricated by a sleeve.

sleeve brick [MATER] Tube-shaped firebrick for lining slag vents.

sleeve burner [ENG] A type of oil burner for domestic heating.

sleeve coupling [DES ENG] A hollow cylinder which fits over the ends of two shafts or pipes, thereby joining them.

sleeve valve [MECH ENG] An admission and exhaust valve on an internal-combustion engine consisting of one or two hollow sleeves that fit around the inside of the cylinder and move with the piston so that their openings align with the inlet and exhaust ports in the cylinder at proper stages in the cycle.

slenderness ratio [AERO ENG] A configuration factor expressing the ratio of a missile's length to its diameter. [CIV ENG] The ratio of the length of a column L to the radius of gyration r about the principal axes of the section.

slewing [ENG] Moving a radar antenna or a sonar transducer rapidly in a horizontal or vertical direction, or both.

slewing mechanism [ENG] Device which permits rapid traverse or change in elevation of a weapon or instrument.

slice [MIN ENG] **1.** A thin broad piece cut off, as a portion of ore cut from a pillar or face. **2.** To remove ore by successive slices.

slice bar [ENG] A broad, flat steel blade used for chipping and scraping.

slice drift [MIN ENG] In sublevel caving, the crosscuts driven between every other slice from 18 to 36 feet (5.5 to 11.0 meters) apart.

slicing method [MIN ENG] Removal of a horizontal layer from a massive ore body.

slickens [MIN ENG] The light earth removed by sluicing in hydraulic mining.

slide [ENG] **1.** A sloping chute with a flat bed. **2.** A sliding mechanism. [MECH ENG] The main reciprocating member of a mechanical press, guided in a press frame, to which the punch or upper die is fastened. [MIN ENG] **1.** An upright rail fixed in a shaft with corresponding grooves for steadying the cages. **2.** A trough used to guide and to support rods in a tripod when drilling an angle hole. Also known as rod slide.

slide conveyor [ENG] A slanted, gravity slide for the forward downward movement of flowable solids, slurries, liquids, or small objects.

slide gate [CIV ENG] A crest gate which has high frictional resistance to opening because it slides on its bearings in opening and closing.

slide projector *See* optical lantern.

slide rail *See* guard rail.

slider coupling [MECH ENG] A device for connecting shafts that are laterally misaligned. Also known as double-slider coupling; Oldham coupling.

slide rest [MECH ENG] An adjustable slide for holding a cutting tool, as on an engine lathe.

slide-rule dial [ENG] A dial in which a pointer moves in a straight line over long straight scales resembling the scales of a slide rule.

slide valve [MECH ENG] A sliding mechanism to cover and uncover ports for the admission of fluid, as in some steam engines.

sliding-block linkage [MECH ENG] A mechanism in which a crank and sliding block serve to convert rotary motion into translation, or vice versa.

sliding-chain conveyor [MECH ENG] A conveying machine to handle cases, cans, pipes, or similar products on the plain or modified links of a set of parallel chains.

sliding fit [DES ENG] A fit between two parts that slide together.

sliding gear [DES ENG] A change gear in which speed changes are made by sliding gears along their axes, so as to place them in or out of mesh.

sliding-gear transmission [MECH ENG] A transmission system utilizing a pair of sliding gears.

sliding pair [MECH ENG] Two adjacent links, one of which is constrained to move in a particular path with respect to the other; the lower, or closed, pair is completely constrained by the design of the links of the pair.

sliding way [CIV ENG] One of the timbers which form the upper part of the cradle supporting a ship during its construction, and which slide over the ground ways with the ship when it is launched.

slime [ENG] Liquid slurry of very fine solids with slime- or mudlike appearance. Also known as mud; pulp; sludge.

slim hole [ENG] A drill hole of the smallest practicable size, drilled with less-than normal-diameter tools, used primarily as a seismic shothole and for structure tests and sometimes for stratigraphic tests. [PETRO ENG] A diamond-drill borehole having a diameter of 5 inches (12.7 centimeters) or less.

sling [ENG] A length of rope, wire rope, or chain used for attaching a load to a crane hook.

sling psychrometer [ENG] A psychrometer in which the wet- and dry-bulb thermometers are mounted upon a frame connected to a handle at one end by means of a bearing or a length of chain; the psychrometer may be whirled in the air for the simultaneous measurement of wet- and dry-bulb temperatures.

sling thermometer [ENG] A thermometer mounted upon a frame connected to a handle at one end by means of a bearing or length of chain, so that the thermometer may be whirled by hand.

slip [CIV ENG] A narrow body of water between two piers. [MATER] A suspension of fine clay in water with a creamy consistency, used in the casting process and in decorating ceramic ware. Also known as slurry.

slip additive [MATER] A plastics modifier that acts as an internal lubricant by exuding to the surface of the plastic during and immediately after processing to reduce friction and improve slip.

slip casting [ENG] A process in the manufacture of shaped refractories, cermets, and other materials in which the slip is poured into porous plaster molds.

slip clay [MATER] Clay containing a high percentage of fluxing impurities; easily fusible and used in clayware to produce a natural glaze.

slip form [CIV ENG] A narrow section of formwork that can be easily removed as concrete placing progresses.

slip forming [ENG] A plastics-sheet forming technique in which some of the sheet is allowed to slip through the mechanically operated clamping rings during stretch-forming operations.

slip friction clutch [MECH ENG] A friction clutch designed to slip when too much power is applied to it.

slip joint [CIV ENG] **1.** Contraction joint between two adjoining wall sections, or at the horizontal bearing of beams, slabs, or precast units, consisting of a vertical tongue fitted into a groove which allows independent movement of the two sections. **2.** A telescoping joint between two parts. [ENG] **1.** A method of laying-up plastic veneers in flexible-bag molding, wherein edges are beveled and allowed to overlap part or all of the scarfed area. **2.** A mechanical union that allows limited endwise movement of two solid items for example, pipe, rod, or duct with relation to each other.

slippage [ENG] The leakage of fluid between the plunger and the bore of a pump piston. Also known as slippage loss. [PETRO ENG] The movement of gas past or through a liquid-phase reservoir front; this movement occurs instead of driving the liquid forward; it can exist in gas-drive reservoirs or in gas-lift oil-well bores.

slippage loss [ENG] **1.** Unintentional movement between the faces of two solid objects. **2.** *See* slippage.

slipper brake [MECH ENG] **1.** A plate placed against a moving part to slow or stop it. **2.** A plate applied to the wheel of a vehicle or to the track roadway to slow or stop the vehicle.

slipping [MIN ENG] Enlarging an excavation by breaking one or more walls.

slipping cut [MIN ENG] A drill-hole pattern used in a wide tunnel face, in which each successive vertical line of shots breaks to the face made by the previous round, so that the relieving cut moves across the end being blasted. Also known as slabbing cut; swing cut.

slip plane [ENG] A plane visible by reflected light in a transparent material; caused

by poor welding and shrinkage during cooling.

slip ratio [MECH ENG] For a screw propeller, relates the actual advance to the theoretic advance determined by pitch and spin.

slips [ENG] A wedge-shaped steel collar fabricated in two sections, designed to hold a string of casing between various portions of the drilling operation.

slip tongue [ENG] A pole on a horse-drawn wagon that is fastened by slipping it between two plates connected to the fore-carriage.

slipway [CIV ENG] The space in a shipyard where a foundation for launching ways and keel blocks exists and which is occupied by a ship while under construction.

slip-weld hanger [PETRO ENG] A type of hanger used to suspend the lower strings of an oil-well casing pipe.

slit [DES ENG] A long, narrow opening through which radiation or streams of particles enter or leave certain instruments.

slitter [MECH ENG] A synchronized feeder-knife variation of a rotary cutter; used for precision cutting of sheet material, such as metal, rubber, plastics, or paper, into strips.

slitting [MECH ENG] The passing of sheet or strip material (metal, plastic, paper, or cloth) through rotary knives.

sliver [MATER] A piece of propellant grain of triangular cross section which remains unburned when the web of multiperforated grains has been burned through. [MET] A thin, elongated fragment of metal that has been rolled onto the surface of the parent metal and is attached by only one end.

slope control [MET] Electronic production of a change in the welding current within set limits and a selected interval of time.

slope conveyor [MECH ENG] A troughed belt conveyor used for transporting material on steep grades.

slope course [ENG] A proving ground facility consisting of a large mound of earth with various sloping sides on which are roads having different grades; this slope course is used to measure the slope performance of military and other vehicles, including maximum speed on various grades, the most suitable gear for best performance, traction, and the holding ability of brakes.

slope deviation [AERO ENG] The difference between planned and actual slopes of aircraft travel, expressed in either angular or linear measurement.

slope engineer [MIN ENG] In anthracite and bituminous coal mining, one who operates a hoisting engine to haul loaded and empty mine cars along a haulage road in a mine.

slope mine [MIN ENG] A mine opened by a slope or an incline.

slope stake [MIN ENG] Stake set at the point where the finished side slope of an excavation or embankment meets the original grade.

slosh test [ENG] A test to determine the ability of the control system of a liquid-propelled missile to withstand or overcome the dynamic movement of the liquid within its fuel tanks.

slot [AERO ENG] **1.** An air gap between a wing and the length of a slat or certain other auxiliary airfoils, the gap providing space for airflow or room for the auxiliary airfoil to be depressed in such a manner as to make for smooth air passage on the surface. **2.** Any of certain narrow apertures made through a wing to improve aerodynamic characteristics. [DES ENG] A narrow, vertical opening. [MIN ENG] To hole; to undercut or channel.

slot distributor [ENG] A long, narrow discharge opening (slot) in a pipe or conduit; used for the extrusion of sheet material, such as plastics.

slot dozing [ENG] A method of moving large quantities of material with a bulldozer using the same path for each trip so that the spillage from the sides of the blade builds up along each side; afterward all material pushed into the slot is retained in front of the blade.

slot extrusion [ENG] A method of extruding plastics-film sheet in which the molten thermoplastic compound is forced through a straight slot.

slotted-head screw [DES ENG] A screw fastener with a single groove across the diameter of the head.

slotted nut [DES ENG] A regular hexagon nut with slots cut across the flats of the hexagon so that a cotter pin or safety wire can hold it in place.

slotter [MECH ENG] A machine tool used for making a mortise or shaping the sides of an aperture.

slotting [MECH ENG] Cutting a mortise or a similar narrow aperture in a material using a machine with a vertically reciprocating tool.

slotting machine [MECH ENG] A vertically reciprocating planing machine, used for making mortises and for shaping the sides of openings.

slot washer [DES ENG] **1.** A lock washer with an indentation on its edge through which a nail or screw can be driven to hold it in place. **2.** A washer with a slot extending from its edge to the center hole to allow the washer to be removed without first removing the bolt.

slot weld [MET] Similar to plug weld, but the hole is elongated and may extend to the edge of a member without closing.

slough [ENG] The fragments of rocky material from the wall of a borehole.

slow-curing liquid asphaltic material [MATER] An asphalt cement blended with slow-volatilizing gas oil. Also known as SC asphalt.

slow igniter cord [ENG] An igniter cord made with a central copper wire around which is extruded a plastic incendiary material with an iron wire embedded to give greater strength; the whole is enclosed in a thin extruded plastic coating.

slow match [ENG] A match or fuse that burns at a known slow rate; used for igniting explosive charges.

slow sand filter [CIV ENG] A bed of fine sand 20–48 inches (151–122 centimeters) deep through which water, being made suitable for human consumption and other purposes, is passed at a fairly low rate, 2,500,000 to 10,000,000 gallons per acre (23,000 to 94,000 cubic meters per hectare); an underdrain system of graded gravel and perforated pipes carries the water from the filters to the point of discharge.

sludge [CIV ENG] *See* sewage sludge. [ENG] **1.** Mud from a drill hole in boring. **2.** Sediment in a steam boiler. **3.** A precipitate from oils, such as the products from crankcase oils in engines. **4.** *See* slime.

sludge acid [MATER] The residue from the sulfuric-acid treatment of petroleum lubricants. Also known as spent acid; waste acid.

sludge assay [MIN ENG] The chemical assaying of drill cuttings for a specific metal or group of metals.

sludge barrel *See* calyx.

sludge box [MIN ENG] A wooden box in which sludge settles from the mud flush.

sludge bucket *See* calyx.

sludge pit *See* slushpit.

sludge pond *See* slushpit.

sludge pump [MECH ENG] *See* sand pump. [MIN ENG] A short iron pipe or tube fitted with a valve at the lower end, with which the sludge is extracted from a borehole.

sluff [ENG] The mud cake detached from the wall of a borehole. [MIN ENG] The falling of decomposed, soft rocks from the roof or walls of mine openings.

slug [MET] **1.** A small, roughly shaped piece of metal for subsequent processing, as by forging or extruding. **2.** The piece of material produced by piercing a hole in a sheet. [MIN ENG] To inject a borehole with cement, slurry, or various liquids containing shredded materials in an attempt to restore lost circulation by sealing off the openings in the borehole-wall rocks.

slug bit *See* insert bit.

slugging [MET] Adding a separate piece of material to a weld joint, resulting in a joint which does not meet specifications.

sluice [CIV ENG] **1.** A passage fitted with a vertical sliding gate or valve to regulate the flow of water in a channel or lock. **2.** A body of water retained by a floodgate. **3.** A channel serving to drain surplus water.

sluice box [MIN ENG] A long, inclined trough or launder with riffles in the bottom that provide a lodging place for heavy minerals in ore concentration.

sluice gate [CIV ENG] The vertical slide gate of a sluice.

sluice tender [MIN ENG] In metal mining, a laborer who tends sluice boxes.

sluicing [MIN ENG] **1.** Washing auriferous earth through sluices provided with riffles and other gold-saving appliances. **2.** Separation of minerals in a flowing stream of water. **3.** Moving earth, sand, gravel, or other rock or mineral materials by flowing water.

sluicing pond *See* scouring basin.

slump test [ENG] Determining the consistency of concrete by filling a conical mold with a sample of concrete, then inverting it over a flat plate and removing the mold; the amount by which the concrete drops below the mold height is measured and this represents the slump.

slurry [MATER] **1.** A semiliquid refractory material, such as clay, used to repair furnace refractories. **2.** A free-flowing, pumpable suspension of fine solid material in liquid. **3.** An emulsion of a sulfonated soluble oil in water used to cool and lubricate metal during cutting operations. **4.** A plastic mixture of portland cement and water pumped into an oil well; after hardening, it provides support for the casing and a seal for the well bore. **5.** *See* slip.

slurry blasting agent [MATER] A dense, insensitive, high-velocity explosive of great power and very high water resistance used principally for blasting hard rock or where blastholes are wet.

slurrying [ENG] The formation of a mud or a suspension from a liquid and nonsoluble solid particles.

slurry mining [MIN ENG] The hydraulic breakdown of a subsurface ore matrix with drill-hole equipment, and the eduction of the resulting slurry to the surface for processing.

slurry preforming [ENG] The preparation of reinforced plastics preforms by wet-processing techniques; similar to pulp molding.

slurry truck [ENG] A mobile unit that transports dry blasting ingredients, and mixes them in required proportions for introduction as explosive slurry into blastholes.

slush casting [MET] Producing a hollow casting without a core in the mold by rotating a liquid alloy in a hollow metal mold until a solid layer chills onto the mold, and then pouring off the remaining liquid.

slusher [ENG] A method for the application of vitreous enamel slip to ware by dashing it on the ware to cover all its parts, excess then being removed by shaking the ware.

slushing compound [MATER] A temporary, corrosion-protective coating for metals; made of nondrying oil, grease, or other similar material.

slushing grease [MATER] A special grade of grease used as a metal coating to prevent corrosion.

slushing oil [MATER] A nondrying oil which is strongly adhesive to metal and is applied to metal surfaces to minimize corrosion.

slush molding [ENG] A thermoplastic casting in which a liquid resin is poured into a hot, hollow mold where a viscous skin forms; excess slush is drained off, the mold is cooled, and the molded product is stripped out.

slushpit [ENG] An excavation or diked area to hold water, mud, sludge, and other discharged matter from an oil well. Also known as mud pit; sludge pit; sludge pond.

slush pump [PETRO ENG] A pump used to circulate the drilling fluid during rotary drilling.

small-diameter blasthole [ENG] A blast hole 1½ to 3 inches (3.8 to 7.6 centimeters) in diameter, in low-face quarries.

small-lot storage [IND ENG] Generally, a quantity of less than one pallet stack, stacked to maximum storage height; thus, the term refers to a lot consisting of from one container to two or more pallet loads, but is not of sufficient quantity to form a complete pallet column.

smalt [MATER] A blue glass made by fusing silica and potash with cobalt oxides; used as a pigment for glass, ceramics, paints, and dyes.

smelter [MET] A furnace used for smelting.

smelting [MET] The heating of ore mixtures accompanied by a chemical change resulting in liquid metal.

Smithell's burner [ENG] Two concentric tubes that can be added to a bunsen burner to separate the inner and outer flame cones.

smith forging [MET] Manual forging of small, hot metal parts with flat or simple-shaped dies, as with a hammer and anvil.

Smith-McIntyre sampler [MECH ENG] A device for taking samples of sediment from the ocean bottom; the digging and hoisting mechanisms are independent: the digging bucket is forced into the sediment before the hoisting action occurs.

smoke [ENG] Dispersions of finely divided (0.01–5.0 micrometers) solids or liquids in a gaseous medium.

smokebox [MECH ENG] A chamber external to a boiler for trapping the unburned products of combustion.

smoke chamber [ENG] That area in a fireplace directly above the smoke shelf.

smoke detector [ENG] A photoelectric system for an alarm when smoke in a chimney or other location exceeds a predetermined density.

smokeless powder [MATER] Nitrocellulose containing 13.1 percent nitrogen with small amounts of stabilizers (amines) and plasticizers usually present, as well as various modifying agents (nitrotoluene and nitroglycerin salts); used in ammunition.

smoke point [ENG] The maximum flame height in millimeters at which kerosine will burn without smoking, tested under standard conditions; used as a measure of the burning cleanliness of jet fuel and kerosine.

smoke shelf [ENG] A horizontal surface directly behind the throat of a fireplace to prevent downdrafts.

smokestack [ENG] A chimney for the discharge of flue gases from a furnace operation such as in a steam boiler, powerhouse, heating plant, ship, locomotive, or foundry.

smoke test [ENG] A test used on kerosine to determine the highest point to which the flame can be turned before smoking occurs.

smoke washer [ENG] A device for removing particles from smoke by forcing it through a spray of water.

S Monel [MET] An alloy similar to H Monel but containing 4% silicon.

smooth blasting [ENG] Blasting to ensure even faces without cracks in the rock.

smooth drilling [ENG] Drilling in a rock formation in which a fast rotation of the drill stem, a fast rate of penetration, and a high recovery of core can be achieved with vibration-free rotation of the drill stem.

smoothing [ENG] Making a level, or continuously even, surface.

smoothing mill [MECH ENG] A revolving stone wheel used to cut and bevel glass or stone.

smoothing plane [DES ENG] A finely set hand tool, usually 5.5–10 inches (14.0–25.4 centimeters) long, for finishing small areas on wood.

smother kiln [ENG] A kiln into which smoke can be introduced for blackening pottery.

smudge oil [MATER] A dark petroleum distillate or gas oil that is burned in citrus fruit-growing areas to prevent frost damage.

smudging [ENG] A frost-preventive measure used in orchards; properly, it means the production of heavy smoke, supposed to prevent radiational cooling, but it is generally applied to both heating and smoke production.

smut [MET] A reaction product left on the surface of a metal after pickling.

snagging [MECH ENG] Removing surplus metal or large surface defects by using a grinding wheel.

snake [MET] **1.** A twisted and bent hod rod formed before subsequent rolling operations. **2.** A flexible mandrel used to prevent collapse of a shaped piece during bending operations.

snake hole [ENG] **1.** A blasting hole bored directly under a boulder. **2.** A drill hole used in quarrying or bench blasting.

snaking [ENG] Towing a load with a long cable.

snap-back forming [ENG] A plastic-sheet-forming technique in which an extended, heated, plastic sheet is allowed to contract over a form shaped to the desired final contour.

snapback method *See* repetitive time method.

snap fastener [DES ENG] A fastener consisting of a ball on one edge of an article that fits in a socket on an opposed edge, and used to hold edges together, such as those of a garment.

snap flask [MET] A foundry flask having its sides latched on one corner to allow removal of the flask from around the sand mold.

snap gage [DES ENG] A device with two flat, parallel surfaces spaced to control one limit of tolerance of an outside diameter or a length.

snap hook *See* spring hook.

snapper [ENG] A device for collecting samples from the ocean bottom, and which closes to prevent the sample from dropping out as it is raised to the surface.

snap ring [DES ENG] A form of spring used as a fastener; the ring is elastically deformed, put in place, and allowed to snap back toward its unstressed position into a groove or recess.

snatch block [DES ENG] A pulley frame or sheave with an eye through which lashing can be passed to fasten it to a scaffold or pole.

snatch plate [ENG] A thick steel plate through which a hole about one-sixteenth of an inch larger than the outside diameter of the drill rod on which it is to be used is drilled; the plate is slipped over the drill rod and one edge is fastened to a securely anchored chain, and if rods must be pulled because high-pressure water is encountered, the eccentric pull of the chain causes the outside of the rods to be gripped and held against the pressure of water; the rod is moved a short distance out of the hole each time the plate is tapped.

S-N diagram [ENG] In fatigue testing, a graphic representation of the relationship of stress S and the number of cycles N before failure of the material.

snifter valve [ENG] A valve on a pump that allows air to enter or escape, and accumulated water to be released.

snorkel [ENG] Any tube which supplies air for an underwater operation, whether it be for material or personnel.

snow bin [ENG] A box for measuring the amount of snowfall; a type of snow gage.

snow blower [MECH ENG] A machine that removes snow from a road surface or pavement using a screw-type blade to push the snow into the machine and from which it is ejected at some distance.

snowbreak [CIV ENG] Any barrier designed to shelter an object or area from snow.

snow fence [CIV ENG] An open-slatted board fence usually 4 to 10 feet (1.2 to 3.0 meters) high, placed about 50 feet (15 meters) on the windward side of a railroad track or highway; the fence serves to disrupt the flow of the wind so that the snow is deposited close to the fence on the leeward side, leaving a comparatively clear, protected strip parallel to the fence and slightly farther downwind.

snowflake *See* flake.

snow load [CIV ENG] The unit weight factor considered in the design of a flat or pitched roof for the probable amount of snow lying upon it.

snow mat [ENG] A device used to mark the surface between old and new snow, consisting of a piece of white duck 28 inches (71 centimeters) square, having in each corner triangular pockets in which are inserted slats placed diagonally to keep the mat taut and flat.

snow-melting system [CIV ENG] A system of pipes containing a circulating nonfreezing liquid or electric-heating cables, embedded beneath the surface of a road, walkway, or other area to be protected from snow accumulation.

snow pillow [ENG] A device used to record the changing weight of the snow cover at a point, consisting of a fluid-filled bladder lying on the ground with a pressure transducer or a vertical pipe and float connected to it.

snowplow [MECH ENG] A device for clearing away snow, as from a road or railway track.

snow resistograph [ENG] An instrument for recording a hardness profile of a snow cover by recording the force required to move a blade up through the snow.

snow sampler [ENG] A hollow tube for collecting a sample of snow in place. Also known as snow tube.

snow scale *See* snow stake.

snowshed [CIV ENG] A structure to protect an exposed area as a road or rail line from snow.

snow stake [ENG] A wood scale, calibrated in inches, used in regions of deep snow to measure its depth; it is bolted to a wood post or angle iron set in the ground. Also known as snow scale.

snow tube *See* snow sampler.

snub [MIN ENG] **1.** To increase the height of an undercut by means of explosives or otherwise. **2.** To check the descent of a car by the turn of a rope around a post.

snubber [MECH ENG] A mechanical device consisting essentially of a drum, spring, and friction band, connected between axle and frame, in order to slow the recoil of the spring and reduce jolting.

Snyder sampler [ENG] A mechanical device for obtaining small representative quantities from a moving stream of pulverized or granulated solids; it consists of a cast-iron plate revolving in a vertical plane on a horizontal axis with an inclined sample spout; the material to be sampled comes to the sampler by way of an inclined chute whenever the sample spout comes in line with the moving stream.

soak cleaning [MET] Cleaning the surface of a metal by immersion in a cleaning solution without electrolysis.

soaking [MET] Heating an alloy, usually an ingot, to a temperature not far below its melting temperature and holding it there for a long time to eliminate segregation that occurred on solidification.

soaking pit [MET] A high-temperature, gas-fired, tightly covered, refractory-lined hole or pit into which a hot metal ingot (with liquid interior) is held at a fixed temperature until needed for rolling into sheet or other forms.

soap [MATER] **1.** A particular type of detergent, in which the water-solubilizing group is a carboxylate, COO—, and the positive ion is usually sodium, Na^+, or potassium, K^+. **2.** A soap compound mixed with a fragrance and other ingredients and then cast into soap bars of different shapes.

soap bubble test [ENG] A leak test in which a soap solution is applied to the surface of the vessel under internal pressure test; soap bubbles form if the tracer gas leaks from the vessel.

soap builder [MATER] A substance mixed with soap to modify the alkali content, to add water-softening ability, or to improve otherwise the cleaning properties; examples are rosin and sodium phosphate.

soar [AERO ENG] To fly without loss of altitude, using no motive power other than updrafts in the atmosphere.

socket-head screw [DES ENG] A screw fastener with a geometric recess in the head into which an appropriate wrench is inserted for driving and turning, with consequent improved nontamperability.

socket wrench [DES ENG] A wrench with a socket to fit the head of a bolt or a nut.

soda-acid extinguisher [ENG] A fire-extinguisher from which water is expelled at a high rate by the generation of carbon dioxide, the result of mixing (when the extinguisher is tilted) of sulfuric acid and sodium bicarbonate.

soda blasting powder *See* B blasting powder.

soda lime [MATER] A mixture of sodium or potassium hydroxide with calcium oxide; granules are used to absorb water vapor and carbon dioxide gas.

soda-lime glass [MATER] Glass made by fusion of sand with sodium carbonate, or sodium sulfate and lime, or limestone; used for window glass.

sodium lead alloy [MATER] A highly toxic, explosionprone alloy of lead and sodium; contains 10% sodium and 90% lead when used to make tetraethyllead, and 2% sodium and 98% lead when used as a deoxidizer and homogenizer in lead-containing nonferrous alloys; reacts with moisture, acids, and oxidizing agents.

sodium nitrate gelignites [MATER] A group of explosives, modifications of blasting gelatin in which varying percentages of nitroglycerin are replaced by sodium nitrate and combustible material; characterized by plastic consistency, high densities, medium velocity of detonation, good resistance to the effects of water, and fume characteristics which are suitable for underground workings.

soffit [CIV ENG] The underside of a horizontal structural member, such as a beam or a slab.

softening agent [MATER] **1.** A substance that is added to another substance to increase softness; for example, stearic acid added to plastics, fat-liquoring agents to leather, and fatty alcohol to fabrics. **2.** A chemical that softens hard water by removing or trapping calcium and magnesium ions.

soft ground [MIN ENG] **1.** A mineral deposit which can be mined without drilling and shooting hard rock. **2.** The rock about underground openings that does not stand well and requires heavy timbering.

soft hammer [ENG] A hammer having a head made of a soft material, such as copper, lead, rawhide, or plastic; used to prevent damage to a finished surface.

soft-iron ammeter [ENG] An ammeter in which current in a coil causes two pieces of magnetic material within the coil, one fixed and one attached to a pointer, to become similarly magnetized and to repel each other, moving the pointer; used for alternating-current measurement.

soft landing [AERO ENG] The act of landing on the surface of a planet or moon without damage to any portion of the vehicle or payload, except possibly the landing gear.

soft missile base [CIV ENG] A missile-launching base that is not protected against a nuclear explosion.

soft patch [ENG] A patch in a crack in a vessel such as a steam boiler consisting of a soft material inserted in the crack and covered by a metal plate bolted or riveted to the vessel.

soft phosphate [MATER] Powdery, impure tricalcium phosphate separated in fertilizer manufacture from rock and pebble phosphates.

soft rock [MIN ENG] Rock that can be removed by air-operated hammers, but cannot be handled economically by a pick.

soft solder [MET] Solder composed of an alloy of lead and tin. Also known as low melting solder.

soft soldering [MET] Soldering with a soft solder.

soft wood [MATER] Wood from a coniferous tree.

soil-cement [MATER] A compacted mixture of soil, cement, and water used as a base course or surface for roads and airport paving.

soil mechanics [ENG] The application of the laws of solid and fluid mechanics to soils and similar granular materials as a basis for design, construction, and maintenance of stable foundations and earth structures.

soil pipe [CIV ENG] A vertical cast-iron or plastic pipe for carrying sewage from a building into the soil drain.

soil stabilizer [MATER] A chemical that alters the engineering property of a natural soil; used to stabilize soil slopes, to prepare for building foundations, and to prevent erosion.

soil stack [BUILD] The main vertical pipe into which flows the waste water from all fixtures in a structure.

soil thermograph [ENG] A remote-recording thermograph whose sensing element may be buried at various depths in the earth.

soil thermometer [ENG] A thermometer used to measure the temperature of the soil, usually the mercury-in-glass thermometer. Also known as earth thermometer.

solar attachment [ENG] A device for determining the true meridian directly from the sun; used an an attachment on a surveyor's transit or compass.

solar engine [MECH ENG] An engine which converts thermal energy from the sun into electrical, mechanical, or refrigeration energy; may be used as a method of spacecraft propulsion, either directly by photon pressure on huge solar sails, or indirectly from solar cells or from a reflector-boiler combination used to heat a fluid.

solar furnace [ENG] An image furnace in which high temperatures are produced by focusing solar radiation.

solar heat exchanger drive [AERO ENG] A proposed method of spacecraft propulsion in which solar radiation is focused on an area occupied by a boiler to heat a working fluid that is expelled to produce thrust directly. Abbreviated SHED.

solar heating [MECH ENG] The conversion of solar radiation into heat for technological, comfort-heating, and cooking purposes.

solar heat storage [ENG] The storage of solar energy for later use; usually accomplished by the heating of water or fusing a salt, although sand and gravel have been used as storage media.

solar house [BUILD] A house with large expanses of glass designed to catch solar radiation for heating.

solarimeter [ENG] 1. A type of pyranometer consisting of a Moll thermopile shielded from the wind by a bell glass. 2. See pyranometer.

solar magnetograph [ENG] An instrument that utilizes the Zeeman effect to directly measure the strength and polarity of the complex patterns of magnetic fields at the sun's surface; comprises a telescope, a differential analyzer, a spectrograph, and a photoelectric or photographic means of differencing and recording.

solar power [MECH ENG] The conversion of the energy of the sun's radiation to useful work.

solar probe [AERO ENG] A space probe whose trajectory passes near the sun so that instruments on board may detect and transmit back to earth data about the sun.

solar propulsion [AERO ENG] Spacecraft propulsion with a system composed of a type of solar engine.

solar rocket [AERO ENG] A rocket designed to carry instruments to measure and transmit parameters of the sun.

solar sail [AERO ENG] A surface of a highly polished material upon which solar light radiation exerts a pressure. Also known as photon sail.

solar satellite [AERO ENG] A space vehicle designed to enter into orbit about the sun. Also known as sun satellite.

solar turboelectric drive [AERO ENG] A proposed method of spacecraft propulsion in which solar radiation is focused on an area occupied by a boiler to heat a working fluid that drives a turbine generator system, producing electrical energy. Abbreviated STED.

solder [MET] 1. To join by means of solder. 2. An alloy, such as of zinc and copper, or of tin and lead, used when melted to join metallic surfaces.

solder brazing [MET] Brazing by means of a relatively high-melting solder.

solder glass [MATER] A special glass having a relatively low softening point (below 500°C); used to join two pieces of higher-melting glass without softening and deforming them.

soldering embrittlement [MET] The reduction in mechanical properties of a metal due to the local penetration of solder along grain boundaries.

soldering flux [MET] A chemical substance which aids the flow of solder and serves to remove and prevent the formation of oxides on the pieces to be united.

soldering gun [ENG] A soldering iron shaped like a gun.

soldering iron [ENG] A rod of copper with a handle on one end and pointed or wedge-shaped at the other end, and used for applying heat in soldering.

soldering pencil [ENG] A small soldering iron, about the size and weight of a standard lead pencil, used for soldering or unsoldering joints on printed wiring boards.

sole [BUILD] The horizontal member beneath the studs in a framed building.

solenoid brake [MECH ENG] A device that retards or arrests rotational motion by means of the magnetic resistance of a solenoid.

solenoid valve [MECH ENG] A valve actuated by a solenoid, for controlling the flow of gases or liquids in pipes.

solepiece [CIV ENG] One of two steel plates, port and starboard, whose forward parts are bolted to the ground ways supporting a ship about to be launched, while their aft parts are attached to the sliding ways; at the start of the launch, they are cut simultaneously with burning torches to release the ship. Also known as soleplate.

soleplate [BUILD] The plate on which stud bases butt in a stud partition. [CIV ENG] See solepiece. [ENG] **1.** The supporting base of a machine. **2.** A plate on which a bearing can be attached and, if necessary, adjusted slightly.

solid asphalt [MATER] Asphalt with a penetration number of less than 10 under specified test conditions.

solid box [MECH ENG] A solid, unadjustable ring bearing lined with babbitt metal, used on light machinery.

solid car [MIN ENG] A mine car equipped with a swivel coupling and generally used with a rotary dump.

solid coupling [MECH ENG] A flanged-face or a compression-type coupling used to connect two shafts to make a permanent joint and usually designed to be capable of transmitting the full load capacity of the shaft; a solid coupling has no flexibility.

solid crib timbering [MIN ENG] Shaft timbering with cribs laid solidly upon one another.

solid cutter [DES ENG] A cutter made of a single piece of material.

solid die [DES ENG] A one-piece screw-cutting tool with internal threads.

solid drilling [ENG] In diamond drilling, using a bit that grinds the whole face, without preserving a core for sampling.

solid explosive [MATER] An explosive employed in the form of a powder, a light-running granulated mass, or as solid sticks.

solidification shrinkage [MET] Volume contraction of a metal during solidification.

solid injection system [MECH ENG] A fuel injection system for a diesel engine in which a pump forces fuel through a fuel line and an atomizing nozzle into the combustion chamber.

solid-phase welding [MET] A welding method in which the weld is consummated by pressure or by heat and pressure without fusion.

solid-piled [MATER] Pertaining to plywood which is fresh from clamps or a hot press, and which is piled onto a solid flat base without stickers and weighted down until it reaches its normal temperature and moisture content. Also known as bulked-down; dead-piled.

solid propellant [MATER] A rocket propellant in solid form, usually containing both fuel and oxidizer combined or mixed, and formed into a monolithic (not powdered or granulated) grain. Also known as solid rocket fuel; solid rocket propellant.

solid propellant binder [MATER] The ingredient component of a propellant that is the agent for holding all the other ingredients together; contributes most to the physical or mechanical properties of the grain.

solid-propellant rocket engine [AERO ENG] A rocket engine fueled with a solid propellant; such motors consist essentially of a combustion chamber containing the propellant, and a nozzle for the exhaust jet.

solid rocket [AERO ENG] A rocket that is propelled by a solid-propellant rocket engine.

solid rocket fuel See solid propellant.

solid rocket propellant See solid propellant.

solid shafting [MECH ENG] A solid round bar that supports a roller and wheel of a machine.

solid shank tool [ENG] A cutting tool in which the shank and cutting edges are machined from one piece.

solid state [ENG] Pertaining to a circuit, device, or system that depends on some combination of electrical, magnetic, and optical phenomena within a solid that is usually a crystalline semiconductor material.

solid-state welding [MET] Welding processes which coalesce materials at temperatures below the melting point of the base metal by methods such as welding or diffusion welding without the use of filler metal.

solid stowing [MIN ENG] The complete filling of the waste area behind a longwall face with stone and dirt.

solid-web girder [CIV ENG] A beam, such as a box girder, having a web consisting of a plate or other solid section but not a lattice.

soluble castor oil See Turkey red oil.

soluble cutting oil [MATER] A petroleum oil containing an emulsifying agent to make it mix easily with water; used as a coolant for metal-cutting tools.

soluble oil [MATER] An oil that readily forms a stable emulsion or colloidal suspension in water. Also known as emulsifying oil.

soluble starch [MATER] A group of water-soluble polymers formed from starch, such as the starches derived from corn or potato, by acetylation, acid hydrolysis, chlorination, or by action of enzymes to form starch acetates, ethers, and esters;

used as textile sizing agents, emulsifying agents, and paper coatings.

solution gas [PETRO ENG] Gaseous reservoir hydrocarbons dissolved in liquid reservoir hydrocarbons because of the prevailing pressures in the reservoir. Also known as dissolved gas.

solution gas drive *See* internal gas drive.

solution-gas reservoir [PETRO ENG] Oil reservoir initially at or above the bubble-point pressure of the gas-oil mixture, and produced primarily by the expansion of the oil and its dissolved gas. Also known as dissolved-gas reservoir.

solution heat treatment [MET] Heating and holding an alloy at a temperature at which one (or more) constituent enters into solid solution, then cooling the alloy rapidly to prevent the constituent from precipitating.

solution mining [MIN ENG] The extraction of soluble minerals from subsurface strata by injection of fluids, and the controlled removal of mineral-laden solutions.

solution porosity [PETRO ENG] A generic designation for reservoir-rock porosity created by solution action; some examples are crystalline limestone and dolomite, porous cap rock, and honeycombed anhydrite.

solvent molding [ENG] A process to form thermoplastic articles by dipping a mold into a solution or dispersion of the resin and drawing off (evaporating) the solvent to leave a plastic film adhering to the mold.

solvent naphtha [MATER] Refined petroleum naphtha of restricted boiling range; used as a solvent and paint thinner and in dry cleaning; Stoddard solvent is a special grade of solvent naphtha.

solvent-refined oil [MATER] A lubricating oil that has been solvent-treated during refining, such as most motor, aircraft, diesel-engine, steam-turbine, and other high-quality oils.

sonar [ENG] **1.** A system that uses underwater sound, at sonic or ultrasonic frequencies, to detect and locate objects in the sea, or for communication; the commonest type is echo-ranging sonar, other versions are passive sonar, scanning sonar, and searchlight sonar. Derived from sound navigation and ranging. **2.** *See* sonar set.

sonar beacon [ENG ACOUS] An underwater beacon that transmits sonic or ultrasonic signals for the purpose of providing bearing information; it may have receiving facilities that permit triggering an external source.

sonar boomer transducer [ENG ACOUS] A sonar transducer that generates a large pressure wave in the surrounding water when a capacitor bank discharges into a flat, epoxy-encapsulated coil, creating opposed magnetic fields from the coil and from eddy currents in an adjacent aluminum disk, which cause the disk to be driven away from the coils with great force.

sonar capsule [ENG ACOUS] A capsule that reflects high-frequency sound waves; the sonar capsule, if attached to a reentry body, may be used to locate the reentry body.

sonar dome [ENG] A streamlined, watertight enclosure that provides protection for a sonar transducer, sonar projector, or hydrophone and associated equipment, while offering minimum interference to sound transmission and reception.

sonar projector [ENG ACOUS] An electromechanical device used under water to convert electrical energy to sound energy; a crystal or magnetostriction transducer is usually used for this purpose.

sonar set [ENG] A complete assembly of sonar equipment for detecting and ranging or for communication. Also known as sonar.

sonar target [ENG ACOUS] An object which reflects a sufficient amount of a sonar signal to produce a detectable echo signal at the sonar equipment.

sonar transducer [ENG ACOUS] A transducer used under water to convert electrical energy to sound energy and sound energy to electrical energy.

sonar transmission [ENG ACOUS] The process by which underwater sound signals generated by a sonar set travel through the water.

sonar window [ENG ACOUS] The portion of a sonar dome or sonar transducer that passes sound waves at sonar frequencies with little attenuation while providing mechanical protection for the transducer.

sonde [ENG] An instrument used to obtain weather data during ascent and descent through the atmosphere, in a form suitable for telemetering to a ground station by radio, as in a radiosonde.

sonic altimeter [ENG] An instrument for determining the height of an aircraft above the earth by measuring the time taken for sound waves to travel from the aircraft to the surface of the earth and back to the aircraft again.

sonic anemometer [ENG] An anemometer which measures wind speed by means of the properties of wind-borne sound waves; it operates on the principle that the propagation velocity of a sound wave in a moving medium is equal to the velocity of sound with respect to the medium plus the velocity of the medium.

sonic barrier [AERO ENG] A popular term for the large increase in drag that acts upon an aircraft approaching acoustic velocity; the point at which the speed of sound is attained and existing subsonic and supersonic flow theories are rather indefinite. Also known as sound barrier.

sonic chemical analyzer [ENG] A device to characterize the composition of a gas, liquid, or solid by the attenuation or change in the velocity of sound waves through a sample; the effect is related to molecular structure and intermolecular interactions.

sonic cleaning [ENG] Cleaning of contaminated materials by the action of intense sound in the liquid in which the material is immersed.

sonic depth finder [ENG] A sonar-type instrument used to measure ocean depth and to locate underwater objects; a sound pulse is transmitted vertically downward by a piezoelectric or magnetostriction transducer mounted on the hull of the ship; the time required for the pulse to return after reflection is measured electronically. Also known as echo sounder.

sonic drilling [MECH ENG] The process of cutting or shaping materials with an abrasive slurry driven by a reciprocating tool attached to an audio-frequency electromechanical transducer and vibrating at sonic frequency.

sonic flaw detection [ENG] The process of locating imperfections in solid materials by observing internal reflections or a variation in transmission through the materials as a function of sound-path location.

sonic liquid-level meter [ENG] A meter that detects a liquid level by sonic-reflection techniques.

sonic pump [PETRO ENG] A type of lifting pump used in a shallow oil well to pump out the crude; consists of a string of tubing equipped with a check valve at each point, and mechanical means on the surface to vibrate the tubing string vertically; creates a harmonic condition that results in several hundred strokes per minute, with the strokes being a small fraction of an inch.

sonic sifter [MECH ENG] A high-speed vibrating apparatus used in particle size analysis.

sonic sounding [ENG] Determining the depth of the ocean bottom by measuring the time for an echo to return to a shipboard sound source.

sonic thermometer [ENG] A thermometer based upon the principle that the velocity of a sound wave is a function of the temperature of the medium through which it passes.

sonic well logging [ENG] A well logging technique that uses a pulse-echo system to measure the distance between the instrument and a sound-reflecting surface; used to measure the size of cavities around brine wells, and capacities of underground liquefied petroleum gas storage chambers.

sonobuoy [ENG] An acoustic receiver and radio transmitter mounted in a buoy that can be dropped from an aircraft by parachute to pick up underwater sounds of a submarine and transmit them to the aircraft; to track a submarine, several buoys are dropped in a pattern that includes the known or suspected location of the submarine, with each buoy transmitting an identifiable signal; an electronic computer then determines the location of the submarine by comparison of the received signals and triangulation of the resulting time-delay data. Also known as radio sonobuoy.

sonograph [ENG] **1.** An instrument for recording sound or seismic vibrations. **2.** An instrument for converting sounds into seismic vibrations.

Sonolog [PETRO ENG] An acoustical device used for sound-reflection logging of oil well boreholes to determine the fluid level in a pumping well.

sonometer [ENG] An instrument for measuring rock stress; a piano wire is stretched between two bolts in the rock, and any change of pitch after destressing is observed and used to indicate stress.

sonoscan [ENG] A type of acoustic microscope in which an unfocused acoustic beam passes through the object and produces deformations in a liquid-solid interface that are sensed by a laser beam reflected from the surface.

soot [MATER] Impure black carbon with oily compounds obtained from the incomplete combustion of resinous materials, oils, wood, or coal.

soot blower [ENG] A system of steam or air jets used to maintain cleanliness, efficiency, and capacity of heat-transfer surfaces by the periodic removal of ash and slag from the heat-absorbing surfaces.

sorbent [MATER] A material, compound, or system that can provide a sorption function, such as adsorption, absorption, or desorption.

Sorel cement *See* oxychloride cement.

sorption pumping [ENG] A technique used to reduce the pressure of gas in an atmosphere; the gas is adsorbed on a granular sorbent material such as a molecular sieve in a metal container; when this sorbent-filled container is immersed in liquid nitrogen, the gas is sorbed.

sortie [AERO ENG] An operational flight by one aircraft.

sortie number [ENG] A reference used to identify the images taken by all the sensors during one air reconnaissance sortie.

sorting table [ENG] Any horizontal conveyor where operators, along its side, sort bulk material, packages, or objects from the conveyor.

sound analyzer [ENG] An instrument which measures the amount of sound energy in various frequency bands; it generally consists of a set of fixed electrical filters or a tunable electrical filter, along with associated amplifiers and a meter which indicates the filter output.

sound barrier *See* sonic barrier.

sound effects [ENG ACOUS] Mechanical devices or recordings used to provide lifelike imitations of various sounds.

sound exclusion [PETRO ENG] Several techniques used to prevent a borehole from sloughing in drilling a well through a reservoir rock that has an unconsolidated nature, similar to beach sand; the borehole can be lined with a screen, the sand consolidated with a binding material, or prepack gravel liner can be used.

sound film [ENG ACOUS] Motion picture film having a sound track along one side for reproduction of the sounds that are to accompany the film.

sound filmstrip [ENG ACOUS] A filmstrip that has accompanying sound on a separate disk or tape, which is manually or automatically synchronized with projection of the pictures in the strip.

sound gate [ENG ACOUS] The gate through which film passes in a sound-film projector for conversion of the sound track into audio-frequency signals that can be amplified and reproduced.

sound head [ENG ACOUS] **1.** The section of a sound motion picture projector that converts the photographic or magnetic sound track to audible sound signals. **2.** In a sonar system, the cylindrical container for the transmitting projector and the receiving hydrophone.

sounding [ENG] **1.** Determining the depth of a body of water by an echo sounder or sounding line. **2.** Measuring the depth of bedrock by driving a steel rod into the soil. **3.** Any penetration of the natural environment for scientific observation. [MIN ENG] **1.** Knocking on a mine roof to see whether it is sound or safe to work under. **2.** Subsurface investigation by observing the penetration resistance of the subsurface material without drilling holes, by driving a rod into the ground or by using a penetrometer.

sounding balloon [ENG] A small free balloon used for carrying radiosonde equipment aloft.

sounding device [PETRO ENG] An acoustical device used to measure the liquid level in a wellbore; for example, a Sonolog or an Echometer.

sounding lead [ENG] A lead used for determining the depth of water.

sounding line [ENG] The line attached to a sounding lead. Also known as lead line.

sounding machine [ENG] An instrument for measuring the depth of water, consisting essentially of a reel of wire; to one end of this wire there is attached a weight which carries a device for measuring and recording the depth; a crank or motor reels in the wire.

sounding pole [ENG] A pole or rod used for sounding in shallow water, and usually marked to indicate various depths.

sounding rocket [AERO ENG] A rocket that carries aloft equipment for making observations of or from the upper atmosphere.

sounding sextant *See* hydrographic sextant.

sounding wire [ENG] A wire used with a sounding machine in determining depth of water.

sound-level meter [ENG] An instrument used to measure noise and sound levels in a specified manner; the meter may be calibrated in decibels or volume units and includes a microphone, an amplifier, an output meter, and frequency-weighting networks.

sound locator [ENG ACOUS] A device formerly used to detect aircraft in flight by sound, consisting of four horns, or sound collectors (two for azimuth detection and two for elevation), together with their associated mechanisms and controls, which enabled the listening operator to determine the position and angular velocity of an aircraft.

sound navigation and ranging *See* sonar.

sound-powered telephone [ENG ACOUS] A telephone operating entirely on current generated by the speaker's voice, with no external power supply; sound waves cause a diaphragm to move a coil back and forth between the poles of a powerful but small permanent magnet, generating the required audio-frequency voltage in the coil.

sound production [ENG ACOUS] Conversion of energy from mechanical or electrical into acoustical form, as in a siren or loudspeaker.

soundproofing *See* damping.

sound ranging [ENG ACOUS] Determining the location of a gun or other sound source by measuring the travel time of the sound wave to microphones at three or more different known positions.

sound reception [ENG ACOUS] Conversion of acoustical energy into another form, usually electrical, as in a microphone.

sound recording [ENG ACOUS] The process of recording sound signals so they may be reproduced at any subsequent time, as on a phonograph disk, motion picture sound track, or magnetic tape.

sound-reinforcement system [ENG ACOUS] An electronic means for augmenting the sound output of a speaker, singer, or musical instrument in cases where it is either too weak to be heard above the general noise or too reverberant; basic elements of such a system are microphones, amplifiers, volume controls, and loudspeakers. Also known as public address system.

sound-reproducing system [ENG ACOUS] A combination of transducing devices and associated equipment for picking up sound at one location and time and reproducing it at the same or some other location and at the same or some later time. Also known as audio system; reproducing system; sound system.

sound reproduction [ENG ACOUS] The use of a combination of transducing devices and associated equipment to pick up sound at one point and reproduce it either at the same point or at some other point, at the same time or at some subsequent time.

sound spectrograph [ENG ACOUS] An instrument that records and analyzes the spectral composition of audible sound.

sound speed [ENG] The speed of sound motion picture film, standardized at 24 frames per second (silent film speed is 18 frames per second).

soundstripe [ENG ACOUS] A longitudinal stripe of magnetic material placed on some motion picture films for recording a magnetic sound track.

sound system *See* sound-reproducing system.

sound track [ENG ACOUS] A narrow band, usually along the margin of a sound film, that carries the sound record; it may be a variable-width or variable-density optical track or a magnetic track.

sound transducer *See* electroacoustic transducer.

sour corrosion [PETRO ENG] Corrosion occurring in oil or gas wells where there is an iron sulfide corrosion product, and hydrogen sulfide is present in the produced reservoir fluid.

sour crude [MATER] Crude oil containing an abnormally large amount of sulfur compounds that, upon refining, liberate corrosive sulfur compounds; opposite to sweet crude.

sour dirt [PETRO ENG] Sulfate-impregnated soil or soil characterized by escaping sulfur dioxide or hydrogen sulfide; considered an indicator of oil in the area. Also known as copper dirt.

sour gas [MATER] Natural gas that contains corrosive, sulfur-bearing compounds, such as hydrogen sulfide and mercaptans.

southbound node *See* descending node.

sow [MET] **1.** A mold of larger size than a pig. **2.** A channel that conducts molten metal to molds in a pig bed.

sow block [MET] In forging, a removable block set into the hammer anvil to reduce wear on the anvil.

soya bean oil *See* soybean oil.

soybean lecithin *See* lecithin.

soybean oil [MATER] A pale yellow, fixed drying oil produced by solvent extraction from soybeans; soluble in alcohol, chloroform, and ether; used for soap manufacture, cattle feeds, and printing inks, and in margarine, salad dressing, and high-protein foods. Also known as Chinese bean oil; soya bean oil; soy oil.

soy lecithin *See* lecithin.

soy oil *See* soybean oil.

Soyuz Program [AERO ENG] A crewed space-flight program begun in 1967 by the Soviet Union.

space capsule [AERO ENG] A container, manned or unmanned, used for carrying out an experiment or operation in space.

spacecraft [AERO ENG] Devices, manned and unmanned, which are designed to be placed into an orbit about the earth or into a trajectory to another celestial body. Also known as space ship; space vehicle.

spacecraft ground instrumentation [ENG] Instrumentation located on the earth for monitoring, tracking, and communicating with manned spacecraft, satellites, and space probes. Also known as ground instrumentation.

spacecraft launching [AERO ENG] The setting into motion of a space vehicle with sufficient force to cause it to leave the earth's atmosphere.

spacecraft propulsion [AERO ENG] The use of rocket engines to accelerate space vehicles.

spacecraft tracking [ENG] The determination of the positions and velocities of spacecraft through radio and optical means.

space detection and tracking system [ENG] System capable of detecting and tracking space vehicles from the earth, and reporting the orbital characteristics of these vehicles to a central control facility. Abbreviated SPADATS.

spaced loading [ENG] Loading shot holes so that cartridges are separated by open spacers which do not prevent the concussion from one charge from reaching the next.

space flight [AERO ENG] Travel beyond the earth's sensible atmosphere; space flight may be an orbital flight about the earth or it may be a more extended flight beyond the earth into space.

space-flight trajectory [AERO ENG] The track or path taken by a spacecraft.

space frame [BUILD] A three-dimensional steel building frame which is stable against wind loads.

space lattice [BUILD] A space frame built of lattice girders.

space mission [AERO ENG] A journey by a vehicle, manned or unmanned, beyond the earth's atmosphere, usually for the purpose of collecting scientific data.

spaceport [AERO ENG] An installation used to test and launch spacecraft.

space power system [AERO ENG] An on-board assemblage of equipment to generate and distribute electrical energy on satellites and spacecraft.

space probe [AERO ENG] An instrumented vehicle, the payload of a rocket-launching system designed specifically for flight missions to other planets or the moon and into deep space, as distinguished from earth-orbiting satellites.

spacer [ENG] **1.** A piece of metal wire twisted at one end to form a guard to keep the explosive in a shothole in place and twisted at the other end to form a guard to hold the tamping in its place. **2.** A piece of wood doweling interposed between charges to extend the column of explosive. **3.** A device for holding two members at a given distance from each other. Also known as spacer block. **4.** The tapered section of a pug joining the barrel to the die; clay is compressed in this section before it issues through the die.

spacer block *See* spacer.

space reconnaissance [AERO ENG] Reconnaissance of the surface of a planet from a space ship or satellite.

space research [AERO ENG] Research involving studies of all aspects of environmental conditions beyond the atmosphere of the earth.

spacer strip [MET] A strip or bar of metal placed in the root of a weld joint, prepared for a groove weld, to serve as backing and maintain root opening during welding.

space satellite [AERO ENG] A vehicle, manned or unmanned, for orbiting the earth.

space shuttle [AERO ENG] A spacecraft designed to travel from the earth to a space station and to return to earth.

space simulator [AERO ENG] **1.** Any device which simulates one or more parameters of the space environment and which is used to test space systems or components. **2.** Specifically, a closed chamber capable of reproducing approximately the vacuum and normal environments of space.

space station [AERO ENG] An autonomous, permanent facility in space for the conduct of scientific and technological research, earth-oriented applications, and astronomical observations.

space suit [ENG] A pressure suit for wear in space or at very low ambient pressures within the atmosphere, designed to permit the wearer to leave the protection of a pressurized cabin.

space technology [AERO ENG] The systematic application of engineering and scientific disciplines to the exploration and utilization of outer space.

Space Tracking and Data Acquisition Network [ENG] A network of ground stations operated by the National Aeronautics and Space Administration, which tracks, commands, and receives telemetry for United States and foreign unmanned satellites. Abbreviated STADAN.

space vehicle *See* spacecraft.

space walk [AERO ENG] The movement of an astronaut outside the protected environment of a spacecraft during a space flight; the astronaut wears a spacesuit.

spacing clamp [PETRO ENG] A clamp for maintaining the rod string in the correct pumping position while the well is in the final stages of being fitted to the pump.

spackling [ENG] The process of repairing a part of a plaster wall or mural by cleaning out the defective spot and then patching it with a plastering material.

SPADATS *See* space detection and tracking system.

spade [DES ENG] A shovellike implement with a flat oblong blade; used for turning soil by pushing against the blade with the foot.

spade bolt [DES ENG] A bolt having a spade-shaped flattened head with a transverse hole, used to fasten shielded coils, capacitors, and other components to a chassis.

spade drill [DES ENG] A drill consisting of three main parts: a cutting blade, a blade holder or shank, and a device, such as a screw, which fastens the blade to the holder; used for cutting holes over 1 inch (2.54 centimeters) in diameter.

spade lug [DES ENG] An open-ended flat termination for a wire lead, easily slipped under a terminal nut.

spall [ENG] **1.** To reduce irregular stone blocks to an approximate size by chipping with a hammer. **2.** To break off thin chips from, and parallel to, the surface of a material, such as a metal or rock. [MIN ENG] To break ore.

span [AERO ENG] **1.** The dimension of a craft measured between lateral extremities; the measure of this dimension. **2.** Specifically, the dimension of an airfoil from tip to tip measured in a straight line. [ENG] A structural dimension measured between certain extremities.

spandrel [BUILD] The part of a wall between the sill of a window and the head of the window below it.

spandrel wall [BUILD] A wall on the outer surface of a vault to fill the spandrels.

Spanish lavender oil *See* spike oil.

Spanish spike oil *See* spike oil.

spanner [DES ENG] A wrench with a semicircular head having a projection or hole at one end. [ENG] **1.** A horizontal brace. **2.** An artificial horizon attachment for a sextant.

spar [AERO ENG] A principal spanwise member of the structural framework of an airplane wing, aileron, stabilizer, and such; it may be of one-piece design or a fabricated section. [MIN ENG] A small clay vein in a coal seam.

spare part [ENG] In supply usage, any part, component, or subassembly kept in reserve for the maintenance and repair of major items of equipment.

spare parts list [ENG] List approved by designated authorities, indicating the total quantities of spare parts, tools, and equipment necessary for the maintenance of a specified number of major items for a definite period of time.

spark arrester [ENG] **1.** An apparatus that prevents sparks from escaping from a chimney. **2.** A device that reduces or eliminates electric sparks at a point where a circuit is opened and closed.

spark coil leak detector [ENG] A coil similar to a Tesla coil which detects leaks in a vacuum system by jumping a spark between the leak hole and the core of the coil.

spark-ignition engine [MECH ENG] An internal combustion engine in which an electrical discharge ignites the explosive mixture of fuel and air.

spark knock [MECH ENG] The knock produced in an internal combustion engine precedes the arrival of the piston at the top dead-center position.

spark lead [MECH ENG] The amount by which the spark precedes the arrival of the pis-ton at its top (compression) dead-center position in the cylinder of an internal combustion engine.

sparkle metal [MET] A crude mixture of sulfides containing 74% copper produced by the smelting of copper ore.

spark machining [MET] Cutting metal by repetitive sparking between a tool (the cathode) and the workpiece (the anode).

spark-over-initiated discharge machining [MECH ENG] An electromachining process in which a potential is impressed between the tool (cathode) and workpiece (anode) which are separated by a dielectric material; a heavy discharge current flows through the ionized path when the applied potential is sufficient to cause rupture of the dielectric.

sparkproof [ENG] **1.** Treated with a material to prevent ignition or damage by sparks. **2.** Generating no sparks.

spark recorder [ENG] Recorder in which the recording paper passes through a spark gap formed by a metal plate underneath and a moving metal pointer above the paper; sparks from an induction coil pass through the paper periodically, burning small holes that form the record trace.

spar varnish [MATER] A flammable varnish made of drying oils, resins, thinners, and driers to provide a durable, water-resistant coating for outside or other severe service.

spatter [MET] Particles of metal expelled during arc or gas welding.

spatter dash [CIV ENG] **1.** A finish put on stucco by dashing a mortar and sand mixture against it. **2.** Paint spattered on a different-colored ground coat.

speaker *See* loudspeaker.

spear [DES ENG] A rodlike fishing tool having a barbed-hook end, used to recover rope, wire line, and other materials from a borehole.

spearmint oil [MATER] A colorless to yellowish essential oil obtained from spearmint with characteristic taste and scent; soluble in alcohol, ether, and chloroform; used as a flavor and a source of carvone.

special cargo [IND ENG] Cargo which requires special handling or protection, such as pyrotechnics, detonators, watches, and precision instruments.

special flight [AERO ENG] An air transport flight, other than a scheduled service, set up to move a specific load.

special gelatin [MATER] The brand name of a series of ammonia-gelatin-type dynamites used in open-pit mining, underground metal mining, quarrying, and construction.

special-purpose item [ENG] In supply usage, any item designed to fill a special requirement, and having a limited application; for example, a wrench or other tool designed to be used for one particular model of a piece of machinery.

special-purpose vehicle [ENG] A vehicle having a special chassis, or a general-purpose chassis incorporating major modifications, designed to fill a specialized requirement; all tractors (except truck tractors) and tracklaying vehicles, regardless of design, size, or intended purpose, are classified as special-purpose vehicles.

specification [ENG] An organized listing of basic requirements for materials of construction, product compositions, dimensions, or test conditions; a number of organizations publish standards (for example, ASME, API, ASTM), and many companies have their own specifications. Also known as specs. [IND ENG] A quantitative description of the required characteristics of a device, machine, structure, product, or process.

specific fuel consumption [MECH ENG] The weight flow rate of fuel required to produce a unit of power or thrust, for example, pounds per horsepower-hour. Abbreviated SFC. Also known as specific propellant consumption.

specific gravity bottle [ENG] A small bottle or flask used to measure the specific gravities of liquids; the bottle is weighed when it is filled with the liquid whose specific gravity is to be determined, when filled with a reference liquid, and when empty. Also known as density bottle; relative density bottle.

specific gravity hydrometer [ENG] A hydrometer which indicates the specific gravity of a liquid, with reference to water at a particular temperature.

specific impulse [AERO ENG] A performance parameter of a rocket propellant, expressed in seconds, equal to the thrust in pounds divided by the weight flow rate in pounds per second. Also known as specific thrust.

specific productivity index [PETRO ENG] Barrels per day of oil produced per pound decline in bottom-hole pressure per foot of effective reservoir thickness.

specific propellant consumption *See* specific fuel consumption.

specific speed [MECH ENG] A number, N_s, used to predict the performance of centrifugal and axial pumps or hydraulic turbines: for pumps, $N_s = N \sqrt{Q}/H^{3/4}$; for turbines, $N_s = N \sqrt{P}/H^{5/4}$, where N_s is specific speed, N is the rotational speed in revolutions per minute, Q is the rate of flow in gallons per minute, H is head in feet, and P is shaft horsepower.

specific thrust *See* specific impulse.

specs *See* specification.

spectral density *See* frequency spectrum.

spectral hygrometer [ENG] A hygrometer which determines the amount of precipitable moisture in a given region of the atmosphere by measuring the attenuation of radiant energy caused by the absorption bands of water vapor; the instrument consists of a collimated energy source, separated by the region under investigation and a detector which is sensitive to those frequencies that correspond to the absorption bands of water vapor.

spectral pyrometer *See* narrow-band pyrometer.

spectrum analyzer [ENG] Test instrument used to show the distribution of energy contained in the frequencies emitted by a pulse magnetron; also used to measure the Q of resonant cavities and lines, and to measure the cold impedance of a magnetron.

speculum alloy [MET] A brilliant white, hard, brittle alloy composed of copper and tin in a 2:1 proportion and sometimes with additions of other elements.

speech amplifier [ENG ACOUS] An audio-frequency amplifier designed specifically for amplification of speech frequencies, as for public-address equipment and radiotelephone systems.

speech clipper [ENG ACOUS] A clipper used to limit the peaks of speech-frequency signals, as required for increasing the average modulation percentage of a radiotelephone or amateur radio transmitter.

speech coil *See* voice coil.

speed cone [MECH ENG] A cone-shaped pulley, or a pulley composed of a series of pulleys of increasing diameter forming a stepped cone.

speed density metering [AERO ENG] A type of aircraft carburetion in which the fuel feed is regulated by the parameters of engine feed and intake manifold pressure.

speed-in [PETRO ENG] To start drilling by making a hole.

speed lathe [MECH ENG] A light, pulley-driven lathe, usually without a carriage or back gears, used for work in which the tool is controlled by hand.

speed of travel [MET] The speed at which a weld is made along its longitudinal axis; measured in inches or spots per minute.

speedometer [ENG] An instrument that indicates the speed of travel of a vehicle in miles per hour, kilometers per hour, or knots.

speed reducer [MECH ENG] A train of gears placed between a motor and the machinery which it will drive, to reduce the speed with which power is transmitted.

speiss [MET] A mixture of impure metal arsenides and antimonides resulting from the smelting of certain ores such as cobalt and lead.

spelter [MET] A commercially pure grade of zinc used in galvanizing; contains lead or iron as impurities.

spelter solder [MET] Brass composed of equal parts of copper and zinc; used in brazing as a filler metal. Also known as brazing brass.

spent acid *See* sludge acid.

spent iron sponge [MATER] Iron sponge saturated with sulfur; prone to spontaneous heating. Also known as spent oxide.

spent liquor [MATER] The liquid effluent from the digestion of wood during pulping; contains wood chemicals (for example, lignin) and spent digestant (caustic, sulfite, or sulfate, depending on the process used).

spent oxide *See* spent iron sponge.

spermaceti [MATER] A white, crystalline, oily (waxy) solid that separates from sperm oil; soluble in ether, chloroform, and carbon disulfide, insoluble in water; melts at 42 to 50°C; used in ointments, emulsions, candles, soaps, and cosmetics; and for linen finishing. Also known as spermaceti wax.

spermaceti wax *See* spermaceti.

sperm oil [MATER] A combustible, yellowish oil found in the head cavities and blubber of the sperm whale; soluble in ether, chloroform, and benzene; used as a lubricant for precision machinery, for rustproofing metals, and in transmission fluids.

spherical powder [MATER] A powder consisting of globular-shaped particles.

spherical separator [PETRO ENG] A gas-oil separator in the form of a spherical vessel.

spheroidal graphite cast iron *See* nodular cast iron.

spheroidized carbides [MET] Globular forms of carbide, as formed in spheroidized steel.

spheroidized steel [MET] Steel that has been heat-treated to produce a spheroidized carbide structure.

spheroidizing [MET] Heating steels just below Ae_1 until the shape of cementite particles becomes relatively spherical.

spherometer [ENG] A device used to measure the curvature of a spherical surface.

spider [ENG] **1.** The part of an ejector mechanism which operates ejector pins in a molding press. **2.** In extrusion, the membranes which support a mandrel within the head-die assembly. [ENG ACOUS] A highly flexible perforated or corrugated disk used to center the voice coil of a dynamic loudspeaker with respect to the pole piece without appreciably hindering in-and-out motion of the voice coil and its attached diaphragm. [PETRO ENG] A steel block with a tapered opening which permits passage of pipe during movement into or from a well; designed to hold pipe suspended in the well when the slips are placed in the tapered opening and in contact with the pipe.

spiegeleisen [MET] An iron alloy containing 15–30% manganese and 5% carbon used in steelmaking.

spike [DES ENG] A large nail, especially one longer than 3 inches (7.6 centimeters), and often of square section.

spike microphone [ENG ACOUS] A device for clandestine aural surveillance in which the sensor is a spike driven into the wall of the target area and mechanically coupled to the diaphragm of a microphone on the other side of the wall.

spike oil [MATER] A pale yellow essential oil extracted by distillation from the flowers of *Lavandula latifolia*; soluble in fixed oils and propylene glycol; used in soap and as an alcohol denaturant and flavoring agent. Also known as lavender spike oil; Spanish lavender oil; Spanish spike oil.

spile [MIN ENG] **1.** A temporary lagging driven ahead on levels in loose ground. **2.** A short piece of plank sharpened flatwise and used for driving into watery stratums as sheet piling to assist in checking the flow of water.

spiling *See* forepoling.

spill [ENG] The accidental release of some material, such as nuclear material or oil, from a container.

spill box [CIV ENG] A device such as a flume that maintains a constant head on a measuring weir or orifice.

spill pit *See* runoff pit.

spillway [CIV ENG] A passage in or about a dam or other hydraulic structure for escape of surplus water.

spillway apron [CIV ENG] A concrete or timber floor at the bottom of a spillway to prevent soil erosion from heavy or turbulent flow.

spillway channel [CIV ENG] An outlet channel from a spillway.

spillway dam *See* overflow dam.

spillway gate [CIV ENG] A gate for regulating the flow from a reservoir.

spindle [DES ENG] A short, slender or tapered shaft.

spinner [ENG] **1.** Automatically rotatable radar antenna, together with directly associated equipment. **2.** Part of a mechanical scanner which rotates about an axis, generally restricted to cases where the speed of rotation is relatively high.

spinneret [ENG] An extrusion die with many holes through which plastic melt is forced to form filaments.

spinning [ENG] The extrusion of a spinning solution (such as molten plastic) through a spinneret. [MECH ENG] Shaping and finishing sheet metal by rotating the workpiece over a mandrel and working it with a round-ended tool. Also known as metal spinning.

spinning machine [MECH ENG] **1.** A machine that winds insulation on electric wire. **2.** A machine that shapes metal hollow ware.

spin rocket [AERO ENG] A small rocket that imparts spin to a larger rocket vehicle or spacecraft.

spin stabilization [AERO ENG] Directional stability of a spacecraft obtained by the action of gyroscopic forces which result from spinning the body about its axis of symmetry.

spin welding [ENG] Fusion of two objects (for example, plastics) by forcing them together while one of the pair is spinning; frictional heat melts the interface, spinning is stopped, and the bodies are held together until they are frozen in place (welded).

spiral bevel gear [DES ENG] Bevel gear with curved, oblique teeth to provide gradual engagement and bring more teeth together at a given time than an equivalent straight bevel gear.

spiral chute [DES ENG] A gravity chute in the form of a continuous helical trough spiraled around a column for conveying materials to a lower level.

spiral conveyor *See* screw conveyor.

spiral cutterhead [MIN ENG] A rotary digging device which dislodges and feeds alluvial sand or gravel to the intake of a suction dredge.

spiral flow tank [CIV ENG] An aeration tank of the activated sludge process into which air is diffused in a spiral helical movement guided by baffles and proper location of diffusers.

spiral flow test [ENG] The determination of the flow properties of a thermoplastic resin by measuring the length and weight of resin flowing along the path of a spiral cavity.

spiral gage *See* spiral pressure gage.

spiral gear [MECH ENG] A helical gear that transmits power from one shaft to another, nonparallel shaft.

spiral-jaw clutch [MECH ENG] A modification of the square-jaw clutch permitting gradual meshing of the mating faces, which have a helical section.

spiral mold cooling [ENG] Cooling an injection mold by passing a liquid through a spiral cavity in the body of the mold.

spiral pipe [DES ENG] Strong, lightweight steel pipe with a single continuous welded helical seam from end to end.

spiral pressure gage [ENG] A device for measurement of pressures; a hollow tube spiral receives the system pressure which deforms (unwinds) the spiral in direct relation to the pressure in the tube. Also known as spiral gage.

spiral ramp system [MIN ENG] Development of a mine by driving moderately inclined haulageways from the surface to underground ore horizons.

spiral scanning [ENG] Scanning in which the direction of maximum radiation describes a portion of a spiral; the rotation is always in one direction; used with some types of radar antennas.

spiral spring [DES ENG] A spring bar or wire wound in an Archimedes spiral in a plane; each end is fastened to the force-applying link of the mechanism.

spiral thermometer [ENG] A temperature-measurement device consisting of a bimetal spiral that winds tighter or opens with changes in temperature.

spiral-tube heat exchanger [ENG] A countercurrent heat-exchange device made of a group of concentric spirally wound coils, generally connected by manifolds; used for cryogenic exchange in air-separation plants.

spiral welded pipe [DES ENG] A steel pipe made of long strips of steel plate fitted together to form helical seams, which are welded.

spirit level *See* level.

spirit stain [MATER] A dye dissolved in methylated spirits; used to stain wood surfaces.

spirit thermometer [ENG] A temperature-measurement device consisting of a closed capillary tube with a liquid (for example, alcohol) reservoir bulb at the bottom; as the bulb is heated, the liquid expands up into the capillary tubing, indicating the temperature of the bulb.

spirit varnish [MATER] An artificial varnish consisting of resin, asphalt, or a cellulose ester dissolved in a volatile solvent.

Spiroid gear [DES ENG] A trade name of the Illinois Tool Works, it resembles a hypoid-type bevel gear but performs like a worm mesh; used to connect skew shafts.

spit [ENG] To light a fuse.

spitted fuse [ENG] A slow-burning fuse which has been cut open at the lighting end for ease of ignition.

spitting rock [ENG] A rock mass under stress that breaks and ejects small fragments with considerable velocity.

splashdown [AERO ENG] **1.** The landing of a spacecraft or missile on water. **2.** The moment of impact of a spacecraft on water.

splash lubrication [ENG] An engine-lubrication system in which the connecting-rod bearings dip into troughs of oil, splashing the oil onto the cylinder and piston rods.

splice [ENG] To unite two parts, such as rope or wire, to form a continuous length.

splice plate [CIV ENG] A plate for joining the web plates or the flanges of girders.

splicing tape [MATER] A pressure-sensitive nonmagnetic tape used for splicing magnetic tape and motion picture film; it has a hard adhesive that will not ooze and gum up the equipment or cause adjacent layers of tape or film on the reel to stick together.

spline [DES ENG] One of a number of equally spaced keys cut integral with a shaft, or similarly, keyways in a hubbed part; the mated pair permits the transmission of rotation or translatory motion along the axis of the shaft. [ENG] A strip of wood, metal, or plastic.

spline broach [MECH ENG] A broach for cutting straight-sided splines, or multiple keyways in holes.

splined shaft [DES ENG] A shaft with longitudinal gearlike ridges along its interior or exterior surface.

split [MIN ENG] **1.** To divide the air current into separate circuits to ventilate more than one section of the mine. **2.** Any division or branch of the ventilating current.

split-altitude profile [AERO ENG] Flight profile at two separate altitudes.

split barrel [DES ENG] A core barrel that is split lengthwise so that it can be taken apart and the sample removed.

split-barrel sampler [DES ENG] A drive-type soil sampler with a split barrel.

split bearing [DES ENG] A shaft bearing composed of two pieces bolted together.

split cavity [ENG] A cavity, such as in a mold, made in sections.

split die [MET] **1.** A screw-thread die made in one piece with a longitudinal slit connecting the outside to the central hole which allows size adjustment. **2.** *See* segment die.

split flap [AERO ENG] A hinged plate forming the rear upper or lower portion of an airfoil; the lower portion may be deflected downward to give increased lift and drag; the upper portion may be raised over a portion of the wing for the purpose of lateral control.

split link [DES ENG] A metal link in the shape of a two-turn helix pressed together.

splitnut [ENG] A nut cut axially into halves to allow for rapid engagement (closed) or disengagement (open).

split pin [DES ENG] A pin with a split at one end so that it can spread to hold it in place.

split-ring core lifter [DES ENG] A hardened steel ring having an open slit, an outside taper, and an inside or outside serrated surface; in its expanded state it allows the core to pass through it freely, but when the drill string is lifted, the outside taper surface slides downward into the bevel of the bit or reaming shell, causing the ring to contract and grip tightly the core which it surrounds. Also known as core catcher; core gripper; core lifter; ring lifter; split-ring lifter; spring lifter.

split-ring lifter *See* split-ring core lifter.

split-ring mold [ENG] A plastics mold in which a split-cavity block is assembled in a chase to permit the forming of undercuts in a molded piece.

split-ring piston packing [MECH ENG] A metal ring mounted on a piston to prevent leakage along the cylinder wall.

split shovel [DES ENG] A shovel containing parallel troughs separated by slots; used for sampling ground ore.

splitter vanes [ENG] A group of curved, parallel vanes located in a sharp (for example, miter) bend of a gas conduit; the vane shape and its location help guide the moving gas around the bend.

splitting [MIN ENG] **1.** Lamina of mica with a maximum thickness of 0.0012 inch (30 micrometers), split from blocks and thins. **2.** One of a pair of horizontal level headings driven through a pillar, in pillar workings, in order to mine the pillar coal.

split transducer [ENG] A directional transducer with electroacoustic transducing elements which are divided and arranged so that there is an electrical separation of each division.

SP logging *See* spontaneous-potential well logging.

spoil [MIN ENG] **1.** The overburden or non-ore material from a coal mine. **2.** A stratum of coal and dirt mixed.

spoil bank [MIN ENG] **1.** In surface mining, the accumulation of overburden. **2.** The place where spoil is deposited. Also known as spoil heap.

spoil dam [MIN ENG] An earthen dike forming a depression, in which returns from a borehole can be collected and retained.

spoiler [AERO ENG] A plate, series of plates, comb, tube, bar, or other device that projects into the airstream about a body to break up or spoil the smoothness of the flow, especially such a device that projects from the upper surface of an airfoil, giving an increased drag and a decreased lift.

spoil heap *See* spoil bank.

spoke [DES ENG] A bar or rod radiating from the center of a wheel.

spokeshave [ENG] A small tool for planing convex or concave surfaces.

sponge gold *See* cake of gold.

sponge grease [MATER] Fibrous, spongy, soda-base grease.

sponge iron [MET] Iron in porous or powder form made without fusion by heating iron ore in a reducing gas or with charcoal.

sponge metal [MET] Any porous metal made by decomposition or reduction of a compound without melting.

sponge rubber *See* rubber sponge.

spontaneous-potential well logging [ENG] The recording of the natural electrochemical and electrokinetic potential between two electrodes, one above the other, lowered into a drill hole; used to detect permeable beds and their boundaries. Also known as SP logging.

spool [MECH ENG] **1.** The drum of a hoist. **2.** The movable part of a slide-type hydraulic valve.

spool-type roller conveyor [MECH ENG] A type of roller conveyor in which the rolls are of conical or tapered shape with the diameter at the ends of the roll larger than that at the center.

spoon [DES ENG] A slender rod with a cup-shaped projection at right angles to the rod, used for scraping drillings out of a borehole. [MIN ENG] An instrument in which earth or pulp may be delicately tested by washing to detect gold or amalgam.

spot check [IND ENG] A check or inspection of certain steps in an operation, process, or the like, of certain parts of a piece of equipment or of a representative lot of completed parts or articles; the steps or parts inspected would normally be only a small percentage of the total.

spot drilling [MECH ENG] Drilling a small hole or indentation in the surface of a material to serve as a centering guide in later machining operations.

spot facing [MECH ENG] A finished circular surface around the top of a hole to seat a bolthead or washer, or to allow flush mounting of mating parts.

spot gluing [ENG] Applying heat to a glued assembly by dielectric heating to make the glue set in spots that are more or less regularly distributed.

spot hover [AERO ENG] To remain stationary relative to a point on the ground while airborne.

spotting [ENG] Fitting one part of a die to another part by applying an oil color to the surface of the finished part and bringing this against the surface of the intended mating part, the high spots being marked by the transferred color. [MIN ENG] Bringing mine cars or surface wagons to the correct spot for loading, discharging, or any other purpose.

spotting hoist [MIN ENG] A small haulage engine used for bringing mine cars into the correct position under a loading chute or feeder or some other point.

spotty ore [MIN ENG] Ore in which the valuable material is concentrated irregularly as small particles.

spot welding [MET] Resistance welding in which fusion is localized in small circular areas; sometimes also accomplished by various arc-welding processes.

spouting [ENG] A term used in the feeding or ejection of powdered or granulated solids by means of vertical or slanted discharge spouts.

sprag [ENG] A stake used as a brake for a vehicle by inserting it through the spokes of a wheel or digging it into the ground at an angle. [MIN ENG] A prop supporting the roof or ore in a mine.

spragger [MIN ENG] In coal mining, a laborer who rides trains of cars and controls their free movement down gently sloping inclines by throwing switches and by poking sprags between the wheel spokes to stop them.

sprag road [MIN ENG] A road so steep that sprags must be used on the wheels of ore cars during descent.

spray [ENG] A mechanically produced dispersion of liquid into a gas stream; as drops are large, the spray is unstable and the liquid will fall free of the gas stream when velocity decreases.

spray chamber [MECH ENG] A compartment in an air conditioner where humidification is conducted.

spray dryer [MECH ENG] A machine for drying an atomized mist by direct contact with hot gases.

sprayed metal mold [ENG] A plastics mold made by spraying molten metal onto a master form until a shell of predetermined thickness is achieved; the shell is then removed and backed up with plaster, cement, or casting resin; used primarily in plastic sheet forming.

sprayer plate [ENG] A rotating flat-faced or dished metal plate used in an oil burner to enhance atomization.

spray gun [MECH ENG] An apparatus shaped like a gun which delivers an atomized mist of liquid.

spray nozzle [MECH ENG] A device in which a liquid is subdivided to form a stream (mist) of small drops.

spray oil [MATER] A low-viscosity petroleum oil similar to lubricating oil; used to combat pests that attack trees and shrubbery.

spray painting [ENG] Applying a fine, even coat of paint by means of a spray nozzle.

spray pond [ENG] An arrangement for cooling large quantities of water in open reservoirs or ponds; nozzles spray a portion of the water into the air for the evaporative cooling effect.

spray probe [ENG] A device which detects a jet spray of tracer gas in vacuum testing for leaks.

spray quenching [MET] Rapid cooling in a spray of water or oil.

spray torch [ENG] In thermal spraying, a device used for the application of self-fluxing alloys; molten metal is propelled against the substrate by a stream of air and gas.

spray transfer [MET] In arc welding, transfer of filler metal across the arc to the workpiece in the form of droplets.

spray-up [ENG] A term for a number of techniques in which a spray gun is used as the processing tool; for example, in reinforced plastics manufacture, fibrous glass and resin can simultaneously be spray-deposited into a mold or onto a form.

spread [ENG] **1.** The layout of geophone groups from which data from a single shot are recorded simultaneously. **2.** *See* sensitivity.

spreader [CIV ENG] A wood or steel member inserted temporarily between form walls to keep them apart. [MECH ENG] **1.** A tool used in sharpening machine drill bits. **2.** A machine which spreads dumped material with its blades. [MIN ENG] **1.** A horizontal timber below the cap of a set, used to stiffen the legs, and to support the brattice when there are two air courses in the same gangway. **2.** A piece of timber stretched across a shaft as a temporary support of the walls.

spreader beam [ENG] A rigid beam hanging from a crane hook and fitted with a number of ropes at different points along its length; employed for such purposes as lifting reinforced concrete piles or large sheets of glass.

spreader stoker [MECH ENG] A coal-burning system where mechanical feeders and distributing devices form a thin fuel bed on a traveling grate, intermittent-cleaning dump grate, or reciprocating continuous-cleaning grate.

spread footing [CIV ENG] A wide, shallow footing usually made of reinforced concrete.

Sprengel pump [MECH ENG] An air pump that exhausts by trapping gases between drops of mercury in a tube.

spring [ENG] To enlarge the bottom of a drill hole by small charges of a high explosive in order to make room for the full charge; to chamber a drill hole. [MECH ENG] An elastic, stressed, stored-energy machine element that, when released, will recover its basic form or position. Also known as mechanical spring.

springback [MET] **1.** Return of a metal part to its original shape after release of stress. **2.** The degree to which a metal returns to its original shape after forming operations. **3.** In flash, upset or pressure welding, the deflection in the welding machine caused by the upset pressure.

spring balance [ENG] An instrument which measures force by determining the extension of a helical spring.

spring bolt [DES ENG] A bolt which must be retracted by pressure and which is shot into place by a spring when the pressure is released.

spring box mold [ENG] A compression mold with a spacing fork that is removed after partial compression.

spring brass [MET] Common brass containing 70–72% copper which has been cold-worked to make it stiffer.

spring calipers [ENG] Calipers in which tension against the adjusting nut is maintained by a circular spring.

spring clip [DES ENG] **1.** A U-shaped fastener used to attach a leaf spring to the axle of a vehicle. **2.** A clip that grips an inserted part under spring pressure; used for electrical connections.

spring collet [DES ENG] A bushing that surrounds and holds the end of the work in a machine tool; the bushing is slotted and tapered, and when the collet is slipped over it, the slot tends to close and the bushing thereby grips the work.

spring cotter [DES ENG] A cotter made of an elastic metal that has been bent double to form a split pin.

spring coupling [MECH ENG] A flexible coupling with resilient parts.

spring die [DES ENG] An adjustable die consisting of a hollow cylinder with internal cutting teeth, used for cutting screw threads.

spring faucet [ENG] A faucet that is kept closed by a spring; force must be exerted to open it, and it closes when the force is removed.

spring gravimeter [ENG] An instrument for making relative measurements of gravity; the elongation s of the spring may be considered proportional to gravity g, $s = (1/k)g$, and the basic formula for relative measurements is $g_2 - g_1 = k(s_2 - s_1)$.

spring hammer [MECH ENG] A machine-driven hammer actuated by a compressed spring or by compressed air.

spring hook [DES ENG] A hook closed at the end by a spring snap. Also known as snap hook.

spring-joint caliper [DES ENG] An outside or inside caliper having a heavy spring joining the legs together at the top; legs are opened and closed by a knurled nut.

spring lifter *See* split-ring core lifter.

spring-lifter case *See* lifter case.

spring-loaded meter [ENG] A variable-area flowmeter in which the force on an obstruction in a tapered tube created by the fluid flowing past the obstruction is balanced by the force of a spring to which the obstruction is attached, and the resulting differential pressure is used to determine the flow rate.

spring-loaded regulator [MECH ENG] A pressure-regulator valve for pressure vessels or flow systems; the regulator is preloaded by a calibrated spring to open (or close) at the upper (or lower) limit of a preset pressure range.

spring pin [MECH ENG] An iron rod which is mounted between spring and axle on a locomotive, and which maintains a regulated pressure on the axle.

spring scale [ENG] A scale that utilizes the deflection of a spring to measure the load.

spring shackle [ENG] A shackle for supporting the end of a spring, permitting the spring to vary in length as it deflects.

spring steel [MET] Carbon or low-alloy steel which can be processed to give it the hardness and yield strength needed in springs.

spring stop-nut locking fastener [DES ENG] A locking fastener that functions by a spring action clamping down on the bolt.

spring switch [CIV ENG] A railroad switch that contains a spring to return it to the running position after it has been thrown over by trailing wheels moving on the diverging route.

spring temper [MET] **1.** A steel temper characterized by an increased upper limit of elasticity; obtained by hardening and tempering in the usual way, then reheating until the steel turns blue. **2.** A similar temper in brass obtained by cold rolling.

sprinkler system [ENG] A fire-protection system of pipes and outlets in a building, mine, or other enclosure for delivering a fire extinguishing liquid or gas, usually automatically by the action of heat on the sprinkler head. Also known as fire sprinkling system.

sprocket [DES ENG] A tooth on the periphery of a wheel or cylinder to engage in the links of a chain, the perforations of a motion picture film, or other similar device.

sprocket chain [MECH ENG] A continuous chain which meshes with the teeth of a sprocket and thus can transmit mechanical power from one sprocket to another.

sprocket hole [ENG] One of a series of perforations at the edge of a motion picture film, paper tape, or roll of continuous stationery, which are engaged by the teeth of a sprocket wheel to drive the material through some device.

sprocket wheel [DES ENG] A wheel with teeth or cogs, used for a chain drive or to engage the blocks on a cable.

spruce oil [MATER] A combustible, colorless to yellow essential oil, soluble in fixed and mineral oils; derived from spruce needles and branches; the main components are bornyl acetate, cadinene, and pinene; used in flavors and veterinary liniments, and as an odorant for soaps and cosmetics. Also known as hemlock oil.

spruce sulfite extract [MATER] A paper-manufacture byproduct from the sulfite-pulping process; used as a foundry core binder and road binder, and in tanning.

sprue [ENG] **1.** A feed opening or vertical channel through which molten material, such as metal or plastic, is poured in an injection or transfer mold. **2.** A slug of material that solidifies in the channel.

sprue bushing [ENG] A steel insert in an injection mold which contains the sprue hole and has a seat for the injection cylinder nozzle.

sprue gate [ENG] A passageway for the flow of molten resin from the nozzle to the mold cavity.

sprue puller [ENG] A pin with a Z-shaped slot to pull the sprue out of the sprue bushing in an injection mold.

sprung axle [MECH ENG] A supporting member for carrying the rear wheels of an automobile.

sprung weight [MECH ENG] The weight of a vehicle which is carried by the springs, including the frame, radiator, engine, clutch, transmission, body, load, and so forth.

spud [DES ENG] **1.** A diamond-point drill bit. **2.** An offset type of fishing tool used to clear a space around tools stuck in a borehole. **3.** Any of various spade- or chisel-shaped tools or mechanical devices. **4.** *See* grouser. [MIN ENG] A nail, resembling a horseshoe nail, with a hole in the head, driven into mine timbering or into a wooden plug inserted in the rock to mark a surveying station.

spudded-in [MIN ENG] A borehole that has been started and has reached bedrock or in which the standpipe has been set.

spun glass [MATER] Blown glass made of fine threads of glass.

spur dike *See* groin.

spur gear [DES ENG] A toothed wheel with radial teeth parallel to the axis.

spur pile *See* batter pile.

Sputnik program [AERO ENG] A series of Soviet earth-orbiting satellites; Sputnik I, launched on October 4, 1957, was the first artificial satellite.

sputter-ion pump *See* getter-ion pump.

square-edged orifice [ENG] An orifice plate with straight-through edges for the hole through which fluid flows; used to measure fluid flow in fluid conduits by means of differential pressure drop across the orifice.

square engine [MECH ENG] An engine in which the stroke is equal to the cylinder bore.

square groove weld [MET] A groove weld in which the joint edges are square.

square-head bolt [DES ENG] A cylindrical threaded fastener with a square head.

square-jaw clutch [MECH ENG] A type of positive clutch consisting of two or more jaws of square section which mesh together when they are aligned.

square key [DES ENG] A machine key of square, usually uniform, but sometimes tapered, cross section.

square mesh [DES ENG] A wire-cloth textile mesh count that is the same in both directions.

square-nose bit *See* flat-face bit.

square set [MIN ENG] A set of timbers composed of a cap, girt, and post which meet so as to form a solid 90° angle; they are so framed at the intersection as to form a compression joint, and join with three other similar sets.

square-set block caving [MIN ENG] Block caving in which the ore is extracted through drifts supported by square sets.

square-set stopes [MIN ENG] The use of square-set timbering to support the ground as ore is extracted.

square thread [DES ENG] A screw thread having a square cross section; the width of the thread is equal to the pitch or distance between threads.

square wheel [DES ENG] A wheel with a flat spot on its rim.

squaring shear [MECH ENG] A machine tool consisting of one fixed cutting blade and another mounted on a reciprocating crosshead; used for cutting sheet metal or plate.

squeegee [DES ENG] A device consisting of a handle with a blade of rubber or leather set transversely at one end and used for spreading, pushing, or wiping liquids off or across a surface.

squeeze [ENG] **1.** To inject a grout into a borehole under high pressure. **2.** The plastic movement of a soft rock in the walls of a borehole or mine working that reduced the diameter of the opening. [MIN ENG] **1.** The settling, without breaking, of the roof over a considerable area of working. Also known as creep; nip; pinch. **2.** The gradual upheaval of the floor of a mine due to the weight of the overlying strata. **3.** The sections in coal seams that have become constricted by the squeezing in of the overlying or underlying rock.

squeeze roll [MECH ENG] A roller designed to exert pressure on material passing between it and a similar roller.

squeeze time [MET] In resistance welding, the time between the initial application of the current and of the pressure.

squib [ENG] A small tube filled with fine-grained black powder; upon the lighting and burning of the ignition match, the squib assumes a rocket effect and darts back into the hole to ignite the powder charge.

squirt can [ENG] An oil can with a flexible bottom and a tapered spout; pressure applied to the bottom forces oil out the spout.

squirt gun [ENG] A device with a bulb and nozzle; when the bulb is pressed, liquid squirts from the nozzle.

SRA-size [ENG] One of a series of sizes to which untrimmed paper is manufactured; for reels of paper the standard sizes are 450, 640, 900, and 1280 millimeters; for sheets of paper the sizes are SRA0, 900 × 1280 mm; SRA1, 640 × 900 mm; and SRA2, 450 × 640 mm; SRA sizes correspond to A sizes when trimmed.

SR cylinder oil [MATER] A viscous, unfiltered lubrication oil, generally made from reduced petroleum crudes that have had the lighter lubricant fractions removed by direct steam heating; used to lubricate steam engine cylinders and valves. Also known as steam-refined cylinder oil.

SSP *See* static spontaneous potential.

SST *See* supersonic transport.

stab [ENG] In a drilling operation, to insert the threaded end of a pipe joint into the collar of the joint already placed in the hole and to rotate it slowly to engage the threads before screwing up.

stabilator [AERO ENG] A one-piece horizontal tail that is swept back and movable; movement is controlled by motion of the pilot's control stick; usually used in supersonic aircraft.

stability [ENG] The property of a body, as an aircraft, rocket, or ship, to maintain its attitude or to resist displacement, and, if displaced, to develop forces and moments tending to restore the original condition. [MATER] Of a fuel, the capability to retain its characteristics in an adverse environment, for example, extreme temperature.

stability augmentation system [AERO ENG] Automatic control devices which supplement a pilot's manipulation of the aircraft controls and are used to modify inherent aircraft handling qualities. Abbreviated SAS. Also known as stability augmentors.

stability augmentors See stability augmentation system.

stability test [ENG] Accelerated test to determine the probable suitability of an explosive material for long-term storage.

stabilization [ENG] Maintenance of a desired orientation independent of the roll and pitch of a ship or aircraft.

stabilized flight [AERO ENG] Maintenance of desired orientation in flight.

stabilized gasoline [MATER] Gasoline from which the "wild" or low-boiling (high vapor pressure, volatile) hydrocarbons have been removed by stabilization.

stabilized platform See stable platform.

stabilizer [AERO ENG] Any airfoil or any combination of airfoils considered as a single unit, the primary function of which is to give stability to an aircraft or missile. [ENG] **1.** A hardened, splined bushing, sometimes freely rotating, slightly larger than the outer diameter of a core barrel and mounted directly above the core barrel back head. Also known as ferrule; fluted coupling. **2.** A tool located near the bit in the drilling assembly to modify the deviation angle in a well by controlling the location of the contact point between the hole and the drill collars. [MATER] **1.** Any powdered or liquid additive used as an agent in soil stabilization. **2.** An ingredient used in the formulation of some compounded plastics to maintain the physical and chemical properties at their initial values throughout the processing and service life of the material, for example, heat and RV stabilizers.

stabilizing treatment [MET] Any of various treatments intended to promote dimensional stability of a metal part or stabilize the structure of an alloy.

stable-base film [MATER] A particular type of film having high stability in regard to shrinkage and stretching.

stable element [ENG] Any instrument or device, such as a gyroscope, used to stabilize a radar antenna, turret, or other piece of equipment mounted on an aircraft or ship.

stable platform [AERO ENG] A gimbal-mounted platform, usually containing gyros and accelerometers, the purpose of which is to maintain a desired orientation in inertial space independent of craft motion. Also known as stabilized platform.

stable vertical [ENG] Vertical alignment of any device or instrument maintained during motion of the mount.

stack [BUILD] The portion of a chimney rising above the roof. [ENG] **1.** To stand and rack drill rods in a drill tripod or derrick. **2.** Any structure or part thereof that contains a flue or flues for the discharge of gases. **3.** One or more filter cartridges mounted on a single column. **4.** Tall, vertical conduit (such as smokestack, flue) for venting of combustion or evaporation products or gaseous process wastes. **5.** The exhaust pipe of an internal combustion engine. [MET] The cone-shaped section of a blast furnace or cupola above the hearth and melting zone and extending to the throat.

stack cutting [MET] Cutting a stack of metal plates with a single cut using a stream of oxygen.

stacked-beam radar [ENG] Three-dimensional radar system that derives elevation by emitting narrow beams stacked vertically to cover a vertical segment, azimuth information from horizontal scanning of the beam, and range information from echo-return time.

stack effect [MECH ENG] The pressure difference between the confined hot gas in a chimney or stack and the cool outside air surrounding the outlet.

stacker [MECH ENG] A machine for lifting merchandise on a platform or fork and arranging it in tiers; operated by hand, or electric or hydraulic mechanisms.

stacker-reclaimer [MECH ENG] Equipment which transports and builds up material stockpiles, and recovers and transports material to processing plants.

stack gas [ENG] Gas passed through a chimney.

stacking [AERO ENG] The holding pattern of aircraft awaiting their turn to approach and land at an airport.

stack pollutants [ENG] Smokestack emissions subject to Environmental Protection Agency standards regulations, including sulfur oxides, particulates, nitrogen oxides, hydrocarbons, carbon monoxide, and photochemical oxidants.

stack welding [MET] Simultaneous spot welding of stacked plates.

stactometer See stalagmometer.

STADAN See Space Tracking and Data Acquisition Network.

stadia [ENG] A surveying instrument consisting of a telescope with special horizontal parallel lines or wires, used in connection with a vertical graduated rod.

stadia hairs [ENG] Two horizontal lines in the reticule of a theodolite arranged symmetrically above and below the line of sight. Also known as stadia wires.

stadia rod [ENG] A graduated rod used with a stadia to measure the distance from the observation point to the rod by observation of the length of rod subtended by the distance between the stadia hairs.

stadia tables [ENG] Mathematical tables from which may be found, without computation, the horizontal and vertical components of a reading made with a transit and stadia rod.

stadia wires *See* stadia hairs.

stadimeter [ENG] An instrument for determining the distance to an object, but its height must be known; the angle, subtended by the object's bottom and top as measured at the observer's position, is proportional to the object's height; the instrument is graduated directly in distance.

staff bead [BUILD] **1.** A bead between a wooden frame and adjacent masonry. **2.** A molded or beaded angle of wood or metal set into the corner of plaster walls.

staff gage [ENG] A graduated scale placed in a position so that the stage of a stream may be read directly therefrom; a type of river gage.

stage [AERO ENG] A self-propelled separable element of a rocket vehicle or spacecraft. [MIN ENG] **1.** A certain length of underground roadway worked by one horse. **2.** A narrow thin dike, especially one where the material of which the dike is composed is soft. **3.** A platform on which mine cars stand.

stage acidizing [PETRO ENG] An oil-reservoir acid treatment (acidizing) in which the formation is treated with two or more separate stages of acid, instead of a single large treatment.

staged crew [AERO ENG] An aircrew specifically positioned at an intermediate airfield to take over aircraft operating on an air route, thus relieving a complementary crew of flying fatigue and speeding up the traffic rate of the aircraft concerned.

stage loader *See* feeder conveyor.

stage separation [PETRO ENG] A system for gas-oil separation of well fluids by a series of stages instead of a single operation.

staggered blastholes [MIN ENG] Two rows of holes staggered to a triangular pattern to distribute the burden when shot-firing in thick coal seams.

staggered-intermittent fillet welding [MET] Making a line of intermittent fillet welds on each side of a joint in a manner such that the increments on one side are not opposite to those on the other side.

staggered-line drive [PETRO ENG] In the placement and drilling of water-injection wells for water flood recovery of oil, the staggered (versus in-line) areal arrangement of the injection wells.

stagger-tooth cutter [MECH ENG] Side-milling cutter with successive teeth having alternating helix angles.

staging [AERO ENG] The process or operation during the flight of a rocket vehicle whereby a full stage or half stage is disengaged from the remaining body and made free to decelerate or be propelled along its own flightpath.

stain [MATER] **1.** A nonprotective coloring matter used on wood surfaces; imparts color without obscuring the wood grains. **2.** Any colored, organic compound used to stain tissues, cells, cell components, cell contents, or other biological substrates for microscopic examination.

stained glass [ENG] Glass colored by any of several means and assembled to produce a varicolored mosaic or representation.

stainless alloy [MET] Any of a large and complex group of corrosion-resistant iron-chromium alloys (containing 10% or more chromium), sometimes containing other elements, such as nickel, silicon, molybdenum, tungsten, and niobium. Commonly known as stainless steel (SS).

stainless-clad steel [MET] Steel clad on one or two sides with a stainless steel to provide a surface that is corrosion-resistant and attractive.

stainless iron *See* ferritic stainless steel.

stainless steel *See* stainless alloy.

stair [CIV ENG] A series of steps between levels or from floor to floor in a building.

stairway [CIV ENG] One or more flights of stairs connected by landings.

stake [ENG] **1.** To fasten back or prop open with a piece of chain or otherwise the valves or clacks of a water barrel in order that the water may run back into the sump when necessary. **2.** A pointed piece of wood driven into the ground to mark a boundary, survey station, or elevation.

staking [ENG] Joining two parts together by fitting a projection on one part against a mating feature in the other part and then causing plastic flow at the joint.

staking out [ENG] Driving stakes into the earth to indicate the foundation location of a structure to be built.

stalagmometer [ENG] An instrument for measuring the size of drops suspended from a capillary tube, used in the drop-weight method. Also known as stactometer.

stall [AERO ENG] **1.** The action or behavior of an airplane (or one of its airfoils) when by the separation of the airflow, as in the case of insufficient airspeed or of an excessive angle of attack, the airplane or airfoil tends to drop; the condition existing during this behavior. **2.** A flight performance in which an airplane is made to lose flying speed and to drop by pointing the nose steeply upward. **3.** An act or instance of stalling.

stall flutter [AERO ENG] A type of dynamic instability that takes place when the separation of flow around an airfoil occurs during the whole or part of each cycle of a flutter motion.

stalling angle of attack *See* critical angle of attack.

stalling Mach number [AERO ENG] The Mach number of an aircraft when the coefficient of lift of the aerodynamic surfaces is the maximum obtainable for the pressure altitude, true airspeed, and angle of attack under which the craft is operated.

stall warning indicator [AERO ENG] A device that determines the critical angle of attack for a given aircraft; usually operates from vane sensors, airflow sensors, tabs on leading edges of wings, and computing devices such as accelerometers or airspeed indicators.

stalogometer [ENG] A device for measuring surface tension; a drop is suspended from a tube of known radius, and the radius of the drop is measured at the instant the drop detaches itself from the tube.

stamp battery [MIN ENG] A machine for crushing very strong ores or rocks; consists essentially of a crushing member (gravity stamp) which is dropped on a die, the ore being crushed in water between the shoe and the die.

stamp copper [MIN ENG] A copper-bearing rock that is stamped and washed before it is smelted.

stamper [ENG ACOUS] A negative, generally made of metal by electroforming, used for molding phonograph records.

stamp head [MIN ENG] A heavy and nearly cylindrical cast-iron head fixed on the lower end of the stamp rod, shank, or lifter to give weight in stamping the ore.

stamping [MECH ENG] Almost any press operation including blanking, shearing, hot or cold forming, drawing, bending, and coining. [MIN ENG] Reducing to the desired fineness in a stamp mill; the grain is usually not so fine as that produced by grinding in pans.

stamping mill [MIN ENG] A machine in which ore is finely crushed by descending pestles (stamps), usually operated by hydraulic power. Also known as crushing mill.

stand-alone fuel [MATER] A fuel that is burned directly rather than being blended with another fuel.

standard depth-pressure recorder [PETRO ENG] A device for the measurement of pressures at the bottom of a well bore (that is, bottom-hole pressure); a spring-restrained piston moves a recording stylus on a pressure-sealed chart.

standard elemental time [IND ENG] A standard time for individual work elements.

standard fit [DES ENG] A fit whose allowance and tolerance are standardized.

standard gage [CIV ENG] A railroad gage measuring 4 feet 8½ inches (1.4351 meters). [DES ENG] A highly accurate gage used only as a standard for working gages.

standard gold [MET] A gold alloy containing 10% copper; at one time used for legal coinage in the United States.

standard hole [DES ENG] A hole with zero allowance plus a specified tolerance; fit allowance is provided for by the shaft in the hole.

standard hour [IND ENG] The quantity of output required of an operator to meet an hourly production quota. Also known as allowed hour.

standard hour plan [IND ENG] A wage incentive plan in which standard work times are expressed as standard hours and the worker is paid for standard hours instead of the actual work hours.

standardization [DES ENG] The adoption of generally accepted uniform procedures, dimensions, materials, or parts that directly affect the design of a product or a facility. [ENG] The process of establishing by common agreement engineering criteria, terms, principles, practices, materials, items, processes, and equipment parts and components.

standardized product [DES ENG] A product that conforms to specifications resulting from the same technical requirements.

standard leak [ENG] Tracer gas allowed to enter a leak detector at a controlled rate in order to facilitate calibration and adjustment of the detector.

standard load [DES ENG] A load which has been preplanned as to dimensions, weight, and balance, and designated by a number or some classification.

standard output [IND ENG] The reciprocal of standard time.

standard performance [IND ENG] The performance of an individual or of a group on meeting standard output.

standard rate turn [AERO ENG] A turn in an aircraft in which heading changes at the rate of 3° per second.

standard shaft [DES ENG] A shaft with zero allowance minus a specified tolerance.

standard time [IND ENG] A unit time value for completion of a work task as determined by the proper application of the appropriate work-measurement techniques. Also known as direct labor standard; output standard; production standard; time standard.

standard ton *See* ton.

standard white kerosine *See* export kerosine.

standard wire rope [DES ENG] Wire rope made of six wire strands laid around a sisal core. Also known as hemp-core cable.

standing derrick *See* gin pole.

standing ground [MIN ENG] Ground that will stand firm without timbering.

standing operating procedure [AERO ENG] A set of instructions covering those features of operations which lend themselves to a definite or standardized procedure without loss of effectiveness; the procedure is applicable unless prescribed otherwise in a particular case; thus, the flexibility necessary in special situations is retained.

standing valve [PETRO ENG] A sucker-rod-pump (oil well) discharge valve that remains stationary during the pumping cycle, in contrast to a traveling valve.

standpipe [ENG] A vertical tube filled with a material, for example, water, or in petroleum refinery catalytic cracking, a catalyst to serve as a seal between high- and low-pressure parts of the process equipment.

staple [DES ENG] A U-shaped loop of wire with points at both ends; used as a fastener.

stapler [ENG] **1.** A device for inserting wire staples into paper or wood. **2.** A hammer for inserting staples.

star drill [DES ENG] A tool with a star-shaped point, used for drilling in stone or masonry.

Starfighter [AERO ENG] A United States supersonic, single-engine, turbojet-powered, tactical and air superiority fighter; the tactical version employs cannons or nuclear weapons for attack against surface targets, and is capable of providing close support for ground forces; the interceptor version employs Sidewinders or cannons.

star grain [MATER] A rocket propellant grain with a cavity of star-shaped cross section.

Starlifter [AERO ENG] A United States large cargo transport powered by four turbofan engines, capable of intercontinental range with heavy payloads and airdrops.

starter [ENG] A drill used for making the upper part of a hole, the remainder of the hole being made with a drill of smaller gage, known as a follower.

starting barrel [ENG] A short (12 to 24 inches or 30 to 60 centimeters) core barrel used to begin coring operations when the distance between the drill chuck and the bottom of the hole or to the rock surface in which a borehole is to be collared is too short to permit use of a full 5- or 10-foot-long (1.5- or 3.0-meter-long) core barrel.

starting mix [MATER] In pyrotechnic devices, an easily ignited mixture which transmits flame from an initiating device to a less readily ignitable composition.

starting resistance [MECH ENG] The force needed to produce an oil film on the journal bearings of a train when it is at a standstill.

starting sheet [MET] A thin sheet of metal used as the initial cathode in electrowinning or electrorefining.

starting taper [DES ENG] A slight end taper on a reamer to aid in starting.

start-to-leak pressure [MECH ENG] The amount of inlet pressure at which the first bubble occurs at the outlet of a safety relief valve with a resilient disk when the valve is subjected to an air test under a water seal.

start-up curve [IND ENG] A learning curve applied to a job for the purpose of adjusting work times that are longer than the standard because of the introduction of new jobs or new workers.

starved joint [ENG] A glued joint containing insufficient or inadequate adhesive. Also known as hungry joint.

static firing [AERO ENG] The firing of a rocket engine in a hold-down position to measure thrust and to accomplish other tests.

static gearing ratio [AERO ENG] The ratio of the control-surface deflection in degrees to angular displacement of the missile which caused the deflection of the control surface.

static grizzly [MIN ENG] A grizzly in the form of a stationary bar screen which either allows suitable pieces of rock or ore to pass over and unwanted small sizes to drop through, or rejects oversize pieces while allowing suitable material to drop through.

static line [AERO ENG] A line attached to a parachute pack and to a strop or anchor cable in an aircraft so that when the load is dropped the parachute is deployed automatically.

static pressure tap *See* pressure tap.

static-pressure tube [ENG] A smooth tube with a rounded nose that has radial holes in the portion behind the nose and is used to measure the static pressure within the flow of a fluid.

static SP *See* static spontaneous potential.

static spontaneous potential [PETRO ENG] Theoretical maximum spontaneous potential current that can be measured in a down-hole, mud log in clean sand. Abbreviated SSP; static SP.

static test [AERO ENG] In particular, a test of a rocket or other device in a stationary or hold-down position, either to verify structural design criteria, structural integrity, and the effects of limit loads, or to measure the thrust of a rocket engine. [ENG] A measurement taken under conditions where neither the stimulus nor the environmental conditions fluctuate.

static tube [ENG] A device used to measure the static (not kinetic or total) pressure in a stream of fluid; consists of a perforated, tapered tube that is placed parallel to the flow, and has a branch tube that is connected to a manometer.

static work [IND ENG] Manual work performed with no significant motion.

station [ENG] Any predetermined point or area on the seas or oceans which is patrolled by naval vessels. [MIN ENG] **1.** An enlargement in a mining shaft or gallery on any level used for a landing at any desired place and also for receiving loaded mine cars that are to be sent to the surface. **2.** An opening into a level which heads out of the side of an inclined plane; the point at which a surveying instrument is planted or observations are made.

stationary cone classifier [MECH ENG] In a pulverizer directly feeding a coal furnace, a device which returns oversize coal to the pulverizing zone.

stationary engine [MECH ENG] A permanently placed engine, as in a power house, factory, or mine.

stationary orbit [AERO ENG] A circular, equatorial orbit in which the satellite revolves about the primary body at the angular rate at which the primary body rotates on its axis; from the primary body, the satellite thus appears to be stationary over a point on the primary body; a stationary orbit must be synchronous, but the reverse need not be true.

stationary satellite [AERO ENG] A satellite in a stationary orbit.

station pole [CIV ENG] One of various rods used in surveying to mark stations, to sight points and lines; or to measure elevation with respect to the transit.

station time [AERO ENG] Time at which crews, passengers, and cargo are to be on board air transport and ready for the flight.

statistical quality control [IND ENG] The use of statistical techniques as a means of controlling the quality of a product or process.

stator [MECH ENG] A stationary machine part in or about which a rotor turns.

statoscope [ENG] **1.** A barometer that records small variations in atmospheric pressure. **2.** An instrument that indicates small changes in an aircraft's altitude.

statuary bronze [MET] Special bronze alloys used for casting statues and other ornamental objects; a typical bronze for statuary work contains 90% copper, 6% tin, 3% zinc, and 1% lead.

stave [DES ENG] **1.** A rung of a ladder. **2.** Any of the narrow wooden strips or metal plates placed edge to edge to form the sides, top, or lining of a vessel or structure, such as a barrel.

stay [ENG] In a structure, a tensile member which holds other members of the structure rigidly in position.

staybolt [DES ENG] A bolt with a thread along the entire length of the shaft; used to attach machine parts that are under pressure to separate.

stayed-cable bridge [CIV ENG] A modified cantilever bridge consisting of girders or trusses cantilevered both ways from a central tower and supported by inclined cables attached to the tower at the top or sometimes at several levels.

Staypak [MATER] Wood made more dense by pressure and heat, with the introduction of added resin.

stay time [AERO ENG] In rocket engine usage, the average value of the time spent by each gas molecule or atom within the chamber volume.

STD recorder See salinity-temperature-depth recorder.

steadite [MET] A hard structural constituent of cast iron consisting of the eutectic of ferrite and iron phosphide (Fe_3P); composition of the eutectic is 10.2% phosphorus and 89.8% iron; melts at 1049°C (1920°F).

steady pin [ENG] **1.** A retaining device such as a dowel, pin, or key that prevents a pulley from turning on its axis. **2.** A guide pin used to lift a cope or pattern.

steady-state model [PETRO ENG] Electric or electrolytic analogs of a reservoir formation used to study the steady-state flow of fluids through porous media; includes gel, blotter, liquid, potentiometric, and similar models.

steam accumulator [MECH ENG] A pressure vessel in which water is heated by steam during off-peak demand periods and regenerated as steam when needed.

steam atomizing oil burner [ENG] A burner which has two supply lines, one for oil and the other for a jet of steam which assists in the atomization process.

steam attemperation [MECH ENG] The control of the maximum temperature of superheated steam by water injection or submerged cooling.

steam bending [ENG] Forming wooden members to a desired shape by pressure after first softening by heat and moisture.

steam boiler [MECH ENG] A pressurized system in which water is vaporized to steam by heat transferred from a source of higher temperature, usually the products of combustion from burning fuels. Also known as steam generator.

steam bronze [MET] A leaded tin-bronze containing 88% copper, 6% tin, 4.5% zinc, 1.5% lead; used for steam valve bodies, gears, and bearings.

steam calorimeter See throttling calorimeter.

steam cock [ENG] A valve for the passage of steam.

steam condenser [MECH ENG] A device to maintain vacuum conditions on the exhaust of a steam prime mover by transfer of heat to circulating water or air at the lowest ambient temperature.

steam cure [ENG] To cure concrete or mortar in water vapor at an elevated temperature, at either atmospheric or high pressure.

steam drive [MECH ENG] Any device which uses power generated by the pressure of expanding steam to move a machine or a machine part.

steam dryer [MECH ENG] A device for separating liquid from vapor in a steam supply system.

steam emulsion test [ENG] A test used for measuring the ability of oil and water to separate, especially for steam-turbine oil; after emulsification and separation, the time required for the emulsion to be reduced to 3 milliliters or less is recorded at 5-minute intervals.

steam engine [MECH ENG] A thermodynamic device for the conversion of heat in steam into work, generally in the form of a positive displacement, piston and cylinder mechanism.

steam engine indicator [ENG] An instrument that plots the steam pressure in an engine cylinder as a function of piston displacement.

steam gage [ENG] A device for measuring steam pressure.

steam-generating furnace See boiler furnace.

steam generator See steam boiler.

steam hammer [MECH ENG] A forging hammer in which the ram is raised, lowered, and operated by a steam cylinder.

steam-hammer oil See tempering oil.

steam-heated evaporator [MECH ENG] A structure using condensing steam as a heat source on one side of a heat-exchange surface to evaporate liquid from the other side.

steam heating [MECH ENG] A system that used steam as the medium for a comfort or process heating operation.

steam jacket [MECH ENG] A casing applied to the cylinders and heads of a steam engine, or other space, to keep the surfaces hot and dry.

steam jet [ENG] A blast of steam issuing from a nozzle. [MIN ENG] A system of ventilating a mine by means of a number of jets of steam at high pressure kept constantly blowing off from a series of pipes in the bottom of the upcast shaft.

steam-jet cycle [MECH ENG] A refrigeration cycle in which water is used as the refrigerant; high-velocity steam jets provide a high vacuum in the evaporator, causing the water to boil at low temperature and at the same time compressing the flashed vapor up to the condenser pressure level.

steam-jet ejector [MECH ENG] A fluid acceleration vacuum pump or compressor using the high velocity of a steam jet for entrainment.

steam locomotive [MECH ENG] A railway propulsion power plant using steam generally in a reciprocating, noncondensing engine.

steam loop [ENG] Two vertical pipes connected by a horizontal one, used to condense boiler steam so that it can be returned to the boiler without a pump or injector.

steam molding [ENG] The use of steam, either directly on the material or indirectly on the mold surfaces, as a heat source to mold parts from preexpanded polystyrene beads.

steam nozzle [MECH ENG] A streamlined flow structure in which heat energy of steam is converted to the kinetic form.

steam pump [MECH ENG] A pump driven by steam acting on the coupled piston rod and plunger.

steam-refined asphalt [MATER] Petroleum-derived asphalt that has been refined in the presence of steam during the distillation of crude oil.

steam-refined cylinder oil See SR cylinder oil.

steam-refined stock [MATER] Unfiltered petroleum products distilled with heat applied in the form of steam, for example, for gear oils, lubricating oils, and cylinder oils.

steam reheater [MECH ENG] A steam boiler component in which heat is added to intermediate-pressure steam, which has given up some of its energy in expansion through the high-pressure turbine.

steam rig [PETRO ENG] A drilling rig which is powered by a number of portable boilers.

steam roller [MECH ENG] A road roller driven by a steam engine.

steam separator [MECH ENG] A device for separating a mixture of the liquid and vapor phases of water. Also known as steam purifier.

steam shovel [MECH ENG] A power shovel operated by steam.

steam superheater [MECH ENG] A boiler component in which sensible heat is added to the steam after it has been evaporated from the liquid phase.

steam tracing [ENG] A steam-carrying heater (such as tubing or piping) next to or twisted around a process-fluid or instrument-air line; used to keep liquids from solidifying or condensing.

steam trap [MECH ENG] A device which drains and removes condensate automatically from steam lines.

steam-tube dryer [MECH ENG] Rotary dryer with steam-heated tubes running the full length of the cylinder and rotating with the dryer shell.

steam turbine [MECH ENG] A prime mover for the conversion of heat energy of steam into work on a revolving shaft, utilizing fluid acceleration principles in jet and vane machinery.

steam valve [ENG] A valve used to regulate the flow of steam.

steam washer [ENG] A device for removing contaminants, such as silica, from the steam produced in a boiler.

Steckel rolling [MET] Cold metal rolling in which the strip is pulled through idler rolls by front tension only; the direction is reversed until the desired thickness is attained.

STED *See* solar turboelectric drive.

steel [MET] An iron base alloy, malleable under proper conditions, containing up to about 2% carbon.

steel bronze [MET] A hardened bronze consisting of 92% copper and 8% tin; used as a substitute for steel in guns.

steel-cable conveyor belt [DES ENG] A rubber conveyor belt in which the carcass is composed of a single plane of steel cables.

steel-clad rope [DES ENG] A wire rope made from flat strips of steel wound helically around each of the six strands composing the rope.

steel converter [MET] A retort in which cast iron is converted to steel; an example is the Bessemer converter.

steel emery [MET] An abrasive material composed of chilled iron produced by forcing iron through a steam jet; used in tumbling barrels and for grinding stones.

Steelflex coupling [MECH ENG] A flexible coupling made with two grooved steel hubs keyed to their respective shafts and connected by a specially tempered alloy-steel member called the grid.

steel foil [MET] A very thin sheet of steel, the thickness of which is measured in thousandths of an inch.

steel jack [MIN ENG] A screw jack suitable in mechanical mining; used for legs or upright timbers.

steelmaking [MET] Any of various processes for making steel from pig iron.

steel sets [MIN ENG] Steel beam used in main entries of coal mines and in shafts of metal mines; I beams for caps and H beams for posts or wall plates.

steel tunnel support [MIN ENG] One of the tunnel support systems made of steel; five types are continuous rib; rib and post; rib and wall plate; rib, wall plate, and post; and full circle rib.

steel wool [MET] Fine steel threads matted into a mass.

steelyard [ENG] A weighing device with a counterbalanced arm supporting the load to be weighed on the short end.

steen [CIV ENG] To line an excavation such as a cellar or well with stone, cement, or similar material without the use of mortar.

steering arm [MECH ENG] An arm that transmits turning motion from the steering wheel of an automotive vehicle to the drag link.

steering brake [MECH ENG] Means of turning, stopping, or holding a tracked vehicle by braking the tracks individually.

steering gear [MECH ENG] The mechanism, including gear train and linkage, for the directional control of a vehicle or ship.

steering wheel [MECH ENG] A hand-operated wheel for controlling the direction of the wheels of an automotive vehicle or of the rudder of a ship.

stellite [MET] A hard, wear- and corrosion-resistant family of nonferrous alloys of cobalt (20–65%), chromium (11–32%), and tungsten (2–5%); resistance to softening is exceptionally high at high temperature.

stem [ENG] **1.** The heavy iron rod acting as the connecting link between the bit and the balance of the string of tools on a churn rod. **2.** To insert packing or tamping material in a shothole.

stem bag [MIN ENG] A fire-resisting paper bag filled with dry sand for stemming shotholes.

stemming rod [ENG] A nonmetallic rod used to push explosive cartridges into position in a shothole and to ram tight the stemming.

stem-winding [MECH ENG] Pertaining to a timepiece that is wound by an internal mechanism turned by an external knob and stem (the winding button of a watch).

stench *See* odorant.

stenometer [ENG] An instrument for measuring distances; employs a telescope in which two target images a known distance apart are superimposed by turning a micrometer screw.

step [ENG] A small offset on a piece of core or in a drill hole resulting from a sudden sidewise deviation of the bit a it enters a hard, tilted stratum or rock underlying a softer rock. [MIN ENG] The portion of a longwall face at right angles to the line of the face formed when a place is worked in front of or behind an adjoining place.

step aeration [CIV ENG] An activated sludge process in which the settled sewage is introduced into the aeration tank at more than one point.

step bearing [MECH ENG] A device which supports the bottom end of a vertical shaft. Also known as pivot bearing.

step block [ENG] A metal block, usually of steel or cast iron, with integral stepped sections to allow application of clamps when securing a workpiece to a machine tool table.

step brazing [MET] Brazing consecutive joints at sequentially lower temperatures to maintain the integrity of preceding joints.

step-climb profile [AERO ENG] The aircraft climbs a specified number of feet whenever its weight reaches a predetermined amount, thus stepping to an optimum altitude as gross weight decreases.

step gage [DES ENG] **1.** A plug gage containing several cylindrical gages of increasing diameter mounted on the same axis. **2.** A gage consisting of a body in which a blade slides perpendicularly; used to measure steps and shoulders.

step-out well [PETRO ENG] A well drilled at a later time over remote, undeveloped portions of a partially developed continuous reservoir rock. Also known as delayed development well.

stepped cone pulley [DES ENG] A one-piece pulley with several diameters to engage transmission belts and thereby provide different speed ratios.

stepped footing [CIV ENG] A widening at the bottom of a wall consisting of a series of steps in the proportion of one horizontal to two vertical units.

stepped gear wheel [DES ENG] A gear wheel containing two or more sets of teeth on the same rim, with adjacent sets slightly displaced to form a series of steps.

stepped screw [DES ENG] A screw from which sectors have been removed, the remaining screw surfaces forming steps.

step pulley [MECH ENG] A series of pulleys of various diameters combined in a single concentric unit and used to vary the velocity ratio of shafts. Also known as cone pulley.

step rocket *See* multistage rocket.

step soldering [MET] Soldering consecutive joints at sequentially lower temperatures to maintain the integrity of preceding joints.

stereo *See* stereophonic; stereo sound system.

stereo amplifier [ENG ACOUS] An audio-frequency amplifier having two or more channels, as required for use in a stereo sound system.

stereomicrometer [ENG] An instrument attached to an optical instrument (such as a telescope) to measure small angles.

stereophonic [ENG ACOUS] Pertaining to three-dimensional pickup or reproduction of sound, as achieved by using two or more separate audio channels. Also known as stereo.

stereophonics [ENG ACOUS] The study of reproducing or reinforcing sound in such a way as to produce the sensation that the sound is coming from sources whose spatial distribution is similar to that of the original sound sources.

stereophonic sound system *See* stereo sound system.

stereo pickup [ENG ACOUS] A phonograph pickup designed for use with standard single-groove two-channel stereo records; the pickup cartridge has a single stylus that actuates two elements, one responding to stylus motion at 45° to the right of vertical and the other responding to stylus motion at 45° to the left of vertical.

stereoplanigraph [ENG] An instrument for drawing topographic maps from observations of stereoscopic aerial photographs with a stereocomparator.

stereo preamplifier [ENG ACOUS] An audio-frequency preamplifier having two channels, used in a stereo sound system.

stereo record [ENG ACOUS] A single-groove disk record having V-shaped grooves at 45° to the vertical; each groove wall has one of the two recorded channels.

stereo recorded tape [ENG ACOUS] Recorded magnetic tape having two separate recordings, one for each channel of a stereo sound system.

stereo sound system [ENG ACOUS] A sound reproducing system in which a stereo pickup, stereo tape recorder, stereo tuner, or stereo microphone system feeds two independent audio channels, each of which terminates in one or more loudspeakers arranged to give listeners the same audio perspective that they would get at the original sound source. Also known as stereo; stereophonic sound system.

stereo tape recorder [ENG ACOUS] A magnetic-tape recorder having two stacked playback heads, used for reproduction of stereo recorded tape.

stereo tuner [ENG ACOUS] A tuner having provisions for receiving both channels of a stereo broadcast.

sterhydraulic [MECH ENG] Pertaining to a hydraulic press in which motion or pressure is produced by the introduction of a solid body into a cylinder filled with liquid.

sterilizer [ENG] An apparatus for sterilizing by dry heat, steam, or water.

sterling silver [MET] A silver alloy having a defined standard of purity of 92.5% silver and the remaining 7.5% usually of copper.

stern attack [AERO ENG] In air intercept, an attack by an interceptor aircraft which terminates with a heading crossing angle of 45° or less.

sterrometal [MET] Hard brass containing a small amount of iron and manganese; used for hydraulic cylinders and marine castings.

Stetefeldt furnace [MET] A furnace for desulfurizing and chloridizing silver ores; the ores are powdered, mixed with salt, and dropped through a hot atmosphere.

stick [ENG] **1.** A rigid bar hinged to the boom of a dipper or pull shovel and fastened to the bucket. **2.** A long slender tool bonded with an abrasive for honing or sharpening tools and for dressing of wheels.

stick gage [ENG] A suitably divided vertical rod, or stick, anchored in an open vessel so that the magnitude of rise and fall of the liquid level may be observed directly.

stick shellac [MATER] Shellac in the form of a solid stick and in a variety of colors; used for filling imperfections in wood.

stiffened-shell fuselage *See* semimonocoque.

stiffener [CIV ENG] A steel angle or plate attached to a slender beam to prevent its buckling by increasing its stiffness.

stiffleg derrick [MECH ENG] A derrick consisting of a mast held in the vertical position by a fixed tripod of steel or timber legs. Also known as derrick crane; Scotch derrick.

stile [BUILD] The upright outside framing piece of a window or door.

Stiles method [PETRO ENG] A technique for computing oil recovery by waterflood methods, taking into account the distribution of varying permeability stratums throughout the reservoir.

stilling basin [ENG] A depressed area in a channel or reservoir that is deep enough to reduce the velocity of the flow. Also known as stilling box.

stilling box *See* stilling basin.

stillingia oil [MATER] A combustible, toxic, pale-yellow drying oil with a linseed-oil scent and a mustard taste, derived from tallow tree seeds; used in lubricants, candles, textile dressing, and soap. Also known as tallow-seed oil.

Stillson wrench [DES ENG] A trademark for an adjustable pipe wrench consisting of an L-shaped jaw in a sleeve, the sleeve being pivoted to a handle; pressure on the handle increases the grip of the jaw.

still wax *See* wax tailings.

stimulation treatment [PETRO ENG] One of the techniques to increase (stimulate) oil- or gas-reservoir production, such as acidizing, fracturing, controlled underground explosions, or various cleaning techniques. Also known as well stimulation.

stinger [PETRO ENG] A support which is attached to the stern of a pipe-laying barge and which controls the bending of the pipe as it leaves the barge to enter the water.

stink *See* gob stink.

stinkdamp [MIN ENG] The hydrogen sulfide that occurs in mines.

stir-in resin [MATER] A vinyl resin that does not require grinding in order to disperse in a plastisol or organisol.

Stirling engine [MECH ENG] An engine in which work is performed by the expansion of a gas at high temperature; heat for the expansion is supplied through the wall of the piston cylinder.

stirrup [CIV ENG] In concrete construction, a U-shaped bar which is anchored perpendicular to the longitudinal steel as reinforcement to resist shear. [MIN ENG] **1.** A piece of steel hung from a gallows frame to engage the endgate hooks when a mine car is tilted over; used at dumps. **2.** A screw joint suspended from the brakestaff of a spring pole, by which the boring rods are adjusted to the depth of the borehole. Also known as temper screw.

stitch bonding [ENG] A method of making wire connections between two or more points on an integrated circuit by using impulse welding or heat and pressure while feeding the connecting wire through a hole in the center of the welding electrode.

stitching [ENG] Progressive welding of thermoplastic materials (resins) by successive applications of two small, mechanically operated, radio-frequency-heated electrodes; the mechanism is similar to that of a normal sewing machine.

stitch rivet [ENG] One of a series of rivets joining the parallel elements of a structural member so that they act as a unit.

stitch welding [MET] A series of spaced spot resistance welds.

stock [IND ENG] **1.** A product or material kept in storage until needed for use or transferred to some ultimate point for use, for example, crude oil tankage or paper-pulp feed. **2.** Designation of a particular material, such as bright stock or naphtha stock.

stock accounting [IND ENG] The establishment and maintenance of formal records of material in stock reflecting such information as quantities, values, or condition.

stockage objective [AERO ENG] The maximum quantities of material to be maintained on hand to sustain current operations; it will consist of the sum of stocks represented by the operating level and the safety level.

stock control [IND ENG] Process of maintaining inventory data on the quantity, location, and condition of supplies and equipment due in, on hand, and due out, to determine quantities of material and equipment available or required for issue and to facilitate distribution and management of material.

stock coordination [IND ENG] A supply management function exercised usually at department level which controls the assignment of material cognizance for items or categories of material to inventory managers.

stocking cutter [MECH ENG] **1.** A gear cutter having side rake or curved edges to rough out the gear-tooth spaces before they are formed by the regular gear cutter. **2.** A concave gear cutter ganged beside a regular gear cutter and used to finish the periphery of a gear blank by milling ahead of the regular cutter.

stock number [IND ENG] Number assigned to an item, principally to identify that item for storage and issue purposes.

stockpile [ENG] A reserve stock of material, equipment, raw material, or other supplies.

stock rail [CIV ENG] The fixed rail in a track, against which the switch rail operates.

stock record account [IND ENG] A basic record showing by item the receipt and issuance of property, the balances on hand, and such other identifying or stock control data as may be required by proper authority.

Stoddard solvent [MATER] A petroleum naphtha product with a comparatively narrow boiling range; used mostly for dry cleaning.

stoker [MECH ENG] A mechanical means, as used in a furnace, for feeding coal, removing refuse, controlling air supply, and mixing with combustibles for efficient burning.

Stokes number 2 [ENG] A dimensionless number used in the calibration of rotameters, equal to $1.042\ m_f g \rho\ (1 - \rho/\rho_f) R^3 / \mu^2$, where ρ and μ are the density and dynamic viscosity of the fluid respectively, m_f and ρ_f are the mass and density of the float respectively, and R is the ratio of the radius of the tube to the radius of the float. Symbol St_2.

stoking [MET] Presintering or sintering a metal powder in such a way as to advance the compacts through the furnace at a fixed rate. Also known as continuous sintering.

STOL *See* short takeoff and landing.

STOL aircraft [AERO ENG] Heavier-than-air craft that cannot take off and land vertically, but can operate within areas substantially more confined than those normally required by aircraft of the same size. Derived from short takeoff and landing aircraft.

stoneboat [MIN ENG] A flat runnerless sled for transporting heavy material.

stone dust [MIN ENG] Inert dust spread on roadways in coal mines as a defense against the danger of coal-dust explosions; effective because the stone dust absorbs heat.

stone-dust barrier [MIN ENG] A device erected in mine roadways to arrest explosions; consists of trays or Vee troughs loaded with stone dust, which are upset or overturned by the pressure wave in front of an explosion and the flame, producing a dense cloud of inert dust which blankets the flame and stops further propagation of the explosion.

stone gobber [MIN ENG] In bituminous coal mining, one who removes stone and other refuse from coal mine floors and dumps the refuse into mine cars for disposal.

stoneware [MATER] Vitrified ware with impermeable surface; used for corrosive materials in the laboratory and for some industrial operations.

stonework [CIV ENG] A structure or the part of a structure built of stone.

Stoney gate [CIV ENG] A crest gate which moves along a series of rollers traveling vertically in grooves in masonry piers, independently of the gate and piers.

stop and stay *See* absolute stop.

stop bead [BUILD] A molding on the pulley stile of a window frame; forms one side of the groove for the inner sash.

stop cock [ENG] A small valve for stopping or regulating the flow of a fluid through a pipe.

stope [MIN ENG] **1.** To excavate ore in a vein by driving horizontally upon it a series of workings, one immediately over the other, or vice versa; each horizontal working is called a stope because when a number of them are in progress, each working face under attack assumes the shape of a flight of stairs. **2.** Any subterranean extraction of ore except that which is incidentally performed in sinking shafts or driving levels for the purpose of opening the mine.

stope assay plan [MIN ENG] A plan that details assay value of ore exposures in a stope.

stope board [MIN ENG] A timber staging on the floor of a stope for setting a rock drill; the stage is tilted so that the bottom holes can be drilled in the same inclined direction.

stope fillings [MIN ENG] Broken waste material or low-grade matter from a lode or vein used to fill stopes on abandonment.

stope hoist [MIN ENG] A small, portable, compressed-air hoist for operating a scraper-loader or for pulling heavy timbers into position, often used in narrow stopes.

stope pillar [MIN ENG] An ore column left in place to support the stope.

stoper *See* stoping drill.

stoping drill [MIN ENG] A small air or electric drill, usually mounted on an extensible column, for working stopes, raises, and narrow workings. Also known as stoper.

stoping ground [MIN ENG] Part of an ore body opened by drifts and raises, and ready for breaking down.

stoplog [CIV ENG] A log, plank, or steel or concrete beam that fits into a groove or rack between walls or piers to prevent the flow of water through an opening in a dam, conduit, or other channel.

stop nut [DES ENG] **1.** An adjustable nut that restricts the travel of an adjusting screw. **2.** A nut with a compressible insert that binds it so that a lock washer is not needed.

stopoff *See* resist.

stopping [MIN ENG] A brattice, or more commonly, a masonry or brick wall built across old headings, chutes, or airways to confine the ventilating current to certain passages, or to lock up the gas in old workings, or to smother a mine fire.

stopping off [MET] **1.** Local deposition of a protective coating, such as copper or fireclay, to prevent carburization, decarburization, or nitriding during heat treatment. **2.** Filling up a portion of the mold cavity to keep out molten metal. **3.** Applying a nonconducting layer to avoid electrodeposition in certain areas.

stop valve [ENG] A valve that can be opened or closed to regulate or stop the flow of fluid in a pipe.

storable propellant [MATER] A propellant capable of being placed and kept in a tank without benefit of special measures for temperature or pressure control.

storage battery locomotive [MIN ENG] An underground locomotive powered by storage batteries.

storage calorifier *See* cylinder.

storage reservoir *See* impounding reservoir.

stored-energy welding [MET] Welding by means of energy accumulated electrostatically, electrochemically, or electromagnetically at a low rate.

storm choke [PETRO ENG] A device installed in an oil-well tubing string below the surface to shut in the well when the flow reaches a predetermined rate; provides an automatic shutoff in case Christmastree or control valves are damaged. Also known as tubing safety valve.

storm drain [CIV ENG] A drain which conducts storm surface, or wash water, or drainage after a heavy rain from a building to a storm or a combined sewer. Also known as storm sewer.

storm sewage [CIV ENG] Refuse liquids and waste carried by sewers during or following a period of heavy rainfall.

storm sewer *See* storm drain.

storm window [BUILD] A sash placed on the outside of an ordinary window to give added protection from the weather.

Storrow whirling hygrometer [ENG] A hygrometer in which the two thermometers are mounted side by side on a brass frame and fitted with a loose handle so that it can be whirled in the atmosphere to be tested; the instrument is whirled at some 200 revolutions per minute for about 1 minute and the readings on the wet- and dry-bulb thermometers are recorded; used in conjunction with Glaisher's or Marvin's hygrometrical tables.

story [BUILD] The space between two floors or between a floor and the roof.

stove [ENG] A chamber within which a fuel-air mixture is burned to provide heat, the heat itself being radiated outward from the chamber; used for space heating, process-fluid heating, and steel blast furnaces.

stove bolt [DES ENG] A coarsely-threaded bolt with a slotted head, which with a square nut is used to join metal parts.

stovepipe [ENG] Large-diameter pipe made of sheet steel.

stovewood [MATER] Firewood sawed into short lengths for use in a stove.

stoving *See* baking.

stowboard [MIN ENG] A mine heading used for storing waste.

straddle milling [MECH ENG] Face milling of two parallel vertical surfaces of a workpiece simultaneously by using two side-milling cutters.

straight beam [ENG] In ultrasonic testing, a longitudinal wave emitted from an ultrasonic search unit in a wavetrain which travels perpendicularly to the test surface.

straight bevel gear [DES ENG] A simple form of bevel gear having straight teeth which, if extended inward, would come together at the intersection of the shaft axes.

straight dynamite [MATER] Any of the powerful, quick-acting dynamites composed of nitroglycerin, a combustible such as wood meal, sodium nitrate, and an antacid such as calcium or magnesium carbonate; made in 15 to 60% strength, the percentage representing the proportion of nitroglycerin in the dynamite.

straightedge [DES ENG] A strip of wood, plastic, or metal with one or more long edges made straight with a desired degree of accuracy.

straightening vanes [ENG] Horizontal vanes mounted on the inside of fluid conduits to reduce the swirling or turbulent flow ahead of the orifice or the venturi meters.

straight filing [ENG] Filing by pushing a file in a straight line across the work.

straight-line mechanism [MECH ENG] A linkage so proportioned and constrained that some point on it describes over part of its motion a straight or nearly straight line.

straight piecework system *See* one-hundred-percent premium plan.

straight polarity [MET] Arrangement of an arc welding circuit in which the electrode is connected to the negative terminal.

straight proportional system *See* one-hundred-percent premium plan.

straight-run gasoline [MATER] Gasoline comprised of only natural ingredients from crude oil or natural-gas liquids; for example, no cracked, polymerized, alkylated, reformed, or visbroken stock.

straight-run pitch [MATER] A petroleum pitch that is run directly during distillation to the desired consistency without any compounding or aftertreatment.

straight-run stock *See* virgin stock.

straight strap clamp [DES ENG] A clamp made of flat stock with an elongated slot for convenient positioning; held in place by a T bolt and nut.

straight-tube boiler [MECH ENG] A water-tube boiler in which all the tubes are devoid of curvature and therefore require suitable connecting devices to complete the circulatory system. Also known as header-type boiler.

straight turning [MECH ENG] Work turned in a lathe so that the diameter is constant over the length of the workpiece.

straightway pump [MECH ENG] A pump with suction and discharge valves arranged to give a direct flow of fluid.

straight wheel [DES ENG] A grinding wheel whose sides or face are straight and not in any way changed from a cylindrical form.

strain aging [MET] Change of mechanical properties of a metal by aging induced by plastic deformation.

strain bursts [MIN ENG] Rock bursts in which there is spitting, flaking, and sudden fracturing at the face, indicating increased pressure at the site.

strainer [ENG] A porous or screen medium used ahead of equipment to filter out harmful solid objects and particles from a fluid stream; used for example, in river-water intakes for process plants or to remove decomposition products from the circulating fluid in a hydraulic system.

strain foil [ENG] A strain gage produced from thin foil by photoetching techniques; may be applied to curved surfaces, has low transverse sensitivity, exhibits negligible hysteresis under cycling loads, and creeps little under sustained loads.

strain gage [ENG] A device which uses the change of electrical resistance of a wire under strain to measure pressure.

strain gage accelerometer [ENG] Any accelerometer whose operation depends on the fact that the resistance in a wire changes when it is strained; these devices are classified as bonded or unbonded.

strain gage bridge [ENG] A bridge arrangement of four strain gages, cemented to a stressed part in such a way that two gages show increases in resistance and two show decreases when the part is stressed; the change in output voltage under stress is thus much higher than that for a single gage.

strain hardening [MET] Increasing the hardness and tensile strength of a metal by cold plastic deformation.

strain relief method [MIN ENG] A method for determining absolute strain and stress within rock in place by boring a smooth hole in the rock and inserting a gage capable of measuring diametral deformation, and overcoring the hole with a large coring bit; the change in the diameter of the hole when the rock cylinder is free to expand is a function of the original stress in the rock and its elastic modulus.

strain restoration method [MIN ENG] A method for determining absolute strain and stress within rock in place by the installation of strain gages on the rock surface, cutting of a slot in the rock between the strain gages so that the surface rock is free to expand, installation of a flat jack in the slot, and application of hydraulic pressure to the flat jack until the rock is restored to its original state of strain; the original stress in the rock is presumed to be equal to the final pressure in the flat jack.

strain seismograph   [ENG] A seismograph that detects secular strains related to tectonic processes and tidal yielding of the solid earth; also detects strains associated with propagating seismic waves.

strain seismometer   [ENG] A seismometer that measures relative displacement of two points in order to detect deformation of the ground.

strake   [MIN ENG] A relatively wide trough set at a slope and covered with a blanket or corduroy for catching comparatively coarse gold and any valuable mineral.

strand   [ENG] **1.** One of a number of steel wires twisted together to form a wire rope or cable or an electrical conductor. **2.** A thread, yarn, string, rope, wire, or cable of specified length. **3.** One of the fibers or filaments twisted or laid together into yarn, thread, rope, or cordage.

strand burner   [ENG] A device that determines the rate at which a propellant burns at various pressures by using a propellant strand.

stranded caisson   *See* box caisson.

stranger   [AERO ENG] In air intercept, an unidentified aircraft, bearing, distance, and altitude as indicated relative to an aircraft.

strap bolt   [DES ENG] **1.** A bolt with a hook or flat extension instead of a head. **2.** A bolt with a flat center portion and which can be bent into a U shape.

strap hammer   [MECH ENG] A heavy hammer controlled and operated by a belt drive in which the head is slung from a strap, usually of leather.

strap hinge   [DES ENG] A hinge fastened to a door and the adjacent wall by a long hinge.

strapping   [PETRO ENG] A petroleum industry procedure in which storage tanks are strapped (measured) on their outside with steel measuring tapes to calculate the volumetric capacity of the tank for increments of height.

strapping table   [PETRO ENG] A tabular record of tank volume versus height so that taped (strapped) measurements of liquid depth can be converted into liquid volumes.

Strasbourg turpentine   [MATER] A balsam from the European white fir; a heavy, thick material, it is sometimes used in painting mediums and glazes, but an excessive amount causes smearing and slow drying of the paint.

strategic airlift   [AERO ENG] The continuous, sustained air movement of units, personnel, and materiel in support of all U.S. Department of Defense agencies between area commands.

strategic material   [IND ENG] A material needed for the industrial support of a war effort.

strategic transport aircraft   [AERO ENG] Aircraft designed primarily for the carriage of personnel or cargo over long distances.

Stratofortress   [AERO ENG] A United States all-weather, intercontinental, strategic heavy bomber powered by eight turbojet engines; it is capable of delivering nuclear and nonnuclear bombs, air-to-surface missiles, and decoys; its range is extended by in-flight refueling. Designated B-52.

Stratofreighter   [AERO ENG] A United States strategic, aerial tanker-freighter powered by four reciprocating engines; it is equipped for inflight refueling of bombers and fighters. Designated KC-97.

Stratojet   [AERO ENG] A United States all-weather strategic medium bomber; it is powered by six turbojet engines, has intercontinental range through in-flight refueling, and is capable of delivering nuclear and nonnuclear bombs. Designated B-47.

Stratotanker   [AERO ENG] A United States multipurpose aerial tanker-transport powered by four turbojet engines; and equipped for high speed, high-altitude refueling of bombers and fighters. Designated KC-135.

straw oil   [MATER] A straw-colored petroleum paraffin oil; used for many process applications.

stray current corrosion   [MET] Corrosion of metals caused by a stray current.

stray line   [ENG] An ungraduated portion of the line connected to a current pole, used so that the pole will acquire the speed of the current before a measurement is begun.

stream gage   *See* river gage.

streamlining   [DES ENG] The contouring of a body to reduce its resistance to motion through a fluid.

stream takeoff   [AERO ENG] Aircraft taking off in tail/column formation.

street   [CIV ENG] A paved road for vehicular traffic in an urban area.

street elbow   [DES ENG] A pipe elbow with an internal thread at one end and an external thread at the other.

stremmatograph   [ENG] An instrument for measuring longitudinal stress in rails as trains pass over.

stress amplitude   [MECH ENG] One half the algebraic difference between the maximum and minimum stress in one fatigue test cycle.

stress corrosion [MET] Corrosion that is accelerated by stress, applied or residual, in a metal.

stress-corrosion cracking [MET] Failure by cracking under the conjoint action of a constant tensile stress, which is applied to residual, in certain chemical environments specific to the metal.

stressed skin construction [CIV ENG] A type of construction in which the outer skin and the framework interact, thus contributing to the flexural strength of the unit.

stress raiser [MET] A notch, hole, or other discontinuity in contour or structure which causes localized stress concentration.

stress relief cracking [MET] Cracking between metal grains in the heat-affected zone of a weldment during exposure to high temperatures.

stress relieving [MET] Low-temperature heating to reduce residual stress.

stretch [PETRO ENG] The increase in length of oil-well casing or tubing when freely suspended in fluid mediums.

stretcher [CIV ENG] A brick or block that is laid with its length paralleling the wall. [MIN ENG] A bar used for roof support on roadways and which is either wedged against or pocketed into the sides of the roadway without support of legs or struts.

stretcher bar [MIN ENG] A single screw column, capable of holding one machine drill; used in small drifts.

stretcher leveling [MET] Removing warp and distortion from a piece of metal by gripping it at each end and subjecting it to stress beyond the yield strength. Also known as patent leveling.

stretcher strains *See* Lüders' lines.

stretch former [MECH ENG] A machine used to stretch form materials, such as metals and plastics.

stretch forming [MECH ENG] Shaping metals and plastics by applying tension to stretch the heated sheet or part, wrapping it around a die, and then cooling it. Also known as wrap forming.

stretch out [IND ENG] A reduction in the delivery rate specified for a program without a reduction in the total quantity to be delivered.

striding compass [ENG] A compass mounted on a theodolite for orientation.

strike [MET] **1.** A very thin, initially electroplated film or the plating solution with which to deposit such a film. **2.** A local crater in a metal surface due to accidental contact with the welding electrode.

strike board [MIN ENG] A board at the top of a shaft from which the bucket is tipped; used in shaft sinking; formerly, the beam or plank at the shaft top on which the baskets were landed. Also known as strike tree.

strike-off board [ENG] A straight-edge board used to remove excess, freshly placed plaster, concrete, or mortar from a surface.

strike plate [DES ENG] A metal plate or box which is set in a door jamb and is either pierced or recessed to receive the bolt or latch of a lock.

strike plating [MET] Applying a thin electroplated film prior to depositing the principal electroplate.

strike tree *See* strike board.

striking hammer [ENG] A hammer used to strike a rock drill.

string [ENG] A piece of pipe, casing, or other down-hole drilling equipment coupled together and lowered into a borehole.

string bead [MET] A continuous weld bead made without appreciable transverse oscillation.

string course [BUILD] A horizontal band of masonry, generally narrower than other courses and sometimes projecting, extending across the facade of a structure and in some instances encircling pillars or engaged columns. Also known as belt course.

string electrometer [ENG] An electrometer in which a conducting fiber is stretched midway between two oppositely charged metal plates; the electrostatic field between the plates displaces the fiber laterally in proportion to the voltage between the plates.

stringer [CIV ENG] **1.** A long horizontal member used to support a floor or to connect uprights in a frame. **2.** An inclined member supporting the treads and risers of a staircase. [MET] An elongated mass of microconstituents or foreign material in wrought metal oriented in the direction of working.

string galvanometer [ENG] A galvanometer consisting of a silver-plated quartz fiber under tension in a magnetic field, used to measure oscillating currents. Also known as Einthoven galvanometer.

stringing [PETRO ENG] The connecting of lengths of pipe end to end (tubing or casing) to make a string long enough to reach to the desired depth in a well bore.

string milling [MECH ENG] A milling method in which parts are placed in a row and milled consecutively.

string shot [PETRO ENG] An oil-well stimulation technique in which a string of explosive (for example, Prima Cord) is hung opposite to the producing zone down a wellbore and detonated; used to remove deposits (gypsum, mud, or paraffin) from the formation face.

strip [ENG] To remove insulation from a wire. [MIN ENG] To remove coal, stone, or other material from a quarry or from a working that is near the surface of the earth.

strip-borer drill [MECH ENG] An electric or diesel skid- or caterpillar-mounted drill used at quarry or opencast sites to drill 3- to 6-inch-diameter (8- to 15-centimeter-diameter), horizontal blast holes up to 100 feet (30 meters) in length, without the use of flush water.

strip-chart recorder [ENG] A recorder in which one or more writing pens or other recording devices trace changes in a measured variable on the surface of a strip chart that is moved at constant speed by a time-clock motor.

strip mine [MIN ENG] An opencut mine in which the overburden is removed from a coal bed before the coal is taken out.

strip mining [MIN ENG] The mining of coal by surface mining methods.

stripper [ENG] A hand or motorized tool used to remove insulation from wires. [PETRO ENG] A well from which oil production is quite small.

stripper plate [ENG] In plastics molding, a plate that strips a molded article free of core pins or force plugs.

stripper punch [MET] In powder metallurgy, a device used as the bottom or top of the die cavity which can be pushed into the die to eject the formed compact.

stripper rubber [PETRO ENG] A pressure-actuated seal used to control gas pressure in the casing-tubing annulus of low-pressure wells while inserting (running) or withdrawing (pulling) tubing.

stripping [MET] Removing a coating from the surface of a metal.

stripping area [MIN ENG] In stripping operations, an area encompassing the pay material, its bottom depth, the thickness of the layer of waste, the slope of the natural ground surface, and the steepness of the safe slope of cuts.

stripping a shaft [MIN ENG] **1.** Removing the timber from an abandoned shaft. **2.** Trimming or squaring the sides of a shaft.

stripping ratio [MIN ENG] The unit amount of spoil or waste that must be removed to gain access to a similar unit amount of ore or mineral material.

stripping shovel [MIN ENG] A shovel with an especially long boom and stick, enabling it to reach further and pile higher.

strip pit [MIN ENG] **1.** A coal or other mine worked by stripping. **2.** An open-pit mine.

strip printer [ENG] A device that prints computer, telegraph, or industrial output information along a narrow paper tape which resembles a ticker tape.

stroboscope [ENG] An instrument for making moving bodies visible intermittently, either by illuminating the object with brilliant flashes of light or by imposing an intermittent shutter between the viewer and the object; a high-speed vibration can be made visible by adjusting the strobe frequency close to the vibration frequency.

stroboscopic disk [ENG] A printed disk having a number of concentric rings each containing a different number of dark and light segments; when the disk is placed on a phonograph turntable or rotating shaft and illuminated at a known frequency by a flashing discharge tube, speed can be determined by noting which pattern appears to stand still or to rotate slowly.

stroboscopic tachometer [ENG] A stroboscope having a scale that reads in flashes per minute or in revolutions per minute; the speed of a rotating device is measured by directing the stroboscopic lamp on the device, adjusting the flashing rate until the device appears to be stationary, then reading the speed directly on the scale of the instrument.

stroke [MECH ENG] The linear movement, in either direction, of a reciprocating mechanical part.

stroke-bore ratio [MECH ENG] The ratio of the distance traveled by a piston in a cylinder to the diameter of the cylinder.

stromeyerite [MIN ENG] CuAgS A metallic-gray orthorhombic mineral with a blue tarnish composed of silver copper sulfide occurring in compact masses.

struck capacity [MIN ENG] The volume of water a mine car, tram, hoppet, or wagon would hold if the conveyance were of watertight construction.

struck joint [CIV ENG] A mortar joint in brickwork formed by pressing the trowel in at the lower edge, so that a recess is formed at the bottom of the joint; suitable only for interior work.

structural adhesive [MATER] An adhesive capable of bearing loads of considerable magnitude; a structural adhesive will not fail when a bonded joint prepared from the thickness of metal, or other material typical for that industry, is stressed to its yield point.

structural analysis [ENG] The determination of stresses and strains in a given structure.

structural clay tile [MATER] Hollow burned-clay masonry unit with parallel cells used as facing tile, load-bearing tile, partition tile, fireproofing tile, furring tile, floor tile, and header tile.

structural connection [CIV ENG] A means of joining the individual members of a structure to form a complete assembly.

structural drill [MECH ENG] A highly mobile diamond- or rotary-drill rig complete with hydraulically controlled derrick mounted on a truck, designed primarily for rapidly drilling holes to determine the structure in subsurface strata or for use as a shallow, slim-hole producer or seismograph drill.

structural drilling [ENG] Drilling done specifically to obtain detailed information delineating the location of folds, domes, faults, and other subsurface structural features indiscernible by studying strata exposed at the surface.

structural engineering [CIV ENG] A branch of civil engineering dealing with the design of structures such as buildings, dams, and bridges.

structural foam [MATER] A type of cellular plastic with a dense outer skin surrounding a foamed core.

structural riveting [ENG] Riveting structural members by using punched holes.

structural shape [MET] A piece of metal of a standard design used in construction.

structural steel [MET] Steel used in engineering structures, usually manufactured by either open-hearth or the electric-furnace process.

structural tile [MATER] A hollow clay product which may be load-bearing or non-load-bearing; used for facing, flooring, or partitions.

structural wall *See* bearing wall.

structural weight *See* construction weight.

structure [AERO ENG] The construction or makeup of an airplane, spacecraft, or missile, including that of the fuselage, wings, empennage, nacelles, and landing gear, but not that of the power plant, furnishings, or equipment. [CIV ENG] Something, as a bridge or a building, that is built or constructed and designed to sustain a load.

structure number [DES ENG] A number, generally from 0 to 15, indicating the spacing of abrasive grains in a grinding wheel relative to their grit size.

strut [AERO ENG] A bar supporting the wing or landing gear of an airplane. [CIV ENG] A long structural member of timber or metal, or a bar designed to resist pressure in the direction of its length. [ENG] **1.** A brace or supporting piece. **2.** A diagonal brace between two legs of a drill tripod or derrick. [MIN ENG] A vertical-compression member in a structure or in an underground timber set.

Stuart windmill *See* Fales-Stuart windmill.

stub axle [MECH ENG] An axle carrying only one wheel.

stub entry [MIN ENG] A short, narrow entry turned from another entry and driven into the solid coal, but not connected with other mine workings.

stub mortise [ENG] A mortise which passes through only part of a timber.

Stubs gage [DES ENG] A number system for denoting the thickness of steel wire and drills.

stub switch [ENG] A pair of short switch rails, held only at or near one end and free to move at the other end; used in mining and to some extent on narrow-gage industrial tramways.

stub tenon [ENG] A tenon that fits into a stub mortise.

stub tube [MECH ENG] A short tube welded to a boiler or pressure vessel to provide for the attachment of additional parts.

stucco [MATER] A smooth plasterlike material applied to the outside wall or other exterior surface of a building or structure.

stud [BUILD] One of the vertical members in the walls of a framed building to which wallboards, lathing, or paneling is nailed or fastened.

stud [DES ENG] **1.** A rivet, boss, or nail with a large, ornamental head. **2.** A short rod or bolt threaded at both ends without a head.

stud wall [BUILD] A wall formed with timbers; studs are usually spaced 12–16 inches (30–41 centimeters) on center.

stud welding [MET] Arc-welding using the heat of an electric arc produced between a metal stud and another part, and then bringing the parts together under pressure.

stuffing [ENG] A method of sealing the mechanical joint between two metal surfaces; packing (stuffing) material is inserted within the seal area container (the stuffing or packing box), and compressed to a liquid-proof seal by a threaded packing ring follower. Also known as packing.

stuffing box [ENG] A packed, pressure-tight joint for a rod that moves through a hole, to reduce or eliminate fluid leakage.

stuffing nut [ENG] A nut for adjusting a stuffing box.

stull [MIN ENG] A platform laid on timbers, braced across a working from side to side, to support workers or to carry ore or waste.

stull piece [MIN ENG] **1.** A piece of timber placed slanting over the back of a level to prevent rock falling into the level from the stopes above. **2.** Timbers bracing the platform of a stull.

stull stoping [MIN ENG] Stull timbers placed between the foot and hanging walls, which constitute the only artificial support provided during the excavation of a stope.

stump [MIN ENG] A small pillar of coal left between the gangway or airway and the breasts to protect these passages; any small pillar.

stupp [MIN ENG] A black residue from distilled mercury ore, consisting of soot, hydrocarbons, mercury and mercury compounds, and ore dust.

style See gnomon.

stylus [ENG ACOUS] The portion of a phonograph pickup that follows the modulations of a record groove and transmits the resulting mechanical motions to the transducer element of the pickup for conversion to corresponding audio-frequency signals. Also known as needle; phonograph needle; reproducing stylus.

styrene-butadiene rubber [MATER] The most common type of synthetic rubber, made by the copolymerization of styrene and butadiene monomers; used in tires, footwear, adhesives, and sealants. Also known as SBR.

styrene-rubber plastic [MATER] A plastic-rubber mixture consisting of at least 50% of a styrene plastic combined with rubber and various compounding ingredients.

subaqueous mining [MIN ENG] Surface mining in which the mined material is removed from the bed of a natural body of water.

subassembly [ENG] A structural unit, which, though manufactured separately, was designed for incorporation with other parts in the final assembly of a finished product.

subatmospheric heating system [MECH ENG] A system which regulates steam flow into the main throttle valve under automatic thermostatic control and maintains a fixed vacuum differential between supply and return by means of a differential controller and a vacuum pump.

subbottom depth recorder [ENG] A compact seismic instrument which can provide continuous soundings of strata beneath the ocean bottom utilizing the low-frequency output of an intense electrical spark discharge source in water.

subboundary structure [MET] A network of low-angle grain boundaries of less than one degree within the main crystals of a metal.

subcomponent [DES ENG] A part of a component having characteristics of the component.

subcontract [ENG] A contract made with a third party by one who has contracted to perform work or service for whole or part performance of that work or service.

subcontractor [ENG] A manufacturer or organization that receives a contract from a prime contractor for a portion of the work on a project.

subdrainage [CIV ENG] Natural or artificial removal of water from beneath a lined conduit.

subdrilling [ENG] Refers to the breaking of the base in which boreholes are drilled 1 foot (0.3 meter) or several feet below the level of the quarry floor.

subfloor [BUILD] The rough floor which rests on the floor joists and on which the finished floor is laid.

subgrade [CIV ENG] The soil or rock leveled off to support the foundation of a structure.

subgrain [MET] The portion of a metal crystal or grain with an orientation that differs slightly from the orientation of neighboring portions of the same crystal.

sublevel [MIN ENG] An intermediate level opened a short distance below the main level; or in the caving system of mining, a 15–20-foot (4.6–6.1-meter) level below the top of the ore body, preliminary to caving the ore between it and the level above.

sublevel caving [MIN ENG] A stoping method in which relatively thin blocks of ore are caused to cave by successively undermining small panels.

sublevel drive [MIN ENG] A drive often made in a section which divides the deposit into narrower panels and zones.

sublevel stoping [MIN ENG] A mining method involving overhand, underhand, and shrinkage stoping; the characteristic feature is the use of sublevels which are worked simultaneously, the lowest on a given block being farthest advanced and the sublevels above following one another at short intervals.

subliming rocket propellant [MATER] A propellant characterized by sublimation of the material at the heated surface.

submarine blast [ENG] A charge of high explosives fired in boreholes drilled in the rock underwater for dislodging dangerous projections and for deepening channels.

submarine gate [ENG] An edge gate with the opening from the runner into the mold positioned below the printing line or mold surface.

submarine mine [MIN ENG] A mine for the extraction of minerals or ores under the sea.

submarine oscillator [ENG ACOUS] A large, electrically operated diaphragm horn which produces a powerful sound for signaling through water.

submarine pipeline [ENG] A pipeline installed under water, resting on the bed of the waterway; frequently used for petroleum or natural gas transport across rivers, lakes, or bays.

submarine sentry [ENG] A form of underwater kite towed at a predetermined constant depth in search of elevations of the bottom; the kite rises to the surface upon encountering an obstruction.

submarine wave recorder [ENG] An instrument for measuring the changing water height above a hovering submarine by measuring the time required for sound emitted by an inverted echo sounder on the submarine to travel to the surface and return.

submerged-arc furnace [MET] An arc-heating furnace in which the arcs may be completely submerged under the charge or in the molten bath under the charge.

submerged-arc welding [MET] Arc welding with a bare metal electrode, the arc and tip of the electrode being shielded by a blanket of granular, fusible material.

submerged-combustion evaporator [ENG] A liquid-evaporation device in which heat is provided by combustion gases bubbling up through the liquid; the burner is submerged in the body of the liquid.

submerged-combustion heater [ENG] A combustion device in which fuel and combustion air are mixed and ignited below the surface of a liquid; used in heaters and evaporators where absorption of the combustion products will not be detrimental.

submerged weir [CIV ENG] A dam which, when in use, has the downstream water level at an elevation equal to or higher than the crest of the dam.

submersible pump [MECH ENG] A pump and its electric motor together in a protective housing which permits the unit to operate under water.

suboptimization [SYS ENG] The process of fulfilling or optimizing some chosen objective which is an integral part of a broader objective; usually the broad objective and lower-level objective are different.

subscale [MET] Oxidation occurring within a metal instead of on the surface.

subsidence [MIN ENG] A sinking down of a part of the earth's crust due to underground excavations.

subsidence break [MIN ENG] A fracture in the rocks overlying a coal seam or mineral deposit resulting from mining operations.

subsidiary conduit [CIV ENG] Terminating branch of an underground conduit run extending from a manhole or handhole to a nearby building, handhole, or pole.

subsidiary transport [MIN ENG] The conveying or haulage of coal or mineral from the working faces to a junction or loading point.

subsieve analysis [MET] Analysis by size distribution of metal powder particles all of which pass through a standard 44-micrometer sieve.

subsieve fraction [MET] The fraction of particles of a metal powder which pass through a standard 44-micrometer sieve.

subsonic flight [AERO ENG] Movement of a vehicle through the atmosphere at a speed appreciably below that of sound waves; extends from zero (hovering) to a speed about 85% of sonic speed corresponding to ambient temperature.

subsonic inlet [ENG] An entrance or orifice for the admission of fluid flowing at speeds less than the speed of sound in the fluid.

subsonic nozzle [ENG] A nozzle through which a fluid flows at speed less than the speed of sound in the fluid.

substantive dye *See* direct dye.

substation [ENG] An intermediate compression station to repressure a fluid being transported by pipeline over a long distance. [MIN ENG] A subsidiary station for the conversion of power to the type, usually direct current, and voltage needed for mining equipment and fed into the mine power system.

substitution solid solution [MET] A solid alloy having the atoms of the solute located at some lattice of points of the solvent.

substrate [ENG] Basic surface on which a material adheres, for example, paint or laminate.

substructure [CIV ENG] The part of a structure which is below ground.

subsurface waste disposal [ENG] A waste disposal method for manufacturing wastes in porous underground rock formations.

subsystem [ENG] A major part of a system which itself has the characteristics of a system, usually consisting of several components.

subtense bar [ENG] The horizontal bar of fixed length in the subtense technique of distance measurement method.

subtense technique [CIV ENG] A distance measuring technique in which the transit angle subtended by the subtense bar enables the computation of the transit-to-bar distance.

subway [CIV ENG] An underground passage.

successive fracture treatment [PETRO ENG] A second or third fracturing operation of an oil well in an oil reservoir to fracture a new part or zone.

sucker rod [PETRO ENG] A connecting rod between a down-hole oil-well pump and the lifting or pumping device on the surface.

sucker-rod pump [PETRO ENG] A cylinder-piston-type pump used to displace oil into the oil-well tubing string, and to the surface.

suction anemometer [ENG] An anemometer consisting of an inverted tube which is half-filled with water that measures the change in water level caused by the wind's force.

suction boundary layer control [AERO ENG] A technique that is used in addition to purely geometric means to control boundary layer flow; it consists of sucking away the retarded flow in the lower regions of the boundary through slots or perforations in the surface.

suction cup [ENG] A cup, often of flexible material such as rubber, in which a partial vacuum is created when it is inverted on a surface; the vacuum tends to hold the cup in place.

suction-cutter dredger [MECH ENG] A dredger in which rotary blades dislodge the material to be excavated, which is then removed by suction as in a sand-pump dredger.

suction head *See* suction lift.

suction lift [MECH ENG] The head, in feet, that a pump must provide on the inlet side to raise the liquid from the supply well to the level of the pump. Also known as suction head.

suction line [ENG] A pipe or tubing feeding into the inlet of a fluid impelling device (for example, pump, compressor, or blower), consequently under suction.

suction pump [MECH ENG] A pump that raises water by the force of atmospheric pressure pushing it into a partial vacuum under the valved piston, which retreats on the upstroke.

suction stroke [MECH ENG] The piston stroke that draws a fresh charge into the cylinder of a pump, compressor, or internal combustion engine.

suede [MATER] Leather with a velvet finish on the flesh side of the skin; calfskin is the commonest suede leather. Also known as napped leather.

sugarcane wax [MATER] Hard, tan to dark-green wax extracted from sugarcane; melts at 77°C; used in polishes, lubricants, and food wrappers.

sulfate paper [MATER] Paper made by the sulfate process, and which cannot be bleached as white as soda or sulfite paper; strong papers, such as kraft paper, are made from unbleached sulfate materials.

sulfite paper [MATER] Paper made from sulfite pulp.

sulfite pulp [MATER] Wood chips digested with a solution of magnesium, ammonium, or calcium disulfite, with free sulfur dioxide present; used to make paper and paper products from spruce and other coniferous woods.

sulfite waste liquor [MATER] Waste reactants and other impurities from the sulfite pulping of wood; used as a foaming and emulsifying agent, in adhesives and tanning, and for road construction.

sulfonated castor oil *See* Turkey red oil.

sulfonated oil [MATER] Mineral or vegetable oil treated with sulfuric acid to make a water-soluble (emulsifiable) form; used as lubricants, emulsifiers, defoamers, and softeners.

sulfur cement [MATER] Cement used for connecting iron parts; made of equal parts of sulfur and pitch.

sulfur dome [MET] An inverted container containing a high concentration of sulfur dioxide gas, used in die casting to cover a pot of molten magnesium to prevent burning.

sulfur hexameter [ENG] An instrument used to measure or to continuously monitor the amount of sulfur hexafluoride present in a waveguide or other device in which this gas is used as a dielectric.

sulfurized oil [MATER] Any of various mineral oils and fatty oils containing active sulfur to increase film strength and load-carrying ability; used generally as cutting fluids.

Sullivan angle compressor [MECH ENG] A two-stage compressor in which the low-pressure cylinder is horizontal and the high-pressure cylinder is vertical; a compact compressor driven by a belt, or directly connected to an electric motor or diesel engine.

Sulzer two-cycle engine [MECH ENG] An internal combustion engine utilizing the Sulzer Co. system for the effective scavenging and charging of the two-cycle diesel engine.

sumac wax *See* Japan wax.

sumbul oil [MATER] Oil from the root of a muskroot (*Ferula sumbul*); formerly used in medicine as an antispasmodic.

sumi ink *See* India ink.

summer black oil *See* tempering oil.

summer savory oil *See* savory oil.

sump [ENG] A pit or tank which receives and temporarily stores drainage at the lowest point of a circulating or drainage system.

sump fuse [ENG] A fuse used for underwater blasting.

sump pump [MECH ENG] A small, single-stage vertical pump used to drain shallow pits or sumps.

sun-and-planet motion [MECH ENG] A train of two wheels moving epicyclically with a small wheel rotating a wheel on the central axis.

sunflower oil [MATER] A combustible, pale-yellow, semidrying oil with a pleasant scent, expressed from the seeds of the common sunflower; soluble in alcohol, ether, and carbon disulfide; consists mostly of mixed triglycerides of fatty acids; used for resins, soaps, edible oils, and margarines.

sun gear See central gear.

sunk [MIN ENG] Drilled downward, as a shaft.

sunshine integrator [ENG] An instrument for determining the duration of sunshine (daylight) in any locality.

sunshine recorder [ENG] An instrument designed to record the duration of sunshine without regard to intensity at a given location; sunshine recorders may be classified in two groups according to the method by which the time scale is obtained: in one group the time scale is obtained from the motion of the sun in the manner of a sun dial, in the second group the time scale is supplied by a chronograph.

sun-synchronous orbit [AERO ENG] An earth orbit of a spacecraft so that the craft is always in the same direction relative to that of the sun; as a result, the spacecraft passes over the equator at the same spots at the same times.

superalloy [MET] A thermally resistant alloy for use at elevated temperatures where high stresses and oxidation are encountered.

supercalendered finish [MATER] A shiny, smooth-surface paper obtained by passing the paper between alternating fiber-filled and steel rolls with the application of steam and pressure.

supercalendering [ENG] A calendering process that uses both steam and high pressure to give calendered material, for example, paper, a high-density finish.

supercentrifuge [MECH ENG] A centrifuge built to operate at faster speeds than an ordinary centrifuge.

supercharge method [ENG] A method for measuring the knock-limited power, under supercharge rich-mixture conditions, of fuels for use in spark-ignition aircraft engines.

supercharger [MECH ENG] An air pump or blower in the intake system of an internal combustion engine used to increase the weight of air charge and consequent power output from a given engine size.

supercharging [MECH ENG] A method of introducing air for combustion into the cylinder of an internal combustion engine at a pressure in excess of that which can be obtained by natural aspiration.

supercobalt drill [DES ENG] A drill made of 8% cobalt highspeed steel; used for drilling work-hardened stainless steels, silicon chrome, and certain chrome-nickel alloy steels.

superconducting alloy [MET] An alloy capable of exhibiting superconductivity, such as an alloy of niobium and zirconium or an alloy of lead and bismuth.

superconducting gyroscope See cryogenic gyroscope.

superconducting metal [MET] A metal capable of exhibiting superconductivity.

supercritical wing [AERO ENG] A wing developed to permit subsonic aircraft to maintain an efficient cruise speed very close to the speed of sound; the middle portion of the wing is relatively flat with substantial downward curvature near the trailing edge; in addition, the forward region of the lower surface is more curved than that of the upper surface with a substantial cusp of the rearward portion of the lower surface.

superficial Rockwell hardness test [MET] A test to determine surface hardness of thin sheet material which applies relatively light loads producing minimal penetration and damage.

superfines [MET] The portion of a metal powder composed of particles smaller than 10 micrometers.

superfinishing [MET] Fine honing of a metal surface with abrasive stones.

superheater [MECH ENG] A component of a steam-generating unit in which steam, after it has left the boiler drum, is heated above its saturation temperature.

superhighway [CIV ENG] A broad highway, such as an expressway, freeway, turnpike, for high-speed traffic.

superimposed back pressure [MECH ENG] The static pressure at the outlet of an operating pressure relief device, resulting from pressure in the discharge system.

superjacent roadway system See Hirschback method.

supernatant liquor [ENG] The liquid above settled solids, as in a gravity separator.

superphosphate [MATER] The most important phosphorus fertilizer, derived by action of sulfuric acid on phosphate rock (mostly tribasic calcium phosphate) to produce a mix of gypsum and monobasic calcium phosphate.

superplasticity [MET] The unusual ability of some metals and alloys to elongate uniformly by thousands of percent at elevated temperatures, much like hot polymers or glasses.

Super Sabre [AERO ENG] A United States supersonic, single-engine, turbojet-powered, tactical and air superiority fighter capable of delivering either nuclear or nonnuclear bombs, rockets, and Bullpup missiles against surface targets, or cannons and Sidewinder missiles against airborne targets; it is capable of providing close support for ground forces, and it can be refueled in flight. Designated F-100.

supersonic aircraft [AERO ENG] Aircraft capable of supersonic speeds.

supersonic airfoil [AERO ENG] An airfoil designed to produce lift at supersonic speeds.

supersonic compressor [MECH ENG] A compressor in which a supersonic velocity is imparted to the fluid relative to the rotor blades, the stator blades, or both, producing oblique shock waves over the blades to obtain a high-pressure rise.

supersonic diffuser [MECH ENG] A diffuser designed to reduce the velocity and to increase the pressure of fluid moving at supersonic velocities.

supersonic inlet [AERO ENG] An inlet of a jet engine at which single, double, or multiple shock waves form.

supersonic nozzle *See* convergent-divergent nozzle.

supersonic transport [AERO ENG] A transport plane capable of flying at speeds higher than the speed of sound. Abbreviated SST.

superstructure [CIV ENG] The part of a structure that is raised on the foundation.

supervisory control [ENG] A control panel or room showing key readings or indicators (temperature, pressure, or flow rate) from an entire operating area, allowing visual supervision and control of the overall operation.

supervisory controlled manipulation [ENG] A form of remote manipulation in which a computer enables the operator to teach the manipulator motion patterns to be remembered and repeated later.

supplied-air respirator [ENG] An atmospheric-supplying device which provides the wearer with respirable air from a source outside the contaminated area; only those with manual or motor-operated blowers are approved for immediately harmful or oxygen-deficient atmospheres.

supply control [IND ENG] The process by which an item of supply is controlled within the supply system, including requisitioning 'receipt, storage, stock control, shipment, disposition, identification, and accounting.

support base [ENG] A place from which logistic support is provided for a group of launch complexes and their control center.

suppressed-zero instrument [ENG] An indicating or recording instrument in which the zero position is below the lower end of the scale markings.

surcharge [CIV ENG] The load supported above the level of the top of a retaining wall.

surcharged wall [CIV ENG] A retaining wall with an embankment on the top.

surface [ENG] The outer part (skin with a thickness of zero) of a body; can apply to structures, to micrometer-sized particles, or to extended-surface zeolites.

surface-active agent [MATER] A soluble compound that reduces the surface tension of liquids, or reduces interfacial tension between two liquids or a liquid and a solid. Also known as surfactant.

surface air leakage [MIN ENG] The amount of surface air entering the fan through the casing at the top of the upcast shaft, the air-lock doors, and fan-drift walls.

surface analyzer [ENG] An instrument that measures or records irregularities in a surface by moving the stylus of a crystal pickup or similar device over the surface, amplifying the resulting voltage, and feeding the output voltage to an indicator or recorder that shows the surface irregularities magnified as much as 50,000 times.

surface area [ENG] Measurement of the extent of the area (without allowance for thickness) covered by a surface.

surface carburetor [MECH ENG] A carburetor in which air is passed over the surface of gasoline to charge it with fuel.

surface combustion [ENG] Combustion brought about near the surface of a heated refractory material by forcing a mixture of air and combustible gases through it or through a hole in it, or having the gas impinge directly upon it; used in muffles, crucibles, and certain types of boiler furnaces.

surface condenser [MECH ENG] A heat-transfer device used to condense a vapor, usually steam under vacuum, by absorbing its latent heat in cooling fluid, ordinarily water.

surface drilling [MIN ENG] Boreholes collared at the surface of the earth, as opposed to boreholes collared in mine workings or underwater.

surface-effect ship [MECH ENG] A transportation device with fixed side walls, which is supported by low-pressure, low-velocity air and operates on water only.

surface finish [ENG] The surface roughness of a component after final treatment, measured by a surface profile.

surface gage [DES ENG] **1.** A scribing tool in an adjustable stand, used to mark off castings and to test the flatness of surfaces. **2.** A gage for determining the distances of points on a surface from a reference plane.

surface grinder [MECH ENG] A grinding machine that produces a plane surface.

surface hardening [MET] Hardening the surface of steel by one of several processes, such as carburizing, carbonitriding, nitriding, flame or induction hardening, and surface working.

surface ignition [ENG] The initiation of a flame in the combustion chamber of an automobile engine by any hot surface other than the spark discharge.

surface lift [MIN ENG] In the freezing method of shaft sinking, freezing and heaving of the surface around the shaft due to the formation of ice and the variation of temperature.

surface mining [MIN ENG] Mining at or near the surface; includes placer mining, mining in open glory-hole or milling pits, mining and removing ore from opencuts by hand or with mechanical excavating and transportation equipment, and the removal of capping or overburden to uncover the ores.

surface pipe [PETRO ENG] The string of casing first set in a well, usually to shut off shallow, fresh-water sands from contamination by deeper, saline waters.

surface plate [DES ENG] A plate having a very accurate plane surface used for testing other surfaces or to provide a true surface for accurately measuring and locating testing fixtures.

surface rights [MIN ENG] **1.** Ownership of the surface land only, mineral rights being reserved. **2.** Ownership of the surface land plus mineral rights. **3.** The right of a mineral owner or an oil and gas lessee to use as much surface land as may be reasonably necessary for the conduct of operations under the lease.

surface rolling [MET] A cold-rolling process for hardening the surface of a metal.

surface roughness [ENG] The closely spaced unevenness of a solid surface (pits and projections) that results in friction for solid-solid movement or for fluid flow across the solid surface.

surface-set bit [DES ENG] A bit containing a single layer of diamonds set so that the diamonds protrude on the surface of the crown. Also known as single-layer bit.

surface sizing *See* sizing treatment.

surface thermometer [ENG] A thermometer, mounted in a bucket, used to measure the temperature of the sea surface.

surface treating [ENG] Any method of treating a material (metal, polymer, or wood) so as to alter the surface, rendering it receptive to inks, paints, lacquers, adhesives, and various other treatments, or resistant to weather or chemical attack.

surface vibrator [MECH ENG] A vibrating device used on the surface of a pavement or flat slab to consolidate the concrete.

surface waterproofing [ENG] Waterproofing concrete by painting a waterproofing liquid on the surface.

surfacing [MET] Depositing filler metal on a metal surface by welding or spraying.

surfacing mat *See* overlay.

surfactant *See* surface-active agent.

surfactant flooding *See* micellar flooding.

surge [ENG] **1.** An upheaval of fluid in a processing system, frequently causing a carryover (puking) of liquid through the vapor lines. **2.** The peak system pressure. **3.** An unstable pressure buildup in a plastic extruder leading to variable throughput and waviness of the hollow plastic tube.

surge bunker [MIN ENG] A large-capacity storage hopper, installed near the pit bottom or at the input end of a processing plant to provide uniform bulk deliveries.

surge column [PETRO ENG] A large-sized pipe of sufficient height to provide a static head able to absorb the surging liquid discharge of the process tank to which it is connected. Also known as boot.

surge drum *See* accumulator.

surge header *See* accumulator.

surge tank [ENG] **1.** A standpipe or storage reservoir at the downstream end of a closed aqueduct or feeder pipe, as for a water wheel, to absorb sudden rises of pressure and to furnish water quickly during a drop in pressure. Also known as surge drum. **2.** An open tank to which the top of a surge pipe is connected so as to avoid loss of water during a pressure surge. [MIN ENG] In pumping of ore pulps, a relatively small tank which maintains a steady loading of the pump.

surging [ENG] Motion of a ship that alternately moves forward and aft, usually when moored.

surveillance [ENG] Systematic observation of air, surface, or subsurface areas or volumes by visual, electronic, photographic, or other means, for intelligence or other purposes.

surveillance satellite [AERO ENG] A satellite whose function is to make systematic observations of the earth, usually by photographic means, for military intelligence or for other purposes.

survey [ENG] **1.** The process of determining accurately the position, extent, contour, and so on, of an area, usually for the purpose of preparing a chart. **2.** The information so obtained.

surveying altimeter [ENG] A barometric-type instrument consisting of a pressure-sensitive element which contracts or expands in proportion to atmospheric pressure, connected through a linkage to a pointer; its dial is graduated in units of linear measurement (feet or meters) to indicate differences of elevation only.

surveying sextant *See* hydrographic sextant.

Surveyor program [AERO ENG] A program in which unmanned spacecraft made soft landings on the moon to take photographs, analyze samples of soil, and obtain other data that could be transmitted back to earth for guidance in planning manned landings.

surveyor's compass [ENG] An instrument used to measure horizontal angles in surveying.

surveyor's cross [ENG] An instrument for setting out right angles in surveying; consists of two bars at right angles with sights at each end.

surveyor's level [ENG] A telescope and spirit level mounted on a tripod, rotating vertically and having leveling screws for adjustment.

surveyor's measure [ENG] A system of measurement used in surveying having the engineer's, or Gunter's chain, as a unit.

survivor curve [IND ENG] A curve showing the percentage of a group of machines or facilities surviving at a given age.

Surwell clinograph [ENG] A directional surveying instrument which records photographically the direction and magnitude of well deviations from the vertical; powered by batteries, it contains a box level gage (indicating vertical deviation), a gyroscopic compass (indicating azimuth direction) and a watch and a dial thermometer, so that a simultaneous record of amount and direction of deviation, temperature, and time can be made on 16-millimeter film.

susceptometer [ENG] An instrument that measures paramagnetic, diamagnetic, or ferromagnetic susceptibility.

suspended acoustical ceiling [BUILD] An acoustical ceiling which is suspended from either the roof or a higher ceiling.

suspended ceiling [BUILD] The suspension of the furring members beneath the structural members of a ceiling.

suspended formwork [CIV ENG] Formwork suspended from supports for the floor being cast.

suspended span [CIV ENG] A simple span supported from the free ends of cantilevers.

suspended tray conveyor [MECH ENG] A vertical conveyor having pendant trays or other carriers on one or more endless chains.

suspended tubbing [MIN ENG] A permanent method of lining a circular shaft, in which the tubbing (German type) is temporarily suspended from the next wedging curb above and for which no temporary supports are required; slurry is run in behind the tubbing by means of a funnel passing through the holes provided in the segments.

suspension [ENG] A fine wire or coil spring that supports the moving element of a meter. [MIN ENG] The bolting of rock to secure fragments or sections, such as small slabs barred down after blasting blocks of rock broken by fracture or joint patterns, which may subsequently loosen and fall.

suspension bridge [CIV ENG] A fixed bridge consisting of either a roadway or a truss suspended from two cables which pass over two towers and are anchored by backstays to a firm foundation.

suspension cable [ENG] A freely hanging cable; may carry mainly its own weight or a uniformly distributed load.

suspension roast *See* flash roast.

suspension roof [BUILD] A roof that is supported by steel cables.

suspension system [MECH ENG] A system of springs, shock absorbers, and other devices supporting the upper part of a motor vehicle on its running gear.

sustainer rocket engine [AERO ENG] A rocket engine that maintains the velocity of a rocket vehicle once it has achieved its programmed velocity by use of a booster or other engine.

Sutro weir [CIV ENG] A dam with at least one curved side and horizontal crest, so formed that the head above the crest is directly proportional to the discharge.

swab [MIN ENG] **1.** A pistonlike device provided with a rubber cap ring that is used to clean out debris inside a borehole or casing. **2.** The act of cleaning the inside of a tubular object with a swab. [PETRO ENG] In petroleum drilling, to pull the drill string so rapidly that the drill mud is sucked up and overflows the collar of the borehole, thus leaving an undesirably empty borehole.

swage bolt [DES ENG] A bolt having indentations with which it can be gripped in masonry.

swaging [MET] Tapering a rod or tube or reducing its diameter by any of several methods, such as forging, squeezing, or hammering.

swallow buoy *See* swallow float.

swallow float [ENG] A tubular buoy used to measure current velocities; it can be adjusted to be neutrally buoyant and to drift at a selected density level while being tracked by shipboard listening devices. Also known as neutrally buoyant float; swallow buoy.

swamp buggy [MECH ENG] A wheeled vehicle that runs on land, mud, or through shallow water; used especially in swamps.

swamper [MIN ENG] **1.** A rear brakeman in a metal mine. **2.** A laborer who assists in hauling ore and rock, coupling and uncoupling cars, throwing switches, and loading and unloading carriers.

swarf [ENG] Chips, shavings, and other fine particles removed from the workpiece by grinding tools.

swash-plate pump [MECH ENG] A rotary pump in which the angle between the drive shaft and the plunger-carrying body is varied.

sway brace [CIV ENG] One or a pair of diagonal members designed to resist horizontal forces, such as wind.

sway frame [CIV ENG] A unit in the system of members of a bridge that provides bracing against side sway; consists of two diagonals, the verticals, the floor beam, and the bottom strut.

sweat [MET] Exudate of low-melting-point constituents from a metal on solidification.

sweat cooling [AERO ENG] A technique for cooling combustion chambers or aerodynamically heated surfaces by forcing a coolant through a porous wall, resulting in film cooling at the interface. Also known as transpiration cooling.

sweated wax [MATER] A white, moisture-free petroleum wax with the oil removed by a sweating process, in which the unrefined wax is heated in shallow pans.

sweating out [MET] Bringing small globules of low-melting constituents to the surface of an alloy during heat treatment (as lead out of bronze).

sweat soldering [MET] Soldering two parts by precoating with solder and merging them by application of heat.

sweep [MET] A profile pattern used to form molds for symmetrical articles made by sweep casting.

sweepback [AERO ENG] **1.** The backward slant of a leading or trailing edge of an airfoil. **2.** The amount of this slant, expressed as the angle between a line perpendicular to the plane of symmetry and a reference line in the airfoil.

sweep-out pattern [PETRO ENG] The areal pattern of water advance in a petroleum reservoir, as for a waterflood operation.

sweet basil oil See basil oil.

sweet bay oil See volatile laurel oil.

sweet corrosion [PETRO ENG] Corrosion occurring in oil or gas wells where there is no iron-sulfide corrosion product, and

there is no odor of hydrogen sulfide in the produced reservoir fluid.

sweet crudes [MATER] Crude petroleum oil containing little sulfur.

sweet gas [MATER] A petroleum natural gas containing no corrosive components, such as hydrogen sulfide and mercaptans.

sweet oil See olive oil.

sweet orange oil [MATER] A sweet, yellow, mild essential oil expressed from the peel of an orange, *Citrus aurantium*; soluble in glacial acetic acid, somewhat in alcohol; used in flavors, perfumes, and medicine. Also known as orange oil.

sweet roast See dead roast.

swell See horseback.

swell diameter [AERO ENG] In a body of revolution having an ogival portion, such as a projectile, the swell diameter is in the diameter of the maximum transverse section of the geometrical ogive.

sweptback wing [AERO ENG] An airplane wing on which both the leading and trailing edges have sweepback, the trailing edge forming an acute angle with the longitudinal axis of the airplane aft of the root. Also known as swept wing.

swept wing See sweptback wing.

swing [ENG] **1.** The arc or curve described by the point of a pick or mandril when being used. **2.** Rotation of the superstructure of a power shovel on the vertical shaft in the mounting. **3.** To rotate a revolving shovel on its base.

swing-around trajectory [AERO ENG] A planetary round-trip trajectory which requires minimal propulsion at the destination planet, but instead uses the planet's gravitational field to effect the bulk of the necessary orbit change to return to earth.

swing bridge [CIV ENG] A movable bridge that pivots in a horizontal plane about a center pier.

swing cut See slipping cut.

swing-frame grinder [MECH ENG] A grinding machine hanging by a chain so that it may swing in all directions for surface grinding heavy work.

swing-hammer crusher [MIN ENG] A rotary crusher with rotating hammers that break up ore by impelling it against breaker plates.

swinging a claim [MIN ENG] The adjustment of the boundaries of a mining claim to more nearly conform to the strike of the vein.

swinging load [ENG] The load in pressure equipment which changes at frequent intervals.

swing joint [DES ENG] A pipe joint in which the parts may be rotated relative to each other.

swing pipe [ENG] A discharge pipe whose intake end can be raised or lowered on a tank.

swing shift [IND ENG] Working arrangement in a three-shift, continuously run plant with working hours changed at regular intervals; during a swing shift the morning shift becomes the afternoon shift, while the afternoon shift becomes the morning shift of the next day, with only an 8-hour break on the first day of change.

swirl flowmeter *See* vortex precession flowmeter.

Swiss pattern file [DES ENG] A type of fine file used for precision filing of jewelry, instrument parts, and dies.

switch [CIV ENG] **1.** A device for enabling a railway car to pass from one track to another. **2.** The junction of two tracks.

switch angle [CIV ENG] The angle between the switch and stock rails of a railroad track, measured at the point of juncture between the gage lines.

switchback [MIN ENG] A zigzag arrangement of railroad tracks by means of which a train can reach a higher or lower level by a succession of easy grades.

switchblade knife [DES ENG] A knife in which the blade is restrained by a spring and swings open when released by a pushbutton.

switching device [ENG] An electrical or mechanical device or mechanism, which can bring another device or circuit into an operating or nonoperating state. Also known as switching mechanism.

switching mechanism *See* switching device.

switchman [MIN ENG] A laborer who throws switches of mine tracks in a coal mine.

switch plate [MIN ENG] An iron plate on tramroads in mines to change the direction of movement.

swivel [DES ENG] A part that oscillates freely on a headed bolt or pin. [PETRO ENG] A short piece of casing having one end belled over a heavy ring, and having a large hole through both walls, the other end being threaded.

swivel block [DES ENG] A block with a swivel attached to its hook or shackle permitting it to revolve.

swivel coupling [MECH ENG] A coupling that gives complete rotary freedom to a deflecting wedge-setting assembly.

swivel head [MECH ENG] The assembly of a spindle, chuck, feed nut, and feed gears on a diamond-drill machine that surrounds, rotates, and advances the drill rods and drilling stem; on a hydraulic-feed drill the feed gears are replaced by a hydraulically actuated piston assembly.

swivel hook [DES ENG] A hook with a swivel connection to its base or eye.

swivel joint [DES ENG] A joint with a packed swivel that allows one part to move relative to the other.

swivel neck *See* water swivel.

swivel pin *See* kingpin.

sylvester [MIN ENG] A hand-operated device for withdrawing supports from the waste or old workings in a mine by means of a long chain which allows the device to be positioned at a safe distance from the support to be extracted.

symballophone [ENG] A double stethoscope for the comparison and lateralization of sounds; permits the use of the acute function of the two ears to compare intensity and varying quality of sounds arising in the body or mechanical devices.

symon fault *See* horseback.

Symon's cone crusher [MIN ENG] A modified gyratory crusher used in secondary ore crushing that consists of a downward-flaring bowl within which is gyrated a conical crushing head; the main shaft is gyrated by means of a long eccentric which is driven by bevel gears.

Symon's disk crusher [MIN ENG] A mill in which the crushing is done between two cup-shaped plates that revolve on shafts set at a small angle to each other.

sympathetic detonation [ENG] Explosion caused by the transmission of a detonation wave through any medium from another explosion.

synchromesh [MECH ENG] An automobile transmission device that minimizes clashing; acts as a friction clutch, bringing gears approximately to correct speed just before meshing.

synchronization [ENG] The maintenance of one operation in step with another, as in keeping the electron beam of a television picture tube in step with the electron beam of the television camera tube at the transmitter. Also known as sync.

synchronization indicator [ENG] An indicator that presents visually the relationship between two varying quantities or moving objects.

synchronized shifting [MECH ENG] Changing speed gears, with the gears being brought to the same speed before the change can be made.

synchronous [ENG] In step or in phase, as applied to two or more circuits, devices, or machines.

synchronous orbit [AERO ENG] **1.** An orbit in which a satellite makes a limited number of equatorial crossing points which are then repeated in synchronism with some defined reference (usually earth or sun). **2.** Commonly, the equatorial, circular, 24-hour case in which the satellite appears to hover over a specific point of the earth.

synchronous satellite *See* geostationary satellite.

synchronous timing [MET] Regulating the welding-transformer primary current in spot, seam, or projection welding so that the following conditions prevail: the first half-cycle is initiated at the proper time in relation to the voltage to ensure a balanced current wave, each succeeding half-cycle is essentially the same as the first, and the last half-cycle is of the opposite polarity to the first.

synchroscope [ENG] An instrument for indicating whether two periodic quantities are synchronous; the indicator may be a rotating-pointer device or a cathode-ray oscilloscope providing a rotating pattern; the position of the rotating pointer is a measure of the instantaneous phase difference between the quantities.

synchro-shutter [ENG] A camera shutter with a circuit that flashes a light the instant the shutter opens.

Syncom [AERO ENG] One of a series of communication satellites placed in synchronous equatorial orbit; used for relaying television and radio communications over long distances.

syndet *See* synthetic detergent.

synergic curve [AERO ENG] A curve plotted for the ascent of a rocket, space-air vehicle, or space vehicle, calculated to give optimum fuel economy and optimum velocity.

synergism [MATER] An action where the total effect of two active components in a mixture is greater than the sum of their individual effects, for example, a mixture volume that is greater than the sum of the individual volumes, or in resin formulation, the use of two or more stabilizers, where the combination improves polymer stability more than expected from the additive effect of the stabilizers, a material that causes such an effect is known as a synergist.

syntactic foam [MATER] A cellular polymer made by dispersing rigid, microscopic particles in a fluid polymer and then curing it.

syntactic semigroup [SYS ENG] For a sequential machine, the set of all transformations performed by all input sequences.

syntectic [MET] Isothermal, reversible conversion of a solid phase into two conjugate liquid phases on applying heat.

synthetic aperture [ENG] A method of increasing the ability of an imaging system, such as radar or acoustical holography, to resolve small details of an object, in which a receiver of large size (or aperture) is in effect synthesized by the mo-

tion of a smaller receiver and the proper correlation of the detected signals.

synthetic-aperture radar [ENG] A radar system in which an aircraft moving along a very straight path emits microwave pulses continuously at a frequency constant enough to be coherent for a period during which the aircraft may have traveled about 1 kilometer; all echoes returned during this period can then be processed as if a single antenna as long as the flight path had been used.

synthetic crude [MATER] The total liquid, multicomponent hydrocarbon mixture resulting from a process involving molecular rearrangement of charge stock, as from oil shale or synthesis gas. Also known as synthetic oil.

synthetic detergent [MATER] A liquid or solid material able to dissolve oily materials and disperse them (or emulsify them) in water. Also known as syndet.

synthetic gem [MATER] A precious or semi-precious stone made by artificial processes, for example, synthetic diamonds made by extreme heat and pressure on carbon, used industrially; and synthetic rubies made by high-temperature crystallization of aluminum oxide, used in laser equipment.

synthetic graphite [MATER] Graphitic crystalline material made by the high-temperature and pressure processing of carbon.

synthetic lubricant [MATER] Any of a group of products, some of them based on petroleum, used as lubricants where heat, chemical resistance, and other requirements can be better met than with straight petroleum products.

synthetic mica [MATER] A fluorphlogopite mica made artificially by heating a large batch of raw material in an electric resistance furnace and allowing the mica to crystallize from the melt during controlled slow cooling.

synthetic oil *See* synthetic crude.

synthetic quartz [MATER] A quartz crystal grown commercially at high temperature and pressure around a seed of quartz suspended in a solution which contains scraps of natural quartz crystals.

synthetic rubber [MATER] Synthetic products whose properties are similar to those of natural rubber, including elasticity and ability to be vulcanized; usually produced by the polymerization or copolymerization of petroleum-derived olefinic or other unsaturated compounds.

system [ENG] A combination of several pieces of equipment integrated to perform a specific function; thus a fire control system may include a tracking radar, computer, and gun.

systematic error   [ENG] An error due to some known physical law by which it might be predicted; these errors produced by the same cause affect the mean in the same sense, and do not tend to balance each other but rather give a definite bias to the mean.

systematic sampling   [MIN ENG] Extracting samples at evenly spaced periods or in fixed quantities from a unit of coal.

systematic support   [MIN ENG] The regular setting of timber or steel supports at fixed intervals irrespective of the condition of the roof and sides.

system effectiveness   [ENG] A measure of the extent to which a system may be expected to achieve a set of specific mission requirements expressed as a function of availability, dependability, and capability.

system engineering   *See* systems engineering.

system life cycle   [ENG] The continuum of phases through which a system passes from conception through disposition.

system optimization   *See* optimization.

system reliability   [ENG] The probability that a system will accurately perform its specified task under stated environmental conditions.

system safety   [ENG] The optimum degree of safety within the constraints of operational effectiveness, time, and cost, attained through specific application of system safety engineering throughout all phases of a system.

system safety engineering   [ENG] An element of systems management involving the application of scientific and engineering principles for the timely identification of hazards, and initiation of those actions necessary to prevent or control hazards within the system.

systems analysis   [ENG] The analysis of an activity, procedure, method, technique, or business to determine what must be accomplished and how the necessary operations may best be accomplished.

systems engineering   [ENG] The design of a complex interrelation of many elements (a system) to maximize an agreed-upon measure of system performance, taking into consideration all of the elements related in any way to the system, including utilization of manpower as well as the characteristics of each of the system's components. Also known as system engineering.

systems implementation test   [ENG] The test program that exercises the complete system in its actual environment to determine its capabilities and limitations; this test also demonstrates that the system is functionally operative, and is compatible with the other subsystems and supporting elements required for its operational employment.

systems test   [ENG] A test of an entire interconnected set of components for the purpose of determining proper functions and interconnections.

T

tab-card cutter [DES ENG] A device for die-cutting card stock to uniform tabulating-card size.

table [MECH ENG] That part of a grinding machine which directly or indirectly supports the work being ground. [MIN ENG] **1.** In placer mining, a wide, shallow sluice box designed to recover gold or other valuable material from screened gravel. **2.** A platform or plate on which coal is screened and picked.

table flotation [MIN ENG] A flotation process in which a slurry of ore is fed to a shaking table where flotable particles become glomerules, held together by minute air bubbles and edge adhesion; the glomerules roll across the table and are discharged nearly opposite the feed end; the process is helped by jets of low-pressure air.

tableting [ENG] A punch-and-die procedure for the compaction of powdered or granular solids; used for pharmaceuticals, food products, fireworks, vitamins, and dyes.

tabling [MIN ENG] Separation of two materials of different densities by passing a dilute suspension over a slightly inclined table having a reciprocal horizontal motion or shake with a slow forward motion and a fast return.

tachometer [ENG] An instrument that measures the revolutions per minute or the angular speed of a rotating shaft.

tack [DES ENG] A small, sharp-pointed nail with a broad flat head. [MATER] Adhesive stickiness, such as occurs on the surface of a varnish or ink that has not completely dried. Also known as tackiness.

tack coat [CIV ENG] A thin layer of bitumen, road tar, or emulsion laid on a road to enhance adhesion of the course above it.

tackiness See tack.

tackiness agent [MATER] An additive used to impart adhesive properties to otherwise nonadhesive materials, such as oils and greases.

tacking [MET] Making small, isolated tack welds.

tackle [MECH ENG] Any arrangement of ropes and pulleys to gain a mechanical advantage.

tack range [ENG] The length of time during which an adhesive will remain in the tacky-dry condition after application to an adherent.

tack weld [MET] A joint between two pieces of metal made by welding at isolated points.

tactical aircraft shelter [CIV ENG] A shelter to house fighter-type aircraft and to provide protection to the aircraft from attack by conventional weapons, or damage from high winds or other elemental hazards.

tactical air-direction center [AERO ENG] An air operations installation under the overall control of the tactical air-control center, from which aircraft and air warning service functions of tactical air operations in an area of responsibility are directed.

tactical air force [AERO ENG] An air force charged with carrying out tactical air operations in coordination with ground or naval forces.

tactical airlift [AERO ENG] That airlift which provides the immediate and responsive air movement and delivery of combat troops and supplies directly into objective areas through air landing, extraction, airdrop, or other delivery techniques; and the air logistic support of all theater forces, including those engaged in combat operations, to meet specific theater objectives and requirements.

tactical air observer [AERO ENG] An officer trained as an air observer whose function is to observe from airborne aircraft and report on movement and disposition of friendly and enemy forces, on terrain and weather and hydrography, and to execute other missions as directed.

tactical air operations [AERO ENG] An air operation involving the employment of air power in coordination with ground or naval forces to gain and maintain air superiority, to prevent movement of enemy forces into and within the objective area and to seek out and destroy these forces and their supporting installations, and to join with ground or naval forces in operations within the objective area in order to assist directly in attainment of their immediate objective.

tactical air reconnaissance [AERO ENG] The use of air vehicles to obtain information concerning terrain, weather, and the disposition, composition, movement, installations, lines of communications, and electronic and communication emissions of enemy forces.

tactical air support [AERO ENG] Air operations carried out in coordination with surface forces which directly assist the land or naval battle.

tactical air transport [AERO ENG] The use of air transport in direct support of airborne assaults, carriage of air-transported forces, tactical air supply, evacuation of casualties from forward airdromes, and clandestine operations.

tactical control radar [ENG] Antiaircraft artillery radar which has essentially the same inherent capabilities as the target acquisition radar (physically it may be the same type of set) but whose function is chiefly that of providing tactical information for the control of elements of the antiaircraft artillery defenses in battle.

tactical range recorder [ENG] A sonar device in surface ships used to plot the time-range coordinates of submarines and determine firing of depth charges.

tactical transport aircraft [AERO ENG] Aircraft designed primarily for carrying personnel or cargo over short or medium distances.

taffrail log [ENG] A log consisting essentially of a rotator towed through the water by a braided log line attached to a distance-registering device usually secured at the taffrail, the railing at the stern. Also known as patent log.

Tag-Robinson colorimeter [ENG] A laboratory device used to determine the color shades of lubricating and other oils; the color, reported as a number, is determined by varying the thickness of a column of oil until its color matches that of a standard color glass.

tail [AERO ENG] **1.** The rear part of a body, as of an aircraft or a rocket. **2.** The tail surfaces of an aircraft or a rocket.

tail assembly *See* empennage.

tail fin [AERO ENG] A fin at the rear of a rocket or other body.

tailgate [CIV ENG] The downstream gate of a canal lock. [ENG] A hinged gate at the rear of a vehicle or railroad freight car that can be let down for convenience in loading.

tailing *See* residual oil.

tailings [ENG] The lighter particles which pass over a sieve in milling, crushing, or purifying operations. [MIN ENG] **1.** The parts, or a part, of any incoherent or fluid material separated as refuse, or separately treated as inferior in quality or value. **2.** The decomposed outcrop of a vein or bed. **3.** The refuse material resulting from processing ground ore.

tailings settling tank [MIN ENG] A vessel in which solids are removed from the tailings effluent in mineral processing plants.

tail out [PETRO ENG] To place sucker rods on a rack as they are pulled from a well during oil production.

tail pulley [MECH ENG] A pulley at the tail of the belt conveyor opposite the normal discharge end; may be a drive pulley or an idler pulley.

tailrace [ENG] A channel for carrying water away from a turbine, waterwheel, or other industrial application. [MIN ENG] A channel for conveying mine trailings.

tail rope [MIN ENG] **1.** The rope which passes around the return sheave in main-and-tail haulage or a scraper loader layout. **2.** The rope that is used to draw the empty cars back into a mine in a tail-rope system. **3.** A counterbalance rope attached beneath the cage when the cages are hoisted in balance. **4.** A hemp rope used for moving pumps in shafts.

tail-rope system [MIN ENG] Haulage by a hoisting engine and two separate drums in which the main rope is attached to the front end of a trip of cars, and the tail rope is attached to the rear end of the trip.

tail sheave [MIN ENG] The return sheave for an endless rope or the tail rope of the main-and-tail-rope system, placed at the far end of a haulageway.

tailstock [MECH ENG] A part of a lathe that holds the end of the work not being shaped, allowing it to rotate freely.

tail surface [AERO ENG] A stabilizing or control surface in the tail of an aircraft or missile.

tail track system [MIN ENG] A form of track layout for car or trip loading in which the track can be extended down the heading, turned right or left, or turned back, U-fashion, in an adjacent heading.

tail warning radar [ENG] Radar installed in the tail of an aircraft to warn the pilot that an aircraft is approaching from the rear.

Tainter gate [CIV ENG] A spillway gate whose face is a section of a cylinder; rotates about a horizontal axis on the downstream end of the gate and can be closed under its own weight. Also known as radial gate.

takeoff [AERO ENG] Ascent of an aircraft or rocket at any angle, as the action of a rocket vehicle departing from its launch pad or the action of an aircraft as it becomes airborne.

takeoff assist [AERO ENG] **1.** The extra thrust given to an airplane or missile during takeoff through the use of a rocket motor or other device. **2.** The device used in such a takeoff.

takeoff weight [AERO ENG] The weight of an aircraft or rocket vehicle ready for takeoff, including the weight of the vehicle, the fuel, and the payload.

takeup [MECH ENG] A tensioning device in a belt-conveyor system for taking up slack of loose parts.

takeup pulley [MECH ENG] An adjustable idler pulley to accommodate changes in the length of a conveyor belt to maintain proper belt tension.

takeup reel [ENG] The reel that accumulates magnetic tape after it is recorded or played by a tape recorder.

talking [MIN ENG] A series of small bumps or cracking noises within the mine walls, indicating that the rock is beginning to yield to stresses.

talk-listen switch [ENG ACOUS] A switch provided on intercommunication units to permit using the loudspeaker as a microphone when desired.

tall oil [MATER] A yellow-black, malodorous, resinous admixture of rosin, fatty acids, sterols, high-molecular-weight alcohols, and other materials, derived from wood-pulping waste liquors; used in paint drying oils, alkyd resins, linoleum, soaps, lubricants, and greases. Also known as liquid rosin; tallol.

tallol See tall oil.

tallow [MATER] Animal fat with carbon chains containing 16–18 carbons, derived from cattle, sheep, and horses; used for soaps, leather dressings, candles, food, and greases, and as a chemical intermediate.

tallow-seed oil See stillingia oil.

tally [MIN ENG] A mark, number, or tin ticket placed by the miner on each car of coal or ore that is sent from the work place, thus facilitating a count or tally of all filled cars.

tamp [ENG] To tightly pack a drilled hole with clay or other stemming material after the charge has been placed.

tamping bag [ENG] A bag filled with stemming material such as sand for use in horizontal and upward sloping shotholes.

tamping bar [ENG] A piece of wood for pushing explosive cartridges or forcing the stemming into shotholes.

tamping plug [ENG] A plug of iron or wood used instead of tamping material to close up a loaded blasthole.

tanbark [MATER] The fibrous portion of ground oak or hemlock bark; it is burned in a mixture with other fuels to maintain combustion; also used on a running track for horses.

tandem [AERO ENG] The fore and aft configuration used in boosted missiles, long-range ballistic missiles, and satellite vehicles; stages are stacked together in series and are discarded at burnout of the propellant for each stage.

tandem-drive conveyor [MECH ENG] A conveyor having the conveyor belt in contact with two drive pulleys, both driven with the same motor.

tandem hoisting [MIN ENG] Hoisting in a deep shaft with two skips running in one shaft; the lower skip is suspended from the tail rope of the upper skip.

tandem mill [MET] A rolling mill consisting of two or more stands in succession, synchronized so that the metal passes directly from one to another.

tandem roller [MECH ENG] A steam- or gasoline-driven road roller in which the weight is divided between heavy metal rolls, of dissimilar diameter, one behind the other.

tang [ENG] **1.** The part of a file that fits into a handle. **2.** The end of a drill shank which allows transmission of torque from the drill press spindle to the body of the drill.

tangent bending [MET] In a single piece of metal, forming a series of identical bends with parallel axes.

tangent galvanometer [ENG] A galvanometer in which a small compass is mounted horizontally in the center of a large vertical coil of wire; the current through the coil is proportional to the tangent of the angle of deflection of the compass needle from its normal position parallel to the magnetic field of the earth.

tangential helical-flow turbine See helical-flow turbine.

tangent offset [ENG] In surveying, a method of plotting traverse lines; angles are laid out by linear measurement, using a constant times the natural tangent of the angle.

tangent screw [ENG] A screw providing tangential movement along an arc, such as the screw which provides the final angular adjustment of a marine sextant during an observation.

tangent welding [MET] Arc welding in which two or more electrodes are in a plane parallel to the line of travel.

tangerine oil See mandarin oil.

tank [ENG] A large container for holding, storing, or transporting a liquid.

tankage [ENG] Contents of a storage tank. [MATER] Slaughter-house entrails and scraps used as fertilizer.

tank balloon [ENG] An air- and vapor-tight flexible container fitted to the breather pipe of a gasoline storage tank to receive gasoline vapors; as the tank cools, the vapors return to the tank.

tank battery [PETRO ENG] A grouping of interconnected storage tanks situated to receive the output of one or more oil wells.

tank car [ENG] Railroad car onto which is mounted a cylindrical, horizontal tank designed for the transport of liquids, chemicals, gases, meltable solids, slurries, emulsions, or fluidizable solids.

tank farm [PETRO ENG] An area in which a number of large-capacity storage tanks are located, generally used for crude oil or petroleum products.

tank gage [ENG] A device used to measure the contents of a liquid storage tank; can be manual or automatic.

tank scale [ENG] A counterweighted suspension or platform weighing mechanism for tanks, hoppers, and similar solids or liquids containers.

tank switch [PETRO ENG] An automatic control of lease tanks in oil fields, including controls of lines to fill tanks and for pipeline runs; can be electrically or pneumatically actuated.

tank truck [ENG] A truck body onto which is mounted a cylindrical, horizontal tank, designed for the transport of liquids, chemicals, gases, meltable solids, slurries, emulsions, or fluidizable solids.

tanning [ENG] A process of preserving animal hides by chemical treatment (using vegetable tannins, metallic sulfates, and sulfurized phenol compounds, or syntans) to make them immune to bacterial attack, and subsequent treatment with fats and greases to make them pliable.

tanning agent [MATER] Any one of the tannins used to treat skins and hides to preserve the hide substance and to protect it from decay.

tanning extract [MATER] Tannin-rich liquor extracted from woods and pulps; used in leather tanning.

tantalum [MET] A lustrous, platinum-gray ductile metal used in making dental and surgical tools, pen points, and electronic equipment.

tantiron [MET] An iron alloy containing silicon, carbon, manganese, phosphorus, and sulfur; used for chemical equipment where resistance to acids is needed.

tap [DES ENG] **1.** A plug of accurate thread, form, and dimensions on which cutting edges are formed; it is screwed into a hole to cut an internal thread. **2.** A threaded cone-shaped fishing tool. [ENG] A small, threaded hole drilled into a pipe or process vessel; used as connection points for sampling devices, instruments, or controls. [MET] **1.** A quantity of molten metal run out from a furnace at one time. **2.** To remove excess slag from the floor of a pot furnace. [MIN ENG] To intersect with a borehole and withdraw or drain the contained liquid, as water from a water-bearing formation or from underground workings.

tap bolt [DES ENG] A bolt with a head that can be screwed into a hole and held in place without a nut. Also known as tap screw.

tap density [MET] The apparent density of a volume of metal powder obtained when its receptacle is tapped or vibrated.

tap drill [MECH ENG] A drill used to make a hole of a precise size for tapping.

tape [ENG] A graduated steel ribbon used, instead of a chain, in surveying.

tape cartridge [ENG ACOUS] A cartridge that holds a length of magnetic tape in such a way that the cartridge can be slipped into a tape recorder and played without threading the tape; in stereophonic usage, usually refers to an eight-track continuous-loop cartridge, which is larger than a cassette. Also known as cartridge.

tape-controlled machine [MECH ENG] A machine tool whose movements are automatically controlled by means of a magnetic or punched tape.

tape correction [ENG] A quantity applied to a taped distance to eliminate or reduce errors due to the physical condition of the tape and the manner in which it is used.

tape deck [ENG ACOUS] A tape-recording mechanism that is mounted on a motor board, including the tape transport, electronics, and controls, but no power amplifier or loudspeaker.

tape drive *See* tape transport.

tape-float liquid-level gage [ENG] A liquid-level measurement by a float connected by a flexible tape to a rotating member, in turn connected to an indicator mechanism.

tape gage [ENG] A box- or float-type tide gage which consists essentially of a float attached to a tape and counterpoise; the float operates in a vertical box or pipe which dampens out short-period wind waves while admitting the slower tidal movement; for the standard installation, the tape is graduated with numbers increasing toward the float and is arranged with pulleys and counterpoise to pass up and down over a fixed reading mark as the tide rises and falls.

tape loop [ENG ACOUS] A length of magnetic tape having the ends spliced together to form an endless loop; used in message repeater units and in some types of tape cartridges to eliminate the need for rewinding the tape.

tape player [ENG ACOUS] A machine designed only for playback of recorded magnetic tapes.

taper [AERO ENG] An airfoil feature in which either the thickness or the chord length or both decrease from the root to the tip.

taper bit [DES ENG] A long, cone-shaped noncoring bit used in drilling blastholes and in wedging and reaming operations.

tape recorder [ENG ACOUS] A device that records audio signals and other information on magnetic tape by selective magnetization of iron oxide particles that form a thin film on the tape; a recorder usually also includes provisions for playing back the recorded material.

tape recording [ENG ACOUS] The record made on a magnetic tape by a tape recorder.

tapered core bit [DES ENG] A core bit having a conical diamond-inset crown surface tapering from a borehole size at the bit face to the next larger borehole size at its upper, shank, or reaming-shell end.

tapered joint [DES ENG] A firm, leakproof connection between two pieces of pipe having the thread formed with a slightly tapering diameter.

tapered pipeline [PETRO ENG] A changing of the pressure grade, either by change of wall thickness or material, of pipeline sections as working pressure is lessened.

tapered thread [DES ENG] A screw thread cut on the surface of a tapered part; it may be either a pine or box thread, or a V-, Acme, or square-screw thread.

tapered wheel [DES ENG] A flat-face grinding wheel with greater thickness at the hub than at the face.

taper-in-thickness ratio [AERO ENG] A gradual change in the thickness ratio along the wing span with the chord remaining constant.

taper key [DES ENG] A rectangular machine key that is slightly tapered along its length.

taper pin [DES ENG] A small, tapered self-holding peg or nail used to connect parts together.

taper pipe thread See pipe thread.

taper plug gage [DES ENG] An internal gage in the shape of a frustrum of a cone used to measure internal tapers.

taper reamer [DES ENG] A reamer whose fluted portion tapers toward the front end.

taper ring gage [DES ENG] An external gage having a conical internal contour; used to measure external tapers.

taper-rolling bearing [MECH ENG] A roller bearing capable of sustaining end thrust by means of tapered rollers and coned races.

taper shank [DES ENG] A cone-shaped part on a tool that fits into a tapered sleeve on a driving member.

taper tap [DES ENG] A threaded cone-shaped tool for cutting internal screw threads.

tape speed [ENG ACOUS] The speed at which magnetic tape moves past the recording head in a tape recorder; standard speeds are $15/16$, $1\frac{7}{8}$, $3\frac{3}{4}$, $7\frac{1}{2}$, 15, and 30 inches per second (2.38125, 4.7625, 9.525, 19.05, 38.1, and 76.2 centimeters per second); faster speeds give improved high-frequency response under given conditions.

tape transport [ENG ACOUS] The mechanism of a tape recorder that holds the tape reels, drives the tape past the heads, and controls various modes of operation. Also known as tape drive.

taphole [MET] A hole in a furnace or ladle through which molten metal is tapped.

taping [ENG] The process of measuring distances with a surveyor's tape.

tappet [MECH ENG] A lever or oscillating member moved by a cam and intended to tap or touch another part, such as a push rod or valve system.

tappet rod [MECH ENG] A rod carrying a tappet or tappets, as one for opening or closing the valves in a steam or an internal combustion engine.

tapping [MECH ENG] Forming an internal screw thread in a hole or other part by means of a tap. [MET] Opening the pouring hole of a melting furnace to remove molten metal.

tap wrench [ENG] A tool used to clamp taps during tapping operations.

tar [MATER] A viscous material composed of complex, high-molecular-weight compounds derived from the distillation of petroleum or the destructive distillation of wood or coal.

tar acid [MATER] A mixture of phenols (phenols, cresols, and xylenols) found in tars and tar distillates; toxic, combustible, and soluble in alcohol and coal-tar hydrocarbons; used as a wood preservative and an insecticide for farm animals and also to make disinfectants.

tar distillate [MATER] A petroleum product produced by a tar still to which is charged the tarlike bottoms from continuous crude stills, pressure stills, cracking coils, or other petroleum refinery equipment.

target [ENG] **1.** The sliding weight on a leveling rod used in surveying to enable the staffman to read the line of collimation. **2.** The point that a borehole or an exploratory work is intended to reach.

3. In radar and sonar, any object capable of reflecting the transmitted beam.

target acquisition [AERO ENG] The process of optically, manually, mechanically, or electronically orienting a tracking system in direction and range to lock on a target.

target acquisition radar [ENG] An antiaircraft artillery radar, normally of lesser range capabilities but of greater inherent accuracy than that of surveillance radar, whose normal function is to acquire aerial targets either by independent search or on direction of the surveillance radar, and to transfer these targets to tracking radars.

target approach point [AERO ENG] In air transport operations, a navigational checkpoint over which the final turn-in to the drop zone–landing zone is made.

target drone [AERO ENG] A pilotless aircraft controlled by radio from the ground or from a mother ship and used exclusively as a target for antiaircraft weapons.

target pattern [AERO ENG] The flight path of aircraft during the attack phase.

target-type flowmeter [ENG] A fluid-flow measurement device with a small circular target suspended centrally in the flow conduit; the target transmits force to a force-balance transmitter by means of a pivoted bar.

tariff [IND ENG] A government-imposed duty on imported or exported goods.

tarnish [MET] Discoloration of a metal surface due to the formation of a thin film of oxide, sulfide, or some other corrosion product.

tar paper [MATER] Heavy construction paper coated or impregnated with tar.

tarpaulin [MATER] A sheet of waterproof canvas or other material; used to cover and protect construction materials and equipment, athletic fields, vehicles, or other exposed objects.

tarragon oil *See* estragon oil.

tarring [ENG] The coating of piles for permanent underground work with prepared acid-free tar.

taut-band ammeter [ENG] A modification of the permanent-magnet movable-coil ammeter in which the jeweled bearings and control springs are replaced by a taut metallic band rigidly held at the ends; the coil is firmly attached to the band, and restoring torque is supplied by twisting of the band.

taut-line cableway [MECH ENG] A cableway whose operation is limited to the distance between two towers, usually 3000 feet (914 meters) apart, has only one carrier, and the traction cable is reeved at the carrier so that loads can be raised and lowered; the towers can be mounted on trucks or crawlers, and the machine shifted across a wide area.

tawing [ENG] A tanning process in which alum is used as a partial tannage, supplementing or replacing chrome.

taxi channel [CIV ENG] A defined path, on a water airport, intended for the use of taxiing aircraft.

taxiway [CIV ENG] A specially prepared or designated path on an airport for taxiing aircraft.

Taylor process [MET] A process for making extremely fine wire by stretching wire in a glass tube at elevated temperatures, or drawing wire through a bead of molten glass and then through dies.

T beam [CIV ENG] A metal beam or bar with a T-shaped cross section.

T bolt [DES ENG] A bolt with a T-shaped head, made to fit into a T-shaped slot in a drill swivel head or in the bed of a machine.

teakwood [MATER] The strong, durable, yellowish-brown wood obtained from the teak tree, *Tectona grandis*.

tear down [ENG] **1.** To disassemble a drilling rig preparatory to moving it to another drill site. **2.** To disassemble a machine or change the jigs and fixtures.

tear-down time [IND ENG] The downtime of a machine following a given work order which usually involves removing parts such as jigs and fixtures and which must be completely finished before setting up for the next order.

teardrop balloon [AERO ENG] A sounding balloon which, when operationally inflated, resembles an inverted teardrop; this shape was determined primarily by aerodynamic considerations of the problem of obtaining maximum stable rates of a balloon ascension.

tear gas [MATER] A substance (usually liquid) which, when atomized and of a certain concentration, causes temporary but intense eye irritation and a blinding flow of tears; chloroacetophenone is a common tear gas. Also known as lacrimator.

technetium [MET] Silver-gray metal with a high melting point, slightly magnetic.

technical characteristics [ENG] Those characteristics of equipment which pertain primarily to the engineering principles involved in producing equipment possessing desired characteristics; for example, for electronic equipment; technical characteristics include such items as circuitry, and types and arrangement of components.

technical cohesive strength [MET] Fracture stress in a notched tensile test.

technical evaluation [ENG] The study and investigation to determine the technical suitability of material, equipment, or a system.

technical information [ENG] Information, including scientific information, which relates to research, development, engineering, testing, evaluation, production, operation, use, and maintenance of equipment.

technical inspection [ENG] Inspection of equipment to determine whether it is serviceable for continued use or needs repairs.

technical maintenance [ENG] A category of maintenance that includes the replacement of unserviceable major parts, assemblies, or subassemblies, and the precision adjustment, testing, and alignment of internal components.

technical representative [IND ENG] A person who represents one or more manufacturers in an area and who gives technical advice on the application, installation, operation, and maintenance of their products, in addition to selling the products.

technical specifications [ENG] A detailed description of technical requirements stated in terms suitable to form the basis for the actual design-development and production processes of an item having the qualities specified in the operational characteristics.

technical white oil *See* white oil.

tectonics [CIV ENG] **1.** The science and art of construction with regard to use and design. **2.** Design relating to crustal deformations of the earth.

tectonometer [ENG] An apparatus, including a microammeter, used on the surface to obtain knowledge of the structure of the underlying rocks.

tee [ENG] Shaped like the letter T.

tee joint [ENG] A joint in which members meet at right angles, forming a T.

teel oil *See* sesame oil.

teeming [MET] Pouring molten metal, usually a ferrous metal, into an ingot mold from a furnace or ladle.

Tego [MATER] A thin film phenol-formaldehyde adhesive, used with thin veneers.

telegraph buoy [ENG] A buoy used to mark the position of a submarine telegraph cable.

telemeteorograph [ENG] Any meteorological instrument, such as a radiosonde, in which the recording instrument is located at some distance from the measuring apparatus; for example, a meteorological telemeter.

telemeteorography [ENG] The science of the design, construction, and operation of various types of telemeteorographs.

telemeter [ENG] **1.** The complete measuring, transmitting, and receiving apparatus for indicating or recording the value of a quantity at a distance. Also known as telemetering system. **2.** To transmit the value of a measured quantity to a remote point.

telemetering [ENG] Transmitting the readings of instruments to a remote location by means of wires, radio waves, or other means. Also known as remote metering; telemetry.

telemetering wave buoy [ENG] A buoy assembly that transmits a radio signal that varies in frequency proportional to the vertical acceleration experienced by the buoy, thereby conveying information about the buoy's vertical motion as it rides the waves.

telemetry *See* telemetering.

telephone *See* telephone set.

telephone dial [ENG] A switch operated by a finger wheel, used to make and break a pair of contacts the required number of times for setting up a telephone circuit to the party being called.

telephone receiver [ENG ACOUS] The portion of a telephone set that converts the audio-frequency current variations of a telephone line into sound waves, by the motion of a diaphragm activated by a magnet whose field is varied by the electrical impulses that come over the telephone wire.

telephone set [ENG ACOUS] An assembly including a telephone transmitter, a telephone receiver, and associated switching and signaling devices. Also known as telephone.

telephone transmitter [ENG ACOUS] The microphone used in a telephone set to convert speech into audio-frequency electric signals.

telephotometer [ENG] A photometer that measures the received intensity of a distant light source.

telepsychrometer [ENG] A psychrometer in which the wet- and dry-bulb thermal elements are located at a distance from the indicating elements.

telerecording bathythermometer [ENG] A device which transmits measurements of sea water depth and temperature over a wire to a ship, where a graph of temperature versus depth is recorded.

telescope [ENG] Any device that collects radiation, which may be in the form of electromagnetic or particle radiation, from a limited direction in space.

telescopic alidade [ENG] An alidade used with a plane table, consisting of a telescope mounted on a straightedge ruler, fitted with a level bubble, scale, and vernier to measure angles, and calibrated to measure distances.

telescopic derrick [ENG] A drill derrick divided into two or more sections, with the uppermost sections nesting successively into the lower sections.

telescopic loading trough [MIN ENG] A shaker conveyor trough of two sections, one nested in the other, used near the face for advancing the trough line without the necessity of adding either a standard or a short length of pan after each cut.

telescopic tripod [ENG] A drill or surveyor's tripod each leg of which is a series of two or more closely fitted nesting tubes, which can be locked rigidly together in an extended position to form a long leg or nested one within the other for easy transport.

telescoping gage [DES ENG] An adjustable internal gage with a telescoping plunger that expands under spring tension in the hole to be measured; it is locked into position to allow measurement after being withdrawn from the hole.

telescoping valve [MECH ENG] A valve, with sliding, telescoping members, to regulate water flow in a pipe line with minimum disturbance to stream lines.

telethermometer [ENG] A temperature-measuring system in which the heat-sensitive element is located at a distance from the indicating element.

telethermoscope [ENG] A temperature telemeter, frequently used in a weather station to indicate the temperature at the instrument shelter located outside.

television film scanner [ENG] A motion picture projector adapted for use with a television camera tube to televise 24-frame-per-second motion picture film at the 30-frame-per-second rate required for television.

television satellite [AERO ENG] An orbiting satellite that relays television signals between ground stations.

television tower [ENG] A tall metal structure used as a television transmitting antenna, or used with another such structure to support a television transmitting antenna wire.

telford pavement [CIV ENG] A road pavement having a firm foundation of large stones and stone fragments, and a smooth hard-rolled surface of small stones.

telltale [ENG] A marker on the outside of a tank that indicates on an exterior scale the amount of fluid inside the tank.

telltale float [CIV ENG] A water-level indicator in a reservoir.

tellurometer [ENG] A microwave instrument used in surveying to measure distance; the time for a radio wave to travel from one observation point to the other and return is measured and converted into distance by phase comparison, much as in radar.

telpher [MECH ENG] An electric hoist hanging from and driven by a wheeled cab rolling on a single overhead rail or a rope.

Telsmith breaker [MECH ENG] A type of gyratory crusher, often used for primary crushing; consists of a spindle mounted in a long eccentric sleeve which rotates to impart a gyratory motion to the crushing head, but gives a parallel stroke, that is, the axis of the spindle describes a cylinder rather than a cone, as in the suspended spindle gyratory.

Telstar satellite [AERO ENG] A spherical active repeater communications satellite, first launched on July 10, 1962.

temper [ENG] **1.** To moisten and mix clay, plaster, or mortar to the proper consistency for use. **2.** To anneal or toughen glass. [MET] **1.** The hardness and strength of a rolled metal. **2.** The nominal carbon content of steel. **3.** To soften hardened steel or cast iron by reheating to some temperature below the eutectoid temperature. **4.** An alloy added to pure tin to make the finest pewter.

tempera [MATER] **1.** An opaque watercolor paint consisting of pigment ground in water and mixed with egg yolk. **2.** A poster paint that uses glue or gum as a binder.

temperature-actuated pressure relief valve [MECH ENG] A pressure relief valve which operates when subjected to increased external or internal temperature.

temperature-chlorinity-depth recorder [ENG] An instrument in which an underwater unit suspended from a cable records temperature, chlorinity, and depth sequentially on a single-pen strip recorder, each quantity being recorded for several seconds at a time.

temperature control [ENG] A control used to maintain the temperature of an oven, furnace, or other enclosed space within desired limits.

temperature error [ENG] That instrument error due to nonstandard temperature of the instrument.

temperature-indicating compound [MATER] A temperature-sensitive material with a predetermined melting point; used to indicate when a predesignated temperature is reached in such processes as heat treating, welding, and molding.

temperature log [PETRO ENG] A continuous record of temperature versus depth in an oil-well borehole.

temperature profile recorder [ENG] A portable instrument for measuring temperature as a function of depth in shallow water, particularly in lakes, in which a thermistor element transmits data over an electrical cable to a recording drum and depth is measured by the amount of wire paid out.

temperature sensor [ENG] A device designed to respond to temperature stimulation.

temperature survey [PETRO ENG] An analysis of temperature changes or differences down an oil-well borehole, based on a temperature log; used to locate the top of casing cement, lost circulation zones, or gas entry zones.

temperature transducer [ENG] A device in an automatic temperature-control system that converts the temperature into some other quantity such as mechanical movement, pressure, or electric voltage; this signal is processed in a controller, and is applied to an actuator which controls the heat of the system.

temper brittleness [MET] A brittle state resulting when certain low-alloy steels are slowly cooled in a range of 600–300°C or reheated in this range after quenching from 600°C.

tempering [MET] Heat treatment of hardened steels to temperatures below the transformation temperature range, usually to improve toughness.

tempering air [ENG] Low-temperature air added to a heated airstream to regulate the stream temperature.

tempering oil [MATER] A high-viscosity neutral petroleum oil, such as a steam cylinder stock, used for the drawing or tempering of steel. Also known as steam-hammer oil; summer black oil.

temper screw See stirrup.

temper time [MET] In resistance welding, that part of the postweld interval during which the current is suitable for tempering or heat treatment.

tempilstick [MATER] A crayon that, when applied to a surface, indicates when the surface temperature exceeds a given value by changing color.

template [ENG] **1.** A two-dimensional representation of a machine or other equipment used for building layout design. **2.** A guide or a pattern used in manufacturing items. Also spelled templet.

templet See template.

tendon [CIV ENG] A steel bar or wire that is tensioned, anchored to formed concrete, and allowed to regain its initial length to induce compressive stress in the concrete before use.

tenon [ENG] A tonguelike projection from the end of a framing member which is made to fit into a mortise.

tensile bar [ENG] A molded, cast, or machined specimen of specified cross-sectional dimensions used to determine the tensile properties of a material by use of a calibrated pull test. Also known as tensile specimen; test specimen.

tensile specimen See tensile bar.

tensile test [ENG] A test in which a specimen is subjected to increasing longitudinal pulling stress until fracture occurs.

tension jack [MIN ENG] A type of jack with a jackscrew for wedging against the mine roof and a ratchet device for applying tension on a chain that is attached to the tail or foot section of a belt conveyor, and used to restore the proper tension to the belt.

tension packer [PETRO ENG] A device to pressure-seal the annular space between an oil-well casing and tubing, held in place by tension against an upward push; a type of production packer.

tension pulley [MECH ENG] A pulley around which an endless rope passes mounted on a trolley or other movable bearing so that the slack of the rope can be readily taken up by the pull of the weights.

tension-type hanger [PETRO ENG] A type of tubing hanger for multiple-completion oil wells to allow for the varying lengths of tubing strings.

teraohmmeter [ENG] An ohmmeter having a teraohm range for measuring extremely high insulation resistance values.

terminal clearance capacity [ENG] The amount of cargo or personnel that can be moved through and out of a terminal on a daily basis.

terminal operations [ENG] The reception, processing, and staging of passengers; the receipt, transit storage, and marshaling of cargo; the loading and unloading of ships or aircraft; and the manifesting and forwarding of cargo and passengers to destination.

terminal pressure [ENG] A pressure drop across a unit when the maximum allowable pressure drop is reached, as for a filter press.

terminal unit [MECH ENG] In an air-conditioning system, a unit at the end of a branch duct through which air is transferred or delivered to the conditioned space.

termite shield [BUILD] A strip of metal, usually galvanized iron, bent down at the edges and placed between the foundation of a house and a timber floor, around pipes, and other places where termites can pass.

ternary alloy [MET] An alloy composed of three principal elements.

terne [MET] A lead alloy having a composition of 10–20% tin and 80–90% lead; used to coat iron or steel surfaces.

terneplate [MATER] A sheet of iron or steel coated with a lead-tin alloy; used chiefly for roofing.

terra alba [MATER] Pure white uncalcined gypsum, used as a filler in paper and paints and as a nutrient in growing yeast.

terrace [BUILD] **1.** A flat roof. **2.** A colonnaded promenade. **3.** An open platform extending from a building, usually at ground level.

terra-cotta [MATER] A brownish-orange clay used in the production of high-quality earthenware, vases, and statuettes, and for tile floors and roofing.

terrain clearance indicator See absolute altimeter.

terrain profile recorder See airborne profile recorder.

terrain sensing [ENG] The gathering and recording of information about terrain surfaces without actual contact with the object or area being investigated; in particular, the use of photography, radar, and infrared sensing in airplanes and artificial satellites.

terra japonica See gambir.

terrazzo [MATER] A mosaic flooring surface made by embedding marble or granite chips in mortar, allowing the mortar to harden, and then grinding and polishing the surface.

tertiary air [MECH ENG] Combustion air added to primary and secondary air.

tertiary creep [MET] Creep strain occurring at an accelerating rate leading to fracture.

tertiary crushing [MIN ENG] **1.** The preliminary breaking down of run-of-mine ore and sometimes coal. **2.** The third stage in crushing, following primary and secondary crushing.

tertiary grinding [MIN ENG] The third-stage grinding in a ball mill when a particularly fine grinding of ore is needed.

tessara [MATER] A small rectangular piece of ceramic tile, stone, or other material used in a mosaic design.

test [IND ENG] A procedure in which the performance of a product is measured under various conditions. [PETRO ENG] A procedure for the analysis of current, potential and ultimate product flow, and pressure-decline properties of various types of petroleum reservoirs.

test-bed [AERO ENG] A base, mount, or frame within or upon which a piece of equipment, especially an engine, is secured for testing.

test chamber [AERO ENG] The test section of a wind tunnel. [ENG] A place, section, or room having special characteristics where a person or object is subjected to experimental procedures, as an altitude chamber.

test firing [AERO ENG] The firing of a rocket engine, either live or static, with the purpose of making controlled observations of the engine or of an engine component.

test flight [AERO ENG] A flight to make controlled observations of the operation or performance of an aircraft or a rocket, or of a component of an aircraft or rocket.

test hole [MIN ENG] A drill hole or shallow excavation to assess an ore body or to obtain rock samples to determine their structural and physical characteristics.

test oscillator See signal generator.

test pile [CIV ENG] A pile equipped with a platform on which a load of sand or pig iron is placed in order to determine the load a pile can support (usually twice the working load) without settling.

test pit [CIV ENG] An open excavation used to obtain soil samples in foundation studies.

test section [AERO ENG] The section of a wind tunnel where objects are tested to determine their aerodynamic characteristics.

test specimen See tensile bar.

test stand [AERO ENG] A stationary platform or table, together with any testing apparatus attached thereto, for testing or proving engines and instruments.

test well [PETRO ENG] A well to determine the presence of petroleum oil and its potential commercial value in terms of abundance and accessibility.

tetrytol [MATER] A high-explosive mixture of tetryl and trinitrotoluene (TNT) in any of several proportions which permit melt loading.

Texas tower [ENG] A radar tower built in the sea offshore, to serve as part of an early-warning radar network; it resembles offshore oil derricks in the Gulf of Mexico.

textile [MATER] A material made of natural or man-made fibers and used for the manufacture of items such as clothing and furniture fittings.

textile oil [MATER] A specially compounded oil used to condition raw textile fibers, yarn, and fabric for manufacturing and finishing operations.

textile softener [MATER] A chemical that attaches molecularly to textile fibers with the polar (charged) end of the cation oriented toward the fiber and the fatty tail exposed to give a feeling of softness to the fabric.

text-to-speech synthesizer [ENG ACOUS] A voice response system that provides an automatic means to take a specification of any English text at the input and generate a natural and intelligible acoustic speech signal at the output by using complex sets of rules for predicting the needed phonemic states directly from the input message and dictionary pronunciations.

thallium [MET] Bluish-white metal with tinlike malleability, but a little softer; used in alloys.

thaw house [ENG] A small building that is designed for thawing frozen dynamite and which is capacious enough for a supply of thawed dynamite for a day's work.

thawing [ENG] Warming dynamite, to reduce risk of premature explosion. [MIN ENG] Working permanently frozen ground by pumping water at a temperature of from 50 to 60°F (10 to 15.5°C) through pipes down into the frozen gravel.

thaw pipe [MIN ENG] A string of pipe drilled into a string of drill rods that is frozen in a borehole in permafrost, through which water is circulated to thaw the ice and free the drill rods.

theobroma oil *See* cocoa butter.

theoretical air [ENG] The amount of air theoretically required for complete combustion.

theoretical relieving capacity [MECH ENG] The capacity of a theoretically perfect nozzle calculated in volumetric or gravimetric units.

Therberg system [IND ENG] A system of categorizing hand movements that is used in the standard motion-and-time analysis technique.

therblig *See* elemental motion.

therblig chart [IND ENG] An operation chart with the suboperations divided into basic motions, all designated with appropriate symbols.

thermal ammeter *See* hot-wire ammeter.

thermal analysis [MET] Determining transformations in a metal by observing the temperature-time relationship during uniform cooling or heating; phase transformations are indicated by irregularities in a smooth curve.

thermal-arrest calorimeter [ENG] A vacuum device for measurement of heats of fusion; a sample is frozen under vacuum and allowed to melt as the calorimeter warms to room temperature.

thermal barrier [AERO ENG] A limit to the speed of airplanes and rockets in the atmosphere imposed by heat from friction between the aircraft and the air, which weakens and eventually melts the surface of the aircraft. Also known as heat barrier.

thermal bulb [ENG] A device for measurement of temperature; the liquid in a bulb expands with increasing temperature, pressuring a spiral Bourdon-type tube element and causing it to deform (unwind) in direct relation to the temperature in the bulb.

thermal compressor [MECH ENG] A steam-jet ejector designed to compress steam at pressures above atmospheric.

thermal conductivity gage [ENG] A pressure measurement device for high-vacuum systems; an electrically heated wire is exposed to the gas under pressure, the thermal conductivity of which changes with changes in the system pressure.

thermal cutting [MET] A group of processes to sever metals by melting or by chemical reaction of oxygen with the metal at elevated temperatures.

thermal detector *See* bolometer.

thermal flame safeguard [MECH ENG] A thermocouple located in the pilot flame of a burner; if the pilot flame is extinguished, an elective circuit is interrupted and the fuel supply is shut off.

thermal gasoline [MATER] Gasoline produced in petroleum refineries by thermal processes, for example, thermal cracking and thermal reforming.

thermal instrument [ENG] An instrument that depends on the heating effect of an electric current, such as a thermocouple or hot-wire instrument.

thermal-loss meter *See* heat-loss flowmeter.

thermal mapper *See* line scanner.

thermal microphone [ENG ACOUS] Microphone depending for its action on the variation in the resistance of an electrically heated conductor that is being alternately increased and decreased in temperature by sound waves.

thermal power plant [ENG] A facility to produce electric energy from thermal energy released by combustion of a fuel or consumption of a fissionable material.

thermal probe [ENG] An instrument which measures the heat flow from ocean bottom sediment. [MECH ENG] A calorimeter in a boiler furnace which measures heat absorption rates.

thermal relief [ENG] A valve or other device that is preset to open when pressure becomes excessive due to increased temperature of the system.

thermal spraying [MET] Spraying finely divided particles of powder or droplets of atomized metal wire or rod for coating a substrate.

thermal telephone receiver [ENG ACOUS] A thermophone used as a telephone receiver.

thermal transducer [ENG] Any device which converts energy from some form other than heat energy into heat energy; an example is the absorbing film used in the thermal pulse method.

thermal wattmeter [ENG] A wattmeter in which thermocouples are used to measure the heating produced when a current is passed through a resistance.

thermic boring [ENG] Boring holes into concrete by means of a high temperature, produced by a steel lance packed with steel wool which is ignited and kept burning by oxyacetylene or other gas.

thermit *See* thermite.

thermite [MATER] A fire-hazardous mixture of ferric oxide and powdered aluminum; upon ignition by a magnesium ribbon, it reaches a temperature of 4000°F (2200°C), sufficient to soften steel; used for industrial purposes or as an incendiary bomb. Also spelled thermit.

thermit process [MET] An exothermic reaction when heating finely divided aluminum on a metal oxide causing reduction of the oxide.

thermit welding [MET] Welding with molten iron which is obtained by igniting aluminum and an iron oxide in a crucible, whereby the aluminum floats to the top of the molten metal and is poured off.

thermoammeter [ENG] An ammeter that is actuated by the voltage generated in a thermocouple through which is sent the current to be measured; used chiefly for measuring radio-frequency currents. Also known as electrothermal ammeter; thermocouple ammeter.

thermocompression bonding [ENG] Use of a combination of heat and pressure to make connections, as when attaching beads to integrated-circuit chips; examples include wedge bonding and ball bonding.

thermocompression evaporator [MECH ENG] A system to reduce the energy requirements for evaporation by compressing the vapor from a single-effect evaporator so that the vapor can be used as the heating medium in the same evaporator.

thermocouple [ENG] A device consisting basically of two dissimilar conductors joined together at their ends; the thermoelectric voltage developed between the two junctions is proportional to the temperature difference between the junctions, so the device can be used to measure the temperature of one of the junctions when the other is held at a fixed, known temperature, or to convert radiant energy into electric energy.

thermocouple ammeter *See* thermoammeter.

thermocouple pyrometer *See* thermoelectric pyrometer.

thermocouple vacuum gage [ENG] A vacuum gage that depends for its operation on the thermal conduction of the gas present; pressure is measured as a function of the voltage of a thermocouple whose measuring junction is in thermal contact with a heater that carries a con-

stant current; ordinarily, used over a pressure range of 10^{-1} to 10^{-3} millimeter of mercury.

thermoelectric cooler [ENG] An electronic heat pump based on the Peltier effect, involving the absorption of heat when current is sent through a junction of two dissimilar metals; it can be mounted within the housing of a device to prevent overheating or to maintain a constant temperature.

thermoelectric cooling [ENG] Cooling of a chamber based on the Peltier effect; an electric current is sent through a thermocouple whose cold junction is thermally coupled to the cooled chamber, while the hot junction dissipates heat to the surroundings. Also known as thermoelectric refrigeration.

thermoelectric heating [ENG] Heating based on the Peltier effect, involving a device which is in principle the same as that used in thermoelectric cooling except that the current is reversed.

thermoelectric laws [ENG] Basic relationships used in the design and application of thermocouples for temperature measurement; for example, the law of the homogeneous circuit, the law of intermediate metals, and the law of successive or intermediate temperatures.

thermoelectric pyrometer [ENG] An instrument which uses one or more thermocouples to measure high temperatures, usually in the range between 800 and 2400°F (425 and 1315°C). Also known as thermocouple pyrometer.

thermoelectric refrigeration *See* thermoelectric cooling.

thermoelectric series [MET] A series of metals arranged in order of their thermoelectric voltage-generating ratings with respect to some reference metal, such as lead.

thermoelectric thermometer [ENG] A type of electrical thermometer consisting of two thermocouples which are series-connected with a potentiometer and a constant-temperature bath; one couple, called the reference junction, is placed in a constant-temperature bath, while the other is used as the measuring junction.

thermoforming [ENG] Forming of thermoplastic sheet by heating it and then pulling it down onto a mold surface to shape it.

thermogalvanic corrosion [MET] Corrosion associated with the passage of an electric current in which the anode and cathode are at different temperatures, the anode usually being the colder of the two.

thermogalvanometer [ENG] Instrument for measuring small high-frequency currents by their heating effect, generally consisting of a direct-current galvanometer connected to a thermocouple that is heated by a filament carrying the current to be measured.

thermograd probe [ENG] An instrument that makes a record of temperature versus depth as it is lowered to the ocean floor, and measures heat flow through the ocean floor.

thermogram [ENG] The recording made by a thermograph.

thermograph [ENG] An instrument that senses, measures, and records the temperature of the atmosphere. Also known as recording thermometer.

thermograph correction card [ENG] A table for quick and accurate correction of the reading of a thermograph to that of the more accurate dry-bulb thermometer at the same time and place.

thermography [ENG] A method of measuring surface temperature by using luminescent materials: the two main types are contact thermography and projection thermography.

thermointegrator [ENG] An apparatus, used in studying soil temperatures, for measuring the total supply of heat during a given period; it consists of a long nickel coil (inserted into the soil by an attached rod) forming a 100-ohm resistance thermometer and a 6-volt battery, the current used being recorded on a galvanometer; a mercury thermometer can be used.

thermojet [AERO ENG] Air-duct-type engine in which air is scooped up from the surrounding atmosphere, compressed, heated by combustion, and then expanded and discharged at high velocity.

thermometal [MET] A bimetallic strip which, on temperature change, deflects because of differences in the coefficients of expansion of the two bonded metals.

thermometer [ENG] An instrument that measures temperature.

thermometer anemometer [ENG] An anemometer consisting of two thermometers, one with an electric heating element connected to the bulb; the heated bulb cools in an airstream, and the difference in temperature as registered by the heated and unheated thermometers can be translated into air velocity by a conversion chart.

thermometer-bulb liquid-level meter [ENG] Detection of liquid level by temperature measurement changes using an immersed bulb-type thermometer.

thermometer frame [ENG] A frame designed to hold two or more reversing thermometers; such a frame is often attached directly to a Nansen bottle.

thermometer screen *See* instrument shelter.

thermometer shelter *See* instrument shelter.

thermometer support [ENG] A device used to hold liquid-in-glass maximum and minimum thermometers in the proper recording position inside an instrument shelter, and to permit them to be read and reset.

thermophone [ENG ACOUS] An electroacoustic transducer in which sound waves having an accurately known strength are produced by the expansion and contraction of the air adjacent to a strip of conducting material, whose temperature varies in response to a current input that is the sum of a steady current and a sinusoidal current; used chiefly for calibrating microphones.

thermopile [ENG] An array of thermocouples connected either in series to give higher voltage output or in parallel to give higher current output, used for measuring temperature or radiant energy or for converting radiant energy into electric power.

thermoplastic insulation [MATER] Electrical insulation made of a thermoplastic material.

thermoplastic resin [MATER] A material with a linear macromolecular structure that will repeatedly soften when heated and harden when cooled; for example, styrene, acrylics, cellulosics, polyethylenes, vinyls, nylons, and fluorocarbons.

thermoregulator [ENG] A high-accuracy or high-sensitivity thermostat; one type consists of a mercury-in-glass thermometer with sealed-in electrodes, in which the rising and falling column of mercury makes and breaks an electric circuit.

thermoscreen *See* instrument shelter.

thermosetting resin [MATER] A plastic that solidifies when first heated under pressure, and which cannot be remelted or remolded without destroying its original characteristics; examples are epoxies, malamines, phenolics, and ureas.

thermosiphon [MECH ENG] A closed system of tubes connected to a water-cooled engine which permit natural circulation and cooling of the liquid by utilizing the difference in density of the hot and cool portions.

thermostat [ENG] An instrument which measures changes in temperature and directly or indirectly controls sources of heating and cooling to maintain a desired temperature. Also known as thermorelay.

thermovoltmeter [ENG] A voltmeter in which a current from the voltage source is passed through a resistor and a fine vacuum-enclosed platinum heater wire; a thermocouple, attached to the midpoint of the heater, generates a voltage of a few millivolts, and this voltage is measured by a direct-current millivoltmeter.

thickened fuel *See* gelatinized gasoline.

thickened oil [MATER] Any oil to which a thickening agent has been added to increase viscosity or produce thixotropic properties; grease is a thickened oil.

thickener [ENG] A nonfilter device for the removal of liquid from a liquid-solids slurry to give a dewatered (thickened) solids product; can be by gravity settling or centrifugation.

thickening [MIN ENG] Concentrating dilute slime pulp into a pulp containing a smaller percentage of moisture by rejecting the free liquid.

thickness gage [ENG] A gage for measuring the thickness of a sheet of material, the thickness of an object, or the thickness of a coating; examples include penetration-type and backscattering radioactive thickness gages and ultrasonic thickness gages.

thickness ratio [AERO ENG] The ratio of the maximum thickness of an airfoil section to the length of its chord.

thief [PETRO ENG] In the petroleum industry, a device that permits the taking of samples from a predetermined location in the liquid body to be sampled.

thinner [MATER] A liquid used to thin paint, varnish, cement, or other material to a desired consistency.

thin-plate orifice [ENG] A thin-metal orifice sheet used in fluid-flow measurement in fluid conduits by means of differential pressure drop across the orifice.

thioindigo [MATER] A group of sulfur dyes made by treating the appropriate organic compound with sodium sulfide; colors are fast to washing and light.

third rail [CIV ENG] The electrified metal rail which carries current to the motor of an electric locomotive or other railway car.

thirling *See* holing.

Thoma cavitation coefficient [MECH ENG] The equation for measuring cavitation in a hydraulic turbine installation, relating vapor pressure, barometric pressure, runner setting, tail water, and head.

Thomas converter [MET] A basic Bessemer converter; that is, one in which air is forced upward through holes in the bottom of the steel container which has a basic lining, usually dolomite, and which employs a basic slag.

Thomas meter [ENG] An instrument used to determine the rate of flow of a gas by measuring the rise in the gas temperature produced by a known amount of heat.

thorium [MET] A heavy malleable metal that changes from silvery-white to dark gray or black in air; potential source of nuclear energy; used in manufacture of sunlamps.

thoroughfare [CIV ENG] **1.** An important, unobstructed public street or highway. **2.** A street going through from one street to another. **3.** An inland waterway for passage of ships usually not between two bodies of water.

thread [DES ENG] A continuous helical rib, as on a screw or pipe. [MIN ENG] A more or less straight line of stall faces, having no cuttings, loose ends, fast ends, or steps.

thread contour [DES ENG] The shape of thread design as observed in a cross section along the major axis, for example, square or round.

thread cutter [MECH ENG] A tool used to cut screw threads on a pipe, screw, or bolt.

thread gage [DES ENG] A design gage used to measure screw threads.

threading die [MECH ENG] A die which may be solid, adjustable, or spring adjustable, or a self-opening die head, used to produce an external thread on a part.

threading machine [MECH ENG] A tool used to cut or form threads inside or outside a cylinder or cone.

thread plug [ENG] Mold part which shapes an internal thread onto a molded article; must be unscrewed from the finished piece.

thread plug gage [DES ENG] A thread gage used to measure female screw threads.

thread protector [ENG] A short-threaded ring to screw onto a pipe or into a coupling to protect the threads while the pipe is being handled or transported. Also known as pipe-thread protector.

thread rating [ENG] The maximum internal working pressure allowable for threaded pipe or tubing joints; important for pressure systems, chemical processes, and oil-well systems.

thread ring gage [DES ENG] A thread gage used to measure male screw threads.

three-jaw chuck [DES ENG] A drill chuck having three serrated-face movable jaws that can grip and hold fast an inserted drill rod.

three-piece set [MIN ENG] A set of timber consisting of a cap and its two supportive posts.

three-point bending [MET] Bending a piece of metal by placing the specimen on two supports and then applying a load on it between the supported ends.

three-point problem [ENG] The problem of locating the horizontal position of a point of observation from the two observed horizontal angles subtended by three known sides of a triangle.

three-quarters hard [MET] A temper designation for various nonferrous metals, such as aluminum, copper, and magnesium alloys, expressing degree of hardness achieved by mechanical working.

three-shift cyclic mining [MIN ENG] A system of cyclic mining on a longwall conveyor face, with coal cutting on one shift, hand filling and conveying on the next, and ripping, packing, and advancement of the face conveyor on the third shift.

threshold [BUILD] A piece of stone, wood, or metal that lies under an outside door. [ENG] The least value of a current, voltage, or other quantity that produces the minimum detectable response in an instrument or system.

threshold speed [ENG] The minimum speed of current at which a particular current meter will measure at its rated reliability.

thribble [PETRO ENG] In drilling operations, a stand of pipe comprising three joints, each about 30 feet (9.4 meters) long.

throat [DES ENG] The narrowest portion of a constricted duct, as in a diffuser or a venturi tube; specifically, a nozzle throat. [ENG] **1.** The smaller end of a horn or tapered waveguide. **2.** The area in a fireplace that forms the passageway from the firebox to the smoke chamber.

throatable [DES ENG] Of a nozzle, designed to allow a change in the velocity of the exhaust stream by changing the size and shape of the throat of the nozzle.

throat balls *See* niter balls.

throat depth [MET] The distance from the center line of the electrodes or platens of a resistance welding machine to the nearest point of interference for flat work.

throat microphone [ENG ACOUS] A contact microphone that is strapped to the throat of a speaker and reacts directly to throat vibrations rather than to the sound waves they produce.

throat of fillet weld [MET] The thinnest part of a fillet weld, or the shortest distance from the root of a fillet weld to its face.

throat velocity *See* critical velocity.

throttle valve [MECH ENG] A choking device to regulate flow of a liquid, for example, in a pipeline, to an engine or turbine, from a pump or compressor.

throttling [AERO ENG] The varying of the thrust of a rocket engine during powered flight.

throttling calorimeter [ENG] An instrument utilizing the principle of constant enthalpy expansion for the measurement of the moisture content of steam; steam drawn from a steampipe through sampling nozzles enters the calorimeter through a throttling orifice and moves into a well-insulated expansion chamber in which its temperature is measured. Also known as steam calorimeter.

through arch [CIV ENG] An arch bridge from which the roadway is suspended as distinct from one which carries the roadway on top.

through bridge [CIV ENG] A bridge that carries the deck within the height of the superstructure.

through-feed centerless grinding [MECH ENG] A metal cutting process by which the external surface of a cylindrical workpiece of uniform diameter is ground by passing the workpiece between a grinding and regulating wheel.

throughput [MIN ENG] The quantity of ore or other material passed through a mill or a section of a mill in a given time or at a given rate.

through street [CIV ENG] A street at which all cross traffic is required to stop before crossing or entering. Also known as throughway.

through transmission [ENG] An ultrasonic testing method in which mechanical vibrations are transmitted into one end of the workpiece and received at the other end.

throughway *See* expressway; through street.

through weld [MET] A long weld made through the unbroken surface of one member to the other member in a lap or tree joint.

throwing power [MET] The ability of an electroplating solution to deposit metal uniformly on an irregularly shaped cathode.

throwout [MECH ENG] In automotive vehicles, the mechanism or assemblage of mechanisms by which the driven and driving plates of a clutch are separated.

throw-out spiral *See* lead-out groove.

thrust [MIN ENG] **1.** A crushing of coal pillars caused by excess weight of the superincumbent rocks, the floor being harder than the roof. **2.** The ruins of the fallen roof, after pillars and stalls have been removed.

thrust [MECH ENG] The weight or pressure applied to a bit to make it cut.

thrust augmentation [AERO ENG] The increasing of the thrust of an engine or power plant, especially of a jet engine and usually for a short period of time, over the thrust normally developed.

thrust augmenter [AERO ENG] Any contrivance used for thrust augmentation, as a venturi used in a rocket.

thrust axis [AERO ENG] A line or axis through an aircraft or a rocket, along which the thrust acts; an axis through the longitudinal center of a jet or rocket engine, along which the thrust of the engine acts. Also known as axis of thrust; center of thrust.

thrust bearing [MECH ENG] A bearing which sustains axial loads and prevents axial movement of a loaded shaft.

thrust coefficient *See* nozzle thrust coefficient.

thruster [AERO ENG] A control jet employed in spacecraft; an example would be one utilizing hydrogen peroxide.

thrust horsepower [AERO ENG] **1.** The force-velocity equivalent of the thrust developed by a jet or rocket engine. **2.** The thrust of an engine-propeller combination expressed in horsepower; it differs from the shaft horsepower of the engine by the amount the propeller efficiency varies from 100%.

thrust load [MECH ENG] A load or pressure parallel to or in the direction of the shaft of a vehicle.

thrust meter [ENG] An instrument for measuring static thrust, especially of a jet engine or rocket.

Thrustor [MECH ENG] Trademark for a hydraulic device for applying a controllable force, as to a brake.

thrust output [AERO ENG] The net thrust delivered by a jet engine, rocket engine, or rocket motor.

thrust-pound [AERO ENG] A unit of measurement for the thrust produced by a jet engine or rocket.

thrust power [AERO ENG] The power usefully expended on thrust, equal to the thrust (or net thrust) times airspeed.

thrust reverser [AERO ENG] A device or apparatus for reversing thrust, especially of a jet engine.

thrust section [AERO ENG] A section in a rocket vehicle that houses or incorporates the combustion chamber or chambers and nozzles.

thrust terminator [AERO ENG] A device for ending the thrust in a rocket engine, either through propellant cutoff (in the case of a liquid) or through diverting the flow of gases from the nozzle.

thrust-weight ratio [AERO ENG] A quantity used to evaluate engine performance, obtained by dividing the thrust output by the engine weight less fuel.

thrust yoke [MECH ENG] The part connecting the piston rods of the feed mechanism on a hydraulically driven diamond-drill swivel head to the thrust block, which forms the connecting link between the yoke and the drive rod, by means of which link the longitudinal movements of the feed mechanism are transmitted to the swivel-head drive rod. Also known as back end.

thuja oil [MATER] An essential oil from white cedar leaves; pale-yellow, combustible oil with a camphor aroma, soluble in alcohol, ether, chloroform, carbon disulfide, and fixed oils; used in medicine, perfumery, and flavorings. Also known as arbor vitae oil.

thumbscrew [DES ENG] A screw with a head flattened in the same axis as the shaft so that it can be gripped and turned by the thumb and forefinger.

thump [ENG ACOUS] Low-frequency transient disturbance in a system or transducer characterized audibly by the vocal imitation of the word.

Thunderchief [AERO ENG] A United States supersonic, single-engine, turbojet-powered tactical fighter capable of delivering nuclear weapons as well as nonnuclear bombs and rockets; an all-weather attack fighter, it is also capable of close support for ground forces, and its range can be extended by in-flight refueling; it is equipped with the Sidewinder missile. Designated F-105.

Thunderstreak [AERO ENG] A United States one-man fighter also used for reconnaissance; it has a range of over 2000 miles (3200 kilometers) and a speed of over 600 miles (970 kilometers) per hour; the bomb load is 6000 pounds (2700 kilograms) of conventional or nuclear bombs, incendiary gel, or rockets, and there are six .50-caliber machine guns. Designated F-84.

thyme oil [MATER] An essential oil found in the flowers of the thymes *Thymus vulgaris* or *T. zygis*, a colorless to reddish-brown liquid with a sharp taste and pleasant aroma, soluble in alcohol, slightly soluble in water; used in medicine, perfumery, cosmetics, flavoring, and soap.

tidal lock *See* entrance lock.

tidal quay [CIV ENG] A quay in an open harbor or basin with sufficient depth to enable ships lying alongside to remain afloat at any state of the tide.

tide gage [ENG] A device for measuring the height of a tide; may be observed visually or may consist of an elaborate recording instrument.

tide gate [CIV ENG] **1.** A restricted passage through which water runs with great speed due to tidal action. **2.** An opening through which water may flow freely when the tide sets in one direction, but which closes automatically and prevents the water from flowing in the other direction when the direction of flow is reversed.

tide indicator [ENG] That part of a tide gage which indicates the height of tide at any time; the indicator may be in the immediate vicinity of the tidal water or at some distance from it.

tide lock *See* entrance lock.

tide machine [ENG] An instrument that computes, sometimes for years in advance, the times and heights of high and low waters at a reference station by mechanically summing the harmonic constituents of which the tide is composed.

tide pole [ENG] A graduated spar used for measuring the rise and fall of the tide. Also known as tide staff.

tide staff *See* tide pole.

tie [CIV ENG] One of the transverse supports to which railroad rails are fastened to keep them to line, gage, and grade. [ENG] A beam, post, rod, or angle to hold two pieces together; a tension member in a construction. [MIN ENG] A support for the roof in coal mines.

tieback [MIN ENG] **1.** A beam serving a purpose similar to that of a fend-off beam, but fixed at the opposite side of the shaft or inclined road. **2.** The wire ropes or stay rods that are sometimes used on the side of the tower opposite the hoisting engine, either in place of or to reinforce the engine braces.

tie bar [CIV ENG] **1.** A bar used as a tie rod. **2.** A rod connecting two switch rails on a railway to hold them to gage.

tied arch [CIV ENG] An arch having the horizontal reaction component provided by a tie between the skewbacks of the arch ends.

tied concrete column [CIV ENG] A concrete column reinforced with longitudinal bars and horizontal ties.

tie-down diagram [ENG] A drawing indicating the prescribed method of securing a particular item of cargo within a specific type of vehicle.

tie-down point [ENG] An attachment point provided on or within a vehicle.

tie-down point pattern [ENG] The pattern of tie-down points within a vehicle.

tie plate [CIV ENG] A metal plate between a rail and a tie to hold the rail in place and reduce wear on the tie. [MECH ENG] A plate used in a furnace to connect tie rods.

tier building [CIV ENG] A multistory skeleton frame building.

tie rod [CIV ENG] A structural member used as a brace to take tensile loads. [ENG] A round or square iron rod passing through or over a furnace and connected with buckstays to assist in binding the furnace together. [MECH ENG] A rod used as a mechanical or structural support between elements of a machine. [MIN ENG] Vertical rods mounted in overlying horizontal shaft timbers.

Tiger [AERO ENG] A United States single-engine, single-seat, supersonic jet fighter designed for operating from aircraft carriers for the interception and destruction of enemy aircraft, and the support of troops ashore; armament consists of Sidewinders, cannons, and rocket packs. Designated F-11.

tight [ENG] **1.** Unbroken, crack-free, and solid rock in which a naked hole will stand without caving. **2.** A borehole made impermeable to water by cementation or casing. [MECH ENG] **1.** Inadequate clearance or the barest minimum of clearance between working parts. **2.** The absence of leaks in a pressure system.

tight fit [DES ENG] A fit between mating parts with slight negative allowance, requiring light to moderate force to assemble.

TIG welding *See* tungsten–inert gas welding.

tile [MATER] **1.** A piece of fired clay, stone, concrete, or other material used ornamentally to cover roofs, floors, or walls. **2.** A hollow building unit made of burned clay or other material.

tilt [AERO ENG] The inclination of an aircraft, winged missile, or the like from the horizontal, measured by reference to the lateral axis or to the longitudinal axis.

tilting dozer [MECH ENG] A bulldozer whose blade can be pivoted on a horizontal center pin to cut low on either side.

tilting idlers [MECH ENG] An arrangement of idler rollers in which the top set is mounted on vertical arms which pivot on spindles set low down on the frame of the roller stool.

tilting mixer [MECH ENG] A small-batch mixer consisting of a rotating drum which can be tilted to discharge the contents; used for concrete or mortar.

tilting-type boxcar unloader [CIV ENG] A mechanism that is used to unload material such as grain from a boxcar; the car, with its door open, is held by end clamps on the specialized piece of track and tilted 15% from the vertical and then tilted endwise 40% to the horizontal to discharge the material at one end of the car, and 40% in the opposite direction to discharge the material from the opposite end.

tiltmeter [ENG] An instrument used to measure small changes in the tilt of the earth's surface, usually in relation to a liquid-level surface or to the rest position of a pendulum.

tilt mold [MET] A mold that rotates from a horizontal to a vertical position during filling to reduce agitation and risk of dross entrapment.

tilt/rotate code [ENG] A code that instructs a "golf ball" printing element which angle of tilt and rotation is needed to print a given character.

tilt rotor [AERO ENG] An assembly of rapidly rotating blades on a vertical takeoff and landing aircraft, whose plane of rotation can be continuously varied from the horizontal to the vertical, permitting performance as helicopter blades or as propeller blades.

timber [MATER] Wood used for building, carpentry, or joinery.

timbered stope [MIN ENG] A stope made of square-set timbering or any of its variations.

timbering [MIN ENG] The timber structure used for supporting the faces of an excavation during the progress of construction.

timbering machine [MIN ENG] An electrically driven machine to raise and hold timber in place while the supporting posts are being set, the posts having been cut to desired length previously by the machine's power-driven saw.

timber mat [MIN ENG] Broken timber forming the roof of an ore deposit that is being extracted by a caving method, such as top slicing.

timber packer See pack builder.

timber puller [MIN ENG] A machine used to remove the timber supports in a mine.

timber trolley [MIN ENG] A carriage consisting of a timber or steel base, mounted on wheels, with U-shaped arms.

timber truck [MIN ENG] Any truck or car used for hauling timber inside of a mine.

time and material contract [IND ENG] A contract providing for the procurement of supplies or services on the basis of direct labor hours at specified fixed hourly rates (which rates include direct and indirect labor, overhead, and profit), and material at cost.

time and motion study [IND ENG] Observation, analysis, and measurement of the steps in the performance of a job to determine a standard time for each performance. Also known as time-motion study.

time break [ENG] A distinctive mark shown on an exploration seismogram to indicate the exact detonation time of an explosive energy source.

time-change component [ENG] A component which because of design limitations or safety is specified to be rebuilt or overhauled after a specified period of operation (for example, an engine or propeller of an airplane).

time curve See time front.

time-distance graph [AERO ENG] A graph used to determine the ground distance for air-route legs of a specified time interval; time-distance relationships are often simplified by considering air, wind, and ground distances for flight legs of 1-hour duration.

time formula [IND ENG] A formula to determine the standard time of an operation as a function of one or more variables in the operation.

time front [AERO ENG] A locus of points representing the maximum ground distances from a departure point that can be covered by an aircraft in a prescribed time interval. Also known as hour-out line; time curve.

time fuse [ENG] A fuse which contains a graduated time element to regulate the time interval after which the fuse will function.

time-motion study See time and motion study.

time of set [MATER] The time required for freshly mixed concrete to stiffen (initial set, about 1 hour) or to attain a minimum specified hardness (final set, about 10 hours); actual times vary with the type of cement used.

time quenching [MET] Interrupted quenching in which the time in the quenching medium is controlled.

timer [ENG] **1.** A device for automatically starting or stopping a machine or other device. **2.** See interval timer. [MECH ENG] A device that controls timing of the ignition spark of an internal combustion engine at the correct time.

time separation [AERO ENG] The time interval between adjacent aircraft flying approximately the same path.

time standard See standard time.

time study [IND ENG] A work measurement technique, generally using a stopwatch or other timing device, to record the actual elapsed time for performance of a task, adjusted for any observed variance from normal effort or pace, unavoidable or machine delays, rest periods, and personal needs.

time switch [ENG] A clock-controlled switch used to open or close a circuit at one or more predetermined times.

timing [MECH ENG] Adjustment in the relative position of the valves and crankshaft of an automobile engine in order to produce the largest effective output of power.

timing belt [DES ENG] A power transmission belt with evenly spaced teeth on the bottom side which mesh with grooves cut on the periphery of the pulley to produce a positive, no-slip, constant-speed drive. Also known as cogged belt; synchronous belt. [MECH ENG] A positive drive belt that has axial cogs molded on the underside of the belt which fit into grooves on the pulley; prevents slip, and makes accurate timing possible; combines the advantages of belt drives with those of chains and gears. Also known as positive drive belt.

timing belt pulley [MECH ENG] A pulley that is similar to an uncrowned flat-belt pulley, except that the grooves for the belt's teeth are cut in the pulley's face parallel to the axis.

timing gears [MECH ENG] The gear train of reciprocating engine mechanisms for relating camshaft speed to crankshaft speed.

Timken film strength [ENG] A test used on a gear lubricant to determine the amount of pressure the film of oil can withstand before rupturing.

Timken wear test [ENG] A test used on a gear lubricant to determine its abrasive effect on gear metals.

tin [MET] A lustrous silver-white ductile, malleable metal used in alloys, for solder, terneplate, and tinplate.

tin bronze [MET] A tin-copper alloy.

tincture [MATER] A dilute solution (aqueous or aqueous alcoholic) of a drug or chemical; more dilute than fluid extracts, less volatile than spirits.

tinfoil [MET] Foil made of tin or a tin alloy.

tinned wire [MET] Copper wire that has been coated during manufacture with a layer of tin or solder to prevent corrosion and simplify soldering of connections.

tinner's rivet [DES ENG] A special-purpose rivet that has a flat head, used in sheet metal work.

tinning [MET] **1.** Covering or preserving with tin. **2.** A protective coating of tin.

tin pest [MET] Transformation of tin to a brittle, gray variety occurring spontaneously at temperatures below 0°C.

tinplate [MET] Thin sheet iron or steel coated with tin.

tin sweat [MET] Exudation of tin-rich low-melting-point material from a tin-bronze surface as a result of inverse segregation in bronze casting, or overheating of the alloy.

tip [DES ENG] A piece of material secured to and differing from a cutter tooth or blade.

tipped bit [DES ENG] A drill bit in which the cutting edge is made of especially hard material.

tipped solid cutters [DES ENG] Cutters made of one material and having tips or cutting edges of another material bonded in place.

tipper [MIN ENG] An apparatus for emptying coal or ore cars by turning them upside down and then righting them, with a minimum of manual labor.

tipping-bucket rain gage [ENG] A type of recording rain gage; the precipitation collected by the receiver empties into one side of a chamber which is partitioned transversely at its center and is balanced bistably upon a horizontal axis; when a predetermined amount of water has been collected, the chamber tips, spilling out the water and placing the other half of the chamber under the receiver; each tip of the bucket is recorded on a chronograph, and the record obtained indicates the amount and rate of rainfall.

tipple [MIN ENG] **1.** The place where the mine cars are tipped and emptied of their coal. **2.** The tracks, trestles, and screens at the entrance to a colliery, where coal is screened and loaded.

tire [ENG] A continuous metal ring, or pneumatic rubber and fabric cushion, encircling and fitting the rim of a wheel.

tire iron [DES ENG] A single metal bar having bladelike ends of various shapes to insert between the rim and the bead of a pneumatic tire to remove or replace the tire.

Tiros satellite [AERO ENG] Television infrared observation satellite; a meteorological satellite that takes television pictures of cloud cover, using radiation sensors and cameras; it stores and transmits this information on ground command.

tirrill burner [ENG] A modification of the bunsen burner which allows greater flexibility in the adjustment of the air-gas mixture.

tissue paper [MATER] Extremely lightweight paper, available in a great many colors and used in craft projects and in collage painting.

titanium [MET] A lustrous, silvery-gray, strong, light metal that is hard and brittle when cold, malleable when heated, and ductile when pure; used in the pure state or in alloys for aircraft and chemical-plate metals, for surgical instruments, and in cermets, and metal-ceramic brazing.

TNT–ammonium nitrate explosive [MATER] An explosive containing ammonium nitrate sensitized with trinitrotoluene; a proportion of aluminum powder or calcium silicide may be added to increase power and sensitiveness.

to-and-fro ropeway See jig back.

Tocco process [MET] A patented process for local hardening of steel by applying a high-frequency current to the area for a few seconds or until heated to the desired depth, then water-spraying the surface.

toe [CIV ENG] The part of a base of a dam or retaining wall on the side opposite to the retained material. [MET] The junction between the face of a weld and the base metal. [MIN ENG] **1.** The burden of material between the bottom of the bore-hole and the free face. **2.** The bottom of the borehole. **3.** A spurn, or small pillar of coal. **4.** The base of a bank in an open-pit mine.

toe crack [MET] A crack in the base metal at the toe of a weld.

toe cut [ENG] In underground blasting, the cut obtained by the use of toe holes.

toe hole [ENG] A blasting hole, usually drilled horizontally or at a slight inclination into the base of a bank, bench, or slope of a quarry or open-pit mine.

toe-in [MECH ENG] The degree (usually expressed in fractions of an inch) to which the forward part of the front wheels of an automobile are closer together than the rear part, measured at hub height with the wheels in the normal "straight ahead" position of the steering gear.

toenailing [ENG] The technique of driving a nail at an angle to join two pieces of lumber.

toe-out [MECH ENG] The outward inclination of the wheels of an automobile at the front on turns due to setting the steering arms at an angle.

toe-to-toe drilling [ENG] The drilling of vertical large-diameter blasting holes in quarries and opencast pits.

toggle [MECH ENG] A form of jointed mechanism for the amplification of forces.

toggle bolt [DES ENG] A bolt having a nut with a pair of pivotal wings that close against a spring; wings open after emergence through a hole or passage in a thin or hollow wall to fasten the unit securely.

toggle press [MECH ENG] A mechanical press in which a toggle mechanism actuates the slide.

tolerance [DES ENG] The permissible variations in the dimensions of machine parts. [ENG] A permissible deviation from a specified value, expressed in actual values or more often as a percentage of the nominal value.

tolerance chart [DES ENG] A chart indicating graphically the sequence in which dimensions must be produced on a part so that the finished product will meet the prescribed tolerance limits.

tolerance limits [DES ENG] The extreme values (upper and lower) that are permitted by the tolerance.

tolerance unit [DES ENG] A unit of length used to express the degree of tolerance allowed in fitting cylinders into cylindrical holes, equal, in micrometers, to $0.45 D^{1/3} + 0.001 D$, where D is the cylinder diameter in millimeters.

ton [IND ENG] A unit of volume of sea freight, equal to 40 cubic feet. Also known as freight ton; measurement ton; shipping ton. [MECH ENG] A unit of refrigerating capacity, that is, of rate of heat flow, equal to the rate of extraction of latent heat when one short ton of ice of specific latent heat 144 international table British thermal units per pound is produced from water at the same temperature in 24 hours; equal to 200 British thermal units per minute, or to approximately 3516.85 watts. Also known as standard ton.

tong hold [MET] The end of a forging billet that is gripped by the operator's tongs; it is removed at the end of the forging operation.

tongs [DES ENG] Any of various devices for holding, handling, or lifting materials and consisting of two legs joined eccentrically by a pivot or spring.

tongue and groove [DES ENG] A joint in which a projecting rib on the edge of one board fits into a groove in the edge of another board.

ton-kilometer [MIN ENG] A unit of measurement equal to the weight in tons of material transported in a mine multiplied by the number of kilometers driven.

ton-mile [CIV ENG] In railroading, a standard measure of traffic, based on the rate of carriage per mile of each passenger or ton of freight.

tool [ENG] Any device, instrument, or machine for the performance of an operation, for example, a hammer, saw, lathe, twist drill, drill press, grinder, planer, or screwdriver. [IND ENG] To equip a factory or industry for production by designing, making, and integrating machines, machine tools, and special dies, jigs, and instruments, so as to achieve manufacture and assembly of products on a volume basis at minimum cost.

tool bit [ENG] A piece of high-strength metal, usually steel, ground to make single-point cutting tools for metal-cutting operations.

toolbox [ENG] A box to hold tools.

tool-check system [IND ENG] A system for temporary issue of tools in which the employee is issued a number of small metal checks stamped with the same number; a check is surrendered for each tool obtained from the crib.

tool design [DES ENG] The division of mechanical design concerned with the design of tools.

tool-dresser [MECH ENG] A tool-stone-grade diamond inset in a metal shank and used to trim or form the face of a grinding wheel.

tool extractor [ENG] An implement for grasping and withdrawing drilling tools when broken, detached, or lost in a borehole.

toolhead [MECH ENG] The adjustable tool-carrying part of a machine tool.

tool joint [ENG] A coupling element for a drill pipe; designed to support the weight of the drill stem and the strain of frequent use, and to provide a leakproof seal.

toolmaker's vise See universal vise.

tool nipper [MIN ENG] A person whose duty it is to carry powder, drills, and tools to the various levels of the mine and to bring dull tools and drills to the surface.

tool post [MECH ENG] A device to clamp and position a tool holder on a machine tool.

tool steel [MET] Any of various steels capable of being hardened sufficiently so as to be a suitable material for making cutting tools.

tooth [DES ENG] **1.** One of the regular projections on the edge or face of a gear wheel. **2.** An angular projection on a tool or other implement, such as a rake, saw, or comb.

tooth point [DES ENG] The chamfered cutting edge of the blade of a face mill.

top and bottom process [MET] A process in which sodium sulfide is added to molten copper-nickel sulfide to form a two-layer melt, with the bulk of the nickel in the bottom layer.

top-benching [MIN ENG] The method by which the bench is removed from above, as with a dragline.

top-blown rotary converter [MET] A rotary converter used for making nickel and steel; oxygen and other gases are fed to the furnace by a lance at the elevated end of the converter to permit formation of a metal product without oxidation.

top cager [MIN ENG] A person at the top of a mine shaft who superintends the lowering and raising of the cage, and, at most mines, the removing of loaded cars from the placing of empty cars in the cage.

top cut [MIN ENG] A machine cut made in the coal at or near the top of the working face in a mine.

top dead center [MECH ENG] The dead-center position of an engine piston and its crankshaft arm when at the top or outer end of its stroke.

top hooker See lander.

top lander See lander.

topographic survey [ENG] A survey that determines ground relief and location of natural and man-made features thereon.

topped crude [MATER] A residual product remaining after the removal by distillation or other means of an appreciable quantity of the more volatile components of crude petroleum.

topping governor See limit governor.

topside sounder [AERO ENG] A satellite designed to measure ion concentration in the ionosphere from above the ionosphere.

top slicing [MIN ENG] A method of stoping in which the ore is extracted by excavating a series of horizontal (sometimes inclined) timbered slices alongside each other, beginning at the top of the ore body and working progressively downward.

top slicing and cover caving [MIN ENG] A mining method that entails the working of the ore body from the top down in successive horizontal slices that may follow one another sequentially or simultaneously; the overburden or cover is caved after mining a unit.

torch [ENG] A gas burner used for brazing, cutting, or welding.

toromatic transmission [MECH ENG] A semiautomatic transmission; it contains a compound planetary gear train with a torque converter.

torpedo [ENG] An encased explosive charge slid, lowered, or dropped into a borehole and exploded to clear the hole of obstructions or to open communications with an oil or water supply. Also known as bullet.

torque arm [MECH ENG] In automotive vehicles, an arm to take the torque of the rear axle.

torque-coil magnetometer [ENG] A magnetometer that depends for its operation on the torque developed by a known current in a coil that can turn in the field to be measured.

torque converter [MECH ENG] A device for changing the torque speed or mechanical advantage between an input shaft and an output shaft.

torque-load characteristic [ENG] For electric motors, the armature torque developed versus the load on the motor at constant speed.

torquemeter [ENG] An instrument to measure torque.

torque reaction [MECH ENG] On a shaft-driven vehicle, the reaction between the bevel pinion with its shaft (which is supported in the rear axle housing) and the bevel ring gear (which is fastened to the differential housing) that tends to rotate the axle housing around the axle instead of rotating the axle shafts alone.

torque-tube flowmeter [ENG] A liquid-flow measurement device in which a flexible torque tube transmits bellows motion (caused by differential pressure from the liquid flow through the pipe) to the recording pen arm.

torque-type viscometer [ENG] A device that measures liquid viscosity by the torque needed to rotate a vertical paddle submerged in the liquid; used for both Newtonian and non-Newtonian liquids and for suspensions.

torque-winding diagram [MECH ENG] A diagram showing how the winding load on a winch drum varies and is used to decide the method of balancing needed; made by plotting the turning moment in pounds per foot on the vertical axis against time, or revolutions or depth on the horizontal axis.

torque wrench [ENG] **1.** A hand or power tool used to turn a nut on a bolt that can be adjusted to deliver a predetermined amount of force to the bolt when tightening the nut. **2.** A wrench that measures torque while being turned.

Torricellian barometer *See* mercury barometer.

torsiometer [MECH ENG] An instrument which measures power transmitted by a rotating shaft; consists of angular scales mounted around the shaft from which twist of the loaded shaft is determined. Also known as torsionmeter.

torsion balance [ENG] An instrument, consisting essentially of a straight vertical torsion wire whose upper end is fixed while a horizontal beam is suspended from the lower end; used to measure minute gravitational, electrostatic, or magnetic forces.

torsion bar [MECH ENG] A spring flexed by twisting about its axis; found in the spring suspension of truck and passenger car wheels, in production machines where space limitations are critical, and in high-speed mechanisms where inertia forces must be minimized.

torsion damper [MECH ENG] A damper used on automobile internal combustion engines to reduce torsional vibration.

torsion galvanometer [ENG] A galvanometer in which the force between the fixed and moving systems is measured by the angle through which the supporting head of the moving system must be rotated to bring the moving system back to its zero position.

torsion hygrometer [ENG] A hygrometer in which the rotation of the hygrometric element is a function of the humidity; such hygrometers are constructed by taking a substance whose length is a function of the humidity and twisting or spiraling it under tension in such a manner that a change in length will cause a further rotation of the element.

torsionmeter *See* torsiometer.

torsion-string galvanometer [ENG] A sensitive galvanometer in which the moving system is suspended by two parallel fibers that tend to twist around each other.

total air [ENG] The actual quantity of air supplied for combustion of fuel in a boiler, expressed as a percentage of theoretical air.

total carbon [MET] The sum of free and combined carbon in a ferrous alloy, especially steel.

total cyanide [MET] Total amount of cyanide contained in an electroplating bath, including both simple and complex ions.

total impulse [AERO ENG] The product of the thrust and the time over which the thrust is produced, expressed in pounds (force)–seconds; used especially in reference to a rocket motor or a rocket engine.

total lift [AERO ENG] The upward force produced by the gas in a balloon; it is equal to the sum of the free lift, the weight of the balloon, and the weight of auxiliary equipment carried by the balloon.

total pressure [MIN ENG] The total ventilating pressure in a mine, usually measured in the fan drift.

total radiation pyrometer [ENG] A pyrometer which focuses heat radiation emitted by a hot object on a detector (usually a thermopile or other thermal type detector), and which responds to a broad band of radiation, limited only by absorption of the focusing lens, or window and mirror.

touch feedback [ENG] A type of force feedback in which servos provide the manipulator fingers with a sense of resistance when an object is grasped, so that the operator does not crush the object.

tough pitch copper [MET] Copper refined in a reverberatory furnace to adjust the oxygen content to 0.2–0.5%.

tour *See* shift.

towbar [ENG] An element which connects to a vehicle that is not equipped with an integral drawbar, for the purpose of towing or moving the vehicle.

tower bolt *See* barrel bolt.

tower crane [CIV ENG] A crane mounted on top of a tower which is sometimes incorporated in the frame of a building.

tower excavator [MIN ENG] A cableway excavator designed specifically for levee work but which is used extensively in the stripping of overburden, spoil, or waste in surface mining: basically, it is a Sauerman-type excavator with towers either fixed or movable, and when the head tower is located on the spoil pile and the tail tower on the unexcavated wall, pits of almost unlimited width can be dug.

tower loader [MIN ENG] A front-end loader whose bucket is lifted along tracks on a more or less vertical tower.

towing tank *See* model basin.

trace [ENG] The record made by a recording device, such as a seismometer or electrocardiograph.

tracer [ENG] A thread of contrasting color woven into the insulation of a wire for identification purposes.

tracer gas [ENG] In vacuum testing for leaks, a gas emitting through a leak in a pressure system and subsequently conducted into the detector.

tracer milling [MECH ENG] Cutting a duplicate of a three-dimensional form by using a mastic form to direct the tracer-controlled cutter.

tracing distortion [ENG ACOUS] The nonlinear distortion introduced in the reproduction of a mechanical recording because the curve traced by the motion of the reproducing stylus is not an exact replica of the modulated groove.

tracing paper [MATER] Thin paper used both for tracing and original drawings, and of various types and surfaces.

track [AERO ENG] The actual line of movement of an aircraft or a rocket over the surface of the earth; it is the projection of the history of the flight path on the surface. Also known as flight track. [ENG] **1.** The groove cut in a rock by a diamond inset in the crown of a bit. **2.** A pair of parallel metal rails for a railway, railroad, tramway, or for any wheeled vehicle. [MECH ENG] **1.** The slide or rack on which a diamond-drill swivel head can be moved to positions above and clear of the collar of a borehole. **2.** A crawler mechanism for earth-moving equipment.

track cable [ENG] Steel wire rope, usually a locked-coil rope which supports the wheels of the carriers of a cableway.

track cable scraper [MIN ENG] A type of excavator that uses a bottomless scraper bucket which conveys its load over the ground and is operated by a two-drum hoist which controls a track cable that spans the working area and a haulage cable that leads to the front of the bucket.

Tracker [AERO ENG] A United States twin-reciprocating-engine, antisubmarine aircraft capable of operating from carriers, and designed primarily for the detection, location, and destruction of submarines. Designated S-2.

track gage [CIV ENG] The width between the rails of a railroad track; in the United States the standard gage is 4 feet 8½ inches.

track haulage [MIN ENG] Movement or transportation of excavated or mined materials in cars or trucks that run on rails.

track hopper [ENG] A hopper-shaped receiver mounted beside or below railroad tracks, into which railroad boxcars or bottom-dump cars are discharged; used for solid materials.

tracking [ENG] **1.** A motion given to the major lobe of a radar or radio antenna such that some preassigned moving target in space is always within the major lobe. **2.** The process of following the movements of an object; may be accomplished by keeping the reticle of an optical sys-

tem or a radar beam on the object, by plotting its bearing and distance at frequent intervals, or by a combination of techniques. [ENG ACOUS] **1.** The following of a groove by a phonograph needle. **2.** Maintaining the same ratio of loudness in the two channels of a stereophonic sound system at all settings of the ganged volume control.

tracking error [ENG ACOUS] Deviation of the vibration axis of a phonograph pickup from tangency with a groove; true tangency is possible for only one groove when the pickup arm is pivoted; the longer the pickup arm, the less is the tracking error.

tracking jitter [ENG] Minor variations in the pointing of an automatic tracking radar.

tracking network [ENG] A group of tracking stations whose operations are coordinated in tracking objects through the atmosphere or space.

tracking radar [ENG] Radar used to monitor the flight and obtain geophysical data from space probes, satellites, and high-altitude rockets.

tracking station [ENG] A radio, radar, or other station set up to track an object moving through the atmosphere or space.

tracking system [ENG] General name for apparatus, such as tracking radar, used in following and recording the position of objects in the sky.

trackless mine [MIN ENG] A mine in which rubber-tired vehicles are used for haulage and transport.

trackless tunneling [MIN ENG] Tunneling by means of loaders mounted on caterpillars.

track made good [AERO ENG] The actual path of an aircraft over the surface of the earth, or its graphic representation.

track shifter [ENG] A machine or appliance used to shift a railway track laterally.

traction meter [ENG] A load-sensing device placed between a locomotive and the car immediately behind it to measure pulling force exerted by the locomotive.

traction tube [ENG] A device for measuring the minimum water velocities capable of moving various sizes of sand grains; it consists of a horizontal glass tube half-filled with sand.

tractor [MECH ENG] **1.** An automotive vehicle having four wheels or a caterpillar tread used for pulling agricultural or construction implements. **2.** The front pulling section of a semitrailer. Also known as truck-tractor.

tractor drill [MECH ENG] A drill having a crawler mounting to support the feed-guide bar on an extendable arm.

tractor gate [CIV ENG] A type of outlet control gate used to release water from a reservoir; there are two types, roller and wheel.

tractor loader [MECH ENG] A tractor equipped with a tipping bucket which can be used to dig and elevate soil and rock fragments to dump at truck height. Also known as shovel dozer; tractor shovel.

tractor shovel *See* tractor loader.

traffic [ENG] The passage or flow of vehicles, pedestrians, ships, or planes along defined routes such as highways, sidewalks, sea lanes, or air lanes.

trafficability [CIV ENG] Capability of terrain to bear traffic, or the extent to which the terrain will permit continued movement of any or all types of traffic.

traffic control [ENG] Control of the movement of vehicles, such as airplanes, trains, and automobiles, and the regulatory mechanisms and systems used to exert or enforce control.

traffic density [CIV ENG] The average number of vehicles that occupy 1 mile or 1 kilometer of road space, expressed in vehicles per mile or per kilometer.

traffic engineering [CIV ENG] The determination of the required capacity and layout of highway and street facilities that can safely and economically serve vehicular movement between given points.

traffic flow [CIV ENG] The total number of vehicles passing a given point in a given time, expressed as vehicles per hour.

traffic noise [ENG] The general disturbance in sonar transmissions which is due to ships but is not associated with a specific vessel.

traffic pattern [AERO ENG] The traffic flow that is prescribed for aircraft landing at, taxiing on, and taking off from an airport; the usual components of a traffic pattern are upwind leg, crosswind leg, downwind leg, base leg, and final approach.

traffic recorder [ENG] A mechanical counter or recorder used to determine traffic movements (hourly variations and total daily volumes of traffic at a point) on an existing route; the air-impulse counter, magnetic detector, photoelectric counter, and radar detector are used.

traffic signal [CIV ENG] With the exception of traffic signs, any power-operated device for regulating, directing, or warning motorists or pedestrians.

tragacanth [MATER] The gummy exudate produced by certain Asiatic species of *Astragalus*; consists of a soluble portion containing uronic acid and arabinose, and an insoluble portion that absorbs water and swells to make a stiff opalescent mucilage.

T rail [CIV ENG] A rail shaped like a T in cross section due to a wide head, web, and flanged base.

trail angle [AERO ENG] The angle at an aircraft between the vertical and the line of sight to an object over which the aircraft has passed.

trailer [MECH ENG] The section of a semitrailer that is pulled by the tractor.

trail formation [AERO ENG] Aircraft flying singly or in elements in such manner that each aircraft or element is in line behind the preceding aircraft or element. [ENG] Vehicles proceeding one behind the other at designated intervals. Also known as column formation.

trailing edge [AERO ENG] The rear section of a multipiece airfoil, usually that portion aft of the rear spar.

trailing-edge tab [AERO ENG] One of the devices on the aircraft elevator that reduce or eliminate hinge movements required to deflect the elevator during flight.

train [ENG] To aim or direct a radar antenna in azimuth.

training aid [ENG] Any item which is developed or procured primarily to assist in training and the process of learning.

training wall [CIV ENG] A wall built along the bank of a river or estuary parallel to the direction of flow to direct and confine the flow.

train shed [CIV ENG] **1.** A structure to protect trains from weather. **2.** The part of a railroad station that covers the tracks.

trajectory-measuring system [ENG] A system used to provide information on the spatial position of an object at discrete time intervals throughout a portion of the trajectory or flight path.

trammel [ENG] A device consisting of a bar, each of whose ends is constrained to move along one of two perpendicular lines; used in drawing ellipses and in the Rowland mounting.

tramming [MIN ENG] Pushing tubs, mine cars, or trams by hand.

tramp metal [MIN ENG] Unwanted metal which finds its way into the mill ore stream.

tramp metal detector [MIN ENG] A sensing device which detects presence of unwanted metal in an ore stream, and sounds an alarm or removes the metal.

tramway [MECH ENG] An overhead rail, rope, or cable on which wheeled cars run to convey a load.

Trancor [MET] A commercial magnetic alloy composed of 3.5% silicon and 96.5% iron.

transcription [ENG ACOUS] A 16-inch-diameter (40.6-centimeter-diameter), 33-1/3-rpm disk recording of a complete radio program, made especially for broadcast purposes. Also known as electrical transcription.

transcrystalline [MET] Across the crystals of a metal; used of cracks in metals. Also known as intracrystalline; transgranular.

transducer [ENG] Any device or element which converts an input signal into an output signal of a different form; examples include the microphone, phonograph pickup, loudspeaker, barometer, photoelectric cell, automobile horn, doorbell, and underwater sound transducer.

transfer [MIN ENG] A vertical or inclined connection between two or more levels, used as an ore pass.

transfer caliper [DES ENG] A caliper having one leg which can be opened (or closed) to remove the instrument from the piece being measured; used to measure inside recesses or over projections.

transfer car [MIN ENG] A quarry car provided with transverse tracks, on which the gang car may be conveyed to or from the saw gang.

transfer chamber [ENG] In plastics processing, a vessel in which thermosetting plastic is softened by heat and pressure before being placed in a closed mold for final curing.

transfer chute [ENG] A chute used at a transfer point in a conveyor system; the chute is designed with a curved base or some other feature so that the load is discharged in a centralized stream and in the same direction as the receiving conveyor.

transfer constant [ENG] A transducer rating, equal to one-half the natural logarithm of the complex ratio of the product of the voltage and current entering a transducer to that leaving the transducer when the latter is terminated in its image impedance; alternatively, the product may be that of force and velocity or pressure and volume velocity; the real part of the transfer constant is the image attenuation constant, and the imaginary part is the image phase constant. Also known as transfer factor.

transfer ellipse *See* transfer orbit.

transfer factor *See* transfer constant.

transfer molding [ENG] Molding of thermosetting materials in which the plastic is softened by heat and pressure in a transfer chamber, then forced at high pressure through suitable sprues, runners, and gates into a closed mold for final curing.

transfer orbit [AERO ENG] In interplanetary travel, an elliptical trajectory tangent to the orbits of both the departure planet

and the target planet. Also known as transfer ellipse.

transfer ratio [ENG] From one point to another in a transducer at a specified frequency, the complex ratio of the generalized force or velocity at the second point to the generalized force or velocity applied at the first point; the generalized force or velocity includes not only mechanical quantities, but also other analogous quantities such as acoustical and electrical; the electrical quantities are usually electromotive force and current.

transferred arc [MET] In plasma arc welding, an arc established between the electrode and the workpiece.

transformation temperature [MET] **1.** The temperature at which a change in phase occurs in a metal during heating or cooling. **2.** The maximum or minimum temperature of a transformation temperature range.

transformation temperature ranges [MET] The ranges of temperatures within which austenite forms during heating and transforms during cooling.

transformer oil [MATER] A high-quality insulating oil in which windings of large power transformers are sometimes immersed to provide high dielectric strength, high insulation resistance, high flash point, freedom from moisture, and freedom from oxidization.

transgranular *See* transcrystalline.

transit [ENG] **1.** A surveying instrument with the telescope mounted so that it can measure horizontal and vertical angles. Also known as transit theodolite. **2.** To reverse the direction of the telescope of a transit by rotating 180° about its horizontal axis. Also known as plunge.

transit circle [ENG] A type of astronomical transit instrument having a micrometer eyepiece that has an extra pair of moving wires perpendicular to the vertical set to measure the zenith distance or declination of the celestial object in conjunction with readings taken from a large, accurately calibrated circle attached to the horizontal axis. Also known as meridian circle; meridian transit.

transit declinometer [ENG] A type of declinometer; a surveyor's transit, built to exacting specifications with respect to freedom from traces of magnetic impurities and quality of the compass needle, has a 17-power telescope for sighting on a mark and for making solar and stellar observations to determine true directions.

Transite pipe *See* asbestos-cement pipe.

transitional fit [DES ENG] A fit with varying clearances due to specified tolerances on the shaft and sleeve or hole.

transition altitude [AERO ENG] The altitude in the vicinity of an aerodrome at or below which the vertical position of an aircraft is controlled by reference to true altitude.

transition flow [AERO ENG] A flow of fluid about an airfoil that is changing from laminar flow to turbulent flow.

transition frequency [ENG ACOUS] The frequency corresponding to the intersection of the asymptotes to the constant-amplitude and constant-velocity portions of the frequency-response curve for a disk recording; this curve is plotted with output-voltage ratio in decibels as the ordinate, and the logarithm of the frequency as the abscissa. Also known as crossover frequency; turnover frequency.

transition lattice [MET] An unstable, intermediate configuration formed in a metal lattice during solid-state reactions such as precipitation or transformation.

transition temperature [MET] The temperature at which a fracture changes from tough to brittle in various tests, such as notched-bar impact test.

transit mix [MATER] Concrete or mortar mixed in a rotating cylinder en route to or at the construction site.

Transit satellite [AERO ENG] One of a system of passive, low-orbiting satellites which provide high-accuracy fixes using the Doppler technique several times a day at every point on earth, for navigation and geodesy.

transit survey [ENG] A ground surveying method in which a transit instrument is set up at a control point and oriented, and directions and distances to observed points are recorded.

transit theodolite See transit.

transmission [MECH ENG] The gearing system by which power is transmitted from the engine to the live axle in an automobile. Also known as gearbox.

transmission dynamometer [ENG] A device for measuring torque and power (without loss) between a propulsion power plant and the driven mechanism, for example, wheels or propellers.

transmission oil [MATER] A lubricant especially compounded for automobile transmissions.

transmission tower [ENG] A concrete, metal, or timber structure used to carry a transmission line.

transmissometer [ENG] An instrument for measuring the extinction coefficient of the atmosphere and for the determination of visual range. Also known as hazemeter; transmittance meter.

transmittance meter See transmissometer.

transobuoy [ENG] A free-floating or moored automatic weather station developed for the purpose of providing weather reports from the open oceans; it transmits barometric pressure, air temperature, seawater temperature, and wind speed and direction.

transom [BUILD] A window above a door.

transonic flight [AERO ENG] Flight of vehicles at speeds near the speed of sound (660 miles per hour or 1060 kilometers per hour, at 35,000 feet or 10,700 meters altitude), characterized by great increase in drag, decrease in lift at any altitude, and abrupt changes in the moments acting on the aircraft; the vehicle may shake or buffet.

transonic wind tunnel [ENG] A type of high-speed wind tunnel capable of testing the effects of airflow past an object at speeds near the speed of sound, Mach 0.7 to 1.4; sonic speed occurs where the cross section of the tunnel is at a minimum, that is, where the test object is located.

transosonde [ENG] The flight of a constant-level balloon, whose trajectory is determined by tracking with radio-direction-finding equipment; thus, it is a form of upper-air, quasi-horizontal sounding.

transpiration cooling See sweat cooling.

transport [ENG] Conveyance equipment such as vehicular transport, hydraulic transport, and conveyor-belt setups.

transportation emergency [ENG] A situation which is created by a shortage of normal transportation capability and of a magnitude sufficient to frustrate movement requirements, and which requires extraordinary action by the designated authority to ensure continued movement.

transportation engineering [ENG] That branch of engineering relating to the movement of goods and people; major types of transportation are highway, water, rail, subway, air, and pipeline.

transportation priorities [ENG] Indicators assigned to eligible traffic which establish its movement precedence; appropriate priority systems apply to the movement of traffic by sea and air.

transportation problem [IND ENG] A programming problem that is concerned with the optimal pattern of the distribution of goods from several points of origin to several different destinations, with the specified requirements at each destination.

transport capacity [ENG] The number of persons or the tonnage (or volume) of equipment which can be carried by a vehicle under given conditions.

transport case [ENG] A moistureproof nonconductive wood, plastic, or fabric container used to transport safely small quantities of dynamite sticks to and from blasting sites.

transporter crane [MECH ENG] A long lattice girder supported by two lattice towers which may be either fixed or moved along rails laid at right angles to the girder; a crab with a hoist suspended from it travels along the girder.

transport network [ENG] The complete system of the routes pertaining to all means of transport available in a particular area, made up of the network particular to each means of transport.

transport vehicle [MECH ENG] Vehicle primarily intended for personnel and cargo carrying.

transverse gallery [MIN ENG] An auxiliary crosscut made in thick deposits across the ore body in order to divide it into sections along the strike.

transverse magnetization [ENG ACOUS] Magnetization of a magnetic recording medium in a direction perpendicular to the line of travel and parallel to the greatest cross-sectional dimension.

transverse stability [ENG] The ability of a ship or aircraft to recover an upright position after waves or wind roll it to one side.

trap [AERO ENG] That part of a rocket motor that keeps the propellant grain in place. [CIV ENG] A bend or dip in a soil drain which is always full of water, providing a water seal to prevent odors from entering the building. [MECH ENG] A device which reduces the effect of the vapor pressure of oil or mercury on the high-vacuum side of a diffusion pump.

trapdoor [BUILD] A hinged, sliding, or lifting door to cover an opening in a roof, ceiling, or floor.

trapezoidal excavator [MECH ENG] A digging machine which removes earth in a trapezoidal cross-section pattern for canals and ditches.

trapped-air process [ENG] A procedure for the blow-mold forming of closed plastic objects; the bottom pinch is conventional and, after blowing, sliding pinchers close off the top to form a sealed-air, inflated product.

trapped fuel [ENG] The fuel in an engine or fuel system that is not in the fuel tanks.

trash screen [CIV ENG] A screen placed in a waterway to prevent the passage of trash.

Trauzl test [ENG] A test to determine the relative disruptive power of explosives, in which a standard quantity of explosive (10 grams) is placed in a cavity in a lead block and exploded; the resulting volume of cavity in the block is compared with the volume produced under the same conditions by a standard explosive, usually trinitrotoluene (TNT).

traveling block [MECH ENG] The movable unit, consisting of sheaves, frame, clevis, and hook, connected to, and hoisted or lowered with, the load in a block-and-tackle system. Also known as floating block; running block.

traveling compartment [MIN ENG] The section of a mine shaft used for raising and lowering the miners.

traveling detector [ENG] Radio-frequency probe which incorporates a detector used to measure the standing-wave ratio in a slotted-line section.

traveling gantry crane [ENG] A type of hoisting machine with a bridgelike structure spanning the area over which it operates and running along tracks at ground level.

traveling-grate stoker [MECH ENG] A type of furnace stoker; coal feeds by gravity into a hopper located on top of one end of a moving (traveling) grate; as the grate passes under the hopper, it carries a bed of fresh coal toward the furnace.

traveling road [MIN ENG] A roadway used by miners for walking to and from the face, that is, from the shaft bottom or main entry to the workings.

traveling valve [PETRO ENG] A sucker-rod-pump (oil well) discharge valve that moves with the plunger of a stationary-barrel-type pump, and with the barrel of a traveling-barrel-type pump; contrasted with a standing valve.

traverse [ENG] **1.** A survey consisting of a set of connecting lines of known length, meeting each other at measured angles. **2.** Movement to right or left on a pivot or mount, as of a gun, launcher, or radar antenna.

traversing mechanism [ENG] Mechanism by which a gun or other device can be turned in a horizontal plane.

trawl [ENG] A baglike net whose mouth is kept open by boards or by a leading diving vane or depressor at the foot of the opening and a spreader bar at the top; towed by a ship at specified depths for catching forms of marine life.

tray elevator [MECH ENG] A device for lifting drums, barrels, or boxes; a parallel pair of vertical-mounted continuous chains turn over upper and lower drive gears, and spaced trays on the chains cradle and lift the objects to be moved.

tread [CIV ENG] **1.** The horizontal part of a step in a staircase. **2.** The distance between two successive risers in a staircase. [ENG] The part of a wheel or tire that bears on the road or rail.

tree [MET] A projecting treelike aggregate of crystals formed at areas of high local current density in electroplating.

tremie [ENG] An apparatus for placing concrete underwater, consisting of a large metal tube with a hopper at the top end and a valve arrangement at the bottom, submerged end.

trencher *See* trench excavator.

trench excavator [MECH ENG] A digging machine, usually on crawler tracks, and having either a movable wheel or a continuous chain on which buckets are mounted. Also known as bucket-ladder excavator; ditcher; trencher; trenching machine.

trenching machine *See* trench excavator.

trench sampling [MIN ENG] A slight refinement of grab sampling in which the ore material to be sampled is spread out flat and channeled in one direction with a shovel, and the material for the sample is taken at regular intervals along the channel.

trennschaukel apparatus [ENG] An instrument for determining the thermal diffusion factors of gases and gas mixtures, consisting of 20 suitably interconnected tubes whose top ends are maintained at the same temperature and whose bottom ends are maintained at the same temperature, with the temperature of the top ends greater than that of the bottom ends.

trepanning tool [MECH ENG] A cutting tool in the form of a circular tube, having teeth on the end; the workpiece or tube, or both, are rotated and the tube is fed axially into the workpiece, leaving behind a narrow grooved surface in the workpiece.

trestle [CIV ENG] A series of short bridge spans supported by a braced tower. [ENG] **1.** A movable support usually with legs that spread diagonally. **2.** A braced structure of timber, reinforced concrete, or steel spanning a land depression to carry a road or railroad.

trestle bent [CIV ENG] A transverse frame that supports the ends of the stringers in adjoining spans of a trestle.

trial batch [ENG] A batch of concrete mixed to determine the water-cement ratio that will produce the required slump and compressive strength; from a trial batch, one can also compute the yield, cement factor, and required quantities of each material.

trial pit [MIN ENG] A shallow hole, 2 to 3 feet (60 to 90 centimeters) in diameter, put down to test shallow minerals or to establish the nature and thickness of superficial deposits and depth to bedrock.

trial shots [ENG] The experimental shots and rounds fired in a sinking pit, tunnel, opencast, or quarry to determine the best drill-hole pattern to use.

triangle cut [MIN ENG] A zigzag arrangement of drill holes permitting larger openings to be obtained as the drill holes can break out between the preceding row of holes.

triangle equation *See* angle equation.

triangular method [MIN ENG] A method of ore reserve estimation based on the assumption that a linear relationship exists between the grade difference and the distance between all drill holes.

triangular-notch weir [CIV ENG] A measuring weir with a V-shaped notch for measuring small flows. Also known as V-notch weir.

triangulation [ENG] A surveying method for measuring a large area of land by establishing a base line from which a network of triangles is built up; in a series, each triangle has at least one side common with each adjacent triangle.

triangulation mark [ENG] A bronze disk set in the ground to identify a point whose latitude and longitude have been determined by triangulation.

trickling filter [CIV ENG] A bed of broken rock or other coarse aggregate onto which sewage or industrial waste is sprayed intermittently and allowed to trickle through, leaving organic matter on the surface of the rocks, where it is oxidized and removed by biological growths.

tricone bit [ENG] A rock bit with three toothed, conical cutters, each of which is mounted on friction-reducing bearings.

tricycle landing gear [AERO ENG] A landing-gear arrangement that places the nose gear well forward of the center of gravity on the fuselage and the two main gears slightly aft of the center of gravity, with sufficient distance between them to provide stability against rolling over during a yawed landing in a crosswind, or during ground maneuvers.

trifilter hydrophotometer [ENG] An instrument that uses red, green, and blue filters to measure the transparency of the water at three wavelengths.

trigonal lattice *See* rhombohedral lattice.

trigonometric leveling [ENG] A method of determining the difference of elevation between two points, by using the principles of triangulation and trigonometric calculations.

trilateration [ENG] The measurement of a series of distances between points on the surface of the earth, for the purpose of establishing relative positions of the points in surveying.

trim [AERO ENG] The orientation of an aircraft relative to the airstream, as indicated by the amount of control pressure required to maintain a given flight performance.

trimmer [BUILD] One of the single or double joists or rafters that go around an opening in the framing type of construction. [MIN ENG] **1.** A piece of bent wire used to regulate the size of the flame of a safety lamp without removing the top of the lamp. **2.** A worker who arranges coal in the hold of a vessel (miner, ship) as the coal is discharged into it from bins. **3.** A person who cleans miners' lamps. **4.** An apparatus for trimming a pile of coal into a regular form (as a cone or prism). **5.** *See* rib hole.

trimmer conveyor [MECH ENG] A self-contained, lightweight portable conveyor, usually of the belt type, for use in unloading and delivering bulk materials from trucks to domestic storage places, and for trimming bulk materials in bins or piles.

trimming [MET] Removing irregular edges from a drawn part, or parting-line flash from a forging, or gates, risers, and fins from a casting.

Trinidad asphalt [MATER] Natural asphaltic material found in Trinidad; contains about 47% bitumen and 28% clay, and the remainder is water.

trip [ENG] To release a lever or set free a mechanism. [MIN ENG] **1.** The line of cars hauled by mules or by motor, or run on a slope, plane, or sprag road. **2.** An automatic arrangement for dumping cars.

trip change [MIN ENG] The period during which the loaded cars are taken away and empties are brought back.

trip hammer [MECH ENG] A large power hammer whose head is tripped and falls by cam or lever action.

trip lamp [MIN ENG] A removable self-contained mine lamp, designed for marking the rear end of a train (trip) of mine cars.

triple-base propellant [MATER] A propellant with three principal active ingredients, such as nitrocellulose, nitroglycerin, and nitroguanidine.

triple entry [MIN ENG] A system of opening a mine by driving three parallel entries as main entries.

triple superphosphate [MATER] A phosphatic fertilizer produced by the reaction of phosphate rock with phosphoric acid so as to give higher concentrations of calcium phosphate than for ordinary superphosphate.

triple thread [DES ENG] A multiple screw thread having three threads or starts equally spaced around the periphery; the lead is three times the pitch.

triplex chain block [MECH ENG] A geared hoist using an epicyclic train.

tripod [DES ENG] An adjustable, collapsible three-legged support, as for a camera or surveying instrument.

tripod drill [MECH ENG] A reciprocating rock drill mounted on three legs and driven by steam or compressed air; the drill steel is removed and a longer drill inserted about every 2 feet (60 centimeters).

tripper [CIV ENG] A device activated by a passing train to work a signal or switch or to apply brakes. [MECH ENG] A device that snubs a conveyor belt causing the load to be discharged.

tripping [MIN ENG] **1.** The process of pulling or lowering drill-string equipment in a borehole. **2.** To open a latch or locking device, thereby allowing a door or gate to open to empty the contents of a skip or bailer.

trip rider [MIN ENG] A rider who throws switches, gives signals, and makes couplings. Also known as rope rider.

trip spear [ENG] A fishing tool intended to recover lost casing; if the casing is found to be immovable, the hold is broken by operating the trip release.

tritonal [MATER] An explosive composed of 80% trinitrotoluene (TNT) and 20% powdered aluminum; can be melt-loaded, and is used in bombs for its blast effect.

triton block [MATER] Block of pressed trinitrotoluene (TNT) used for demolition purposes.

trolley [MECH ENG] **1.** A wheeled car running on an overhead track, rail, or ropeway. **2.** An electric streetcar.

trolley locomotive [MECH ENG] A locomotive operated by electricity drawn from overhead conductors by means of a trolley pole.

trommel [MIN ENG] **1.** A revolving cylindrical screen used to grade coarsely crushed ore; the ore is fed into the trommel at one end, the fine material drops through the holes, and the coarse is delivered at the other end. Also known as trommel screen. **2.** To separate coal into various sizes by passing it through a revolving screen.

trommel screen *See* trommel.

tropical finish [ENG] A finish that is applied to electronic equipment to resist the high relative humidity, fungus, and insects encountered in tropical climates.

tropicalize [ENG] To prepare electronic equipment for use in a tropical climate by applying a coating that resists moisture and fungi.

troughed belt conveyor [MECH ENG] A belt conveyor with the conveyor belt edges elevated on the carrying run to form a trough by conforming to the shape of the troughed carrying idlers or other supporting surface.

troughed roller conveyor [MECH ENG] A roller conveyor having two rows of rolls set at an angle to form a trough over which objects are conveyed.

troughing idler [MECH ENG] A belt idler having two or more rolls arranged to turn up the edges of the belt so as to form the belt into a trough.

troughing rolls [MECH ENG] The rolls of a troughing idler that are so mounted on an incline as to elevate each edge of the belt into a trough.

trough washer [MIN ENG] A sloping wooden trough, 1½ to 2 feet wide, 8 to 12 feet long, and 1 foot deep (about 50 by 300 by 30 centimeters), open at the tail end but closed at the head end; it is used to float adhering clay or fine stuff from the coarser portions of an ore or coal.

trowel [DES ENG] Any of various hand tools consisting of a wide, flat or curved blade with a short wooden handle; used by gardeners, plasterers, and bricklayers.

Trube's correlation [PETRO ENG] An empirical correlation (based on pseudocritical properties) for compressibilities of undersaturated oil-reservoir fluids.

truck [MECH ENG] A self-propelled wheeled vehicle, designed primarily to transport goods and heavy equipment; it may be used to tow trailers or other mobile equipment. [MIN ENG] See barney.

truck crane [MECH ENG] A crane carried on the bed of a motortruck.

truck-mounted drill rig [MECH ENG] A drilling rig mounted on a lorry or caterpillar tracks.

truck-tractor See tractor.

true airspeed [AERO ENG] The actual speed of an aircraft relative to the air through which it flies, that is, the calibrated airspeed corrected for temperature, density, or compressibility.

true-airspeed indicator [AERO ENG] An instrument for measuring true airspeed. Also known as true-airspeed meter.

true-airspeed meter See true-airspeed indicator.

true porcelain See porcelain.

true rake [MECH ENG] The angle, measured in degrees, between a plane containing a tooth face and the axial plane through the tooth point in the direction of chip flow.

true width [MIN ENG] The width of thickness of a vein or stratum as measured perpendicular to or normal to the dip and the strike; the true width is always the least width.

truing [MECH ENG] **1.** Cutting a grinding wheel to make its surface run concentric with the axis. **2.** Aligning a wheel to be concentric and in one plane.

truncation error [ENG] The error resulting from the analysis of a partial set of data in place of a complete or infinite set.

trunk buoy [ENG] A mooring buoy having a pendant extending through an opening in the buoy, with the ship's anchor chain or mooring line being secured to this pendant.

trunk roadway [MIN ENG] The main developing heading from the pit bottom and is usually driven along the strike of the coal seam.

trunk sewer [CIV ENG] A sewer receiving sewage from many tributaries serving a large territory.

trunnion [DES ENG] **1.** Either of two opposite pivots, journals, or gudgeons, usually cylindrical and horizontal, projecting one from each side of a piece of ordnance, the cylinder of an oscillating engine, a molding flask, or a converter, and supported by bearings to provide a means of swiveling or turning. **2.** A pin or pivot usually mounted on bearings for rotating or tilting something.

truss [CIV ENG] A frame, generally of steel, timber, concrete, or a light alloy, built from members in tension and compression.

truss bridge [CIV ENG] A fixed bridge consisting of members vertically arranged in a triangular pattern.

trussed beam [CIV ENG] A beam stiffened by a steel tie rod to reduce its deflection.

trussed rafter [BUILD] A triangulated beam in a trussed roof.

truss rod [CIV ENG] A rod attached to the ends of a trussed beam which transmits the strain due to downward pressure.

trypan blue [MATER] An acid disazo dye of the benzopurpurine series used as a vital stain.

try square [ENG] An instrument consisting of two straightedges secured at right angles to each other, used for laying off right angles and testing whether work is square.

Tschudi engine [MECH ENG] A cat-and-mouse engine in which the pistons, which are sections of a torus, travel around a toroidal cylinder; motion of the pistons is controlled by two cams which bear against rollers attached to the rotors.

T slot [DES ENG] A recessed slot, in the form of an inverted T, in the table of a machine tool, to receive the square head of a T-slot bolt.

tubbing [MIN ENG] The watertight cast-iron lining of a circular shaft built up of segments with the space outside the tubbing grouted to add strength and to improve watertightness.

tube [ENG] A long cylindrical body with a hollow center used especially to convey fluid.

tube cleaner [MECH ENG] A device equipped with cutters or brushes used to clean tubes in heat transfer equipment.

tube core [AERO ENG] One type of sandwich configuration used in structural materials in aircraft; aluminum, steel, and titanium have been used for face materials with cores of wood, rubber, plastics, steel, and aluminum in the form of tubes.

tube door [MECH ENG] A door in a boiler furnace wall which facilitates the removal or installation of tubes.

tube hole [ENG] A hole in a tube sheet through which a tube is passed prior to sealing.

tube mill [MECH ENG] A revolving cylinder used for fine pulverization of ore, rock, and other such materials; the material, mixed with water, is fed into the chamber from one end, and passes out the other end as slime.

tube plug [ENG] A solid plug inserted into the end of a tube in a tube sheet.

tubercle [MET] A mound of corrosive products on the surface of a metal that is subjected to local corrosive attack.

tuberculation [MET] Corrosive attack with formation of tubercles.

tube reducing [MET] Reducing the diameter and wall thickness of tubing by means of a mandrel and rolls.

tube seat [ENG] The surface of the tube hole in a tube sheet which contacts the tube.

tube sheet [ENG] A mounting plate for elements of a larger item of equipment; for example, filter cartridges, or tubes for heat exchangers, coolers, or boilers.

tube shield [ENG] A shield designed to be placed around an electron tube.

tube socket [ENG] A socket designed to accommodate electrically and mechanically the terminals of an electron tube.

tube stock [MET] Semifinished metal tubing.

tube turbining [MECH ENG] Cleaning tubes by passing a power-driven rotary device through them.

tube voltmeter *See* vacuum-tube voltmeter.

tubing [ENG] Material in the form of a tube, most often seamless.

tubing hanger *See* hanger.

tubing head [PETRO ENG] A spool-type unit or housing attached to the top flange on the uppermost oil-well-casing head to support the tubing string and to seal the annulus between the tubing string and the production casing string.

tubing-head adapter flange [PETRO ENG] An intermediate flange used in oil wells to connect the top tubing-head flange to the master valve (Christmas tree) and to provide support for the tubing.

tubing pump [PETRO ENG] A type of oil-well, sucker-rod pump in which the pump barrel is attached to the tubing string, and lowered into the well bore with the tubing.

tubing safety valve *See* storm choke.

tuck-and-pat pointing *See* tuck pointing.

tuck joint pointing *See* tuck pointing.

tuck pointing [BUILD] The finishing of old masonry joints in which the joints are first cleaned out and then filled with fine mortar which projects slightly or has a fillet of putty or lime. Also known as tuck-and-pat pointing; tuck joint pointing.

tugger [MIN ENG] A small portable pneumatic or electric hoist mounted on a column and used in a mine.

Tukon tester [ENG] A device that uses a diamond (Knoop) indenter applying average loads of 1 to 2000 grams to determine microhardness of a metal.

tumble-plating process [MET] A method of zinc-coating small metal parts by first applying zinc powder with an adhesive, then tumbling with glass beads to roll out the powder into a continuous coat.

tumbler [ENG] **1.** A device in a lock cylinder that must be moved to a particular position, as by a key, before the bolt can be thrown. **2.** A device or mechanism in which objects are tumbled.

tumbler feeder *See* drum feeder.

tumbler gears [MECH ENG] Idler gears interposed between spindle and stud gears in a lathe gear train; used to reverse rotation of lead screw or feed rod.

tumbling [AERO ENG] An attitude situation in which the vehicle continues on its flight, but turns end over end about its center of mass. [ENG] A surface-finishing operation for small articles in which irregularities are removed or surfaces are polished by tumbling them together in a barrel, along with wooden pegs, sawdust, and polishing compounds. [MECH ENG] Loss of control in a two-frame free gyroscope, occurring when both frames of reference become coplanar.

tumbling mill [MECH ENG] A grinding and pulverizing machine consisting of a shell or drum rotating on a horizontal axis.

tundish [MET] A funnel or pouring basin used for transferring a stream of molten metal.

tuned-reed frequency meter *See* vibrating-reed frequency meter.

tung oil [MATER] A yellow, combustible drying oil extracted from the seed of the tung tree; soluble in ether, chloroform, carbon disulfide, and oils; used in formulations for paints, varnishes, varnish driers, paper waterproofing, and linoleum. Also known as China wood oil.

tungsten [MET] A hard, brittle, ductile, heavy gray-white metal used in the pure form chiefly for electrical purposes and with other substances in dentistry, pen points, x-ray-tube targets, phonograph needles, and high-speed tool metal, and as a radioactive shield.

tungsten–inert gas welding [MET] Welding in which an arc plasma from a nonconsumable tungsten electrode radiates heat onto the work surface, to create a weld puddle in a protective atmosphere provided by a flow of inert shielding gas; heat must then travel by conduction from this puddle to melt the desired depth of weld. Abbreviated TIG welding.

tungsten steel [MET] Steel containing tungsten with other alloys; formerly used for cutting and forging tools but replaced by high-speed steel.

tuning fork [ENG] A U-shaped bar for hard steel, fused quartz, or other elastic material that vibrates at a definite natural frequency when struck or when set in motion by electromagnetic means; used as a frequency standard.

tunnel [ENG] A long, narrow, horizontal or nearly horizontal underground passage that is open to the atmosphere at both ends; used for aqueducts and sewers, carrying railroad and vehicular traffic, various underground installations, and mining.

tunnel-bearing grease [MATER] Lubricating grease for the main engine and propeller shaft (in the shaft tunnel) of ships.

tunnel blasting [ENG] A method of heavy blasting in which a heading is driven into the rock and afterward filled with explosives in large quantities, similar to a borehole, on a large scale, except that the heading is usually divided in two parts on the same level at right angles to the first heading, forming in plan a T, the ends of which are filled with explosives and the intermediate parts filled with inert material like an ordinary borehole.

tunnel borer [MECH ENG] Any boring machine for making a tunnel; often a ram armed with cutting faces operated by compressed air.

tunnel carriage [MECH ENG] A machine used for rapid tunneling, consisting of a combined drill carriage and manifold for water and air so that immediately the carriage is at the face, drilling may commence with no lost time for connecting up or waiting for drill steels; the air is supplied at pressures of 95 to 100 pounds per square inch (6.55 to 6.89 × 10^5 newtons per square meter).

tunnel liner [CIV ENG] Any of various materials, especially timber, concrete, and cast iron, applied to the inner surface of a vehicular or railroad tunnel. [MIN ENG] The timber, brick, concrete, or steel supports erected in a mine tunnel to maintain dimensions and safe working conditions.

tunnel set [MIN ENG] Timbers 6 to 8 inches (15 to 20 centimeters) in diameter and of sufficient height to support the roof of the tunnel.

tunnel system [MIN ENG] A method of mining in which tunnels or drifts are extended at regular intervals from the floor of the pit into the ore body.

turbine [MECH ENG] A fluid acceleration machine for generating rotary mechanical power from the energy in a stream of fluid.

turbine propulsion [MECH ENG] Propulsion of a vehicle or vessel by means of a steam or gas turbine.

turbine pump *See* regenerative pump.

turbining [MECH ENG] The removal of scale or other foreign material from the internal surface of a metallic cylinder.

turboblower [MECH ENG] A centrifugal or axial-flow compressor.

turbodrill [PETRO ENG] A rotary tool used in drilling oil or gas wells in which the bit is rotated by a turbine motor inside the well.

turbofan [AERO ENG] An air-breathing jet engine in which additional propulsive thrust is gained by extending a portion of the compressor or turbine blades outside the inner engine case.

turbojet [AERO ENG] A jet engine incorporating a turbine-driven air compressor to take in and compress the air for the combustion of fuel (or for heating by a nuclear reactor), the gases of combustion (or the heated air) being used both to rotate the turbine and to create a thrust-producing jet.

turbosupercharger [MECH ENG] A centrifugal air compressor, gas-turbine driven, usually used to increase induction system pressure in an internal combustion reciprocating engine.

turbulent burner [ENG] An atomizing burner which mixes fuel and air to produce agitated flow.

Turkey red oil [MATER] Sulfonated castor oil, soluble in water; autoignites at 833°F (445°C); used in textiles, leather, and paper coatings, for manufacture of soaps, and as an alizarin dye assistant. Also known as soluble castor oil; sulfonated castor oil.

Turkish geranium oil *See* palmarosa oil.

Turk's head rolls [MET] A group of four idler rolls, arranged in a square or rectangular pattern, through which strip metal can be drawn to form angled sections.

turmeric [MATER] An orange-red or reddish-brown dye obtained from the rhizome of turmeric.

turnaround [ENG] The length of time between arriving at a point and departing from that point; it is used in this sense for the turnaround of vehicles, ships in ports, and aircraft.

turnaround cycle [ENG] A term used in conjunction with vehicles, ships, and aircraft, and comprising the following: loading time at home, time to and from destination, unloading and loading time at destination, unloading time at home, planned maintenance time, and, where applicable, time awaiting facilities.

turnbuckle [DES ENG] A sleeve with a thread at one end and a swivel at the other, or with threads of opposite hands at each end so that by turning the sleeve connected rods or wire rope will be drawn together and tightened.

turning [MECH ENG] Shaping a member on a lathe.

turning bar *See* chimney bar.

turning basin [CIV ENG] An open area at the end of a canal or narrow waterway to allow boats to turn around.

turning-block linkage [MECH ENG] A variation of the sliding-block mechanical linkage in which the short link is fixed and the frame is free to rotate. Also known as the Wentworth quick-return motion.

turning error *See* northerly turning error.

turning table [ENG] In plastics molding, a rotating table or wheel carrying various molds in a multimold, single-parison blow-molding operation.

turnkey contract [ENG] A contract in which an independent agent undertakes to furnish for a fixed price all materials and labor, and to do all the work needed to complete a project.

turnout [ENG] **1.** A contrivance consisting of a switch, a frog, and two guardrails for passing from one track to another. **2.** The branching off of one rail track from another. **3.** A siding. [MIN ENG] To shovel coal toward the track for more convenient loading.

turnover cartridge [ENG ACOUS] A phonograph pickup having two styli and a pivoted mounting that places in playing position the correct stylus for a particular record speed.

turnover frequency *See* transition frequency.

turnpike [CIV ENG] A toll expressway.

turntable [ENG ACOUS] The rotating platform on which a disk record is placed for recording or playback.

turntable rumble [ENG ACOUS] Low-frequency vibration that is mechanically transmitted to a recording or reproducing turntable and superimposed on the reproduction. Also known as rumble.

turpentine [MATER] An essential oil produced by steam distillation of pine woods and from gum turpentine; used as a solvent and a thinner for paints and varnishes.

turret coal cutter [MIN ENG] A coal cutter in which the horizontal jib can be adjusted

vertically to cut at different levels in the seam; for example, an overcut.

turret lathe [MECH ENG] A semiautomatic lathe differing from the engine lathe in having the tailstock replaced with a multisided, indexing tool holder or turret designed to hold several tools.

turtle oil [MATER] The oil derived from the muscles and genital glands of the giant sea turtle; melts at 25°C; used in cosmetics.

tusche [MATER] A liquid lithographic ink that can be used with pen or brush on lithographic stones or metal plates; it is also used in the silk-screen process in a glue-resist system of pattern making; an extremely greasy material.

tuyere [MET] An opening in the shell and refractory lining of a furnace through which air is forced.

Twaddell scale [ENG] A scale for specific gravity of solutions that is the first two digits to the right of the decimal point multiplied by two; for example, a specific gravity of 1.4202 is equal to 84.04°Tw.

tweeter [ENG ACOUS] A loudspeaker designed to handle only the higher audio frequencies, usually those well above 3000 hertz; generally used in conjunction with a crossover network and a woofer.

twin band [MET] A line on a polished or etched surface representing the section through crystal twins.

twin-cable ropeway [MECH ENG] An aerial ropeway which has parallel track cables with carriers running in opposite directions; both rows of carriers are pulled by the same traction rope.

twine [MATER] A strong string made up of two or more strands twisted together.

twin entry [MIN ENG] A pair of parallel entries, one of which is an intake air course and the other the return air course; rooms can be worked from both entries.

twin-geared press [MECH ENG] A crank press having the drive gears attached to both ends of the crankshaft.

twist [DES ENG] In a fiber, rope, yarn, or cord, the turns about its axis per unit length; usually expressed as TPI (turns per inch).

twist drill [DES ENG] A tool having one or more helical grooves, extending from the point to the smooth part of the shank, for ejecting cuttings and admitting a coolant.

two-cycle engine [MECH ENG] A reciprocating internal combustion engine that requires two piston strokes or one revolution to complete a cycle.

two-level mold [ENG] Placement of one cavity of a plastics mold above another instead of alongside it; reduces clamping force needed.

two-lip end mill [MECH ENG] An end-milling cutter having two cutting edges and straight or helical flutes.

two-part adhesive [MATER] A glue supplied in two parts, a resin and an accelerator, which are mixed only just before application.

two-piece set [MIN ENG] A set of timbers consisting of a cap and a single post.

two-point press [MECH ENG] A mechanical press in which the slide is actuated at two points.

two-position propeller [AERO ENG] An airplane propeller whose blades are limited to two angles, one for take off and climb and the other for cruising.

two-sided sampling plans [IND ENG] Any sampling plan whereby the acceptability of material is determined against upper and lower limits.

two-stage hoisting [MIN ENG] Deep shaft hoisting with two winders, one at the surface, and the other at mid-depth in the shaft.

two-stroke cycle [MECH ENG] An internal combustion engine cycle completed in two strokes of the piston.

two-tone diaphone [ENG ACOUS] A diaphone producing blasts of two tones, the second tone being of a lower pitch than the first tone.

two-way slab [CIV ENG] A concrete slab supported by beams along all four edges and reinforced with steel bars arranged perpendicularly.

tyfon *See* typhon.

Tyler Standard screen scale [ENG] A scale for classifying particles in which the particle size in micrometers is correlated with the meshes per inch of a screen.

Tyndallization [ENG] Heat sterilization by steaming the food or medium for a few minutes at atmospheric pressure on three or four successive occasions, separated by 12- to 18-hour intervals of incubation at a temperature favorable for bacterial growth.

type metal [MET] Any of various low-melting-point alloys, composed mainly of lead (50–90%), antimony (2–30%), and tin (2–20%), used for casting printers' type.

typhon [ENG ACOUS] A diaphragm horn which operates under the influence of compressed air or steam. Also spelled tyfon.

U

U-bend die [MECH ENG] A die with a square or rectangular cross section which provides two edges over which metal can be drawn.

U blades [DES ENG] Curved bulldozer blades designed to increase moving capacity of tractor equipment.

U bolt [DES ENG] A U-shaped bolt with threads at the ends of both arms to receive nuts.

udometer *See* rain gage.

Ugine Sejournet process *See* Sejournet process.

ullage [ENG] The amount that a container, such as a fuel tank, lacks of being full.

ullage rocket [AERO ENG] A small rocket used in space to impart an acceleration to a tank system to ensure that the liquid propellants collect in the tank in such a manner as to flow properly into the pumps or thrust chamber.

ultimate elongation [MET] The percentage of permanent deformation remaining after tensile rupture.

ultimate-load design [DES ENG] Design of a beam that is proportioned to carry at ultimate capacity the design load multiplied by a safety factor. Also known as limit-load design; plastic design; ultimate-strength design.

ultimate recovery [PETRO ENG] Estimated total (ultimate) recovery of hydrocarbon fluids expected from a reservoir during its productive lifetime.

ultimate set [ENG] The ratio of the length of a specimen plate or bar before testing to the length at the moment of fracture; usually expressed as a percentage.

ultimate-strength design *See* ultimate-load design.

ultracentrifuge [ENG] A laboratory centrifuge that develops centrifugal fields of more than 100,000 times gravity.

ultramicrobalance [ENG] A differential weighing device with accuracies better than 1 microgram; used for analytical weighings in microanalysis.

ultramicrotome [ENG] A microtome which uses a glass or diamond knife, allowing sections of cells to be cut 300 nanometers in thickness.

ultrasonic atomizer [MECH ENG] An atomizer in which liquid is fed to, or caused to flow over, a surface which vibrates at an ultrasonic frequency; uniform drops may be produced at low feed rates.

ultrasonic bonding [MET] Bonding of two identical or dissimilar metals by mechanical pressure combined with a wiping motion produced by ultrasonic vibration.

ultrasonic cleaning [ENG] A method used to clean debris and swarf from surfaces by immersion in a solvent in which ultrasonic vibrations are excited.

ultrasonic delay line [ENG ACOUS] A delay line in which use is made of the propagation time of sound through a medium such as fused quartz, barium titanate, or mercury to obtain a time delay of a signal. Also known as ultrasonic storage cell.

ultrasonic depth finder [ENG] A direct-reading instrument which employs frequencies above the audible range to determine the depth of water; it measures the time interval between the emission of an ultrasonic signal and the return of its echo from the bottom.

ultrasonic drill [MECH ENG] A drill in which a magnetostrictive transducer is attached to a tapered cone serving as a velocity transformer; with an appropriate tool at the end of the transformer, practically any shape of hole can be drilled in hard, brittle materials such as tungsten carbide and gems.

ultrasonic drilling [MECH ENG] A vibration drilling method in which ultrasonic vibrations are generated by the compression and extension of a core of electrostrictive or magnetostrictive material in a rapidly alternating electric or magnetic field.

ultrasonic flaw detector [ENG ACOUS] An ultrasonic generator and detector used together, much as in radar, to determine the distance to a wave-reflecting internal crack or other flaw in a solid object.

ultrasonic generator [ENG ACOUS] A generator consisting of an oscillator driving an electroacoustic transducer, used to produce acoustic waves above about 20 kilohertz.

ultrasonic imaging device [ENG ACOUS] An imaging device in which a wave is generated by a transducer external to the body; the reflected wave is detected by the same transducer.

ultrasonic leak detector [ENG] An instrument which detects ultrasonic energy resulting from the transition from laminar to turbulent flow of a gas passing through an orifice.

ultrasonic machining [MECH ENG] The removal of material by abrasive bombardment and crushing in which a flat-ended tool of soft alloy steel is made to vibrate at a frequency of about 20,000 hertz and an amplitude of 0.001–0.003 inch (0.0254–0.0762 millimeter) while a fine abrasive of silicon carbide, aluminum oxide, or boron carbide is carried by a liquid between tool and work.

ultrasonic metal inspection [MET] The application of ultrasonic vibrations to materials for detection of flaws, pits, or wall thickness.

ultrasonic sealing [ENG] A method for sealing plastic film by using localized heat developed by vibratory mechanical pressure at ultrasonic frequencies.

ultrasonic soldering [MET] A method used for aluminum soldering by ultrasonically vibrating the soldering iron to disrupt the oxide film on the metal.

ultrasonic testing [ENG] A nondestructive test method that employs high-frequency mechanical vibration energy to detect and locate structural discontinuities or differences and to measure thickness of a variety of materials.

ultrasonic thickness gage [ENG] A thickness gage in which the time of travel of an ultrasonic beam through a sheet of material is used as a measure of the thickness of the material.

ultrasonic transducer [ENG ACOUS] A transducer that converts alternating-current energy above 20 kilohertz to mechanical vibrations of the same frequency; it is generally either magnetostrictive or piezoelectric.

ultrasonic transmitter [ENG ACOUS] A device used to track seals, fish, and other aquatic animals: the device is fastened to the outside of the animal or fed to it, and has a loudspeaker which is made to vibrate at an ultrasonic frequency, propagating ultrasonic waves through the water to a special microphone or hydrophone.

ultrasonic welding [MET] A nonfusion welding process in which the atomic movement required for coalescence is stimulated by ultrasonic vibrations.

ultrasonoscope [ENG] An instrument that displays an echosonogram on an oscilloscope; usually has auxiliary output to a chart-recording instrument.

ultraspeed welding See commutator-controlled welding.

ultraviolet absorber fixative [MATER] A protective fixative that includes a material to filter ultraviolet light from the sun and from artificial light; it helps to keep colors from fading.

umbilical connections [AERO ENG] Electrical and mechanical connections to a launch vehicle prior to lift off; the umbilical tower adjacent to the vehicle on the launch pad supports these connections which supply electrical power, control signals, data links, propellant loading, high pressure gas transfer, and air conditioning.

umbilical cord [AERO ENG] Any of the servicing electrical or fluid lines between the ground or a tower and an uprighted rocket vehicle before the launch. Also known as umbilical.

umbilical tower [AERO ENG] A vertical structure supporting the umbilical cords running into a rocket in launching position.

umpire [MIN ENG] An assay made by a third party to settle the difference in assays made by the purchaser and the seller of ore.

unavoidable delay [IND ENG] Any delay in a task, the occurrence of which is outside the control or responsibility of the worker.

unavoidable-delay allowance [IND ENG] An adjustment of standard time to allow for unavoidable delays in a task.

unbalanced cutter chain [MIN ENG] A cutter chain which carries more picks along the bottom line than along the top line.

unbalanced hoisting [MIN ENG] The method of hoisting in small one-compartment shafts with only one cage in operation, as opposed to balanced winding.

unbalanced shothole [MIN ENG] A shothole in which the explosive charge breaks down the coal at the back of the machine cut while leaving the front portion standing or in large blocks.

unbonded strain gage [ENG] A type of strain gage that consists of a grid of fine wires strung under slight tension between a stationary frame and a movable armature; pressure applied to the bellows or to the diaphragm sensing element moves the armature with respect to the frame, increasing tension in one half of the filaments and decreasing tension in the rest.

uncage [ENG] To release the caging mechanism of a gyroscope, that is, the mechanism that erects the gyroscope or locks it in position.

uncharged demolition target [ENG] A demolition target which has been prepared to receive the demolition agent, the necessary quantities of which have been calculated, packaged, and stored in a safe place.

unconfined explosion [ENG] Explosion occurring in the open air where the (atmospheric) pressure is constant.

uncouple [ENG] To unscrew or disengage.

unctuous [MATER] Greasy, oily, or soapy to the touch.

underbead crack [MET] A crack in the heat-affected zone of a weldment which usually does not reach the base-metal surface.

undercarriage [AERO ENG] The landing gear assembly for an aircraft.

undercast [MIN ENG] An air crossing in which one airway is deflected to pass under the other airway.

underchain haulage [MIN ENG] Haulage in which the chains are placed beneath the mine car at certain intervals with suitable hooks that thrust against the car axle.

undercoater [MATER] A type of solvent-thinned paint that is formulated to have good adhesion to a substrate and furnish a good base for additional coats of paint, and having a relatively small amount of pigment.

undercooling [MET] Cooling a metal below the transformation temperature without obtaining the transformation.

undercut [ENG] Underside recess either cut or molded into an object so as to leave a topside lip or protuberance. [MET] An unfilled groove melted into the base metal at the toe of a weld. [MIN ENG] To cut below or in the lower part of a coal bed by chipping away the coal with a pick or mining machine; cutting is usually done on the level of the floor of the mine, extending laterally the entire face and 5 or 6 feet (1.5 or 1.8 meters) into the material.

underdraft [MET] Downward curving of a metal part on leaving the rolls because of higher speed of the upper roll.

underdrain [CIV ENG] A subsurface drain with holes into which water flows when the water table reaches the drain level.

underdrive press [MECH ENG] A mechanical press having the driving mechanism located within or under the bed.

underfeed stoker [ENG] A coal-burning system in which green coal is fed from beneath the burning fuel bed.

underfill [MET] A depression on the face of the weld which falls below the surface of the adjacent base metal.

underfloor raceway [BUILD] A raceway for electric wires which runs beneath the floor.

underground [ENG] Situated, done, or operating beneath the surface of the ground.

underground gasification See gasification.

underground glory-hole method [MIN ENG] A mining method used in large deposits with a very strong roof: the deposit is divided by levels and on every level chutes are raised to the next level; mining starts from the mouth of the chutes in such a way as to develop a funnel-shaped excavation (mill, or glory) with slopes so steep that the broken ore falls into the chutes and thus to the cars on the lower level; a sufficiently strong pillar is left for protection at the higher level. Also known as underground milling.

underground milling See underground glory-hole method.

underhand stoping [MIN ENG] Mining downward or from upper to lower level; the stope may start below the floor of a level and be extended by successive horizontal slices, either worked sequentially or simultaneously in a series of steps; the stope may be left as an open stope or supported by stulls or pillars.

underhand work [MIN ENG] Picking or drilling downward.

underhead crack [MET] A subsurface crack in the heat-affected zone of the base metal near a weld.

underhole [MIN ENG] To mine out a portion of the bottom of a seam, by pick or powder, thus leaving the top unsupported and ready to be blown down by shots, broken down by wedges, or mined with a pick or bar.

underhung crane [MECH ENG] An overhead traveling crane in which the end trucks carry the bridge suspended below the rails.

underlay shaft [MIN ENG] A shaft sunk in the footwall and following the dip of a vein. Also known as footwall shaft; underlier.

underlier See underlay shaft.

undermine [MIN ENG] To excavate the earth beneath, especially for the purpose of causing to fall; to form a mine under.

underpinning [CIV ENG] **1.** Permanent supports replacing or reinforcing the older supports beneath a wall or a column. **2.** Braced props temporarily supporting a structure. [MIN ENG] Building up the wall of a mine shaft to join that above it.

underplate [DES ENG] An unfinished plate which forms part of an armored front for a mortise lock, and which is fastened to the case.

underream [ENG] To enlarge a drill hole below the casing.

undersea mining [MIN ENG] The working of economic deposits (usually coal) situated in strata or rocks below the seabed.

undershot wheel [MECH ENG] A water wheel operated by the impact of flowing water against blades attached around the periphery of the wheel, the blades being partly or totally submerged in the moving stream of water.

undersize [ENG] That part of a crushed material (for example, ore) which passes through a screen.

understressing [MET] Repeated stressing below the fatigue limit or below the final applied stress; can improve fatigue properties as a result of strain-aging effects.

underwater sound projector [ENG ACOUS] A transducer used to produce sound waves in water.

underwater transducer [ENG ACOUS] A device used for the generation or reception of underwater sounds.

underway bottom sampler *See* underway sampler.

underway sampler [ENG] A device for collecting samples of sediment on the ocean bottom, consisting of a cup in a hollow tube; on striking the bottom, the cup scoops up a small sample which is forced into the tube which is then closed with a lid, and the device is hoisted to the surface. Also known as scoopfish; underway bottom sampler.

undisturbed [ENG] Pertaining to a sample of material, as of soil, subjected to so little disturbance that it is suitable for determinations of strength, consolidation, permeability characteristics, and other properties of the material in place.

unfinished bolt [DES ENG] One of three degrees of finish in which standard hexagon wrench-head bolts and nuts are available; only the thread is finished.

unidirectional hydrophone [ENG ACOUS] A hydrophone mainly sensitive to sound that is incident from a single solid angle of one hemisphere or less.

unidirectional microphone [ENG ACOUS] A microphone that is responsive predominantly to sound incident from one hemisphere, without picking up sounds from the sides or rear.

unified screw thread [DES ENG] Three series of threads: coarse (UNC), fine (UNF), and extra fine (UNEF); a ¼-inch (0.006 millimeter) diameter thread in the UNC series has 20 threads per inch, while in the UNF series it has 28.

unifilar suspension [ENG] The suspension of a body from a single thread, wire, or strip.

uniflow engine [MECH ENG] A steam engine in which steam enters the cylinder through valves at one end and escapes through openings uncovered by the piston as it completes its stroke.

uniform click track [ENG ACOUS] A click track with regularly spaced clicks.

uniform corrosion [MET] Corrosion which takes place uniformly over the entire exposed surfaces.

uniform mat [CIV ENG] A type of foundation mat, consisting of a reinforced concrete slab of constant thickness, supporting walls, and columns; it is thick, rigid, and strong.

unilateral tolerance method [DES ENG] Method of dimensioning and tolerancing wherein the tolerance is taken as plus or minus from an explicitly stated dimension; the dimension represents the size or location which is nearest the critical condition (that is maximum material condition), and the tolerance is applied either in a plus or minus direction, but not in both directions, in such a way that the permissible variation in size or location is away from the critical condition.

union [DES ENG] A screwed or flanged pipe coupling usually in the form of a ring fitting around the outside of the joint.

union joint [DES ENG] A threaded assembly used for the joining of ends of lengths of installed pipe or tubing where rotation of neither length is feasible.

union shop [IND ENG] An establishment in which union membership is not a requirement for original employment but becomes mandatory after a specified period of time.

Uniontown method [ENG] A road-test method to determine the knock characteristics of motor fuels.

unit [ENG] An assembly or device capable of independent operation, such as a radio receiver, cathode-ray oscilloscope, or computer subassembly that performs some inclusive operation or function.

unitary air conditioner [MECH ENG] A small self-contained electrical unit enclosing a motor-driven refrigeration compressor, evaporative cooling coil, air-cooled condenser, filters, fans, and controls.

unit assembly [IND ENG] Assemblage of machine parts which constitutes a complete auxiliary part of an end item, and which performs a specific auxiliary function, and which may be removed from the parent item without itself being disassembled.

unit cell [MIN ENG] In flotation, a single cell.

unit construction [BUILD] An assembly comprising two or more walls, plus floor and ceiling construction, ready for shipping to a building site.

unit cost [IND ENG] Cost allocated to a specified unit of a product; computed as the cost over a period of time divided by the number of units produced.

unit die [MET] A die block having more than one cavity insert and allowing several different castings to be made.

United States standard dry seal thread [DES ENG] A modified pipe thread used for pressure-tight connections that are to be assembled without lubricant or sealer in refrigeration pipes, automotive and aircraft fuel-line fittings, and gas and chemical shells.

unit heater [MECH ENG] A heater consisting of a fan for circulating air over a heat-exchange surface, all enclosed in a common casing.

unitized body [ENG] An automotive body that has the body and frame in one unit; side members are designed on the principle of a bridge truss to gain stiffness, and sheet metal of the body is stressed so that it carries some of the load.

unitized cargo [IND ENG] Grouped cargo carried aboard a ship in pallets, containers, wheeled vehicles, and barges or lighters.

unitized load [IND ENG] A single item or a number of items packaged, packed, or arranged in a specified manner and capable of being handled as a unit; unitization may be accomplished by placing the item or items in a container or by banding them securely together. Also known as unit load.

unit load *See* unitized load.

unit mold [ENG] A simple plastics mold composed of a simple cavity without further mold devices; used to produce sample containers having shapes difficult to blow-mold.

unit of coal [MIN ENG] The quantity of coal from which the sample is taken and which the sample represents.

unit of issue [IND ENG] In reference to special storage, the quantity of an item, such as each number, dozen, gallon, pair, pound, ream, set, or yard.

unit power [MET] A unit describing machinability of a metal; the power needed to remove a unit volume in unit time, usually expressed as horsepower per cubic inch per minute.

unit procurement cost [IND ENG] The net basic cost paid or estimated to be paid for a unit of a particular item including, where applicable, the cost of government-furnished property and the cost of manufacturing operations performed at government-owned facilities.

unit train [MIN ENG] A system for delivering coal in which a string of cars, with distinctive markings and loaded to full visible capacity, is operated without service

frills or stops along the way for cars to be cut in and out.

universal chuck [ENG] A self-centering chuck whose jaws move in unison when a scroll plate is rotated.

universal grinding machine [MECH ENG] A grinding machine having a swivel table and headstock, and a wheel head that can be rotated on its base.

universal instrument *See* altazimuth.

universal joint [MECH ENG] A linkage that transmits rotation between two shafts whose axes are coplanar but not coinciding.

universal mill [MET] A rolling mill having both horizontal and vertical sets of rolls.

universal output transformer [ENG ACOUS] An output transformer having a number of taps on its winding, to permit its use between the audio-frequency output stage and the loudspeaker of practically any radio receiver by proper choice of connections.

universal vise [ENG] A vise which has two or three swivel settings so that the workpiece can be set at a compound angle. Also known as toolmaker's vise.

unloader [MECH ENG] A power device for removing bulk materials from railway freight cars or highway trucks; in the case of railway cars, the car structure may aid the unloader; a transitional device between interplant transportation means and intraplant handling equipment.

unloading conveyor [MECH ENG] Any of several types of portable conveyors adapted for unloading bulk materials, packages, or objects from conveyances.

unproductive development [MIN ENG] The drifts, tunnels, and crosscuts driven in stone, preparatory to opening out production faces in a coal seam or ore body.

unprotected reversing thermometer [ENG] A reversing thermometer for sea-water temperature which is not protected against hydrostatic pressure.

unproven area [MIN ENG] An area in which it has not been established by drilling operations whether oil or gas may be found in commercial quantities.

unreserved minerals [MIN ENG] Minerals which belong to the owner of the land on which or in which they are located.

unscheduled maintenance [IND ENG] Those unpredictable maintenance requirements that had not been previously planned or programmed but require prompt attention and must be added to, integrated with, or substituted for previously scheduled workloads.

unscrambler [IND ENG] A part of a feeding and packaging line that aids in arranging cartons for the filling machines; there are rotary, straight-line, and walking-beam types.

Unsin engine [MECH ENG] A type of rotary engine in which the trochoidal rotors of eccentric-rotor engines are replaced with two circular rotors, one of which has a single gear tooth upon which gas pressure acts, and the second rotor has a slot which accepts the gear tooth.

unsprung axle [MECH ENG] A rear axle in an automobile in which the housing carries the right and left rear-axle shafts and the wheels are mounted at the outer end of each shaft.

unsprung weight [MECH ENG] The weight of the various parts of a vehicle that are not carried on the springs such as wheels, axles, brakes, and so forth.

unwater [ENG] To remove or draw off water; to drain.

upcast [MIN ENG] **1.** The opening through which the return air ascends and is removed from the mine; the opposite of downcast or intake. **2.** An upward current of air passing through a shaft. **3.** Material that has been thrown up, as by digging.

up-Doppler [ENG ACOUS] The sonar situation wherein the target is moving toward the transducer, so the frequency of the echo is greater than the frequency of the reverberations received immediately after the end of the outgoing ping; opposite of down-Doppler.

updraft carburetor [MECH ENG] For a gasoline engine, a fuel-air mixing device in which both the fuel jet and the airflow are upward.

updraft furnace [MECH ENG] A furnace in which volumes of air are supplied from below the fuel bed or supply.

upgrade [MIN ENG] **1.** To increase the commercial value of a coal or mineral product by appropriate treatment. **2.** To increase the quality rating of diamonds beyond or above the rating implied by their particular classification.

uphole [MIN ENG] A borehole collared in an underground working place and drilled in a direction pointed above the horizontal plane of the drill-machine swivel head.

uplift pressure [CIV ENG] Pressure in an upward direction against the bottom of a structure, as a dam, a road slab, or a basement floor.

upmilling [MECH ENG] Milling a workpiece by rotating the cutter against the direction of feed of the workpiece.

upper punch [MET] In powder metallurgy, the member of the die assembly that moves downward into the die body to transmit pressure to the metal powder in the cavity.

upright [CIV ENG] A vertical structural member, post, or stake.

upset [ENG] To increase the diameter of a rock drill by blunting the end. [MET] A localized increase in the cross-sectional area of a metal during working, caused by the application of pressure; enables a head to be formed on fasteners such as bolts. [MIN ENG] **1.** A narrow heading connecting two levels in inclined coal. **2.** A capsized or broken skip.

upsetting test [MET] A test used to identify the role of variables in forging, which demonstrates that the force to forge is a function of the strength of the material, coefficient of friction, and ratio of the lateral to thickness dimensions of the workpiece.

upset welding [MET] Pressure butt welding in which heat is generated by resistance to the passage of current across the joint.

upslope time [MET] In resistance welding, the time associated with an increase in current when slope control is used.

upstream face [CIV ENG] The side of a dam nearer the source of water.

uptake [ENG] A large pipe for exhaust gases from a boiler furnace that runs upward to a chimney or smokestack.

uranium [MET] A dense, silvery, ductile, strongly electropositive metal.

urbanization [CIV ENG] The state of being or becoming a community with urban characteristics.

urban renewal [CIV ENG] Redevelopment and revitalization of a deteriorated urban community.

usable iron ore [MET] A steel industry term for high-grade iron ore, concentrates, or agglomerates which can be used in blast furnaces or other processing plants.

U-shaped abutment [CIV ENG] A bridge abutment with wings perpendicular to the face which act as counterforts; a very stable abutment, often used for architectural effect.

utilidor [CIV ENG] An insulated, heated conduit built below the ground surface or supported above the ground surface to protect the contained water, steam, sewage, and fire lines from freezing.

utility [ENG] One of the nonprocess (support) facilities for a manufacturing plant; usually considered as facilities for steam, cooling water, deionized water, electric power, refrigeration, compressed and instrument air, and effluent treatment.

utilization rate [AERO ENG] The amount of flying time produced in a specific period expressed in hours per period per aircraft. Also known as flying hour rate.

U-tube manometer [ENG] A manometer consisting of a U-shaped glass tube partly filled with a liquid of known specific gravity; when the legs of the manometer are connected to separate sources of pressure, the liquid rises in one leg and drops in the other; the difference between the levels is proportional to the difference in pressures and inversely proportional to the liquid's specific gravity. Also known as liquid-column gage.

uviol glass [MATER] A type of glass that is highly transparent to ultraviolet radiation.

V

vacuum brake [MECH ENG] A form of air brake which operates by maintaining low pressure in the actuating cylinder; braking action is produced by opening one side of the cylinder to the atmosphere so that atmospheric pressure, aided in some designs by gravity, applies the brake.

vacuum brazing [MET] Brazing utilizing a chamber at subatmospheric pressure.

vacuum breaker [ENG] A device used to relieve a vacuum formed in a water supply line to prevent backflow. Also known as backflow preventer.

vacuum casting [MET] Metal casting in a vacuum.

vacuum cleaner [MECH ENG] An electrically powered mechanical appliance for the dry removal of dust and loose dirt from rugs, fabrics, and other surfaces.

vacuum concrete [CIV ENG] Concrete poured into a framework that is fitted with a vacuum mat to remove water not required for setting of the cement; in this framework, concrete attains its 28-day strength in 10 days and has a 25% higher crushing strength.

vacuum degassing [MET] A process for removing gases from a metal either by melting or heating the solid metal in a vacuum.

vacuum deposition [MET] Deposition of a thin coating of metal by condensation on a cool work surface in a vacuum.

vacuum drying [ENG] The removal of liquid from a solid material in a vacuum system; used to lower temperatures needed for evaporation to avoid heat damage to sensitive material.

vacuum evaporation [ENG] Deposition of thin films of metal or other materials on a substrate, usually through openings in a mask, by evaporation from a boiling source in a hard vacuum.

vacuum evaporator [ENG] A vacuum device used to evaporate metals and spectrographic carbon to coat (replicate) a specimen for electron spectroscopic analysis or for electron microscopy.

vacuum filter [ENG] A filter device into which a liquid-solid slurry is fed to the high-pressure side of a filter medium, with liquid pulled through to the low-pressure side of the medium and a cake of solids forming on the outside of the medium.

vacuum filtration [ENG] The separation of solids from liquids by passing the mixture through a vacuum filter.

vacuum forming [ENG] Plastic-sheet forming in which the sheet is clamped to a stationary frame, then heated and drawn down into a mold by vacuum.

vacuum freeze dryer [ENG] A type of indirect batch dryer used to dry materials that would be destroyed by the loss of volatile ingredients or by drying temperatures above the freezing point.

vacuum fusion [MET] A technique for determining the oxygen, hydrogen, and sometimes nitrogen content of metals; can be applied to a wide variety of metals with the exception of alkali and alkaline earth metals.

vacuum gage [ENG] A device that indicates the absolute gas pressure in a vacuum system.

vacuum heating [MECH ENG] A two-pipe steam heating system in which a vacuum pump is used to maintain a suction in the return piping, thus creating a positive return flow of air and condensate.

vacuum mat [CIV ENG] A rigid flat metal screen faced by a linen filter, the back of which is kept under partial vacuum; used to suck out surplus air and water from poured concrete to produce a dense, well-shrunk concrete.

vacuum measurement [ENG] The determination of a fluid pressure less in magnitude than the pressure of the atmosphere.

vacuum metallurgy [MET] The melting, shaping, and treating of metals and alloys under reduced pressure that ranges from subatmospheric pressure to ultra-high vacuum.

vacuum pencil [ENG] A pencillike length of tubing connected to a small vacuum pump, for picking up semiconductor slices or chips during fabrication of solid-state devices.

vacuum plating [MET] Producing a surface film of metal on a heated surface, often in a vacuum, either by decomposition of the vapor of a compound at the work surface, or by direct reaction between the work surface and the vapor. Also known as vapor deposition.

vacuum pump [MECH ENG] A compressor for exhausting air and noncondensable gases from a space that is to be maintained at subatmospheric pressure.

vacuum relief valve [ENG] A pressure relief device which is designed to allow fluid to enter a pressure vessel in order to avoid extreme internal vacuum.

vacuum shelf dryer [ENG] A type of indirect batch dryer which generally consists of a vacuum-tight cubical or cylindrical chamber of cast-iron or steel plate, heated supporting shelves inside the chamber, a vacuum source, and a condenser; used extensively for drying pharmaceuticals, temperature-sensitive or easily oxidizable materials, and small batches of high-cost products where any product loss must be avoided.

vacuum support [MECH ENG] That portion of a rupture disk device which prevents deformation of the disk resulting from vacuum or rapid pressure change.

vacuum-tube voltmeter [ENG] Any of several types of instrument in which vacuum tubes, acting as amplifiers or rectifiers, are used in circuits for the measurement of alternating-current or direct-current voltage. Abbreviated VTVM. Also known as tube voltmeter.

valerian oil [MATER] A combustible, yellow to brown liquid with a penetrating aroma; soluble in alcohol, acetone, and other organic solvents; derived from the roots and rhizome of the garden heliotrope (*Valeriana officinalis*), the main components being pinene, camphene, borneol, and esters; used in medicine, flavors, and industrial odorants and to perfume tobacco.

valley [BUILD] An inside angle formed where two sloping sides intersect.

valley rafter [BUILD] A part of the roof frame that extends diagonally from an inside corner plate to the ridge board at the intersection of two roof surfaces.

value analysis *See* value engineering.

value control *See* value engineering.

value engineering [IND ENG] The systematic application of recognized techniques which identify the function of a product or service, and provide the necessary function reliably at lowest overall cost. Also known as value analysis; value control.

value theory [SYS ENG] A concept normally associated with decision theory; it strives to evaluate relative utilities of simple and mixed parameters which can be used to describe outcomes.

valve [MECH ENG] A device used to regulate the flow of fluids in piping systems and machinery.

valve follower [MECH ENG] A linkage between the cam and the push rod of a valve train.

valve guide [MECH ENG] A channel which supports the stem of a poppet valve for maintenance of alignment.

valve head [MECH ENG] The disk part of a poppet valve that gives a tight closure on the valve seat.

valve-in-head engine *See* overhead-valve engine.

valve lifter [MECH ENG] A device for opening the valve of a cylinder as in an internal combustion engine.

valve seat [DES ENG] The circular metal ring on which the valve head of a poppet valve rests when closed.

valve stem [MECH ENG] The rod by means of which the disk or plug is moved to open and close a valve.

valve train [MECH ENG] The valves and valve-operating mechanism for the control of fluid flow to and from a piston-cylinder machine, for example, steam, diesel, or gasoline engine.

van [MIN ENG] **1.** A test of the value of an ore, made by washing (vanning) a small quantity, after powdering it, on the point of a shovel. **2.** To separate, as ore from veinstone, by washing it on the point of a shovel. **3.** A shovel used in ore dressing.

vanadium [MET] A silvery-white, ductile metal resistant to corrosion; used in alloy steels and as an x-ray target.

vanadium steel [MET] A low-alloy steel containing 0.10–0.15% vanadium.

Van Dorn sampler [ENG] A sediment sampler that consists of a Plexiglas cylinder closed at both ends by rubber force cups; in the armed position the cups are pulled outside the cylinder and restrained by a releasing mechanism, and after the sample is taken, a length of surgical rubber tubing connecting the cups is sufficiently prestressed to permit the force cups to retain the sample in the cylinder.

vane [AERO ENG] A device that projects ahead of an aircraft to sense gusts or other actions of the air so as to create impulses or signals that are transmitted to the control system to stabilize the aircraft. [MECH ENG] A flat or curved surface exposed to a flow of fluid so as to be forced to move or to rotate about an axis, to rechannel the flow, or to act as the impeller; for example, in a steam turbine, propeller fan, or hydraulic turbine.

vane anemometer [ENG] A portable instrument used to measure low wind speeds and airspeeds in large ducts; consists of a number of vanes radiating from a common shaft and set to rotate when facing the wind.

vane motor rotary actuator [MECH ENG] A type of rotary motor actuator which consists of a rotor with several spring-loaded sliding vanes in an elliptical chamber; hydraulic fluid enters the chamber and forces the vanes before it as it moves to the outlets.

vane-type instrument [ENG] A measuring instrument utilizing the force of repulsion between fixed and movable magnetized iron vanes, or the force existing between a coil and a pivoted vane-shaped piece of soft iron, to move the indicating pointer.

Vanguard satellite [AERO ENG] One of three artificial satellites launched by the United States in 1958 and 1959, using a modified Viking rocket, as a part of the International Geophysical Year program; *Vanguard 1* was the first spacecraft to use solar cells.

vanner [MIN ENG] A machine for dressing ore; the name is given to various patented devices in which the peculiar motions of the shovel in the miner's hands in the operation of making a van are, or are supposed to be, successfully imitated. Also known as vanning machine.

vanning machine *See* vanner.

vapor barrier [CIV ENG] A layer of material applied to the inner (warm) surface of a concrete wall or floor to prevent absorption and condensation of moisture.

vapor blasting [MET] Cleaning the surface of a metal with a fine abrasive suspended in water and propelled at high speed by air or steam. Also known as liquid honing; vapor honing.

vapor-compression cycle [MECH ENG] A refrigeration cycle in which refrigerant is circulated through a machine which allows for successive boiling (or vaporization) of liquid refrigerant as it passes through an expansion valve, thereby producing a cooling effect in its surroundings, followed by compression of vapor to liquid.

vapor degreasing [ENG] A type of cleaning procedure, for metals to remove grease, oils, and lightly attached solids; a solvent such as trichloroethylene is boiled, and its vapors are condensed on the metal surfaces.

vapor deposition *See* vacuum plating.

vapor-filled thermometer [ENG] A gas- or vapor-filled temperature measurement device that moves or distorts in response to temperature-induced pressure changes from the expansion or contraction of the sealed, vapor-containing chamber.

vapor honing *See* vapor blasting.

vaporimeter [ENG] An instrument used to measure a substance's vapor pressure, especially that of an alcoholic liquid, in order to determine its alcohol content.

vaporization cooling [ENG] Cooling by volatilization of a nonflammable liquid having a low boiling point and high dielectric strength; the liquid is flowed or sprayed on hot electronic equipment in an enclosure where it vaporizes, carrying the heat to the enclosure walls, radiators, or heat exchanger. Also known as evaporative cooling.

vapor-pressure thermometer [ENG] A thermometer in which the vapor pressure of a homogeneous substance is measured and from which the temperature can be determined; used mostly for low-temperature measurements.

vapor-recovery unit [ENG] **1.** A device or system to catch vaporized materials (usually fuels or solvents) as they are vented. **2.** In petroleum refining, a process unit to which gases and vaporized gasoline from various processing operations are charged, separated, and recovered for further use.

vapor volume equivalent [PETRO ENG] The volume of vapor to which a specified amount of liquid would be equivalent at designated standard conditions (for example, 14.65 psia and 60°F, or 15.5°C); used in the petroleum industry to calculate the specific gravity of fluids from gas-condensate wells.

vara [CIV ENG] A surveyors' unit of length equal to $33\frac{1}{3}$ inches (84.7 centimeters).

variable-area exhaust nozzle [AERO ENG] On a jet engine, an exhaust nozzle of which the exhaust exit opening can be varied in area by means of some mechanical device, permitting variation in the jet velocity.

variable-area meter [ENG] A flowmeter that works on the principle of a variable restrictor in the flowing stream being forced by the fluid to a position to allow the required flow-through.

variable-area track [ENG ACOUS] A sound track divided laterally into opaque and transparent areas; a sharp line of demarcation between these areas corresponds to the waveform of the recorded signal.

variable click track [ENG ACOUS] A click track with irregularly spaced clicks.

variable costs [IND ENG] Costs which vary directly with the number of units produced; direct labor and material are examples.

variable-cycle engine [AERO ENG] A type of gas turbine jet engine whose cycle parameters, such as pressure ratio, temperature, gas flow paths, and air-handling characteristics, can be varied between those of a turbojet and a turbofan, enabling it to combine the advantages of both.

variable-density sound track [ENG ACOUS] A constant-width sound track in which the average light transmission varies along the longitudinal axis in proportion to some characteristic of the applied signal.

variable-depth sonar [ENG] Sonar in which the projector and receiving transducer are mounted in a watertight pod that can be lowered below a vessel to an optimum depth for minimizing thermal effects when detecting underwater targets.

variable geometry aircraft [AERO ENG] Aircraft with variable profile geometry, such as variable sweep wings.

variable-inductance accelerometer [ENG] An accelerometer consisting of a differential transformer with three coils and a mass which passes through the coils and is suspended from springs; the center coil is excited from an external alternating-current power source, and two end coils connected in series opposition are used to produce an ac output which is proportional to the displacement of the mass.

variable-pitch propeller [ENG] A controllable-pitch propeller whose blade angle may be adjusted to any angle between the low and high pitch limits.

variable radio-frequency radiosonde [ENG] A radiosonde whose carrier frequency is modulated by the magnitude of the meteorological variables being sensed.

variable-reluctance microphone *See* magnetic microphone.

variable-reluctance pickup [ENG ACOUS] A phonograph pickup that depends for its operation on variations in the reluctance of a magnetic circuit due to the movements of an iron stylus assembly that is a part of the magnetic circuit. Also known as magnetic cartridge; magnetic pickup; reluctance pickup.

variable-resistance accelerometer [ENG] Any accelerometer which operates on the principle that electrical resistance of any conductor is a function of its dimensions; when the dimensions of the conductor are varied mechanically, as constant current flows through it, the voltage across it varies as a function of this mechanical excitation; examples include the strain-gage accelerometer, and an acce-

lerometer making use of a slide-wire potentiometer.

variable-speed drive [MECH ENG] A mechanism transmitting motion from one shaft to another that allows the velocity ratio of the shafts to be varied continuously.

variety [SYS ENG] The logarithm (usually to base 2) of the number of discriminations that an observer or a sensing system can make relative to a system.

variograph [ENG] A recording variometer.

variometer [ENG] A geomagnetic device for detecting and indicating changes in one of the components of the terrestrial magnetic field vector, usually magnetic declination, the horizontal intensity component, or the vertical intensity component.

varmeter [ENG] An instrument for measuring reactive power in vars. Also known as reactive volt-ampere meter.

varnish [MATER] A transparent surface coating which is applied as a liquid and then changes to a hard solid; all varnishes are solutions of resinous materials in a solvent.

varnish makers' and painters' naphtha [MATER] A petroleum naphtha that has a narrow boiling range and is used mainly as a thinner in paint and varnish. Abbreviated VM & P naphtha.

varnish paper *See* insulating paper.

vat dye [MATER] One of the dyes that are easily reduced to a soluble and colorless form in which they easily impregnate fibers; subsequent oxidation produces the final color; examples are indigo and indanthrene blue.

vat printing assistant [MATER] The carrier for the dye in the printing of fabrics with vat dyes; a mixture of gums and reducing and wetting agents to assist in penetrating the fabric.

V belt [DES ENG] An endless power-transmission belt with a trapezoidal cross section which runs in a pulley with a V-shaped groove; it transmits higher torque at less width and tension than a flat belt. [MECH ENG] A belt, usually endless, with a trapezoidal cross section which runs in a pulley with a V-shaped groove, with the top surface of the belt approximately flush with the top of the pulley.

V-bend die [MECH ENG] A die with a triangular cross-sectional opening to provide two edges over which bending is accomplished.

V block [ENG] A square or rectangular steel block having a 90° V groove through the center, and sometimes provided with clamps to secure round workpieces.

V-bucket carrier [MECH ENG] A conveyor consisting of two strands of roller chain separated by V-shaped steel buckets; used for elevating and conveying nonabrasive materials, such as coal.

V cut [ENG] In mining and tunneling, a cut where the material blasted out in plan is like the letter V; usually consists of six or eight holes drilled into the face, half of which form an acute angle with the other half.

vectopluviometer [ENG] A rain gage or array of rain gages designed to measure the inclination and direction of falling rain; vectopluviometers may be constructed in the fashion of a wind vane so that the receiver always faces the wind, or they may consist of four or more receivers arranged to point in cardinal directions.

vector impedance meter [ENG] An instrument that not only determines the ratio between voltage and current, to give the magnitude of impedance, but also determines the phase difference between these quantities, to give the phase angle of impedance.

vector steering [AERO ENG] A steering method for rockets and spacecraft wherein one or more thrust chambers are gimbal-mounted so that the direction of the thrust force (thrust vector) may be tilted in relation to the center of gravity of the vehicle to produce a turning movement.

vector voltmeter [ENG] A two-channel high-frequency sampling voltmeter that measures phase as well as voltage of two input signals of the same frequency.

vee path [ENG] In ultrasonic testing, the path of an angle beam from an ultrasonic search unit in which the waves are reflected off the opposite surface of the test piece and returned to the examination surface in a manner which has the appearance of the letter V.

Vegard's Law [MET] Linear relation between lattice parameters and composition of solid solution alloys expressed as atomic percentage.

vegetable black [MATER] Carbon made by the incomplete combustion or destructive distillation of vegetable matter, for example, wood.

vegetable dye [MATER] Any colorant that is obtained from a vegetable source; for example, indigo or madder.

vegetable fat [MATER] A semisolid vegetable oil, used chiefly for food; for example, Suari fat, ucuhuba tallow, Mahuba fat, gamboge butter (gurgi, murga), Sierra Leone butter (kamga, lamy), and Mafura tallow.

vegetable glue [MATER] Mostly starch- or dextrine-based glues mixed with gums, resins, or antioxidants; tapioca paste is the most common; used on cheaper plywoods, postage stamps, envelopes, and labels.

vegetable ivory [MATER] A material from the ivory nut, a seed of the palm *Phytelephas macrocarpa*, which grows in tropical America; the nut has a white color and fine texture and is used to make buttons and similar small articles.

vegetable oil [MATER] An edible, mixed glyceride oil derived from plants (fruit, leaves, and seeds), including cottonseed, linseed, tung, and peanut; used in food oils, shortenings, soaps, and medicine, and as a paint drying oil.

vegetable parchment [MATER] A paperlike material made from a base of cotton rags or alpha cellulose called waterleaf, and containing no sizing or filling materials; used for documents and food packaging.

vegetable tanning [ENG] Leather tanning using plant extracts, such as tannic acid.

vegetable wax [MATER] A waxy substance of vegetable origin, composed of fatty acids in combination with higher alcohols (instead of glycerin, as in fats and oils); includes Japan wax, jojoba oil, candelilla, and carnauba wax.

vehicle [AERO ENG] **1.** A structure, machine, or device, such as an aircraft or rocket, designed to carry a burden through air or space. **2.** More restrictively, a rocket vehicle. [MECH ENG] A self-propelled wheeled machine that transports people or goods on or off roads; automobiles and trucks are examples.

vehicle control system [AERO ENG] A system, incorporating control surfaces or other devices, which adjusts and maintains the altitude and heading, and sometimes speed, of a vehicle in accordance with signals received from a guidance system. Also known as flight control system.

vehicle mass ratio [AERO ENG] The ratio of the final mass of a vehicle after all propellant has been used, to the initial mass.

veining [MET] Lines in a polished and etched metal surface marking slight imperfections in structure of an otherwise single grain.

vellum [MATER] A high-grade paper made to resemble genuine parchment.

velocimeter [ENG] An instrument for measuring the speed of sound in water; two transducers transmit acoustic pulses back and forth over a path of fixed length, each transducer immediately initiating a pulse upon receiving the previous one; the number of pulses occurring in a unit time is measured.

velocity-head tachometer [ENG] A type of tachometer in which the device whose speed is to be measured drives a pump or blower, producing a fluid flow, which is converted to a pressure.

velocity hydrophone [ENG ACOUS] A hydrophone in which the electric output essentially matches the instantaneous particle velocity in the impressed sound wave.

velocity microphone [ENG ACOUS] A microphone whose electric output depends on the velocity of the air particles that form a sound wave; examples are a hot-wire microphone and a ribbon microphone.

velocity ratio [MECH ENG] The ratio of the velocity given to the effort or input of a machine to the velocity acquired by the load or output.

velocity-type flowmeter [ENG] A turbine-type fluid-flow measurement device in which the fluid flow actuates the movement of a wheel or turbine-type impeller, giving a volume-time reading. Also known as current meter; rotating meter.

velour paper [MATER] A paper with a velvetlike finish, produced by flocking the surface with fine bits of rayon, nylon, cotton, or wool; it is sometimes embossed in various patterns.

veneer [MATER] **1.** A thin sheet of wood of uniform thickness used for facing furniture or, bonded to make plywood. **2.** A facing, as of brick or marble, on the outside of a wall.

Venera space program [AERO ENG] A series of unmanned space vehicle flights to probe the conditions near and on the planet Venus; the program was initiated by the Soviet Union.

vent [ENG] **1.** A small passage made with a needle through stemming, for admitting a squib to enable the charge to be lighted. **2.** A hole, extending up through the bearing at the top of the core-barrel inner tube, which allows the water and air in the upper part of the inner tube to escape into the borehole. **3.** A small hole in the upper end of a core-barrel inner tube that allows water and air in the inner tube to escape into the annular space between the inner and outer barrels. **4.** An opening provided for the discharge of pressure or the release of pressure from tanks, vessels, reactors, processing equipment, and so on. [MET] A small opening in a casting mold to allow for the escape of gases.

vented baffle *See* reflex baffle.

ventilation [ENG] Provision for the movement, circulation, and quality control of air in an enclosed space.

ventilator [ENG] A device with an adjustable aperture for regulating the flow of fresh or stagnant air. [MECH ENG] A mechanical apparatus for producing a current of air, as a blowing or exhaust fan.

vent stack [BUILD] The portion of a soil stack above the highest fixture.

venturi flume [ENG] An open flume with a constricted flow which causes a drop in the hydraulic grade line; used in flow measurement.

venturi meter [ENG] An instrument for efficiently measuring fluid flow rate in a piping system; a nozzle section increases velocity and is followed by an expanding section for recovery of kinetic energy.

venturi tube [ENG] A constriction that is placed in a pipe and causes a drop in pressure as fluid flows through it, consisting essentially of a short straight pipe section or throat between two tapered sections; it can be used to measure fluid flow rate (a venturi meter), or to draw fuel into the main flow stream, as in a carburetor.

Venus probe [AERO ENG] A probe for exploring and reporting on conditions on or about the planet Venus, such as Pioneer and Mariner probes of the United States, and Venera probes of the Soviet Union.

veratria *See* veratrine.

veratrine [MATER] An alakaloid mixture from the seeds of sabadilla (*Schoenocaulon officinale*); which is toxic, colorless, soluble in alcohol and ether, very slightly soluble in water, and melts at about 150°C; used in medicine. Also known as veratria.

verbena oil [MATER] A volatile oil from lemon verbena leaves; contains 30% citral; used to make perfumes.

verge [BUILD] The edge of a sloping roof which projects over a gable.

vergeboard [BUILD] One of the boards utilized as the finish of the eaves on the gable end of a structure.

vernier [ENG] A short, auxiliary scale which slides along the main instrument scale to permit accurate fractional reading of the least main division of the main scale.

vernier caliper [ENG] A caliper rule with an attached vernier scale.

vernier dial [ENG] A tuning dial in which each complete rotation of the control knob causes only a fraction of a revolution of the main shaft, permitting fine and accurate adjustment.

vernier engine [AERO ENG] A rocket engine of small thrust used primarily to obtain a fine adjustment in the velocity and trajectory of a rocket vehicle just after the thrust cutoff of the last sustainer engine, and used secondarily to add thrust to a booster or sustainer engine. Also known as vernier rocket.

vernier rocket *See* vernier engine.

vertical band saw [MECH ENG] A band saw whose blade operates in the vertical plane; ideal for contour cutting.

vertical boiler [MECH ENG] A fire-tube boiler having vertical tubes between top head and tube sheet, connected to the top of an internal furnace.

vertical boring mill [MECH ENG] A large type of boring machine in which a rotating workpiece is fastened to a horizontal table, which resembles a four-jaw independent chuck with extra radial T slots, and the tool has a traverse motion.

vertical broaching machine [MECH ENG] A broaching machine having the broach mounted in the vertical plane.

vertical compliance [ENG ACOUS] The ability of a stylus to move freely in a vertical direction while in the groove of a phonograph record.

vertical conveyor [MECH ENG] A materials-handling machine designed to move or transport bulk materials or packages upward or downward.

vertical-current recorder [ENG] An instrument which records the vertical electric current in the atmosphere.

vertical curve [CIV ENG] A curve inserted between two lengths of a road or railway which are at different slopes.

vertical-face breakwater [CIV ENG] A breakwater whose mound of rubble does not rise above the water, but is surmounted by a vertical-face superstructure of masonry or concrete; may be built without mound rubble, provided sea bed is firm.

vertical field balance [ENG] An instrument that measures the vertical component of the magnetic field by means of the torque that the field component exerts on a horizontal permanent magnet.

vertical firing [MECH ENG] The discharge of fuel and air perpendicular to the burner in a furnace.

vertical force instrument See heeling adjuster.

vertical guide idlers [MECH ENG] Idler rollers about 3 inches (8 centimeters) in diameter so placed as to make contact with the edge of the belt conveyor should it run too much to one side.

vertical gyro [AERO ENG] A two-degree-of-freedom gyro with provision for maintaining its spin axis vertical; output signals are produced by gimbal angular displacements which correspond to components of the angular displacements of the base about two orthogonal axes; used in aircraft to measure both bank angle and pitch attitude.

vertical intensity variometer [ENG] A variometer employing a large permanent magnet and equipped with very fine steel knife-edges or pivots resting on agate planes or saddles and balanced so that its magnetic axis is horizontal. Also known as Z variometer.

vertical-lift bridge [CIV ENG] A movable bridge with a span that rises on towers, lifted by steel ropes.

vertical-lift gate [CIV ENG] A dam spillway gate of which the movable parts are raised and lowered vertically to regulate water flow.

vertical obstacle sonar [ENG] An active sonar used to determine heights of objects in the path of a submersible vehicle; its beam sweeps along a vertical plane, about 30° above and below the direction of the vehicle's motion. Abbreviated VOS.

vertical-position welding [MET] Welding in which the weld axis is essentially vertical.

vertical recording [ENG ACOUS] A type of disk recording in which the groove modulation is perpendicular to the surface of the recording medium, so the cutting stylus moves up and down rather than from side to side during recording. Also known as hill-and-dale recording.

vertical retort process [MET] A zinc-smelting method using a vertical retort of silicon carbide brick. Also known as New Jersey retort process.

vertical scale [DES ENG] The ratio of the vertical dimensions of a laboratory model to those of the natural prototype; usually exaggerated in relation to the horizontal scale.

vertical seismograph [ENG] An instrument that records the vertical component of the ground motion during an earthquake.

vertical separation [AERO ENG] A specified vertical distance measured in terms of space between aircraft in flight at different altitudes or flight levels.

vertical separator [PETRO ENG] A gas-oil separator in the form of a vertical cylindrical tank.

vertical speed indicator See rate-of-climb indicator.

vertical tail [AERO ENG] A part of the tail assembly of an aircraft; consists of a fin (a symmetrical airfoil in line with the center line of the fuselage) fixed to the fuselage or body and a rudder which is movable by the pilot.

vertical takeoff and landing [AERO ENG] A flight technique in which an aircraft rises directly into the air and settles vertically onto the ground. Abbreviated VTOL.

vertical turbine pump See deep-well pump.

vertical turret lathe [DES ENG] Similar in principle to the horizontal turret lathe but capable of handling heavier, bulkier workpieces; it is constructed with a rotary, horizontal worktable whose diameter (30–74 inches, or 76–188 centimeters) normally designates the capacity of the machine; a crossrail mounted above the worktable carries a turret, which indexes in a vertical plane with tools that may be fed either across or downward.

vessel [ENG] A container or structural envelope in which materials are processed, treated, or stored; for example, pressure vessels, reactor vessels, agitator vessels, and storage vessels (tanks).

vestibule [BUILD] A hall or chamber between the outer door and the interior, or rooms, of a building.

vestibule school [IND ENG] A school organized by an industrial concern to train new employees in specific tasks or prepare employees for promotion.

vestibule training [IND ENG] A procedure used in operator training in which the training location is separate from the main productive areas of the plant; includes student carrels, lecture rooms, and in many instances the same type of equipment that the trainee will use in the work station.

vetiver oil [MATER] A combustible, viscous essential oil from partially dried roots of *Vetiveria zizanioides* (East Indian grass), with a violet scent; soluble in fixed oils; used in perfumery. Also known as cuscus oil; vetivert.

vetivert *See* vetiver oil.

V flume [MIN ENG] A V-shaped flume, supported by trestlework, and used by miners for bringing down timber and wood from the high mountains; the flume water is also used for mining purposes.

VFR *See* visual flight rules.

VFR between layers [AERO ENG] A flight condition wherein an aircraft is operated under modified visual flight rules while in flight between two layers of clouds or obscuring phenomena, each of which constitutes a ceiling.

VFR flight *See* visual flight.

VFR on top [AERO ENG] A flight condition wherein an aircraft is operated under modified visual flight rules while in flight above a layer of clouds or an obscuring phenomenon sufficient to constitute a ceiling.

VFR terminal minimums [AERO ENG] A set of operational weather limits at an airport, that is, the minimum conditions of ceiling and visibility under which visual flight rules may be used.

V guide [MECH ENG] A V-shaped groove serving to guide a wedge-shaped sliding machine element.

viaduct [CIV ENG] A bridge structure supported on high towers with short masonry or reinforced concrete arched spans.

vibrating coring tube [ENG] A sediment corer made to vibrate in order to eliminate the resistance of compacted ocean floor sediments, sands, and gravel.

vibrating feeder [MECH ENG] A feeder for bulk materials (pulverized or granulated solids), which are moved by the vibration of a slightly slanted, flat vibrating surface.

vibrating grizzlies [MECH ENG] Bar grizzlies mounted on eccentrics so that the entire assembly is given a forward and backward movement at a speed of some 100 strokes a minute.

vibrating needle [ENG] A magnetic needle used in compass adjustment to find the relative intensity of the horizontal components of the earth's magnetic field and the magnetic field at the compass location.

vibrating pebble mill [MECH ENG] A size-reduction device in which feed is ground by the action of vibrating, moving pebbles.

vibrating-reed electrometer [ENG] An instrument using a vibrating capacitor to measure a small charge, often in combination with an ionization chamber.

vibrating-reed frequency meter [ENG] A frequency meter consisting of steel reeds having different and known natural frequencies, all excited by an electromagnet carrying the alternating current whose frequency is to be measured. Also known as Frahm frequency meter; reed frequency meter; tuned-reed frequency meter.

vibrating-reed magnetometer [ENG] An instrument that measures magnetic fields by noting their effect on the vibration of reeds excited by an alternating magnetic field.

vibrating-reed tachometer [ENG] A tachometer consisting of a group of reeds of different lengths, each having a specific natural frequency of vibration; observation of the vibrating reed when in contact with a moving mechanical device indicates the frequency of vibration for the device.

vibrating screen [MECH ENG] A sizing screen which is vibrated by solenoid or magnetostriction, or mechanically by eccentrics or unbalanced spinning weights.

vibrating screen classifier [MECH ENG] A classifier whose screening surface is hung by rods and springs, and moves by means of electric vibrators.

vibrating wire transducer [ENG] A device for measuring ocean depth, consisting of a very fine tungsten wire stretched in a magnetic field so that it vibrates at a frequency that depends on the tension in the wire, and thereby on pressure and depth.

vibration damping [MECH ENG] The processes and techniques used for converting the mechanical vibrational energy of solids into heat energy.

vibration drilling [MECH ENG] Drilling in which a frequency of vibration in the range of 100 to 20,000 hertz is used to fracture rock.

vibration galvanometer [ENG] An alternating-current galvanometer in which the natural oscillation frequency of the moving element is equal to the frequency of the current being measured.

vibration isolation [ENG] The isolation, in structures, of those vibrations or motions that are classified as mechanical vibration; involves the control of the supporting structure, the placement and arrangement of isolators, and control of the internal construction of the equipment to be protected.

vibration machine [MECH ENG] A device for subjecting a system to controlled and reproducible mechanical vibration. Also known as shake table.

vibration meter See vibrometer.

vibration puddling [CIV ENG] A technique used to achieve proper consolidation of concrete; vibrating machines may be drawn vertically through the cement, or used on the surface, or placed against the form holding the concrete in place. Also known as mechanical puddling.

vibration separation [MECH ENG] Classification or separation of grains of solids in which separation through a screen is expedited by vibration or oscillatory movement of the screening mediums.

vibrator [MECH ENG] An instrument which produces mechanical oscillations.

vibratory centrifuge [MECH ENG] A high-speed rotating device to remove moisture from pulverized coal or other solids.

vibratory equipment [MECH ENG] Reciprocating or oscillating devices which move, shake, dump, compact, settle, tamp, pack, screen, or feed solids or slurries in process.

vibratory hammer [MECH ENG] A type of pile hammer which uses electrically activated eccentric cams to vibrate piles into place.

vibroenergy separator [MECH ENG] A screen-type device for classification or separation of grains of solids by a combination of gyratory motion and auxiliary vibration caused by balls bouncing against the lower surface of the screen cloth.

vibrograph [ENG] An instrument that provides a complete oscillographic record of a mechanical vibration; in one form a moving stylus records the motion being measured on a moving paper or film.

vibrometer [ENG] An instrument designed to measure the amplitude of a vibration. Also known as vibration meter.

Vicat needle [ENG] An apparatus used to determine the setting time of cement by measuring the pressure of a special needle against the cement surface.

Vickers hardness test See diamond pyramid hardness test.

Victaulic coupling [DES ENG] A development in which a groove is cut around each end of pipe instead of the usual threads; two ends of pipe are then lined up and a rubber ring is fitted around the joint; two semicircular bands, forming a sleeve, are placed around the ring and are drawn together with two bolts, which have a ridge on both edges to fit into the groove of the pipe; as the bolts are tightened, the rubber ring is compressed, making a watertight joint, while the ridges fitting in the grooves make it strong mechanically.

Vigilante [AERO ENG] A United States supersonic, twin-engine, turbojet tactical, all-weather attack aircraft designed to operate from aircraft carriers, and capable of delivering nuclear or nonnuclear weapons; it has electronic countermeasures equipment, long-range radar, automatic pilot guidance features, inflight refueling capabilities, and a crew of pilot and bombardier. Designated A-5.

Viking spacecraft [AERO ENG] A series of two United States spacecraft, each consisting of one module which remained in orbit around Mars and another which landed on the planet's surface in 1976.

vinasse [MATER] Residue from the fermentation of molasses or grapes; used as a fertilizer and a source of potassium salts. Also known as schlempe.

vinegar [MATER] The product of the incomplete oxidation to acetic acid of ethyl alcohol produced by a primary fermentation of vegetable materials; contains not less than 4 grams of acetic acid per gallon; used in preparation of pickled fruits and vegetables and in salad dressing.

virgin stock [MATER] A petroleum-derived liquid stream processed from natural (virgin) crude oil; it contains no cracked or otherwise chemically modified material. Also known as straight-run stock.

Virmel engine [MECH ENG] A cat-and-mouse engine that employs vanelike pistons whose motion is controlled by a gear-and-crank system; each set of pistons stops and restarts when a chamber reaches the spark plug.

virtual leak [ENG] The semblance of the vacuum system leak caused by a gradual desorptive release of gas at a rate which cannot be accurately predicted.

virtual PPI reflectoscope [ENG] A device for superimposing a virtual image of a chart on a plan position indicator (PPI) pattern; the chart is usually prepared with white lines on a black background to the scale of the plan position indicator range scale.

viscometer [ENG] An instrument designed to measure the viscosity of a fluid.

viscometer gage [ENG] A vacuum gage in which the gas pressure is determined from the viscosity of the gas.

viscometry [ENG] A branch of rheology; the study of the behavior of fluids under conditions of internal shear; the technology of measuring viscosities of fluids.

viscous damping [MECH ENG] A method of converting mechanical vibrational energy of a body into heat energy, in which a piston is attached to the body and is arranged to move through liquid or air in a cylinder or bellows that is attached to a support.

viscous-drag gas-density meter [ENG] A device to measure gas-mixture densities; driven impellers in sample and standard chambers create measurable turbulences (drags) against respective nonrotating impellers.

viscous fillers [MECH ENG] A packaging machine that fills viscous product into cartons; there are two basic types, straight-line and rotary plunger; the former operates intermittently on a given number of containers, while the latter fills and discharges containers continuously.

viscous impingement filter [ENG] A filter made up of a relatively loosely arranged medium, such that the airstream is forced to change direction frequently as it passes through the filter medium; the medium usually consists of spun-glass fibers, metal screens, or layers of crimped expanded metal whose surfaces are coated with a tacky oil.

viscous neutral oil [MATER] The bottoms from reducing, by distillation, of a petroleum neutral oil fraction; such oils are frequently blended with bright stock to make finished oils of various viscosities.

vise [DES ENG] A tool consisting of two jaws for holding a workpiece; opened and closed by a screw, lever, or cam mechanism.

visibility meter [ENG] An instrument for making direct measurements of visual range in the atmosphere or of the physical characteristics of the atmosphere which determine the visual range.

visual comparator *See* optical comparator.

visual flight [AERO ENG] An aircraft flight occurring under conditions which allow navigation by visual reference to the earth's surface at a safe altitude and with sufficient horizontal visibility, and operating under visual flight rules. Also known as VFR flight.

visual flight rules [AERO ENG] A set of regulations set down by the U.S. Civil Aeronautics Board (in Civil Air Regulations) to govern the operational control of aircraft during visual flight. Abbreviated VFR.

vitreous enamel [MATER] A glass coating applied to a metal by covering the surface with a powdered glass frit and heating until fusion occurs. Also known as porcelain enamel.

vitreous silica *See* silica glass.

vitrified-clay pipe [MATER] A pipe, made of clay treated in a kiln to induce vitrification, with the surface glazed for water-tightness; used for drainage.

vitrified wheel [DES ENG] A grinding wheel with a glassy or porcelanic bond.

vixen file [DES ENG] A flat file with curved teeth; used for filing soft metals.

V jewels [DES ENG] Jewel bearings used in conjunction with a conical pivot, the bearing surface being a small radius located at the apex of a conical recess; found primarily in electric measuring instruments.

VM & P naphtha *See* varnish makers' and painters' naphtha.

V-notch weir *See* triangular-notch weir.

voice coil [ENG ACOUS] The coil that is attached to the diaphragm of a moving-coil loudspeaker and moves through the air gap between the fixed pole pieces due to interaction of the fixed magnetic field with that associated with the audio-frequency current flowing through the voice coil. Also known as loudspeaker voice coil; speech coil (British usage).

voice print [ENG ACOUS] A voice spectrograph that has individually distinctive patterns of voice characteristics that can be used to identify one person's voice from other voice patterns.

voice response [ENG ACOUS] The process of generating an acoustic speech signal that communicates an intended message, such that a machine can respond to a request for information by talking to a human user.

void channels [ENG] The open passages of a porous or packed medium through which liquid or gas can flow.

volatile laurel oil [MATER] A bright-yellow liquid with an aromatic aroma, distilled from the leaves or berries of the laurel, *Laurus nobilis*; main components are cineole and pinene; soluble in alcohol, ether, benzene, and chloroform; used in perfumes, flavors, and medicine. Also known as sweet bay oil.

volatile-oil reservoir [PETRO ENG] A type of bubble-point oil reservoir in which the temperature is high and the liquid density is low (leading to a volatilized oil situation), reducing the amount of producible liquids.

volley [ENG] A round of holes fired at any one time.

voltmeter [ENG] An instrument for the measurement of potential difference between two points, in volts or in related smaller or larger units.

voltmeter-ammeter [ENG] A voltmeter and an ammeter combined in a single case but having separate terminals.

volt-ohm-milliammeter [ENG] A test instrument having a number of different ranges for measuring voltage, current, and resistance. Also known as circuit analyzer; multimeter; multiple-purpose tester.

volume [ENG ACOUS] The magnitude of a complex audio-frequency current as measured in volume units on a standard volume indicator.

volume compressor [ENG ACOUS] An audio-frequency circuit that limits the volume range of a radio program at the transmitter, to permit using a higher average percent modulation without risk of overmodulation; also used when making disk recordings, to permit a closer groove spacing without overcutting. Also known as automatic volume compressor.

volume control [ENG ACOUS] A potentiometer used to vary the loudness of a reproduced sound by varying the audio-frequency signal voltage at the input of the audio amplifier.

volume control system [ENG ACOUS] An electronic system that regulates the signal amplification or limits the output of a circuit, such as a volume compressor or a volume expander.

volume expander [ENG ACOUS] An audio-frequency control circuit sometimes used to increase the volume range of a radio program or recording by making weak sounds weaker and loud sounds louder; the expander counteracts volume compression at the transmitter or recording studio. Also known as automatic volume expander.

volume indicator [ENG ACOUS] A standardized instrument for indicating the volume of a complex electric wave such as that corresponding to speech or music; the reading in volume units is equal to the number of decibels above a reference level which is realized when the instrument is connected across a 600-ohm resistor that is dissipating a power of 1 milliwatt at 100 hertz. Also known as volume unit meter.

volume meter [ENG] Any flowmeter in which the actual flow is determined by the measurement of a phenomenon associated with the flow.

volume range [ENG ACOUS] The difference, expressed in decibels, between the maximum and minimum volumes of a complex audio-frequency signal occurring over a specified period of time.

volumetric efficiency [MECH ENG] In describing an engine or gas compressor, the ratio of volume of working substance actually admitted, measured at a specified temperature and pressure, to the full piston displacement volume; for a liquid-fuel engine, such as a diesel engine, volumetric efficiency is the ratio of the volume of air drawn into a cylinder to the piston displacement.

volumetric performance [PETRO ENG] The volume production of gas and oil from a reservoir; usually expressed as gas-oil ratio.

volumetric radar [ENG] Radar capable of producing three-dimensional position data on a multiplicity of targets.

volume unit [ENG ACOUS] A unit for expressing the audio-frequency power level of a complex electric wave, such as that corresponding to speech or music; the power level in volume units is equal to the number of decibels above a reference level of 1 milliwatt as measured with a standard volume indicator. Abbreviated VU.

volume unit meter *See* volume indicator.

volute [DES ENG] A spiral casing for a centrifugal pump or a fan designed so that speed will be converted to pressure without shock.

volute pump [MECH ENG] A centrifugal pump housed in a spiral casing.

vomiting gas [MATER] Any one of a group of toxic gases, such as adamsite, that causes coughing, sneezing, sometimes vomiting, and other effects.

von Arx current meter [ENG] A type of current-measuring device using electromagnetic induction to determine speed and, in some models, direction of deep-sea currents.

vortex amplifier [ENG] A fluidic device in which the supply flow is introduced at the circumference of a shallow cylindrical chamber; the vortex field developed can substantially reduce or throttle flow; used in fluidic diodes, throttles, pressure amplifiers, and a rate sensor.

vortex burner [ENG] Combustion device in which the combustion air is fed tangentially into the burner, creating a spin (vortex) to mix it with the fuel as it is injected.

vortex cage meter [ENG] In flow measurement, a type of quantity meter which exerts only a slight retardation on the flowing fluid; the elements rotate at a speed that is linear with fluid velocity; revolutions are counted either by coupling to a local mounted counter or by a proximity detector for remote transmission.

vortex precession flowmeter [ENG] An instrument for measuring gas flows from the rate of precession of vortices generated by a fixed set of radial vanes placed in the flow. Also known as swirl flowmeter.

vortex-shedding meter [ENG] A flowmeter in which fluid velocity is determined from the frequency at which vortices are generated by an obstruction in the flow.

vortex thermometer [ENG] A thermometer, used in aircraft, which automatically corrects for adiabatic and frictional temperature rises by imparting a rotary motion to the air passing the thermal sensing element.

VOS *See* vertical obstacle sonar.

Voskhod program [AERO ENG] The multimanned spaceflight program of the Soviet Union which began with the flight of *Voskhod 1*, October 12, 1964.

Vostok spacecraft [AERO ENG] One of a series of manned artificial satellites launched by the Soviet Union; *Vostok 1* launched on April 12, 1961, was the first manned spacecraft.

votator [MECH ENG] Efficient heat-exchange units for chilling and mechanically working a continuous stream of emulsion; used in food industries in preparation of margarine.

VTOL *See* vertical takeoff and landing.

VTVM *See* vacuum-tube voltmeter.

VU *See* volume unit.

vulcanized fiber [MATER] A laminated plastic made by chemically treating layers of 100% rag-content paper to gelatinize the paper and fuse the layers into a solid mass; when dried under pressure, it forms a hard, tough material having good electrical properties along with mechanical strength and dimensional stability.

vulcan power [MATER] A high explosive composed of 30% nitroglycerin, 52.5% sodium nitrate, 10.5% charcoal, and 7% sulfur.

W

wafer [ENG] A flat element for a process unit, as in a series of stacked filter elements.

wage curve [IND ENG] A graphic representation of the relationship between wage rates and point values for key jobs.

wage incentive plan [IND ENG] A wage system which provides additional pay for qualitative and quantitative performance which exceeds standard or normal levels. Also known as incentive wage system.

wagon drill [MECH ENG] **1.** A vertically mounted, pneumatic, percussive-type rock drill supported on a three- or four-wheeled wagon. **2.** A wheel-mounted diamond drill machine.

Wahl correlation [PETRO ENG] A pressure-volume-temperature (PVT) correlation used to estimate the total oil recovery from a solution-gas-drive oil reservoir; it is based on assumed PVT data, and may be in error.

waist [ENG] The center portion of a vessel or container that has a smaller cross section than the adjacent areas.

waiting line [IND ENG] A line formed by units waiting for service. Also known as queue.

waiting time *See* idle time.

waler [CIV ENG] A horizontal reinforcement utilized to keep newly poured concrete forms from bulging outward. Also spelled whaler.

walking beam [MECH ENG] A lever that oscillates on a pivot and transmits power in a manner producing a reciprocating or reversible motion; used in rock drilling and oil well pumping.

walking dragline [MECH ENG] A large-capacity dragline built with moving feet; disks 20 feet (6 meters) in diameter support the excavator while working.

walking machine [MECH ENG] A machine designed to carry its operator over various types of terrain; the operator sits on a platform carried on four mechanical legs, and movements of his arms control the front legs of the machine while move-ments of his legs control the rear legs of the machine.

walking props *See* self-advancing supports.

wall [ENG] A vertical structure or member forming an enclosure or defining a space. [MIN ENG] **1.** The side of a level or drift. **2.** The country rock bounding a vein laterally. **3.** The face of a longwall working or stall, commonly called coal wall.

wall anchor [BUILD] A steel strap fastened to the end of every second or third common joist and built into the brickwork of a wall to provide lateral support. Also known as joist anchor.

wallboard [MATER] Panels of various materials for surfacing ceilings and walls, including asbestos cement sheet, plywood, gypsum plasterboard, and laminated plastics.

wall cake [PETRO ENG] In drilling operations, the solid deposit along the hole wall due to filtration of the fluid part of the mud into the formation.

wall crane [MECH ENG] A jib crane mounted on a wall.

waller *See* pack builder.

Walley engine [MECH ENG] A multirotor engine employing four approximately elliptical rotors that turn in the same clockwise sense, leading to excessively high rubbing velocities.

wall off [ENG] To seal cracks or crevices in the wall of a borehole with cement, mud cake, compacted cuttings, or casing.

wall plate [BUILD] A piece of timber laid flat along the tip of the wall; it supports the rafters.

wallplate [MIN ENG] A horizontal timber supported by posts resting on sills and extending lengthwise on each side of the tunnel; roof supports rest on the wall-plates.

wall ratio [DES ENG] Ratio of the outside radius of a gun, a tube, or jacket to the inside radius; or ratio of the corresponding diameters.

wall tie [BUILD] A rigid, corrosion-resistant metal tie fitted into the bed joints across the cavity of a cavity wall.

Walter engine [MECH ENG] A multirotor rotary engine that uses two different-sized elliptical rotors.

wandering sequence [MET] A welding sequence in which the increments of weld bead are longitudinally deposited in a random fashion.

Wankel engine [MECH ENG] An eccentric-rotor-type internal combustion engine with only two primary moving parts, the rotor and the eccentric shaft; the rotor moves in one direction around the trochoidal chamber containing peripheral intake and exhaust ports.

warehouse [IND ENG] A building used for storing merchandise and commodities.

warm-air heating [MECH ENG] Heating by circulating warm air; system contains a direct-fired furnace surrounded by a bonnet through which air circulates to be heated.

warning agent *See* odorant.

Warren truss [CIV ENG] A truss having only sloping members between the top and bottom horizontal members.

wash [AERO ENG] The stream of air or other fluid sent backward by a jet engine or a propeller. [ENG] **1.** To clean cuttings or other fragmental rock materials out of a borehole by the jetting and buoyant action of a copious flow of water or a mud-laden liquid. **2.** The erosion of core or drill string equipment by the action of a rapidly flowing stream of water or mud-laden drill-circulation liquid. [MET] **1.** A coating applied to the face of a mold prior to casting. **2.** A sand expansion defect on the surface finish of a casting due to radiation from the metal rising in the mold and causing increased volume and shear of the interface sand on the upper layers.

washability [MIN ENG] Coal properties determining the amenability of a coal to improvement in quality by cleaning.

washboard course [ENG] A test course for vehicles consisting of a series of waves or convolutions having arbitrary amplitude and frequency; a common type is the so-called sine-wave course.

wash-built terrace *See* alluvial terrace.

washer [DES ENG] A flattened, ring-shaped device used to improve the tightness of a screw fastener.

washing plant [MIN ENG] A plant where slimes are removed from relatively coarse ore by washing, tumbling, or scrubbing.

Washita stone [MATER] A relatively porous and not very dense oilstone, used chiefly for whetstones and for sharpening coarse tools.

wash metal [MET] Molten metal used to clean out a furnace, ladle, or other container.

wash oil *See* absorption oil.

washout [ENG] **1.** An overlarge well bore caused by the solvent and erosional action of drilling fluid. **2.** A fluid-cut opening resulting from leaking fluid. [MIN ENG] *See* horseback.

wash primer [MET] A synthetic vehicle primer containing phosphoric acid and zinc chromate; used as a corrosion-inhibiting first paint coat on metals.

waste [ENG] **1.** Rubbish from a building. **2.** Dirty water from mining, industrial, and domestic use. **3.** The amount of excavated material exceeding fill. [MIN ENG] **1.** The barren rock in a mine. **2.** The refuse from ore dressing and smelting plants. **3.** The fine coal made in mining and preparing coal for market.

waste acid *See* sludge acid.

waste bank [MIN ENG] A bank made of earth excavated during the digging of a ditch and laid parallel to it.

waste filling [MIN ENG] Material used for support in heavy ground and in large stopes to prevent failure of rock walls and to minimize or control subsidence and to make it possible to extract pillars of ore left in the earlier stages of mining; material used for filling includes waste rock sorted out in the stopes or mined from rock walls, milltailing, sand and gravel, smelter slag, and rock from surface open cuts or quarries.

waste heat [ENG] Sensible heat in gases not subject to combustion and used for processes downstream in a system.

waste lubrication [ENG] A method in which a lubricant is delivered to a bearing surface by the wicking action of cloth waste or yarn.

waste pipe [CIV ENG] A pipe to carry waste water from a basin, bath, or sink in a building.

waste raise [MIN ENG] An excavation in the mine in which barren rock and other material is broken up for use as filling at a stope.

waste rock [MIN ENG] Valueless rock that must be fractured and removed in order to gain access to or upgrade ore. Also known as muck; mullock.

water-base mud [PETRO ENG] Oil-well drilling mud in which the liquid component is water, into which are mixed the thickeners and other additives.

water-base paint [MATER] Paint in which the vehicle or binder is dissolved in water or in which the vehicle or binder is dispersed as an emulsion; an example of the dispersion type is latex paint. Also known as water-thinned paint.

water block [PETRO ENG] The tendency of accumulated water-oil emulsion around the lower (producing) end of an oil well borehole to block the movement of formation fluids through the formation and toward the borehole.

water brake [ENG] An absorption dynamometer for measuring power output of an engine shaft; the mechanical energy is converted to heat in a centrifugal pump, with a free casing where turning moment is measured.

water break [MET] A break in the continuity of the film of water on the surface of a metal withdrawn from an aqueous bath.

water calorimeter [ENG] A calorimeter that measures radio-frequency power in terms of the rise in temperature of water in which the r-f energy is absorbed.

watercolor [MATER] A pigment ground in a solution of gum arabic, water, and plasticizer, such as glycerin; the glycerin film retards drying in the tube and prevents brittleness in the paint film.

watercolor paper [MATER] A special drawing paper with a surface texture suitable to accept watercolors; the better grades can withstand the harsh scraping that is sometimes necessary to produce highlights; for permanent painting, the paper should be 100% rag, not wood pulp.

water column [MECH ENG] A tubular column located at the steam and water space of a boiler to which protective devices such as gage cocks, water gage, and level alarms are attached.

water-cooled condenser [MECH ENG] A steam condenser which is for the maintenance of vacuum, and in which water is the heat-receiving fluid.

water-cooled furnace [MECH ENG] A fuel-fired furnace containing tubes in which water is circulated to limit heat loss to the surroundings, control furnace temperature, and generate steam.

water cooling [ENG] Cooling in which the primary coolant is water.

water curb *See* garland.

water-drive reservoir [PETRO ENG] An oil or gas reservoir in which pressure is maintained to a greater or lesser extent by an influx of water as the oil or gas is removed.

waterflooding *See* flooding.

water-flow pyrheliometer [ENG] An absolute pyrheliometer, in which the radiation-sensing element is a blackened, water calorimeter; it consists of a cylinder, blackened on the interior, and surrounded by a special chamber through which water flows at a constant rate; the temperatures of the incoming and outgoing water, which are monitored continuously by thermometers, are used to compute the intensity of the radiation.

water gage [ENG] A gage glass with attached fittings which indicates water level in a vessel.

water garland *See* garland.

water gas [MATER] A mixture of carbon monoxide and methane produced by passing steam through deep beds of incandescent coal; used for industrial heating and as a gas engine fuel.

water-gas coke [MATER] Coke which is used in the manufacture of water gas, and which should have a low ash content, a softening temperature of about 2500°F (1370°C), a low sulfur content, and a size larger than 2 inches (5 centimeters).

water heater [MECH ENG] A tank for heating and storing hot water for domestic use.

water influx [PETRO ENG] **1.** The incursion of water (natural or injected) into oil- or gas-bearing formations. **2.** One of the mechanisms of oil production in which the water movement (drive) displaces and moves the reservoir fluids toward the well borehole.

water jacket [ENG] A casing for circulation of cooling water.

water knockout drum [PETRO ENG] A device for removal of water from oil well fluids (gas, or gas with oil). Also known as water knockout trap; water knockout vessel.

water knockout trap *See* water knockout drum.

water knockout vessel *See* water knockout drum.

water leg [ENG] The vertical area of a vessel or accessory to a vessel for the collection of water. Also known as sump.

water main [CIV ENG] The water pipe in a street from which water is delivered to individual service pipes supplying domestic property.

water meter [ENG] An instrument for measuring the amount of water passing a specified point in a piping system.

water paint [MATER] A paint in which the vehicle or binder is dissolved in water; examples are calcimine in which the vehicle is glue, and casein paints in which the vehicle is casein.

water path [ENG] In ultrasonic testing, distance from an ultrasonic search unit to the test piece in an immersion or water column examination.

waterproof [ENG] Impervious to water.

waterproof grease [MATER] A viscous lubricating material that does not dissolve in water and that resists being washed out of bearings or gears; it usually has a low content of oil and metallic soaps of aluminum, barium, calcium, or strontium.

waterproofing agent [MATER] A substance used to make textiles, paper, wood, and other porous or absorbent materials impervious to penetration by water.

water-pump lubricant [MATER] A lubricating grease suitable for the types of automotive water pumps that require grease lubrication.

water purification [CIV ENG] Any of several processes in which undesirable impurities in water are removed or neutralized; for example, chlorination, filtration, primary treatment, ion exchange, and distillation.

water repellent [MATER] Chemicals used to treat textiles, leather, paper, or wood to make them resistant (but not proof) to wetting by water; includes various types of resins, aluminum of zirconium acetates, or latexes.

water right [ENG] The right to use water for mining, agricultural, or other purposes.

water ring See garland.

water sample [ENG] A portion of water brought up from a depth to determine its composition.

water seal [ENG] A seal formed by water to prevent the passage of gas.

water-sealed holder [ENG] A low-pressure gas holder which consists of cylindrical sections or lifts telescoping into a pit or tank filled with water; the inside section is closed in on top.

waterspout [ENG] A pipe or orifice through which water is discharged or by which it is conveyed.

water-supply engineering [CIV ENG] A branch of civil engineering concerned with the development of sources of supply, transmission, distribution, and treatment of water.

water swivel [DES ENG] A device connecting the water hose to the drill-rod string and designed to permit the drill string to be rotated in the borehole while water is pumped into it to create the circulation needed to cool the bit and remove the cuttings produced. Also known as gooseneck; swivel neck.

water-thinned paint See water-base paint.

water tower [CIV ENG] A tower or standpipe for storing water in areas where ordinary water pressure is inadequate for distribution to consumers.

water treatment [CIV ENG] Purification of water to make it suitable for drinking or for any other use.

water-tube boiler [MECH ENG] A steam boiler in which water circulates within tubes and heat is applied from outside the tubes to generate steam.

water tunnel [AERO ENG] A device similar to a wind tunnel, but using water as the working fluid instead of air or other gas. [CIV ENG] A tunnel to transport water in a water-supply system.

waterwall [MECH ENG] The side of a boiler furnace consisting of water-carrying tubes which absorb radiant heat and thereby prevent excessively high furnace temperatures.

waterway [CIV ENG] A channel for the escape or passage of water.

water well [CIV ENG] A well sunk to extract water from a zone of saturation.

waterwheel [MECH ENG] A vertical wheel on a horizontal shaft that is made to revolve by the action or weight of water on or in containers attached to the rim.

water-white kerosine [MATER] Kerosine or refined oil from the crude still before it is treated or rerun; has the whitest (nearest to colorless) of the three standard kerosine colors, namely, water white, prime white, and standard white.

waterworks [CIV ENG] The whole system of supply and treatment utilized in acquisition and distribution of water to consumers.

watt-hour meter [ENG] A meter that measures and registers the integral, with respect to time, of the active power of the circuit in which it is connected; the unit of measurement is usually the kilowatt-hour.

wattle gum [MATER] A gum arabic extracted from the Australian and East African trees of the genus Acacia, as A. dealbata, and other species (called mimosa in Kenya); contains 65% tannin.

wattmeter [ENG] An instrument that measures electric power in watts ordinarily.

wave gage [ENG] A device for measuring the height and period of waves.

wavemeter [ENG] A device for measuring the geometrical spacing between successive surfaces of equal phase in an electromagnetic wave.

wave microphone [ENG ACOUS] Any microphone whose directivity depends upon some type of wave interference, such as a line microphone or a reflector microphone.

wave motor [MECH ENG] A motor that depends on the lifting power of sea waves to develop its usable energy.

wave shaper [ENG] Of explosives, an insert or core of inert material or of explosives having different detonation rates, used for changing the shape of the detonation wave.

wave soldering See flow soldering.

wave trap [CIV ENG] A device used to reduce the size of waves from sea or swell entering a harbor before they penetrate as far as the quayage; usually in the form of diverging breakwaters, or small projecting breakwaters situated close within the entrance.

wax [MATER] Any of a group of substances resembling beeswax in appearance and character, and in general distinguished by their composition of esters and higher alcohols, and by their freedom from fatty acids.

wax distillate [MATER] A neutral distillate from distillation of crude oil that contains a high percentage of crystallizable paraffin wax; used as a primary base for paraffin wax and neutral lubricating oils.

waxed paper [MATER] Paper that is treated or coated with wax to make it waterproof and greaseproof; used for wrapping.

wax original [ENG ACOUS] An original recording made on a wax surface and used to make a master. Also known as wax master.

wax stain [MATER] A semitransparent pigment mixed with beeswax, and thinned with turpentine.

wax tailings [MATER] A sticky, pitchlike substance with dark-brown color, the last volatile product distilling off an oil charge before it is coked; used as wood preservative and in manufacture of roofing paper. Also known as petroleum tailings; still wax.

ways [CIV ENG] **1.** The tracks and sliding timbers used in launching a vessel. **2.** The building slip or space upon which the sliding timbers or ways, supporting a vessel to be launched, travel. [MECH ENG] Bearing surfaces used to guide and support moving parts of machine tools; may be flat, V-shaped, or dovetailed.

weak ground [MIN ENG] Roof and walls of underground excavations which would be in danger of collapse unless suitably supported.

wear [ENG] Deterioration of a surface due to material removal caused by relative motion between it and another part.

wearing course [CIV ENG] The top layer of surfacing on a road.

weathercocking [AERO ENG] The aerodynamic action causing alignment of the longitudinal axis of a rocket with the relative wind after launch. Also known as weather vaning.

weathercock stability *See* directional stability.

weathered crude [MATER] Crude petroleum that, owing to evaporation and other natural causes during storage and handling, has lost an appreciable quantity of its more volatile components.

weather observation radar *See* weather radar.

weatherometer [ENG] A device used to subject articles and finishes to accelerated weathering conditions; for example, a rich ultraviolet source, water spray, or salt water.

weatherproof [ENG] Able to withstand exposure to weather without damage.

weather radar [ENG] Generally, any radar which is suitable or can be used for the detection of precipitation or clouds. Also known as weather observation radar.

weather resistance [ENG] The ability of a material, paint, film, or the like to withstand the effects of wind, rain, or sun and to retain its appearance and integrity.

weather strip [BUILD] A piece of material, such as wood or rubber, applied to the joints of a window or door to stop drafts.

weather vaning *See* weathercocking.

weather window [PETRO ENG] That part of the year when the weather is suitable for operations, such as pipelaying or platform installation, which cannot be undertaken in adverse sea conditions.

web [CIV ENG] The vertical strip connecting the upper and lower flanges of a rail or girder. [MATER] In a grain of propellant, the minimum thickness of the grain between any two adjacent surfaces. [MECH ENG] For twist drills and reamers, the central portion of the tool body that joins the loads. [MET] In forging, the thin section of metal remaining at the bottom of a depression or at the location of the punches.

wedge [DES ENG] A piece of resistant material whose two major surfaces make an acute angle. [ENG] In ultrasonic testing, a device which directs waves of ultrasonic energy into the test piece at an angle.

wedge bit [DES ENG] A tapered-nose noncoring bit, used to ream out the borehole alongside the steel deflecting wedge in hole-deflection operations. Also known as bull-nose bit; wedge reaming bit; wedging bit.

wedge bonding [ENG] A type of thermocompression bonding in which a wedge-shaped tool is used to press a small section of the lead wire onto the bonding pad of an integrated circuit.

wedge core lifter [MECH ENG] A core-gripping device consisting of a series of three or more serrated-face, tapered wedges contained in slotted and tapered recesses cut into the inner surface of a lifter case or sleeve; the case is threaded to the inner tube of a core barrel, and as the core enters the inner tube, it lifts the wedges up along the case taper; when the barrel is raised, the wedges are pulled tight, gripping the core.

wedge photometer [ENG] A photometer in which the luminous flux density of light from two sources is made equal by pushing into the beam from the brighter source a wedge of absorbing material; the wedge has a scale indicating how much it reduces the flux density, so that the luminous intensities of the sources may be compared.

wedge reaming bit *See* wedge bit.

wedging [ENG] **1.** A method used in quarrying to obtain large, regular blocks of building stones; a row of holes is drilled, either by hand or by pneumatic drills, close to each other so that a longitudinal crevice is formed into which a gently sloping steel wedge is driven, and the block of stone can be detached without shattering. **2.** The act of changing the course of a borehole by using a deflecting wedge. **3.** The lodging of two or more wedge-shaped pieces of core inside a core barrel, and therefore blocking it. **4.** The material, moss, or wood used to render the shaft lining tight.

wedging bit *See* wedge bit.

weep hole [CIV ENG] A hole in a wood sill, retaining wall, or other structure to allow accumulated water to escape.

weeping core [PETRO ENG] A core cut that is covered with tearlike drops of fluid when it is brought to the surface; usually indicative of a deposit that will produce little oil.

weighing rain gage [ENG] A type of recording rain gage, consisting of a receiver in the shape of a funnel which empties into a bucket mounted upon a weighing mechanism; the weight of the catch is recorded, on a clock-driven chart, as inches of precipitation; used at climatological stations.

weight and balance sheet [AERO ENG] A sheet which records the distribution of weight in an aircraft and shows the center of gravity of an aircraft at takeoff and landing.

weight barometer [ENG] A mercury barometer which measures atmospheric pressure by weighing the mercury in the column or the cistern.

weighting [ENG] The artificial adjustment of measurements to account for factors that, in the normal use of the device, would otherwise be different from conditions during the measurements.

weighting network [ENG ACOUS] One of three or more circuits in a sound-level meter designed to adjust its response; the A and B weighting networks provide responses approximating the 40- and 70-phon equal loudness contours, respectively, and the C weighting network provides a flat response up to 8000 hertz.

weight-loaded regulator [ENG] A pressure-regulator valve for pressure vessels or flow systems; the regulator is preloaded by counterbalancing weights to open (or close) at the upper (or lower) limit of a preset pressure range.

weir [CIV ENG] A dam in a waterway over which water flows, serving to regulate water level or measure flow.

weir tank [PETRO ENG] A type of oil-field storage tank with high- and low-level weir boxes and liquid-level controls for metering the liquid content of the tank.

weld [MET] A union made between two metals by welding.

weldability [MET] Suitability of a metal to be welded under specified conditions.

weld bead [MET] A deposit of filler metal from a single welding pass. Also known as bead.

weldbonding [MET] A process for joining metals in which adhesive, typically an epoxy paste, is applied to the parts, which are then clamped together, spot-welded, and put into an oven (250°F, or 121°C, for 1 hour) to cure the adhesive.

weld decay [MET] Intercrystalline corrosion of austenitic stainless steels near welded areas; caused by chromium carbide precipitation along grain boundaries of alloy subject to prolonged heating in the temperature range 400–850°C.

weld delay time [MET] Delay of the time current in spot, seam, or projection welding with respect to starting the forge delay timer used to synchronize pressure and heat.

welder [MET] **1.** A machine used in welding. Also known as welding machine. **2.** A person who performs a welding operation.

weld gage [ENG] A device used to check the shape and size of welds.

welding [MET] Joining two metals by applying heat to melt and fuse them, with or without filler metal.

welding cycle [MET] The complete sequence of events involved in making a resistance weld.

welding electrode [MET] **1.** In arc welding, the current-carrying rod or rods used to strike an arc between rod and work. **2.** In resistance welding, the component of a machine through which current and pressure are applied to the work.

welding force *See* electrode force.

welding ground *See* work lead.

welding rod [MET] Filler metal in the form of a rod or heavy wire.

welding schedule [MET] A record of all welding machine settings plus identification of the machine needed to produce a weld for a given material of a given size and finish.

welding sequence [MET] The order for welding component parts of a weldment or structure.

welding stress [MET] Residual stress resulting from localized heating and cooling during welding.

welding tip [ENG] A replaceable nozzle for a gas torch used in welding. [MET] An electrode used in spot or projection welding.

welding torch [ENG] A gas-mixing and burning tool for the welding of metal.

weld interval [MET] The total heat and cool times for making one multiple-impulse weld.

weld-interval timer [ENG] A device used to control weld interval.

weld line [MET] The junction of the weld metal and base metal, or the junction of base-metal parts when filler metal is not used.

weld mark *See* flow line.

weldment [ENG] An assembly or structure whose component parts are joined by welding.

weld metal [MET] The metal constituting the fused zone in spot, seam, or projection welding.

weld time [MET] The time that the welding current is applied to the work in single-impulse and flash welding.

Welge method [PETRO ENG] A method of calculation of the anticipated oil-recovery performance of a gas-cap-drive oil reservoir.

well [BUILD] An open shaft in a building, extending vertically through floors to accommodate stairs or an elevator. [ENG] A hole dug into the earth to reach a supply of water, oil, brine, or gas.

wellbore hydraulics [PETRO ENG] A branch of oil production engineering that deals with the motion of fluids (oil, gas, or water) in wellbore tubing or casing, or the annulus between tubing and casing.

well completion [PETRO ENG] The final sealing off of a drilled well (after drilling apparatus is removed from the borehole) with valving, safety, and flow-control devices.

well conditioning [PETRO ENG] **1.** Preparation of a well for sampling procedures by control of production rate and associated pressure drawdown. **2.** Removal of accumulated scale, wax, mud, and sand from the inner surfaces of a wellbore, or breakage of water blocks to increase production of oil or gas.

well core [ENG] A sample of rock penetrated in a well or other borehole obtained by use of a hollow bit that cuts a circular channel around a central column or core.

well drill [MECH ENG] A drill, usually a churn drill, used to drill water wells.

wellhead [CIV ENG] The top of a well.

wellhole [MIN ENG] **1.** A large-diameter vertical hole used in quarries and opencast pits for taking heavy explosive charges in blasting. **2.** The sump, or portion of a shaft below the place where skips are caged at the bottom of the shaft, in which water collects.

well injectivity [PETRO ENG] The ability of an injection well (water or gas) to receive injected fluid; can be negatively influenced by formation plugging, borehole scale, or liquid blocking around the lower end of the borehole.

well logging [ENG] The technique of analyzing and recording the character of a formation penetrated by a drill hole in mineral exploration and exploitation work.

well performance [PETRO ENG] The measurement of a well's production of oil or gas as related to the well's anticipated productive capacity, pressure drop, or flow rate.

wellpoint [CIV ENG] A component of a wellpoint system consisting of a perforated pipe about 4 feet (1.2 meters) long and about 2 inches (5 centimeters) in diameter, equipped with a ball valve, a screen, and a jetting tip.

wellpoint system [CIV ENG] A method of keeping an excavated area dry by intercepting the flow of groundwater with pipe wells located around the excavation area.

well shooting [ENG] The firing of a charge of nitroglycerin, or other high explosive, in the bottom of a well for the purpose of increasing the flow of water, oil, or gas.

well spacing [PETRO ENG] Areal location and interrelationship between producing oil or gas wells in an oil field; calculated for the maximum ultimate production from a given reservoir.

well stimulation *See* stimulation treatment.

well-type manometer [ENG] A type of double-leg, glass-tube manometer; one leg has a relatively small diameter, and the second leg is a reservoir; the level of the liquid in the reservoir does not change appreciably with change of pressure; a mercury barometer is a common example.

Wentworth quick-return motion *See* turning-block linkage.

West African cedar *See* Sapele mahogany.

Westphal balance [ENG] A direct-reading instrument for determining the densities of solids and liquids; a plummet of known mass and volume is immersed in the liquid whose density is to be measured or, alternatively, a sample of the solid whose density is to be measured is immersed in a liquid of known density, and the loss in weight is measured, using a balance with movable weights.

wet and dry bulb thermometer *See* psychrometer.

wet assay [MIN ENG] The determination of the quantity of a desired constituent in ores, metallurgical residues, and alloys by the use of the processes of solution, flotation, or other liquid means.

wet blasting [ENG] Shot firing in wet holes. [MET] Liquid honing in which an impeller wheel drives the liquid suspension.

wet-bulb thermometer [ENG] A thermometer having the bulb covered with a cloth, usually muslin or cambric, saturated with water.

wet-cell caplight [MIN ENG] A rechargeable head lamp; the batteries are worn on the belt.

wet classifier [ENG] A device for the separation of solid particles in a mixture of solids and liquid into fractions, according to particle size or density by methods other than screening; operates by the difference in the settling rate between coarse and fine or heavy and light particles in a tank-confined liquid.

wet collector *See* scrubber.

wet cooling tower [MECH ENG] A structure in which water is cooled by atomization into a stream of air; heat is lost through evaporation. Also known as evaporative cooling tower.

wet corrosion [MET] Corrosion caused by exposure to aqueous solutions.

wet drawing [MET] Drawing in which the dies and blocks are completely immersed in the lubricant.

wet drill [MECH ENG] A percussive drill with a water feed either through the machine or by means of a water swivel, to suppress the dust produced when drilling.

wet emplacement [AERO ENG] A launch emplacement that provides a deluge of water for cooling the flame bucket, the rocket engines, and other equipment during the launch of a missile.

wet engine [MECH ENG] An engine with its oil, liquid coolant (if any), and trapped fuel inside.

wet gas [MATER] Natural gas produced along with crude petroleum in oil fields or from gas-condensate fields; in addition to methane, it contains ethane, propane, butanes, and some higher hydrocarbons, such as pentane and hexane.

wet grinding [MECH ENG] **1.** The milling of materials in water or other liquid. **2.** The practice of applying a coolant to the work and the wheel to facilitate the grinding process.

wet hole [ENG] A borehole that traverses a water-bearing formation from which the flow of water is great enough to keep the hole almost full of water.

wet mill [MECH ENG] **1.** A grinder in which the solid material to be ground is mixed with liquid. **2.** A mill in which the grinding energy is developed by a fast-flowing liquid stream; for example, a jet pulverizer.

wet sleeve [MECH ENG] A cylinder liner which is exposed to the coolant over 70% or more of its surface.

wet slip [CIV ENG] An opening between two wharves or piers where dock trials are usually conducted, and the final fitting out is done.

wet strength [MATER] **1.** The strength of a material saturated with water. **2.** The ability to withstand water (as for paper products) with a wet-strength additive or resin finish.

wet-strength paper [MATER] Paper with increased water resistance due to processing and interlocking of fibers, as well as impregnation with small amounts of resins (for example, melamine or urea formaldehyde). Also known as wet-strong paper.

wet tabling [MIN ENG] A tabling process in which a pulp of two or more minerals flows across an inclined, riffled plane surface, is shaken longwise, and is water-washed crosswise.

wet-test meter [ENG] A device to measure gas flow by counting the revolutions of a shaft upon which water-sealed, gas-carrying cups of fixed capacity are mounted.

wetting [MET] Spreading liquid filler metal or flux on a solid base metal.

wetting phase [PETRO ENG] In a two-phase oil reservoir system (oil and water), one phase (water) will wet the pore surfaces of the reservoir formation, the other (oil) will not.

wet well [MECH ENG] A chamber which is used for collecting liquid, and to which the suction pipe of a pump is attached.

whale oil [MATER] A combustible, nontoxic, yellow-brown fixed oil obtained from whale blubber; soluble in alcohol, ether, chloroform, carbon disulfide, and benzene; used as a lubricant, illuminant, and leather dressing, and in soapmaking and fat manufacture. Also known as blubber oil.

whaler *See* waler.

wharf [CIV ENG] A structure of open construction built parallel to the shoreline; used by vessels to receive and discharge passengers and cargo.

wheat germ oil [MATER] A light-yellow oil extracted from wheat germ; used as a source for vitamin E, as a dietary supplement, and in medicine.

wheel [DES ENG] A circular frame with a hub at the center for attachment to an axle, about which it may revolve and bear a load.

wheelbarrow [ENG] A small, hand-pushed vehicle with a single wheel and axle between the front ends of two shafts that support a boxlike body and serve as handles at the rear. Also known as barrow.

wheel base [DES ENG] The distance in the direction of travel from front to rear wheels of a vehicle, measured between centers of ground contact under each wheel.

wheel-bearing lubricant [MATER] A lubricating grease with the character, structure, and consistency needed to make it suitable for use in antifriction wheel bearings.

wheel dresser [ENG] A tool for cleaning, resharpening, and restoring the mechanical accuracy of the cutting faces of grinding wheels.

wheeled crane [MECH ENG] A self-propelled crane that rides on a rubber-tired chassis with power for transportation provided by the same engine that is used for hoisting.

wheel load capacity [CIVIL ENG] The capacity of airfield runways, taxiways, parking areas, or roadways to bear the pressures exerted by aircraft or vehicles in a gross weight static configuration.

wheel sleeve [DES ENG] A flange used as an adapter on precision grinding machines where the hole in the wheel is larger than the machine arbor.

whetstone [MATER] Any hard, fine-grained, naturally occurring, usually siliceous rock suitable for sharpening cutting instruments.

whipstock [PETRO ENG] A long wedge dropped or placed in a petroleum well in order to deflect the drill from some obstruction.

white carbon black [MATER] A white silica powder made from silicon tetrachloride; used as a replacement for carbon black in rubber compounding.

white cast iron [MET] An extremely hard cast iron, rapidly cooled from the melt; contains about 3% carbon in the form of cementite and fine pearlite.

white cement [MATER] Pure white portland cement, made from raw materials with a low iron content, or by using a reducing flame to fire the clinker.

white coat [BUILD] The finishing coat in plastering.

white cutch *See* gambir.

white damp [MIN ENG] In mining, carbon monoxide (CO); a gas that may be present in the afterdamp of a gas or coal-dust explosion, or in the gases given off by a mine

fire; it is an important constituent of illuminating gas, supports combustion, and is very poisonous.

whiteheart malleable iron [MET] White cast iron malleableized and decarburized by heat treatment in an oxidizing material at 900°C for 100–150 hours; decarburization produces a light-colored fracture, in contrast to blackheart malleable iron, which is not decarburized. Also known as blackheart malleable iron.

white iron [MET] A brittle cast iron whose total carbon content is in the combined forms, and containing little or no graphite; a fresh fracture is white.

white metal [MET] **1.** Any of several white-colored metals and their alloys of relatively low melting points, such as lead, tin, antimony, and zinc. **2.** A copper matte of about 77% copper, obtained from the smelting of sulfide copper ores.

white-metal bearing alloy *See* lead-base babbitt.

white mineral oil [MATER] A highly refined, colorless hydrocarbon oil with low volatility; used as a laxative and in medicine. Also known as liquid petrolatum; paraffinum liquidum.

white oil [MATER] Any of various highly refined, colorless hydrocarbon oils of low volatility and a wide range of viscosities; used for lubrication of food and textile machinery and as medicinal and mineral oils. Also known as technical white oil.

white portland cement [MATER] Finely ground portland cement made from pure calcite limestone and white clay.

whitewash [MATER] A simple mixture of hydrated lime and water, used mostly for painting fences and outbuildings; common whitewash is not water-resistant and rubs off easily.

whiting *See* chalk.

Whitworth screw thread [DES ENG] A British screw thread standardized to form and dimension.

whizzer mill *See* Jeffrey crusher.

whole range point [AERO ENG] The point vertically below an aircraft at the moment of impact of a bomb released from that aircraft, assuming that the aircraft's velocity has remained unchanged.

wicket dam [CIV ENG] A movable dam consisting of a number of rectangular panels of wood or iron hinged to a sill and propped vertically; the prop is hinged and can be tripped to drop the wickets flat on the sill.

wicking [ENG] The flow of solder under the insulation of covered wire.

Wiese formula [ENG] An empirical relationship for motor fuel antiknock values above 100 in relation to performance numbers; basis for the ASTM scale, in which octane numbers above 100 are related to increments of tetraethyllead added to isooctane.

wiggle stick *See* divining rod.

wildcat drilling [MIN ENG] The drilling of boreholes in unproved territory. Also known as cold nosing; wildcatting.

wildcatting *See* wildcat drilling.

Wild fence [ENG] A wooden enclosure about 16 feet square and 8 feet high with a precipitation gage in its center; the function of the fence is to minimize eddies around the gage, and thus ensure a catch which will be representative of the actual rainfall or snowfall.

wild gasoline [MATER] Unstabilized casing-head gasoline.

wildness [MET] A condition that exists when molten metal, during cooling, evolves so much gas that it becomes violently agitated, forcibly ejecting metal from its container.

Wilfley table [MIN ENG] A flat, rectangular surface that can be tilted and shaken about the long axis and has horizontal riffles for imposing restraint in removing minerals from classified sand. Also known as shaking table.

Willans line [MECH ENG] The line (nearly straight) on a graph showing steam consumption (pounds per hour) versus power output (kilowatt or horsepower) for a steam engine or turbine; frequently extended to show total fuel consumed (pounds per hour) for gas turbines, internal combustion engines, and complete power plants.

winch [MECH ENG] A machine having a drum on which to coil a rope, cable, or chain for hauling, pulling, or hoisting.

windbreak [ENG] Any device designed to obstruct wind flow and intended for protection against any ill effects of wind.

wind cone [ENG] A tapered fabric sleeve, shaped like a truncated cone and pivoted at its larger end on a standard, for the purpose of indicating wind direction; since the air enters the fixed end, the small end of the cone points away from the wind. Also known as wind sleeve; wind sock.

wind correction [ENG] Any adjustment which must be made to allow for the effect of wind; especially, the adjustments to correct for the effect on a projectile in flight, on sound received by sound ranging instruments, and on an aircraft flown by dead reckoning navigation.

wind direction indicator [ENG] A device to indicate the direction from which the wind blows; an example is a weather vane.

winder [BUILD] A step, generally wedge-shaped, with a tread that is wider at one end than the other; often used in spiral staircases.

windmill [MECH ENG] Any of various mechanisms, such as a mill, pump, or electric generator, operated by the force of wind against vanes or sails radiating about a horizontal shaft.

windmill anemometer [ENG] A rotation anemometer in which the axis of rotation is horizontal; the instrument has either flat vanes (as in the air meter) or helicoidal vanes (as in the propeller anemometer); the relation between wind speed and angular rotation is almost linear.

window [AERO ENG] An interval of time during which conditions are favorable for launching a spacecraft on a specific mission. [BUILD] An opening in the wall of a building or the body of a vehicle to admit light and usually to permit vision through a transparent or translucent material, usually glass. [MATER] A globular defect in a thermoplastic sheet or film caused by incomplete plasticization; similar to a fisheye.

wind shield *See* rain gage shield.

windshield [ENG] A transparent glass screen that protects the passengers and compartment of a vehicle from wind and rain.

wind sleeve *See* wind cone.

wind sock *See* wind cone.

wind tee [ENG] A weather vane shaped like the letter T or like an airplane, situated on an airport or landing field to indicate the wind direction. Also known as landing tee.

wind triangle [AERO ENG] A vector diagram showing the effect of the wind on the flight of an aircraft; it is composed of the wind direction and wind speed vector, the true heading and true airspeed vector, and the resultant track and ground speed vector.

wind tunnel [ENG] A duct in which the effects of airflow past objects can be determined.

wind-tunnel balance [AERO ENG] A device or apparatus that measures the aerodynamic forces and moments acting upon a body tested in a wind tunnel.

wind-tunnel instrumentation [ENG] Measuring devices used in wind-tunnel tests; in addition to conventional laboratory instruments for fluid flow, thermometry, and mechanical measurements, there are sensing devices capable of precision measurement in the small-scale environment of the test setup.

wind vane [ENG] An instrument used to indicate wind direction, consisting basically of an asymmetrically shaped object mounted at its center of gravity about a vertical axis; the end which offers the greater resistance to the motion of air moves to the downwind position; the direction of the wind is determined by reference to an attached oriented compass rose.

wing [AERO ENG] **1.** A major airfoil. **2.** An airfoil on the side of an airplane's fuselage or cockpit, paired off by one on the other side, the two providing the principal lift for the airplane.

wing assembly [AERO ENG] An aeronautical structure designed to maintain a guided missile in stable flight; it consists of all panels, sections, fastening devices, chords, spars, plumbing accessories, and electrical components necessary for a complete wing assembly.

wing axis [AERO ENG] The locus of the aerodynamic centers of all the wing sections of an airplane.

wing dam *See* groin.

wingless abutment [CIV ENG] A straight-sided bridge abutment designed to resist pressure in back and provide a bridge seat.

wing loading [AERO ENG] A measure of the load carried by an airplane wing per unit of wing area; commonly used units are pounds per square foot and kilograms per square meter.

wing nut [DES ENG] An internally threaded fastener with wings to permit it to be tightened or loosened by finger pressure only. Also known as butterfly nut.

wing panel [AERO ENG] That portion of a multipiece wing section that usually lies between the front and rear spars; it may be designed to include either the leading edge or the trailing edge as an integral part, but never both, and excludes control surfaces.

wing profile [AERO ENG] The outline of a wing section.

wing rib [AERO ENG] A chordwise member of the wing structure of an airplane, used to give the wing section its form and to transmit the load from the fabric to the spars.

wing section *See* airfoil profile.

wing structure [AERO ENG] In an aircraft, the combination of outside fairing panels that provide the aerodynamic lifting surfaces and the inside supporting members that transmit the lifting force to the fuselage; the primary load-carrying portion of a wing is a box beam (the prime box) made up usually of two or more vertical webs, plus a major portion of the upper and lower skins of the wing, which serve as chords of the beam.

wing-tip rake [AERO ENG] The shape of the wing when the tip edge is straight in plan but not parallel to the plane of symmetry; the amount of rake is measured by the acute angle between the straight portion of the wing tip and the plane of symme-

try; the rake is positive when the trailing edge is longer than the leading edge.

winning [MIN ENG] **1.** A new mine opening. **2.** The portion of a coal field laid out for working. **3.** Mining.

winnowing gold [MIN ENG] Tossing up dry powdered auriferous material in air, and catching the heavier particles not blown away.

winterization [ENG] The preparation of equipment for operation in conditions of winter weather; this applies to preparation not only for cold temperatures, but also for snow, ice, and strong winds.

winze [MIN ENG] A vertical or inclined opening or excavation connecting two levels in a mine, differing from a raise only in construction; a winze is sunk underhand, and a raise is put up overhand.

wiped joint [MET] A joint wherein filler metal is applied in liquid form, and the joint is wiped mechanically to distribute the metal.

wiping effect [MET] Activation of a metal surface by mechanically rubbing or wiping to enhance the formation of a conversion coating.

wire [MET] A thin, flexible, continuous length of metal, usually of circular cross section.

wire cloth [DES ENG] Screen composed of wire crimped or woven into a pattern of squares or rectangles.

wire-cut brick [MATER] A brick cut from clay shaped by extrusion before burning; the long bar of extruded clay is cut into bricks by a set of wires 9 inches (23 centimeters) apart.

wire drag [ENG] An apparatus for surveying rocky underwater areas where normal sounding methods are insufficient to ensure the discovery of all existing submerged obstructions, small shoals, or rocks above a given depth or for determining the least depth of an area; it consists essentially of a buoyed wire towed at the desired depth by two launches.

wiredrawing [MET] Pulling a metal rod or wire through a die to reduce its cross section.

wire-fabric reinforcing [CIV ENG] Reinforcing concrete or mortar with a welded wire fabric.

wire flame spray gun [ENG] A device which utilizes the heat from a gas flame and material in the form of wire or rod to perform a flame-spraying operation.

wire gage [DES ENG] **1.** A gage for measuring the diameter of wire or thickness of sheet metal. **2.** A standard series of sizes arbitrarily indicated by numbers, to which the diameter of wire or the thickness of sheet metal is usually made, and which is used in describing the size or thickness.

wire glass [MATER] Sheet glass with woven wire mesh embedded in the center of the sheet; used in building construction for windows, doors, floors, and skylights.

wire insulation [MATER] A flexible insulation used to cover an electric wire.

wire line [DES ENG] **1.** Any cable or rope made of steel wires twisted together to form the strands. **2.** A steel wire rope ⁵⁄₁₆ inch or less in diameter. [PETRO ENG] A line or cable used to lower and raise devices and gages in oil well boreholes; used for logging instruments and bottom-hole pressure gages.

wire-line coring [PETRO ENG] A method for obtaining samples of reservoir rocks during the drilling phase of oil wells.

wire nail [DES ENG] A nail made of wire and having a circular cross section.

wire recorder [ENG ACOUS] A magnetic recorder that utilizes a round stainless steel wire about 0.004 inch (0.01 centimeter) in diameter instead of magnetic tape.

wire recording [ENG ACOUS] Magnetic recording by use of a magnetized wire.

wire rod [MET] A metal rod used in wiredrawing.

wire rope [ENG] A rope formed of twisted strands of wire.

wire saw [MECH ENG] A machine employing one- or three-strand wire cable, up to 16,000 feet (4900 meters) long, running over a pulley as a belt; used in quarries to cut rock by abrasion.

wiresonde [ENG] An atmospheric sounding instrument which is supported by a captive balloon and used to obtain temperature and humidity data from the ground level to a height of a few kilometers; height is determined by means of a sensitive altimeter, or from the amount of cable released and the angle which the cable makes with the ground, and the information is telemetered to the ground through a wire cable.

wire stripper [ENG] A hand-operated tool or special machine designed to cut and remove the insulation for a predetermined distance from the end of an insulated wire, without damaging the solid or stranded wire inside.

wire tack [DES ENG] A tack made from wire stock.

wire train [ENG] An assembly that normally consists of an extruder, a crosshead and die, a means of cooling, and feed and take-up spools for the wire; used to coat wire with resin.

wireway [ENG] A trough which is lined with sheet metal and has hinged covers, designed to house electrical conductors or cables.

wire weight gage [ENG] A river gage in which a weight suspended on a wire is lowered to the water surface from a bridge or other overhead structure to measure the distance from a point of known elevation on the bridge to the water surface; the distance is usually measured by counting the number of revolutions of a drum required to lower the weight, and a counter is provided which reads the water stage directly.

witch hazel [MATER] A water extract from the dried leaves of the witch hazel shrub (*Hamamelis virginiana*); a solution of 14% alcohol with 1% witch hazel extract is commonly known as witch hazel; used as a tonic and sedative.

wobble wheel roller [MECH ENG] A roller with freely suspended pneumatic tires used in soil stabilization.

Wollaston wire [ENG] An extremely fine platinum wire, produced by enclosing a platinum wire in a silver sheath, drawing them together, and using acid to dissolve away the silver; used in electroscopes, microfuses, and hot-wire instruments.

wood [MATER] Lumber or timber obtained from trees.

wood filler [MATER] A paste designed to fill the pores of open-grained woods such as ash, chestnut, mahogany, and oak.

wood flour [MET] Pulverized wood used in casting molds to furnish a reducing atmosphere, help overcome expansion of sand, increase flowability of the metal, and improve casting finish.

wood physics [MATER] The area of wood science concerned with the physical and mechanical properties of wood and the factors which affect them.

wood preservative [MATER] A material used to coat wood to kill insects and fungi, but not usually classed as an insecticide; coal tar creosote and its derivatives are the most widely used wood preservatives.

wood pulp *See* pulp.

Woodruff key [DES ENG] A self-aligning machine key made by a side-milling cutter in the form of a segment of a disk.

wood screw [DES ENG] A threaded fastener with a pointed shank, a slotted or recessed head, and a sharp tapered thread of relatively coarse pitch for use only in wood.

Wood's glass [MATER] A type of glass that has a high transmission factor for ultraviolet radiation but is relatively opaque to visible radiation.

Wood's metal [MET] A fusible alloy of the Cerro Corporation that contains 50% bismuth, 25% lead, 12.5% tin, and 12.5% cadmium, and melts at 158°F (70–72°C); used for automatic sprinkler plugs.

woodstave pipe [DES ENG] A pipe made of narrow strips of wood placed side by side and banded with wire, metal collars, and inserted joints, used largely for municipal water supply, outfall sewers, and mining irrigation.

woody structure [MET] A fibrous appearance in a fracture, particularly found in wrought iron and extruded aluminum alloys, usually associated with elongated inclusions or grains.

woofer [ENG ACOUS] A large loudspeaker designed to reproduce low audio frequencies at relatively high power levels; usually used in combination with a crossover network and a high-frequency loudspeaker called a tweeter.

wool fat *See* wool grease.

wool grease [MATER] A highly complex mixture of wax ester, alcohols, and fatty acids coating the surface of sheep wool fibers and obtained by scouring the wool with soap or synthetic detergent; used in the manufacture of lanolin and its derivatives, for dressing leather, and in lubricating and slushing oils, soaps, and ointments. Formerly known as degras. Also known as wool fat; wool oil; wool wax.

wool oil *See* wool grease.

wool wax *See* wool grease.

word concatenation system [ENG ACOUS] The simplest form of voice response system, which retrieves previously spoken versions of words or phrases and carefully forms them into a sequence without pauses, to approximate normally spoken word sequences.

workability [MATER] The ease with which concrete can be placed.

work angle [MET] In arc welding, the angle in a plane normal to the weld axis between the electrode and one member of the joint.

work cycle [IND ENG] A sequence of tasks, operations, and processes, or a pattern of manual motions, elements, and activities that is repeated for each unit of work.

worked-out [MIN ENG] Exhausted, referring to a coal seam or ore deposit.

worked penetration [ENG] Penetration of a sample of lubricating grease immediately after it has been brought to a specified temperature and subjected to strokes in a standard grease worker.

work hardening [MET] Increased hardness accompanying plastic deformation of a metal below the recrystallization temperature range.

working [MIN ENG] **1.** The whole strata excavated in working a seam. **2.** Ground or rocks shifting under pressure and producing noise.

working load [ENG] The maximum load that any structural member is designed to support.

working place [MIN ENG] The place in a mine at which coal or ore is being actually mined.

working pressure [ENG] The allowable operating pressure in a pressurized vessel or conduit, usually calculated by ASME (American Society of Mechanical Engineers) or API (American Petroleum Institute) codes.

work lead [MET] The electrical conductor connecting the source of current to the work in arc welding. Also known as ground lead; welding ground.

work sampling [IND ENG] A technique to measure work activity as related to delays consisting of intermittent observations of actual work and delays. Also known as activity sampling; frequency study; ratio delay study.

work standardization [IND ENG] The establishment of uniformity of working conditions, tools, equipment, technical procedures, administrative procedures, workplace arrangements, motion sequences, materials, quality requirements, and similar factors which affect the performance of work.

work task [IND ENG] A specified amount of work, set of responsibilities, or occupation assigned to an individual or to a group.

worm [DES ENG] A shank having at least one complete tooth (thread) around the pitch surface; the driver of a worm gear. [MET] Sweat of molten metal which exudes through the crust of solidifying metal in a casting, and is caused by gas evolution.

worm conveyor *See* screw conveyor.

worm gear [DES ENG] A gear with teeth cut on an angle to be driven by a worm; used to connect nonparallel, nonintersecting shafts.

wormseed oil [MATER] An essential oil distilled from the seeds and leaf stems of the plant *Chenopodium anthelminticum*, which is grown in Maryland, and which contains the alkaloid ascoridole; used in worm treatment of animals. Also known as Baltimore oil.

worm wheel [DES ENG] A gear wheel with curved teeth that meshes with a worm.

wormwood oil *See* absinthe oil.

wove paper [MATER] Paper characterized by a uniform, unlined surface and a soft, smooth finish.

wow [ENG ACOUS] A low-frequency flutter; when caused by an off-center hole in a disk record, occurs once per revolution of the turntable.

wrap-around hanger [PETRO ENG] An oil-well tubing hanger made up of two hinged halves with a resilient sealing element between two steel mandrels.

wrap forming *See* stretch forming.

wrapper sheet [MECH ENG] **1.** The outer plate enclosing the firebox in a fire-tube boiler. **2.** The thinner sheet of a boiler drum having two sheets.

wrecking ball *See* skull cracker.

wrench [ENG] A manual or power tool with adapted or adjustable jaws or sockets either at the end or between the ends of a lever for holding or turning a bolt, pipe, or other object.

wrench-head bolt [DES ENG] A bolt with a square or hexagonal head designed to be gripped between the jaws of a wrench.

Wright system [PETRO ENG] A method for mining oil from partially drained sands that involves drilling a shaft through the productive strata, followed by long, slanting holes drilled radially in all directions from the shaft bottom into the oil sands.

wringing fit [DES ENG] A fit of zero-to-negative allowance.

wrinkling [MET] Waviness around the edges of a drawn metal product.

wrist pin *See* piston pin.

wrought alloy [MET] An alloy that has been mechanically worked after casting.

wrought iron [MET] A commercial iron consisting of slag fibers, primarily iron silicate, embedded in a ferrite matrix.

wye [ENG] A pipe branching off a straight main run at an angle of 45°. Also known as Y; yoke.

wye level *See* Y level.

X

X engine [MECH ENG] An in-line engine with the cylinder banks so arranged around the crankshaft that they resemble the letter X when the engine is viewed from the end.

X frame [DES ENG] An automotive frame which either has side rails bent in at the center of the vehicle, making the overall form that of an X, or has an X-shaped member which joins the side rails with diagonals for added strength and resistance to torsional stresses.

x-ray diffractometer [ENG] An instrument used in x-ray analysis to measure the intensities of the diffracted beams at different angles.

x-ray goniometer [ENG] A scale designed to measure the angle between the incident and refracted beams in x-ray diffraction analysis.

x-ray machine [ENG] The x-ray tube, power supply, and associated equipment required for producing x-ray photographs.

x-ray microscope [ENG] **1.** A device in which an ultra-fine-focus x-ray tube or electron gun produces an electron beam focused to an extremely small image on a transmission-type x-ray target that serves as a vacuum seal; the magnification is by projection; specimens being examined can thus be in air, as also can the photographic film that records the magnified image. **2.** Any of several instruments which utilize x-radiation for chemical analysis and for magnification of 100–1000 diameters; it is based on contact or projection microradiography, reflection x-ray microscopy, or x-ray image spectrography.

x-ray monochromator [ENG] An instrument in which x-rays are diffracted from a crystal to produce a beam having a narrow range of wavelengths.

x-ray telescope [ENG] An instrument designed to detect x-rays emanating from a source outside the earth's atmosphere and to resolve the x-rays into an image; they are carried to high altitudes by balloons, rockets, or space vehicles; although several types of x-ray detector, involving gas counters, scintillation counters, and collimators, have been used, only one, making use of the phenomenon of total external reflection of x-rays from a surface at grazing incidence, is strictly an x-ray telescope.

x-ray thickness gage [ENG] A thickness gage used for measuring and indicating the thickness of moving cold-rolled sheet steel during the rolling process without making contact with the sheet; an x-ray beam directed through the sheet is absorbed in proportion to the thickness of the material and its atomic number.

XY recorder [ENG] A recorder that traces on a chart the relation of two variables, neither of which is time.

Y

Y *See* wye.

yard [CIV ENG] A facility for building and repairing ships.

yardage [MIN ENG] The extra compensation a miner receives in addition to the mining price for working in a narrow place or in deficient coal, usually at a certain price per yard advanced.

yard crane *See* crane truck.

yard maintenance [ENG] A category of maintenance that includes the complete rebuilding of parts, subassemblies, or components.

yaw damper [AERO ENG] A control system or device that reduces the yaw of an aircraft, guided missile, or the like.

yaw indicator [AERO ENG] A device that measures the angular direction of the airflow relative to the longitudinal vertical plane of the aircraft; this may be accomplished by a balanced vane or by a differential pressure sensor that aligns the detector to the airflow, and in so doing transmits the measured angle between the normal axis and the detector as the yaw angle.

Y block [MET] A Y-shaped test casting used to appraise low-shrinkage alloys.

yellow cake [MIN ENG] The final precipitate formed in the milling of uranium ores.

yellow cedar *See* Alaska cedar.

yellow cypress *See* Alaska cedar.

yellow scale [MATER] The commercial name for low-grade paraffin wax.

yellow wax *See* beeswax.

Y factor [PETRO ENG] An empirical relationship of bubble-point data (pressure and formation volume) used to smooth oil reservoir solution-gas/oil-ratio data for graphical presentation.

yield [ENG] Product of a reaction or process as in chemical reactions or food processing.

yielding arches [MIN ENG] Steel arches installed in underground openings as the ground is removed to support loads caused by changing ground movement or by faulted and fractured rock; when the ground load exceeds the design load of the arch as installed, yielding takes places in the joint of the arch, permitting the overburden to settle into a natural arch of its own and thus tending to bring all forces into equilibrium.

yielding floor [MIN ENG] A soft floor which heaves and flows into open spaces when subjected to heavy pressure from packs or pillars.

yielding prop [MIN ENG] A steel prop which is adjustable in length and incorporates a sliding or flexible joint which comes into operation when the roof pressure exceeds a set load or value.

yield-pillar system [MIN ENG] A method of roof control whereby the natural strength of the roof strata is maintained by the relief of pressure in working areas and the controlled transference of load to abutments which are clear of the workings and roadways.

yield temperature [ENG] The temperature at which a fusible plug device melts and is dislodged by its holder and thus relieves pressure in a pressure vessel; it is caused by the melting of the fusible material, which is then forced from its holder.

ylang-ylang oil *See* ilang-ilang oil.

Y level [ENG] A surveyor's level with Y-shaped rests to support the telescope. Also known as wye level.

yoke [DES ENG] A clamp or similar device to embrace and hold two other parts. [ENG] **1.** A bar of wood used to join the necks of draft animals for working together. **2.** *See* wye. [MECH ENG] A slotted crosshead used instead of a connecting rod in some steam engines.

Z

Zanzibar gum [MATER] A combustible, hard, fossil-type copal, which is insoluble in most solvents and melts at about 245°C; used in varnishes.

zein [MATER] A combustible, white to yellowish protein powder derived from corn; insoluble in water, soluble in dilute alcohol; used in inks, fibers, microencapsulation, and coatings for paper and food.

zero adjuster [ENG] A device for adjusting the pointer position of an instrument or meter to read zero when the measured quantity is zero.

zero bevel gears [DES ENG] A special form of bevel gear having curved teeth with a zero-degree spiral angle.

zero defects [IND ENG] A program for improving product quality to the point of perfection, so there will be no failures due to defects in construction.

zero length [AERO ENG] In rocket launchers, zero length indicates that the launcher is designed to hold the rocket in position for launching but not to give it guidance.

zero level [ENG ACOUS] Reference level used for comparing sound or signal intensities; in audio-frequency work, a power of 0.006 watt is generally used as zero level; in sound, the threshold of hearing is generally assumed as the zero level.

zero-lift angle [AERO ENG] The angle of attack of an airfoil when its lift is zero.

zero-lift chord [AERO ENG] A chord taken through the trailing edge of an airfoil in the direction of the relative wind when the airfoil is at zero-lift angle of attack.

zero method *See* null method.

Ziegler catalyst [MATER] A special catalyst developed to produce stereospecific polymers, and derived from a transition-metal halide and a metal hydride or metal alkyl.

zinc [MET] A shiny, bluish-white, lustrous metal that is ductile when pure; used in alloys, metal coatings, electrical fuses, anodes, and dry cells.

zinc baryta white *See* lithopone.

zinc sulfide white *See* lithopone.

zipper [ENG] A generic name for slide fasteners in which two sets of interlocking teeth of the same design provide sturdy and continuous closure for adjacent pieces of textile, leather, and other materials.

zipper conveyor [MECH ENG] A type of conveyor belt with zipperlike teeth that mesh to form a closed tube; used to handle fragile materials.

zirconia brick [MATER] A type of brick containing zirconium oxide, used to line metallurgical furnaces.

zirconium [MET] A hard, lustrous, grayish metal that is strong and ductile; used in alloys, pyrotechnics, welding fluxes, and explosives.

Zond spacecraft [AERO ENG] One of a series of Soviet space probes which have photographed the moon and made observations in interplanetary space.

zone control [ENG] The zoning of a process or building, and the independent heating or temperature controls for each zone.

zone heat [CIV ENG] A central heating system arranged to allow different temperatures to be maintained at the same time in two or more areas of a building.

zone-position indicator [ENG] Auxiliary radar set for indicating the general position of an object to another radar set with a narrower field.

zone purification *See* zone refining.

zone refining [MET] A technique to purify materials in which a narrow molten zone is moved slowly along the complete length of the specimen to bring about impurity segregation, and which depends on differences in composition of the liquid and solid in equilibrium. Also known as zone purification.

zoning [CIV ENG] Designation and reservation under a master plan of land use for light and heavy industry, dwellings, offices, and other buildings; use is enforced by restrictions on types of buildings in each zone.

Z variometer *See* vertical intensity variometer.

Zyglo method [ENG] A procedure for visualizing incipient cracks caused by fatigue failure, in which the part is immersed in a special activated penetrating oil and viewed under black light.